Marine Mussels

Marine Mussels

Ecology, Physiology, Genetics and Culture

Elizabeth Gosling
Galway-Mayo Institute of Technology,
Ireland

WILEY

This edition first published 2022

Registered Offices
John Wiley & Sons, Inc., 111 River Street, Hoboken, NJ 07030, USA
John Wiley & Sons Ltd, The Atrium, Southern Gate, Chichester, West Sussex, PO19 8SQ, UK

Editorial Office
The Atrium, Southern Gate, Chichester, West Sussex, PO19 8SQ, UK

For details of our global editorial offices, customer services, and more information about Wiley products visit us at www.wiley.com.

Wiley also publishes its books in a variety of electronic formats and by print-on-demand. Some content that appears in standard print versions of this book may not be available in other formats.

Library of Congress Cataloging-in-Publication Data

Names: Gosling, E. M., author.
Title: Marine mussels : ecology, physiology, genetics and culture / Elizabeth Gosling, Galway-Mayo Institute of Technology, Ireland.
Description: First edition. | Hoboken, NJ : Wiley, 2021. | Includes bibliographical references and index.
Identifiers: LCCN 2021032409 (print) | LCCN 2021032410 (ebook) | ISBN 9781119293903 (cloth) | ISBN 9781119293965 (adobe pdf) | ISBN 9781119293934 (epub)
Subjects: LCSH: Mytilidae. | Mytilidae–Ecology. | Mussel culture.
Classification: LCC QL430.7.M95 G67 2021 (print) | LCC QL430.7.M95 (ebook) | DDC 594/.4–dc23
LC record available at https://lccn.loc.gov/2021032409
LC ebook record available at https://lccn.loc.gov/2021032410

Cover Design: Wiley
Cover Image: Photo courtesy of Philippe Archambault, Dèpartment de Biologie, Université de Laval, Canada.

Printed and bound by CPI Group (UK) Ltd, Croydon, CR0 4YY

C9781119293903_151121

Contents

Preface

A comprehensive review of the literature on marine mussels was published in 1976,[1] but by the early 1990s a wealth of additional material had appeared, not just in the traditional areas of investigation, but also in the newer fields of nutrition, environmental monitoring, population genetics, toxicology, disease and public health. In view of this, in 1992 I edited a thorough compilation of information on mussels in the form of critical review papers by leading authorities,[2] titled *The Mussel Mytilus*. I was also the author of two of the chapters: 'Systematics and geographic distribution of *Mytilus*' and 'Genetics of *Mytilus*', reviewing not just population genetics but cytogenetics and selective breeding studies. The book, with an extensive bibliography of more than 2000 references, is still considered the standard reference text on *Mytilus*. Although out of print, it continues to have a reasonably good citation rate. For example, individual chapters have citations ranging from 47 to 744, with a total of 2404 citations for the complete volume (Google Scholar, December 2020).

Subsequently, I wrote *Bivalve Molluscs*,[3] a comprehensive book covering all major aspects of this important class of invertebrates, which was published by Blackwell Publishing UK in 2003. It was well received by reviewers and did well in terms of sales, and in 2011 I was invited by Wiley Blackwell to prepare a second edition, *Marine Bivalve Molluscs*,[4] which was published in June 2015. Work on these two volumes entailed a large amount of reading in diverse fields, including biology, ecology, physiology, genetics, culture, disease and public health – an experience that proved invaluable in the preparation of this new volume on marine mussels.

It is now almost 30 years since I edited *The Mussel Mytilus*. In the interim, there have been many significant advances, along with the emergence of new areas of research in marine mussels, one reason for which is the range of new techniques being applied in their biology and ecology. For example, DNA genetic markers play an increasingly important role in disease diagnosis, detection of different strains of pathogenic bacteria (Chapter 11), identification and tagging of larvae (Chapter 5) and the completely new area of bivalve genomics (Chapter 9). Also, isotope ratios and fatty acid signatures are being used to identify dietary food sources in bivalves, while measurements of stable oxygen isotopes in fossil and contemporary bivalve shells are allowing reconstruction of past sea surface temperatures. Trace elemental fingerprinting, based on naturally occurring elements in bivalve shells, is being used to assess larval origins and trajectories, while fluorescence *in situ* hybridisation (FISH) combined with cell sorting can identify and rapidly count and sort larvae of several bivalve species in diverse plankton samples. Immunosensors – devices that use specific biochemical reactions mediated by antibodies to detect chemical compounds – may soon replace liquid chromatography, which has long been the official method for detection and quantification of algal toxins in bivalves (Chapter 12). Also, a wide range of molecular biomarkers are increasingly being used to identify exposure to certain chemicals, provide information on spatial and temporal changes in the concentration of contaminants and indicate environmental quality or occurrence of adverse ecological consequences (Chapter 8). Another significant development has been the increasing use of sophisticated mathematical models to describe and predict growth and bioenergetics of individual animals, to track larval movement and recruitment, to study contaminant bioaccumulation in marine bivalves and to predict geographic range of species in global warming scenarios (Chapters 3, 4, 6 and 8). Models are being employed to describe carrying capacity and environmental effects in mussel culture areas (Chapter 10) and to simulate disease and forecast outbreaks (Chapters 11 and 12). Another area that has changed is the management of culture systems, where the emphasis is now more on the use of legislation and control measures, particularly in Europe, North America, Australia and New Zealand (Chapter 10). This also pertains to the control of diseases in bivalves and to hygiene standards for shellfish and shellfish waters.

This volume does not focus solely on marine mussels in the genus *Mytilus* but also incorporates research on other commercially important marine genera such as *Perna*, *Aulacomya* and *Choromytilus*, as well as noncommercial genera such as *Modiolus*, *Geukensia*, *Brachidontes* and hydrothermal vent *Bathymodiolus* species. It therefore has a broader scope and content than *The Mussel Mytilus* and consequently should not be regarded as a second edition.

Chapter 1 covers the phylogeny and evolution of the Bivalvia and Mytilida. Chapter 2 provides a detailed description of external and internal anatomy. Chapter 3 describes global and local distribution patterns and the physical and biological factors influencing marine mussel distribution and abundance, with a section on climate change and the actual and potential impacts of global warming, ocean acidification and hypoxia on marine mussels. Chapter 4 covers factors influencing filter feeding, together with dietary components and assimilation efficiency, and presents new material on the impacts of mussels on marine ecosystems. Chapter 5 deals with reproduction, larval development and settlement, and has new sections on sperm–egg interaction, larval dispersal, connectivity, identification and abundance estimates. Chapter 6 covers methods of measuring growth, the mussel shell as a marine archive and exogenous and endogenous modulators of growth. The processes of circulation, respiration, excretion and osmoregulation are described in Chapter 7, while Chapter 8 covers types and levels of contaminants in the marine environment, contaminant absorption and assimilation, kinetic modelling, mussel monitoring programmes and biomarkers of pollution. Chapter 9 describes the application of genetic methods, population genetics, global breeding programmes and the relatively new area of bivalve genomics. The fundamentals of mussel aquaculture are dealt with in Chapter 10, focusing on a number of key mussel species for detailed treatment and the effects of mussel culture on the environment; it includes a section on facilitating sustainable aquaculture development. Chapter 11 deals with diseases and parasites, with a large amount of new information on diagnostic methods and the diverse defence mechanisms utilised by mussels. Finally,

Chapter 12 looks at the role of mussels in disease transmission to humans, with sections on production and processing controls, regulation of monitoring and quality control, including the HACCP system.

Notes

1 Bayne, B.L. (ed.) (1976) Marine Mussels: Their Ecology and Physiology. Cambridge University Press, Cambridge.
2 Gosling, E.M. (ed.) (1992) The Mussel Mytilus: Ecology, Physiology, Genetics and Culture. Elsevier Science, Amsterdam.
3 Gosling, E. (2003) Bivalve Molluscs: Biology, Ecology and Culture. Blackwell Publishing, Oxford.
4 Gosling, E. (2015) Marine Bivalve Molluscs, 2nd edition. John Wiley & Sons, Chichester.

Acknowledgements

I benefited greatly from the assistance and knowledge of librarians at the Hardiman Library of the National University of Ireland, Galway.

A book like this is in many ways as good as its illustrations, and I thank those who provided figures: Philippe Archambault, Sara Barrento, Brian Beal, Guisla Boehs, Frank Alberto Ocaña Borrego, Craig Burton, Thomas Carefoot, Noèlia Carrasco, Antonio Checa, Siu Gin Cheung, John Costello, David Cowles, Jeff Davidson, Elizabeth Fly, Gael Force Fusion, Argyll, Scotland, UK, Laas Hiebenthal, Jade Irisarri, Jaafar Kefi, Anja Monika Landes, Ionan Marigomez, Katherine McFarland, Ivona Mladineo, Arthur Morris, Jorge Navarro, Sandra Noel, Jiří Novák, Aida Ovejero, Bernadette Pagoda, Peter Petraitis, David Polo, Guido Poppe, Chris Richardson, Mindy Richlen, Gianluca Sarà, Julia Sigwart, Tore Strohmeier, Tianli Sun, Cindy Lee Van Dover and David Wethey.

Thanks also to the team at John Wiley & Sons, Ltd: Sonali Melwani, Kerry Powell, Rebecca Ralf, Blesy Regulas, and, in particular, Sivasri Chandrasekaran, for their encouragement and feedback on the long course from commissioning to proofs.

Finally, and above all, I wish to thank my son, Marcus Gosling, who provided invaluable technical support and graphics expertise, and my partner, Jim, an exemplar of diligence and support.

1

Phylogeny and Evolution of Marine Mussels

Introduction

The phylum Mollusca is the second largest phylum of animals, with about 130 000 named extant species and 70 000 described fossil species (Haszprunar *et al.* 2008). While most of these are marine, many live in freshwater and terrestrial habitats. Research has indicated that molluscs had a terminal Precambrian origin, with rapid divergence occurring in the Cambrian era some 540–560 million years ago (Stöger *et al.* 2013). All molluscs have a soft body that, with the exception of some groups (see later), is protected by a hard calcium shell. Inside the shell is a heavy fold of tissue called the mantle that encloses the internal organs of the animal. Another feature of the phylum is a large muscular foot that is generally used for locomotion. Although most molluscs share this basic body plan, the group is characterised by a great diversity of form and habit.

Phylogeny of the Phylum Mollusca

Eight living classes (lineages) of molluscs have been recognised, primarily based on clad[1] (phylogenetic) analysis of morphological characters (Haszprunar *et al.* 2008). Aplacopora incorporates two classes, Solenogastres and Caudofoveata; these are worm-shaped, deep-water animals lacking a shell. Polyplacophora, often referred to as chitons, inhabit hard substrates on rocky shores and are characterised by eight dorsal shell plates. Aplacophora and Polyplacophora are grouped in the clade[2] Aculifera, which is regarded as monophyletic (i.e. all taxa in this group share a common ancestor) (Sigwart & Sutton 2007). The remaining five classes are grouped in the clade Conchifera, which is also regarded as a monophyletic group. Monoplacophora live in deep waters and are small and limpet-like, with a single cap-like shell. The class Bivalvia includes laterally compressed animals enclosed in two shell valves, such as clams mussels, oysters and scallops. Scaphopoda, commonly known as tusk shells, live in marine mud and sediments. The class Gastropoda is the largest and most diverse, containing spirally coiled snails, flat-shelled limpets, shell-less sea slugs and terrestrial snails and slugs. Octopus, squid and cuttlefish are in the class Cephalopoda and represent the largest, most organised and specialised of all the molluscs. The Monoplacophora are generally accepted as the earliest extant offshoot of the Conchifera (Haszprunar 2008).

Marine Mussels: Ecology, Physiology, Genetics and Culture, First Edition. Elizabeth Gosling.

Morphological disparity among the lineages has given rise to numerous conflicting phylogenetic hypotheses. Molecular investigations using nuclear ribosomal gene sequences (18S and 28S) offered little resolution (Ponder & Lindberg 2008). More recent studies, using phylogenomic-scale molecular data sets (Kocot *et al.* 2011; Smith *et al.* 2011; Stöger *et al.* 2013), have significantly advanced understanding of molluscan phylogeny by providing well-supported tree topologies and generally congruent results (Telford & Budd 2011; Kocot 2013; Schrödl & Stöger 2014). Probably the most important achievement of these studies is the establishment of the Aculifera and Conchifera groupings.

Several major evolutionary hypotheses and sister relationships have been proposed for the eight living molluscan classes, which are illustrated in Figure 1.1 (Sigwart & Lindberg 2015). Sigwart & Lindberg (2015) assembled a data set of 42 unique published trees describing molluscan interrelationships, which included at least five out of the eight classes. They found that almost 60% of trees based on morphological data (N = 27) were similar, while only 45% of those based on molecular data (N = 15) were similar; the distances separating morphological and molecular trees indicated that they were similar in almost 30% of pairs and different in 20%. It is interesting that there are no studies in which both molecular and morphological data have been simultaneously analysed for all eight classes of Mollusca (Sigwart & Lindberg 2015). Regarding some of the hypotheses illustrated in Figure 1.1, the authors found that support for Cyrtostoma or Diasoma was relatively weak, while the Serialia concept was deserving of due consideration (see also Stöger *et al.* 2013). They found no consensus support for the topology of the morphological Testaria concept. Integration of new molecular techniques with morphological and developmental data using multiple type species from all eight molluscan classes will no doubt continue to deepen our understanding of molluscan phylogeny and evolution (Kocot 2013; Sigwart & Lindberg 2015).

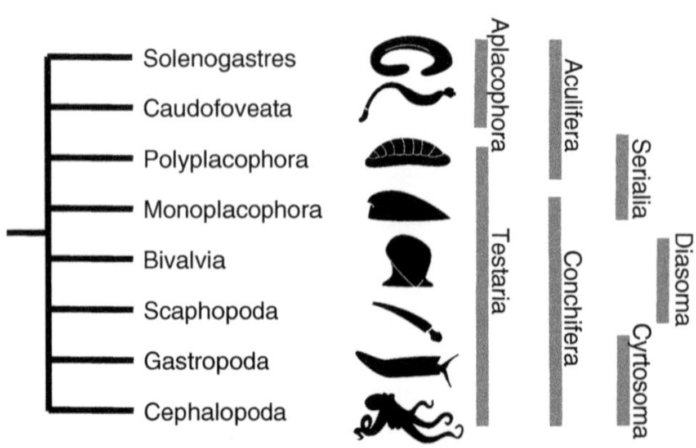

Figure 1.1 Schematic topology of the major evolutionary hypotheses and sister relationships proposed for the eight living classes in Mollusca: **Aculifera** (Solenogastres + Caudofoveata + Polyplacophora), **Aplacophora** (Caudofoveata + Solenogastres), **Conchifera** (Monoplacophora + Bivalvia + Scaphopoda + Gastropoda + Cephalopoda), **Cyrtosoma** (Gastropoda + Cephalopoda; historically also including Monoplacophora), **Diasoma** (Bivalvia + Scaphopoda), **Serialia** (Polyplacophora + Monoplacophora), and **Testaria** (Conchifera + Polyplacophora). Text in bold indicate the seven different hypotheses. *Source:* From Sigwart & Lindberg (2015). Reproduced with permission from Oxford University Press.

Phylogeny and Evolution of Bivalvia

Bivalvia is the second largest class within the Mollusca, with more than 9000 extant species. Individuals are bilaterally symmetrical with a laterally compressed body enclosed in two shell valves. Bivalves are an important component of marine and freshwater ecosystems, with more than 80% of species living in oceanic habitats and exhibiting varied ecologies. Sessile epifaunal bivalves, such as oysters and mussels, attach themselves to hard surfaces, while infaunal bivalves, such as clams, burrow to different depths in sand or sediment on the seafloor or in riverbeds. Other sessile forms bore into hard sediments such as coral and wood. Some species, such as scallops, are free-living and can move through the water by clapping the two shell valves together or dig into the sediment using their muscular foot. Although some bivalves are deposit feeders, the majority feed using greatly enlarged gill surfaces to filter food particles from the surrounding water. However, because of their mode of feeding, they pump large volumes of water and thus have the potential to accumulate contaminants, bacteria, viruses and toxins, frequently posing significant public health risks. Despite this, many species form the basis of valuable aquaculture and fisheries industries worldwide.

Although bivalves, with their strong shells, provide one of the most complete fossil record, their systematics[3] until recently lagged behind that of other animal groups. Historically, there was a heavy reliance on single-character systems (e.g. shell hinge teeth, shell ligament, gill structure, gill ciliation and stomach morphology). This changed in the 1970s with the development of numerical systematics based on simultaneous analysis of multiple character systems. From the early 1990s, gene sequence data became available, and over the past two decades researchers have been increasingly involved in large-scale phylogenetic analyses using shell morphology and anatomy, fossils and, more recently, molecular sequence data. These data sources have made a significant contribution in systematic studies, encompassing all Bivalvia as well as major groups within the class (Carter *et al.* 2000; Harper *et al.* 2000; Giribet & Wheeler 2002 and references therein; Matsumoto 2003; Plazzi & Passamonti 2010; Plazzi *et al.* 2011; Tsubaki *et al.* 2011; Sharma *et al.* 2012; Yuan *et al.* 2012). Recently, Bieler *et al.* (2014) provided a new analysis of bivalve relationships integrating classic and novel morphological characters with a combination of up to nine molecular markers. While their results are largely consistent with many of the previous schemes of bivalve phylogenetics, they made significant progress in resolving previously uncertain relationships, which allowed them to refine higher order bivalve classification (Figure 1.2). Bieler *et al.* (2014) confirmed that the Bivalvia consists of two major clades, Protobranchia and Autobranchia (not shown in Figure 1.2), with the latter dividing into Pteriomorphia and Heterodonta (Heteroconchia). Heterodonta in turn consists of Archiheterodonta, Palaeoheterodonta and Euheterodonta. Protobranchia has been confirmed as a monophyletic group comprising Nuclida, Solemyida and Nuculanida – much in agreement with earlier morphology-based classifications. Protobranchs are primitive, marine, infaunal bivalves that use their large labial palps in deposit feeding – the ctenidia being used solely for respiration, in contrast to other subclasses within Bivalvia. They possess a lecithotrophic larval type, found in other primitive mollusc groups, as well as the respiratory pigment haemocyanin, found in nonbivalve groups such as cephalopods, polyplacophorans and gastropods (reviewed in Zardus 2002). Autobranchia, confirmed as

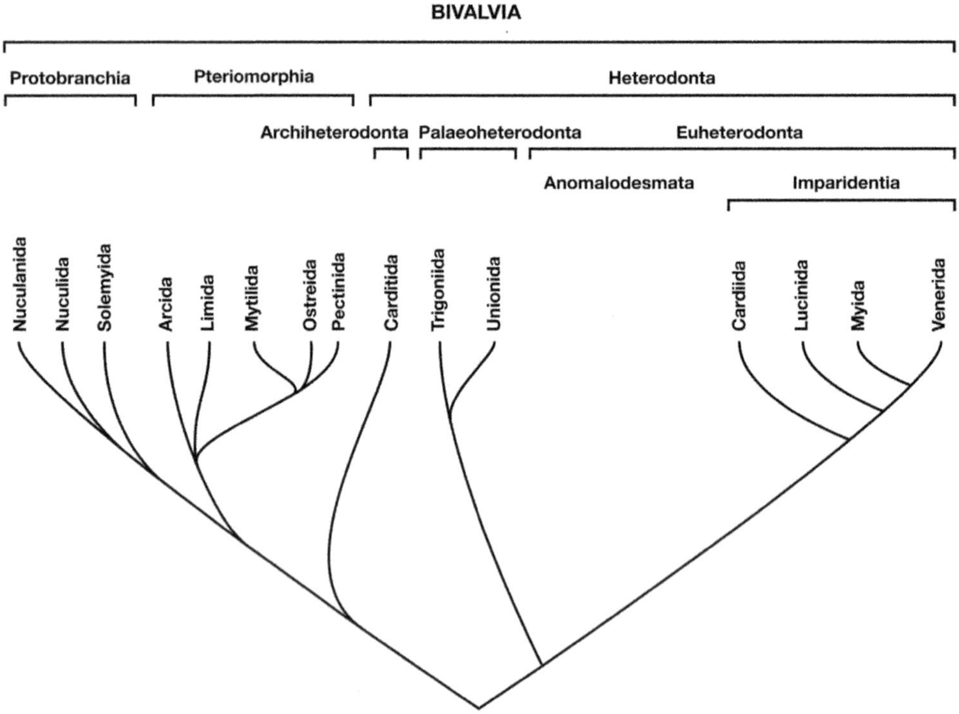

Figure 1.2 Phylogenetic diagram showing hypothesised relationships between the major clades recognised for the living members of the class Bivalvia. While formal endings imply certain ranks in the Linnaean hierarchy, no attempt was made to re-rank all hypothesised clades from Bieler *et al.*'s study (2014) as they considered 'such a step as premature until a denser family-level sampling is presented'. *Source:* Based on Bieler *et al.* (2014).

the monophyletic sister group of Protobranchia, contains all of the remaining bivalve lineages. It is characterised by the presence of enlarged ctenidia with a filtering function and comprises two sister taxa, the Pteriomorphia and Heteroconchia. The former is monophyletic and contains four well-supported marine clades: Arcida (arks), Mytilida (mussels), Ostreida (oysters) and Pectinida + Limida (scallops) (Figure 1.2). A molecular analysis based on 42 mitochondrial genomes of phylogenetic relationships within the Pteriomorphia supports the monophyly of the Pteriomorphia (Sun & Gao 2017), although it includes an additional clade, Pterida, now accepted as Ostreida (http://www.marine species.org). Heterodonta is also confirmed as monophyletic and, as mentioned earlier, is composed of three clades, Palaeoheteterodonta, Archiheterodonta and Euheterodonta. Molecular analyses support a divergence of Archiheterodonta prior to the split of the other two clades, which have traditionally been grouped in the Heteroconchia but are now regarded as sister taxa (Bieler *et al.* 2014). The well-supported Palaeoheteterodonta consists of the marine Trigoniida, remnants of a once diverse group, and the diverse freshwater Unionida (freshwater mussels, pearl mussels). The likewise well-supported clade Archiheterodonta is monophyletic and comprises marine clams Carditida (González & Giribet 2014).

Euheterodonta consists of two well-supported clades, Anomalodesmata, a deep-sea group of marine clams with unusual morphology, and the remaining Heterodonta, to which Bieler *et al.* (2014) have applied the name 'Imparidentia', a sister group to Anomalodesmata. The new clade Imparidentia, the relationships of which are not fully resolved (Bieler *et al.* 2014; González *et al.* 2015), includes some of the best-known bivalves and most of the commercial species, excluding mussels, oysters and scallops, all of which are members of the Pteriomorphia.

Evolution and Adaptive Radiation in Bivalvia

As already mentioned, bivalves, because of their strong shells, provide one of the most complete fossil records of any animal group. The earliest known bivalves are the Early Cambrian (540 mya) *Pojetaia runnegari* from Australia, the Middle Cambrian (510 mya) *Tuarangia gravgaerdensis* from New Zealand and the Upper Cambrian (485 mya) *Fordilla troyensis* (details in Giribet 2008). These were all <1 cm in length and possessed a bivalved shell, a simple ligament, bivalve-like pedal muscle insertions and well-developed adductor muscle insertions (Morton 1992). According to Morton (1996), they were probably surface dwellers that used the foot for feeding and locomotion. Protobranchs closely resembling modern-day representatives (Figure 1.2) were widespread by the early Ordovician (485 mya), as were all extant lineages and feeding types from most other lineages (see later). Protobranchs are clearly separated from the rest of the Bivalvia by a number of distinct morphological traits that are unique to the group (see earlier), but whether these represent an ancestral or a derived condition (i.e. traits that have evolved as a result of the clade's evolution) is not clear (Cope 2000 and references therein). Results from a comprehensive cladistic analysis that incorporated both morphological and molecular characters show that Protobranchia are basal to the entire Bivalvia class (Giribet & Wheeler 2002).

One of the most significant developments in the evolution of modern bivalves from the more primitive form (Figure 1.3a) was moving the site of water intake to the posterior of the animal so that water both enters and exits the mantle cavity posteriorly (Figure 1.3d; see alternative evolutionary pathway in Figure 1.3b). This made it possible for bivalves to penetrate sand or mud 'head first', with the posterior end freely communicating with the water above. Extensions of the mantle to form siphons at the posterior enabled the animals to live deeper and deeper under the surface. As bivalves evolved, plankton in the incoming current were increasingly adopted as a source of food, the ctenidia (gills) replacing the palp processes as the feeding organs (see Chapter 2 for details on mussel morphology). The chief modification of the ctenidia for filtering was the lengthening and folding of individual gill filaments. In addition, many extra filaments were added so that they extended as far forward as the labial palps. Both of these modifications greatly increased the surface area of the ctenidia. It is believed that the triangular-shaped filaments of the primitive bivalve gill progressively changed over evolutionary time to the W-shaped filaments of the modern bivalve gill (see Cannuel *et al.* 2009). A notch at the bottom of each side of the W lines up with similar notches on adjacent filaments to form a food groove that extends the length of the underside of the ctenidia. Yonge (1941) suggested that since the food groove was necessary for nutrition, these notches probably preceded folding of the gill filaments. Changes in

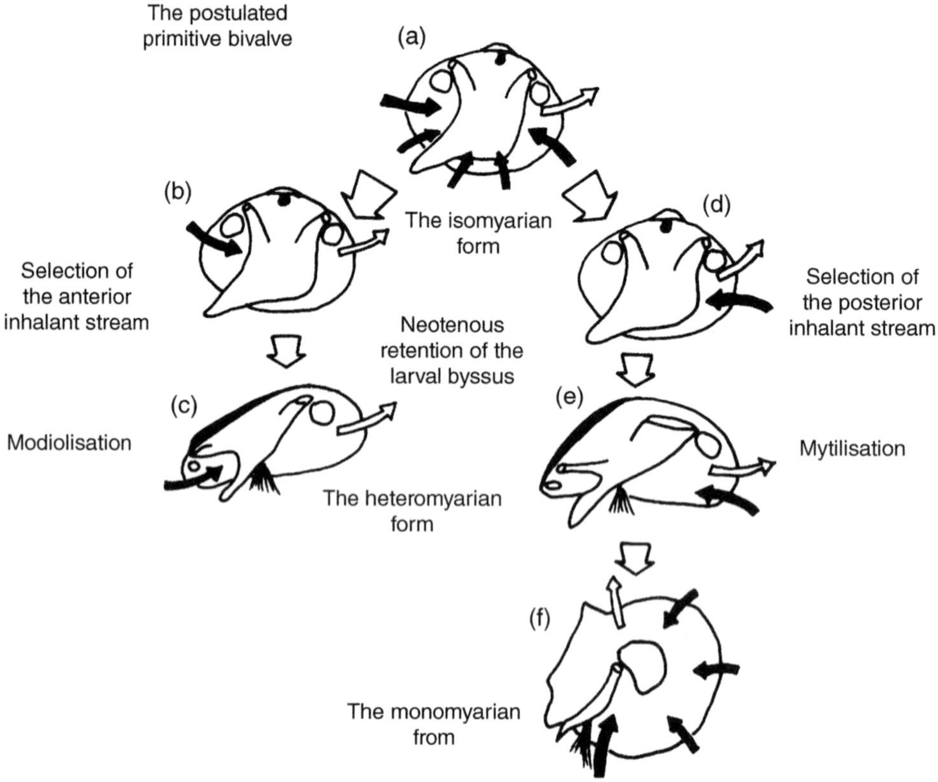

Figure 1.3 Evolution of the heteromyarian form, and ultimately of the monomyarian form, from an isomyarian ancestor. (a) Postulated primitive isomyarian bivalve such as the protobranch *Nucula* or the heterodont *Glycymeris*, with water capable of entering the mantle from anterior and posterior directions. (b) Selection of the anterior inhalant stream by representatives of such groups as the Lucinida (shallow burrowers in tropical mud) can only result in the process of heteromyarianisation, leading to (c) a modioliform shell found in representatives of the Arcida. (d) Selection of the posterior inhalant stream can result in full expression of (e) the heteromyarian form (e.g. in the mussel *Mytilus*) and ultimately (f) the monomyarian form (in pteriomorph oyster and scallop species). See also Figure 1.2. *Source:* From Morton (1992). Reproduced with permission from Elsevier.

both ciliation and water circulation followed. The exploitation of filter feeding led to the first increase in bivalve diversity and body plan divergence, so that by the Ordovician period (~450 mya) all extant higher lineages and feeding types were present and had colonised a wide variety of habitats that had hitherto been inaccessible to their protobranch ancestors (Giribet 2008 and references therein).

An important factor in this diversification was the development of a larval byssal apparatus in the basal Autobranchia; this was absent in the Protobranchia. Byssal threads provided an effective method of attachment on to hard substrates, enabling bivalves to adopt an epibenthic lifestyle in new adaptive niches (Giribet 2008). The byssal apparatus is seen as a persistent post-larval structure that evolved for temporary attachment of the animal to the substrate during the vulnerable stage of metamorphosis. While the byssus is subsequently lost in many bivalve species, in mussels it persists into adult life. In byssally-attached forms, there has been a tendency for the anterior (head) end of the animal to become smaller, with a corresponding enlargement of the posterior end, thereby allowing free access, posteriorly, to the water above (Morton 1992). Accompanying this change, there has also been a reduction of the anterior adductor muscle and an increase in the size of the posterior adductor muscle. This led to the evolution of the heteromyarian form, a pronounced triangular shape brought about through reduction of the anterior and enlargement of the posterior ends of the shell, which is very marked in the Mytilida (Figure 1.3c and e). The heteromyarian form, coupled with secure byssal attachment throughout life, allowed mussels to anchor themselves and live in high population densities in wave-exposed habitats (Morton 1996). These two adaptations were so successful that they have evolved independently among otherwise unrelated lineages (e.g. in freshwater dreissenid mussels), a perfect example of convergence (Seed *et al.* 2000).

The heteromyarian condition has been seen as a stepping stone from the isomyarian form of primitive prosobranchs (Figure 1.3a) toward the monomyarian form and the adoption of a horizontal posture (Figure 1.3f). Monomyarian bivalves, such as the Pectinida (scallops) and Ostreida (oysters), have largely circular shells, no trace of the anterior adductor muscle and a body reorganised around the enlarged and more or less centrally placed posterior muscle. Water enters around two-thirds or more of the rounded margins of the shell. Shell attachment has led to varying degrees of inequality in the size of the two shell valves. In scallops, the shell valves are circular, but they may be concave and similar or the left (uppermost) one may be flat. Like oysters, they also lie in a horizontal position on the substrate. However, scallops, far from being fixed, are active swimming bivalves. In early life, they use byssus threads for attachment to algae, but before they attain a size of 15 mm the majority of species have detached themselves to take up a free-living existence on the seabed.

Cementation is another mode of attachment that evolved during the early Mesozoic era (252 mya). This adaptation arose independently in Pteriomorphia, Heterodonta and Palaeohterodonta, peaking in the Late Triassic and Jurassic periods of the Mesozoic era (252–145 mya) as a possible response to the appearance of many predatory groups (Vermeij 1977; Harper 1991). During the Triassic, another important development occurred when an ancestral unionid (Paleoheterodonta) colonised freshwater environments, thereby gaining access to a bivalve-free ecosystem. Giribet (2008) suggests that this move may have been triggered by the evolution of a novel mode of development using microscopic glochidia larvae with fish as intermediate hosts.

Burrowing into the substrate is the habit most extensively exploited by bivalves. Contact is maintained with the surface by way of siphons that extend from the posterior end of the animal. During the Cenozoic era (up to 60 mya), soft, nutrient-rich sediments on continental margins allowed diversification of shallow burrowing, globular, strongly ribbed forms and deep burrowers with smooth, blade-like shells (Giribet 2008). Cockles (e.g. *Cardium* spp.) are shallow burrowers, while many clam species (e.g. razor clams) burrow as deep as 60 cm. The geoducks (*Panopea*) on the West Coast of the United States are among the deepest burrowers, digging down to a depth of over 1 m, aided by a streamlined shell for fast burrowing and fusion of the mantle edges (apart from a small gape for the large muscular foot), which prevents entry of sediment into the mantle cavity. Many bivalves that burrow deeply (>30 cm) live in permanent burrows, moving deeper as they grow larger. This lifestyle is brought to an extreme in bivalves that bore into hard substrates such as shell, coral, wood and rock and are permanently locked in their burrows, which are therefore inevitably dependent on outside sources of food. Both shallow and deep-sea wood-boring bivalves excavate 'sawdust' as their principal food source with the aid of gill-associated symbiotic bacteria (Voight 2007).

Phylogeny and Evolution in the Mytilida

Pteriomorphia is monophyletic and contains four well-supported marine clades: Arcida, Mytilida, Ostreida and Pectinida + Limida (Figure 1.2). The order Mytilida originated in the Cambrian era some 540–560 million years ago and is composed of a single superfamily, the Mytiloidea, which is considered to be representative of and basal to the Pteriomorphia (Morton 2015 and references therein). Information based on shell morphology and molecular data suggests that Mytiloidea is a monophyletic clade (Morton *et al.* 2016 and references therein). For more than five decades, there has been no consensus on the number of families contained within the Mytiloidea. For example, Boss (1982) recognised only a single family, the Mytilidae, divided into four subfamilies, Mytilinae, Crenellinae, Lithophaginae and Modiolinae. Coan *et al.* (2000), and subsequently Coan & Valentich-Scott (2012), also recognised a single family, but added three additional subfamilies: Bathymodiolinae, Dacrydiinae and Septiferinae. Bieler *et al.* (2010) added one more, Limnoperninae, making a total of eight subfamilies within the Mytilidae. In contrast, others have recognised between three and four families in the Mytiloidea; Carter *et al.* (2011), for example, in a classification including fossil taxa, recognised three families, Mytilidae, Crenellidae and Septiferidae, while Morton (2015) proposed four, the Mytilidae, Crenellidae, Modiolidae and Musculidae – although he suggested that the Septiferinae might be eligible for family status subject to additional morphological and molecular analyses. Bieler & Mikkelsen (2006) initially recognised five families, the Mytilidae, Crenellidae, Musculidae, Modiolidae and Septiferidae, although as we have seen they later recognised only one, the Mytilidae (Bieler *et al.* 2010). Table 1.1 presents a summary of five different taxonomic classifications of extant Mytiloidea, which are only a fraction of the published classifications for this superfamily. The majority of classifications have been heavily reliant on shell features, such as shape and characteristics of the valves, hinges and dentition, with few authors to date using anatomical features (reviewed in Morton 2015) or molecular data.

The Mytilidae is the largest family within the Mytiloida, with subfamily numbers varying between four and eight (Table 1.1). The family consists of 52 genera and 411 species (www.marinespecies.org). Species are recognised as members of the family by the shell form, sculpture, hinge and muscle scars (see Soot-Ryen 1955 for details). Mytiliform species, with a broadly triangular shape and terminal umbones, are placed in the genus *Mytilus* (Figure 1.4A), while those with subterminal umbones with a distinct anterior margin – the modioliform species – are consigned to the genus *Modiolus* (Figure 1.4B). Elongate forms that burrow in rock or mud and have parallel dorsal and ventral shell margins and subterminal umbones are placed in the genus *Lithophaga* (Figure 1.4C), species with pronounced anterior and posterior radiating ribs, except for a smooth median area, are placed in the

Table 1.1 Classification schemes in the superfamily Mytiloidea. Families within the Mytiloidea are in bold, while those in plain text are subfamilies recognized within the individual families. *Source:* From Morton (2015). Reprinted with permission from Taylor & Francis.

Boss 1982[a]	Coan *et al.* 2000[a]	Bieler *et al.* 2010[a,b,c]	Carter *et al.* 2011[a,b,c]	[†]Morton 2015[a]
Mytilidae	**Mytilidae**	**Mytilidae**	**Mytilidae**	**Mytilidae**
Mytilinae	Mytilinae	Mytilinae	Mytilinae	Mytilinae
Crenellinae	Bathymodiolinae	Bathymodiolinae	Arcuatulinae	Brachidontinae
Lithophaginae	Crenellinae	Crenellinae	Bathymodiolinae	Botulinae
Modiolinae	Dacrydiinae	Dacrydiinae	Lithophaginae	Adulinae[§]
	Lithophaginae	Limnoperninae[‡]	Modiolinae	Lithophaginae
	Modiolinae	Lithophaginae	**Crenellidae**	Septiferinae
	Septiferinae	Modiolinae	Crenellinae	**Crenellidae**
		Septiferinae	Musculinae	Crenellinae
			Septiferidae	Dacrydiinae
			Septiferinae	**Modiolidae**
			Limnoperninae	Modiolinae
				Musculidae
				Musculinae
				Arcuatulinae
				Bathymodiolinae
				Limnoperninae

Based on [a]morphological/anatomical characters; [b]molecular data; [c]paleontological data. [†]The World Register of Marine Species (WoRMS; www.marinespecies.org) used this source in their classification of the Mytiloidea. [§]Based on the fossil record, shell morphology, life history traits and spat behavior, Ockelmann & Dinesen (2009) concluded that the Adulinae, a subfamily of mechanical borers into soft rock, should be included in the Mytilinae rather than the Lithophaginae, another subfamily of borers. [‡]A freshwater subfamily.

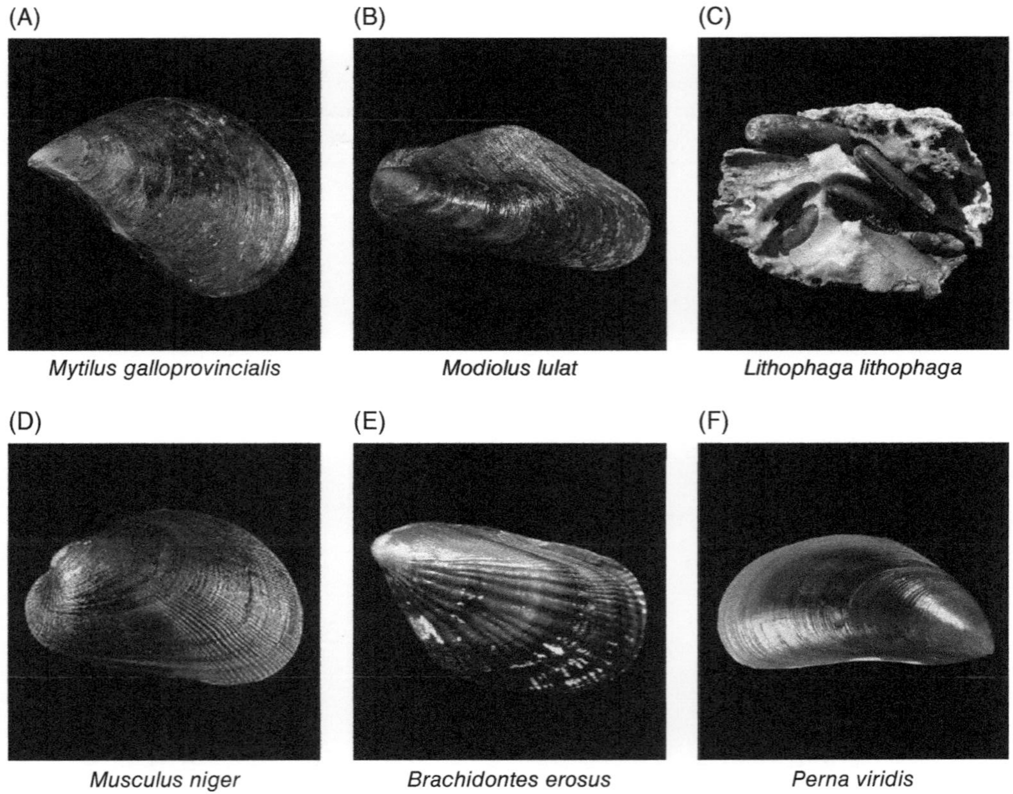

Figure 1.4 Different external shell forms in the family Mytilidae. (A) Mytiliform *Mytilus galloprovincialis*. *Source:* From MgA. Jiří Novák (biolib.cz). (B) Modioliform *Modiolus lulat*. *Source:* From MgA. Jiří Novák (biolib.cz). (C) Elongate coral boring form *Lithophaga lithophaga*. *Source:* From MgA. Jiří Novák (biolib.cz). (D) Shell with pronounced anterior and posterior radiating ribs, as in *Musculus niger*. *Source:* From Guido Poppe © G. & Ph. Poppe (http://www.poppe-images.com/). (E) Radiating sculpture over entire shell surface, as in *Brachidontes erosus*. *Source:* From MgA. Jiří Novák (biolib.cz). (F) *Perna viridis*. *Source:* From Guido Poppe © G. & Ph. Poppe (http://www.poppe-images.com/) (*See colour plate section for colour representation of this figure*).

genus *Musculus* (Figure 1.4D), and species with radiating sculpture over the whole outer shell surface are assigned to the genus *Brachidontes* (Figure 1.4E). Classification solely based on external shell features is problematic as these can be influenced by environmental conditions and similar forms may arise as a result of convergent adaptations to common environmental conditions. For example, a phylogenic study of four subfamilies of Mytilidae found that Crenellinae and Lithophaginae were monophyletic but Mytilinae and Modiolinae were polyphyletic[4] (i.e. they do not share a common ancestor), suggesting that the mytiliform and/or the modioliform body plans have evolved independently in at least two mytilid lineages (Distel 2000). However, in a recent study using species in five subfamilies (Mytilinae, Modiolinae, Bathymodiolinae, Lithophaginae and Crenellinae), Liu *et al.* (2018) found that phylogenetic trees reconstructed with a combination of two mitochondrial (COI and 16S rRNA) and three nuclear (18S and 28S rRNA and histone H3) genes indicated monophyly of the family Mytilidae and the subfamily Bathymodiolinae, polyphyly of the subfamilies Modiolinae and Lithophaginae and paraphyly[5] of the subfamily Mytilinae.

Significant Evolutionary Developments in the Family Mytilidae

As already mentioned, one of the most significant developments in bivalve evolution was the emergence of the heteromyarian form, possibly from a putative isomyarian infaunal ancestor (Figures 1.3 and 1.5A). The form – most pronounced in the Mytilidae – when coupled with secure byssal attachment allowed mussels to anchor themselves and live in high population densities in wave-exposed habitats (Morton 1996). Species of Mytilidae that exhibit varying degrees of heteromyarianism are found in the subfamilies Arcuatulinae, Brachidontinae, Mytilinae, Modiolinae, Musculinae, Septiferinae and Limnoperninae. Infaunal (endobyssate) mytilids inhabit both hard and soft substrates and are less heteromyarian, but more diverse, than epifaunal (epibyssate) rocky shore mytilids. The most significant adaptation in infaunal mytilids is a shell that is long and narrow (modioliform), allowing large numbers of individuals to form dense beds. The following text is heavily reliant on Morton (1992).

The posteriorly elongated shell keeps the inhalant and exhalant orifices in these fixed bivalves above the sediment water interface (Figure 1.5B). Enlargement of the posterior face of the shell and reduction of the anterior has resulted in the development of larger posterior adductor and posterior byssal retractor muscles and reduced anterior adductor and anterior byssal retractor muscles. In *Perna*, the anterior adductor muscle has been lost entirely, making the genus the only monomyarian one that has retained the heteromyarian form (Morton 1987). In addition, there has been an enlargement of the ligament, which probably acts to keep the two shell valves correctly aligned, particularly since another consequence of heteromyarianism is either a large reduction in hinge plate and teeth or their loss. Another adaptive feature, exemplified by species of *Musculista*, *Modiolus* and *Arcuatula*, is a thin, brittle shell that provides sufficient protection for an attached, infaunal lifestyle. Epibyssate mytilids (e.g. *Septifer* and *Brachidontes*) have much thicker sculptured shells that help to withstand the effects of wave action and provide protection against

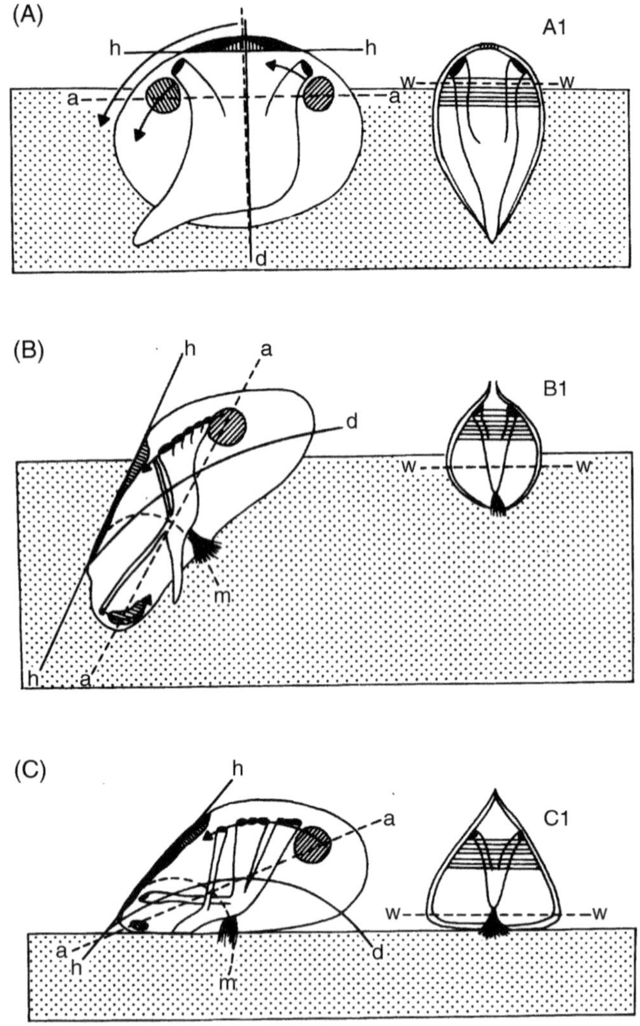

Figure 1.5 Lateral views (A–C) and transverse sections (A1–C1) of (A) an isomyarian, (B) a modioliform, and (C) a mytiliform bivalve; a = axis through the adductor muscles; d = the mid-dorso-ventral axis of the shell; h = the hinge axis; m = the mid-dorsal and foot (byssus) axis; and w = the position of greatest shell width. *Source:* From Morton (1992). Reproduced with permission from Elsevier.

predatory snails, crabs and birds. To enhance stability around the point of byssal attachment, the antero-ventral region of the shell valves is flattened, both laterally and antero-posteriorly. Flattening of the shell provides greater stability on wave exposed shores by opposing an overturning force and lowering the centre of gravity, thereby reducing the effects of drag. Such an adaptation is clearly associated with the full expression of heteromyarianism and thus the successful colonisation of hard subtrates (Figure 1.5C). For some mytilids (e.g. *Mytilus* spp.), however, this adaptation has released them from a colonial life, and some (e.g. *Septifer bilocularis*) are completely solitary. Ventral flattening also resulted in development of the umbones into 'beaks', which, with the progressive assumption of the heteromyarian form, are located in a subterminal position beyond the reduced anterior edge of the shell. There are a few points worth noting. In infaunal habitats, some species (e.g. *Brachiodontes erosus*) possess a cyclindrical shell, but the shell is more triangular and ventrally flattened when the species is epifaunal (Morton 1991). According to Morton (1970), the essential adaptations necessary in the development of the heteromyarian shell probably commenced in infaunal lineages, as a wide range of extant mytilids still possess features that adapt them for an endobyssate mode of life (e.g. a pallial sinus, an elongated modioliform shell and a smooth and often thin shell).

The evolution of hard calcareous coral reefs in the Mesozoic (252–66 mya) was an event that foresaw important phases of molluscan adaptive radiation (Morton 1990). In the family Mytilidae, borers of coral skeletons, rock and even wood constitute an excellent example of such a radiation. The evolution of gastropod predators is regarded as the driving force for coral–bivalve associations (Morton 1990). Exploitation of living coral as a habitat was facilitated by the evolution of larval adaptations, enabling them to penetrate living coral tissue and thereby develop mechanisms to overcome coral defences (Morton 1990). The boring life habit developed independently in both epifaunal and infaunal ancestors and also in at least five other families of clams: Gastrochaenidae, Petricolidae, Pholadidae, Clavagellidae and Tridacnidae (references in Ockelmann & Dinesen 2009). In the Mytilidae there are about 78 species of borers in six genera: *Adula* (7 species), *Botula* (6), *Fungiacava* (1), *Gregariella* (17), *Leiosolenus* (36) and *Lithophaga* (11). The majority of species use chemical means to bore into calcareous rocks or corals, while *Adula* and *Botula*, for example, are mechanical borers in softer clay and chalk (Kleemann 1990). The boring habit was made possible by the elongated shell as seen in *Adula*, *Botula* and *Lithophaga*, which probably evolved from a heteromyarian ancestor like *Modiolus* (Kleemann 1990). The exceptionally long shell ligament in the Mytilidae, with its consequent powerful opening thrust, represents another preconditioning factor that made boring possible. Coevolution between some borers and certain living corals has been suggested (Mokady *et al.* 1994).

In the chemical borer *Lithophaga*, the dorsal extension of the mantle edges was accompanied by increased mucus secretion to aid in cleansing. Protrusion of these tissues, and direct application by them of the mucous, provided the means of chemical boring (Kleemann 1990). Initially, the mucous was believed to be acidic, but histochemical tests have since shown that the secretion is a neutral mucoprotein with calcium-binding ability. This finding indicates that in *Lithophaga* the complexing of calcium by the secretion of the pallial glands is the mechanism of chemical boring into calcareous rock (Jaccarini *et al.* 1968). The boring glands in the *Lithophaga* may be derived from the cocoon-forming cells of a nest-building bivalve like *Arcuatula* (Morton 1982). According to Yonge (1955), all boring in the Mytilidae was initially mechanical. In mechanical borers such as *Botula* and *Adula*, firm attachment is made by byssal threads arranged in a large anterior and a small posterior group. These threads are much more numerous but less localised than in chemical borers. Noncalcareous rock is abraded by the dorsal surfaces of the valves, which are deeply eroded. The posterior byssal retractors and the exceptionally long ligament provide the force needed for boring. The former deepens the boring by forcing the anterior end of the shell against the rock, while the latter widens it by forcing apart the shell valves.

The result of boring activity is referred to as bioerosion (Warme 1975), which includes both bioabrasion and biocorrosion. Species of *Lithophaga* bore into dead and live coral and are most abundant subtidally, with some attacking reef corals to their lower depth limits. The siphonal openings of *Lithophaga* typically have a keyhole-like appearance on coral surfaces and the circular holes penetrate vertically into the coral skeleton, from 1 to 10 cm deep depending upon the species. Population densities in productive equatorial eastern Pacific waters range from 500 to 10 000 individuals/m^2 (Scott *et al.* 1988), which can lead to rapid reef erosion. Bioerosion also renders coral more susceptible to physical erosion (see review by Glynn & Manzello 2015).

An unusual example of adaptive radiation in the Mytilidae is one in which mussels have colonised chemosynthetic habitats such as hydrothermal and cold-seep vents in deep oceanic waters (Figure 1.6). These are places where chemical-rich fluids exude from the seafloor, often providing the energy to sustain rich communities in very harsh environments. The habitats differ from one another in the underlying conditions that form and drive them. While mussels dominate many vents and cold seeps worldwide, they are also common in other sulphide-rich habitats such as whale carcasses and sunken wood (Duperron 2010) – habitats that are regarded as stepping stones to deep-sea vents (Distel *et al.* 2000). Mussels thrive in such habitats because of the symbiotic bacteria housed in their gills. Some of these symbionts harness energy by oxidising methane to carbon dioxide, while others do so by oxidising hydrogen sulphide to sulphate (Orphan & Hoehler 2011). The sulphide-oxidising bacteria can also use hydrogen as an energy source, oxidising it to water (Petersen *et al.* 2011 and references therein). Mussels absorb nutrients synthesised by the bacteria, while they also filter-feed from surrounding waters. Genera such as *Bathymodiolus* and *Tamu* are dominant at hydrothermal vents and cold seeps, while small-sized mussels in the genera *Idas*, *Adipicola* and *Benthomodiolus* inhabit sunken wood, vegetal debris and bones on the ocean

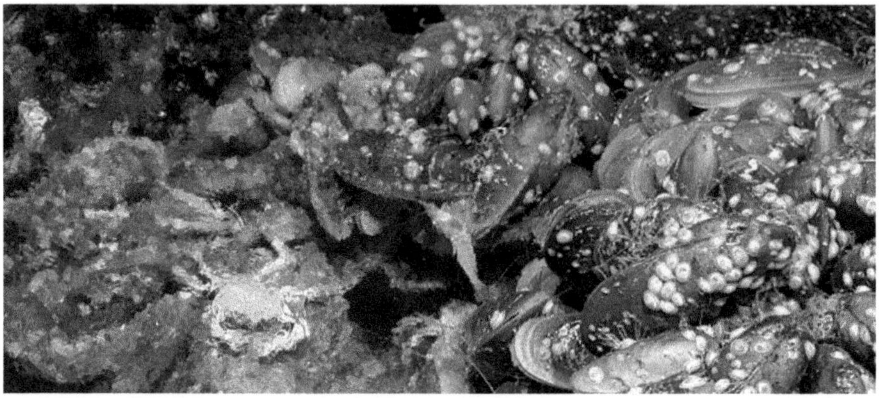

Figure 1.6 Vent mussels and associated fauna are bathed in hydrothermal fluids at the Wideawake vent field on the Mid-Atlantic Ridge. Photo taken by the remotely operated, deep-sea submersible MARUM-QUEST. *Source:* From Orphan & Hoehler (2011). Reproduced with permission from Nature Publishing Group.

floor. This decomposing organic matter produces sulphide, which is used by mussels through symbionts similar to those of vent and seep mussels (Duperron *et al.* 2009). Another genus, *Giganticus*, inhabits warm water seeps near active volcanoes in the Southern Hemisphere. In addition to a diversity of habitats, many mussels live in a dual symbiosis with both sulphur- and methane-oxidising bacteria. Indeed, as many as six distinct symbionts have been recorded in the gills of a single species of *Idas*, thereby allowing greater flexibility to the host in the use of substrates (Duperron *et al.* 2008). Mussels from chemosynthesis-based ecosystems have several features in common, including a modioliform shell, reduced labial palps and digestive system, greatly enlarged gills and planktotrophic larvae with large dispersal abilities (see Duperron 2010 for details).

Although these mussels were initially regarded as archaic, molecular age estimates and fossil records suggest that most of the modern vent and seep animals appeared during a short time interval between the Late Mesozoic and the Early Cenozoic, less than 100 mya (Van Dover *et al.* 2002). The evolutionary history of deep-sea symbiotic mussels has been investigated over the past two decades using a range of mitochondrial and nuclear DNA markers. Distel *et al.* (2000) were the first to propose the stepping stone hypothesis (i.e. that vent and seep mussels derived from ancestors associated with wood and whale fall ecosystems). Results from subsequent phylogenetic studies have strongly supported this proposal (Samadi *et al.* 2007; Miyazaki *et al.* 2010; Lorion *et al.* 2013; Thubaut *et al.* 2013). Rather than a single step toward colonisation of vents and seeps, there were multiple habitat shifts from organic substrates to vents and cold seeps, always in the one direction (Thubaut *et al.* 2013). Cold seeps seem to be an intermediate habitat for the colonisation of deep vents (Thubaut *et al.* 2013). Ancestors of *Bathymodiolus* mussels were shallow species that acquired the ability to associate with bacteria, most likely sulphur oxidisers (Duperron 2010). For mussels, the acquisition of such symbionts is seen as a prerequisite for their adaptation to, and successful radiation within, chemosynthetic environments (Lorion *et al.* 2013). Subsequent acquisition of methanotrophic symbionts allowed the colonisation of new niches within the vent and seep habitats, resulting in a second wave of diversification (Lorion *et al.* 2013). To date, there is no general consensus on the phylogenetic relationships of deep-sea *Bathymodiolus* mussels and their mytilid relatives (see Jones *et al.* 2006; Fujita *et al.* 2009; Miyazaki *et al.* 2010; Thubaut *et al.* 2013; Oliver 2015).

Notes

1 A system of classification based on the phylogenetic relationships and evolutionary history of groups of organisms, rather than purely on shared features.

2 A monophyletic group of organisms believed to comprise all the evolutionary descendants of a common ancestor.

3 The classification of organisms and the evolutionary relationships among them.

4 A polyphyletic taxon is defined as one that does not include the common ancestor of all members of the taxon.

5 A paraphyletic taxon is one that includes the most recent common ancestor but not all of its descendants.

References

Bieler, R. & Mikkelsen, P.A. (2006) Bivalvia – a look at the branches. *Zoological Journal of the Linnean Society*, **148**, 223–235.

Bieler, R., Carter, J.G. & Coan, E.V. (2010) Classification of bivalve families. In: *Nomenclator of Bivalve Families* (eds P. Bouchet & J.-P. Rocroi), pp. 113–133. Conch Books, Hackenheim.

Bieler, R., Mikkelsen, P.M., Collins, T.M., Glover, E.A., González, V.L., Graf, D.L. *et al.* (2014) Investigating the bivalve tree of life – an exemplar-based approach combining molecular and novel morphological characters. *Invertebrate Systematics*, **28**, 32–115.

Boss, K.J. (1982) Mollusca. In: *Synopsis and Classification of Living Organisms*, 1st edn (ed S.P. Parker), pp. 945–1166. McGraw-Hill, New York.

Cannuel, R., Beninger, P.G., McCombie, H. & Boudry, P. (2009) Gill development and its functional and evolutionary implications in the blue mussel *Mytilus edulis* (Bivalvia: Mytilidae). *Biological Bulletin*, **217**, 173–188.

Carter, J.G., Campbell, D.C. & Campbell, M.R. (2000) Cladistic perspectives on early bivalve evolution. In: *The Evolutionary Biology of the Bivalvia* (eds E.M. Harper, J.D. Taylor & J.A. Crame), pp. 47–79. Geological Society, London.

Carter, J.G., Altaba, C.R., Anderson, L.C., Araujo, R., Biakov, A.S., Bogan, A.E. *et al.* (2011) A synoptical classification of the Bivalvia (Mollusca). *Paleontological Contributions*, **4**, 1–47.

Coan, E.V. & Valentich-Scott, P. (2012) *Bivalve Seashells of Tropical West America. Marine Bivalve Mollusks from Baja California to Northern Peru*. Santa Barbara Museum of Natural History, Santa Barbara, CA.

Coan, E.V., Valentich-Scott, P. & Bernard, F.R. (2000) *Bivalve Seashells of Western North America. Bivalve Mollusks from Arctic Alaska to Baja California*. Santa Barbara Museum of Natural History, Santa Barbara, CA.

Cope, J.C.W. (2000) A new look at early bivalve phylogeny. In: *The Evolutionary Biology of the Bivalvia* (eds E.M. Harper, J.D. Taylor & J.A. Crame), pp. 81–95. Geological Society, London.

Distel, D.L. (2000) Phylogenetic relationships among Mytilidae (Bivalvia): 18S rRNA data suggest convergence in mytilid body plans. *Molecular Phylogenetics and Evolution*,**15**, 25–33.

Distel, D.L., Baco, A.R., Chuang, E., Morrill, W., Cavanaugh, C. & Smith, C.R. (2000) Marine ecology: do mussels take wooden steps to deep-sea vents? *Nature*, **403**, 725–726.

Duperron, S. (2010) The diversity of deep-sea mussels and their bacterial symbioses. In: *The Vent and Seep Biota* (ed S. Kiel), pp. 137–167. Springer, New York.

Duperron, S., Halary, S., Lorion, J., Sibuet, M. & Gaill F. (2008) Unexpected co-occurrence of six bacterial symbionts in the gills of the cold seep mussel *Idas* sp. (Bivalvia: Mytilidae). *Environmental Microbiology*, **10**, 433–445.

Duperron, S., Lorion, J., Samadi, S., Gros, O. & Gaill, F. (2009) Symbioses between deep-sea mussels (Mytilidae: Bathymodiolinae) and chemosynthetic bacteria: diversity, function and evolution. *Comptes Rendus Biologies*, **332**, 298–310.

Fujita, Y., Matsumoto, H., Fujiwara, Y., Hashimoto, J., Galkin, S.V., Ueshima, R. *et al.* (2009) Phylogenetic relationships of deep-sea *Bathymodiolus* mussels to their mytilid relatives from sunken whale carcasses and wood. *Venus*, **67**, 123–134.

Giribet, G. (2008) Bivalvia. In: *Phylogeny and Evolution of the Mollusca* (eds W. Ponder & D.R. Lindberg), pp. 105–141. University of California Press, Berkeley, CA.

Giribet, G. & Wheeler, W.C. (2002) On bivalve phylogeny: a high-level analysis of the Bivalvia (Mollusca) based on combined morphology and DNA sequence data. *Invertebrate Biology*, **121**, 271–324.

Glynn, P.W. & Manzello, D.P. (2015) Bioerosion and coral reef growth: a dynamic balance. In: *Coral Reefs in the Anthropocene* (ed C. Birkeland), pp. 67–97. Springer, New York.

González, V.L. & Giribet, G. (2014) A multilocus phylogeny of archiheterodont bivalves (Mollusca, Bivalvia, Archiheterodonta). *Zoologica Scripta*, **44**, 41–58.

González, V.L., Andrade, S.C.S., Bieler, R., Collins, T.M., Dunn, C.W., Mikkelsen, P.M. *et al.* (2015) A phylogenetic backbone for Bivalvia: an RNA-seq approach. *Proceedings of the Royal Society B*, **282**, 20142332.

Harper, E.M. (1991) The role of predation in the evolution of cementation in bivalves. *Palaeontology*, **34**, 455–460.

Harper, E.M., Taylor, J.D. & Crame, J.A. (eds) (2000) *The Evolutionary Biology of the Bivalvia*. Geological Society London, Special Publications, **177**.

Haszprunar, G. (2008) Monoplacophora (Tryblidia) In: *Phylogeny and Evolution of the Mollusca* (eds W. Ponder & D. R. Lindberg), pp.97–104. University of California Press, Berkeley, CA.

Haszprunar, G., Schander, C. & Halanych, K.M. (2008) Relationships of higher miolluscan taxa. In: *Phylogeny and Evolution of the Mollusca* (eds W. Ponder & D.R. Lindberg), pp. 19–32. University of California Press, Berkeley, CA.

Jaccarini, V., Bannister, W.H. & Micallef, H. (1968) The pallial glands and rock boring in *Lithophaga lithophaga* (Lamellibranchia, Mytilidae). *Journal of Zoology*, **154**, 397–401.

Jones, W.J., Won, Y.J., Maas, P.A.Y., Smith, P.J., Lutz, R.A. & Vrijenhoek, R.C. (2006) Evolution of habitat use by deep-sea mussels. *Marine Biology*, **148**, 841–851.

Kleemann, K. (1990) Evolution of chemically boring Mytilidae (Bivalvia). In: *The Bivalvia. Proceedings of a Memorial Symposium in Honour of Sir Charles Maurice Yonge, Edinburgh, 1986* (ed B. Morton), pp. 111–124. Hong Kong University Press, Hong Kong.

Kocot, K.M. (2013) Recent advances and unanswered questions in deep molluscan phylogenetics. *American Malacological Bulletin*, **31**, 195–208.

Kocot, K.M., Cannon, J.T., Todt, C., Citarella, M.R., Kohn, A.B., Meyer, A. *et al.* (2011) Phylogenomics reveals deep molluscan relationships. *Nature*, **477**, 452–457.

Liu, J., Liu, H. & Zhang, H. (2018) Phylogeny and evolutionary radiation of the marine mussels (Bivalvia: Mytilidae) based on mitochondrial and nuclear genes. *Molecular Phylogenetics and Evolution*, **126**, 233–240.

Lorion, J., Kiel, S., Faure, B., Kawato, M., Ho, S.Y.M., Marshall, B. *et al.* (2013) Adaptive radiation of chemosymbiotic deep-sea mussels. *Proceedings of the Royal Society B*, *280*, 1–9.

Matsumoto, M. (2003) Phylogenetic analysis of the subclass Pteriomorphia (Bivalvia) from mtDNA COI sequences. *Molecular Phylogenetics and Evolution*, **27**, 429–440.

Miyazaki, J-I., de Oliveira Martins, D., Fujita, Y., Matsumoto, H. & Fujiwara, Y. (2010) Evolutionary process of deep-sea *Bathymodiolus* mussels. *PLOS One*, **5**, e10363.

Mokady, O., Rozenblatt, S., Graur, D. & Loya, Y. (1994) Coral-host specificity of Red Sea *Lithophaga* bivalves: interspecific and intraspecific variation in 12S mitochondrial ribosomal RNA. *Molecular Marine Biology and Biotechnology*, **3**, 158–164.

Morton, B. (1970) The evolution of the heteromyarian condition in the Dreissenacea (Bivalvia). *Palaeontology*, **13**, 563–572.

Morton, B. (1982) The mode of life and functional morphology of *Gregariella coralliophaga* (Gmelin 1791) (Bivalvia: Mytilidae) with a discussion on the evolution of the boring Lithophaginae and adaptive radiation in the Mytilidae. In: *Proceedings of the First International Marine Biological Workshop: the Marine Flora and Fauna of Hong Kong and southern China, Hong Kong, 1980* (eds B. Morton & C.K. Tseng), pp. 875–895. Hong Kong University Press, Hong Kong.

Morton, B. (1987) The functional morphology of the organs of the mantle cavity of *Perna viridis* (Linneaus, 1758) (Bivalvia: Mytilacea). *American Malacological Bulletin*, **5**, 159–164.

Morton, B. (1990) Bivalvia. In: *Encyclopedia Brittanica. Vol. 24*, pp. 321–327. Encyclopedia Brittanica Inc., Chicago, IL.

Morton, B. (1991) The biology of *Brachidontes erosus* (Bivalvia: Mytilidae) in Princess Royal Harbour, Albany, Western Australia. In: *Proceedings of the Third International Marine Biological Workshop: The Marine Flora and Fauna of Albany, Western Australia, 1988* (eds F.E. Wells, D.I. Walker, H. Kirkham & R. Lethbridge), pp. 693–712. Western Australian Museum, Perth, WA.

Morton, B. (1992) The evolution and success of the heteromyarian form in the Mytiloida. In: *The Mussel Mytilus: Ecology, Physiology, Genetics and Culture* (ed E.M. Gosling), pp. 21–52. Elsevier Science Publishers B.V., Amsterdam.

Morton, B. (1996) The evolutionary history of the Bivalvia. In: *Origin and Evolutionary Radiation of the Mollusca* (ed J.D. Taylor), pp. 337–359. Oxford University Press, Oxford.

Morton, B. (2015) Evolution and adaptive radiation in the Mytiloidea (Bivalvia): clues from the pericardial–posterior byssal retractor musculature complex. *Molluscan Research*, **35**, 227–245.

Morton, B., Dinesen, G.E. & Ockelmann, K.W. (2016) Functional morphology, biology and sexual strategy of the circumboreal, adventitious crypt-building, *Crenella decussata* (Bivalvia: Mytiloidea: Crenellidae). *Journal of the Marine Biological Association of the UK*, **96**, 1597–1616.

Ockelmann, K.W. & Dinesen, G.E. (2009) Systematic relationship of the genus *Adula* and its descent from a *Mytilus*-like ancestor (Bivalvia, Mytilidae, Mytilinae). *Steenstrupia*, **30**, 141–152.

Oliver, P.G. (2015) Description and morphology of the 'Juan de Fuca vent mussel', *Benthomodiolus erebus* sp. n. (Bivalvia, Mytilidae, Bathymodiolinae): 'Phylogenetically basal but morphologically advanced'. *Zoosystematics and Evolution*, **91**, 151–165.

Orphan, V.J. & Hoehler, T.M. (2011) Hydrogen for dinner. *Nature*, **476**, 154–155.

Petersen, J.M., Zielinski, F.U., Pape, T., Seifert, R., Moraru, C., Amann, R. *et al.* (2011) Hydrogen is an energy source for hydrothermal vent symbioses. *Nature*, **476**, 176–180.

Plazzi, F. & Passamonti, M. (2010) Towards a molecular phylogeny of mollusks: bivalves' early evolution as revealed by mitochondrial genes. *Molecular Phylogenetics and Evolution*, **57**, 641–657.

Plazzi, F., Ceregato, A., Taviani, M. & Passamonti, M. (2011) A molecular phylogeny of bivalve mollusks: ancient radiations and divergences as revealed by mitochondrial genes. *PLOS One*, **6**, e27147.

Ponder, W. & Lindberg, D.R. (2008) Molluscan evolution and phylogeny an introduction. In: *Phylogeny and Evolution of the Mollusca* (eds W. Ponder & D.R. Lindberg), pp. 1–17. University of California Press, Berekely, CA.

Samadi, S., Quéméré, E., Lorion, J., Tillier, A., von Cosel, R. & Lopez, P. *et al.* (2007) Molecular phylogeny in mytilids supports the wooden steps to deep-sea vents hypothesis. *Comptes Rendus Biologies*, **330**, 446–456.

Schrödl, M. & Stöger, I. (2014) A review on deep molluscan phylogeny: old markers, integrative approaches, persistent problems. *Journal of Natural History*, **48**, 2773–2804.

Scott, P.J.B., Risk, M.J. & Carriquiry, J.D. (1988) El Niño, bioerosion and the survival of east Pacific reefs. In: *Proceedings of the 6th International Coral Reef Symposium, Townsville Australia* (eds J.H. Choat, D. Barnes, M.A. Borowitzka, J.C. Coll, P.J. Davies, P. Flood, *et al.*), pp. 517–520. International Coral Reef Symposium, Townsville, QLD.

Seed, R., Richardson, C.A. & Smith, K. (2000) Marine mussels, their evolutionary success, ecological significance and use as chronometers of environmental change. In: *The Evolutionary Biology of the Bivalvia* (eds E.M. Harper, J.D. Taylor & J.A. Crame), pp. 465–478. Geological Society, London.

Sharma, P.P., González, V.L., Kawauchi, G.Y., Andrade, S.C.S., Guzmán, A., Collins, T.M. *et al.* (2012) Phylogenetic analysis of four nuclear protein-encoding genes largely corroborates the traditional classification of Bivalvia (Mollusca). *Molecular Phylogenetics and Evolution*, **65**, 64–74.

Sigwart, J.D. & Sutton, M.D. (2007) Deep molluscan phylogeny: synthesis of palaeontological and neontological data. *Proceedings of the Royal Society B*, **274**, 2413–2419.

Sigwart, J.D. & Lindberg, D.R. (2015) Consensus and confusion in molluscan trees: evaluating morphological and molecular phylogenies. *Systematic Biology*, **64**, 384–395.

Smith, S.A., Wilson, N.G., Goetz, F.E., Feehery, C., Andrade, S.C.S., Rouse, G.W. *et al.* (2011) Resolving the evolutionary relationships of molluscs with phylogenomic tools. *Nature*, **480**, 364–369.

Soot-Ryen, T. (1955) A report on the family Mytilidae (Pelecypoda). *Allan Hancock Pacific Expeditions*, **20**, 1–175.

Stöger, I., Sigwart, J.D., Kano, Y., Knebelsberger, T. & Marshall, B.A. (2013) The continuing debate on deep molluscan phylogeny: evidence for Serialia (Mollusca, Monoplacophora + Polyplacophora). *BioMed Research International*, **2013**, art. 407072.

Sun, W. & Gao, L. (2017) Phylogeny and comparative genomic analysis of Pteriomorphia (Mollusca: Bivalvia) based on complete mitochondrial genomes. *Marine Biology Research*, **13**, 255–268.

Telford, M.J. & Budd, G.E. (2011) Invertebrate evolution: bringing order to the molluscan chaos. *Current Biology*, **21**, R964–R966.

Thubaut, J., Puillandre, N., Faure, B., Cruaud, C. & Samadi, S. (2013) The contrasted evolutionary fates of deep-sea chemosynthetic mussels (Bivalvia, Bathymodiolinae). *Ecology and Evolution*, **3**, 4748–4766.

Tsubaki, R., Kameda, Y. & Kato, M. (2011) Pattern and process of diversification in an ecologically diverse epifaunal bivalve group Pterioidea (Pteriomorphia, Bivalvia). *Molecular Phylogenetics and Evolution*, **58**, 97–104.

Van Dover, C.L., German, C.R., Speer, K.G., Parson, L.M. & Vrijenhoek, R.C. (2002) Evolution and biogeography of deep-sea vent and seep invertebrates. *Science*, **295**, 1253–1257.

Vermeij, G.J. (1977) The Mesozoic marine revolution: evidence from snails, predators and grazers. *Paleobiology*, **3**, 245–258.

Voight, J.R. (2007) Experimental deep-sea deployments reveal diverse Northeast Pacific wood-boring bivalves of Xylophagainae (Myoida: Pholadidae). *Journal of Molluscan Studies*, **73**, 377–391.

Warme, J.E. (1975) Borings as trace fossils, and the process of marine erosion. In: *The Study of Trace Fossils* (ed R.W. Frey), pp. 181–227. Springer, New York.

Yuan, Y., Li, Q., Yu, H. & Kong, L. (2012) The complete mitochondrial genomes of six heterodont bivalves (Tellinoidea and Solenoidea): variable gene arrangements and phylogenetic implications. *PLOS One*, **7**, e32353.

Yonge, C.M. (1941) The Protobranchiata Mollusca: a functional interpretation of their structure and evolution. *Philosophical Transactions of the Royal Society B*, **230**, 79–147.

Yonge, C.M. (1955) Adaptation to rock boring in *Botula* and *Lithophaga* (Lamellibranchia, Mytilidae) with a discussion on the evolution of this habit. *Quarterly Journal of Microscopical Science*, **96**, 383–410.

Zardus, J.D. (2002) Protobranch bivalves. *Advances in Marine Biology*, **42**, 1–65.

2

Functional Morphology

Introduction

In this chapter, the approach is to use much studied representative species of marine mussels to describe the general morphology and functions of the shell, mantle, foot, gill, alimentary canal, gonad, heart, kidney and nervous tissue. Additional information on their particular roles in feeding, reproduction, circulation, excretion, osmoregulation and contaminant accumulation is presented in Chapters 4, 5, 7 and 8.

Shell

Mussels have two shell valves that are hinged dorsally and connected by an elastic ligament. Adductor muscles hold the valves together, and relaxation of the ligament and contraction of these muscles open and close the shell, respectively. In *Mytilus* spp., the two shell valves are similar in size, and are roughly triangular in shape (Figure 2.1a). The valves are hinged together at the anterior by means of a ligament. This area of the shell is called the umbo. A series of interlocking teeth and sockets along the hinge line prevent the valves from sliding against one another. The interior of the shell is white with a broad border of purple or dark blue, while in *Perna viridis* the shell's interior has a pale blue sheen. The border is called the pallial line and is the part of the shell along which the mantle is attached when empty shells are examined (Figure 2.1b). On the inside of each valve are two muscle scars, the attachment points for the large posterior adductor muscle and the much-reduced anterior adductor muscle. All *Perna* species lack an anterior adductor muscle, so the shell anterior has only one muscle scar. Anterior and posterior retractor muscles are also attached to the inner shell; these control the movement of the foot (see later). The foot in turn secretes a byssus, a bundle of tough threads of tanned protein. These threads emerge through the ventral part of the shell and serve as mooring lines for attachment of the mussel to the substrate and to other mussels. Details of shell characteristics in selected mussel species are presented in Table 2.1. In summary, most mussel species have an equivalve, oblong, dark-coloured shell with a smooth or ribbed surface marked with concentric growth lines; the inner shell is often iridescent white to purple, with two adductor scars and a variable number of hinge teeth.

Marine Mussels: Ecology, Physiology, Genetics and Culture, First Edition. Elizabeth Gosling.

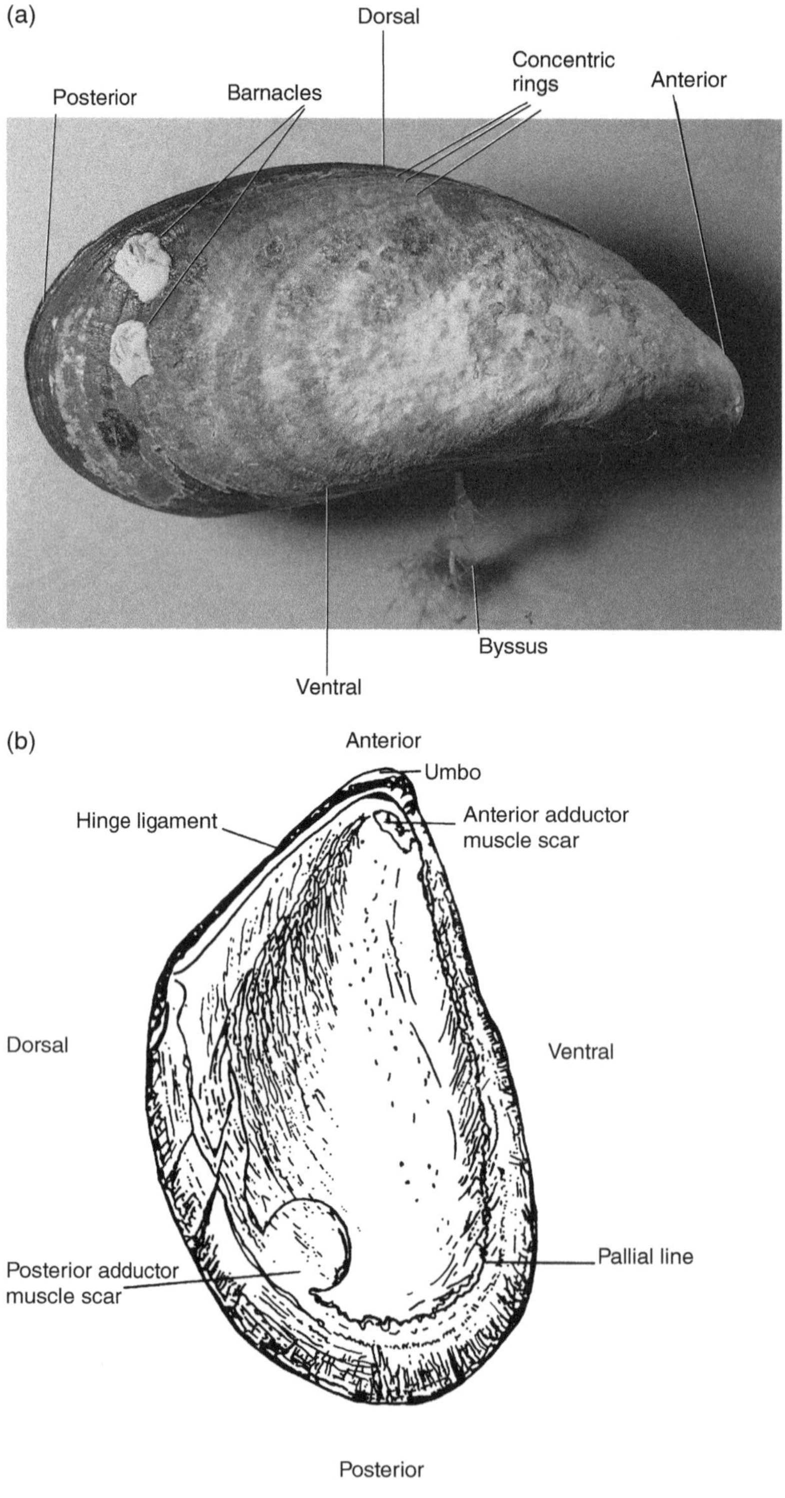

Figure 2.1 (a) External and (b) internal shell features of the mussel *Mytilus edulis*. *Source:* Photograph and permission to reproduce from Craig Burton.

Table 2.1 Details of shell characteristics in selected mussel species. (www.fao.org/fishery/species/search/en). *Sources:* Adapted from information from species databases: Centre for Agriculture and Biosciences International (www.cabi.org), Encyclopedia of Life (eol.org), Invasive Species Specialist Group (www.issg.org) Sanctuary Integrated Monitoring Network (SIMoN; www.sanctuarymonitoring.org) and FAO Aquatic Species Fact Lists (www.fao.org/fishery/species/search/en).

Species	Shell exterior	Shell interior	Maximum shell length
Arcuatula senhousia	Olive-green to brown with darker radial lines or zig-zag markings; umbones subterminal	Lustrous purple-grey; no hinge teeth	35 mm
Aulacomya atra	Brown ribbed eqivalve shell, which darkens to black with age; pointed umbo; concentric growth grooves	Nacreous; hinge with a single tooth that fits in the groove of the opposed shell	200 mm
Bathymodiolus childressi	Thin, fragile equivalve shell; dark brown to straw-yellow	Nacreous, off-white colour; no hinge teeth in individuals >11 mm shell length; muscle scars and pallial line indistinct	120 mm
Geukensia demissa	Solid, oblong shell with blunt umbones; glossy yellow or brown-black with grooved or ribbed surface	Iridescent, white with purple tints; no hinge teeth	100 mm
Modiolus modiolus	Purple or dark blue; annual growth lines clear with fine sculpturing of concentric grooves and ridges	White with a broad pallial line	220 mm
Mytella charruana	Mostly dark brown to black exhibiting a wavy dark pattern on a lighter background; concentric growth rings evident on shell surface	Iridescent purple	40 mm
Mytilus californianus	Brown to dark purple, bluish or black and often eroded; teardrop shaped inequivalve shell with several strong concentric grooves and weak radial lines	Blue-gray; may be slightly iridescent	250 mm
Mytilus galloprovincialis	Black-violet colour; equivalve shell, nearly quadrangular, with one valve pointed with slightly bent umbo and other rounded	Mantle edge blue or purple	120 mm
Perna viridis	Equivalve shell; vivid green to dark brownish-green near the dorsal edge and olive-green near the umbo, which has a downturned beak	Surface iridescent blue-green hue; three interlocking hinge teeth; wavy pallial line and large kidney-shaped adductor muscle, both diagnostic for the species	150 mm

Structure

The main component of the shell is calcium carbonate, formed by the deposition of crystals of this salt in an organic matrix composed of polysaccharide β-chitin, a hydrophobic silk protein and a complex assemblage of hydrophilic proteins, many rich in aspartic acid (Addadi *et al.* 2006). The mineral component represents more than 95% of the shell weight, while the organic matrix represents 1–5%. Despite this, the organic component increases the fracture toughness of the shell by two or three orders of magnitude (Kamat *et al.* 2000). Three layers make up the shell: (1) a thin outer periostracum of conchiolin, a tough insoluble protein forming the organic matrix of the shell, within which calcium carbonate is deposited – this layer is often a different colour to the underlying shell layers; (2) a middle prismatic layer composed of fibres of aragonite or calcite – both crystalline forms of calcium carbonate (Figure 2.2); and (3) an inner nacreous layer composed of tablets of aragonite arranged in layers separated by sheets of an organic matrix of elastic biopolymers composed of chitin and silk-like proteins – the nacre is either of a dull texture or iridescent mother-of-pearl, depending on the species.

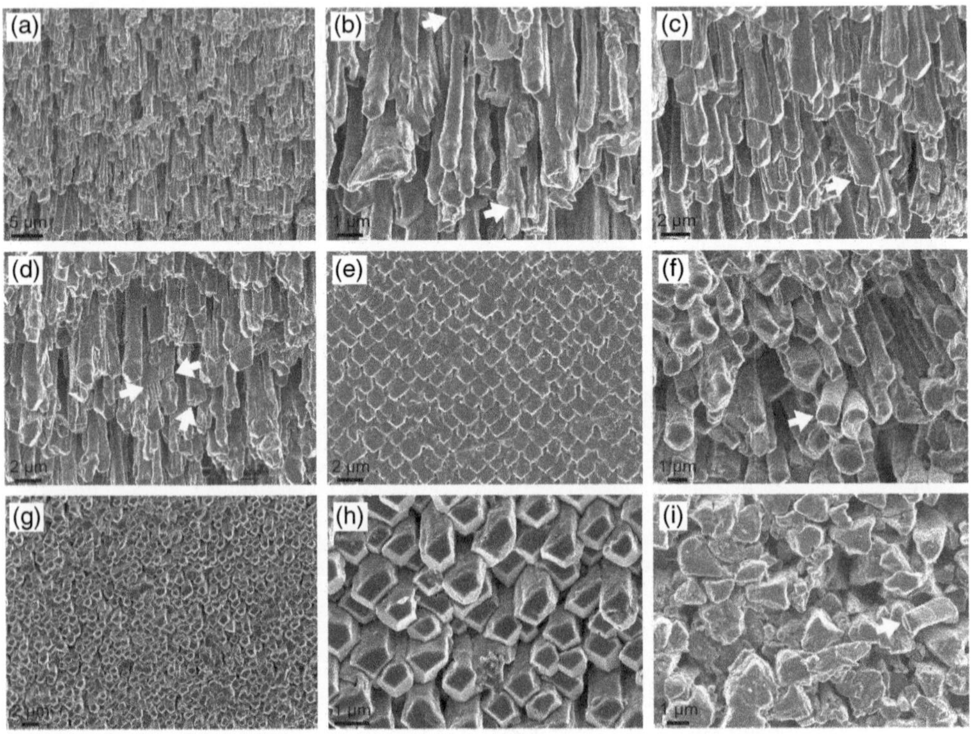

Figure 2.2 Calcitic fibrous layers of adult mytilids. (a–d) Fibrous layer of *Bathymodiolus azoricus*. (a) General view showing high irregularity in size of fibres. (b–d) Details of fibres. The arrows point to (b) newly formed, (c) twisting or (b,d) splitting fibres. (e,f) *Choromytilus chorus*. (e) General view of the layer, showing high regularity in size and distribution. (f) Detail of the same layer close to the shell margin. The arrow points to bending fibres. (g,h) *Mytilus edulis platensis* (now accepted as *M. platensis* by World Register of Marine Species, WoRMS; www.marinespecies.org). General view and detail of the fibrous layer. Note the even orientation of rhombohedral faces in (h). (i) *Mytilus californianus*. Fibres are characterised by irregular outlines and endings. The arrow indicates a bent fibre. *Source:* From Checa et al. (2014). Reproduced with permission from Schweizerbart Science Publishers.

A range of techniques, such as scanning electron microscopy (SEM), transmission electron microscopy (TEM), X-ray diffraction (XRD), electron backscatter diffraction (EBSD) and atomic force microscopy (AFM), have played an important role in clarifying the microstructure of the different shell layers (Chateigner *et al.* 2000; Furuhashi *et al.* 2009; Checa *et al.* 2014; Nakamura Filho *et al.* 2014), while analytical methods such as secondary ion mass spectrometry (SIMS; Shirai *et al.* 2008), electron probe microanalysis (EPMA; Jacob *et al.* 2008) and laser ablation inductively coupled plasma mass spectrometry (LA-ICP-MS; Jacob *et al.* 2008) have been used to elucidate chemical composition of the shell in several species. In addition, the NanoSIMS microprobe allows trace element imaging and quantification in bivalve shells (Shirai *et al.* 2008).

The construction of the shell begins very early in larval development. An area of the ectoderm thickens in the dorsal region of the developing embryo. The area invaginates to become a shell gland, which forms a groove, which eventually becomes the future ligament between the two shell valves (Marin & Luquet 2004). The peripheral cells of the shell gland produce an extracellular lamella, the future periostracum, which will serve as a scaffold for the developing shell. Subsequently, the shell gland everts and the shell field spreads by flattening of the cells and mitotic divisions, thus becoming the calcifying mantle. Between the periostracum and the cells of the shell field, primary mineralisation takes place. The first larval shell is the prodissoconch I stage, followed by the prodissoconch II stage and then the dissoconch stage after metamorphosis (details in Chapter 5). At the prodissoconch I stage, the mineral produced is usually amorphous calcium carbonate, followed by either aragonite or calcite at prodissoconch II (Marin & Luquet 2004). The sequence can vary. For example, in the oyster, *Ostrea edulis*, the first mineral deposited is calcite, followed by aragonite at the prodissoconch II stage; at the dissoconch stage, the fraction of calcite rapidly increases and that of aragonite decreases (Medaković *et al.* 1997). By the time the juvenile stage is reached, the shell is heavily calcified and has different pigmentation and more conspicuous sculpturing than the larval shell.

Shell formation in juveniles and adults involves three separate elements: the mantle and its outer epithelium, the periostracum and the interface between the outer epithelium, the periostracum and the growing shell (Marin *et al.* 2012). Distinct areas of the outer epithelium secrete prismatic and nacre layers. Also, the epithelium contains membrane pumps and channels for extruding the inorganic precursors of calcium carbonate such as calcium (Ca^{2+}) and bicarbonate (HCO_3^-). Calcium and bicarbonate ions are taken up from food, from water filtration activity or by passive diffusion over the body surface; bicarbonate may also come from hydration of metabolic carbon dioxide.

The outer mantle epithelium is not in direct contact with the shell but is separated from it by the minute fluid-filled extrapallial space (ES). Calcium and bicarbonate ions are transported in the haemolymph to the calcifying epithelium, where they are stored as granules and pumped into the ES when needed. In addition to these precursor ions, the fluid contains other inorganic ions, minor elements, proteins and glycosaminoglycans (GAGs) (Marin *et al.* 2012 and references therein). As the extrapallial fluid is supersaturated, macromolecules, especially acidic proteins and GAGs, maintain calcium in solution by inhibiting the precipitation of calcium carbonate and by allowing it to precipitate where needed.

The edge of the mantle has three folds (Figure 2.3). The ridge between the outer and middle fold is called the periostracal groove; here, specialised cells secrete the periostracum and prismatic layers, while the inner nacreous layer is secreted by the general mantle

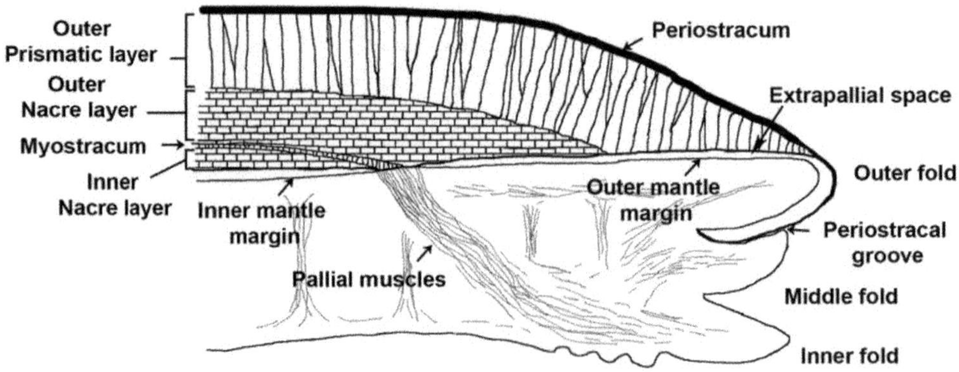

Figure 2.3 Schematic representation of the mytilid shell margin. *Source:* After Taylor *et al.* (1969). From Génio *et al.* (2012). Reproduced with permission from Elsevier.

surface. The myostracum is a very thin mineral layer covering the nacre, where muscles attach to the shell. The shell grows in circumference by the addition of material from the edge of the mantle and grows in thickness by deposition from the general mantle surface. The functions of the periostracum are to act as a first support for calcium carbonate crystals, to protect the shell and to serve as a seal of the ES in such a way that supersaturation conditions – a prerequisite for crystal formation – can be attained (Marin & Luquet 2004). In adults, the periostracum is often much reduced due to mechanical abrasion, fouling organisms, parasites or disease.

While the organic fraction (matrix) of the shell is small (1–5%), it is comprised of a mixture of extracellular macromolecules that play an important role in the mineralisation process. A wide variety of shell matrix proteins (SMPs) have been identified in molluscs, including mussels, but to date their specific functions have not been elucidated (Marin *et al.* 2012; Gao *et al.* 2015; Yarra *et al.* 2016). However, several candidate genes coding for nacreous and prismatic layer proteins have been identified in bivalves (Inoue *et al.* 2010; Jackson *et al.* 2010; Kinoshita *et al.* 2011; Hüning *et al.* 2016; Bjärnmark *et al.* 2016 and references therein). A wide range of enzyme proteins are expressed during the formation of the shell, including carbonic anhydrase, alkaline phosphatase, DOPA-oxidase (tyrosinase)/peroxidase and chitin synthase. After proteins, polysaccharides are the next most important component of the organic matrix; for example, chitin is a long-chain insoluble polymer that plays a key role by defining the interlamellar matrix between nacre tablets (Addadi *et al.* 2006). Soluble acidic polysaccharides, many of which are bound to protein, are also present in the matrix, but their characterisation in mussels is still in its infancy.

External Characteristics

Shell colour in mussels is controlled by one (Gantsevich & Tyunnikova 2005) or several (Innes & Haley 1977; Newkirk 1980) genes, and also varies depending on the age and location of the animal. In the intertidal zone, *M. edulis* has a blue-black and heavy shell, while in the sublittoral region, where mussels are continuously submerged, the shell is thin and brown with dark brown to purple radial markings. In *Perna perna*, the shell is red-maroon with irregular patches of light brown and green. Juvenile green mussels, *P. viridis*, have bright green or blue-green shells, but older individuals tend to have more brown (Siddall 1980).

Presence of concentric rings on the shell exterior has been extensively used in age determination. In scallops and clams, these rings are annual in origin and therefore can be used as a reliable estimate of age, but in mussels there are few geographic locations where they provide an accurate estimate (Lutz 1976). Age, however, can be determined by examining annual growth bands in sections of the prismatic and inner nacreous shell layers or in other parts of the shell umbo, hinge plate, pallial line scar or posterior adductor muscle scar (see Chapter 6).

The principal parameters used in shell measurement are: height, the maximum distance from the hinge to the shell margin; length, the widest part of the shell at 90° to the height; and width, the thickest part of the two shell valves (Figure 2.4). Under optimal conditions, such as in the sublittoral zone, *M. edulis* and the Mediterranean mussel, *M. galloprovincialis*, attain a shell length of 100–130 mm, whereas in marginal conditions, such as the high intertidal zone on an exposed shore, mussels may measure as little as 20–30 mm, even after 15–20 years (Seed 1976). This is not however a universal pattern. In South Africa, the native mussel *Perna perna* is largest on more exposed shores whereas the invasive mussel *M. galloprovincialis* is largest at intermediate levels of shore exposure (McQuaid *et al.* 2000; Hammond & Griffiths 2004). Shell shape is also very variable in these two mussel species. The shells of densely packed mussels have higher length to height ratios than those from less crowded conditions. This is most extreme in older mussels and ensures that they can more readily exploit posterior feeding currents, since they are effectively elevated above younger mussels in the same clump (Seed & Suchanek 1992). Density also has a negative effect on shell thickness in the intertidal mussel *Perumytilus purpuratus* (Briones *et al.* 2014).

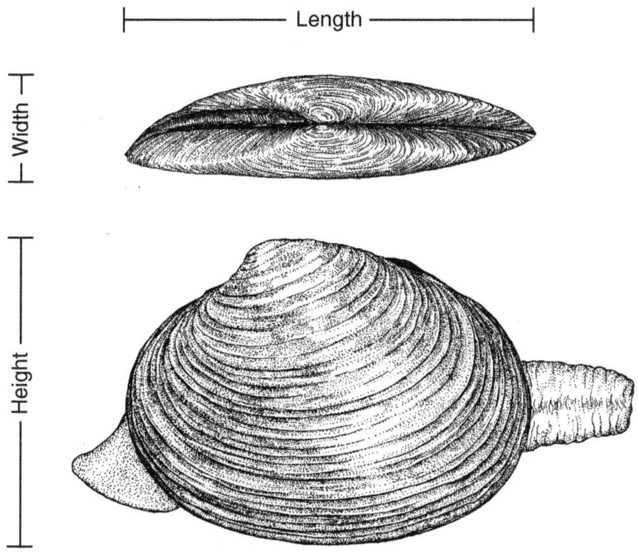

Figure 2.4 The convention used for the main external shell parameters in bivalves. *Source:* Sandra Noel, http://www.noeldesigninterp.com. Reproduced with permission.

Shell morphology can also be correlated with wave exposure; on the west coast of Canada, both juvenile and adult *M. trossulus* at wave-exposed sites show a lower shell height to width ratio and a thicker shell than mussels from sheltered locations (Akester & Martel 2000). Shell shape, as well as internal features of the shell, have also been used to differentiate both within and between various mytilid species (Innes & Bates 1999; Aguirre *et al.* 2006; Krapivka *et al.* 2007; Gardner & Thompson 2009; Valladares *et al.* 2010; Bonel *et al.* 2013; Van der Molen *et al.* 2013; Lajus *et al.* 2015; Katolikova *et al.* 2016). Figure 2.5 illustrates 18 shell characters which when used in combination give good separation of *M. edulis, M. galloprovincialis* and *M. trossulus* in the Northern Hemisphere (McDonald *et al.* 1991). See Chapter 6 for more on the methods used in shell shape analysis.

The shells of mussels, and other bivalves, are increasingly being used as potential bio-archives of proxies for physicochemical changes in their marine habitats. In marine mussels, isotopes ($\delta^{18}O$, $\delta^{13}C$) and rare earth element values, along with Mg/Ca and Sr/C ratios in shells, have been used to reconstruct changes in surface water salinity (Hahn *et al.* 2012), temperature (Cusack *et al.* 2008; Bau *et al.* 2010; Ford *et al.* 2010) and freshwater inputs to the ocean (McConnaughey & Gillikin 2008) (see further details in Chapter 3). This type of

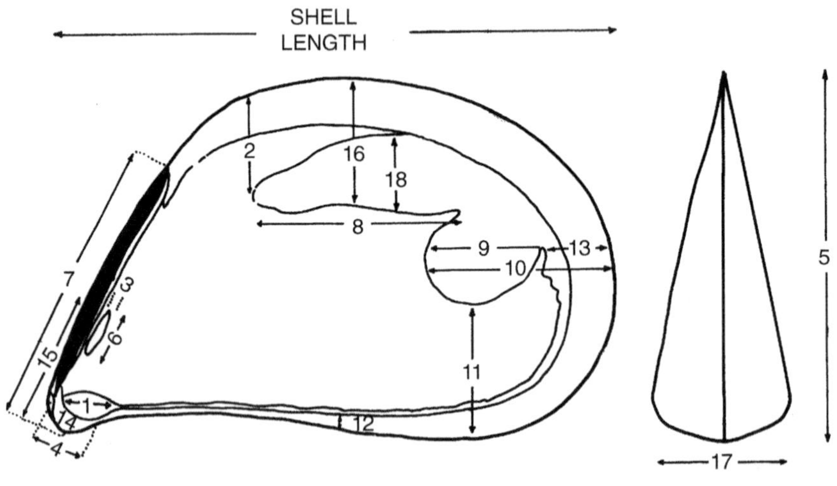

Figure 2.5 Shell morphological characters used to distinguish between different *Mytilus* taxa. (1) Length of the anterior adductor muscle scar. (2) Distance between the anterior end of the posterior retractor muscle scar and the dorsal shell margin. (3) Width of the anterior adductor muscle scar. (4) Length of the hinge plate. (5) Shell height. (6) Length of the anterior retractor muscle scar. (7) Distance between the umbo and the posterior end of the ligament. (8) Length of the posterior retractor muscle scar. (9) Length of the posterior adductor muscle scar. (10) Distance between the anterior edge of the posterior adductor muscle scar and the posterior shell margin. (11) Distance between the ventral edge of the posterior adductor muscle scar and the ventral shell margin. (12) Distance between the pallial line and the ventral shell margin midway along the shell. (13) Distance between the posterior edge of the posterior adductor muscle scar and the posterior shell margin. (14) Number of major teeth on the hinge plate, excluding any small crenulations, which may be present, especially on the posterior ventral face of the hinge plate. (15) Distance between the umbo and the posterior end of the anterior retractor scar. (16) Distance between the ventral edge of the posterior retractor muscle scar and the dorsal shell margin. (17) Shell width. (18) Width of the posterior retractor muscle scar. *Source:* From McDonald *et al.* (1991). Reproduced with permission from Springer.

research is particularly relevant in the light of observed and potential climate change impacts on marine organisms (see Chapter 3). Another use of shells is in trace elemental fingerprinting, which assesses larval origins or trajectories based on the elemental composition of their shells, which in turn reflects the chemistry of the water in which they were formed (see Chapter 5). The technique has also been used to identify population sources and thus redistribution potentials in adult *M. edulis* (Sorte *et al.* 2013).

Finally, mussel shells are extensively used to assess environmental contamination (Bellotto & Miekeley 2007; Pereira *et al.* 2012; see Chapter 8). Radionucleotides (e.g. uranium) and metals such as Cu, Cd, Cr, Pb, U, V and Zn will be highly concentrated in contaminated shells (Widdows & Donkin 1992; Boisson *et al.* 1998; Avelar *et al.* 2000; Richardson *et al.* 2001).

Mantle

Structure

In bivalves, the mantle consists of two lobes of tissue that completely enclose the animal within the shell (Figure 2.6). Between the mantle and the internal organs is a capacious mantle cavity. Unlike in other marine bivalves, the mantle in mussels contains most of the gonad. Gametes proliferate within the mantle and are carried along ciliated channels to paired gonoducts that discharge through the exalant opening of the mantle (see Chapter 5). The colour of the mantle varies from a creamy white to pink, brown or orange, depending on the stage of gametogenesis. After mussels have released their gametes, the mantle is thin and transparent. The mantle consists of connective tissue with haemolymph ('blood') vessels, nerves and muscles that are particularly well developed near the mantle margins. The mantle edge is usually darkly pigmented, which may give protection from the harmful effects of solar radiation. Cilia on the inner surface of the mantle play an important role in directing particles on to the gills and in deflecting heavier material along rejection tracts toward the inhalant opening, the entry point on the mantle for incoming water (see Chapter 4). Periodically, the rejected material is discharged by sudden and forceful closure of the shell valves; this is sufficient to blow the rejected material out of the mantle cavity through the inhalant opening.

Function

The mantle plays a crucial role in the formation of the shell, a process that has already been covered in some detail. Also, the mantle is the site of gametogenesis and the main location for the storage of nutrient reserves, especially glycogen. In *M. edulis*, reserves are laid down in summer and utilised in autumn and winter in the formation of gametes (see Chapter 5). For a full discussion of energy metabolism in the mantle and other tissues, see de Zwann & Mathieu (1992).

The mantle is also involved in pearl formation. Mussels (*Mytilus* spp.) produce pearls in response to infection by the larva of a small parasitic flatworm (*Gymnophallus* spp.). If the larva gets between the mantle epithelium and the shell, the bivalve, in self-defence, encapsulates it with a pearly (nacreous) coat produced by the outermost layer of the mantle.

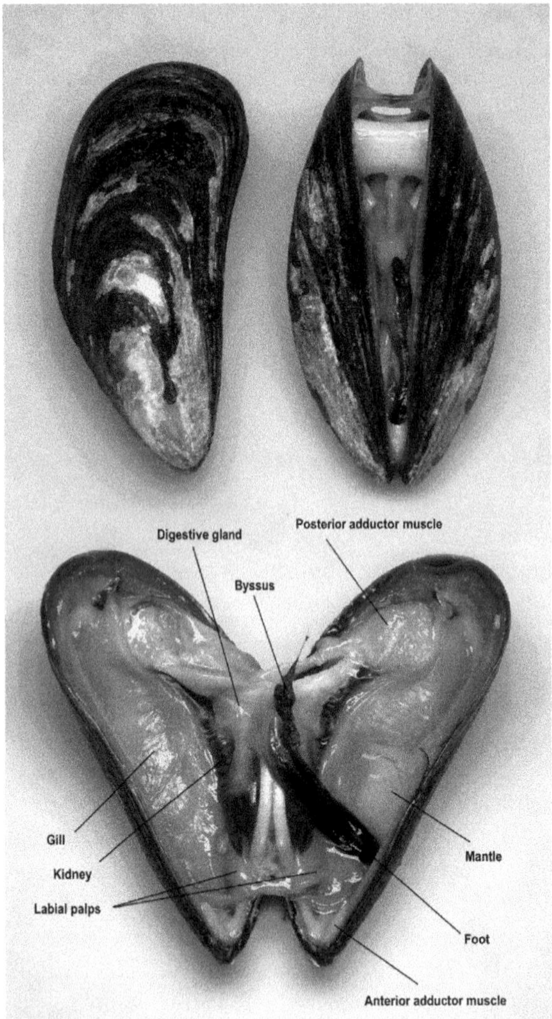

Figure 2.6 Inner anatomy of *Mytilus edulis*. The white posterior adductor muscle is visible in the upper image but has been cut in the lower image to allow the valves to open fully. *Source:* Photograph by Rainer Zenz. (*See colour plate section for colour representation of this figure*).

Pearling can be extremely damaging as it affects the potential for the development of growing mussels for the live – most profitable – market (Wilcox 2013). The problem can be eliminated by avoiding areas where pearl formation occurs or by growing mussels on ropes and marketing them before any pearls reach a detectable size (Morse & Rice 2010). The mantle is also host to various nonpathogenic viruses, potentially pathogenic protozoans, commensal cnidarians and parasitic flatworms. The parasitic flatworm, *Proctoeces maculatus*, seriously reduces glycogen energy reserves in heavily infected mussels. This can lead to disturbances of gametogenesis and possible castration and death (Bower 2009). Additional information on diseases, parasites and pests of mussel mantle is presented in Chapter 11.

The mantle margins are thrown into three folds (Figure 2.3): the outer one, next to the shell, is concerned with shell secretion (see earlier); the middle one has a sensory function; and the inner one is muscular and controls water flow in the mantle cavity. The ES separates the mantle from the shell, except in the regions of muscle attachment. As already

Figure 2.7 Exhalant (white and smooth) and inhalant (fringed with tentacles) openings in the mantle of the mussel *Mytilus edulis*. *Source:* Photo courtesy of John Costelloe, Aquafact International Services Ltd., Galway, Ireland.

seen, the calcareous and organic materials for shell secretion are deposited into this space. The mantle is attached to the shell by pallial muscle fibres in the inner fold; the line of attachment, the pallial line, runs in a semicircle a short distance from the edge of the shell. In mussels, the exhalant opening is small, smooth and conical, and the inhalant aperture is wider and fringed by sensory papillae (Figure 2.7). The middle fold has assumed a sensory role in the evolution of the bivalve form from the ancestral mollusc – a change that involved the loss of the head and associated sense organs. The middle fold is frequently drawn out into short tentacles that contain tactile and chemoreceptor cells. Both of these cell types play an important role in predator detection and avoidance. Ocelli, which are sensitive to sudden changes in light intensity, may also be present on the middle fold. In mussels, these 'eyes' are simple invaginations lined with pigment cells and filled with a mucoid substance or 'lens', whereas in scallops they are highly developed, with a cornea, lens and retina that produce a low-contrast image (Colicchia *et al.* 2009). The inner mantle fold, or velum, the largest of the three folds, has small sensory tentacles or papillae that usually fringe the fold. There is also a large muscular component, especially on the inhalant opening. The velum plays an important role in controlling the flow of water into and out of the mantle cavity.

Gills

Filter feeding is believed to have evolved in some group of early protobranch molluscs, giving rise to the Autobranchia, the dominant subclass of modern bivalves. These feed by filtering the incoming current as a source of food, the gills having replaced the palps as the feeding organs. One important development in the evolution of filter feeding was movement of the site of water intake to the posterior of the animal (see Chapter 1).

Structure

The gills, often referred to as ctenidia, are two large, curtain-like structures that are suspended from the gill axis, which is fused along the dorsal margin of the mantle (Figure 2.8A). Within the gill axis are the branchial nerves and afferent and efferent branchial

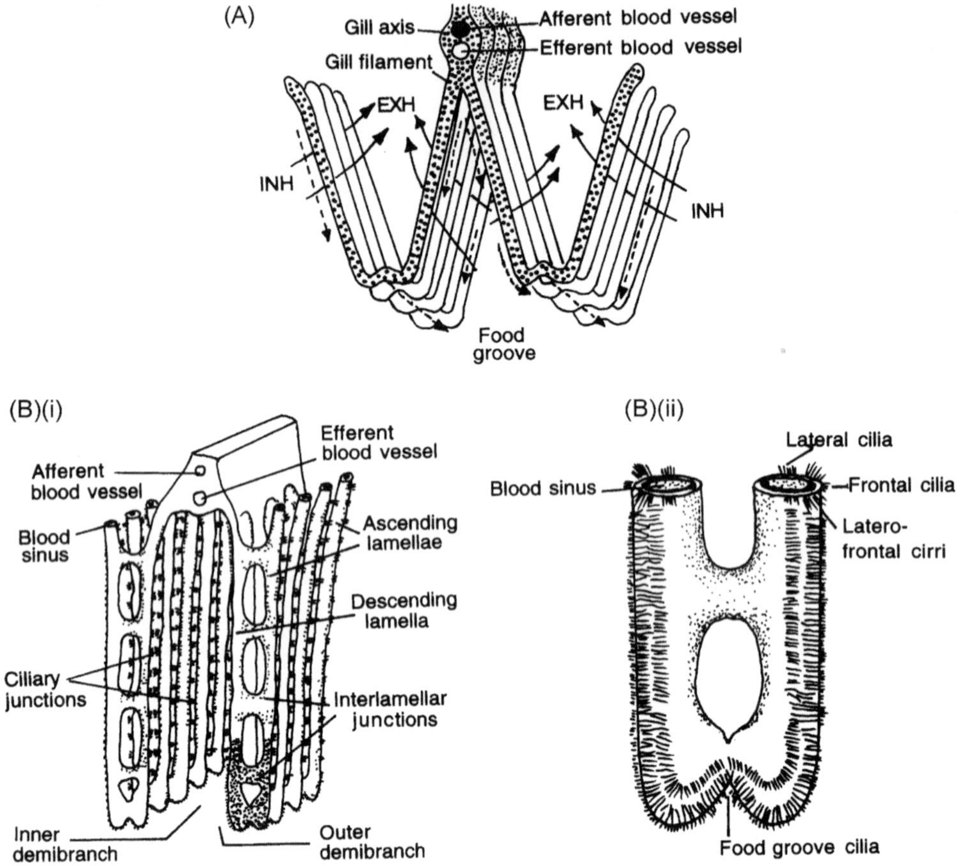

Figure 2.8 (A) Section of a lamellibranch gill showing the ctenidial axis and four W-shaped filaments. For greater clarity, the descending and ascending lamellae of each demibranch have been separated. Solid arrows indicate direction of water flow through the filaments from inhalant (INH) to exhalant (EXH) chambers and broken arrows indicate path of particle transport to the food grooves. (B) (i) section of a fillibranch gill in the mussel, *Mytilus edulis*. Adjacent filaments are joined together by ciliary junctions. (ii) Transverse section through one fillibranch gill filament (shaded in Bi), showing pattern of ciliation. *Source:* (A) From Barnes *et al.* (1993), with permission from John Wiley & Sons; (B) From Pechenik (2010). Reproduced with permission from the McGraw–Hill Companies.

haemolymph (blood) vessels. Each gill is made up of numerous W-shaped (or double-V) ciliated filaments and an internal skeletal rod rich in collagen strengthens each filament. Each V is known as a demibranch and each arm is called a lamella, giving an inner descending and outer ascending lamella (Figure 2.8Bi). In the space between the descending and ascending lamellae is the exhalant chamber, connected to the exhalant area of the mantle edge; the space ventral to the filaments is the inhalant chamber, connected to the inhalant area of the mantle edge (Figure 2.8B). In mussels, the gills follow the curvature of the shell margin, with the maximum possible surface exposed to the inhalant water flow (Figure 2.6).

When the individual filaments are free or loosely attached to one another through interlocking clumps of cilia, this is known as the fillibranch condition (Figure 2.8Bi); it is seen in mussels and scallops. In more advanced bivalves, neighbouring filaments are joined to each other at regular intervals by tissue connections, leaving narrow openings or ostia between them. This gill type, termed eulamellibranch, is a solid structure and is found in the majority of bivalves. In oysters, tissue connections are less extensive than in most eulamellibranch species, so the gills are often referred to as pseudoeulamellibranch. Also, when filaments are similar, as in mussels, the gill is termed homorhabdic, and when there are different types of filaments, through folding of the gill area, it is called heterorhabdic. The filibranch homorhabdic gill in adult mussels is regarded as the ancestral condition from which the other gill types evolved (Beninger & Dufour 2000).

Functions

Cilia on the gill filaments have specific arrangements and functions (Figure 2.8Bii). Lateral cilia are set along the sides of the filaments in fillibranch gills and in the ostia of eulamellibranch gills. These cilia are responsible for drawing water into the mantle cavity and passing it through the gill filaments or the ostia, and then upward to the exhalant chamber and on to the exhalant opening. Lying between the lateral and frontal cilia (see later) are the large feather-like latero-frontal cilia, which are unique to bivalves. When the incoming current hits the gill surface, these cilia flick particles from the water and convey them to the frontal cilia. The frontal cilia, which are abundantly distributed on the free outer surface of the gill facing the incoming current, convey particles aggregated in mucous – secreted by the filaments – downward toward the ciliated food grooves on the ventral side of each lamella. The movement of cilia is under nervous control. Each gill axis is supplied with a branchial nerve from a visceral ganglion, which subdivides to innervate individual groups of filaments. The general architecture and fine structure of the gill vary little from one mussel species to the next, even when rock (e.g. *Lithophaga lithophaga*; Akşit & Falakali Mutaf 2014) and sediment (e.g. *Mytella falcate*; David & Fontanetti 2005) burrowing species are considered. See Chapter 4 for a detailed description of the role of the gill in water pumping and particle capture.

In bivalves, the gills have a respiratory as well as a feeding role. Their large surface area and rich haemolymph supply make them well suited for gas exchange. Deoxygenated haemolymph is carried from the kidneys to the gills by way of the afferent gill vein. Each filament receives a small branch of this vein. The filaments are essentially hollow tubes within which the haemolymph circulates. Gas exchange takes place across the thin walls of the filaments. The oxygenated haemolymph from each filament is collected into the efferent gill vein, which goes to the kidney and on to the heart. It is likely that gas exchange also occurs over the general mantle surface.

The gills perform an additional function in hydrothermal vents mussels, which depend almost entirely on endosymbiont chemosynthetic bacteria in the gill filaments as an energy source. The bacteria use the energy obtained from the oxidation of reduced sulphur compounds and methane from hydrothermal fluid for the fixation of the CO_2 required for primary production (Duperron *et al.* 2016 and references therein; see also Chapter 4).

Due to their dominant role in ingestion and respiration, the gills are among the main target organs in the bioaccumulation of pesticides, soluble heavy metals and hydrocarbons. Complex mixtures of heavy metals and polycyclic aromatic hydrocarbons (PAHs) cause morphological changes in the gill epithelium of *Mytella falcata*, leading to an increase in the number of gill mucous cells, haemocytes and cell turnover processes. These are possible mechanisms to compensate for cell injury and prevent entry of pollutants from gill filaments into the entire organism (David & Fontanetti 2005; David *et al.* 2008). Exposure to mercury over a 24-day period caused an initial deterioration in neural and epithelial cells, increased interstitial cell oedema and reduced ciliation in *Perna perna*. However, after a metal-free recovery period, gill filament morphology returned to near normal (Gregory *et al.* 2002). This is not unexpected, as metallothioneins (MTs) – metal-binding, heat-stable, low-molecular-weight proteins – play an important role in detoxifying trace metals in bivalves and are widely distributed in gill and digestive gland tissue. Consequently, mussel MT levels are increasingly being used as a biomarker of heavy metal contamination in coastal ecosystems (Khati *et al.* 2012; see Chapter 8).

Foot

The foot first appears when bivalve larvae are about 200 μm in length, and becomes functional in crawling and attachment at ~260 μm shell length. This is the pediveliger stage of development, which immediately precedes settlement and metamorphosis (see Figure 5.10). The ciliated foot is proportionately very large and sock shaped, and is made up of layers of circular and longitudinal muscles surrounding a capacious haemolymph space. A byssal duct opens at the 'heel' of the foot, and a byssal pedal groove extends forward along the 'sole' from this opening. The groove is embedded in secretory tissue, which produces the different byssal thread components (see later). While swimming, the foot is fully extended, and periodically the velum (larval swimming organ) is withdrawn and the larva sinks to the bottom and begins to crawl. If the substrate is unsuitable (i.e. does not stimulate the secretion of byssus), the foot is withdrawn and the larva once again swims off (Lutz & Kennish 1992). This cycle can be repeated many times over a period of a few days. In *Mytilus*, when a suitable substrate is found, the larva continues to crawl for some time, gradually ceases movement, protrudes the foot and quickly secretes a single byssal thread. In the newly attached mussel larva, this thread can be repeatedly broken and reformed before final settlement takes place. As the mussel grows in length, more and more attachment threads are secreted; this is not surprising, as larger individuals are subject to greater mechanical stress than smaller ones. To resist dislodgement, mussels cluster their threads in the direction of applied forces (e.g. facing ebb and flow of tide). The adhesive in mussel larvae differs from that of adults, resembling the mucous secreted by other benthic marine species at the larval stage (Petrone *et al.* 2008). The green crenella, *Musculus discors*, is unusual in that the byssus threads that are used to fix it to the substrate are woven into a nest or cage surrounding the shell, similar to a ball of twine. Eggs in mucous strings are retained within this nest, which may incorporate a variety of macroalgae (Merrill & Turner 1963).

Byssus Composition

There are four distinct regions in the mussel byssus: root, stem, thread and plaque (Figure 2.9). The root is embedded in the muscular tissue at the base of the foot. The stem is divided into sections, each with a thread attached; each thread ends in a plaque at which

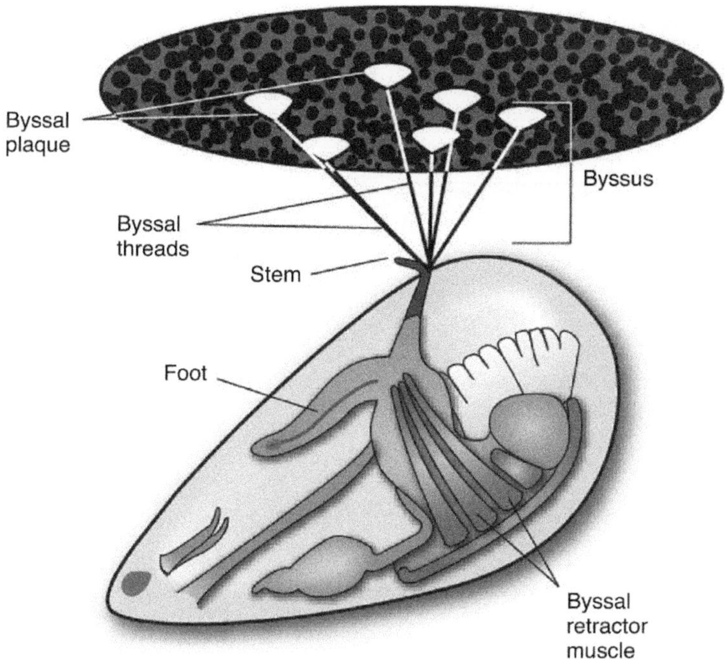

Figure 2.9 Anatomy of the byssus in *Mytilus edulis*. *Source:* From Silverman & Roberto (2007). Reproduced with permission from Springer Nature.

attachment to the substrate takes place. In *M. edulis* and *M. galloprovincialis*, byssal threads are a few centimetres in length and less than 0.1 mm in diameter (Waite *et al.* 2002). Each thread is further divided into a smooth and stiff distal region and a soft and wavy proximal region (Figure 2.10A); in terms of mechanical properties, the latter resembles soft rubber (elasticity of 200%) and the former behaves like rigid nylon with a high tensile strength (Young's modulus = 500 MPa) (Waite *et al.* 2002). The overall mechanical properties of threads reflect those of elastomers, withstanding significant deformations without rupture and returning to their original state when the stress is removed (Hagenau *et al.* 2009). Each thread has a flexible, collagenous inner core covered by a tough, durable cuticle (Holten-Andersen *et al.* 2009), which has a thickness of 2–5 μm and is composed of the protein mefp-1, rich in dihydroxyphenylalanine (DOPA) residues. Metal crosslinking of DOPA residues further enhances the toughness of the byssal threads (Arnold *et al.* 2010). The collagenous proteins that make up the core are known as preCOLs (prepepsinised collagens). The proximal region of the thread contains preCOL-P, a protein with remarkable extensibility and toughness due to the coiled nature of the collagen fibres (Figure 2.10B). The distal region of the thread contains preCOL-D, arranged in straight bundles which provide the stiff properties of the thread–plaque interface. A third protein, preCOL-NG, is distributed throughout the core and is believed to mediate the function between the elastic proximal and stiff distal regions of the thread (Figure 2.11). PreCols extend into the thread–plaque interface. Another component of the core is the thread matrix proteins (TMPs; Figure 2.11), which are distributed throughout the thread and provide a viscoelastic matrix around the collagen fibres (Figure 2.10A). Apparently, TMPs lubricate the fibres and help in reforming

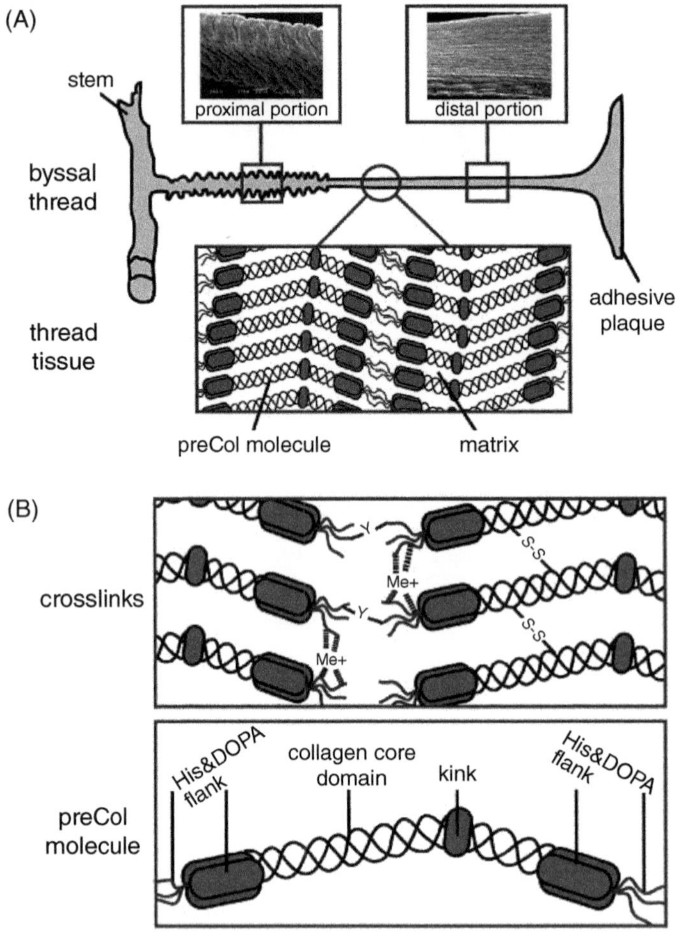

Figure 2.10 Model of the hierarchical arrangement of a mussel byssal thread. (A) Originating from a stem, individual threads show a gradual transition from an elastic proximal to a stiff distal portion ending in an adhesive plaque. (B) Triple helix arrangement of the underlying collagen proteins (preCOLs). Crosslinking for lateral and longitudinal assembly of triple helices is accomplished through lateral His–metal chelate complexes (Me^{+}), disulphide bridges (S-S) or linearly after oxidising two individual DOPA side chains. Deviations from a repeat motif in preCOL-D can lead to kinks in the characteristically rod-like collagen morphology. *Source:* From Hagenau *et al.* (2009). Reproduced with permission from John Wiley & Sons.

byssal threads following deformation from tensile loads (Sagert & Waite 2009). The cuticle, which covers the thread and plaque regions, is about fivefold harder than the thread core. Six different families of foot proteins have been classified from *M. edulis* byssal threads: Mefp-1, Mefp-2, and so on (Figure 2.11). One distinguishing characteristic of all Mefps is the presence of the post-translational modification of tyrosine to 3,4-dihydroxyphenyl-l-alanine (DOPA) as well as organic ions, notably iron (Fe^{3+}). Although DOPA is found throughout the byssus, it is present in higher amounts in the Mefps found at the plaque–substrate interface and is believed to be an important component in wet adhesion (Anderson *et al.* 2010). The plaque has a solid foam structure and is commonly only ~0.15 mm in diameter where it

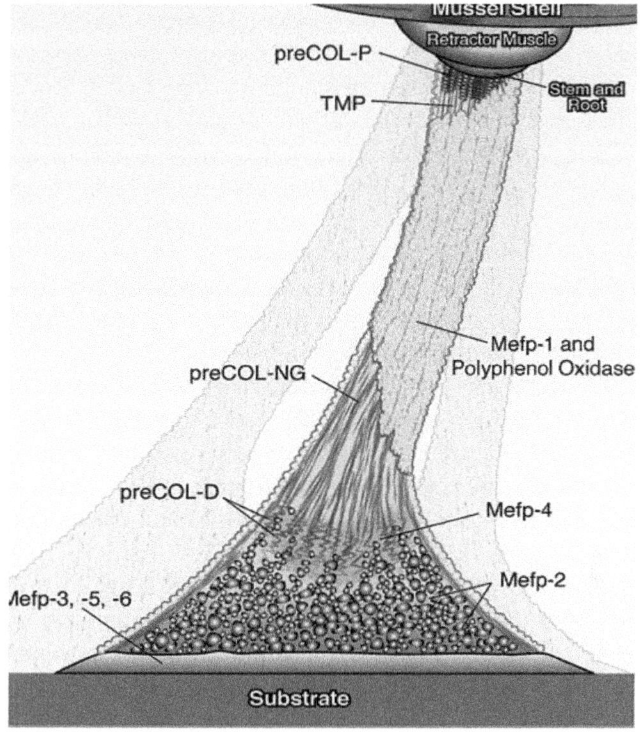

Figure 2.11 Localisation of adhesive proteins in the byssal thread and plaque of *Mytilus*. *Source:* From Silverman & Roberto (2011). Reproduced with permission from Springer Nature.

meets the thread and ~2–3 mm at the substrate interface. The six foot proteins are utilised in plaque formation: fp-3 and fp-5 are adhesive and adhere to the substrate; fp-6 is cohesive and binds the former two proteins together; fp-2 is also cohesive and binds most of the plaque matrix; fp-4 is cohesive and connects the plaque to the thread; and fp-1 is cohesive and forms the plaque and thread outer sheath. Fifteen additional foot proteins have been identified through transcriptomic analysis of the foot of the mussel *M. californianus* (DeMartini *et al.* 2017). The discovery of these new Mfps sets the groundwork for future biochemical investigations to build a more complete model of byssus structure and function in bioadhesion. Mussels concentrate metals like iron, zinc, copper and manganese more than 100 000-fold higher in the adhesive plaque compared to their quantities in the sea, which suggests that metal ions play an important role in adhesion (Bandara *et al.* 2013; Merten 2013). How the mussel processes all these proteins and transition metals into a high-strength, robust, mechanical adhesive is still not understood (Ornes 2013). For additional information on the fine structure of byssal thread and plaque, see Silverman & Roberto (2011).

Attachment

To prepare a surface for attachment, the mussel uses the distal tip of the foot to scrub it, removing weakly attached micro fouling and dirt particles (Wiegemann 2005). The foot is then placed firmly on the surface, forming an airtight and watertight seal. Adhesive proteins are secreted and/or released into the ventral pedal groove, and within 3–10 min (Waite 1992) the foot is lifted to reveal a single thread that connects the mussel and its shell to the plaque attached to the surface (Silvermann & Roberto 2011). Using the same process, additional threads are added to the byssus, thus tethering the mussel to the substrate.

Adhesion is a surface physico-chemical process. It is achieved through a combination of adsorption, mechanical interlocking and molecular diffusion across an interface. The primer proteins fp-3 and fp-5, which connect the plaque to the surface, have unusually high DOPA contents (>20%). Both primers are highly hydroxylated and therefore have the potential to form numerous hydrogen bonds (Wiegemann 2005). The primers are also very low-molecular-weight proteins, which likely causes them to have greater mobility to dissolve into the interstitial areas of a surface and bond by mechanical interlocking. DOPA not only mediates adhesion to the surface but is also able to form strong hydrogen bonds with hydrophilic polymers, as well as strong complexes with metal ions, metal oxide and silicon oxide present in mineral surfaces (Wiegemann 2005). Also, histidine-rich domains in preCOLs form crosslinks with metal ions such as Zn^{2+} and Cu^{2+} (Harrington & Waite 2007; Figure 2.10B). These bonds are pH sensitive, which in the face of ocean acidification in a climate change scenario could have implications for mussel attachment in suspension culture and for intertidal communities anchored by mussels (O'Donnell *et al.* 2013; Carrington *et al.* 2015; see also Chapter 3). Hydrogen bonds and complex formation contribute to the cohesive strength of the adhesive plaque. George *et al.* (2018) examined the effect of seawater temperature, salinity and dissolved oxygen concentration in *M. trossulus*, using tensile testing, atomic force miscroscopy (AFM) and amino acid compositional analysis. High temperature (30°C) and hyposalinity (1 psu) had no effect on adhesion strength, while incubation in hypoxia (0.9 mg l^{-1}) caused plaques to have a mottled colouration and to prematurely peel from substrates, leading to a 51% decrease in adhesion strength. AFM imaging of the plaque cuticle found that plaques cured in hypoxia had regions of lower stiffness throughout, indicative of reductions in DOPA crosslinking between adhesive proteins. A better understanding of the dynamics of plaque curing could aid in the design of better synthetic adhesives, particularly in medicine, where adhesion must take place within wet body cavities (see later).

Various abiotic and biotic factors also influence byssal thread formation and strength of attachment. Carrington *et al.* (2008) examined the effect of water flow on byssal production in four species, *M. trossulus*, *M. galloprovincialis*, *M. californianus* and *Modiolus modiolus*, and found that for all four, thread formation decreased with flow rates above ~25 cm/s, with the critical flow threshold estimated at 50 cm/s. But how can mussels persist on shores with rates of flow considerably higher than this? Apparently, living as they do in dense beds modulates flow, thereby creating microhabitats that are conducive to thread production. Temperature is another factor that affects thread production. *Mytella charruana* is native to Central/South America, but has been introduced along the south-eastern Atlantic coast of the United States. When water temperature was manipulated to near lethal temperatures 6–9°C for this species – thread production ceased at 10 °C (Brodsky 2011). However, *Geukensia demissa* collected from the same area produced some threads at 10 °C and showed no difference in mean thread production between 13 and 23 °C, while *M. charruana* had significantly fewer threads at 13 than at 23 °C. These data suggest that *M. charruana* may experience difficulty surviving in the wild at 10 °C for extended periods of time, which could have implications for its survival and future spread as an alien species. Garner & Litvaitis (2013a) have shown that *M. edulis* can increase the strength (see later), number and attachment sites of byssal threads in response to waterborne cues from an array of predators and injured conspecifics. Also, the same authors found that more threads with greater attachment strength were produced when mussels were fouled with epibionts (Garner & Litvaitis 2013b). Epibionts increase the chance of dislodgement due to an increase in hydrodynamic forces exerted on mussels: specifically, a higher drag-induced loading.

Wave action is probably the factor that has been most often cited as influencing mussel attachment (references in Garner & Litvaitis 2013b). Babarro & Carrington (2011) compared byssus tenacity (attachment) and associated features in mussels at an exposed and a sheltered site in the Ría de Vigo (NW Spain). They found that mussels inhabiting the rougher outer Ría secreted stronger and stiffer threads and had a higher potential to form crosslinks or metal chelation in the byssal collagen in order to gain structural integrity when needed (Figure 2.10B). Their results from reciprocal transplants indicated that mussels have the potential to change byssus diameter and mechanical properties in order to increase strength in stressful abiotic conditions, and can reallocate energy for vital activities such as gonadal and soft tissue growth in more benign environments (see also Babarro & Carrington 2013). However, Moeser & Carrington (2006) and Moeser *et al.* (2006) suggest that seasonal variations in material properties of the byssus play an even more significant role than wave action in determining mussel attachment strength. They found that thread strength and extensibility increase after autumn and winter, leading to the strongest attachment occurring during the spring, at which point energetic resources switch their focus toward gamete production (see also Zardi *et al.* 2007). This shift in energetic allocation, combined with increased thread decay, decreases attachment strength throughout the summer, leading to the weakest attachment strength occurring in the autumn. Hawkins & Bayne (1985) estimate that byssus production can consume up to 8% of a mussel's monthly energy expenditure.

The mechanical properties of byssal threads vary depending on the species. An examination of the material and structural properties of the threads of *M. californianus*, *M. galloprovincialis* and *M. trossulus* indicated that while the material properties of the threads were similar among species, the distal portion of the threads of *M. californianus* extended further before breaking, leading to a stronger attachment strength (8–17% increase) relative to the other two species. This may be a factor in the domination of *M. californianus* on wave-exposed shores on the Pacific coast of North America (Bell & Gosline 1996). The preCOLs in this species are more divergent from those of the other two than the preCOLs of *M. galloprovincialis* and *M. trossulus* are from each other. The single most influential factor in the tensile superiority of *M. californianus* is the greater abundance of silk-like alanine-rich sequences in the flanking domains of preCOLs (Harrington & Waite 2007). See Bell & Gosline (1997), Lucas *et al.* (2002), Brazee & Carrington (2006), Pearce & LaBarbera (2009), Bouhlel *et al.* (2017), George *et al.* (2018) and Newcomb *et al.* (2019) and references therein for additional comparative studies on mytilid thread properties.

As seen earlier, marine mussels use a natural adhesive to adhere to a wide variety of substrates in an aqueous environment, and to date there are no synthetic glues that are as strong, versatile and unaffected by water as mussel glue. It is not surprising therefore that mussel adhesive proteins are attractive targets for biomimetic technology, which entails using designs from nature to solve problems in engineering, materials science, medicine and other fields. So far, more than a dozen adhesive proteins have been identified and characterised, and recombinant DNA technology has been used to obtain them in large amounts for conventional adhesion tests and practical applications. Two commercial mussel adhesive products are available on the market: Cell-Tak, a naturally extracted adhesive consisting of fp-1 and fp-2, and MAP, which contains only fp-1 (Cha *et al.* 2008). Alternatively, the exceptional adhesive properties exhibited by the native proteins can be captured in synthetic polymer systems (see Lee *et al.* 2011 for review). These have potential use as coatings for a wide range of organic and inorganic materials. For example, they are used for adhesion and sealing in foetal membrane rupture, corneal tissue sutures, surgical repair of nerves and cancer drug delivery (Kaushik *et al.* 2015). They are also used in antifouling coatings (Lee *et al.* 2011), to create hydrogels for drug delivery (Lee & Konst 2014) and to anchor nanoparticles on to a variety of surfaces (Zhu & Pan 2014).

Labial Palps

Each gill terminates within a pair of triangular-shaped palps that are situated on either side of the mouth (Figure 2.6) and extend posteriorly about one-third of the length of the mantle cavity (Morton 1992). The inner surface of each palp faces the gill and is folded into numerous ridges and grooves that carry a complicated series of ciliary tracts. The outer surfaces of the palps are smooth, and between the inner and outer surfaces there is muscular connective tissue (see Figures 4.15 and 4.16).

The main function of the labial palps is to continually remove material from the food tracts on the gills in order to prevent gill saturation. In dense suspensions, sorting and rejection tracts on the palps channel most of the filtered material away from the mouth and deposit it as pseudofaeces so that the animal can continue to filter and ingest at an optimum rate. The pseudofaeces is carried along rejectory tracts on the mantle to the inhalant opening and periodically forcefully ejected through it. When the ingestive capacity is not exceeded, particles from the gill move along acceptance tracts on the labial palps toward the mouth (see Chapter 4).

Alimentary Canal

Stomach and Digestive Gland

The mouth is ciliated and leads into a narrow ciliated oesophagus. Ciliary movement helps to propel material toward the anterior part of the stomach. Indeed, this method of moving material is found throughout the length of the alimentary canal, primarily because it lacks a muscular wall. The stomach is large and oval-shaped and lies completely embedded in the digestive gland, which opens into it by several ducts. A semi-transparent gelatinous, tapering rod, the crystalline style, originates in a style sac at the posterior end and projects forward and dorsally across the cavity of the stomach to rest against the gastric shield, a thickened area of the stomach wall (Figure 2.12; details in Morton 1992). The projecting

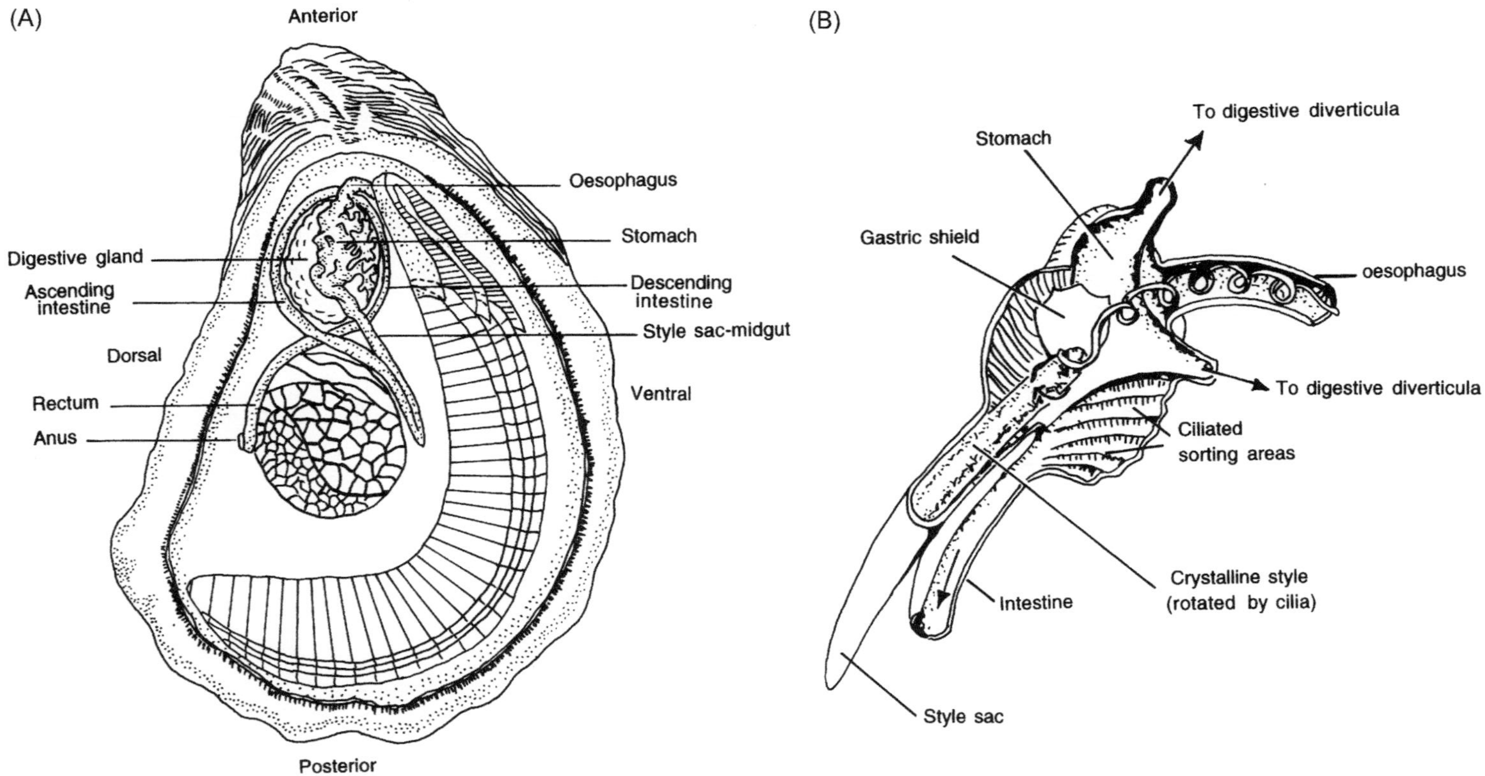

Figure 2.12 (A) The bivalve digestive system. Redrawn from Langdon & Newell (1996), after Galtsoff (1964). Reprinted with permission from Maryland Sea Grant. (B) Bivalve stomach showing rotation of crystalline style and winding of food string. Rejectory groove on floor of stomach not shown. *Source:* From Pechenik (1991). Reproduced with permission from the McGraw-Hill Companies.

anterior end of the style is rotated against the gastric shield by the style sac cilia, and in the process the style end is abraded and dissolved, releasing a range of carbohydrate-, fat- and protein-splitting enzymes in the process (see Chapter 4). This loss is made good by continual additions by the style sac to the base of the style. Some digestive enzymes are actually synthesised in the digestive gland and transported to the style sac, where they are secreted (Sakamoto *et al.* 2008). Rotation of the style also aids in pulling a food-laden mucous strand through the mouth into the stomach.

The length of the style is correlated with shell length; in *M. edulis* and *M. galloprovincialis*, the length is approximately 50% of shell length (Alyakrinskaya 2001). Style length changes with the season, with maximum length in spring when food intake is high. Also, the length exhibits a tidal cycle, with maximum length when the animal is submerged and feeding; in the absence of water, the style shortens by approximately 25% in *M. edulis* (Alyakrinskaya 2001). While the style has an ephemeral existence in most bivalves, in *Perna canaliculus* it is a firm, robust, permanent structure. The greater proportion of high-molecular-weight mucin-like proteins (>500 kDa) in the *P. canaliculus* style suggests that these proteins may be important in the formation of the hard, permanent nature of the style, perhaps through interaction with medium-molecular-weight proteins (MacKenzie & Marshall 2014).

The style has additional functions in the digestive process. The low pH of the stomach facilitates the dislodgement of particles from the mucous string. These particles are then mixed with the other contents of the stomach, including the liberated enzymes from the style. The rotation of the style helps the mixing process. While all the mixing and extracellular digestion is taking place, the stomach contents come under the influence of ciliary tracts that cover all areas of the stomach except those occupied by the gastric shield. These ciliated tracts have fine ridges and grooves and act as sorting areas in much the same way as the labial palps. Finer particles and digested matter are kept in suspension by cilia at the crests of the ridges, and this material is continually swept toward the digestive gland duct openings. Larger particles segregate out and are channelled into the intestine along a deep rejectory groove on the floor of the stomach (see Chapter 4).

The digestive gland, which is brown or black and consists of blind-ending tubules that connect to the stomach by several ciliated ducts, is the major site of intracellular digestion. The epithelium of the tubules is composed of two cell types, digestive cells and basophil (secretory) cells. The former are the most abundant type and are responsible for intracellular digestion of food. Digestion takes place within large vesicles called lysosomes that contain hydrolytic enzymes. The end products of digestion are released directly into the haemolymph system and waste products are contained in residual bodies within the digestive cells. The cells eventually rupture and the waste material enclosed in excretory spheres is swept along the ciliated secondary and primary ducts of the digestive gland toward the stomach, and ultimately to the intestine. The intestine terminates in an anus, and faeces in the form of faecal pellets are swept away through the exhalant opening. The basophil cells display a highly developed rough endoplasmic reticulum and numerous secretory granules, and carry out extensive protein synthesis and probably secrete digestive enzymes (Dimitriadis *et al.* 2004; Beninger & Le Pennec 2016). A more detailed description of the morphology and role of these two cell types is presented in Chapter 4.

Gonads

Mussels are dioecious (i.e. the sexes are separate), and there are usually equal numbers of males and females. The gonads extend throughout most parts of the body, except the gills, muscles and foot. Most of the gonad is in the mantle (Figure 2.6), thus accounting for its unusual thickness in mussels. The colour of the reproductive tissue is not a good indicator of the sex of an individual. Each gonad consists of a converging system of ducts leading to a gonopore on the tip of the genital papilla, which is located in the exhalant portion of the mantle cavity. Gametes are shed through the exhalant opening of the mantle and fertilisation takes place in the water column. After mussels have released their gametes, the mantle is thin and transparent. See Chapter 5 for a full description of mussel reproduction.

Heart and Haemolymph Vessels

The heart lies in the mid-dorsal region of the body, close to the hinge line of the shell, in a space called the pericardial cavity, which surrounds the heart dorsally and a portion of the intestine ventrally. The wall of the cavity, a thin, translucent layer, is called the **pericardium**. The heart consists of a single, muscular ventricle and two thin-walled auricles. Haemolymph – a colourless fluid of haemocytes and plasma (cell-free haemolymph) – flows from the auricles into the ventricle, which contracts to drive it into a single vessel, the anterior aorta. The aorta divides into many arteries, the most important of which are the pallial arteries from the posterior aorta, which supply the mantle with haemolymph, and the visceral arteries (gastro-intestinal, hepatic and terminal) from the anterior aorta, which supply the stomach, intestine, shell muscles and foot (Figure 2.13). While a posterior aorta is present in *Guekensia* spp., it is absent in *Mytilus* spp. The arteries break up into a network

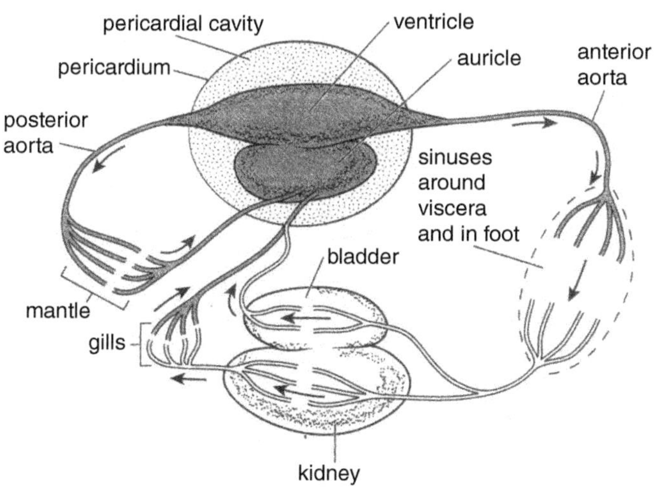

Figure 2.13 Circulatory system of a typical bivalve. The shaded areas indicate the route of oxygenated haemolymph. While the bivalve heart has two auricles, only one of these is illustrated. *Source:* From Pechenik (2010). Reproduced with permission from the McGraw-Hill Companies.

of vessels in all tissues and then join to form veins, which empty into three extensive spaces, the pallial, pedal and median ventral sinuses. The circulatory system is therefore an open system with haemolymph in the sinuses bathing the tissues directly. One consequence of an open circulatory system is that the animal is continually exposed to fluctuations in environmental factors, such as temperature, salinity and contaminants. From the sinuses, haemolymph is carried to the kidneys for purification. In *Mytilus*, some of the haemolymph from the kidney network enters the gills, discharging into the afferent gill vein, which gives off a branch to each gill filament, descending on one side and ascending on the other. The ascending vessels join to form an efferent gill vein that passes back to the kidneys. The haemolymph from the kidneys returns to the auricles of the heart. In other bivalves (e.g. *Pecten* spp.), haemolymph from the gills does not return to the kidney but flows directly from the gills to the heart (Figure 2.13). In all bivalves, there are well-developed circulatory pathways through the mantle, which therefore serves as an additional site of oxygenation. See Field (1922) for a very detailed description of the arterial and venous systems in *M. edulis*.

Haemolymph plays a number of important roles in bivalve physiology. These include gas exchange, osmoregulation (see Chapter 7), nutrient distribution, waste elimination and internal defence (see Chapter 11). Because haemolymph constitutes 40–60% of the fresh tissue weight, it also serves as a fluid skeleton, giving temporary rigidity to such organs as the labial palps, foot and mantle edges. The haemolymph contains cells called haemocytes, which float in a colourless plasma. Most bivalves lack circulating respiratory pigments, probably because their sedentary lifestyle and large exposed surfaces (for oxygen uptake) preclude the need for such pigments. However, haemocyanin, the typical molluscan respiratory pigment, is found in some protobranch bivalves, while haemoglobin has been reported in several bivalve families (references in Giribet 2008). Haemocytes are not confined to the haemolymph system but move freely out of the sinuses into surrounding connective tissue, the mantle cavity and gut lumen. Therefore, it is not surprising that these cells play an important role in physiological processes such as nutrient digestion and transport, excretion, tissue repair, heavy metal metabolism and internal defence. See Chapter 7 for details on haemocyte types and their functions.

Excretory Organs

There are two types of excretory organs in bivalves, the pericardial glands and the paired nephridia or kidneys. In *Mytilus*, the reddish-brown elongate kidneys lie ventral to the pericardial cavity surrounding the heart and dorsal to the gill axis, and in fact extend the complete length of the gill axis from the labial palps to the posterior adductor muscle (Figure 2.6). One arm of each kidney is glandular and opens into the pericardium, and the other end is a thin-walled bladder that opens through a nephridiopore and empties into the exhalant chamber of the mantle cavity. See Pirie & George (1979) for a more detailed description of the excretory system in *M. edulis*.

The brown-coloured pericardial glands, sometimes referred to as Keber's organs, develop from the epithelial lining of the pericardium and come to lie over the auricular walls of the heart. Waste accumulates in certain cells of the pericardial glands and is periodically discharged into the pericardial cavity, and from there is eliminated via the kidneys. Other cells of the pericardial glands are involved in filtering the haemolymph, the first stage of urine formation. The filtrate then flows to the glandular part of the kidney, where the processes of secretion and reabsorption of ions occur. The end result is urine that has a high concentration of ammonia and smaller amounts of amino acids and creatine. Most aquatic invertebrates excrete ammonia as the end product of protein metabolism. Ammonia is highly toxic but its small molecular size and high solubility in water ensure that it very rapidly diffuses away from the animal.

While the kidneys and pericardial glands are the major excretory organs, excretory products are probably also lost across the general body surface and particularly across the gills (see Chapter 7 for details on excretion and osmoregulation). The kidney also plays a very important role in the storage and elimination of radionuclides and heavy metals such as silver, cobalt, mercury, manganese, lead and zinc (Metian *et al.* 2011 and references therein; Pouil *et al.* 2015). In scallops, Metian *et al.* (2009) have shown that several of these metals are sequestered in renal concretions, mostly of calcium carbonate, before being eliminated in the urine.

Nerves and Sensory Receptors

The nervous system of mussels is fundamentally simple. It is bilaterally symmetrical and consists of three pairs of ganglia and several pairs of nerves (Figure 2.14). The cerebral ganglia are joined by a short commissure dorsal to the oesophagus. From each cerebral ganglion, two pairs of nerve cords extend to the posterior of the animal. One pair extends directly back to the visceral ganglion, which is located on the surface of the posterior adductor muscle. The second pair extends posteriorly and ventrally to the pedal ganglia in

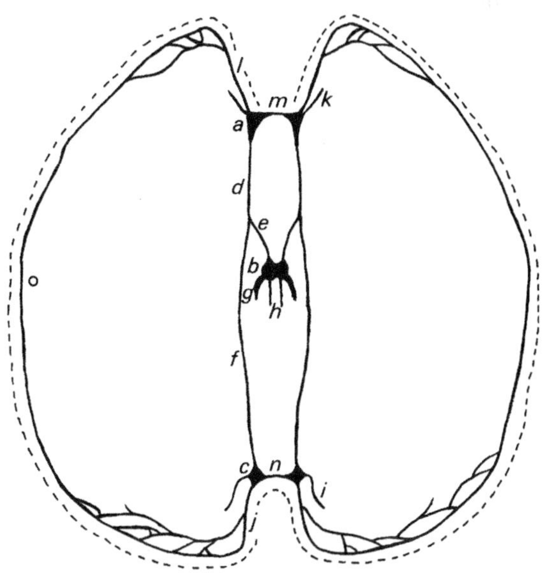

Figure 2.14 Schematic representation of the nervous system in the mussel *Mytilus edulis*. Dashed line indicates the outline of the two shell valves. (a) Cerebral ganglia. (b) Pedal ganglia. (c) Visceral ganglia. (d) Cerebrovisceral pedal connective. (e) Cerebropedal connective. (f) Cerebrovisceral connective. (g) Pedal nerve. (h) Byssal retractor nerve. (i) Branchial nerve. (j) Posterior pallial nerve. (k) Anterior pallial nerve. (l) Buccal nerve. (m) Cerebral commissure. (n) Visceral commissure. (o) Circumpallial nerve. Ganglia are small, their longest dimension never exceeding 2 mm, but each contains numerous neurosecretory cells. *Source:* From de Zwann & Mathieu (1992). Reproduced with permission from Elsevier.

the foot. The cerebral ganglia innervate the palps, anterior adductor muscle and part of the mantle, as well as the sensory organs (see later). The pedal ganglia control the foot. The visceral ganglia control a large area: gills, heart, pericardium, kidney, digestive tract, gonad, posterior adductor muscle, part or all of the mantle, siphons and pallial sense organs. See Field (1922) for more detail on individual ganglia and nerves in *M. edulis.*

The ganglia also have a major neurosecretory role in bivalves. Several different types of neurosecretory cells have been identified in mussels, and most of these are located in the cerebral ganglia (de Zwann & Mathieu 1992). These cells produce peptides that are released into the circulatory system. At least four different neuropeptides have been identified that mediate reproductive-related events in bivalves (Morishita *et al.* 2010). Insulin-related peptides have also been identified in neurosecretory cells of several bivalve species and have been shown to be involved in growth regulation by stimulating protein synthesis in mantle edge cells involved in shell and soft tissue growth. More details on the role of neurohormones in gametogenesis and growth are presented in Chapters 5 and 6, respectively.

During the evolution of bivalves, with loss of a distinct head, most of the sense organs withdrew from the anterior end and came to lie at the edge of the mantle. Most sensory receptors are located on the middle fold of the mantle. This fold is thick and bears a large number of pallial tentacles, their length and number varying with the species (Figure 2.7). The tentacles are covered in epithelial tactile cells that are sensitive to touch. A slight tactile stimulus elicits local contraction of the mantle or siphon musculature. This is a reflex action and is not under the control of the central nervous system (CNS). A strong stimulus produces a coordinated retraction of the whole animal into its shell. This more general and clearly adaptive type of contraction is under the control of the visceral ganglion. The mantle also contains sensory cells (chemoreceptors) that are stimulated by waterborne chemicals. These cells are capable of detecting the presence of gametes in the water column and provide a powerful chemical stimulus for ripe mussels to release their gametes, thereby enhancing the chances of fertilisation. Chemoreceptors also help juvenile mussels to select a suitable substrate for settlement (see Chapter 5).

Ocelli, which can detect sudden changes in light intensity, may also be present on the middle fold of the mantle or siphons. These may take the form of invaginated eyecups lined with pigmented sensory cells and filled with a mucoid substance that acts as a 'lens', or they may be very well developed structures that produce a low-contrast image as in scallops. Research shows that swimming scallops tend to have better vision than sessile scallops, and this suggests that mobile scallops may visually detect preferred habitats (Speiser & Johnsen 2008a). Also, during shell gaping, when eyes on the mantle edges are exposed, scallops may visually detect the size and speed of moving particles and use this information to help identify favourable feeding conditions (Speiser & Johnsen 2008b). For more information on the visual physiology of scallops, see Speiser & Wilkens (2016).

Sensory receptors called osphradia are well known in gastropods, but in bivalves they are difficult to detect, either because of their small size or because they are in fact absent in some species. These receptors, usually paired, consist of a pigmented epidermal patch of sensory and secretory cells located in the gill axis and enervated by the branchial nerve. There has been some debate as to the role of these structures in bivalves (reviewed by Haszprunar 1987), but the general consensus is that they have a dual function: reception of chemical spawning cues by the sensory cells and synchronisation of gamete emission by the secretory cells through the release of serotonin, a powerful stimulant for spawning in bivalves (Beninger *et al.* 1995). While these sensory structures are no doubt real anatomical features, Lindberg & Sigwart (2015) question the assumption that there is a single molluscan osphradium and propose that at least two distinct classes of epithelial sensory structures have been identified as such.

Another type of sensory receptor is the statocyst, which lies in the foot near the pedal ganglia and is innervated by the cerebral ganglia. The structure of the statocyst, while varying between invertebrate species, typically consists of a dense mass known as the statolith or of multiple smaller statoconia, in a fluid-filled chamber lined by sensory hair cells (Budelmann 1988). The solid concretion(s) interact with the cilia and convey information to the mussel on its orientation in space. Anthropogenic activities in the oceans, such as pile driving, involve direct contact with the seabed, creating radiating particle motion waves. Sensitivity of *M. edulis* to substrate-borne vibration has been quantified by exposure of the species to vibration under controlled conditions (Roberts *et al.* 2015). The results indicate that vibration is likely to impact overall fitness in this species due to disruption of natural valve gape periodicity. It is likely that vibration travelling through the mussel body may stimulate movement of the statocyst system, as observed in other invertebrates (Roberts *et al.* 2015 and references therein).

Finally, another sensory receptor, the abdominal sense organ (ASO), first described in scallops but now reported from at least 19 bivalve families (Haszprunar 1983), is a small piece of tissue situated on the adductor muscle near the anus. The receptor epithelium contains ciliated sensory cells and mucocytes, the former making up 90% of the total cell number (details in Zhadan *et al.* 2004). While some believe that the ASO functions in the regulation of water flow within the pallial cavity (Beninger & Le Pennec 2006), others have suggested that it may be involved in the perception of waterborne vibrations – similar in some respects to the acoustic lateral line system in fish – and Zhadan (2005) has indeed shown that the scallops *Mizuhopecten yessoensis* and *Chlamys swifti* are sensitive to ultrasonic vibrations in the range 30–1000 Hz. This type of sensory input could give warning of the approach of a predator.

References

Addadi, L., Joester, D., Nudelman, F. & Weiner, S. (2006) Mollusk shell formation: a source of new concepts for understanding biomineralization processes. *Chemistr – A European Journal*, **12**, 980–987.

Aguirre, M.L., Perez, S.I. & Sirch, Y.N. (2006) Morphological variability of *Brachidontes* Swainson (Bivalvia, Mytilidae) in the marine Quaternary of Argentina (SW Atlantic). *Palaeogeography, Palaeoclimatology, Palaeoecology*, **239**, 100–125.

Akester, R.J. & Martel, A.L. (2000) Shell shape, dysodont tooth morphology, and hinge-ligament thickness in the bay mussel *Mytilus trossulus* correlate with wave exposure. *Canadian Journal of Zoology*, **78**, 240–253.

Akşit, D. & Falakali Mutaf, B. (2014) The gill morphology of the date mussel *Lithophaga lithophaga* (Bivalvia: Mytilidae). *Turkish Journal of Zoology*, **38**, 61–67.

Alyakrinskaya, I.O. (2001) The dimensions, characteristics and functions of the crystalline style of mollusks. *Biological Bulletin*, **28**, 523–535.

Anderson, T.H., Yu, J., Estrada, A., Hammer, M.U., Waite, J.H. & Israelachvili, J.N. (2010) The contribution of DOPA to substrate–peptide adhesion and internal cohesion of mussel-inspired synthetic peptide films. *Advanced Functional Materials*, **20**, 4196–4205.

Arnold, A. A., Byette, F., Séguin-Heine, M.O., LeBlanc,A., Sleno, L., Tremblay, R. *et al.* (2010) Solid-state NMR structure determination of whole anchoring threads from the blue mussel *Mytilus edulis*. *Biomacromolecules*, **14**, 132–141.

Avelar, W.E.P., Mantelatto, F.L.M., Tomazelli, A.C., Silva, D.M.L., Shuhama, T. & Lopes, J.L.C. (2000) The marine mussel *Perna perna* (Mollusca, Bivalvia, Mytilidae) as an indicator of contamination by heavy metals in the Ubatuba Bay, São Paulo, Brazil. *Water, Air, and Soil Pollution*, **118**, 65–72.

Babarro, J.M.F. & Carrington, E. (2011) Byssus secretion of *Mytilus galloprovincialis*: effect of site at macro- and micro-geographical scales within Ría de Vigo (NW Spain). *Marine Ecology Progress Series*, **435**, 125–140.

Babarro, J.M.F. & Carrington, E. (2013) Attachment strength of the mussel *Mytilus galloprovincialis*: effect of habitat and body size. *Journal of Experimental Marine Biology and Ecology*, **443**, 188–196.

Bandara, N., Zeng, H. & Wu, J. (2013) Marine mussel adhesion: biochemistry, mechanisms, and biomimetics. *Journal of Adhesion Science and Technology*, **27**, 2139–2162.

Barnes, R.S.K., Callow, P. & Olive, P.J.W. (1993) *The Invertebrates: A New Synthesis*, 2nd edn. Blackwell Scientific Publications, Oxford.

Bau, M., Balan, S., Schmidt, K. & Koschinsky, A. (2010) Rare earth elements in mussel shells of the Mytilidae family as tracers for hidden and fossil high-temperature hydrothermal systems. *Earth and Planetary Science Letters*, **299**, 310–316.

Bell, E.C. & Gosline, J.M. (1996) Mechanical design of mussel byssus: material yield enhances attachment strength. *Journal of Experimental Biology*, **199**, 1005–1017.

Bell, E.C. & Gosline, J.M. (1997) Strategies for life in flow: tenacity, morphometry, and probability of dislodgment of two *Mytilus* species. *Marine Ecology Progress Series*, **159**, 197–208.

Bellotto, V.R. & Miekeley, N. (2007) Trace metals in mussel shells and corresponding soft tissue samples: a validation experiment for the use of *Perna perna* shells in pollution monitoring. *Analytical and Bioanalytical Chemistry*, **389**, 769–776.

Beninger, P.G. & Dufour, S.C. (2000) Evolutionary trajectories of a redundant feature: lessons from bivalve gill abfrontal cilia and mucocyte distributions. In: *The Evolutionary Biology of the Bivalvia* (eds E.M. Harper, J.D. Taylor & J.A. Crame), pp. 273–278. Geological Society, London.

Beninger, P.G. & Le Pennec, M. (2006) Structure and function in scallops. In: *Scallops: Biology, Ecology and Aquaculture*, 2nd edn (eds S.E. Shumway & G.J. Parsons), pp. 123–227. Elsevier Science Publishers B.V., Amsterdam.

Beninger, P.G. & Le Pennec, M. (2016) Scallop structure and function. In: *Scallops: Biology, Ecology Aquaculture and Fisheries*, 3rd edn. (eds S.E. Shumway & G.J. Parsons), pp. 85–159. Elsevier Science Publishers B.V., Amsterdam.

Beninger, P.G., Donval, A. & Le Pennec, M. (1995) The osphradium in *Placopocten magellanicus* and *Pecten maximus* (Bivalvia, Pectinidae): histology, ultrastructure, and implications for spawning synchronization. *Marine Biology*, **123**, 121–129.

Bjärnmark, N.A., Yarra, T., Churcher, A.M., Felix, R.C., Clark, M.S. & Power, D.M. (2016) Transcriptomics provides insight into *Mytilus galloprovincialis* (Mollusca: Bivalvia) mantle function and its role in biomineralisation. *Marine Genomics*, **27**, 37–45.

Boisson, F., Cotret, O. & Fowler, S.W. (1998) Bioaccumulation and retention of lead in the mussel *Mytilus galloprovincialis* following uptake from seawater. *Science of the Total Environment*, **222**, 55–61.

Bonel, N., Solari, L.C. & Lorda, J. (2013) Differences in density, shell allometry and growth between two populations of *Limnoperna fortunei* (Mytilidae) from the Río De La Plata Basin, Argentina. *Malacologia*, **56**, 43–58.

Bouhlel, Z., Genard, B., Ibrahim, N., Carrington, E., Babarro, J.M.F., Lok, A. *et al.* (2017) Interspecies comparison of the mechanical properties and biochemical composition of byssal threads. *Journal of Experimental Biology*, **220**, 984–994.

Bower, S. M. (2009) Synopsis of infectious diseases and parasites of commercially exploited shellfish. Available at: https://www.dfo-mpo.gc.ca/science/aah-saa/diseases-maladies/index-eng.html (accessed 28 May 2021).

Brazee, S.L. & Carrington, E. (2006) Interspecific comparison of the mechanical properties of mussel byssus. *Biological Bulletin*, **211**, 263–274.

Briones, C., Rivadeneira, M.M., Fernández, M. & Guiñez, R. (2014) Geographical variation of shell thickness in the mussel *Perumytilus purpuratus* along the Southeast Pacific coast. *Biological Bulletin*, **227**, 221–231.

Brodsky, S. (2011) Cold temperature effects on byssal thread production by the native mussel *Geukensia demissa* versus the non-native mussel *Mytella charruana*. *The University of Central Florida Undergraduate Research Journal*, **5**, 1–10.

Budelmann, B.U. (1988) Morphological diversity of equilibrium receptor systems in aquatic invertebrates. In: *Sensory Biology of Aquatic Animals* (eds J. Atema, R.R. Fay, A.N. Popper & W.N. Tavolga), pp. 757–782. Springer-Verlag, New York.

Carrington, E., Moeser, G.M., Thompson, S.B., Coutts, L.C. & Craig, C.A. (2008) Mussel attachment on rocky shores: the effect of flow on byssus production. *Integrative and Comparative Biology*, **48**, 801–807.

Carrington, E., Waite, J.H., Sarà, G. & Sebens, K.P. (2015) Mussels as a model system for integrative ecomechanics. *Annual Review of Marine Science*, **7**, 443–469.

Cha, H.J., Hwang, D.S. & Lim, S. (2008) Development of bioadhesives from marine mussels. *Biotechnology Journal*, **3**, 631–638.

Chateigner, D., Hedegaard, C. & Wenk, H.-R. (2000) Mollusc shell microstructures and crystallographic textures. *Journal of Structural Geology*, **22**, 1723–1735.

Checa, A.G., Pina, C.M., Osuna-Mascaró, A.J., Rodríguez-Navarro, A.B. & Harper, E.M. (2014) Crystalline organization of the fibrous prismatic calcitic layer of the Mediterranean mussel *Mytilus galloprovincialis*. *European Journal of Mineralogy*, **26**, 495–505.

Colicchia, G., Waltner, C., Hopf, M. & Wiesner, H. (2009) The scallop's eye – a concave mirror in the context of biology. *Physics Education*, **44**, 175–179.

Cusack, M., Parkinson, D., Freer, A., Pérez-Huerta, A., Fallick, A.E. & Curry, G.B. (2008) Oxygen isotope composition in *Modiolus modiolus* aragonite in the context of biological and crystallographic control. *Mineralogical Magazine*, **72**, 569–577.

David, J.A.O. & Fontanetti, C.S. (2005) Surface morphology of *Mytella falcata* gill filaments from three regions of the Santos Estuary. *Brazilian Journal of Morphological Sciences*, **22**, 203–210.

David, J.A.O., Salaroli, R.B. & Fontanetti, C.S. (2008) The significance of changes in *Mytella falcata* (Orbigny, 1842) gill filaments chronically exposed to polluted environments. *Micron*, **39**, 1293–1299.

DeMartini, D.G., Errico, J.M., Sjoestroem, S., Fenster, A. & Waite, JH. (2017) A cohort of new adhesive proteins identified from transcriptomic analysis of mussel foot glands. *Journal of the Royal Society Interface*, **14**, 20170151.

de Zwann, A. & Mathieu, M. (1992) Cellular biochemistry and endocrinology. In: *The Mussel Mytilus: Ecology, Physiology, Genetics and Culture* (ed E.M. Gosling), pp. 223–307. Elsevier Science Publishers B.V., Amsterdam.

Dimitriadis, V.K., Domouhtsidou, G.P. & Cajaraville, M.P. (2004) Cytochemical and histochemical aspects of the digestive gland cells of the mussel *Mytilus galloprovincialis* (L.) in relation to function. *Journal of Molecular Histology*, **35**, 501–509.

Duperron, S., Quiles, A., Szafranski, K.M., Léger, N. & Shillito, B. (2016) Estimating symbiont abundances and gill surface areas in specimens of the hydrothermal vent mussel *Bathymodiolus puteoserpentis* maintained in pressure vessels. *Frontiers in Marine Science*, **3**, article 16.

Field, I.A. (1922) Biology and economic value of the sea mussel *Mytilus edulis*. *Bulletin of the US Bureau of Fisheries*, **38**, 127–259.

Ford, H. L., Schellenberg, S.A., Becker, B.J., Deutschman, D.L., Dyck, K.A. & Koch, P.L. (2010) Evaluating the skeletal chemistry of *Mytilus californianus* as a temperature proxy: effects of microenvironment and ontogeny. *Paleoceanography*, **25**, 1–14.

Furuhashi, T., Schwarzinger, C., Miksik, I., Smrz, M. & Beran, A. (2009) Molluscan shell evolution with review of shell calcification hypothesis. *Comparative Biochemistry and Physiology Part-B*, **154**, 351–371.

Galtsoff, P.S. (1964) The American oyster *Crassostrea virginica* Gmelin. *Fishery Bulletin*, **64**, 1–480.

Gantsevich, M.M. & Tyunnikova, A.V. (2005) The genetics of shell pigmentation of the Mediterranean mussel *Mytilus galloprovincialis* Lamarck, 1819 (Bivalvia, Mytilida). *Doklady Biological Sciences*, **404**, 370–371.

Gao, P., Liao, Z., Wang, X.X., Bao, L.F., Fan, M.H., Li, X.M. *et al.* (2015) Layer-by-layer proteomic analysis of *Mytilus galloprovincialis* shell. *PLoS ONE*, **10**, e0133913.

Gardner, J.P.A. & Thompson, R.J. (2009) Influence of genotype and geography on shell shape and morphometric trait variation among North Atlantic blue mussel (*Mytilus* spp.) populations. *Biological Journal of the Linnean Society*, **96**, 875–897.

Garner, Y.L. & Litvaitis, M.K. (2013a) Effects of injured conspecifics and predators on byssogenesis, attachment strength and movement in the blue mussel, *Mytilus edulis*. *Journal of Experimental Marine Biology and Ecology*, **448**, 136–140.

Garner, Y.L. & Litvaitis, M.K. (2013b) Effects of wave exposure, temperature and epibiont fouling on byssal thread production and growth in the blue mussel, *Mytilus edulis*, in the Gulf of Maine. *Journal of Experimental Marine Biology and Ecology*, **446**, 52–56.

Génio, L., Kiel, S., Cunha, M.R., Grahame, J. & Little, C.T.S. (2012) Shell microstructures of mussels (Bivalvia: Mytilidae: Bathymodiolinae) from deep sea chemosynthetic sites: do they have a phylogenetic significance? *Deep Sea Research I*, **64**, 86–103.

George, M.N., Pedigo, B. & Carrington, E. (2018) Hypoxia weakens mussel attachment by interrupting DOPA cross-linking during adhesive plaque curing. *Journal of the Royal Society Interface*, **15**, 20180489.

Giribet, G. (2008) Bivalvia. In: *Phylogeny and Evolution of the Mollusca* (eds W.F Ponder & D.R. Lindberg), pp 105–141. University of California Press, Berkeley, CA.

Gregory, M.A., Marshall, D.J., George, R.C., Anandraj, A. & McClurg, T.P. (2002) Correlations between metal uptake in the soft tissue of *Perna perna* and gill filament pathology after exposure to mercury. *Marine Pollution Bulletin*, **45**, 114–125.

Hagenau, A., Scheidt, H.A., Serpell, L., Huster, D. & Scheibel, T. (2009) Structural analysis of proteinaceous components in byssal threads of the mussel *Mytilus galloprovincialis*. *Macromolecular Bioscience*, **9**, 162–168.

Hahn, S., Rodolfo-Metalpa, R., Griesshaber, E., Schmahl, W.W., Buhl, D., Hall-Spencer, J.M. *et al.* (2012) Marine bivalve shell geochemistry and ultrastructure from modern low pH environments: environmental effect versus experimental bias. *Biogeosciences*, **9**, 1897–1914.

Hammond, W. & Griffiths, C.L. (2004) Influence of wave exposure on South African mussel beds and their associated infaunal communities. *Marine Biology*, **144**, 547–552.

Harrington, M.J. & Waite, J.H. (2007) Holdfast heroics: comparing the molecular and mechanical properties of *Mytilus californianus* byssal threads. *Journal of Experimental Biology*, **210**, 4307–4318.

Haszprunar, G. (1983) Comparative analysis of the abdominal sense organs of Pteriomorpha (Bivalvia). *Journal of Molluscan Studies, Supplement*, **12A**, 47–50.

Haszprunar, G. (1987) The fine morphology of the osphradial sense organs of the Mollusca. III. Placophora and Bivalvia. *Philosophical Transactions of the Royal Society B*, **315**, 37–61.

Hawkins, A.J.S. & Bayne, B.L. (1985) Seasonal variation in the relative utilization of carbon and nitrogen by the mussel *Mytilus edulis*: budgets, conversion efficiencies and maintenance requirements. *Marine Ecology Progress Series*, **25**, 181–188.

Holten-Andersen, N., Zhao, H. & Waite, J.H. (2009) Stiff coatings on compliant biofibers: the cuticle of *Mytilus californianus* byssal threads. *Biochemistry*, **48**, 2752–2759.

Hüning, A.K., Lange, S.M., Ramesh, K., Jacob, D.E., Jackson, D.J., Panknin, U. *et al.* (2016) A shell regeneration assay to identify biomineralization candidate genes in mytilid mussels. *Marine Genomics*, **27**, 57–67.

Innes, D.J. & Haley, L.E. (1977) Inheritance of a shell-color polymorphism in the mussel. *Journal of Heredity*, **68**, 203–204.

Innes, D.J. & Bates, J.A. (1999) Morphological variation of *Mytilus edulis* and *Mytilus trossulus* in eastern Newfoundland. *Marine Biology*, **133**, 691–699.

Inoue, N., Ishibashi, R., Ishikawa, T., Atsumi, T., Aoki, H. & Komaru, A. (2010) Gene expression patterns and pearl formation in the Japanese pearl oyster (*Pinctada fucata*): a comparison of gene expression patterns between the pearl sac and mantle tissues. *Aquaculture*, **308**, S68–S74.

Jackson, D.J., McDougall, C., Woodcroft, B., Moase, P., Rose, R.A., Kube, M. *et al.* (2010) Parallel evolution of nacre building gene sets in molluscs. *Molecular Biology and Evolution*, **27**, 591–608.

Jacob, D.E., Soldati, A.L., Wirth, R., Huth, J., Wehrmeister, U. & Hofmeister, W. (2008) Nanostructure, composition and mechanisms of bivalve shell growth. *Geochimica et Cosmochimica Acta*, **72**, 5401–5415.

Kamat, S., Su, X., Ballarini, R. & Heuer, A.H. (2000) Structural basis for the fracture toughness of the shell of the conch *Strombus gigas*. *Nature*, **405**, 1036–1040.

Katolikova, M., Khaitov, V., Väinölä, R., Gantsevich, M. & Strelkov, P. (2016) Genetic, ecological and morphological distinctness of the blue mussels *Mytilus trossulus* Gould and *M. edulis* L. in the White Sea. *PLoS ONE*, **11**, e0152963.

Kaushik, N.K., Kaushik, N., Pardeshi, S., Sharma, J.G., Lee, S.H. & Choi, E.H. (2015) Biomedical and clinical importance of mussel-inspired polymers and materials. *Marine Drugs*, **13**, 6792–6817.

Khati, W., Ouali, K., Mouneyrac, C. & Banaoui, A. (2012) Metallothioneins in aquatic invertebrates: their role in metal detoxification and their use in biomonitoring. *Energy Procedia*, **18**, 784–794.

Kinoshita, S., Wang, N., Inoue, H., Maeyama, K., Okamoto, K., Nagai, K. *et al.* (2011) Deep sequencing of ESTs from nacreous and prismatic layer producing tissues and a screen for novel shell formation-related genes in the pearl oyster. *PLoS ONE*, **6**, e21238.

Krapivka, S., Toro, J.E., Alcapán, A.C., Astorga, M., Presa, P., Pérez, M. *et al.* (2007) Shell-shape variation along the latitudinal range of the Chilean blue mussel *Mytilus chilensis* (Hupe 1854). *Aquaculture Research*, **38**, 1770–1777.

Lajus, D., Katolikova, M., Strelkov, P. & Hummel, H. (2015) Fluctuating and directional asymmetry of the blue mussel (*Mytilus edulis*): improving methods of morphological analysis to explore species performance at the northern border of its range. *Symmetry*, **7**, 488–514.

Langdon, C.J. & Newell, R.I.E. (1996) Digestion and nutrition in larvae and adults. In: *The Eastern Oyster Crassostrea virginica* (eds V.S. Kennedy, R.I.E. Newell & A.F. Eble), pp. 231–269. Maryland Sea Grant, College Park, MD.

Lee, B.P. & Konst, S. (2014) Novel hydrogel actuator inspired by reversible mussel adhesive protein chemistry. *Advanced Materials*, **26**, 3415–3419.

Lee, B.P., Messersmith, P.B., Israelachvili, J.N. & Waite, J.H. (2011) Mussel-inspired adhesives and coatings. *Annual Review of Materials Research*, **41**, 99–132.

Lindberg, D.R. & Sigwart, J.D. (2015) What is the molluscan osphradium? A reconsideration of homology. *Zoologischer Anzeiger*, **256**, 14–21.

Lucas, J.M., Vaccaro, E. & Waite, J.H. (2002) A molecular, morphometric and mechanical comparison of the structural elements of byssus from *Mytilus edulis* and *Mytilus galloprovincialis*. *Journal of Experimental Biology*, **205**, 1807–1817.

Lutz, R.A. (1976) Annual growth patterns in the inner shell layer of *Mytilus edulis*. *Journal of the Marine Biological Association of the UK*, **56**, 723–731.

Lutz, R.A. & Kennish, M.J. (1992) Ecology and morphology of larval and early postlarval mussels. In: *The Mussel Mytilus: Ecology, Physiology, Genetics and Culture* (ed E.M. Gosling), pp. 53–85. Elsevier Science Publishers B.V., Amsterdam.

MacKenzie, A.L. & Marshall, C. (2014) Proteins in the crystalline styles of the marine mussels *Perna canaliculus* Gmelin and *Mytilus galloprovincialis* Lamarck. *Journal of Shellfish Research*, **33**, 673–685.

Marin, F. & Luquet, G. (2004) Molluscan shell proteins. *Comptes Rendus Palevol*, **3**, 469–492.

Marin, F., Le Roy, N. & Marie, B. (2012) The formation and mineralization of mollusk shell. *Frontiers in Bioscience (Scholar Edn)*, **4**, 1099–1125.

McConnaughey, T.A. & Gillikin, D.P. (2008) Carbon isotopes in mollusk shell carbonates. *Geo-Marine Letters*, **28**, 287–299.

McDonald, J.H., Seed, R., Koehn, R.K. (1991) Allozymes and morphometric characters of three species of *Mytilus* in the Northern and Southern Hemispheres. *Marine Biology*, **111**, 323–333.

McQuaid, C.D., Lindsay, J.R. & Lindsay, T.L. (2000) Interactive effects of wave exposure and tidal height on population structure of the mussel *Perna perna*. *Marine Biology*, **137**, 925–932.

Medaković, D., Popović, S., Gržeta, B., Plazonić, M. & Hrs-Brenko, M. (1997) X-ray diffraction study of calcification processes in embryos and larvae of the brooding oyster *Ostrea edulis*. *Marine Biology*, **129**, 615–623.

Merrill, A.S. & Turner, R.D. (1963) Nest building in the bivalve genera *Musculus* and *Lima*. *Veliger*, **6**, 55–59.

Merten, B.J. (2013) Review of mussel adhesion mechanism and scoping study. Technical memorandum no. MERL-2013-43. Materials Engineering and Research Laboratory, Technical Service Center, Bureau of Reclamation, US Department of the Interior, Denver, CO.

Metian, M., Warnau, M., Hédouin, L. & Bustamante, P. (2009) Bioaccumulation of essential metals (Co, Mn and Zn) in the king scallop *Pecten maximus*: seawater, food and sediment exposures. *Marine Biology*, **156**, 2063–2075.

Metian, M., Warnau, M., Teyssié, J-L. & Bustamante, P. (2011) Characterization of ^{241}Am and ^{134}Cs bioaccumulation in the king scallop *Pecten maximus*: investigation via three exposure pathways. *Journal of Environmental Radioactivity*, **102**, 543–550.

Moeser, G.M. & Carrington, E. (2006) Seasonal variation in mussel byssal thread mechanics. *Journal of Experimental Biology*, **209**, 1996–2003.

Moeser, G.M., Leba, H. & Carrington, E. (2006) Seasonal influence of wave action on thread production in *Mytilus edulis*. *Journal of Experimental Biology*, **209**, 881–890.

Morishita, F., Furukawa, Y., Matsushima, O. & Minakata, H. (2010) Regulatory actions of neuropeptides and peptide hormones on the reproduction of molluscs. *Canadian Journal of Zoology*, **88**, 825–845.

Morse, D. & Rice, M.A. (2010) Mussel aquaculture in the Northeast. NRAC Publication No. 211–2010. University of Maryland, Baltimore, MD.

Morton, B. (1992) The evolution and success of the heteromyarian form in the Mytiloida. In: *The Mussel Mytilus: Ecology, Physiology, Genetics and Culture* (ed E.M. Gosling), pp. 21–52. Elsevier Science Publishers B.V., Amsterdam.

Nakamura Filho, A., Corrêa de Almeida, A., Espinoza Riera, H., Ferreira de Araújo, J.L., Pinto Gouveia, V.J. & de Carvalho, M.D. (2014) Polymorphism of $CaCO_3$ and microstructure of the shell of a Brazilian invasive mollusc (*Limnoperna fortunei*). *Materials Research*, **17**, 15–22.

Newcomb, L A, George, M.N., O'Donnell, M.J. & Carrington, E. (2019) Only as strong as the weakest link: structural analysis of the combined effects of elevated temperature and pCO2 on mussel attachment. *Conservation Physiology*, **7**, coz068.

Newkirk, G.F. (1980) Genetics of shell colour in *Mytilus edulis* L. and the association of growth rate with shell color. *Journal of Experimental Marine Biology and Ecology*, **47**, 89–94.

O'Donnell, M.J., George, M.N. & Carrington, E. (2013) Mussel byssus attachment weakened by ocean acidification. *Nature Climate Change*, **3**, 587–590.

Ornes, S. (2013) Mussels' sticky feet lead to applications. *Proceedings of the National Academy of Sciences*, **110**, 16697–16699.

Pearce, T. & LaBarbera, M. (2009) A comparative study of the mechanical properties of Mytilid byssal threads. *Journal of Experimental Biology*, **212**, 1442–1448.

Pechenik, J.A. (1991) *Biology of the Invertebrates*, 2nd edn. Wm. C. Brown (WCB) Publishers, Dubuque, Iowa.

Pechenik, J.A (2010) *Biology of the Invertebrates*, 6th edn. McGraw-Hill, New York.

Pereira, E.R., Soares, B.M., Vieira, J.P., Mai, A.C.G., Picoloto, R.S., Muller, E.I. *et al.* (2012) Assessment of inorganic contaminants in golden mussel (*Limnoperna fortunei*) in Southern Brazil. *Journal of the Brazilian Chemical Society*, **23**, 846–853.

Petrone, L., Ragg, N.L.C. & McQuillan, A. J. (2008) *In situ* infrared spectroscopic investigation of *Perna canaliculus* mussel larvae primary settlement. *Biofouling*, **24**, 405–413.

Pirie, B.J.S. & George, S.G. (1979) Ultrastructure of the heart and excretory system of *Mytilus edulis* (L.). *Journal of the Marine Biological Association of the UK*, **59**, 819–829.

Pouil, S., Bustamante, P., Warnau, M., Oberhansli, F., Teyssié, J-L. & Metian, M. (2015) Delineation of ^{134}Cs uptake pathways (seawater and food) in the variegated scallop *Mimachlamys varia*. *Journal of Environmental Radioactivity*, **148**, 74–79.

Richardson, C.A., Chenery, S.R.N. & Cook, J.M. (2001) Assessing the history of trace metal (Cu, Zn, Pb) contamination in the North Sea through laser ablation – ICP-MS of horse mussel *Modiolus modiolus* shells. *Marine Ecology Progress Series*, **211**, 157–167.

Roberts, L., Cheesman, S., Breithaupt, T. & Elliott, M. (2015) Sensitivity of the mussel *Mytilus edulis* to substrate-borne vibration in relation to anthropogenically generated noise. *Marine Ecology Progress Series*, **538**, 185–195.

Sagert, J. & Waite, H. (2009) Hyperunstable matrix proteins in the byssus of *Mytilus galloprovincialis*. *Journal of Experimental Biology*, **212**, 2224–2236.

Sakamoto, K., Uji, S., Kurokawa, T. & Toyohara, H. (2008) Immunohistochemical, *in situ* hybridization and biochemical studies on endogenous cellulose of *Corbicula japonica*. *Comparative Biochemistry and Physiolgy Part-B*, **150**, 216–221.

Seed, R. (1976) Ecology. In: *Marine Mussels: Their Ecology and Physiology* (ed B.L. Bayne), pp. 13–65. Cambridge University Press, Cambridge.

Seed, R. & Suchanek, T.H. (1992) Population and community ecology of *Mytilus*. In: *The Mussel Mytilus: Ecology, Physiology, Genetics and Culture* (ed E.M. Gosling), pp. 87–169. Elsevier Science Publishers B.V., Amsterdam.

Shirai, K., Takahata, N., Yamamoto, H., Omata, T., Sasaki, T. & Sano Y. (2008) Novel analytical approach to bivalve shell biogeochemistry: a case study of hydrothermal mussel shell. *Geochemical Journal*, **42**, 413–420.

Siddall, S.E. (1980) A classification of the genus *Perna* (Mytilidae). *Bulletin of Marine Science*, **30**, 858–870.

Silverman, H.G. & Roberto, F.F. (2007) Understanding marine mussel adhesion. *Marine Biotechnology*, **9**, 661–681.

Silverman, H.G. & Roberto, F.F. (2011) Byssus formation in *Mytilus*. In: *Biological Adhesive Systems: From Nature to Technical and Medical Application* (eds J. von Byern & I. Grunwald), pp. 273–283. Springer-Verlag, Vienna.

Sorte, C.J.B., Etter, R.J., Spackman, R., Boyle, E.E. & Hannigan, R.E. (2013) Elemental fingerprinting of mussel shells to predict population sources and redistribution potential in the Gulf of Maine. *PLoS ONE*, **8**, e80868.

Speiser, D.I. & Johnsen, S. (2008a) Comparative morphology of the concave mirror eyes of scallops (Pectinoidea). *American Malacological Bulletin*, **26**, 27–33.

Speiser, D.I. & Johnsen, S. (2008b) Scallops visually respond to the size and speed of virtual particles. *Journal of Experimental Biology*, **211**, 2066–2070.

Speiser, D.I. & Wilkens, L.A. (2016) Neurobiology and behavior of the scallop. In: *Scallops: Biology, Ecology Aquaculture and Fisheries*, 3rd edn. (eds S.E. Shumway & G.J. Parsons), pp. 219–251. Elsevier Science Publishers B.V., Amsterdam.

Taylor, J.D., Kennedy, W.J. & Hall, A. (1969) The shell structure and mineralogy of the Bivalvia. Introduction, Nuculacea-Trigonacea. *Bulletin of the British Museum of Natural History (Zoology)*, **S3**, 1–125.

Valladares, A., Manríquez, G. & Suárez-Isla, B.A. (2010) Shell shape variation in populations of *Mytilus chilensis* (Hupe1854) from southern Chile: a geometric morphometric approach. *Marine Biology*, **157**, 2731–2738.

Van der Molen, S., Márquez, F., Idaszkin, Y.A. & Adami, M. (2013) Use of shell-shape to discriminate between *Brachidontes rodriguezii* and *Brachidontes purpuratus* species (Mytilidae) in the transition zone of their distributions (south-western Atlantic). *Journal of the Marine Biological Association of the UK*, **93**, 803–808.

Waite, J.H. (1992) The formation of mussel byssus: anatomy of a natural manufacturing process. In: *Results and Problems in Cell Differentiation*, Vol. 19 (ed S.T. Case), pp. 27–54. Springer-Verlag, Berlin.

Waite, J.H., Vaccaro, E., Sun, C. & Lucas, J.M. (2002) Elastomeric gradients: a hedge against stress concentration in marine holdfast? *Philosophical Transactions of the Royal Society B*, **357**, 143–153.

Widdows, J. & Donkin, P. (1992) Mussels and environmental contaminants: bioaccummmulation and physiological aspects. In: *The Mussel Mytilus: Ecology, Physiology, Genetics and Culture* (ed E.M. Gosling), pp. 383–424. Elsevier Science Publishers B.V., Amsterdam.

Wiegemann, M. (2005) Adhesion in blue mussels (*Mytilus edulis*) and barnacles (genus *Balanus*): mechanisms and technical applications. *Aquatic Sciences*, **67**, 166–176.

Wilcox, R. (2013) *Review on the incidence of pearling in the blue mussel, Mytilus edulis: with relevance to the mussel beds of the Walney Channel, Barrow-in-Furness, Cumbria, England.* Cumbria Wildlife Trust, Kendal.

Yarra, T., Gharbi, K., Blaxter, M., Peck, L.S. & Clark, M.S. (2016) Characterization of the mantle transcriptome in bivalves: *Pecten maximus*, *Mytilus edulis* and *Crassostrea gigas*. *Marine Genomics*, **27**, 9–15.

Zardi, G.I., McQuaid, C.D. & Nicastro, K.R. (2007) Balancing survival and reproduction: seasonality of wave action, attachment strength and reproductive output in indigenous *Perna perna* and invasive *Mytilus galloprovincialis* mussels. *Marine Ecology Progress Series*, **334**, 155–163.

Zhadan, P.M. (2005) Directional sensitivity of the Japanese scallop *Mizuhopecten yessoensis* and Swift scallop *Chlamys swifti* to water-borne vibrations. *Russian Journal of Marine Biology*, **31**, 28–35.

Zhadan, P.M., Sizov, A.V. & Dautov, S.S. (2004) Ultrastructure of the abdominal sense organ of the scallop *Mizuhopecten yessoensis* (Jay). *Cell Tissue Research*, **318**, 617–629.

Zhu, Q. & Pan, Q. (2014) Mussel-inspired direct immobilization of nanoparticles and application for oil-water separation. *ACS Nano*, 8, 1402–1409.

3

Ecology of Marine Mussels

Introduction

Ecology is the study of how organisms interact with one another and with their physical environment. The distribution and abundance of organisms on Earth is shaped by both biotic (living organism-related) and abiotic (physical) factors. Ecology is studied at many levels, including at the level of the organism, population, ecosystem and biosphere. At the organismal level, ecologists study adaptations, beneficial features arising by natural selection, that allow organisms to live in specific habitats. These adaptations can be morphological, physiological or behavioural. At the population level, the size, density and structure of populations and how they change over time are studied. An ecosystem consists of all the organisms in an area, the community and the abiotic factors that influence that community. Ecosystem ecologists often focus on flow of energy and recycling of nutrients (reviewed in Dame 2012; Riisgård 2017). Finally, the biosphere is the planet Earth, viewed as an ecological system. Ecologists working at this level study global patterns, species distribution, interactions among ecosystems and phenomena that affect the entire globe, such as climate change (OpenStax College 2013).

The environment embodies everything outside the organism that impinges on it, including physical factors (e.g. temperature, salinity and light) and biological factors (e.g. predators, competitors and parasites). An organism's response to these factors influences its distribution and abundance on both a local and a regional scale. In this chapter, the effects of temperature and salinity – probably the two most important physical factors governing the distribution of marine organisms – will be described. Their effects on other aspects of mussel biology, such as feeding, reproduction, growth and respiration, and osmotic regulation, are covered in Chapters 4, 5, 6 and 7, respectively. Other limiting factors, such as aerial exposure, oxygen concentration, currents and substrate type, will be briefly reviewed in this chapter. In addition, effects of biological factors such as predation and competition will be considered. Pathogens and parasites (see Chapter 11) also influence the ecology of bivalve populations. Finally, the potential and observed impacts of climate change on marine ecosystems will be discussed.

Before describing the major factors that influence mussel distribution patterns, information on global and local ranges of representative marine mussel species is presented in the following section.

Marine Mussels: Ecology, Physiology, Genetics and Culture, First Edition. Elizabeth Gosling.

Global and Local Distribution Patterns

The World Register of Marine Species (WoRMS; www.marinespecies.org) is a database that provides an authoritative and comprehensive list of names of marine organisms. The content of the registry is edited and maintained on an ongoing basis by scientific specialists on each group of organisms. It provides information on the marine global distribution of individual species in specific geographical units, such as North Sea, Baltic Sea, Gulf of Mexico and Mediterranean Sea. Published sources occurrence records are also included. The registry is linked to numerous websites, including the Encyclopedia of Life (www.eol.org), a free online collaborative encyclopedia intended to document all of the 1.3 million species that have been identified and described on Earth to date. It is compiled from existing databases and from contributions by experts and non-experts throughout the world. It also provides distribution maps and records for individual species.

Global Distribution

Mussels in the genus *Mytilus* are a dominant component of rocky shore communities in cooler waters of the Northern and Southern Hemispheres. As described in Chapters 5 and 9, *Mytilus edulis* has a wide distribution in the Northern Hemisphere, occurring in European waters from Spitsbergen (74–81 °N; Berge *et al.* 2005) to western France, and on the Atlantic coast of North America from the Canadian Maritimes (New Brunswick, Nova Scotia and Prince Edward Island), along the coast of Maine (Hayhurst & Rawson 2009), to Cape Hatteras in North Carolina. In the subarctic and Arctic, *M. edulis* is the most abundant *Mytilus* species, being recorded in Arctic regions of Russia, along the Norwegian coast, in Iceland and on the west Greenland coast (Mathiesen *et al.* 2017 and references therein). Like *M. edulis*, *Mytilus trossulus* inhabits cool water regions of the Northern Hemisphere such as the coasts of the North Pacific, eastern Canada, the Baltic Sea, Scotland, Iceland, the Barents Sea, the White Sea and Norwegian fjords (Zbawicka *et al.* 2012; Brooks & Farmen 2013). It is also found from Alaska south to Monterey Bay, California (~38 °N). *Mytilus galloprovincialis* occurs in the Mediterranean Sea, the British Isles, continental Europe and on Moroccan and Mediterranean Tunisian shores (Lourenço *et al.* 2015), and it has been introduced, intentionally, for aquaculture into parts of the NE Pacific (e.g. Washington, USA and British Columbia, Canada; Wonham 2004). It is also found from Mexico north to Humboldt Bay, California (~38 °N). What seems like a stable hybrid zone between indigenous *M. trossulus* and invasive *M. galloprovincialis* extends between San Diego (~32 °N) and Humboldt Bay (Wonham 2004 and references therein). The species has also been reported in SW Greenland, Svalbard Island to the east of Greenland and in the Pechora Sea, NW Russia (Mathiesen *et al.* 2017). Some of the present distribution of *M. galloprovincialis* is due to accidental introductions into Hawaii (Apte *et al.* 2000), Japan (Wilkins *et al.* 1983), Hong Kong (Lee & Morton 1985), South Korea (Daguin & Borsa 2000), SE China (McDonald *et al.* 1990), southern California (Geller 1999), southern Chile (Toro *et al.* 2005; Larraín *et al.* 2018), the south and west coastlines of South Africa (McQuaid *et al.* 2015 and references therein) and Australia and New Zealand (Westfall & Gardner 2010; Ab Rahim *et al.* 2016 and references therein). In Australia, *M. galloprovincialis* is distributed throughout Tasmania, in Perth in Western Australia and along the mainland coast of Victoria, extending as far north as Port

Stephens in New South Wales. In New Zealand, its distribution includes the South and North Islands and all offshore islands, spanning approximately 35–52 °S latitude (1800 km). Other *Mytilus* species have a more restricted range. For example, the Californian mussel, *M. californianus*, is confined to the Pacific coast of North America, occurring from northern Mexico to the Aleutian Islands of Alaska (Addison *et al.* 2008), a distance of ~6000 km, while the Chilean mussel (*Mytilus chilensis*) is distributed along approximately 1900 km of the Chilean coast from the Arauco Gulf (35 °S) in the north to Cape Horn (55 °S) in the south (Araneda *et al.* 2016 and references therein).

Species in the genus *Perna* have a subtropical to tropical distribution in the Southern Hemisphere. The brown mussel, *Perna perna*, is found throughout South Africa and on the Atlantic coasts of Namibia and Andsteina; in the Cape Verde Islands and from Mauritania northward; on the Indian Ocean coast of Mozambique and the west coast of Madagascar; extending through the Gulf of Aden into the Red Sea; along the Atlantic coast of South America from Rio de la Plata, Argentina to Recife, Brazil; and on to the Caribbean shores of Venezuela (Wood *et al.* 2007). The mussel has invaded the western Gulf of Mexico, including Texas and northern Mexico (Hicks & McMahon 2002). The green mussel, *Perna viridis*, has an extensive Indo-Pacific distribution from the Persian Gulf through India and South East Asia as far north as Japan, Korea and southern China and as far east as Papua New Guinea (Wood *et al.* 2007; McDonald 2012). The species is a recent invader in the Caribbean Sea, including the subtropical southeastern United States (SE-US) region (Baker *et al.* 2007). The green-lipped or New Zealand mussel, *Perna canaliculus*, is endemic to New Zealand – with one exception, when very small numbers were reported at Port Adelaide in Southern Australia in 1996. There is no evidence that the species has since become established in the area (Wiltshire *et al.* 2010).

Modiolus modiolus, the horse mussel, is found on the European coasts of the Atlantic Ocean. It has been recorded from the White Sea and Norway, off the Faroes and Iceland, south to the Bay of Biscay, in the Mediterranean Sea, occasionally in North Africa and from Labrador to North Carolina in the Atlantic and from the Bering Sea south to Japan and to California in the Pacific (Tyler-Walters 2007). The bearded horse mussel, *Modiolus barbatus*, occurs from the British Isles south to Mauritania, West Africa and is also found in the Mediterranean Sea and along the Croatian coastline of the Adriatic Sea (Peharda *et al.* 2007). The eared horse mussel, *Modiolus auriculatus*, inhabits the western Indian Ocean, the Red Sea, the eastern Mediterranean, the North Atlantic Ocean, New Zealand, Mozambique and South Africa (www.marinespecies.org). The ribbed mussel, *Geukensia demissa*, occurs along the Atlantic coast of North America from Cape Cod, Massachusetts to the Gulf of Maine, Florida and the Gulf of Mexico (Franz 2001). In the late 1880s, it was introduced into San Francisco Bay on the Pacific West Coast of California, and since then has extended its range as far as Texas (GISD 2015) and Venezuela (Honig 2013 and references therein).

The charru mussel, *Mytella charruana*, is native to Central and South America, with a distribution on the Pacific coast from Mexico to Ecuador and on the South Atlantic coast from Columbia to Argentina (Yuan *et al.* 2016b). Since 2004, it has been found along the SE-US from South Carolina to Florida. The introduced SE-US population originated from multiple source populations, with the Panama region the primary one (Calazans *et al.* 2018). The mangrove mussel, *Mytella bicolor*, previously known as *Mytella guyanensis*, is found in intertidal zones of mangrove and estuarine regions from western Mexico to Peru and from Venezuela to Brazil (Introíni *et al.* 2010). The pharaonic mussel, *Brachidontes pharaonis*, is native to the Indian Ocean but has colonised the eastern part of the Mediterranean Sea via the Suez Canal, becoming abundant on the coasts of Israel, Egypt, Syria, Lebanon, southern Turkey, Greece, Croatia and Sicily (CABI 2020). The scorched mussel, *Brachidontes exustus*, is a complex of five species in the Gulf of Mexico and the Caribbean Sea, plus two little differentiated species on the Pacific coast of Central America (see Chapter 9). The purple mussel, *Perumytilus purpuratus*, once known as *Brachidontes purpuratus*, is distributed from Ecuador to the Strait of Magellan, the southern tip of South America, and also extends along the Atlantic coast up to La Lobería on the northern shore of Golfo San Matias, Argentina (Prada & Castilla 2006). The dwarf mussel, *Semimytilus algosus*, is native to Chile but is also found as an introduced species on the west coast of South Africa, extending some 500 km from Groenriviersmond in the north to Bloubergstrand in the south (de Greef *et al.* 2013). The Magellan or ribbed mussel, *Aulacomya atra*, is native to South American coasts, from southern Brazil to Tierra del Fuego on the Atlantic coast and from there to El Callao, Peru on the Pacific coast (Pérez *et al.* 2013). The species is also found on the coasts of New Zealand, from Namibia to Port Alfred in South Africa and in the Kerguelen Islands in the Indian Ocean. Tierra del Fuego, Argentina is the southernmost limit of the species distribution. *A. atra* has been introduced into the North Sea and to UK coasts (www.marinespecies.org). Like *Aulacomya atra*, *Choromytilus meridionalis*, the black mussel, is native to the west coasts of Namibia and the west and south coasts of South Africa (Grant *et al.* 1984). The Asian date mussel, *Arcuatula* (= *Musculista*) *senhousia*, is native to the Malay Peninsula and northward to Siberia. It has invaded estuaries in the Mediterranean, Australia and New Zealand and the NE Pacific from San Diego, California to British Columbia, Canada. Such globally widespread introductions were probably through ships' ballast water, by transport of planktonic larvae or adults rafting on debris or boat hulls, or via oyster mariculture (details in Asif & Krug 2012). The green crenella, *Musculus discors*, is common around most of the British Isles from Shetland to the Channel Isles, but also has a wider global distribution. It is found from the Arctic Circle south through the Bering Sea to Japan or to the Puget Sound in the Pacific, and south to New York in the Atlantic, including Norway, Sweden, Denmark, The Netherlands, Belgium, France, the North Sea, the western Baltic Sea and the Mediterranean Sea (www.marinespecies.org).

Bathymodiolus spp. are found in hydrothermal vents (hot springs) in the deep ocean. These typically form along the mid-ocean ridges, such as the East Pacific Rise and the Mid-Atlantic Ridge (MAR), located where two tectonic plates spread apart and new magma rises and cools to form new crust and volcanic mountain chains. Species such as *Bathymodiolus azoricus* and *B. puteoserpentis* dominate hydrothermal vent sites and cold seeps (vents with cooler, weaker flows), while small-sized mussels in the genera *Idas*, *Adipicola* and *Benthymodiolus* inhabit sunken wood, vegetal debris and bones on the ocean floor. Another genus, *Giganticus*, inhabits warm water seeps near active volcanoes in the Southern Hemisphere (see Chapters 1 and 9).

Local Distribution

On a local scale, *Mytilus* mussels dominate the intertidal to subtidal regions of rocky shores. *M. edulis* has the widest distribution pattern in the genus, extending from high intertidal to shallow subtidal (1–5 m) and even deeper waters, from estuarine to fully marine conditions, and from sheltered to extremely wave-exposed shores. On rocky shores of open coasts, the species attaches to rock surfaces and crevices, and in sheltered harbours and estuaries it occurs on rocks and piers, often as dense masses of one or more layers, with individuals bound together by byssus threads. Young mussels colonise spaces within the bed, increasing spatial complexity and providing numerous niches for other organisms (Seed 1996). At exposed sites, the mussel prefers gently sloping, slow-draining platforms to steep rock faces (Seed 1976). On such shores, dense beds of small mussels, often several layers thick, blanket the rocks (Figure 3.1). Local distributions on the upper and lower shores can be very erratic on a seasonal and annual basis, due to irregular cycles of settlement, competition, predation, wave action, exposure to air and desiccation, and detachment during storms (see later). Death of underlying individuals may detach the mussel bed from the substratum, leaving the bed vulnerable to tidal scour and wave action (Seed & Suchanek 1992). On sheltered shores and in estuaries, fewer but larger mussels are found (Figure 3.2). In these habitats, infaunal beds may occur on gravel or even quite sandy areas, although it is likely that some harder substratum embedded within more sandy areas is required. Subtidal populations on pier pilings, oil platforms and culture ropes enjoy a virtually predator-free environment, and this, combined with continuous submersion, allows *M. edulis* to reach a large size in a relatively short period of time (e.g. 50 mm shell length in six to eight months; Page & Hubbard 1987). In its native range, *M. galloprovincialis* is found from exposed rocky outer coasts to sandy bottoms (Ceccherelli & Rossi 1984). As an invader, it typically requires rocky coastlines with a high rate of water flow. The mussel is able to outcompete and displace native mussels and becomes the dominant mussel species in certain localities. For example, survivorship and tolerance to air exposure and desiccation in *M. galloprovincialis* are reported to be double those of any of the indigenous species in South Africa (Branch & Steffani 2004). Experiments in which mussels were held for 42 weeks at the high tide mark, where they experienced seven days of continuous exposure to air, resulted in 92% survivorship in *M. galloprovincialis* compared to 78% for *Perna perna*, 37–46% for *Choromytilus meridionalis* and only 0–10% for *Aulacomya atra*. Also, *M. galloprovincialis* grows faster than native mussels and has a reproductive output between 20 and 200% greater than that of indigenous species (Van Erkom Schurink & Griffiths 1993).

Figure 3.1 Layers of small mussels (*Mytilus*) on barnacle-covered rocks on a semi-exposed shore at Clogherhead, Louth, Ireland.

Figure 3.2 A cluster of mussels (*Mytilus*) on a sheltered shore at Lough Hyne, Cork, Ireland. Mussels in this habitat can reach shell lengths of 50–60 mm or greater.

In New Zealand, *P. canaliculus* is distributed widely throughout the two main islands, but it is more common in the warmer northern parts of the country. Dense beds of up to 100 individuals m^{-2} can be found in northern coastal areas, which can include rocky reefs, wharf piles and soft bottom habitats. Intertidal (mid-littoral[1]) populations are limited by aerial exposure, while subtidal (down to 50 m) populations are limited by predation pressure. Environmental parameters, such as temperature and salinity, are strong determinants of the distribution of this species. While its temperature range is from 5.3 °C in the south to 27 °C in the north, this mussel appears to prefer the warmer waters of bays and estuaries in northern coastal areas. A wide range of salinities are tolerated by the species, although the optimum range is 30–35 psu (Alfaro *et al.* 2011 and references therein). *P. viridis* mussels generally inhabit intertidal, subtidal and estuarine environments with high salinity. They form a thick carpetlike growth (up to 35 000 individuals m^{-2}) on rocky surfaces and submerged structures like wharves, pilings, breakwaters and buoys, and they play a very important role in rocky shore ecosystems. The species colonises submerged rocks, wood, concrete, metal, old submerged logs, boats, PVC pipes, ropes, muddy sea bottoms and even seagrass beds and aerial roots of mangroves. *P. viridis* tolerates a temperature range of 15–32.5 °C without much problem. For example, the species can survive a temperature of 39 °C for about 200 min (Rajagopal *et al.* 1995), which is considerably greater than the thermal tolerance of *P. perna* (see later). Also, *P. viridis* thrives well at winter water temperatures as low as 12 °C and has a higher degree of adaptability to salinity changes than *P. perna* (Rajagopal *et al.* 2006).

P. perna, a native of tropical and subtropical rocky shores, colonises jetties, navigation buoys, petroleum platforms, wrecks and other artificial hard substrata. Hicks & McMahon

(2002) found that the mussel's long-term, incipient lower and upper thermal limits were 7.5–30 °C, similar to the seasonal ambient water temperature range of 10–30 °C reported for other populations worldwide. These narrow incipient thermal limits, a low capacity for temperature acclimation and a poor freeze resistance may account for the mussel's restriction to subtidal and lower eulittoral zones of cooler subtropical rocky shores. Adult salinity tolerance ranges from 19 to 44 psu, which is lower than that reported for a nonindigenous Texas population (15–55 psu) (GISD 2019).

The larvae of the ribbed mussel, *Geukensia demissa*, settle on subtidal oyster reefs, in intertidal salt marshes and on artificial structures in these habitats. Sometimes, they attach to one another in aggregations or to clumps of hollow grass stems (*Spartina alterniflora*) in low marshes. They are most abundant at the lowest shore levels within salt marshes, and occur in small numbers in the high marsh zone above the average high watermark (GISD 2015). The highest water temperature tolerated is 56 °C, and salinities are tolerated from nearly freshwater up to 70 psu, which is twice as salty as the open ocean (Cohen 2005).

Modiolus modiolus occurs, partly buried, in soft sediments or coarse grounds or attached to hard substrata, forming clumps or extensive beds or reefs. The mussel may be found on the lower shore in rock pools or in laminarian holdfasts but is more commonly found subtidally to a depth of ~280 m (Tyler-Walters 2007). Dense populations of very young *M. modiolus* do occasionally occur subtidally in estuaries, although the species is more poorly adapted to fluctuating salinity than other mussel species (Bayne 1976). *M. barbatus* is found among rocks and stones on the lower shore, extending into the sublittoral to depths of around 100 m. *Modiolus auriculatus* is found attached to rocks and in rock crevices in the littoral and sublittoral zones to a depth of 25 m. It also inhabits sandy mudflats and sometimes intertidal reefs in rock pools and crevices. Abdel-Razak *et al.* (2014) examined the distribution of the species over the intertidal zone, coastal fringing reefs, offshore reefs and lagoons along the Red Sea coast of Egypt. They identified several diverse habitats: flat and rocky with a few sandy depressions in both intertidal and subtidal zones, with moderate vegetation of macro algae and sea grass beds and dead and live small coral patches; mangrove trees and tidal zones that reach to about 100 m; and a substrate consisting of small gravel pieces mixed with sand, extending down to ~70 m, with a dense vegetation of macro algae and sea grasses.

In the Mediterranean, *Brachidontes pharaonis* is confined to high-temperature and high-salinity habitats, where it has established very high densities, completely covering rock surfaces at intertidal sites (Safriel *et al.* 1980; Sarà *et al.* 2008). Adult mussels show large temperature tolerances (9–31 °C) and occur at salinities from 35 to 53 psu. Lower winter temperatures limit their physiological activity (Galil 2006). *B. exustus* was believed to commonly inhabit rock pilings, seawalls and wharf pilings in the intertidal zone and was most abundant in the lower intertidal (Seed 1980). However, DNA barcoding has since revealed that *B. exustus* is actually a complex of five species (see earlier and Chapter 9). *Perumytilus purpuratus* dominates the mid and mid to high intertidal zones on rocky shores, forming dense three-dimensional matrices that serve as a microhabitat for a wide variety of organisms. Consequently, these mussels have been defined as ecosystem bioengineers that play an important role in regional biodiversity (Pérez *et al.* 2008). Prado & Castilla (2006) tested the hypothesis that factors determining the habitat structural complexity and environmental heterogeneity of *P. purpuratus* matrices have significant effects on the associated macrofaunal community. They found 92 invertebrate taxa in *P. purpuratus* matrices. The number of layers (stratification) in the matrix had a significant effect on evenness (degree of similarity in abundance among the species): the greater the stratification, the lower the evenness index. Sediment retention by matrices in sheltered sectors had a significant effect on evenness: greater sediment retention resulted in lower evenness. Sediment retention also determined significant differences in macrofaunal assemblages. In matrices without sediment retention, mussel layering and the presence of algae on the shells of *P. purpuratus* determined significant macrofaunal differences. As a dominant competitor, *P. purpuratus* plays a major role in intertidal rocky shores where it is present, structuring communities and determining local biodiversity. *Semimytilus algosus* is another bioengineer mussel, cohabiting with *P. purpuratus* on most Chilean rocky shores. However, unlike *P. purpuratus* (see earlier), *S. algosus* occurs in the low intertidal zone and is a weak competitor of *P. purpuratus*, and *S. algosus* post-settlers show high mobility to relocate in the intertidal, which may serve not only to allow this mussel to reach the optimal physiological position in the intertidal, but also to reduce competition interaction (Brante *et al.* 2019).

Arcuatula senhousia inhabits littoral and sublittoral rock and other hard substrata, and also hard and soft substrates in the intertidal and shallow subtidal zones to 20 m depth. The mussel may settle on hard surfaces, but prefers to settle gregariously on soft substrates, burrowing until only the hind part of the shell protrudes, after which the species secretes fibrous threads that attach to sediment particles to form a kind of nest or byssal cocoon around the shell. Gregarious settlement results in the formation of large colonies, raised several centimetres above the surface of the sand or mud (NIMPIS 2002). In some areas of California, the mussel has been reported to reach densities of 15 000 m^{-2}. The byssus mats formed by the mussel may smother native biota and can make the underlying seabed anoxic. *A. senhousia* is tolerant of low salinity and low oxygen levels. In San Francisco Bay, it has been collected at salinities of 17–33 psu and temperatures of 17–24 °C, and in southern California at 35–37 psu and 25–27 °C (Cohen 2005). *Aulacomya atra* is well known for its presence in deep water. For example, in Punta Arenas Cove, Chile, natural banks of the species are found at depths of 15 to >30 m, where temperatures vary from 12 to 16 °C. The mussel is also commonly found in shallow nearshore kelp-bed communities and algae holdfasts in Central Chile, subantarctic and Antarctic waters, South Africa and cold and deep waters of the Kerguelen Islands (Caza *et al.* 2016). In South Africa, where *A. atra* is indigenous, its existence is compromised in many areas by the presence of alien species, such as *M. galloprovincialis*, which has a faster growth rate and a higher resistance to exposure to air on exposed and semi-exposed shores compared to *A. atra*, which is predominantly a subtidal species (see later on interspecific competition). *Choromytilus meridionalis* is most abundant in the lower intertidal and subtidal zones and is geographically restricted to areas of cool upwelled water on the west and south coasts of South Africa. The mussel co-occurs with *P. perna* on the southern coast of S. Africa. The two species exhibit a degree of habitat separation, with *P. perna* inhabiting sand-free rocks and *C. meridionalis* inhabiting rocks subject to frequent sand cover (Marshall & McQuaid 1993).

Musculus discors is distributed from the shallow subtidal to depths up to 50 m in the British Isles. The mussel is found in scattered, gregarious clumps growing epiphytically on the holdfasts of seaweeds and among assemblages of attached animals growing on hard substrata. It occasionally forms extensive, dense aggregations covering upward-facing rock surfaces (Tyler-Walters 2001). Once attached, adults weave a 'nest' of several thousand fine byssus threads around their shells, so that they are suspended in a network of threads, like a 'ball of twine'. The nest completely encloses the adult so that the mussel is only visible when its valves are open and it is feeding with siphons extended (Merrill & Turner 1963; see Chapter 5).

The charru mussel, *Mytella charruana*, is an epifaunal tropical (6–36 °C; Brodsky *et al.* 2009) and subtropical mussel. Its indigenous habitat is typically shallow lagoons and mudflats in bays, where it reaches high densities (de Oliveira *et al.* 2005 and references therein). Mussels are found tightly attached to a variety of naturally submerged and intertidal substrates, such as muddy sand, rocks, oyster shells, water intake pipes, wood pilings, driftwood floating docks and boat hulls. Juvenile and adult individuals are euryhaline, surviving at a wide range of salinities (2–40 psu; Yuan *et al.* 2010) at 20 °C, although the salinity tolerance range narrows as the temperature decreases or increases (Yuan *et al.* 2016b). *Mytella guyanensis* is a widespread euryhaline bivalve that occurs half-buried in the intertidal zone of mangroves and estuaries from Baja California, Mexico to Santa Catarina, Brazil (de Oliveira *et al.* 2005).

Factors Affecting Geographic Distribution

Temperature and salinity not only set limits on the spatial distribution of bivalves but also affect every aspect of biology, including feeding, reproduction, growth, respiration, osmoregulation and parasite-disease interactions (see details in Chapters 4, 5, 6, 7, 11 and 12). When it comes to distribution on a large geographic scale, it is generally recognised that temperature plays a more important role than salinity. The synergistic effect of temperature and salinity, acting in concert with other environmental variables such as water depth, substrate type, wave action, food availability, water turbidity and the occurrence of competitors, predators and disease, can have more profound consequences than either factor acting alone. Geographical distribution is also governed by hydrographic barriers to larval dispersal, such as oceanic currents, confluences, gyres and surface water stratification.

Temperature, Salinity and Hydrographic Factors

Most marine bivalves live within a temperature range from −3 to 44°C (Vernberg & Vernberg 1972). Within this range, the degree of temperature tolerance is species-specific, and within individual species early embryos and larva have a narrower temperature tolerance than adults (see Chapter 5). In addition, the temperature required for spawning is invariably higher than the minimum temperature required for growth. All of these factors set limits on the natural distribution of individual species on both regional and local scales. A few relevant examples from marine mussels will elucidate this.

As described earlier, the distribution of *M. edulis* on the Atlantic coast of North America extends from the Canadian Maritimes southward to Cape Hatteras in North Carolina, the historical southern limit for this species. It is believed that the northward-moving warm water Gulf Stream meets the southward-moving cold Labrador Current in the region of

Cape Hatteras (35.25 °N), thus providing a temperature barrier for the distribution and survival of mussel larvae south of this point. Field experimental data, combined with a modelling approach, indicate that the southern limit of *M. edulis* is mediated by intolerance to high summer temperatures (Jones *et al.* 2009). Since 1960, seasonal air and water temperatures have increased along the eastern US seaboard, and south of Lewes, Delaware (38.8 °N) summer sea surface temperature (SST) increases have exceeded the upper lethal limits (32 °C) of *M. edulis* (Jones *et al.* 2010), resulting in geographic contraction of the southern, equatorward range edge of *M. edulis*, shifting it approximately 350 km north of the previous limit at Cape Hatteras. At the southern part of the range, high water and air temperatures cause mass mortality events, while along the more northerly portion mortality is caused by high temperatures during aerial exposure. Ultimately, water temperatures in excess of thermal tolerances have caused contraction of the mussel's biogeographic range.

Two *Mytilus* conspecifics, *M. trossulus* and *M. galloprovincialis*, co-occur on the West Coast of the United States, and both have exhibited changes in biogeographic ranges over several decades. The range of *M. trossulus* once extended from Baja California north to Alaska. However, when the Mediterranean mussel, *M. galloprovincialis*, was introduced into southern California in the first half of the 20th century, the invasive mussel was able to completely displace the native *M. trossulus* throughout the southern portion of its range. Fields *et al.* (2006) investigated whether biochemical adaptation to temperature might potentially play a role in invasion success. An examination of cytosolic malate dehydrogenase (cMDH), an enzyme known to exhibit distinct patterns of temperature adaptation in kinetic properties, showed that a minor change in structure permits *M. galloprovincialis* cMDH to function at warmer temperatures and may be a part of a broad suite of molecular adaptations that has allowed this species to displace its congener throughout the warmer part of *M. trossulus*' original range in North America, and also in Japan. In a subsequent study, Lockwood & Somero (2011a) analysed a variety of physiological, biochemical and molecular systems, including cardiac function, enzymatic activity and gene expression, in the same two species. In all comparisons, *M. galloprovincialis* was more warm-adapted than *M. trossulus*. Higher activities of enzymes involved in ATP generation show that the native *M. trossulus* is better adapted to colder conditions than *M. galloprovincialis*. The higher thermal tolerance of the invasive species is likely due in part to its enhanced ability to induce changes in the expression of particular genes and proteins in response to acute heat stress. Taken together, these data predict that *M. galloprovincialis* will continue to be the dominant blue mussel species along the warmer range of the California Current,[2] yet its northward spread is limited. In support of this, Hilbish *et al.* (2010) reported that from 1995 to 1999 the poleward movement of *M. galloprovincialis* showed a reversal concomitant with a cooling phase of the Pacific Decadal Oscillation (PDO),[3] an important driver of climate. *M. galloprovincialis* has declined in abundance over the northern third of its geographic range (~540 km), and has become rare or absent across the northern 200 km of the range it previously colonised during its initial invasion. The distribution of the native species *M. trossulus* has, however, remained unchanged over the same time period. The difference in SST between warm and cold phases of the PDO is small (~1 °C), but Hilbish *et al.* (2010) deduced that even this minor decrease during the cold phase of the PDO might be enough to retard larval development in *M. galloprovincialis*, such that recruitment is handicapped in northern waters.

The Asian green mussel, *P. viridis*, has been introduced into coastal waters of Florida; it was also reported in coastal Georgia in 2003 and has spread as far north as South Carolina (Benson 2019). Its current distribution is assumed to be limited by low temperatures during winter. To test this, Urian *et al.* (2011) analysed lethal and sublethal effects of cold water and air temperatures in two size classes of the mussel from Florida in an effort to determine the effects of current and forecasted temperatures on the potential for range expansion. Mussels were exposed to water temperatures of 14, 10, 7 and 3 °C for up to 30 days, or to air temperatures of 14, 7, 0 and −10 °C for two hours. Mortality was significantly increased at all water and air temperatures ≤14 °C. No differences in mortality rates were observed between small (15–45 mm) and large (75–105 mm) size classes except after exposure to 7 °C air, in which small mussels had higher mortality. The temperature threshold for survival in this population appears to be between 10 and 14 °C, suggesting that under present conditions *P. viridis* may already be at the northern edge of its potential range in the United States, unless with global climate change northerly flowing currents permit range expansion beyond the coastal waters of South Carolina.

Beside rising air and water temperatures, global climate change may also entail increases in precipitation, and estuarine species in particular may be exposed to increasing hypoosmotic stress due to decreasing salinities impacting on species' distribution ranges (Somero 2012). Indeed, increased precipitation leading to reduced salinity in estuarine habitats may in some cases be more important in governing local distributions than changes in temperatures (Somero 2012; Braby & Somero 2006a,b). The open ocean has surface salinities between 33 and 37 psu, with an average of 35 psu. In contrast, estuaries and bays are subject to pronounced salinity fluctuations because of evaporation, rainfall and inflow from rivers. Many mussels, in particular *Mytilus* spp., are euryhaline (i.e. they can tolerate an extremely wide range of salinity, 4–40 psu, in their natural environment). In the northern Baltic, *M. trossulus* is living at the margin of its salinity tolerance (4.5 psu), and although dwarfed by the low-salinity conditions, the species is very abundant in this area (Westerbom *et al.* 2002; see Chapter 7 for salinity tolerance values in marine mussels).

Tomanek *et al.* (2012) examined the proteomic responses to hyposaline stress in *M. trossulus* and *M. galloprovincialis*, whose ranges overlap on the west coast of N. America (see earlier). Mussels were exposed to short time periods (4 hr) of hyposaline stress, followed by a recovery period to mimic conditions typical for bays and coastal areas experiencing heavy freshwater input, with a quick return to full salinity with incoming tides and mixing with full-strength seawater. The differences in protein abundances in gill tissue suggested that *M. trossulus* was able to respond to a greater hyposaline challenge (24.5 psu) than *M. galloprovincialis* (29.8 psu). These differences, in a scenario of reduced coastal salinities, may enable *M. trossulus* to cope with greater hyposaline stress and outcompete *M. galloprovincialis* in the southern part of the *M. trossulus* range, thereby preventing *M. galloprovincialis* from expanding northward. Interestingly, Lockwood & Somero (2011b), in a transcriptomic analysis of gill tissue of *Mytilus* specimens from the same experiment (but limited to 29.8 and 35.0 psu), found no major differences in salinity stress tolerance between the two congeners at the level of transcriptional regulation.

P. viridis and *Mytella charruana* are two species of non-native marine mussels that have recently invaded Florida, United States; as they are now well established, they are potential causes of concern along SE-US coasts. Both species have been documented on intertidal oyster reefs in Florida (Spinuzzi *et al.* 2013), where they have a negative impact on the commercially important eastern oyster *Crassostrea virginica* (Yuan *et al.* 2016a). The simultaneous effects of salinity and temperature on the two species were investigated to get a better understanding of their respective invasion potentials (Yuan *et al.* 2016b). Survival at three salinity ranges (5–9, 20–22.5 and 35–40 psu) in both cold and warm water was estimated for juveniles and adults of both species. Yuan *et al.* (2016b) found that *P. viridis* can survive at a wide range of temperatures (9–35 °C) when the salinity is 35–37 psu, but as salinity decreases, the thermal survival range for *P. viridis* becomes narrower. The data for *M. charruana* indicate that juvenile and adult individuals can survive at a wide range of salinities (5–40 psu) at 20 °C, but the salinity tolerance range narrows as the temperature decreases or increases. Also, temperature rapidly impacts survival of both species (within hours), while salinity impacts are more gradual (days to weeks). As the species are tolerant of many salinity–temperature combinations found in SE-US, they should be able to both persist in their present invaded range and expand into new estuaries and bays along the Atlantic coastline of Florida and west within the Gulf of Mexico. Freezing weather, however, could impede expansions into colder waters. Given that both species have a detrimental effect on *C. virginica* and possibly on many other native species, the abiotic variables could be used to help predict successful introductions and future biogeographic expansions.

In 2004, Berge *et al.* (2005) were the first to report the presence of settled mussels (*M. edulis*), for the first time since the age of the Vikings (793–1066 AD), in the high Arctic Archipelago of Svalbard – a Norwegian archipelago between mainland Norway and the North Pole. Their data indicated that most mussels settled at a single site as spat in 2002 and that larvae were transported by the West Spitsbergen Current northward from the Norwegian coast to Svalbard the same year. The extension of the mussel's distribution range was made possible by the unusually high northward mass transport of warm Atlantic water resulting in elevated sea surface temperatures in the North Atlantic and along the west coast of Svalbard. This was suggested to be a one-time event as all mussels were of similar age. Recently, Leopold *et al.* (2019) carried out a detailed field survey with scientific divers to map the *Mytilus* spp. distribution along the west coast of Svalbard, with a focus on the west coast of Spitsbergen, the largest island of the Svalbard archipelago, where strong Atlantification (transition from cold, fresh Arctic waters to a warm, salty Atlantic regime) has been documented over the last few decades. Low densities (<0.5 individuals m^{-2}) of live *Mytilus* were recorded at 21 locations on the west coast of Svalbard and Spitsbergen. Using genetic markers, the authors found that these were mainly *M. edulis*, although some samples were classified as *M. galloprovincalis* or as hybrids of both *M. edulis*/*M. galloprovincalis* and *M. edulis*/*M. trossulus*. However, Leopold *et al.* (2019) regarded the three species as one complex, referring to them as '*Mytilus* spp.' throughout their publication. The oldest mussel collected settled around the year 2000, indicating, despite low mussel densities at all sites, that settlement of blue mussels is a recurring event, with evidence of recruitment every year since 2000. The presence of *Mytilus* on Svalbard is the probable outcome of at least two distinct dispersal vectors: natural larval advection[4] by ocean currents and human introduction by ship traffic. Based on the estimated year of settlement of mussels found in non-harbour locations, Leopold *et al.* (2019) concluded that the settlement of *Mytilus* spp. is an annual event, rather than the unique one suggested by Berge *et al.* (2005), which implies that current environmental conditions are favourable for continued persistence of Svalbard's *Mytilus* spp. populations.

The coast of South Africa comprises three broad biogeographic regions: the cool-temperate western coast, the warm-temperate southern coast and the subtropical eastern coast; these are aligned with two globally important currents and characterised by a clear decrease of upwelling intensity and frequency and by primary production from west to east (McQuaid *et al.* 2015 and references therein). The two currents are the upwelling-driven Benguela Current (BC) on the southwestern coast and the warm, temperate Agulhas Current (AC), which follows the eastern and southern coasts (Figure 3.3). Upwelling is believed to have strong effects on the advection of larvae between coastal waters and the intertidal. In southern Africa, the indigenous mussel, *P. perna*, dominates the subtropical and warm-temperate bioregions from central Mozambique to the Cape of Good Hope, west of Cape Agulhas on the southwest coast of S. Africa. The species is absent from there to central Namibia due to the cold, upwelling waters of the Benguela system – a distributional gap of more than 1000 km. From there, *P. perna* extends northward along the west coast of Africa to the Mediterranean Sea, as far as the Gulf of Tunis (Van Erkom Schurink &

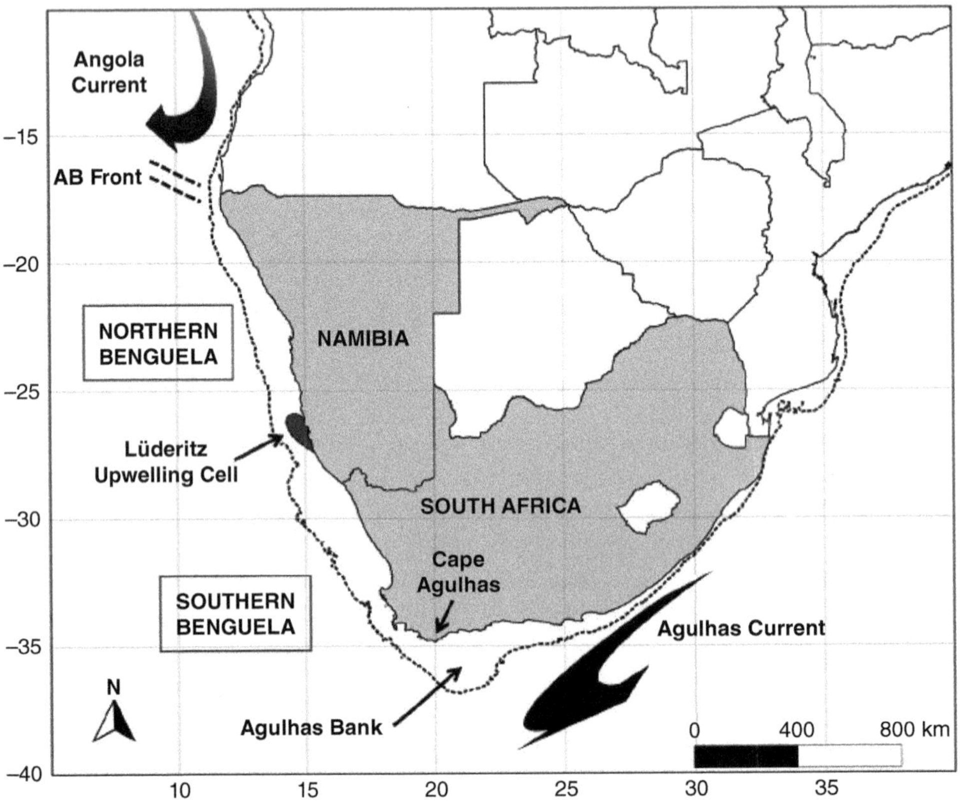

Figure 3.3 Map of the Benguela Current region bordering Namibia and South Africa, showing the 500 m depth contour (dashed line) and the approximate locations of the Lüderitz upwelling cell (which separates the northern and southern Benguela Current subsystems) and the Agulhas Current. *Source:* From Roux *et al.* (2013). Reproduced with permission from the Bulletin of Marine Science.

Griffiths 1990; Zardi *et al.* 2015). The invasive *M. galloprovincialis* is considered to have arrived on the western coast of South Africa via shipping in the 1970s (Grant & Cherry 1985). The species subsequently spread both north and south, with its northerly spread being more rapid under the influence of the north-flowing BC, which flows from the Cape of Good Hope to Lüderitz, after which it is deflected away from the coast to the northwest. The northern limit of *M. galloprovincialis* on the west coast is in central Namibia, where conditions may represent the limit of the species' temperature tolerance (Zardi *et al.* 2007a). The second distributional limit for *M. galloprovincialis* occurs on the SE southern African coast. There, the warm AC, which flows to the southwest, diverges away from the coast, thereby reducing SST significantly. And because warm waters tend to have fewer nutrients than cold waters, the distributional limits of *M. galloprovincialis* may also be affected by unfavourable trophic conditions beyond these boundaries, in addition to prohibitive thermal conditions (Assis *et al.* 2015).

There is increasing modelling and experimental evidence that pronounced alterations to oceanographic features (dominant currents and upwelling systems) due to climatic change are rearranging species' distributions globally (Lourenço *et al.* 2017). Researchers at the University of Miami found that the Indian Ocean's warm Agulhas Current is getting wider rather than strengthening (Beal & Elipot 2016). Their findings, which have important implications for global climate change, suggest that intensifying winds in the region may be increasing the turbulence of the current, rather than increasing its flow rate. In the case of the Benguela Current, global climate change has led to a rise in the temperature of this cold current. Not only has there been a rise in water temperature, but the water has become increasingly saline (Global Climate 2019). An El Niño[5] effect has already been detected. The outcome of these changes from the Benguela El Niño is expected to have a dramatic effect on marine life on the southwest coast of Africa. How exactly it will impact global distributions of marine mussels remains to be seen.

Factors Affecting Local Distribution

Marine mussels on a local scale dominate the intertidal to subtidal regions of rocky shores (see earlier). Mussels extend from high intertidal to shallow subtidal (1–5 m) and even deeper waters, from estuarine to fully marine conditions and from sheltered to extremely wave-exposed shores. On rocky shores of open coasts, mussels attach to the rock surface and crevices, and in sheltered harbours and estuaries they attach on rocks and piers, often occurring as dense masses of one or more layers, with individuals bound together by byssus threads. Young mussels colonise spaces within the bed, increasing spatial complexity and providing numerous niches for other organisms. It comes as no surprise that temperature and salinity, which play a key role in geographic distribution, are also important in determining species distribution on a local scale. For subtidal species, additional factors such as water depth, substrate type, currents, turbidity and predation and competition play an important role. Upper distributional limits in the intertidal zone are believed to be determined primarily by desiccation and temperature stress, while lower limits are determined by biological factors, particularly predators. Anthropogenic factors such as waterborne pollutants, introduced species and disease can also be significant in determining local distribution patterns.

Rocky shore ecosystems are governed by the tidal movement of water, which creates zonation patterns from high to low tide. The area above the spring high-tide mark is called the supralittoral (splash) zone, which is regularly splashed but not submerged by seawater, except during storms with high tides (Purcell 2018). The area around the high-tide mark is known as the intertidal fringe; organisms in this sparse habitat include anemones, barnacles, crabs, lichens, limpets, mussels, periwinkles and whelks, with very little vegetation. Below this zone is the intertidal, with three zones: high, mid and low, based on the average exposure. Each of these zones has its own characteristic animal and plant community. The high intertidal area is only flooded during high tides; common organisms are anemones, barnacles, chitons, crabs, isopods, mussels, sea stars and snails. This zone can also contain rock pools inhabited by small fish and larger seaweeds. The mid intertidal zone is covered and uncovered twice a day with saltwater from the tides. Therefore, temperatures are less extreme due to shorter direct exposure to the sun, and salinity is only marginally higher than ocean levels. However, wave action in the mid intertidal zone is generally more extreme than in the high tide and splash zones. Typical organisms are snails, sponges, sea stars, barnacles, mussels, crabs and brown and green algae. The low intertidal is only exposed on unusually low tides. Common organisms in this region are brown and green seaweed, crabs, hydroids, mussels, limpets, sea cucumber, sea urchins, sea stars, shrimp and snails. These organisms are not well adapted to long periods of dryness or to extreme temperatures. Below the low-tide mark is the littoral zone, which can again be divided into three zones based on areas of tidal action, from shallow to deep: the supralittoral zone, the eulittoral zone and the sublittoral zone. The littoral zones are much more stable than the intertidal zones. Different types of oysters, star fish, sea urchins, coral, crabs and anemones live in the littoral zones, some of which are significant predators of marine mussels (see later).

Physical Factors

Temperature

Organisms in the rocky intertidal zone have to cope with being out of water at regular intervals. For those in the high intertidal, emersion times are longest, and consequently these individuals are often subjected to temperature extremes and desiccation. Upper distribution limits for a species are set by its ability to tolerate such extremes via a well-coordinated set of physiological, behavioural, biochemical and molecular adaptive responses (McQuaid *et al.* 2015).

Considering the enormous amount known about the ecology and physiology of mussels, the main focus in this section will be on *Mytilus* species and the impacts of thermal stress on them in the rocky intertidal zone – a habitat that has been described as among the most physically harsh environments on Earth (Tomanek & Helmuth 2002), but which has emerged as a model system for investigating the ecological impacts of global climate change (Zhou *et al.* 2018; see later). When mussels are submerged during high tide, their body temperature is close to that of the surrounding water. However, during aerial exposure at low tide, body temperatures are driven by the interactions of climatic factors such as solar radiation, cloud cover, wind speed, relative humidity and air and ground temperatures, which drive the flux of heat into and out of the mussels' bodies (Helmuth & Hofmann 2001; Helmuth 2002 and references therein). Consequently, body temperatures can be considerably higher (≥20 °C) than those experienced during submersion and can vary substantially from surrounding air and substrate temperatures (Helmuth 1999). It is important to note that the rate of heat transfer between an organism and its environment is determined to some extent by the size and morphology of the organism, and can be strongly affected by characteristics such as colour and material properties (Helmuth 2002). Also, changes in shell coloration, presence or absence of the periostracum and algal growth on the shell are all likely to modify a mussel's reflectivity and hence its body temperature (Helmuth 1999). Consequently, organisms exposed to identical environmental conditions can experience quite different body temperatures (Helmuth 2002). For example, barnacles keep a relatively large proportion of their total surface area in contact with the underlying substrate and display body temperatures that are tightly coupled with ground temperature. In contrast, mussels are predicted to have body temperatures that are largely decoupled from the temperature of the underlying substrate, at least while living in beds of conspecifics, and are often several degrees warmer than the surrounding air. Daily variations in temperature may also be extremely different between individuals that are just a few metres apart but located on surfaces facing different directions, or in or out of crevices or other shaded areas (Helmuth *et al.* 2010).

Until recently, measuring body temperatures in the intertidal zone was inefficient or unfeasible at large spatial or temporal scales, as temperature loggers were large and expensive and required frequent servicing (Lima *et al.* 2011). In addition, the intertidal zone, with its rapidly varying temperatures, tremendous wave forces and hard substrata, often caused loss of sensors and data loggers. In the early 2000s, however, advances in technology made it possible to measure the body temperatures of intertidal organisms over long time scales, because commercially available temperature loggers were sufficiently small and robust to be deployed for long periods of time. As wtih the organism being monitored, the colour, morphology and mass of a temperature logger can significantly affect the temperature that it records. Therefore, biomimetic sensors/loggers (robomussels) that are specifically matched (modified) to the thermal characteristics of the mussels being measured have been developed. These can provide realistic measurements of body temperature in the field, over a broad variety of temporal and spatial scales (Fitzhenry *et al.* 2004). Robomussels have the shape, size and colour of actual mussels (Figure 3.4), with miniature built-in sensors that track temperature inside the mussel beds. The fact that they are self-contained, robust and inconspicuous makes it unlikely that they will be intentionally destroyed. Robomussels are made from a polyester resin with a black colouring, which is poured into moulds created from *Mytilus* shells. A TidbiT logger (Figure 3.4) is embedded in the polyester resin, which after hardening is smoothed and shaped to resemble a real mussel. The logger battery lasts for more than five years and the memory stores ~42 000 12-bit temperature measurements (enough for more than two and a half years) with a sampling frequency of 30 min (can be set from 1 s to 18 hr). The loggers have an accuracy of ±0.20 °C at 0–50 °C (details in Fitzhenry *et al.* 2004; Lima *et al.* 2011). In the field, robomussels are usually deployed on hard rock in intact mussel beds using epoxy putty, taking care to ensure that they are completely surrounded by other mussels, since loggers deployed as solitary

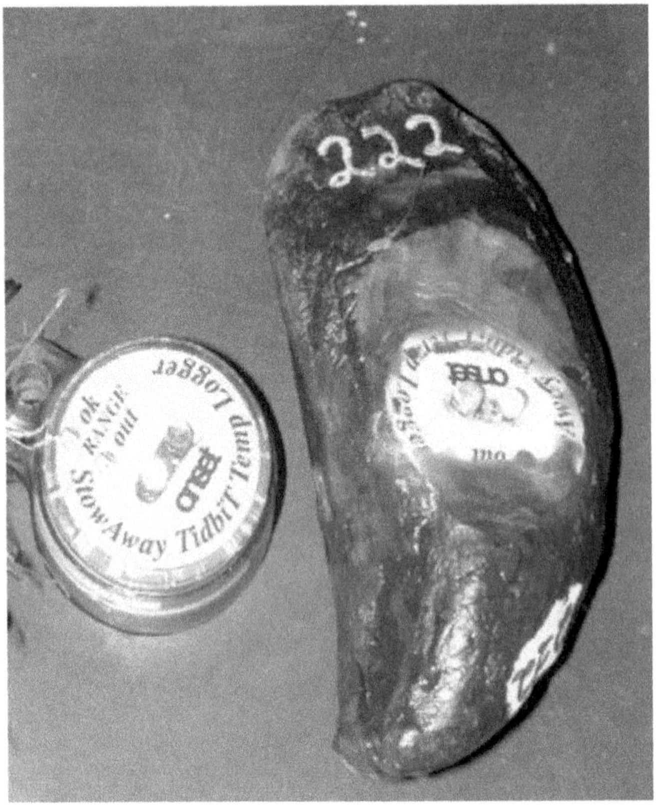

Figure 3.4 A'robomussel' (right) molded from polyester resin and containing a TidbiT logger thermally matched to the characteristics of a living mussel. On the left is an unmodified Onset TidbiT logger. *Source:* From Fitzhenry *et al.* (2004). Reproduced with permission from Springer Nature.

individuals tend to yield abnormally high readings. Logger programming and data retrieval is done through the instrument's LEDs, which are exposed on the outside of the robomussel. An optical USB interface allows users to offload data in seconds. Biomimetic loggers have also been used in animals such as limpets, barnacles, snails and seastars. See Helmuth (2002) and Fitzhenry *et al.* (2004) for the potential pitfalls of logger use, Lima *et al.* (2011) for different types of biomimetic data loggers and Judge *et al.* (2018) for recent advances in temperature logging in intertidal systems.

In the first study in which robomussels were deployed, Helmuth & Hofmann (2001) continuously monitored temperatures at a location in central California for a period of two years, using loggers designed to mimic the thermal characteristics of *M. californianus*. Model mussel temperatures were recorded on both a horizontal (Figure 3.5) and a vertical, north-facing microsite and in an adjacent tidepool. Mussels at each microsite were periodically measured for levels of heat shock proteins (Hsp 70), a measure of thermal stress (Fields *et al.* 2012). The role of Hsps in the stress response is to assist misfolded proteins to attain or regain their native state and prevent heat-damaged protein accumulation in cells, thereby

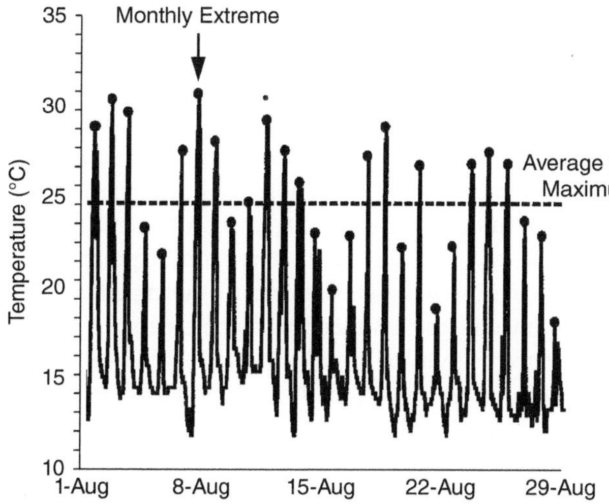

Figure 3.5 Example of fluctuations in temperature experienced over one month (August 1999) at a horizontal microsite. Daily maxima were calculated from temperature data collected every 5–10 min. The highest daily maximum was recorded as the monthly extreme ('acute') high temperature at each site. The average of the daily maxima was calculated as a measure of 'chronic' high-temperature exposure. Similarly, average daily minima and monthly minima were calculated. *Source:* From Helmuth & Hofmann (2001). Reprinted with permission from The University of Chicago Press.

preventing the formation of cytotoxic aggregates. Other factors that have a bearing on an individual's response to thermal stress include: size, shape, mass and colour (Helmuth 2002), vertical position in the sediment (Jost & Helmuth 2007), geographic location (Fields *et al.* 2012), timing of low tide (Mislan *et al.* 2009) and interaction with stressors such as anoxia (Veldhuizen-Tsoerkan *et al.* 1991), salinity fluctuations (Podlipaeva & Berger 2012), hypoxic conditions (Anestis *et al.* 2010), reduced food supply (Schneider *et al.* 2010), pollutants (Mahmood *et al.* 2014; Banni *et al.* 2017) and pathogens (Cellura *et al.* 2007). In Helmuth & Hofmann (2001), mussel temperatures were consistently higher on the horizontal surface than on the vertical one, and differences between these surfaces were reflected in the amount of Hsp70 produced. Temperatures recorded in the tidepool were not as high as those on the exposed horizontal microsite but in general exceeded those recorded on the aerially exposed vertical face. One interesting finding of the study was that seasonal peaks in extreme (acute) high temperatures did not always coincide with peaks in average daily maxima (chronic high temperatures). A subsequent study showed that thermal stress may depend not only on exposure to temperature extremes but also on the thermal history of the temperature signal (Helmuth 2002). For example, on days when daily maximum temperatures increase gradually, mussels may be able to acclimate better to temperature extremes compared to days when temperature maxima are preceded by cool days. Helmuth (2002) also found through deployment of loggers at multiple sites that each site had a unique thermal signature due to interactions of terrestrial climate, tidal cycle and wave exposure, but also significant within-site differences due to tidal height and substratum angle.

In a more recent study, Olabarria *et al.* (2016) investigated the effect of a heatwave on the physiological and behavioural responses in monospecific or mixed aggregations of the invasive mussel *Xenostrobus securis*, which has successfully colonised the inner part of the Galician Rias Baixas (NW Spain), where it co-occurs with the commercially important mussel *M. galloprovincialis* (see Chapter 10). In a mesocosm experiment, mussels were exposed to simulated tidal cycles and similar temperature conditions to those experienced in the field during a heatwave that occurred in the summer of 2013, when field robomussels registered temperatures up to 44.5 °C at low tide. In monospecific aggregations, *M. galloprovincialis* was more vulnerable than *X. securis* to heat exposure during emersion. However, in mixed aggregations, the presence of the invader was associated with lower mortality in *M. galloprovincialis*. The greater sensitivity of *M. galloprovincialis* to heat exposure was reflected in a higher mortality level, greater induction of Hsp70 protein and higher rates of respiration and gaping activity (opening of shell valves to permit evaporative cooling), which were accompanied by a lower heart rate (bradycardia). It appears that the invader enhanced the physiological performance of *M. galloprovincialis*, highlighting the importance of species interactions in regulating responses to environmental stress. In a complementary study, Nicastro *et al.* (2012) found that under conditions of heat stress, aggregations of the gaping mussel *P. perna* exhibited lower mortality rates than isolated occurences or aggregations of the nongaping mussel *M. galloprovincialis*, because the gaping behaviour of *P. perna* ameliorated stressful environmental conditions of mussels through evaporative cooling (see Chapter 7 for details on gaping). The drawback of this response is an increased risk of desiccation. Large mussels have greater amounts of water available in their tissues than small ones (Kennedy 1976; Helmuth 1998), which provides greater protection from desiccation and enables them to use evaporative cooling for longer periods. To test this, laboratory experiments were conducted to determine the sensitivity of mussels (*M. trossulus*) to the full range of temperatures and desiccation levels experienced in the field (Jenewein & Gosselin 2013). Mussels (1–2 mm shell length) experienced a threshold of heat tolerance at 34 °C and a threshold of desiccation tolerance at vapour pressure deficit levels of 1.01 kPa. Extended periods of temperatures reaching or exceeding lethal levels for newly settled *M. trossulus* occur relatively rarely in Barkley Sound, British Columbia, Canada, the study mussel collection site, which has a consistently high *M. trossulus* settlement. Extended periods of temperatures reaching or exceeding lethal levels for newly settled *M. trossulus* occurred relatively rarely at this site, whereas lethal levels of desiccation occurred often during the recruitment season and were usually sustained for several hours, indicating that desiccation appears to be a substantially greater threat to recently settled *M. trossulus* than heat. Mussels became highly tolerant to desiccation when they reached a size of 2–3 mm shell length. This closely corresponds to the size at which juvenile *M. trossulus* relocate from protective filamentous algal habitat to adult habitat, suggesting ontogenetic shifts in habitat use by juvenile *M. trossulus* are a response to changing sensitivity to desiccation. In a scenario of global warming, survival of newly settled mussels, and thus possibly the persistence of mussel populations, will likely depend even more upon the persistence of protective microhabitats created by filamentous and fucoid algae.

Body temperature during aerial exposure is driven by multiple interacting climatic factors, and extremes during low tide far exceed those during submersion (Helmuth 2002).

Physiological effects of thermal stress in both low- and high-tide conditions were compared between *M. galloprovincialis* and *M. trossulus* under laboratory conditions (Schneider 2008). To simulate the natural range of tidal thermal stress, mussel populations of the two species were established in seawater tanks that mimicked a daily tidal cycle by filling and draining at different times. During high tide, mussels were completely immersed in water, while during low tide, they were exposed to one of three aerial temperature treatments: 20 °C (cool), 25 °C (warm) or 30 °C (hot). A subtidal control group, which remained underwater during the entire tidal cycle, was also established. Separate experiments were carried out in two different water temperatures, 18 and 12 °C, which represented those pertaining where the species co-occur in California (Braby & Somero 2006a) and found during the summer along the west coast of the United States. In 18 °C water, there was no effect of the aerial treatments on growth or survival in either species. In contrast, in 12 °C water, aerial exposure affected the survival and growth of both species. Growth and survival rates of *M. galloprovincialis* were higher in all conditions than those of *M. trossulus*, especially in the 18 °C water experiments and in the aerial exposure treatments of the 12 °C water experiment. *M. galloprovincialis* seems to be warm-adapted with regard to both low- and high-tide thermal stress. These results suggest that *M. galloprovincialis* will spread, at least compared to *M. trossulus*, as water and air temperatures increase in a global warming scenario.

M. californianus has a large distributional range (Washington to southern California, United States) and experiences a correspondingly wide geographical range in temperature. Therefore, one might ask to what extent its temperature regime can be predicted from latitudinal position, or even from a few temperature measurements taken at a site in air or water over a relatively short period of time. To investigate this, Helmuth *et al.* (2006) used long-term (five years) high-frequency temperature data from biomimetic sensors deployed at 10 locations along the West Coast of the United States (Figure 3.6). Their objective was to test the hypothesis that local modifying factors, such as the timing of low tide in summer, can lead to large-scale geographic mosaics of body temperature. The results indicated that patterns of body temperature during aerial exposure at low tide varied in physiologically meaningful and often counterintuitive ways, over large sections of the species' geographic range. An evaluation of spatial correlations among sites to explore how body temperatures change along the latitudinal gradient showed that 'hot spots' and 'cold spots' exist where temperatures are hotter or colder than expected based on latitude. Four major hot spots and four cold spots were identified along the entire geographic gradient, with at least one hot spot and one cold spot occurring in regions as climatically distinct as Washington/Oregon, Central California and Southern California. This pattern was driven by differences in the timing of low tides among regions. At northern latitudes, low tides in the summer occur at midday and expose intertidal organisms to high aerial temperatures, while at southern sites low tides during summer occur mostly at night and organisms are thus often submerged during hot days. Temporal autocorrelation analysis of year-to-year consistency and temporal predictability in the mussel body temperatures revealed that southern mussels experience higher levels of predictability in thermal signals compared to northern animals. Helmuth *et al.* (2006) also explored the role of wave splash at a subset of sites and found that while average daily temperature extremes varied between sites with different levels of wave splash, yearly extreme temperatures were often similar, as were patterns of

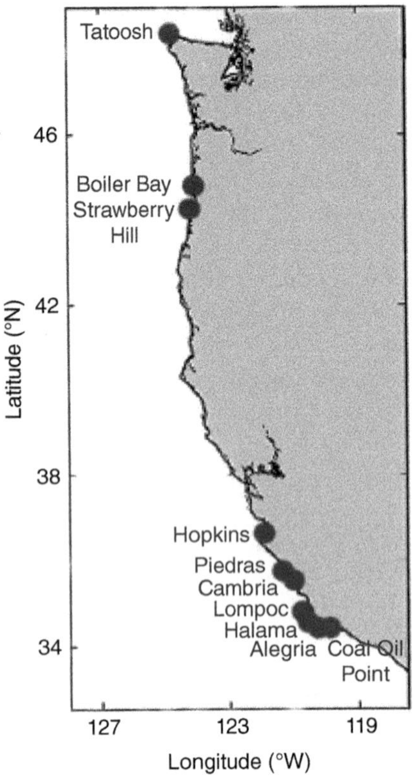

Figure 3.6 Map of the 10 deployment sites used by Helmuth *et al.* (2006) along the United States Pacific coast. *Source:* From Helmuth *et al.* (2006). Reproduced with permission from John Wiley and Sons.

predictability. In summary, rather than simple latitudinal gradients, these intertidal mussels experience a complex thermal mosaic, with many potential variables affecting the thermal state of an intertidal organism. The results of this study highlight the importance of quantitatively assessing biogeographic patterns in temperatures and other environmental variables at scales relevant to the organisms themselves and of forecasting the impacts of changes in climate across species ranges.

Between 1998 and 2016, Helmuth *et al.* (2016) further deployed biomimetic sensors that approximated the thermal characteristics of intertidal mussels at 71 sites worldwide. Biomimetic loggers (robomussels) were programmed to record at intervals of 10–30 min and were left in the field for periods up to nine months before they were removed for downloading and replaced with another logger. The loggers were used at multiple intertidal elevations to estimate temperatures of the mussels *M. californianus* (west coast of North America), *M. edulis* and *Geukensia demissa* (east coast of North America), *M. chilensis* (Chile), *P. perna* (South Africa) and *P. canaliculus* (New Zealand). Unmodified ('off-the-shelf') commercial loggers were also deployed on rock surfaces at multiple sites (Australia, Ireland, Mexico, Scotland, United Kingdom, United Staes), which recorded temperatures relevant to barnacles, newly settled mussels and other organisms that are sufficiently small that their temperatures mirror those of the underlying rock. Comparisons against direct measurements of mussel tissue temperature indicated errors of ~2.0–2.5 °C during daily fluctuations that often exceeded 15–20 °C. Geographic patterns in thermal stress, based on biomimetic logger measurements, were generally far more complex than expected, based only on 'habitat-level' measurements of air temperature or SST. This unique data set provides an opportunity to link physiological measurements with spatially- and temporally-explicit field observations of body temperature in intertidal mussels.

In cold conditions, mobile intertidal species can hide in rock crevices or migrate to deeper water to avoid freezing. But sessile bivalves, often exposed to subzero temperatures during winter, do not have such protection. In northeastern Canada, temperatures can drop to −35°C in winter. Mussels (*M. trossulus*) survive such low temperatures even when their tissue temperatures are as low as −10°C (Williams 1970), with large adults surviving laboratory conditions of −16 °C for 24 hours (Bourget 1983). As much as 60% of their extracellular fluid (ECF) is frozen at this temperature. The unfrozen ECF becomes more concentrated with solutes, and this process draws water by osmosis out of cells, thus lowering the intracellular freezing point. The high osmotic concentration of the ECF places an osmotic stress on the cells that can damage membranes and enzymes. This damage can be minimised through the production of cryoprotectants (e.g. the amino acids glycine, alanine and proline), as well as the end products of anaerobic metabolism (e.g. lactate, succinate and strombine (Loomis *et al.* 1988; Loomis & Zinser 2001) and glucose (Gionet *et al.* 2009)). Calcium also acts as a cryoprotectant in the mussel *G. demissa* by binding to cell membranes and reducing cell damage during freezing, either through physical stabilisation of the membrane against mechanical disruption caused by cell shrinkage or by prevention of the denaturation of membrane compounds (Ansart & Vernon 2003). Another mechanism to avoid intracellular ice formation – an invariably lethal process – is the production of ice-nucleating proteins, which are secreted into the ECF and act to induce and control extracellular ice formation. These proteins reduce undercooling from the range −15 to −20 °C to the range of −5 to −10 °C. *G. demissa* is a freeze-tolerant saltmarsh mussel that is regularly exposed to subzero temperatures for extended periods during low tides. The species' cold tolerance varies seasonally, ranging from a lower lethal temperature of −10 °C in the summer to one of −13 °C in the winter. Although it lacks ice-nucleating proteins, it utilises at least one strain of ice-nucleating bacteria, *Pseudomonas fulva*, from seawater. These bacteria, which are found in the gills of *G. demissa*, could perform the same function as hemolymph ice-nucleating proteins by limiting ice formation to extracellular compartments (Loomis & Zinser 2001).

A decrease in temperature usually reduces membrane fluidity, which can lead to membrane dysfunction. Ectotherms typically counteract this temperature effect by remodelling membrane lipids via changes in phospholipid head groups, fatty acid composition and cholesterol content. Lipid dynamics and the physiological responses of cold-adapted *M. edulis* and the warmer-water oyster, *Crassostrea virginica*, were compared during simulated overwintering and onward to spring conditions (Pernet *et al.* 2007; see also Fokina *et al.* 2018). To simulate overwintering, bivalves were held at 0, 4 and 9 °C for three months. They were then gradually brought up to 20 °C and held at that temperature for five weeks, to simulate spring–summer conditions. Major differences were observed in triglyceride (TAG) metabolism during overwintering. TAGs are the main constituents of natural fats and oils. Mussels used digestive gland TAG stores for energy metabolism or reproductive processes during the winter, whereas oysters did not accumulate large TAG stores prior to overwintering. Mussel TAG contained high levels of 20:5n−3 compared to levels in oysters, which may help to counteract the effect of low temperature by reducing the melting point of the TAG, thereby increasing the availability of storage fats at low temperature. The physiological responses of mussels to temperature change and acclimation to high and low temperatures are dealt with in more detail in Chapter 7.

Salinity

On fully marine shores, mussels experience a salinity of about 35 psu most of the time. *M. edulis* is likely to encounter hyper-saline conditions in tide pools, crevices and sediments exposed on hot, breezy days, where evaporation can increase salinity to values as high as 42 psu (Tyler-Walters 2008). In contrast, large amounts of rainfall dilute standing water on the shore, decreasing its salinity. But since many mussels, in particular *Mytilus* spp., are euryhaline, they can tolerate an extremely wide range of salinity (4–40 psu) in their natural environment (see Chapter 7 for details on salinity tolerances in marine mussels). Mussels in subarctic Norway commonly occur in shallow intertidal pools. By living in pools, they are insulated against low air temperatures but exposed to high salinities beneath overlying ice. They avoid exposing their tissues to salinities as high as 65 psu because a shell valve closure response to low temperature operates at about −1.5 °C, before ice sheets form and bottom water salinities rise (Davenport & Carrion-Cotrina 1981). The increase in salinity appears to take several hours, but the return to normal seawater salinities occurs in just a few minutes when the ice is melted and the pools are flushed out by the relatively warm (3–5 °C) water of the incoming tide. In estuarine waters, mean salinity decreases and salinity variation increases with distance upstream, and both these factors have deleterious effects on bivalve distribution, with the result that species diversity is significantly less in estuaries than on fully marine shores.

For marine species in general, salinity is one of the most important abiotic factors contributing to distribution and affecting many physiological rates (Dame 2012; see Chapter 7). To understand the invasion potential of *Mytella charruana*, newly introduced to the southeastern United States, Yuan *et al.* (2010) examined salinity thresholds for this species, which inhabits colonising rocky substrates in estuaries. Their study addressed the following questions: (1) In what range of salinities can this species survive if it is slowly adjusted to test salinities? (2) In what range of salinities can this species survive when it experiences rapid changes of salinity? and (3) In what range of salinities can this species survive when it experiences temporary, rapid changes of salinity (6 hr duration). They tested survival in salinities ranging from freshwater to hypersaline conditions (0–45 psu) and determined whether mussel size affected experimental results. All experiments examined survivorship of mussels by increasing or decreasing the salinity from the field value under laboratory conditions. Mortality in each tank was recorded daily for 43 days for the gradual adjustment trials and for 12 days for permanent and six-hour shock trials. Large mussels (20–54 mm) survived best in salinities from 2 to 23 psu, with 100% mortality at 0 psu and 45 psu with gradual adjustment. Small mussels (3–19 mm) survived in a wider range of salinities (2–40 psu) with gradual adjustment to new salinities. However, survival of both large and small mussels was significantly lower in permanent shock trials at salinity extremes. Six-hour shock trials had no effect on survival at any of the test salinities (0–45 psu) for both large and small *M. charruana*. Acclimation experiments on Philippine charru mussels (Rice *et al.* 2016) provided very similar results, except that significant size-based differences in salinity tolerances were not detected. Both studies indicated that charru mussels have the capacity to invade a wide variety of saline environments with significant freshwater or marine input, particularly when they have an opportunity to acclimate to gradual changes in salinity with significant freshwater or marine input, and possibly surviving wide salinity swings in Philippine estuaries associated with the rapid onset of abundant monsoonal rains. See also the study (described earlier) of Yuan *et al.* (2016b), who investigated the simultaneous effects of salinity and temperature on two invasive species, *M. charruana* and *Perna viridis*, in order to obtain a better understanding of their respective invasion potentials.

Habitat salinity may also determine the outcome of competition between native and invasive mussel species (see Tomanek *et al.* 2012 and Lockwood & Somero 2011b, described earlier, and Sarà & de Pirro 2011, described in Chapter 7).

Wave Exposure

Wave exposure, through both wave force and changes in immersion patterns, has a powerful influence on patterns of zonation and abundance on rocky shores. Lack of tolerance to high wave forces may limit species composition in this habitat, but because intertidal zonation patterns are driven by emersion time, wave action tends to extend biological zones vertically (upshore) by effectively decreasing the frequency and duration of emersion. High wave action will therefore cause a point on the shore to behave as if it is effectively lower than its still water tidal height. Harley & Helmuth (2003) described a method for measuring effective shore level (ESL), a metric that combines the influence of wave splash and tidal regime on patterns of emersion and immersion. They identified immersion events as sharp drops in temperature recorded by submersible dataloggers and compared the tide height at the time of the temperature drop to the wave height recorded by an offshore buoy. Using this method, Gilman *et al.* (2006) compared the effects of wave action on immersion patterns within multiple sites on the Pacific coast of North America. They deployed miniature temperature loggers at fixed intertidal heights at each site and recorded temperatures at intervals of 5–15 min for periods of up to five years, and developed a programme that can be used to identify temperature drops in a datalogger time series and calculate daily and monthly summary statistics of temperature. They concluded that ESL parameters provide a useful metric for comparing the effects of wave action on immersion patterns within sites. A sensor has now been designed to measure wave forces and wave heights on a time and size scale appropriate for organisms inhabiting the surf zone of rocky intertidal shores. The instrument is capable of continuous deployments for a duration of several months. The sensor/logger package consists of a three-axis force transducer connected to a resin model of a mussel (*M. edulis*), a custom-made datalogger in a waterproof housing and their respective communication cables (details in Lima *et al.* 2011).

Bivalves on wave-exposed shores are primarily mussels (*Mytilus* and *Modiolus*), often forming dense, single-species zones. Mussels on these shores are subjected to high water velocities from breaking waves, often as high as 25 m s^{-1} when waves interact with the local topography of a shore (Denny *et al.* 2003). One might expect that the higher the waves, the larger the forces imposed on intertidal organisms. However, while Helmuth & Denny (2003) found that 146 out of 221 microsites at a moderately exposed shore in central California showed a positive correlation between force and wave height; at some of these microsites force increased nonlinearly toward a statistically defined limit, making it difficult to classify a shore in regard to its exposure.

One might also expect that when higher water velocities are encountered, mussels will produce more byssal threads to increase strength of attachment. However, this is not the case, as shown by Moeser *et al.* (2006), who reported that in flume experiments mussels (*M. edulis*) significantly reduced thread production at water velocities above 15 cm s^{-1} (but see contrary findings in Zardi *et al.* 2007b). Similar findings were reported in a subsequent study on four mussel species, *M. trossulus*, *M. galloprovincialis*, *M. californianus* and *Modiolus modiolus* by Carrington *et al.* (2008). For all four species, velocities above 20 cm s^{-1} visibly hindered the mussel's ability to extend its foot beyond the margin of the shell, a posture that must be held for several minutes in order to mold and attach a new thread to the substrate. Even in the exposed shore species, *M. californianus*, velocities above 30 cm s^{-1} precluded thread formation. Altogether, this study established 50 cm s^{-1} as a reasonable threshold limiting byssal thread production in solitary mussels. The authors found that flow was greatly ameliorated within mussel aggregations, ranging from 0.1 to 10% of free-stream velocity, thus explaining why mussels can persist on shores with water flows in excess of their physiological limits.

Wave action also has a controlling influence on mussel bed communities by causing dislodgement through lift and drag, especially when mussel beds are dense and firmly packed, as on the majority of wave-exposed shores (Figure 3.1). When dislodgement occurs, new space is created for colonisation. The risk of dislodgement in *M. edulis* increases with flow speed and mussel size and decreases with mussel tenacity or attachment strength; the latter varies twofold during the year, but this cycle is not aligned precisely with seasonal patterns of wave velocity (Carrington 2002). In a subsequent study, mussels (*M. galloprovincialis*) on rocky shores were found to allocate resources to reduce risk of dislodgement (smaller, thicker shell, stronger byssal threads) instead of promoting growth and reproduction (Babarro & Carrington 2013).

Denny (1995 and references therein) outlined a method for using information regarding the long-term 'waviness' of the ocean to predict the rate of disturbance for individual organisms. This approach potentially provides a mechanism for 'predicting the weather' at arbitrary sites on the shore and can thus serve as a mechanistic link between wave climate and the rate of physical disturbance of individuals in wave-swept populations. The rate of disturbance can then be used to predict the practical limits to a species' distribution. Several studies have used this mechanistic model to predict frequency and severity of mussel dislodgement (Hunt & Scheibling 2001; Schneider *et al.* 2005; Carrington *et al.* 2009; Cole & Denny 2014). To illustrate, using inputs of wave height and mussel attachment strength on wave-exposed shores on Rhode Island, United States (Bass Rock and Black Point), Carrington *et al.* (2009) found that when averaged over three years, the predictions of the mechanistic model matched observed patterns, with minimal dislodgment from November to July and peak dislodgment in the hurricane season between August and October (Figures 3.7 and 3.8). Overall, the mechanistic model accounted for over 80% of the variation in mean monthly wave dislodgment during their study (linear regression, $r^2 = 0.85$, $P < 0.001$, $N = 12$, observed $= 0.08 + 0.574 \times$ predicted).

Mussel aggregations provide a habitat for a diverse assemblage of organisms, some of which attach directly to the mussel shells (epibionts). Three-dimensional epibionts

Figure 3.7 Photographs of a quadrat from Black Point, Rhode Island Sound, United States, taken on (top) October 3 and (bottom) October 21, 2001. Mussels (*M. edulis*) were dislodged by waves between the two censuses, as evidenced by the large gap of bare rock in the latter image. Scale bar 5.0 cm. *Source:* From Carrington *et al.* (2009). Reproduced with permission from John Wiley and Sons.

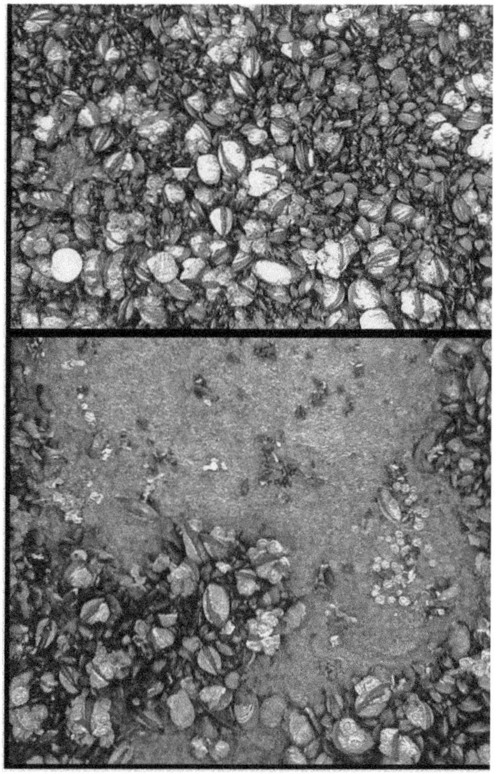

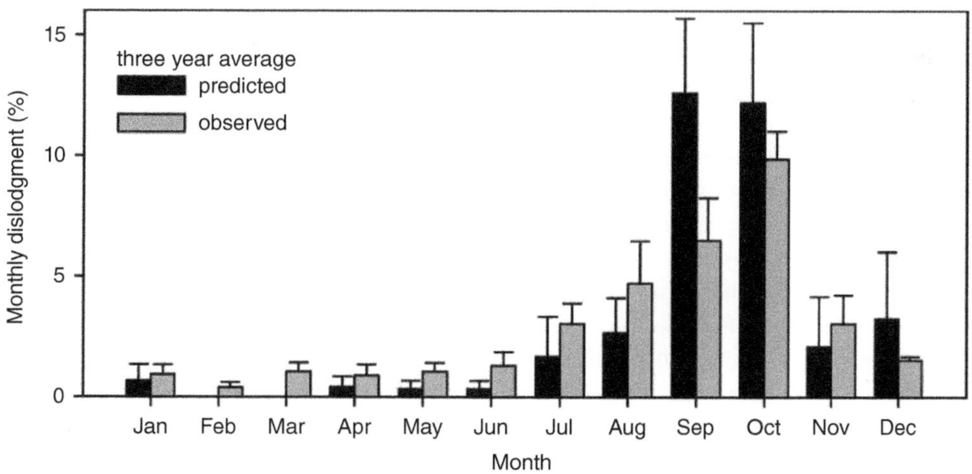

Figure 3.8 Summary of mussel dislodgment in Rhode Island, 2001–2003. Bars are means (± SE) of monthly values of predicted (black) and observed (gray) mortality. *Source:* From Carrington *et al.* (2009). Reproduced with permission from John Wiley and Sons.

(e.g. kelp, barnacles and slipper limpets) increase the height of the mussel and thus the probability it will be dislodged, as compared to epibionts that form thin layers (e.g. encrusting bryozoans, colonial tunicates and sheet-like sponges). The increased chance of dislodgement of mussels covered with 3D epibionts is due to increased hydrodynamic forces exerted on the mussel, specifically faster flow velocities that lead to higher drag-induced loading (Garner & Litvaitis 2017 and references therein; O'Connor *et al.* 2006). When mussels were artificially fouled, attachment strength increased by 7%, or by 17% during winter (Garner & Litvaitis 2013). Attachment strength and position within mussel beds were measured in two mussel species, *M. edulis* and *M. galloprovincialis* (Schneider *et al.* 2005). Results indicated that *M. edulis* moved more frequently and more quickly to the exterior of simulated mussel beds than did *M. galloprovincialis*. Together with measurements of attachment strength in the field, Schneider *et al.* (2005) used a wave force model to examine the probability of dislodgement for each species under a range of water velocities. By preferentially moving to the exterior of beds, the results suggested that *M. edulis* experiences higher dislodgement rates due to exposure to large hydrodynamic forces than does *M. galloprovincialis*. Because of lower attachment strengths, *M. edulis* is also predicted to have higher mortality rates than *M. galloprovincialis* in interior portions of the bed. Thus, the authors concluded that differential movement behaviour may contribute to the differential genotype-specific mortality rates observed in the *Mytilus* spp. hybrid zone in southwest England (see Chapter 9) and is an example of behaviour potentially modifying rates of exogenous selection in an intertidal hybrid zone. However, a genetic survey of 20 sites on Irish coasts has indicated no apparent advantage for either *M. galloprovincialis* or *M. edulis* genotypes at exposed shore locations (Gosling *et al.* 2008).

Substrate

In wave-exposed areas, *Mytilus* requires a hard and stable substratum such as rocks or large boulders on which to form beds, while in sheltered areas infaunal beds may occur on gravel or even in quite sandy areas. Mussel larvae settle on a wide variety of substrates (e.g. rocks and ridged, hard surfaces, filamentous macroalgae, hydroids, eelgrass, byssus of conspecific adults, artificial substrates such as polypropylene fibrous ropes or artificial seaweed; see details on the mechanism of byssus attachment to substrates and larval settlement cues in Chapters 2 and 5, respectively). The capability of mussels to attach byssal threads to the substratum and to form dense aggregations has permitted them to colonise both hard- and soft-bottom habitats, where they attach to one another because little suitable attachment substratum is available (Aguilera *et al.* 2017). Specifically, shells of recently dead mussels, for example, are common in mussel beds, yet they do not offer the same hold as the shells of living conspecifics. However, shells (empty, alive or fragmented) do offer substrata for attachment of epibionts, provide refuge from predation, alleviate physical or physiological stress and control transport of solutes and particles in the benthic environment (Gutiérrez *et al.* 2003).

In field studies, Aguilera *et al.* (2017) examined the attachment strength of the mussels *Perumytilus purpuratus*, growing on Pacific hard- and soft-bottom shores in Chile, and *M. edulis*, growing on an Atlantic rocky shore in France and a sedimentary shore in the North Sea (Germany). They found that the attachment strength of *P. purpuratus* on hard bottoms was substantially higher than that on soft bottoms even though mussels produced more byssus in the latter habitat. In contrast, *M. edulis* showed only a slightly reduced attachment strength on soft compared to hard bottoms. In laboratory experiments, the authors examined the mussel substratum selectivity of both bivalve species from soft and hard bottoms by offering living versus dead, barnacle-fouled versus unfouled and firmly attached versus loose conspecifics. In these experiments, *P. purpuratus* from both habitats showed a significant preference for living conspecifics, while *M. edulis* had no preference for particular conspecifics – except those that preferred fouled over clean mussels from soft-bottom habitats. In summary, their results confirmed that marine mussels show active substratum choice for byssus attachment, which depends on mussel species and habitat type. Also, their results suggest that mussels are adapted to a particular habitat type, with *P. purpuratus* showing lower adaptation to soft-bottom areas and *M. edulis* showing equal preference for both substrate types. In a previous study, Thiel & Ullrich (2002) described the fauna associated with hard-bottom mussel (*P. purpuratus*) beds at eight sites along the Pacific coast of Chile, with the aim of elucidating the functional role of mussel beds in hard- and soft-bottom environments. At all sites, the associated fauna was dominated by suspension feeding organisms (cirripeds, spionid and sabellid polychaetes), followed by grazing crustaceans and gastropods. Predators and scavengers also reached high abundances, while deposit- and detritus-feeding organisms were of minor importance. The majority of organisms associated with these hard-bottom mussel beds feed on resources obtained from the water column, or grow on the mussels rather than on materials (faeces, pseudofaeces) deposited by the mussels. This is in contrast to the fauna associated with mussel beds on soft bottoms, which comprise many species feeding on material accumulated by mussels and deposited within the mussel bed, indicating that mussels on hard bottoms primarily provide substratum for associated fauna, while mussels on soft bottoms provide both substratum and food resources.

Growth and mortality rates and inducible defence characters on medium-sized *M. edulis* (18–22 mm shell length) exposed to shore crab (*Carcinus maenas*) predation were examined on three different substrate types in combined field and laboratory experiments (Frandsen & Dolmer 2002). The substrate types used were: a smooth substrate, classified as simple, that structurally resembles sand; unbroken shells, classified as complex; and live *M. edulis*, classified as complex. High complexity and heterogeneity of a substrate is believed to reduce predation pressure by increasing the number of spatial refuges (references in Frandsen & Dolmer 2002). The experiments showed that crab predation was significantly higher (one-way ANOVA, $P < 0.001$) on the smooth substrate compared to the two more complex substrates, with no significant difference in predation between the complex types. However, increased intraspecific competition for food on the complex substrates resulted in significantly lower growth rates of the mussels. Inducible defence characters were also influenced by substrate type. Mussels were more affected by predators on the structurally simple substrate, where they developed thicker shells and a significantly ($P < 0.01$) larger posterior adductor muscle, both of which are defence responses that cause predators to take much longer to open the mussels (Freeman 2007). Finally, interspecific substrate preferences have been described in Gilg *et al.* (2010, ch. 5) and Katolikova *et al.* (2016, ch. 9).

Disturbance

The lower limit in mussel beds is typically controlled by the interaction between recruitment and sea star predation, and the upper limit by desiccation (Robles & Desharnais 2002). But within the mussel bed itself, the ability of mussels to dominate space is limited primarily by the dislodgement of individuals by waves (see earlier). Dislodgment opens patches (gaps) of bare substratum in the bed, temporarily providing space for fugitive species, which are eventually snuffed out by reinvasion of the bed (Denny & Gaylord 2010). This process, referred to as 'disturbance', has been defined as 'the displacement, damage or death of organisms caused by an external physical force or condition or incidentally by a biological entity' (Sousa 2007, p. 186). Most of the information on the effects of, and recovery from, disturbance comes from studies of *Mytilus* beds and their associated flora and fauna on exposed shore sites on the Pacific and NE Atlantic coasts of North America (Seed & Suchanek 1992; Svane & Ompi 1993; Wootton 1993; Beukema & Cadee 1996; Carroll & Highsmith 1996; Hunt & Scheibling 1998, 2001; Bertness *et al.* 2002; Guichard *et al.* 2003; Sousa 2007, 2012; Calcagno *et al.* 2012).

Apart from wave action, common physical agents that initiate gap formation include impact or abrasion by waveborne objects such as cobbles, logs or ice; extremes of air or water temperatures; desiccation; fouling by brown algae and barnacles; hummocking; abrasion by suspended sand; and burial under deposited sand (Sousa 2007). Wave and log damage occurs mostly during the winter and is typically responsible for removing 1–5% of *M. californianus* cover per month on exposed shores (Paine & Levin 1981). The initial size of disturbance gaps can range from single mussel size to areas as large as 60 m^2. Subsequent enlargement of the gap (as much as 5000%) may occur, especially during winter months, primarily due to weaker byssal thread attachments (Witman & Suchanek 1984). While some mussel beds remain uniform and flat, others form elevated hummocks consisting of small groups of 10–20 mussels, which become detached from the rock and are forced upward above the bed surface. Mussels (*Brachidontes rodriguezii*) in hummocks show lower attachment strength than those in the single-layered matrix (Gutiérrez *et al.* 2015). Accordingly, wave conditions associated with the passage of cold fronts (i.e. transition zones from warm air to cold air accompanied by moderate to strong winds and wave action, with seven-day average recurrence times based on historical weather data) caused detectable mussel dislodgment in a high proportion of hummocks but have virtually no impact on single-layered areas. Biological disturbances that disrupt the matrix of *Mytilus* beds are predators such as crabs and sea stars and epizoism by algal fronds; these usually occur on the scale of individual or a few mussels, but the sea star *Picaster* forms larger gaps in *M. californianus* beds (Seed & Suchanek 1992).

Patch structure and dynamics will probably differ among habitat types because of environmental differences influencing the intensity and outcome of biological interactions. Hunt & Scheibling (2001) compared the dynamics of natural and experimentally constructed mussel patches (*M. trossulus* and *M. edulis*) in two intertidal habitats, tidepools and emergent rock, over a time series (5, 10, 15 months) and among seasons (three successive five-month intervals). In tidepools, mussels naturally occurred in small patches (median <25 cm^2), while mussels on emergent rock formed extensive beds with decimetre-scale gaps, but these beds started as small patches following disturbance such as ice scour. For their experimental patches (15 cm^2) in both habitats, the authors assessed the relative importance of physical (wave disturbance) and biological (predation, growth, recruitment and immigration) processes in determining patch size and structure. Individual experimental and natural patches varied greatly in size over time, but mean patch area remained relatively constant. Mean size of individuals in experimental patches decreased due to the loss of larger mussels, while numbers increased due to recruitment. Wave disturbance appeared to be more important than predation in determining patch structure and dynamics, although losses due to either process did not differ consistently between habitats over time. Growth rates were low (≤0.4mm per month), but were greater in tidepools than on emergent rock, whereas recruitment and immigration rates generally did not differ between habitats. Although each process contributed to changes in patch size and structure, overall they did not result in a marked divergence in mean patch area or biomass between tidepools and emergent rock over the 15-month experiment. This study highlights how integrative approaches, with monitoring of patches and of individuals within patches, can provide detailed mechanistic understanding of patch structure and dynamics.

Minchinton *et al.* (1997) monitored the recovery of intertidal algae and sessile macrofauna after a rare occurrence of scouring sea ice denuded the intertidal area of an exposed rocky shore in Nova Scotia, Canada. They found that barnacle cover was restored soon after the ice scour and macroalgae cover in about two years, but that it took considerably longer (four to six years) for the mussel beds (*M. edulis* and/or *M. trossulus*) to recover. Similar results have been reported by Brosnan & Crumrine (1994), who trampled 250 experimental plots in the intertidal zone for a period of one year and then monitored recovery. The algal–barnacle community recovered in the year following trampling, but mussel beds (*M. alifornianus*) and their associated species had not recovered within two years following cessation of trampling, and over this period no mussel recruitment was recorded. Two trampling studies (Beauchamp & Gowing 1982; Goldstein 1992) were performed 14 years apart at the same intertidal site, Santa Cruz on the west coast of California, with areas differing in intensity of trampling. There was little evidence of impact or change in either studies, although similar studies elsewhere have documented clear trampling effects (Ghazanshahi *et al.* 1983; Durán & Castilla 1989; Underwood & Kennelly 1990; Keough *et al.* 1993; Keough & Quinn 1998; Clark *et al.* 2002), as have others on the California coast (Beauchamp & Gowing 1982; Ghazanshahi *et al.* 1983; Goldstein 1992; Addessi 1994). In 1995, Van De Werfhorst & Pearse (2007) carried out a new study at the same Santa Cruz sites as Beauchamp & Gowing (1982) and Goldstein (1992), using a different strategy sampling regime, and were able to demonstrate trampling effects. Mussel bed cover, with the associated higher number of species within it, decreased with increased human trampling, while the frequency of bare rock areas increased. Consistent with the previous studies, the two brown algal species that were abundant on the less trampled sites were completely absent at the most trampled. These results, in conjunction with those of the earlier studies, indicate that rocky intertidal assemblages can be resilient to human trampling, and that sampling scale and design are important in evaluating and monitoring trampling impacts (Van De Werfhorst & Pearse 2007). Micheli *et al.* (2016) evaluated the separate and combined influences of disturbance from storm waves and disturbance associated with human trampling of rocky shores by conducting an experiment mimicking controlled levels of trampling at sites with different wave exposures and before and after a major storm event in central California, United States. Their results show that trampling and storm waves affected the same taxa and had comparable and additive effects on rocky shore assemblages. Both disturbance types caused significant reduction in percent cover of mussels (*M. californianus*) and erect macroalgae, and resulted in significant reorganisation of assemblages associated with these habitat-forming taxa. A single extreme storm event caused similar per cent cover losses of mussels and erect macroalgae as 6–12 months of trampling, which combined additively rather than synergistically. Mussel beds in wave-exposed sites were more vulnerable to trampling impacts than algal beds at protected sites. Mussels and erect macroalgae recovered within five years after trampling stopped. These results suggest that impacts from local human use can be reversed in relatively short periods. See Mendez *et al.* (2019) for a similar study on *Brachidontes* spp. in Patagonia, Argentina.

Waves are not only associated with hydrodynamic stress but can also carry heavy loads of sand, periodically disturbing intertidal shores through sand burial or sand scour (reviewed in McQuaid *et al.* 2015). Acting as an agent of disturbance, sand removes plant tissue, epiphytes or invertebrates with poor attachment to the rock surface through scouring and decreases light, oxygen and substratum available to organisms through burial. It can lead to temporary species impoverishment by selective elimination of maladapted species, although in the longer term it may also enhance species richness by increasing habitat heterogeneity, allowing within-shore coexistence of sand-intolerant species and those associated with sand deposits (McQuaid & Dower 1990).

Unexpectedly, sand stress strongly affects the survival of *M. galloprovincialis* and *P. perna* individuals but is not related to their physiological tolerances and does not explain their vertical zonation. When buried under sand, *P. perna* mortality rates are higher than those of *M. galloprovincialis* in both laboratory and field experiments, yet it is *P. perna* that dominates the low shore where sand inundation is recurrent (Zardi *et al.* 2006). Although both species accumulate sediments within the shell valves while still alive and sand buried, the quantities are much greater for *P. perna*, causing intense visible damage and clogging of the gills, which explains its higher mortality rates. Presumably, the accumulation of sand within the shell of *P. perna* is linked to its gaping behaviour. Wave and sand stress vary also in time, altering the timing and mortality rates of the two mussel species (Zardi *et al.* 2008). During periods of high sand accumulation in mussel beds, the indigenous species has increased mortality rates that are higher than those of *M. galloprovincialis*, while the pattern is reversed during winter, when wave action is high (Zardi *et al.* 2008). When sand stress is high, the less stable secondary substratum of sand and shell fragments weakens the attachment strength of mussels living within a bed. Consequently, the indigenous species loses its advantage in attachment strength over the invasive species, and this results in a seasonal shift in the competitive balance between the two.

There has been considerable interest in the ecology of patches, especially concerning the successional[6] events leading to their recolonisation. In the case of *M. californianus*, the patches are initially colonised by diatoms, then by ephemeral algae such as *Ulva* and *Porphyra* spp. and then by perennial algae. These successional stages are followed by acorn barnacles, *Balanus glandula* and *Semibalanus cariosus*, then goose barnacles, *Pollicipes polymerus* and sometimes *M. trossulus*, and then by *M. californianus*. The latter needs secondary space (as opposed to primary space, i.e. rock surface), such as substrates like algae, barnacles and byssus threads, for larval settlement, but later becomes competitively dominant over all other sessile species. These events take about five years, depending upon degree of wave exposure (Dayton 1971; Paine & Levin 1981; Wootton 2002).

The mechanism driving these successional events is generally thought to be competition with larger, later-colonising species assuming competitive dominance over smaller, early-succession species, until finally sea mussels dominate. *M. californianus* exhibits competitive dominance: small barnacles are smothered, larger barnacles such as *Semibalanus cariosus* are overgrown and abraded and goose-neck barnacles, *Pollicipes polymerus*, are slowly crushed to death. Both *M. californianus* and *Semibalanus cariosus* have refuge in size from predation by whelks *Nucella* spp., and thus could potentially monopolise all space were it not for predation by sea stars, *Pisaster ochraceus*, from lower positions on the shore and for creation of new colonisable spaces by log damage. Studies on the Olympic Peninsula, Washington, United States also show that by consuming the early successional stages, predators such as whelks and birds and herbivores such as chitons and limpets actually accelerate succession (Suchanek 1981; Paine 1984).

The importance of large-scale ocean currents in the global distribution of bivalves has already been dealt with in this chapter. Locally, areas with strong currents usually provide favourable feeding conditions for bivalves. However, very strong currents can have an inhibitory effect on feeding and consequently growth. Also, strong currents may prevent larval settlement and byssal attachment of spat, ultimately resulting in local variability in recruitment.

Fishing methods can affect bivalve abundance directly by causing significant mortality and indirectly by causing shell damage. In Spain, mussel seed from intertidal exposed rocky shores is the method most used by farmers to seed ropes in mussel culture areas (Peteiro *et al.* 2007). This practice, while legal, must have a detrimental effect on mussel beds and their community structure, although to date there is no documented evidence of damage.

Biological Factors

Just as humans greatly appreciate the delicate flavour of bivalves, so also do a whole range of other organisms from groups as diverse as fish, birds, mammals, crustaceans, echinoderms, flatworms and even other molluscs. Bivalves are usually preyed upon by several of these groups, which operate at specific times of the year, and which generally focus on the smaller size classes. Predators are probably the single most important source of natural mortality in bivalve molluscs and have the potential to influence population size structure, overall abundance and local distribution patterns.

In this section, the main predators of mussels will be dealt with, along with major pests, fouling organisms and competitors. Most of the following information comes from studies in intertidal and shallow water environments.

Predators, Biofouling and Competitors

Gastropods are significant predators of mussels worldwide. The dogwhelk, *Nucella lapillus*, is widely distributed on exposed shores in northern Europe and on the east coast of North America, where it feeds extensively on barnacles and small mussels. Predation is often seasonal, with whelks remaining aggregated in pools and crevices over the wintertime.

However, numbers on mussel beds on the low and mid shore start to increase in the spring, and densities as high as 300 whelks m^{-2} have been recorded over the summer months in NE England (Seed 1969). Profitability (energy assimilated from a food item relative to handling time) for dogwhelks feeding on mussels increases with prey size (Hughes & Dunkin 1984). Yet whelks prefer mussels smaller than the largest available. Hughes (1986) suggests that dogwhelks choose mussels with the maximum average profitability in the face of competition from other dogwhelks, which are attracted to the predator by olfactory stimuli from the damaged prey. Marples *et al.* (2018) recently suggested that profitability is a function of the current state of both the predator and the prey individuals, and that it should be considered to be an attribute of a particular encounter, in contrast to its present usage as an attribute of a prey species. The whelk uses the radula, a modified toothed chitinous structure, to drill a small hole through the thinnest part of the shell around the umbone or adductor muscle insertion regions (Seed 1976), or through the shell area overlying the glycogen-rich digestive gland (Hughes & Dunkin 1984). Prior to drilling, the whelk softens the area using a secretion from the foot. The proboscis is inserted through the hole and the flesh of the prey is rasped away by the radula, converted into a 'soup' and devoured. Alternatively, a more efficient mechanism that is used by experienced whelks involves insertion of the proboscis through the valve gape and induction of muscular paralysis by injection of toxins. Feeding only occurs when conditions are conducive to such an activity, and during these times the dogwhelk consumes large quantities of food, so that the gut is always kept as full as possible. This allows shelter until more food is required, when foraging resumes. If waves are large or there is an excessive risk of water loss, the dogwhelk will remain inactive in sheltered locations for long periods. Feeding rates (drilling plus ingestion times) peak during the summer, but as water temperatures fall through the autumn, the time needed for ingestion lengthens, more than tripling the total handling time (Miller 2013). Prey handling time per mussel is generally in the range of two to three days, which agrees well with results from laboratory experiments showing that an adult whelk can consume about two mussels (1–3 cm shell length) per week during the summer (Seed 1969). Although this level of consumption may appear small, the high density of foraging whelks makes a serious impact on mussel coverage on exposed shores. For example, on rocky intertidal sites in Alaska, United States, where *Nucella lima* occurs at densities of >100 m^{-2}, Carrol & Highsmith (1996) estimated that the whelk can eliminate 60–90% of mussels (*M. trossulus*) at a given site in one season. Preference for mussels, as opposed to barnacles, appears to be fixed in early life: adult whelks transferred from sites with no mussel cover to those with a high coverage of mussels largely ignored the mussels, preferring to feast on barnacles (Wieters & Navarette 1998). Mussels 'fight' back by ensnaring and immobilising whelks in their byssus threads. An ensnared whelk can be overturned, thus arresting the drilling process (Figure 3.9). Whatever the fate of the mussel, the whelk, once ensnared, is trapped and exposed to predation (Petraitis 1987; Davenport *et al.* 1998; Farrell & Crowe 2007; Chiu *et al.* 2011). Sherker *et al.* (2017) were the first to use organisms raised in the field (Atlantic Canada), rather than in the lab, to demonstrate that predator-induced morphological responses in bivalve prey hinder predation. During the spring and summer of 2016, they ran a field experiment that manipulated dogwhelk presence to test their nonconsumptive effects on mussel traits. Dogwhelk cues elicited thickening at the lip, centre and base of mussel shells while simultaneously limiting shell length

Figure 3.9 Photograph of a whelk, *Nucella lapillus*, flipped on its back by a mussel, *Mytilus edulis*. Mussels attach a number of byssus threads to the body whorl of the predator gastropod, then retract them, flipping the predator over and immobilising it, and thus exposing it to crab predation. *Source:* Photo courtesy of P. Petraitis, Department of Biology, University of Pennsylvania, Philadelphia, Pennsylvania, USA. (*See colour plate section for colour representation of this figure*).

growth. As shell mass was unaffected by dogwhelk presence, a trade-off between shell thickening and elongation was revealed. Thickening was most pronounced at the thinnest parts of the shell. Using the field-raised organisms, a lab experiment showed that dogwhelks took, on average, 55% longer to drill and consume mussels previously exposed to dogwhelk cues than mussels grown without such a cue exposure. Dogwhelks drilled at the thinnest parts of the shell, but nevertheless the consumed cue-exposed mussels had thicker shells at the borehole than the consumed mussels not previously exposed to cues, which likely explains the observed difference in handling time. As handling time normally decreases predation success, this study indicates that the plastic structural modifications in mussels triggered by dogwhelk cues in the field hinder predation by these drilling predators.

On the West Coast of the United States, *Nucella canaliculata* and *N. emarginata* are major predators of mussels but seem to favour the thinner-shelled species, *M. galloprovincialis*, to the thicker-shelled and less nutritious *M. californianus* (Suchanek 1981). Other predatory gastropods such as *Ocenebra poulsoni*, *Acanthina sopirata*, *Ceratostoma nuttalli* and *Jaton festivus* also feed on *Mytilus* (Shaw *et al.* 1988), while various birds, crustaceans and fish feed on *N. lapillus* (details in Crothers 1985). Chemical cues are commonly used by prey to evaluate risk. Large *et al.* (2010) conducted a study to investigate the nature of cues used by prey hunted by generalist predators. Using *Nucella* and a suite of its potential predators as a model system, they explored how (1) predator type, (2) predator diet and (3) injured conspecifics and heterospecifics influence *Nucella* behaviour. Taking laboratory flumes, they found that *N. lapillus* responded only to the invasive green crab, *Carcinus maenas* – the predator it most frequently encounters – and not to rock crabs (*Cancer irroratus*) or Jonah crabs (*Cancer borealis*), which are sympatric predators but do not frequently encounter *N. lapillus* because they are primarily subtidal. Predator diet did not affect whelk responses to risk, although starved predator response was not significantly different from controls. Since green crabs are generalist predators, diet cues do not reflect predation risk, and thus altering behaviour as a function of predator diet would not likely benefit *N. lapillus*. However, the whelk did react to injured conspecifics – a strategy that may allow it to recognise threats when predators are difficult to detect. *N. lapillus* did not react to injured heterospecifics including *M. edulis* and herbivorous snails, *Littorina littorea*, suggesting that they respond to chemical cues unique to their species, allowing them to minimise costs associated with predator avoidance. The ability of prey to detect and respond to predator signals varies with environmental conditions (Large *et al.* 2011 and references therein).

Sea stars are also important predators that influence the distribution and abundance of mussels on the lower shore and in the sublittoral zone. Sea stars predate on mussels and other bivalves either by using force or by secreting an anaesthetic from their stomach that numbs the mussel and causes it to gape. They then extrude their stomach through their mouth into the shell opening and digest the prey. Sea stars with short or inflexible arms (e.g. *Astropecten* or *Luidia* spp.) usually swallow their prey whole while digestion occurs in the stomach, before egesting shell pieces through the mouth. Semmens *et al.* (2013) have reported a novel neurophysin-associated neuropeptide that triggers stomach eversion and retraction, which may provide a basis for development of nonpeptidic small molecule agonists or antagonists that mimic or block the effects of neuropeptides; this could be used for chemical control of starfish feeding.

There are numerous studies of the sea star–bivalve interaction on rocky shores, primarily because of the ease of observation and manipulation of this pairing. Such studies have led to a greater understanding of the causes of zonation, and have provided additional evidence on size and spatial refuges in bivalve populations (Dame 2012 and references therein). On the Pacific coast of the United States, the mussel *M. californianus* exists as a well-defined band on rocky intertidal shores. The sea star, *P. ochraceus*, is a major predator of this mussel (Figure 3.10). In a series of classic studies, removal of sea stars over a 10-year period produced marked changes in zonation patterns, with a notable downward shift in the lower limit of mussel distribution of about 2 m through redistribution of adults and

Figure 3.10 The sea star *Pisaster ochraceus* predating on mussels, *Mytilus californianus*, on the US Pacific coast. *Source:* David Cowles, http://rosario.wallawalla.edu/inverts. Reproduced with permission. (*See colour plate section for colour representation of this figure*).

normal settlement of mussel larvae (Paine 1974). Consequently, the starfish has been regarded as a 'keystone' predator, one which through its feeding activities exerts a disproportionate influence on community structure, in this case setting the lower limits of mussel distribution. In a keystone predator-dominated system, other invertebrate predators have minor effects on community structure, but in the absence of the keystone predator, such species may adopt a major role in the altered system (Navarrete & Menge 1996; see also Menge 1992; Menge *et al.* 1994; Naverette *et al.* 2000; Gosnell & Gaines 2012).

Until recently, the accepted explanation for the distinct zonation patterns on wave-exposed rocky shores was that dense populations of sedentary organisms, such as mussels, form in static prey refuges above the reach of natural predators. Robles *et al.* (2009) showed that prey refuges are not in fact static. On the West Coast of the United States, experimental alteration of starfish (*P. ochraceus*) densities caused the downward extension of the lower boundaries of the mussel *M. californianus*, while experimental increases in sea star densities caused the upward recession of the lower boundary well into the zone presumed to be a spatial refuge for mussels from predation. As small mussel prey are depleted by sea stars over time, larger mussels are attacked, including those that would otherwise represent the lower boundary of their distribution. Based partly on what is known of *Pisaster*'s feeding behaviour, moving into the higher reaches of the mussel bed at high tide, then retreating to the lower parts of the bed at low tide, one might predict that the predators are more susceptible to stress than their prey. This was tested in Oregon by transfering mid-tidal-level mussels in April to high and low edges of the mussel bed and caging half of them with *Pisaster* (Petes *et al.* 2008). Three caging combinations were employed: mussels with and without sea stars, and sea stars without mussels. At four four-weekly intervals in the summer, both mussels and sea stars were assessed for growth and reproductive status, and tissues were sampled for levels of Hsp70 (see Chapter 9). Results showed that mussels in the high-level cages spawned earlier and had significantly higher levels of Hsp70 than those in the low-level cages. There was no significant difference in growth of mussels at the two levels. Sea stars suffered higher mortality at the upper edge of the mussel bed, while those at the lower edge lost body mass but saw less mortality. Sea stars at either level exhibited no significant changes in the level of Hsp70. The authors concluded that intertidal stress factors, such as desiccation and temperature, affect the motile predator more than its sedentary prey. In a subsequent study, Hayne & Palmer (2013) found that *P. ochraceus* on wave-exposed shores had narrower arms and were lighter per unit arm length than those from sheltered sites. Narrower arms probably reduce both lift and drag in breaking waves, while on protected shores fatter arms may provide more thermal inertia to resist overheating, or more body volume for gametes.

In northern Europe, *Asterias rubens* is a serious predator of *M. edulis*. This sea star aggregates seasonally on mussel beds in large numbers – sometimes as high as 450 m^{-2} – often completely destroying local mussel populations (Dare 1982). In contrast to the results of laboratory-based experiments, *A. rubens* shows no size selectivity when feeding in the field. The solid structure of interconnected mussels forming the bed, however, restricts predation to only those mussels situated at the bed surface, thus providing a refuge from predation for smaller mussels deeper down (Dolmer 1998). Not all mussels are equally susceptible to starfish predation. About 70% of *M. edulis* of North Sea origin were able to resist *A. rubens*, whereas all Baltic mussels (presumably *M. trossulus*) were opened within one hour

(Norberg & Tedengren 1995; see also Lowen *et al.* 2013). *M. edulis* cultured in close vicinity to *A. rubens* were significantly smaller in shell length, height and width but had significantly larger posterior adductor muscles, thicker shells and more meat/shell volume. These morphological changes have an adaptive value in that predator-exposed mussels have a significantly higher survival rate than unexposed mussels (Reimer & Tedengren 1996). Behavioural changes were also evident: predator-exposed mussels in the laboratory formed larger aggregates, migrated less and sought structural refuges more often (Reimer & Tedengren 1997). Also, chemical aspects of epibionts, such as barnacles, algae, sponges and hydrozoans, on mussel (*M. edulis*) shells have the potential to modify the top-down control by starfish (*A. rubens*) by changing or masking prey properties the sea stars cues upon, or by producing their own repellants (Laudien & Wahl 2004). This may explain why, in an earlier study, *A. rubens* was reported to prefer clean subtidal mussels over barnacle-overgrown intertidal ones (Saier 2001). The most popular optimal foraging theory states that predators should maximise prey profitability (i.e. select that prey item that contains the highest energy content per handling time). However, Hummel *et al.* (2011) hypothesised that sea stars (*A. rubens*) do not forage on mussels (*M. edulis*) according to the classical optimal foraging theory, but actively avoid damage that may be caused by, for example, capture of foraging on too-strong mussel shells; hence, they will have a stronger preference for mussels that are smaller than the most profitable ones. They found that *A. rubens* as a predator indeed chooses much smaller blue mussels to forage on than the most profitable ones. Hence, their study does not support the optimal foraging theory. There may be other constraints involved in foraging than just optimising energy intake. For example, predators may also be concerned with preventing potential loss or damage of their foraging body parts.

Sea star wasting disease or starfish wasting syndrome is a disease of starfish and several other echinoderms that appears sporadically, causing mass mortality of those affected. There are around 40 different species of sea stars that have been affected by this disease. It seems to be associated with raised water temperatures in some places but not others. It starts with the emergence of lesions, followed by body fragmentation and death. In 2014, it was suggested that the disease is associated with a densovirus now known as the sea star-associated densovirus (SSaDV), although it is still not fully understood; see details in MARINe (2019).

Crabs (*Cancer*, *Carcinus* and *Pachygrapsus* spp.) are also significant predators of mussels on the lower shore and in the sublittoral zone. Their effect on mussel abundance is seasonal, with reduced predation in winter when crabs migrate offshore. Results from laboratory and field experiments show that crabs employ size-selection of prey, with the upper size limit that can be opened being directly related to the size of the crab (references in Seed 1976). Crabs will almost always choose small-sized prey when offered a range of sizes. It is handling time rather than the energetic costs of handling, estimated as a mere 2% of corresponding gains, that is the basis on which prey are selected (Rovero *et al.* 2000). During the handling period, the crab is at risk from other predators, competitors and even claw damage. Small mussels are therefore particularly vulnerable to predation, since they are easily crushed by most size classes of crab. A mussel must attain a shell length of at least 45 mm before it is relatively safe from crab predation. Once again, mussels show several defence mechanisms. In laboratory experiments, *M. edulis* increased byssus volume in response to waterborne cues from *Cancer pagurus* and *Carcinus maenas* (Cote 1995; Leonard *et al.* 1999). Similar findings were reported when *P. viridis* were exposed to crab

(*Thalamita danae*) that had recently consumed conspecifics (Chiu *et al.* 2011). In addition, mussels subject to heavy predation develop thicker and more robust shells in response not just to crabs but also to the broken shells of other mussels (Leonard *et al.* 1999; see also Freeman & Byers 2006 and comments from Rawson *et al.* 2007 and Freeman & Byers 2007). Similar effects have also been reported in mussels subject to heavy whelk predation (Smith & Jennings 2000), and increased production of byssal threads has been shown in *P. viridis* in response to crab and gastropod predation (Cheung *et al.* 2004a,b, 2006, 2009). A behavioural strategy in response to crab predation has been reported in the Wadden Sea, Germany, where oysters (*Crassostrea gigas*) have invaded native mussel (*M. edulis*) beds (Eschweiler & Christensen 2011). Mussels subjected to direct contact with crabs (*C. maenas*) migrated from the top of the oyster reef to interspaces at the bottom of the reef, where they showed significantly reduced growth rates and conditions to mussels on the top of the reef. Mussels experience a trade-off between survival and food supply, preferring to take refuge from predation even when this decreases growth and condition. In a previous study, Shin *et al.* (2008) investigated the effect of cues from damaged conspecifics and heterospecifics on the induction of refuge seeking and enhancement of byssus production as responses to the crab *Carcinus maenas* in the mussel *Brachidontes variabilis*, hypothesising that the mussel would seek refuge more readily and prefer a smaller refuge. More byssal threads (thicker and longer ones) should also be produced. They found that *B. variabilis* was able to differentiate between the sizes of available refuges and to stay in appropriate ones according to the level of risk they perceived. Staying in a smaller refuge would reduce the chance of the mussels being dislodged and consumed by the crabs. This helps explain why *B. variabilis* tended to stay in smaller refuges when predation risk was high, as simulated by the presence of damaged conspecifics and heterospecifics. In contrast, higher food and oxygen availabilities were found in large refuges, although the predation risk was also higher. Staying in larger refuges would be advantageous to the mussels only when predation risk was low. The preference toward certain sizes of refuge, therefore, should be a trade-off between physiological requirements and the risk of predation.

Aquatic prey encounter an array of threat cues from multiple predators and killed conspecifics, yet the vast majority of induced defences are investigated using cues from single predator species. In most cases, it is unclear if odours from multiple predators will disrupt defences observed in single-predator induction experiments (Freeman *et al.* 2009). Freeman *et al.* (2009) compared the inducible defences of *M. edulis* to waterborne odours from pairwise combinations of three predators representing two attack strategies, the sea star *A. rubens*, which pulls mussel shells open, and the crabs *Carcinus maenas* and *Cancer irroratus*, which crush or peel[7] shells. Mussels increased adductor muscle mass in response to cues from unfed *Asterias* and increased shell thickness in response to unfed *Carcinus*. However, they did not express either predator-specific response when exposed to the combined cues of *Asterias* and *Carcinus*, and did not increase shell thickness when exposed to cues from *Cancer* alone or any pairwise combination of the three predators. Shell closure or 'clamming up' did not occur in response to any predator combination. These results suggest that predator-specific responses to *Asterias* and *Carcinus* are poorly integrated and cannot be expressed simultaneously.

Biotic invasions can result in the displacement of native species. This can alter the availability of native prey and the choices made by native predators. Skein *et al.* (2018)

investigated prey selection by two native South African predators, the lobster *Jasus lalandii* and the starfish *Marthasterias africana*, in response to the invasive mussels *M. galloprovincialis* and *Semimytilus algosus* and the native mussels *Aulacomya atra* and *Choromytilus meridionalis*. As the diets of the two predators are broad, the authors hypothesised that they would consume the most abundant prey, regardless of its native or alien status. Laboratory studies presented predators with varying proportions of native and invasive mussels that represented pre- and post-invasion scenarios. Both predators exhibited preference toward the native mussel *C. meridionalis*, even when it was the least abundant prey. The selection of native species occurred despite mussel parameters (shell strength, adductor muscle size and energy content), suggesting that invasive species would be easier to consume. These findings highlight the potential for facilitation of prey invasions, especially when predators avoid alien prey and select for native comparators that may offer resistance to the invasion through interspecific competition. This study does not present the first observation of native predators failing to select for invasive prey (e.g. López *et al.* 2010; Veiga *et al.* 2011). However, in other cases the avoided invasive prey were suggested to possess physical characteristics that might hinder predation from native predators (Skein *et al.* 2018).

Several bird species are predators of mussels. The main ones in western Europe that feed on mussels at low tide are the oystercatcher, *Haematopus ostralegus*, the common eider duck, *Somateria mollissima*, and the herring gull, *Larus argentatus*, although the latter only feeds on small mussels on newly established beds. Of these, the most significant is the oystercatcher (Figure 3.11). Predation is seasonal, with birds switching in the spring from mussels (and cockles) to deep-living clams such as *Scrobicularia plana* and *Macoma balthica*, and back to surface bivalves in autumn in order to maximise intake rate (Zwarts *et al.* 1996a). The birds cannot survive if their diet is restricted to one or two prey species; they need to switch between three or four, and have to roam over feeding areas measuring at least some tens of km^2. Oystercatchers use one of two general techniques when preying

Figure 3.11 The American oystercatcher, *Haematopus palliatus*, a significant predator of bivalves, eating a surf clam (*Spisula solidissima*) at Nickerson Beach, Long Island, New York. *Source:* ©Arthur Morris, www.birdsasart.com. (*See colour plate section for colour representation of this figure*).

on *M. edulis*: they either 'stab' their bills between the gaping valves of the mussel or 'hammer' through the shell on the dorsal or ventral side (Zieritz *et al.* 2012 and references therein). Stabbers and dorsal hammerers open the mussel in situ, but ventral hammerers usually tear the mussel off from the substrate and carry it to a suitably firm patch, where the shell is broken at the ventral region. They invariably select thin-shelled mussels to hammer through because these are easier to crack than thick-shelled mussels; it is the thickness of the prismatic layer that largely determines the vulnerability of mussel shells (Le Rossignol *et al.* 2011). Wintering oystercatchers feed extensively on *M. edulis* in the estuaries of southern Britain. They show a marked preference for brown-shelled mussels over the commoner black-shelled morph, and show that this enables them to maximise their rate of energy gain over a longer period than a single foraging bout (Nagarajan *et al.* 2002a). The brown and black mussels do not differ in ventral thickness or energy content, which are the main criteria for mussel selection and the most important for short-term optimisation. The brown mussels contain significantly less moisture, so by selecting them, oystercatchers can pack more mussel flesh into their limited oesophageal storage capacity. This enables them to increase their overall consumption during a feeding bout and increases their long-run energy gain rate, to an extent that is large enough to be significant for survival, especially during the short exposure of the mussel beds in winter. All birds show size selection within the prey species; this is because flesh content increases more steeply with prey size than handling time (Zwarts *et al.* 1996b). Oystercatchers are highly selective toward mussels of between 35 and 50 mm shell length, and fewer than 5% of mussels taken are below 35 mm or above 50 mm (Nagarajan *et al.* 2002b). The oystercatchers select ventrally thin-shelled mussels, especially if the length is more than 35 mm. Removal of the largest mussels may reduce protected refuge for younger mussels, but may also allow younger mussels to grow at a faster rate – although, as already mentioned, gulls preferentially prey on small mussels (Goss-Custard *et al.* 1996).

Different diving duck species, including eiders (*Somateria* spp.), scoters (*Melanitta* spp.) and scaups (*Aythya* spp.), also predate extensively on mussels. In the case of eiders, mussels often constitute as much as 60% of their diet (Nehls & Ruth 1994). Eiders select mussels of smaller than optimal size, because this minimises shell ingestion, even though larger available prey would provide greater net energy gain per prey item (Bustnes & Erikstadk 1990; Hamilton *et al.* 1999). In the process of zoning in on their prey, the ducks may remove whole mussel clumps, thus causing mussel mortality over and above that produced by direct predation (Raffaelli *et al.* 1990).

Numerous studies have been undertaken to provide actual data on the impact of oystercatchers and other bird predators on commercial mussel beds. Analysis of faeces or regurgitated pellets provides information on the sizes of mussels selected by bird species. In the laboratory, the ash-free dry weights (AFDW) of different sizes of mussels are determined in order to calculate the biomass eliminated (Hilgerloh 1999). According to exclosure experiments on the Danish tidal flats, oystercatchers and eiders eliminated 116 g AFDW m^{-2} (Egerrup & Høegh Laursen 1992), while on the Dutch tidal flats, oystercatchers alone eliminated 48 g AFDW m^{-2}. Eiders consumed 360 g AFDW m^{-2} on the west cost of Denmark. In a regional study in the Danish Wadden Sea, 300 g AFDW m^{-2} emerged as a potential consumption, assuming that 100% of the food of all three predator species (eiders, oystercatchers and herring gulls) consisted of mussels and that eiders did not feed on mussels that lived submerged or on hard substrates (Faldborg *et al.* 1994). In the Wadden Sea of Lower Saxony, the same three bird predators consumed a total of 32 and 71 g AFDW m^{-2} in two different years (Hilgerloh 1997). On a newly settled mussel bed in the same area, oystercatchers and herring gulls consumed 71 g AFDW m^{-2} in five to six months (Hilgerloh *et al.* 1997). Predation pressure differs according to the available size classes and the predatory bird species, as birds are size selective.

It is well established that sea ducks feed in mussel aquaculture sites, but whether they are able to identify those mussels as being of higher quality or are only attracted by farms because of better accessibility of mussels (i.e. higher densities and convenient suspension in the water column) is not known. Varennes *et al.* (2015) showed that when detectability is controlled, eiders still choose the cultivated mussels. Preferences for cultivated mussels and their foraging advantages have important implications for sea ducks and habitat management.

Other birds that feed on mussels include knots, *Calidris* spp. (Alerstam *et al* 1992), and crows, *Corvus* spp. (Berrow *et al.* 1992a). Indeed, crows are significant predators of mussels in the intertidal zone and show several interesting adaptations. They frequently cache mussels during low tide, and recover them during high tide some two to three days later. This behaviour is believed to be a response to short-term, daily fluctuations in food availability (Berrow *et al.* 1992a). In order to break them open, the crows drop mussels and other hard-shelled prey on to hard surfaces such as roads or rocky shores (Berrow *et al.* 1992b). This behaviour peaks during October–February and usually involves only large-sized mussels, no doubt an adaptation to food shortages in winter.

Mussel farms, with their very high densities of small, thin-shelled mussels, can be foraging hot spots for diving ducks, particularly during spring and autumn when birds are building up their energy reserves for reproduction, migration or overwintering. For example, in spring 2011 in Baie de Chaleurs, Quebec, Canada, all mussel growers were severely hit by scoter predation, losing almost all their collectors and over 30% of their one- to two-year-old mussels on ropes (Varennes *et al.* 2013). Acoustic and visual deterrents have been tried with little success (Dionne *et al.* 2006). Provincial aquaculture authorities in Prince Edward Island, Canada proposed the use of protective socking material as a potential solution to the problem of diving ducks. The material consists of the standard polypropylene sock with a biodegradable protective layer stitched around it (Figure 3.12). When mussels are put into socks and hung in the water, they start migrating toward the outside of the sock in order to filter feed properly, making them more vulnerable to predation by diving ducks. The purpose of the second layer, with its smaller mesh openings, is to prevent their migrating outside the sock, keeping them between the layers until the bays freeze up in winter and the predation threat is over. Currently, in Canada, the United States and Europe, the only effective method that provides a complete and long-term control of bird predation in culture facilities is the use of exclusion nets deployed around longlines and rafts of suspended mussel ropes (Varennes *et al.* 2013).

On the west coast of North America, the sea otter (*Enhydra lutris*) is an important predator of *M. californianus*. This species removes large clumps of mussels, which it sorts and consumes on the sea surface by pounding them on a flat stone on its chest or against other mussels. So, although sea otters are selective in terms of the size of prey they consume, they have profound effects on all size classes of mussel (Seed & Suchanek 1992). In addition,

Figure 3.12 Socking used for protection of mussels from predatory birds. (a) Regular socking materials used are the GDI-4S, GDI-5M and GDI-7L (Go-Deep International Inc., Fredericton, New Brunswick, Canada), for small, medium and large mussel seeds, respectively. These socks are made of interwoven, flattened polypropylene strands. (b) Protective socking material composed of regular (GDI) mussel socks with a biodegradable loop-knitted sleeve in a 50:50 cotton:polyester blend sewn over the polypropylene strands, giving it its protective layer. *Source:* From Dionne *et al.* (2006). Reproduced with permission from Springer Nature.

otters have substantial, indirect effects on the biomass of mussel bed-associated communities. The total biomass of species associated with mussel beds was found to be more than three times higher where otters were absent (Singh *et al.* 2013). Other predators of mussels include sea urchins (*Strongylocentrotus droebachiensis*), lobsters (*Panulirus interruptus* and *Homarus americanus*), flatfish (*Platichthys flesus*, *Pleuronectes platessa* and *Limanda limanda*) and seals, walruses and turtles (see Seed & Suchanek 1992 for references).

The most common pests of bottom-dwelling mussels are shell-burrowing sponges (*Cliona* spp.), polychaetes (*Polydora* spp.) and pea crabs (*Pinnotheres* spp.). The detrimental effects of pea crabs and boring polychaetes are described in Chapter 11.

Several management measures that prevent predation in mussel culture are described in Kammermans & Capelle (2019). In bouchot culture (Dardignac-Corbeil 1975), crabs (*Carcinus meanas*, *Maja brachydactyla*) that predate on mussels can be prevented by placing a sheet around the bouchots. Predation by birds (e.g. gulls or diving ducks) on mussels on bouchots can be reduced by using nylon threads to prevent their landing. When sea stars and mollusc-drilling snails (*Nucella lapillus*) are present in high densities and predation levels are high, they need to be manually removed.

Predation may exert a top-down limitation on production, especially in bottom culture, since mussel plots are accessible for benthic predators as well as for fish and birds. Intertidal mussels are preyed upon by shore crabs and birds (oystercatchers, herring gulls), while subtidal mussels are preyed upon by shore crabs, sea stars and diving ducks. The number of sea stars on culture plots is reduced by freshwater treatment and there is a selective fishery on sea stars with sea star mops in The Netherlands, the United Kingdom, Germany, Ireland and purse-seines (large walls of netting deployed around an entire area or school of fish) in Denmark (Petersen *et al.* 2016). Freshwater treatment is applied before seeding when mussels are in the vessels' hold; the process consists of the joint exposure of mussels and associated sea stars to freshwater for several hours. Mussels will keep their shells shut, while sea stars are unable to protect themselves against osmotic stress and will not survive. Sea star mops are made of fuzzy rope entwined around small chains that are towed over the mussel plots ensnaring the sea stars, thereby enabling removal. Calderwood *et al.* (2016) estimated the efficiency of sea star removal by mops in Belfast Lough in Northern Ireland and found a large variation in the catch efficiency (4–78%), while the mean sea star reduction when applying this method was 27% (±SE 3.2.)

Studying the effect of exclusion of shore crabs in newly formed intertidal mussel beds on a scale of 800 m^2, Davies *et al.* (1980) found a 400–500% increase in yield over a period of two years. Rope or net culture of mussels has the advantage over bottom culture in that benthic predators cannot reach the mussels directly. Predation by mobile predators on mussels in raft or longline culture is therefore limited to diving birds and fish. However, predators with pelagic larvae (e.g. sea stars) commonly settle in long-line farms. Ducks such as eider ducks that primarily feed on mussels can cause extensive damage to longline mussel cultures (Dunthorn 1971; Žydelis *et al.* 2009). In Maine, United States, mussels are protected by nets placed around the mussel rafts (Newell & Richardson 2014). Mussel ropes and nets are very attractive for a range of fish species. Due to a dramatic decline in Croatian shellfish production, Šegvić-Bubić *et al.* (2011) investigated the most abundant fish species at a mussel farm situated along the eastern coast of the Adriatic Sea. Over a period of two years, the most abundant species observed were gilthead seabream (*Sparus aurata*), sand smelt (*Atherina hepsetus*) and mullet (Mugilidae species) in summer and autumn, *A. hepsetus* and bogue (*Boops boops*) in winter and the sand smelt (*A. hepsetus*), mullet (Mugilidae species) and saddled bream (*Oblada melanura*) in spring. At control locations, characterised by significantly lower fish assemblage abundances, the most abundant species were *A. hepsetus* and mullet (Mugilidae species) in summer, *A. hepsetus* and *O. melanura* in autumn, *A. hepsetus* and picarel (*Spicara flexuosa*) in winter and *A. hepsetus* and bogue

(*B. boops*) in spring. Gilthead sea bream was extremely abundant at the mussel farm, with 5936 individuals censused in 155 of 192 fish counts (80%) and a maximum abundance of 285 individuals per 5000 m^3. Stomach content analysis confirmed the presence of *M. galloprovincialis* as the dominant prey. During a single month, monitoring of 423 ropes revealed that approximately 828 kg of mussels with an average shell length of 34.3 ± 2.54 mm were destroyed within the first week of mussel deposition into the sea, highlighting the degree of fish predation on mussels at this farm. Šegvić-Bubić *et al.* (2011) suggest that shifting farm concessions toward locations with greater depths may reduce fish predation.

Bivalves provide an excellent substrate for the settlement of many fouling organisms. Biofouling appears to be a significant cause of mortality in intertidal mussels, mainly due to dislodgement caused by the increased weight, especially from barnacles and seaweed. Fouling is a particular problem in suspended mussel culture, and almost 100 invertebrate species, including gastropods, crustaceans, bivalves, polychaetes, ascidians, sponges and hydroids, have been identified on mussel ropes (Hickman 1992). These organisms cause reduced growth and productivity through competition for space, but are not a major cause of mortality in suspended culture. See Chapter 10 for details on biofouling.

Mussels are the most prominent competitors for space in mid- to low-shore areas on gently sloping rocky shores, but on steeper shores they tend to be replaced by barnacles or algae. Generally, where two mussel species coexist, there is competition but rarely elimination of one by the other. There are a few notable exceptions. *M. galloprovincialis* was accidentally introduced on the west coast of South Africa in the late 1970s, where it outcompetes the indigenous mussels *Aulacomya atra* and *Choromytilus meridionalis* by reason of its superior reproductive output, faster growth rate and greater tolerance to desiccation (Hockey & van Erkom Schurink 1992; Figure 3.13), while it exhibits partial habitat segregation with the local mussel, *P. perna*, on the south coast (Rius & McQuaid 2006). Because of weaker attachment strength in *M. galloprovincialis*, however, the species will be largely excluded from open coast sites, where wave action is generally stronger, although its greater capacity for exploitation competition through recolonisation will allow it to outcompete *P. perna* in more sheltered areas (especially in bays) that are periodically disturbed by storms (Erlandsson *et al.* 2006). Currently, South Africa is experiencing a second mussel invasion, with detection of the Chilean *Semimytilus algosus* in 2009 (de Greef *et al.* 2013). Both invasive species are now much more abundant intertidally than either of the indigenous mussels, *Aulacomya atra* or *Choromytilus meridionalis*, which have become largely confined to sublittoral and sand-inundated habitats, respectively. The two invasive mussels display strong spatial segregation, with *M. galloprovincialis* dominating the midshore and *S. algosus* blanketing the lower shore. Alexander *et al.* (2015) predict that *S. algosus* will become established along the south coast of South Africa and that *M. galloprovincialis* will maintain dominance on the south and west coasts.

The mussels *S. algosus* and *P. purpuratus* cohabit most of the Chilean rocky shores, with the former inhabiting the low intertidal zone and the latter dominating the mid and mid to high zones. Field and laboratory experiments show that *S. algosus* is a weak competitor with respect to *P. purpuratus*, and post-settlers present high mobility to relocate in the intertidal (Brante *et al.* 2019). Under this scenario, Brante *et al.* (2019) evaluated the dispersal behaviour of juveniles and adults of *S. algosus* as a potential response to competition

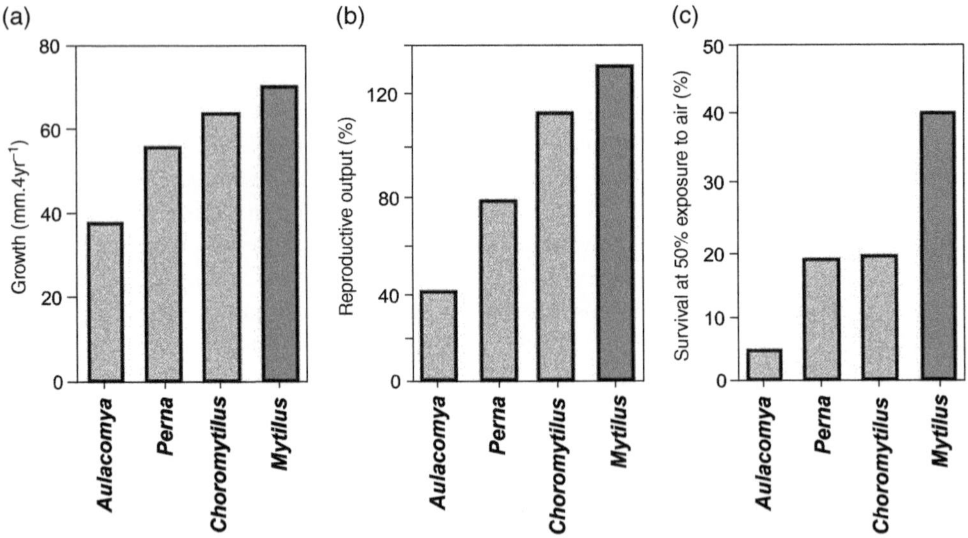

Figure 3.13 Performance of three indigenous South African mussels, *Aulacomya atra*, *Perna perna* and *Choromytilus meridionalis*, relative to *Mytilus galloprovincialis*. (a) Growth rate (mm in the first four years). (b) Total annual reproductive output as a percentage of body mass. (c) Survival rate after 24 months at a shore height experiencing 50% exposure to air per tidal cycle. *Source:* From Branch & Steffani (2004). Reproduced with permission from Elsevier.

with *P. purpuratus*. They also measured the attachment strength of *S. algosus* in the presence of its competitor as a measure of its escape response ability. Their results show that the presence of *P. purpuratus* increased the movement activity of juveniles and adults of *S. algosus* and decreased their attachment strength. Field experiments carried out with marked individuals on a Chilean rocky shore showed that *S. algosus* exhibits higher local dispersion in the zone where *P. purpuratus* is present. Mussels' high dispersal ability throughout the whole benthic phase may serve not only to reach the optimal physiological position in the intertidal, but also to reduce interspecific competition.

Another example of interspecific competition is the contribution of *M. galloprovincialis* to the displacement of *M. trossulus* along much of its historic range in southern California (Shinen & Morgan 2009). Other examples are those between *M. californianus* and the sea palm, *Postelsia palmaeformis*, on the Pacific coast (Dayton 1973; Blanchette 1996). The sea palm grows quickly and may overgrow *M. californianus*, eventually causing mussels to be torn free in the waves. This creates the space necessary for settlement of spores and recolonisation by *Postelsia*. Then there is the competition between *M. galloprovincialis* and the large indigenous limpet, *Scutellastra argenvillei*, on the west coast of South Africa, where the mussel is capable of forming dense, almost monospecific stands low on the shore. A survey indicated that at wave-exposed locations, the abundance of *M. galloprovincialis* changes with exposure (Steffani & Branch 2003). At such locations, the mussel covered up to 90% of the primary substratum, whereas in semi-exposed situations it was never abundant. As the cover of *M. galloprovincialis* increased, the abundance and size of *S. argenvillei* on rock declined, becoming confined to patches within a matrix of mussel beds. Both species were absent from sheltered shores and diminished where wave action was extreme. Comparisons with previous surveys indicated that exposed sites now largely covered by the alien mussel were once dominated by dense populations of the limpet. Therefore, the results of this survey provide circumstantial correlative evidence of a competitive interaction between *M. galloprovincialis* and *S. argenvillei*, and suggest that wave action mediates the strength of this interaction. The presence of mussel beds provides a novel settlement and living substratum for recruits and juveniles of *S. argenvillei*, albeit at much lower densities than in limpet patches. Adult limpets are virtually excluded from the mussel beds owing to their large size, which indicates the unsuitability of this habitat as a replacement substratum after competitive exclusion from primary rock space.

The American slipper limpet, *Crepidula fornicate*, was unintentionally introduced to Europe in the 1870s with oysters imported for farming purposes. Since the limpet is a filter feeder, trophic competition and associated negative effects when epizootic on bivalves have been assumed (Figure 3.14). Thielges *et al.* (2005) were the first to experimentally test the effects of *C. fornicata* on survival and growth of its major basibiont, *M. edulis*. In two field experiments (each lasting two weeks), epigrowth by *C. fornicat*a resulted in a four- to eight-fold reduction in survival of mussels, equivalent to a mortality of 28 and 30%, respectively. Shell growth in surviving mussels with attached *C. fornicata* was three to five times lower compared to unfouled mussels, but similar to that with artificial limpets. As a causative agent, interference competition in the form of changes in small-scale hydrodynamics due to *C. fornicata* stacks was suggested. This could result in a high energy expenditure for byssus production of mussels. In general, interference and not exploitation competition is suggested to be the major impact of epizootic *C. fornicata* on its basibionts (organisms that are host to epibionts) in Europe.

Figure 3.14 Stack of four American slipper limpets, *Crepidula fornicata*, attached to the mussel *Mytilus edulis*. *Source:* From Thieltges (2005). Reproduced with permission from Inter-Research.

Typically, however, intraspecific competition for space is a more serious problem than interspecific competition, in that heavy spat fall of mussels on to adult beds can cause the underlying mussels to suffocate, thus loosening the entire population from the rock surface (Seed 1976). See Chapter 5 for information on interspecific and intraspecific gamete competition in *Mytilus* taxa.

Climate Change and Potential and Observed Impacts on Marine Mussels

Climate in a narrow sense is usually defined as the average weather, or more rigorously as the statistical description in terms of the mean and variability of relevant quantities over a period of time ranging from months to thousands or millions of years. Climate encompasses patterns of temperature, precipitation, humidity, wind and seasons. The classical period for averaging these variables is 30 years, as defined by the World Meteorological Organization (https://public.wmo.int/en). Our climate is now changing at a rate faster than any seen in the last 2000 years (www.ecology.wa.gov). This is due to rising levels of carbon dioxide and other heat-trapping (greenhouse) gases in the atmosphere. The primary greenhouse gases in Earth's atmosphere are water vapour (H_2O), carbon dioxide (CO_2), nitrous oxide (N_2O), methane (CH_4) and ozone (O_3). These act like a blanket around Earth, causing it to warm – a phenomenon referred to as 'global warming'. According to an ongoing temperature analysis conducted by scientists at NASA's Goddard Institute for Space Studies (GISS), the average global temperature on Earth has increased by about 0.8 °C since 1880. Two-thirds of the warming has occurred since 1975, at a rate of roughly 0.15–0.20 °C per decade (Figure 3.15). Wide-ranging impacts of rising temperatures include rising sea levels, melting snow and ice, more extreme heat events, fires and drought and more extreme storms, rainfall and floods. This rise in global average temperature is attributed to an increase in greenhouse gas emissions. The link between global temperatures and greenhouse gas concentrations, especially of CO_2, has held throughout Earth's history (Lacis *et al.* 2010). CO_2 concentrations in the atmosphere are now well over 400 ppm, their highest levels in over 800 000 years (Figure 3.16). Globally, we emit over 36 billion tonnes of CO_2 per year, and this continues to increase. There are large differences – more than 100-fold – in per capita CO_2 emissions between countries (Richie & Roser 2020). For example, the United States has contributed most to global CO_2 emissions to date, accounting for 25% of cumulative emissions. This is followed by the EU-28 (22%), China (13%), Russia (6%) and Japan (4%). There are large inequalities in CO_2 emissions: the world's poorest countries have contributed less than 1% of emissions, but will be the most vulnerable to climate change impacts. Rising levels of atmospheric CO_2 is increasing ocean acidity, which has been predicted to double over the next 100 years in an uncontrolled emission scenario (Fauville *et al.* 2013). Approximately one-third of the anthropogenic carbon added to the atmosphere is absorbed by the oceans. Uptake of atmospheric CO_2 results in a decrease in ocean water pH, an effect referred to as 'ocean acidification' (OA; see later).

The Intergovernmental Panel on Climate Change (IPCC) is the UN body responsible for assessing the science related to climate change. It was established by the United Nations Environment Programme and the World Meteorological Organization (WMO) in 1988 to

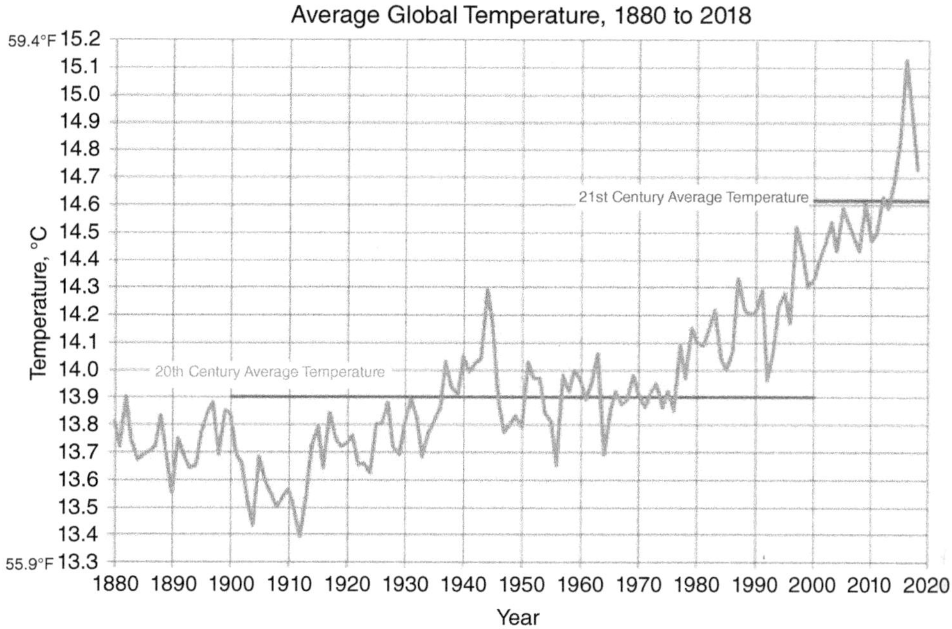

Figure 3.15 Global average temperature for the period 1880–2018. While global average temperatures vary from year to year, the overall trend from 1880 to the present is one of increased temperature. *Source:* Data from National Centers for Environmental Information. https://www.ncdc.noaa.gov/cag/global/time-series/globe/land_ocean/1/1/1880-2020.

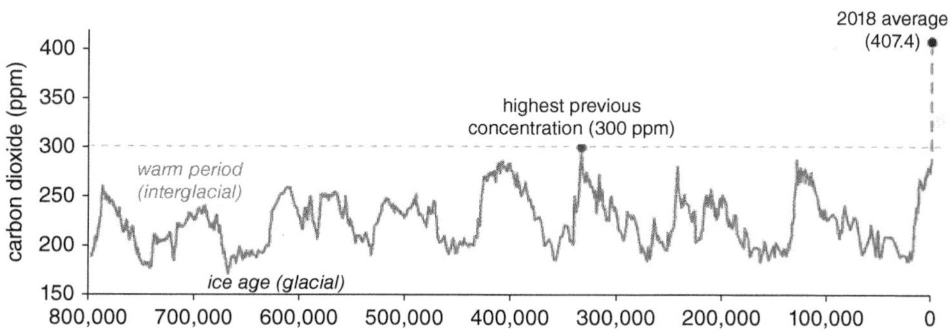

Figure 3.16 Global atmospheric carbon dioxide (CO_2) concentrations in parts per million (ppm) for the past 800 000 years. The peaks and valleys track ice ages (low CO_2) and warmer interglacials (higher CO_2). During these cycles, CO_2 was never higher than 300 ppm. In 2018, it reached 407.4 ppm. On the geologic time scale, the increase (dashed line) looks virtually instantaneous. *Source:* Data from Lindsey (2020).

provide policymakers with regular scientific assessments concerning climate change, its implications and its potential future risks, as well as to put forward adaptation and mitigation strategies. The IPCC's reports are comprehensive and balanced assessments of the state of knowledge on topics related to climate change. There are different types of reports, but all go through a rigorous process of scoping, drafting and review to ensure the highest quality. To date, the IPCC has 195 member states. The Paris Agreement is an agreement within the United Nations Framework Convention on Climate Change (UNFCCC) dealing with greenhouse gas emissions mitigation, adaptation and finance. As of November 2019, 190 of 195 IPCC member countries have signed the Agreement. Its long-term temperature goal is to keep the increase in global average temperature to well below 2 °C above pre-industrial levels and to pursue efforts to limit the increase to 1.5 °C, recognising that this would substantially reduce the risks and impacts of climate change. The Paris Agreement was intended to become fully effective in 2020. Predictions of global environmental conditions for the end of the century coupled with ever-increasing experimental evidence suggest wide-ranging impacts of future OA and warming scenarios on marine life (Poloczanska *et al.* 2016). Humans, through activities such as burning fossil fuels, deforestation, industrial processes and some agricultural practices, are largely responsible.

So, what are the consequences of climate change for marine ecosystems? Mora *et al.* (2013), using global climate models, have shown that in the next 100 years, the entire world's ocean surface will be simultaneously impacted by varying intensities of ocean warming, acidification, oxygen depletion or shortfalls in productivity. In contrast, only a very small fraction of the world's ocean surface, mostly in polar regions, will experience increased oxygenation and productivity, and almost nowhere will there be cooling or pH increase. From a compiled list of 32 marine habitats and biodiversity hot spots, Mora *et al.* (2013) found that all would experience simultaneous exposure to changes in multiple biogeochemical parameters, which will demand multiple physiological adjustments from marine biota. However, regional-scale differences in response to climate change can often be more relevant than global averages. For example, a study of SST change in 63 global large marine ecosystems (LMEs) over a 50-year period (1957–2006) revealed strong regional variation, with the Subarctic Gyre, European Seas and East Asian Seas warming at two to four times the global mean rate (Belkin 2009). The Subarctic Gyre warming is likely caused by natural variability in relation to the North Atlantic Oscillation, a climatic phenomenon which varies over time but has no particular periodicity. The most rapid warming was observed in the land-locked or semi-enclosed European and East Asian Seas (Baltic and North Seas, Black Sea, Japan Sea/East Sea and East China Sea), and also over the Newfoundland–Labrador Shelf. The proximity of the European and East Asian Seas to major industrial/population agglomerations suggests a possible direct anthropogenic effect. In a comparable study, Alexander *et al.* (2018) examined changes in SSTs in 18 LMEs adjacent to North America, Europe and the Arctic Ocean. The annual SST trends over 1976–2099 in all 18 were positive, ranging from 0.05 to 0.5 °C per decade. SST changes by the end of the 21st century will primarily be due to a positive shift in the mean, with only modest changes in the variability in most LMEs, resulting in a substantial increase in warm extremes and decrease in cold extremes. The shift in the mean is so large that in many regions SSTs during 2070–2099 will always be warmer than the warmest year during 1976–2005. The SST trends are generally stronger in summer than in winter, which amplifies the seasonal cycle of SST over the 21st century. In the Arctic, the mean SST and its variability increase substantially during summer, when it is ice-free, but not during winter, when a thin layer of ice reforms and SSTs remain near the freezing point. While basin-wide changes in the ocean are expected (Alexander *et al.* 2018), it is critical to examine temperature changes along continental margins, which supply more than 75% of the world's marine fish catch. Lima & Wethey (2012) explored global and monthly warming patterns along 19 276 coastal locations between 1982 and 2010. They demonstrated that 46% of the coastlines had experienced a significant decrease in the frequency of extremely cold events, while extremely hot days were becoming more common in 38%. They further showed that the onset of the warm season was advancing significantly earlier in the year in 36% of the temperate coastal regions. More importantly, it is now possible to analyse local patterns within the global context, which is useful for a broad array of scientific fields, policymakers and the general public. Li *et al.* (2019) similarly conducted a global analysis of SST at 26 locations in Chinese coastal waters.

Climate Warming

Latitudinal distributions of many organisms are limited by temperature. One major response is a shift in distribution, usually poleward (Root *et al.* 2003). Physiological processes that set thermal tolerance limits are thought to determine, or at least contribute to, some of the shifts that have been observed (Tomanek 2008 and references therein). As already mentioned, seasonal air and water temperatures since 1960 have increased along the eastern US seaboard, and south of Lewes, Delaware (38.8 °N) summer SST increases have exceeded the upper lethal limits (32 °C) of *M. edulis* (Jones *et al.* 2010), resulting in geographic contraction of its southern, equatorward range edge approximately 350 km north, or ~7.5 km per year (Somero 2012). At the southern part of the range, high water and air temperatures cause mass mortality events, while along the more northerly portion, mortality is caused by high temperatures during aerial exposure. Ultimately, water temperatures in excess of thermal tolerances have caused contraction of the mussel's biogeographic range (Jones *et al.* 2010).

Range shifts vary greatly between species, and the distributions of *Mytilus* populations all over the world are responding differently to climate change. For example, Harley *et al.* (2011) compared distributions of *M. californianus* from 2009 to 2010 to a historical data set from 1957–1958. Sampling sites were believed to be within ~30 m of the original survey sites. The 52 years separating the two sampling intervals span a period of climatic warming. During the latter half of the 20th century, maximum air temperatures near the eastern and western ends of the Strait of Juan de Fuca near Washington state increased by ~0.2 and ~0.13 °C per decade, respectively, and mean annual water temperatures along the southern and western Vancouver Island coast warmed by ~0.08–0.11 °C per decade. Average daily maximum air temperatures during the summer, which are particularly relevant to thermal stress experienced in the intertidal zone, have warmed even more rapidly. At Victoria, on the southern end of Vancouver Island, summer average daily maxima have risen approximately linearly at a rate of 0.654 °C per decade since 1950, corresponding to an increase of 3.48 °C over 52 years, which has caused a 51% decrease in the vertical distribution of *M. californianus*, providing strong evidence that this change is linked to global warming. Associated with this decrease are local extinctions of *M. californianus* at 13% of the sites resurveyed between 2009 and 2010. Historical data for *M. trossulus* were only available for one site (former vertical range = 49 cm). By 2010, *M. trossulus* had been completely eliminated from that site, with the exception of three small juveniles found under a single rock.

From 1995 to 1999, the poleward movement of *M. galloprovincialis* showed a reversal concomitant with a cooling phase of the PDO (Hilbish *et al.* 2010). *M. galloprovincialis* has declined in abundance over the northern third of its geographic range (~540 km) and has become rare or absent across the northern 200 km of the range it previously colonised during its initial invasion. The distribution of the native species *M. trossulus* has, however, remained unchanged over the same time period. The difference in SST between warm and cold phases of the PDO is small (~1 °C), but Hilbish *et al.* (2010) deduced that even this minor decrease in temperature during the cold phase of the PDO may be enough to retard larval development in *M. galloprovincialis*, such that recruitment is handicapped in northern waters.

Geographic ranges of species are determined by climatic conditions, and this forms the basis of a number of models that use associations between aspects of climate and species' occurrences to estimate the conditions that are suitable for maintaining viable populations (Araújo & Peterson 2012). The most frequently used climate change model is species distribution modelling (SDM), also known as environmental (or ecological) niche modelling, habitat modelling, predictive habitat distribution modelling and range mapping. SDM uses computer algorithms to predict the distribution of a species across geographic space and time using environmental data (Elith & Leathwick 2009). It can explain how environmental conditions influence the occurrence or abundance of a species, and can provide models (ecological forecasting) of a species' future distribution under climate change, a species' past distribution (in order to assess evolutionary relationships) or the potential future distribution of an invasive species. There are two main types of SDM: correlative SDMs, also known as climate envelope models or bioclimatic models, which model the observed distribution of a species as a function of environmental conditions; and mechanistic SDMs, also known as process-based models or biophysical models, which use independently derived information about a species' physiology to develop a model of the environmental conditions under which it can exist (Kearney & Porter 2009). In empirical modelling, a number of climatic variables, such as maximum and minimum temperatures and precipitation, are measured for many different locations and statistically compared to the occurrence of the focal species at these locations (Jeschke & Strayer 2008). Most bioclimatic models do not explicitly consider biotic interactions (e.g. predation and competition between species) or limitations to dispersal, and assume that species lack sufficient plasticity to adapt to environments beyond those currently occupied (Jeschke & Strayer 2008; Fuller *et al.* 2010). The ecological effects of climate change can only be predicted if there is an understanding of the physiology of the species in its natural environment. Evidence based on temperature indicates that species' distributions are driven both by short-term exposure to lethal conditions and by repeated or longer-term exposures leading to energetic failures. The relationship between physiological performance (e.g. scope for growth, SFG) and body temperature can be depicted as a thermal performance curve (Figure 3.17). Woodin *et al.* (2013) define the difference between lethal and sublethal exposure limits as the transient event margin (TEM), the range of environmental conditions below the short-term lethal limit that a species may endure on a transient basis but which would lead to mortality over longer time spans. The magnitude of the disparity (TEM) between performance and tolerance temperature thresholds relative to environmental variance

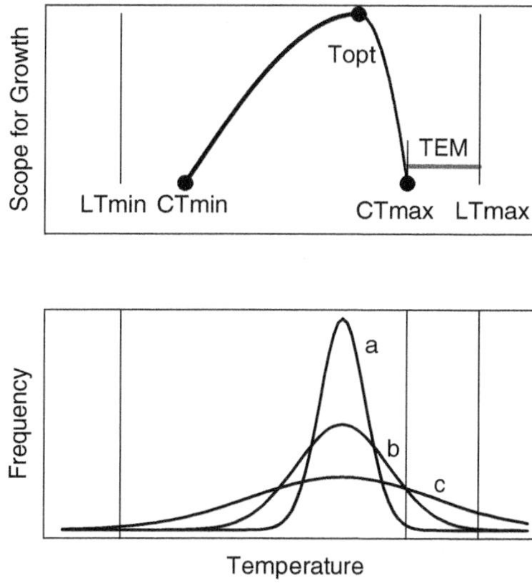

Figure 3.17 (a) Relation between performance as scope for growth (SFG) and temperature. Lethal limits are LTmin and LTmax, performance limits are critical minimum temperature (CTmin) and critical maximum temperature (CTmax), the body temperature at any point in space and time is Tbody and the optimal temperature is Topt. The upper transient event margin (TEM) = LTmax – CTmax. Thermal response curves vary from species to species but all typically display some optimum temperature (Topt) at which performance is maximised, as well as critical minima (CTmin) and critical maxima (CTmax) beyond which mortality and/or reproductive failure occur (Helmuth *et al.* 2010). (b) Temperature versus frequency for environments a, b and c. TEM from (a) is expressed on yearly environmental variance curves for three environments, all with the same mean environmental temperature but differing in variance. From Woodin *et al.* (2013). *Source:* From Woodin *et al.* (2013). Reprinted with permission from John Wiley & Sons Ltd. CC BY 3.0.

determines the likelihood of failure of bioclimatic model predictions (Woodin *et al.* 2013). To define TEM, both CT_{max} SFG and LT_{max} data are necessary, but in the majority of data sets only LT_{max} estimates are available. Jones *et al.* (2010) used a mechanistic biogeographic model for *M. edulis* on the Atlantic coast of N. America based on LT_{max}. Physiological limits were compared against environmental temperatures and the biogeographic distribution of the species was accurately predicted. The hindcast using the model also successfully predicted the historical changes in distribution over half a decade described earlier. However, when the same model applied to Europe, it failed to predict the distribution of *M. edulis*; in reality, its distribution in Europe was 50% less than that predicted. When an energetics model was applied to Europe, the predicted distribution was close to the actual distribution of the species on western European coasts. This highlights the need for caution in applying the same model across large geographic distances (see details in Woodin *et al.* 2013).

Closely related species with different physiological tolerances and distributions make ideal systems for documenting range shifts in response to a changing climate. The closely related species *M. edulis*, *M. trossulus* and *M. galloprovincialis* have distinct biogeographical ranges that are correlated with SST. Fly & Hilbish (2013) determined the SFG of these three species over a range of temperatures (5–30 °C) to determine whether energetics could predict their distributions. SFG represents energy available for growth and/or reproduction above that necessary for maintenance requirements. The results showed, as expected, reasonable correlation with the temperature regimes of habitats presently occupied, with *M. trossulus*, a boreal species, exhibiting positive SFG up to 17 °C, *M. edulis*, a cold temperate species, up to 23 °C, and *M. galloprovincialis*, a warm-water species, up to 30 °C. In the spring, temperatures of peak performance (T_{opt}) for both *M. trossulus* and *M. edulis* occurred at ~15 °C, and both species exhibited negative SFG at 25 °C, which was the T_{opt} for *M. galloprovincialis*. Overall, SFG was lower in summer than in autumn and winter. *M. trossulus* only showed positive SFG at 10 °C, while *M. edulis* and *M. galloprovincialis* maintained positive SFG at all temperatures above 5 °C, with T_{opt}s at 20 and 25 °C, respectively (see Figure 6.15). The warm end of each species' range correlated positively with the seawater temperature at which that species' SFG became negative (i.e. when metabolic expenditure exceeds energy acquisition) in summer and autumn. Energetics at cold temperatures did not predict the cold end of the species' ranges, as there was no clear SFG advantage to explain the dominance of *M. trossulus* in cold habitats. As SST continues to warm due to climate change, the energetics of these three species provide a basis for developing mechanistic models[8] predicting future distribution and productivity changes in mussel populations (Fly & Hilbish 2013).

Montalto *et al.* (2016) used three interconnected models (climatic, biophysical and energetics) to estimate changes in growth, reproduction and mortality risk by 2050 for three commercially and ecologically important mussels (*Brachidontes pharaonis*, *Mytilaster minimus* and *M. galloprovincialis*) at 51 sites in the Mediterranean Sea. Helmuth *et al.* (2014) have cautioned that to focus solely on range limits may lead to failure to notice highly significant impacts that are both occurring now and will likely occur in the future. All three species inhabit intertidal habitats, and thus are affected by changes in both terrestrial and marine environments. They differ from one another in their thermal optima, performance breadth and lethal temperatures. Montalto *et al.* (2016) predicted highly variable responses (both positive and negative) in the timing of reproductive maturity and the risk of lethality among the species and sites that do not conform to simple latitudinal gradients, which would be undetectable by methods focused only on lethal limits and/or range boundaries. Also, results strongly suggest that the three species will likely experience changes in the timing of reproductive maturity, and that in intertidal zones thermal stress may cause changes to the risk of mortality. However, these consequences are highly species- and site-specific in ways that do not always conform to simple latitudinal gradients, and in some cases reveal potential benefits for Mediterranean Sea ecosystems.

Challenges with managing bivalve culture may become more pronounced as the uncertainty associated with climate change makes it difficult to predict future production levels. Steeves *et al.* (2018) coupled a previously established high-resolution climate model with two dynamic energy budget (DEB) models (see Chapter 6) to explore the future growth and distribution of two economically and ecologically important species, the oyster *Crassostrea virginica* and the mussel *M. edulis*, along the Atlantic coast of Canada. SST data were extracted from the climate model and used as a forcing variable in the bioenergetic models. This approach was applied across three distinct time periods: the past (1986–1990), the present (2016–2020) and the future (2046–2050), thus permitting a comparison of bivalve performance under different temporal scenarios. Results showed that growth in the future is variable both spatially and interspecifically. Modelling outcomes suggested that warming ocean temperatures will cause an increase in growth rates of both species as a result of their ectothermic nature. Since the thermal tolerance of *C. virginica* is higher than that of *M. edulis*, oysters will generally out-perform mussels (Gosling 2015). The predicted effects of temperature on bivalve physiology also provided insight into vulnerabilities, such as mortality, under future SST scenarios. This information is useful for adapting future management strategies for both farmed and wild shellfish. Although the study focused on a geographically specific area, the approach of coupling bioenergetic and climate models is valid for species and environments across the globe.

Apart from the impact on mussel biogeographic distribution and physiological traits already discussed, there is evidence that global warming can impact on: interspecific (Petes *et al.* 2007) and predator–prey (Freitas *et al.* 2007; Broitman *et al.* 2009; Menge *et al.* 2008; Harley 2011; Kordas *et al.* 2011; Monaco *et al.* 2016; Torossian *et al.* 2020) interactions; reproductive timing and larval dispersal (Carson *et al.* 2010); the immune response (Matozzo & Marin 2011; Beaudry *et al.* 2016; Hernroth & Baden 2018); byssal thread strength (Newcomb *et al.* 2019); developmental instability (Nishizaki *et al.* 2015); species invasion success (US EPA 2008; Occhipinti-Ambrogi & Galil 2010; Firth *et al.* 2011; Somero 2010, 2011; Thyrring *et al.* 2017); mussel culture (Guyondet *et al.* 2015; Steeves *et al.* 2018; Silva *et al.* 2017; Filgueira *et al.* 2016; Callaway *et al.* 2012); and radiation-induced damage (Dallas *et al.* 2016).

Ocean Acidification

Human activities such as the burning of oil, coal and gas, as well as deforestation, are the primary cause of the increased CO_2 concentrations in the atmosphere. The global average atmospheric CO_2 in 2018 was 407.4 ppm, with a range of uncertainty of $\pm$0.1 ppm. Carbon dioxide levels today are higher than at any point in at least the past 800 000 years (NOAA Global Climate Report 2019). Human-induced rise in atmospheric CO_2 concentration has been linked to changes in seawater carbonate chemistry and a decrease in ocean pH, a process known as ocean acidification (OA) (Ventura *et al.* 2016; see Chapter 6). When carbon dioxide diffuses into the ocean, it reacts with water to create carbonic acid (H_2CO_3), most of which quickly dissociates into a hydrogen ion (H^+) and bicarbonate (HCO_3^-), which can further dissociate into carbonate (CO_3^{2-}) and hydrogen ions (Figure 3.18). Some of the carbonate ions already in the ocean combine with some of the hydrogen ions to form further bicarbonate, thereby reducing the availability of carbonate ions, which are necessary for marine calcifying organisms such as corals, foraminiferans, echinoderms, crustaceans and molluscs to produce their $CaCO_3$ shells and skeletons (Fabry *et al.* 2008).

Since the Industrial Revolution, a time span of less than 250 years, the pH of surface oceans has dropped by 0.1 pH units, representing a ~30% increase in hydrogen ion concentration relative to the preindustrial value (Guinotte & Fabry 2008). The pH scale is logarithmic, and consequently each whole unit decrease in pH is equal to a 10-fold increase in acidity. The pH of the oceans is projected to drop another 0.3–0.4 pH units by 2100 (Mackenzie *et al.* 2014). The rate of this change is cause for concern, as many marine organisms – particularly those that calcify (see earlier) – may not be able to adapt quickly enough to survive. The degree of change will depend on whether we adopt a concerted rapid CO_2 mitigation effort or a 'business-as-usual' attitude (Mora *et al.*2013). The lowest surface ocean pH is found in the equatorial regions from 20 °N to 20 °S, especially in the eastern Pacific (Jiang *et al.* 2019). The Arctic Ocean shows the largest spatial variability, followed by the Southern Ocean, which surrounds Antarctica. The globally and annually averaged surface ocean pHs in the Atlantic, Pacific and Indian Oceans (60 °N to 60 °S) are

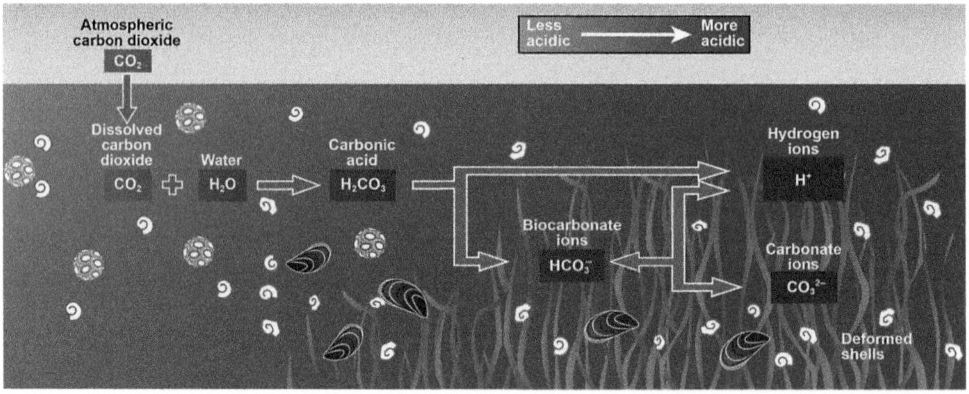

Figure 3.18 Schematic diagram of ocean acidification (OA). The reaction between dissolved carbon dioxide (CO_2) and water results in an increase in the concentration of hydrogen ions (H^+); additional changes include an increase in bicarbonate ions (HCO3−) and a great decrease in carbonate ions ($CO_3{}^{2-}$). These ions will modify the carbonate saturation state, leading to acidification. *Source:* From Birchenough *et al.* (2017). Open Government Licence v3.0.

very similar at 8.07 ± 0.02, 8.06 ± 0.03 and 8.07 ± 0.02, respectively, with a global average of 8.07 ± 0.02 between 60 °N to 60 °S. The most rapid acidification is presently occurring in the Arctic Ocean (see Qi *et al.* 2017).

Dupont & Pörtner (2013) reviewed a selection of papers covering a range of experimental approaches used to investigate the impact of OA on marine species and ecosystems. They found that while the vast majority of studies (>90%) on the potential effects of OA were laboratory-based, there were a few field studies using natural gradients or CO_2-rich environments and large-scale field studies using mesocosms that could provide insights into the short- and long-term responses at the ecosystem level. Since their metastudy, there has been an enormous increase in the number of studies examining the effects of OA on a range of marine species, particularly the calcifying ones. The following paragraphs provide a selection of such studies, with the emphasis on those dealing with marine mussels.

To the best of my knowledge, all studies on the potential effects of OA on marine mussels are laboratory-based. Impacts of pH on *M. edulis* and their larvae have been widely investigated. Thomsen & Melzner (2010) analysed the impacts of pH on metabolism in adult *M. edulis*. They observed reduced shell growth under severe acidification, and suggested this may be a result of synergistic effects of increased cellular energy demand and nitrogen loss. In a study on the effects of OA on early life stages of *M. edulis*, Gazeau *et al.* (2010) showed negative impacts of increasing seawater acidity on parameters such as hatching rates and D-veliger shell growth, while Kapsenberg *et al.* (2018) observed abnormal soft tissue in *Mytilus* D-larvae. When early life stages of *M. edulis* were exposed to pH 7.6, Bechmann *et al.* (2011) observed no significant effect on fertilisation success, development time, shell abnormality or feeding when compared to pH 8.1. In a coupled field and laboratory study, Thomsen *et al.* (2013) examined the annual pCO_2 (p = partial pressure) variability in Kiel Fjord, Western Baltic Sea and the combined effects of elevated pCO_2 and food availability on juvenile *M. edulis* growth and calcification. In the laboratory experiment, mussel growth and calcification were found to chiefly depend on food supply, with only minor impacts of pCO_2 up to 3350 μatm. In the field growth experiment in Kiel Fjord, a brackish and CO_2-enriched habitat, the authors found seven times higher growth and calcification rates of *M. edulis* at a high-CO_2 inner fjord field station (mean pCO_2 ~1000 μatm) in comparison to a low-CO_2 outer fjord station (~600 μatm). High mussel productivity at the inner fjord site was enabled by higher particulate organic carbon concentrations. Thomsen *et al.* (2013) concluded that benthic stages of *M. edulis* tolerate high ambient pCO_2 when food supply is abundant and that energy availability needs to be considered to predict species vulnerability to OA.

In a subsequent study, Ventura *et al.* (2016) investigated the effects of a wide range of seawater pHs on different physiological parameters of *M. edulis* developing larvae in order to find the physiological tipping point beyond which they are no longer capable of carrying out the functions necessary to their survival and recruitment into the adult population. The results confirmed that decreasing seawater pH and decreasing calcium carbonate saturation state (Ω) increase larval mortality rate and the percentage of abnormally developing larvae. Virtually no larvae reared at average pH 7.16 were able to feed or reach the D-shell stage, and their development appeared to be arrested at the trochophore stage. However, larvae were capable of reaching the D-shell stage under milder acidification (pH ≈ 7.35, 7.6, 7.85), including in undersaturated seawater with aragonite saturation state[9] (Ωa) as low as 0.54 ± 0.01 (mean ± SEM), with a tipping point for normal development identified at pH_T 7.765. Also, the growth rate of normally developing larvae was not affected by lower pH_T despite potential increased energy costs associated with compensatory calcification in response to increased shell dissolution. Altogether, the results for OA impacts on mussel larvae suggest an average pH of 7.16 is beyond their physiological tolerance threshold and indicate a shift in energy allocation toward growth in some individuals, thus revealing potential OA resilience.

Navarro *et al.* (2013) evaluated the impact of medium-term (70 days) exposure to elevated pCO_2 levels (380, 750 and 1200 ppm) on physiological processes of juvenile *M. chilensis*. SFG was reduced by 13% at 750 ppm and 28% at 1200 ppm CO_2 compared to the control treatment, 380 ppm (see Figure 6.21). This could represent a significant loss in annual production for commercial operations. A reduction in growth to 55% was also reported for juvenile and adult *M. edulis* maintained under long-term moderate CO_2 levels (Michaelidis *et al.* 2005). Subsequently, Navarro *et al.* (2016) examined the combined effect of temperature (12 and 16 °C) and elevated pCO_2 levels (380, 750 and 1000 μatm) on juvenile *M. chilensis*. They found, as in their previous study, that SFG was significantly lower at the highest pCO_2 concentration compared with the control, and mussels exposed to 700 μatm did not show a significantly different SFG from the other two treatments. SFG was significantly higher at 16 °C than at 12 °C, which may be because these temperatures are within the thermal tolerance of *M. chilensis* in southern Chile. When strong pCO_2 stress was coupled to food limitation, each significantly decreased shell length growth, and both significantly influenced the magnitude of inner shell surface dissolution in *M. edulis* (Melzner *et al.* 2011). In contrast, Range *et al.* (2012) found that even under extreme levels of CO_2-induced acidification, juvenile *M. galloprovincialis* can continue to calcify and grow.

Several studies have examined the combined effects of warming and OA on mussel growth. In one, adult mussels (*M. galloprovincialis*) were exposed to low and high pCO_2

and 12, 14, 16, 18, 20 and 24 °C for one month (Kroeker *et al.* 2014). Although high pCO_2 significantly reduced mussel growth at 14 °C, this effect gradually lessened with increasing temperature, illustrating how warming can moderate the effects of OA in this species. Similar results have also been reported for adult *M. galloprovincialis* (Gazeau *et al.* 2014) and juveniles of *M. edulis* (Hiebenthal *et al.* 2013) and *M. coruscus* (Wang *et al.* 2015). However, Vihtakari *et al.* (2013) found that increasing temperature might have a larger impact on sperm motility and very early larval stages in *M. galloprovincialis* than OA at the levels predicted for the end of the century.

In mussels, elevated pCO_2 impacts on larval and broodstock feeding (Diaz *et al.* 2018), byssal attachment (O'Donnell *et al.* 2013), sea star and gastropod predation (Kepell *et al.* 2015; Sadler *et al.* 2018), anti-predator defense strategies (Kong *et al.* 2019), levels of predation vulnerability (Kroeker *et al.* 2016), cellular signaling pathways involved in the immune response (Bibby *et al.* 2008), immune parameters of haemocytes (Wu *et al.* 2018), host–pathogen interactions (Asplund *et al.* 2014), antimicrobial activity (Hernroth *et al.* 2016) and expression of genes involved in energy and protein metabolism (Hüning *et al.* 2013).

As already mentioned, OA is altering the oceanic carbonate saturation state and threatening the survival of marine calcifying organisms. Production of their calcium carbonate exoskeletons is dependent not only on the environmental seawater carbonate chemistry but also on the ability to produce biominerals through proteins. A detailed description of the production and structure of mussel shells is given in Chapter 2. Fitzer *et al.* (2014) examined the responses of *M. edulis* to four pCO_2 concentrations (380, 550, 750 and 1000 µatm pCO_2) over a six-month incubation period. These concentrations represent future OA scenarios leading up to the year 2100. Mussels were also exposed to combined increases in CO_2 and temperature (ambient + 2 °C) relating to future projected climate change. They were examined for shell structural and crystallographic orientation, growth, calcite and aragonite (crystalline forms of calcium carbonate) thickness and carbonic anhydrase concentration. The aim of the investigation was to determine the presence of any OA 'tipping' point or threshold which, once reached, might cause calcifiers to experience difficulties in maintaining control of biomineralisation and producing structurally sound shell growth. In a related study, Fitzer *et al.* (2015) found that OA resulted in rounder, flatter mussel shells with thinner aragonite layers likely to be more vulnerable to fracture under changing environments and predation. These changes in shape could present a compensatory mechanism to enhance protection against predators and changing environments. Mussels employ transient phases of amorphous calcium carbonate (ACC) in the construction of crystalline shells. Fitzer *et al.* (2016) investigated the influence of OA on ACC formation in the shells of adult *M. edulis*. Their results demonstrated that OA induces more ACC formation and less crystallographic control in mussels, suggesting that ACC is used as a repair mechanism to combat shell damage under OA. However, the resultant reduced crystallographic control in mussels raises concerns for shell protective function under predation and changing environments.

Gaylord *et al.* (2011) tested effects of OA on the larval stage of *M. californianus*, a critical community member on rocky shores of the northeastern Pacific. Larvae were cultured for eight days in seawater containing CO_2 concentrations associated with a present-day global-mean atmospheric concentration (~380 ppm) contrasted against a 'fossil fuel-intensive' projection (970 ppm) and a more optimistic prediction (540 ppm). The authors analysed the strength, size and thickness of larval shells and found that acidification of the seawater had a strong impact on shell strength. The shells of five-day-old larvae raised in 970 ppm CO_2 were 20% weaker than those of larvae reared at the current CO_2 level, while the shells of larvae reared at 540 ppm CO_2 were only 13% weaker than those of control individuals. They also found that after eight days at 970 ppm CO_2, the shells were up to 15% thinner and 5% smaller, and the body mass of the mussels within the shells were as much as 33% smaller than those of mussels grown at modern levels. Potential ecological consequences of OA are larvae weakened by rising CO_2 levels, slow development, vulnerability to predation, susceptibility to stress and risk of desiccation, ultimately altering mussel survivorship and thereby overall community dynamics.

The following is a succinct summary of the current OA scenario:

> Over the next decades, it is likely that ocean acidification will pose serious consequences for many marine and estuarine shelled molluscs. A comparison of the available literature to date suggests that while detrimental effects on adults remain uncertain, the most sensitive life-history stage seems to be the larvae, with a large majority of studies on this critical stage of development revealing negative effects. Despite these obvious trends, our current understanding of the biological consequences of an acidifying ocean over the next century is still dominated by large uncertainties. This is because the majority of studies done to date have measured single-species responses on one stage in the life cycle, without considering the synergistic effects of other stressors (i.e. temperature, hypoxia, food concentration) and have not considered the potential for species to adapt, nor the underlying mechanisms responsible for adaptation or acclimation. In order to fully understand the consequences of ocean acidification at the population and ecosystem level, multi-generational and multi-stressor experiments on multiple species from geographically distinct locations are needed to assess the adaptive capacity of shelled mollusc species and the potential winners and losers in an acidifying ocean over the next century.
>
> *(Gazeau et al. 2013, p. 2239)*

The high economic value of bivalves for aquaculture has stimulated a number of studies to estimate their adaptation potential to future oceanic conditions (Parker *et al.* 2012, 2013, 2015; Sunday *et al.* 2011, 2014; Thomsen *et al.* 2017; Vargas *et al.* 2017). To illustrate, Thomsen *et al.* (2017) recorded the successful settlement of wild mussel larvae (*M. edulis*) in a periodically CO_2-enriched habitat. The larval fitness of the population originating from the enriched habitat was compared to the response of a population from a nonenriched one. The high CO_2-adapted population showed higher fitness under elevated pCO_2 than the nonadapted cohort, demonstrating, for the first time, an evolutionary response of a natural mussel population to OA. To assess the rate of adaptation, the authors performed a selection experiment over three generations. Tolerance to CO_2 differed substantially between the families within the F1 generation, and survival was drastically decreased in the highest – yet, realistic – pCO_2 treatment. Selection of CO_2-tolerant F1 mussels resulted in higher calcification performance of F1 larvae during early shell formation but did not improve overall survival. Hence, the results reveal significant short-term selective responses of traits directly affected by OA and long-term adaptation potential in a key bivalve species. Because immediate response to selection did not directly translate into increased fitness, multigenerational studies need to take into consideration the multivariate nature of selection acting in natural habitats. Combinations of short-term selection with long-term adaptation in populations from CO_2-enriched versus nonenriched natural habitats represent promising approaches toward estimating the adaptive potential of organisms facing global change.

Hypoxia

Another serious consequence of global warming that has gained attention only in the last two decades is the decrease in dissolved O_2 content of the world's oceans (Keeling *et al.* 2010; Levin & Le Bris 2015). Global ocean deoxygenation is a direct effect of warming. A warmer ocean is more stratified because warm water is less dense than cold water, and strong density gradients reduce vertical mixing. The combined effects of reduced oxygen solubility in warmer water and increased thermal stratification create widespread oxygen reduction, termed 'hypoxia' or 'deoxygenation'. In coastal and inland waters, nutrient input from wastewater treatment plants, runoff from farmland, urban and suburban areas and air pollution from cars and so forth exacerbates the problem, because nutrients can lead to the proliferation of primary production and consequently enhance reduction of dissolved oxygen (DO) by microbes (Jewett & Romanou 2017). Increased stratification leads to reduced mixing of oxygen into the ocean interior and accounts for up to 85% of global ocean oxygen loss (Helm *et al.* 2011). Effects of ocean temperature change and stratification on oxygen loss are strongest in intermediate or mode waters (nearly vertically homogeneous) at bathyal depths (generally 200–3000 m), and also nearshore and in the open ocean; these changes are especially evident in tropical and subtropical waters globally, in the eastern Pacific and in the Southern Ocean (Jewett & Romanou 2017).

Hypoxic zones are areas in the ocean where the oxygen concentration is so low that animals can suffocate and die, and as a result are often called 'dead zones' (Diaz 2013). The largest hypoxic zone in the United States, and the second largest worldwide, occurs in the northern Gulf of Mexico adjacent to the Mississippi River. Diaz & Rosenberg (2008) assembled a global database of over 400 dead zones, a number that has not increased much since (415 in 2020). They predicted that global warming could cause dead zones to grow by a factor of 10 or more by the year 2100. Notable dead zones include the Gulf of Mexico, the Atlantic coast of North America, the east China Sea and, in European waters, the Adriatic Sea, the German Bight, the Baltic Sea and parts of the Black Sea. However, it has been demonstrated that decreased nutrient loading strongly decreases the probability of hypoxic events (Mee *et al.* 2005; Kemp *et al.* 2009). Several mitigation programmes are currently underway to reduce nutrient loading into the Gulf of Mexico, Chesapeake Bay on the Atlantic coast of the United States, and the Baltic and Black Seas.

There is growing awareness that low-oxygen regions of the ocean are also acidified as a result of warming-enhanced biological respiration. While the effects of hypoxia on marine mussels are well documented (see Chapter 7), the combined effects of low oxygen and acidification were until recently largely unknown. This is now an area of active research in bivalves (Meire *et al.* 2013; Melzner *et al.* 2013) and marine mussels (Gu *et al.* 2019; Sui *et al.* 2017). As seen earlier, OA can pose negative effects for the physiological energetics of mussels (Navarro *et al.* 2013). Therefore, it may interact with hypoxia synergistically to cause much stronger effects than those of each in isolation. To test this, Gu *et al.* (2019) evaluated the interactive effects of elevated CO_2 and hypoxia on physiological energetics in *M. edulis*. Adult mussels from the East China Sea were exposed to three pH levels (8.1, 7.7, 7.3) at two dissolved oxygen (DO) levels (6 and 2 mg L^{-1}). Clearance rate, absorption efficiency, respiration rate, excretion rate, SFG and O:N ratio were measured during a14-day exposure time. After exposure, all parameters except excretion rate were significantly reduced under low-pH and hypoxic conditions, while excretion rate was significantly increased. Additive effects of low pH and hypoxia were evident for all parameters, and low pH appeared to elicit a stronger effect than hypoxia (2 mg L^{-1}). Overall, the study showed that hypoxia can aggravate the effects of acidification and that its impacts on physiology and SFG may have ecological and economic ramifications in the short term.

Sui *et al.* (2017) investigated the combined effects of hypoxia and OA on the defence responses of adult *M. coruscus* in the East China Sea, one of the largest coastal hypoxic zones observed in the world (Chen *et al.* 2007). Mussels were exposed to three pH/pCO_2 levels (7.3/2800 μatm, 7.7/1020 μatm and 8.1/376 μatm) at two DO concentrations (6 and 2 mg L^{-1}) for 72 hr. Results showed that byssus thread parameters (e.g. number, diameter, attachment strength, plaque area) were reduced by low DO, and shell-closing strength was significantly weaker under both hypoxia and low-pH conditions. Expression patterns of genes related to mussel byssus proteins were affected by hypoxia. Generally, hypoxia reduced expressions of some of these and increased expression of others. Overall, both hypoxia and low pH induced negative effects on mussel defense responses, with hypoxia being the main driver of change. In addition, significant interactive effects between pH and DO were observed on shell-closing strength. Depressed byssus attachment strength and shell-closing strength may enable predators and water currents to remove mussels from the substrate more easily. In a similar study, the effects of hypoxia and salinity on antipredator responses were examined in juvenile *P. viridis* from Hong Kong coastal waters (Wang *et al.* 2012). In hypoxic and low-salinity conditions, *P. viridis* produced fewer byssal threads and had a thinner shell and adductor muscle, indicating that hypoxia and low salinity are severe environmental stressors for self-defence of mussels, and that their interactive effects further increase the predation risk.

Growing human pressures, including climate change, are having profound and diverse consequences for marine ecosystems (Doney *et al.* 2012). Rising atmospheric carbon dioxide is one of the most critical problems, because its effects are globally pervasive and irreversible on ecological time scales. The primary direct consequences are increasing ocean temperatures and acidity. Also, both warming and altered ocean circulation act to reduce subsurface oxygen concentrations. Furthermore, climbing temperatures create a multitude of additional changes, such as rising sea level, increased ocean stratification, decreased sea-ice extent and altered patterns of ocean circulation, precipitation, freshwater input and extreme weather events (reviewed in Doney *et al.* 2012; Howard *et al.* 2013; USGCRP 2017). In recent decades, the rates of change have been rapid, perhaps exceeding the current and potential future tolerances of many organisms to adapt. In addition, the rates of physical and chemical change in marine ecosystems will almost certainly accelerate over the next several decades in the absence of immediate and dramatic efforts toward climate mitigation.

Notes

1 The littoral zone extends from the high water mark, which is rarely inundated, to shoreline areas, which are permanently submerged.
2 A Pacific Ocean current that moves southward along the western coast of North America, beginning off southern British Columbia and ending off southern Baja California Peninsula.
3 The PDO is detected as warm or cool surface waters in the Pacific Ocean north of 20 °N and, depending on the phase of the oscillation, may regionally ameliorate or counter global climate trends (Stenseth *et al.* 2003).
4 The horizontal movement of a mass of fluid (e.g. an ocean current); also, transport (e.g. of larvae or pollutants) by such movement.
5 An abnormal weather pattern caused by the warming of the Pacific Ocean near the equator, off the coast of South America.
6 Ecological succession is the process of change in the species structure of an ecological community over time.
7 Crabs peel bivalve shells by inserting a large dactyl molar, a movable finger of the claw, into the aperture of the shell and progressively chipping away at its lip.
8 Mechanistic species distribution models use independently derived information about a species' physiology to develop a model of the environmental conditions under which the species can exist (Kearney & Porter 2009).
9 Aragonite saturation state is commonly used to track ocean acidification because it is a measure of carbonate ion concentration. Aragonite is one of the more soluble forms of calcium carbonate and is widely used by marine calcifiers. When aragonite saturation state falls below 3, these organisms become stressed, and when saturation state is less than 1, shells and other aragonite structures begin to dissolve.

References

Abdel Razek, F.A., Abu-Zaid, M.M., Helal, A.M., Abdel-Gaid, S.E. & Aziz, T.A. (2014) The distribution of eared horse mussel, *Modiolus auriculatus* (Krauss 1848) along the Red Sea coast of Egypt. *Egyptian Journal of Aquatic Research*, **40**, 469–472.

Ab Rahim, E.S., Nguyen, T.T.T., Ingram, B., Rigino, C., Weston, K.J. & Sherman, C.D.H. (2016) Species composition and hybridisation of mussel species (Bivalvia: Mytilidae) in Australia. *Marine and Freshwater Research*, **67**, 1955–1963.

Addessi, L. (1994) Human disturbance and long-term changes on a rocky intertidal community. *Ecological Applications*, **4**, 786–797.

Addison, J., Ort, B.S., Mesa, K.A. & Pogson, G.H. (2008) Range-wide genetic homogeneity in the California sea mussel (*Mytilus californianus*): a comparison of allozymes, nuclear DNA markers, and mitochondrial DNA sequences. *Molecular Ecology*, **17**, 4222–4232.

Aguilera, M. A., Thiel, M., Ullrich, N., Luna-Jorquera, G. & Buschbaum, C. (2017) Selective byssus attachment behavior of mytilid mussels from hard- and soft-bottom coastal systems. *Journal of Experimental Marine Biology and Ecology*, **497**, 61–70.

Alerstam, T., Gudmundsson, G.A. & Johannesson, K. (1992) Resources for long-distance migration intertidal exploitation of *Littorina* and *Mytilus* by knots *Calidris canutus* in Iceland. *Oikos*, **65**, 179–189.

Alexander, M.E., Adams, R., Dick, J.T.A. & Robinson, T.B. (2015) Forecasting invasions: resource use by mussels informs invasion patterns along the South African coast. *Marine Biology*, **162**, 2493–2500.

Alexander, M.A., Scott, J.D., Friedland, K.D., Mills, K.E., Nye, J.A. Pershing, A.J. *et al.* (2018) Projected sea surface temperatures over the 21st century: changes in the mean, variability and extremes for large marine ecosystem regions of Northern Oceans. *Elementa: Science of the Anthropocene*, **6**, 9.

Alfaro, A.C., Jeffs, A.G., Gardner, J.P.A., Bollard Breen, B.A. & Wilkin, J. (2011) Green-lipped mussels in GLM 9. New Zealand Fisheries Assessment Report 2011/48.

Anesti, A. Pörtner, H.O. & Michaelidis, B. (2010) Anaerobic metabolic patterns related to stress responses in hypoxia exposed mussels *Mytilus galloprovincialis*. *Journal of Experimental Marine Biology and Ecology*, **394**, 123–133.

Ansart, A. & Vernon, P. (2003) Cold hardiness in molluscs. *Acta Oecologica*, **24**, 95–102.

Apte, S., Holland, B. S., Godwin, L.S. & Gardner, J.P.A. (2000) Jumping ship: a stepping stone event mediating transfer of a non-indigenous species via a potentially unsuitable environment. *Biological Invasions*, **2**, 75–79.

Araneda, C., Larraín, M.A., Hecht, B. & Narum, S. (2016) Adaptive genetic variation distinguishes Chilean blue mussels (*Mytilus chilensis*) from different marine environments. *Ecology and Evolution*, **6**, 3632–3644.

Araújo, M.B. & Peterson, A.T. (2012) Uses and misuses of bioclimatic envelope modeling. *Ecology*, **93**, 1527–1539.

Asif, J.H. & Krug, P.J. (2012) Lineage distribution and barriers to gene flow among populations of the globally invasive marine mussel *Musculista senhousia*. *Biological Invasions*, **14**, 1431–1444.

Asplund, M.E, Baden, S.P., Russ, S., Ellis, R.P., Gong, N. & Hernroth, B.E. (2014) Ocean acidification and host–pathogen interactions: blue mussels, *Mytilus edulis*, encountering *Vibrio tubiashii*. *Environmental Microbiology*, **16**, 1029–1039.

Assis, J., Zupan, M., Nicastro, K.R., Zardi, G.I., McQuaid, C.D. & Serrão, E.A. (2015) Oceanographic conditions limit the spread of a marine invader along southern African shores. *PLoS ONE*, **10**, e0128124.

Babarro, J.M.F. & Carrington, E. (2013) Attachment strength of the mussel *Mytilus galloprovincialis*: effect of habitat and body size. *Journal of Experimental Marine Biology and Ecology*, **443**, 188–196.

Baker, P., Fajans, J.S., Arnold, W.S., Ingrao, D.A., Marelli, D.C. & Baker, S.M. (2007) Range and dispersal of a tropical marine invader, the Asian green mussel, *Perna viridis*, in subtropical waters of the southeastern United States. *Journal of Shellfish Research*, **26**, 345–355.

Banni, M., Sforzini, S., Arlt, V. M., Barranger, A., Dallas, L. J., Oliveri, C. *et al.* (2017) Assessing the impact of benzo[a]pyrene on marine mussels: application of a novel targeted low density microarray complementing classical biomarker responses. *PLoS ONE*, **12**, e0178460.

Bayne, B.L. (1976) *Marine Mussels: Their Ecology and Physiology*. Cambridge University Press, Cambridge.

Beal, L. & Elipot, S. (2016) Broadening not strengthening of the Agulhas Current since the early 1990s. *Nature*, **540** (7634), 570–573.

Beauchamp, K. A. & M. M. Gowing (1982) A quantitative assessment of human trampling effects on a rocky intertidal community. *Marine Environmental Research*, **7**, 279–293.

Beaudry, A., Fortier, M., Masson, S., Auffret, M., Brousseau, P. & Fournier, M. (2016) Effect of temperature on immunocompetence of the blue mussel (*Mytilus edulis*). *Journal of Xenobiotics*, **6**, 5889.

Bechmann, R.K., Taban, I.C., Westerlund, S., Godal, B.F., Arnberg, M., Vingen, S. *et al.* (2011) Effects of ocean acidification on early life stages of shrimp (*Pandalus borealis*) and mussel (*Mytilus edulis*). *Journal of Toxicology and Environmental Health, Part A*, **74**, 424–438.

Belkin, I.M. (2009) Rapid warming of Large Marine Ecosystems. *Progress in Oceanography*, **81**, 207–213.

Benson, A.J. (2019) *Perna viridis Linneaus 1758: US Geological Survey*, Nonindigenous Aquatic Species Database, Gainesville, FL. Available at: https://nas.er.usgs.gov/queries/FactSheet.aspx?speciesID=110 (accessed 28 May 2021).

Berge, J., Johnsen, G., Nilsen, F., Gulliksen, B. & Slagstad, D. (2005) Ocean temperature oscillations enable reappearance of blue mussels *Mytilus edulis* in Svalbard after a 1000 year absence. *Marine Ecology Progress Series*, **303**, 167–175.

Berrow, S.D., Kelly, T.C. & Myers, A.A. (1992a) The mussel catching behaviour of hooded crows *Corvus corone cornix*. *Bird Study*, **39**, 115–119.

Berrow, S.D., Kelly, T.C. & Myers, A.A. (1992b) The diet of coastal breeding hooded crows (*Corvus corone cornix*). *Ecography*, **15**, 337–346.

Bertness, M.D., Trussell, G.C., Ewanchuk, P.J. & Silliman, B.R. (2002) Do alternate stable community states exist in the Gulf of Maine rocky intertidal zone? *Ecology*, **83**, 3434–3448.

Beukema, J.J. & Cadee, G. (1996) Consequences of the sudden removal of nearly all mussels and cockles from the Dutch Wadden Sea. *Marine Ecology*, **17**, 279–289.

Bibby, R., Widdicombe, S., Parry, H., Spicer, J. & Pipe, R. (2008) Effects of ocean acidification on the immune response of the blue mussel *Mytilus edulis*. *Aquatic Biology*, **2**, 67–74.

Birchenough, S., Williamson, P. & Turley, C. (2017) *Future of the Sea: Ocean Acidification*. Government of United Kingdom, Office for Science, London.

Blanchette, C.A. (1996) Seasonal patterns of disturbance influence recruitment of the sea palm, *Postelsia palmaeformis*. *Journal of Experimental Marine Biology and Ecology*, **197**, 1–14.

Bourget, E. (1983) Seasonal variations of cold tolerance in intertidal mollusks and relation to environmental conditions in the St. Lawrence Estuary. *Canadian Journal of Zoology*, **61**, 1193–1201.

Braby, C.E. & Somero, G.N. (2006a) Ecological gradients and relative abundance of native (*Mytilus trossulus*) and invasive (*Mytilus galloprovincialis*) blue mussels in the California hybrid zone. *Marine Biology*, **148**, 1249–1262.

Braby, C.E. & Somero, G.N. (2006b) Following the heart: temperature and salinity effects on heart rate in native and invasive species of the blue mussels (genus *Mytilus*). *Journal of Experimental Biology*, **209**, 2554–2566.

Branch, G.M. & Steffani, N.C. (2004) Can we predict the effects of alien species? A case-history of the invasion of South Africa by *Mytilus galloprovincialis* (Lamarck). *Journal of Experimental Marine Biology and Ecology*, **300**, 189–215.

Brante, A., Rieraa, R. & Cartes, V. (2019) Post-settlement movement as response to interspecific competition between the bioengineer mussels *Semimytilus algosus* and *Perumytilus purpuratus*. *Journal of Sea Research*, **154**, 101809.

Brodsky, S., Walters L., Hoffman E. & Schneider K. (2009) Thermal tolerances of the invasive mussel *Mytella charruana* [Abstract]. Presented at the Society of Integrative and Comparative Biology Conference, Boston, MA.

Broitman, B.R., Szathmary, P.L., Mislan, K.A.S., Blanchette, C.A. & Helmuth, B. (2009) Predator-prey interactions under climate change: the importance of habitat vs body temperature. *Oikos*, **118**, 219–224.

Brooks, S.J. & Farmen, E. (2013) The distribution of the mussel *Mytilus* species along the Norwegian coast. *Journal of Shellfish Research*, **32**, 265–270.

Brosnan, D.M. & Crumrine, L.L. (1994) Effects of human trampling on marine rocky shore communities. *Journal of Experimental Marine Biology and Ecology*, **177**, 79–97.

Bustnes, J.O. & Erikstadk. E. (1990) Size selection of common mussels, *Mytilus edulis*, by common eiders, *Somateria mollissima*: energy maximization or shell weight minimization? *Canadian Journal of Zoology*, **68**, 2280–2283.

CABI (2020) *Invasive Species Compendium*. CAB International, Wallingford.

Calazans, C.S.H., Walters, L.J., Fernandes, F.C., Ferreira, C.E.L. & Hoffman, E.A. (2018) Genetic structure provides insights into the geographic origins and temporal change in the invasive charru mussel (Sururu) in the southeastern United States. *PLoS ONE*, **13**, e0195159.

Calcagno, J.A., Curelovich, J.N., Fernandez, V.M., Thatje, S. & Lovrich, G.A. (2012) Effects of physical disturbance on a sub-Antarctic middle intertidal bivalve assemblage. *Marine Biology Research*, **8**, 937–953.

Calderwood J, O'Connor N.E. & Roberts, D. (2016) Efficiency of starfish mopping in reducing predation on cultivated benthic mussels (*Mytilus edulis* Linnaeus). *Aquaculture*, **452**, 88–96.

Callaway, R., Shinn, A.P., Grenfell, S.E., Bron, J.E., Burnell, G. & Cook, E.J. (2012) Review of climate change impacts on marine aquaculture in the UK and Ireland. *Aquatic Conservation Marine and Freshwater Ecosystems*, **22**, 389–421.

Carrington, E. (2002) Seasonal variation in the attachment strength of blue mussels: causes and consequences. *Limnology and Oceanography*, **47**, 1723–1733.

Carrington, E., Moeser, G.M., Thompson, S.B., Coutts, L.C. & Craig, C.A. (2008) Mussel attachment on rocky shores: the effect of flow on byssus production. *Integrative and Comparative Biology*, **48**, 801–807.

Carrington, E., Moeser, G.M., Dimond, J., Mello, J. J. & Boller, M.L. (2009) Seasonal disturbance in mussel beds: field test of a mechanistic model predicting wave dislodgment. *Limnology and Oceanography*, **54**: 978–986.

Carroll, M.L. & Highsmith, R. (1996) Role of catastrophic disturbance in mediating *Nucella-Mytilus* interactions in the Alaskan rocky intertidal. *Marine Ecology Progress Series*, **138**, 125–133.

Carson, H.S., López-Duarte, P.C., Rasmussen, L., Wang, D. & Levin, L.A. (2010) Reproductive timing alters population connectivity in marine metapopulations. *Current Biology*, **20**, 1926–1931.

Caza, F., Cledon, M. & St-Pierre, Y. (2016) Biomonitoring climate change and pollution in marine ecosystems: a review on *Aulacomya ater*. *Journal of Marine Biology*, **2016**, art. 7183813.

Ceccherelli, V.U. & Rossi, R. (1984) Settlement, growth and production of the mussel *Mytilus galloprovincialis*. *Marine Ecology Progress Series*, **16**, 173–184.

Cellura, C., Toubiana, M., Parrinello, N. & Roch, R. (2007) Specific expression of antimicrobial peptide and HSP70 genes in response to heat-shock and several bacterial challenges in mussels. *Fish ands Shellfish Immunology*, **22**, 340–350.

Chen, C.C., Gong, G.C. & Shiah, F.K. (2007) Hypoxia in the East China Sea: one of the largest coastal low-oxygen areas in the world. *Marine Environmental Research*, **64**, 399–408.

Cheung, S.G., Tong, P.Y., Yip, K.M. & Shin, P.K.S. (2004a) Chemical cues from predators and damaged conspecifics affect byssus production in the green-lipped mussel *Perna viridis*. *Marine and Freshwater Behaviour and Physiology*, **37**, 127–135.

Cheung, S.G., Lam, S., Gao, Q.F., Mak, K.K. & Shin, P.K.S. (2004b) Induced anti-predator responses of the green mussel, *Perna viridis* (L.), on exposure to the predatory gastropod, *Thais clavigera* Küster and the swimming crab, *Thalamita danae* Stimpson. *Marine Biology*, **144**, 675–684.

Cheung, S.G., Luk, K.C. & Shin, P.K.S. (2006) Predator-labeling effect on byssus production in marine mussels *Perna viridis* (L.) and *Brachidontes variabilis* (Krauss). *Journal of Chemical Ecology*, **32**, 1501–1512.

Cheung, S.G., Yang, F.Y., Chiu, J.M.Y., Liu, C.C. & Shin, P.K.S. (2009) Anti-predator behaviour in the green-lipped mussel *Perna viridis*: byssus thread production depends on the mussel's position in clump. *Marine Ecology Progress Series*, **378**, 145–151.

Chiu, J.M.Y., Shin, P.K.S., Yang, F.Y. & Cheung, S.G. (2011) Can a scavenging gastropod with a mussel conspecific diet induce anti-predator defence in the mussel *Perna viridis*? *Journal of Experimental Marine Biology and Ecology*, **401**, 85–88.

Clark, B.M., Hauck, M., Harris, J.M., Salo, K. & Russell, E. (2002) Identification of subsistence fishers, fishing areas, resource use and activities along the South African coast. *African Journal of Marine Science*, **24**, 425–437.

Cohen, A.N. (2005) *Guide to the Exotic Species of San Francisco Bay. Geukensia demissa*. San Francisco Estuary Institute, Oakland, CA.

Cole, A. & Denny, M. (2014) United we fail: group versus individual strength in the California sea mussel, *Mytilus californianus*. *Biological Bulletin*, **227**, 61–67.

Cote, I.M. (1995) Effects of predatory crab effluent on byssus production in mussels. *Journal of Experimental Marine Biology and Ecology*, **188**, 233–241.

Crothers, J.H. (1985) Dog-whelks: an introduction to the biology of *Nucella lapillus* (L.). *Field Studies*, **6**, 291–360.

Daguin, C. & Borsa, P. (2000) Genetic relationships of *Mytilus galloprovincialis* Lamarck populations worldwide: evidence from nuclear-DNA markers. In: *The Evolutionary Biology of the Bivalvia* (eds E.M. Harper, J.D. Taylor & J.A. Crame), pp. 389–397. Geological Society, London.

Dallas, L.J., Bean, T.P., Turner, A., Lyons, B.P. & Jha, A.N. (2016) Exposure to tritiated water at an elevated temperature: Genotoxic and transcriptomic effects in marine mussels (*M. galloprovincialis*). *Journal of Environmental Radioactivity*, **164**, 325–336.

Dame, R.F. (2012) *Ecology of Marine Bivalves: An Ecosystem Approach*, 2nd edn CRC Press, Boca Raton, FL.

Dardignac-Corbeil, M.-J. (1975) La culture des moules sur bouchots. *Science et Pêche*, **244**, 1–10.

Dare, P.J. (1982) Notes on the swarming behaviour and population density of *Asterias rubens* L. (Echinodermata: Asteroidea) feeding on the mussel *Mytilus edulis*. *Journal du Conseil*, **40**, 112–118.

Davenport, J. & Carrion-Cotrina, M. (1981) Responses of the mussel *Mytilus edulis* L. to simulated subarctic tide pool conditions. *Journal of Thermal Biology*, **6**, 257–265.

Davenport, J., Moore, P.G., Magill, S.H. & Fraser, L.A. (1998) Enhanced condition in dogwhelks *Nucella lapillus* (L.) living under mussel hummocks. *Journal of Experimental Marine Biology and Ecology*, **230**, 225–234.

Davies, G.P., Dare, P.J. & Edwards, D.B. (1980) Fenced enclosures for the protection of seed mussels (*Mytilus edulis* L.) from predation by shore crabs (*Carcinus maenas* L.). *Fisheries Research Technical Report*, Issue 56.

Dayton, P.K. (1971) Competition, disturbance, and community organization: the provision and subsequent utilization of space in a rocky intertidal community. *Ecological Monographs*, **41**, 351–389.

Dayton, P.K. (1973) Dispersion, dispersal, and persistence of the annual intertidal alga, *Postelsia palmaeformis* Ruprecht. *Ecology*, **54**, 433–438.

Denny, M.W. (1995) Predicting physical disturbance: mechanistic approaches to the study of survivorship on wave-swept shores. *Ecolgical Monographs*, **65**, 371–418.

Denny, M.W. & Gaylord, B. (2010) Marine ecomechanics. *Annual Review of Marine Science*, **2**, 89–114.

Denny, M.W., Miller, L.P., Stokes, M.D., Hunt, H. & Helmuth, B.S.T. (2003) Extreme water velocities: topographical amplification of wave-induced flow in the surf zone of rocky shores. *Limnology and Oceanography*, **48**, 1–8.

Diaz, R. (2013) What are dead zones? [video file]. Virginia Institute of Marine Science, October 3. Retrieved from https://www.youtube.com/watch?v=OJEWP0MlI5c.

Diaz, R.J. & Rosenberg, R. (2008) Spreading dead zones and consequences for marine ecosystems. *Science*, **321**, 926–929.

Diaz, R., Lardies, M.A., Tapia, F.J., Tarifeño, E. & Vargas, C.A. (2018) Transgenerational effects of $p\mathrm{CO}_2$-driven ocean acidification on adult mussels *Mytilus chilensis* modulate physiological response to multiple stressors in larvae. *Frontiers in Physiology*, **9**, 1349.

Dionne, M., Lauzon-Guay, J.-S., Hamilton, D.J. & Barbeau, M.A. (2006) Protective socking material for cultivated mussels: a potential non-disruptive deterrent to reduce losses to diving ducks. *Aquaculure International*, **14**, 595–613.

Dolmer, P. (1998) The interactions between bed structure of *Mytilus edulis* L. and the predator *Asterias rubens* L. *Journal of Experimental Marine Biology and Ecology*, **228**, 137–150.

Doney, S.C., Ruckelshaus, M., Duffy, J.E., Barry, J.P., Chan, F., English, C.A. *et al.* (2012) Climate change impacts on marine ecosystems. *Annual Review of Marine Science*, **4**, 11–37.

Dunthorn, A (1971) The predation of cultivated mussels by eiders. *Bird Study*, **18**, 107–112.

Dupont, S. & Pörtner, H.-O. (2013) A snapshot of ocean acidification research. *Marine Biology*, **160**, 1765–1771.

Durán, L. R. & Castilla. J.C. (1989) Variation and persistence of the middle rocky intertidal community of central Chile, with and without human harvesting. *Marine Biology*, **103**, 555–562.

Egerrup, M. & Hoegh Laursen, M.L. (1992) Aspects of predation in intertidal blue mussels (*Mytilus edulis* L.) in the Danish Wadden Sea. CM/ICES K 25, 1–20.

Elith, J. & Leathwick, J.R. (2009) Species distribution models: ecological explanation and prediction across space and time. *Annual Review of Ecology, Evolution, and Systematics*, **40**, 677–697.

Erlandsson, J., Purba, P. & McQuaid, C.D. (2006) Re-colonisation rate differs between co-existing indigenous and invasive intertidal mussels following major disturbance. *Marine Ecology Progress Series*, **320**, 169–176.

Eschweiler, N. & Christensen H.T. (2011) Trade-off between increased survival and reduced growth for blue mussels living on Pacific oyster reefs. *Journal of Experimental Marine Biology and Ecology*, **403**, 90–95.

Fabry, V.J., Seibel, B.A., Feely, R.A. & Orr, J.C. (2008) Impacts of ocean acidification on marine fauna and ecosystem processes. *ICES Journal of Marine Science*, **65**, 414–432.

Faldborg, K., Jensen, K. T. & Maagaard, L. (1994), Dynamics, growth, secondary production and elimination by waterfowl of an intertidal population of *Mytilus edulis* L. *Ophelia, Supplement* **6**, 187–200.

Farrell, E.D. & Crowe, T.P. (2007) The use of byssus threads by *Mytilus edulis* as an active defence against *Nucella lapillus*. *Journal of the Marine Biological Association of the UK*, **87**, 559–564.

Fauville, G., Säljö, R. & Dupont, S. (2013) Impact of ocean acidification on marine ecosystems: educational challenges and innovations. *Marine Biology*, **160**, 1863–1874.

Fields, P.A., Rudomin, E.L. & Somero, G.N. (2006) Temperature sensitivities of cytosolic malate dehydrogenases from native and invasive species of marine mussels (genus *Mytilus*): sequence–function linkages and correlations with biogeographic distribution. *Journal of Experimental Biology*, **209**, 656–667.

Fields, P.A., Cox, K.M. & Karch, K.R. (2012) Latitudinal variation in protein expression after heat stress in the salt marsh mussel *Geukensia demissa*. *Integrative and Comparative Biology*, **52**, 636–647.

Filgueira, R., Guyondet, T., Comeau, L.A. & Tremblay, R. (2016) Bivalve aquaculture-environment interactions in the context of climate change. *Global Change Biology*, **22**, 3901–3913.

Firth, L.B., Knights, A.M. & Bell, S.S. (2011) Air temperature and winter mortality: implications for the persistence of the invasive mussel, *Perna viridis* in the intertidal zone of the south-eastern United States. *Journal of Experimental Marine Biology and Ecology*, **400**, 250–256.

Fitzer, S.C., Phoenix, V.R., Cusack, M. & Kamenos, N.A. (2014) Ocean acidification impacts mussel control on biomineralisation. *Scientific Reports*, **4**, 6218.

Fitzer, S.C., Vittert, L., Bowman, A., Kamenos, N.A., Phoenix, V.R. & Cusack, M. (2015) Ocean acidification and temperature increase impact mussel shell shape and thickness: problematic for protection? *Ecology and Evolution*, **5**, 4875–4884.

Fitzer, S.C., Chung, P., Maccherozzi, F., Dhesi, S.S., Kamenos, N.A., Phoenix, V.R. *et al.* (2016) Biomineral shell formation under ocean acidification: a shift from order to chaos. *Scientific Reports*, **6**, 21076.

Fitzhenry, T., Halpin, P.M. & Helmuth, B. (2004) Testing the effects of wave exposure, site, and behavior on intertidal mussel body temperatures: applications and limits of temperature logger design. *Marine Biology*, **145**, 339–349.

Fly, E.K. & Hilbish, T.J. (2013) Physiological energetics and biogeographic range limits of three congeneric mussel species. *Oecologia*, **172**, 35–46.

Fokina, N.N., TRuokolainen, T.R. & Nemova, N.N. (2018) The effect of intertidal habitat on seasonal lipid composition changes in blue mussels, *Mytilus edulis* L., from the White Sea. *Polar Record*, **54**, 133–151.

Frandsen, R.P. & Dolmer, P. (2002) Effects of substrate type on growth and mortality of blue mussels (*Mytilus edulis*) exposed to the predator *Carcinus maenas*. *Marine Biology*, **141**, 253–262.

Franz D.R. (2001) Recruitment, survivorship, and age structure of a New York ribbed mussel population (*Geukensia demissa*) in relation to shore level–a nine year study. *Estuaries*, **24**, 319–327.

Freeman, A.S. (2007) Specificity of induced defenses in *Mytilus edulis* and asymmetrical predator deterrence. *Marine Ecology Progress Series*, **334**, 145–153.

Freeman, A.S & Byers, J.E. (2006) Divergent induced responses to an invasive predator in marine mussel populations. *Science*, **313**, 831–833.

Freeman, A.S & Byers, J.E. (2007) Response to comment on 'divergent induced responses to an invasive predator in marine mussel populations'. *Science*, **316**, pp.53.

Freeman, A.S., Meszaros, J. & Byers, J.E. (2009) Poor phenotypic integration of blue mussel inducible defences in environments with multiple predators. *Oikos*, **118**, 758–766.

Freitas, V., Campos, J., Fonds, M. & Van der Veer, H.W. (2007) Potential impact of temperature change on epibenthic predator–bivalve prey interactions in temperate estuaries. *Journal of Thermal Biology*, **32**, 328–340.

Fuller, A., Dawson, T., Helmuth, B., Hetem, R.S., Mitchell, D. & Maloney, S.K. (2010) Physiological mechanisms in coping with climate change. *Physiological and Biochemical Zoology*, **83**, 713–720.

Galil, B. (2006) *Brachidontes pharaonic*. DAISIE-Delivering Alien Invasive Species Inventories for Europe.

Garner, Y.L. & Litvaitis, M.K. (2013) Effects of wave exposure, temperature and epibiont fouling on byssal thread production and growth in the blue mussel, *Mytilus edulis*, in the Gulf of Maine. *Journal of Experimental Marine Biology and Ecology*, **446**, 52–56.

Garner, Y.L. & Litvaitis, M.K. (2017) Effects of artificial epibionts on byssogenesis, attachment strength, and movement in two size classes of the blue mussel, *Mytilus edulis*. *Invertebrate Biology*, **136**, 15–21.

Gaylord, B., Hill, T.M., Sanford, E., Lenz, E.A., Jacobs, L.A., Sato, K.N. *et al.* (2011) Functional impacts of ocean acidification in an ecologically critical foundation species. *Journal of Experimental Biology*, **214**, 2586–2594.

Gazeau, F., Gattuso, J.-P., Dawber, C., Pronker, A.E., Peene, F., Peene, J. *et al.* (2010) Effect of ocean acidification on the early life stages of the blue mussel *Mytilus edulis*. *Biogeosciences*, **7**, 2927–2947.

Gazeau, F., Parker, L.M., Cormeau, S., Gattuso, J.-P., O'Connor, W.A., Martin, S. *et al.* (2013) Impacts of ocean acidification on marine shelled molluscs. *Marine Biology*, **160**, 2207–2245.

Gazeau, F., Alliouane, S., Bock, C., Bramanti, L., López Correa, M., Gentile, M. *et al.* (2014) Impact of ocean acidification and warming on the Mediterranean mussel (*Mytilus galloprovincialis*). *Frontiers in Marine Science*, **1**, *art.* 62.

Geller J.B. (1999) Decline of a native mussel masked by sibling species invasion. *Conservation Biology*, **13**, 661–664.

Ghazanshahi, J., Huchel, T.D. & Devinny, J.S. (1983) Alteration of southern California rocky shore ecosystems by public recreational use. *Journal of Environmental Management*, **16**, 379–394.

Gilg, M.R., Hoffman, E.A., Schneider, K.R., Ryabinov, J., El-Khoury, C. & Walters, L.J. (2010) Recruitment preferences of non-native mussels: interaction between marine invasions and land-use changes. *Journal of Molluscan Studies*, **76**, 333–339.

Gilman, S.E., Harley, C.D.G., Strickland, D.C., Vanderstraeten, O., O'Donnell, M.J. & Helmuth, B. (2006) Evaluation of effective shore level as a method of characterizing intertidal wave exposure regimes. *Limnology Oceanography-Methods*, **4**, 448–457.

Gionet, C., Mayrand, E. & Landry, T. (2009) The effect of energy reserves and cryoprotectants on overwintering mortality in *Mercenaria mercenaria* notata (Say 1822) at two tidal levels. *Aquaculture International*, **17**, 589–605.

GISD (2015) Species profile: *Geukensia demissa*. Available at: http://www.iucngisd.org/gisd/species.php?sc=1153 (accessed 28 May 2021).

GISD (2019) Species profile: *Perna perna*. Available at: http://www.iucngisd.org/gisd/species.php?sc=742 (accessed 28 May 2021).

Global Climate (2019) Benguela Current. Available at: https://www.climate-policy-watcher.org/global-climate-2/benguela-current.html (accessed 28 May 2021).

Goldstein, D. S. (1992) Long term ecological stability within a disturbed intertidal region. Senior thesis, Dept. of Biology, University of California, Santa Cruz, CA.

Gosling, E.M. (2015) *Marine Bivalve Molluscs*, 2nd edn. Wiley Blackwell, Oxford.

Gosling, E.M., Doherty, S.D. & Howley, N. (2008) Genetic characterization of hybrid mussel (*Mytilus*) populations on Irish coasts. *Journal of the Marine Biological Association UK*, **88**, 341–346.

Goss-Custard, J.D., McGrorty, S. & Durell, S.E.A. (1996) The effect of oystercatchers *Haematopus ostralegus* on shellfish populations. *Ardea*, **84A**, 453–468.

Gosnell, J.S. & Gaines, S.D. (2012) Keystone intimidators in the intertidal: non-consumptive effects of a keystone sea star regulate feeding and growth in whelks. *Marine Ecology Progress Series*, **450**, 107–114.

Grant, W.S., Cherry, M.I. & Lombard, A.T. (1984) A cryptic species of *Mytilus* (Mollusca: Bivalvia) on the west coast of South Africa, *South African Journal of Marine Science*, **2**, 149–162.

Grant, W.S. & Cherry, M.I. (1985) *Mytilus galloprovincialis* Lmk. in South Africa. *Journal of Experimental Marine Biology and Ecology*, **90**, 179–191.

de Greef, K.C., Griffiths, C.L. & Zeeman, Z. (2013) Deja vu? A second mytilid mussel, *Semimytilus algosus*, invades South Africa's west coast. *African Journal of Marine Science*, **35**, 307–313.

Gu, H., Shang, Y., Clements, J., Dupont, S., Wang, T., Wei, S. *et al.* (2019) Hypoxia aggravates the effects of ocean acidification on the physiological energetics of the blue mussel *Mytilus edulis*. *Marine Pollution Bulletin*, **149**, 110538.

Guichard, F., Halpin, P.M., Allison, G.W., Lubchenco, J. & Menge, B.A. (2003) Mussel disturbance dynamics: signatures of oceanographic forcing from local interactions. *American Naturalist*, **161**, 889–904.

Guinotte, J.M. & Fabry, V.J. (2008) Ocean acidification and its potential effects on marine ecosystems. *Annals of the New York Academy of Sciences*, **1134**, 320–342.

Gutiérrez, J.L., Jones, C.G., Strayer, D.L. & Iribarne, O.O. (2003) Mollusks as ecosystem engineers: the role of shell production in aquatic habitats. *Oikos*, **101**, 79–90.

Gutiérrez, J.L., Palomo, M.G., Bagur, M., Arribas, L.P. & Soria, S.A. (2015) Wave action limits crowding in an intertidal mussel. *Marine Ecology Progress Series*, **518**, 153–163.

Guyondet, T., Comeau, L.A., Bacher, C., Grant, J., Rosland, R., Sonier, R. *et al.* (2015) Climate change influences carrying capacity in a coastal embayment dedicated to shellfish aquaculture. *Estuaries and Coasts*, **38**, 1593–1618.

Hamilton, D.J., Nudds, T.D. & Neate, J. (1999) Size-selective predation of blue mussels (*Mytilus edulis*) by common eiders (*Somateria mollissima*) under controlled field conditions. *Auk*, **116**, 403–416.

Harley, C.D.G. (2011) Climate change, keystone predation, and biodiversity loss. *Science*, **334**, 1124–1127.

Harley, C.D.G. & Helmuth, B.S.T. (2003) Local- and regional-scale effects of wave exposure, thermal stress, and absolute versus effective shore level on patterns of intertidal zonation. *Limnology and Oceanography-Methods*, **48**, 1498–1508.

Hayhurst, S. & Rawson, P.D. (2009) Species-specific variation in larval survival and patterns of distribution for the blue mussels *Mytilus edulis* and *Mytilus trossulus* in the Gulf of Maine. *Journal of Molluscan Studies*, **75**, 215–222.

Hayne, K.J.R. & Palmer, A.R. (2013) Intertidal sea stars (*Pisaster ochraceus*) alter body shape in response to wave action. *The Journal of Experimental Biology*, **216**, 1717–1725.

Helm, K.P., Bindoff, N.L. & Church, J.A. (2011) Observed decreases in oxygen content of the global ocean. *Geophysical Research Letters*, **38**, L23602.

Helmuth, B. (1998) Intertidal mussel microclimates: predicting the body temperature of a sessile invertebrate. *Ecological Monographs*, **68**, 51–74.

Helmuth, B. (1999) Thermal biology of rocky intertidal mussels: quantifying body temperatures using climatological data. *Ecology*, **80**, 15–34.

Helmuth, B. (2002) How do we measure the environment? Linking intertidal thermal physiology and ecology through biophysics. *Integrative and Comparative Biology*, **42**, 837–845.

Helmuth, B.S.T. & Hofmann, G.E. (2001) Microhabitats, thermal heterogeneity and patterns of physiological stress in the rocky intertidal zone. *Biological Bulletin*, **201**, 374–384.

Helmuth, B., Broitman, B.R., Blanchette, C.A., Gilman, S., Halpin, P., Harley, C.D.G. *et al.* (2006) Mosaic patterns of thermal stress in the rocky intertidal zone: implications for climate change. *Ecological Monographs*, **76**, 461–479.

Helmuth, B., Broitman, B.R., Yamane, L., Gilman, S.E., Mach, K., Mislan, K.A.S. *et al.* (2010) Organismal climatology: analyzing environmental variability at scales relevant to physiological stress. *Journal of Experimental Biology*, **213**, 995–1003.

Helmuth, B., Russell, B.D., Connell, S.D., Dong, Y., Harley, C.D.G., Lima F.P *et al.* (2014) Beyond long-term averages: making biological sense of a rapidly changing world. *Advances in Climate Change Research*, **1**, 6.

Helmuth, B. *et al.* (2016) Long-term, high frequency *in situ* measurements of intertidal mussel bed temperatures using biomimetic sensors. *Scientific Data*, **3**, 160087.

Helmuth, B. & Denny, M.W. (2003) Predicting wave exposure in the rocky intertidal zone: do bigger waves always lead to larger forces? *Limnology and Oceanography*, **48**, 1338–1345.

Hernroth, B., Baden, S., Tassidis, H., Hörnaeus, K., Guillemant, J., Bergström, Lind. S. *et al.* (2016) Impact of ocean acidification on antimicrobial activity in gills of the blue mussel (*Mytilus edulis*). *Fish and Shellfish Immunology*, **55**, 452–459.

Hernroth, B.E. & Baden, S.P. (2018) Alteration of host-pathogen interactions in the wake of climate change–increasing risk for shellfish associated infections? *Environmental Research*, **161**, 425–438.

Hicks, D.W. & McMahon, R.F. (2002) Temperature acclimation of upper and lower thermal limits and freeze resistance in the nonindigenous brown mussel, *Perna perna* (L.), from the Gulf of Mexico. *Marine Biology*, **140**, 1167–1179.

Hickman, R.W. (1992) Mussel Cultivation. In: *The Mussel Mytilus: Ecology, Physiology, Genetics and Culture* (ed E.M. Gosling), pp. 465–510. Elsevier Science Publishers B.V., Amsterdam.

Hiebenthal, C., Philipp, E.E.R., Eisenhauer, A. & Wahl, M. (2013) Effects of seawater pCO_2 and temperature on shell growth, shell stability, condition and cellular stress of Western Baltic Sea *Mytilus edulis* (L.) and *Arctica islandica* (L.). *Marine Biology*, **160**, 2073–2087.

Hilbish, T.J., Brannock, P.M., Jones, K.R., Smith, A.B., Bullock, B.N. & Wethey, D.S. (2010) Historical changes in the distributions of invasive and endemic marine invertebrates are contrary to global warming predictions: the effects of decadal climate oscillations. *Journal of Biogeography*, **37**, 423–431.

Hilgerloh, G. (1997) Predation by birds on blue mussel *Mytilus edulis* beds of the tidal flats of Spiekeroog (Southern North-Sea). *Marine Ecology Progress Series*, **146**, 61–72.

Hilgerloh, G (1999) Predation pressure by birds on mussels. In: *Proceedings of the 22nd International Ornithological Congress held in Durban, South Africa, 1998* (eds N.J. Adams & R.H. Slotow), pp. 2345–2352. BirdLife South Africa, Johannesburg.

Hilgerloh, G., Herlyn, M. & Michaelis, H. (1997) The influence of predation by herring gulls *Larus argentatus* and oystercatchers *Haematopus ostralegus* on a newly established mussel *Mytilus edulis* bed in autumn and winter. *Helgoländer Meeresunters*, **51**, 173–189.

Hockey, P.A.R. & Van Erkom Schurink, C. (1992) The invasive biology of the mussel *Mytilus galloprovincialis* on the southern African coast. *Transactions of the Royal Society of South Africa*, **48**, 123–139.

Honig, A. (2013) *Population ecology of the ribbed mussel in southeastern Louisiana*. MS thesis, Louisiana State University, Baton Rouge, LA.

Howard, J., Auer, C., Beard, R., Bond, N., Boyer, T., Brown, D. *et al.* (2013) Climate-driven physical and chemical changes in marine ecosystems In: *Oceans and Marine Resources in a Changing Climate* (eds R. Griffis & J. Howard), pp. 7–39. Island Press, Washington, DC.

Hughes, R.N. (1986) *A Functional Biology of Marine Gastropods*. Crook Helm, London.

Hughes, R.N. & Dunkin, S. de B. (1984) Behaviour components of prey selection by dogwhelks, *Nucella lapillus* (L.) feeding on mussels, *Mytilus edulis* L. in the laboratory. *Journal of Experimental Marine Biology and Ecology*, **77**, 45–68.

Hummel, C., Honkoop, P. & van der Meer, J. (2011) Small is profitable: no support for the optimal foraging theory in sea stars *Asterias rubens* foraging on the blue edible mussel *Mytilus edulis*. *Estuarine, Coastal and Shelf Science*, **94**, 89–92.

Hüning, A.K., Melzner, F., Thomsen, J., Gutowska, M.A., Krämer, L., Frickenhaus, S. *et al.* (2013) Impacts of seawater acidification on mantle gene expression patterns of the Baltic Sea blue mussel: implications for shell formation and energy metabolism. *Marine Biology*, **160**, 1845–1861.

Hunt, H.L. & Scheibling, R.E. (1998) Spatial and temporal variability of patterns of colonization by mussels (*Mytilus trossulus, M. edulis*) on a wave-exposed rocky shore. *Marine Ecology Progress Series*, **167**, 155–169.

Hunt, H.L. & Scheibling, R.E. (2001) Predicting wave dislodgement of mussels: variation in attachment strength with body size, habitat and season. *Marine Ecology Progress Series*, **213**, 157–164.

Introíni, G.O., Maester, F.M., Pereira Leite, F.P. & Recco-Pimentel, S.M. (2010) Sperm ultrastructure of *Mytella* (Bivalvia) populations from distinct habitats along the northern coast of São Paulo State, *Brazil. Biocell*, **34**, 103–111.

Jiang, l.Q., Carter, B.R., Feely, R.A., Lauvset, S.K. & Olsen, A. (2019) Surface ocean pH and buffer capacity: past, present and future. *Scientific Reports*, **9**, 18624.

Jenewein, B.T. & Gosselin, L.A. (2013) Ontogenetic shift in stress tolerance thresholds of *Mytilus trossulus*: effects of desiccation and heat on juvenile mortality. *Marine Ecology Progress Series*, **481**, 147–159.

Jeschke, J.M. & Strayer, D.L. (2008) Usefulness of bioclimatic models for studying climate change and invasive species. *Annals of the New York Academy of Sciences*, **1134**, 1–24.

Jewett, L. & Romanou, A. (2017) Ocean acidification and other ocean changes. In: *Climate Science Special Report: Fourth National Climate Assessment, Volume I* (eds D.J. Wuebbles, D.W. Fahey, K.A. Hibbard, D.J. Dokken, B.C. Stewart & T.K. Maycock), pp. 364–392. US Global Change Research Program, Washington, DC.

Jones, S.J., Mieszkowska, N. & Wethey, D.S. (2009) Linking thermal tolerances and biogeography: *Mytilus edulis* (l.) at its southern limit on the east coast of the United States. *Biological Bulletin*, **217**, 73–85.

Jones, S.J., Lima, F.P. & Wethey, D.S. (2010) Rising environmental temperatures and biogeography: poleward range contraction of the blue mussel, *Mytilus edulis* L., in the western Atlantic. *Journal of Biogeography*, **37**, 2243–2259.

Jost, J. & Helmuth, B. (2007) Morphological and ecological determinants of body temperature of *Geukensia demissa*, the Atlantic ribbed mussel, and their effects on mussel mortality. *Biological Bulletin*, **213**, 141–151.

Judge, R., Choi, F. & Helmuth, B. (2018) Recent advances in data logging for intertidal ecology. *Frontiers in Ecology and Evolution*, **6**, 213.

Kamermans, P. & Capelle, J.J. (2019) Provisioning of mussel seed and its efficient use in culture. In: *Goods and Services of Marine Bivalves* (eds A.C. Smaal, J.G. Ferreira, J. Grant, J.K. Petersen & O. Strand), pp. 27–49. SpringerOpen, New York.

Kapsenberg, L., Miglioli, A., Bitter, M.C., Tambutte, E., Dumollard, R. & Gattuso, J.-P. (2018) Ocean pH fluctuations affect mussel larvae at key developmental transitions. *Proceedings of the Royal Society London B*, **285**, 20182381.

Katolikova, M., Khaitov, V., Väinölä, R., Gantsevich, M. & Strelkov, P. (2016) Genetic, ecological and morphological distinctness of the blue mussels *Mytilus trossulus* Gould and *M. edulis* L. in the White Sea. *PLoS ONE*, **11**, e0152963.

Kearney, M. & Porter, W. (2009) Mechanistic niche modelling: combining physiological and spatial data to predict species' ranges. *Ecology Letters*, **12**, 334–350.

Keeling, R.F., Körtzinger, A. & Gruber, N. (2010) Ocean deoxygenation in a warming world. *Annual Review of Marine Science*, **2**, 199–229.

Kemp, W.M., Testa, J.M., Conley, D.J., Gilbert, D. & Hagy, J.D. (2009) Temporal responses of coastal hypoxia to nutrient loading and physical controls. *Biogeosciences*, **6**, 2985–3008.

Kennedy, V. S. (1976) Desiccation, higher temperatures and upper intertidal limits of three species of sea mussels (mollusca: Bivalvia) in New Zealand. *Marine Biology*, **35**, 127–137.

Keough, M. J. Quinn G.P. & King, A. (1993) Correlations between human collecting and intertidal mollusc populations on rocky shores. *Conservation Biology*, **7**, 378–390.

Keough, M. J. & Quinn, G.P. (1998) Effects of periodic disturbances from trampling on rocky intertidal algal beds. *Ecological Applications*, **8**, 141–161.

Keppel, E.A., Scrosati, R.A. & Courtenay, S.C. (2015) Interactive effects of ocean acidification and warming on subtidal mussels and sea stars from Atlantic Canada. *Marine Biology Research*, **11**, 337–348.

Kong, H., Clements, J.C., Dupont, S., Wang, T., Huang, X., Shang, Y. *et al.* (2019) Seawater acidification and temperature modulate anti-predator defenses in two co-existing *Mytilus* species. *Marine Pollution Bulletin*, **145**, 118–125.

Kordas, R.L., Harley, C.D.G. & O'Connor, M.I. (2011) Community ecology in a warming world: the influence of temperature on interspecific interactions in marine systems. *Journal of Experimental Marine Biology and Ecology*, **400**, 218–226.

Kroeker, K.J., Gaylord, B., Hill, T.M., Hosfelt, J.D., Miller, S.H. & Sanford, E. (2014) The role of temperature in determining species' vulnerability to ocean acidification: a case study using *Mytilus galloprovincialis*. *PLoS ONE*, **9**, e100353.

Kroeker, K.J., Sanford, E., Rose, J.M., Blanchette, C.A., Chan, F., Chavez, F.P. *et al.* (2016) Interacting environmental mosaics drive geographic variation in mussel performance and predation vulnerability. *Ecology Letters*, **19**, 771–779.

Lacis, A.A., Schmidt, G.A., Rind, D. & Ruedy, R.A. (2010) Atmospheric CO_2: principal control knob governing Earth's temperature. *Science*, **330**, 356–359.

Large, S.I. & Smee, D.L. (2010) Type and nature of cues used by *Nucella lapillus* to evaluate predation risk. *Journal of Experimental Marine Biology and Ecology*, **396**, 10–17.

Large, S.I., Smee, D.L. & Trussell, G.C. (2011) Environmental conditions influence the frequency of prey responses to predation risk. *Marine Ecology Progress Series*, **422**, 41–49.

Larraín, M.A., Zbawicka, M., Araneda, C., Gardner, J.P.A. & Wenne, R. (2018) Native and invasive taxa on the Pacific coast of South America: impacts on aquaculture, traceability and biodiversity of blue mussels (*Mytilus* spp.). *Evolutionary Applications*, **11**, 298–311.

Laudien, J. & Wahl, M. (2004) Associational resistance of fouled blue mussels (*Mytilus edulis*) against starfish (*Asterias rubens*) predation: relative importance of structural and chemical properties of the epibionts. *Helgoland Marine Research*, **58**, 162–167.

Lee, S.Y. & Morton, B.S. (1985) The introduction of the Mediterranean mussel *Mytilus galloprovincialis* into Hong Kong. *Malacological Review*, **18**, 107–109.

Leonard, G.H., Bertness, M.D. & Yund, P.O. (1999) Crab predation, waterborne cues, and inducible defenses in the blue mussel, *Mytilus edulis*. *Ecology*, **80**, 1–14.

Leopold, P., Renaud, P.E., Ambrose Jr., W.G. & Berge, J. (2019) High Arctic *Mytilus* spp.: occurrence, distribution and history of dispersal. *Polar Biology*, **42**, 237–244.

Le Rossignol, A.P., Buckingham, S.G., Lea, S.E.G. & Nagarajan, R. (2011) Breaking down the mussel (*Mytilus edulis*) shell: which layers affect, oystercatchers' (*Haematopus ostralegus*) prey selection? *Journal of Experimental Marine Biology and Ecology*, **405**, 87–92.

Levin, L.A. & Le Bris, N. (2015) The deep ocean under climate change. *Science*, **350**, 766–768.

Li, Y., von Storch, H., Wang, Q., Zhou, Q. & Shengquan, T. (2019) Testing the validity of regional detail in global analyses of sea surface temperature – the case of Chinese coastal waters. *Ocean Science*, **15**, 1455–1467.

Lima, F.P. & Wethey, D.S. (2012) Three decades of high-resolution coastal sea surface temperatures reveal more than warming. *Nature Communications*, **3**, art. 704.

Lima, F.P., Burnett, N.P., Helmuth, B., Kish, N., Aveni-Deforge, K. & Wethey, D.S. (2011) Monitoring the intertidal environment with biomimetic devices. In: *Biomimetic Based Applications* (ed M. Cavrak), pp. 499–522. InTech, London.

Lindsey, R. (2020) Climate change: atmospheric carbon dioxide. Available at: https://www.climate.gov/news-features/understanding-climate/climate-change-atmospheric-carbon-dioxide (accessed 28 May 2021).

Lockwood, B.L. & Somero, G.N. (2011a) Invasive and native blue mussels (genus *Mytilus*) on the California coast: the role of physiology in a biological invasion. *Journal of Experimental Marine Biology and Ecology*, **400**, 167–174.

Lockwood, B.L. & Somero, G.N. (2011b) Transcriptomic responses to salinity stress in invasive and native blue mussels (genus *Mytilus*). *Molecular Ecology*, **20**, 517–529.

Loomis, S.H. &, Zinser, M. (2001) Isolation and identification of an ice-nucleating bacterium from the gills of the intertidal bivalve mollusc *Geukensia demissa*. *Journal of Experimental Marine Biology and Ecology*, **261**, 225–235.

Loomis, S.H., Carpenter, J.F. & Crowe, J.H. (1988) Identification of strombine and taurine as cryoproctectants in the intertidal bivalve *Mytilus edulis*. *Biochimica et Biophysica Acta*, **943**, 113–118.

López, M.S., Coutinho, R., Ferreira, C.E & Rilov, G. (2010) Predator–prey interactions in a bioinvasion scenario: differential predation by native predators on two exotic rocky intertidal bivalves. *Marine Ecology Progress Series*, **403**, 101–112.

Lourenço, C.R., Nicastro, K.R., Serrão, E.A., Castilho, R. & Zardi. G.I. (2015) Behind the mask: cryptic genetic diversity of *Mytilus galloprovincialis* along southern European and northern African shores. *Journal of Molluscan Studies*, **81**, 380–387.

Lourenço, C.R., Nicastro, K.R., McQuaid, C.D., Chefaoui, R.M., Assis, J., Taleb, M.Z. *et al.* (2017) Evidence for rangewide panmixia despite multiple barriers to dispersal in a marine mussel. *Scientific Reports*, **7**, 10279.

Lowen, J.B., Innes, D.J. & Thompson, R.J. (2013) Predator-induced defences differ between sympatric *Mytilus edulis* and *M. trossulus*. *Marine Ecology Progress Series*, **475**, 135–143.

Mackenzie, C.L., Ormondroyd. G.A., Curling, S.F., Ball, R.J., Whiteley. N.M. & Malham, S.K. (2014) Ocean warming, more than acidification, reduces shell strength in a commercial shellfish species during food limitation. *PLoS ONE*, **9**, e86764.

Mahmood, K., Jadoon, S., Mahmood, Q., Irshad, M. & and Hussain, J. (2014) Synergistic effects of toxic elements on heat shock proteins. *BioMed Research International*, **2014**, art. 564.

MARINe (2019) Sea star wasting syndrome. Available at: https://marine.ucsc.edu/data-products/sea-star-wasting/index.html (accessed 28 May 2021).

Marples, N.M., Speed, M.P. & Thomas, R.J. (2018) An individual-based profitability spectrum for understanding interactions between predators and their prey. *Biological Journal of the Linnean Society*, **125**, 1–13.

Marshall, D.J. & McQuaid, C.D. (1993) Differential physiological and behavioural responses of the intertidal mussels, *Choromytilus meridionalis* (Kr.) and *Perna perna* L., to exposure to hypoxia and air: a basis for spatial separation. *Journal of Experimental Marine Biology and Ecology*, **171**, 225–237.

Mathiesen, S.S., Thyrring, J., Hemmer-Hansen, J., Berge, J., Sukhotin, A., Leopold, P. *et al.* (2017) Genetic diversity and connectivity within *Mytilus* spp. in the subarctic and Arctic. *Evolutionary Applications*, **10**, 39–55.

Matozzo, V. & Marin, M.G. (2011) Bivalve immune responses and climate changes: is there a relationship? *Invertebrate Survival Journal*, **8**, 70–77.

McDonald, J.H., Koehn, R.K., Balakirev, E.S., Manchenko, G.P., Pudovkin, A.I., Sergiyevskii, S.O. *et al.* (1990) Species identity of the common mussel inhabiting Asiatic coats of the Pacific Ocean. *Soviet Journal of Marine Biology*, **16**, 10–18.

McDonald, J.I. (2012) Detection of the tropical mussel species *Perna viridis* in temperate Western Australia: possible association between spawning and a marine heat pulse. *Aquatic Invasions*, **7**, 483–490.

McQuaid, C. & Dower, K. (1990) Enhancement of habitat heterogeneity and species richness on rocky shores inundated by sand. *Oecologia*, **84**, 142–144.

McQuaid, C.D., Porri, F., Nicastro, K.R. & Zardi, G.I. (2015) Simple, scale-dependent patterns emerge from very complex effects-an example from the intertidal mussels *Mytilus galloprovincialis* and *Perna perna*. In: *Oceanography and Marine Biology: An Annual Review* (eds R.N. Hughes, D.J. Hughes, I.P. Smith & A.C. Dale), pp. 127–156. Taylor & Francis, Abingdon.

Mee, L.D., Friedrich, J. & Gomoiu, M.T. (2005) Restoring the Black Sea in times of uncertainty. *Oceanography*, **18**, 100–111.

Meire, L., Soetaert, K.E.R. & Meysman, F.J.R. (2013) Impact of global change on coastal oxygen dynamics and risk of hypoxia. *Biogeosciences*, **10**, 2633–2653.

Melzner, F., Stange, P., Trübenbach, K., Thomsen, J., Casties, I., Panknin, U. *et al.* (2011) Food supply and seawater pCO_2 impact calcification and internal shell dissolution in the blue mussel *Mytilus edulis*. *PLoS ONE*, **6**, e24223.

Melzner, F., Thomsen, J., Koeve, W., Oschlies, A., Gutowska, M.A., Bange, H.W. *et al.* (2013) Future ocean acidification will be amplified by hypoxia in coastal habitats. *Marine Biology*, **160**, 1875–8188.

Mendez, M.M., Livore, J.P. & Bigatti, G. (2019) Interaction of natural and anthropogenic stressors on rocky shores: community resistance to trampling. *Marine Ecology Progress Series*, **631**, 117–126.

Menge, B.A. (1992) Community regulation: under what conditions are bottom-up factors important on rocky shores? *Ecology*, **73**, 755–765.

Menge, B.A. Berlow, E.L., Blanchette, C., Navarrete, S.A. & Yamada, S.B. (1994) The keystone species concept: variation in interaction strength in a rocky intertidal habitat. *Ecological Monographs*, **64**, 249–286.

Menge, B.A., Chan, F. & Lubchenco, J. (2008) Response of a rocky intertidal ecosystem engineer and community dominant to climate change. *Ecology Letters*, **11**, 151–162.

Merrill, A.S. & Turner, R.D. (1963) Nest building in the bivalve genera *Musculus* and *Lima*. *Veliger*, **6**, 55–59.

Michaelidis, B., Ouzounis, C., Paleras, A. & Pörtner, H.O. (2005) Effects of long-term moderate hypercapnia on acid–base balance and growth rate in marine mussels *Mytilus galloprovincialis*. *Marine Ecology Progress Series*, **293**, 109–118.

Micheli F, Heiman, K.W., Kappel, C.V., Martone, R.G., Sethi, S.A., Osio, G.C. *et al.* (2016) Combined impacts of natural and human disturbances on rocky shore communities. *Ocean and Coastal Management*, **126**, 42–50.

Miller, L.P. (2013) The effect of water temperature on drilling and ingestion rates of the dogwhelk *Nucella lapillus* feeding on *Mytilus edulis* mussels in the laboratory. *Marine Biology*, **160**, 1489–1496.

Minchinton, T.E., Scheibling, R.E. & Hunt, H.L. (1997) Recovery of an intertidal assemblage following a rare occurrence of scouring by sea-ice in Nova-Scotia, *Canada. Botanica Marina*, **40**, 139–148.

Mislan, K.A.S., Wethey, D.S. & Helmuth, B. (2009) When to worry about the weather: role of tidal cycle in determining patterns of risk in intertidal ecosystems. *Global Change Biology*, **15**, 3056–3065.

Moeser, G.M., Leba, H. & Carrington, E. (2006) Seasonal influence of wave action on thread production in *Mytilus edulis*. *Journal of Experimental Biology*, **209**, 881–890.

Monaco, C.J., Wethey, D.S. & Helmuth, B. (2016) Thermal sensitivity and the role of behavior in driving an intertidal predator–prey interaction. *Ecological Monographs*, **86**, 429–447.

Montalto, V., Helmuth, B., Rut, P.M., Dell'Aquila, A., Rinaldi, A. & Sarà, G. (2016) A mechanistic approach reveals non linear effects of climate warming on mussels throughout the Mediterranean Sea. *Climatic Change*, **139**, 293–306.

Mora, C., Wei, C-L., Rollo, A., Amaro, T., Baco, A.R., Billett, D. *et al.* (2013) Biotic and human vulnerability to projected changes in ocean biogeochemistry over the 21st century. *PLoS ONE*, **11**, e1001682.

Nagarajan, R., Goss-Custard, J.D. & Lea, S.E.G. (2002a) Oystercatchers use colour preference to achieve longer-term optimality. *Proceeding of the Royal Society London B*, **269**, 523–528.

Nagarajan, R., Lea, S.E.G. & Goss-Custard, J.D. (2002b) Reevaluation of patterns of mussel (*Mytilus edulis*) selection by European Oystercatchers (*Haematopus ostralegus*). *Canadian Journal of Zoology*, **80**, 846–853.

Navarrete, S.A. & Menge, B. (1996) Keystone predation and interaction strength: interactive effects of predators on their main prey. *Ecolgical Monographs*, **66**, 409–429.

Navarrete, S.A., Menge, B.A. & Bryon, A. (2000) Species interactions in intertidal food webs: prey or predation regulation of intermediate predators? *Ecology*, **81**, 2264–2277.

Navarro, J.M., Torres, R., Acuña, K., Duarte, C., Manriquez, P.H., Lardies, M. *et al.* (2013) Impact of medium-term exposure to elevated pCO_2 levels on the physiological energetics of the mussel *Mytilus chilensis*. *Chemosphere*, **90**, 1242–1248.

Navarro, J.M., Duarte, C., Manríquez, P.H., Lardies, M.A., Torres, R., Acuña, K. *et al.* (2016) Ocean warming and elevated carbon dioxide: multiple stressor impacts on juvenile mussels from southern Chile. *ICES Journal of Marine Science*, **73**, 764–771.

Nehls, G. & Ruth, M. (1994) Eiders, mussels, and fisheries in the Wadden Sea: continuous conflicts or relaxed relations. *Ophelia*, **56**, 263–278.

Newcomb, L.A., George, M.N., O'Donnell, M.J. & Carrington, E. (2019) Only as strong as the weakest link: structural analysis of the combined effects of elevated temperature and pCO_2 on mussel attachment. *Conservation Physiology*, **7**, coz068.

Newell, C.R. & Richardson, J. (2014) The effects of ambient and aquaculture structure hydrodynamics on the food supply and demand of mussel rafts. *Journal of Shellfish Research*, **33**, 257–272.

Nicastro, K.R., Zardi, G.I., McQuaid, C.D., Pearson, G.A. & Serrão, E.A. (2012) Love thy neighbour: group properties of gaping behaviour in mussel aggregations. *PLoS ONE*, **7**, e47382.

NIMPIS (2002) *Musculista senhousia* species summary. National Introduced Marine Pest Information System (eds C.L. Hewitt, R.B Martin, C. Sliwa, F.R. McEnnulty, F.R. Murphy, N.E. Jones *et al.*). Available at: http://crimp.marine.csiro.au/nimpis (accessed 28 May 2021).

Nishizaki, M.T., Barron, S. & Carew, E. (2015) Thermal stress increases fluctuating asymmetry in marine mussels: environmental variation and developmental instability. *Ecosphere*, **6**, art. 85.

NOAA Global Climate Report (2019) Available at: https://www.ncdc.noaa.gov/sotc/global/201907 (accessed 28 May 2021).

Norberg, J. & Tedengren, M. (1995) Attack behavior and predatory success of *Asterias rubens* L. related to differences in size and morphology of the prey mussel *Mytilus edulis* L. *Journal of Experimental Marine Biology and Ecology*, **186**, 207–220.

Occhipinti-Ambrogi, A. & Galil, B. (2010) Marine alien species as an aspect of global change. *Advances in Oceanography and Limnology*, **1**, 199–218.

O'Connor, N.E., Crowe, T.P. & McGrath, D. (2006) Effects of epibiotic algae on the survival, biomass and recruitment of mussels, *Mytilus* L. (Bivalvia: Mollusca). *Journal of Experimental Marine Biology and Ecology*, **328**, 265–276.

O'Donnell, M.J., George, M.N. & Carrington, E. (2013) Mussel byssus attachment weakened by ocean acidification. *Nature Climate Change*, **3**, 587–390

Olabarria, C., Gestoso, I., Lima, F.P., Vázquez, E., Comeau, L.A., Gomes, F. *et al.* (2016) Response of two mytilids to a heatwave: the complex interplay of physiology, behaviour and ecological interactions. *PLoS ONE*, **11**, e0164330.

de Oliveira, M.E., Russo, C.A., Lazoski, C., Vianna, P.R. & Solé-Cava, A.M. (2005) Genetic variation and population structure of two species of neo-tropical mud-mussels (*Mytella* spp). *Genetics and Molecular Research*, **4**, 197–202.

OpenStax College (2013) Biology. Available at: http://cnx.org/content/col11448/latest (accessed 28 May 2021).

Page, H.M. & Hubbard, D.M. (1987) Temporal and spatial patterns of growth in mussels *Mytilus edulis* on an offshore platform: relationships to water temperature and food availability. *Journal of Experimental Marine Biology and Ecology*, **111**, 159–179.

Paine, R.T. (1974) Intertidal community structure. *Oecologia*, **15**, 93–120.

Paine, R.T. (1984) Ecological determinism in the competition for space. *Ecology*, **65**, 1339–1348.

Paine, R.T. & Levin, S.A. (1981) Intertidal landscapes: disturbance and the dynamics of pattern. *Ecological Monographs*, **51**, 145–178.

Parker, L.M., Ross, P.M., O'Connor, W.A., Borysko, L., Raftos, D.A. & Pörtner, H-O. (2012) Adult exposure influences offspring response to ocean acidification in oysters. *Global Change Biology*, **18**, 82–92.

Parker, L.M., Ross, P.M., O'Connor, W.A., Pörtner, H-O., Scanes, E. & Wright, J.M. (2013) Predicting the response of molluscs to the impact of ocean acidification. *Biology*, **2**, 651–692.

Parker, L.M., O'Connor, W.A., Raftos, D.A., Pörtner, H-O. & Ross, P.M. (2015) Persistence of positive carryover effects in the oyster, *Saccostrea glomerata*, following transgenerational exposure to ocean acidification. *PLoS ONE*, **10**, e0132276.

Peharda, M., Richardson, C.A., Mladineo, I., Šestanović, S., Popović, Z., Bolotin, J. *et al.* (2007) Age, growth and population structure of *Modiolus barbatus* from the Adriatic. *Marine Biology*, **151**, 629–638.

Pérez, A.F., Boy, C.C., Curelovich, J., Pérez-Barros, P. & Calcagno, J.A. (2013) Relationship between energy allocation and gametogenesis in *Aulacomya atra* (Bivalvia: Mytilidae) in a sub-Antarctic environment. *Revista de Biología Marina y Oceanografía*, **48**, 459–469.

Pérez, M., Guıñez R., Llavona A., Toro J.E., Astorga M. & Presa P. (2008) Development of microsatellite markers for the ecosystem bioengineer mussel *Perumytilus purpuratus* and cross-priming testing in six Mytilinae genera. *Molecular Ecology Resources*, **8**, 449–451.

Pernet, F., Tremblay, R., Comeau, L. & Guderley, H. (2007) Temperature adaptation in two bivalve species from different thermal habitats: energetics and remodelling of membrane lipids. *Journal of Experimental Biology*, **210**, 2999–3014.

Peteiro, L.G., Filgueira, R., Labarta, U. & Fernández-Reiriz, M.J. (2007) Settlement and recruitment patterns of *Mytilus galloprovincialis* L. in the Ría de Ares-Betanzos (NW Spain) in the years 2004/2005. *Aquaculture Research*, **38**, 957–964.

Petersen, A.J.K, Gislason, H., Fitridge, I., Saurel, C., Degel, H. & Nielsen, C.F. (2016) Fiskeri efter søstjerner i Limfjorden. *Fagligt grundlag for en forvaltningsplan*. Institut for Akvatiske Ressourcer, Danmarks Tekniske Universitet.

Petes, L.E., Menge, B.A. & Murphy, G.D. (2007) Environmental stress decreases survival, growth, and reproduction in New Zealand mussels. *Journal of Experimental Marine Biology and Ecology*, **351**, 83–91.

Petes, L.E., Mouchka, M.E., Milston-Clements, R.H., Momoda, T.S. & Menge, B.A. (2008) Effects of environmental stress on intertidal mussels and their sea star predators. *Oecologia*, **156**, 671–680.

Petraitis, P.S. (1987) Immobilisation of the predatory gastropod, *Nucella lapillus*, by its prey *Mytilus edulis*. *Biological Bulletin*, **172**, 307–314.

Podlipaeva, Y. I & Berger, V.Ya (2012) *Cell and Tissue Biology*, **6**, 498–502.

Poloczanska, E.S., Burrows, M.T., Brown, C.J., Garcia Molinos, J., Halpern, B.S., Hoegh-Guldberg, O. *et al.* (2016) Responses of marine organisms to climate change across oceans. *Frontiers in Marine Science*, **3**, *art.* 62.

Prado, L. & Castilla, J.C. (2006) The bioengineer *Perumytilus purpuratus* (Mollusca: Bivalvia) in central Chile: biodiversity, habitat structural complexity and environmental heterogeneity. *Journal of the Marine Biological Association of the UK*, **86**, 417–421.

Purcell, A. (2018) Rocky shore. Available at: https://basicbiology.net/environment/marine/rocky-shore (accessed 28 May 2021).

Qi, D., Chen, L., Chen, B., Gao, Z., Zhong, W., Feely, R.A. *et al.* (2017) Increase in acidifying water in the western Arctic Ocean. *Nature Climate Change*, **7**, 195–199.

Raffaelli, D., Falcy, V. & Galbraith, C. (1990) Eider predation and the dynamics of mussel bed communities. In: *Trophic Relationships in the Marine Environment* (eds M. Barnes & R.N. Gibson), pp. 89–103. Aberdeen University Press, Aberdeen.

Range, P., Piló, D., Ben-Hamadou, R., Chícharo, M.A., Matias, D., Joaquim, S. *et al.* (2012) Seawater acidification by CO_2 in a coastal lagoon environment: effects on life history traits of juvenile mussels *Mytilus galloprovincialis*. *Journal of Experimental Marine Biology and Ecology*, **424–425**, 89–98.

Rice, M.A., Rawson, P.D., Salinas, A.D. & Rosario, W.R. (2016) Identification and salinity tolerance of the western hemisphere mussel *Mytella charruana* (D'Orbigny, 1842) in the Philippines. *Journal of Shellfish Research*, **35**, 865–873.

Rajagopal, S., Venugopalan, V.P., Azariah J. & Nair K.V.K. (1995) Response of green mussel *Perna viridis* (L.) to heat treatment in relation to power plant biofouling control. *Biofouling*, **8**, 313–330.

Rajagopal, S., Venugopalan, V.P., van der Velde, G. & Jenner, H.A. (2006) Greening of the coasts: a review of the *Perna viridis* success story. *Aquatic Ecology*, **40**, 273–297.

Rawson, P.D., Yund, P.O. & Lindsay, S.M. (2007) Comment on 'divergent induced responses to an invasive predator in marine mussel population'. *Science*, **316**, 53b.

Reimer, O. & Tedengren, M. (1996) Phenotypical improvement of morphological defenses in the mussel *Mytilus edulis* induced by exposure to the predator *Asterias rubens. Oikos*, **75**, 383–390.

Reimer, O. & Tedengren, M. (1997) Predator-induced changes in byssal attachment, aggregation and migration in the blue mussel, *Mytilus edulis. Marine and Freshwater Behaviour and Physiology*, **30**, 251–266.

Ritchie, H. & Roser, M. (2020) CO_2 and greenhouse gas emissions. Available at: https://ourworldindata.org/co2-and-other-greenhouse-gas-emissions (accessed 28 May 2021).

Riisgård, H.U. (2017) *General Ecology – Outline of Contemporary Ecology for University Students*, 1st edn. Bookboon, London.

Rius, M. & McQuaid, C.D. (2006) Wave action and competitive interaction between the invasive mussel *Mytilus galloprovincialis* and the indigenous *Perna perna* in South Africa. *Marine Biology*, **150**, 69–78.

Robles C. & Desharnais R. (2002) History and development of a paradigm of predation in rocky intertidal communities. *Ecology*, **83**, 1521–1536.

Robles, C.D., Desharnais, R.A., Garza, C., Donahue, M.J. & Martinez, C.A. (2009) Complex equilibria in the maintenance of boundaries: experiments with mussel beds. *Ecology*, **90**, 985–995.

Root, T.L., Price, J.T., Hall, K.R., Schneider, S.H., Rosenzweig, C. & Pounds, J.A. (2003) Fingerprints of global warming on wild animals and plants. *Nature*, **421**, 57–60.

Roux, J-P., van der Lingen, C.D., Gibbons, M.J., Moroff, N.E., Shannon, L.J., Smith, A.D.M. *et al.* (2013) Jellyfication of marine ecosystems as a likely consequence of overfishing small pelagic fishes: lessons from the Benguela. *Bulletin of Marine Science*, **89**, 249–284.

Rovero, F., Hughes, R.N. & Chelazzi, G. (2000) When time is of the essence: choosing a currency for prey-handling costs. *Journal of Animal Ecology*, **69**, 683–689.

Sadler, D.E., Lemasson, A.J. & Knights, A.M. (2018) The effects of elevated CO_2 on shell properties and susceptibility to predation in mussels *Mytilus edulis. Marine Environmental Research*, **139**, 162–168.

Safriel, U.N., Gilboa, A. & Felsenburg, T. (1980) Distribution of rocky intertidal mussels in the Red Sea coasts of Sinai, the Suez Canal and the Mediterranean coast of Israel, with special reference to recent colonizers. *Journal of Biogeography*, **7**, 39–62.

Saier, B. (2001) Direct and indirect effects of seastars *Asterias rubens* on mussel beds (*Mytilus edulis*) in the Wadden Sea. *Journal of Sea Research*, **46**, 29–42.

Sarà, G. & de Pirro, M. (2011) Heart beat rate adaptations to varying salinity of two intertidal Mediterranean bivalves: the invasive *Brachidontes pharaonis* and the native *Mytilaster minimus. Italian Journal of Zoology*, **78**, 193–197.

Sarà, G., Romano, C., Mazzola, A. (2008) A new Lessepsian species in the western Mediterranean (*Brachidontes pharaonis* Bivalvia: Mytilidae): density, resource allocation and biomass. *Marine Biodiversity Records*, **1**, e8.

Schneider, K.R., Wethey, D.S., Helmuth, B.S.T. & Hilbish, T.J. (2005) Implications of movement behavior on mussel dislodgement: exogenous selection in a *Mytilus* spp. hybrid zone. *Marine Biology*, **146**, 333–343.

Schneider, K.R. (2008) Heat stress in the intertidal: comparing survival and growth of an invasive and native mussel under a variety of thermal conditions. *Biological Bulletin*, **215**, 253–264.

Schneider, K.R., Van Thiel, L.E. & Helmuth, B. (2010) Interactive effects of food availability and aerial body temperature on the survival of two intertidal *Mytilus* species. *Journal of Thermal Biology*, **35**, 161–166.

Seed, R. (1969) The ecology of *Mytilus edulis* L. (Lamellibranchiata) on exposed shores. 2. Growth and mortality. *Oecologia*, **3**, 317–350.

Seed, R. (1976) Ecology. In: *Marine Mussels: Their Ecology and Physiology* (ed B.L. Bayne), pp. 13–65. Cambridge University Press, Cambridge.

Seed, R. (1980) A note on the relationship between shell shape and life habits in *Geukensia demissa* and *Brachidontes exustus* (Mollusca: Bivalvia). *Journal of Molluscan Studies*, **46**, 293–299.

Seed, R. (1996) Patterns of biodiversity in the macro-invertebrate fauna associated with mussel patches on rocky shores. *Journal of the Marine Biological Association of the UK*, **76**, 203–210.

Seed, R. & Suchanek, T.H. (1992) Population and community ecology of *Mytilus*. *The Mussel Mytilus: Ecology, Physiology, Genetics and Culture* (ed. E.M. Gosling), pp. 87–169. Elsevier Science Publishers B.V., Amsterdam.

Šegvić-Bubić T, Grubišić L, Karaman N, Tičina V, Jelavić KM & Katavić, I. (2011) Damages on mussel farms potentially caused by fish predation – self service on the ropes? *Aquaculture*, **319**, 497–504.

Semmens, D.C., Dane, R.E., Pancholi, M.R., Slade, S.E., Scrivens, J.H. & Elphick, M.R. (2013) Discovery of a novel neurophysin-associated neuropeptide that triggers cardiac stomach contraction and retraction in starfish. *Journal of Experimental Biology*, **216**, 4047–4053.

Shaw, W.N., Hassler, T.J. & Moran, D.P. (1988) Species profiles: life histories and environmental requirements of coastal fishes and invertebrates (Pacfic southwest): California Sea mussel and Bay mussel. US Fisheries and Wildlife Services Biology Report 82(11.84)TR-EL-82-4.

Sherker, Z.T., Ellrich, J.A. & Scrosati, R.A. (2017) Predator-induced shell plasticity in mussels hinders predation by drilling snails. *Marine Ecology Progress Series*, **573**, 167–175.

Shin, P.K.S., Liu, C.C., Liu, Z.X. & Cheung, S.G. (2008) Marine mussels *Brachidontes variabilis* selected smaller places of refuge and enhanced byssus production upon exposure to conspecific and heterospecific cues. *Journal of Experimental Marine Biology and Ecology*, **361**, 16–20.

Shinen, J.S. & Morgan, S.G. (2009) Mechanisms of invasion resistance: competition among intertidal mussels promotes establishment of invasive species and displacement of native species. *Marine Ecology Progress Series*, **383**, 187–197.

Silva, A.F., Sousa, M.C., Bernardes, C. & Dias, J.M. (2017) Will climate change endanger the current mussel production in the Rias Baixas (Galicia, Spain)? *Journal of Aquaculture and Fisheries*, **1**, 001.

Singh, G.G., Markel, R.W., Martone, R.G., Salomon, A.K., Harley, C.D.G. & Chan K.M.A. (2013) Sea otters homogenize mussel beds and reduce habitat provisioning in a rocky intertidal ecosystem. *PLoS ONE*, **8**, e65435.

Skein, L., Robinson,T.B. & Alexander, M.E. (2018) Impacts of mussel invasions on the prey preference of two native predators. *Behavioural Ecology*, **29**, 353–359.

Smith, L.D. & Jennings, J.A. (2000) Induced defensive responses by the bivalve *Mytilus edulis* to predators with different attack modes. *Marine Biology*, **136**, 461–469.

Somero, G.N. (2010) The physiology of climate change: how potentials for acclimatization and genetic adaptation will determine 'winners' and 'losers'. *Journal of Experimental Biology*, **213**, 912–920.

Somero, G.N. (2011) Comparative physiology: a 'crystal ball' for predicting consequences of global change. *American Journal of Physiology – Regulatory, Integrative and Comparative Physiology*, **301**, R1–R14.

Somero, G.N. (2012) The physiology of global change: linking patterns to mechanisms. *Annual Review of Marine Science*, **4**, 39–61.

Sousa W.P. (2007) Disturbance In: *Encyclopedia of Tidepools and Rocky Shores* (eds M.W. Denny & S. Gaines), pp. 186–193. University of California Press, Berkeley, CA.

Sousa, W.P (2012) Disturbance and patch dynamics on rocky intertidal shores. In: *The Ecology of Natural Disturbance and Patch Dynamics* (eds S.T.A. Pickett & P.S. White), pp. 101–124. Academic Press, Cambridge, MA.

Spinuzzi, S.S., Schneider, K.R., Walters, L.J., Yuan, W.S. & Hoffman, E.A. (2013 Tracking the distribution of non-native marine invertebrates (*Mytella charruana*, *Perna viridis* and *Megabalanus coccopoma*) along the south-eastern USA. *Marine Biodiversity Records*, **6**, art. e55.

Steeves, L.E., Filgueira, R., Guyondet, T., Chassé, J. & Comeau, L. (2018) Past, present, and future: performance of two bivalve species under changing environmental conditions. *Frontiers in Marine Science*, **5**, art. 184.

Steffani, C.N. & Branch, G.M. (2003) Spatial comparisons of populations of an indigenous limpet *Scutellastra argenvillei* and an alien mussel *Mytilus galloprovincialis* along a gradient of wave energy. *African Journal of Marine Science*, **25**, 195–212.

Stenseth, N.C., Ottersen, G., Hurrell, J.W., Mysterud, A., Lima, M., Chan, K-S. *et al.* (2003) Studying climate effects on ecology through the use of climate indices: the North Atlantic Oscillation, El Niño Southern Oscillation and beyond. *Proceedings of the Royal Society London B*, **270**, 2087–2096.

Suchanek, T.H. (1981) The role of disturbance in the evolution of life history strategies in the intertidal mussels *Mytilus edulis* and *Mytilus californianus*. *Oecologia*, **50**, 143–152.

Sui, Y., Liu, Y., Zhao, X., Dupont, S., Hu, M. & Wu, F. (2017) Defense responses to short-term hypoxia and seawater acidification in the thick shell mussel *Mytilus coruscus*. *Frontiers in Physiology*, **8**, art. 145.

Sunday, J.M., Crim, R.N., Harley, C.D.G. & Hart, M.W. (2011) Quantifying rates of evolutionary adaptation in response to ocean acidification. *PLoS ONE*, **6**, e22881.

Sunday, J.M., Calosi, P. Dupont, S., Munday, P.L., Stillman, J.H. & Reusch, T.B.H. (2014) Evolution in an acidifying ocean. *Trends in Ecology and Evolution*, **29**, 117–125.

Svane, I. & Ompi, M. (1993) Patch dynamics in beds of the blue mussel *Mytilus edulis* L: effects of site, patch size, and position within patch. *Ophelia*, **37**, 187–202.

Thiel, M. & Ullrich, N. (2002) Hard rock versus soft bottom: the fauna associated with intertidal mussel beds on hard bottoms along the coast of Chile, and considerations on the functional role of mussel beds. *Helgoland Marine Research*, **56**, 21–30.

Thieltges, D.W. (2005) Impact of an invader: epizootic American slipper limpet *Crepidula fornicata* reduces survival and growth in European mussels. *Marine Ecology Progress Series*, **286**, 13–19.

Thomsen, J. & Melzner, F. (2010) Moderate seawater acidification does not elicit long-term metabolic depression in the blue mussel *Mytilus edulis*. *Marine Biology*, **157**, 2667–2676.

Thomsen, J., Casties, I., Pansch, C., Körtzinger, A. & Melzner, F. (2013) Food availability outweighs ocean acidification effects in juvenile *Mytilus edulis*: laboratory and field experiments. *Global Change Biology*, **19**, 1017–1027.

Thomsen, J., Stapp, L.S., Haynert, K., Schade, H., Danelli, M., Lannig, G. *et al.* (2017) Naturally acidified habitat selects for ocean acidification-tolerant mussels. *Science Advances*, **3**, e1602411.

Thyrring, J., Blicher, M.E., Sørensen, J.G. Wegeberg, S. & Sejr, M.K. (2017) Rising air temperatures will increase intertidal mussel abundance in the Arctic. *Marine Ecology Progress Series*, **584**, 91–104.

Tomanek, L. (2008) The importance of physiological limits in determining biogeographical range shifts due to global climate change: the heat-shock response. *Physiological and Biochemical Zoology*, **81**, 709–717.

Tomanek, L. & Helmuth, B. (2002) Physiological ecology of rocky intertidal organisms: a synergy of concepts. *Integrative and Comparative Biology*, **42**, 771–775.

Tomanek, L., Zuzow, M.J., Hitt, L., Serafini, L. & Valenzuela, J.J. (2012) Proteomics of hyposaline stress in blue mussel congeners (genus *Mytilus*): implications for biogeographic range limits in response to climate change. *Journal of Experimental Biology*, **215**, 3905–3916.

Toro, J.E., Ojeda, J.A., Vergara, A.M., Castro, G.C. & Alcapán, A.C. (2005) Molecular characterization of the Chilean blue mussel (*Mytilus chilensis* Hupé 1854) demonstrates evidence for the occurrence of *Mytilus galloprovincialis* in southern Chile. *Journal of Shellfish Research*, **24**, 1117–1121.

Torossian, J.L., Hosek, K.E., Donelan, S.C., Trussell, G.C., Helmuth, B. S. & Zippay, M.L. (2020) Physiological and biochemical responses to acute environmental stress and predation risk in the blue mussel, *Mytilus edulis*. *Journal of Sea Research*, **159**, 101891.

Tyler-Walters, H. (2001) Green crenella (*Musculus discors*). In: *Marine Life Information Network: Biology and Sensitivity Key Information Reviews* (eds H. Tyler-Walters & K. Hiscock). Marine Biological Association of the United Kingdom, Plymouth. Available at: https://www.marlin.ac.uk/species/detail/1645 (accessed 28 May 2021).

Tyler-Walters, H. (2007) Horse mussel (*Modiolus modiolus*). In: *Marine Life Information Network: Biology and Sensitivity Key Information Reviews* (eds H. Tyler-Walters & K. Hiscock). Marine Biological Association of the United Kingdom, Plymouth. Available at: https://www.marlin.ac.uk/species/detail/1532 (accessed 28 May 2021).

Tyler-Walters, H. (2008) Common mussel (*Mytilus edulis*). In: *Marine Life Information Network: Biology and Sensitivity Key Information Reviews* (eds H. Tyler-Walters & K. Hiscock). Marine Biological Association of the United Kingdom, Plymout. Available at: https://www.marlin.ac.uk/species/detail/1421 (accessed 28 May 2021).

Underwood, A. J. & Kennelly, S.J. (1990) Pilot studies for designs of surveys of human disturbance of intertidal habitats in New South Wales. *Australian Journal of Marine and Freshwater Research*, **1**: 165–173.

Urian, A.G., Hatle, J.D. & Gilg, M.R. (2011) Thermal constraints for range expansion of the invasive green mussel, *Perna viridis*, in the Southeastern United States. *Journal of Experimental Zoology Part A: Ecological Genetics and Physiology*, **315**, 12–21.

US EPA (2008) Effects of climate change for aquatic invasive species and implications for management and research. National Center for Environmental Assessment, Washington, DC; EPA/600/R-08/014.

USGCRP (2017) *Climate Science Special Report: Fourth National Climate Assessment*, Volume I (eds D.J. Wuebbles, D.W. Fahey, K.A. Hibbard, D.J. Dokken, B.C. Stewart & T.K. Maycock). US Global Change Research Program, Washington, DC.

Van De Werfhorst, L.C. & Pearse, J.S. (2007) Trampling in the rocky intertidal of central California: a follow-up study. *Bulletin of Marine Science*, **81**, 245–254.

Van Erkom Schurink, C. & Griffiths, C.L. (1990) Marine mussels of southern Africa–their distribution, standing stocks, exploitation and culture. *Journal of Shellfish Research*, **9**, 75–85.

Van Erkom Schurink, C. & Griffiths, C.L. (1993) Factors affecting relative growth in four South African mussel species. *Aquaculture*, **109**, 257–273.

Varennes, E., Hanssen, S.A., Bonardelli, J. & Guillemette, M. (2013) Sea duck predation in mussel farms: the best nets for excluding common eiders safely and efficiently. *Aquaculture Environment Interactions*, **4**, 31–39.

Varennes, E., Hanssen, S.A., Bonardelli, J.C. & Guillemette, M. (2015) A large molluscivore bird (common eider, *Somateria mollissima*) is able to discriminate quality of blue mussels (*Mytilus edulis*) based on size and provenance. *Canadian Journal of Zoology*, **93**, 655–663.

Vargas, C.A., Lagos, N.A., Lardies, M.A., Duarte, C., Manríquez, P.H., Aguilera, V.M. *et al.* (2017) Species-specific responses to ocean acidification should account for local adaptation and adaptive plasticity. *Nature Ecology and Evolution*, **1**, 0084.

Veiga, P., Rubal, M., Arenas, F., Incera, M., Olabarria, C. & Sousa-Pinto, I. (2011) Does *Carcinus maenas* facilitate the invasion of *Xenostrobus securis*? *Journal of Experimental Marine Biology and Ecology*, **406**, 14–20.

Veldhuizen-Tsoerkan, M.B., Holwerda, D.A., de Bont, A.M.T., Smaal, A.C. & Zandee, D.I. (1991) A field study on stress indices in the sea mussel, *Mytilus edulis*: application of the 'stress approach' in biomonitoring. *Archives of Environmental Contamination and Toxicology*, **21**, 497–504.

Ventura, A., Schulz, S. & Dupont, S. (2016) Maintained larval growth in mussel larvae exposed to acidified undersaturated seawater. *Scientific Reports*, **6**, 23728.

Vernberg, F. J. & Vernberg, W.B. (1972) *Environmental Physiology of Marine Animals*. Springer–Verlag, New York.

Vihtakari, M., Hendriks, I.E., Holding, J., Renaud, P.E., Duarte, C.M. & Havenhand, J.N. (2013) Effects of ocean acidification and warming on sperm activity and early life stages of the Mediterranean mussel (*Mytilus galloprovincialis*). *Water*, **5**, 1890–1915.

Wang, Y., Hu, M., Cheung, S.G., Shin, P.K.S., Lu, W. & Li, J. (2012) Chronic hypoxia and low salinity impair anti-predatory responses of the green-lipped mussel *Perna viridis*. *Marine Environmental Research*, **77**, 84–89.

Wang, Y., Li, L., Hu, M. & Lu, W. (2015) Physiological energetics of the thick shell mussel *Mytilus coruscus* exposed to seawater acidification and thermal stress. *Science of the Total Environment*, **514**, 261–272.

Westerbom, M., Kilpi, M. & Mustonen, M. (2002) Blue mussels, *Mytilus edulis*, at the edge of the range: population structure, growth and biomass along a salinity gradient in the north-eastern Baltic Sea. *Marine Biology*, **140**, 991–999.

Westfall, K.M. & Gardner, J.P.A, (2010) Genetic diversity of Southern hemisphere blue mussels (Bivalvia: Mytilidae) and the identification of non-indigenous taxa. *Biological Journal of the Linnean Society*, **101**, 898–909.

Wieters, E.A. & Navarrete, S.A. (1998) Spatial variability in prey preferences of the intertidal whelks *Nucella canaliculata* and *Nucella emarginata*. *Journal of Experimental Marine Biology and Ecology*, **222**, 133–148.

Wilkins, N.P., Fujino, K. & Gosling, E.M. (1983) The Mediterranean mussel *Mytilus galloprovincialis* Lmk in Japan. *Biological Journal of the Linnean Society*, **20**, 365–374.

Williams, R.J. (1970) Freezing tolerance in *Mytilus edulis*. *Comparative Biochemistry and Physiology*, **35**, 145–161.

Wiltshire, K., Rowling, K. & Deveney, M. (2010) Introduced marine species in South Australia: a review of records and distribution mapping. South Australian Research and Development Institute (Aquatic Sciences), Adelaide. SARDI Publication No. F2010/000305-1. SARDI Research Report Series No. 468, 232pp.

Witman, J.D. & Suchanek, T.H. (1984) Mussels in flow: drag and dislodgement by epizoans. *Marine Ecology Progress Series*, **16**, 259–268.

Wonham, M.J. (2004) Mini-review: distribution of the Mediterranean mussel *Mytilus galloprovincialis* (Bivalvia: Mytilidae) and hybrids in the northeast Pacific. *Journal of Shellfish Research*, **23**, 535–543.

Wood, A.R., Apte, S., MacAvoy, E.S. & Gardner, J.P.A. (2007) A molecular phylogeny of the marine mussel genus *Perna* (Bivalvia: Mytilidae) based on nuclear (ITS1&2) and mitochondrial (COI) DNA sequences. *Molecular Phylogenetics and Evolution*, **44**, 685–698.

Woodin, S.A., Hilbish, T.J., Helmuth, B., Jones, S.J. & Wethey, D.S. (2013) Climate change, species distribution models, and physiological performance metrics: predicting when biogeographic models are likely to fail. *Ecology and Evolution*, **3**, 3334–3346.

Wootton, J.T. (1993) Size-dependent competition: effects on the dynamics versus the end-point of mussel bed succession. *Ecology*, **74**, 195–206.

Wootton J.T. (2002) Mechanisms of successional dynamics: consumers and the rise and fall of species dominance. *Ecological Research*, **17**, 249–260.

Wu, F., Xie, Z., Lan, Y., Dupont, S., Sun M, Cui, S. *et al.* (2018) Short-term exposure of *Mytilus coruscus* to decreased pH and salinity change impacts immune parameters of their haemocytes. *Frontiers in Physiology*, **9**, art. 166.

Yuan, W., Walters L.J., Schneider K.R. & Hoffman E.A. (2010) Exploring the survival threshold: a study of salinity tolerance of the nonnative mussel *Mytella charruana*. *Journal of Shellfish Research*, **29**, 415–422.

Yuan, W.S., Hoffman, E.A. & Walters, L.J. (2016a) Effects of nonnative invertebrates on two life stages of the native eastern oyster *Crassostrea virginica*. *Biological Invasions*, **18**, 689–701.

Yuan, W.S., Walters, L.J., Brodsky, S.A., Schneider, K.R. & Hoffman, E.A. (2016b) Synergistic effects of salinity and temperature on the survival of two nonnative bivalve molluscs, *Perna viridis* (Linnaeus 1758) and *Mytella charruana* (d'Orbigny 1846). *Journal of Marine Biology*, **2016**, art. 9261309.

Zardi G.I., Nicastro, K.R., Porri, F. & McQuaid, C.D. (2006) Sand stress as a non-determinant of habitat segregation of indigenous (*Perna perna*) and invasive (*Mytilus galloprovincialis*) mussels in South Africa. *Marine Biology*, **148**, 1031–1038.

Zardi, G.I., McQuaid, C. Teske, P.R. & Barker, N. P. (2007a) Unexpected genetic structure of mussel populations in South Africa: indigenous *Perna perna* and invasive *Mytilus galloprovincialis*. *Marine Ecology Progress Series*, **337**, 135–144.

Zardi, G.I., McQuaid, C.D. & Nicastro, K.R. (2007b) Balancing survival and reproduction: seasonality of wave action, attachment strength and reproductive output in indigenous *Perna perna* and invasive *Mytilus galloprovincialis* mussels. *Marine Ecology Progress Series*, **334**, 155–163.

Zardi, G. I., Nicastro, K.R., McQuaid, C.D. & Erlandsson, J. (2008) Sand and wave induced mortality in invasive (*Mytilus galloprovincialis*) and indigenous (*Perna perna*) mussels. *Marine Biology*, **153**, 853–858.

Zardi, G.I. Nicastro, K.R., McQuaid, C.D., Castilho, R., Costa, J., Serrão, E.A. *et al.* (2015) Intraspecific genetic lineages of a marine mussel show behavioural divergence and spatial segregation over a tropical/subtropical biogeographic transition. *BMC Evolutionary Biology*, **15**, art. 100.

Zbawicka, M., Drywa, A., Śmietanka, B. & Wenne, R. (2012) Identification and validation of novel SNP markers in European populations of marine *Mytilus* mussels. *Marine Biology*, **159**, 1347–1362.

Zhou, X., Ji, X., Wang, B., Cheng, Y., Ma, Z., Choi, F. *et al.* (2018) Pido: Predictive Delay Optimization for Intertidal Wireless Sensor Networks. *Sensors*, **18**, 1464.

Zieritz, A., Clucas, G., Axtmann, L. & Aldridge, D.C. (2012) Shell ecophenotype in the blue mussel (*Mytilus edulis*) determines the spatial pattern in foraging behaviour of an oystercatcher (*Haematopus ostralegus*) population. *Marine Biology*, **159**, 863–872.

Zwarts, L., Wanink, J.H. & Ens, B.J. (1996a) Predicting seasonal and annual fluctuations in the local exploitation of different prey by oystercatchers *Haematopus ostralegus*: a 10-year study in the Wadden Sea. *Ardea*, **84A**, 401–440.

Zwarts, L., Ens, B.J., Goss-Custard, J.D., Hulscher, J.B. & Durell, S.E.A. (1996b) Causes of variation in prey profitability and its consequences for the intake rate of the oystercatcher *Haematopus ostralegus*. *Ardea*, **84A**, 229–268.

Żydelis, R., Esler, D., Kirk, M. & Boyd W.S. (2009) Effects of off-bottom shellfish aquaculture on winter habitat use by molluscivorous sea ducks. *Aquatic Conservation: Marine and Freshwater Ecosystems*, **19**, 34–42.

4

Particle Processing and Nutrient Utilisation

Introduction

As already mentioned, suspension or filter feeding is believed to have evolved in some group of early protobranch molluscs, giving rise to the Autobranchia, the dominant subclass of modern bivalves. These feed by filtering the incoming current as a source of food, the gills (ctenidia) having replaced the labial palps as the feeding organs (see Chapter 1). The chief modification of the ctenidia for filtering was the lengthening and folding of individual gill filaments. In addition, many extra filaments were added so that they extended as far forward as the labial palps. Both of these modifications greatly increased the surface area of the ctenidia. It is believed that the triangular-shaped filaments of the primitive bivalve gill progressively changed over evolutionary time to the W-shaped filaments of the modern bivalve gill (see Cannuel *et al.* 2009). The ctenidia, with their different ciliary tracts, remove suspended particles from the water pumped through the mantle cavity. For a full description and illustration of gill structure and ciliation, see Chapter 2.

The present chapter focuses on feeding, digestion and absorption in mussels, the various mechanisms they employ to control the quantity and quality of their diet and how their mode of feeding can influence marine ecosystem processes.

Filtration and Clearance Rates

Filtration rate (FR), often referred to as pumping rate, is the volume of water flowing through the gills per unit of time. Clearance rate (CR) is defined as the volume of water cleared of suspended particles of a certain size per unit of time (Riisgård 2001a). If the particles are 100% retained by the gill, which is the case for particles ≥4 μm in mussels (Rosa *et al.* 2015), then the clearance rate is representative of the filtration rate. If the retention efficiency is less than 100%, CR < FR.

Estimation of Filtration Rate

Riisgård (2001a) has reviewed the methods used to measure filtration rate in bivalves, with the emphasis on marine mussels. The review accentuates the advantages and disadvantages of the different methods, and examples of shortcomings are mentioned to warn against pitfalls. Because of their suitability in physiological investigations, direct methods, where exhaled water from the bivalve is physically separated from the surrounding water and is collected and measured, have been widely used to estimate filtration rates in bivalves. Reliable filtration rates in *Mytilus edulis* were first obtained by means of the direct technique developed by Famme *et al.* (1986); a version of the experimental set-up is illustrated in Figure 4.1. A mussel is placed in a tank that is divided into two chambers C_1 and C_2 by a membrane with a slit to accommodate the mussel and to separate its inhalant and exhalant siphons. A shunt connects the two chambers, which is closed at the start of the experiment. As the animal pumps water from C_1 to C_2, the water level in C_2 is monitored with a laser beam striking a mirror fixed to a floating ping-pong ball. The mirror reflects the laser beam on to a scale set at a distance of about 10 m from the mirror. A deflection of 1 cm on the scale represents a 0.1 mm change in the water level in C_2. When the mussel pumps water from C_1 to C_2, a pressure difference develops, which may be fixed at any selected value by adjusting a calibrated peristaltic pump that pumps water back from C_2 to C_1. The use of this method in bivalves is limited to species which have a well-defined exhalant siphon. Famme *et al.* (1986) showed that high pumping rates were correlated with wide-open shell valves and extended and diverging mantle edges. Also, the filtration rates measured agreed with the filtration rates measured on the same set of animals with the clearance and suction methods described later.

Indirect methods are used to determine clearance rates. An example is the indirect clearance method, whereby a bivalve is placed in a known volume of water containing a suspension of particles, usually unicellular algae, for a set period of time. Filtration results in a

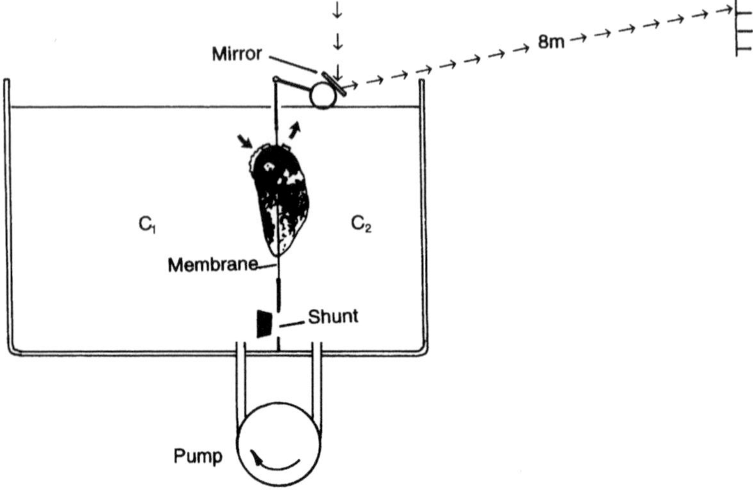

Figure 4.1 Experimental set-up for direct measurement of filtration rates in bivalves. *Source:* From Famme *et al.* (1986). Reproduced with permission from Springer Nature.

decrease in particle concentration over time. This reduction is tracked either by measuring the fluorescence intensity of chlorophyll *a* in water pumped through the apparatus (Pleissner *et al.* 2013) or by taking water samples at fixed time intervals and measuring particle concentration using an electronic particle counter (Riisgård *et al.* 2011a). A disadvantage of the method is that particle concentration declines during the experiment, but this can be overcome by adding new particles to maintain concentration within a certain range (Riisgård 2001a). The filtration rate (*F*) is determined from the exponential decrease in algal concentration as a function of time using the usual clearance formula (Riisgård *et al.* 2011b; Pleissner *et al.* 2013):

$$F_{CR} = \left(V/\text{tn}\right)\ln\left(C_0/C_t\right) = V/n \times b$$

where C_0 and C_t are initial and terminal particle concentrations, *V* is the water volume in the aquarium, *n* is the number of individuals and *b* is the slope of the regression line in a semi-log plot for the reduction of algal concentration with time in a well-mixed aquarium (Figure 4.2). Hence:

$$F_{CR} = V \times b/n$$

Another indirect method, the flow-through chamber (FTC) method, has been widely used despite the fact that its erroneous use gives incorrect values of filtration rates, hence leading to conflicting data in the literature. Clearance rate (ml min^{-1}) is determined by using the equation:

$$\text{CR} = \text{Fl} \times \left(C_i - C_o\right)/C_i$$

where Fl = flow rate (volume of water flowing through the chamber per unit of time) and C_i and C_o = concentration of suspended algal cells at inlet and outlet of the chamber, respectively (Figure 4.3). The size and shape (i.e. the geometry) of the FTC, as well as the

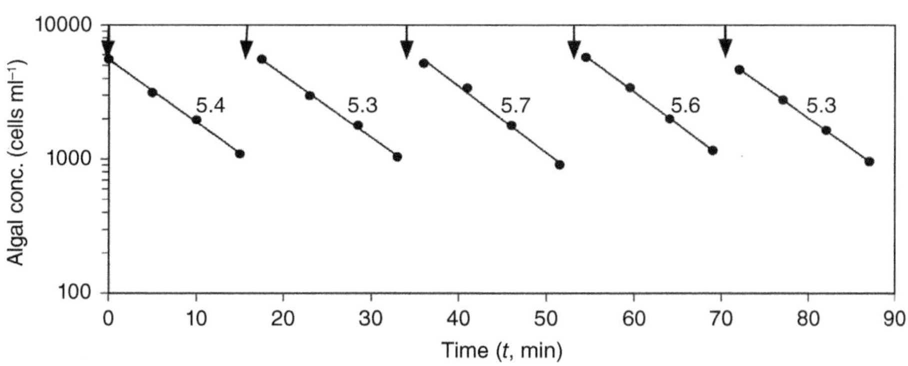

Figure 4.2 Exponential decrease in algal cell concentration as a function of time (*t*, min). in fully open and actively filtering *Mytilus edulis* during a clearance experiment in an aquarium with well-mixed seawater. Regression lines and estimated filtration rates (l h^{-1} ind^{-1}) are shown. Arrows indicate addition of algal cells. Low algal concentrations, 1000–4000 *Rhodomonas* sp. cells ml^{-1}, equivalent to about 1–5 μg chl *a* l^{-1}, were used. *Source:* From Kittner & Riisgård (2005). CC BY 4.0.

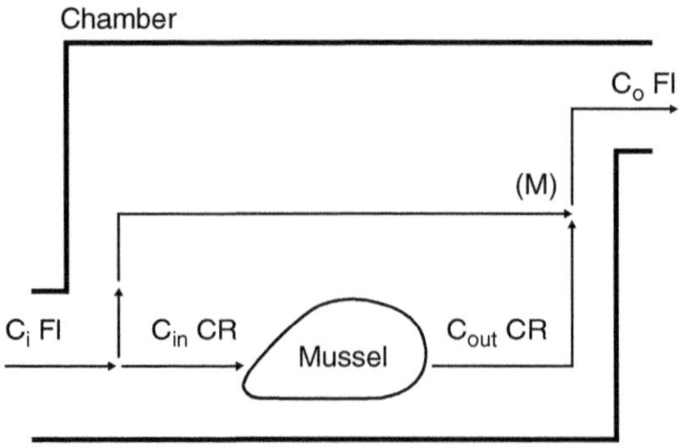

Figure 4.3 Diagram of the flow-through chamber (FTC) method. (M), point of perfect mixing; C_i and C_o, concentration of chamber inflow and outflow, respectively; C_{in} (= C_i) and C_{out} (= 0 for 100% retention), concentration of inhalant and exhalant flows, respectively; Fl and CR, chamber through-flow and clearance rate, respectively. *Source:* From Larsen & Riisgård (2011).

flow rate, are of great importance when using this method. The requirements for its suitable use are that: the food crossing the chamber is completely available to the bivalve, otherwise the available food will be less than the theoretical value; the geometry of the chamber minimises water recirculation to prevent dilution of the incoming food concentration; and the algal cells are 100% efficiently retained by the gills (e.g. algal diameter >4 μm for mussels), to reduce underestimation of the measurement (Filgueira *et al.* 2006). The suction method, however, is superior to the FTC method because the 'geometry' problems of the chamber are eliminated (details in Møhlenberg & Riisgård 1979; Riisgård 2001a; Riisgård & Seerup 2003).

In field experiments, CR measurements cannot be easily used without disturbing the bivalves. The biodeposit method may be a useful field supplement to laboratory methods. Hawkins *et al.* (1996) calculated CR in *M. edulis* from direct collection of biodeposits according to the ratio CR = (inorganic matter egested as faeces and pseudofaeces)/(total inorganic matter in seawater). The method is easy to set up and has been used successfully in several field studies (details in Riisgard 2001a). Hansen *et al.* (2011) designed an open-top chamber for measuring ambient community clearance rate on undisturbed bivalve populations in the field (Figure 4.4). The chamber was pressed 5–10 cm down in the sediment on the mussel bed. It held approximately a 30–40 cm water column, equal to a volume of 43–77 l. It was provided with an air lift connected to a SCUBA pressure tank, generating a continuous and gentle water circulation, which ensured a complete mixture of suspended particles, and thereby a maximum filtration by the bivalves. An in situ fluorometer was mounted to record plant pigment reduction due to mussel clearance in real time. To calibrate the in situ fluorometer, triplicate water samples were obtained initially in each of the bivalve filtration measurements. The water samples were filtrated, extracted and later analysed for plant pigment concentration on a laboratory spectrophotometer. In ambient natural seawater, Hansen *et al.* (2011) found that an assemblage of bivalves (mussels and oysters) only realised 10% of their theoretical filtration capacity, but when cultivated algal cells were added to the experimental chamber this soon increased the filtration activity to 42% of the theoretical (i.e. maximum filtration rate measured in the laboratory using cultivated algae by Møhlenberg & Riisgård 1979). Recently, Lüskow & Riisgård (2018), who slightly modified the design of the open-top chamber set-up, found that the in situ measured filtration rates are comparable to filtration rates obtained in laboratory studies using different methods and controlled diets of cultivated algal cells. Furthermore, they found

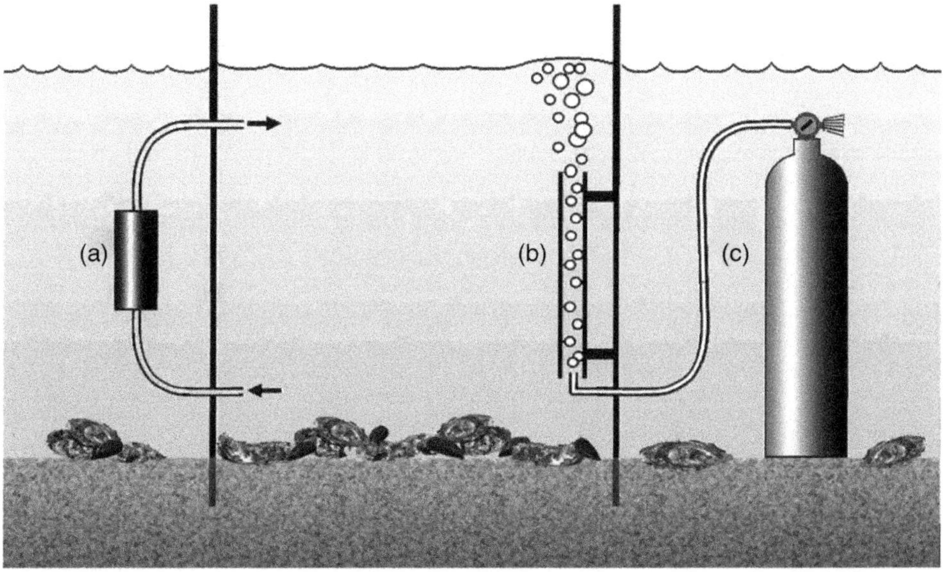

Figure 4.4 Schematic side-view of the open-top chamber for real-time in situ measurements of mussel bed clearance rate. (a) CTD (conductivity, temperature and depth; the primary tool for determining essential physical properties of sea water), equipped with a flow-through fluorometer and water pump. (b) Air lift. (c) SCUBA pressure tank equipped with an adjustable first-stage pressure regulator (see details in text). *Source:* From Hansen *et al.* (2011). Reproduced with permission from John Wiley and Sons.

that the retention efficiency (see later) of free-living bacteria was between 22.2 and 29.9%, in good agreement with values from laboratory studies. As Bayne (2004) so rightly stated, it is by such field-orientated studies, using natural particulates, that we can best expect to understand the full complexities of bivalve feeding behaviour.

Details on other indirect methods are in Riisgård (2001a). For example, the photoaquarium method is based on a complicated automatic recording apparatus that maintains a constant algal concentration. The method is sophisticated, but also rather difficult to use as compared to another indirect method, the steady-state method, where a constant algal concentration is maintained over longer periods (up to 47 days) by means of a dosing pump that can be adjusted to balance the increased clearance of growing bivalves (Riisgård & Randløv 1981). The video observation method deduces the filtration rate from particle tracking by means of microscope-video recording with subsequent image analysis. The replacement method (Coughlan & Ansell 1964), which is based on replacement of the inhalant current by a metered stream of colored water, may be used on bivalves remaining in their natural positions. This also applies to the thermistor method, based on the measurement of velocity of exhalant streams by a thermistor probe, which, however, requires very accurate positioning and temperature control (Meyhöfer 1985). The recent development of the VacuSIP technique, which allows in situ sampling of inhalant and exhalant currents from undisturbed suspension feeders, may provide more reliable estimates of clearance rates (Morganti *et al.* 2016).

Riisgård (2001a) presented a very comprehensive and stimulating review of this and seven other methods, highlighting methodological shortcomings and possible misinterpretations of data and suggesting guidelines to aid in obtaining reliable clearance rate data. This was later followed by a comparison of three of these methods (Peterson 2004), which in turn led to a series of comments and debate (Bayne 2004; Petersen 2004). More recently, Filgueira *et al.* (2006) proposed a way of designing experimental chambers and developed a protocol for evaluating them for CR measurement using the FTC method. A similar approach has been applied for the FTC and steady-state methods (Larsen & Riisgård 2011), clearance and biodeposit methods (Navarro & Vlasco 2003), FTC, clearance and biodeposit methods (Petersen 2004) and clearance and particle image velocity methods (Frank *et al.* 2008).

Factors Influencing Filtration Rate

Size

In bivalves, filtration rate and feeding capacity scale closely to gill area and size. The allometric relationship between filtration rate (F; l h^{-1}) and tissue dry weight (W, g) is described by the equation

$$F = \mathrm{a}W^{b}$$

where the constants *a* and *b* are fitted parameters, with *a* representing the filtration rate of an individual of unit body weight (or length) and *b* the power to which increase in size raises that rate. As experimentally verified by Jones *et al.* (1992), Riisgård (2001a) and Riisgård *et al.* (2014a), the filtration rate (F) and gill area (G) in *M. edulis* are nearly proportional to the square of the shell length (L) of the mussel (i.e. $F \approx L^2$ and $G \approx L^2$). Likewise, the body dry weight (W) may be nearly proportional to L^3 – or, reversed, L is proportional to $W^{1/3}$, so that $F \approx (W^{1/3})^2 = W^{2/3} = W^{0.67}$. For veliger and post-metamorphosed larvae (*M. edulis*), filtration rate has been reported to scale with weight (b =1.03, Riisgård *et al.* 1980). The high scaling factor was significantly different from exponents obtained for larger *M. edulis* (b = 0.66, Møhlenberg & Riisgård 1979; b = 0.72, Riisgård & Møhlenberg 1979). A rapid nonisometric development of the gills in the early post-metamorphic stages could account for the high b exponent, which declines when gill growth becomes isometric (Riisgård *et al.* 1980). The relationship between shell length and body size is not constant, but varies from population to population during the year due to factors such as spawning in spring, rapid growth during summer and starvation during winter (Riisgård *et al.* 2014a). Weight is expected to fluctuate not only within a year, but also between years, affecting the relation with clearance rate. In other studies, it had previously been concluded that gill area generally scales well with length, and that therefore clearance rate estimates based on length can be considered the actual clearance rates (e.g. Filgueira *et al.* 2008; Riisgård *et al.* 2014a).

Table 4.1 shows allometric equations for mussel filtration rate (F l h^{-1}) versus weight (W g body dry wt) and length (L mm shell length), measured in 12 studies using eight different methods. There is considerable variation in filtration rates both within and between species. While some of this variation represents genuine differences, some could be attributed to the differing experimental conditions employed in the different studies.

Table 4.1 Shell length (*L*), body dry weight (*W*) and filtration rate (*F*) of mussel species measured in various studies by means of different methods. Estimated allometric power-law equations for *F* versus *W* and *L* are shown. FTC, flow-through chamber; PIV, particle image velocimetry; FCA, fluorometer-controlled apparatus.

Equation	Species	Method	Reference
$F = 7.37W^{0.72}$	*Mytilus edulis*	Photoaquarium	Riisgård & Møhlenberg (1979)
$F = 7.45W^{0.66}$	*M. edulis*	Suction	Møhlenberg & Riisgård (1979)
$F = 6.00W^{0.75}$	*Modiolus modiolus*	Suction	Møhlenberg & Riisgård (1979)
$F = 5.37W^{0.60}$	*Choromytilus meridionalis*	Clearance	Griffiths (1980)
$F = 0.0012L^{2.14}$	*M. edulis*	Suction	Kiørboe & Møhlenberg (1981)
$F = 8.85W^{0.66}$	*Perna perna*	Clearance	Berry & Schleyer (1983)
$F = 0.0027L^{1.86}$	*Perna perna*	Clearance	Berry & Schleyer (1983)
$F = 6.15W^{0.83}$	*Geukensia demissa*	Clearance	Riisgård (1988)
$F = 1.78W^{0.70}$	*M. edulis*	Impeller	Jones *et al.* (1992)
$F = 0.0002L^{2.19}$	*M. edulis*	Impeller	Jones *et al.* (1992)
$F = 5.80W^{0.60} / 5.02W^{0.50}$	*Mytilus galloprovincialis*	FTC	Filgueira *et al.* (2008)
$F = 0.0035/0.0039L^{1.72}$	*Mytilus galloprovincialis*	FTC	Filgueira *et al.* (2008)
$F = 0.0008L^{2.26}$	*M. edulis*	PIV	Riisgård *et al.* (2011b)
$F = 0.0024L^{2.01}$	*M. edulis*	FCA	Pleissner *et al.* (2013)
$F = 6.77W^{0.68}$	*M. edulis*	Clearance	Riisgård *et al.* (2014a)
$F = 0.0014L^{2.09}$	*M. edulis*	Clearance	Riisgård *et al.* (2014a)
$F = 0.74W^{0.67}$; $F = 0.0004L^{2.00}$	*M. edulis*	Fluorescence probe	Jacobs *et al.* (2015)

However, in *M. edulis*, filtration rates ($F\,\mathrm{l\,h^{-1}}$) as a function of weight (W) were often quite similar (e.g. 7.37, 7.45 and 6.77 ($\mathrm{l\,h^{-1}}$)), even when the photoaquarium, suction and clearance methods, respectively, were used. Also, rates were often very similar between species when the same or different methods were used (Table 4.1). The filtration rate $F = 1.78W^{0.70}$ for *M. edulis* (Jones *et al.* 1992) is low – about four times lower than values shown in Table 4.1 – which may indicate that the method was inadequate in measuring filtration rate (Riisgård 2001a). In Table 4.1, filtration rates, as a function of length, vary quite a lot, ranging between 0.0002 and 0.0037, with an average rate of 0.0018 ($\mathrm{l\,h^{-1}}$). Filtration rates in Jacobs *et al.* (2015) were among the lowest recorded. Their study was performed on densely populated pieces of rope or large numbers per water volume. This may have resulted in refiltration of water, leading to lower rates compared to the maximums determined in studies performed under controlled laboratory conditions. Finally, b values for W and L were 0.687 and 2.03, respectively – both very close to the theoretically expected 2/3 (0.66) and 2.0, respectively (Riisgård *et al.* 2014a).

Particle Concentration

When filter-feeding bivalves experience very low algal concentrations, this eventually leads to reduced siphon-opening and valve-gape, or complete closure, along with cessation of filtration activity, which lasts until the algal concentration is elevated above a lower threshold level (Riisgård *et al.* 2003). In controlled laboratory studies, using video-camera recording of valve-gape responses of *M. edulis* to presence or absence of algal cells in the ambient water, Riisgard *et al.* (2013a and references therein) revealed that the critical algal concentration (C_{min}) below which the mussel closes its valves is between about 0.9 (Riisgård *et al.* 2006) and 0.5 μg chl *a* $\mathrm{l^{-1}}$ (Maire *et al.* 2007; Pascoe *et al.* 2009). The upper algal concentration below which it exploits its filtration capacity over an extended period of time has been identified to be between about 5000 and 8000 (*Rhodomonas salina*) cells $\mathrm{ml^{-1}}$, equivalent to 6.3 and 10.0 μg chl *a* $\mathrm{l^{-1}}$, respectively (Riisgård *et al.* 2011a). In a subsequent study, food saturation levels (C_{sat}) were found to be ~6000–7000 *R. salina* cells $\mathrm{ml^{-1}}$, equivalent to 7.5–8.8 μg chl *a* $\mathrm{l^{-1}}$, respectively (Pleissner *et al.* 2013a). Chlorophyll *a* concentration is used as a proxy for algal biomass, because it is coloured and is specific to and shared among all primary producers (Huot *et al.* 2007).

Laboratory studies of valve-gape responses of mussels (*M. edulis*) to presence or absence of algal cells have shown that valve opening and closing responses are strongly influenced by the preceding feeding conditions (Riisgård *et al.* 2006; Figure 4.5). Field studies have demonstrated that phytoplankton concentrations higher than ~1.0 μg chl *a* $\mathrm{l^{-1}}$ stimulate mussels to keep their valves wide open, while lower near-bed concentrations cause valve closure, or reduced valve gape, which represents a physiological adaptation (see later) to presumably frequent situations with scanty, suspended phytoplankton in the near-bottom water (Riisgård 2001b; Riisgård *et al.* 2003).

As particle concentration increases, the rate of pseudofaeces (material cleared from suspension but rejected before ingestion) production rises. Subtracting the rate of pseudofaeces production from filtration rate gives the true ingestion rate (Velasco & Navarro 2002). Video-microscope recordings of pseudofaeces production in *M. edulis* showed that the trigger threshold concentration for formation of pseudofaeces was about 12 000 cells $\mathrm{ml^{-1}}$ (Riisgård *et al.* 2011a). The production of pseudofaeces increases with the inorganic

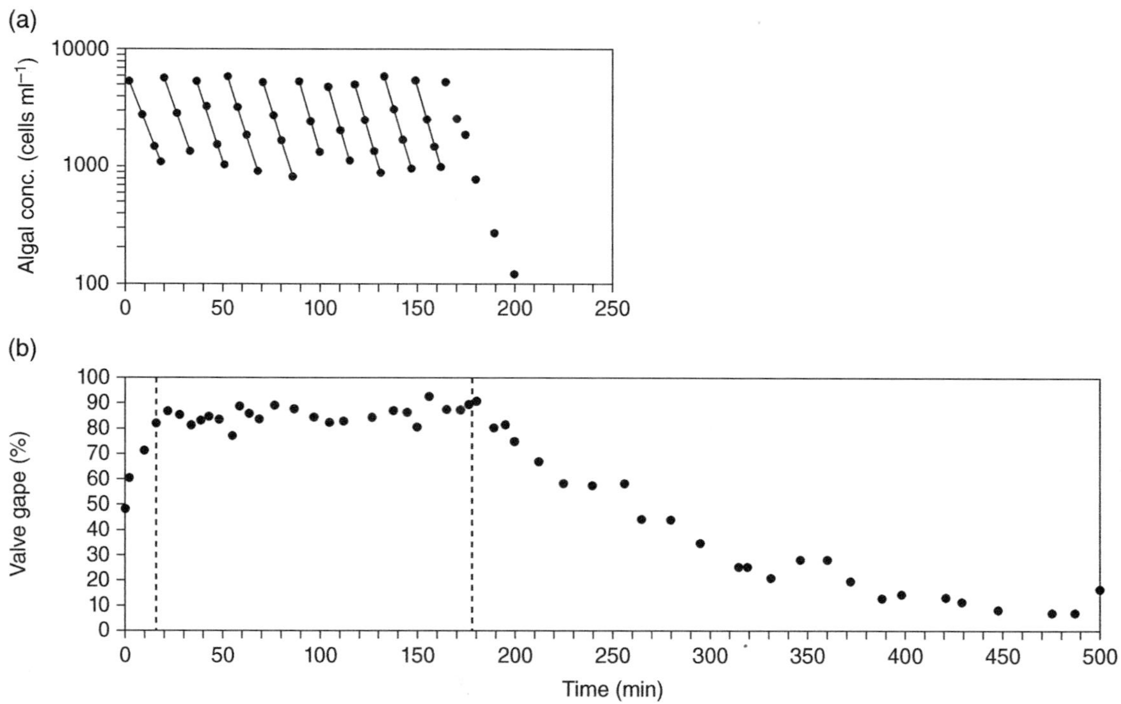

Figure 4.5 (a) Exponential decrease in algal cell concentration after repeated additions of algal suspension to a well-mixed aquarium with actively grazing *Mytilus edulis*. Slope of regression lines represent filtration rates. (b) Mean valve gape of the mussels. Opening response (0–16 min), constant filtration rate and maximal valve gape (16–178 min), closing response (178–315 min). *Source:* From Riisgård *et al.* (2006). Reproduced with permission from the National Shellfisheries Association-BioONE.

particulate content of the water, or when the ingestive capacity of the bivalve is exceeded by the food particles captured.

Many studies have also examined filtration rate in bivalves in relation to quality of the seston (i.e. plankton or other organic matter; Velasco & Navarro 2002 and references therein). Mussels (*M. edulis*) were fed four concentrations of seston (0.99–10.3 mg total seston l^{-1}), comprising three levels of organic content (71.9, 63.6 and 40.8%), made up from natural silt and two species of cultured phytoplankton (Bayne *et al.* 1989, 1993). Two of the seston concentrations were below, and two above, the threshold at which pseudofaeces were produced. Mussels fed at a high concentration of seston of low organic content increased their filtration rate, rejected a higher proportion of filtered material as pseudofaeces (see earlier) and increased the efficiency with which they selected filtered matter of higher organic content for ingestion (see later). This resulted in a constancy of the relationship between ingestion rate and the concentration of particulate organic material (POM), regardless of differences in seston organic content. When mussels (*Geukensia demissa*) were moved from high-plankton, low-inorganic waters to high-inorganic, low-plankton waters, they adapted to the new, poorer-quality seston within six days of submersion, with a doubling of their filtration rate (Galimany *et al.* 2015).

Temperature

The effect of temperature on bivalve filtration rates has been well documented (Kittner & Riisgård 2005 and references therein). Filtration rates were investigated in two species, *M. edulis* and *Crassostrea virginica*, that were collected in the Gulf of St Lawrence, eastern Canada, the northernmost distribution of the oyster species (Comeau *et al.* 2008). Initially, the bivalves were acclimated at 9 °C for a 63-day period, after which time six individuals of each species were randomly distributed among six experimental tanks with a closed system, and initially inoculated with 10 000 algal cell ml^{-1}. They were maintained in cold water (0, 4 or 9 °C) over a 63-day period. For *M. edulis*, the tank mean CR fell significantly at 0 °C, whereas for *C. virginica*, it was similar at all investigated temperatures (Figure 4.6). A species effect was present (i.e. tank mean CR was significantly lower for *C. virginica* than for *M. edulis*), except at 0 °C, where the two means were similar. This was mediated by temperature (species × temperature, $P = 0.016$). For tank max CR, species and temperature main effects were highly significant ($P < 0.01$) without interaction (species × temperature, $P = 0.146$). Tank max CR was substantially lower, by 66%, on average for *C. virginica* than for *M. edulis*, and fell significantly at 0 °C (compared to 9 °C) for both species. Thus, *C. virginica* cleared fewer particles than did *M. edulis* within the range of investigated temperatures, and both species exhibited reduced feeding at 0 °C. Unlike *C. virginica*, *M. edulis* became progressively more tolerant to low temperatures over the 63-day period (data not shown), which suggests that *C. virginica* is disadvantaged in terms of grazing on seasonal phytoplankton blooms, including toxic (domoic acid) blooms (see Chapter 12), which have historically occurred at low temperatures (<9 °C) in eastern Canadian waters.

In a much earlier study, Widdows & Bayne (1971) found that, at between 5 and 20 °C, the immediate response to a reduction in temperature was a decrease in FR in mussels (*M. edulis*) that were acclimated to 15 °C, while an increase in temperature resulted in a rise in FR. However, above 20 °C, mussels acclimated to 15 °C responded by reducing FR. Between 5 and 20 °C, there was complete acclimation of FR over a period of two

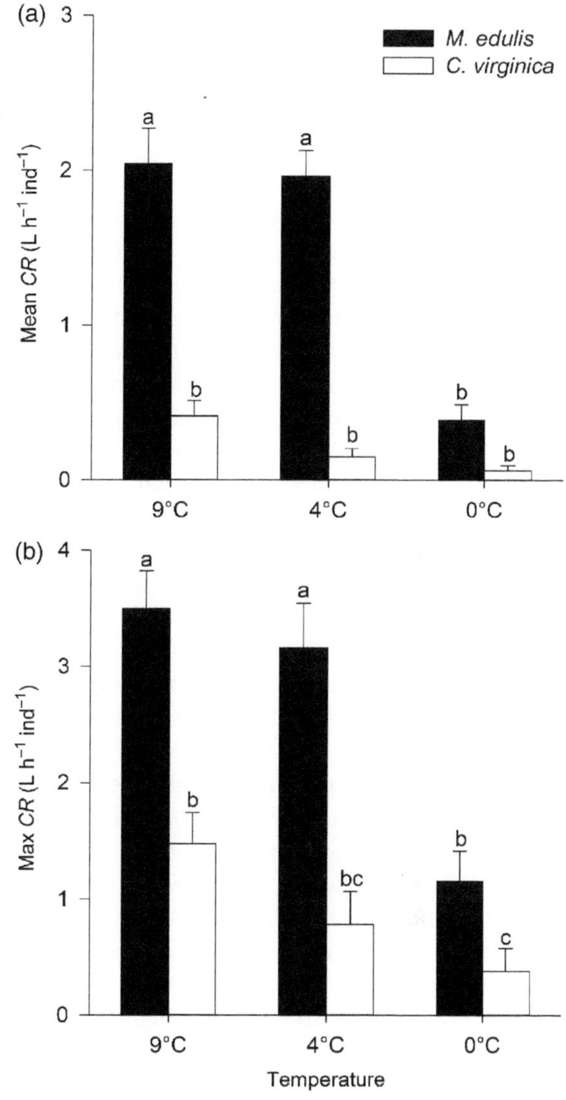

Figure 4.6 (a) Mean and (b) max clearance rate (CR) as a function of species (*Mytilus edulis*, *Crassostrea virginica*) and temperature (0, 4 and 9 °C). Error bars represent one standard error of the mean. Different letters indicate significant differences ($P < 0.05$). *Source:* From Comeau *et al.* (2008). © 2008 Fisheries and Oceans, Canada.

weeks, but only partial acclimation above 25 °C (Bayne *et al.* 1976 and references therein). Widdows (1976) suggested that *M. edulis* has compensatory mechanisms that allow mussels to acclimate FR to both cyclic and constant temperatures. In addition, Widdows (1978) proposed that the feeding activity of *M. edulis* is maintained at a relatively constant level throughout the year as a result of the acclimation of filtration rate to temperature. On the basis of the results of these studies, Kittner & Riisgård (2005) examined the effect of temperature on FR in *M. edulis*, placing special emphasis on possible temperature compensation. But they found no evidence of temperature acclimation in this species, a result that is in agreement with similar studies on *M. edulis* (Jorgensen *et al.* 1990 and references therein) and other bivalves, including the clams *Hiatella arctica* (Petersen *et al.* 2003) and *Mya arenaria* (Riisgård & Seerup 2003).

Salinity

Mussels are able to tolerate a wide range of salinities, from 4–5 psu to fully marine conditions (details in Chapter 7). In an experimental study, Pleissner *et al.* (2013) investigated the acute effect of changing salinity on filtration rate of mussels (*M. edulis*) collected from the Kerteminde Fjord, Denmark. After an acclimation period of one week, mussels were transferred to an experimental aquarium in which four experiments (I, II, III and IV) using three groups (1, 2 and 3) of mussels were carried out (Table 4.2). All experiments were performed at 12.0 ± 0.3 °C and 21.8 ± 1.2 psu. FR was unaffected by decreasing salinity down to about 10 psu, after which mussels began to close their valves, followed by an immediate cessation of filtration activity. However, when salinity was subsequently increased, normal FR was soon restored. The changing rate of FR (ΔF) was strongly influenced by the salinity changing rate (ΔS); that is, the faster the salinity was either reduced or increased, the more pronounced the change in FR (Table 4.2; Figure 4.7). Another example examined the effect of different salinities (10, 15, 25 and 35 psu) on clearance rates in the green mussel *Perna viridis*, a recent invader to Florida, United Staes, and the eastern oyster *Crassostrea virginica* (McFarland *et al.* 2013). At 25 and 35 psu, clearance rates in mussels were approximately double those of oysters, but unlike in oysters, mussel clearance rates decreased by an order of magnitude at 10 and 15 psu, and mussels increasingly closed their valves over time, so that within 120 hours of exposure all were either closed or dead (Figure 4.8). The chief concern about *P. viridis* is that it might compete with native bivalves for food and space. The results suggest, however, that this recent invader could be salinity-limited, providing *C. virginica* with a refuge from competition in estuaries that experience acute periods of low salinity.

Riisgård *et al.* (2013b) investigated the effect of salinity on the filtration rate of *M. edulis* from the brackish (mean salinity 17 psu) Great Belt, Denmark and of dwarfed (<25 mm shell length) *M. edulis* from the low-saline (mean salinity 6.5 psu) Central Baltic Sea, Askö, Sweden. They measured the effect of long-term (weeks) constant ambient salinities between 5 and 30 psu on the FR of *M. edulis* (30–33 mm shell length) from the Great Belt. At salinities 10–30 psu, the FRs did not vary a lot, but at 5 psu FRs were significantly lowered. In dwarfed *M. edulis*, the maximum filtration rates (F, ml min^{-1} ind.$^{-1}$) as a function of shell length (L, mm) and dry weight of soft parts (W, mg) were $F = 0.003L^{2.71}$ and

Table 4.2 Variation in filtration rates (ΔF) of mussels at different salinity changing rates (ΔS). Experiments were performed at a mean algal concentration (3770 ± 690 *Rhodomonas salina* cells ml^{-1}) with three groups of mussels (*Mytilus edulis*). The number of mussels used in each experiment was 20 and the mean shell length was 29.4 ± 0.4 mm. *Source:* From Pleissner *et al.* (2013). CC BY 3.0.

Experiment	Group	$-\Delta S$ (psu h^{-1})	$+\Delta S$ (psu h^{-1})	$-\Delta F$ (ml min^{-1} h^{-1})	$+\Delta F$ (ml min^{-1} h^{-1})
I	2	− 0.69	—	− 1.41	—
II	1	− 0.69	0.96	−2.99	4.17
III	1	− 2.20	1.82	−2.09	3.17
IV	3	− 0.18	0.55	−0.35	2.84

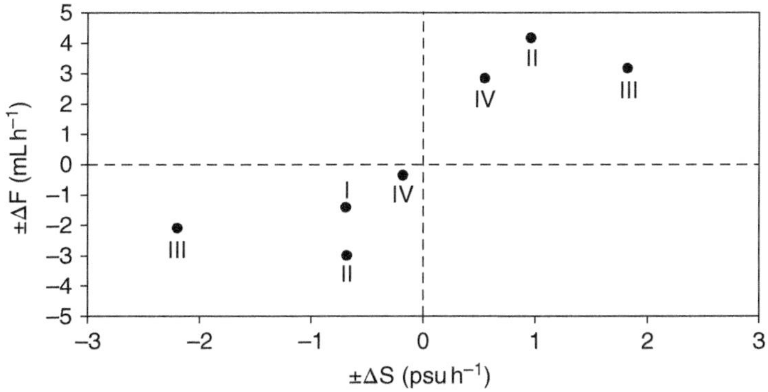

Figure 4.7 Filtration changing rates (Δ*F*) of mussels (*Mytilus edulis*) as a function of salinity changing rates (Δ*S*) in all experiments based on data in Table 4.2. *Source:* From Pleissner *et al.* (2013). CC BY 3.0.

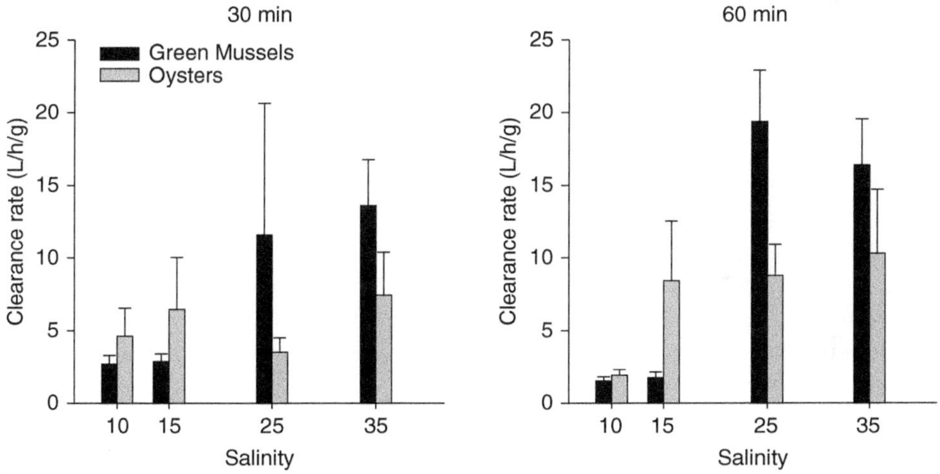

Figure 4.8 Mean clearance rates of five pooled replicates for green mussels (*Perna viridis*) and oysters (*Crassostrea virginica*) at 30 and 60 mins. Bars represent standard errors. *Source:* From McFarland *et al.* (2013). CC BY 2.0.

$F = 0.478W^{0.92}$, respectively, which indicates that FRs of these mussels are comparable to those of Great Belt mussels of similar size exposed to salinities >10 psu. When dwarfed mussels, acclimatised to 20 psu in the laboratory, were exposed to 6.5 psu, this caused a substantial reduction in FR, but after about two days the previous high filtration rate was regained at 6.5 psu, and, furthermore, a similar pattern was observed when the 6.5 psu-exposed mussels were subsequently re-exposed to 20 psu. The ease with which Baltic Sea mussels can adjust their FR back and forth between 6.5 and 20 psu contrasts remarkably with the inability of Great Belt mussels to do. However, this difference may be genetically determined as mussels have been identified, using a range of molecular markers, as primarily *M. edulis* in the Great Belt region, while those from the Central Baltic Sea are *M. trossulus* (Zbawicka *et al.* 2014; Stuckas *et al.* 2017). The genetic structure of *Mytilus* populations will be discussed in Chapter 9.

Current Velocity

Marine bivalves orientate to maximise their filtration capacities and exploit their food supply. Velocity can directly affect the filtration rates of mussels facing into unidirectional flows. As current speed increases from 10 to 30 cm s^{-1}, mussels decrease their filtration rates (Newell *et al.* 2001 and references therein). At low velocities (0–5 cm s^{-1}), filtration rates are flow-limited indirectly due to the effects of reductions in ambient particle concentrations from local refiltration (Jonsson *et al.* 2005). An increase in current speed compensates for the effects of particle depletion by replacing the seston-depleted water with a new source of seston. However, at higher velocities, filtration rates of bivalves with siphons facing into the flow are directly inhibited by hydrodynamic effects on the functioning of the bivalve pump due to pressure differentials between inhalant and exhalant openings (Newell *et al.* 2001).

A number of studies on the clearance rate of different *Mytilus* species as a function of current velocity have given different results (reviewed in Nielsen & Vismann 2014). For example, the CR of *M. trossulus* (Ackerman & Nishizaki 2004) and *M. galloprovincialis* (Denis *et al.* 1999) correlates positively with current velocity in the range of 0.01–0.18 m s^{-1} and 0.05–0.25 m s^{-1}, respectively. Other studies show either no effect on CR at current velocities in the range between the lower and upper tolerance limit (Widdows *et al.* 2002) or negative correlation to all experimentally applied current velocities (Wildish & Miyares 1990; Newell *et al.* 2001). The number of mussels used seems to be important for the outcome of the study. With this in mind, Nielsen & Vismann (2014) used either three or twenty mussels (*M. edulis*) to clarify the effect of water current velocities on CR, their hypothesis being that if mussels are allowed to form clumps, they can withstand greater current velocities than individuals. They examined the effect of increasing current velocity (0.05–1.4 m sec^{-1}) on CR. The CR of 20 mussels was unaffected by current velocities up to 1.4 m sec^{-1}, whereas experiments with three mussels showed that CR decreased progressively at current velocities greater than 0.2 m sec^{-1}, and reached 0 litres h^{-1} $indiv^{-1}$ at velocities >0.6 m sec^{-1} (Figure 4.9). The constant feeding at all current velocities for 20 mussels was most likely a result of a combination of reduced current velocity within the aggregated mussels, a current velocity-dependent change of siphon orientation and attachment of mussels at current velocities less than 1.2 m sec^{-1}, whereas with three mussels, they were detached at ≥0.4 m sec^{-1} and their valves were generally closed. These results emphasise the importance of discriminating between the individual level and the level of mussel aggregations in the study of current velocity effects on clearance, especially when the study is made on a bivalve species that form dense beds in nature. Average CRs (1.6 ± 0.3–3.2 ± 0.2 l h^{-1} $indiv^{-1}$) for *M. edulis* were in agreement with CRs measured in other studies using similar sized *M. edulis* (~30 mm shell length) and a similar food species and concentration (Riisgård 1991; Clausen & Riisgård 1996; Pleissner *et al.* 2013), an indication that mussels in general were filtering under optimal conditions when they were active.

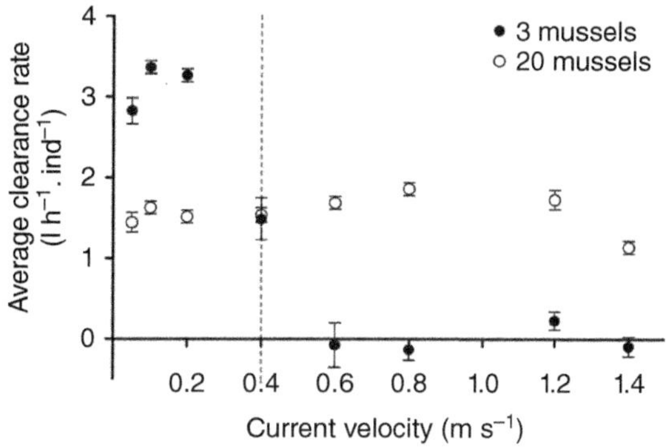

Figure 4.9 Average clearance rate ± SE (l h^{-1} ind^{-1}) for *Mytilus edulis* at eight different current velocities (m s^{-1}) measured with either three or twenty mussels present in the tank and 3000 *Rhodomonas salina* cells ml^{-1}. The vertical dotted line indicates the shift between experiments measured at low and high current velocities. *Source:* From Nielsen & Vismann (2014).

Mussel FR is not only influenced by current velocity but also by currents created by the mussels themselves. There is little known about the detailed fluid mechanics of flow near a mussel that is generated by the flow through the exhalant siphon appearing as a well-structured jet. This flow in conjunction with a possible imposed external flow from currents or other mussels will influence the grazing impact and thus the concentration of food particles reaching the inhalant aperture, thereby affecting feeding conditions for mussels, as in the situation of refiltration, whereby part of the once filtered and exhaled water reaches the inhalant aperture of either the same or other mussels (Riisgård *et al.* 2011b and references therein; Uslu & Pekkan 2016). See later for more information on inhalant and exhalant flow in mussels.

Seawater oxygen concentration is another factor that affects FR in mussels. Tang & Riisgård (2018) found that FR of *M. edulis* was independent of variations in oxygen concentrations above 2.0 mg O_2 /l. Below this concentration, FR, as well as respiration rate (RR), decreased rapidly and simultaneously, and mussels responded by gradually closing their valves. When mussels that had been exposed to 0 mg O_2 /l were subsequently exposed to oxygen concentration above 2 mg O_2 /l, they reopened their valves, and normal FR and RR were resumed. These observations support the hypothesis that *M. edulis* is able to save energy during starvation periods by reducing its shell-opening degree, whereby the FR and thus RR become strongly reduced (Riisgård & Larsen 2015).

Tidal height effects may influence FR of mussels. For example, the ribbed mussel, *Geukensia demissa* is mainly found in the intertidal zone of salt marshes on the Atlantic coast of the United States. It has been proposed that the cultivation and harvest of ribbed mussels could be used to bioremediate the eutrophication of coastal waters. To accomplish this, *G. demissa* would be grown in suspension culture underwater, a habitat quite different from its natural habitat, in which it is exposed to a tidal regime of submersion and emersion. Galimany *et al.* (2013) quantified the filter-feeding activities of *G. demissa* collected from either an intertidal location or a group of mussels from the same population kept submerged in pearl nets for two months to adapt to constant submersion. Mussel filtration was quantified under laboratory conditions with ultra-filtered seawater with added cultured phytoplankton. Mussels taken from the intertidal population had significantly higher filtration rates than the submerged population initially, but after three days of submersion this difference disappeared. Moreover, all experimental *G. demissa* had higher filtration rates during natural low tide than during natural high tide. These results indicate a potential for *G. demissa* to be grown in suspension culture for nutrient bioextraction purposes.

Factors responsible for significant reductions in FR include toxic algae (May *et al.* 2010, *Perna viridis*; Navarro & Contreras 2010, *Mytilus chilensis*; Galimany *et al.* 2014, *Geukensia demissa*), hydrocarbons (Culbertson *et al.* 2008, *Geukensia demissa*), heavy metals (Gregory *et al.* 2002; *Perna perna*), chlorine (Rajagopal *et al.* 2003, *P. perna*), synthetic surfactants (Ostroumov & Widdows 2006, *Mytilus* spp.) and loess, a flocculent clay that is sprayed in the vicinity of aquaculture sites to mitigate the effect of harmful algal blooms (Shumway *et al.* 2003, *Mytilus* spp.).

Control of Filtration Rate

It is still a matter of considerable discussion as to whether filtration rate in suspension-feeding bivalves is physiologically regulated (Bayne 1998; Hawkins *et al.* 1998b, 1999) or if it should be regarded as a basically autonomous process (Jørgensen *et al.* 1988; Riisgård 2001b; Jørgensen 1990, 1996; Maire *et al.* 2007; MacDonald & Ward 2009). The most widely held belief, mainly based on evidence from laboratory studies on *M. edulis* (reviewed in Riisgård *et al.* 2011a), is that the process *is* under physiological control, and that mussels regulate pumping rate according to both the availability and the composition of suspended seston. The alternative less popular view, first mooted by Jørgensen in 1966, is that accumulation of food particles on the gills is not regulated according to the nutritional needs of the individual, but is an automatic process basically determined by the capacity of the pump, the concentration of food particles in the water and the efficiency with which these particles are retained on the gills. Riisgård *et al.* (2003, 2011a, 2015) have argued that these different points of view are mainly due to inconsistencies in methodological measurements of filtration, and that most of the reported experiments have been conducted in laboratories using high algal concentrations that are much higher than those encountered by mussels in the field. Also, Riisgård & Larsen (2007) maintain that the effect of temperature on filtration rate is purely mechanical, controlled by viscosity, a parameter inversely related to water temperature. The controversy regarding physiological versus autonomous regulation of filtration rate still continues to the present day.

Energetic Costs of Filtration

In order to meet their food requirements, suspension-feeding bivalves must filter large volumes of water on a daily basis. Because of life in a nutritionally dilute environment, they possess low-energy (i.e. low-pressure) pumps that continuously pump the surrounding water through gills that are efficient in retaining small food particles (Riisgård & Larsen 1995). There are few published estimates of the metabolic cost of pumping. Several studies have inferred high-energy costs in mussel species. For example, in *M. edulis* and *M. californianus*, Bayne & Newell (1983) estimated that 24% of an ingested ration of algae cells represented the cost of feeding, of which digestion and assimilation accounted for only 4–6%. These very high costs of water processing were regarded as 'sound energetic grounds for the very variable rates of filtration that occur in these and other bivalves, the majority of the energetic costs of activity being eliminated by a reduction in filtration when available ration is insufficient to meet the costs of feeding'. The results from numerous studies on mussels and other bivalves infer high energy costs of water processing (references in Jørgensen 1990). However, low values (1–3% of total metabolic expenditure) have also been reported for bivalve species, including *Mytilus* (Bernard & Noakes 1990; Jørgensen 1990; Hawkins & Bayne 1992; Riisgård & Larsen 1995). The energy costs of filter feeding may be evaluated using the ratio of useful pumping power (*P*) and total metabolic power expenditure *Rt* (Riisgård & Larsen 2000). This ratio, referred to as the overall pump efficiency, *P*/*Rt*, was estimated for various suspension feeders, including bivalves. Useful pump work constitutes from 0.3 to 4% of the total metabolic energy expenditure, with a value of just over 1% for *M. edulis.* Other measures of efficiency other than overall pump efficiency may be appropriate when estimating the energy costs of filter feeding. If the costs of powering bands of lateral cilia and their associated cells (R_p) are considered instead, then a minimum efficiency would be $R_p/Rt = 6.6\%$, rather than the 1% calculated for *M. edulis.* Furthermore, if the ratio gill to metabolic rate $R_g/Rt = 19\%$ is used, it is clear that the energetic costs of filter feeding by means of large specialised gills are considerable, since such structures are expensive to maintain irrespective of whether the mussel is pumping water or not (Riisgård *et al.* 2015).

Particle Processing on the Gills and Labial Palps and in the Stomach

Particle processing mechanisms in suspension-feeding bivalves can be divided into the following components: encounter, capture, transport, selection and ingestion. The development of a number of research tools or approaches has deepened our understanding of the overall process (Silverman *et al.* 2000). Several complementary techniques have made this possible. Video endoscopy (Figure 4.10) uses an optical insertion tube inserted into the mantle that permits direct in vivo observations of whole intact structures (e.g. gills and labial palps) in relatively undisturbed animals (Beninger *et al.* 1992; Ward *et al.* 1993, 1996; Richoux & Thompson 2001). Combined with fluid sampling and subsequent chemical and histochemical analyses, histology and mucocyte (mucus-secreting cell) mapping, endoscopy has been responsible for a major leap forward in our understanding of mechanisms of particle transport, particle selection and ingestion, albeit the technique does have several limitations (see Beninger 2000; Silverman *et al.* 2000). Scanning electron micrographs (SEMs) have also played an important role in surveying mantle rejection pathways for pseudofaeces transport in *M. edulis* and other bivalves (Beninger *et al.* 1999). Another technique, confocal laser scanning microscopy (CLSM), has the ability to provide high-resolution images from selected depths. Images are taken point-by-point and reconstructed with a computer to produce 3D reconstructions of complex objects. CLSM, in conjunction with high-speed video recording, has been used to simultaneously visualise particles and the spatial relationships, structures and movements of cilia and cirri on living gill filaments in *M. edulis* (Silverman *et al.* 1999). CLSM has also been combined with tissue reflectance and autofluorescence to visualise mucus position and mucociliary transport in excised

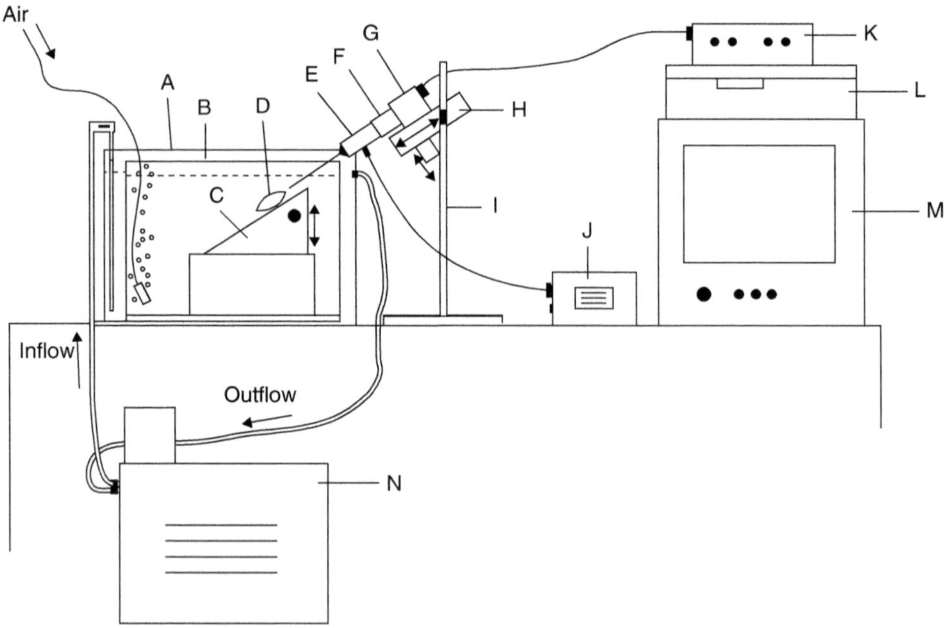

Figure 4.10 Video endoscopy set-up used in particle concentration and temperature experiments with *Mytilus edulis*. (A) Outer tank containing coolant water and connected with temperature control unit. (B) Inner experimental chamber containing seawater and algal suspension. (C) Adjustable plastic stand. (D) Mussel. (E) Endoscope. (F) Zoom adaptor. (G) Camera head. (H) Micromanipulator. (I) Support stand. (J) Light source. (K) Camera controls. (L) Hi 8 video recorder. (M) Monitor. (N) Temperature control unit. *Source:* From Richoux & Thompson (2001).

living gill tissue from *M. edulis* (Beninger *et al.* 1997). Finally, water flow inside the bivalve shell can be measured by phase-contrast magnetic resonance imaging (MRI). Seo *et al.* (2014) used the technique to map the flow in the mantle cavity of a live *M. galloprovincialis*, measuring a velocity of more than 5 cm s^{-1} entering the exhalant siphon.

In *Mytilus*, the adult gill/palp feeding mechanism develops once the larval feeding mechanism is lost, and becomes functional within 48 hours of metamorphosis (see Chapter 5). Studying the development of gills (see Chapter 1) has led to new insights into the evolution of these structures and the phylogenetic relationships among bivalve taxa (Cannuel & Beninger 2006; Beninger & Decottignies 2008; Giribet 2008; Cannuel *et al.* 2009). As seen in Chapter 2, the simplest gill structure, found only in mussels, is the homorhabdic filibranch type, in which individual gill filaments are free or loosely attached to one another through interlocking clumps of cilia (see Figure 2.8B). Other gill types that possess a somewhat more complex structure are: plicate heterorhabdic filibranch (scallops), eulamellibranch (clams) and plicate heterorhabdic pseudolamellibranch (oysters), as described in Chapter 2 and in Gosling (2015, ch. 4). In the following sections, we look at particle processing on the gills, labial palps and stomach as they apply to marine mussel species.

As introduced in Chapter 2, cilia on the gill filaments have specific arrangements and functions (see Figure 2.8Bii). Lateral cilia (lc), set along the sides of the filaments, produce the water current between adjacent filaments (Figure 4.11). This water passes ventrally

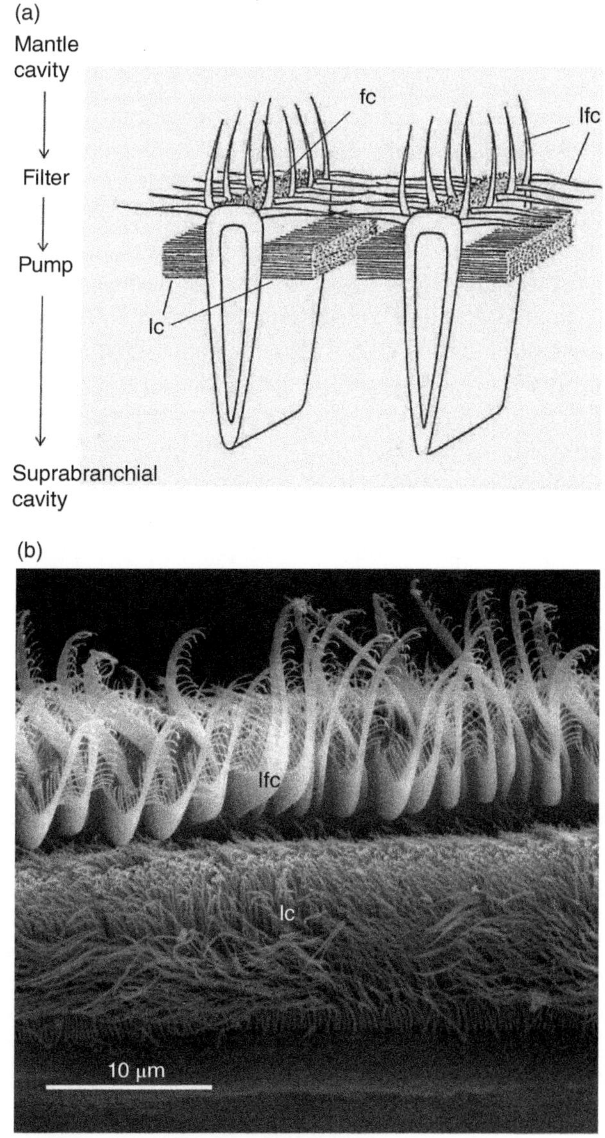

Figure 4.11 (a) Cross-section of two gill filaments of *Mytilus edulis.* lfc, laterofrontal cirri (alternating cirri beat synchronously and differ half a phase with the two adjacent ones; effective stroke is upward); lc, lateral cilia (effective stroke downward); fc, frontal cilia. *Source:* Dral, A.D.G. (1967) The movement of the latero-frontal cilia and mechanism of particle retention in the mussel (*Mytilus edulis* L.). *Netherlands Journal of Sea Research*, **3**, 391–422. Reproduced with permission from Elsevier. (b) Scanning electron micrograph (SEM) of an *M. edulis* single gill filament with a band of water-pumping lateral cilia (lc) and a row of particle-capturing laterofrontal cirri (lfc). The lateral cilia are inactive because the nerve has been cut. *Source:* From Riisgård *et al.* (2015).

into the inhalant part of the mantle cavity and then through the gill filaments to the exhalant chamber above them to leave the mussel as an exhalant jet with an essentially constant velocity of 8 cm s^{-1}, in mussels of shell length 16–82.6 mm (Riisgård *et al.* 2011b). The beat of the lateral cilia is downward, but involves a sideways movement (metachronal wave) in the recovery stroke (Aiello & Sleigh 1972; details in Riisgård *et al.* 2015). Lying between the lateral and frontal cilia are the large feather-like latero-frontal cilia (lfc), which are unique to bivalves. When the incoming current hits the gill surface of *M. edulis*, the actively beating lfc act as a filter mechanism called 'cirri trapping' (Riisgård & Larsen 2010) that ensures that suspended food particles in the water are captured and transferred to the frontal surface of the gill filaments. When the lfc are inactivated by the neurotransmitter serotonin (5-hydroxytryptamine; 10^{-5}–10^{-4}M concentrations), capture efficiency (CE) in *M. edulis* drops significantly but does not fall to zero, suggesting that other mechanisms in addition to cirral trapping are involved in particle capture (Ward *et al.* 1998a). The frontal cilia (fc), which are abundantly distributed on the free outer surface of the gill facing the incoming current, convey particles aggregated in mucous – secreted by the filaments – downward toward the ciliated food grooves on the ventral side of each lamella. There, the material is incorporated into mucus strings, which are transported inside the grooves toward the labial palps (details later). In the ventral gill particle grooves, material is never in suspension, even at low particle concentration (Ward *et al.* 1993). This mechanism of particle processing is seen in mussels but also in eulamellibranchs (clams); that is, in the vast majority of bivalve species. Between the lfc and fc is a row of pro-latero-frontal cilia (p-lfc), the function of which remains speculative. Riisgård *et al.* (2015) suggest that the cilia may transfer arrested particles from the lfc to the frontal surface of the gill filament, and likewise that the bending of the lfc in the recovery stroke may release captured particles, although this possibility has not yet been supported by observations. Cilia on the abfrontal gill surface, facing the excurrent flow, are too sparse to function in water pumping in either *M. edulis* or *Modiolus modiolus* (Dufour & Beninger 2001). *M. edulis* had the highest density of simple abfrontal cilia (1–3 μm), which are evenly distributed throughout the gill, on the posterior side of the filaments only, while some longer composite cilia are either interspersed between simple cilia or present on the anterior side (Figure 4.12a–c). *M. modiolus* also has both types of cilia, but their densities and positions are more variable (Figure 4.12d–f). Also, on the abfrontal surface, mucocyte densities are high, but there is no particle transport in this area. The secretion of mucous may be adaptive if it reduces friction between water and the gill epithelium in a manner similar to the reduction produced by fish epithelial mucous (discussed in Dufour & Beninger 2001). The gills are not mere sieves but are able to control the rate at which suspended material is removed from the surrounding water, either by altering the angle of ciliary beat or by altering gill dimensions (interfilament width or, in the case of eulamellibranchs, cross-sectional area of water tubes) through contraction of gill muscles (Medler & Silverman 2001).

Selective Particle Capture on the Gills

Particle capture is the first step in suspension feeding and is the result of two associated processes: particle encounter and particle retention (Ward *et al.* 1998a). Encounter efficiency relates to the proportion of particles that come into contact with the capture unit (i.e. gill filaments), whereas retention efficiency is the proportion of encountered particles that are actually retained on the filaments. Many studies have used the term 'retention efficiency' to describe particle CE. But unless in vivo techniques are used to differentiate the number of particles that are encountered from those that are actually retained, retention efficiency cannot be determined. Hence, Rosa *et al.* (2015) have recommended that future studies using in situ techniques should use the term 'capture efficiency' (CE) in place of 'retention efficiency'.

Bivalves in coastal ecosystems are exposed to large amounts of suspended material that includes both nutritious and non-nutritious particles. In order to efficiently process the

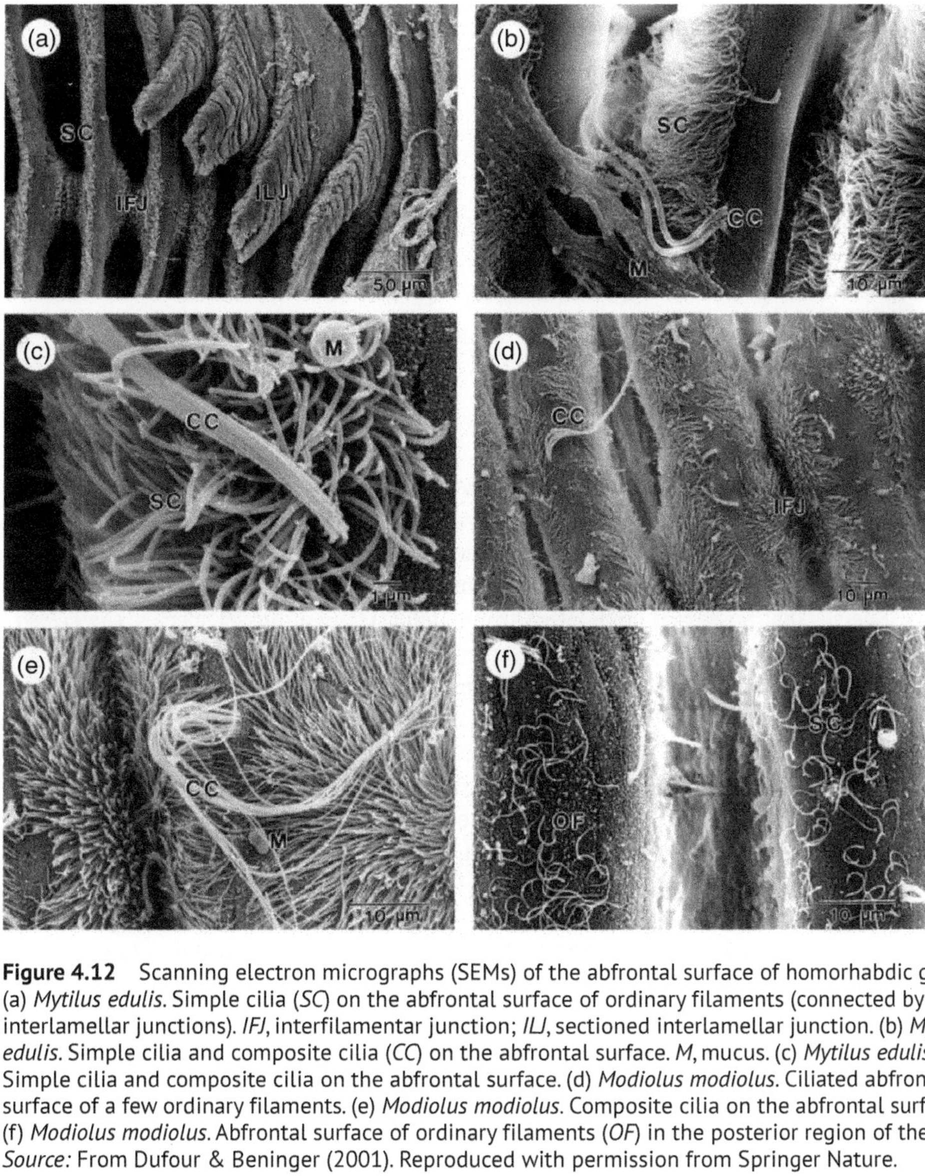

Figure 4.12 Scanning electron micrographs (SEMs) of the abfrontal surface of homorhabdic gills. (a) *Mytilus edulis*. Simple cilia (*SC*) on the abfrontal surface of ordinary filaments (connected by interlamellar junctions). *IFJ*, interfilamentar junction; *ILJ*, sectioned interlamellar junction. (b) *Mytilus edulis*. Simple cilia and composite cilia (*CC*) on the abfrontal surface. *M*, mucus. (c) *Mytilus edulis*. Simple cilia and composite cilia on the abfrontal surface. (d) *Modiolus modiolus*. Ciliated abfrontal surface of a few ordinary filaments. (e) *Modiolus modiolus*. Composite cilia on the abfrontal surface. (f) *Modiolus modiolus*. Abfrontal surface of ordinary filaments (*OF*) in the posterior region of the gill. *Source:* From Dufour & Beninger (2001). Reproduced with permission from Springer Nature.

complex mixture of material that they encounter, bivalves have evolved capabilities for selective feeding that allow them to reject some of the captured material in pseudofaeces (Bacon *et al.* 1998). CE in relation to particle size has been determined in a wide variety of bivalves using both direct and indirect methods (see Ward & Shumway 2004 and references therein). In general, CE has been shown to be strongly dependent on particle size, increasing asymptotically to a maximum of about 100% for particles ca. ≥4 µm in diameter (Figure 4.13). Møhlenberg and Riisgård (1978) examined the CE of 13 species of bivalves fed mixtures of natural particles and cultured phytoplankton. They found that CE decreased with particle

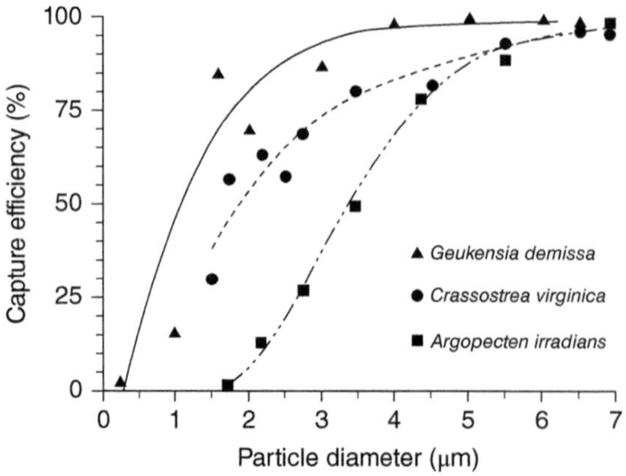

Figure 4.13 Relationship between capture efficiency (CE) and particle diameter in three species of bivalves. *Source:* From Ward & Shumway (2004).

size, although the degree to which efficiency changed and the particle diameter at which inefficient CE began were species-specific. All bivalves captured particles greater than ~6 μm in diameter with almost 100% efficiency, and some species could capture 1 μm particles with an efficiency of 60–70%. Riisgård (1988), in a similar study, examined CE in six species of bivalves. Species that possessed large latero-frontal cilia (i.e. two mussel species, *Geukensia demissa* and *Brachidontes exustus*, and two clam species, *Spisula solidissima* and *Mercenaria mercenaria*) captured all particles >4 μm in diameter with an efficiency of 100%, whereas those with small (or no) latero-frontal cilia (the oyster *Crassostrea virginica* and the scallop *Argopecten irradians*) retained 4 μm particles with an efficiency of about 75–85%, suggesting that in both of these studies, CE is dependent, in part, on the morphology of the latero-frontal cilia of the gill. Mussels can retain motile flagellates as small as 1–2 μm diameter, but can also remove bacteria (0.3–1.0 μm) from suspension, albeit with low efficiency. The mussel *Geukensia demissa*, for example, retains particles as small as 0.4–0.6 μm with 86% of its efficiency for larger particle capture. This ability is related to the narrow space (~0.5 μm) between the latero-frontal cilia (Wright *et al.* 1982). The mussel *Perna perna* also shows high CE for particles of this size, and therefore both of these species seem specifically adapted to exploit the bacterial size-fraction. While bacterial cells are much higher in numerical concentration than phytoplankton cells, the former are not as good a source of carbon as the latter (Newell & Field 1983). The cold seep mussel, *Bathymodiolus childressi*, can efficiently capture and utilise ultraplankton (i.e. any plankton organism <2 μm in size, including bacteria). Pile & Young (1999) found that CE significantly increased with increasing mussel length, with 27 mm mussels capturing about 20% of bacteria and 84 mm mussels capturing about 80%. It appears that the species supplements its symbiotically acquired nitrogen by feeding selectively on nitrogen-rich bacterioplankton. In contrast, Yahel *et al.* (2009) reported size-independent capture of naturally occurring phytoplankton and bacterial species in the coral-boring, tropical mytilid *Lithophaga simplex*. Using flow cytometry to examine the types of particles entering the inhalant siphon and comparing these with particles exiting the exhalent siphon, they found that some species were captured in higher proportions than others, despite the overlap in mean cell size distribution. They concluded that CE is not strictly size-dependent in this species, and probably involves other cell attributes such as cell-surface properties (see Rosa *et al.* 2015) and motility.

Strohmeier *et al.* (2012) have shown that mussels (*M. edulis*) can modulate CE in relation to seasonal shifts in the composition of particles. During May and June, when abundant particle volumes included most sizes between 3 and 35 μm, mussels captured most effectively (80–100%) particles larger than 20–30 μm. In contrast, in late July and August, as the seston particle spectra shifted toward the dominance of smaller particle volumes, there was a decrease in CE maxima to 14–64 and 12–86% for particles between 7 and 15 μm, respectively. These data clearly indicate that *M. edulis* has the capacity to substantially control CE mechanisms, likely in response to exogenous dietary factors. However, in a subsequent study, results for CE of microspheres unequivocally showed that differently sized particles ≥4 μm were captured by *M. edulis* with close to 100% efficiency regardless of season (Rosa *et al.* 2015). The authors suggested that the purported changes in CE reported by Strohmeier *et al.* (2012) may be a result of instrument artifacts or other factors associated with the seston, and not of behavioral or physiological responses of the mussel.

CE can also depend on particle shape (Bayne *et al.* 1977), as well as qualitative factors such as nutrient status (Defossez & Hawkins 1997), particle stickiness or molecular excretions (e.g. living versus dead diatoms; Beninger *et al.* 2008), lectin (mucopolysaccharide) secretion on the gills (see later) and particle surface charge and wettability (Hernroth *et al.* 2000; Rosa *et al.* 2015, 2017a). The CE of *M. edulis* for the bacterium *Salmonella typhimurium* increased when the surface charge of the bacterium (~1 μm) was experimentally lowered (Hernroth *et al.* 2000). CE experiments were used to examine the effects of several physicochemical surface properties on particle–mucus interactions in *M. edulis* and the scallop *Argopecten irradians*, both of which have fillibranch gills, although *M. edulis* has well developed latero-frontal cilia, capable of capturing ≥3.2 μm particles with ~90% efficiency (Vahl 1972), while the scallop has shorter latero-frontal cilia and captures 3.3 μm particles with an efficiency of only 50% (Palmer & Williams 1990). Rosa *et al.* (2017b) tested the null hypothesis that different physicochemical properties of particles have no effect on particle adhesion and capture by these two species of bivalves. Mucous from the gills of the two species was isolated separately, adsorbed on to glass microscope slides and incubated with polystyrene microspheres (10 μm) that were either untreated or covalently bound to different neoglycoproteins (NGP; N-acetyl-glucosamine, D-mannose) or bovine serum albumin (BSA). Untreated microspheres and spheres with covalently bound NGP and BSA (ranging from 2 to 10 μm in diameter) were directly delivered to the incurrent aperture of the two bivalves, and the excurrent water was sampled and the CEs calculated. Results demonstrated that the presence of epiparticulate BSA and NGP can significantly change the surface characteristics of microspheres, but the differences have minimal effect on adhesion to mucus or on CE. A notable exception was a significant reduction in CE of D-mannose-treated 2 and 3 μm-diameter microspheres by mussels compared to the control spheres of the same size. Overall, results suggested that mussels and scallops can efficiently capture particles with a range of physicochemical surface properties, especially those larger than ~4 μm (Figures 4.14 and 4.15). Also, the observed differences in adhesion and CE of the treated microspheres by the two bivalves demonstrated the importance of species-specific interactions. These different responses could be a consequence of differences in the latero-frontal cilia and specific capture mechanisms (Ward *et al.* 1991; Beninger *et al.* 1992). Furthermore, the specific mucous constituents are different in these two species (Beninger *et al.* 1993), which could also affect adhesion.

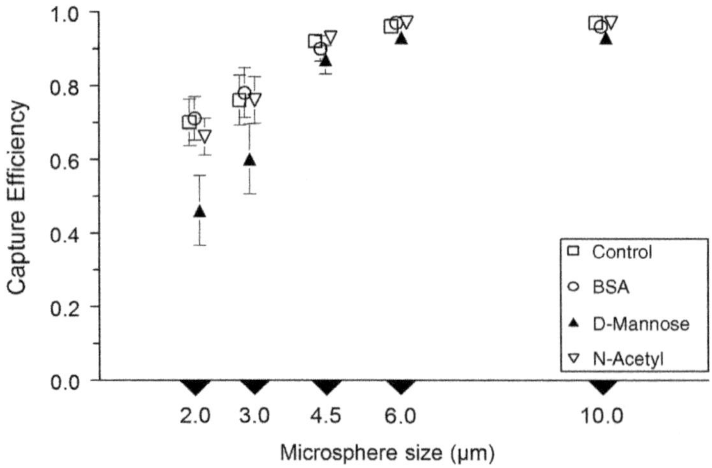

Figure 4.14 Capture efficiencies (CEs) of microspheres delivered to *Mytilus edulis*. Microspheres were either untreated (control) or treated with bovine serum albumin (BSA) or one of two neoglycoproteins (NGPs) (NGP with D-mannose or N-acetyl-glucosamine). There was a significant treatment effect on CE, with the 2 and 3 μm D-Mannose-coupled microspheres being captured at a significantly lower efficiency than the other spheres of the same size ($P < 0.05$). Data presented as means ± SE, N = 8–13. *Source:* From Rosa *et al.* (2017a).

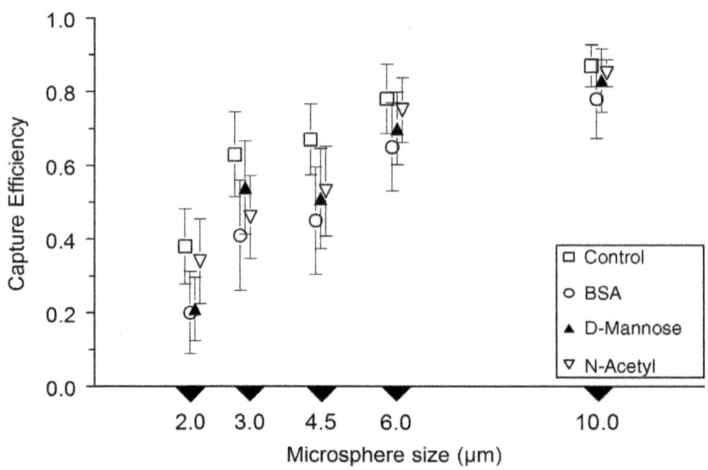

Figure 4.15 Capture efficiencies (CEs) of microspheres delivered to the scallop *Argopecten irradians*. Microspheres were either untreated (control) or treated with bovine serum albumin (BSA) or one of two neoglycoproteins (NGPs) (NGP with D-mannose or N-acetyl-glucosamine). There was no significant effect of microsphere treatment on CE ($P > 0.05$). CE rates for the scallops were significantly lower than those calculated for mussels (Figure 4.14), indicating a significant species effect on particle CE ($P < 0.05$). Data presented as means ± SE, N = 6–10. *Source:* From Rosa *et al.* (2017a).

Pre-ingestive Particle Processing on the Gills and Labial Palps

Two of the major problems faced by suspension-feeding bivalves are the regulation of the volume of material processed and ingested (ingestion volume regulation) and the sorting of edible seston from that which is inorganic, of low nutritional value or even toxic (selection) (Beninger *et al.* 1999). Both processes culminate in rejection, which results in the formation of pseudofaeces. Elimination of pseudofaeces is the final rejection step in the cascade of particle processing events that begins with capture on the gill. The rejected particles are wrapped in mucous and then expelled without having passed through the digestive tract. The production and voidance of pseudofaeces is an important physiological function in suspension-feeding bivalves, and contributes significantly to bivalve biodeposits (see later) in the aquatic ecosystem (Beninger *et al.* 1999 and references therein). Strong selection of particles at the post-capture stage in the feeding process should enhance the quality of material ingested, thereby increasing the efficiency of postingestive particle sorting in the stomach (Ward & Shumway 2004).

Role of Mucus

Mucus plays a key role in all aspects of particle processing, and different types are used for different functions on gills, labial palps, mouth and mantle (Beninger & St-Jean 1997a). In homorhabdic fillibranchs, mucocytes are present on both the frontal and abfrontal surfaces of gill filaments, but not in the ventral groove of *M. edulis* (Beninger *et al.* 1993; Beninger & St-Jean 1997a; Dufour & Beninger 2001). Generally, high-viscosity mucus, secreted by acid mucopolysaccharide (AMPS) mucocytes, is used for particle transport on surfaces exposed directly to inhalant flow (e.g. in the ventral groove, on the abfrontal surfaces of filaments) or leading to areas exposed directly to flow, on rejection pathways and in initial transport on the gill for a subsequent 'decision' – pertinent only to mussels and clams that have a single ventral particle groove. Transport in these cases is counter to the current flow, and so highly viscous mucus greatly facilitates particle transport. 'Counter' is defined as an angle approximately 0–90° with the current flow. Lower-viscosity mucus, secreted by mixed mucopolysaccharide (MMPS) mucocytes, is used when particle transport is on an enclosed or semi-enclosed surface, such as the ridged areas of the labial palps (see later); that is, for particles destined for ingestion and moving with the current flow. Neutral MPS mucocytes are also found in areas where low-viscosity mucous is important, such as the ridged areas of the labial palps (Beninger & St-Jean 1997a,b). Although mucus is an important element in suspension feeding, there are no data on the energetic cost of its production. However, the costs are unlikely to be high since mucus produced in feeding is either reabsorbed or ingested, and the animal is generally in adequate energy balance to support the costs of mucus produced for rejection (Beninger *et al.* 1993).

Labial Palps and Pseudofaeces Production and Transport

The principal structures involved in the production of pseudofaeces are the labial palps and the food grooves on the gill. The labial palps are paired, fleshy leaf-shaped structures on either side of the mouth. The inner surface of each palp faces the gill and is folded into numerous ridges that carry a complicated series of ciliary tracts and various types of mucocytes (Beninger *et al.* 1995). The outer surfaces are smooth, and between the inner and outer surfaces there is muscular connective tissue. The palps may reach posteriorly for some one-third to three-quarters of the length of the gills, depending on the species (Garrido *et al.* 2012). Palp length increases with shell length in both *M. galloprovincialis* and *P. perna*, although the former has significantly ($P < 0.001$) longer palps over the entire size range (25–67 mm) tested (Zardi *et al.* 2006). Individual bivalve species, including mussels, can alter the size of the labial palps (and gills), which may be an adaptation to seston quality and quantity (Sunde 2013).

Bivalve gill type dictates the nature of particle processing on the labial palps (see Gosling 2015). On the mussel homorhabdic filibranch gill, all captured particles move ventrally into the ciliated ventral particle groove, even when particle concentrations are

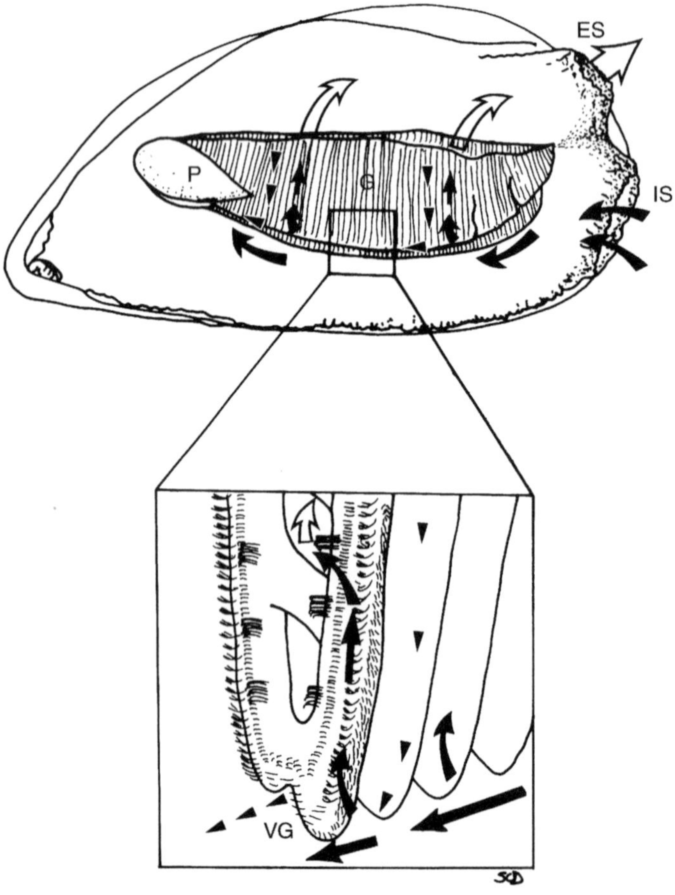

Figure 4.16 Principal pathway of current flow and particle transport on the homorhabdic filibranch gill of *Mytilus edulis*. Water enters the animal through the inhalant siphon (IS). The frontal surface of the gill (G) is exposed to an antero-posterior flow at the ventral margin and to a ventro-dorsal flow on the rest of the frontal surface (solid arrows). Water exits the gill from the abfrontal region, out through the exhalant siphon (ES; open arrows). Arrowheads represent particle transport to the ventral particle groove (VG); P is the labial palps. *Source:* From Beninger & St-Jean (1997a). Reproduced with permission from Springer Nature.

high, and proceed anteriorly embedded in a viscous mucus cord toward the palps (Figure 4.16). The ventral groove is the principal route in *M. edulis* by which food material is carried to the labial palps for sorting and then ingestion (Ward *et al.* 1993). Material is never observed in suspension in the ventral grooves, even at the lowest particle concentration. The dorsal canal of the gill is not involved in particle accumulation and transport. Little if any selection takes place on the gill, possibly due to its simple, nonplicate (i.e. nonpleated) structure (Ward *et al.* 1998b). The outer and inner palps enclose the anterior gill region (Figure 4.17). The outer demibranch is applied to the ridged surface of the outer palp and the inner demibranch to the ridged surface of the inner palp. The gill rests against a specialised region of the palp called the dorsal fold (Figure 4.18). This is a smooth, non-ridged flap of densely ciliated tissue that covers approximately half of the ridged surface of each palp. The dense ciliation of the dorsal fold correlates with the particle transport function of this surface (Beninger *et al.* 1995). Its ventral margin is unattached to the underlying ridges, forming an antero-posterior ciliated tract called the palp particle groove. This part of the dorsal fold is contractile, allowing it to cover or expose areas of the ridged palp surface. When the ingestive/handling capacity of the gill is not exceeded, mucus-particle cords are directed into the palp particle groove, where they are transformed into a slurry by the mechanical action of the palp crests and dorsal fold, and probably also the addition of low-viscosity neutral mucopolysaccharides present on the palp crests (Beninger & St-Jean 1997b). The resulting slurry is amenable to both ingestion and particle selection; selected particles continue their anteriorward movement in the palp particle groove, while rejected particles are transported ventrally in the AMPS-rich crest troughs to join the rejection pathway along the palp ventral margin. The mucus-particle cord is subjected to mechanical stress through kneading of palp ridges in the palp particle groove and to biochemical stress through mucus secretion. Consequently, the cord disintegrates and the material becomes less viscous and more flocculent. Some of this material travels anteriorly, presumably for ingestion, while some enters the palp troughs, exiting at the palp ventral margin for eventual rejection as pseudofaeces. When the ingestive/handling capacity is overloaded, the dorsal fold retracts, thus preventing the mucus cord from entering the palp particle groove. Instead, the cord is moved to the ventral margins of the palps for rejection. Thus, the dorsal fold, which is unique to homorhabdic filibranchs, determines whether material from the gill is destined for ingestion or rejection. Video-microscope recordings of pseudofaeces production in *M. edulis* revealed that the trigger threshold concentration for formation of pseudofaeces was about 12 000 cells ml^{-1} (Riisgård *et al.* 2011a). The mucous particle cord may be transferred directly from the gill ventral groove to the mantle wall without the participation of the palps, in which case no selection or ingestion takes place (Garrido *et al.* 2012).

As seen earlier, the main function of the labial palps is to continually remove material from the food tracts on the gills in order to prevent gill saturation. In dense suspensions, sorting and rejection tracts on the palps channel most of the filtered material away from the mouth and deposit it as pseudofaeces so that the animal can continue to filter and ingest at an optimum rate. In homorhabdic fillibranches, pseudofaeces is transported along well-defined rejection tracts on the ciliated mantle surface to the inhalant opening, and periodic, sudden and forceful closure of the shell valves ensures that it is carried away from the animal in the exhalant current. In mussels, the tracts are characterised by composite cilia

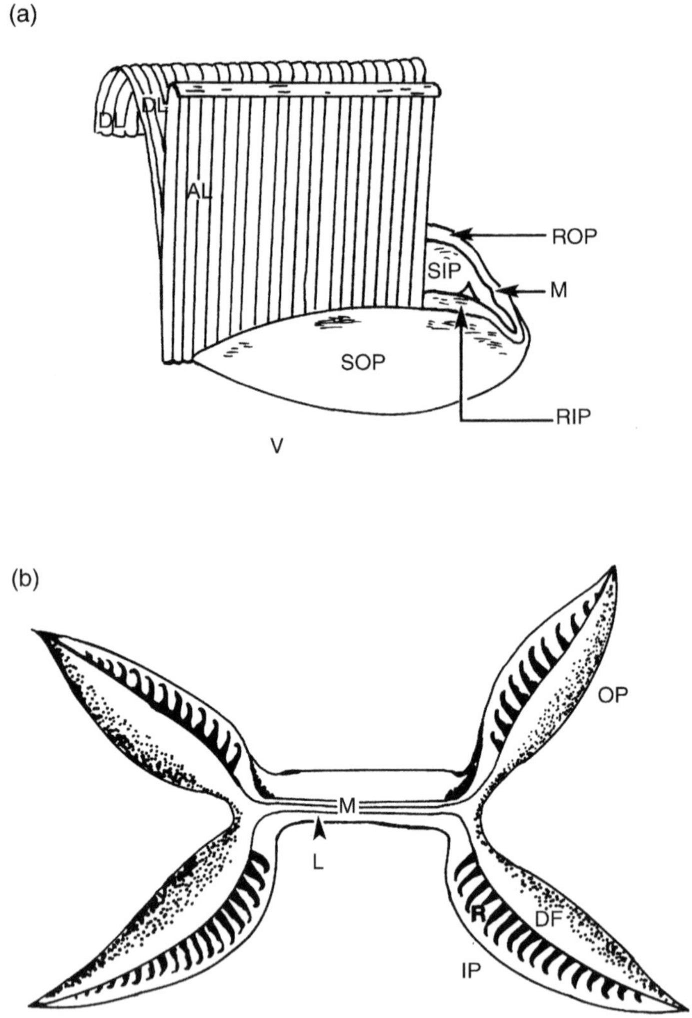

Figure 4.17 (a) Schematic showing the natural position and anatomical relations of the gill, labial palps and mouth in *Mytilus edulis*. AL, ascending lamella of gill; D, dorsal; DL, descending lamella; M, mouth; RIP, ridged surface of inner palp; ROP, ridged surface of outer palp; SIP, smooth surface of inner palp; SOP, smooth surface of outer palp; V, ventral. (b) General aspect of labial palps spread apart and gills removed, with ridged surface only visible. DF, dorsal fold; IP, inner palp; L, lip; M, mouth; OP, outer palp; R, ridges. *Source:* From Beninger *et al.* (1995). Reproduced with permission from Springer Nature.

that are extraordinary long (20 μm) compared to the short, simple cilia (SSC; 5 μm) on the general mantle surface (Beninger & Venoit 1999; Beninger *et al.* 1999). These long cilia are grouped into closely adhering tufts, hence the name. They effectively elevate and isolate the pseudofaeces from the rest of the mantle surface. In addition, high-viscosity mucus secreted in the tracts firmly binds the rejected material and anchors it to the cilia, thus ensuring counter-current transport to the inhalant siphon. The mucociliary transport of

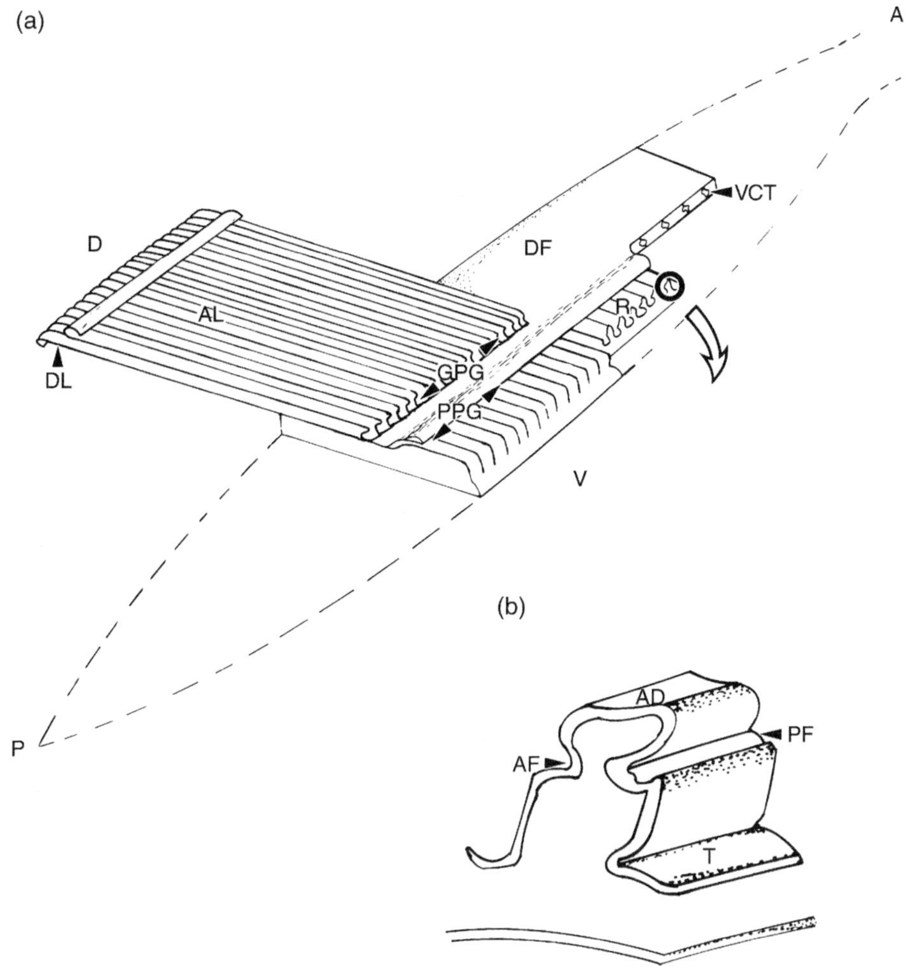

Figure 4.18 Schematic sections showing anatomical relationships and gross internal anatomy of gill and labial palp. (a) Drawing of the ridged surface of a single labial palp in *Mytilus edulis*. A, anterior; AL, ascending lamella of gill; D, dorsal; DF, dorsal fold; DL, descending lamella of gill; GPG, gill particle groove; P, posterior; PPG, palp particle groove; R, ridges; V, ventral; VCT, vestigial ciliated tracts (derived from the fusion of the epithelium of the palp crests with that of the dorsal fold). (b) Schematic drawing of a single palp ridge. AD, apical depression; AF, anterior fold; PF, posterior fold; T, trough. *Source:* From Beninger *et al.* (1995). Reproduced with permission from Springer Nature.

the pseudofeces probably conforms to the two-layer model proposed by numerous researchers and, subsequently, by Beninger *et al.* (1997). In all species with mantle rejection tracts, the tracts are positioned away from the gill ventral particle groove (Beninger *et al.* 1999). This is to ensure that the pseudofaeces travelling in an antero-posterior direction does not entangle with the mucus strings moving in the opposite direction along the ventral gill particle groove. In *M. edulis*, where the gill extends almost to the mantle edge, the tracts are situated along the middle of the mantle.

Particle Selection Efficiency

As seen earlier, the first step in particle processing is preferential capture by the gill, resulting in nonretention of some particles. This step is followed by selection and differential transport by the gill, leading to some particles being rejected as pseudofaeces. The subsequent step in mussels, discussed in this section, is selection by the labial palps, resulting in preferential ingestion of some particles and production of pseudofaeces (Ward & Shumway 2004).

While results from numerous studies show that bivalves can sort and ingest certain types of particles preferentially over others (see later), little was known until the 1990s about the mechanisms of particle sorting and the contribution of the main feeding organs (gill, labial palps) to the selection process. We now know from video endoscopy and histological data (Ward *et al.* 1993; Beninger & St-Jean 1997b) that particles delivered to the labial palps can be handled individually, because the mode of transport is a mucus slurry, not a string. Particles delivered to the palps in a mucous string can still be handled individually, because the action of opposing palp lamellae reduces the cohesiveness of mucus, allowing embedded particles to be dispersed. The observed mechanical fluidisation of mucus by the palps explains how selection of individual particles is possible. Finally, after being processed by the labial palps, particles are ingested in a mucus slurry, not in a continuous mucous string as was traditionally thought (reviewed in Ward & Shumway 2004).

Bivalve food consists of a variety of suspended particles such as bacteria, phytoplankton, microzooplankton and detritus, but also dissolved organic matter (DOM) such as amino acids and sugars (see later). However, this nutritious material is held in suspension with silt, which is usually present in much higher concentrations than the food particles. To compensate for the dilution of organic material in suspension, mussels are able to preferentially select nutritive particles such as algae and reject particles of poor nutritive value in their pseudofaeces (Bayne *et al.* 1993; Hawkins *et al.* 1996, 1998a,b). Selection efficiency is calculated as

$$\mathrm{SE} = 100\,x\left[1-\left(P/F\right)\right]$$

where P is the organic content of the pseudofaeces and F is the organic content (fraction) of the seston (Bayne *et al.* 1993). Kiørboe & Møhlenberg (1981) determined the efficiency with which 10 species of filter-feeding bivalves, including mussels (*M. edulis*), from temperate latitudes separated the diatom *Phaeodactylum tricornutum* from natural silt when their concentrations were 10–30 cells μl^{-1} and 10–20 mg l^{-1}, respectively. All species were able to ingest the diatom selectively and reject the silt. Selection efficiency (SE) was expressed as the ratio of chlorophyll *a* in the suspension offered and pseudofaeces rejected by undisturbed filtering bivalves. This ratio varied considerably between species and was significantly correlated with the size of the labial palps and turbidity of the environment. For example, *M. edulis* from the Wadden Sea, NW Netherlands, had relatively larger palps and a higher SE than mussels from the Øresund, a strait between Sweden and Denmark; the authors suggested that this could be an adaptation to high particulate load in the Wadden Sea.

Bivalves are also capable of discriminating between similar-sized algae cells in their diet. Algal food particles can be distinguished quantitatively, even when they are of similar size, by the detection of their fluorescent pigments using a flow cytometer. Cucci *et al.* (1985) presented *M. edulis* a mixed diet of three different species of nearly equal dimensions: the diatom *Phaeodactylum tricornutum*, the dinoflagellate *Prorocentrum* sp. and the cryptomonad flagellate *Chroomonas salina*. The mussel preferentially rejected the dinoflagellate and diatom species in favour of the cryptomonad flagellate. Subsequently, Bougrier *et al.* (1997) reported that *M. edulis* can selectively ingest the phytoflagellate *Tetraselmis suecica* (8.3 μm) in preference to the diatom *Skeletonema costatum* (9.0 μm) and the phytoflagellate *Pavlova lutheri* (4.6 μm).

The efficiency with which bivalves select and preferentially ingest organic matter increases with the organic quality of the seston to a maximum at about 40% organic content, and then decreases again at higher qualities. Food quality is usually measured in terms of particulate organic material (POM) or chlorophyll content (Ward & Shumway 2004). A similar trend was reported when *M. edulis* was fed a diet comprising three levels of organic content (72, 64 and 41%) made up from natural silt and two cultured algal species, *Tetraselmis suecica* and *Phaeodactylum tricornutum* (Bayne *et al.* 1993). At the lower organic content (41%), filtration rates were increased, the proportion of filtered material rejected increased and there was an increase in the efficiency with which organic material was selected for ingestion. At the higher organic contents (64 and 72%), SE decreased from the maximum 40% to approximately 30 and 17%, respectively. Ward & Shumway (2004) have suggested that the decrease in SE in response to increasing organic content may be due to limitations of the selection process in bivalves. They contend that as the proportion of non-organic particles in the seston decreases with increasing organic content, selectively rejecting the few poor-quality particles may be difficult. This, taken together with the reduction in the amount of rejected material that occurs with increasing organic content, causes a drop in SE. The decrease in organic SE in response to decreasing seston organics (below ~40%) may also reflect limitations of the selection process, because of a less heterogeneous mix at low organic content. Alternatively, proportionally more pseudofaeces is produced when bivalves are exposed to seston of low quality, and mucous represents the main organic component of this rejected material. The presence of this mucous could be responsible for the decrease observed in SE when the quality of seston is low (Urrutia *et al.* 1996). A later study did indeed show this to be the case in the cockle *Cerastoderma edule* (Urrutia *et al.* 2001).

The mechanism of selection is at least partially mediated by interactions between lectins in the mucus covering the gills and labial palps and carbohydrates associated with the surface of microalgal cells (Pales Espinosa *et al.* 2009). To date, specific microalgal-binding lectins have been identified in the mucous produced by feeding organs in *M. edulis* (Pales Espinosa *et al.* 2010a,b) and the oyster *Crassostrea virginica* (Pales Espinosa *et al.* 2009). The physic-chemical surface properties of particles, such as charge and wettability, may also contribute to the particle selection process (Rosa *et al.* 2013, 2017b). See Chapter 11 on the role of lectins in humoral defence. Carbohydrate–lectin interactions are involved in mediating particle sorting in *M. edulis* (Pales Espinosa *et al.* 2010a) and the oyster *Crassostrea virginica* (Pales Espinosa *et al.* 2009). A mucosal C-type lectin (called MeML) in *M. edulis* was expressed in mucocytes lining the epithelium of pallial organs (gills, labial palps and mantle) and intestine (Figure 4.19), and its expression was significantly upregulated following starvation. The MeML transcript was not detected in other tissues,

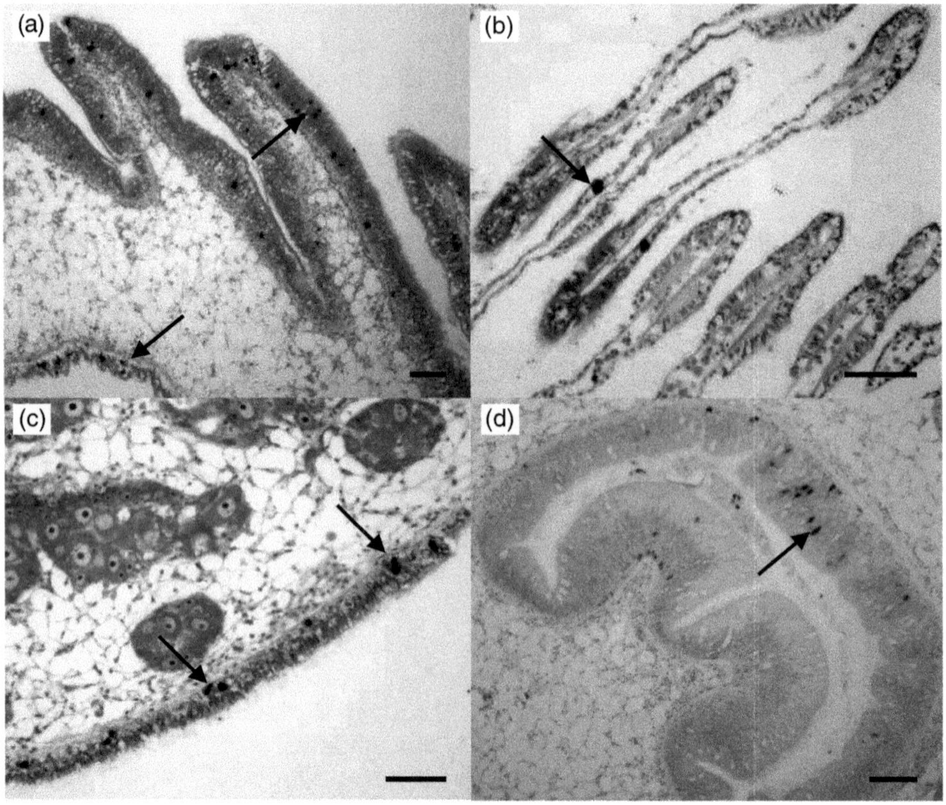

Figure 4.19 In situ hybridisation localisation of MeM, a C-type lectin, in (a) labial palps, (b) gills, (c) mantle and (d) intestine of *Mytilus edulis*. The arrows indicate positive cells for MeML transcripts. Scale bar = 50 µm. *Source:* From Pales Espinosa *et al.* (2010b). Reproduced with permission from Elsevier.

including hemocytes. Subsequently, Pales Espinosa & Allam (2013) demonstrated that the expression of MeML is regulated to favour the selection of high-quality food. Pales Espinosa *et al.* (2016) tested whether some carbohydrates contributed more than others in directing particle selection in *M. edulis* and *C. virginica* when several marine microalgae were presented to them in feeding experiments. Both species preferentially selected microalgae rich in glucose/mannose residues, indicating that cell surface carbohydrates are good predictors for particle fate.

Postingestive Particle Processing in the Stomach

Particles in suspension enter the mouth from an oral groove at the base of the labial palps and travel down the oesophagus into the stomach where the final opportunity for particle selection takes place. The coordinated ciliary beat of the stomach and intestinal epithelia cause the crystalline style, which projects across the floor of the stomach, to rotate against the gastric shield (details in Chapter 2). The rotating style, together with the gastric shield teeth, grinds and crushes particulate material, thereby breaking apart large particles and particle aggregates. Enzymes liberated from the dissolving head of the style mix with this particle slurry, which is then subjected to ridged sorting areas of the stomach and stomach pouches (Ward & Shumway 2004). Finer particles and digested matter are kept in suspension by cilia at the crests of the ridges, and this material is continually swept toward the digestive gland, which opens into the stomach by several ducts. Lighter organic particles and particle fragments enter the ducts of the digestive diverticula and are absorbed in the digestive diverticula by endocytosis or pinocytosis for intracellular degradation in digestive cells aided by the lysosomal system (Moore *et al.* 2007; see later). Larger and more dense material such as sand grains is segregated out and channeled into the intestine along a deep rejection groove on the floor of the stomach and transported to the mid-gut, where it mixes with other undigested material to be incorporated into faecal pellets, which are ejected through the anus and onward to the exhalant opening.

Compared to the process of selective particle capture and preingestive particle sorting, described earlier, there is much less published information on postingestive selection in bivalves, and even less in mussels (see Gosling 2015). The process may occur either by the retention of some particles longer than others in the stomach so that extracellular digestion has more time to act, or by direction of some particles to the digestive gland for intracellular digestion. In both cases, gut retention time (GRT) can be used to indicate whether sorting has occurred (reviewed by Ward & Shumway 2004). The assumption is that bivalves retain larger, less dense particles longer (long GRT) than smaller, denser particles (short GRT) because organic matter tends to be larger and lighter than inorganic particles, which contain little nutritive value (Doyle *et al.* 2016 and references therein). Thus, the separation of particles based on size and density increases digestive efficiency and reduces digestive investment in material with little to no nutritive value. Several studies with filter-feeding bivalves generally show good corrrelation of GPT[1] and assimilation efficiency (AE; see later). This parameter is critical in affecting AE because it determines the time for which food particles can be retained within the gut. Longer retention may enable food materials to be subjected to more rigorous digestion (i.e. intracellular digestion) and more efficient absorption (Wang & Fisher 1996).

Several studies regarding GRT have found that *M. edulis* retains natural food particles for a period of approximately 2.5 hr (Bayne *et al.* 1989), whereas the oyster *Crassostrea virginica* was found to retain natural food particles for approximately 9 hr depending on feeding rate and tidal cycle (Ward & Shumway 2004 and references therein). One approach to studying postingestive processes is to feed the bivalves different types of particles and then determine the time period over which they are egested to calculate a GRT. For example, three bivalve species (*M. edulis*, the cockle *Cerastoderma edule* and the clam *Venerupis pullastra*, now known as *V. corrugata*) were fed a mixture of equal numbers of the alga *Phaeodactylum tricornutum* (29 ± 4.5 μm diameter) and alumina particles (7.5–31.5 ± μm, with a mode of 17.5). The bivalves egested a predominance of alumina after one hour, but after a further hour a greater proportion of *P. tricornutum* was egested (Foster-Smith 1975). Therefore, although alumina was equally as acceptable as *P. tricornutum* by the gills, it was selectively rejected in the gut.

Alimentary Canal and Digestive Process

The system of organs concerned with the ingestion and digestion of food and elimination of faeces consists of the mouth, a short oesophagus, stomach, crystalline style sac, digestive diverticula, midgut, rectum and anus. With the exception of a short section of the rectum, the entire alimentary canal lies within the visceral mass, a cluster of internal organs, and is completely immobilised by the surrounding connective tissue (see Figure 2.12). Food is moved from the mouth toward the anus exclusively by the strong ciliary motion of the epithelial lining of the system, which is well supplied with mucocytes that secrete both acid and neutral mucopolysaccharides, even when the animal in not feeding (Beninger *et al.* 1991). The oesophagus does not have a digestive function, merely serving to propel material along ciliated tracts toward the stomach. The stomach, digestive gland and, to a much lesser extent, intestine are the main organs of the digestive system. From the stomach, there are two pathways for particle processing: an intestinal path and a glandular path. Material destined for the intestine is subjected to extracellular digestion during gut passage, while particles processed via the glandular pathway pass from the stomach to the digestive gland and are subjected to both extracellular digestion in the stomach and intracellular digestion in the digestive gland.

The Stomach and Extracellular Digestion

The stomach is a flattened, oval-shaped sac with several outgrowths or pouches, into which the oesophagus opens at the anterior end and from which the midgut leaves at the posterior end (see Figure 2.12B). Detailed information on stomach structure in fillibranch bivalves is provided in Purchon (1957). The principal organ of extracellular digestion is the noncellular, semi-transparent, gelatinous crystalline style in the stomach. This is synthesised in a part of the digestive gland called the style sac, which is probably a modified part of the midgut (Giribet 2008). The style grows and thus projects forward and dorsally across the cavity of the stomach to rest against the gastric shield, a chitinous, thickened area of the stomach wall. The projecting anterior end of the style is rotated against the gastric shield by the style sac cilia, and in the process the anterior and central parts of the style are abraded and dissolved, thereby releasing a variety of digestive enzymes into the stomach. The stomach wall also releases enzymes into the stomach. Rotation of the style aids in pulling a food-laden mucous strand through the mouth into the stomach, where the material is stirred and mixed with enzymes. While all the mixing and extracellular digestion is taking place, the stomach contents come under the influence of ciliary tracts, which cover all areas of the stomach except those occupied by the gastric shield. These ciliated tracts have fine ridges and grooves and act as sorting areas in much the same way as the labial palps. Finer particles and digested matter are kept in suspension by cilia at the crests of the ridges, and this material is continually swept toward the digestive gland duct openings. Larger particles segregate out and are channeled into the intestine along a deep rejectory groove on the floor of the stomach.

The style is largely composed of a suite of high-molecular-weight (>800 kDa and ~95–170 kDa) glycolproteins and medium-molecular weight (40–50 kDa) protein duplexes (Mackenzie & Marshall 2014). Mackenzie & Marshall (2014) suggest that these 40–50 kDa proteins play a role in the formation of the protein gel that establishes the structural integrity of the style in *M. galloprovincialis* and *Perna canaliculus*. The style sac secretes a variety of digestive enzymes (e.g. amylase, cellulase, agarase, laminarase, protease and lipase), which are incorporated into the crystalline style as it is being secreted (Alyakrinskaya *et al.* 2001; Tizon *et al.* 2013). These enzymes make up only a minor proportion of total protein content of the styles. Total carbohydrates make up a small proportion of the mass of the style. In bivalves, irrespective of species, the ratio of protein to carbohydrate is ~3:1 (Xu 2002).

The crystalline style is not a permanent structure in the stomach. Under 'normal' conditions, style length, as a percentage of mussel shell length (mm) (*M. edulis*, *M. galloprovincialis*), is ~50% (Alyakrinskaya *et al.* 2001). The size of the style has been reported to vary systematically with the tidal cycle, with the maximum size corresponding to the time when there is most food in the stomach (i.e. at high tide, when the animal is feeding) (Alyakrinskaya *et al.* 2001). Digestive enzyme activities extracted from the crystalline style of the clam *Pholas orientalis* exhibited a semi-diurnal pattern. In a 24-hour period, two peaks of digestive activities were generally observed, suggesting that the pattern might be related to their feeding rhythmicity and influenced by tide (Tizon *et al.* 2013). However, during low tide, when drying in the littoral zone for about seven hours, the style can shorten in *M. edulis* by approximately 25%. The crystalline style may be easily dissolved under the influence of unfavorable factors, and the degree and the rate of its dissolution can greatly vary (Alyakrinskaya *et al.* 2001 and references therein). In *M. galloprovincialis*, for example, the style was dissolved in water polluted with oil, in deoxygenated water and in deoxygenated water with a significant admixture of hydrogen sulfide. In *Modiolus* sp., absence of food caused complete dissolution of the style within 24 hours.

Apart from the crystalline style, there is a paucity of information on enzymes produced by the stomach in marine mussels. However, the following enzymes have been reported from the stomach wall of bivalves (Reid 1968; Mathers 1973a,b; additional references in Bayne *et al.* 1976): esterases, which are important in lipid digestion; acid and alkaline phosphatases, which are believed to play a role in absorption and phagocytosis of material from the stomach; and endopeptidases (e.g. trypsin), which break down proteins. High to moderate activity of carbohydrate-splitting enzymes such as α-amylase, α- and β-glucosidase, β-galactosidase, maltase, chitinase and cellulases has also been reported. Chitin is a structural component in diatoms and cellulose is a common component of algal cell walls. Not surprisingly, the activity of cellulases is particularly high in bivalve stomachs. While the stomach wall is important in digestive enzyme secretion, it is also likely that some enzymes from the digestive gland enter the lumen of the stomach and act extracellularly.

The Digestive Gland and Intracellular Digestion

The stomach is surrounded by a mass of dark tissue known as digestive diverticula, which is made up of a large number of blind tubules emptying into several large ducts that lead to the interior of the stomach. Within these ducts, there is a continuous flow: materials enter the digestive gland for intracellular digestion and absorption and wastes leave *en route* to the stomach and intestine (Figure 4.20). The cells that form the interior of a tubule are large and well vacuolated, with large nuclei at their base (Galtsoff 1964).

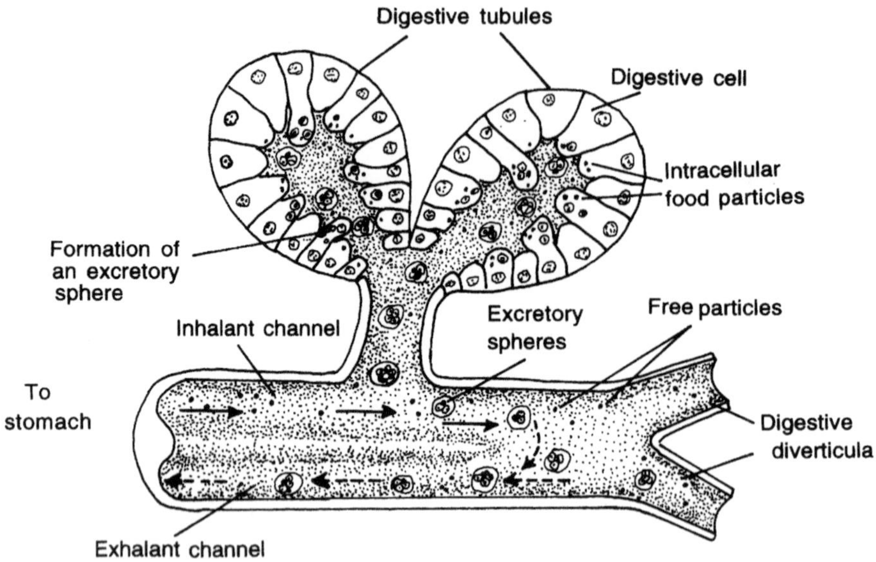

Figure 4.20 Section of the digestive gland showing absorption and intracellular digestion of material coming from the stomach (solid arrows) and outward movement of wastes (broken arrows). *Source:* From Owen (1955). Reproduced with permission from Company of Biologists Ltd.

The tubules consist of two cell types: digestive cells and basophil secretory cells. The former are round and smooth cells (5–30 μm) with intracellular organelles such as lysosomes, secondary lysosomes, lipid granules and residual bodies (Birmelin *et al.* 1999; Dimitriadis *et al.* 2004; Figure 4.21). In addition, the lysosomal content is characterised by the presence of acid hydrolases dealing with the intracellular digestion of nutrients. Digestive cells take up material by pinocytosis (uptake of extracellular fluid into a cell) and digest it within vacuoles. The end products of digestion are released directly into the system and the waste is retained in membrane-bound residual bodies, which are later released into the lumen of the tubules. The digestive cells eventually rupture and are replaced by new cells. The released waste enclosed in excretory spheres is swept toward the stomach and ultimately to the intestine. At high seston concentrations, the digestive tubules can reach saturation, giving rise to a greater proportion of material being transported directly to the intestine without being subjected to intracellular digestion, and ultimately resulting in reduced absorption efficiency (AE; see later). This waste may also contain digestive enzymes, which can be utilised by the stomach for extracellular digestion (Mathers 1972). Basophilic cells (5–20 μm) are like digestive cells in that they also have microvilli on the free surface. The cells display a highly developed rough endoplasmic reticulum (RER) and numerous secretory granules (Figure 4.22) and are considered mainly as digestive enzyme-secreting cells, since the presence of enzymes in their secretory granules has been reported (Dimitriadis *et al.* 2004; Beninger & Le Pennec 2016). In addition, the presence of pinocytotic-coated vesicles in basophilic cells suggests the occurrence of endocytic processes.

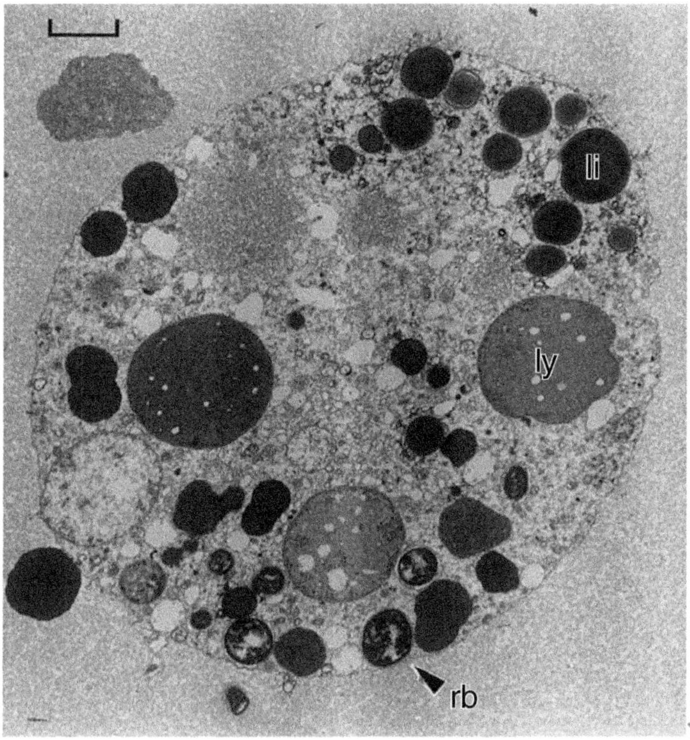

Figure 4.21 Electron micrograph of a cultured digestive cell from the digestive gland of *Mytilus edulis*. li, lipid granules; ly, lysosome; rb, residual body. Scale bar = 2 μm. *Source:* From Birmelin *et al.* (1999). Reproduced with permission from Springer.

Digestive gland cells have a finite life span, after which they disintegrate and are replaced. The results from various studies suggest that there is a well-defined rhythm to the cycle of changes taking place in the digestive tubules. In mussels, these have been linked with tidal periodicity (Langton 1977; Connor *et al.* 2016), salinity (Pipe & Moore 1985), natural light (Robson *et al.* 2010) and food availability (Thompson *et al.* 1974; Langton 1975; Wong & Cheung 2001). To determine whether cell proliferation in the digestive gland of mussels follows a circadian pattern, the distribution of proliferating cells in the digestive gland of subtidal *M. galloprovincialis* was investigated by bromodeoxyuridine (BrdU) immunohistochemistry (Zaldibar *et al.* 2004). BrdU is an analogue of thymidine. During the S-phase of the cell cycle, when DNA is replicated, it is incorporated into the DNA of proliferative cells. Mussels were exposed to 4 mg BrdU/l seawater continuously. Starting at six hours after treatment, samples were collected every two hours for two days and BrdU labelling was estimated by direct counting using a light microscope, with values being noted per thousand BrdU-positive cells. A BrdU-positive reaction was observed in the nuclei of digestive, basophilic, duct and stomach cells and in haemocytes. Cell renewal, produced by autologous division of mature digestive gland cells, was synchronised following a circatidal pattern (i.e. BrdU labelling increased during low tide

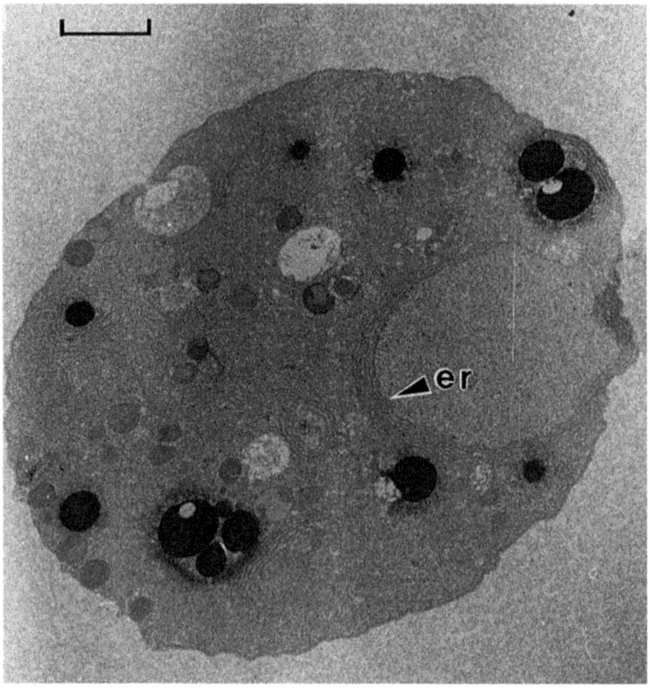

Figure 4.22 Electron micrograph of a cultured basophilic (secretory) cell from the digestive gland of *Mytilus edulis*. er, endoplasmc reticulum. Scale bar = 2 µm. *Source:* From Birmelin *et al.* (1999). Reproduced with permission from Springer.

and decreased during high tide; Figure 4.23). Epithelial cell renewal in the stomach also appeared to be synchronised. Therefore, tidal rhythms may be a perfect Zeitgeber[2] for rhythmic processes in cell proliferation within the digestive system (Zaldibar *et al.* 2004). Also, marked seasonal differences in BrdU labeling were recorded in the same species, with much higher proliferating activity in summer than in autumn and winter (Zaldibar *et al.* 2008). This is because summer is the period when increased somatic growth or enhanced cell turnover takes place, and also because more digestion activity causes digestive cells to be exhausted more rapidly. In contrast, the digestive gland epithelium of mussels from intertidal and subtidal populations differed not only in the level but also in the pattern of variation of BrdU labelling, which in intertidal mussels appeared to be modulated by photoperiod and tide – unlike in subtidal mussels, in which variations followed a circatidal pattern. Size did not have a great impact on the proliferative pattern, since both small and large mussels showed a very similar pattern.

The stomach, digestive gland and sometimes the intestine produces a wide variety of digestive enzymes, such as polysaccharide- and sugar-splitting enzymes, fat-digesting enzymes (e.g. lipase) and protein/peptide-digesting enzymes (e.g. proteases, trypsin and chymotrypsin) (Table 4.3). The digestive gland also plays an important role in the storage of metabolic reserves, such as lipids, which are used as an energy source during the process

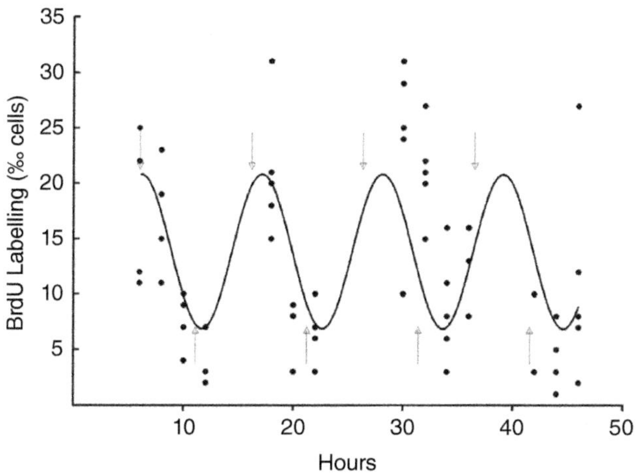

Figure 4.23 Nonlinear regression of BrdU labelling values (‰ cells) recorded in the digestive alveolotubular units against experimental time (in hours). Best fitness corresponded to a sigmoideal function: $Y = 13.838 + 6.978 \sin[(2\pi X/10.988) + 4.324]$; $r^2 = 0.338$, where all regression and correlation coefficients were significant ($P < 0.05$). The distance between peaks and between troughs is approximately 12 hours (upper arrows high tide, lower arrows low tide).
Source: From Zaldibar *et al.* (2004). Reproduced with permission from Springer Nature.

Table 4.3 Carbohydrate-, fat- and protein-splitting enzymes in the digestive gland of mussels.

Species	Enzyme	Substrate	Reference
Perna viridis	Amylase	Starch, glycogen	Teo & Sabapathy (1990) Sabapathy & Teo (1992)
Mytilus californianus	Cellulases	Cellulose	Connor *et al.* (2016)
P. viridis			Teo & Sabapathy (1990)
Mytilus edulis			Xu (2002)
Crenomytilus grayanus *M. californianus*	1,3;1,6-β-D-glucans (laminarinase)	Laminarin (in brown algae)	Vehkova *et al.* (2015) Connor *et al.* (2016)
P. viridis	Maltase	Maltose	Teo & Sabathy (1990); Teo *et al.* (1990)
	Sucrase	Sucrose	
	trehalase	Trehalose	Teo & Lim (1991, 1993)
P. viridis	α-glucosidase	Sucrose	Teo & Sabapathy (1990)
Mytilus trossulus	Protease	Peptides	Kreeger (1993)
M. californianus	Trypsin	Peptides	Connor *et al.* (2016)
P. viridis	Chymotrypsin-like	Protein, peptides	Teo & Sabapathy (1990)
P. viridis	Lipase	Fat molecules	Teo & Sabapathy (1990)

of gametogenesis and during periods of physiological stress (Miller *et al.* 2014 and references therein).

Because of their pronounced endocytotic nature, digestive gland cells are the major environmental interface for the uptake of metal and organic contaminants, especially those associated with natural particulates that are filtered from seawater (Moore & Allen 2002). A range of biomarkers measured at the molecular or cellular level have been identified, and these serve as early-warning tools in environmental quality assessment (Raftopoulou & Dimitriadis 2012; Patetsini *et al.* 2013; Hamza-Chaffai 2014; Franzellitti *et al.* 2015). The most used biomarkers in digestive gland tissue are metallothionein induction (Viarengo *et al.* 1999; Mićović *et al.* 2009; Khati *et al.* 2012), lysosomal alterations (Moore *et al.* 2004, 2007; Izagirre *et al.* 2009), peroxisome proliferation (Cajaraville & Ortiz-Zarragoitia 2006) and heat-shock proteins (Mícovíc *et al.* 2009). Metallothioneins bind to heavy metals and are likely to be involved in detoxification processes. Their synthesis in response to specific toxic metals has been well documented for marine bivalves (reviewed in Amiard *et al.* 2006; Khati *et al.* 2012). Lysosomes are ubiquitous cytoplasmic organelles that contain hydrolytic enzymes which degrade macromolecules to low-molecular-weight products. They accumulate many toxic metals and organic chemical contaminants that provoke loss of membrane integrity, increase in lysosomal size and accumulation of lipid and the pigment lipofuscin. These responses have been used routinely to test for the effects of toxic contaminants in both experimental investigations and environmental impact assessments (Moore *et al.* 2004; Izagirre *et al.* 2009 and references therein). Peroxisomes are cytoplasmic organelles involved in the metabolism of lipids and reactive oxygen free radicals, and peroxisome proliferation is used to assess levels of organic contaminants (e.g. polycyclic aromatic hydrocarbons, PAHs and polychlorinated biphenyls, PCBs) in coastal ecosystems (Cajaraville & Ortiz-Zarragoitia 2006). Heat-shock proteins play a key role in cellular protection against environmental stress, with different forms induced by elevated temperatures and a great variety of chemical stressors (e.g. heavy metals) found in seawater or sediment (Mićović *et al.* 2009 and references therein; Mahmood *et al.* 2014). More detailed information on the use of mussel biomarkers in the assessment of ecosystem health is presented in Chapter 8. The formation of granulocytomas (inflammatory lesions) has also been reported in digestive gland tissue (Usheva *et al.* 2006; Garmendia *et al.* 2011; Rocha *et al.* 2016; see Chapter 11).

The Intestine

The intestine leaves the posterior end of the stomach and loops through the visceral mass before eventually reaching the anus. It is composed of several regions of approximately equal length, of which the last is the rectum, which passes through the pericardial cavity. The rectum ends at the anus, located in the exhalant chamber above the posterior adductor muscle. Rejected particles from the stomach, as well as waste material from the digestive gland, pass into the coiled intestine. Such particles are not necessarily lost, since wandering haemocytes that migrate freely into the gut from the haemolymph stream may phagocytose them, digest them and transport them around the body, thereby playing unique roles in intracellular digestion and nutrient transport. The intestine is also active in extracellular digestion. Waste is formed into faecal pellets, which are voided through the anus and swept away through the exhalant opening. Mathers (1973a,b) reported the following enzymes from the mid-gut wall – the first part of the intestine – of the oysters *Ostrea edulis* and *Crassostrea angulata*: α-amylase, maltase, trehalase, cellobiase and various glucosidases. Esterases, acid and alkaline phosphatases, chitinase and endopeptidases have also been reported from the mid-gut of a range of bivalves (Reid 1968).

Mussel Nutrition

Suspended particulate matter (seston) in oceanic and coastal waters is composed of numerous types of particles originating from various sources. Mussels derive their nutrition from the organic component of seston, which includes living cells such as phytoplankton of a wide range of sizes and palatability, bacteria, DOM (especially dissolved free amino acids) and a range of zooplankton of different sizes, as well as dead organic material such as detritus derived from vascular plants and macroalgae (reviewed in Arapov *et al.* 2010). So, while it is generally assumed that mussels rely on phytoplankton from the water column as their main energy source, the other food sources mentioned may represent a valuable supplement when phytoplankton abundance is low. However, studies have shown that there is spatial, seasonal and tidal variation in the concentration of seston components in coastal waters, particularly in shallow water systems such as estuaries (Huang *et al.* 2003a,b and references therein), making it difficult to characterise the natural diet of suspension-feeding bivalves. Despite such variability, bivalves maintain relatively constant rates of nutrient acquisition by various compensatory adjustments (discussed later). The following sections deal with the standard methods used in bivalve diet analysis. For information on larval diets, see Chapters 6 and 10.

Gut Content Analysis

This method is the traditional and simplest way to examine what bivalves are ingesting. The gut, usually the stomach, is dissected from each individual and fixed in 4–5% formaldehyde, the standard preservative. Tan & Ransangan (2015) have recently shown that a 1% Lugol's solution can be used as a safe alternative to formaldehyde. Stomach contents are identified and quantified under an inverted microscope. The method has been used to characterise and quantify particulate food, including mussel larvae and post-larvae, ingested by the mussel *P. canaliculus* during different times of the year (Alfaro 2006). Adult mussels of different sizes (45–105 mm in shell length) were collected for gut content analysis from three intertidal sites at Ninety Mile Beach, northern New Zealand about 30 minutes after being submerged by the incoming tide. The preserved gut contents were classified as bivalve larvae (<300 and >300 μm), phytoplankton cells (2–20 μm), microzooplankton (20–200 μm) or mesozooplankton (>200 μm). Larger intertidal mussels tended to have more food particles in their stomachs than smaller mussels within all three populations. For all populations, mussel larvae and juveniles contributed about 70% of the food particle consumption during the spawning peak in August, while phytoplankton (i.e. dinoflagellates, *Gymnodinium* and *Ceratium* species, and diatoms, *Chaetoceros*, *Navicula* and *Nitzschia* species) and other zooplankton (i.e. calanoid, cyclopoid and harpacticoid copepods, mysids, zoea and nauplia) constituted the majority of the food source (about 99%) in March, during gametogenesis. Mussels from Scott Point consumed more phytoplankton, microzooplankton and mesozooplankton than those from Tonatona Beach, which consumed more than mussels from The Bluff (Figure 4.24). Mean (± SE) phytoplankton cells were 8.3 ± 1.1, 7.7 ± 0.9 and 0.3 ± 0.1 for Scott Point, Tonatona Beach and The Bluff, respectively. For microzooplankton and mesozooplankton, respectively, the means (± SE) were 6.5 ± 0.9 and 0.8 ± 0.2 for Scott Point, 8.4 ± 1.3 and 1.4 ± 0.3 for Tonatona Beach and 0.3 ± 0.1 and 0.0 ± 0.0 for The Bluff. Phytoplankton and zooplankton numbers/stomach were more abundant in March 2001, followed by August and October 2000. Very few plankton were consumed by mussels from any site in December 2000. All two-way ANOVAs for phytoplankton, microzooplankton and mesozooplankton indicated statistically significant differences among sites and dates, with a significant interaction between site and date.

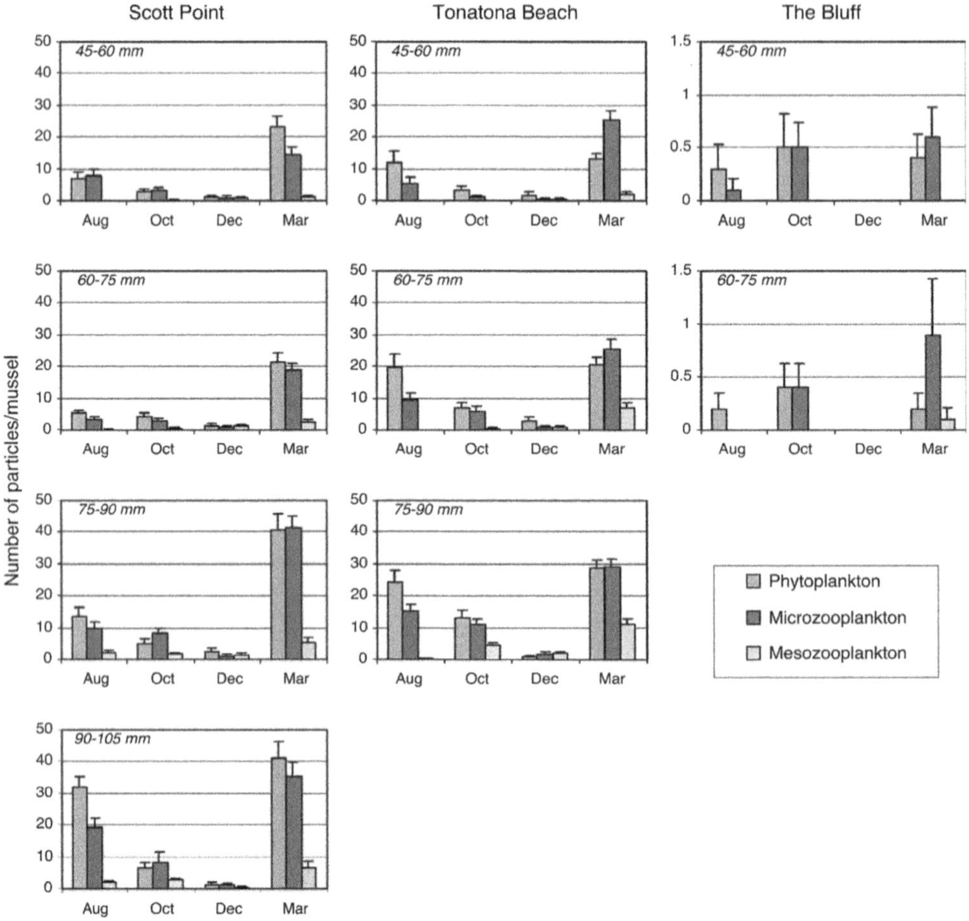

Figure 4.24 Mean (± SE) number of phytoplankton, microzooplankton (<200 μm) and mesozooplankton (>200 μm) in the stomachs of mussels (45–105 mm shell length) from Scott Point, Tonatona Beach and The Bluff, Ninety Mile Beach, northern New Zealand, during August, October and December 2000 and March 2001. *Source:* From Alfaro (2006). Reproduced with permission from Elsevier.

In a separate study, Rouillon *et al.* (2005) collected seasonal data on the phytoplankton composition of seston and stomach contents of *M. edulis* from an estuarine mud flat and a rocky open shore. The objectives were to determine the extent to which spatial differences in phytoplankton composition reflect the particular conditions of both sites and to ascertain the degree of similarity between the stomach contents of mussels and the microalgal composition of the water column at the two sites. Individuals and water samples were collected monthly from November 2001 to December 2002. Stomach contents showed a lower average number of species (N = 6) at both sites and a greater proportion of tychopelagic[3] forms compared with the water samples (N = 20 and 24 in the estuary and open shore, respectively). Maximum phytoplankton density in water samples, mostly due to diatoms, occurred in the May to October period. At the marine site, the stomach contents of mussels displayed two peaks of phytoplankton concentration, in May (caused by the dinoflagellate *Ensiculifera* sp.) and in July (caused by the diatoms *Pseudo-nitzschia pungens* and *Licmophora* sp.). In estuarine mussels, a single peak of abundance was recorded in the month of May, mainly produced by *Ensiculifera* sp. The most notable finding from the study was the increased abundance of dinoflagellates in mussel stomach contents relative to the corresponding seawater samples. Rouillon *et al.* (2005) suggest that dinoflagellates could be ingested in preference to diatoms through differential retention on the gills and/or selection during the pseudofaeces-forming process. Alternatively, dinoflagellates may be far more resistant to extracellular digestion, remaining longer within the gut and being preferentially voided with the faeces (Rouillon *et al.* 2005).

As mentioned earlier, bivalve larvae can constitute a significant proportion of mussel gut contents. Lehane & Davenport (2004) investigated the ingestion of bivalve larvae by *M. edulis*. Laboratory experiments revealed that 90% of bivalve larvae offered to mussels were ingested and apparently fully digested. The larval shell offered no protection against digestive processes, resulting in high larval mortality once inside the stomach. Stomach content analysis (September 2001–January 2003) showed that farmed mussels ingested bivalve larvae year-round, with the exception of March 2002. Numbers of ingested larvae were highest in October 2001 and May 2002, which coincide with the known spawning times of farmed mussels in Ireland. Mussels ingested a large size range of bivalve larvae, suggesting that all stages of the bivalve life cycle are vulnerable to predation. Given the high biomass of mussels present in mussel farms, filtration by adult bivalves significantly reduces numbers of bivalve larvae in nearby waters. The phenomenon has also been reported in other bivalves, including mussels (Alfaro 2006; Troost *et al.* 2008, 2009; Davenport *et al.* 2011; Peharda *et al.* 2012; Ordóñez *et al.* 2013). Besides bivalve larvae, other zooplankton (e.g. copepods) and crustacean, ascidian and gastropod larvae are ingested by adult filter-feeding bivalves (Lehane & Davenport 2002; Nielsen & Maar 2007; Davenport *et al.* 2011; Peharda *et al.* 2012). Also, Webb *et al.* (2013) showed that *Mytilus* spp. and other bivalves ingest sea lice larvae, regardless of phytoplankton presence or absence, and that large individuals consume a significantly greater proportion of larvae than small individuals. They suggested that bivalves could act as biological agents in the control of parasitic sea lice at salmon farms.

Raby *et al.* (1997) used epifluorescence microscopy to analyse the stomach contents of mussel and other bivalve veliger larvae collected from the Baie des Chaleurs, western Gulf of St Lawrence, Canada in order to document food-particle sizes, compare feeding among taxa and compare diets with the in situ phytoplankton community. The stomach contents of the veligers sampled in waters characterised by microbial components were dominated by cyanobacteria (<2 μm) and small autotrophic flagellates (<5 μm). Approximately 50% of all veligers examined contained cells measuring between 5 and 15 μm, whereas only 3% contained algal cells measuring between 15 and 25 μm. The differences in the size ranges of ingested algal cells among similar-sized larvae of different species suggest that veligers actively select food particles. The dominance of <5 μm phytoplankton in the Baie des Chaleurs and in stomach contents suggests that veliger larvae may be an important export pathway for carbon produced by ultraplankton (planktonic organisms <2 μm in size).

Stable Isotope Analysis

Gut content analysis only provides an instant snapshot of the percentages of various materials ingested, all of which may not be assimilated. Also, isotope analysis can be particularly useful in revealing the food sources of bivalves, because the recognition of gut content is difficult – sometimes even impossible – due to the small size of food particles. The application of stable isotopes may overcome some limitations because evaluation of food sources is based on assimilated instead of ingested food and allows us to trace long-term (weeks to months) dietary patterns in bivalves in the field (Gao *et al.* 2006 and references therein). In isotope analysis, the ratios of ^{13}C to ^{12}C and ^{15}N to ^{14}N are expressed in δ notation with respect to deviations from standard reference material. The ratio of stable isotopes of nitrogen ($\delta\,^{15}N$) can be used to estimate trophic position (e.g. intertidal versus subtidal), because the $\delta\,^{15}N$ of a consumer is typically enriched by 3–4% relative to its diet, whereas the ratio of carbon isotopes ($\delta\,^{13}C$), used to identify primary food sources, changes little as carbon moves through food webs (Bustamante & Branch 1996). Focusing first on food sources, Bustamante & Branch (1996) used isotopes ratios (^{13}C, ^{12}C and $^{15}N : {}^{14}N$) to examine the degree to which filter feeders (mussels *M. galloprovincialis* and *Aulacomya*) depend trophically on kelp detritus or pelagic phytoplankton and how much intertidal grazers (limpets, *Patella* spp.) depend on subtidal kelps or intertidal micro- and macroalgae. The filter feeders used kelp-derived detritus as their main source of organic carbon (50%) and nitrogen (65%). Similarly, two abundant and highly specialised grazers used subtidal kelps as their main source of carbon and nitrogen (>69%), although other more generalised intertidal grazers relied mainly on the epilithic algal production. Kelp-derived detritus represented more than 65% of POM, being consistently high all year round and during both high and low tides, although it was particularly high on exposed shores. Phytoplankton was seldom found, and contributed no more than 6% of total intertidal POM. Therefore, kelp-derived detritus represents a steady source of food for filter feeders, upon which they can rely all year round.

In contrast, $\delta^{13}C$ and $\delta^{15}N$ were measured in *M. edulis*, the clams *Macoma balthica* and *Scrobicularia plana* and food sources collected in the salt marsh of the Alguillon Bay, France (Riera *et al.* 1999). The results showed that, although the cordgrass *Spartina anglica* was dominant, it did not contribute significantly to the carbon and nitrogen requirements of the bivalves. In fact, all three preferentially fed on a mixture of benthic diatoms and marine phytoplankton, but in different proportions – estimated through isotope mixing models, which are increasingly being used to quantify consumer diets (reviewed in Phillips *et al.* 2014). The isotopic position of *M. edulis* suggested that this suspension feeder subsists primarily on phytoplankton (64.6%), but also consistently feeds on benthic diatoms (35.4 %). The deposit feeder *M. balthica* used a higher proportion of benthic diatoms (75.8%) compared to phytoplankton (24.2%), while *S. plana* also preferentially used benthic diatoms (64.5%), but used a higher proportion of phytoplankton (35.5%) than did *M. balthica.*

Since stable isotope analysis typically classifies carbon sources into broad functional groups (phytoplankton versus benthic microalgae), the method has fairly recently been combined with polymerase chain reaction-denaturing gradient gel electrophoresis (PCR-DGGE), a method that has the potential to identify ingested organisms to the genus or species level. Using both short-term PCR-DGGE ingestion data and long-term assimilation data, Maloy *et al.* (2013) examined dietary patterns in suspension-feeding *Mytilus* spp. from seven spatially discrete locations within a semi-enclosed marine bay in Northern Ireland. Both methods detected variations on a lough-wide scale, although PCR-DGGE provided greater taxonomic specificity. The combined approach provided a more complete picture of the trophic relationships than either approach alone. However, in the same study, organisms identified in the gut contents of mussels did not necessarily reflect the overall isotopic values, presumably primarily explained by the temporal lag between the approaches.

Isotope studies have also shown that spatial, temporal and hydrographic factors can influence the relative contributions of different food sources in bivalves' diet. Factors include tides (Sarà *et al.* 2007), nearshore–offshore hydrography (Hill *et al.* 2006), upwelling (Allan *et al.* 2010), spatial and temporal scales (Maloy *et al.* 2009, 2013), seasonal changes (Pernet *et al.* 2012) and, as seen earlier, shore type (Bustmante & Branch 1996). To illustrate, Hill *et al.* (2006) investigated the effects of biogeographic and temporal variation in coastal hydrography on food resources of intertidal mussels (*Perna perna* and *M. galloprovincialis*) at 12 sites along the southern African coastline. In their study, $\delta^{13}C$ and $\delta^{15}N$ isotopes of suspended particulate matter, mussels and macroalgae along the southern African coastline were explored. Suspended particulate matter exhibited overall trends of nearshore ^{13}C depletion from southwest to northeast along the coastline, and from nearshore (0 km) to offshore (10 km) waters, in both cases suggesting a shift from a nearshore signature strongly influenced by macroalgal detritus to one more representative of oceanic phytoplankton. It was possible, using discriminant analysis, to categorise mussel populations – with one exception – into four geographic groups, on the basis of both carbon and nitrogen signatures. A linear mixing model indicated that mussels along the entire coastline generally demonstrated more than 50% dependence on nearshore carbon and nitrogen, emphasising the importance of nearshore primary production to intertidal consumers.

Isotopic studies have also shown variable degrees of competition and resource partitioning within and between species. To take an example, the temporal dynamics of carbon and nitrogen isotope ratios of co-occurring suspension feeders in two shellfish culture areas in northern France were investigated over two years to evaluate interspecific trophic partitioning and the relative contributions of different organic matter sources to the diet of benthic suspension feeders (Lefebvre *et al.* 2009). Oysters (*Crassostrea gigas*), mussels (*Mytilus edulis*) and cockles (*Cerastoderma edule*) and two nonbivalves, limpets (*Crepidula fornicata*) and polychaete worms (*Lanice conchilega*), were sampled from an estuarine environment, while oysters and mussels, limpets and worms (*Sabellaria alveolata*) were sampled in a marine one. As already mentioned, phytoplankton (POM) represents the major food source in marine systems for suspension feeders, while in estuarine systems other organic matter sources (OMS) such as riverine particulate material, microphytobenthos, macroalgal detritus and bacteria significantly contribute (references in Lefebvre *et al.* 2009). In the marine system, whatever the year, all species – with the exception of *S. alveolata* – relied strongly on phytoplankton, especially during phytoplankton blooms (May, July), but in winter, when phytoplankton is less abundant, they switched to OMS. In the estuarine system, OMS was the major dietary item for all species. But is there competition for food sources between these co-occurring suspension feeders? The authors suggested that particle capture and selection may be important in limiting competition and may explain why ecosystems can often support large numbers of suspension feeders.

Fatty Acid Profile Analysis

Fatty acid (FA) profiling is another potential trophic marker technique for tracing food sources. The simultaneous use of stable isotopes and FAs as trophic markers may mutually confirm the information obtained from single-marker techniques alone (Gao *et al.* 2006; Allan *et al.* 2010; Redmond *et al.* 2010). Bivalves obtain their lipid requirements from their diet or by synthesis of lipids from dietary protein and carbohydrate precursors. Dietary lipids are broken down into their constituent FAs and incorporated into the consumer's tissues largely unchanged, and therefore generally reflect the FA profiles of the food consumed and provide useful information on trophic sources (Pernet *et al.* 2012 and references therein). While most of the different food types that are available to bivalves have many FAs in common, there are some specific FAs, groups of FAs and FA ratios that serve as biomarkers of food sources such as marine bacteria, diatoms, dinoflagellates, constituents of phytoplankton, vascular plants and farm wastes, all of which show different combinations of specific FAs. For example, high concentrations of 20:5ω3[4] and 16:1ω7, a ratio of 16:1ω7/16:0 close to 1 and the presence of C16 polyunsaturated fatty acid (PUFA) (mainly 16:4ω1) are all diatom markers. The 16:1ω7/16:0 ratio is used to discriminate between diatom versus dinoflagellate dietary input, because 16:1ω7 is abundant in diatoms, while 16:0 is common in dinoflagellates. Another ratio with the potential to differentiate between these two food sources is 20:5ω3/22:6ω3 (Prato *et al.* 2010; Irisarri *et al.* 2014 and references therein). Budge & Parrish (1999) found that two *Pseudo-nitzschia* toxic species both displayed similar FA compositions typical of diatoms, including 16:1ω7, 16:2ω4 and eicosapentaenoic acid (EPA; 20:5ω3) as the major unsaturated FA. Also, 16:4ω1 occurred in both species and therefore could be used as a signature compound to differentiate toxic *Pseudo-nitzschia* from other diatoms. The sum of branched-chain and odd FAs (15:0–17:0 + 18:1ω7) is used to estimate bacterial dietary input (Budge *et al.* 2001), while 20:1ω9 is indicative of a carnivorous diet (Virtue *et al.* 2000). The FAs 18:2ω6, 20:4ω6 and 22:6ω3 are all indicative of heterotrophic flagellates and ciliates, which are the sources of PUFAs in the microbial food loop – an important component of the classical food chain (Zhukova & Kharlamenko 1999). The FAs 22:6ω3 and 18:1ω9 are abundant in lipids of zooplankton and thus provide an indicator of carnivorous organisms, such as copepods (Silina & Zhukova 2007 and references therein). The FAs 18:1ω9, 18:2ω6 and 16:ω 4 in *M. edulis* can be used in tracing the assimilation of salmon pellets into mussels within 28 days of exposure (Redmond *et al.* 2010), suggesting that mussels can perform bioremediating services when integrated with fish farming (see also Gao *et al.* 2006; Peharda *et al.* 2007; Handå *et al.* 2012). It now appears that the lipid profile of *M. edulis* from the White Sea, Russia is modified in response to various environmental factors in order to protect cell membranes, maintain or recover their homeostasis and replenish the cell's energy and metabolic resources, thus securing the mussels' adaptation to changes in environmental conditions (Fokina *et al.* 2017).

Irisarri *et al.* (2014) traced the food sources of mussels (*M. galloprovincialis*) cultured in NW Spain using FA biomarkers. The FA profile of seston, mussel mantle, digestive gland and faeces was analysed during five seasons. The principal FAs in the mussels' organs were 16:0, 16:1ω7, EPA (20:5ω3) and DHA (docosahexaenoic acid) [22:6ω3], while 16:0 predominated in the faeces. Seasonal fluctuations in the seston composition were mirrored in the FA signature of mussel organs and faeces, although the digestive gland had the closest resemblance to the seston FA profile (Figure 4.25). In general, diatom- and bacteria-derived biomarkers predominated in mussels' organs and faeces during the upwelling period (spring–summer), while dinoflagellates dominated during downwelling (autumn–winter). The higher concentration of EPA and DHA in both organs and the faeces compared with the seston suggested a preferential accumulation of these ω3 FA in the mussels' tissues. The authors also compared the FA profiles of mussels cultured in a raft close to fish cages of red sea bream (*Pagellus bogaraveo*) with those reared in a raft distant from the net-pens. Because some of the mussel rafts were located proximate to a fish aquaculture site, a second aim of the study was to assess whether mussels cultured in a raft near fish cages were assimilating any uneaten fish feed as part of their diet. Insights into the effectiveness of the utilisation of fish feed by mussels will be of great importance for the implementation of integrated mussel–fish systems worldwide – but especially for Galicia, a region where mussel farming is the principal aquaculture activity.

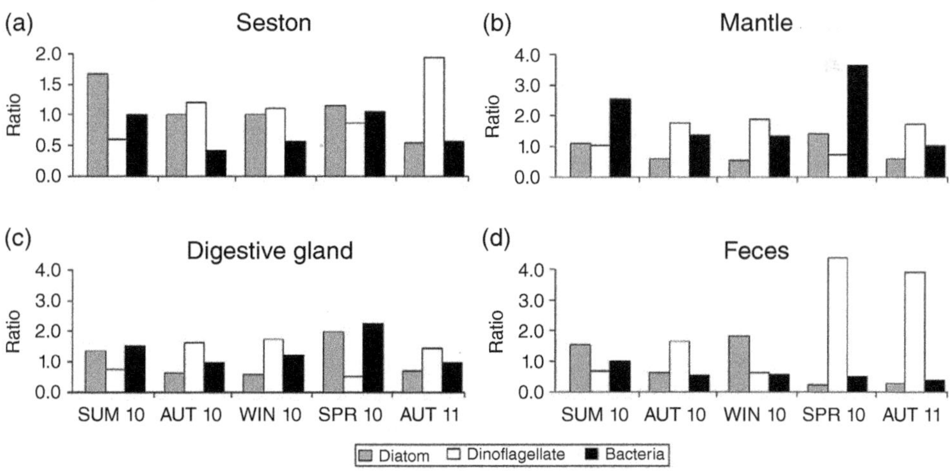

Figure 4.25 Seasonal variations of the principal fatty acid (FA) ratios typical of diatoms (20:5ω3/22:6ω3>1), dinoflagellates (22:6ω3/20:5ω3>1) and bacteria (18:1ω7/18:1ω9>1), measured for the (a) seston, (b) mantle, (c) digestive gland and (d) faeces of the mussel *Mytilus galloprovincialis*. *Source:* From Irisarri *et al.* (2014). Reproduced with permission from Elsevier.

FA analysis has also been used to assess the influence of spatial (Xu & Yang 2007; Allan *et al.* 2010) and seasonal (Colaço *et al.* 2009; Prato *et al.* 2010; Ezgeta-Balić *et al.* 2012; Dernekbaşi *et al.* 2015; Fernández-Reiriz *et al.* 2015a) variability in the diet of mussels, with a combination of FA profile analysis and stable isotope analysis used in some studies (Xu & Yang 2007; Allan *et al.* 2010). To assess the influence of spatial variability, the food sources of bivalves from two habitats – intertidal mussels (*M. galloprovincialis*) and oysters (*Crassostrea gigas*) and subtidal scallops (*Chlamys farreri*) – in Jiaozhou Bay, Qingdao, China were determined by FA analysis (Xu & Yang 2007). Intertidal bivalves exhibited a high level of Chlorophyta FA markers, Σ18:2ω6 and 18:3ω3, which indicates the importance of the local *Ulva pertusa* seaweed bed as a food source for the intertidal species. These also exhibited significant levels of the flagellate marker 22:6ω3. In scallop tissue, there were significant levels of diatom markers, as well as those specific to bacteria and terrestrial materials, suggesting there were substantial bacterial and terrestrial inputs into the food of the species. Dernekbaşi *et al.* (2015) investigated FA composition in *M. galloprovincialis* cultured from October to September at Sinop on the Black Sea. Seasonal variation of lipids was significantly different throughout the study period, with the highest value in winter (1.90 ± 0.10% wet weight) and the lowest value in spring (1.45 ± 0.22%) ($P < 0.05$). PUFAs were dominant among total lipids in spring (56%) and winter (56%), respectively, while in the summer and autumn their values were lower than values, representing 53 and 51% of total lipids, respectively.

Additional Food Sources: Picoplankton, Bacteria and DOM

Picoplankton is composed of organisms, chiefly bacteria and small phytoplankton, measuring between 0.2 and 2 μm. As seen earlier in relation to bacteria, most bivalves cannot efficiently capture and ingest particles in this size range. However, picoplankton can be embedded in large organic aggregates (~0.5 mm in diameter) that are tens to hundreds of times larger than picoplankton size (Waite *et al.* 2000). Aggregates, often referred to as 'marine snow', are made up of organic matter (plankton, diatoms, faecal matter) and inorganic particles (e.g. sand). They are held together by transparent extracellular polysaccharides (TEP), a sugary mucous exuded as waste products by phytoplankton, bacteria and a variety of suspension feeders, including bivalves (Heinonen *et al.* 2007; Li *et al.* 2008). The coupling of inorganic and organic particles, including TEP, creates microhabitats and provides substrates for enriched microbial communities (Kach & Ward 2008). The aggregates, which are formed in the upper water column, eventually sink to the ocean floor, thus providing a primary source of energy for deep-water organisms.

Kach & Ward (2008) have shown that suspension-feeding bivalves can ingest picoplankton-sized particles bound in marine aggregates, which suggests that picoplankton presented in this way may be an important food source. The increasing use of nanotechnology means that there is an increasing risk of nanoparticles (NPs; 1–100 nm) entering the aquatic environment, with potential toxicological consequences for animals and plants. The first evidence of the impact of polystyrene (PS) nanoparticles on filter feeders showed that marine aggregates of nano-sized PS facilitate food ingestion and are translocated from the gut to the circulatory system, where they are retained for more than a month – a finding that has implications for bivalves and their transfer to higher trophic levels, including humans (Ward & Kach 2009; reviewed in Canesi *et al.* 2012). In *M. edulis*, PS NPs caused a reduction in filtration and an increase in production of pseudofaeces (Wegner *et al.* 2012). To date, however, there is no information on the effects at a cellular level in marine organisms, although there is recent evidence that aggregates can be transporters of pathogens into bivalves (Kramer *et al.* 2016). In addition, there is growing evidence to suggest that microplastic fragments represent a potential reservoir of pathogens and antimicrobial resistance bacteria on these distinct artificial matrices that differ markedly from natural particles.

Bacterioplankton is a characteristic attribute of marine as well as fresh waters and represents a substantial amount of organic matter in these systems. Consequently, there is increasing interest in the role of bacteria in bivalve nutrition. For example, probiotics – beneficial bacteria that override pathogens by producing inhibitory substances or prevent pathogenic colonisation in the host – are increasingly used as dietary supplements in bivalve hatcheries (see Chapter 10). Bacterial abundance in aquatic systems usually ranges from 10^5 to 10^7 cells ml^{-1}, and averages in the euphotic zone – the layer of seawater that receives enough sunlight for photosynthesis to occur – are 0.5–2 g C m^{-2} across a range of oceanic systems. In the marine environment, the production of heterotrophic bacterioplankton is in the order of 10–60% of the corresponding phytoplankton production (references in Nevejan *et al.* 2018). Cavallo *et al.* (2009) estimated the abundance of heterotrophic bacteria over an annual cycle in seawater and mussel (*M. galloprovincialis*) samples collected from three locations in the Northern Ionian Sea, Italy. Throughout the year, bacterial concentrations were sometimes higher in the mussel samples compared to the corresponding seawater. Among the bacteria, *Aeromonas* prevailed in both water (16%) and mussel (43%) samples. *Lucibacterium* (3%) and *Vibrio* (1%) were present only in mussel samples. Different percentages of other genera such as *Moraxella* (water 7%, mussels 5%), *Pseudomonas* (12, 6), *Alcaligenes* (4, 4), *Acinetobacter* (5, 2) *Flavobacterium* (4, 2) *Chromobacterium* (2, 4) *Photobacterium* (4, 8) and *Flexibacter* (2, 1) were present in both seawater and mussel samples, highlighting the abundance and diversity in mussels and the surrounding seawater in the Northern Ionian Sea, an important location for mussel culture.

As most bacteria are smaller than 1 μm, CE on the gills is generally low. For example, in experiments using free, unattached bacteria, the oyster *Crassostrea gigas* had a CE of 5%, compared with 16% for the mussel *Geukensia demissa* (Langdon & Newell 1990). Unattached bacteria only meet ~1–10% of the carbon requirements of suspension-feeding bivalves, but in eutrophic estuaries, stagnant water and marshes, where bacterial abundances can reach ~1×10^7cells ml^{-1}, Langdon & Newell (1990) showed that during the spring period, when metabolic activity was high, bacteria satisfied 5.5 and 31% of metabolic demand for carbon and 26.7 and 70.6% of metabolic demand for nitrogen in *C. gigas* and *G. demissa*, respectively. Kreeger & Newell (1996) found that the proportion of available carbon ingested by *G. demissa* did not differ significantly among bacteria (39%), heterotrophic flagellates (58%) or the unicellular algae *Isochrysis galbana* (66%), which was used as a reference diet. In contrast, *M. edulis* ingested a significantly lower proportion of the small bacteria (19%) than of the larger heterotrophic flagellates (58%). Only *G. demissa* was capable of assimilating a considerable amount of carbon from bacteria (42%) as opposed to *M. edulis* (21%). But it is only in certain environments such as eutrophic estuaries, kelp beds and marshes that bacterial concentrations are sufficiently large to contribute significantly to the nutrition of suspension-feeding bivalves. In open coastal and oceanic waters, bacteria are generally less abundant, and therefore are unlikely to be nutritionally significant for bivalves (Langdon & Newell 1990). They probably only meet <4% of a bivalve's nutritional needs (Prieur *et al.* 1990). Because the C:N ratio of bacteria is lower than that of phytoplankton (3.5–3.7 compared with 6.6), bacteria represent a relatively more important nitrogen resource than do phytoplankton, especially during upwelling phases (Seiderer & Newell 1985). Marine bacteria also excrete various substances, including amino acids, carbohydrates and B vitamins. Therefore, DOM from bacteria may also represent a substantial resource of food and possibly growth factors for those bivalves that are capable of directly absorbing it (Stewart 1979; Prieur *et al.* 1990). See Nevejan *et al.* (2018) for a comprehensive review of the role of bacteria as a direct food source for aquaculture animals of commercial importance.

Marine DOM is a complex mixture of thousands of different molecules and represents one of the largest pools of carbon in the global biosphere. DOM consists of soluble organic materials derived from the partial decomposition of organic materials, including soil organic matter and plant residues, and of soluble particles released by living organisms, including bacteria, algae and plants. Because of its complexity, studies of bivalve uptake of DOM have mainly focussed on uptake of relatively simple dissolved free amino acids (DFAA). HPLC analysis of seawater samples from the Wadden Sea showed that total concentrations of DFAA varied seasonally between 1528 ± 668 nM in October 1984 and 1198 ± 597 nM in July 1984, although proportions of individual amino acids were remarkably constant (Siebers & Winkler 1984). Uptake by mussels (*M. edulis*) varied between 29 and 66%, with 78% of the uptake resulting from the five most concentrated amino acids: serine, alanine, glycine/threonine, ornithine and aspartic acid. Mussels could deplete the seawater of individual amino acids down to concentrations below 10 nM and of total amino acids to levels about 0.4 nM. They took up between 30 and 50% of the amino acids dissolved in coastal water and obtained nutritional benefits equivalent to 5–38% of metabolic rate. Manahan *et al.* (1982), on the other hand, demonstrated that *M. edulis* rapidly removed up to 94% of naturally occurring amino acids from seawater – a considerably higher value – and estimated that uptake of amino acids at ambient concentrations in seawater could meet 34% of the metabolic requirements in this species (Manahan *et al.* 1983), although Hawkins & Bayne (1992) suggest a percentage closer to the values of Siebers & Winkler (1984). Transport systems in bivalves (e.g. *M. californianus*) are well adapted for the net accumulation of amino acids from seawater (Wright *et al.* 1989). Total uptake depends on concentration and composition of DFAA (Manahan *et al.* 1983), external salinity and sodium concentration (Wright *et al.* 1989) and the flow rate of seawater across each mussel (Siebers & Winkler 1984). A more recent study has examined the uptake of radiolabelled glucose and two amino acids in the mussel *P. viridis*, and found that mussels accumulate carbon predominantly from POM (phytoplankton, zooplankton, bacteria and detritus), with less than 0.2% coming from DOM (Pan & Wang 2004). While this figure is low, DOM could serve as a potential energy source when other sources of nutrition are least abundant. It may be that bacteria take up DOM more efficiently than bivalves, thus reducing its concentration to a low level. It should be kept in mind that all of these studies have been conducted in the laboratory, and to my knowledge no study has been undertaken to test DFAA uptake in the field.

A small number of bacterial species have developed an endosymbiont relationship with bivalve hosts, particularly mussel and clam species. The host provides access to reduced compounds such as hydrogen, sulphide and methane, as well as oxygen and carbon dioxide, which the symbionts need for their chemosynthetic metabolism, while in return the symbionts provide their hosts with nutrition (Wentrup *et al.* 2014 and references therein). The best-studied species are mussels in the genus *Bathymodiolus*, which are found in hydrothermal vents in the Pacific and Atlantic Oceans. The symbionts are intracellular, residing within the epithelial cells of the gill. Bathymodiolid mussels exploit several food resources: suspended POM and DOM (Riou *et al.* 2010) and nutrient transfer through leakage of metabolites from the symbiont to the host, not excluding lysosomal resorption of dead endosymbionts (Kádár *et al.* 2008). Also, monounsaturated (n-8) FAs, which are diagnostic of methanotrophic endosymbionts, have been detected in gill, posterior adductor and mantle tissues of *Bathymodiolus* spp., although levels were highest in gill tissues, where the endosymbionts were found (Pond *et al.* 1998). With increasing reliance on endosymbionts as a carbon source, there is a progressive reduction in the size of the labial palps, stomach and digestive gland until the digestive system disappears altogether.

Absorption Efficiency

AE compares the organic matter in the food and the faeces, and represents how efficiently bivalves absorb the organic matter from the seston (Widdows & Johnson 1988). It is functionally interrelated with gut capacity, the residence time for food in the gut and the ingestion rate. Altering the rate of ingestion and gut residence time provides a bivalve with a powerful means of adapting to a variable food source (Bayne & Newell 1983). Another mechanism by which AE can be enhanced is increased synthesis of digestive enzymes. This has the effect of both increasing the rates of enzymatic breakdown of ingested material and creating more space in the gut for suchmaterial (Bayne *et al.* 1993; Wong & Cheung 2001). Physiological changes that result in increased AE are believed to occur over a period of days, whereas changes in filtration rate and rejection of particles as pseudofaeces occur over much shorter time scales (Bayne *et al.* 1993).

The most commonly used method for measuring AE, proposed by Conover (1966), is based on the assumption that the inorganic matter contained in food can be used as an inert tracer for the digestive processes. AE is estimated using the following equation:

$$\mathrm{AE} = (F - E) / [(1 - E) * F]$$

where F is the ash-free dry weight : dry weight ratio of the food and E is the ash-free dry weight : dry weight ratio of the faeces. Details on methods of determining AE in mussels can be found in Ren *et al.* (2006), Navarro *et al.* (2009), Ezgeta-Balić *et al.* (2011), Irisarri *et al.* (2013), Srisunont & Babel (2015) and Saggese *et al.* (2016).

The influence of food type on AE is fundamental to the prediction of production, and is of particular relevance in bivalve culture operations. Numerous studies have investigated the effects of seston composition, quantity and quality and of temporal and spatial variability on AE in mussels. For example, Ren *et al.* (2006) investigated how mussels respond to different food diets. AEs were measured in *P. canaliculus* using nine different ^{14}C-labeled phytoplankton species (three diatoms, three flagellates and three dinoflagellates) found in a mussel-farming ecosystem in Marlborough Sounds, New Zealand. AE varied with algal species but was significantly higher when mussels were fed dinoflagellates (85%) compared with diatoms (62%) and flagellates (78%) (Figure 4.26); however, the difference in AEs between flagellates and dinoflagellates was not significant ($P > 0.08$). Gut passage time (GPT) showed a large variation between food types, ranging from 6.9 to 50.9 hr (Figure 4.26). No significant differences in GPT were detected when the phytoplankton species were grouped into diatom, flagellate and dinoflagellate taxa ($P > 0.71$). The cell walls of diatoms are rigid and resistant to enzymatic digestion and physical breakdown, and diatoms also contain significant amounts of inorganic matter. Therefore, there would be little benefit in withholding diatoms in the gut for more efficient digestion. In a similar study, Fernández-Reiriz *et al.* (2015b) examined AE in cultured *M. galloprovincialis* fed three different mono-algal diets (*Rhodomonas lens*, *Isochrysis galbana* clone T-ISO and *Tetraselmis suecica*). AE was significantly higher for *Rhodomonas* (69.6%) than for *Tetraselmis* (38.4%) or T-ISO (23.6%) diets, which could be explained by the greater proportion of refractive nondigestible material contained in *Tetraselmis* and T-ISO diets, together with the low digestibility of the cell wall of *Tetraselmis*. The *Rhodomonas* diet showed the highest protein content, which was reflected in the highest ingestion and absorption of proteins compared with the other diets, indicating a distinct advantage to mussels during the seasonal growth cycle, especially in terms of gametogenesis, shell formation and byssogenesis. Other studies have investigated the effects of seston concentration, quality and spatial variability on AE. For example, Bayne *et al.* (1993) fed mussels (*M. edulis*) with four seston concentrations (between 0.99 and 10.3 mg total seston l^{-1}), comprising three levels of organic content (71.9, 63.6 and 40.8%), made up from natural silt and two phytoplankton species. When fed at a high concentration of seston of low organic content, the mussels rejected a higher proportion of filtered material as pseudofaeces and increased the efficiency with which filtered matter of higher organic content was selected for ingestion, suggesting active regulation of AE. Also, in an earlier study, Bayne & Hawkins (1990) observed that elevated rates of ingestion were accompanied by unchanging GPT so that total gut content increased, helping to maintain organic gut content independent of changes in either food quality or food quantity. Similar findings have been reported for the scallop *Placopecten magellanicus* (Cranford *et al.* 1998) and the oysters *Crassostrea gigas* and *Saccostrea glomerata* (Bayne 2002 and references therein).

Other factors beside diet influence AE. For example, habitat has been shown to have an effect. A comparison of cultured and intertidal mussels (*M. galloprovincialis*) showed that higher values of AE were systematically recorded for cultivated mussels, which may be related to habitat differences in terms of cycles of food availability and air exposure (Labarta *et al.* 1997). In contrast, Charles & Newell (1997), examining the effect of increasing tidal

Figure 4.26 Variation in absorption efficiency (AE) and gut passage time (GPT) of the greenshell mussel *Perna canaliculus* fed with nine phytoplankton species. The first three are diatom species, the next three dinoflagellates and the last three flagellates. Error bars are ±1 SE. *Source:* From Ren *et al.* (2006). Reproduced with permission from the National Shellfisheries Association-BioONE.

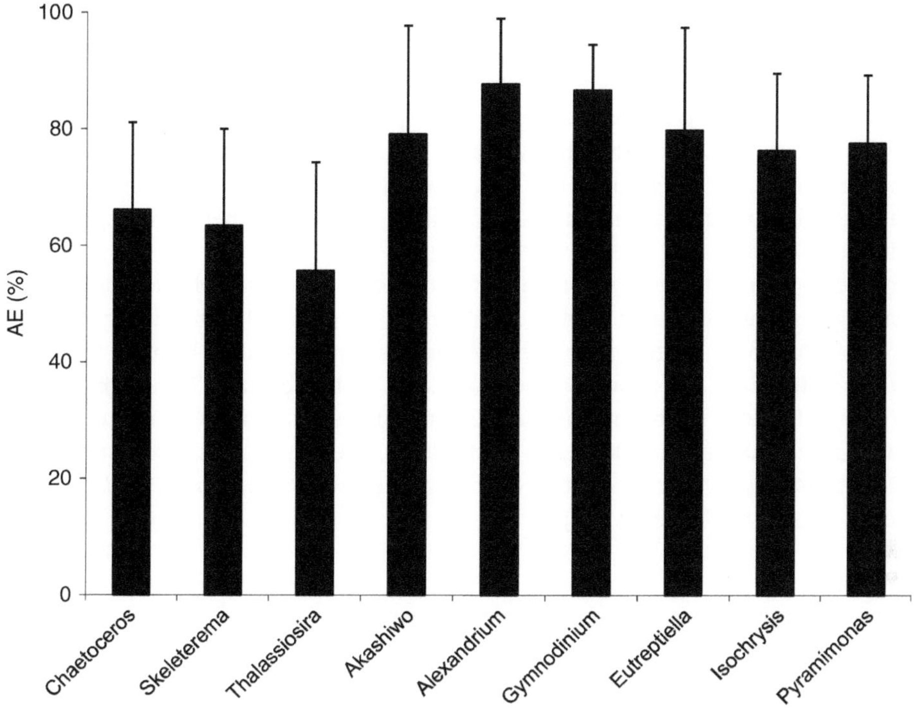
100
80
60
40
20
0
AE (%)
Chaetoceros
Skeleterema
Thalassiosira
Akashiwo
Alexandrium
Gymnodinium
Eutreptiella
Isochrysis
Pyramimonas

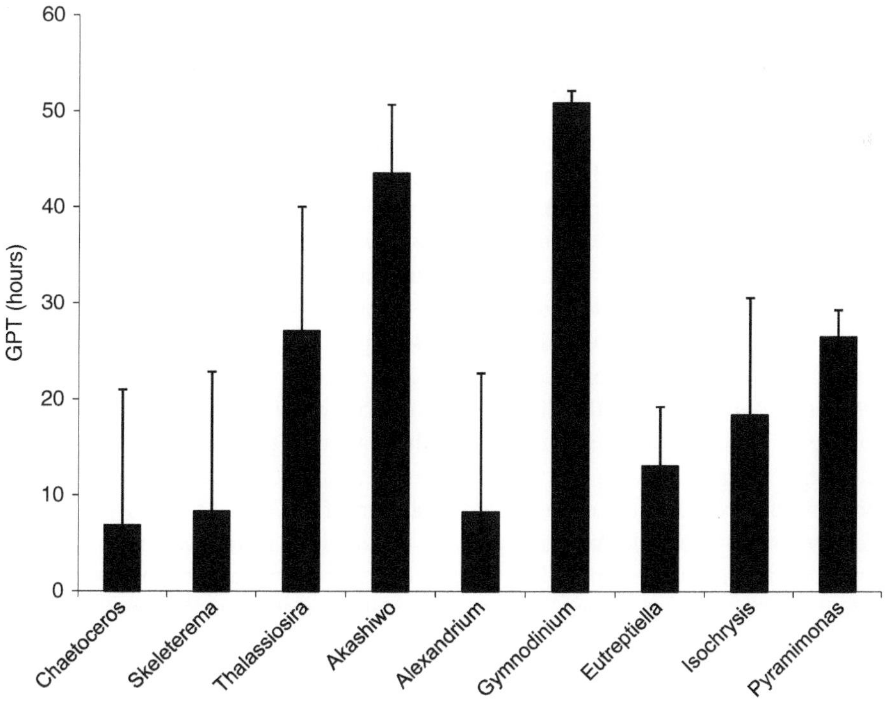
60
50
40
30
20
10
0
GPT (hours)
Chaetoceros
Skeleterema
Thalassiosira
Akashiwo
Alexandrium
Gymnodinium
Eutreptiella
Isochrysis
Pyramimonas

exposure on the mussel *Geukensia demissa*, found that those periodically emersed had AE values between 22 and 26%, which were significantly higher than for those continuously immersed (16%). The authors concluded that mussels compensate for reduced feeding time by exhibiting a higher AE. AE is also affected by atmospheric gas concentrations in oceanic waters. Wang *et al.* (2011) found that in *P. viridis*, AE decreased with decreasing dissolved oxygen concentration. In *M. chilensis*, AE was reduced at high pCO_2 levels; in addition, oxygen uptake fell significantly under these conditions, indicating a metabolic depression (Navarro *et al.* 2013).

Dynamic energy budget (DEB) theory aims to quantify the energetic framework of an individual organism as a dynamic model (details in Chapter 6), and is increasingly being used to reproduce the sequential passage of food in suspension-feeding bivalves through the processes of filtration, selection, rejection, ingestion and absorption. Those wishing to research how DEB theory has been used to model these processes in mussels might consult the following: Beadman *et al.* (2002), Bos *et al.* (2006), Frechette (2012) Frechette *et al.* (2016), Ren (2009), Rosland *et al.* (2009), Saraiva *et al.* (2011, 2012), Sebens *et al.* (2016), Picoche *et al.* (2014) and Sarà *et al.* (2013).

Impacts of Mussels on Marine Ecosystems

The different ways in which suspension-feeding bivalves influence marine ecosystem processes have been thoroughly reviewed by Dame (2012). This section will focus only on the effects that mussels have on their own habitats through feeding, biodeposition of sediments and cycling of nutrients.

Bivalve Feeding and the Plankton Community

Suspension-feeding bivalves have a direct impact on phytoplankton communities in coastal waters. Their large grazing potential causes depletion of phytoplankton biomass above the seabed, but also increases phytoplankton productivity by recycling inorganic nutrients from suspended POM, thereby exerting both top-down (grazing effects) and bottom-up (nutrient recycling) control on phytoplankton communities (Nielsen & Maar 2007 and references therein). To access the scale and magnitude of phytoplankton depletion, most studies have focused on how cultured bivalves control phytoplankton: an effect that has relevance to the assessment of ecological carrying capacity (see Chapter 10).

Maar *et al.* (2008) investigated the distributions of both chl *a* and zooplankton across a cultivation raft of mussels (*M. galloprovincialis*) during summer in the Ría de Vigo, NW Spain. They observed a significant average depletion of 57% for chl *a* and of 26–77% for different zooplankton groups at the edge and downstream of the raft. At the lowest current speeds, removal rates of zooplankton nauplii and copepodites were lower than for chl *a*, indicating that zooplankton escaped filtration. In contrast, removal rates of zooplankton >150 μm increased with current speed, and at the highest current speeds exceeded those of chl *a*, stressing the importance of including heterotrophic plankton in estimates of potential mussel production. In a related study using the same species and location, Froján *et al.* (2014) showed that mussel culture exerted a top-down control over the microbial plankton population, modifying its structure by consuming microplankton and nanoplankton without affecting picoplankton. Also, an excess of ammonium, probably excreted by mussels, and a lower autotrophic carbon : chlrophyll *a* ratio suggested that mussel culture could also exert a bottom-up-like control on the phytoplankton that escapes mussel consumption in culture zones. Other examples of top-down control are provided by Grant *et al.* (2008), Ogilvie *et al.* (2003), Petersen *et al.* (2008), Trottet *et al.* (2008), Lonsale *et al.* (2009) and Dame (2012), with the most dramatic examples found in areas following rapid population growth of invasive bivalve species (McKindsey *et al.* 2006).

Benthic mussels, which take their food not from a homogeneous medium but from the near-bed water layer, also control phytoplankton biomass at the scale of individual bivalve beds or even whole water systems. The distribution of phytoplankton was measured in vertical profiles at three stations in Limfjorden, Denmark within and above *M. edulis* beds (−5 to 200 cm above the bed surface; Dolmer 2000). The density of phytoplankton was 2690–10 600 cells cm^{-3} at 200 cm above the bottom (~1.4–3.2 mg chl *a* m^{-3}), declining toward the bottom to about 1000 cells cm^{-3} at two of the stations; this demonstrates that the near-bed phytoplankton concentrations observed in the field often remained below the concentration level at which mussels reduce their maximal filtration activity. Using the same species and location, Nielsen & Maar (2007) compared the plankton community above a mussel bed with that above a bare sand bed. The vertical distributions of phytoplankton, bacteria, protozoa, benthic plankton and copepods were significantly different above the mussel and sand beds, showing that mussels also impact the zooplankton community indirectly through reduction in the abundance of their plankton prey and directly by preying on other zooplanktons and thereby eliminating pelagic competitors. The mussel bed also recycled nutrients back to the water column as waste products, showing that apart from removing plankton from the water column above them, they recycle nutrients back to the water column as waste products. Consequently, in shallow-water ecosystems like the Limfjorden, mussels have the potential to simultaneously regulate the pelagic food web by performing both bottom-up and top-down control. See Petersen *et al.* (2013) for a related study.

Another effect of bivalve feeding is that turbidity is reduced through removal of phytoplankton and other particles from the water column (Figure 4.27). This has the knock-on effect of increasing light penetration to the sediment, which can potentially enhance the growth of seagrasses and benthic algae; the former provide a heterogenous habitat for sessile and mobile fauna, while the latter are an important food source for benthic herbivores, which in turn are a food item for fish (Newell *et al.* 2002; Newell 2004). One downside of increased transmission of light at the sediment surface is enhanced growth of macroalgal species, some of which may restrict water flow and cause sediment hypoxia when they decay (Rafaelli *et al.* 1998).

As seen earlier, mussels also have the potential to influence zooplankton communities. They do so directly by consuming zooplankton and by competing with zooplankton for phytoplankton resources. Bivalves may out-compete zooplankton because they overwinter as adults and are able to start feeding as soon as water temperatures increase in the spring. In contrast, temperate copepods – a dominant component of zooplankton – have to rebuild their population in the spring, and since most of the phytoplankton is then being grazed by bivalves, copepod populations can diminish, as their fecundity relies on this food source (Newell 2004).

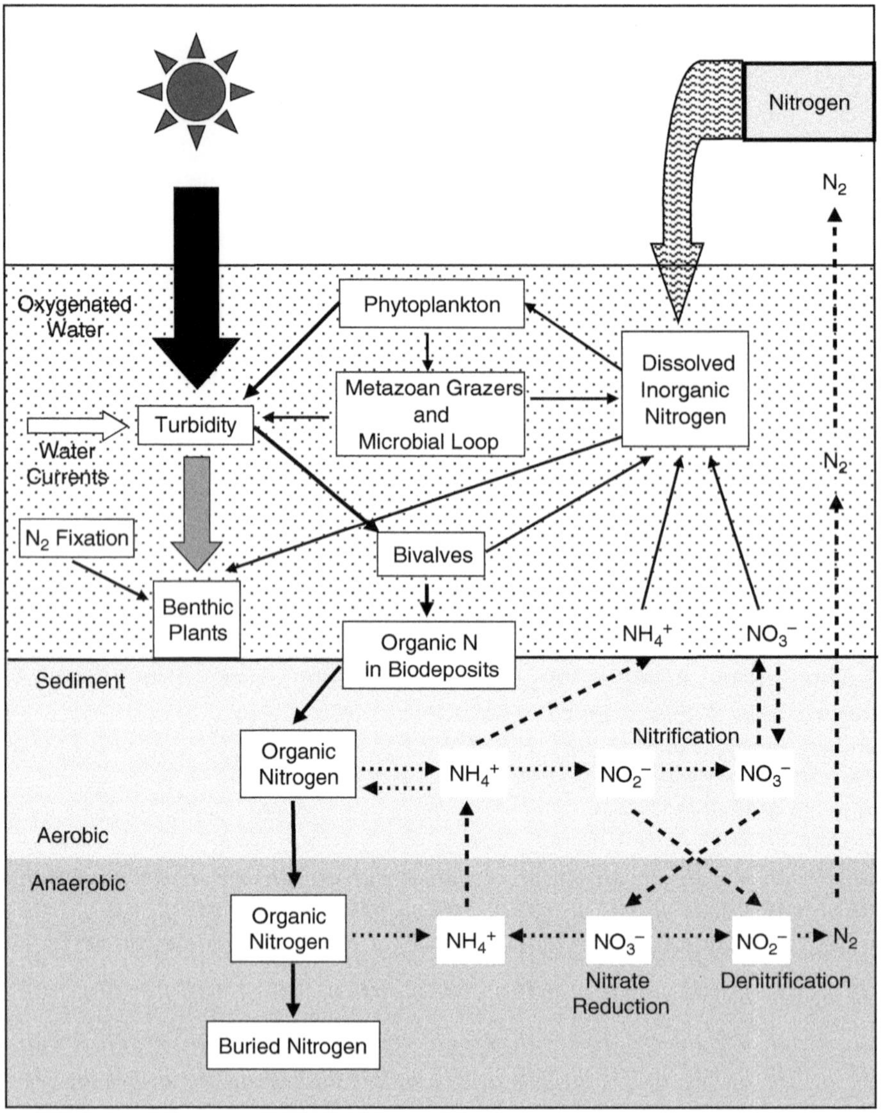

Figure 4.27 Nitrogen (N) cycle in a bivalve suspension feeder-dominated ecosystem. Bivalves remove organic and inorganic particles from the water column and transfer undigested particulate matter in their biodeposits to the sediment surface. Burial of N throughout the sediment leads to removal of N from the water column. Within the aerobic sediment layers, microbial-mediated nitrification occurs, which when linked to denitrification in the underlying anaerobic sediment leads to the further loss of N as N_2 gas. Bivalve grazing on phytoplankton leads to a reduction in turbidity, which enhances the growth of benthic plants (see text for more detail). Solid lines indicate movement of materials; dashed lines indicate diffusion of materials; dotted lines indicate microbial-mediated reactions. *Source:* From Newell *et al.* (2002). Reproduced with permission from John Wiley & Sons.

Biodeposition and Mussels

Waste products from bivalves are excreted in dissolved form or released as faeces and pseudofaeces, usually referred to as biodeposits (Giles & Pildich 2004). The high variation in assimilation efficiency shown by mussels (see earlier), as well as the substantial amounts of total organic matter (TOM) in pseudofaeces (4–20%; Zamora & Jeffs 2012), means that large amounts of undigested POM are transferred to the sediment surface as biodeposits. Their accumulation beneath suspended mussel farms is determined by four factors: the rate of biodeposit production, initial dispersal (i.e. the transport of biodeposits in the water column until their first contact with the seabed), the redistribution of biodeposits on the sediment surface and the rate of biodeposit decay (reviewed in Giles *et al.* 2009). Biodeposit production in suspended mussel culture systems may be considerable and can vary depending on species, size and diet (reviewed in Weise *et al.* 2009). For example, a single raft of *M. galloprovincialis* in the Galician rías, Spain may produce up to 129–190 kg (dry weight)/day^{-1} of biodeposits, which translates into an annual input per raft of 69 t of sediment, 5220 kg of carbon and 620 kg of nitrogen to the seabed (Otero *et al.* 2009 and references therein). Many other examples can be found in Jaramillo *et al.* (1992), Callier *et al.* (2006), Giles *et al.* 2009, McKindsey *et al.* (2011) and references therein and Zuñiga *et al.* (2014). Analysis of mussel biodeposits shows that they are relatively rich in dissolved C, N and P (van Broekhoven *et al.* 2015) and can be of great nutritional value (C:N ratios between 4.8 and 8.5; references in McKindsey *et al.* 2011). Biodeposits may gradually be incorporated into the sediment or resuspended, depending on the velocity of bottom water currents (references in Newell 2004). The quality of organic and other matter in biodeposits influences the rate at which it breaks down, as more labile components may be released to the environment rapidly (Carlsson *et al.* 2010).

Increasing levels of organic input into sediment can have a dramatic effect on benthic infaunal communities. For example, soft-sediment communities, which are typically dominated by large filter-feeders, are replaced first by smaller and greater numbers of deposit feeders, then by polychaetes, then by nematodes (McKindsey *et al.* 2006 and references therein). The physical structure provided by shells falling from suspended mussel operations may also provide a habitat for species normally associated with hard-bottom communities, thereby often increasing local benthic diversity (Grant *et al.* 1995). The magnitude of the effect on infaunal communities depends on farm size, stocking density, age of operation and site-specific characteristics. When the decay of biodeposits consumes oxygen at a rate greater than that of renewal by water exchange, this can create anoxic conditions and induce sulphate reduction to occur near the surface. The development of anoxic sediments can lead to increases in hydrogen sulphide to levels toxic to benthic organisms (Diaz & Rosenberg 1995). In contrast to the preceding scenario, however, Dittman (1990) reported that biodeposition from mussel beds in the North Sea actually led to an enhanced and more diverse invertebrate assemblage. See McKindsey *et al.* (2006) for a detailed discussion on the direct and indirect effects of biodeposition on benthic assemblages.

Biodeposition by mussels and associated organisms affects the biogeochemistry in the sediment by increasing oxygen consumption and nutrient fluxes at the water–sediment (benthic) interface. The elements C, N and P in biodeposits (see earlier) are regenerated and subsequently recycled back to the water column to support further phytoplankton production. The majority of studies on the effects of bivalves on nutrient recycling have focused on nitrogen because this is the most common nutrient-limiting biological production in marine and estuarine systems (National Research Council 2000). Hence, this section – heavily reliant on information in Newell *et al.* (2002), Newell (2004) and Dame (2012) – will focus on nitrogen, which undergoes numerous transformations and state changes as it cycles through the environment (Dame 2012). Bivalves, including mussels, obtain their nitrogen not just from phytoplankton but also from N-rich bacteria and flagellates. When bivalve biodeposits settle to the sediment surface, any remaining particulate organic nitrogen (PON) undergoes bacterial degradation, which can lead to some NH_4^+ being regenerated to the water column (Figure 4.27). This form of nitrogen is preferentially taken up by photoplankton. This is an example of a positive feedback loop, with both bivalves and phytoplankton benefitting from the process (Dame 2012). Measured rates of NH_4^+ flux can be high, ranging from ~1 to 5 mmol N m^{-2} h^{-1} in bivalve populations. Nitrogen that is not bacterially metabolised becomes buried in the accumulating sediments. If there is sufficient oxygen in the sediments then aerobic denitrifying bacteria can oxidise nitrogen compounds to NO_2^- and NO_3^-, some of which diffuses out of the sediment into the water column's pool of dissolved inorganic nitrogen and some of which diffuses down into the underlying anaerobic sediments. Here, denitrifying bacteria reduce NO_2^- and NO_3^- to N_2 gas. This gas, which is in a form unavailable for phytoplankton production, passes to the atmosphere. Where there is an active benthic microalgal community, it limits nutrient release to the water column by assimilating inorganic nitrogen released by bacterial decomposition. But when nitrogen requirements exceed that supplied from the water column and regenerated from the sediment, the plants can convert N_2 to NH_4^+, a process called nitrogen fixation.

Those wishing to explore the regeneration of other biologically essential elements such as carbon, phosphorous and silicon would be well advised to read the very comprehensive review in Dame (2012), plus Prins & Smaal (1994), Cranford *et al.* (2007), Bracken *et al.* (2012), Carlsson *et al.* (2012), Konrad (2014), Filgueira *et al.* (2015) and van Broekhoven *et al.* (2015).

Notes

1 Gut passage/transit time (GPT/GTT) is defined as the minimum time for an organic particle to pass through the digestive tract of an individual after ingestion (Galimany *et al.* 2013).
2 Zeitgeber is a rhythmically occurring natural phenomenon which acts as a cue in the regulation of the body's circadian rhythms.
3 Tychopelagic organisms are normally benthic but have been carried up into the water column by chance factors.
4 The shorthand notation of a fatty acid is A:BωX, where A represents the number of carbon atoms, B the number of double bonds and X the position of the double bond closest to the terminal methyl group.

References

Ackerman, J. & Nishizaki, M. (2004) The effect of velocity on the suspension feeding and growth of the marine mussels *Mytilus trossulus* and *M. californianus*: implications for niche separation. *Journal of Marine Systems*, **49**, 195–207.

Aiello, E. & Sleigh, M.A. (1972) The metachronal wave of lateral cilia of *Mytilus edulis*. *Journal of Cell Biology*, **54**, 493–506.

Alfaro, A.C. (2006) Evidence of cannibalism and bentho-pelagic coupling within the life cycle of the mussel, *Perna canaliculus*. *Journal of Experimental Marine Biology and Ecology*, **329**, 206–217.

Allan, E.L., Ambrose, S.T., Richoux, N.B. & Froneman, P.W. (2010) Determining spatial changes in the diet of nearshore suspension-feeders along the South African coastline: stable isotope and fatty acid signatures. *Estuarine, Coastal and Shelf Science*, **87**, 463–471.

Alyakrinskaya, I.O. (2001) The dimensions, characteristics and functions of the crystalline style of mollusks. *Biological Bulletin*, **28**, 523–535.

Amiard, J-C., Amiard-Triquet, C., Barka, S., Pellerin, J. & Rainbow, P.S. (2006) Metallothioneins in aquatic invertebrates: their role in metal detoxification and their use as biomarkers. *Aquatic Toxicology*, **76**, 160–202.

Arapov, J., Ezgeta–Balić, D., Peharda, M. & Ninčević Gladan, Ž. (2010) Bivalve feeding–how and what they eat? *Croatian Journal of Fisheries*, **68**, 105–116.

Bacon, G.S., MacDonald, B.A. & Ward, J.E. (1998) Physiological responses of infaunal (*Mya arenaria*) and epifaunal (*Placopecten magellanicus*) bivalves to variations in the concentration and quality of suspended particles. *I. Feeding activity and selection. Journal of Experimental Marine Biology and Ecology*, **219**, 105–125.

Bayne, B.L. (1998) The physiology of suspension feeding by bivalve mollusks: an introduction to the Plymouth 'TROPHEE' workship. *Journal of Experimental Marine Biology and Ecology*, **219**, 1–19.

Bayne, B.L. (2002) A physiological comparison between Pacific oysters *Crassostrea gigas* and Sydney Rock oysters *Saccostrea glomerata*: food, feeding and growth in a shared estuarine habitat. *Marine Ecology Progress Series*, **232**, 163–178.

Bayne, B.L. (2004) Comparisons of measurements of clearance rates in bivalve molluscs. *Marine Ecology Progress Series*, **276**, 305–306.

Bayne, B.L. & Newell, R.C. (1983) Physiological energetics of marine molluscs. In: *The Mollusca*, Vol. 4, *Physiology*, Part 1 (eds A.S.M. Saleuddin & K.M. Wilbur), pp. 407–515. Academic Press, New York.

Bayne, B.L. & Hawkins, A.J.S. (1990) Filter-feeding in bivalve molluscs: controls on energy balance. In: *Animal Nutrition and Transport Processes, Vol. 1. Nutrition in Wild and Domestic Animals* (ed J. Mellinger), pp. 70–83. Karger, Basel.

Bayne, B.L., Thompson, R.J. & Widdows, J. (1976) Physiology: I. In: *Marine Mussels: Their Ecology and Physiology* (ed B.L. Bayne), pp. 121–206. Cambridge University Press, Cambridge.

Bayne, B.L, Widdows, J. & Newell, R.I.E. (1977) Physiological measurements on estuarine bivalve molluscs in the field. In: *Biology of Benthic Organisms* (eds B.F. Keegan, P.J.S. Boaden & P.O. Céidigh), pp. 57–68. Pergamon Press, Oxford.

Bayne, B.L., Hawkins, A.J.S., Navarro, E. & Iglesias, I.P. (1989) Effects of seston concentration on feeding, digestion and growth in the mussel *Mytilus edulis*. *Marine Ecology Progress Series*, **55**, 47–54.

Bayne, B.L., Iglesias, J.I.P., Hawkins, A.J.S., Navarro, E., Heral, M. & Deslous Paoli, J. M. (1993) Feeding behaviour of the mussel, *Mytilus edulis*: responses to variations in quantity and organic content of the seston. *Journal of the Marine Biological Association of the UK*, **73**, 813–829.

Beadman, H.A., Willows, R.I. & Kaiser, M.J. (2002) Potential applications of mussel modeling. *Helgoland Marine Research*, **56**, 76–85.

Beninger, P.G. (2000) Limits and constraints: a comment on premises and methods in recent studies of particle capture mechanisms in bivalves. *Limnology and Oceanography*, **45**, 1196–1199.

Beninger, P.G. & St–Jean, S.D. (1997a) The role of mucus in particle processing by suspension-feeding marine bivalves: unifying principles. *Marine Biology*, **129**, 389–397.

Beninger, P.G. & St–Jean, S.D. (1997b) Particle processing on the labial palps of *Mytilus edulis* and *Placopecten magellanicus* (Mollusca: Bivalvia). *Marine Ecology Progress Series*, **147**, 117–127.

Beninger, P.G. & Venoit, A. (1999) The oyster proves the rule: mechanisms of pseudofaeces transport and rejection on the mantle of *Crassostrea virginica* and *C. gigas*. *Marine Ecology Progress Series*, **190**, 179–188.

Beninger, P.G. & Decottignies, P. (2008) Worth a second look: gill structure in *Hemipecten forbesianus* (Adams & Reeve, 1849) and taxonomic implications for the Pectinidae. *Journal of Molluscan Studies*, **74**, 137–142.

Beninger, P.G. & Le Pennec, M. (2016) Scallop structure and function. In: *Scallops: Biology, Ecology Aquaculture and Fisheries*, 3rd edn (eds S.E. Shumway & G.J. Parsons), pp. 85–159. Elsevier Science Publishers B.V., Amsterdam.

Beninger, P.G., Le Pennec, M. & Donval, A. (1991) Mode of particle ingestion in five species of suspension-feeding bivalve mollusks. *Marine Biology*, **108**, 255–261.

Beninger, P.G., Ward, J.E., MacDonald, B.A. & Thompson, R.J. (1992) Gill function and particle transport in *Placopecten magellanicus* (Mollusca, Bivalvia) as revealed using video endoscopy. *Marine Biology*, **114**, 281–288.

Beninger, P.G., St-Jean, S.D, Poussart, Y. & Ward, J.E. (1993) Gill function and mucocyte distribution in *Placopecten magellanicus* and *Mytilus edulis* (Mollusca, Bivalvia): the role of mucus in particle transport. *Marine Ecology Progress Series*, **98**, 275–282.

Beninger, P.G., St–Jean, S.D. & Poussart, Y. (1995) Labial palps of the blue mussel *Mytilus edulis* (Bivalvia: Mytilidae). *Marine Biology*, **123**, 293–303.

Beninger, P.G., Lynn, J.W., Dietz, T.H. & Silverman, H. (1997) Mucociliary transport in living tissue: the two-layer model confirmed in the mussel *Mytilus edulis* L. *Biological Bulletin*, **193**, 4–7.

Beninger, P.G., Veniot, A. & Poussart, Y. (1999) Principles of pseudofaeces rejection on the bivalve mantle: integration in particle processing. *Marine Ecology Progress Series*, **178**, 259–269.

Beninger, P.G., Valdizan, A., Cognie, B., Guiheneuf, F. & Decottignies, P. (2008) Wanted: alive and not dead: functioning diatom status is a quality cue for the suspension-feeder *Crassostrea gigas*. *Journal of Plankton Research*, **30**, 689–697.

Bernard, F.R. & Noakes, D.J. (1990) Pumping rates, water pressure and oxygen use in eight species of marine bivalve molluscs from British Columbia. *Canaian Journal of Fisheries and Aquatic Sciences*, **47**, 1302–1306.

Berry, P.F. & Schleyer, M.H. (1983) The brown mussel *Perna perna* on the Natal coast, South Africa: utilization of available food and energy budget. *Marine Ecology Progress Series*, **13**, 201–210.

Birmelin, C., Pipe, R.K., Goldfarb, P.S. & Livingstone, D.R. (1999) Primary cell-culture of the digestive gland of the marine mussel *Mytilus edulis*: a time-course study of antioxidant- and biotransformation-enzyme activity and ultrastructural changes. *Marine Biology*, **135**, 65–75.

Bos, O.G., Hendriks, I.E., Strasser, M., Dolmer, P. & Kamermans, P. (2006) Estimation of food limitation of bivalve larvae in coastal waters of north-western Europe. *Journal of Sea Research*, **55**, 191–206.

Bougrier, S., Hawkins, A.J.S. Héral, M. (1997) Preingestive selection of different microalgal mixtures in *Crassostrea gigas* and *Mytilus edulis*, analysed by flow cytometry. *Aquaculture*, **150**, 123–134.

Bowley, J., Baker-Austin, C., Porter, A., Hartnell, R. & and Lewis, C. (2021) Oceanic hitchhikers-assessing pathogen risks from marine microplastics. *Trends in Microbiology*, **29**, 107–116.

Bracken, M.E.S., Menge, B.A., Foley, M.M., Sorte, C.J.B., Lubchenco, J. & Schiel, D.R. (2012) Mussel selectivity for high-quality food drives carbon inputs into open-coast intertidal ecosystems. *Marine Ecology Progress Series*, **459**, 53–62.

van Broekhoven, W., Jansen, H., Verdegem, M., Struyf, E., Troost, K., Lindeboom, H. *et al.* (2015) Nutrient regeneration from feces and pseudofeces of mussel *Mytilus edulis* spat. *Marine Ecology Progress Series*, **534**, 107–120.

Budge, S.M. & Parrish, C.C. (1999) Lipid class and fatty acid composition of *Pseudo-nitzschia multiseries* and *Pseudo-nitzschia pungens* and effects of lipolytic enzyme deactivation. *Phytochemistry*, **52**,561–566.

Budge, S.M., Parrish, C.C. & Mckenzie, C.H. (2001) Fatty acid composition of phytoplankton, settling particulate matter and sediments at a sheltered bivalve aquaculture site. *Marine Chemistry*, **76**, 285–303.

Bustamante, R.H. & Branch, G.M. (1996) The dependence of intertidal consumers on kelp-derived organic matter on the west coast of South Africa. *Journal of Experimental Marine Biology and Ecology*, **196**, 1–28.

Cajaraville, M.P. & Ortiz-Zarragoitia, M. (2006) Specificity of the peroxisome proliferation response in mussels exposed to environmental pollutants. *Aquatic Toxicology*, **78S**, S117–123.

Callier, M.D., Weise, A.M., McKindsey, C.W. & Desrosiers, G. (2006) Sedimentation rates in a suspended mussel farm (Great-Entry Lagoon, Canada): biodeposit production and dispersion. *Marine Ecology Progress Series*, **322**, 129–141.

Canesi, L., Ciacci, C., Fabbri, R., Marcomini, A., Pojana, G. & Gallo, G.C. (2012) Bivalve molluscs as a unique target group for nanoparticle toxicity. *Marine Environmental Research*, **76**, 16–21.

Cannuel, R. & Beninger, P.G. (2006) Gill development, functional and evolutionary implications in the Pacific oyster *Crassostrea gigas* (Bivalvia: Ostreidae). *Marine Biology*, **149**, 547–563.

Cannuel, R., Beninger, P.G. McCombie, H. & Boudry, P. (2009) Gill development and its functional and evolutionary implications in the blue mussel *Mytilus edulis* (Bivalvia: Mytilidae). *Biological Bulletin*, **217**, 173–188.

Carlsson, M.S., Glud, R.N. & Petersen, J.K. (2010) Degradation of mussel (*Mytilus edulis*) fecal pellets released from hanging long-lines upon sinking and after settling at the sediment. *Canadian Journal of Fisheries and Aquatic Sciences*, **67**, 1376–1387.

Carlsson, M.S., Engström, P., Lindahl, O., Ljungqvist, L., Petersen, J.K., Svanberg, I. *et al.* (2012) Effects of mussel farms on the benthic nitrogen cycle on the Swedish west coast. *Marine Ecology Progress Series*, **2**, 177–191.

Cavallo, R.A., Acquaviva, M.I. & Stabili, L. (2009) Culturable heterotrophic bacteria in seawater and *Mytilus galloprovincialis* from a Mediterranean area (Northern Ionian Sea – Italy). *Environmental Monitoring and Assessment*, **149**, 465–475.

Charles, F. & Newell, R.I.E. (1997) Digestive physiology of the ribbed mussel *Geukensia demissa* (Dillwyn) held at different tidal heights. *Journal of Experimental Marine Biology and Ecology*, **209**, 201–213.

Clausen, I. & Riisgård, H.U. (1996) Growth, filtration and respiration in the mussel *Mytilus edulis*: no evidence for physiological regulation of the filter-pump to nutritional needs. *Marine Ecology Progress Series*, **141**, 37–45.

Colaço, A., Prieto, C., Martins, A., Figueiredo, M., Lafon, V., Monteiro, M. *et al.* (2009) Seasonal variations in lipid composition of the hydrothermal vent mussel *Bathymodiolus azoricus* from the Menez Gwen vent field, *Marine Environmental Research*, **67**, 146–152.

Comeau, L.A., Pernet, F., Tremblay, R., Bates, S.S. & LeBlanc, A. (2008) Comparison of eastern oyster (*Crassostrea virginica*) and blue mussel (*Mytilus edulis*) filtration rates at low temperatures. *Canadian Technical Report of Fisheries and Aquatic Sciences*, **2810**, vii + 17pp.

Connor, K.M., Sung, A., Garcia, N.S., Gracey, A.Y. & German, D.P. (2016) Modulation of digestive physiology and biochemistry in *Mytilus californianus* in response to feeding level acclimation and microhabitat. *Biology Open*, **5**, 1200–1210.

Conover, R. (1966) Assimilation of organic matter by zooplankton. *Limnology and Oceanography*, **11**, 338–354.

Coughlan, J. & Ansell, A.D. (1964) A direct method for determining the pumping rate of siphonate bivalves. *ICES Journal of Marine Science*, **29**, 205–213.

Cranford, P.J., Emerson, C.W., Hargrave, B.T. & Milligan T.G. (1998) *In situ* feeding and absorption responses of sea scallops *Placopecten magellanicus* (Gmelin) to storm-induced changes in the quantity and composition of the seston. *Journal of Experimental Marine Biology and Ecology*, **219**, 45–70.

Cranford, P.J., Strain, P.M., Dowd, M., Hargrave, B.T., Grant, J. & Archambault, M.-C. (2007) Influence of mussel aquaculture on nitrogen dynamics in a nutrient enriched coastal embayment. *Marine Ecology Progress Series*, **S347**, 61–78.

Cucci, T.L., Shumway, S.E., Newell, R.C., Selvin, R., Guillard, R.L. & Yentsch, C.M. (1985) Flow cytometry: a new method for characterization of differential ingestion, digestion and egestion by suspension feeders. *Marine Ecology Progress Series*, **24**, 201–204.

Culbertson, J.B., Valiela, I., Olsen, Y.S. & Reddy, C.M. (2008) Effect of field exposure to 38-year-old residual petroleum hydrocarbons on growth, condition index, and filtration rate of the ribbed mussel, *Geukensia demissa*. *Environmental Pollution*, **154**, 312–319.

Dame, R.F. (2012) *Ecology of Marine Bivalves: An Ecosystem Approach*, 2nd edn. CRC Press, Boca Raton, FL.

Davenport, J., Ezgeta-Balić, D., Peharda, M., Skejić Ninčević-Gladan, Z. & Matijević, S. (2011) Size-differential feeding in *Pinna nobilis* L. (Mollusca: Bivalvia): exploitation of detritus, phytoplankton and zooplankton. *Estuarine, Coastal Shelf Science*, **92**, 246–254.

Defossez, J.–M. & Hawkins, A.J.S. (1997) Selective feeding in shellfish: size-dependent rejection of large particles within pseudofaeces from *Mytilus edulis*, *Ruditapes philippinarum* and *Tapes decussatus*. *Marine Biology*, **129**, 139–147.

Denis, L., Alliot, E. & Grzebyk. D. (1999) Clearance rate responses of Mediterranean mussels, *Mytilus galloprovincialis*, to variations in the flow, water temperature, food quality and quantity. *Aquatic Living Resources*, **12**, 279–288.

Dernekbaşı S, Öksüz A, Çelik M.Y., Karayücel İ. & Karayücel, S. (2015) The fatty acid composition of cultured mussels (*Mytilus galloprovincialis* Lamarck 1819) in offshore longline system in the Black Sea. *Journal of Aquaculture and Marine Biology*, **2**, 00049.

Diaz, R.J. & Rosenberg, R. (1995) Marine benthic hypoxia, a review of its ecological effects and the behavioral responses of benthic macrofauna. *Oceanography and Marine Biology – An Annual Review*, **33**, 245–303.

Dimitriadis, V.K., Domouhtsidou, G.P. & Cajaraville, M.P. (2004) Cytochemical and histochemical aspects of the digestive gland cells of the mussel *Mytilus galloprovincialis* (L.) in relation to function. *Journal of Molecular Histology*, **35**, 501–509.

Dittmann, S. (1990) Mussel beds–amensalism or amelioration for intertidal fauna? *Helgolander Meeresuntersuchungen*, **44**, 335–352.

Dolmer, P. (2000) Algal concentration profiles above mussel beds. *Journal of Sea Research*, **43**, 113–119.

Doyle, J.J., Ward, J.E. & Mason, R. (2016) Exposure of bivalve shellfish to titania nanoparticles under an environmental-spill scenario: encounter, ingestion and egestion. *Journal of the Marine Biological Association of the UK*, **96**, 137–149.

Dral, A.D.G. (1967) The movement of the latero-frontal cilia and mechanism of particle retention in the mussel (*Mytilus edulis* L.). *Netherlands Journal of Sea Research*, **3**, 391–422.

Dufour, S.C. & Beninger, P.G. (2001) A functional interpretation of cilia and mucocyte distributions on the abfrontal surface of bivalve gills. *Marine Biology*, **138**, 295–309.

Ezgeta-Balić, D., Rinaldi, A., Peharda, M., Prusina, I., Montalto, V., Niceta, N. et. al. (2011) An energy budget for the subtidal bivalve *Modiolus barbatus* (Mollusca) at different temperatures. *Marine Environmental Research*, **71**, 79–85.

Ezgeta-Balić, D., Najdek, M. Peharda, M. & Blažina, M. (2012) Seasonal fatty acid profile analysis to trace origin of food sources of four commercially important bivalves. *Aquaculture*, **334–337**, 89–100.

Famme, P., Riisgård, H.U. & Jørgensen, C.B. (1986) On direct measurements of pumping rates in the mussel *Mytilus edulis*. *Marine Biology*, **92**, 323–327.

Fernández-Reiriz, M-J., Garrido, J-L. & Irisarri, J. (2015a) Fatty acid composition in *Mytilus galloprovincialis* organs: trophic interactions, sexual differences and differential anatomical distribution. *Marine Ecology Progress Series*, **528**, 221–234.

Fernández-Reiriz, M.J., Irisarri, J. & Labarta, U. (2015b) Feeding behaviour and differential absorption of nutrients in mussel *Mytilus galloprovincialis*: responses to three microalgae diets. *Aquaculture*, **446**, 42–47.

Filgueira, R., Labarta, U. & Fernández-Reiriz, M.J. (2006) Flow-through chamber method for clearance rate measurements in bivalves: design and validation of individual chambers and mesocosm. *Limnology and Oceanography-Methods*, **4**, 284–292.

Filgueira, R., Labarta, U. & Fernández-Reiriz, M.J. (2008) Effect of condition index on allometric relationships of clearance rate in *Mytilus galloprovincialis* Lamarck, 1819. *Revista de Biología Marina y Oceanografía*, **43**, 391–398.

Filgueira, R., Byron, C.J., Comeau, L.A., Costa-Pierce, B., Cranford, P.J., Ferreira, J.G. *et al.* (2015) An integrated ecosystem approach for assessing the potential role of cultivated bivalve shells as part of the carbon trading system. *Marine Ecology Progress Series*, **518**, 281–287.

Fokina, N.N., Ruokolainen, T.R. & Nemova, N.N. (2017) Lipid composition modifications in the blue mussels (*Mytilus edulis* L.) from the White Sea. In: *Organismal and Molecular Malacology* (ed. S. Ray), pp. 143–159. InTech, London.

Foster-Smith, R.L. (1975) The effect of concentration of suspension and inert material on the assimilation of algae by three bivalves. *Journal of the Marine Biological Association of the UK*, **55**, 411–418.

Frank, D. M., Ward, J. E., Shumway, S. E., Holohan, B. A. & Gray, C. (2008) Application of particle image velocimetry to the study of suspension feeding in marine invertebrates. *Marine and Freshwater Behaviour and Physiology*, **41**, 11–18.

Franzellitti, S., Buratti, S., Du, B., Haddad, S.P., Chambliss, C.K., Brooks, B.W. *et al.* (2015) A multibiomarker approach to explore interactive effects of propranolol and fluoxetine in marine mussels. *Environmental Pollution*, **205**, 60–69.

Fréchette, M. (2012) A model of clearance rate regulation in mussels. *Journal of Sea Research*, **73**, 32–40.

Fréchette, M., Urquiza, J.M., Daigle, G. & Rioux-Gagnon, D. (2016) Clearance rate regulation inmussels: adding the effect of seston level to a model of internal state-based regulation. *Journal of Experimental Marine Biology and Ecology*, **475**, 1–10.

Froján, M., Arbones, B., Zúñiga, D., Castro, C.G. & Figueiras, F.G. (2014) Microbial plankton community in the Ría de Vigo (NW Iberian upwelling system): impact of the culture of *Mytilus galloprovincialis*. *Marine Ecology Progress Series*, **498**, 43–54.

Galimany, E., Alix, J.H., Dixon, M.S. & Wikfors, G.H. (2013) Short communication: adaptability of the feeding behavior of intertidal ribbed mussels (*Geukensia demissa*) to constant submersion. *Aquaculture International*, **21**, 1009–1015.

Galimany, E., Rose, J.M., Alix, J.H., Dixon, M.S. & Wikfors, G.H. (2014) Responses of the ribbed mussel, *Geukensia demissa*, to the harmful algae *Aureococcus anophagefferens* and *Heterosigma akashiwo*. *Journal of Molluscan Studies*, **80**, 123–130.

Galimany, E., Rose, J.M., Dixon, M.S. and Wikfors, G.H. (2015) Transplant experiment to evaluate the feeding behavior of the Atlantic ribbed mussel, *Geukensia demissa*, moved to a high inorganic seston area. *Marine and Freshwater Research*, **66**, 220–222.

Galtsoff, P.S. (1964) The American oyster *Crassostrea virginica* Gmelin. *Fishery Bulletin*, **64**, 1–480.

Gao, Q.F., Shin, P.K.S., Lin, G.H., Chen, S.P. & Cheung, S.G. (2006) Stable isotope and fatty acid evidence for uptake of organic waste by green-lipped mussels *Perna viridis* in a polyculture fish farm system. *Marine Ecology Progress Series*, **317**, 273–283.

Garmendia, L., Soto, M., Vicario, U., Kim, Y., Cajaraville, M.P. & Marigómez, I. (2011) Application of a battery of biomarkers in mussel digestive gland to assess long-term effects of the Prestige oil spill in Galicia and Bay of Biscay: tissue-level biomarkers and histopathology. *Journal of Environmental Monitoring*, **13**, 915–932.

Garrido, M.V., Chaparro, O.R., Thompson, R.J., Garrido, O. & Navarro, J.M. (2012) Particle sorting and formation and elimination of pseudofaeces in the bivalves *Mulinia edulis* (siphonate) and *Mytilus chilensis* (asiphonate). *Marine Biology*, **159**, 987–1000.

Giles, H. & Pilditch, C.A. (2004) Effects of diet on sinking rates and erosion thresholds of mussel *Perna canaliculus* biodeposits. *Marine Ecology Progress Series*, **282**, 205–219.

Giles, H., Broekhuizen, N., Bryan, K.R. & Pilditch, C.A. (2009) Modelling the dispersal of biodeposits from mussel farms: the importance of simulating biodeposit erosion and decay. *Aquaculture*, **291**, 168–178.

Giribet, G. (2008) Bivalvia. In: *Phylogeny and Evolution of the Mollusca.* (eds W.F Ponder & D.R. Lindberg), pp 105–141. University of California Press, USA.

Gosling, E.M. (2015) *Marine Bivalve Molluscs*, 2nd edn. Wiley Blackwell, Hoboken, NJ.

Grant, J., Hatcher, A., Scott, D.B., Pocklington, P., Schafer, C.T. & Winters, G.V. (1995) A multidisciplinary approach to evaluating impacts of shellfish aquaculture on benthic communities. *Estuaries*, **18**, 124–144.

Grant, J., Bacher, C., Cranford, P.J., Guyondet, T. & Carreau, M. (2008) A spatially explicit ecosystem model of seston depletion in dense mussel culture. *Journal of Marine Science*, **73**, 155–168.

Gregory, M.A., Marshall, D.J., George, R.C., Anandraj, A. & McClurg, T.P. (2002) Correlations between metal uptake in the soft tissue of *Perna perna* and gill filament pathology after exposure to mercury. *Marine Pollution Bulletin*, **45**, 114–125.

Griffiths, R.J. (1980) Filtration, respiration and assimilation in the black mussel *Choromytilus meridionalis. Marine Ecology Progress Series*, **3**, 63–70.

Hamza-Chaffai, A. (2014) Usefulness of bioindicators and biomarkers in pollution biomonitoring, *International Journal of Biotechnology for Wellness Industries*, **3**, 19–26.

Handå, A., Min, H., Wang, X., Broch, O.J., Reitan, K.I., Reinertsen, H. *et al.* (2012) Incorporation of fish feed and growth of blue mussels (*Mytilus edulis*) in close proximity to salmon (*Salmo salar*) aquaculture: implications for integrated multi-trophic aquaculture in Norwegian coastal waters. *Aquaculture*, **356–357**, 328–341.

Hansen, B.W., Dolmer, P. & Vismann, B. (2011) In situ method for measurements of community clearance rate on shallow water bivalve populations. *Limnology and Oceanography-Methods*, **9**, 454–459.

Hawkins, A.J.S. & Bayne, B.L. (1992) Physiological interrelations and the regulation of production. In: *The Mussel Mytilus: Ecology, Physiology, Genetics and Culture* (ed E.M. Gosling), pp. 171–222. Elsevier Science Publishers B.V., Amsterdam.

Hawkins, A.J.S., Smith, R.F.M., Bayne, B.L. & Héral, M. (1996) Novel observations underlying the fast growth of suspension-feeding shellfish in turbid environments: *Mytilus edulis. Marine Ecology Progress Series*, **131**, 179–190.

Hawkins, A.J.S., Smith, R.F.M., Tan, S.H. & Yasin, Z.B. (1998a) Suspension-feeding behaviour in tropical bivalve molluscs, *Perna viridis*, *Crassostrea belcheri*, *Crassostrea iredalei*, *Saccostrea cucculata* and *Pinctada margarifera*. *Marine Ecology Progress Series*, **166**, 173–185.

Hawkins, A.J.S., Bayne, B.L., Bougier, S., Héral, M., Iglesias, J.I.P., Navarro, E. *et al.* (1998b) Some general relationships in comparing the feeding physiology of suspension-feeding bivalve mollusks. *Journal of Experimental Marine Biology and Ecology*, **219**, 87–103.

Hawkins, A.J.S., James, M.R., Hickman, R.W., Hatton, S. & Weatherhead, M. (1999) Modelling of suspension feeding and growth in the green-lipped mussel *Perna canaliculus* exposed to natural and experimental variations of seston availability in the Marlborough Sounds, New Zealand. *Marine Ecology Progress Series*, **191**, 217–232.

Heinonen, K.B., Ward, J.E. & Holohan, B.A. (2007) Production of transparent exopolymer particles (TEP) by benthic suspension feeders in coastal systems. *Journal of Experimental Marine Biology and Ecology*, **341**, 184–195.

Hernroth, B., Larsson, A. & Edebo, L. (2000) Influence on uptake, distribution and elimination of *Salmonella typhimurium* in the blue mussel, *Mytilus edulis*, by the cell surface properties of the bacteria. *Journal of Shellfish Research*, **19**, 167–174.

Hill, J.M., McQuaid, C.D. & Kaehler, S. (2006) Biogeographic and nearshore–offshore trends in isotope ratios of intertidal mussels and their food sources around the coast of southern Africa. *Marine Ecology Progress Series*, **318**, 63–73.

Huang, S.C., Kreeger, D.A. & Newell, R.I.E. (2003a) Tidal and seasonal variations in the quantity and composition of seston in a North American, mid-Atlantic salt marsh. *Estuarine, Coastal Shelf Science*, **56**, 547–560.

Huang, S.C., Kreeger, D.A. & Newell, R.I.E. (2003b) Seston available as a food resource for the ribbed mussel (*Geukensia demissa*) in a North American, mid-Atlantic saltmarsh. *Estuarine, Coastal Shelf Science*, **56**, 561–571.

Huot, Y., Babin, M., Bruyant, F., Grob, C., Twardowski, M.S. & Claustre, H. (2007) Does chlorophyll *a* provide the best index of phytoplankton biomass for primary productivity studies? *Biogeosciences Discussions*, **4**, 707–745.

Irisarri, J., Fernández-Reiriz, M.J., Robinson, S.M.C., Cranford, P.J. & Labarta, U. (2013) Absorption efficiency of mussels *Mytilus edulis* and *Mytilus galloprovincialis* cultured under Integrated Multi-Trophic Aquaculture conditions in the Bay of Fundy (Canada) and Ría Ares-Betanzos (Spain). *Aquaculture*, **388–391**, 182–192.

Irisarri, J., Fernández-Reiriz, M.J., De Troch, M. & Labarta, U. (2014) Fatty acids as tracers of trophic interactions between seston, mussels and biodeposits in a coastal embayment of mussel rafts in the proximity of fish cages. *Comparative Biochemistry and Physiology Part-B*, **172–173**, 105–115.

Izagirre, U., Ruiz, P. & Marigómez, I. (2009) Time-course study of the early lysosomal responses to pollutants in mussel digestive cells using acid phosphatase as lysosomal marker enzyme. *Comparative Biochemistry and Physiology Part-C*, **149**, 587–597.

Jacobs, P., Troost, K., Riegman, R. & van der Meer, J. (2015) Length- and weight-dependent clearance rates of juvenile mussels (*Mytilus edulis*) on various planktonic prey items. *Helgoland Marine Research*, **69**,101–112.

Jaramillo, E., Bertran, C. & Bravo, A. (1992) Mussel biodeposition in an estuary in southern Chile. *Marine Ecology Progress Series*, **82**, 85–94.

Jones, H.D., Richards, O.G. & Southern, T.A. (1992) Gill dimensions, water pumping rate and body size in the mussel *Mytilus edulis*. *Journal of Experimental Marine Biology and Ecology*, **155**, 213–237.

Jonsson, P.R., Petersen, J.K., Karlsson, O., Loo, L-O. & Nilsson, S. (2005) Particle depletion above experimental bivalve beds: *in situ* measurements and numerical modeling of bivalve filtration in the boundary layer. *Limnology and Oceanography*, **50**, 1989–1998.

Jørgensen, C.B. (1990) *Bivalve Filter Feeding: Hydrodynamics, Bioenergetics, Physiology and Ecology*. Olsen & Olsen, Fredensborg, Denmark.

Jørgensen, C.B. (1996) Bivalve filter feeding revisited. *Marine Ecology Progress Series*, **142**, 287–302.

Jørgensen, C.B., Larsen, P.S., Møhlenberg, F. & Riisgård, H.U. (1988) The mussel pump: properties and modelling. *Marine Ecology Progress Series*, **45**, 205–216.

Jørgensen, C.B., Larsen, P.S. & Riisgård, HU (1990) Effects of temperature on the mussel pump. *Marine Ecology Progress Series*, **64**, 89–97.

Kach, D.J. & Ward, J.E. (2008) The role of marine aggregates in the ingestion of picoplankton-size particles by suspension-feeding molluscs. *Marine Biology*, **153**, 797–805.

Kádár, E., Davis, S.A. & Lobo-da-Cunha, A. (2008) Cytoenzymatic investigation of intracellular digestion in the symbiont-bearing hydrothermal bivalve *Bathymodiolus azoricus*. *Marine Biology*, **153**, 995–1004.

Khati, W., Ouali, K., Mouneyrac, C. & Banaoui, A. (2012) Metallothioneins in aquatic invertebrates: their role in metal detoxification and their use in biomonitoring. *Energy Procedia*, **18**, 784–794.

Kiørboe, T.F. & Møhlenberg, F. (1981) Particle selection in suspension-feeding bivalves. *Marine Ecology Progress Series*, **5**, 291–296.

Kittner, C. & Riisgård, H.U. (2005) Effect of temperature on filtration rate in the mussel *Mytilus edulis*: no evidence for temperature compensation. *Marine Ecology Progress Series*, **305**, 147–152.

Konrad, C.P. (2014) Approaches for evaluating the effects of bivalve filter feeding on nutrient dynamics in Puget Sound, Washington: U.S. Geological Survey Scientific Investigations Report 2013–5237, 22pp.

Kramer, A.M., Ward, J.E., Dobbs, F.C., Pierce, M.L. & Drake, J.M. (2016) The contribution of marine aggregate-associated bacteria to the accumulation of pathogenic bacteria in oysters: an agent-based model. *Ecology and Evolution*, **6**, 7397–7408.

Kreeger, D.A. (1993) Seasonal patterns in the utilization of dietary protein by the mussel *Mytilus trossulus*. *Marine Ecology Progress Series*, **95**, 215–232.

Kreeger, D.A. & Newell, R.I.E. (1996) Ingestion and assimilation of carbon from cellulolytic bacteria and heterotrophic flagellates by the mussels *Geukensia demissa* and *Mytilus edulis* (Bivalvia, Mollusca). *Marine Ecology Progress Series*, **11**, 205–214.

Labarta, U., Fernández-Reiriz, M.J. & Babarro, J.M.F. (1997) Differences in physiological energetics between intertidal and raft cultivated mussels *Mytilus galloprovincialis*. *Marine Ecology Progress Series*, **152**, 165–173.

Langton, R.W. (1975) Synchrony in the digestive diverticula of *Mytilus edulis* L. *Journal of the Marine Biological Association of the UK*, **55**, 221–229.

Langton, R.W. (1977) Digestive rhythms in the mussel *Mytilus edulis*. *Marine Biology*, **41**, 53–58.

Langdon, C.J. & Newell, R.I.E. (1990) Utilization of detritus and bacteria as food sources by two bivalve suspension feeders, the oyster, *Crassostrea virginica* and the mussel *Geukensia demissa*. *Marine Ecology Progress Series*, **8**, 299–310.

Larsen, P.S. & Riisgård, H.U. (2011) Validation of the flow-through chamber (FTC) and steady-state (SS) methods for clearance rate measurements in bivalves. *Biology Open*, **1**, 6–11.

Lefebvre, S., Marin-Leal, J.C.M., Dubois, S., Orvain, F., Blin, J.L., Bataille, M.P. *et al.* (2009) Seasonal dynamics of trophic relationships among co-occurring suspension-feeders in two shellfish culture dominated ecosystems. *Estuarine, Coastal and Shelf Science*, **82**, 415–425.

Lehane, C. & Davenport, J. (2002) Ingestion of mesozooplankton by three species of bivalve; *Mytilus edulis*, *Cerastoderma edule* and *Aequipecten opercularis*. *Journal of the Marine Biological Association of the UK*, **82**, 615–619.

Lehane, C. & Davenport, J. (2004) Ingestion of bivalve larvae by *Mytilus edulis*: experimental and field demonstrations of larviphagy in farmed blue mussels. *Marine Biology*, **145**, 101–107.

Li, B., Ward, J.E. & Holohan, B.A. (2008) Transparent exopolymer particles (TEP) from marine suspension feeders enhance particle aggregation. *Marine Ecology Progress Series*, **357**, 67–77.

Lonsdale, D.J., Cerrato, R.M., Holland, R., Mass, A., Holt, L., Schaffner, R.A. *et al.* (2009) Influence of suspension-feeding bivalves on the pelagic food webs of shallow, coastal embayments. *Aquatic Biology*, **6**, 263–279.

Lüskow, F. & Riisgrd, H.U. (2018) In Situ filtration rates of blue mussels (Mytilus edulis) measured by an open-top chamber method. *Open Journal of Marine Science*, **8**, 395–406.

Maar, M., Nielsen, T.G. & Petersen, J.K. (2008) Depletion of plankton in a raft culture of *Mytilus galloprovincialis* in Ría de Vigo, NW Spain. II. Zooplankton. *Aquatic Biology*, **4**, 127–141.

MacDonald, B.A. & Ward, J.E. (2009) Feeding activity of scallops and mussels measured simultaneously in the field: repeated measures sampling and implications for modeling. *Journal of Experimental Marine Biology and Ecology*, **371**, 42–50.

Mackenzie, A. L. & Marshall, C. (2014) Proteins in the crystalline styles of the marine mussels *Perna canaliculus* Gmelin and *Mytilus galloprovincialis* Lamarck. *Journal of Shellfish Research*, **33**, 673–685.

Mahmood, K., Jadoon, S., Mahmood, O., Irshad, M. & Hussain, J. (2014) Synergistic effects of toxic elements on heat shock proteins. *BioMed Research International*, **2014**, art. 564136.

Maire, O., Amouroux, J-M., Duchêne, J-C. & Grémare, A. (2007) Relationship between filtration activity and food availability in the Mediterranean mussel *Mytilus galloprovincialis*. *Marine Biology*, **152**, 1293–1307.

Maloy. A.P., Culloty, S.C. & Slater, J.W. (2009) Use of PCR-DGGE to investigate the trophic ecology of marine suspension feeding bivalves. *Marine Ecology Progress Series*, **381**, 109–118.

Maloy, A.P., Culloty, S.C., Slater, J.W., Nelle, P. & Harrod, C. (2013) Identifying trophic variation in a marine suspension feeder: DNA- and stable isotope-based dietary analysis in *Mytilus* spp. *Marine Biology*, **160**, 479–490.

Manahan, D.T., Wright, S.H., Stephens, G.C. & Rice, M.A. (1982) Transport of dissolved amino acids by the mussel *Mytilus edulis*: demonstration of net uptake from natural seawater. *Science*, **215**, 1253–1255.

Manahan, D.T., Wright, S.H. & Stephens, G.C. (1983) Simultaneous determination of net uptake of sixteen amino acids by a marine bivalve. *American Journal of Physiology*, **244**, 832–838.

Mathers, N.F. (1972) The tracing of a natural algal food labeled with a carbon^{14} isotope through the digestive tract of *Ostrea edulis*. *Proceedings of the Malacological Society London*, **40**, 115–124.

Mathers, N.F. (1973a) Carbohydrate digestion in *Ostrea edulis* L. *Proceedings of the Malacological Society London*, **40**, 359–367.

Mathers, N.F. (1973b) A comparative histochemical survey of enzymes associated with the processes of digestion in *Ostrea edulis* and *Crassostrea angulata* (Mollusca: Bivalvia). *Journal of Zoology*, **169**, 169–179.

May, S.P., Burkholder, J.M., Shumway, S.E. Hégaret, H., Wikfors, G.H. & Frank, D. (2010) Effects of the toxic dinoflagellate *Alexandrium monilatum* on survival, grazing and behavioral response of three ecologically important bivalve molluscs. *Harmful Algae*, **9**, 281–293.

McFarland, K., Donaghy, L. & Volety, A.K. (2013) Effect of acute salinity changes on hemolymph osmolality and clearance rate of the non-native mussel, *Perna viridis*, and the native oyster, *Crassostrea virginica*, in Southwest Florida. *Aquatic Invasions*, **8**, 299–310.

McKindsey, C., Anderson, M.R., Barnes, P., Courtenay, S. Landry, T. & Skinner, M. (2006) Effects of shellfish aquaculture on fish habitat. Canadian Science Advisory Secretariat Research Document. 2006/011.

McKindsey, C.W., Archambault, P., Callier, M.D. & Olivier, F. (2011) Influence of suspended and off-bottom mussel culture on the sea bottom and benthic habitats: a review. *Canadian Journal of Zoology*, **89**, 622–646.

Medler, S. & Silverman, H. (2001) Muscular alteration of gill geometry *in vitro*: implications for bivalve pumping processes. *Biological Bulletin*, **200**, 77–86.

Meyhöfer, E. (1985) Comparative pumping rates in suspension-feeding bivalves. *Marine Biology*, **85**, 137–142.

Mićović, V., Bulog, A., Kučič, N., Jakovac, H. & Radošević-Stašić, B. (2009) Metallothioneins and heat shock proteins 70 in marine mussels as sensors of environmental pollution in Northern Adriatic Sea. *Environmental Toxicology and Pharmacology*, **28**, 439–447.

Miller, M.R. Pearce, L. & Bettjeman, B.I. (2014) Detailed distribution of lipids in Greenshell™ Mussel (*Perna canaliculus*). *Nutrients*, **6**, 1454–1474.

Møhlenberg, F. & Riisgård, H.U. (1978) Efficiency of particle retention in 13 species of suspension feeding bivalves. *Ophelia*, **17**, 239–246.

Møhlenberg, F. & Riisgård, H.U. (1979) Filtration rate, using a new indirect technique, in thirteen species of suspension-feeding bivalves. *Marine Biology*, **54**, 143–148.

Moore, M.N. & Allen, J.I. (2002) A computational model of the digestive gland epithelial cell of marine mussels and its simulated responses to oil-derived aromatic hydrocarbons. *Marine Environmental Research*, **54**, 579–584.

Moore, M.N., Lowe, D.M. & Koehler, A. (2004) Measuring lysosomal membrane stability. *ICES Techniques in Marine Environmental Sciences*, No. 36. ICES, Copenhagen.

Moore, M.N., Viarengo, A., Donkin, P. & Hawkins, A.J.S. (2007) Autophagic and lysosomal reactions to stress in the hepatopancreas of blue mussels. *Aquatic Toxicology*, **84**, 80–91.

Morganti, T., Yahel, G., Ribes, M. & Coma, R. (2016) VacuSIP, an improved InEx method for *in situ* measurement of particulate and dissolved compounds processed by active suspension feeders. *Journal of Visualized Experiments*, 114, e54221.

National Research Council. 2000. *Clean Coastal Waters: Understanding and Reducing the Effects of Nutrient Pollution*. The National Academies Press, Washington, DC.

Navarro, J.M. & Velasco, L.A. (2003) Comparison of two methods for measuring filtration rate in filter-feeding bivalves. *Journal of the Marine Biological Association of the UK*, **83**, 553–558.

Navarro, J.M. & Contreras, A.M. (2010) An integrative response by *Mytilus chilensis* to the toxic dinoflagellate *Alexandrium catenella*. *Marine Biology*, **157**,1967–1974.

Navarro, E., Méndez, S., Ibarrola, I. & Urrutia, M.B. (2009) Comparative utilization of phytoplankton and vascular plant detritus by the cockle *Cerastoderma edule*: digestive responses during diet acclimation. *Aquatic Biology*, **6**, 247–262.

Navarro, J.M., Torres, R., Acuña, K., Duarte, C., Manriquez, P.H., Lardies, M. *et al.* (2013) Impact of medium-term exposure to elevated pCO_2 levels on the physiological energetics of the mussel *Mytilus chilensis*. *Chemosphere*, **90**, 1242–1248.

Nevejan, N., De Schryver, P., Wille, M., Dierckens, K., Baruah, K & Van Stappen, G. (2018) Bacteria as food in aquaculture: do they make a difference? *Reviews in Aquaculture*, **10**, 180–212.

Newell, C.R., Wildish, D.J. & MacDonald, B.A. (2001) The effects of velocity and seston concentration on the exhalant siphon area, valve gape and filtration rate of the mussel *Mytilus edulis. Journal of Experimental Marine Biology and Ecology*, **262**, 91–111.

Newell, R.C. & Field, J.G. (1983) The contribution of bacteria and detritus to carbon and nitrogen flow in a benthic community. *Marine Biology Letters*, **4**, 23–36.

Newell, R.I.E. (2004) Ecosystem influences of natural and cultivated populations of suspension-feeding bivalve molluscs: a review. *Journal of Shellfish Research*, **23**, 51–61.

Newell, R.I.E., Cornwell, J.C. & Owens, M.S. (2002) Influence of simulated bivalve biodeposition and microphytobenthos on sediment nitrogen dynamics, a laboratory study. *Limnology and Oceanography*, **47**, 1367–1379.

Nielsen, P. & Vismann, B. (2014) Clearance rate of *Mytilus edulis* (L.) as a function of current velocity and mussel aggregation. *Journal of Shellfish Research*, **33**, 457–463.

Nielsen, T.G. & Maar, M. (2007) Effects of a blue mussel *Mytilus edulis* bed on vertical distribution and composition of the pelagic food web. *Marine Ecology Progress Series*, **339**, 185–198.

Ogilvie, S., Ross, A.H., James, M.R. & Schiel, D.R. (2003) *In situ* enclosure experiments on the influence of cultured mussels (*Perna canaliculus*) on phytoplankton at times of high and low ambient nitrogen. *Journal of Experimental Marine Biology and Ecology*, **295**, 23–39.

Ordóñez, V., Rius, M., McQuaid, C.D., Pineda, M.C., Pascual, M. & Turon, X. (2013) Early biotic interactions among introduced and native benthic species reveal cryptic predation and shifts in larval behavior. *Marine Ecology Progress Series*, **488**, 65–79.

Ostroumov, S.A. & Widdows, J. (2006) Inhibition of mussel suspension feeding by surfactants of three classes. *Hydrobiologia*, **556**, 381–386.

Otero, X.L., Calvo de Anta, R.M. & Macías, F. (2009) Iron geochemistry under mussel rafts in the Galician ria system (Galicia-NW Spain). *Estuarine, Coastal and Shelf Science*, **81**, 83–93.

Owen, G. (1955) Observations on the stomach and digestive diverticula of the Lamellibranchia. I. The Anisomyaria and Eulamellibranchia. *Quarterly Journal of Microcopical Science*, **96**, 517–537.

Pales Espinosa, E. & Allam, B. (2013) Food quality and season affect gene expression of the mucosal lectin MeML and particle sorting in the blue mussel *Mytilus edulis. Marine Biology*, **160**, 1441–1450.

Pales Espinosa, E., Perrigault, M., Ward, J.E., Shumway, S.E. & Allam, B. (2009) Lectins associated with the feeding organs of the oyster *Crassostrea virginica* can mediate particle selection. *Biological Bulletin*, **217**, 130–141.

Pales Espinosa, E., Hassan, D., Ward, J.E., Shumway, S.E. & Allam, B. (2010a) Role of epicellular molecules in the selection of particles by the blue mussel, *Mytilus edulis. Biological Bulletin*, **219**, 50–60.

Pales Espinosa, E., Perrigault, M. & Allam, B. (2010b) Identification and molecular characterization of a mucosal lectin (MeML) from the blue mussel *Mytilus edulis* and its potential role in particle capture. *Comparative Biochemistry and Physiology, Part-A*, **156**, 495–501.

Pales Espinosa, E., Cerrato, R.M., Wikfors, G.H. & Allam, B. (2016) Modeling food choice in the two suspension-feeding bivalves, *Crassostrea virginica* and *Mytilus edulis. Marine Biology*, **163**, 40.

Palmer, R. E. & Williams, L.G. (1990) Effect of particle concentration on filtration efficiency of the bay scallop *Argopecten irradians* and the oyster *Crassostrea virginica*. *Ophelia*, **19**, 163–174.

Pan, J-F. & Wang, W-X. (2004) Differential uptake of dissolved and particulate organic carbon by the marine mussel *Perna viridis*. *Limnology and Oceanography*, **49**, 1980–1991.

Pascoe, P.L., Parry, H.E. & Hawkins, A.J.S. (2009) Observations on the measurement and interpretation of clearance rate variations in suspension-feeding bivalve shellfish. *Aquatic Biology*, **6**, 181–190.

Patetsini, E., Dimitriadis, V.K. & Kaloyianni, M. (2013) Biomarkers in marine mussels, *Mytilus galloprovincialis*, exposed to environmentally relevant levels of the pesticides, chlorpyrifos and penoxsulam. *Aquatic Toxicology*, **126**, 338–345.

Peharda, M., Zupan, I., Bavcevic, L., Frankic, A. & Klanjscek, T. (2007) Growth and condition index of mussel *Mytilus galloprovincialis* in experimental integrated aquaculture. *Aquaculture Research*, **38**, 1714–1720.

Peharda, M., Ezgeta-Balić, D., Davenport, J., Bojanić, N., Vidjak, O. & Ninčević-Gladan, Z. (2012) Differential ingestion of zooplankton by four species of bivalves (Mollusca) in the Mali Ston Bay, Croatia. *Marine Biology*, **159**, 881–895.

Pernet, F., Malet, N., Pastoureaud, A., Vaquer, A., Quéré, C. & Dubroca, L. (2012) Marine diatoms sustain growth of bivalves in a Mediterranean lagoon. *Journal of Sea Research*, **68**, 20–32.

Petersen, J.K. (2004) Methods for measurement of bivalve clearance rate – hope for common understanding. *Marine Ecology Progress Series*, **276**, 309–310.

Petersen, J.K., Sejr, M.K. & Larsen, J.E.N. (2003) Clearance rates in the Arctic bivalves *Hiatella arctica* and *Mya* sp. *Polar Biology*, **26**, 334–341.

Petersen, J.K., Nielsen, T.K., van Duren, L. & Maar, M. (2008) Depletion of plankton in a raft culture of *Mytilus galloprovincialis* in Ría de Vigo, NW Spain. I. Phytoplankton. *Aquatic Biology*, **4**, 113–125.

Petersen, J.K., Maar, M., Ysebaert, T. & Herman, P.M.J. (2013) Near-bed gradients in particles and nutrients above a mussel bed in the Limfjorden: influence of physical mixing and mussel filtration. *Marine Ecology Progress Series*, **490**, 137–146.

Pile, A.J. & Young, C.M. (1999) Plankton availability and retention efficiencies of cold-seep symbiotic mussels. *Limnology and Oceanography*, **44**, 1833–1839.

Phillips, D.L., Inger, R., Bearhop, S., Jackson, A.L., Moore, J.W., Parnell, A.C. *et al.* (2014) Best practices for use of stable isotope mixing models in food-web studies. *Canadian Journal of Zoology*, **92**, 823–835.

Picoche, C., Le Gendre, R., Flye-Sainte-Marie, J., Françoise, S., Maheux, F., Simon, B. *et al.* (2014) Towards the determination of *Mytilus edulis* food preferences using the Dynamic Energy Budget (DEB) theory. *PLoS ONE*, **9**, e109796.

Pipe, R.K. & Moore, M.N. (1985) Ultrastructural changes in the lysosomal±vacuolar system in digestive cells of *Mytilus edulis* as a response to increased salinity. *Marine Biology*, **87**, 157–163.

Pleissner, D., Lundgreen, K., Lüskow, F. & Riisgård, H.U. (2013) Fluorometer controlled apparatus designed for long-duration algal-feeding experiments and environmental effect studies with mussels. *Journal of Marine Biology*, **2013**, art. 401961.

Pond, D.W., Bell, M.V., Dixon, D.R., Fallick, A.E., Segonzac, M & Sargent, J.R. (1998) Bacterial endosymbionts containing methanotrophic and thiotrophic acids in hydrothermal vent mussels. *Applied and Environmental Microbiology*, **64**, 370–375.

Prato, E., Danieli, A., Maffia, M. & Biandolino, F. (2010) Lipid and fatty acid compositions of *Mytilus galloprovincialis* cultured in the Mar Grande of Taranto (southern Italy): feeding strategies and trophic relationships. *Zoological Studies*, **49**, 211–219.

Prieur, D., Nicolas, J.L., Plusquellec, A. & Vigneulle, M. (1990) Interactions between bivalve mollusks and bacteria in the marine environment. *Oceanography and Marine Ecology*, **28**, 277–352.

Prins, T.C. & Smaal, A.C. (1994) The role of the blue mussel *Mytilus edulis* in the cycling of nutrients in the Oosterschelde estuary (The Netherlands). In: *The Oosterschelde estuary (The Netherlands): A Case Study of a Changing Ecosystem* (eds P.H. Nienhuis and A.C Smaal), pp. 413–429. Kluwer, Dordrecht.

Purchon, R.D. (1957) The stomach in the Filibranchia and Pseudolamellibranchia. *Proceedings of the Zoological Society London*, **129**, 27–60.

Raby, D., Mingelbier, M., Dodson, J.J., Klein, B., Lagadeuc, Y. & Legendre, L. (1997) Food-particle size and selection by bivalve larvae in a temperate embayment. *Marine Biology*, **127**, 665–672.

Rafaelli, D.G., Raven, J.A. & Poole, L.J. (1998) Ecological impact of green macroalgal blooms. *Oceanography and Marine Biology-An Annual Review*, **36**, 97–125.

Raftopoulou, E.K. & Dimitriadis, V.K. (2012) Aspects of the digestive gland cells of the mussel *Mytilus galloprovincialis*, in relation to lysosomal enzymes, lipofuscin presence and shell size: contribution in the assessment of marine pollution biomarkers. *Marine Pollution Bulletin*, **64**, 182–188.

Rajagopal, S., Venugopalan, V.P., van der Velde, G. & Jenner, H.A, (2003) Response of fouling brown mussel, *Perna perna* (L.), to chlorine. *Archives of Environmental Contamination and Toxicology*, **44**, 369–376.

Redmond, K.J., Magnesen, T., Kupka Hansen, P., Strand, O. & Meier, S. (2010) Stable isotopes and fatty acids as tracers of the assimilation of salmon fish feed in blue mussels (*Mytilus edulis*). *Aquaculture*, **298**, 202–210.

Reid, R.G.B. (1968) The distribution of digestive tract enzymes in lamellibranchiate bivalves. *Comparative Biochemistry and Physiology*, **24**, 727–744.

Ren, J.S. (2009) Effect of food quality on energy uptake. *Journal of Sea Research*, **62**, 72–74.

Ren, J.S., Ross, A.H. & Hayden, B.J. (2006) Comparison of assimilation efficiency on diets of nine phytoplankton species of the greenshell mussel *Perna canaliculus*. *Journal of Shellfish Research*, **25**, 887–892.

Richoux, N.B & Thompson, R.J. (2001) Regulation of particle transport within the ventral groove of the mussel *Mytilus edulis* gill in response to environmental conditions. *Journal of Experimental Marine Biology and Ecology*, **260**, 199–215.

Riera, P., Stal, L., Nieuwenhuize, J., Richard, P., Blanchard, G. & Gentil, F. (1999) Determination of food sources for benthic invertebrates in a salt marsh (Aiguillon Bay, France) by carbon and nitrogen stable isotopes: importance of locally produced sources. *Marine Ecology Progress Series*, **187**, 301–307.

Riisgård, H.U. (1988) Efficiency of particle retension and filtration rate in 6 species of northeast American bivalves. *Marine Ecology Progress Series*, **45**, 217–223.

Riisgård, H.U. (1991) Filtration rate and growth in the blue mussel. *Mytilus edulis* Linneaus, 1758: dependence on algal concentration. *Journal of Shellfish Research*, **10**, 29–35.

Riisgård, H.U. (2001a) On measurement of filtration rates in bivalves–the stony road to reliable data: review and interpretation. *Marine Ecology Progress Series*, **211**, 275–291.

Riisgård, H.U. (2001b) Physiological regulation versus autonomous filtration in filter-feeding bivalves: starting points for progress. *Ophelia*, **54**, 193–209.

Riisgård, H.U. & Larsen, P.S. (1995) Filter-feeding in marine macroinvertebrates: pump characteristics, modelling and energy cost. *Biological Reviews*, **70**, 67–106.

Riisgard, H.U. & Larsen, P.S. (2000) Comparative ecophysiology of active zoobenthic filter feeding, essence of current knowledge. *Journal of Sea Research*, **44**, 169–193.

Riisgård, H.U. & Larsen, P.S. (2007) Viscosity of seawater controls beat frequency of water-pumping cilia and filtration rate of mussels *Mytilus edulis*. *Marine Ecology Progress Series*, **343**, 141–150.

Riisgård, H. U. & Larsen, P. S. (2010) Particle-capture mechanisms in marine suspension-feeding invertebrates. *Marine Ecology Progress Series*, **418**, 255–293.

Riisgård, H. U. & Larsen, P. S. (2015) Physiologically regulated valve-closure makes mussels long-term starvation survivors: test of hypothesis. *Journal of Molluscan Studies*, **81**, 303–307.

Riisgård, H.U. & Møhlenberg, F. (1979) An improved automatic recording apparatus for determining the filtration rate of *Mytilus edulis* as a function of size and algal concentration. *Marine Biology*, **52**, 61–67.

Riisgård, H.U. & Randløv, A. (1981) Energy budgets, growth and filtration rates in *Mytilus edulis* at different algal concentration. *Marine Biology*, **61**, 227–234.

Riisgård, H.U. & Seerup, D.F. (2003) Filtration rates in soft clam, *Mya arenaria*: effects of temperature and body size. *Sarsia*, **88**, 425–428.

Riisgård, H.U., Randløv, A. & Kristensen, P.S. (1980) Rates of water processing, oxygen consumption and efficiency of particle retention in veligers and young post-metamorphic *Mytilus edulis*. *Ophelia*, **19**, 37–46.

Riisgård, H.U. Kittner, C. & Seerup, D.F. (2003) Regulation of opening state and filtration rate in filter-feeding bivalves (*Cardium edule, Mytilus edulis, Mya arenaria*) in response to low algal concentration. *Journal of Experimental Marine Biology and Ecology*, **84**, 105–127.

Riisgård, H.U., Lassen, J. & Kittner, C. (2006) Valve-gape response times in mussels (*Mytilus edulis*)-effects of laboratory preceding-feeding conditions and *in situ* tidally induced variation in phytoplankton biomass. *Journal of Shellfish Research*, **25**, 901–911.

Riisgård, H.U., Egede, P.P. & Saavedra, I.B. (2011a) Feeding behavior of the mussel, *Mytilus edulis*: new observations, with a minireview of current knowledge. *Journal of Marine Biology*, **2011**, art. 312459.

Riisgård, H.U., Jørgensen, B.H., Lundgreen, K., Storti, F., Walther, J.H., Meyer, K.E. *et al.* (2011b) The exhalant jet of mussels *Mytilus edulis*. *Marine Ecology Progress Series*, **437**, 147–164.

Riisgård, H.U., Pleissner, D., Lundgreen, K. & Larsen, P.S. (2013a) Growth of mussels *Mytilus edulis* at algal (*Rhodomonas salina*) concentrations below and above saturation levels for reduced filtration rate. *Marine Biology Research*, **9**, 1005–1017.

Riisgård, H.U., Lüskow, F., Pleissner, D., Lundgreen, K. & López, M.Á.P. (2013b) Effect of salinity on filtration rates of mussels *Mytilus edulis* with special emphasis on dwarfed mussels from the low-saline Central Baltic Sea. *Helgoland Marine Research*, **167**, 591–598.

Riisgård, H.U., Larsen. P.S. & Pleissner, D. (2014) Allometric equations for maximum filtration rate in blue mussels *Mytilus edulis* and importance of condition index. *Helgoland Marine Research*, **68**, 193–198.

Riisgård, H.U., Funch, P. & Larsen, P.S. (2015) The mussel filter–pump–present understanding, with a re-examination of gill preparations. *Acta Zoologica*, **96**, 273–282.

Riou, V., Colaço, A., Bouillon, S., Khripounoff, A., Dando, P., Mangion, P. *et al.* (2010) Mixotrophy in the deep sea: a dual endosymbiotic hydrothermal mytilid assimilates dissolved and particulate organic matter. *Marine Ecology Progress Series*, **405**, 187–201.

Robson, A., Garcia de Leaniz. C., Wilson, R. & Halsey. L. (2010) Effect of anthropogenic feeding regimes on activity rhythms of laboratory mussels exposed to natural light. *Hydrobiologia*, **655**, 197–204.

Rocha, T.L., Teixeira Sabóia-Morais, S.M. & Bebianno, M.J. (2016) Histopathological assessment and inflammatory response in the digestive gland of marine mussel *Mytilus galloprovincialis* exposed to cadmium-based quantum dots. *Aquatic Toxicology*, **177**, 306–315.

Rosa, M., Ward, J.E., Shumway, S.E., Wikfors, G.H., Pales Espinosa, E. & Allam, B. (2013) Effects of particle surface properties on feeding selectivity in the eastern oyster *Crassostrea virginica* and the blue mussel *Mytilus edulis*. *Journal of Experimental Marine Biology and Ecology*, **446**, 320–327.

Rosa, M., Ward, J.E., Ouvrard, M., Holohan, B.A., Pales Espinosa, E., Shumway, S.E. *et al.* (2015) Examining the physiological plasticity of particle capture by the blue mussel, *Mytilus edulis* (L.): confounding factors and potential artifacts with studies utilizing natural seston. *Journal of Experimental Marine Biology and Ecology*, **473**, 207–217.

Rosa, M., Ward, J.E., Frink, A. & Shumway, S.E. (2017a) Effects of surface properties on particle capture by two species of suspension-feeding bivalve molluscs. *American Malacological Bulletin*, **35**, 181–188.

Rosa, M., Ward, J.E., Holohan, B.A., Shumway, S.E. & Wikfors, G.H. (2017b) Physicochemical surface properties of microalgae and their combined effects on particle selection by suspension-feeding bivalve molluscs. *Journal of Experimental Marine Biology and Ecology*, **486**, 59–68.

Rosland, R., Strand, Ø., Alunno-Bruscia, M., Bacher, C. & Strohmeier, T. (2009) Applying Dynamic Energy Budget (DEB) theory to simulate growth and bio-energetics of blue mussels under low seston conditions. *Journal of Sea Research*, **62**, 49–61.

Rouillon, G., Guerra Rivas, J., Ochoa, N. & Navarro, E. (2005) Phytoplankton composition of the stomach contents of the mussel *Mytilus edulis* L. from two populations: comparison with its food supply. *Journal of Shellfish Research*, **24**, 5–14.

Sabapathy, U. & Teo, L.H. (1992) A kinetic study of the α-amylase from the digestive gland of *Perna viridis* L. *Comparative Biochemistry and Physiology*, **101B**, 73–77.

Saggese, I., Sarà, G. & Dondero, F. (2016) Silver nanoparticles affect functional bioenergetic traits in the invasive Red Sea mussel *Brachidontes pharaonis*. *BioMed Research International*, **2016**, art. 1872351.

Sarà, G., De Pirro, M., Romano, C., Rumolo, P., Sprovieri, M. & Mazzola, A. (2007) Sources of organic matter for intertidal consumers on *Ascophyllum*-shores (SW Iceland): a multi-stable isotope approach. *Helgoland Marine Research*, **61**, 297–302.

Saraiva, S., van der Meer, J., Kooijman, S.A.L.M. & Sousa, T. (2011) Modelling feeding processes in bivalves: a mechanistic approach. *Ecological Modelling*, **222**, 514–523.

Saraiva, S., van der Meer, J., Kooijman, S.A.L.M., Witbaard, R., Philippart, C.J.M., Hippler, D. *et al.* (2012) Validation of a Dynamic Energy Budget (DEB) model for the blue mussel *Mytilus edulis*. *Marine Ecology Progress Series*, **463**, 141–158.

Sebens, K., Sarà, G. & Nishizaki, M. (2016) Energetics, particle capture, and growth dynamics of benthic suspension feeders. In: *Marine Animal Forests* (eds S. Rossi, L. Bramanti, A. Gori & C.O. Saco del Valle), pp. 813–854. Springer International Publishing, Switzerland.

Seiderer, L.J. & Newell, R.C. (1985) Relative significance of phytoplankton, bacteria and plant detritus as carbon and nitrogen resources for the kelp bed filter-feeder *Choromytilus mendionalis*. *Marine Ecology Progress Series*, **22**, 127–139.

Seo, E., Ohishi, K., Maruyama, T., Imaizumi-Ohashi, Y., Murakami, M. & Seo, Y. (2014) Magnetic resonance imaging analysis of water flow in the mantle cavity of live *Mytilus galloprovincialis*. *Journal of Experimental Biology*, **217**, 2277–2287.

Shumway, S.E., Frank, D.M., Ewart, L.M. & Ward, J.E. (2003) Effect of yellow loess on clearance rate in seven species of benthic, filter-feeding invertebrates. *Aquaculture Research*, **34**, 1391–1402.

Siebers, D. & Winkler, A. (1984) Amino-acid uptake by mussels, *Mytilus edulis*, from natural sea water in a flow-through system. *Helgolander Meeresunters*, **38**, 189–199.

Silina, A.V. & Zhukova, N.V. (2007) Growth variability and feeding of scallop *Patinopecten yessoensis* on different bottom sediments: evidence from fatty acid analysis. *Journal of Experimental Marine Biology and Ecology*, **348**, 46–59.

Silverman, H., Lynn, J.W., Beninger, P. G. & Dietz, T.H. (1999) The role of latero-frontal cirri in particle capture by the gills of *Mytilus edulis*. *Biological Bulletin*, **197**, 368–376.

Silverman, H., Lynn, J.W. & Dietz, T.H. (2000) *in vitro* studies of particle capture and transport in suspension-feeding bivalves. *Limnology and Oceanography*, **45**, 1199–1203.

Srisunont, C. & Babel, S. (2015) Uptake, release, and absorption of nutrients into the marine environment by the green mussel (*Perna viridis*). *Marine Pollution Bulletin*, **97**, 285–293.

Stewart, M.G. (1979) Absorption of dissolved organic nutrients by marine invertebrates. *Oceanography and Marine Biology-An Annual Review*, **17**, 163–192.

Strohmeier, T., Strand, O., Alunno-Bruscia, M., Duinker, A. & Cranford, P.J. (2012) Variability in particle retention efficiency by the mussel *Mytilus edulis*. *Journal of Experimental Marine Biology and Ecology*, **412**, 96–102.

Stuckas, H., Knöbel, L., Schade, H., Breusing, C., Hinrichsen, H-H., Bartel, M. *et al.* (2017) Combining hydrodynamic modelling with genetics: can passive larval drift shape the genetic structure of Baltic *Mytilus* populations? *Molecular Ecology*, **26**, 2765–2782.

Sunde, B.K. (2013) Gill and labial palp areas in blue mussels (*Mytilus edulis)* at sites with different food quantity. MSc thesis, University of Bergen, Norway.

Tan K.S. & Ransangan J. (2015) Cold Lugol's solution as an alternative of formaldehyde-base preservative in preserving stomach content of shellfish. *International Journal of Marine Science*, **5**, 1–4.

Tang, B. & H.U. Riisgård (2018) Relationship between oxygen concentration, respiration and filtration rate in blue mussel *Mytilus edulis*. *Chinese Journal of Oceanology and Limnology*, **36**, 395–404.

Teo, L.H. & Sabapathy, U. (1990) Preliminary report on the digestive enzymes present in the digestive gland of *Perna viridis*. *Marine Biology*, **106**, 403–407.

Teo, L.H. & Lim, E.H. (1991) Some properties of the trehalase from the digestive gland of the green mussel, *Perna viridis* L. *Comparative Biochemistry and Physiology*, **99B**, 489–494.

Teo, L.H. & Lim, E.H. (1993) Effects of storage temperature, tris buffer and divalent cations on the activity of trehalase of *Perna viridis* L. *Journal of the Singapore National Academy of Science*, **20 & 21**, 61–66.

Teo, L.H., Lateef, Z. & Ip, Y.K. (1990) Some properties of the sucrase from the digestive gland of the green mussel *Perna viridis*. *Comparative Biochemistry and Physiology*, **96B**, 47–51.

Thompson, R.J., Ratcliffe N.A. & Bayne, B.L. (1974) Effects of starvation on structure and function in the digestive gland of the mussel (*Mytilus edulis* L.). *Journal of the Marine Biological Association of the UK*, **54**, 699–713.

Tizon, R.U., Serrano Jr., A.E. & Traifalgar, R.F. (2013) Influence of photoperiod on digestive enzyme activities of the angelwing clam (*Pholas orientalis*). *ELBA Bioflux*, **5**, 61–68.

Troost, K., Kamermans, P. & Wolff, W.J. (2008) Larviphagy in native bivalves and an introduced oyster. *Journal of Sea Research*, **60**, 157–163.

Troost, K., Gelderman, E., Kamermans, P., Smaal, A.C. & Wolff, W.J. (2009) Effects of an increasing filter feeder stock on larval abundance in the Oosterschelde estuary (SW Netherlands). *Journal of Sea Research*, **61**, 153–164.

Trottet, A., Roy, S., Tamigneaux, E., Lovejoy, C. & Tremblay, R. (2008) Impact of suspended mussels (*Mytilus edulis* L.) on plankton communities in a Magdalen Islands lagoon (Québec, Canada): A mesocosm approach. *Journal of Experimental Marine Biology and Ecology*, **365**, 103–115.

Urrutia, M.B., Iglesias, J.I.P., Navarro, E. & Prou, J. (1996) Feeding and absorption in *Cerastoderma edule* under environmental conditions in the Bay of Marennes-Oleron (western France). *Journal of the Marine Biological Association of the UK*, **76**, 431–150.

Urrutia, M.B., Nararro, E., Ibarrola, I. & Iglesias, J.I.P. (2001) Preingestive selection processes in the cockle *Cerastoderma edule*: mucus production related to rejection of pseudofaeces. *Marine Ecology Progress Series*, **209**, 177–187.

Usheva, L.N., Vaschenko, M.A. & Durkina, V.B. (2006) Histopathology of the digestive gland of the bivalve mollusk *Crenomytilus grayanus* (Dunker, 1853) from southwestern Peter the Great Bay, Sea of Japan. *Russian Journal of Marine Biology*, **32**, 166–172.

Uslu, F.E. & Pekkan, K. (2016) *Mytilus galloprovincialis* as a smart micro-pump. *Biology Open*, **5**, 1493– 1499.

Vahl, O. (1972) Efficiency of particle retention in *Mytilus edulis* L. *Ophelia*, **10**, 17–25.

Vekhova, E.E., Kusaykin, M.I. & Kiselev, K.V. (2015) Activity of 1,3-β-D-glucanases and expression of *CG-GLU1* in the Mussel *Crenomytilus grayanus* (Bivalvia) in three bays of the Sea of Japan. *Achievements in the Life Sciences*, **9**, 69–77.

Velasco, L.A. & Navarro, J.M. (2002) Feeding physiology of infaunal (*Mulinia edulis*) and epifaunal (*Mytilus chilensis*) bivalves under a wide range of concentrations and qualities of seston. *Marine Ecology Progress Series*, **240**, 143–155.

Viarengo, A., Burlando, B., Dondero, F., Marrò, A. & Fabbri, R. (1999) Metallothionein as a tool in biomonitoring programmes. *Biomarkers*, **4**, 455–466.

Virtue, P., Mayzaud, P., Albessard, E. & Nichols, P. (2000) Use of fatty acids as dietary indicators in northern krill, *Meganyctiphanes norvegica*, from northeastern Atlantic, Kattegat, and Mediterranean waters. *Canadian Journal of Fisheries and Aquatic Sciences*, **57**, 104–114.

Waite, A.M., Safi, K.A., Hall, J.A. & Nodder, S.D. (2000) Mass sedimentation of picoplankton embedded in organic aggregates. *Limnology and Oceanography*, **45**, 87–97.

Wang, W-X & Fisher, N.S. (1996) Assimilation of trace elements and carbon by the mussel *Mytilus edulis*: effects of food composition. *Limnology and Oceanography*, **41**, 197–207.

Wang, Y., Hu, M., Wong, W.H., Shin, P.K.S. & Cheung, S.G. (2011) The combined effects of oxygen availability and salinity on physiological responses and scope for growth in the green-lipped mussel *Perna viridis*. *Marine Pollution Bulletin*, **63**, 255–261.

Ward, J.E. & Shumway, S.E. (2004) Separating the grain from the chaff: particle selection in suspension- and deposit-feeding bivalves. *Journal of Experimental Marine Biology and Ecology*, **300**, 83–130.

Ward, J.E. & Kach, D.J. (2009) Marine aggregates facilitate ingestion of nanoparticles by suspension-feeding bivalves. *Marine Environmental Research*, **68**, 137–142.

Ward, J.E., Beninger, P.G., MacDonald, B.A. & Thompson, R.J. (1991) Direct observations of feeding structures and mechanisms in bivalve mollusks using endoscopic examination and video image-analysis. *Marine Biology*, **111**, 287–291.

Ward, J.E., MacDonald, B.A. & Thompson, R.J. (1993) Mechanisms of suspension feeding in bivalves: resolution of current controversies by means of endoscopy. *Limnolology and Oceanography*, **38**, 265–272.

Ward, J.E., MacDonald, B.A., Thompson, R.J. & Beninger P.G. (1996) Biodynamics of suspension-feeding in adult bivalve molluscs: particle capture, processing, and fate. *Invertebrate Biology*, **115**, 218–231.

Ward, J.E., Sanford, L.P., Newell, R.I.E. & MacDonald, B.A. (1998a) A new explanation of particle capture in suspension-feeding bivalve molluscs. *Limnology and Oceanography*, **43**, 741–752.

Ward, J.E., Levinton, J.S., Shumway, S.E. & Cucci, T. (1998b) Particle sorting in bivalves: *in vivo* determination of the pallial organs of selection. *Marine Biology*, **131**, 283–292.

Webb, J.L., Vandenbor, J., Pirie, B., Robinson, S.M.C., Cross, S.F., Jones, S.R.M. *et al.* (2013) Effects of temperature, diet, and bivalve size on the ingestion of sea lice (*Lepeophtheirus salmonis*) larvae by various filter-feeding shellfish. *Aquaculture*, **406–407**, 9–17.

Wegner, A., Besseling, E., Foekema, E.M., Kamermans, P. & Koelmans, A.A. (2012) Effects of nanopolystyrene on the feeding behavior of the blue mussel (*Mytilus edulis* L.). *Environmental Toxicology and Chemistry*, **31**, 2490–2497.

Weise, A.M., Cromey, C.J., Callier, M.D., Archambault, P., Chamberlain, J. & McKindsey, C.W. (2009) Shellfish-DEPOMOD: modelling the biodeposition from suspended shellfish aquaculture and assessing benthic effects. *Aquaculture*, **288**, 239–253.

Wentrup, C., Wendeberg, A., Schimak, M., Borowski, C. & Dubilier, N. (2014) Forever competent: deep-sea bivalves are colonized by their chemosynthetic symbionts throughout their lifetime. *Environmental Microbiology*, **16**, 3699–3713.

Widdows, J. (1976) Physiological adaptation of *Mytilus edulis* to cyclic temperatures. *Journal of Comparative Physiology*, **105**, 115–128.

Widdows, J. (1978) Combined effects of body size, food concentration and season on the physiology of *Mytilus edulis*. *Journal of the Marine Biological Association of the UK*, **58**, 109–124.

Widdows, J. & Bayne, B.L. (1971) Temperature acclimation of *Mytilus edulis* with reference to its energy budget. *Journal of the Marine Biological Association of the UK*, **51**, 827–843.

Widdows, J. & Johnson, D. (1988) Physiological energetics of *Mytilus edulis*: scope for growth. *Marine Ecology Progress Series*, **46**, 113–121.

Widdows, J., Lucas, J., Brinsley, M., Salkeld, P. & Staff, F. (2002) Investigation of the effects of current velocity on mussel feeding and mussel bed stability using an annular flume. *Helgoland Marine Research*, **56**, 3–12.

Wildish, D. & M. Miyares. 1990. Filtration-rate of blue mussels as a function of flow velocity: preliminary experiments. *Journal of Experimental Marine Biology and Ecology*, **42**, 213–219.

Wong, W.H. & Cheung, S.G. (2001) Feeding rhythms of the green-lipped mussel, *Perna viridis* (Linnaeus, 1758) (Bivalvia: Mytilidae) during spring and neap tidal cycles. *Journal of Experimental Marine Biology and Ecology*, **257**, 13–36.
Wright, R.T., Collins, R.B., Ersing, C.P. & Pearson, D. (1982) Field and laboratory measurements of bivalve filtration of natural marine bacterioplankton. *Limnology and Oceanography*, **27**, 91–98.
Wright, S.H., Moon, D.A.A. & Silva, A.L. (1989) Extracellular Na^+ and the control of amino acid fluxes in the integumental epithelium of a marine bivalve. *Journal of Experimental Biology*, **142**, 293–310.
Xu, B. (2002) Endoglucanase and mannanase from blue mussel, *Mytilus edulis*: purification, characterization, gene and three dimensional structure. PhD thesis, Uppsala University, Sweden.
Xu, Q. & Yang, H. (2007) Food sources of three bivalves living in two habitats of Jiaozhou Bay (Qingdao, China): indicated by lipid biomarkers and stable isotope analysis. *Journal of Shellfish Research*, **26**, 561–567.
Yahel, G., Dominique, M., Beninger, P.G., Eckstein, S. & Genin, A. (2009) *In situ* evidence for pre-capture qualitative selection in the tropical bivalve *Lithophaga simplex*. *Aquatic Biology*, **6**, 235–246.
Zaldibar, B., Cancio, I. & Marigómez, I. (2004) Circatidal variation in epithelial cell proliferation in the mussel digestive gland and stomach. *Cell Tissue Research*, **318**, 395–402.
Zaldibar, B., Cancio, I. & Marigómez, I. (2008) Epithelial cell renewal in the digestive gland and stomach of mussels: season, age and tidal regime related variations. *Histology and Histopatholgy*, **23**, 281–290.
Zamora, L.N. & Jeffs, A.G. (2012) The ability of the deposit-feeding sea cucumber *Australostichopus mollis* to use natural variation in the biodeposits beneath mussel farms. *Aquaculture*, **326–329**, 116–122.
Zardi, G.I., Nicastro, K.R., Porri, F. & McQuaid, C.D. (2006) Sand stress as a non-determinant of habitat segregation of indigenous (*Perna perna*) and invasive (*Mytilus galloprovincialis*) mussels in South Africa. *Marine Biology*, **148**, 1031–1038.
Zbawicka, M., Sanko, T., Strand, J. & Wenne, R. (2014) New SNP markers reveal largely concordant clinal variation across the hybrid zone between *Mytilus* spp. in the Baltic Sea. *Aquatic Biology*, **21**, 25–36.
Zhukova, N.V. & Kharlamenko, V.I. (1999) Sources of essential fatty acids in the marine microbial loop. *Aquatic Microbial Ecology*, **17**, 153–157.
Zúñiga, D., Castro, C.G., Aguiar, E., Labarta, U., Figueiras, F.G. & Fernández-Reiriz, M.J. (2014) Biodeposit contribution to natural sedimentation in a suspended *Mytilus galloprovincialis* Lmk mussel farm in a Galician Ría (NW Iberian Peninsula). *Aquaculture*, **432**, 311–320.

5

Reproduction, Larval Development, Dispersal and Recruitment

Introduction

The reproductive system in bivalves is extremely simple, and mussels are no exception (see Chapter 2). The paired gonads, testes and ovaries are made up of branching tubules, and gametes are budded off their epithelial lining. The tubules unite to form ducts, which lead into larger ducts that eventually terminate in a short gonoduct. In primitive bivalves, the gonoducts open into the kidneys, and eggs and sperm exit through the kidney opening (nephridiopore) into the mantle cavity. In mussels and most other bivalves, the gonoducts are no longer associated with the kidneys but open through independent gonopores into the mantle cavity close to the nephridiopore. Gametes are shed through the exhalant opening of the mantle, and fertilisation takes place in the water column. The fertilised egg goes through a number of larval stages, culminating after several weeks of dispersal in attachment once a suitable substrate is chosen. The larva then undergoes a number of morphological changes to achieve a sedentary lifestyle.

Sex Identification

Mussels are dioecious (i.e. the sexes are separate), and there are usually equal numbers of males and females. But triploids, which show no histological evidence of spawning, have a highly skewed incidence of males (99–100%) in *Mytilus galloprovincialis* (Kiyomoto *et al.* 1996) and *M. edulis* (Brake *et al.* 2004). Also, sex ratio in the charru mussel, *Mytella charruana*, collected from different locations and maintained in the laboratory, changed toward a male-bias under starvation conditions within a short period of time (Stenyakina *et al.* 2010). The green crenella, *Musculus discors*, is unusual among marine mussels in that it is a protandrous hermaphrodite: one-year-olds are functionally male, but they become functional females by the third year of life. Other unusual aspects of this species are internal fertilisation, production of large eggs laid within a nest of adult byssal threads, direct development with no pelagic phase and the departure of juveniles from the nest as free-living crawl-aways (De-Bastos & Tyler-Walters 2001). Small numbers of hermaphrodites have been reported in samples of *Perna viridis* (Lee 1988; Al-Barwani *et al.* 2013) and *Perumytilus purpuratus* (Montenegro Villalobos *et al.* 2010), albeit at low

Marine Mussels: Ecology, Physiology, Genetics and Culture, First Edition. Elizabeth Gosling.

frequencies (≤1%). There may be a link between hermaphrodite occurrence in *P. purpuratus* and environmental pollutants, although the authors provided no supporting data for this. However, a high incidence (26%) of hermaphroditism/intersex individuals was reported in *M. galloprovincialis* from a polluted site in the Bay of Biscay, France (Ortiz-Zarragoitia & Cajaraville 2010). In addition, maturing male gametes and immature oocytes have been reported in the gonad of *Bathymodiolus elongatus*, a deep-sea hydrothermal vent mussel (Le Pennec & Beninger 1997). In a recent study, Oyarzún *et al.* (2020) confirmed trioecy (males, females and hermaphrodites) in the marine mussel *Semimytilus algosus*, previously described as a simultaneous hermaphrodite. Their report was the first in the phylum Mollusca. They examined the frequencies of females, males and hermaphrodites in seven natural populations along a latitudinal gradient of 2500 km in Chile. Hermaphrodites (~95.3%), females (~3.6%) and males (~1.1%) were found in all seven populations. The consistent sex ratios across all populations suggested that the sexual system is stable in space. Phylogenetic analysis (mtDNA) of members of the family Mytilidae indicates that this species alone has developed trioecy from dioecious ancestors. As unisex animals were the least common in the wild, it is likely that they – and especially the males – might be in the process of being lost in the species, and that trioecy might be an intermediate step toward gynodioecy (co-occurrence of females and hermaphrodites) or hermaphrodism.

Although sex cannot be differentiated on external characters, field biologists have used gonad tissue colouration as an indicator of sex in intertidal mussels: individuals with white gonadal tissue are identified as males, and those with orange tissue as females. However, Petes *et al.* (2008a) have shown that gonad colour is not a reliable indicator of sex. When mussels (*M. californianus*) were sampled from different intertidal heights at different locations on the Oregon coast, United States, the gonads of both male and female mussels higher up on the shore were more likely to be orange than those lower down. The authors suggested that environmental stress, most apparent in the higher intertidal zone, leads to greater deposition of carotenoid pigments in the gonads, irrespective of sex. Carotenoids can act as antioxidants and may have a protective function in protecting gametes from damaging oxidative stress experienced during aerial exposure (Petes *et al.* 2008b). The standard method for distinguishing the sexes is to fix gonadal tissue and then use histological processing and light microscope examination (see later). Jabber & Davis (1987) have described a simple colorimetric test for sex determination in mussels that is based on the substantial difference in lipid concentrations between eggs and sperm. Also, Mikhailov *et al.* (1995) have identified a 'male-associated polypeptide' (MAP39) that is virtually absent in females. Both these methods, while cheaper and less time-consuming than histological processing, involve sacrificing the animals. A preparation of squashes from fine-needle aspirates of mantle tissue is preferable because it can be done on live mussels (Burton *et al.* 1996). A novel reverse transcriptase polymerase chain reaction (RT-PCR) method provides a simple diagnostic based on the presence or absence of gender-specific transcripts[1] in the mantle. The targets for this method are vitelline coat lysin (VCL) and vitelline envelope receptor for lysin (VERL), which play key roles in fertilisation: the sperm is able to enter the egg by the sperm-located VCL binding to the egg-located VERL. Of all techniques, RT-PCR is currently the most accurate and robust for absolute gender determination, and unlike the other methods, it is applicable to both ripe and spent *Mytilus* species (Hines *et al.* 2007; Sedik *et al.* 2010; Anantharaman & Craft; Fraser *et al.* 2016). The technique has been used to track the course of gametogenesis in *M. edulis* (Anantharaman & Craft 2012).

Doubly Uniparental Inheritance and Sex Determination

In virtually all animals, mitochondrial DNA (mtDNA) – in contrast to nuclear DNA – is transmitted maternally. This mode of inheritance is responsible for offspring homoplasy: a state in which all mitochondrial genomes[2] are genetically identical. This may be a mechanism that has evolved to avoid competition between different mtDNAs (maternal and paternal) within a cell (see Sharpley *et al.* 2012; Breton & Stewart 2015). Bivalves are unique in that they are the only organisms that have a different inheritance system, known as doubly uniparental inheritance (DUI). First discovered in the *Mytilus edulis* species complex (Skibinski *et al.* 1994a,b; Zouros *et al.* 1994a,b), DUI has now been identified in more than 40 bivalve species, both marine and freshwater (Kyriakou *et al.* 2015). Interestingly, there is no evidence for paternally-inherited mtDNA in the mussel *Perna perna* (Uliano-Silva *et al.* 2016). Species with DUI possess two mitochondrial lineages that are transmitted in a sex-specific manner, one through the eggs (F genome) and the other through the sperm (M genome). Female mussels inherit their F genome only from their mother, but they transmit it to both sons and daughters. Male zygotes inherit their mtDNA from both parents, but they sort the mixture of mitochondrial genomes so that the M genome inherited from their father becomes established in the germ line (Breton *et al.* 2007). While heteroplasmy is the initial state in embryos, females shift from this state to being essentially homoplasmic[3] F-containing adults; traces of the M genome can sometimes be found in female somatic tissues (Dalziel & Stewart 2002; Batista *et al.* 2011) through sperm mtDNA leakage (Kyriakou *et al.* 2010). On the other hand, adult males contain predominantly the F genome in all tissues except the gonad, where the M genome predominates. DUI, therefore, indicates that there is an active mechanism to ensure that the M type will be the exclusive form present in spermatozoa, and a further mechanism to prevent the presence of the M type in eggs, and hence its transmission (Batista *et al.* 2011).

Another unusual aspect of DUI that has only been demonstrated in *Mytilus* is that some females produce female-biased offspring whereas others produce male-biased offspring, regardless of what male they mate with. The coupling of sex and mtDNA transmission in mussels raises the possibility that sex-ratio determination may be under the control of the mtDNA of the female parent. However, this is not the case: it is the mother's nuclear rather than mtDNA genotype that exercises control. The initial model proposed by Zouros (2000) has been discussed or modified by subsequent authors such as Kenchington *et al.* (2002, 2009), Cao *et al.* (2004), Cogswell *et al.* (2006), Breton *et al.* (2007), Passamonti & Ghiselli (2009), Ghiselli *et al.* (2012), Yusa *et al.* (2013), Diz *et al.* (2013), Zouros (2013) and Breton *et al.* (2017). According to the model, described in Passamonti & Ghiselli (2009) and Diz *et al.* (2013), three genetic factors are hypothesised to be involved. The W factor is located on the outer surface of the sperm mitochondrion and is recognised by an egg nuclear-encoded cytoplasmic

X factor produced during spermatogenesis that labels sperm mitochondria. The X factor, produced during oogenesis, recognises W, leading to the elimination of sperm mitochondria. This is assumed to be the same mechanism for sperm mitochondrial elimination in most strict maternal inheritance (SMI) species (Kenchington *et al.* 2002). Finally, the Z factor, which is coded by a nuclear gene and has two alleles, one active (Z) and one inactive (z), works to suppress the X factor, preventing the elimination (or dispersion) of sperm mitochondria in fertilised eggs. Consequently, ZZ females will produce the active factor only, while zz will not, and Zz heterozygotes will produce intermediate quantities of Z. If we assume that the retention of sperm mitochondria somehow induces the masculinisation of the embryo, then the zz female will produce a female-biased offspring and the ZZ a male-biased one, with the heterozygous Zz producing a 1:1 sex ratio offspring.

It is considered that DUI was established in the common ancestor of Autolamellibranchia, and perhaps of all bivalve molluscs. If this is the case, it has a very long history, which may go as far back as 460 million years (Boyle & Etter 2013; Zouros 2013). It appears that DUI was subsequently lost in a number of lineages (e.g. families Ostreoidea, oysters, and Pectinoidea, scallops), but the exact mechanism whereby this might have occurred is presently unknown. The first complete maternal and paternal mitochondrial genomes were sequenced from the Mediterranean mussel, *Mytilus galloprovincialis* (Mizi *et al.* 2005). Since then, F and M genomes from eight other mussel species, one clam species and four freshwater unionid clam species have been sequenced (references in Zouros 2013). In *Mytilus*, while the two genomes differ at the control region (CR), gene content and arrangement are the same in conspecific F and M genomes, but the primary sequence has diverged from 20 to 40%, depending on the species (Breton *et al.* 2006; Zbawicka *et al.* 2010; Smietanka *et al.* 2010). The nucleotide *p* distance[4] varies from species to species: ≈22% in *M. edulis*/*M. galloprovincialis* (the smallest known for fully sequenced conspecific F and M genomes), 24% in *M. trossulus* and 39% in *M. californianus*.

Synonymous substitutions accumulate faster in the M than the F genome, and nonsynonymous ones even faster. Expression studies indicate that the M genome is active only at spermatogenesis. These observations suggest that the M genome is under a more relaxed selective constraint than the F due to its reduced biological role (discussed in Zouros 2013). However, Ghiselli *et al.* (2013) argue that the function of the M genome in gamete production and functionality is an extremely important one and a relaxed selection would affect gamete fitness and hence male fertility and overall population viability. For those wishing to explore the topic of DUI in bivalves in more detail, Zouros (2013) presents a very comprehensive review and synthesis of the literature, proposes a model that incorporates published findings and suggests future work priorities.

Gametogenesis

Mussels undergo an annual reproductive cycle which involves a period of gametogenesis followed by a single, an extended or even several spawning events, which in turn are followed by a period of gonad reconstitution (see later).

Ultrastructural studies of the testis show that different members of the family Mytilidae (e.g. species of *Mytilus*, *Musculus*, *Bathymodiolus*, *Perna*) have similar patterns of spermatogenesis and sperm structure (Longo & Dornfield 1967; Franzén 1983; Bernard & Hodgson 1985; Reunov & Hodgson 1994; Le Pennec & Beninger 1997; Kádár *et al.* 2006). The testis consists of branching acini within a supporting matrix of adipogranular (ADG) cells packed with glycogen particles and fat droplets that serve as a nutrient source for developing sperm. Each acinus contains germ cells at different stages of development. In spermatogenesis, primary spermatogonia undergo repeated mitotic divisions to produce secondary spermatogonia, whcih undergo meiosis to become spermatocytes. These produce spermatids, which differentiate into flagellated spermatazoa. Spermatagonia, the largest cells (6.5–7.5 μm diameter), are located nearest the inner wall of the acinus, spermatocytes (5–6.5 μm) and spherical spermatids (~2.5 μm) are found between the spermatagonia and the acinus lumen, while mature sperm are almost exclusively confined to the central lumen (Eckelbarger & Young 1999). Mature sperm comprise three distinct regions: head, midpiece and flagellum. In *Perna viridis*, the head (4 μm long) is capped by a short, conical nucleated acrosome (about 2 μm long) (Reunov *et al.* 1999). The midpiece contains four or more mitochondria, which power sperm motility, while the flagellum possesses the classical 9 + 2 microtubular structure and can be as long as 50–60 μm in some *Mytilus* species (Oyarzún *et al.* 2014; Figure 5.1). Sperm structure in mussels and most bivalves is of the primitive type, with head and flagellum adapted for external fertilisation (Le Pennec & Beninger 1997).

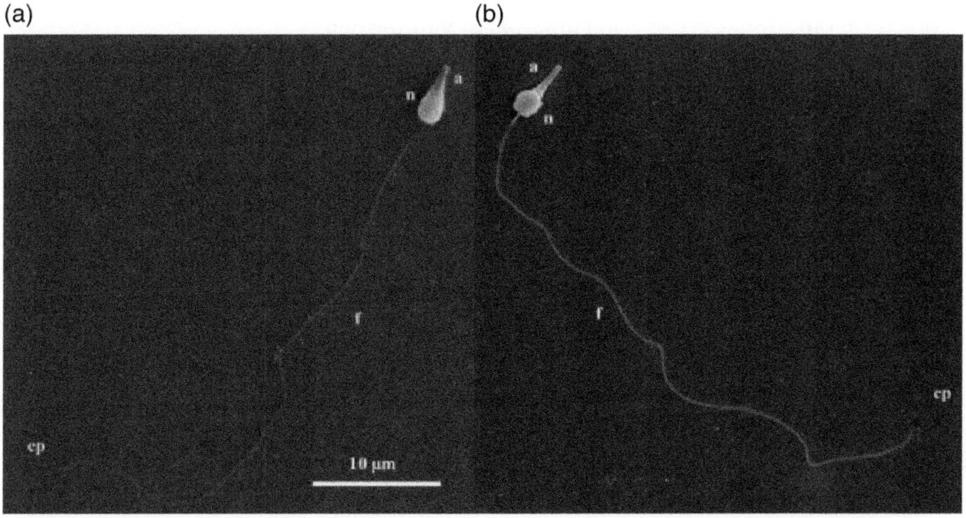

Figure 5.1 Scanning electron micrographs (SEMs) of the mature sperm of (a) *Mytilus chilensis* and (b) *Mytilus galloprovincialis*. a, acrosome; n, nucleus; f, flagellum; ep, end piece or tail. *Source:* From Oyarzún *et al.* (2014). Reproduced with permissions from American Journal of Aquatic Research.

Sperm ultrastructure has been used in systematic studies to differentiate closely related species, especially in the family Mytilidae (Bernard & Hodgson 1985; Garrido & Gallardo 1996; Kafanov & Drozdov 1998; Healy *et al.* 2000; Introíni *et al.* 2004; Introíni *et al.* 2010; Briones *et al.* 2012; Oyarzún *et al.* 2014; Torroglosa & Giménez 2015). Between species, the acrosome appears to be the most variable organelle in both shape and size (Garrido & Gallardo 1996). For example, when sperm ultrastructure was compared in *Mytilus chilensis* and *M. galloprovincialis*, differences in acrosomal length and shape of the acrosomal nucleus were found. These are now considered species-specific taxonomic traits in the genus *Mytilus*; thus, sperm ultrastructure can be used as an additional trait to validate the taxonomic status of *M. chilensis* (Oyarzún *et al.* 2014). Intraspecific sperm polymorphism has also between reported in the green mussel, *Perna viridis* (Reunov *et al.* 1999), the Chilean mussel, *Perumytilus purpuratus* (Briones *et al.* 2012) and in both *M. trossulus* and *Crenomytilus grayanus* (Vekhova *et al.* 2013; Figure 5.2). Kafanov & Drozdov (1998) have used sperm ultrastructure to redefine three subfamilies within the Mytilidae, mainly based on nuclear morphology and the presence or absence of an axial rod (a bundle of filaments that extends from the tip of the acrosome to the nucleus).

Unlike for spermatogenesis, there is a paucity of ultrastructural studies on oogenesis in marine mussel species. Oogenesis follows a similar pattern to spermatogenesis in that primary oogonia undergo repeated mitosis to give secondary oogonia (5–7 µm diameter), which enter meiosis as primary oocytes but are then arrested at the prophase stage of meiosis I. The remaining meiotic stages are completed at fertilisation. The oocytes then enter a period of vitellogenesis, a process that involves the accumulation in the oocyte of lipid globules, vitellogenin (a precursor protein of egg yolk secreted by oocytes) and cortical granules, which are later exocytosed as part of egg envelope synthesis (details in Pipe 1987a; Eckelbarger & Young 1999; Agnese *et al.* 2013). Follicle cells containing vitellogenin, mitochondria, lysosomal-like inclusions and extensive rough endoplasmic reticulum (RER) are closely applied to pre-vitellogenic and vitellogenic oocytes by filaments in the cell membranes (Eckelbarger & Young 1999; Figures 5.3 and 5.4). The role of cortical granules is unknown in bivalves, but in mammals they prevent polyspermy after fertilisation (Sun 2003). In large oocytes (80 µm diameter in the methane seep mussel, *Bathymodiolus childressi*), the envelope is ~0.5 µm thick and consists of evenly spaced parallel microvilli penetrating a homogeneous dense vitelline layer. Lysis (atresia) and resorption of oocytes is usually observed at the end of vitellogenesis, but previtellogenic oocytes may also be affected due to the presence of hydrolytic enzymes released from lysed oocytes. Extremes of temperature, desiccation, low levels of nutrition or exposure to environmental contaminants may instigate lysis (Pipe 1987a; McDowell *et al.* 1999; Ortiz-Zarragoitia & Cajaraville 2010; Baussant *et al.* 2011; Ruiz *et al.* 2011; Cuevas *et al.* 2015). Ultimately, lysed products provide metabolic substrates for energy production at a time when energy reserves in adductor muscle are at their lowest (Beninger & Le Pennec 2006). In a recent paper, Beninger (2017) reported that oocyte atresia may occur at several points in a reproductive cycle, and that its recognition and quantification are therefore essential to the proper interpretation of reproductive processes such as fecundity, reproductive effort (RE) and recruitment in marine bivalves.

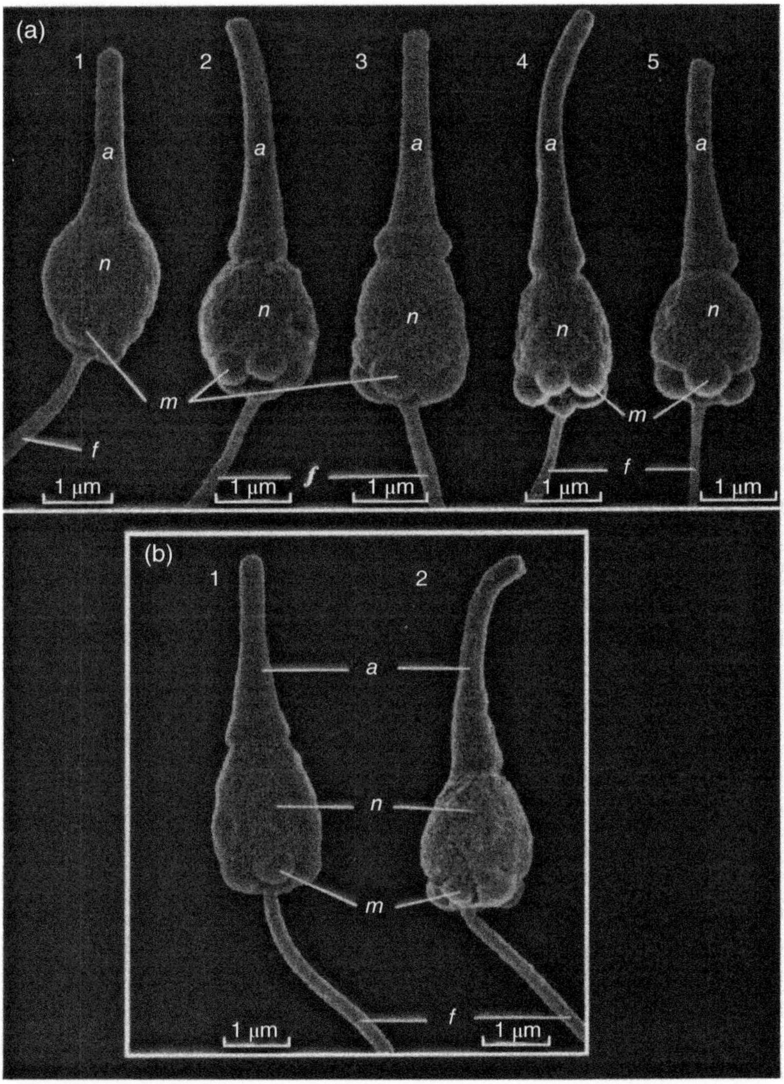

Figure 5.2 Diversity of morphological patterns of spermatazoa in (a) *Mytilus trossulus* and (b) *Crenomytilus grayanus*. *a*, acrosome; *n*, nucleus; *m*, mitochondrion; *f*, flagellum. Note that variant 3 in (a) *M. trossulus* is identical to variant 1 in (b) *C. grayanus*. *Source:* From Vekhova *et al.* (2013). Reproduced with permission from Springer.

Reproductive Cycles

Methods of Assessment

The principle method for assessing the reproductive cycle in mussels is based on histological preparations of gonad. Usually, animals for analysis are collected over a one- to two-year period at monthly or bimonthly intervals. About 10–20 individuals are randomly

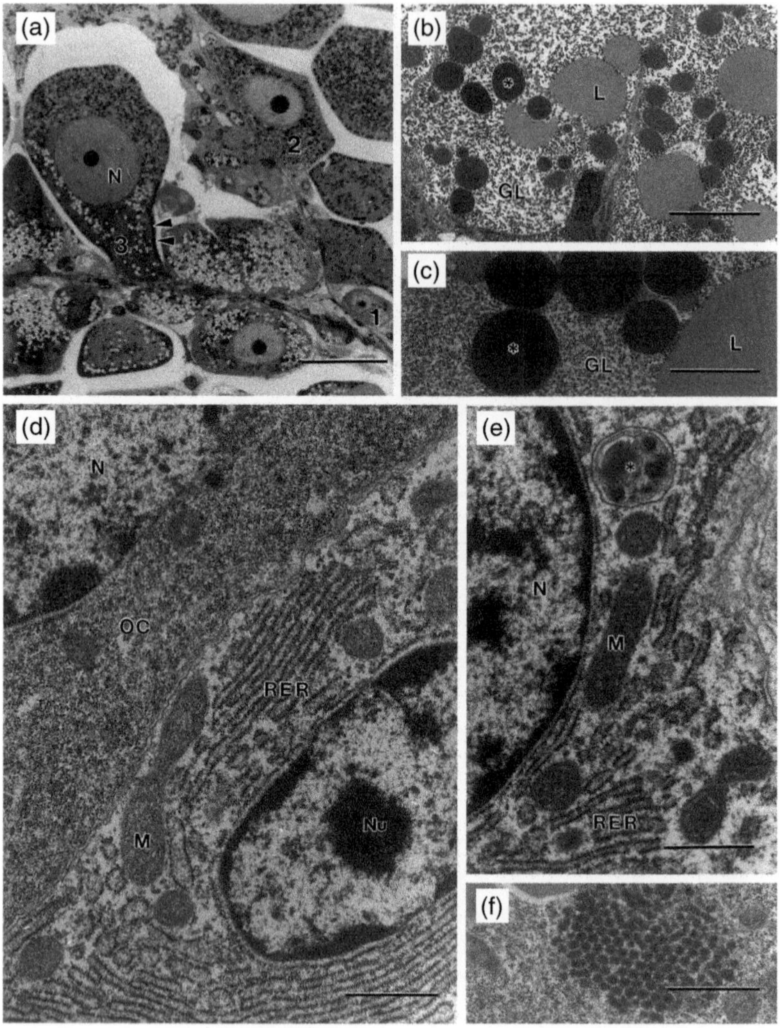

Figure 5.3 Ultrastructure of the ovary of the methane seep mussel, *Bathymodiolus childressi*. (a) Light microscope transverse section through the ovary, showing developing oocytes within acini. Note oocytes in (1) early, (2) middle and (3) late stages of vitellogenesis. Arrowheads indicate stalk-like connection of oocyte to ovary wall. N, nucleus. Scale bar = 40 μm. (b) Adipogranular (ADG) cell, showing lipid droplets (L), electron dense granules (*) and glycogen (GL). Scale bar =10 μm. (c) Higher magnification of ADG cells, showing lipid (L), electron dense granules (*) and glycogen (GL). Scale bar = 1.0 μm. (d) Previtellogenic oocyte (OC) in close association with a follicle cell containing extensive arrays of rough endoplasmic reticulum (RER). N, nucleus; Nu, nucleolus; M, mitochondrion. Scale bar = 2.0 μm. (e) Follicle cell containing nucleus (N), mitochondrion (M), rough endoplasmic reticulum (RER) and lysosomal-like inclusion (*). Scale bar = 2.0 μm. (f) Nuage-like granules in ooplasm of previtellogenic oocyte. Scale bar = 2.0 μm. *Source:* From Eckelbarger & Young (1999). Reproduced with permission from Springer.

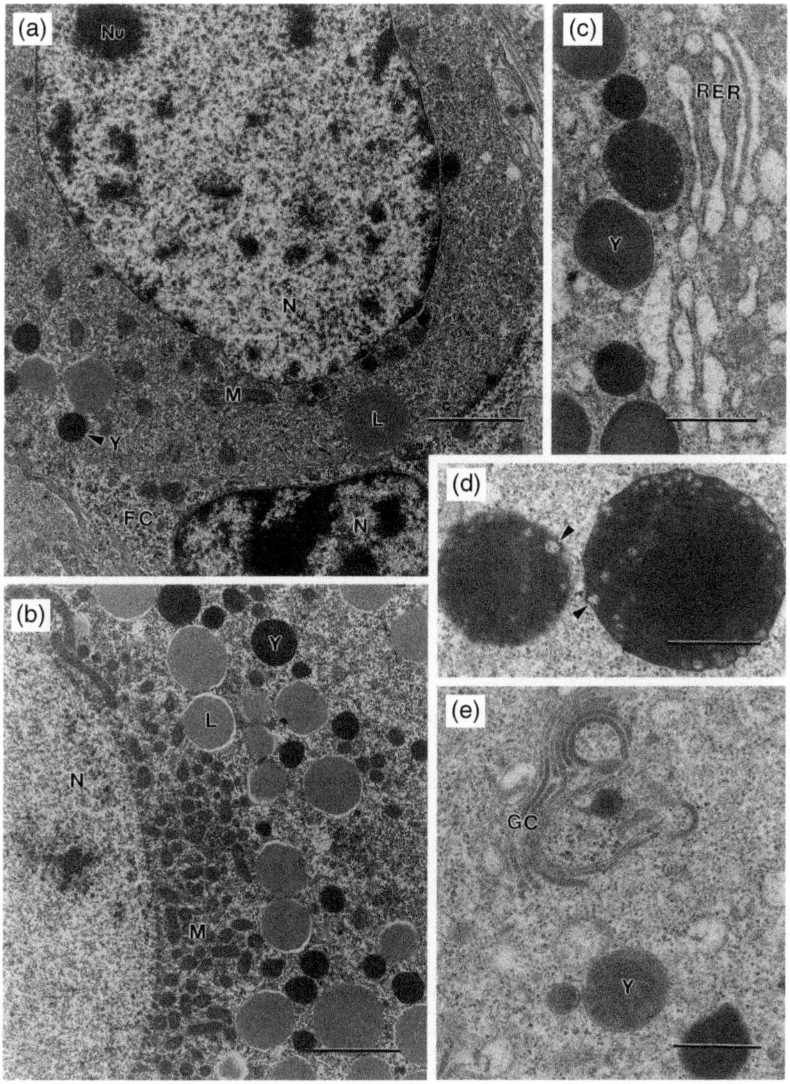

Figure 5.4 Ultrastructure of the ovary of the methane seep mussel, *Bathymodiolus childressi*. (a) Early vitellogenic oocyte containing large nucleus (N), mitochondria (M), lipid droplets (L) and yolk granules (Y). FC, follicle cell; Nu, nucleolus. Scale bar = 4.0 μm. (b) Perinuclear region of early vitellogenic oocyte, showing band of mitochondria (M) and adjacent lipid (L) and yolk granules (Y). N, nucleus. Scale bar = 2.0 μm. (c) Dilated cisternae of rough endoplasmic reticulum (RER) and adjacent yolk bodies (Y). Scale bar = 1.0 μm. (d) High magnification of yolk granules, showing vacuole-like inclusions (arrows). Scale bar = 1.5 μm. (e) Golgi complex (GC) and adjacent nascent yolk granules (Y). Scale bar = 2.0 μm. *Source:* From Eckelbarger & Young (1999). Reproduced with permission from Springer.

sampled, and back in the laboratory shell and soft tissue measurements may be taken before a small section (2–3 mm^3) from the middle area of the mantle lobe is dissected and fixed in a standard fixative such as 10% formalin. The fixed gonadal tissues are subsequently dehydrated in ascending concentrations of alcohol, washed in xylene and embedded in paraffin. This process is generally carried out in an automatic tissue processor. Sections (5–7 μm thickness) are prepared from each individual and stained with haematoxylin–eosin or an equivalent stain. For more detailed information, see Howard *et al.* (2004) and Kim *et al.* (2006). The slides are examined microscopically to estimate the proportion of developing, ripe, spawning and spent individuals in each sample. By multiplying the number of individuals at each development stage by the numerical ranking of that stage and dividing the resulting value by the total number of mussels in the sample, a mean gonad index (GI) is arrived at for each sampling interval. The GI increases during gametogenesis and decreases during spawning. Sections can also be used for gender determination, provided that gametes are present in the gonad (mantle). Qualitative gonad classification schemes, too numerous to mention, abound in the literature. A scheme developed for one particular mussel species is often used or adapted for another species (Chipperfield 1953; Lubet 1959; Wilson & Seed 1974; Brousseau 1982; Shafee 1989; Jaramillo & Navarro 1995; Toro *et al.* 2002; Barber *et al.* 2005; Mladineo *et al.* 2007; Oyarzún *et al.* 2010). One such scheme, which has been used for species of *Mytilus*, *Aulacomya* and *Perna*, is as follows (Wilson & Seed 1974):

Stage 0. Resting stage. Sex indeterminate. No follicles in mantle tissue. Genital canals obscured by densely packed connective tissue.

Stage 1. Early development. First signs of gametogenesis, with small groups of germinal cells scattered in mantle. Spermatogonia and oogonia line the follicle wall, but sex determination may be difficult early in this stage.

Stage 2. Later development. Increase in mantle area occupied by follicles. In male follicles, there is a wide band of spermatogonia, spermatocytes and spermatids, with some spermatozoa free in the lumen. In female follicles, oocytes are growing and accumulating yolk. Young oocytes are still attached to the follicle wall, but some mature oocytes are free in the lumen.

Stage 3A. Morphologically ripe. Mantle is filled with follicles, occupying almost the entire gonad area and displacing connective tissue. In males, there may be a narrow band of early gametogenic products around the edge of the follicles; the latter are distended with spermatozoa, usually arranged in lamellae, although this arrangement may be lost with the formation of a sperm 'plug'. In females, most oocytes are well developed and assume polygonal shapes by being tightly packed in the follicles.

Stage 3B. Spawning. The lamellar arrangement of spermatozoa is disappearing and follicles are less densely packed. Remaining ova are spherical, as spawning reduces pressure in the follicles. There is a general reduction in mantle area occupied by genital tissue.

Stage 3C. Redevelopment. Renewed gametogenesis may occur. In male follicles, a band of early gametogenic stages is apparent. In the female, young attached oocytes are abundant on follicle walls. This stage can be confused with Stage 2, but when full redevelopment has occurred (the equivalent of Stage 3A), the packed follicles do not occupy all the mantle area available, so more connective tissue is evident than before initial spawning (Stage 3B).

Stage 3D. Recently spent. Follicles collapse and degenerate. Amoebocytes attack unspawned gametes, often resulting in the lumen of follicles filling with a mass of cellular debris. The animal returns to the resting state.

Gonad micrographs can also be used to assess ADG tissue via a qualitative index, as described by Bignell *et al.* (2008), based on a score of 1–5 to indicate the 'absent', 'present', 'scattered', 'frequent' or 'abundant' status, respectively, of this cell type.

Gonad classification schemes tend to be subjective because they are based on the qualitative characterisation of overall morphology of the gonads and not on quantitative analysis of the gametes. One such quantitative method expresses the dry weight of the gonad as a percentage of the total soft-tissue dry weight for each individual, male and female, to give a mean gonadosomatic index (GSI) for each sampling interval. Both wet and dry tissue weights can be used, but the latter are preferable because water content varies seasonally and between different tissues. Seasonal changes in GSI reflect accumulation and release of gonad material, as well as utilisation of stored energy reserves. Jaafar Kefi *et al.* (2014) analysed temporal changes in GSI in males and females of the date mussel, *Lithophaga lithophaga*, over a 14-month period and found a gradual increase from September to July in females and from September to August in males, followed by a drop, particularly in females. This decrease in GSI was due to a reduction in gonad weight as a result of spawning, which in this species occurs from July to early September (Table 5.1). In addition to tracking gonad development, the index is increasingly being used as a gross morphological biomarker of mussel reproductive health in contaminated waters (Kopecka *et al.* 2006; Schiedek *et al.* 2006; Saavedra *et al.* 2012; Kristan *et al.* 2014; González-Fernández *et al.* 2015; Peters & Granek 2016).

Another quantitative technique is stereology (Weibel *et al.* 1966), which is used to measure changes in the volume fractions of different components within the gonad (oocytes, nutritive storage cells, connective tissue) from point counts on test grids applied to histological sections of gonad. An important requirement of the technique is that the sections examined are representative of the whole gonad (Lowe *et al.* 1982). The proportion of gonadal tissue comprising follicles that contain developing or ripe gametes is called the gamete volume fraction (GVF). Stereology has been used in mussels to estimate the fractional area of the mantle composed of gametes (%GVF) (Toro *et al.* 2002; Doherty *et al.* 2009). To illustrate, Doherty *et al.* (2009) examined tissue sections with an Olympus CX41 compound microscope interfaced with a PC and Olympus DP-Soft Image Analysis Software. For each slide, a test grid of 30 points measuring 180 × 165 mm was applied to 10 fields of view (400×). The percentage of the mantle volume occupied by gametes was calculated from the relationship between the number of pixels occupied by the gametes and the total pixels in the field for a number of individuals for each sampling interval.

GVF varies between 0% for a reproductively inactive individual and 100% for an individual in peak reproductive condition, and provides a measure of the relative maturity of the gonad (Toro *et al.* 2002). As gametogenesis proceeds, there is an increase in the volume of the gonad and a concomitant increase in oocyte size. Indeed, Lango-Reynoso *et al.* (2000) proposed a quantitative reproductive scale for oysters based on oocyte diameter, which they suggested would complement the traditional qualitative method of staging. Image analysis has been used to measure oocyte area and/or diameter in mussels

Table 5.1 Spawning periods in a selection of mussel species. See text for information on additional species.

Species	Locality	Spawning period	Reference
Arcuatula senhousia	Adriatic Sea, Italy	September–November/December	Mistri (2002); Sgro *et al.* (2002)
Aulacomya atra	Chiloé Island, Chile	June–January	Jaramillo & Navarro (1995)
	Antofagasta Region, Chile	August–September; October–November	Avendaño & Cantillánez (2013)
	Tierra del Fuego, Argentina	August–November	Pérez *et al.* (2013)
	Table Bay, South Africa	December–February/March; April–July/August; September–October	van Erkom Schurink & Griffiths (1991)
Aulacomya maoriana	South Island, New Zealand	August–September	Kennedy (1977)
Brachidontes exustus	Tampa, Florida	April	Barber *et al.* (2005)
Choromytilus meridionalis	Table Bay, South Africa	Summer (November–February/March); midwinter (May–August/September)	van Erkom Schurink & Griffiths (1991)
Lithophaga lithophaga	Bizerte Bay, northern Tunisia	July, August and early September	Jaafar Kefi *et al.* (2014)
Mytilaster lineatus	Balaklava Bay, Black Sea	June	Shalaeva & Lisitskaya (2004)
Mytilus chilensis	Chaihuín, southern Chile (40°S)	October and March	Oyarzún *et al.* (2011)
	Bahía Yal, southern Chile (44°S)	March–June	
	Falkland Islands	January–February	Gray *et al.* (1997)
Mytilus galloprovincialis	Oran, Algeria	November–February, March–June	Rouabhi *et al.* (2019)
Perumytilus purpuratus	Taltal, northern Chile (25.5°S)	December, March, July and October	Oyarzún *et al.* (2010)
	Huasco, northern Chile (28.5°S)	December, February and September	
Septifer virgatus	Cape d'Aguilar, Hong Kong	February –March; September–December	Morton (1995)

(Doherty *et al.* 2009). In scallops (*Pecten maximus*), stereology has been used to measure oocyte volume fraction (OVF) – the per cent ovary volume occupied by oocytes – and to provide different size frequency distributions of developing oocytes (Pazos *et al.* 1996). Enzyme-linked immunosorbent assay (ELISA) has been employed to quantify eggs, and ultimately to estimate reproductive output (RO; fecundity), in oyster and clam species (see Long *et al.* 2008). More recently, magnetic resonance imaging (MRI) has been used to assess seasonal changes in gonad development of oyster and scallop species (Smith & Reddy 2012). To date, however, neither ELISA nor MRI has been employed in assessing gametogenesis in mussels.

All methods of assessing gametogenesis have their advantages and disadvantages, and therefore the optimum strategy is to employ at least one qualitative (histology) and one quantitative (GSI, GVF, gonad weight, oocyte area or OVF) method.

Annual Cycles

As we have seen, the reproductive cycle in mussels and other bivalves involves several stages: growth and ripening of gametes, spawning and gonad redevelopment. Understandably, there is enormous variation in the timing and duration of each stage both between species and between different populations of the same species. In this section, a few appropriate examples will be described in order to illustrate the breadth of variation in mussels. The spawning periods for a wide range of species not covered in the text are presented in Table 5.1.

Mytilus Species

Mussels in the genus *Mytilus* are widely distributed throughout the cooler waters of the Northern and Southern Hemispheres (see Chapter 3). *Mytilus edulis*, which can mature in its first year (Duinker *et al.* 2008), usually commences gonad redevelopment in October or November in the Northern Hemisphere, with the process proceeding over the winter months so that by February the gonads are ripe. Partial spawning occurs in the spring and a second period of gametogenesis may occur over the summer months, culminating in spawning in early autumn (Sunila *et al.* 2004). When feeding conditions are favourable, such as on the low shore, the number of gametes spawned in spring and autumn is comparable. This is not the case under suboptimal conditions (e.g. the high shore), when fewer gametes are shed in the autumn. There are a number of variations on this pattern; for example, some populations exhibit a single short spawning period of a few weeks, while others (e.g. at Strangford Lough, Ireland) have a spawning period from March to September that is sometimes extended into the winter months (Seed & Brown 1975). Also, in western Iceland, just inside the Arctic Circle, *M. edulis* spawns over a period from late June/July until November; food supply is highest at the onset of spawning and remains relatively high until October (Thorarinsdóttir & Gunnarsson 2003). Even within a single location, there are differences in spawning patterns. For example, in Killary Harbour on the west coast of Ireland, wild mussels partially spawn in early spring but spawn completely in summer, while rope-cultured mussels in the harbour spawn twice over the summer months (Rodhouse *et al.* 1986).

The reproductive cycle of *Mytilus galloprovincialis* differs somewhat from that of *M. edulis*. For example, on mussel rafts in the Ria de Vigo, NW Spain, redevelopment of the gonad takes place during late autumn and early winter, and some small spawnings may occur over that period. The peak of the spawning season occurs in spring, coinciding with an increase in temperature and phytoplankton in the area, after which fast gametogenesis begins again, restoring the spawning condition, such that several random spawns may take place until early summer (Cáceres-Martínez & Figueras 1998; Ruiz *et al.* 2011 and references therein). This cycle, while similar to that reported for *M. galloprovincialis* in the Black Sea (Ciocan 2002), differs from that observed for this species in the Venice Lagoon, Italy, where several spawns occur between September and May, with a peak at the end of winter (January–February) and a quiescent period between July and August (Da Ros *et al.* 1985 and references therein).

In the Southern Hemisphere, on the west coast of South Africa, *M. galloprovincialis* spawns over two extended periods, once during summer (October–February) and once in winter (April/May– July) (van Erkom Schurink & Griffiths 1991). It is interesting to note that where the indigenous mussel *Perna perna* and the invasive mussel *M. galloprovincialis* are found together on the South African south coast, the former spawns in summer and early winter and the latter in summer and autumn (see earlier), although both are exposed to the same environmental factors (Zardi *et al.* 2007). Zardi *et al.* (2007) suggest that species-specific endogenous factors or responses to environmental conditions play a key role in regulating gametogenesis and spawning. In the Northern Hemisphere, along the west coast of North America, *M. galloprovincialis* also spawns over a relatively restricted period (October–February), whereas in the same region *M. californianus* dribbles gametes continuously throughout the year, although there may be summer peaks depending on the area (Seed & Suchanek 1992). However, a study at a single exposed shore site on the Pacific coast of Baja California, Mexico showed that *M. galloprovincialis* spawned in early winter while *M. californianus* spawned from midwinter to spring, suggesting a different response to endogenous and/or exogenous factors regulating reproduction in these two species (Curiel-Ramírez & Cáceres-Martínez 2004). In Trinity Bay, Newfoundland, Canada, *M. edulis* and *M. trossulus* exhibit asynchronous spawning activity, with the former spawning over a short period (two to three weeks in July) and the latter over an extended period (late spring to early autumn) – perhaps a possible explanation for the low frequency of hybrids between the two species in this region (Toro *et al.* 2002). However, in a subsequent study, Maloy *et al.* (2003) found no difference in spawning periods (late July) between the two species in Cobscook Bay, eastern Maine, United States. In SW England, where *M. edulis* and *M. galloprovincialis* are sympatric at some sites, the taxa show asynchronous spawning periods and divergent energy storage strategies. The former spawned as a group in June and July 1993 and the latter spawned asynchronously, beginning in June and finishing by August 1993 (Secor *et al.* 2001).

Not all cohabiting mussel species show different spawning cycles; for example, *Perna viridis* and *Brachidontes exustus* on the coast of Florida, United States (Barber *et al.* 2005), *M. edulis* and *Mytilus trossulus* in Maine, United States (Maloy *et al.* 2003) and *M. edulis* and *M. galloprovincialis* on the west coast of Ireland (Doherty *et al.* 2009) all share cycles.

Perna Species

As we have just seen, on the south coast of South Africa the brown mussel, *Perna perna*, spawns in summer (January–March) and early winter (June), at the highest and lowest temperatures recorded in the region (Zardi *et al.* 2007). At another location on this coast, a rather variable pattern is evident, with a brief but intense winter spawning that can occur between April and August accompanied by either one extended or two discrete spring or summer events (van Erkom Schurink & Griffiths 1991). On the South African Transkei coast, Lasiak (1986) reported an extended spawning season from mid-autumn to early spring (April to September), with minor events from December to February – yet again, a different pattern to those already mentioned. In the coastal waters of Yemen (Gulf of Aden) in the Northern Hemisphere, two major spawning events occurred in late summer (SW monsoon, August–September) and early spring (NE monsoon, February–March) (Sokołowski *et al.* 2010). The first coincided with low water temperature and high salinity due to upwelling of colder and nutrient-rich water masses during the SW monsoon, while the second occurred when water temperature, salinity and food availability were relatively low, suggesting that gametes were probably released at the expense of endogenous energy reserves. On the Atlantic coast of Morocco, Benomar *et al.* (2006) found that spawning at Cap Ghir occurred during three periods: January–May, in July and in December; but at Anza, a polluted site nearby, the main period of spawning was February–March, followed by a second spawning in May–August and a third during September–October in males but not in females.

The native range of the warm-water green-lipped mussel, *Perna viridis*, is widely distributed in Asia-Pacific and Indo-Pacific regions, but it has also been introduced into parts of coastal Australia, Japan, the Caribbean and North and South America. In tropical countries, *P. viridis* spawns year-round, with peak spawning coinciding with the monsoon seasons twice a year (e.g. March–April and October–November in Malaysia) (Soon & Ransangan 2014 and references therein). But this is not always the case, as evidenced by Rajagopal *et al.* (1998), who found that peak spawning periods on the east coast of India (May–June and October–November) did not correlate with monsoon periods (July–September and October–December). See Rajagopal *et al.* (2006) for additional information on the considerable variability in reproductive behaviour of this species in tropical locations. An unusual heat pulse in the temperate waters of Western Australia is believed to have been the trigger that initiated spawning in *P. viridis*, an invasive species from South East Asia, emphasising the need to prepare for more invasions of this kind as a consequence of predicted climate change (McDonald 2012).

The green-lipped mussel, *Perna canaliculus*, endemic to New Zealand, has a prolonged spawning season from June to December in North Island populations (Alfaro *et al.* 2001, 2003). However, in the Marlborough Sounds (northern South Island), the species has two distinct spawning periods, early summer and autumn–spring, which has been attributed to water temperature differences between the two islands (Alfaro *et al.* 2011a and references therein). In addition, at Ninety Mile Beach at the tip of North Island, there are variations in reproductive activity over relatively small spatial scales (e.g. between individual intertidal populations and between intertidal and adjacent subtidal populations) (Alfaro *et al.* 2003).

Geukensia Species

The ribbed mussel, *Geukensia demissa*, is native to intertidal salt marshes along the Atlantic coast of North America from the Gulf of Maine to Florida and the Gulf of Mexico

(Franz 2001), but in the late 1880s it was introduced into San Francisco Bay on the Pacific West Coast and since then it has extended its range as far south as Venezuela (Honig 2013 and references therein). Within North Inlet Estuary, South Carolina, United States, peak spawning occurs in the low intertidal between June and August, with high intertidal populations spawning only in September, which suggests that the length of submersion and potential feeding time exert significant influence on the timing of the reproductive cycle (Borrero 1987). In Venezuela, the species has an extended reproductive season, with two distinct peaks of spawning activity that coincide with the onset of the rainy season in July and of the dry season in December (Báez *et al.* 2005). While marsh habitats in the tropics experience little temperature variation throughout the year, there are large fluctuations in ambient salinity during the onset of the tropical rainy season in midsummer and again during the extended dry season from midwinter through spring, thus suggesting that salinity, rather than temperature, plays an important role in the timing and extent of gametogenesis in this species (Baez *et al.* 2005). Salinity also modulates the progression of gametogenesis in *Geukensia granosissima* in coastal Louisiana, United States (Honig *et al.* 2014).

Modiolus Species

The horse mussel, *Modiolus modiolus*, is found on the Atlantic and Pacific coasts of North America as well as on Europeans coasts of the Atlantic Ocean. Populations in Strangford Lough, Northern Ireland showed a slow release of gametes throughout the year, with no pronounced annual cycle (Brown & Seed 1977). However, a small intertidal population located in nearby Belfast Lough exhibited a much more seasonal cycle, with spawning mainly occurring during the autumn and winter, suggesting that localised environmental factors are exceedingly important in controlling the annual reproductive cycle of this species (Brown & Seed 1977). When widely separated populations from western Sweden, western and northern Norway and Northern Ireland were examined, those from western Sweden and western Norway showed a well-defined spawning period in June, followed by redevelopment and a further gamete release. However, northern Norway mussels spawned in July, and subsequent redevelopment was not accompanied by further spawning, indicating that spawning is more dependent than gametogenesis on the attainment of a limited range of minimum temperatures (Brown 1984).

In the eared horse mussel, *Modiolus auriculatus*, at Hurghada, Egypt on the Red Sea coast, the main spawning period extended between March and May, but some spawning continued until June and July, and even when most individuals were spent, there was still a small portion of mussels that were still ripe in August and September – a spawning period of more than six months' duration (Abdel Razek *et al.* 2014). A similar but less prolonged spawning cycle has been reported for the horse bearded mussel, *M. barbatus*, in the Mali Ston Bay area of Croatia, where spawning occurs between June and August, with a further small spawning in October (Mladineo *et al.* 2007).

Mytella Species

The mangrove mussel, *Mytella bicolor*, previously known as *Mytella guyanensis*, is found in intertidal zones of mangrove and estuarine regions from western Mexico to Peru and from Venezuela to Brazil (Introíni *et al.* 2010). On the Pacific coast of Costa Rica, this species spawns year-round, but with two heavy spawning periods from August through November and from February through April (Cruz & Villalobos 1993). This pattern of year-round spawning is characteristic of other tropical species already mentioned. A similar regime was observed in the same species on the northern coast of Brazil, except that spawning peaks coincided with periods of lower rainfall and higher salinity (i.e the dry season) (Gomes *et al.* 2010). The native distribution of the closely related charru mussel, *Mytella charruana*, extends from South and Central America northward through Mexico. In two coastal lagoons along the Pacific coast of Mexico, the spawning period occurred annually from July to October, a much more restricted period than that reported for the more southerly distributed species *M. bicolor* (Cárdenas & Aranda 2000).

Bathymodiolus Species

These species are found in hydrothermal and cold seep vents in deep oceanic waters. They have chemosynthetic bacteria in their gills that oxidise hydrogen sulphide, and they absorb nutrients synthesised by these bacteria while also filter-feeding from surrounding waters (see details in Chapter 3). Despite the apparent continuous supply of energy at seeps and vents, the reproductive patterns of species occurring there do not differ from those of mussel species found in nonchemosynthetic environments (Tyler & Young 1999).

Bathymodiolus azoricus dominates hydrothermal vent communities in the Mid-Atlantic Ridge (MAR) near the Azores region. Dixon *et al.* (2006) were the first to provide evidence of an annual reproductive cycle in this species, with spawning taking place in December–January, a period that coincides with a peak in surface-water primary production that has the potential to act as a food source for both adult mussels and dispersing larvae. Gametogenesis was also investigated in the cold seep mussel *Bathymodiolus childressi* at three sites surrounding the Brine Pool on the continental slope of the northern Gulf of Mexico (Tyler *et al.* 2007). The periods of reproductive development and spawning (October–February) were synchronised across all sites, and were similar to those of *B. azoricus*; this periodicity was apparently correlated with surface production, which peaks during the winter months. A single spawning event (October–December) was also reported for the cold seep mussel *Idas modiolaeformis*, from the eastern Mediterranean Sea (Gaudron *et al.* 2012). The major phytoplankton bloom in this region occurs from October to March, and likely increases food availability for the larvae released during the spawning period (Krom *et al.* 2010).

Factors Controlling Reproduction

The control of reproduction involves a complex interplay between exogenous factors like temperature, phytoplankton density and quality, salinity, photoperiod and tidal rhythm, and endogenous factors such as neuropeptides, steroids, structural proteins and genotype. The role of these factors in the initiation and duration of gametogenesis – a process that extends over several months – most likely differs from that in the timing and synchronisation of spawning, a much briefer event. However, because the two processes, gametogenesis and spawning, occur sequentially, they are considered together in the following section.

Exogenous Regulation of Gametogenesis and Spawning

Water temperature, which varies with season and latitude, is the single exogenous factor that is most often cited as playing a dominant synchronising role in gametogenesis in mussels (Bayne 1975; Seed & Suchanek 1992; Cáceres-Martínez & Figueras 1998; Rajagopal *et al.* 1998, 2006; Le Pennec & Beninger 2000 and references therein; Ciocan 2002; Barber *et al.* 2005; Suárez *et al.* 2005; Mladineo *et al.* 2007; Zardi *et al.* 2007; Alfaro *et al.* 2011a; Abdel Razek *et al.* 2014; Jaafar Kefi *et al.* 2014; Sreedevi *et al.* 2014). However, many decades ago, Barnes (1957) showed that synchronisation of spawning in epibenthic communities was particularly tuned to the main phytoplankton outburst, and only indirectly to temperature. This is not surprising since the process is energy-demanding and therefore dependent on available food supply, stored energy reserves or both, and is supported by the findings of several studies (Kautsky 1982; Suárez *et al.* 2005; Dixon *et al.* 2006; Tyler *et al.* 2007; Gomes *et al.* 2010; Soon & Ransangan 2014). To illustrate, in their classic study, Newell *et al.* (1982) examined the reproductive condition of several populations of *M. edulis* spread over a distance of 700 km along the East Coast of the United States and found no discernible latitudinal trends in reproduction. Two populations, at the same latitude and experiencing the same temperature regime, exhibited the greatest temporal difference in gametogenesis, with summer reproductive maxima separated by a three-month interval. These differences were attributed to temporal and quantitative variations in the energy content of the available food between the two habitats. So, it would appear that in environments where variations in patterns of phytoplankton production are not large (e.g. the tropics), it is likely that the reproductive cycle of a species will be rather constant from year to year. But in habitats where there is year-to-year variation in food abundance (e.g. brackish water environments), the reproductive cycle could be expected to vary (Newell 1989). These results emphasise the importance of food supply in the control of gametogenesis, but also highlight the complex interplay between food supply and temperature. Food quality is also an important consideration in gonad maturation, particularly for hatchery broodstock (see Chapter 10). Phytoplankton, at concentrations found in nature, actually stimulates spawning in species of *Mytilus* (Smith & Strehlow 1983; Starr *et al.* 1990). It is the direct contact with the phytoplankton cells, or detection of substances released by them, that results in spawning. The advantage of phytoplankton as a spawning cue is that it integrates various environmental parameters indicating favourable conditions for larval success (Starr *et al.* 1990).

Mussels in the high intertidal zone are subject to the stress of both decreased food availability and increased exposure to aerial temperature and desiccation on a daily basis. Petes *et al.* (2008b) have shown that mussels (*M. californianus*) inhabiting or transplanted to the high intertidal invest less relative energy in reproduction and spawn all their gametes in the early summer, while low intertidal mussels continuously spawn small batches of gametes throughout the year. In addition, as seen earlier, the gonads of high intertidal mussels accumulate high concentrations of carotenoid pigments, a potentially costly defence against damaging oxidative stress during exposure. Alfaro *et al.* (2003) reported similar results for *P. viridis*, where sub-tidal mussels had consistently higher gonad indices and higher percentages of ripe and spawning individuals than intertidal mussels. Borrero (1987) actually measured the difference in potential feeding time between 10 high-tidal and 11 low-tidal populations of *Geukensia demissa*, and found there was a twofold difference, accompanied by a delay of about two months, in the reproductive activity of the high-tidal mussels. Another consequence of reduced feeding time in this species is that body growth rates are depressed at the higher shore level and mussels become sexually mature a year or more later than do lower-level mussels of the same age (Franz 1996). This pattern is different from that in some populations of *M. edulis*, where depressed growth rates induce rather than delay sexual maturity (Kautsky 1982).

Salinity fluctuations are another factor, albeit a minor one, influencing the gametogenic cycle. In the section on annual cycles, information has already been presented on the effects of fluctuating salinity on mussel gametogenesis in tropical and subtropical waters; see *Geukensia demissa* (Báez *et al.* 2005), *G. granosissima* (Honig *et al.* 2014), *Lithophaga lithophaga* (Jaafar Kefi *et al.* 2014), *P. perna* (Sokołowski *et al.* 2010) and *P. viridis* (Rajagopal *et al.* 2006).

In addition to other exogenous factors that are potentially important in the initiation of spawning events (e.g. temperate, food availability, salinity), the presence of gametes in the water column is thought to provide a powerful chemical stimulus to other ripe individuals, thereby enhancing the chances of fertilisation. Physical stimulation by jarring or scraping the shell or by pulling or cutting the byssus threads can stimulate spawning in *M. edulis*. Seed & Suchanek (1992) suggest that these are the types of environmental cues received during storms that could also signal the presence of storm-generated patches of bare rock on to which mussel spat can settle. See Chapter 10 for a description of chemical, biological and physical stimuli used to induce spawning in hatchery broodstock.

Endogenous Regulation of Gametogenesis

In bivalves, the reproductive cycle is controlled by interactions taking place between ganglion neurosecretions and steroids. The activity of neurosecretory cells, which are mostly located in the cerebral ganglia of the central nervous system (CNS; see Chapter 2), is low during the resting phase of gametogenesis, progressively increases in synchrony with the developing gonad and reaches a maximum just before spawning. This pattern is mirrored by the activity of aspartate transcarbamylase (ACTase), a key enzyme involved in nucleotide synthesis; its activity is closely correlated with active cell multiplication in *Mytilus edulis* (de Zwaan & Mathieu 1992). Neurosecretory cells secrete a variety of neuropeptides that exert various physiological effects on the gonad. One of these is the peptide hormone gonadotropin-releasing hormone (GnRH), which in humans regulates reproduction indirectly by stimulating secretion of follicle stimulating hormone (FSH) or luteinising hormone (LH) from the anterior pituitary gland. In *M. edulis* and the Pacific oyster, *Crassostrea gigas*, GnRH augments incorporation of [^{3}H]-thymidine, a marker of DNA synthesis, into dispersed cells of the testes, suggesting that GnRH stimulates cell division in the gonad (Pazos & Mathieu 1999). It is not clear whether GnRH acts directly on gonadal cells or not. In view of the commercial importance of bivalves, there is a paucity of information on GnRH, and taking bivalves as a whole, there is less information on its role in mussels compared to oyster and scallop species (see Rodet *et al.* 2005; Morishita *et al.* 2010; Bigot *et al.* 2012; Treen *et al.* 2012; Osada & Treen 2013). Other neuropeptides, such as Ala-Pro-Gly-Trp amide (APGWamide) and Phe-Met-Arg-Phe amide (FMRFamide), also play a putative role in bivalve reproduction (Pani *et al.* 2005; reviewed in Morishita *et al.* 2010). When the APGWamide precursor was cloned in *M. edulis*, it was shown that the precursor encoded authentic APGWamide, as well as other analogue peptides (Favrel & Mathieu 1996). The gene encoding this precursor was expressed in discrete neurons within the three pairs of ganglia of the CNS, the only clearly established endocrine centres in this species. However, the role of these peptides in neural regulation in gonadal activity is yet to be elucidated in marine mussels (Morishita *et al.* 2010). The monamine serotonin (5-hydroxytryptamine, 5-HT) is another molecule that acts as a neurohormone, modulating spawning and oocyte maturation in the gonad (references in Garnerot *et al.* 2006). In a recent study, levels of serotonin and two other monoamines, dopamine (DA) and noradrenaline (NA), measured using high-pressure liquid chromatography with electrochemical detection (HPLC-ED), were seen to vary dramatically in the gonads and cerebral and pedal ganglia of *P. perna* over the course of gametogenesis. When samples were analysed from three locations on the Atlantic coast of Morocco, two of which were contaminated by industrial or domestic waste (Klouche *et al.* 2015), marked modification of monamine tissue content was evident at the two polluted sites, which might explain the poor reproductive performance recorded for *P. perna* from this coastline (see also Benomar *et al.* 2006; Moukrim *et al.* 2008).

A range of endogenous sex steroids (progesterone, androstenedione, testosterone, 5α-dihydrotestosterone, 17β-estradiol, estrone, androsterone and 3α-androstanediol) have been identified in *M. edulis* (Reis-Henriques *et al.* 1990); of these, levels of progesterone and androstenedione were highest, and varied seasonally irrespective of gender. Progesterone levels in *M. edulis* males and females during the annual reproductive cycle were analysed from whole animals and gonads (Reis-Henriques & Coimbra 1990). The highest levels in both males and females were observed in July and October, coinciding with the two periods of intense gametogenesis, suggesting that the steroid, or a derivative, plays a role in gamete proliferation. More recently, annual variations in 17β-estradiol (17β-E_2) levels have been reported in gonad tissue of *M. galloprovincialis*, with the highest concentrations (March–April, September–October) during gonad development stages (Kaloyianni *et al.* 2005). The same steroid was detected in the gonad and haemolymph of *M. edulis*, suggesting a hormonal as well as a reproductive function (Stefano *et al.* 2003). The data also suggest that in pedal ganglia, 17β-E_2 can act via an estrogen cell-surface receptor (see also Zhu *et al.* 2003; Nagasawa *et al.* 2015). Most of the required enzymes for production of sex steroids from cholesterol have also been identified in molluscs (reviewed in Lafont & Mathieu 2007), with four enzyme systems detected in *M. edulis* gonads, albeit with extremely low activities except when estrogens were used as substrates (de Longcamp *et al.* 1974). To summarise: while it is clear that vertebrate-type steroids are present in bivalves, the evidence of their endogenous origin and their involvement in endocrine-regulated gametogenesis, as demonstrated in vertebrates, is rather weak. In a review paper, Scott (2012) considers the strength of the evidence for the presence of steroids in molluscs, for steroid biosynthesis and for the presence of steroid receptors, while a related review (Scott 2013) deals with the strength of evidence for biological actions of vertebrate steroids on molluscs. From an examination of the data from more than 50 studies (1970–2012), the author concluded that while vertebrate-type steroids are indeed present in mollusc tissues, they are most likely sequestered from faeces and urine excreted by vertebrates into the aquatic environment. Molluscs concentrate these steroids much in the same way as they concentrate a wide variety of contaminants (see Chapter 8). He found no convincing evidence for biosynthesis of vertebrate steroids in molluscs. Moreover, genes for key enzymes involved in steroid biosynthesis in vertebrates are missing from the genome of molluscs (Markov *et al.* 2009). Finally, there was no incontrovertible evidence for the presence of specific functional receptors for vertebrate steroids in molluscs (Scott 2012), or to show that sex steroids have an endocrine role in mollusc reproduction (Scott 2013).

Annual Storage Cycle

Gametogenesis is an energy-demanding process. During the annual reproductive cycle, nutrients are stored when food supplies are abundant and gonad activity is minimal. Subsequently, these energy reserves are utilised to meet the energetic requirements of gametogenesis. The main energy reserve for gametogenesis is glygogen, a branched polysaccharide of glucose units, and in *Mytilus edulis* the main storage area is the mantle, where gonad and storage tissue coexist (Berthelin *et al.* 2000). Two types of cells are involved in energy storage in mussels: adipogranular (ADG) cells, which are located only in mantle tissue, and which accumulate lipids, protein and small amounts of glycogen (see Figure 5.3), and vesicular connective tissue (VCT) cells, which are located in both mantle and labial palps, and store large quantities of glycogen. The relative proportion of storage tissue to germinal tissue is variable depending on the stage of gametogenesis; it is usually at a maximum level in summer and at a minimum in autumn and winter, depending on the location. For example, in the mantle of *M. edulis* in SW England, ADG and VCT cells exhibit an annual cycle, which consists of a gradual regression from November to April, during which time nutrients are released from the cells, followed by replenishment from June to October, when the cells regain their reserves (Pipe 1987b). See de Zwann & Mathieu (1992) for a detailed description of the structure and function of the two cell types.

Glycogen metabolism is under the control of two enzymes: glycogen phosphorylase, which removes glucose residues from α (1–4) linkages within glycogen molecules to produce glucose-1-phosphate, and glycogen synthase, which catalyses the transfer of a glucose molecule from UDP-glucose to a terminal branch of a glycogen molecule. See Hanquet *et al.* (2011) for details on glucose transport into VCT cells for glygogen synthesis. Glycogen phosphorylase and glycogen synthase control short-term regulation of glycogen metabolism, while long-term regulation is probably influenced by genes that encode the two enzymes. Bacca *et al.* (2005) have identified the genes *Cg-GPH* and *Cg-GYS* for glycogen phosphorylase and glycogen synthase, respectively, in the oyster *Crassostrea gigas*. Relative levels of *Cg-GYS* transcripts were highest in October, corresponding to glycogen storage and the resting period, while relative levels of *Cg-GPH* transcripts were highest in May, corresponding to mobilisation of the glycogen needed for germ cell maturation in this species.

There is little information on hormonal control of glycogen metabolism in mussels. In vertebrates, the peptide hormone glucagon controls the release of glucose from glycogen storage cells. Up to two decades ago, the existence of molluscan glucagon-like peptides was poorly documented. However, Kellner *et al.* (2002), using an immunocytochemical approach, found evidence for the presence of glucagon-like molecules in neurosecretory cells in the cerebral ganglia of *M. edulis*. Also, using an in vitro bioassay, they investigated the effect of vertebrate glucagon on glucose metabolism and found that the hormone significantly decreased incorporation of ^{14}C-labelled glucose into glycogen.

As mentioned earlier, in addition to glycogen, proteins and lipids are also important energy reserves in bivalve gametogenesis. Annual variations in the amounts of these reserves are tightly coupled with the gametogenic cycle of *Modiolus barbatus* on the Adriatic coast of Croatia (Mladineo *et al.* 2007; Figure 5.5). Early and late stages of gametogenesis were observed between March and April, and peak spawning occurred from June to August. A biochemical analysis of gonad and whole tissue over the annual cycle showed that highest concentrations of total carbohydrates were in May, while highest concentrations of lipids were in May and June. The most significant accumulation of lipids occurred in gonad tissue, while carbohydrates were about equally distributed between gonad and whole tissue. A sudden decline in the lipid content of gonad tissue was observed in July, during peak spawning (see earlier), followed by an extended period of low lipid concentrations (July–December). Carbohydrate content started to decline in June, with the decline more rapid in gonad than in whole tissue. Protein content was at its lowest in both gonad and whole tissue in spring and at its highest during summer and autumn. Spawning in this species results in a reduction in energy stores of both carbohydrates and lipids. A rapid

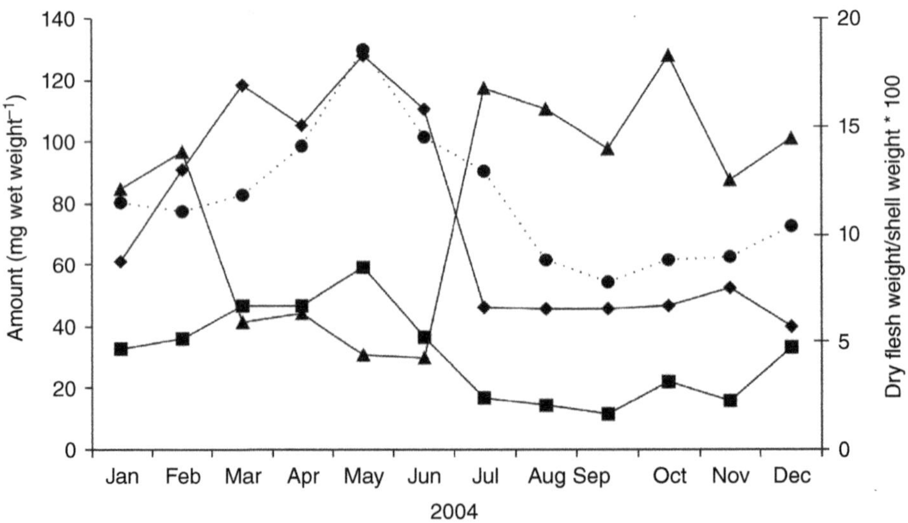

Figure 5.5 Protein (filled triangle), lipid (filled diamond) and carbohydrate (filled square) contents and condition index (CI) expressed as ratio of dry flesh weight to shell weight (filled circle) in *Modiolus barbatus* over a 12-month period. The amount of the biochemical components is expressed as milligrams of the respective component in one gram of gonad wet weight. *Source:* From Mladineo *et al.* (2007). Reproduced with permission from Springer Nature.

decline in gonad lipid content during spawning may be explained by a loss of lipids stored in ripe eggs, while the earlier onset and somewhat slower reduction in carbohydrate content may be due to its increased utilisation in the final stages of maturation. Production of mature oocytes requires a store of lipids to produce yolk (Benomar *et al.* 2010). The lowest values of protein content coincide with maxima of carbohydrate and lipid contents, and the highest protein values are attained when these products are utilised (i.e. over the spring and early summer months). This period of high food abundance is when there is a build up of protein reserves, which can serve as an energy reserve for subsequent gonad and somatic development. For further information on annual storage cycles in other mussel species, see de Zwann & Mathieu (1992), Okumuş & Sterling (1998), Benomar *et al.* (2006, 2010) and Sokołowski *et al.* (2010).

Condition indices (CIs) are also useful measurements of the nutritive status of bivalves. On a seasonal level, CI increases during the period of energy storage and gametogenesis and declines with the main spawning event. Indices are generally expressed as either wet or dry tissue weight (g) divided by some parameter of the shell, such as dry shell weight (g) or length (mm), internal shell capacity (g) or volume (ml). Crosby & Gale (1990) evaluated three different CI formulae – from which all others are derived – and concluded that the formula CI = [dry soft tissue weight (g)] × [1000/internal shell cavity capacity (g)] of Hawkins *et al.* (1987) has less measuring errors and a lower coefficient of variation, is the easiest and fastest to use, and serves best as an index of current bivalve nutritive status and recent stress. Shell cavity capacity is determined by subtracting dry shell weight (g) in air from the total whole live weight (g) in air of a cleaned bivalve. A sample size of 50–100 individuals is recommended, to allow for individual variation (Quayle & Newkirk 1989). In a more recent study, nine frequently used CI methods were applied to compare the productivity of *P. perna* yield in three coastal bays in Brazil, as well as seasonality effects on CI values (Galvao *et al.* 2015). The CI = [soft tissue wet weight (g)] × [shell length (mm) × $100]^{-1}$ formula (Kagley *et al.* 2003) indicated the best harvest time and the most favourable area in which to develop mussel farming.

Condition varies both seasonally and spatially (Peharda *et al.* 2007), being influenced by food levels (Orban *et al.* 2002; Bodin *et al.* 2004; Yildiz *et al.* 2006; Lander *et al.* 2012), mussel size (Hickman & Illingworth 1980), mussel density (Barbarro *et al.* 2003), air exposure (Narváez *et al.* 2009; Knights 2012), local environmental conditions (Okumuş & Stirling 1998; Smith 2000; Oliver *et al.* 2008) and parasitism (Olivas-Valdez & Cáceres-Martínez 2002; Brenner 2010). CI has proved to be a useful biomarker, since mussels exposed to contaminants expend part of their energy budget in detoxification processes to maintain homeostasis, to the detriment of growth and body condition (Andral *et al.* 2004; Pampanin *et al.* 2005; Mubiana *et al.* 2006; Culbertson *et al.* 2008; Rajkumar 2009 and references therein; Yap & Al-Barwani 2012; Gomiero *et al.* 2015). In the date mussel, *Lithophaga lithophaga*, Jaafar Kefi *et al.* (2014) reported that CI values were relatively stable in both males and females over a 12-month period (Figure 5.6). At their study site in Tunisia, seawater temperatures ranged from 11.5 °C in January to 35 °C in August, with an annual average of 22 °C. The consistent CI values are probably a consequence of relatively high temperatures with a high density of nutrients in the water column. It is surprising that there was no obvious drop in CI values over the spawning period (July–September) of this mussel – unlike the situation for GSI values, especially for females in the same study (Figure 5.7).

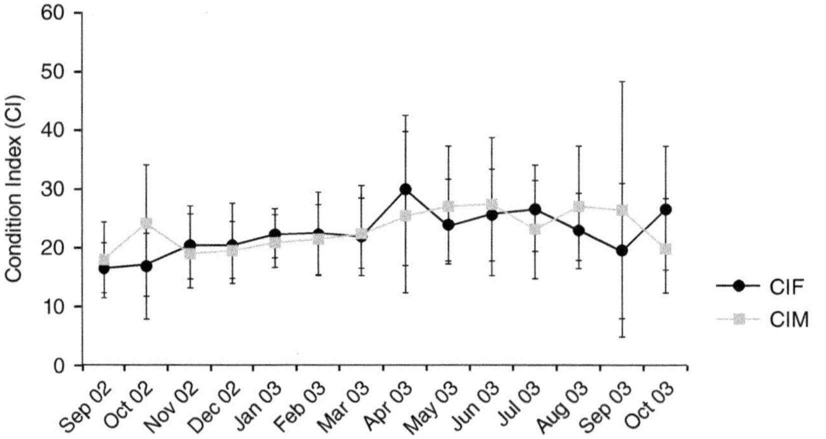

Figure 5.6 Monthly variations in the condition index (CI) of male and female date mussels, *Lithophaga lithophaga*, sampled from Bizerte Bay, northern Tunisia, over a 14-month period. *Source:* From Jaafar Kefi *et al.* (2014). Reproduced with permission from Springer Nature.

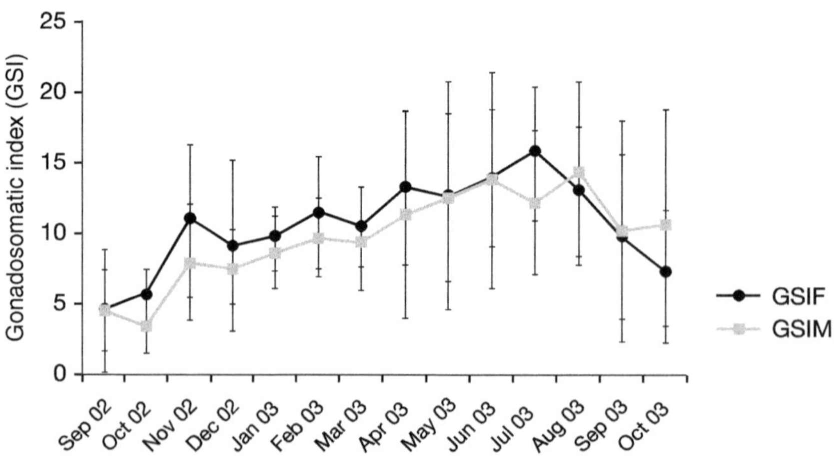

Figure 5.7 Monthly variations in the gonadosomatic index (GSI) of male and female date mussels, *Lithophaga lithophaga*, sampled from Bizerte Bay, northern Tunisia over a 14-month period. *Source:* From Jaafar Kefi *et al.* (2014). Reproduced with permission from Springer Nature.

Reproductive Effort and Fecundity

Young mussels grow fast, converting all available energy into somatic growth; but with increasing size there is a gradual shift from somatic growth into reproduction, so that in the largest animals most production (>90%) is channelled into gamete synthesis (Rodhouse *et al.* 1986). Paine (1976) has suggested that this may be the result of selective pressure for early increase in size as a refuge from predation, coupled with selection for high fecundity (egg production) in larger individuals.

The term 'reproductive effort' (RE) is used to refer to the level of energy allocated to reproduction and is commonly expressed as a proportion of total production:

$$RE = \left[P_g / \left(P_g + P_s\right)\right] \times 100$$

where *Pg* is gamete production and *Ps* is somatic production over the same time period, usually one year. See Bayne *et al.* (1983), Thompson (1984) and Hicks *et al.* (2001) for methods of measuring RE in mussels. Generally, RE increases nonlinearly in planktotrophic bivalves that reproduce multiple times over the adult life span. Bayne & Worrall (1980) reported an age-related increase in RE in *M. edulis*, and also observed an increase in gamete production (*Pg*) per unit body weight, which means that with an increase in body size an increasing proportion of body calories is allocated to reproduction (Bayne *et al.* 1983). For example, RE was estimated to be 10–17, 39 and 64% of total production in the first, second and third year of life in a Moroccan population of *P. perna* (Shafee 1992). There is substantial variation in RE between mussel species, with high values reported for *Choromytilus meridionalis* (up to 75%), *Aulocomya atra* (36–89%) and *P. perna* (76 and 74% for 1993 and 1992 cohorts, respectively) in the Gulf of Mexico (Hicks *et al.* 2001 and references therein). Medium to low values were observed for *M. edulis* (35%) and *Geukensia demissa* (19%). While most low values are related to young populations, the low value in *G. demissa* has been ascribed to stress due to exposure experienced by this species in the intertidal zone, resulting in a channelling of energy from RE. In contrast, *M. edulis* appears to divert energy from somatic growth, as opposed to gamete production, under stressful environmental conditions (Suchanek 1985). Low values may also occur due to variation in year-to-year environmental conditions. There is also considerable variation in RE between conspecific populations. For example, RE in three-year-old mussels in different populations from southwest England ranged between 43 and 95% (Figure 5.8). At Lynher, mussels received a richer food supply and experienced little environmental stress compared to the population at Cattewater, which was located near a warm water outfall, and was therefore subject to thermal stress during winter and spring (Bayne & Widdows 1978; Bayne *et al.* 1983).

Fecundity, or reproductive output (RO), is the amount of gametogenic material produced by an individual. RO can be measured by first determining the body mass index (BMI mg cm^{-3}), defined as ash-free dry weight (AFDW) divided by cubic shell length, for a subsample of individuals, usually female. Each individual in the remainder of the sample is induced to spawn, each in a separate container, and eggs are collected, counted and often measured (Seed & Suchanek 1992 and references therein; Honkoop & van der Meer 1998; see Bayne *et al.* 1983 for related methods of measuring RO). Another method of RO estimation is to multiply the total mantle dry weight of an individual by the gonad volume fraction (GVF). Toro *et al.* (2002) used this method to estimate RO in sympatric populations of *M. edulis* and *M. trossulus*. The latter had a higher RO (but smaller eggs) than *M. edulis* of similar shell length, while hybrids showed intermediate values of RO. More recently, immunological techniques have been used to quantify RO in several bivalve species, but not in mussels (Park *et al.* 2005; Royer *et al.* 2008).

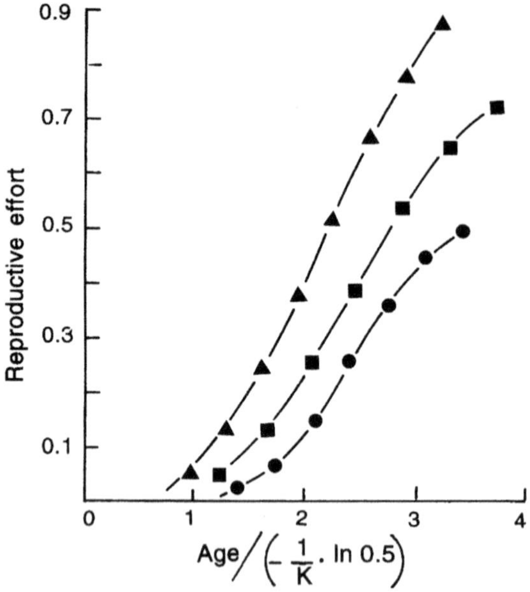

Figure 5.8 Reproductive effort (RE) of the mussel *Mytilus edulis* from three populations, related to age, corrected to exclude differences due to different rates of growth. K is the rate constant in the von Bertalanffy growth equation (see Chapter 6) for each site. Filled triangle, Lynher; filled square, Mothercombe; filled circle, Cattewater (all in SW England). *Source:* From Bayne *et al.* (1983). Reproduced with permission from Springer Nature.

In the mussel *M. edulis*, individual females produce between 7 and 40 million eggs depending on shell length (Thompson 1979; see also references in Sprung 1983; but see Sukhotin & Flyachinskaya 2009 on the effects of ageing on RO in *M. edulis*). In the Baltic, where mussels (*M. trossulus*) are small – dwarfed by their struggle to survive in the low salinity of the brackish waters – maximum fecundity (50–60%) in the largest individuals was of the same magnitude as the maximum found in much larger individuals outside of that region (Kautsky 1982). Kautsky (1982) suggested that the generally wider shells of these mussels may provide the extra volume needed for gonadal tissue. RO is also influenced by environmental factors such as temperature, food supply and tidal exposure. On the island of Helgoland in the North Sea, RO was estimated in a tidal and subtidal population of *M. edulis* over a 15-month period (Sprung 1983). Intertidal mussels had significantly smaller eggs than subtidal individuals, which is in agreement with the results of Bayne *et al.* (1978) and Honkoop & van der Meer (1998) for the same species. Although body mass per unit area of the subtidal population was almost double that of the tidal population, their annual egg outputs were nearly the same (2.3×10^9 eggs m^{-2}). This was achieved by two mechanisms: in the intertidal zone, smaller mussels were already sexually mature (18 mm instead of 30 mm shell length); and individuals of the same shell length had a higher egg output in the intertidal than in the subtidal zone. Sprung (1983) suggests that under favourable conditions, as in the subtidal environment, the period of somatic growth is prolonged, whereas the stressful conditions of the intertidal environment (temperature and salinity fluctuations, reduced feeding time) stimulate mussels to enhance their reproductive period in order to maximise recruitment success.

As seen earlier, the indigenous mussel *P. perna* and the invasive mussel *M. galloprovincialis* coexist on the low shore of the south coast of South Africa. Zardi *et al.* (2007) examined attachment strength and RO in the two species over an 18-month period. Peaks in attachment strength coincided with relatively low gamete output in both, suggesting that mussels cannot afford to invest simultaneously in both processes. The attachment strength of *P. perna* was always significantly higher than that of *M. galloprovincialis*, while the latter had a greater RO. The authors suggest that while this can improve the ability of the invasive species to colonise free space, it may energetically limit its capacity to invade wave-exposed shores (see also Nicastro *et al.* 2010). RO in mussels is rarely examined over meso-scales (i.e. tens to hundreds of kilometres). Phillips (2007), however, examined RO in *M. californianus* from sites spanning a 200 km stretch of coastline in southern and central California, United States. Not surprisingly, there was a positive correlation between RO and mussel size, but the slope of the relationship was on average twice as steep at the southern sites, where mussels had higher growth rates than those at northern sites (Phillips 2005). This spatial pattern may be temperature-driven, since water temperatures were consistently higher at the more southern sites, with summer average water temperature 4 °C warmer than at northern sites over the six-month study period. Feeding rates increase with an increase in temperature, which means that small mussels at the southern sites could grow and reproduce at higher rates.

Fertilisation and Gamete Compatibility

Most marine bivalves are broadcast spawners, simultaneously releasing their eggs and sperm into the water column where fertilisation takes place. In *M. edulis*, males usually spawn first, and the presence of sperm in the water stimulates the females to start spawning (Newell & Thompson 1984). Sperm is discharged in a thin, steady stream through the exhalant opening, while eggs are discharged, more intermittently, in clouds (Helm *et al.* 2004). Male-associated peptide (MAP) is a major protein in the gonad and semen of *M. galloprovincialis* and may be associated with maturation, emission and survival of sperm (Torrado *et al.* 2003). Sperm exploit chemical cues given off by eggs, enabling them to effectively choose between the eggs of different conspecific females (Evans *et al.* 2012). Also, the preferences shown by sperm in egg-choice trials are highly predictive of early embryonic viability (Oliver & Evans 2014).

Species of *Mytilus*, *Crassostrea* and *Dreissena*, the freshwater mussel, have served as valuable models for understanding general mechanisms of fertilisation (Misamore *et al.* 2006 and references therein). The vitelline coat lysin M7 (see later), a protein found in the sperm acrosome of mussels, dissolves the vitelline coat of the egg, allowing fertilisation (Riginos *et al.* 2006). Several carbohydrate-binding proteins (lectins) on the surfaces of eggs and sperm are believed to play a role in sperm–egg binding and fusion (McAnlis *et al.* 2010). The plasma membranes of the sperm and egg fuse and the nucleus and mitochondria of the sperm enter the egg cytoplasm. Only one sperm nucleus joins with one egg nucleus at fertilisation. Polyspermy is prevented by electrical and/or non-electrical changes in the plasma membrane of the egg. In an electrical block, the sperm opens channels in the egg plasma membrane, producing a change in membrane potential, which protects the egg during the first minute after insemination, until the fertilisation envelope is formed (details in Gould & Stephano 2003). Monospermy is ensured after the fast polyspermy block by the collaboration of two strategies: suppression of the acrosome reaction of the supernumerary sperm and blockage of sperm entry at the plasma membrane (Togo *et al.* 1995). It has been suggested that an aminopeptidase-like protease, released into seawater by fertilised eggs, participates in suppression of the acrosome reaction (Togo & Morisawa 1997). Meiosis is completed in the egg once fertilisation has taken place.

Recognition and adhesion between sperm and egg are crucial for fertilisation, and gamete recognition proteins (GRPs) play a central role in this process. The best-studied GRP is the membrane-bound protein M7 lysin, which was first identified in the *M. edulis* species complex (Takagi *et al.* 1994). The protein is located on the acrosomal surface of *Mytilus* sperm and binds to glycoprotein receptors on the egg cell surface, ultimately bonding with the egg vitelline envelope. In fact, M7 lysin is not limited to the acrosome of sperm, but is also expressed in oocytes as well as the somatic tissue of male and female *M. edulis* (Heβ *et al.* 2012). GRPs evolve rapidly and are often subject to positive selection.[5] Gamete incompatibility can result in a mismatch between egg and sperm recognition proteins, and these fertilisation failures are sufficient to reproductively isolate even closely related taxa. As well as maintaining existing species boundaries, gamete-recognition divergence could also contribute to the origin of new species (Springer & Crespi 2007 and references therein). The *M. edulis* species complex (*M. edulis, M. galloprovincialis* and *M. trossulus*) serves as an excellent model to investigate the molecular basis of reproductive isolation. As broadcast spawners, the absence of mating competition or mate choice prior to gamete release means that sexual selection operates solely through gamete-level interactions, which can easily be investigated under laboratory conditions. Another feature of the complex is the ability of species to hybridise in zones of secondary contact, which allows for investigation of molecular evolution and interspecific gene flow of reproductive genes in laboratory crosses of species pairs (Bierne *et al.* 2002; Rawson *et al.* 2003; Miranda *et al.* 2010) or in allopatric and sympatric *Mytilus* populations (Slaughter *et al.* 2008; Heβ *et al.* 2012 and references therein). Despite extensive hybridisation, the ability to retain their integrity through much of their species range suggests the operation of pre- or postzygotic reproductive isolating mechanisms (RIMs; see Chapter 9). One way of testing gamete compatibility between species is to compare conspecific and heterospecific fertilisation success by pair matings with no gamete competition, and pool sperm from both species to allow gamete competition. Klibansky & McCartney (2014) carried out gamete crossing experiments, using *M. edulis* and *M. trossulus* from natural populations in a hybrid zone on the coast of Maine, United States. They found evidence in 5 of 12 crosses for conspecific sperm precedence (CSP), defined as a reduction in heterospecific offspring from competitive fertilisations in vitro compared to that seen in noncompetitive fertilisations. Similar results have been reported for the same (Rawson *et al.* 2003; Slaughter *et al.* 2008; Miranda *et al.* 2010) and different (Bierne *et al.* 2002) pairs of closely related *Mytilus* taxa. The mitochondrial dye MitoTracker has recently been used to track the success of *M. galloprovincialis* sperm from individual males when they compete with those from rivals in egg fertilisation (Lymbery *et al.* 2016).

Studies have demonstrated that positive selection drives divergence of M7 lysin in the *Mytilus* species complex (Riginos & McDonald 2003; Riginos *et al.* 2006; Springer & Crespi 2007; Gilg *et al.* 2013). Gilg *et al.* (2013) reported divergence of this protein in *M. galloprovincialis* in sympatric populations in the east Atlantic, which they suggested was driven by selection pressures following secondary Pleistocene contact between *M. edulis* and *M. galloprovincialis*. This divergence in GRPs could reduce the rate of cross-fertilisation due to gamete incompatibility, thereby acting as an important prezygotic RIM. Lima & McCartney (2013) have shown that different lysins serve as GRPs in different hybrid zones, with M3 lysin likely to play this role in the *M. edulis–M. trossulus* hybrid zone in the west Atlantic (see earlier). There is little evidence that early-acting postzygotic RIMs (e.g. abnormal development or reduced viability of hybrid larvae) play a role in reducing hybridisation. The fact that hybrid zones persist suggests that such zones are being maintained, most likely through highly compatible heterospecific fertilisations that result in few but fertile offspring (Klibansky & McCartney 2014).

Larval Development

There are three main stages in larval development: the blastula, free-swimming larvae (trochophore and veliger) and the pre-settlement pediveliger. The fertilised egg, usually 60–65 μm in diameter, rapidly divides to become a hollow ball of cells, the blastula (Figure 5.9a), which begins to swim once cilia appear some 10–12 hours post-fertilisation (hpf). About 12 hours later, a ciliated trochophore stage (70–75 μm) is reached (24 hpf) (Figure 5.9b). At this stage, although mobile, the trochophore is still reliant on the yolk for nutrients. Also, neuronal development, as demonstrated by neuropeptide peptide FMRFamide and neurohormone serotonin immune reactivity, commences (Voronezhskaya *et al.* 2008). A shell gland begins to secrete the first larval shell, the prodissoconch I, which is thin, translucent and D-shaped in outline (48 hpf). The larva, now an early veliger (75–100 μm shell length; Figure 5.9c) immediately starts to secrete the second larval shell, the prodissoconch II. This shell is secreted by the mantle and exhibits growth lines and other ornamentation. Veliger larvae possess a velum, a circular lobe of tissue bearing a ring of cilia. At 4–10 days post-fertilisation (dpf), the mid veliger larva (100–150 μm) has rudimentary gills, with a well-developed mouth, oesophagus, stomach, intestine and digestive gland (Figure 5.9d). Small particles (1–2 μm diameter) caught by the cilia are swept toward the mouth and onward into the simple gut. At about 15 dpf, the late veliger larva is about 120–250 μm and is characterised by a prominent umbo on the larval shell. Striated muscle velum retractors and the shell anterior adductor muscle develop at this stage (Dyachuk & Odintsova 2009; Figure 5.9e). During the veliger stage, larvae are positively phototrophic and are sensitive to pressure. These responses tend to keep the shelled larvae in the surface waters. Prior to metamorphosis, a pair of pigmented eyespots and an extensible ciliated foot with a byssal gland appears. The larva, now a pediveliger (after 20 dpf; Figure 5.9f) is between 210 and 300 μm in length and is characterised by the features outlined in

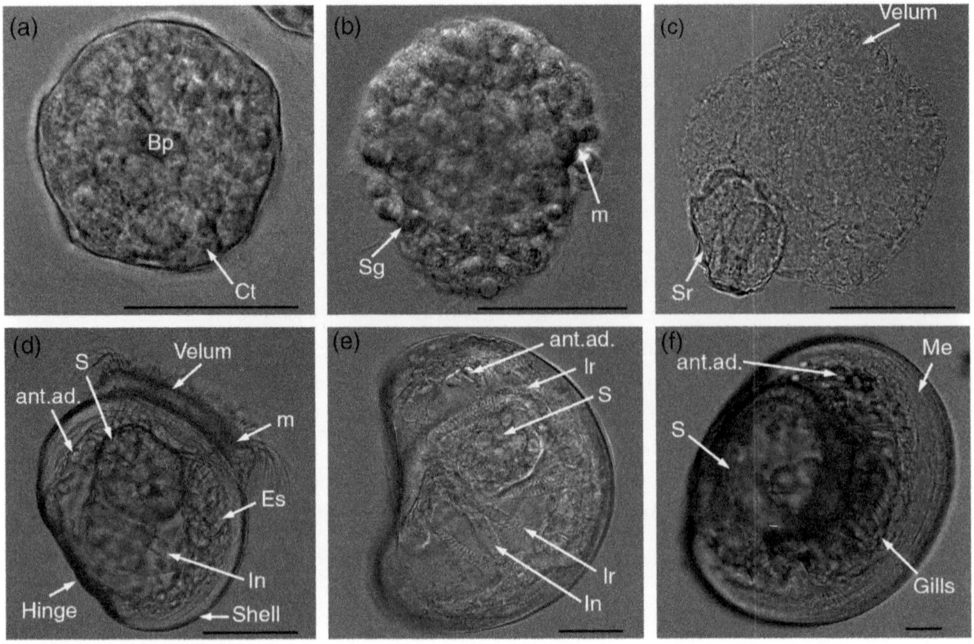

Figure 5.9 Stages in the development of *Mytilus trossulus*. (a) Swimming blastula stage (12 hr post fertilisation (hpf), at 16–17 °C). Bp, blastopore; Ct, ciliary trochoblasts. (b) Late trochophore (36 hpf). m, mouth; Sg, shell gland. (c) Early veliger (48 hpf). Sr, shell rudiment. (d) Veliger (96 hpf). The well-defined velum and the digestive system are visible. ant.ad., anterior adductor; Es, esophagus; In, intestine; m, mouth; S, stomach. (e) Pediveliger (after 20 dpf). Lr, larval striated muscles. (f) Early juvenile (30 dpf). Me, mantle edge. Scale bar = 50 μm. *Source:* From Dyachuk & Odintsova (2009). Reproduced with permission from John Wiley & Sons.

Table 5.2 Morphological features of *Mytilus edulis* pediveliger larvae.

Large velum for swimming and feeding
Ciliated palp that sorts food particles
A few pairs of gill filaments
Mouth, oesophagus, stomach with style sac and large digestive gland, simple intestine
Thin mantle that secretes shell
Foot used in crawling and byssus secretion
Cerebral, pedal and visceral ganglia, sensory pigment spots, pedal statocysts

Sources: Adapted from Bayne (1971) and Lutz & Kennish (1992).

Table 5.2. Pediveligers, in contrast to veligers, are positively geotrophic and insensitive to pressure, which encourages them to descend to the bottom in preparation for settlement. Much of the preceding description is based on Lutz & Kennish (1992) and Dyachuk & Odintsova (2009) for *Mytilus* species in temperate latitudes. Rusk *et al.* (2017) carried out a comprehensive study, incorporating histology, light microscopy, scanning electron microscopy (SEM), immunochemistry and confocal microscopy on embryogenesis and larval development of *P. canaliculus*, a commercially important species endemic to New Zealand. Timing of the different larval stages was similar to that described for *M. trossulus*. However, in species from warmer climes (e.g. the green mussel, *P. viridis*, in the Indo-Pacific region), the length of time taken to reach the individual stages are shorter (see Rajagopal *et al.* 2006).

Factors Affecting Larval Growth

The factors important in larval development are, not surprisingly, those same factors influencing bivalve growth in general, namely temperature, salinity and food ration (see Chapter 6 for details). Almost all information on the effects of these factors on growth comes from larval cultures produced in the laboratory.

Temperature and Salinity

Temperature is probably the most important factor influencing larval growth, as the water temperature in which larvae swim determines metabolic rate (Helm *et al.* 2004). For example, in the green mussel, *P. viridis*, larval growth increased with increasing temperatures from 24 to 31 °C (Nair & Appukuttan 2003). Growth rates (μm day^{-1} ± SD) were 6.01 ± 0.19, 10.44 ± 0.895, 12.08 ± 0.56 and 17.13 ± 0.84 at 24, 27, 29 and 31 °C, respectively (Figure 5.10). Growth rates for days 1–10 were significantly different in larvae reared at 24 °C (3.85 ± 0.57 μm day^{-1}) than in those reared at 27 °C (9.08 ± 0.27 μm day^{-1}). Growth was nearly four times faster at 31 °C (15.9 ± 0.39 μm day^{-1}) than at 24 °C. From days 10 to

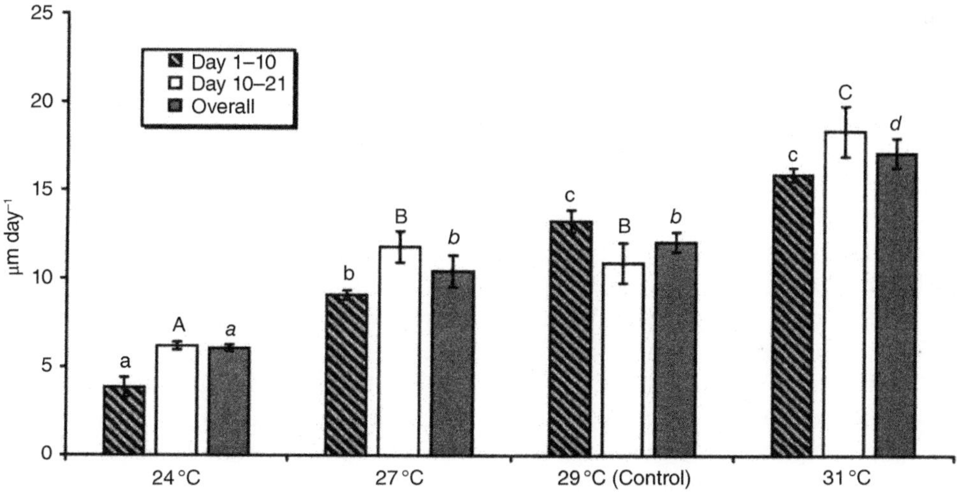

Figure 5.10 Growth rates of *Perna viridis* larvae reared at different temperatures in the first three weeks of larval life. Means with the same subscripts are not significantly different ($P < 0.05$). *Source:* From Nair & Appukuttan (2003). Reproduced with permission from John Wiley & Sons.

21, growth rate almost doubled at 24 °C (6.19 ± 0.22 μm day^{-1}), but it was not as large at 27 °C (11.08 ± 0.87 μm day^{-1}), and even decreased at 29 °C (10.9 ± 11.14 μm day^{-1}). This strong positive correlation between temperature and larval growth has also been reported for other species (Almada-Villela *et al.* 1982; Galley *et al.* 2010). In another study, exposure of larvae (*M. galloprovincialis*) to 17 °C resulted in poor growth, low survival and delayed development (Sánchez-Lazo & Martínez-Pita 2012). Rearing larvae at 20 or 24 °C produced better growth, higher survival rates and faster metamorphosis. These findings have practical implications for hatchery managers, since any factor that reduces growth – thus extending the period of larval development in the water column – will have major effects on mortality and chances of survival to the settlement stage and beyond.

Not surprisingly, larval temperature tolerance is species-specific. For example, when larvae of *M. edulis* and *M. trossulus* were reared separately at constant temperatures of 10, 17 and 24 °C from the D-larval to metamorphosis stage, *M. trossulus* larvae had lower survival than *M. edulis* at 24 °C, while *M. trossulus* showed higher survival at 10 °C (Rayssac *et al.* 2010; see also Hayhurst & Rawson 2009). These survival patterns reflect the distribution of the two species in the wild (see Chapter 3). The species-specific tolerances observed during the larval stages were not observed during post-larval development. This is an unexpected finding, contrasting with that of Beaumont *et al.* (2004), who reported faster growth in *M. galloprovincialis* larvae compared to *M. edulis* and their hybrids irrespective of the temperature tested (10, 14 and 20 °C).

In recent years, several studies have examined the effects of CO_2 and temperature on growth and survival of larvae and juveniles, the rationale being that atmospheric CO_2 concentrations and global temperatures are rising and are expected to continue to rise in the coming decades. Laboratory studies show that larvae exhibit developmental delay, decreased calcification and altered shell ultrastructure in response to decreased pH, a consequence of the high-CO_2 seawater used in the experiments (Kurihara *et al.* 2008; Gazeau *et al.* 2010; Gaylord *et al.* 2011). Possible consequences include an elongated larval life, decreased recruitment and increased vulnerability to predation (discussed in Gaylord *et al.* 2011). However, others have shown that larval shell development and growth depend not on carbon dioxide partial pressure or pH but on the concentration and saturation state of seawater calcium carbonate, a major component of bivalve shells (Bechmann *et al.* 2011; Waldbusser *et al.* 2015; Ventura *et al.* 2016). For information on this complex topic, see Chapter 3 on the potential and actual impacts of climate change on marine bivalves.

Bayne (1965) measured the effect of salinity on larvae from adult *M. edulis* collected from two different locations. Veliger larvae from North Wales did not grow at 19 psu and showed retarded growth at 24 psu, but at 30–32 psu growth was normal. However, larvae from Helsingør in the Øresund, Denmark could grow even at 14–15 psu. The latter site, which is close to the Baltic, is in the transition zone between Kattegat *M. edulis* and Baltic Sea *M. trossulus* populations (Väinölä & Strelkov 2011), which suggests that mussels from Helsingør were probably *M. trossulus*, and not *M. edulis* as described in Bayne (1965). The effect of salinity (5–25 psu) on survival and duration of development from fertilisation to settlement was examined in the same two species, which co-exist in Atlantic Canada (Qiu *et al.* 2002). At 5 and 10 psu, all early stages of both species died, while at 25 psu, >90% of individuals of both species survived through the oocyte to D-veliger stage. However, only

~40% survived through the D-veliger to eyed-veliger stage, and only 30–42% of eyed-veligers metamorphosed successfully. At 15 psu, *M. edulis* did not develop through any of the three stages, in contrast to *M. trossulus*, in which 13% of eyed-veligers metamorphosed into plantigrades. At 20 psu, some *M. edulis* reached plantigrade stage but development time was longer and the percentage at each stage was lower than in *M. trossulus*. This may explain why only *M. trossulus* is found close to several river mouths along the Gaspé Peninsula in Atlantic Canada (Qiu *et al.* 2002).

It is now more usual to look at the combined effects of temperature and salinity on larval growth. Generally speaking, as the limits of salinity tolerance are approached for the larvae of a particular species, the range of temperature tolerance is markedly narrowed. Also, shelled larvae show a greater tolerance to salinity and temperature change than do the embryonic larval stages. The effects of the two factors on larval (*M. edulis*) growth and survival have been investigated (Hrs-Brenko & Calabrese 1969). Optimum growth occurred at 20 °C in salinities from 25–30 psu, with a decrease at both at 25 and 10 °C; the decline was greatest at high (40 psu) and low (20 psu) salinities. Survival at salinities from 15–40 psu was good (≥70%) over the temperature range 5–20 °C, but was drastically reduced at 25 °C, particularly at high and low salinities. In a similar but more recent study, Liang *et al.* (2009) examined the effects of the same two factors on embryonic development and spat survival in *Arcuatula* (*Musculista*) *senhousia*, a mussel native to the northwestern Pacific, but which has now extended its range to Europe, Australia, New Zealand, Korea, Japan, China and parts of the United States (see Chapter 3). Embryos were incubated in water ranging from 0 to 35 °C and from 5 to 40 psu salinity. Optimal conditions for embryo development were 20–25 °C and 25–35 psu, while those for spat survival were considerably higher at 25–35 °C and 30–40 psu. The ability of the species to tolerate low salinity and a wide range of temperatures probably contributes to its success as an invader.

The combined effects of temperature and pressure on embryonic and larval development were analysed in the mussel *M. edulis* (Mestre *et al.* 2009) – a surprising choice, but one based on the premise that the subfamily Mytilinae, to which it belongs, is the closest taxonomic group to the hydrothermal-vent and seep mussels of the subfamily Bathymodiolinae (Distel *et al.* 2000; see Chapter 1). Successful fertilisation under pressure was possible up to 500 atm (50.66 MPa) at 10, 15 and 20 °C. A slower larval development (blastula to D-larval stage) was observed with increasing pressure and decreasing temperature. Pressure narrows the physiological tolerance window in different larval stages and slows down metabolism. From these results, the authors proposed that the theoretical maximum distribution depth for early ontogenetic stages of *M. edulis* is at least 2000 m for the temperatures tested, suggesting the possibility that vent and seep habitats were directly colonised by a cold-eurythermal shallow-water ancestor. But see Distel *et al.* (2000) for an alternative hypothesis.

Food Ration

The planktotrophic eggs of mussels have reserves of lipid, protein and glycogen that fuel the early developmental stages. Once the feeding apparatus is functional, the larvae are totally dependent on the plankton for food. Feeding commences shortly after the development of the shell and velum. Although the diet of mussel larvae in the wild is not known for certain, they probably depend on a ration of phytoplankton cells, heterotrophic flagellates, amino acids, dissolved organic material (DOM) and detritus and bacteria for successful growth and development (Bos *et al.* 2006 and references therein; see also Chapters 4 and 10). A DNA-based approach has been developed to explore qualitative similarities in diets using wild larvae of *Ostrea edulis* and *Mysella* sp. (Maloy *et al.* 2013). Of the sequenced clones, 75% originated from diatoms (Bacillariophyta), 16% from fungi representing the phyla Ascomycota and Basidiomycota and the remaining sequences from flowering plants (Magnoliophyta), single-celled green algae (Prasinophyceae), potential parasites (Ichthyosporea), dinoflagellates (Dinophyceae) and brown algae (Phaeophyceae). No qualitative difference in diet was observed between the two species, even though they belong to different subclasses within the class Bivalvia. There is an impressive literature on the best hatchery foods for larval growth. Mixed algal diets are beneficial, with a combination of two or three high-nutritional-value species, including a suitably sized diatom and a flagellate, providing improved rates of larval growth and development compared to single-species diets (Helm *et al.* 2004; Galley *et al.* 2010; see Chapter 10). However, in the wild, phytoplankton concentrations can be much lower than the concentrations recommended for optimal growth in the laboratory (Bos *et al.* 2006).

Other determinants of larval growth are seawater quality, larval density, contaminants and genetic factors. In northern temperate latitudes, seawater quality in the spring is relatively stable due to the cool temperatures and short day length. However, as spring progresses into early summer, seawater quality can often deteriorate due to intense phytoplankton blooming. Dense concentrations of some algal and bacterial species release toxins that may reduce both growth and survival of bivalve larvae and juveniles, or even lead to mass mortalities in extreme cases (Helm *et al.* 2004). When the abundance of *M. edulis* pelagic larvae was studied quantitatively in the western part of the Dutch Wadden Sea, larvae appeared to have a relatively higher survival at low larval concentrations than at high concentrations (de Vooys 1999). In hatchery production, larvae are normally reared at high densities to increase spat yields. However, the increased food demand and metabolic waste production may impact negatively on water quality, and ultimately result in reduced larval growth (Helm *et al.* 2004). Embryos and larvae of marine organisms are more sensitive to toxic substances than adults, so early life stages of bivalves are routinely used in bioassays to assess seawater quality and sediment elutriates. Larvae are incubated in the test water for a fixed period and at a constant temperature, and parameters such as EC_{50} (the concentration of contaminant that results in 50% abnormal development or retarded growth) and LC_{50} (median lethal concentration) are calculated (His *et al.* 1999). In larval bioassays, growth is significantly retarded by heavy metals such as mercury, lead, copper, cadmium (Beiras & His 1995; Beiras & Albentosa 2004), polycyclic aromatic hydrocarbons (PAHs; Bellas *et al.* 2008) and biocides (Bieras & Bellas 2008). The larval toxicity test has now been adapted for high-throughput screening of emerging contaminants such as endocrine disrupting chemicals (EDCs), brominated and perfluorinated compounds (BFRs and PFCs) and pharmaceuticals (Fabbri *et al.* 2014). Finally, Meyer & Manahan (2010) have identified more than 180 candidate genes for differential larval growth in the oyster *Crassostrea gigas*. Half of these genes are involved in protein metabolism, energy metabolism and regulation of feeding. To date, equivalents have not been identified in marine mussel species.

Larval Dispersal and Population Connectivity

The mobile larval dispersal phase in bivalves is the primary mechanism for connecting and replenishing sedentary adult populations. Variation in larval supply into local populations is believed to be an important determinant in the structure and dynamics of adult populations (Shima *et al.* 2010 and references therein). In planktotrophic (feeding on plankton) marine mussels, the length of larval life (time to reach ~300 μm shell length) varies between three and five weeks, depending on environmental factors such as temperature, salinity and food ration (see earlier). During this planktonic phase, larvae may be transported up to several hundred kilometres by near-shore currents (Alfaro *et al.* 2011a). The dispersal phase is considerably shorter (hours to days) in lecithotrophic (feeding on yolk) larvae and brooded embryos. During the early embryonic phase of development, dispersal is passive, but once feeding starts, vertical dispersal is by active swimming and lateral dispersal is by means of water currents. The main factors influencing vertical distribution of larvae are light, gravity and water flow, and the effects of these vary during the larval life span. Unlike in crustacean zooplankton, there is little evidence that vertical migration in bivalve larvae is a behavioural strategy to avoid predation (see Manuel *et al.* 1996 and references therein). Factors that change over various spatio-temporal scales (e.g. temperature, salinity, tidal phase, season and food availability) are also believed to influence vertical migration patterns (Dobretsov & Miron 2001; Knights *et al.* 2006 and references therein). In a field study in the Irish Sea, Knights *et al.* (2006) investigated the vertical distribution of veliger larvae (*Mytilus*) at a range of depths and tidal states at two sites on two dates. Larvae of different sizes were homogenously distributed throughout the water column during flood tide, whereas they were more densely aggregated in middle and bottom waters during ebb tide. Thus, the relative lack of larvae in the water column during ebb tides compared to flood tides could lead to significant net transport in the direction of flood tides. Larvae are initially photopositive and concentrate in the upper surface waters, but as they approach settlement they become increasingly photonegative and descend into deeper waters in preparation for settlement (Dobretsov & Miron 2001).

Several studies indicate that length of larval life – often referred to as planktonic larval duration (PLD) – and dispersal distance (DD) are positively correlated in marine invertebrates. Many models have estimated DD using PLD coupled with oceanographic models, and they generally assume that larvae are passively dispersed. To test whether dispersal is passive, Shanks (2009) used DD and PD from 67 marine species, including the green mussel, *P. viridis*. In their data, PLD varied from a couple of minutes to several months and DD from less than one metre to hundreds of kilometres. Shanks (2009) plotted PLD against DD and unsurprisingly found a positive correlation between the two. They also plotted the distance larvae would be dispersed if they behaved as passive particles in a steady flow of 10–30 cm s^{-1} or in a more realistic flow field using the Lagrangian passive larval dispersal model of Siegel *et al.* (2003). Overall, they found that all larvae, irrespective of PLD, had DDs many times shorter than those predicted for passive particles. However, the distribution was bimodal: a reasonably tight group represented by species with long PLDs and DDs and a more diffuse and larger group characterised by short DDs and a range of PLDs. There were many species (>60%) with very much shorter dispersal distances (orders of magnitude shorter) than one would predict from their PLDs. They achieve this by remaining close to the bottom, where currents are much reduced. Larvae that have long PLDs (>1 wk) and long DDs (>20 km), as found in bivalves are dispersed into coastal waters, where they remain, close to shore in the coastal boundary layer. Here, currents are typically slower and can alternate back and forth due to changes in wind direction (Largier 2003). All of this limits net alongshore transport of larvae, leading to lower-than-predicted dispersal. Shanks & Brink (2005) consider it makes sense, from an evolutionary perspective, that the larvae of coastal species remain near the coast and their potential settlement sites, rather than allowing themselves to be swept offshore. In summary, larval behaviour plays a critical role in determining dispersal distance (Shanks 2009).

Connectivity – the exchange of individuals among marine populations at the pelagic larval stage – is a hot topic in marine ecology, but before patterns of population connectivity can be fully described we first need to understand the processes that drive dispersal patterns (Cowen & Sponaugle 2009). Coupled models that incorporate both physical dynamics and specific biological traits (e.g. swimming behaviour, food availability, developmental rate, physiological tolerances, mortality) are increasingly being used to properly describe and quantify larval dispersal and assess its role in regulating population connectivity (Hilário *et al.* 2015 and references therein). See Werner *et al.* (2007) and Hinrichsen *et al.* (2011) for an evaluation of coupled biological–physical models. Other methodological approaches to assessing connectivity include applications of physical larval dispersal models (e.g. Lagrangian particle tracking and advection-diffusion models) and natural and artificial tags (see later). A number of key findings have emerged from the modelling of larval dispersal. For example, the interaction of various biophysical mechanisms may counter dispersion to varying extents and enable retention of larvae in their natal habitat; populations may therefore be less open than was originally believed (Warner & Cowen 2002; Levin 2006; Cowen & Sponaugle 2009; Morgan *et al.* 2009; Weidberg *et al.* 2015). Other key insights are the potential importance of biological factors such as larval behaviour (see earlier) and of physical factors such as tidal forces, wind, surface waves, water stratification, turbulence, ocean eddies and coastal topography in influencing larval dispersal (reviewed in Sponaugle *et al.* 2002; Thomas *et al.* 2012). Physical models, in combination with other techniques, have also been used to determine how far larvae travel. To illustrate, Strangford Lough in Northern Ireland is a pilot Marine Protected Area (MPA) where horse mussel (*Modiolus modiolus*) biogenic reefs cover only a fraction of their historical range, with the remaining reefs badly damaged and requiring restoration. Elsäßer *et al.* (2013) used coupled hydrodynamic and particle dispersal models to assess larval dispersion from remnant and translocated populations in order to identify suitable areas for adult *M. modiolus* translocation in the lough. A maximum entropy model (MAXENT) was used to identify whether dispersing larvae could reach habitat suitable for adult *M. modiolus*. From these, they predicted whether translocated mussels would reseed themselves or be able to act as larval sources for nearby reefs. The dispersal models showed that the remnant populations were mainly self-recruiting, with little connectivity between them. The majority of larvae settled near the source, and larval numbers away from the source population were probably too low for successful recruitment. Based on these findings, Elsäßer *et al.* (2013) proposed a combination of total protection of all remaining larval sources and small-scale translocations in each of the identified self-recruiting areas. Gormley *et al.* (2015) have carried out a comparable study on threatened *M. modiolus* bed populations in the Irish Sea. They used biophysical modelling, which involved the development and validation of a three-dimensional (3D) hydrodynamic Stony Brook Parallel Ocean Model (sbPOM) of the Irish Sea, in order to simulate realistic tidal and residual currents throughout the sea. Larval dispersal patterns and genetic structure, using microsatellite DNA markers, were determined for several *M. modiolus* bed populations, including those in Wales (North Pen Llŷn), the Isle of Man (Point of Ayre) and Northern Ireland (Ards Peninsula and Strangford Lough). The authors found moderately significant differentiation between the Northern Ireland populations and those in the Isle of Man and Wales. Simulations of larval dispersal over a 30-day PLD suggest that connectivity over a spatial scale of 150 km is possible between some source and sink populations. The data establish a baseline for underpinning management and conservation of these important and threatened marine habitats in the southern part of the known range of *M. modiolus.*

Knowledge of the magnitude and pathways of larval dispersal can be critical in the design of effective MPAs (Guichard *et al.* 2004; Mace & Morgan 2006; Cowen & Sponaugle 2009; Weersing & Toonen 2009; Pelc *et al.* 2010; Corell *et al.* 2012; Hilário *et al.* 2015). Larval tracking models have also been used to study the spread of newly invasive species such as *P. perna* in the Caribbean and Florida, United States (Gilg *et al.* 2014) and *M. galloprovincialis* on the coasts of South Africa (McQuaid & Phillips 2000). Genetic markers, as seen earlier, are often used to estimate dispersal. The most common metric is Wright's fixation index (F_{ST}), which is a measure of genetic variation among a sample of geographically separated populations. Small differentiation (low F_{ST} values) indicates high gene flow through larval migration between populations. Genetic isolation-by-distance (IBD) models (Palumbi 2003; Won *et al.* 2003; White *et al.* 2010; Selkoe & Toonen 2011; Coolen *et al.* 2019) and direct estimates through parentage analysis, assignment tests (Hogan *et al.* 2012) and clustering analysis using genetic marker data are also useful in estimating larval dispersal distances. See Gagnaire *et al.* (2015) for a good critique of genetic approaches currently being used to assess larval dispersal.

The use of geochemical tags in calcified structures in bivalves is an exciting tool for investigating larval population connectivity (Fodrie *et al.* 2011). Bivalve larval carbonate structures such as shells can be marked artificially using fluorescent dyes such as tetracycline and calcein, elemental tags such as strontium chloride, radioactive isotopes and applied thermal stress marks (Thorrold *et al.* 2002; Moran & Mąrko 2005). Because dilution rates in the sea are too great to yield significant numbers of artificially marked larvae, naturally occurring markers – dyes that mark all larvae exposed to a particular environment – are increasingly being used (Levin 2006). One method is trace elemental fingerprinting, which assesses larval origins or trajectories based on the elemental composition of larval tissues or hard parts, which reflects the chemistry of the water in which the larvae were formed (reviewed in Thorrold *et al.* 2007). To illustrate, Gomes *et al.* (2016) used *M. galloprovincialis* as a model species to investigate connectivity patterns along 120 km of the central Portuguese west coast. Isotopes of eight trace elements (Boron, phosphorus, cobalt, copper, zinc, cerium, lead and uranium) were quantified using laser ablation inductively coupled plasma mass spectrometry (LA-ICP-MS). The authors found three distinctive chemical signatures in larval shells, corresponding to three regions, Estremadura, Cascais and Arrábida, which incorporated two MPAs: Berlengas in the north of the study area and Arrábida in the south. Almost 80% of juveniles were correctly classified to their site of collection. The results demonstrate that the Arrábida MPA was an important source population for the other two regions, and while it showed high rates of self-recruitment, there was limited connectivity with the Berlengas MPA. Such direct measures of demographic connectivity can be a powerful tool to inform policymakers on the conservation and management of ecologically coherent networks of protected areas in coastal marine ecosystems (Gomes *et al.* 2016).

Using the same technique, Becker *et al.* (2007) reared larvae of *M. californianus* and *M. galloprovincialis* in situ at 13 locations spanning 75 km of the Californian coast, United States. Differences in the shell elemental chemistry of the out-planted larva were sufficient to discriminate among the study sites. Using this regional reference map, they compared the expected larval chemistry for each region to the shell chemistry of newly settled recruits, assuming that they had developed in the water column during the period of in situ larval culturing. They then predicted the natal origin of wild-caught juvenile mussels using this chemical reference map. The two species exhibited substantially different connectivity patterns. For *M. californianus*, most individuals (88%) settling at the 13 sites originated in the northern region with high self-recruitment (87%), while there was high larval importation (91%) in the south – a clear indication of asymmetric mixing. *M. galloprovincialis* recruits had more diverse origins from northern, southern and bay areas, but substantial (40%) self-recruitment. These differences are probably a reflection of their different susceptibilities to the prevailing southward surface current just offshore of the region; *M. californianus* larvae are released along the open coast, while *M. galloprovincialis* larvae are mainly released into bays. Overall, the results showed that mussel larvae, previously thought from genetic analysis to be highly dispersed, could be retained within 20–30 km of their natal origin.

Larval Identification and Abundance

Planktonic larval dispersal and recruitment success largely determine distribution patterns of adult bivalves, and in order to distinguish and study these processes in the field, we need tools to identify larvae (Hendricks *et al.* 2005 and references therein). This is particularly important for predicting the time and location of peak spatfall in culture operations based on natural spat collection (see Chapter 10).

Larval identification is a difficult task and requires considerable taxonomic experience because of the small size and similar morphology displayed by many species, particularly during the early larval stages. The technique most commonly used is light microscopy, which examines features of the larval shell such as shape, dimensions and colour patterns, all of which are phenotypically plastic and thus generally inadequate for definitive species identification (Garland & Zimmer 2002; Adams *et al.* 2012). The exception is the hinge apparatus, which is morphologically unique to each species (see Lutz & Kennish 1992 for details). Although SEM of the larval hinge structure has proved useful, the method is costly,

Table 5.3 Characters used to differentiate early and late post-larvae of three species of mussels. Diagnostic characters are marked with an asterisk, but note that hair-like structures are not always present on late post-larvae of *Mytilus galloprovincialis* (i.e. they are diagnostic when present). Umbo: area of shell where the two valves are hinged together. Dissoconch: early juvenile shell. Provincular teeth: the straight part of the larval hinge is called the provinculum, the lateral regions of which bear provincular teeth.

Stage	*Perna perna*	*Mytilus galloprovincialis*	*Choromytilus meridionalis*
Early post-larvae	Umbo narrow, rounded and central	Umbo wide, flat, towards dorsal margin	Umbo wider, rounder and ventral margin less elongate than *Mytilus*
	Red mark on umbo*		
	Hinge reddish*	Hinge rarely pigmented	Hinge often bluish
Late post-larvae	Red zig-zag markings*	Dissoconch growth highly skewed	Ventral margin of early dissoconch straight
	Primary lateral teeth on dorsal hinge*	'Hairs' on dissoconch*	Ventral anterior hinge less prominent than *Perna* or *Mytilus*
			Umbo anterior point of shell*
	22–24 provincular teeth*	30–36 provincular teeth	28–32 provincular teeth

Source: From Bownes *et al.* (2008). Reproduced with permission from Taylor & Francis Ltd.

time-consuming and impractical for processing large numbers of samples. Despite these drawbacks, Bownes *et al.* (2008) were able to distinguish between early and late post-larvae of mussel species using shell morphology and hinge structure (Table 5.3). Improvements in image-analysis technology have lead to the automation of larval shape analysis by computer, reducing the subjectivity of the method (Tiwari & Gallagher 2003; Hendricks *et al.* 2005; Flyachinskaya & Lesin 2006; Goodwin *et al.* 2014).

Molecular methods utilising antibody and DNA probes are increasingly being used to identify larvae to species, regardless of larval development stage (Garland & Zimmer 2002). Monoclonal antibodies that stain all larval stages and permit the identification of *M. galloprovincialis* are being used to identify and monitor abundance (see later) of *M. galloprovincialis* larvae not only in the Galician rías, NW Spain – the biggest mussel-producing region in Europe (Lorenzo-Abalde *et al.* 2005; Pérez *et al.* 2009) – but also in other geographic areas. The development of species-specific DNA probes involves the identification of a nucleotide sequence that is conserved within a species and does not occur in closely related species (see Goffredi *et al.* 2006 and Chapter 9 for details). Probes specific for the 16S and 18S ribosomal (r) RNA genes have proved useful in identifying larvae in mussel species (Bendezu *et al.* 2005; Larsen *et al.* 2005, 2007; Livi *et al.* 2006; Pradillon *et al.* 2007; Jones *et al.* 2008; Harvey *et al.* 2012; Yund *et al.* 2015). Recently, Meistertzheim *et al.* (2017) developed a cheap and user-friendly tool for bivalve species identification through high-resolution melting (HRM) analysis. Based on the polymorphism of the nuclear 18S ribosomal RNA gene among 11 orders of bivalve, the

18S-HRM method allowed for the discrimination of species of the Ostreida, Mytilida and Pectinida orders. In addition, PCR-RFLP analysis (Wang *et al.* 2006; Phillips *et al.* 2008; Adams *et al.* 2010), microsatellites (Morgan & Rodgers 2001; Zhan *et al.* 2008) and a multiplex PCR identification assay based on cytochrome oxidase I (COI) product size (Hare *et al.* 2000) have all been used in larval identification. A proteomic approach has also been used to search for larval proteins specific to *M. galloprovincialis*. Out of 18 spots on gels, six were clearly exclusive to the species (López *et al.* 2005). In hydrothermal vent organisms, including mussels, larval identification is difficult because of the paucity of information on early development stages. Comtet *et al.* (2000) were the first to successfully trap and identify larvae of the vent mussel, *Bathymodiolus azoricus*, using a combination of sediment traps and PCR amplification of the internal transcribed spacer 2 (ITS2) of rDNA.

Knowledge of the temporal and spatial abundance of invertebrate larvae is critical to understanding the dispersal capabilities and recruitment potential of marine and aquatic organisms (Jones *et al.* 2008). The abundance of mussel (*M. edulis*) larvae was estimated at a number of potential farming sites in the Great Belt region of Denmark (Riisgård *et al.* 2015). Water samples were collected from four locations using a small conical zooplankton net (mesh size 80 μm) and an 80 ml collection cylinder. The net was lowered to a certain depth and then slowly hauled to the surface. The sample was transferred to a 100 ml bottle containing Lugol to preserve the larvae. The total number of counted larvae was expressed as individuals per litre of filtered water (V), which was calculated from the circular area of the net ($A = 0.045\ m^2$) and the length (L) of the vertical haul through the water column ($V = A \times L$), which varied depending on the depth of the sampling location (2–6 m). At all locations, there was a pronounced spring peak in larval abundance, apparently closely related to the annual temperature cycle, which synchronises spring spawning. Mussel larvae disappeared at all sites when autumn temperature decreased below ~12 °C. Maximum spring densities ranged between 2.7 and 12.4 individuals per litre, which the authors believe is sufficient for recruitment to future farms sites in the sampling area. Larval densities for the same species at other Danish sites are presented and discussed in Riisgård *et al.* (2015). See Rajagopal *et al.* (2006) for density estimates in other mussel species.

Many probes have been developed to provide a rapid, reliable and high-throughput method for distinguishing mussel larvae (Le Goff-Vitry *et al.* 2007). For example, Henzler *et al.* (2010) used a fluorescence in situ hybridisation (FISH) method, which involves assaying with a fluorescein-labelled species-specific oligonucleotide probe targeted against the 18S rRNA gene in the intact target species; if the target species is present and hybridisation is successful, it will fluoresce green under epifluorescent microscopy. They combined the FISH technique with cell sorting (CS) to identify and rapidly count and sort larvae of several mussel species in diverse plankton samples. Jones *et al.* (2008) developed oligonucleotide probes for mussel (*Mytilus*), barnacle, polychaete and crab larvae occurring in Monterey Bay, California, United States. The probes were printed in arrays on nitrocellulose membranes and deployed in a robotic Environmental Sample Processor (ESP) for near real-time in situ detection and quantification of specific larvae. Also in Monterey Bay, Ryan *et al.* (2014) used a sandwich hybridisation assay (SHA) that quantifies target organism RNA molecules to identify and assess relative abundances of mussel and barnacle larvae from autonomous underwater vehicle (AUV) water samples. Numbers of mussel larvae per litre were estimated using SHA optical absorbance measurements ($\lambda = 450$ nm), which were calibrated using SHA dosage response curves reported by Jones *et al.* (2008) for *Mytilus edulis*.

Short-term fluctuations in larval abundance are influenced by factors such as temperature, salinity, wind velocity, tidal amplitude, chlorophyll *a* levels, larval density and hydrodynamics (Pulfrich 1996; de Vooys 1999; Chícharo & Chícharo 2000 and references therein; Porri *et al.* 2006; Tilberg *et al.* 2012; Yund *et al.* 2015), while seasonal fluctuations are controlled by temperature, the major factor in the timing of bivalve reproduction (Chícharo & Chícharo 2000; Thompson *et al.* 2012; Riisgård *et al.* 2015). Another factor to be considered is predation on larvae, not just by other organisms but also by adult bivalves – a widespread phenomenon called larviphagy, which is believed to reduce numbers of larvae in waters with a high adult bivalve filter feeder biomass (Lehane & Davenport 2004; Porri *et al.* 2008; Troost *et al.* 2009; see Chapter 4).

Settlement, Metamorphosis and Post-Larval Dispersal

Settlement

The act of settlement involves descent from the plankton to the sea bottom, followed by a sequence of swimming and crawling behaviour that culminates in attachment once a suitable substrate is chosen. Morphological change (metamorphosis) then follows, and this heralds the end of the pelagic larval phase and the start of benthic life. Thus, settlement is defined as a process beginning with the onset of a behavioural search for a suitable substrate and ending with metamorphosis. Recruitment, on the other hand, is the process of successful colonisation and implies the lapse of some period of time during which some post-settlement mortality will normally have occurred (Rodríguez *et al.* 1993; Seed & Suchanek 1992). Almost all bivalve larvae become competent to settle in the size range 250–300 μm shell length, although many are able to delay the process until they are considerably larger (see later).

Mussel larvae are capable of discriminating between different substrates, as demonstrated by their behaviour prior to attachment. Not surprisingly, there is considerable interest in elucidating the chemical nature of attractants. It now seems likely that larvae of specific species respond to either a single important cue or to a complex interaction of multiple cues, including positive and negative ones (Morello & Yund 2016 and references therein). Natural cues are often associated with biofilms: thin coats of microorganisms found on solid surfaces in the sea (reviewed by Hadfield 2011). In a series of related studies on *Mytilus coruscus*, J.L. Yang and colleagues have shown, using high-throughput Illumina MiSeq sequencing, that the composition and density of microbial communities in biofilms is influenced by substrata type, which determines attractiveness to settling and metamorphosing larvae (Wang *et al.* 2012; Yang *et al.* 2013a, 2014a,b; Li *et al.* 2014, 2016). However, it appears that it is not the bacteria themselves but their exudates that act as chemical cues for settlement in *P. canaliculus* (Ganesan *et al.* 2012). Interestingly, biofilm exudates actually inhibited settlement in the ribbed mussel, *Aulacomya maoriana* (Alfaro *et al.* 2011b), while larvae of *M. galloprovincialis* and *P. perna* were attracted to filmed rather than bare collectors, and both species avoided conspecific settlers (von der Meden *et al.* 2010). Protozoa, such as ciliates, are also part of the typical biofilm community. In a study on settlement in *M. galloprovincialis*, Shimeta *et al.* (2012) found a 54% reduction in settlement rate on a biofilm of bacteria and ciliates compared to the purely bacterial biofilm, possibly due to direct interactions between larvae and ciliates (physical interactions, interference from ciliates' feeding currents or responses to chemicals from ciliates) or to indirect effects from ciliates altering the bacterial assemblage or its settlement cues.

There is also evidence that other natural chemical cues, such as extracts from macroalgae, may promote larval settlement in some mussel species, such as *P. canaliculus* and *M. galloprovincialis* (Alfaro *et al.* 2006; Yang *et al.* 2007) – although Gribben *et al.* (2011) found that settlement in the former species was significantly higher on rough compared to smooth substrates, with or without algal extracts. It may be that biogenic compounds from marine algae play a much more significant role as repellents to settling larvae and could prove useful as environmentally-friendly antifouling agents (Bhadury & Wright 2004 and references therein).

A wide variety of artificial compounds such as L-3, 4-dihydroxyphenlyalanine (L-DOPA), 5-hydroxytryptamine (serotonin), dopamine, epinephrine, norepinephrine, acetycholine and γ-aminobutyric acid (GABA), as well as amino acids, ammonia (NH_3), potassium chloride (KCl), potassium sulphate (K_2SO_4) and a variety of ions have been shown to be effective in inducing metamorphosis and settlement in some mussel species (Dobretsov & Qian 2003; García-Lavandeira *et al.* 2005; Bao *et al.* 2007; Alfaro *et al.* 2011a; Young *et al.* 2011, 2015; Sánchez-Lazo *et al.* 2012; Yang *et al.* 2015). Exposure of larvae (*P. canaliculus*) to the catecholamines L-DOPA, dopamine (DA) and epinephrine, as well as their precursors L-phenylalanine and L-tyrosine, resulted in maximum settlement induction (50, 60, 51 and ~65%, respectively) at 10^{-5} mol l^{-1} (Young *et al.* 2015). These results suggest that settlement in this species is under endogenous regulation by a catecholaminergic mechanism. Several of these chemicals are used as artificial cues to induce mussel settlement in hatcheries (Kamermans *et al.* 2013).

Mussel larvae settle on a wide variety of substrates, including rocks and ridged, hard surfaces (Christensen *et al.* 2015), filamentous macroalgae (Lasiak & Barnard 1995; Dobretsov & Wahl 2008 and references therein; Yang *et al.* 2007), hydroids (Buchanan & Babcock 1997), eelgrass (Bologna *et al.* 2005), byssus of conspecific adults (Ompi 2010) and artificial substrates such as polypropylene fibrous ropes (Lekang *et al.* 2003) and artificial seaweed (Brenner & Buick 2010). Conditioning substrates can enhance settlement (e.g. submerging collector ropes for long enough to allow biofilm production; Peteiro *et al.* 2007). Gilg *et al.* (2010) investigated whether two species recently introduced to coastal waters of SW United States, *P. viridis* and *Mytella charruana*, settled preferentially on artificial substrates such as plexiglass, wood and rock or on natural substrates such as oysters, green mussels and charru mussels. Their results showed that both species preferentially settled on natural substrates, although it is possible that survival to adulthood was higher on artificial ones.

Studies in the field and in laboratory flumes, as well as the use of mathematical models, have shown that factors such as flow direction and velocity, turbulence intensity and boundary shear stress affect settlement in many benthic invertebrate groups, including bivalves (reviewed in Koehl 2007; Pernet *et al.* 2003). Early settlement patterns of *Perna canaliculus* were investigated within water tanks exposed to three water flow regimes (~1, 5 and 10 cm s^{-1}) and three oxygen concentrations (~6, 9 and 12 mg l^{-1}) (Alfaro 2005). Settlement of larvae increased with increasing water flow, and higher oxygen concentrations appeared to enhance larval settlement. Interestingly, the experiments suggested that exploratory behaviour (i.e. settlement and resettlement) takes place under low and medium water flows but not high ones, highlighting the complexity of settlement and resettlement processes (see later) in this species. Navarette *et al.* (2015) reported that daily settlement over two summers was tightly and positively associated with wave height in two mussels, *Perumytilus purpuratus* and *Semimytilus algosus*, in central Chile. In laboratory flumes, laser-Doppler velocimetry was used to measure the fluctuating water velocities that would be experienced by larvae on spatial scales of hundreds of microns and temporal scales of fractions of a second (Koehl & Reidenbach 2007). Fuchs & DiBaccco (2011) showed that mussel (*M. edulis*) larvae sink in strong turbulence, a behaviour that in energetic coastal waters could improve their chances of being retained near shallow habitats and subsequently settling on to adult mussel beds. Settlement is generally difficult to measure in the field, so it is usually inferred from recruitment data measured some days or even weeks after it has actually occurred (see later).

When a pediveliger encounters a surface while swimming, the velum is retracted and the larva explores the substrate by means of the foot. After a while, it may swim off again, and this cycle of activity may be repeated many times over a period of several days. Once a suitable substrate has been located, however, the larva stops crawling and begins the process of attachment. In mussel, scallop and clam species, attachment is by byssus, while in oysters a drop of 'cement' is squeezed from a gland in the foot and rapidly hardened by the left valve. Predicting the timing of the pediveliger stage is critical to the mussel culture industry, which generally relies on wild seed collection (see Chapter 10).

Metamorphosis and Post-Larval Dispersal

Metamorphosis is a critical phase in the life history of a bivalve, because the ability to move is lost and there is massive reorganisation of body parts to suit a sedentary existence. In all bivalves, rapid morphological changes take place in the attached larva, now termed a 'plantigrade' (Figure 5.11). The extent of morphological reorganisation varies from species to species. In *M. edulis*, within 24–72 hours after the initial secretion of the byssus, the velum degenerates and the labial palps form. The adult gill/palp feeding mechanism develops once the larval feeding mechanism is lost, and becomes functional within 48 hours of metamorphosis. The foot gradually migrates anteriorly and enlarges in the mantle cavity. The ctenidial filaments move forward in the mantle cavity and begin filtration. The filaments also become closely associated with the mouth and labial palps, enhancing the feeding process. Simultaneously, the labial palps become functional, complete with ciliary sorting currents on the palps and division of the posterior mantle cavity into inhalant and exhalant regions, thus fostering the movement of food particles to the mouth (Bayne 1976). After metamorphosis, the adult dissoconch shell is secreted. This shell differs from the larval prodissoconch II shell in pigmentation, ornamentation and mineral content. In addition, a

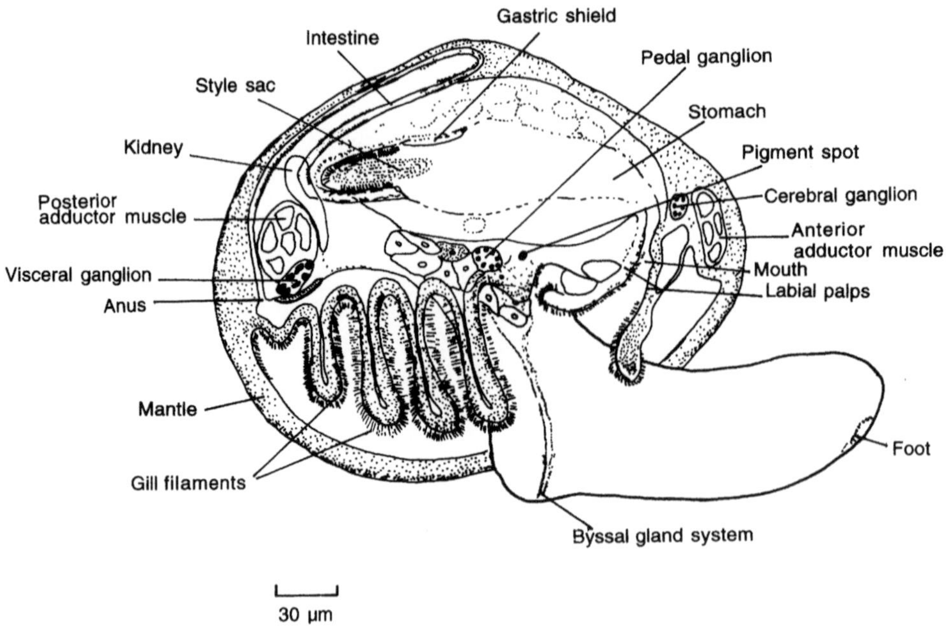

Figure 5.11 Early plantigrade larva of the mussel *Mytilus edulis* immediately after metamorphosis and before secretion of the dissoconch shell has begun. *Source:* From Bayne (1971). Reproduced with permission from Cambridge University Press.

remarkable change in shape takes place, such that the dissoconch shell is more similar to the later adult form. The prodissoconch–dissoconch boundary provides a clear morphological trait that is useful in differentiating true juveniles from metamorphosing post-larvae (Lutz & Kennish 1992). If there is a lack of favourable encounters with a suitable substrate or an absence of appropriate cues, the pediveliger can delay metamorphosis for several weeks (Martel *et al.* 2014 and references therein). Neuroactive compounds, ions and organic solvents are effective inducers of metamorphosis and could prove useful in improving hatchery production of *M. coruscus* (Yang *et al.* 2013b). Also, neurotransmitter blockers that inhibit metamorphosis could be useful tools for antifouling studies (Yang *et al.* 2011).

Many bivalves undergo a post-settlement dispersal phase after metamorphosis (Baker & Mann 1997). Post-metamorphic stages of mussels have the ability to drift up to a size of 2.0–2.5 mm (Sigurdsson *et al.* 1976) for distances on the scale of 1–10^4 m, which probably represents an add-on period of one to two months for pelagic dispersal. Bayne (1964) was the first to propose the hypothesis of primary and secondary settlement in *M. edulis*. Briefly, Bayne proposed that larvae first settle on intertidal filamentous algae, where they metamorphose and grow to about 1 mm in size. They then detach from the algae, re-enter the water column and, using byssal threads, disperse some distance away to eventually settle secondarily on mussel beds, where they grow to juvenile and adult stages. The primary attachment phase away from established mussel beds may have an adaptive value in that it reduces intraspecific competition and prevents the small and vulnerable post-larval stages from entering the strong inhalant currents of larger mussels (Seed & Suchanek 1992).

Many have tested the hypothesis in different mussel species in different regions and, to date, there is weak support for the primary–secondary hypothesis. Overall, it appears that direct settlement on to adult mussel beds is as common as primary settlement on to algae (see McQuaid & Lindsay 2005; Navarrete *et al.* 2015). Navarrete *et al.* (2015) investigated settlement in two intertidal mussels, *Perumytilus purpuratus* and *Semimytilus algosus*, in central Chile. They found that these species do not have a planktonic post-metamorphic stage; the larvae typically arrive in the intertidal zone at a size 80–150 μm larger than the largest planktonic larva, which corresponds to a 3–20-day delay. There were no differences in pediveliger sizes between different algal substrates, mussel beds or artificial collectors, and no evidence that larvae metamorphose in the intertidal and grow in an alternative habit before relocation to mussel beds. On the basis of these results, the authors proposed a new settlement model in which, after metamorphosis beyond the surf zone, the negatively buoyant settlers become semi-benthic and readily sink to the bottom. There, they can be transported onshore through the surf zone by wave-driven near-bed transport. The process of tumbling under the surf may take from a few hours to several days, with larvae arriving at the shoreline in a wide range of sizes at any given time. For some larvae, relocation continues in the intertidal zone for months. Although the model may not apply to *all* mussel species, it could help to explain some of the settlement patterns that have been reported for some species on different shores worldwide (Navarrete *et al.* 2015).

Recruitment

Here, ‘recruitment’ refers to the process of successful colonisation and implies the lapse of some period of time during which some post-settlement mortality will normally have occurred. Mortality is thought to be density-dependent, but it is not known when exactly it occurs or how prevalent it is (Andresen *et al.* 2014). The main causal factors are delay of metamorphosis, biological disturbance, physical disturbance and hydrodynamics, physiological stress, competition and predation, the latter being the best-documented; other possible causes are developmental abnormalities, insufficient energy reserves, disease and parasitism (reviewed in Hunt & Scheibling 1997).

Recruitment in mussels is quantified by counting numbers of spat on mussel beds (Erlandsson & McQuaid 2004; Porri *et al.* 2016), filamentous algae (King *et al.* 1989; Dobretsov & Wahl 2008) or artificial substrates such as plastic mesh, nylon bristle brushes, ropes (polypropylene, nylon, jute) and PVC tubes (Reaugh-Flower *et al.* 2010; Landry & Miron 2011; Honig *et al.* 2015). Tuffy nylon dishwashing pads, made of two strands of artificial plastic filament loosely woven into a 30 cm-long tube that is rolled three times to make an oval pad (approximately 80 × 80 × 60 mm in size, 8.75 g in weight) are increasingly being used as an artificial substrate. They have the advantage over plant material that they possess a relatively constant surface area and textural composition, making it easier to quantify recruitment over fixed time intervals – unlike in natural substrates, where the numbers recorded represent cumulative results from recruitment over an unknown period prior to collection (McGrath *et al.* 1994). Consequently, the use of artificial substrates has facilitated many comparative studies of settlement at spatial scales from tens of metres to between hemispheres, and durations of deployment from days to many years (South 2016 and references therein). One such study carried out over an 18-month period examined spatial scales of variation in mussel (*Brachidontes* spp.) recruitment at rocky intertidal sites on the coast of Argentina (Arribas *et al.* 2015). Recruitment and adult densities (individuals m^{-2}) were compared between two regions ~700 km from each other, two locations (10–20 km apart) within each region and two sites (100–500 m apart) within each location. In total, 32 Tuffy pads (eight per site) were placed in the mid intertidal zone and replaced every 34–120 days. In the laboratory, attached mussels were removed and counted. To avoid potential biases in recruitment estimates due to adult immigration into pads, recruits were defined as individuals measuring <3 mm shell length. Recruitment was also evaluated from cyclindrical core samples of the mussel bed (10 cm diameter, four replicates per site) taken on the same dates as pad collection. Spatial variation in the density of adult mussels was assessed from the core samples taken at the beginning and end of the sampling period. Recruitment was observed to occur all year round (Figure 5.12), although despite continuous recruitment at all sites, peaks were seen in some cases (e.g. March and June 2011 at SE and CC). Between locations, there were considerable differences in the relative magnitude of recruitment estimates on pads versus natural substrate (Figure 5.12), which may be associated with variation across locations in environmental and biotic features that differentially affect the suitability of the substrate for settling larvae. Variance components analysis indicated that most variation in mussel recruitment on pads (71%) and natural substrate (87%) occurred at the scale of locations (i.e. tens of kilometres) (Figure 5.13), which could be due to differences in local larval production and/or oceanographic factors impacting the size of the larval pool and movement of larvae toward the shore. However, Arribas *et al.* (2015) argued that inter-location differences in chlorophyll *a* levels and wave exposure indices were more likely factors. Location also accounted for most spatial variation (82%) in adult densities. There was a close correspondence between the spatial patterns of variation in cumulative recruitment on natural substrates (but not on pads) during the study period and the density of adults at its end, which suggests that differences in mussel abundance along south-western Atlantic rocky shores may be primarily determined by larval recruitment.

The arrival of widely dispersing pelagic larvae and their subsequent recruitment to adult populations are among the most important factors driving metapopulation[6] dynamics, and both factors have a powerful effect on species interactions and community regulation (Pfaff *et al.* 2011 and references therein). It is well documented that recruitment rates of intertidal bivalves can differ spatially by orders of magnitude at multiple scales both within (see example earlier) and between coasts (Rilov *et al.* 2008 and references therein). Determination of the factors underlying this variability has been a major challenge. Several publications have highlighted the relevance of the frequency and variability of upwelling episodes in recruitment dynamics in marine mussels (Peteiro *et al.* 2011 and references therein). Lower recruitment rates have been reported in geographical areas with persistent downwelling or upwelling conditions compared to areas characterised by intermittent upwelling (Peteiro *et al.* 2011 and references therein). Also, temporal and spatial variability in recruitment coincide with seasonal and topographic patterns in coastal upwelling and with differences in wave exposure (Pfaff *et al.* 2011). The surf zone itself may act as a semi-permeable barrier to recruitment of mussel larvae into the intertidal zone. Rilov *et al.* (2008) found that despite temporally variable and sometimes relatively uniform larval abundances,

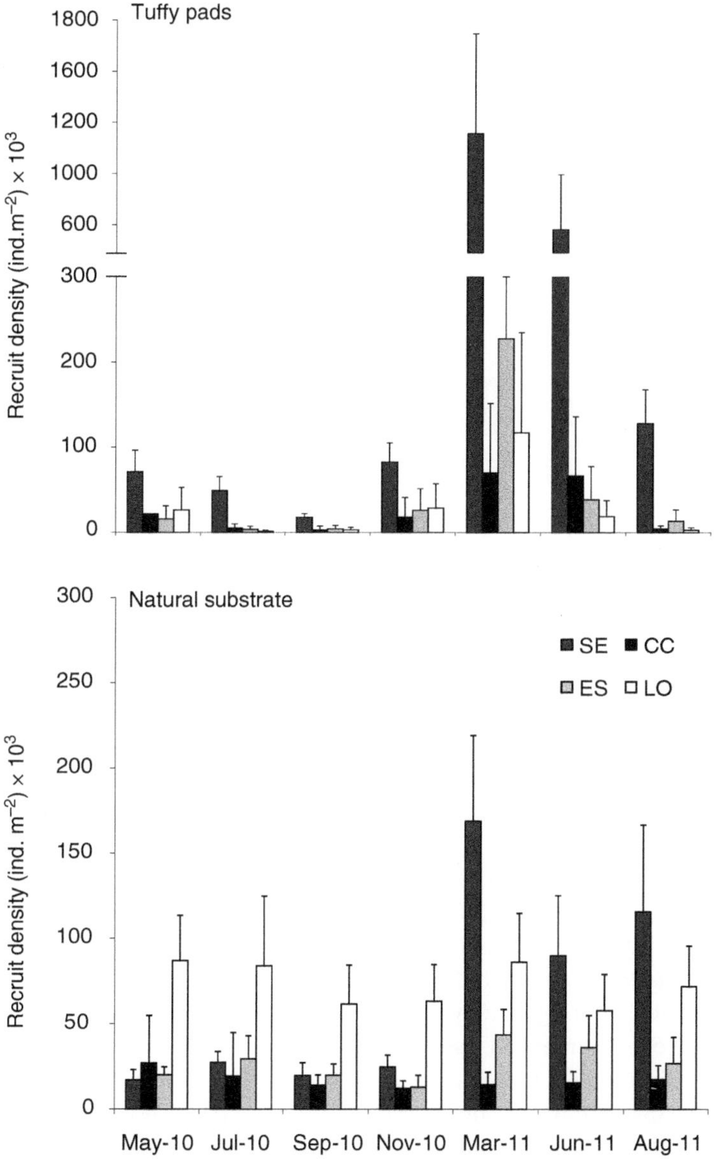

Figure 5.12 Mean (±SD) densities of mussel recruits (indiduals m^{-2}) in Tuffy pads and natural substrate at different sampling locations and dates. SE and CC are Santa Elena and Cabo Corrientes, respectively, in the Buenos Aires Province; ES and LO are El Espigón and La Lobería, respectively, in the Río Negro Province. Note the different scales in the *y* axes of the top and bottom panels. *Source:* From Arribas *et al.* (2015). Reproduced with permission from Elsevier.

recruitment tended to be consistently high at some sites and low at others, suggesting differential transport through the surf zone or differential success in settlement.

At low-shore levels, newly settled recruits are highly vulnerable to fast, mobile predators (e.g. fish, crabs and starfish). Rilov & Shiel (2006) found that in a single day, 80–90% of

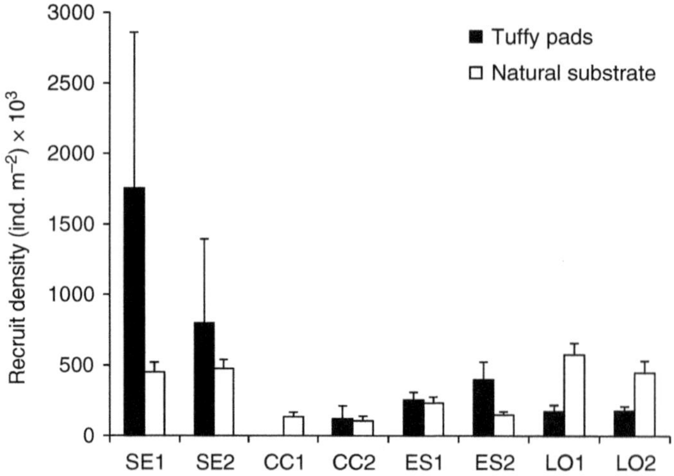

Figure 5.13 Mean (±SD) cumulative mussel recruitment per area unit (individuals m^{-2}) on artificial plastic mesh collectors (Tuffy pads) and natural substrate (mussel beds) at different sampling locations (abbreviations as in Figure 5.12) during an 18-month study period. *Source:* From Arribas *et al.* (2015). Reproduced with permission from Elsevier.

small mussels were removed from experimental tiles on the low shore when unprotected from fish predation, while 20–60% were removed on the mid-shore; in contrast, most fully protected mussels survived for months. Witman *et al.* (2003) found that one year after a large pulse of mussel (*Mytilus edulis*) recruitment, starfish (*Asterias* spp.) recruitment was positively related to the per cent cover of young mussels, suggesting a feedback between food supply (mussels) and consumer (starfish) recruitment, a good example of bottom-up control. See also Marsh (1986), Rilov & Shiel (2011) and Cockrell *et al.* (2015).

Choice of substrate is another factor that influences variation in recruitment. Several studies have shown that adult mussels serve as an important substrate for recruits, thereby reducing post-settlement predation from benthic predators, while at the same time causing recruit mortality through adults' filtering activity (Dolmer & Stenalt 2010; see also Nielsen & Franz 1995). Where adults coexist in mixed species populations (e.g. *M. galloprovincialis* and *P. perna* on the South African coast), recruits of each species are found more frequently on *Mytilus* adults; only large (8–10 mm) recruits of *Perna* responded positively to conspecific *Perna* adults, illustrating how a behavioural trait such as attraction to conspecifics can change ontogenetically within a species (Porri *et al.* 2016).

Finally, several studies have shown that much of the variation in recruitment may be due to factors operating on climatic scales. Using a 14–18-year time series of phytoplankton abundance and mussel recruitment at up to 10 sites along 320 km of the Oregon coast of the United States in a rocky intertidal ecosystem, Menge *et al.* (2009) examined the influence of climatic variation at 3–7-year (El Niño-Southern Oscillation, ENSO), 7–10-year (North Pacific Gyre Oscillation, NPGO) and ≥20-year (Pacific Decadal Oscillation, PDO) time scales. Phytoplankton abundance and mussel (*Mytilus* spp.) recruitment revealed order-of-magnitude increases and seasonal shifts in the 2000s compared to the previous decade, suggesting that increases in larval food may have driven the changes in mussel recruitment. But seasonal shifts were in opposite directions, with phytoplankton blooms occurring earlier and recruitment occurring later in the year. Variation in phytoplankton and recruitment were most strongly associated with climate fluctuation in NPGO and showed weaker associations with ENSO and PDO during the 2000s. In a subsequent study, Menge *et al.* (2011) found that NPGO and upwelling had the strongest associations with recruitment. Thus, much of the variation in recruitment may be due to factors operating on climatic scales. A climate-driven change in larval recruitment rates may have major consequences for the ecology of coastal ecosystems (Harley 2006).

Notes

1 Transcription is the process of making an RNA copy of a gene sequence. This copy, or transcript, called a messenger RNA (mRNA) molecule, leaves the cell nucleus and enters the cytoplasm, where it directs the synthesis of the protein, which it encodes.
2 The genome is the sum total of all genes (nuclear and mtDNA) making up a single individual.
3 Heteroplasmy is the presence of more than one type of mitochondrial DNA within a cell or individual, while homoplasmy refers to a cell whose copies of mitochondrial DNA are all identical.
4 The proportion (p) of nucleotide sites at which the two sequences compared are different.
5 Positive (Darwinian) selection is the process by which new advantageous genetic variants increase in frequency and become fixed in a population.
6 Metapopulations are a group of spatially separated populations of the same species that interact at some level.

References

Abdel Razek, F.A., Abdel-Gaid, S.E., Abu-Zaid, M.M. & Aziz, T.A. (2014) Aspects on the reproduction of eared horse mussel, *Modiolus auriculatus* (Krauss 1848) in Red Sea, Egypt. *Egyptian Journal of Aquatic Research*, **40**, 191–198.

Adams, D.K., Mills, S.W., Shank, T.M. & Mullineaux, L.S. (2010) Expanding dispersal studies at hydrothermal vents through species identification of cryptic larval forms. *Marine Biology*, **157**, 1049–1062.

Adams, D.K., Arellano, S.M. & Govenar, B. (2012) Larval dispersal: vent life in the water column. *Oceanography*, **25**, 256–268.

Agnese, M., Verderame, M., De Meo, E., Prisco, M., Rosati, L., Limatola, E. *et al.* (2013) A network system for vitellogenin synthesis in the mussel *Mytilus galloprovincialis* (L.). *Journal of Cellular Physiology*, **228**, 547–555.

Al-Barwani, S.M., Arshad, A., Amin, S.M.N. & Bujang, J.S. (2013) Incidence of hermaphrodites in green mussel *Perna viridis* along the west coast of Peninsular Malaysia. *Asian Journal of Animal and Veterinary Advances*, **8**, 376–382.

Alfaro, A.C. (2005) Effect of water flow and oxygen concentration on early settlement of the New Zealand green-lipped mussel, *Perna canaliculus*. *Aquaculture*, **246**, 285–294.

Alfaro, A.C., Jeffs, A.G. & Hooker, S.H. (2001) Reproductive behavior of the green-lipped mussel, *Perna canaliculus*, in northern New Zealand. *Bulletin of Marine Science*, **69**, 1095–1108.

Alfaro, A.C., Jeffs, A.G. & Hooker, S.H. (2003) Spatial variability in reproductive behavior of green-lipped mussel populations of northern New Zealand. *Molluscan Research*, **23**, 223–238.

Alfaro, A.C., Copp, B.R., Appleton, D., Kelly, S. & Jeffs, A.G. (2006) Chemical cues promote settlement in larvae of the green-lipped mussel, *Perna canaliculus*. *Aquaculture International*, **14**, 405–412.

Alfaro, A.C., Jeffs, A.G., Gardner, J.P.A., Bollard Breen, B.A. & Wilkin, J. (2011a) Green-lipped mussels in GLM 9. New Zealand Fisheries Assessment Report 2011/48.

Alfaro, A.C., Young, T. & Ganesan, A.M. (2011b) Regulatory effects of mussel (*Aulacomya maoriana* Iredale 1915) larval settlement by neuroactive compounds, amino acids and bacterial biofilms. *Aquaculture*, **322–323**, 158–168.

Almada-Villela, P.C., Davenport, J. & Gruffydd, L.D. (1982) The effects of temperature on the shell growth of young *Mytilus edulis* L. *Journal of Experimental Marine Biology and Ecology*, **59**, 275–288.

Anantharaman, S. & Craft, J.A. (2012) Annual variation in the levels of transcripts of sex-specific genes in the mantle of the common mussel, *Mytilus edulis*. *PLoS ONE*, **7**, e50861.

Andral, B., Stanisiere, J.Y., Sauzade, D., Damier, E., Thebault, H., Galgani, F. *et al.* (2004) Monitoring chemical contamination levels in the Mediterranean based on the use of mussel caging. *Marine Pollution Bulletin*, **49**, 704–712.

Andresen, H., Strasser, M. & van der Meer, J. (2014) Estimation of density-dependent mortality of juvenile bivalves in the Wadden Sea. *PLoS ONE*, **9**, e102491.

Arribas, L.P., Bagur, M., Gutiérrez, J.L. & Palomo, M.G. (2015) Matching spatial scales of variation in mussel recruitment and adult densities across southwestern Atlantic rocky shores. *Journal of Sea Research*, **95**, 16–21.

Avendaño, M. & Cantillánez, M. (2013) Reproductive cycle, collection and early growth of *Aulacomya ater*, Molina 1782 (Bivalvia: Mytilidae) in northern Chile. *Aquaculture Research*, **44**, 1327–1338.

Bacca, H., Huvet, A., Fabioux, C., Daniel, J-Y. & Delaporte, M. (2005) Molecular cloning and seasonal expression of oyster glycogen phosphorylase and glycogen synthase genes. *Comparative Biochemistry and Physiology Part-B*, **140**, 635–646.

Báez, M., García de Severeyn, Y. & Severeyn, H. (2005) Reproductive cycle of *Geukensia demissa* (Bivalvia: Mytilidae) on a beach at Nazaret, El Moján, Zulia State, Venezuela. *Ciencias Marinas*, **31**, 111–118.

Baker, P. & Mann, R. (1997) The postlarval phase of bivalve mollusks: a review of functional ecology and new records of postlarval drifting of Chesapeake Bay bivalves. *Bulletin of Marine Science*, **61**, 409–430.

Bao, W-Y., Satuito C.G., Yang, J-L. & Kitamura, H. (2007) Larval settlement and metamorphosis of the mussel *Mytilus galloprovincialis* in response to biofilms. *Marine Biology*, **150**, 565–574.

Barber, B.J., Fajans, J.S., Baker, S.M. & Baker, P. (2005) Gametogenesis in the non-native green mussel, *Perna viridis*, and the native scorched mussel, *Brachidontes exustus*, in Tampa Bay, Florida. *Journal of Shellfish Research*, **24**, 1087–1095.

Barnes, H. (1957) Process of restoration and synchronization in marine ecology. The spring diatom increase and the spawning of the common barnacle, *Balanus balanoides* L. *Annales Biologiques*, **33**, 67–85.

Batista, F.M., Lallias, D., Taris, N., Guedes-Pinto, H. & Beaumont, A.R. (2011) Relative quantification of the M and F mitochondrial DNA types in the blue mussel *Mytilus edulis* by real-time PCR. *Journal of Molluscan Studies*, **77**, 24–29.

Baussant, T., Ortiz-Zarragoitia, M., Cajaraville, M.P., Bechmann, R.K., Taban, I.C. & Sanni, S. (2011) Effects of chronic exposure to dispersed oil on selected reproductive processes in adult blue mussels (*Mytilus edulis*) and the consequences for the early life stages of their larvae. *Marine Pollution Bulletin*, **62**, 1437–1445.

Bayne, B.L. (1964) *Primary and secondary settlement in Mytilus edulis L. Journal of Animal Ecology*, **33**, 513–523.

Bayne, B.L. (1965) Growth and delay of metamorphosis of the larvae of *Mytilus edulis* (L.). *Ophelia*, **2**, 1–47.

Bayne, B.L. (1971) Some morphological changes that occur at the metamorphosis of the larvae of *Mytilus edulis*. In: *Proceedings of the Fourth European Marine Biology Symposium* (ed. D.J. Crisp), pp. 259–280. Cambridge University Press, Cambridge.

Bayne, B.L. (1975) Reproduction in bivalve molluscs under environmental stress. In: *Physiological Ecology of Estuarine Organisms* (ed F.J. Vernberg), pp. 259–277. University of South Carolina Press, Columbia, SC.

Bayne, B.L. (1976) The biology of mussel larvae. In: *Marine Mussels: Their Ecology and Physiology* (ed B.L. Bayne), pp. 81–120. Cambridge University Press, Cambridge.

Bayne, B.L. & Widdows, J. (1978) The physiological ecology of two populations of *Mytilus edulis* L. *Oecologia*, **37**, 137–162.

Bayne, B.L. & Worrall, C.M. (1980) Growth and production of mussels *Mytilus edulis* from two populations. *Marine Ecology Progress Series*, **3**, 317–328.

Bayne, B.L., Holland, D.L., Moore, M.N., Lowe, D.M. & Widdows, J. (1978) Further studies on the effects of stress in the adults on the egg of *Mytilus edulis*. *Journal of the Marine Biological Association of the UK*, **58**, 825–841.

Bayne, B.L., Salkeld, P.N. & Worrall, C.M. (1983) Reproductive effort and reproductive value in different populations of *Mytilus edulis* L. *Oecologia*, **59**, 18–26.

Beaumont, A.R., Turner, G., Wood, A.R. & Skibinski, D.O.F. (2004) Hybridisations between *Mytilus edulis* and *Mytilus galloprovincialis* and performance of pure species and hybrid veliger larvae at different temperatures. *Journal of Experimental Marine Biology and Ecology*, **302**, 177–188.

Bechmann, R.K., Taban, I.C., Westerlund S., Godal, B.F., Arnberg, M., Vingen, S. *et al.* (2011) Effects of ocean acidification on early life stages of shrimp (*Pandalus borealis*) and mussel (*Mytilus edulis*). *Journal of Toxicology and Environmental Health-Part A*, **74**, 424–438.

Becker, B.J., Levin, L.A., Fodrie, F.J. & McMillan, P.A. (2007) Complex larval connectivity patterns among marine invertebrate populations. *Proceedings of the National Academy of Science*, **104**, 3267–3272.

Beiras, R. & His, E. (1995) Effects of dissolved mercury on embryogenesis, survival, growth and metamorphosis of *Crassostrea gigas* oyster larvae. *Marine Ecology Progress Series*, **113**, 95–103.

Beiras, R. & Albentosa, M. (2004) Inhibition of embryo development of the commercial bivalves *Ruditapes decussatus* and *Mytilus galloprovincialis* by trace metals; implications for the implementation of seawater quality criteria. *Aquaculture*, **230**, 205–213.

Beiras, R. & Bellas, J. (2008) Inhibition of embryo development of the *Mytilus galloprovincialis* marine mussel by organic pollutants; assessment of risk for its extensive culture in the Galician Rias. *Aquaculture*, **277**, 208–212.

Bellas, J., Saco-Álvarez, L., Nieto, O. & Beiras, R. (2008) Ecotoxicological evaluation of polycyclic aromatic hydrocarbons using marine invertebrate embryo–larval bioassays. *Marine Pollution Bulletin*, **57**, 493–502.

Bendezu, I.F., Slater, J.W. & Carney, B.F. (2005) Identification of *Mytilus* spp. and *Pecten maximus* in Irish waters by standard PCR of the 18S rDNA gene and multiplex PCR of the 16S rDNA gene. *Marine Biotechnology*, **7**, 687–696.

Beninger, P.G. (2017) *Caveat observator*: the many faces of pre-spawning atresia in marine bivalve reproductive cycles. *Marine Biology*, **164**,163.

Beninger, P.G. & Le Pennec, M. (2006) Structure and function in scallops. In: *Scallops: Biology, Ecology and Aquaculture, 2nd edn* (eds S.E. Shumway & G.J. Parson), pp. 123–127. Elsevier Science, Boston, MA.

Benomar, S., Bouhaimi, A., El Hamidi, F., Mathieu, M., Ouichou, A. & Moukrim, A. (2006) Cycle de reproduction de la moule africaine *Perna perna* (Mollusca, Bivalvia) dans la baie d'Agadir: impact des rejets d'eaux usées domestiques et industrielles. *Biologie et Santé*, **6**, 25–36.

Benomar, S., Costil, K., El Filali, F., Mathieu, M. & Moukrim, A. (2010) Annual dynamics of glycogen, lipids and proteins during the sexual cycle of *Perna perna* (Mollusca: Bivalvia) from south-western Morocco. *Journal of the Marine Biological Association of the UK*, **90**, 335–346.

Bernard, R.T.F. & Hodgson, A.N. (1985) The fine structure of the sperm and spermatid differentiation in the brown mussel *Perna perna*. *South African Journal of Zoology*, **20**, 5–9.

Berthelin, C., Kellner, K. & Mathieu, M. (2000) Storage metabolism in the Pacific oyster (*Crassostrea gigas*) in relation to summer mortalities and reproductive cycle (West Coast of France). *Comparative Biochemistry & Physiology-Part B*, **125**, 359–369.

Bhadury, P. & Wright. P.C. (2004) Exploitation of marine algae: biogenic compounds for potential antifouling applications. *Planta*, **219**, 561–578.

Bierne, N., David, P., Boudry, P. & Bonhomme, F. (2002) Assortative fertilization and selection at larval stage in the mussels *Mytilus edulis* and *M. galloprovincialis*. *Evolution*, **56**, 292–298.

Bignell, J.P., Dodge, M.J., Feist, S.W., Lyons, B., Martin, P.D., Taylor, N.G.H. *et al.* (2008) Mussel histopathology: effects of season, disease and species. *Aquatic Biology*, **2**, 1–15.

Bigot, L., Zatylny-Gaudin, C., Rodet, F., Bernay, B., Boudry, P. & Favrel, P. (2012) Characterization of GnRH-related peptides from the Pacific oyster *Crassostrea gigas*. *Peptides*, **34**, 303–310.

Bodin, N., Burgeot, T., Stanisière, J. Y., Bocquené, G., Menard, D., Minier, C. *et al.* (2004) Seasonal variations of a battery of biomarkers and physiological indices for the mussel *Mytilus galloprovincialis* transplanted into the northwest Mediterranean Sea. *Comparative Biochemistry & Physiology Part-C* **138**, 411–427.

Bologna, P.A.X., Fetzer, M.L., McDonnell, S. & Moody, E.M. (2005) Assessing the potential benthic–pelagic coupling in episodic blue mussel (*Mytilus edulis*) settlement events within eelgrass (*Zostera marina*) communities. *Journal of Experimental Marine Biology and Ecology*, **316**, 117–131.

Borrero, F. (1987) Tidal height and gametogenesis: reproductive variation among populations of *Geukensia demissa*. *Biological Bulletin*, **173**, 160–168.

Bos, O.G., Hendriks, I.E., Strasser, M., Dolmer, P. & Kamermans, P. (2006) Estimation of food limitation of bivalve larvae in coastal waters of north-western Europe. *Journal of Sea Research*, **55**, 191–206.

Bownes, S., Barker, N.P. & McQuaid, C.D. (2008) Morphological identification of primary settlers and post-larvae of three mussel species from the coast of South Africa. *African Journal of Marine Science*, **30**, 233–240.

Boyle, E.E. & Etter, R. J. (2013) Heteroplasmy in a deep-sea protobranch bivalve suggests an ancient origin of doubly uniparental inheritance of mitochondria in Bivalvia. *Marine Biology*, **160**, 413–422.

Brake, J., Davidson, J. &, Davis, J. (2004) Field observations on growth, gametogenesis, and sex ratio of triploid and diploid *Mytilus edulis*. *Aquaculture*, **236**, 179–191.

Brenner, M. (2010) Assessing the health of blue mussels (*Mytilus edulis*) for offshore site-selection. In: *The Annual Meeting of the European Aquaculture Society, Porto, Portugal, 5–8 October 2010*, pp. 1–23.

Brenner, M. & Buck, B.H. (2010) Attachment properties of blue mussel (*Mytilus edulis* L.) byssus threads on culture-based artificial collector substrates. *Aquacultural Engineering*, **42**, 128–139.

Breton, S. & Stewart, D.T. (2015) Atypical mitochondrial inheritance patterns in eukaryotes. *Genome*, **58**, 423–431.

Breton, S., Burger, G., Stewart, D.T. & Blier, P.U. (2006) Comparative analysis of gender-associated complete mitochondrial genomes in marine mussels *Mytilus* spp. *Genetics*, **172**, 1107–1119.

Breton, S., Doucet-Beaupré, H., Stewart, D.T., Hoeh, W.R. & Blier, P.U. (2007) The unusual system of doubly uniparental inheritance of mtDNA: isn't one enough? *Trends in Genetics*, **23**, 465–474.

Breton, S., Capt, C., Guerra, D. & Stewart, D. (2018) Sex-Determining Mechanisms in Bivalves. In: *Transitions Between Sexual Systems* (ed J. L. Leonard), pp. 165–192. Springer, Cham.

Briones, C., Guiñez, R., Garrido, O., Oyarzún, P.A., Toro, J.E. & Pérez, M. (2012) Sperm polymorphism and genetic divergence in the mussel *Perumytilus purpuratus*. *Marine Biology*, **159**, 1865–1870.

Brousseau, D (1982) Gametogenesis and spawning in a population of *Geukensia demissa* (Pelecypoda: Mytilidae) from Westport, Connecticut. *Veliger*, **24**, 247–251.

Brown, R.A. (1984) Geographical variations in the reproduction of the horse mussel, *Modiolus modiolus* (Mollusca: Bivalvia). *Journal of the Marine Biological Association of the UK*, **64**, 751–770.

Brown, R.A. & Seed, R. (1977) *Modiolus modiolus* (L.). An autoecological study. In: *Biology of Benthic Organisims* (eds B.K. Keegan, P. O'Chéidigh & P.J.S. Boddin), pp. 93–100. Pergamon Press, Oxford.

Buchanan, S. & Babcock, R. (1997) Primary and secondary settlement by the Greenshell mussel *Perna canaliculus*. *Journal of Shellfish Research*, **16**, 71–76.

Burton, S.A., Johnson, G.R. & Davidson, T. J. (1996) Cytologic sexing of marine mussels (*Mytilus edulis*). *Journal of Shellfish Research*, **15**, 345–347.

Cáceres-Martínez, J. & Figueras, A. (1998) Long-term survey on wild and cultured mussels (*Mytilus galloprovincialis* Lmk) reproductive cycles in the Ria de Vigo (NW Spain). *Aquaculture*, **162**, 141–156.

Cao, L., Kenchington, E.L., Zouros, E. & Rodakis, G.C. (2004) Evidence that the large noncoding sequence is the main control region of maternally and paternally transmitted mitochondrial genomes of the marine mussel (*Mytilus* spp.). *Genetics*, **167**, 835–850.

Cárdenas, E.B. & Aranda, D.A. (2000) A review of reproductive patterns of bivalve mollusks from Mexico. *Bulletin of Marine Science*, **66**, 13–27.

Chícharo, L.M.Z. & Chícharo, M.A. (2001) Juvenile recruitment prediction model for *Ruditapes decussatus* (L.) (Bivalvia, Mollusca). *Fisheries Research*, **53**, 219–233.

Chipperfield, P.N.J. (1953) Observations on the breeding and settlement of *Mytilus edulis* (L) in British waters. *Journal of the Marine Biological Association of the UK*, **32**, 449–476.

Christensen, H.T., Dolmer, P., Hansen, B.W., Holmer, M., Kristensen, L.D., Poulsen, L.K. *et al.* (2015) Aggregation and attachment responses of blue mussels, *Mytilus edulis*–impact of substrate composition, time scale and source of mussel seed. *Aquaculture*, **435**, 245–251.

Ciocan, C. (2002) Reproduction of *Mytilus galloprovincialis* (Lmk) in the Romanian waters of the Black Sea. *Cercetari Marine*, **34**, 23–28.

Cockrell, M.L., Bernhardt, J.R. & Leslie, H.A. (2015) Recruitment, abundance, and predation on the blue mussel (*Mytilus edulis*) on northeastern estuarine rocky shores. *Ecosphere*, **61**, art. 18.

Cogswell, A., Kenchington, E.L. & Zouros, E. (2006) Segregation of sperm mitochondria in two and four cell embryos of the blue mussel *Mytilus edulis*: implications for the mechanism of doubly uniparental inheritance of mitochondrial DNA. *Genome*, **49**, 799–807.

Coolen, J.W.P., Boon, A.R., Crooijmans, R., van Pelt, H., lKleissen, F., Gerla, D. *et al.* (2020)1 Marine stepping-stones: connectivity of *Mytilus edulis* populations between offshore energy installations. *Molecular Ecology*, **29**, 686–703.

Comtet, T., Jollivet, D., Khripounoff, A. & Segonzac, M. (2000) Molecular and morphological identification of settlement-stage vent mussel larvae, *Bathymodiolus azoricus* (Bivalvia: Mytilidae), preserved in situ at active vent fields on the Mid-Atlantic Ridge. *Limnology and Oceanography*, **45**, 1655–1660.

Corell, H., Moksnes, P.-O., Engqvist, A., Döös, K. & Jonsson, P.R. (2012) Depth distribution of larvae critically affects their dispersal and the efficiency of marine protected areas. *Marine Ecology Progress Series*, **467**, 29–46.

Cowen, R.K. & Sponaugle, S. (2009) Larval dispersal and marine population connectivity. *Annual Review of Marine Science*, **1**, 443–466.

Crosby, M.P. & Gale, L.D. (1990) A review and evaluation of bivalve condition index methodologies with a suggested standard method. *Journal of Shellfish Research*, **9**, 233–237.

Cruz, R.A. & Villalobos, C.R. (1993) Shell length at sexual maturity and spawning cycle of *Mytella guyanensis* (Bivalvia: Mytilidae) from Costa Rica. *Revista de Biologia Tropical*, **41**, 89–92.

Cuevas, N., Zorita, I., Costa, P.M., Franco, J. & Larreta, J. (2015) Development of histopathological indices in the digestive gland and gonad of mussels: integration with contamination levels and effects of confounding factors. *Aquatic Toxicology*, **162**, 152–164.

Culbertson, J.B., Valiela, I., Olsen, Y.S. & Reddy, C.M. (2008) Effect of field exposure to 38-year-old residual petroleum hydrocarbons on growth, condition index, and filtration rate of the ribbed mussel, *Geukensia demissa*. *Environmental Pollution*, **154**, 312–319.

Curiel-Ramírez, S. & Cáceres-Martínez, J. (2004) Reproductive cycle of coexisting mussels, *Mytilus californianus* and *Mytilus galloprovincialis*, in Baja California, New Mexico. *Journal of Shellfish Research*, **23**, 515–520.

Dalziel, A.C. & Stewart, D.T. (2002) Tissue-specific expression of male-transmitted mitochondrial DNA and its implications for rates of molecular evolution in *Mytilus* mussels (Bivalvia: Mytilidae). *Genome*, **45**, 348–355.

Da Ros, L., Bressan, M. & Marin, M.G. (1985) Reproductive cycle of the mussel (*Mytilus galloprovincialis* Lmk) in Venice Lagoon (NorthAdriatic). *Bolletino di Zoologia*, **52**, 223–229.

De-Bastos, E. & Tyler-Walters, H. (2001) Green crenella (*Musculus discors*). In: *Marine Life Information Network: Biology and Sensitivity Key Information Reviews* (eds H. Tyler-Walters & K. Hiscock). Marine Biological Association of the United Kingdom, Plymouth. Available at: http://www.marlin.ac.uk/species/detail/1645 (accessed 28 May 2021).

de Zwann, A. & Mathieu, M. (1992) Cellular biochemistry and endocrinology. In: *The Mussel Mytilus: Ecology, Physiology, Genetics and Culture* (ed E.M. Gosling), pp. 223–307. Elsevier Science Publishers B.V., Amsterdam.

Distel, D.L., Baco, A.R., Chuang, E., Morrill, W., Cavanaugh, C. & Smith, C.R. (2000) Marine ecology: do mussels take wooden steps to deep-sea vents? *Nature*, **403**, 725–726.

Dixon, D.R., Lowe, D.M., Miller, P.I., Villemin, G.R., Colaço, A., Serrão-Santos, R. *et al.* (2006) Evidence of seasonal reproduction in the Atlantic vent mussel *Bathymodiolus azoricus*, and an apparent link with the timing of photosynthetic primary production. *Journal of the Marine Biological Association of the UK*, **86**, 1363–1371.

Diz, A.P., Dudley, E., Cogswell, A., MacDonald, B.W., Kenchington, E.L.R., Zouros, E. *et al.* (2013) Proteomic analysis of eggs from *Mytilus edulis* females differing in mitochondrial DNA transmission mode. *Molecular & Cellular Proteomics*, **12**, 3068–3080.

Dobretsov, S.V. & Miron, G. (2001) Larval and post-larval vertical distribution of the mussel *Mytilus edulis* in the White Sea. *Marine Ecology Progress Series*, **218**, 179–187.

Dobretsov, S.V. & Qian, P.-Y. (2003) Pharmacological induction of larval settlement and metamorphosis in the blue mussel *Mytilus edulis* L. *Biofouling*, **19**, 57–63.

Dobretsov, S. & Wahl, M. (2008) Larval recruitment of the blue mussel *Mytilus edulis*: the effect of flow and algae. *Journal of Experimental Marine Biology and Ecology*, **355**, 137–144.

Doherty, S.D., Brophy, D. & Gosling, E. (2009) Synchronous reproduction may facilitate introgression in a hybrid mussel (*Mytilus*) population. *Journal of Experimental Marine Biology and Ecology*, **378**, 1–7.

Dolmer, P. & Stenalt, E. (2010) The impact of the adult blue mussel (*Mytilus edulis*) population on settling of conspecific larvae. *Aquaculture International*, **18**, 3–17.

Duinker, A., Håland, L., Hovgaard, P. & Mortensen, S. (2008) Gonad development and spawning in one and two year old mussels (*Mytilus edulis*) from Western Norway. *Journal of the Marine Biological Association of the UK*, **88**, 1465–1473.

Dyachuk, V. & Odintsova, N. (2009) Development of the larval muscle system in the mussel *Mytilus trossulus* (Mollusca, Bivalvia). *Development, Growth and Differentiation*, **51**, 69–79.

Eckelbarger, K.J. & Young, C.M. (1999) Ultrastructure of gametogenesis in a chemosynthetic mytilid bivalve (*Bathymodiolus childressi*) from a bathyal, methane seep environment (northern Gulf of Mexico). *Marine Biology*, **135**, 635–646.

Elsäßer, B., Fariñas-Franco, J.M., Wilson, C.D., Kregting, L. & Roberts, D. (2013) Identifying optimal sites for natural recovery and restoration of impacted biogenic habitats in a special area of conservation using hydrodynamic and habitat suitability modeling. *Journal of Sea Research*, **77**, 11–21.

van Erkom Schurink, C. & Griffiths, C.L. (1991) A comparison of reproductive cycles and reproductive output in 4 southern African mussel species. *Marine Ecology Progress Series*, **76**, 123–134.

Erlandsson, J. & McQuaid, C.D. (2004) Spatial structure of recruitment in the mussel *Perna perna* at local scales: effects of adults, algae and recruit size. *Marine Ecology Progress Series*, **267**, 173–185.

Evans, J.P., Garcia-Gonzalez, F., Almbro, M., Robinson, O. & Fitzpatrick, J.L. (2012) Assessing the potential for egg chemoattractants to mediate sexual selection in a broadcast spawning marine invertebrate. *Proceedings of the Royal Society B*, **279**, 2855–2861.

Fabbri, R., Montagna, M., Balbi, T., Raffo, E., Palumbo, F. & Canesi, L. (2014) Adaptation of the bivalve embryotoxicity assay for the high throughput screening of emerging contaminants in *Mytilus galloprovincialis*. *Marine Environmental Research*, **99**, 1–8.

Favrel, P. & Mathieu, M. (1996) Molecular cloning of a cDNA encoding the precursor of Ala-Pro-Gly Trp amide-related neuropeptides from the bivalve mollusc *Mytilus edulis*. *Neuroscience Letters*, **205**, 210–214.

Flyachinskaya, L.P. & Lesin, P.A. (2006) Using 3D reconstruction method in the investigations of Bivalvia larval development (by the example of *Hiatella arctica* L.). *Proceedings of the Zoological Institute Russian Academy of Sciences*, **310**, 45–50.

Fodrie, F J., Becker, B.J., Levin, L.A., Gruenthal, K. & McMillan, P.A. (2011) Connectivity clues from short-term variability in settlement and geochemical tags of mytilid mussels. *Journal of Sea Research*, **65**, 141–150.

Franz, D.R. (1996) Size and age at first reproduction of the ribbed mussel *Geukensia demissa* (Dillwyn) in relation to shore level in a New York salt marsh. *Journal of Experimental Marine Biology*, **205**, 1–13.

Franz, D.R. 2001. Recruitment, survivorship, and age structure of a New York ribbed mussel population (*Geukensia demissa*) in relation to shore level – a nine-year study. *Estuaries*, **24**, 319–327.

Franzén, A. (1983) Ultrastructural studies of spermatozoa in three bivalve species with notes on evolution of elongated sperm nucleus in primitive spermatozoa. *Gamete Research*, **7**, 199–214.

Fraser, M., Fortier, M., Roumier, P.-H., Parent, L., Brousseau, P., Fournier, M. *et al.* (2016) Sex determination in blue mussels: which method to choose? *Marine Environmental Research*, **120**, 78–85.

Fuchs, H.L. & DiBacco, C. (2011) Mussel larval responses to turbulence are unaltered by larval age or light conditions. *Limnology and Oceanography: Fluids and Environments*, **1**, 120–134.

Gagnaire, P.-A., Broquet, T., Aurelle,D., Viard,F., Souissi, A., Bonhomme, F. *et al.* (2015) Using neutral, selected, and hitchhiker loci to assess connectivity of marine populations in the genomic era. *Evolutionary Applications*, **8**, 769–786.

Galley, T.H., Batista, F.M., Braithwaite, R., King, J. & Beaumont, A.R. (2010) Optimisation of larval culture of the mussel *Mytilus edulis* (L.). *Aquaculture International*, **18**, 315–325.

Galvao, P., Longo, R., Machado Torres, J. P. & Malm, O. (2015) Estimating the potential production of the brown mussel *Perna perna* (Linnaeus, 1758) reared in three tropical bays by different methods of condition indices. *Journal of Marine Biology*, **2015**, art. 948053.

Ganesan, A.M., Alfaro, A.C., Higgins, C.M. & Brooks, J.D. (2012) The effects of bacterial cell suspensions on mussel (*Perna canaliculus*) larval settlement. *Aquaculture*, **350–353**, 143–146.

García-Lavandeira, M., Silva, A., Abad, M., Pazos, A.J., Sánchez, J.L. & Pérez-Parallé, M.L. (2005) Effects of GABA and epinephrine on the settlement and metamorphosis of the larvae of four species of bivalve molluscs. *Journal of Experimental Marine Biology and Ecology*, **316**, 149–156.

Garland, E.D. & Zimmer, C.A. (2002) Techniques for the identification of bivalve larvae. *Marine Ecology Progress Series*, **225**, 299–310.

Garnerot, F., Pellerin, J., Blaise, C. & Mathieu, M. (2006) Immunohistochemical localization of serotonin (5-hydroxytryptamine) in the gonad and digestive gland of *Mya arenaria* (Mollusca: Bivalvia). *General and Comparative Endocrinology*, **149**, 278–284.

Garrido, O. & Gallardo, C.S. (1996) Ultrastructure of sperms in bivalve molluscs of the Mytilidae family. *Invertebrate Reproduction and Development*, **29**, 95–102.

Gaudron, S.M., Demoyencourt, E. & Duperron, S. (2012) Reproductive traits of the cold-seep symbiotic mussel *Idas modiolaeformis*: gametogenesis and larval biology. *Biological Bulletin*, **222**, 6–16.

Gaylord, B., Hill, T.M., Sanford, E., Lenz. E.A., Jacobs, L.A. Sato, K.N. *et al.* (2011) Functional impacts of ocean acidification in an ecologically critical foundation species. *The Journal of Experimental Biology*, **214**, 2586–2594.

Gazeau, F., Gattuso, J.-P., Dawber, C., Pronker, A.E., Peene, F., Peene, J. *et al.* (2010) Effect of ocean acidification on the early life stages of the blue mussel *Mytilus edulis*. *Biogeosciences*, **7**, 2051–2060.

Ghiselli, F., Milani, L., Chang, P. L., Hedgecock, D., Davis, J. P., Nuzhdin, S. V. *et al.* (2012) De novo assembly of the Manila clam *Ruditapes philippinarum* transcriptome provides new insights into expression bias, mitochondrial doubly uniparental inheritance and sex determination. *Molecular Biology and Evolution*, **29**, 771–786.

Ghiselli, F., Milani, L., Guerra, D., Chang, P.L., Breton, S., Nuzhdin, S.V. *et al.* (2013) Structure, transcription, and variability of metazoan mitochondrial genome: perspectives from an unusual mitochondrial inheritance system. *Genome Biology and Evolution*, **5**, 1535–1554.

Gilg, M.R., Hoffman, E.A., Schneider, K.R., Ryabinov, J., El-Khoury, C. & Walters, L.J. (2010) Recruitment preferences of non-native mussels: interaction between marine invasions and land-use changes. *Journal of Molluscan Studies*, **76**, 333–339.

Gilg, M.R., Restrepo, M.C., Walton, R., Brannock, P.M., Hilbish, T.J. & Rodriguez, E. (2013) Geographic variation in allele frequency of the gamete recognition protein M7 lysin throughout a mosaic blue mussel hybrid zone. *Marine Biology*, **160**, 1737–1750.

Gilg, M.R., Howard, R., Turner, R., Middlebrook, M., Abdulnour, M., Lukaj, E. *et al.* (2014) Estimating the dispersal capacity of the introduced green mussel, *Perna viridis* (Linnaeus, 1758), from field collections and oceanographic modeling. *Journal of Experimental Marine Biology and Ecology*, **461**, 233–242.

Goffredi, S.K., Jones, W.J., Scholin, C.A., Marin III, R. & Vrijenhoek, R.C. (2006) Molecular detection of marine invertebrate larvae. *Marine Biotechnology*, **8**, 149–160.

Gomes, C.P., Beasley, C.R., Oliveira Perote, SM., Silva Favacho, A., Sousa da Silva, L., Tagliaro, C.H. *et al.* (2010) Quantitative evaluation of gametogenesis in the mangrove mussel *Mytella guyanensis*. *Ecotropica*, **16**, 125–139.

Gomes, I., Peteiro, L.G., Albuquerque, R., Nolasco, R., Dubert, J., Swearer, S.E. et. al. (2016) Wandering mussels: using natural tags to identify connectivity patterns among Marine Protected Areas. *Marine Ecology Progress Series*, **552**, 159–176.

Gomiero, A., Volpato, E., Nasci, C., Perra, G., Viarengo, A., Dagnino, A. *et al.* (2015) Use of multiple cell and tissue-level biomarkers in mussels collected along two gas fields in the northern Adriatic Sea as a tool for long term environmental monitoring. *Marine Pollution Bulletin*, **93**, 228–244.

González-Fernández, C., Albentosa, M., Campillo, J.A., Viñas, L., Fumega, J., Franco, A. *et al.* (2015) Influence of mussel biological variability on pollution biomarkers. *Environmental Research*, **137**, 14–31.

Goodwin, J.D., North, E.W. & Thompson, C.M. (2014) Evaluating and improving a semi-automated image analysis technique for identifying bivalve larvae. *Limnology and Oceanography: Methods*, **12**, 548–562.

Gormley, K., Mackenzie, C., Robins, P., Coscia, I., Cassidy, A., James, J. *et al.* (2015) Connectivity and dispersal patterns of protected biogenic reefs: implications for the conservation of *Modiolus modiolus* (L.) in the Irish Sea. *PLoS ONE* **10**, e0143337.

Gould, M.C. & Stephano, J.L. (2003) Polyspermy prevention in marine invertebrates. *Microscopic Research and Technology*, **61**, 379–388.

Gray, A.P., Seed, R. & Richardson, C.A. (1997) Reproduction and growth of *Mytilus edulis chilensis* from the Falkland Islands. *Scientia Marina*, **61**, 39–48.

Gribben, P.E., Jeffs, A.G., de Nys, R. & Steinberg, P.D. (2011) Relative importance of natural cues and substrate morphology for settlement of the New Zealand Greenshell™ mussel, *Perna canaliculus*. *Aquaculture*, **319**, 240–246.

Guichard, F., Levin, S.A., Hastings, A. & Siegal, D. (2004) Toward a dynamic metacommunity approach to marine reserve theory. *BioScience*, **54**, 1003–1011.

Hadfield, M.G. (2011) Biofilms and marine invertebrate larvae: what bacteria produce that larvae use to choose settlement sites. *Annual Review of Marine Science*, **3**, 453–70.

Hanquet, A.-C., Jouaux, A., Heude, C., Mathieu, M. & Kellner, K. (2011) A sodium glucose co-transporter (SGLT) for glucose transport into *Crassostrea gigas* vesicular cells: impact of alimentation on its expression. *Aquaculture*, **313**, 123–128.

Hare, M.P., Palumbi, S.R. & Butman, C.A. (2000) Single-step species identification of bivalve larvae using multiplex polymerase chain reaction. *Marine Biology*, **137**, 953–961.

Harley, C.D.G., Hughes, A.R., Hultgren, K. M., Miner, B.G., Sorte, C.J.B., Thornber, C.S. *et al.* (2006) The impacts of climate change in coastal marine systems. *Ecology Letters*, **9**, 228–241.

Harvey, J.B.J., Ryan, J.P., Marin III, R., Preston, C.M., Alvarado, N., Scholin, C.A. *et al.* (2012) Robotic sampling, in situ monitoring and molecular detection of marine zooplankton. *Journal of Experimental Marine Biology and Ecology*, **413**, 60–70.

Hawkins, C.M., Rowell, T.W. & Woo, P. (1987) The importance of cleansing in the calculation of condition index in the soft-shell clam, *Mya arenaria* (L.). *Journal of Shellfish Research*, **6**, 29–36.

Hayhurst, S. & Rawson, P.D. (2009) Species-specific variation in larval survival and patterns of distribution for the blue mussels *Mytilus edulis* and *Mytilus trossulus* in the Gulf of Maine. *Journal of Molluscan Studies*, **75**, 215–222.

Healy, J.M., Keys, J.L. & Daddow, L.Y.M. (2000) Comparative sperm ultrastructure in pteriomorphian bivalves with special reference to phylogenetic and taxonomic implications. In: *The Evolutionary Biology of the Bivalvia* (eds E.M. Harper, J.D. Taylor & J.A. Crame), pp. 169–190. Geological Society, London,

Hendriks, I.E., van Duren, L.A. & Herman, P.M.J. (2005) Image analysis techniques: a tool for the identification of bivalve larvae? *Journal of Sea Research*, **54**, 151–162.

Helm, M.M., Bourne, N. & Lovatelli, A. (2004) *Hatchery Culture of Bivalves. A Practical Manual.* FAO Fisheries Technical Paper No. 471. Food and Agriculture Organization (FAO) of the United Nations, Rome.

Henzler, C.M., Hoaglund, E.A. & Gaines, S.D. (2010) FISH-CS – a rapid method for counting and sorting species of marine zooplankton. *Marine Ecology Progress Series*, **410**, 1–11.

Heβ, A.-K., Bartel, M, Roth, K, Messerschmidt, K., Heilmann, K., Kenchington, E. *et al.* (2012) Expression of M6 and M7 lysin in *Mytilus edulis* is not restricted to sperm, but occurs also in oocytes and somatic tissue of males and females. *Molecular Biology and Evolution*, **79**, 517–524.

Hickman, R.W. & Illingworth, J. (1980) Condition cycle of the green lipped mussel *Perna canaliculus*, in New Zealand. *Marine Biology*, **60**, 27–38.

Hicks, D, W., Tunnell Jr, J.W. & McMahon, R.F. (2001) Population dynamics of the nonindigenous brown mussel *Perna perna* in the Gulf of Mexico compared to other world-wide populations. *Marine Ecology Progress Series*, **211**, 181–192.

Hilário, A., Metaxas, A., Gaudron, S.M., Howell, K.L., Mercier, A., Mestre, N.C. *et al.* (2015) Estimating dispersal distance in the deep sea: challenges and applications to marine reserves. *Frontiers in Marine Science*, **2**, art. 6.

Hines, A., Yeung, W.H., Craft, J., Brown, M., Kennedy, J., Bignell, J. *et al.* (2007) Comparison of histological, genetic, metabolomics, and lipid-based methods for sex determination in marine mussels. *Analytical Biochemistry*, **369**, 175–186.

Hinrichsen, H.-H., Dickey-Collas, M., Huret, M., Peck, M.A. & Vikebø, F.B. (2011) Evaluating the suitability of coupled biophysical models for fishery management. *ICES Journal of Marine Science*, **68**, 1478–1487.

His, E., Beiras, R. & Seaman, M.N.L. (1999) The assessment of marine pollution–bioassays with bivalve embryos and larvae. *Advances in Marine Biology*, **37**, 1–178.

Hogan, J.D., Thiessen, R.J., Sale, P.F. & Heath, D.D. (2012) Local retention, dispersal and fluctuating connectivity among populations of a coral reef fish. *Oecologia*, **168**, 61–71.

Honig, A. (2013) *Population ecology of the ribbed mussel in southeastern Louisiana*. MS thesis, Louisiana State University, Baton Rouge, LA.

Honig, A., La Peyre, M. & Supan, J. (2014) Effects of low and high salinity regimes on seasonal gametogenesis of the ribbed mussel *Geukensia granosissima* in coastal Louisiana, USA. *Sexuality and Early Development in Aquatic Organisms*, **1**, 75–82.

Honig, A., Supan, J. & La Peyre, M. (2015) Population ecology of the gulf ribbed mussel across a salinity gradient: recruitment, growth and density. *Ecosphere*, **6**, art. 226.

Honkoop, P.J.C. & van der Meer, J. (1998) Experimentally induced effects of water temperature and immersion time on reproductive output of bivalves in the Wadden Sea. *Journal of Experimental Marine Biology and Ecology*, **220**, 227–246.

Howard, D.W., Lewis, E.J., Keller, B.J. & Smith, C.S. (2004) Histological techniques for marine bivalve mollusks and crustaceans. *NOAA Technical Memorandum NOS NCCOS* 5.

Hrs-Brenko, M. & Calabrese, A. (1969) The combined effects of salinity and temperature on larvae of the mussel *Mytilus edulis*. *Marine Biology*, **4**, 224–226.

Hunt, H.L. & Scheibling, R.E. (1997) Role of early post-settlement mortality in recruitment of benthic marine invertebrates. *Marine Ecology Progress Series*, **155**, 269–301.

Introíni, G.O., Magalhães, C.A., Aguiar-Jr, O., Quaresma, A.J.C., Lino-Neto, J. & Recco-Pimentel, S.M. (2004) Spermatozoan morphology of *Brachidontes darwinianus* and *Brachidontes solisianus* (Bivalvia, Mytilidae) from the southern Brazilian coast. *Invertebrate Reproduction and Development*, **46**, 149–158.

Introíni, G.O., Maester, F.M., Pereira Leite, F.P. & Recco-Pimentel, S.M. (2010) Sperm ultrastructure of *Mytella* (Bivalvia) populations from distinct habitats along the northern coast of São Paulo State, Brazil. *Biocell*, **34**, 103–111.

Jabbar, A. & Davies, J.I. (1987) A simple and convenient biochemical method for sex identification in the marine mussel, *Mytilus edulis* L. *Journal of Experimental Marine Biology and Ecology*, **107**, 39–44.

Jaafar Kefi, F., Boubaker, S. & Trigui El Menif, N. (2014) Relative growth and reproductive cycle of the date mussel *Lithophaga lithophaga* (Linnaeus, 1758) sampled from the Bizerte Bay (Northern Tunisia). *Helgoland Marine Research*, **68**, 439–450.

Jaramillo, R. & Navarro, J. (1995) Reproductive cycle of the Chilean ribbed mussel *Aulacomya ater* (Molina, 1782). *Journal of Shellfish Research*, **14**, 165–171.

Jones, W.J., Preston, C.M., Marin III, R., Scholin, C.A. & Vrijenhoek, R.C. (2008) A robotic molecular method for *in situ* detection of marine invertebrate larvae. *Molecular Ecology Resources*, **8**, 540–550.

Kádár, E., Lobo-da-Cunha, A., Santos, R.S. & Dando, P. (2006) Spermatogenesis of *Bathymodiolus azoricus* in captivity matching reproductive behaviour at deep-sea hydrothermal vents. *Journal of Experimental Marine Biology and Ecology*, **335**, 19–26.

Kafanov, A.I. & Drozdov, A.L. (1998) Comparative sperm morphology and phylogenetic classification of recent mytiloidea (Bivalvia). *Malacologia*, **39**, 129–139.

Kagley, A.N., Snider, R.G., Krishnakumar, P.K. & Casillas, E. (2003) Assessment of seasonal variability of cytochemical responses to contaminant exposure in the blue mussel *Mytilus edulis* (complex). *Archives of Environmental Contamination and Toxicology*, **44**, 43–52.

Kaloyianni, M., Stamatiou, R. & Dalianis, S, (2005) Zinc and 17β-estradiol induce modifications in Na+/H+exchanger and pyruvate kinase activity through protein kinase C in isolated mantle/gonad cells of *Mytilus galloprovincialis. Comparative Biochemistry & Physiology-Part C*, **141**, 257–266.

Kautsky, N. (1982) Quantitative studies on gonad cycle, fecundity, reproductive output and recruitment in a Baltic *Mytilus edulis* population. *Marine Biology*, **68**, 143–160.

Kellner, K., Heude-Berthelin, C. & Mathieu, M. (2002) Immunocytochemical demonstration of glucagon-like peptides in *Mytilus edulis* cerebral ganglia and an *in vitro* effect of vertebrate glucagon on glycogen metabolism. *Tissue Cell*, **34**, 109–116.

Kamermans P., Galley T., Boudry P., Fuentes J., McCombie H., Batista F. *et al.* (2013) Blue mussel hatchery technology in Europe. In: *Advances in Aquaculture Hatchery Technology* (eds G. Allan & G. Burnell), pp. 339–373. Woodhead Publishing, Cambridge.

Kenchington, E., MacDonald, B., Cao, L., Tsagkarakis, D. & Zouros, E. (2002) Genetics of mother-dependent sex ratio in blue mussels (*Mytilus* spp.) and implications for doubly uniparental inheritance of mitochondrial DNA. *Genetics*, **161**, 1579–1588.

Kenchington, E.L., Hamilton, L., Cogswell, A. & Zouros, E. (2009) Paternal mtDNA and maleness are co-inherited but not causally linked in mytilid mussels. *PLOS One*, **4**, e6976.

Kennedy, V.S. (1977) Reproduction in *Mytilus edulis aoteanus* and *Aulacomya maoriana* (Mollusca: Bivalvia) from Taylors mistake, New Zealand. *New Zealand Journal of Marine and Freshwater Research*, **11**, 255–267.

Kim, Y., Ashton-Alcox, K.A. & Powell, E.N. (2006) Histological techniques for marine bivalve molluscs: Update. Silver Spring, MD. NOAA Technical Memorandum NOS NCCOS 27.

King, P.A., McGrath, D. & Gosling, E.M. (1989) Reproduction and settlement of *Mytilus edulis* on an exposed rocky shore in Galway Bay, west coast of Ireland. *Journal of the Marine Biological Association of the UK*, **69**, 355–365.

Kiyomoto, M., Komaru, A., Scarpa, J., Wada, K.T., Danton, E. & Awaji, M. (1996) Abnormal gametogenesis, male dominant sex ratio, and Sertoli cell morphology in induced triploid mussels, *Mytilus galloprovincialis*. *Zoological Science*, **13**, 393–402.

Klibansky, L.K.J. & McCartney, M.A. (2014) Conspecific sperm precedence is a reproductive barrier between free-spawning marine mussels in the northwest Atlantic *Mytilus* hybrid zone. *PLOS One*, **9**, e108433.

Klouche, M.S., De Deurwaerdère, P., Dellu-Hagedorn. F., Lakhdar-Ghazal, N. & Benomar, S. (2015) *Monoamine content during the reproductive cycle of Perna perna depends on site of origin on the Atlantic Coast of Morocco. Scientific Reports*, **5**, 13715.

Knights, A.M. (2012) Spatial variation in body size and reproductive condition of subtidal mussels: considerations for sustainable management. *Fisheries Research*, **113**, 45–54.

Knights, A.M., Crowe, T.P. & Burnell, G. (2006) Mechanisms of larval transport: vertical distribution of bivalve larvae varies with tidal conditions. *Marine Ecology Progress Series*, **326**, 167–174.

Koehl, M.R.A. (2007) Mini review: Hydrodynamics of larval settlement into fouling communities. *Biofouling*, **23**, 357–368.

Koehl, M.A.R. & Reidenbach, M.A. (2007) Swimming by microscopic organisms in ambient water flow. *Experimental Fluids*, **43**, 755–68.

Kopecka, J., Lehtonen, K.K., Baršiene, J., Broeg, K., Vuorinen, P.J., Gercken, J. *et al.* (2006) Measurements of biomarker levels in flounder (*Platichthys flesus*) and blue mussel (*Mytilus trossulus*) from the Gulf of Gdańsk (southern Baltic). *Marine Pollution Bulletin*, **53**, 406–421.

Kristan, U., Kanduč, T., Osterc, A., Šlejkovec, Z., Ramšak, A. & Stibilj, V. (2014) Assessment of pollution level using *Mytilus galloprovincialis* as a bioindicator species: the case of the Gulf of Trieste. *Marine Pollution Bulletin*, **89**, 455–463.

Krom, M.D., Emeis, K-C. & Van Cappellen, P. (2010) Why is the Eastern Mediterranean phosphorus limited? *Progress in Oceanography*, **85**, 236–244.

Kurihara, H., Takamasa, A., Shoji, K. & Ishimatsu, A. (2008) Effects of elevated pCO_2 on early development in the mussel *Mytilus galloprovincialis*. *Aquatic Biology*, **4**, 225–233.

Kyriakou, E., Zouros, E. & Rodakis, G.C. (2010) The atypical presence of the paternal mitochondrial DNA in somatic tissues of male and female individuals of the blue mussel species *Mytilus galloprovincialis*. *BMC Research Notes*, **3**, 222.

Kyriakou, E., Kravariti, L., Vasilopoulos, T., Zouros, E. & Rodakis, G.C. (2015) A protein binding site in the M mitochondrial genome of *Mytilus galloprovincialis* may be responsible for its paternal transmission. *Gene*, **562**, 83–94.

Lafont, R. & Mathieu, M. (2007) Steroids in aquatic invertebrates. *Ecotoxicology*, **16**, 109–130.

Lander, T.R., Robinson, S.M.C., MacDonald, B.A. & Martin, J.D. (2012) Enhanced growth rates and condition index of blue mussels (*Mytilus edulis*) held at integrated multitrophic aquaculture sites in the Bay of Fundy. *Journal of Shellfish Research*, **31**, 997–1007.

Landry, S. & Miron, G. (2011) Effect of vertical planktonic distribution of competent larvae on spat location in a soft-bottom intertidal zone: a case study for the softshell clam (*Mya arenaria*) and the blue mussel (*Mytilus edulis*). *Journal of Experimental Marine Biology and Ecology*, **410**, 87–100.

Lango–Reynoso, F., Chávez-Villalba, J., Cochard, J.C. & Le Pennec, M. (2000) Oocyte size, a means to evaluate the gametogenic development of the Pacific oyster, *Crassostrea gigas* (Thunberg). *Aquaculture*, **190**, 183–199.

Largier, J.L. (2003) Considerations in estimating larval dispersal distances from oceanographic data. *Ecolological Applications*, **13**, S71–89.

Larsen, J.B., Frischer, M.E., Rasmussen, L.J. & Hansen, B.W. (2005) Single step nested multiplex PCR to differentiate between various bivalve larvae. *Marine Biology*, **146**, 1119–1129.

Larsen, J.B., Frischer, M.E., Ockelmann, K.W., Rasmussen, L.J. & Hansen, B.W. (2007) Temporal occurrence of planktotrophic bivalve larvae identified morphologically and by single step nested multiplex PCR. *Journal of Plankton Research*, **29**, 423–436.

Lasiak, T. (1986) The reproductive cycles of the intertidal bivalves *Crassostrea cucullata* (Born, 1778) and *Pema perna* (Lnnaeus, 1758) from the Transkei coast, southern Africa. *Veliger*, **29**, 226–230.

Lasiak, T.A. & Barnard, T.C.E. (1995) Recruitment of the brown mussel *Perna perna* onto natural substrata: a refutation of the primary/secondary settlement hypothesis. *Marine Ecology Progress Series*, **120**, 147–153.

Lee, S.Y. (1988) The reproductive cycle and sexuality of the green mussel *Perna viridis* (L.) (Bivalvia: Mytilacea) in Victoria Harbour, Hong Kong. *Journal of Molluscan Studies*, **54**, 317–325.

Le Goff-Vitry, M.C., Chipman, A.D. & Comtet, T. (2007) *In situ* hybridization on whole larvae: a novel method for monitoring bivalve larvae. *Marine Ecology Progress Series*, **343**, 161–172.

Lehane, C. & Davenport, J. (2004) Ingestion of bivalve larvae by *Mytilus edulis*: experimental and field demonstrations of larviphagy in farmed blue mussels. *Marine Biology*, **145**, 101–107.

Lekang, O.I., Stevik, T.K. & Bomo, A.M. (2003) Evaluation of different combined collectors used in longlines for blue mussel farming. *Aquacultural Engineering*, **27**, 89–104.

Le Pennec, M. & Beninger, P.G. (1997) Ultrastructural characteristics of spermatogenesis in three species of deep-sea hydrothermal vent mytilids. *Canadian Journal of Zoology*, **75**, 308–316.

Le Pennec, M. & Beninger, P.G. (2000) Reproductive characteristics and strategies of reducing-system bivalves. *Comparative Biochemistry and Physiology Part -C*, **126**, 1–16.

Levin, L.A. (2006) Recent progress in understanding larval dispersal: new directions and digressions. *Integrative and Comparative Biology*, **46**, 282–297.

Li, Y.F., Chen, Y.R., Yang, J.L., Bao, W.Y., Guo, X.P., Liang, X. *et al.* (2014) Effects of substratum type on bacterial community structure in biofilms in relation to settlement of plantigrades of the mussel *Mytilus coruscus*. *International Biodeterioration & Biodegradation*, **96**, 41–49.

Li, Y.F., Guo, X.P., Chen, Y.R., Ding, D.W. & Yang, J.L. (2016) Comparative analysis of biofilm community on different coloured substrata in relation to mussel settlement. *Journal of the Marine Biological Association of the UK*, **97**, 81–89.

Liang, Z.L., Kim, Y.H., Zhang, Z.F., Lim, S.M. & Kang, K.H. (2009) Water temperature and salinity tolerance of embryos and spat of the mussel, *Musculista senhousia*. *Korean Journal of Malacology*, **25**, 179–187.

Lima, T.G. & McCartney, M.A. (2013) Adaptive evolution of M3 lysin–a candidate gamete recognition protein in the *Mytilus edulis* species complex. *Molecular Biology and Evolution*, **30**, 2688–2698.

Livi, S., Cordisco, C., Damiani, C., Romanelli, M. & Crosetti, D. (2006) Identification of bivalve species at an early developmental stage through PCR-SSCP and sequence analysis of partial 18S rDNA. *Marine Biology*, **149**, 1149–1161.

Long, W.C., Bromage, E., Seitz, R.D. & Kaattari, S. (2008) Quantifying fecundity in *Macoma balthica* using an enzyme-linked immunosorbent assay (ELISA). *Aquatic Biology*, **3**, 187–193.

de Longcamp, D., Lubet, P. & Drosdowsky, M. (1974) The *in vitro* biosynthesis of steroids by the gonad of the mussel (*Mytilus edulis*). *General and Comparative Endocrinology*, **22**, 116–127.

Longo, F.J. & Dornfield, E.J. (1967) The fine structure of spermatid differentiation in the mussel *Mytilus edulis*. *Journal of Ultrastructure Research*, **20**, 462–480.

Lorenzo-Abalde, S.L., González-Fernández, A., de Miguel Villegas, E. & Fuentes, J. (2005) Two monoclonal antibodies for the recognition of *Mytilus* spp. larvae: studies on cultured larvae and tests on plankton samples. *Aquaculture*, **250**, 736–747.

Lowe, D., Moore, M.N. & Bayne, B.L. (1982) Aspects of gametogenesis in the marine mussel *Mytilus edulis* L. *Journal of the Marine Biological Association of the UK*, **62**, 133–145.

López J.L., Abalde, S.L. & Fuentes, J. (2005) Proteomic approach to probe for larval proteins of the mussel *Mytilus galloprovincialis*. *Marine Biotechnology*, **7**, 396–404.

Lubet, P. (1959) Recherches sur le cycle sexuel et l'émission des gamètes chez les mytilidés et les pectinidés. *Revue des Travaux de l'Institut des Pêches Maritimes*, **23**, 395–548.

Lutz, R.A. & Kennish, M.J. (1992) Ecology and morphology of larval and early postlarval mussels. In: *The Mussel Mytilus: Ecology, Physiology, Genetics and Culture* (ed E.M. Gosling), pp. 53–85. Elsevier Science Publishers B.V., Amsterdam.

Lymbery, R.A., Kennington, W.J. & Evans, J.P. (2016) Fluorescent sperm offer a method for tracking the real-time success of ejaculates when they compete to fertilise eggs. *Scientific Reports*, **6**, 22689.

Mace, A.J. & Morgan, S.G. (2006) Larval accumulation in the lee of a small headland: implications for the design of marine reserves. *Marine Ecology Progress Series*, **318**, 19–29.

Maloy, A.P., Barber, B.J. & Rawson, P.D. (2003) Gametogenesis in a sympatric population of blue mussels, *Mytilus edulis* and *Mytilus trossulus* from Cobscook Bay (USA). *Journal of Shellfish Research*, **22**, 119–123.

Maloy, A.P., Culloty, S.C. and Slater, J.W. (2013) Dietary analysis of small planktonic consumers: a case study with marine bivalve larvae. *Journal of Plankton Research*, **36**, 271–275.

Manuel, J.L., Gallager, S.M., Pearce, C.M., Manning, D.A. & O'Dor, R.K. (1996) Veligers from different populations of sea scallop *Placopecten magellanicus* have different vertical migration patterns. *Marine Ecology Progress Series*, **142**, 147–163.

Markov, G.V., Tavares, R., Dauphin-Villemant, C., Demeneix, B.A., Baker, M.E. & Laudet, V. (2009) Independent elaboration of steroid hormone signaling pathways in metazoans. *Proceedings of the National Academy of Sciences*, **106**, 11913–11918.

Marsh, C.P. (1986) Rocky intertidal community organization: the impact of avian predators on mussel recruitment. *Ecology*, **67**, 771–786.

Martel, A.L., Tremblay, R., Toupoint, N., Olivier, F. & Myrand, B. (2014) Veliger size at metamorphosis and temporal variability in prodissoconch II morphometry in the blue mussel (*Mytilus edulis*): potential impact on recruitment. *Journal of Shellfish Research*, **33**, 443–455.

McAnlis, K.M.K., Lynn, J.W. & Misamore, M.J. (2010) Lectin labeling of surface carbohydrates on gametes of three bivalves: *Crassostrea virginica*, *Mytilus galloprovincialis*, and *Dreissena bugensis*. *Journal of Shellfish Research*, **29**,193–201.

McDonald, J.I. (2012) Detection of the tropical mussel species *Perna viridis* in temperate Western Australia: possible association between spawning and a marine heat pulse. *Aquatic Invasions*, **7**, 483–490.

McDowell, J.E., Lancaster, B.A., Leavitt, D.F., Rantamaki, P. & Ripley, B. (1999) The effects of lipophilic organic contaminants on reproductive physiology and disease processes in marine bivalve molluscs. *Limnology and Oceanography*, **44**, 903–909.

McGrath, D., King, P.A. & Reidy, M. (1994) Conditioning of artificial substrata and settlement of the marine mussel *Mytilus edulis* L.: a field experiment. *Proceedings of the Royal Irish Academy*, **94B**, 53–56.

McQuaid, C.D. & Phillips, T.E. (2000) Limited wind-driven dispersal of intertidal mussel larvae: *in situ* evidence from the plankton and the spread of the invasive species *Mytilus galloprovincialis* in South Africa. *Marine Ecology Progress Series*, **201**, 211–220.

McQuaid, C.D & Lindsay, J.R. (2005) Interacting effects of wave exposure, tidal height and substratum on spatial variation in densities of mussel *Perna perna* plantigrades. *Marine Ecology Progress Series*, **301**, 173–184.

von der Meden, C.E.O., Porri, F., McQuaid, C.D., Faulkner, K. & Robey, J. (2010) Fine-scale ontogenetic shifts in settlement behaviour of mussels: changing responses to biofilm and conspecific settler presence in *Mytilus galloprovincialis* and *Perna perna*. *Marine Ecology Progress Series*, **411**, 161–171.

Meistertzheim, A-L., Héritier, L. & Lejart, M. (2017) High-Resolution Melting of 18S rDNA sequences (18S-HRM) for discrimination of bivalve species at early juvenile stage: application to a spat survey. *Marine Biology*, **164**, 133.

Menge, B.A., Chan, F., Nielsen, K.J., Di Lorenzo, E. & Lubchenco, J. (2009) Climatic variation alters supply-side ecology: impact of climate patterns on phytoplankton and mussel recruitment. *Ecological Monographs*, **79**, 379–395.

Menge, B.A., Gouhier, T.C., Freidenburg, T. & Lubchenco, J. (2011) Linking long-term, large-scale climatic and environmental variability to patterns of marine invertebrate recruitment: toward explaining 'unexplained' variation. *Journal of Experimental Marine Biology and Ecology*, **400**, 236–249.

Mestre, N.C., Thatje, S. & Tyler, P.A. (2009) The ocean is not deep enough: pressure tolerances during early ontogeny of the blue mussel *Mytilus edulis*. *Proceedings of the Royal Society B*, **276**, 717–726.

Meyer, E. & Manahan, D.T. (2010) Gene expression profiling of genetically determined growth variation in bivalve larvae (*Crassostrea gigas*). *Journal of Experimental Biology*, **213**, 749–758.

Mikhailov, A.T., Torrado, M. & Mendez, J. (1995) Sexual differentiation of reproductive tissue in bivalve molluscs: identification of male associated polypeptide in the mantle of *Mytilus galloprovincialis* Lmk. *The International Journal of Developmental Biology*, **39**, 545–548.

Miranda, M.B.B., Innes, D.J. & Thompson, R.J. (2010) Incomplete reproductive isolation in the blue mussel (*Mytilus edulis* and *M. trossulus*) hybrid zone in the northwest Atlantic: role of gamete interactions and larval viability. *Biological Bulletin*, **218**, 266–281.

Misamore, M.J., Stein, K.K. & Lynn, J.W. (2006) Sperm incorporation and pronuclear development during fertilization in the freshwater bivalve *Dreissena polymorpha*. *Molecular Reproduction and Development*, **73**, 1140–1148.

Mistri M. (2002) Ecological characteristics of the invasive Asian date mussel, *Musculista senhousia*, in the Sacca di Goro (Adriatic Sea, Italy). *Estuaries*, **25**, 431–440.

Mizi A, Zouros E, Moschonas N. & Rodakis, G.C. (2005) The complete maternal and paternal mitochondrial genomes of the Mediterranean mussel *Mytilus galloprovincialis*: implications for the doubly uniparental inheritance mode of mtDNA. *Molecular Biology and Evolution*, **22**, 952–967.

Mladineo, I., Peharda, M., Orhanović, S., Bolotin, J., Pavela-Vranĉić, M. & Treursić, B. (2007) The reproductive cycle, condition index and biochemical composition of the horse-bearded mussel *Modiolus barbatus*. *Helgoland Marine Research*, **61**, 183–192.

Montenegro Villalobos, D., Olivares Paz, P.A. & González, M.T. (2010) Hermaphroditism in marine mussel *Perumytilus purpuratus* (Lamarck, 1819), (Mollusca: Mytilidae). *International Journal of Morphology*, **28**, 569–573.

Moran, A.L. & Marko, P.B. (2005) A simple technique for physical marking of larvae of marine bivalves. *Journal of Shellfish Research*, **24**, 567–571.

Morello, S.L. & Yund, P.O. (2016) Response of competent blue mussel (*Mytilus edulis*) larvae to positive and negative settlement cues. *Journal of Experimental Marine Biology and Ecology*, **480**, 8–16.

Morgan, S.G., Fisher, J.L., Miller, S.H., McAfee, S.T. & Largier, J.L. (2009) Nearshore larval retention in a region of strong upwelling and recruitment limitation. *Ecology*, **90**, 3489–3502.

Morgan, T.S. & Rogers, A.D. (2001) Specificity and sensitivity of microsatellite markers for the identification of larvae. *Marine Biology*, **139**, 967–973.

Morishita, F., Furukawa, Y., Matsushima, O. & Minakata, H. (2010) Regulatory actions of neuropeptides and peptide hormones on the reproduction of molluscs. *Canadian Journal of Zoology*, **88**, 825–845.

Morton, B. (1995) The population dynamics and reproductive cycle of *Septifer virgatus* (Bivalvia: Mytilidae) on an exposed rocky shore in Hong Kong. *Journal of Zoology*, **235**, 485–500.

Moukrim, A., Id Halla, M., Kaaya, A., Bouhaimi, A., Benomar, S. & Mathieu, M. (2008) Pattern of reserve storage of the two mussel species *Perna perna* and *Mytilus galloprovincialis* living on Moroccan coasts: annual variation and effect of pollution. *Iberus*, **26**, 17–28.

Mubiana, V.K. Vercauteren, K. & Blust, R. (2006) The influence of body size, condition index and tidal exposure on the variability in metal bioaccumulation in *Mytilus edulis*. *Environmental Pollution*, **144**, 272–279.

Nagasawa, K., Treen, N., Kondo, R., Otoki, Y., Itoh, N., Rotchell, J.M. *et al.* (2015) Molecular characterization of an estrogen receptor and estrogen-related receptor and their autoregulatory capabilities in two *Mytilus* species. *Gene*, **564**, 153–159.

Nair, M.R. & Appukuttan, K.K. (2003) Effect of temperature on the development, growth, survival and settlement of green mussel *Perna viridis* (Linnaeus, 1758). *Aquaculture Research*, **34**, 1037–1045.

Narváez, M., Freites, L., Mendoza, J. & Guevara, M. (2009) Influence of spat origin and environmental parameters on biochemical composition and biometry of the brown mussel *Perna perna* (Linné, 1758), under culture conditions. *Revista de Biología Marina y Oceanografía*, **44**, 343–355.

Navarrete, S.A., Parragué, M. & Wieters, E.A. (2008) Local and meso-scale patterns of recruitment and abundance of two intertidal crab species that compete for refuges. *Marine Biology*, **155**, 223–232.

Navarrete, S.A., Largier, J.L., Vera, G., Tapia, F.J., Parragué, M., Ramos, E. *et al.* (2015) Tumbling under the surf: wave-modulated settlement of intertidal mussels and the continuous settlement relocation model. *Marine Ecology Progress Series*, **520**, 101–121.

Newell, R.I.E. (1989) *Species profiles: life histories and environmental requirements of coastal fishes and invertebrates (North and Mid-Atlantic) – blue mussel. United States Fish and Wildlife Service Biological Report 82(11.102)*. US Army Corps of Engineers, TR El–82–4.

Newell, R.I.E. & Thompson, R.J. (1984) Reduced clearance rates associated with spawning in the mussel, *Mytilus edulis* (L.) (Bivalvia, Mytilidae). *Marine Biology Letters*, **5**, 21–33.

Newell, R.I.E., Hilbish, T.J., Koehn, R.K. & Newell, C.J. (1982) Temporal variation in the reproductive cycle of *Mytilus edulis* L. (Bivalvia, Mytilidae) from localities on the east coast of the United States. *Biological Bulletin*, **162**, 299–310.

Nicastro, K.R., Zardi, G.I. & McQuaid, C.D. (2010) Differential reproductive investment, attachment strength and mortality of invasive and indigenous mussels across heterogeneous environments. *Biological Invasions*, **12**, 2165–2177.

Nielsen, K.L. & Franz, D.R. (1995) The influence of adult conspecifics and shore level on recruitment of the ribbed mussel *Geukellsia demissa* Dillwyn)). *Journal of Experimental Marine Biology and Ecology*, **188**, 89–98.

Okumuş, I. & Stirling, H.P. (1998) Seasonal variations in the meat weight, condition index and biochemical composition of mussels *Mytilus edulis* L. in suspended culture in two Scottish sea lochs. *Aquaculture*, **159**, 249–261.

Olivas-Valdez & Cáceres-Martínez, J. (2002) Infestation of the blue mussel *Mytilus galloprovincialis* by the copepod *Pseudomyicola spinosus* and its relation to size, density, and condition index of the host. *Journal of Invertebrate Pathology*, **79**, 65–71.

Oliver, L.R., Seed, R. & Reynolds, B. (2008) The effect of high flow events on mussels (*Mytilus edulis*) in the Conwy estuary, North Wales, UK. *Hydrobiologia*, **606**, 117–127.

Oliver, M. & Evans, J.P. (2014) Chemically moderated gamete preferences predict offspring fitness in a broadcast spawning invertebrate. *Proceedings of the Royal Society B*, **281**, 20140148.

Ompi, M. (2010) Settlement behaviour and size of mussel larvae from the family mytilidae (*Brachidontes erosus* (Lamarck, 1819), *Brachidontes rostratus* (Dunker, 1857), *Trichomya hirsutus* (Lamarck, 1819), and *Mytilus galloprovincialis* Lamarck, 1819)[1]. *Journal of Coastal Development*, **13**, 215–227.

Orban, E., Di Lena, G., Nevigato, T., Casini, I., Marzetti, A. & Caproni, R. (2002) Seasonal changes in meat content, condition index and chemical composition of mussels (*Mytilus galloprovincialis*) cultured in two different Italian sites. *Food Chemistry*, **77**, 57–65.

Ortiz-Zarragoitia, M. & Cajaraville, M.P. (2010) Intersex and oocyte atresia in a mussel population from the Biosphere's Reserve of Urdaibai (Bay of Biscay). *Ecotoxicology and Environmental Safety*, **73**, 693–701.

Osada, M. & Treen, N. (2013) Molluscan GnRH associated with reproduction. *General and Comparative Endocrinology*, **181**, 254–258.

Oyarzún, P.A., Toro, J.E., Jaramillo, R., Guiñez, R., Briones, C. & Astorga, M. (2010) Comparative analysis of the gametogenic cycle from *Perumytilus purpuratus* (Bivalvia: Mytilidae), at the localities of Taltal and Huasco, northern Chile. *Revista de Biología Marina y Oceanografía*, **45**, 43–58.

Oyarzún, P.A., Toro, J.E., Jaramillo, R., Guiñez, R., Briones, C. & Astorga, M. (2011) Gonadal cycle of the mussel *Mytilus chilensis* (Bivalvia: Mytilidae) at two localities in southern of Chile. *Latin American Journal of Aquatic Research*, **39**, 512–525.

Oyarzún, P.A., Toro, J.E., Garrido, O., Briones, C. & Guiñez, R. (2014) Differences in sperm ultrastructure between *Mytilus chilensis* and *Mytilus galloprovincialis* (Bivalvia, Mytilidae): could be used as a taxonomic trait? *Latin American Journal of Aquatic Research*, **42**, 172–179.

Oyarzún, P.A., Nuñez, J.J., Toro, J.E. & Gardner, J.P.A. (2020) Trioecy in the marine mussel *Semimytilus algosus* (Mollusca, Bivalvia): stable sex ratios across 22 degrees of a latitudinal gradient. *Frontiers in Marine Science*, **7**, 348.

Paine, R.T. (1976) Size-limited predation: an observational and experimental approach with the *Mytilus-Pisaster* interaction. *Ecology*, **57**, 858–873.

Palumbi, S.R. (2003) Population genetics, demographic connectivity, and the design of marine reserves. *Ecological Applications*, **13**, S146–158.

Pampanin, D.M., Volpato, E., Marangon, I. & Nasci, C. (2005) Physiological measurements from native and transplanted mussel (*Mytilus galloprovincialis*) in the canals of Venice. Survival in air and condition index. *Comparative Biochemistry and PhysiologyPart-B*, **140**, 41–52.

Pani, K., Roghani, A. & Croll, R.P. (2005) Identification of peptide-like substances in the *Placopecten*: possible role in growth and reproduction. *Biogenic Amines*, **19**, 47–67.

Park, K.-I., Choi, J.-W. & Choi, K.-S. (2005) Development of a *Saxidomus purpuratus* (Mollusca: Bivalvia) egg-specific antibody for the quantification of eggs using an enzyme-linked immunosorbent assay. *Journal of Shellfish Research*, **24**, 1079–1086.

Passamonti, M. & Ghiselli, F. (2009) Doubly uniparental inheritance: two mitochondrial genomes, one precious model for organelle DNA inheritance and evolution. *DNA and Cell Biology*, **28**, 79–89.

Pazos, A.J. & Mathieu, M. (1999) Effects of five natural gonadotropin-releasing hormones on cell suspensions of marine bivalve gonad: stimulation of gonial DNA synthesis. *General and Comparative Endocrinology*, **113**, 112–120.

Pazos, A.J., Román, G., Acosta, C.P., Abad, M. & Sánchez, J.L. (1996) Stereological studies on the gametogenic cycle of the scallop, *Pecten maximus*, in suspended culture in Ria de Arousa (Galicia, NW Spain). *Aquaculture*, **142**, 119–135.

Peharda, M., Župan, I., Bavčević, L., Frankić, A. & Klanjšček, T. (2007) Growth and condition index of mussel *Mytilus galloprovincialis* in experimental integrated aquaculture. *Aquaculture Research*, **38**, 1714–1720.

Pelc, R.A., Warner, R.R., Gaines, S.D. & Paris, C.D. (2010) Detecting larval export from marine reserves. *Proceedings of the National Academy of Sciences*, **107**, 18266–18271.

Pérez, A.F., Boy, C.C., Curelovich, J., Pérez-Barros, P. & Calcagno, J.A. (2013) Relationship between energy allocation and gametogenesis in *Aulacomya atra* (Bivalvia: Mytilidae) in a sub-Antarctic environment. *Revista de Biología Marina y Oceanografía*, **48**, 459–469.

Pérez, D., Lorenzo-Abalde, S., González-Fernández, A. & Fuentes, J. (2009) Immunodetection of *Mytilus galloprovincialis* larvae using monoclonal antibodies to monitor larval abundance on the Galician coast: optimization of the method and comparison with identification by morphological traits. *Aquaculture*, **294**, 86–92.

Pernet, F., Tremblay, R. & Bourget, E. (2003) Settlement success, spatial pattern and behavior of mussel larvae *Mytilus* spp. in experimental 'downwelling' systems of varying velocity and turbulence. *Marine Ecology Progress Series*, **260**, 125–140.

Peteiro, L.G., Filgueira, R., Labarta, U. Fernández-Reiriz, M.J., Álvarez-Salgado, X.A., Filgueira, R. *et al.* (2007) Effect of submerged time of collector ropes on the settlement capacity of *Mytilus galloprovincialis* L. *Aquaculture Research*, **38**, 1679–1681.

Peteiro,L.G., Labarta, U., Fernández-Reiriz, M. J., Álvarez-Salgado, X.A, Filgueira, R. & Piedracoba, S. (2011) Influence of intermittent-upwelling on *Mytilus galloprovincialis* settlement patterns in the Ría de Ares-Betanzos. *Marine Ecology Progress Series*, **443**, 111–127.

Peters, J.R. & Granek, E.F. (2016) Long-term exposure to fluoxetine reduces growth and reproductive potential in the dominant rocky intertidal mussel, *Mytilus californianus*. *Science of the Total Environment*, **545–546**, 621–628.

Petes, L.E., Menge, B.A., Chan, F. & Webb, M.A.H. (2008a) Gonadal tissue color is not a reliable indicator of sex in rocky intertidal mussels. *Aquatic Biology*, **3**, 63–70.

Petes, L.E., Menge, B.A. & Harris, A.L. (2008b) Intertidal mussels exhibit energetic trade-offs between reproduction and stress resistance. *Ecological Monographs*, **78**, 387–402.

Pfaff, M.C., Branch, G.M., Wieters, E.A., Branch, R.A. & Broitman, B.R. (2011) Upwelling intensity and wave exposure determine recruitment of intertidal mussels and barnacles in the southern Benguela upwelling region. *Marine Ecology Progress Series*, **425**, 141–152.

Phillips, N.E. (2005) Growth of filter-feeding benthic invertebrates from a region with variable upwelling intensity. *Marine Ecology Progress Series*, **295**, 79–89.

Phillips, N.E. (2007) A spatial gradient in the potential reproductive output of the sea mussel *Mytilus californianus*. *Marine Biology*, **151**, 1543–1550.

Phillips, N.E., Wood, A.R. & Hamilton, J.S. (2008) Molecular species identification of morphologically similar mussel larvae reveals unexpected discrepancy between relative abundance of adults and settlers. *Journal of Experimental Marine Biology and Ecology*, **362**, 90–94.

Pipe, R.K. (1987a) Oogenesis in the marine mussel *Mytilus edulis*: an ultrastructural study. *Marine Biology*, **95**, 405–414.

Pipe, R.K. (1987b) Ultrastructural and cytochemical study on interactions between nutrient storage cells and gametogenesis in the mussel *Mytilus edulis* L. *Marine Biology*, **96**, 519–528.

Porri, F., McQuaid, C.D. & Radloff, S. (2006) Spatio-temporal variability of larval abundance and settlement of *Perna perna*: differential delivery of mussels. *Marine Ecology Progress Series*, **315**, 141–150.

Porri, F., Jordaan, T. & McQuaid, C.D. (2008) Does cannibalism of larvae by adults affect settlement and connectivity of mussel populations? *Estuarine, Coastal and Shelf Science*, **79**, 687–693.

Porri, F., McQuaid, C.D. & Erlandsson, J. (2016) The role of recruitment and behaviour in the formation of mussel-dominated assemblages: an ontogenetic and taxonomic perspective. *Marine Biology*, **163**, 157.

Pradillon, F., Schmidt, A., Peplies, J. & Dubilier, N. (2007) Species identification of marine invertebrate early stages by whole-larvae *in situ* hybridisation of 18S ribosomal RNA. *Marine Ecology Progress Series*, **333**, 103–116.

Pulfrich, A. (1997) Seasonal variation in the occurrence of planktic bivalve larvae in the Schleswig-Holstein Wadden Sea. *Helgoländer Meeresunters*, **51**, 23–39.

Qiu, J.-W., Tremblay, R. & Bourget, E. (2002) Ontogenetic changes in hyposaline tolerance in the mussels *Mytilus edulis* and *M. trossulus*: implications for distribution. *Marine Ecology Progress Series*, **228**, 143–152.

Quayle, D.B. & Newkirk, G.F. (1989) *Farming bivalve molluscs: methods for study and development*. The World Aquaculture Society and the International Development Research Centre, Baton Rouge.

Rajagopal, S., Venugopalan, V.P., Nair, K.V.K., van der Velde, G., Jenner, H.A. & den Hartog, C. (1998) Reproduction, growth rate and culture potential of the green mussel, *Perna viridis* (L.) in Edaiyur backwaters, east coast of India. *Aquaculture*, **162**, 187–202.

Rajagopal, S., Venugopalan, V.P., van der Velde G. & Jenner, H.A. (2006) Greening of the coasts: a review of the *Perna viridis* success story. *Aquatic Ecology*, **40**, 273 –297.

Rajkumar, A. (2009) *Physiological and cytological biomarker studies using Perna perna for marine pollution monitoring*. MSc thesis, University of KwaZulu-Natal, Durban.

Rawson, P.D., Slaughter, C. & Yund, P.O. (2003) Patterns of gamete incompatibility between the blue mussels *Mytilus edulis* and *M. trossulus*. *Marine Biology*, **143**, 317–325.

Rayssac, N., Pernet, F., Lacasse, O. & Tremblay, R. (2010) Temperature effect on survival, growth, and triacylglycerol content during the early ontogeny of *Mytilus edulis* and *M. trossulus*. *Marine Ecology Progress Series*, **417**, 183–191.

Reaugh-Flower, K.E., Branch, G.M., Harris, J.M., McQuaid, C.D., Currie, B., Dye, A. *et al.* (2010) Patterns of mussel recruitment in southern Africa: a caution about using artificial substrata to approximate natural recruitment. *Marine Biology*, **157**, 2177–2185.

Reis-Henriques, M.A. & Coimbra, J. (1990) Variations in the levels of progesterone in *Mytilus edulis* during the annual reproductive cycle. *Comparative Biochemistry and Physiology-Part A*, **95**, 343–348.

Reis-Henriques, M.A., Le Guellec, D., Rémy-Martin, J.P. & Adessi, G.L. (1990) Studies on endogenous steroids from the marine mollusc *Mytilus edulis* L. by gas chromatography and mass spectrometry. *Comparative Biochemistry and Physiology-Part B*, **95**, 303–309.

Reunov, A.A. & Hodgson, A.N. (1994) Ultrastructure of the spermatozoa of five species of South African bivalves (Mollusca), and an examination of early spermatogenesis. *Journal of Morphology*, **219**, 275–283.

Reunov, A.A., Au, D.W.T. & Wu, R.S.S. (1999) Spermatogenesis of the green-lipped mussel *Perna viridis* with dual patterns of acrosome and tail development in spermatids. *Helgoland Marine Research*, **53**, 62–69.

Riginos, C. & McDonald, J.H. (2003) Positive selection on an acrosomal sperm protein, M7 lysin, in three species of the mussel genus *Mytilus*. *Molecular Biology and Evolution*, **20**, 200–207.

Riginos, C., Wang, D. & Abrams, A.J. (2006) Geographic variation and positive selection on M7 lysin, an acrosomal sperm protein in mussels (*Mytilus* spp.). *Molecular Biology and Evolution*, **23**, 1952–1965.

Riisgård, H.U., Lundgreen, K. & Pleissner, D. (2015) Environmental factors and seasonal variation in density of mussel larvae (*Mytilus edulis*) in Danish waters. *Open Journal of Marine Science*, **5**, 280–289.

Rilov, G. & Schiel, D.R. (2006) Trophic linkages across seascapes: subtidal predators limit effective mussel recruitment in rocky intertidal communities. *Marine Ecology Progress Series*, **327**, 83–93.

Rilov, G. & Schiel, D.R. (2011) Community regulation: the relative importance of recruitment and predation Intensity of an intertidal community dominant in a seascape context. *PLOS One*, **6**, e23958.

Rilov, G., Dudas, S.E., Menge, B.A., Grantham, B.A., Lubchenco, J. & Schiel, D.R. (2008) The surf zone: a semi-permeable barrier to onshore recruitment of invertebrate larvae? *Journal of Experimental Marine Biology and Ecology*, **361**, 59–74.

Rodet, F., Lelong, C., Dubos, M.P., Costil, K. & Favrel, P. (2005) Molecular cloning of a molluscan gonadotropin-releasing hormone receptor orthologue specifically expressed in the gonad. *Biochimica et Biophysica Acta*, **1730**, 187–195.

Rodhouse, P.G., McDonald, J.H., Newell, R.I.E. & Koehn, R.K. (1986) Gamete production, somatic growth and multiple-locus enzyme heterozygosity in *Mytilus edulis*. *Marine Biology*, **90**, 209–214.

Rodriguez, S.R., Ojedal, F.P. & Inestrosa, N.C. (1993) Settlement of benthic marine invertebrates. *Marine Ecology Progress Series*, **97**, 193–207.

Rouabhi, Y.L., Grosjean P., Boutiba, Z., Rouane Hacene, O. & Richir, J. (2019) Reproductive cycle and follicle cleaning process of *Mytilus galloprovincialis* (Mollusca: Bivalvia) from a polluted coastal site in Algeria. *Invertebrate Reproduction & Development*, **63**, 255–267.

Royer, J., Seguineau, C., Park, K.-I., Pouvreau, S., Choi, K.-S. & Costil, K. (2008) Gametogenetic cycle and reproductive effort assessed by two methods in 3 age classes of Pacific oysters, *Crassostrea gigas*, reared in Normandy. *Aquaculture*, **277**, 313–320.

Ruiz, Y., Suarez, P., Alonso, A., Longo, E., Villaverde, A. & San Juan, F. (2011) Environmental quality of mussel farms in the Vigo estuary: pollution by PAHs, origin and effects on reproduction. *Environmental Pollution*, **159**, 250–265.

Rusk, A.B., Alfaro, A.C., Young, T., Watts, E. & Adams, S.L. (2017) Investigation of early mussel (*Perna canaliculus*) development using histology, SEM imaging, immunochemistry and confocal microscopy. *Marine Biology Research*, **13**, 314–329.

Ryan, J.P., Harvey, J.B.J., Zhang, Y. & Woodson, C.B. (2014) Distributions of invertebrate larvae and phytoplankton in a coastal upwelling system retention zone and peripheral front. *Journal of Experimental Marine Biology and Ecology*, **459**, 51–60.

Saavedra, L., Leonardi, M., Morin, V. & Quiñones, R.A. (2012) Induction of vitellogenin-like lipoproteins in the mussel *Aulacomya ater* under exposure to 17β-estradiol. *Revista de Biología Marina y Oceanografía*, **47**, 429–438.

Sánchez-Lazo, C. & Martínez-Pita, I. (2012) Effect of temperature on survival, growth and development of *Mytilus galloprovincialis* larvae. *Aquaculture Research*, **43**, 1127–1133.

Schiedek, D., Broeg, K., Baršiene, J., Lehtonen, K.K. Gercken, J., Pfeifer, S. *et al.* (2006) Biomarker responses as indication of contaminant effects in blue mussel (*Mytilus edulis*) and female eelpout (*Zoarces viviparus*) from the southwestern Baltic Sea. *Marine Pollution Bulletin*, **53**, 387–405.

Scott, A.P. (2012) Do mollusks use vertebrate sex steroids as reproductive hormones? Part I: Critical appraisal of the evidence for the presence, biosynthesis and uptake of steroids. *Steroids*, **77**, 1450–1468.

Scott, A.P. (2013) Do mollusks use vertebrate sex steroids as reproductive hormones? II. Critical review of the evidence that steroids have biological effects. *Steroids*, **78**, 268–281.

Secor, C.L., Day, A.J. & Hilbish, T.J. (2001) Factors influencing differential mortality within a marine mussel (*Mytilus* spp.) hybrid population in southwestern England: reproductive effort and parasitism. *Marine Biology*, **138**, 731–739.

Sedik, W.F., Dempsey, K.E., Meng, X. & Craft, J.A. (2010) Temporal expression of sex-specific genes in the mantle of the common mussel (*Mytilus edulis*). *Marine Biology*, **157**, 639–646.

Seed, R. & Brown., R.A. (1975) The influence of reproductive cycle, growth and mortality on population structure in *Modiolus modiolus* (L.), *Cerastoderma edule* (L.), and *Mytilus edulis* (L.) (Mollusca: Bivalvia). In: *Proceedings of the Ninth European Marine Biology Symposium* (ed H. Barnes), pp. 257–274. Aberdeen University Press, Aberdeen.

Seed, R. & Brown, R.A. (1977) A comparison of the reproductive cycles of *Modiolus modiolus* (L.), *Cerastoderma (=Cardium) edule* (L.), and *Mytilus edulis* L. in Strangford Lough, Northern Ireland. *Oecologia*, **30**, 173–188.

Seed, R. & Suchanek, T.H. (1992) Population and community ecology of *Mytilus*. In: *The Mussel Mytilus: Ecology, Physiology, Genetics and Culture* (ed E.M. Gosling), pp. 87–169. Elsevier Science Publishers B.V., Amsterdam.

Selkoe, K.A. & Toonen, R.J. (2011) Marine connectivity: a new look at pelagic larval duration and genetic metrics of dispersal. *Marine Ecology Progress Series*, **436**, 291–305.

Sgro, L., Turolla, E., Rossi, R. & Mistri, M. (2002) Sexual maturation and larval development of the immigrant Asian date mussel, *Musculista senhousia, in a Po River deltaic lagoon. Italian Journal of Zoology*, **69**, 223–228.

Shafee, M. (1989) Reproduction of *Perna picta* (Mollusca: Bivalvia) from the Atlantic coast of Morocco. *Marine Ecology Progress Series*, **53**, 235–245.

Shafee, M.S. (1992) Production estimate of a mussel population *Perna picta* (Born) on the Atlantic coast of Moroccco. *Journal of Experimental Marine Biology and Ecology*, **163**, 183–197.

Shalaeva, E.A. & Lisitskaya, E.V. (2004) Seasonal dynamics of the numbers of larvae of common fouling organisms in Balaklava Bay (Black Sea). *Russian Journal of Marine Biology*, **30**, 371–378.

Shanks, A.L. (2009) Pelagic larval duration and dispersal distance revisited. *Biological Bulletin*, **216**, 373–385.

Shanks, A.L. & Brink, L. (2005) Upwelling, downwelling, and cross-shelf transport of bivalve larvae: test of a hypothesis. *Marine Ecology Progress Series*, **302**, 1–12.

Sharpley, M.S., Marciniak, C., Eckel-Mahan, K., McManus, M., Crimi, M., Waymire, K. *et al.* (2012) Heteroplasmy of mouse mtDNA is genetically unstable and results in altered behavior and cognition. *Cell*, **151**, 333–343.

Shima, J.S., Noonburg, E.G. & Phillips, N.E. (2010) Life history and matrix heterogeneity interact to shape metapopulation connectivity in spatially structured environments. *Ecology*, **91**, 1215–1224.

Shimeta, J., Cutajar, J., Watson, M.G. & Vlamis, T. (2012) Influences of biofilm-associated ciliates on the settlement of marine invertebrate larvae. *Marine Ecology Progress Series*, **449**, 1–12.

Siegel, D.A., Kinlan, B.P., Gaylord, B. & Gaines, S.D. (2003) Lagrangian descriptions of marine larval dispersion. *Marine Ecology Progress Series*, **260**, 83–96.

Sigurdsson, J.B., Titman, C.W. & Davies, P.A. (1976) The dispersal of young post-larval bivalve molluscs by byssus threads. *Nature*, **262**, 386–387.

Skibinski, D.O.F., Gallagher, C. & Beynon, C.M. (1994a) Mitochondrial DNA inheritance. *Nature*, **368**, 817.

Skibinski, D.O.F., Gallagher, C. & Beynon, C.M. (1994b) Sex-limited mitochondrial DNA transmission in the marine mussel *Mytilus edulis*. *Genetics*, **138**, 801–809.

Slaughter, C. & McCartney, M.A. & Yund, P.O. (2008) Comparison of gamete compatibility between two blue mussel species in sympatry and in allopatry. *Biological Bulletin*, **214**, 57–66.

Śmietanka, B., Burzyński, A. & Wenne, R. (2010) Comparative genomics of marine mussels (*Mytilus* spp.) gender associated mtDNA: rapidly evolving atp8. *Journal of Molecular Evolution*, **71**, 385–400.

Smith, E.B., Scott, K.M., Nix, E.R., Korte, C. & Fisher, C.R. (2000) Growth and condition of seep mussels (*Bathymodiolus childressi*) at a Gulf of Mexico brine pool. *Ecology*, **81**, 2392–2403.

Smith, J.R. & Strehlow, D.R. (1983) Algal-induced spawning of the marine mussel *Mytilus californianus* (Conrad). *International Journal of Invertebrate Reproduction*, **6**, 129–133.

Smith, P.T. & Reddy, N. (2012) Application of magnetic resonance imaging (MRI) to study the anatomy and reproductive condition of live Sydney rock oyster, *Saccostrea glomerata* (Gould). *Aquaculture*, **334–337**, 191–198.

Sokołowski, A., Bawazir, A.S., Sokołowska, E. & Wołowicz, M. (2010) Seasonal variation in the reproductive activity, physiological condition and biochemical components of the brown mussel *Perna perna* from the coastal waters of Yemen (Gulf of Aden). *Aquatic Living Resources*, **23**, 177–186.

Soon, T.K. & Ransangan, J. (2014) A review of feeding behavior, growth, reproduction and aquaculture site selection for green-lipped mussel, *Perna viridis*. *Advances in Bioscience and Biotechnology*, **5**, 462–469.

South, P.M. (2016) An experimental assessment of measures of mussel settlement: effects of temporal, procedural and spatial variations. *Journal of Experimental Marine Biology and Ecology*, **482**, 64–74.

Sponaugle, S., Cowen, R.K., Shanks, A., Morgan, S.G., Leis, J.M., Pineda, J. *et al.* (2002) Predicting self-recruitment in marine populations: biophysical correlates and mechanisms. *Bulletin of Marine Science*, **70S**, 341–375.

Springer, S.A. & Crespi, B.J. (2007) Adaptive gamete-recognition divergence in a hybridizing *Mytilus* population. *Evolution*, **61**, 772–783.

Sprung, M. (1983) Reproduction and fecundity of the mussel *Mytilus edulis* at Helgoland (North Sea). *Helgoländer wiss. Meeresunters*, **36**, 243–255.

Sreedevi, P.R., Uthayakumar, V., Jayakumar, R. & Ramasubramanian, R. (2014) Influence of rearing water temperature on induced gonadal development and spawning behaviour of tropical green mussel, *Perna viridis*. *Asian Pacific Journal of Reproduction*, **3**, 204–209.

Starr, M., Himmelman, J.H. & Therriault, J.C. (1990) Direct coupling of marine invertebrate spawning with phytoplankton blooms. *Science*, **247**, 1071–1074.

Stefano, G.B., Cadet, P., Mantione, K., Cho, J.J., Jones, D. & Zhu, W. (2003) Estrogen signaling at the cell surface coupled to nitric oxide release in *Mytilus edulis* nervous system. *Endocrinology*, **144**, 1234–1240.

Stenyakina, A., Walters, L.J., Hoffman, E.A. & Calestani, C. (2010) Food availability and sex reversal in *Mytella charruana*, an introduced bivalve in the southeastern United States. *Molecular Reproduction and Development*, **77**, 222–230.

Suárez, M.P., Alvarez, C., Molist, P. & San Juan, F. (2005) Particular aspects of gonadal cycle and seasonal distribution of gametogenic stages of *Mytilus galloprovincialis* cultured in the estuary of Vigo. *Journal of Shellfish Research*, **24**, 531–540.

Suchanek, T.H. (1985) Mussels and their role in structuring roky shore communities. In: *The Ecology of Rocky Coasts* (eds P.G. Moore & R. Seed), pp. 70–96. Hodder & Stoughton, Sevenoaks.

Sukhotin, A. & Flyachinskaya, L.P. (2009) Aging reduces reproductive success in mussels *Mytilus edulis*. *Mechanisms of Ageing and Development*, **130**, 754–761.

Sun, Q.-T. (2003) Cellular and molecular mechanisms leading to cortical reaction and polyspermy block in mammalian eggs. *Microscopy Research and Technique*, **61**, 342–348.

Sunila, I., Williams, L., Russo, S. & Getchis, T. (2004) Reproduction and pathology of blue mussels, *Mytilus edulis* (L.) in an experimental longline in Long Island Sound, Connecticut. *Journal of Shellfish Research*, **23**, 731–740.

Takagi, T., Nakamura, A., Deguchi, R. & Kyozuka, K. (1994) Isolation, characterization, and primary structure of three major proteins obtained from *Mytilus edulis* sperm. *Journal of Biochemistry*, **116**, 598–605.

Thomas, Y., Garen, P., Bennett, A., Le Pennec, M. & Clavier, J. (2012) Multi-scale distribution and dynamics of bivalve larvae in a deep atoll lagoon (Ahe, French Polynesia). *Marine Pollution Bulletin*, **65**, 453–462.

Thompson, C.M., York, R.H. & Gallager, S.M. (2012) Species-specific abundance of bivalve larvae in relation to biological and physical conditions in a Cape Cod estuary. *Marine Ecology Progress Series*, **469**, 53–69.

Thompson, R.J. (1979) Fecundity and reproductive effort in the blue mussel (*Mytilus edulis*), the sea urchin (*Strongylocentrotus droebachiensis*), and the snow crab (*Chionoecetes opilio*) from populations in Nova Scotia and Newfoundland. *Journal of the Fishery Research Board of Canada*, **36**, 955–964.

Thompson, R.J. (1984) Production, reproductive effort, reproductive value and reproductive cost in a population of the blue mussel *Mytilus edulis* from a subarctic environment. *Marine Ecology Progress Series*, **16**, 249–257.

Thorarinsdóttir, G. & Gunnarsson, K. (2003) Reproductive cycles of *Mytilus edulis* L. on the west and east coasts of Iceland. *Polar Research*, **22**, 217–223.

Thorrold, S.R., Jones, G.P., Holmberg, M.E., Burton, R.S., Swearer, S.E., Neigel, J.E. *et al.* (2002) Quantifying larval retention and connectivity in marine populations with artificial and natural markers. *Bulletin of Marine Science*, **70**, S291–308.

Thorrold, S.R., Zacherl, D.C. & Levin, L.A. (2007) Population connectivity and larval dispersal: using geochemical signatures in calcified structures. *Oceanography*, **20**, 80–89.

Tilburg, C.E., McCartney, M.A. & Yund, P.O. (2012) Across-shelf transport of bivalve larvae: can the interface between a coastal current and inshore waters act as an ecological barrier to larval dispersal? *PLOS One*, **7**, e48960.

Tiwari, S. & Gallagher, S. (2003) Machine learning and multiscale methods in the identification of bivalve larvae. In: *Proceedings of the Ninth IEEE International Conference on Computer Vision. Nice, France 13–16 October 2003, Vol. I*, pp. 494–501. IEEE Computer Society, Piscataway, NJ.

Togo, T. & Morisawa, M. (1997) Aminopeptidase-like protease released from oocytes affects oocyte surfaces and suppresses the acrosome reaction in establishment of polyspermy block in oocytes of the mussel *Mytilus edulis*. *Developmental Biology*, **182**, 219–227.

Togo, T., Osanai, K. & Morisawa, M. (1995) Existence of three mechanisms for blocking polyspermy in oocytes of the mussel *Mytilus edulis*. *Biological Bulletin*, **189**, 330–339.

Toro, J.E., Thompson, R.J. & Innes, D.J. (2002) Reproductive isolation and reproductive output in two sympatric mussel species (*Mytilus edulis, M. trossulus*) and their hybrids from Newfoundland. *Marine Biology*, **141**, 897—909.

Torrado, M., Paz, M., Korochkin, L.I. & Mikhailov, A.T. (2003) Mussel MAP, a major gonad-duct esterase-like protein, is released into sea water as a dual constituent of the seminal fluid and the spermatozoon. *Journal of Experimental Biology*, **206**, 313–326.

Torroglosa, M.E. & Giménez, J. (2015) Sperm ultrastructure in two species of *Brachidontes* (Bivalvia, Mytilidae) from the south-western Atlantic Ocean. *Journal of the Marine Biological Association of the UK*, **95**, 991–998.

Treen, N., Itoh, N., Miura, H., Kikuchi, I., Ueda, T., Takahashi, K.G. *et al.* (2012) Mollusc gonadotropin-releasing hormone directly regulates gonadal functions: a primitive endocrine system controlling reproduction. *General and Comparative Endocrinology*, **176**, 167–172.

Troost, K., Gelderman, E., Kamermans, P., Smaal, A.C. & Wolff, W.C. (2009) Effects of an increasing filter feeder stock on larval abundance in the Oosterschelde estuary (SW Netherlands). *Journal of Sea Research*, **61**, 153–164.

Tyler, P.A. & Young, C.M. (1999) Reproduction and dispersal at vents and cold seeps. *Journal of the Marine Biological Association of the UK*, **79**, 193–208.

Tyler, P., Young, C.M., Dolan, E., Arellano, S.M., Brooke, S.D. & Baker, M. (2007) Gametogenic periodicity in the chemosynthetic cold-seep mussel '*Bathymodiolus*' *childressi*. *Marine Biology*, **150**, 829–840.

Uliano-Silva, M., Americo, J., Schomaker Bastos, A., Furtado, C., de Freitas Rebelo, M. & Prosdocimi, F. (2016) Complete mitochondrial genome of the brown mussel *Perna perna* (Bivalve, Mytilidae). *Mitochondrial DNA Part A*, **27**, 3955–3956.

Väinölä, R. & Strelkov, P. (2011) *Mytilus trossulus* in Northern Europe. *Marine Biology*, **158**, 817–833.

Vekhova, E.E., Reunov, A.A., Alexandrova, Y.N., Reunova Y.A. & Adrianov, A.V. (2013) First data on sperm heteromorphism in the mussels *Crenomytilus grayanus* and *Mytilus trossulus* (Bivalvia) from the Sea of Japan. *Doklady Biological Sciences*, **453**, 358–361.

Ventura, A., Schulz, S. & Dupont, S. (2016) Maintained larval growth in mussel larvae exposed to acidified undersaturated seawater. *Scientific Reports*, **6**, 23728.

de Vooys, C.G.N. (1999) Numbers of larvae and primary plantigrades of the mussel *Mytilus edulis* in the western Dutch Wadden Sea. *Journal of Sea Research*, **41**, 189–201.

Voronezhskaya, E.E., Nezlin, L.P., Odintsova, N.A., Plummer, J.T. & Croll, R.P. (2008) Neuronal development in larval mussel *Mytilus trossulus* (Mollusca: Bivalvia). *Zoomorphology*, **127**, 97–110.

Waldbusser, G.G., Hales, B., Langdon, C.L., Haley, B.A., Schrader, P., Brunner, E.L. *et al.* (2015) Saturation-state sensitivity of marine bivalve larvae to ocean acidification. *Nature Climate Change*, **5**, 273–280.

Wang, C., Bao, W.Y., Gu, Z.Q., Li, Y.F., Liang, X., Ling, Y. *et al.* (2012) Larval settlement and metamorphosis of the mussel *Mytilus coruscus* in response to natural biofilms. *Biofouling*, **28**, 249–256.

Wang, S., Bao, Z., Zhang, L., Li, N., Zhan, A., Guo, W. *et al.* (2006) A new strategy for species identification of planktonic larvae: PCR–RFLP analysis of the internal transcribed spacer region of ribosomal DNA detected by agarose gel electrophoresis or DHPLC. *Journal of Plankton Research*, **28**, 375–384.

Warner, R.R. & Cowen, R.K. (2002) Local retention of production in marine populations: evidence, mechanisms, and consequences. *Bulletin of Marine Science*, **70**, 245–249.

Weersing, K. & Toonen, R.J. (2009) Population genetics, larval dispersal, and connectivity in marine systems. *Marine Ecology Progress Series*, **393**, 1–12.

Weibel, E.R.G., Kistler, S. & Scherle, W.F. (1966) Practical stereological methods for morphometric cytology. *The Journal of Cell Biology*, **30**, 23–38.

Weidberg, N., Porri, F., Von der Meden, C.E.O., Jackson, J.M., Goschen, W. & McQuaid, C.D. (2015) Mechanisms of nearshore retention and offshore export of mussel larvae over the Agulhas Bank. *Journal of Marine Systems*, **144**, 70–80.

Werner, F.E., Cowen, R.K. & Paris, C.B. (2007) Coupled biological and physical models: present capabilities and necessary developments for future studies of population connectivity. *Oceanography*, **20**, 54–69.

White, C., Selkoe, K.A., Watson, J., Siegel, D.A., Zacherl, D.C. & Toonen, R.J. (2010) Ocean currents help explain population genetic structure. *Proceedings of the Royal Society B*, **277**, 1685–1694.

Wilson, J.H. & Seed, R. (1974) Reproduction in *Mytilus edulis* L. (Mollusca: Bivalvia) in Carlingford Lough, Northern Ireland. *Irish Fisheries Investigations*, **15B**, 1–30.

Witman, J.D., Genovese, S.J., Bruno, J.F., McLaughlin, J.W. & Pavlini, B.I. (2003) Massive prey recruitment and the control of rocky subtidal communities on large spatial scales. *Ecological Monographs*, **73**, 441–462.

Won, Y., Young, C.R., Lutz, R.A. & Vrijenhoek, R.C. (2003) Dispersal barriers and isolation among deep-sea mussel populations (Mytilidae: *Bathymodiolus*) from eastern Pacific hydrothermal vents. *Molecular Ecology*, **12**, 169–184.

Yang, J.L., Satuito, C.G., Bao, W.Y. & Kitamura, H. (2007) Larval settlement and metamorphosis of the mussel *Mytilus galloprovincialis* on different macroalgae. *Marine Biology*, **152**, 1121–1132.

Yang, J.L., Li, Y.F., Bao, W.Y., Satuito, C.G. & Kitamura, H. (2011) Larval metamorphosis of the mussel *Mytilus galloprovincialis* Lamarck, 1819 in response to neurotransmitter blockers and tetraethylammonium. *Biofouling*, **27**, 193–199.

Yang, J.L., Shen, P.J., Liang, X., Li, Y.F., Bao, W.Y. & Li, J.L. (2013a) Larval settlement and metamorphosis of the mussel *Mytilus coruscus* in response to monospecific bacterial biofilms. *Biofouling*, **29**, 247–259.

Yang, J.L., Li, S.H., Li, Y.F., Liu, Z.W., Liang, X., Bao, W.Y. *et al.* (2013b) Effects of neuroactive compounds, ions and organic solvents on larval metamorphosis of the mussel *Mytilus coruscus*. *Aquaculture*, **396–399**, 106–112.

Yang, J.L., Zhou, X., Li, Y.F., Guo, X.P., Liang, X. & Li, J.L. (2014a) Plantigrade settlement of the mussel *Mytilus coruscus* in response to natural biofilms on different surfaces. *Journal of the Marine Biological Association of the UK*, **94**, 1639–1649.

Yang, J.L., Li, X., Liang, X., Bao, W.Y., Shen, H.D. & Li, J.L. (2014b) Effects of natural biofilms on settlement of plantigrades of the mussel *Mytilus coruscus*. *Aquaculture*, **424–425**, 228–233.

Yang, J.L., Li, S.H., Bao, W.Y., Yamada, H. & Kitamura, H. (2015) Effect of different ions on larval metamorphosis of the mussel *Mytilus galloprovincialis*. *Aquaculture Research*, **46**, 155–162.

Yap, C.K. & Al-Barwani, S. M. (2012) A comparative study of condition indices and heavy metals in *Perna viridis* populations at Sebatu and Muar, Peninsular Malaysia. *Sains Malaysiana*, **41**, 1063–1069.

Yildiz, H., Palaz, M. & Bulut, M. (2006) Condition indices of Mediterranean mussels (*Mytilus galloprovincialis* L. 1819) growing on suspended ropes in Dardanelles. *Journal of Food Technology*, **4**, 221–224.

Young, T., Alfaro, A.C. & Robertson, J. (2011) Effect of neuroactive compounds on the settlement of mussel (*Perna canaliculus*) larvae. *Aquaculture*, **319**, 277–283.

Young, T., Alfaro, A.C., Sánchez-Lazo, C. & Robertson, J. (2015) Putative involvement of adrenergic receptors in regulation of mussel (*Perna canaliculus*) larval settlement. *Marine Biology Research*, **11**, 655–665.

Yund, P.O., Tilburg, C.E. & McCartney, M.A. (2015) Across-shelf distribution of blue mussel larvae in the northern Gulf of Maine: consequences for population connectivity and a species range boundary. *Royal Society Open Science*, **2**, 150513.

Yusa, Y., Breton, S. & Hoeh, W.R. (2013) Population genetics of sex determination in *Mytilus* mussels: reanalyses and a model. *Journal of Heredity*, **104**, 380–385.

Zardi, G.I., McQuaid, C.D. & Nicastro, K.R. (2007) Balancing survival and reproduction: seasonality of wave action, attachment strength and reproductive output in indigenous *Perna perna* and invasive *Mytilus galloprovincialis* mussels. *Marine Ecology Progress Series*, **334**, 155–163.

Zbawicka, M., Burzyński, A., Skibinski, D. & Wenne, R. (2010) Scottish *Mytilus trossulus* mussels retain ancestral mitochondrial DNA: complete sequences of male and female mtDNA genomes. *Gene*, **456**, 45–53.

Zhan, A., Hu, X., Bao, L., Lu, W., Peng, W., Wang, M. *et al.* (2008) Molecular identification of scallop planktonic larvae using species-specific microsatellites. *Acta Oceanologica Sinica*, **27**, 134–146.

Zhu, W., Mantione, K., Jones, D., Salamon, E., Cho, J.J., Cadet, P. *et al.* (2003) The presence of 17-β estradiol in *Mytilus edulis* gonadal tissues: evidence for estradiol isoforms. *Neuroendocrinology Letters*, **24**, 137–140.

Zouros, E., Oberhouser Ball, A., Saavedra, C. & Freeman, K.R. (1994a) Mitochondrial DNA inheritance. *Nature*, **368**, 817.

Zouros, E., Oberhouser Ball, A., Saavedra, C. & Freeman, K.R. (1994b) An unusual type of mitochondrial DNA inheritance in the blue mussel *Mytilus*. *Proceedings of the National Academy of Science*, **91**, 7463–7467.

Zouros, E. (2000) The exceptional mitochondrial DNA system of the mussel family Mytilidae. *Genes & Genetic Systems*, **75**, 313–318.

Zouros, E. (2013) Biparental inheritance through uniparental transmission: the doubly uniparental inheritance (DUI) of mitochondrial DNA. *Evolutionary Biology*, **40**, 1–31.

6

Growth

Introduction

Growth in bivalves is usually described in terms of an increase in some dimension of the shell valves. In mussels, for example, length – the maximum distance between the anterior and posterior margins of the shell – is the dimension of choice, whereas in clams shell height – the maximum distance between the dorsal (hinge) and ventral shell margins – is used (see Figure 2.4). Growth rate can be measured in one of two ways: either the size of the whole organism is related to age (absolute growth) or the rate of growth of one size variable is related to that of another (allometric growth; Seed & Suchanek 1992). Shell length as an indicator of size is chosen because it is more easily measured than the more appropriate parameter, flesh weight, and also because growth history is recorded in the shell. In most bivalve species, shell and flesh weight increase are uncoupled. For example, in the mussel *Mytilus edulis*, shell growth is rapid during the spring and summer and slows down over the winter, while flesh weight is subject to seasonal fluctuations associated with the reproductive cycle (Kautsky 1982; see also Borrero & Hilbish 1988).

This chapter deals with methods used to measure growth, bioenergetics growth models and exogenous and endogenous modulators of growth.

Measurement of Absolute Growth

Methods used to measure rates of increase in shell growth include length-frequency distribution (LFD) analysis, growth increment analysis (GIA) following mark-and-recapture experiments using different labelling techniques, shell growth-ring analysis, internal growth lines analysis in shell sections, elemental analysis and analysis of stable oxygen isotopes.

Length-Frequency Distributions

The biggest advantage of this method is that it measures growth of a population under undisturbed natural conditions. The accuracy of LFD analyses depends on how well the collected samples reflect the size structure of the actual population. Well-defined age cohorts

Marine Mussels: Ecology, Physiology, Genetics and Culture, First Edition. Elizabeth Gosling.

and large sample sizes are also required (Herrmann *et al.* 2009). Shell lengths are measured from a random sample of individuals, ideally more than 1500 sampled over a minimum six-month period (Pauly 1987), and the data are divided into size frequency classes and plotted in the form of a histogram. Size class intervals are normally small, from 1 to 5% of the size of the largest specimen in the population (Cerrato 1980). Where recruitment is seasonal and the life span is short, and there is little variability in individual growth rates, individual year classes can often be identified as distinct modes (Figure 6.1). By following the position of these modes over time, the mean seasonal growth rate of each year class can be estimated. However, in long-lived species with extended recruitment and variable individual growth rates, it is not possible to use size frequency distributions to measure growth rate due to merging of year classes. This is particularly true for older populations, where year classes overlap to such an extent that each loses its identity because slow growers of one year class overlap with fast growers of the previous one. Several methods are available that can separate an LFD into groups that are assumed to represent separate cohorts. Early methods were graphical, such as the Bhattacharya plot (Sparre & Venema 1992 and references therein), which separates an LFD into a series of normal distributions or pseudo-cohorts. But various programs have superseded these methods, such as Electronic LEngth-Frequency ANalysis (ELEFAN; Pauly 1987), MULTIple Length Frequency ANalysis (MULTIFAN; Fournier *et al.* 1990) and Shepherds's Length Composition Analysis (SLCA; Shepherd 1987). FAO-ICLARM Stock Assessment Tools II (FiSAT II) is another program package developed mainly for the analysis of length-frequency data, but also enabling related analyses of

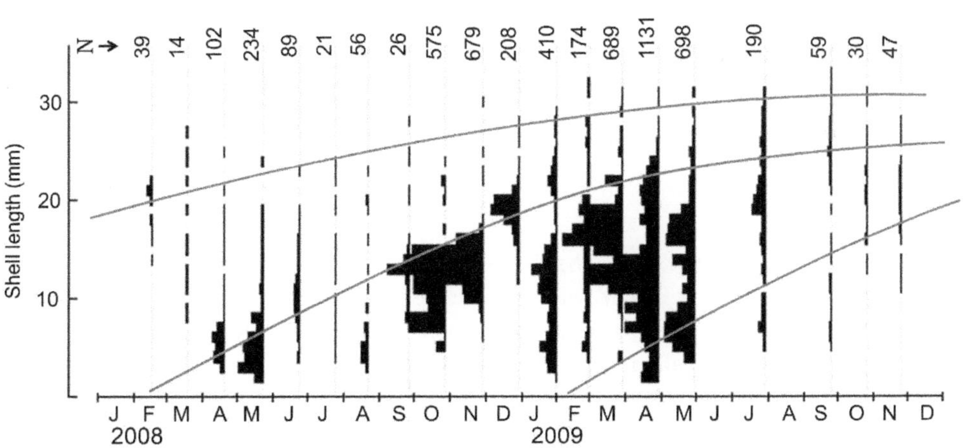

Figure 6.1 Monthly length-frequency data (black histograms; 5471 individuals grouped in 1 mm size classes) and growth curves (grey lines) in the clam *Donax striatus* for the period February 2008 to November 2009. Monthly length-frequency distributions (LFDs) were used for growth estimations, to which a von Bertalanffy growth function (VBGF) was fitted throughout the ELEFAN routine of the FiSAT II program. Preliminary estimates of L_∞ and k were used to identify the VBGF of best fit for the monthly size-frequency data, using the Rn value in ELEFAN as a criterion of goodness-of-fit. The smallest recorded individual measured 2.7 mm and the largest 33.3 mm. The growth parameters used to construct best-fit curves (Rn = 0.15) were: L_∞ = 36.1 mm, k = 0.8/yr and t_0 = 0.2/yr. *Source:* From Ocaña (2015). CC BY-ND 3.0 CR.

size-at-age, catch-at-age, selection and so on (Gayanilo *et al.* 2005). A number of growth models (see later) have been incorporated into these programs.

ELEFAN I is a program package for the analysis of growth, mortality and recruitment from multiple size-frequency samples. The main advantage of this method is that no assumptions have to be made about the number of cohorts or the shape of the underlying distributions of a size-frequency sample. The program first restructures the length distributions by assigning positive values to length classes containing many animals and small or negative values to length classes with few animals. Goodness-of-fit scores are then calculated by summing the values of the length classes through which each growth curve passes. An accurate growth curve will pass through length classes (or modes) with large numbers of animals and thus will accumulate a high goodness-of-fit score. The growth curve with the highest score is considered to be the best estimate. MULTIFAN is a likelihood-based analysis of several size-frequency samples at the same time. In contrast to ELEFAN, MULTIFAN depends on several assumptions (e.g. normal distribution within cohorts), which may not be met by natural data sets. Growth curves can also be fitted in ways that use more information than that contained just in the position of the modes. As in ELEFAN, SLCA is based on the goodness of fit of the location of modes expected from a von Bertalanffy growth curve to the observed modes in the length distribution. However, length distributions are not restructured as in ELEFAN, and the goodness-of-fit criterion in SLCA is similar to that used in time series analysis (Pauly & Arreguín-Sánchez 1995).

Growth Curves

A growth curve can be estimated from the relative position of the modes of an LFD, the assumption being that the modes represent groups spawned one year apart. Growth curves may also be constructed from mark-and-recapture data and analysis of growth checks on the shell (see later). Various equations (length-age models) are used to fit curves to the growth data, but the one which best describes growth in bivalves and in fish is the von Bertalanffy (1938) growth equation:

$$L_t = L_\infty \left(1 - e^{-k(t-t_0)}\right)$$

where Lt is the length at age t, $L\infty$ is the theoretical maximum size attained under specific environmental conditions, k is a growth constant reflecting the rate at which maximum size $L\infty$ is approached and $t0$ is another constant representing time when $Lt = 0$ (Figure 6.2). A commonly used graphical method of estimating $L\infty$ and k is a Ford–Walford plot (Walford 1946), where length at age t is plotted against length at $t + 1$ years. $L\infty$ is given where the line of best fit intercepts the 45^0 angle, or by

$$\frac{y}{1 - slop}$$

where y is the intercept of best-fit line on the y axis. The growth constant is calculated as minus the natural log of the slope (see King 2007 for details).

While the von Bertalanffy growth model (VBGM) is the most studied and most commonly applied among all length-age models, other frequently used alternatives are the generalised VBGM (Pauly 1979), the Gompertz growth model (Gompertz 1825), the logistic

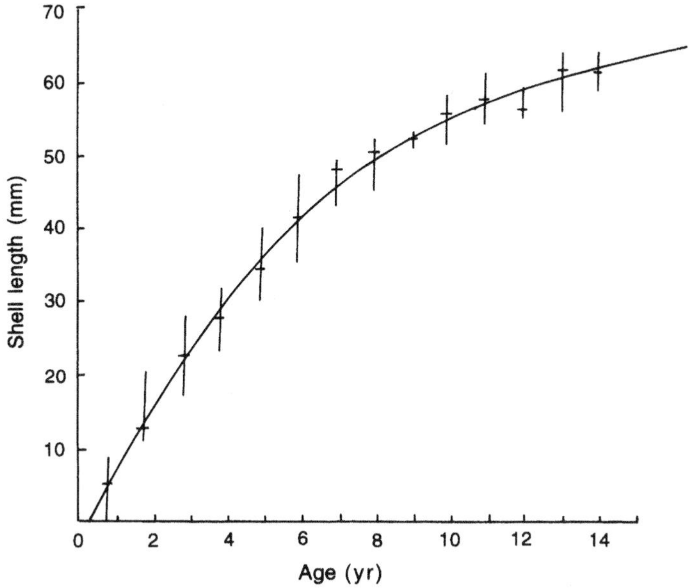

Figure 6.2 Von Bertalanffy growth curve for a sample of mussels *Mytilus edulis* (N = 417) from Dungarvan Harbour, south coast of Ireland. The predicted maximum size (L_∞) for this population is 70.74 mm shell length, achieved after considerably more than 14 years of growth; k = 0.1598 and t_0 = 0.423. The L_∞ value is fairly typical of Irish sheltered shore mussels. Note that, as in most bivalves, growth is rapid during the first years of life and slows down as size increases.
Source: Ottway & Ross (unpublished data).

model (Ricker 1975) and the Schnute–Richards model (Schnute & Richards 1990). The Gompertz model is similar to the VBGM but uses the logarithms of length. The assumption of these models is that growth is determinate; that is, some maximum attainable size exists for any given population. However, in many bivalves, including mussels, growth may be indeterminate (i.e. it continues right through the entire life span; Seed & Suchanek 1992 and references therein), and because of this some workers use polynomial expressions to describe shell and tissue growth in preference to the von Bertalanffy or equivalent growth equations. Neither the VBGM nor the Gompertz model takes into account seasonal variations in growth rate. However, the VBGM has been modified to produce several growth models that allow for seasonal oscillations in length within each growth year (García-Berthou *et al.* 2012 and references therein).

In addition to information on age-specific growth, size frequency distributions also provide information on recruitment patterns and size-selective mortality caused by disturbance, predation or intraspecific competition for space (Tumanda *et al.* 1997; Ramón *et al.* 2007; Calcagno *et al.* 2012; Bagur *et al.* 2013).

Mark-and-Recapture and Growth Increment Analysis

Mark-and-recapture involves marking a number of individuals from a population, returning them to that population, and then recapturing a set number of individuals at a later date. The number of marked individuals relative to the number of captured individuals provides an estimate of the population size at the time of marking and release. Marking in bivalves was originally achieved by filing a notch (1 mm) on the shell (Alfaro *et al.* 2008), painting with coloured ink or applying a metal or plastic tag with adhesive. While these methods were inexpensive and relatively easily applied, it is difficult to detect microscopic shell growth increases on a scale of less than tens of micrometres, since such increases must be measured using a sliding callipers or optical microscopy (Fujikura *et al.* 2003). For this reason, and also because of tag loss and growth checks caused by filing, other methods have come to replace them. For example, fluorescent compounds such as oxytetracycline (OTC), calcein and alizarin red S (ARS) offer a rapid and noninvasive option in mussels (Peharda *et al.* 2007; Bownes & McQuaid 2010; Connor & Robles 2015; Fitzer *et al.* 2016) and other bivalves (Bernhard *et al.* 2015 and references therein). Organisms exposed to fluorochromes incorporate the chemical into growing calcified structures in the form of an internal growth mark (observable under ultraviolet or blue light), which is subsequently used to estimate growth from time of exposure (Kaehler & McQuaid 1999). Of the three, calcein is the one that is most often used as a growth marker, probably because of its low toxicity and reliable marking quality (Figure 6.3). OTC and calcein have been used in combination to double-mark individual shells; specific 'bar codes' can be used to distinguish between different cohorts or spat deployment strategies (Lucas *et al.* 2008). The elements strontium and manganese have also been used to mark shells in species other than mussels (Fujikura *et al.* 2003; Lartaud *et al.* 2010).

To illustrate the use of calcein to measure growth in *Mytilus californianus* on two rocky shores on the west coast of North America, Connor & Robles (2015) injected mussels with fluorescent dye (300 mg/l) seawater solutions. A small hole (approximately 1 mm in diameter) was made on the shell margin off centre to the medial line of the dorso-ventral surface of each mussel. A fine-needle syringe (1.65 mm diameter, 25 mm length) was used to inject the dye solution into the mantle cavity through the hole until excess escaped between the valves; concentrations as high as 650 mg/l do not increase mortality rates of mussels (Kaehler & McQuaid 1999). Mussels were allowed to grow for about 90 days, after which

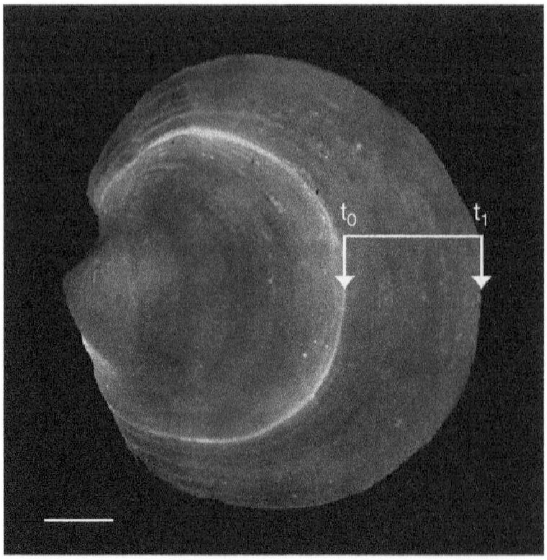

Figure 6.3 Photograph of the right valve of a calcein-marked *Loripes lacteus* shell. The position of the growing edge at the time of calcein administration is given by t_0; the position of the growing edge three months later is given by t_1. Thus, t_0 to t_1 represents shell growth (increase in height). Scale bar: 1 mm. *Source:* Van der Geest *et al.* (2011). Reproduced with permission from Elsevier. (*See colour plate section for colour representation of this figure*).

they were collected and had their shell lengths measured with a digital callipers. To measure growth increment, the shells were cut along the longest axis between the anterior and posterior margins with a lapidary diamond saw and viewed under an epifluorescent microscope for growth past the calcein dye mark secreted at the mantle edge. Mussel growth greater than 2 mm was beyond the field of view and therefore measured using a digital callipers. For more details on the Connor & Robles (2015) study, see later.

Comparing the two methods, LFD and GIA from mark-and-recapture, for the measurement of growth in clam (*Donax*) species from different climate areas, Herrmann *et al.* (2009) found that by carrying out a relatively short-term mark-and-recapture experiment, they could obtain a good growth estimate for temperate bivalves. However, they recommended that both methods should be used to determine growth in tropical species. Since tropical species exhibit continuous spawning events or recruit over a longer period compared to temperate species, LFD analysis may not be a useful method to estimate growth in the former. The decided advantage of GIA is accuracy, as it allows daily growth-rate measurements on a much smaller number of individuals than does LFD analyses. GIA can also be used on scattered populations where it is difficult to sample sufficient individuals for clear cohort detection. On the other hand, the LFD method allows detection of seasonal growth and causes much less disturbance than GIA, with its much shorter sampling intervals. Herrmann *et al.* (2009) further found that GIA was more time-saving compared to LFD, achieving results after only six weeks of sampling time and three months of laboratory time, compared to 25 months of sampling time for LFD analysis. LFD also entailed costs twice as high as those of GIA.

Shell Growth Rings

Age can be determined from growth rings on the external shell, or from growth lines in shell or ligament cross-sections. In scallops and clams, surface rings or lines are usually produced annually during the winter period of suspended shell growth and therefore can be used as a reliable estimate of age (Figure 6.4). Unlike clams and scallops, the majority of mussels, including freshwater species, lack clearly defined surface growth lines, precluding any reliable age estimates (see Sansom *et al.* 2013; Peharda *et al.* 2015). In bivalves with well-defined external growth lines, the shell-secreting mantle edges withdraw from the shell margins during cold periods, and shell deposition slows down or ceases. When conditions are again suitable for shell growth, the mantle margins extend. Thus, old and new regions of the shell are no longer continuous, and an obvious growth ring results (Richardson 2001). Mark-and-recapture studies are used to confirm that growth lines are formed annually. Confirmation that rings are annual is essential before they can be used to construct growth curves. Annual marks can often be difficult to distinguish from spawning or disturbance lines on the shell, which may be caused as a result of storms, dredging, handling, transplanting or predation attempts (reviewed in Richardson 2001). Age determination is often unreliable in species that do not mark a clear ring during the first year of life, and also in older bivalves where growth rings near the border of the shell are so close together that they are difficult or impossible to tell apart.

An alternative ageing method for species where external growth rings are absent or difficult to see is the use of annual growth checks and microgrowth patterns in shell

Figure 6.4 The cockle *Clinocardium nuttallii*, with the most obvious external growth lines indicated, suggesting an individual of three years of age. 1×. *Source:* Photograph courtesy of Thomas Carefoot and A SNAIL'S ODYSSEY website.

sections or other parts of the shell (e.g. umbo, hinge plate, pallial line scar, posterior adductor muscle scar). Reliable estimates of age can be obtained from growth lines within acetate peels. The following method is mainly based on that of Richardson (2001); modifications to it can be found in Grey *et al.* (1997), Millstein & O'Clair (2001), Alfaro *et al.* (2008), Bagur *et al.* (2013) and Peharda *et al.* (2015). Clean, dry shell valves are sectioned whole or embedded in epoxy resin and then sectioned along the major growth axis using a diamond saw. The cut surface is ground on increasingly finer grades of wet and dry paper, polished and etched with a 1% HCl and immediately blotted dry. The etched surface is flooded with acetone, and a sheet of acetate (0.5 mm thick) is firmly applied. After six minutes, the acetate peel is removed, mounted on a glass slide and viewed under the microscope to determine the number of growth lines in the middle prismatic and inner nacreous shell layers. An alternative is fluorescence spectroscopy, which does not require any chemical treatment of the polished shell section and yields results comparable to the acetate peel method (Wanamaker *et al.* 2009). Another method is thin sectioning; although it is more labour-intensive, it tends to produce superior results to acetate peels (Cerrato 1980; Gosselin *et al.* 2013; Sansom *et al.* 2013).

Nacreous lines are formed annually, while fine microgrowth bands within the prismatic layer have a tidal periodicity. The annual lines are thin, translucent and darkly pigmented and first appear at the end of the winter. Between each winter line is a wider area that represents the spring–summer growth increment (Figure 6.5). Counts should be made at or near the umbonal region of the shell, as this is the only region of the nacreous layer that contains a relatively complete record of growth (Lutz 1976). Abrasion of the umbo or poor definition of the first winter ring can result in an underestimation of age, as for example in large, old mussels. However, the method is generally reliable, although it requires more time and expertise than the external shell ring method. It is particularly useful in habitats such as offshore platforms or water intake pipes where conventional methods involving repeated sampling of the population are difficult or impracticable (Richardson *et al.* 1990).

It is now widely accepted that there is a tidal periodicity to microgrowth band formation in intertidal bivalves (Richardson 1989; Richardson *et al.* 1990; Grey *et al.* 1997; Abades *et al.* 2000; Schöne & Giere 2005). For example, Grey *et al.* (1997) showed that there was

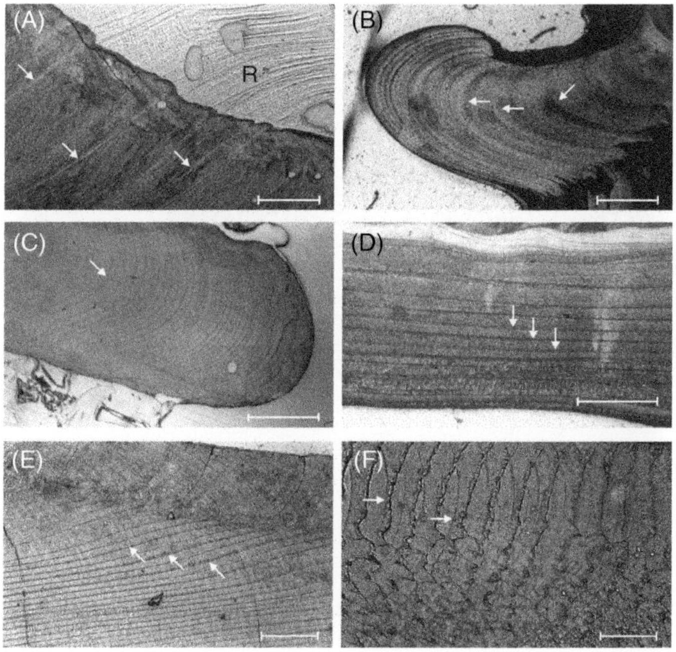

Figure 6.5 Acetate peel replicas of shell sections of (A) the ligament scar in the oyster *Ostrea edulis*. R, resilium (enlargement of the ligament); arrows indicate positions of annual growth lines. (B, C) Umbones of the mussel *Modiolus modiolus* and the clam *Glycymeris glycymeris*, respectively, showing annual growth lines (arrows). (D) Inner nacreous layer of *Modiolus modiolus* and annual growth lines (arrows). (E) Strongly defined growth bands (arrows) alternating with weakly defined bands in the crossed lamellar layer of the cockle *Cerastoderma edule*. (F) Weakly defined endogenous bands in the prismatic layer of the clam *Ruditapes philippinarum*. Scale bar = 100 μm, except (D) and (E) where scale bar = 0.5 mm. *Source:* Photomicrographs courtesy of C. Richardson, Bangor University, Wales, UK.

close agreement between the number of clearly defined bands in the prismatic layer and the predicted number of tidal emersions in a mussel (*Mytilus chilensis*) bed. Results from two experiments showed that the expected numbers of emersions were 12 and 29, while the number of well-defined bands (mean $\pm$ SD) were 9.17 ± 3.76 and 24.33 ± 10.22, respectively. Also, bands were stronger in definition during spring tides, due to the increased period of emersion, and weaker during neap tides, when the mussels were not always uncovered. There was also a narrowing of the tidal bands during autumn and narrow increments in the winter, with wider growth increments subsequently deposited in summer (Figure 6.6). The width and definition of microgrowth bands is influenced by such factors as season, temperature at time of emersion, tidal amplitude, tidal cycle and reproductive stage (Richardson 2001; Seed *et al.* 2000; Abades *et al.* 2008).

The observed patterns of microgrowth bands is thought to result from the sequential alternation of aerobic and anaerobic metabolic pathways occurring during tidal immersion and emersion, respectively, in intertidal bivalves. These periods of aerobic–anaerobic metabolism result in a wide growth increment and a narrow, dark microgrowth band, respectively (Abades *et al.* 2008 and references therein). To illustrate how microgrowth

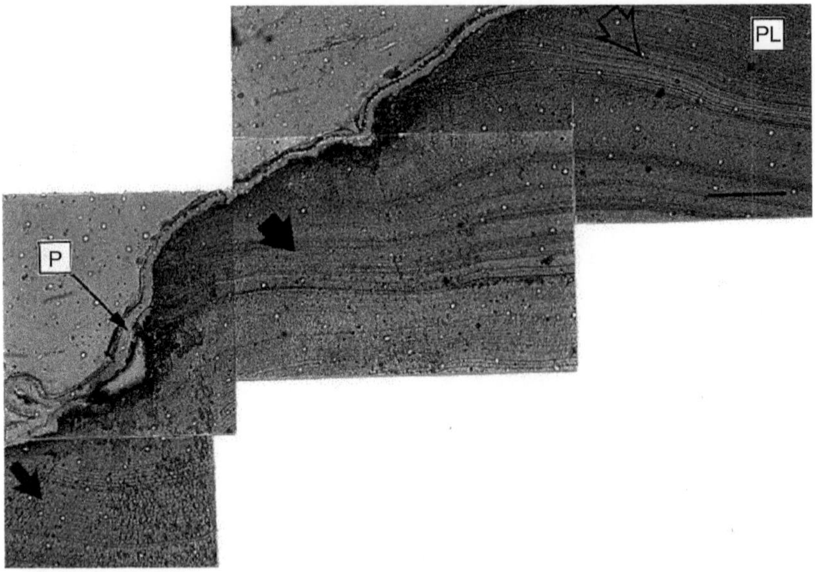

Figure 6.6 Photomicrograph of acetate peel replicas of the mussel *Mytilus edulis chilensis*, collected from the middle of the mussel zone at Camilla Creek, East Falkland, showing a narrowing of the tidal bands during autumn (open arrow), narrow increments in the winter (large filled arrow) and wider growth increments subsequently deposited in summer (small filled arrow). P, periostracum; PL,: prismatic layer. *Source:* From Gray *et al.* (1997). Reproduced with permission from Scientia Marina.

bands can be used to estimate age and annual growth rate in mussel populations, a sample of the largest mussels were collected from the middle of a *M. chilensis* mussel bed at three sites in the Falkland Islands (Gray *et al.* 1997). Acetate peels were prepared and the position of each annual check within each shell was identified from narrowing of the growth bands and marked directly on the peel. The age of each mussel was estimated and the distance between each successive check and the umbo was measured to the nearest 0.1 mm using vernier callipers. Von Bertalanffy growth curves were then fitted to these measurements of size at age from the three populations.

Other methods of measuring absolute growth include optical techniques such as laser diffraction, photographic techniques and the uptake and decay of radionuclides (often referred to as radioisotopes) from seawater into shells. The nuclei of such chemicals are unstable and spontaneously change or decay into a more stable nucleus of another element. Once in the shell and isolated from the seawater source, the radionuclide atoms undergo radioactive decay. The number of atoms remaining is thus a function of the time since the deposition of the shell. In addition, the seasonal variation in the ratio of the oxygen-stable isotopes, ^{18}O and ^{16}O, in shell calcite may be used to validate annual formation of growth rings. The ratio is determined primarily by the thermodynamic equilibrium between the isotopes in calcite and in seawater at the time of shell deposition. Calcite that has been deposited at equilibrium with seawater has a higher $^{18}O/^{16}$ ratio (or $\delta^{18}O$ when expressed relative to a standard) than that of seawater. Also, the difference between

$\delta^{18}O_{calcite}$ and $\delta^{18}O_{seawater}$ increases as temperature decreases (see later). Ratios $\delta^{18}O/\delta^{13}C$ and $\delta^{14}C$ in shell carbonate have also been used to verify annual formation of rings, but to date the majority of studies have focused on clam and scallop species (Vadopalas *et al.* 2011; Chute *et al.* 2012 and references therein), with just two – several decades apart – looking at mussels, *Modiolus modiolus* (Margosian *et al.* 1987) and *Lithophaga lithophaga* (Peharda *et al.* 2015).

The Mussel Shell as a Marine Archive

Bivalves secrete shells composed of successive daily layers of calcium carbonate (see Chapter 2). The interaction between the environment and the bivalve's physiology controls most of the features of shell growth patterns and geochemical composition (Elliot *et al.* 2003). The shells of mussels and other bivalves are thus increasingly being used as potential bioarchives of proxies for physicochemical changes in their marine habitats. Marine bivalves occupy widely distributed oceanic habitats and provide a rich fossil record going back the Triassic era (200–250 mya). Research in this area, often referred to as sclerochronology,[1] is particularly relevant in the light of observed and potential climate change impacts on marine organisms (see Chapter 3). In marine mussels, isotope ($\delta^{18}O$, $\delta^{13}C$) and Mg/Ca and Sr/C ratios in shells have been used to reconstruct changes in surface water salinity (Klein *et al.* 1996; Wanamaker *et al.* 2007; Versteegh *et al.* 2012), changes in sea surface temperature (SST) (Cusack *et al.* 2008; Wanamaker *et al.* 2007, 2008; Ford *et al.* 2010; Peharda *et al.* 2015), primary production variations (Gosselin *et al.* 2013), ocean acidification (OA) events (Hahn *et al.* 2012), changes in carbon dynamics (Pfister *et al.* 2011), hydrothermal venting activity (Schöne & Giere 2005) and freshwater inputs to the ocean (McConnaughey & Gillikin 2008). Freitas *et al.* (2008), however, have shown that, for *M. edulis*, the use of shell Mg/Ca ratios as a temperature proxy is limited, because of intraspecific heterogeneity in Mg incorporation into calcite. Factors such as growth rate and metabolic activity (Klein *et al.* 1996) may also influence minor element ratios. Mussel shells can further serve as bioarchives of rare earth element and yttrium (REY) distribution in their habitat, and may have the potential to be used as pH and temperature proxies (Ponnurangam *et al.* 2016).

The following example illustrates how prismatic calcite $\delta^{18}O$ variations in *M. californianus* can be used to assess the potential of this species as a recorder of mean annual temperature (Ford *et al.* 2010). Individuals were transplanted into high and low intertidal areas of San Diego Harbour, California and left for over a year with in situ recording of ambient temperature and concentrations of $\delta^{18}O_{seawater}$ and Mg/Ca. The prismatic calcite layer of eight variously sized individuals from each intertidal position was serially microsampled and geochemically analysed for $\delta^{18}O_{calcite}$ and Mg/Ca. Observed specimen values were significantly correlated with predicted equilibrium values, but showed ^{18}O enrichments of 0.2–0.5% (Figure 6.7). Average intraspecimen $\delta^{18}O$ values significantly covaried with temperature, whereas Mg/Ca values were significantly correlated with growth rate and intertidal position but not with temperature – a finding previously reported for *M. edulis* (see earlier). The $\delta^{18}O$ data appear to provide a reliable proxy for past temperature events, thus making *M. californianus* 'a valuable source of information for paleoceanographic reconstruction' (Ford *et al.* 2010).

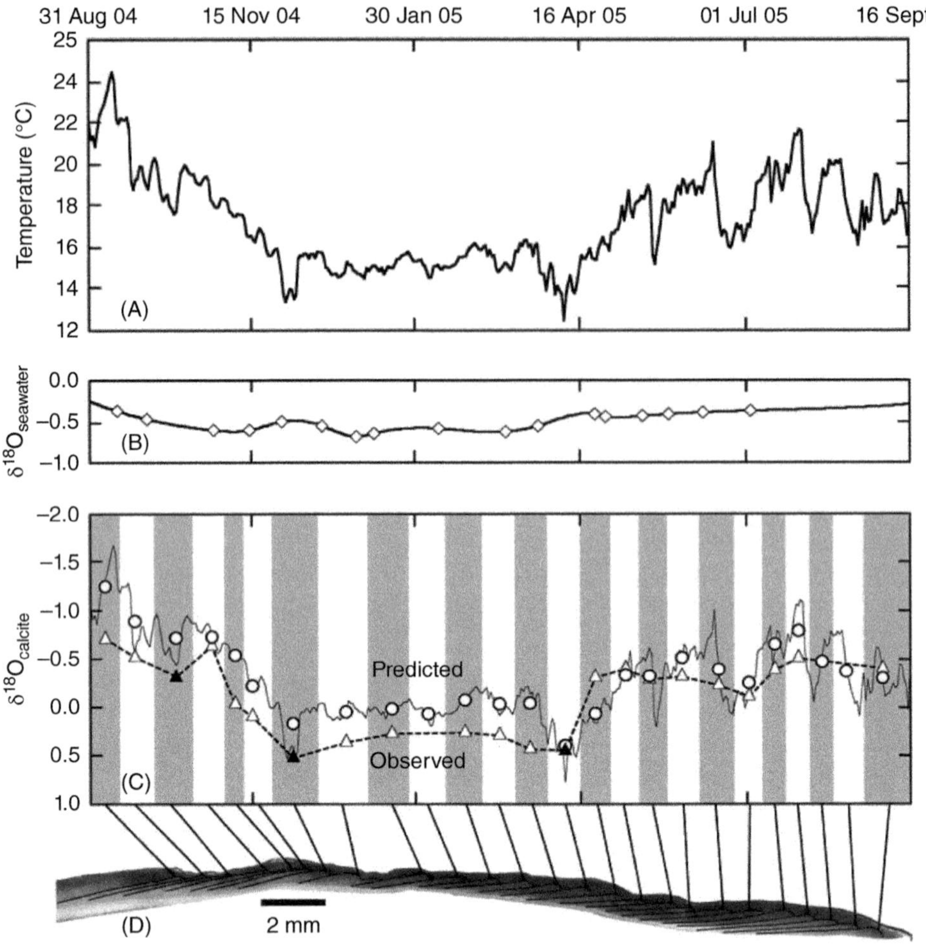

Figure 6.7 Construction of predicted $\delta^{18}O_{\text{equilibrium calcite}}$ record and comparison to observed $\delta^{18}O_{M.\ californianus}$. (A) Average daily submerged water temperature. (B) $\delta^{18}O_{\text{seawater}}$ values. (C) Predicted daily $\delta^{18}O_{\text{equilibrium calcite}}$ values (solid line) and observed $\delta^{18}O_{M.\ californianus}$ (triangles with dashed line, where solid triangles are microsamples used to align observed $\delta^{18}O_{M.\ californianus}$ with the predicted values). (D) Mosaic of thin section images for the specimen prior to the removal of the inner nacreous layer and subsequent microsampling of the prismatic layer. Overlay lines on the specimen define the area of each microsample within the prismatic layer, and each microsample is tied to its corresponding temporal bin (N = 25 microsamples). *Source:* From Ford *et al.* (2010). Reproduced with permission from John Wiley and Sons.

An exciting development in recent years is the recognition of the enormous potential of long-lived marine clams such as *Arctica islandica*, *Glycymeris glycymeris*, *Panopea generosa*, *Laternula elliptica* and *Yoldia eightsi* as high-resolution marine archives. For those wishing to explore this topic in greater depth, see Gosling (2015), Reynolds *et al.* (2016) and volume 465, 2017 of the journal *Palaeogeography, Palaeoclimatology, Palaeoecology*.

Measurement of Allometric Growth

Growth in bivalves is often accompanied by progressive changes in the relative proportions of the shell with increasing body size, referred to as relative or allometric growth. These changes are the result of differential growth vectors operating at different points around the mantle edge; this plays a key role in shell secretion (see Chapter 2). The relationship between any two size variables can be described by the allometric equation

$$y = ax^b$$

when y is one growth variable (e.g. shell length), x is the other variable (e.g. shell height) and a and b are coefficients. The exponent b is the growth coefficient and represents the relative growth rate of the two variables, while a is the value of y when x is unity. To estimate the coefficients of the allometric equation, the data are usually log-transformed and a linear regression is fitted. Logarithmic transformation is generally appropriate because morphological data tend to have log normal structure (Katsanevakis *et al.* 2007 and references therein). In logarithmic form, the equation becomes

$$\log y = \log a + b \log x$$

The slope b and intercept a of the transformed data are estimated by regression analysis. If x and y have the same units of measurement, a value of unity for b indicates that the relative growth of the two variables is identical (isometric) and that geometrical similarity is maintained with increasing size. Values of b greater than unity indicate that y is increasing relatively faster than x (positive allometry), while values of b less than unity indicate the reverse (negative allometry). Different criteria for isometry and allometry apply if the dimensions of the two variables x and y differ (Seed 1980). However, b is not necessarily constant and may change either continuously or abruptly at specific breakpoints (points of discontinuity in slope b); thus, the allometric model may not always be the best one to describe allometric growth (discussed in Katsanevakis *et al.* 2007).

Until relatively recently, most morphological analyses of mussel shells were based on shell linear morphometric characters or their ratios (McDonald *et al.* 1991; Gosling 1992; Karakousis *et al.* 1992; Gardner 1996). One disadvantage of bivariate data is that only two variables are compared at any given time, so that this type of analysis deals only with changing proportions, not with changing shape of shell. This approach was replaced by multivariate morphometrics, which applies multivariate statistical analysis to sets of morphological variables that include linear measurements as well as counts, ratios and angles; see McDonald *et al.* (1991) for an example of its use in mussel shell shape analysis. Statistical analyses typically include principal components analysis (PCA), factor analysis, canonical variates analysis (CVA) and discriminant function analysis. However, linear distance measurements are usually highly correlated with size. Consequently, a lot of effort was spent developing methods for size correction, so that size-free shape variables could be used. Problems persisted, so alternative methods of quantifying and analysing morphological shape were developed (reviewed by Adams *et al.* 2004). Such methods, which capture the geometry of morphological structure and are becoming the standard approach in shell shape analysis, are usually referred to as 'geometric morphological methods'. Examples that have been used in shell shape analysis are outline methods (e.g. Fourier analysis) and landmark methods.

In Fourier analysis, shell outlines are captured by video imagery, and the Fourier series mathematically compartmentalises the shape outline into a series of standard sinusoidal components which, when summated, converge to the empirical shape of the outline. The method, which is capable of discerning rather subtle differences among very similar, nearly oval shapes, has been used for shells of mussel (Ferson *et al.* 1985; Innes & Bates 1999; Krapivka *et al.* 2007; Gardner & Thompson 2009; Van der Molen *et al.* 2013), scallop (Kenchington & Full 1994), oyster (Rajaei *et al.* 2014) and clam (Rufino *et al.* 2006; Costa *et al.* 2008; Márquez & Van der Molen 2011) species.

Landmark-based geometric morphometric methods quantify overall shell shape from the coordinates of locations (landmarks), after the effects of nonshape variation have been mathematically held constant (Adams *et al.* 2004). Landmarks must be homologous between specimens. An important advantage of this approach is that shape information from 2D and 3D discrete landmarks – as well as points along curves and points on anatomical surfaces, referred to as semi-landmarks – can be included in the same analysis. A good example of how the technique has been used is described by Serb *et al.* (2011), who used landmark-based geometric morphometrics to quantify shell shape in seven scallop species displaying a range of behavioural habits. They used 506 3D landmarks and semi-landmarks to quantify shell shape and shell surface texture. A large number of landmarks were chosen, to ensure that the textural information of the shell surface was well quantified by the procedure. Initially, the authors obtained high-resolution surface scans of the left valve of each individual using a surface scanner. The locations of five homologous landmarks and eleven semi-landmarks along the ventral edge of the valve were digitised (Figure 6.8). Digitising 490 semi-landmarks on the surface of each scan quantified the general shell surface. To accomplish this, the authors produced a mesh of 490 relatively evenly spaced surface semi-landmarks on a single specimen and treated this as a template. They then used thin-plate splines, a mapping function for coordinate transformation (Bookstein 1991), to warp the template to a second specimen, using the fixed and edge landmarks as points of correspondence between the template and the specimen. The points on the scan nearest to the 490 points on the template were taken as the surface semi-landmarks for that specimen. This process was repeated on all specimens to obtain surface semi-landmarks for each. Once all specimens were digitised, the authors aligned them using a generalised Procrustes superimposition, whereby shape variables are constructed. From the aligned specimens, a set of Procrustes shape coordinates were obtained, which were used as shape variables in subsequent statistical analyses (details in Serb *et al.* 2011). The authors found that all species had a distinct shell shape – with the exception of the two long-distance swimmers, whose shells were indistinguishable. These two species also displayed reduced morphological variance relative to the other five.

Apart from this study, landmark-based methods have also been used to quantify shell shape in mussels (Aguirre *et al.* 2006; Valladares *et al.* 2010, Scalici *et al.* 2016; Marquez *et al.* 2018), clams (Rufino *et al.* 2006; Moneva *et al.* 2014), oysters (Rajaei *et al.* 2014) and an additional scallop species (Márquez *et al.* 2010). To illustrate how the method has been used in mussels, Marquez *et al.* (2018) investigated the variation in the shell-shape allometric trajectories of two species throughout their distributional range on the SW Atlantic

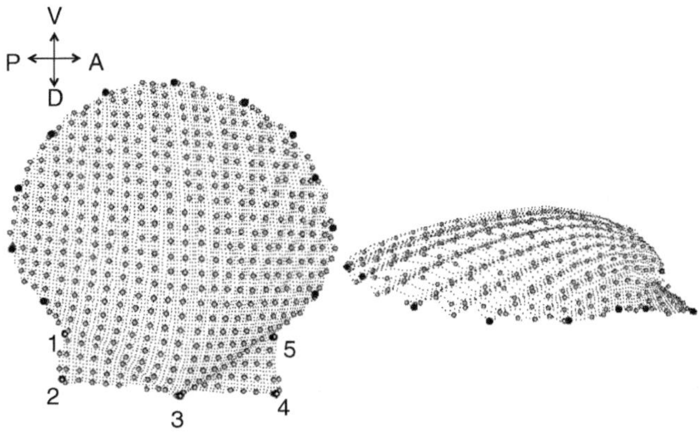

Figure 6.8 3D surface scan of the left valve of a representative scallop, with the positions of 506 landmarks and semi-landmarks indicated. Fixed landmarks are shown as numbered, open circles: (1) ventro-posterior auricle; (2) dorso-posterior auricle; (3) umbo; (4) dorso-anterior auricle; (5) ventro-anterior auricle. Semi-landmarks along the ventral edge of the valve are shown as closed circles. Surface semi-landmarks on the scallop valve are shown as grey circles. Dorso-ventral and antero-posterior axes are provided. *Source:* From Serb *et al.* (2011). Reproduced with permission from Oxford University Press.

coast, from latitude 34 °S to 53 °S in South America. Shells of *Brachidontes rodriguezii* were collected at two Uruguayan and four Argentinean localities, while shells of *Perumytilus purpuratus* were collected at six Argentinean localities. The shell shapes of the specimens were studied using landmark and semi-landmark methods (Figure 6.9). *B. rodriguezii* seems to have a long history in the region and is present in a restricted area, the warm temperate region of the SW Atlantic, while *P. purpuratus* appears to have a recent history and occupies a wide thermal range, involving the cold-temperate regions of the SW Atlantic and SE Pacific and the warm-temperate region of the SE Pacific. Based on this information, Marquez *et al.* (2018) hypothesised that allometric trajectories of *P. purpuratus* shell shape would be more variable and adapted to a more specific habitat than that of *B. rodriguezii*. Graphical evidence showed that the two species differed in allometric growth forms, and allometric shell morphology changes reflected environmental constraints and adaptation. While the intertidal mussel beds of the South American coasts are ecologically similar in appearance, the authors found that the allometric variation of the two species occurred in response to diverging evolutionary processes, such as phenotypic plasticity in *P. purpuratus* and canalisation in *B. rodriguezii*. Canalisation is a measure of the ability of a population to produce the same phenotype regardless of variability in its environment or genotype; it is a form of evolutionary robustness (Willmore *et al.* 2007).

Allometric changes in the shell may simply be associated with the maintenance of physiologically favourable surface area to volume ratios as body size increases (Seed & Suchanek 1992). For example, in the mussel *M. edulis*, differential growth among various shell parameters (height, length, width, weight) is reflected in a gradual change in shell shape, and larger (older) mussels have relatively heavier and more elongate shells, in which width frequently exceeds shell height (Seed 1968). This may be particularly beneficial to

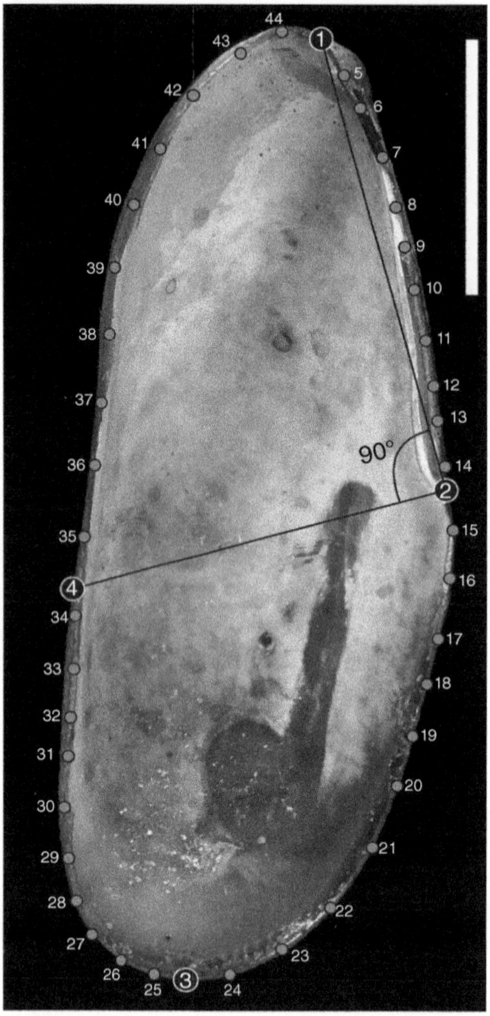

Figure 6.9 Landmark and semi-landmark configurations on the outline of the shell of *Brachidontes rodriguezii*. Landmarks (filled in black circles) are: (1) umbo (anterior end); (2) tip of the posterior hinge ligament; (3) maximum posterior curvature; (4) projection at 90° of the vector formed between landmarks 1 and 2 on the outline of the shell. Semi-landmarks (grey dots) are: (5–14) along the boundary between landmarks 1 and 2; (15–24) between landmarks 2 and 3; (25–34); between landmarks 3 and 4; and (35–44) between landmarks 4 and 1. Scale bar = 1 cm. *Source:* From Marquez *et al.* (2018). Reproduced with permission from Springer Nature.

mussels living in densely crowded conditions, as a wedge-shaped profile effectively elevates the posterior current flow among conspecifics (Yonge & Campbell 1968). Also, highly significant differences in shell shape components were found between wild and cultured mussels (*M. chilensis*) from southern Chile (Valladares *et al.* 2010). There may be a simple ecophysiological explanation. For example, wild populations, facing higher predator pressure and other environmental stress factors, may allocate a higher energy fraction to reinforce shells and promote adductor muscle growth at the expense of somatic growth. In contrast, individuals inhabiting calm aquaculture environments are relatively protected from predators and use most of their assimilated energy in somatic growth. In turn, this growth depends on changes that covary with shell morphological change.

Allometric changes in the shell may reflect adaptive responses to a variety of local changing environmental conditions. In mussels, such factors can include food concentration (Alunno-Bruscia *et al.* 2001), tidal flow and waves (Moschino *et al.* 2015), latitudinal clinal variation (Katsanevakis *et al.* 2007), population density (i.e. crowding, already mentioned; Bonel *et al.* 2013), predation (Valladares *et al.* 2010), substratum structure (Devescovi 2009) and sediment type (Seed 1980). The allometric equation is also employed in physiological investigations and to obtain estimates of seasonal variation in growth or productivity (see later).

Scope for Growth and Dynamic Energy Budget Models

Scope for Growth

Determination of the energy available for growth, based on the physiological analysis of the energy budget rather than on direct measurement of growth itself, has proved to be particularly useful in the detection of responses to environmental change prior to measurable effects on the growth, reproduction or survival of the individual (Dame 2012). Physiological energetics is concerned with the study of energy balance within individuals, in terms not only of the acquisition and expenditure of energy, but also of the efficiency with which it is converted from one form to another (Bayne & Newell 1983). The most common way to assess energy balance in an individual is to measure the various components of the balanced energy equation:

$$C = P + R + F + U$$

where C is total consumption of food energy, P is total production (shell, soft tissue and gametes), R is the respiratory energy expenditure, F is faecal energy loss and U is the loss of the end products of nitrogen metabolism. The absorbed ration (A) is the actual amount of material digested and is represented by $C - F$. The efficiency with which the ingested ration (C) is absorbed, A/C, is called the absorption efficiency.

By rearranging the energy balance equation, we get

$$P = A - (R + E)$$

where excretion $E = F + U$. Production is therefore the difference between energy gains from the absorbed ration (A) and metabolic losses through respiration and excretion ($R + E$). An organism can only allocate energy to growth or reproduction if the energy gain from A exceeds total metabolic losses. This energy, surplus to metabolic demands, is referred to as 'scope for growth' (SFG; Warren & Davis 1967). Many laboratory and field studies show that energy lost via excreta is a small proportion ($<5\%$) of the total energy budget (Widdows 1993). Therefore, for routine monitoring, the energy balance equation may be further simplified to

$$SFG = A - R$$

SFG provides an instantaneous measure of the energy status of an animal, which can range from maximum positive values under optimal conditions to negative values when the animal is severely stressed and utilising body reserves (Widdows & Donkin 1992). SFG was primarily developed for use with suspension-feeding mussels, in particular *Mytilus* spp., and in combination with the analysis of chemical contaminants. It can be measured either in the field using a mobile laboratory or in the laboratory under standardised conditions. Thus, the effect on SFG of parameters such as temperature, salinity, pollutants, food type and ration can be quantified. No significant differences have been found between laboratory and field measurements (Widdows *et al.* 1995). The preferred season for measuring SFG in mussels is the period of maximum growth potential (early summer to early autumn). Reproductive condition can also influence SFG, as respiration and excretion rates are higher when gonads are fully developed (Dame 2012), so sampling during the main spawning season should be avoided. Also, factors such as parasitic infection and contaminant exposure lower SFG estimates. Widdows & Staff (2006) provide details on: collection, preparation, maintenance and feeding of mussels before physiological measurements; standardisation of laboratory conditions; collection of faecal material; and measurement of respiration rate, clearance rate and food absorption efficiency.

Calculation of SFG

The energy consumed or ingested is calculated as follows: C = [maximum clearance rate: $l\ g^{-1}\ h^{-1}$] × [mg POM l^{-1}] × [23 J mg^{-1} POM], where the energy content of particulate organic matter (POM) or algal food is 23 J mg^{-1} ash-free POM (Widdows & Staff 2006). Absorption efficiency (AE) is measured by comparing the proportion of the organic matter in the food and faeces of animals kept in individual experimental chambers (Conover 1966). The energy absorbed (A) is then estimated from (C) × (AE). Oxygen consumption (R) is measured using an oxygen electrode to track the decline in oxygen partial pressure for individual animals maintained in closed glass respirometers held in a temperature-controlled water bath. The energy respired is expressed as: $R = (\mu mol\ O_2\ g^{-1}\ h^{-1}) \times 0.456$, where the heat equivalent of oxygen uptake is 0.456 J $\mu mol^{-1}\ O_2$. All physiological rates are converted to mass-specific rates for a standard individual of 1 g dry weight using the appropriate weight exponent b; in mussels, b has a value of 0.67 (Widdows & Staff 2006). Since all measured physiological responses have been converted into energy equivalents, J g^{-1} h^{-1}, SFG (P) is expressed in these terms. Generally, there is good agreement between growth rates estimated from energy budgets and those measured by more direct methods (references in Bayne 1998). The methods that are used for estimating production in individual animals can equally well be applied to whole populations, or even to communities (Asmus 1987). Data on populations are normally derived by integrating field measurements of population density, age structure and production with laboratory estimates of C, A, R and E of different-sized individuals. The results are generally expressed in terms of ash-free dry weight (AFDW) or energy flux per unit area of habitat, kJ $m^{-2}\ yr^{-1}$. Production, expressed as a percentage of consumption, along with P_g and P_s (gonad and somatic growth, respectively) values for several mussel species, is given in Table 6.1. Values for production vary from as low of 1.5% in *Mytilus edulis* to 15% in *Geukensia demissa*. Such variation reflects interspecific growth differences (somatic and reproductive) and differences in recruitment, immigration, mortality and emigration. Within-species variation in P_g and P_s estimates largely reflects differences in age-structure. This can be seen when individuals of the same species are analysed from the same or different locations (Table 6.1). At both Lynher and Cattewater in southern England, for example, P_s estimates were lowest and P_g values were highest in older mussels (Bayne & Worrall 1980). This is not

Table 6.1 Energy budget parameters for several mussel populations. Somatic growth (*Ps*), reproductive output (*Pg*) and total production (*P*) are each expressed as a percentage of consumption (*C*; kJ m^{-2} yr^{-1}). Dashes indicate that no measurement was made. The enormous variation in *C* mainly reflects differences in population density. L, Lynher, southern England; C, Cattewater, southern England. Figures in brackets indicate number of years of settlement.

Species	C	P_s	P_g	P	Reference
Aulacomya atra	27 499	3.4	5.8	9.2	Griffiths & King (1979)
Choromytilus meridionalis	831 890	1.6	4.8	6.4	Griffiths (1981)
Geukensia demissa	468	12.4	2.5	14.9	Kuenzler (1961)
Mytilus edulis L (3–4)	–	4.30	0.22	4.52	Bayne & Worrall (1980)
L (5–6)	–	5.01	1.75	6.76	
L (7–8)	–	2.21	3.33	5.54	
Mytilus edulis C (3–4)	–	3.67	0.05	3.72	
C (5–6)	–	2.87	0.24	3.11	
C (7–8)	–	1.09	0.39	1.48	

unexpected, as somatic growth is fastest in small mussels (Figure 6.2), and reproductive output (i.e. P_g) is positively correlated with shell length (see Chapter 5). However, overall production (*P*) was consistently higher for all year groups at the Lynher site. Another study, in Killary Harbour, a fjord on the Irish west coast, showed that production of cultured mussels exceeded production of the wild population on the shore by an order of magnitude (Rodhouse *et al.* 1985). However, the wild population, which is dominated by large, older mussels (40–60 mm shell length) with a high reproductive output, makes a major contribution to total gamete production in the inlet.

Dynamic Energy Budget

The dynamic energy budget (DEB) bioenergetics model is another model that has received considerable attention in recent years. A very comprehensive account of the major differences between the SFG and DEB models is provided by Filgueira *et al.* (2011), but can be summarised as follows. The main difference is that in the SFG model, maintenance costs are subtracted from the assimilated products before they are allocated to reserves and somatic tissue, while in the DEB model, assimilated energy is first stored in reserves, which are then utilised to fuel other metabolic processes. Thus, reserves reflect the feeding history of the organisms, and consequently the structural growth dynamics in a DEB model become different from those in the SFG model, particularly in situations with temporal fluctuations in energy supply. Another important assumption of the DEB model is the κ-rule, which implies that a fixed proportion κ of the available energy is allocated to somatic maintenance and growth, with priority for maintenance, while the remaining 1 − κ is allocated to maturation and reproduction/maturity maintenance for juveniles and adults, respectively (Figure 6.10). So, SFG models are forced by environmental conditions such as temperature, food

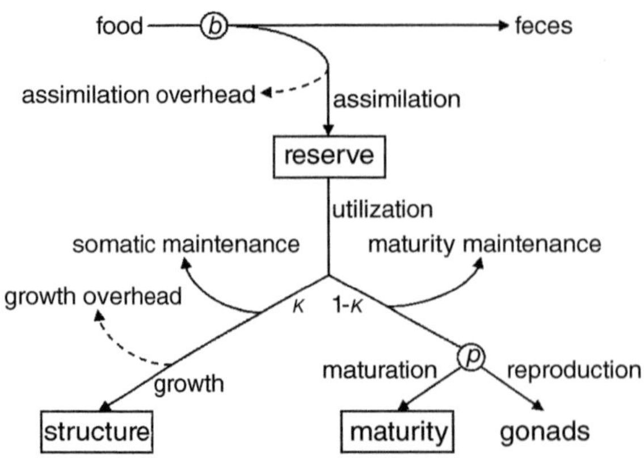

Figure 6.10 Schematic representation of energy flows in the standard dynamic energy budget (DEB) model (the κ-rule model of Kooijman 2010). Commonly tracked state variables are denoted by rectangles. The model assumes that an organism ingests food at a rate dependent on its size and the food density. Energy is extracted from the food and added to the reserves. The rate at which energy becomes available to the organism depends on its size and stored energy density. Provided somatic maintenance requirements are met, a fixed proportion κ of the available energy is allocated to somatic maintenance and growth combined, and the remaining 1 − κ to either maturation (for embryos and juveniles) or reproduction and maturity maintenance (for adults). Growth ceases when this fixed fraction κ just meets somatic maintenance demands. Then, the organism may still reproduce provided that the energy made available exceeds the requirements for somatic and maturity maintenance. Note that nodes *b* and *p* indicate metabolic switches at onset of larval feeding, and at onset of reproduction in mature individuals. *Source:* Based on Jusup *et al.* (2017).

concentration and quality, without taking into account the internal state of the organism, while DEB models describe the rates of energy absorption and utilisation as a function of the organism and the environment (see Duarte *et al.* 2010 and references therein). Although the conceptual basis is different in the two models, each has been used successfully to simulate growth in a variety of bivalve species, including mussels (Larsen *et al.* 2014 and references therein; Filgueria *et al.* 2011; Barillé *et al.* 2011). For additional information on DEB theory, see Kooijman (2010, 2012, 2013), Sousa *et al.* (2008, 2010), Sebens *et al.* (2016) and volume 94, 2014 of the *Journal of Sea Research*.

The following is an example of how the DEB model has been utilised in mussels. While the physiology of *Mytilus californianus* has been reasonably well studied, DEB parameters until recently had not been estimated for this species – a dominant mussel on rocky intertidal shores on the west coast of North America. Matzelle *et al.* (2014) estimated a set of DEB parameters and compared them to parameter values from other bivalve species (Tables 6.2 and 6.3). These estimates were meant to serve as an initial exploration of how this ecologically important species may differ in its response to changing environmental conditions from its better-studied congeners. *M. californianus* exhibited a higher assimilation efficiency κX than *M. edulis*, increasing the fraction of energy in food that is fixed in reserve by 26% (0.72 vs. 0.53 in *M. californianus* and *M. edulis*, respectively), a likely energetic advantage for *M. californianus* under low-food conditions. The somatic

Table 6.2 Dynamic energy budget (DEB) parameters obtained for *Mytilus californianus* at a reference temperature of 20 °C. Maximum size differs between species, so a dimensionless zoom factor *z* is applied and used for comparison of the physical parameters of an organism to a 'reference organism' of maximum length 1 cm. Surface area-specific assimilation rate and energy conductance increases proportionately with length (Kooijman 2010). The relationship between volume and length, which varies significantly between species of different morphologies, is captured using the shape coefficient.

Parameter description	Symbol	Unit
Assimilation efficiency	κ_X	–
Surface area-specific assimilation rate	p_{Am}	J cm^{-2} d^{-1}
Energy conductance	v	cm^{-1}
Allocation fraction to soma	κ	–
Reproduction efficiency	κ_R	–
Volume-specific somatic maintenance rate	$[p_M]$	J cm^{-3} d^{-1}
Specific cost for structure	$[E_G]$	J cm^{-3}
Zoom factor	z	–
Shape coefficient	δ_M	–

Sources: Adapted from Matzelle *et al.* (2014). Notation after Kooijman (2010).

maintenance cost [*pM*] for *M. californianus* fell in the lower range across all bivalve species. The value [*EG*], the amount of reserve energy for structure, was lower in *M. californianus* (3140 J cm^3) than in *M. edulis* (4783 J cm^3), but these species had the highest values in comparison with the other bivalve species in Table 6.3. The κ value for *M. californianus* fell within the high range of the rest of the other species. These results are consistent with the comparatively long life span of *M. californianus*, in that more energy appears to be devoted to somatic maintenance than to reproduction (1 – κ). Overall, the results suggest that, based on the parameter set obtained, *M. californianus* has favourable energetic strategies enabling it to contend with a range of environmental conditions, including prolonged periods of starvation.

DEB models are being increasingly used to simulate physiological responses of cultured mussels reared under varying conditions of food quantity and quality (Cardoso *et al.* 2006; Duarte *et al.* 2010, 2012; Handå *et al.* 2011; Ren 2009; Rosland *et al.* 2009; Troost *et al.* 2010; Thomas *et al.* 2011), temperature (Ren & Ross 2005; Thomas *et al.* 2011), salinity (Maar *et al.* 2015), density (Rosland *et al.* 2011) and contaminants (Muller *et al.* 2014). They have also been used to estimate the carrying capacity of coastal ecosystems for aquaculture (Grant *et al.* 1993; Beadman *et al.* 2002; Guyondet *et al.* 2010). More detailed and complex models are also used, such as ecosystem models for food depletion (Dowd 1997; Duarte *et al.* 2008; Grant *et al.* 2008; Filgueira & Grant 2009; Ren *et al.* 2010) and 2D and 3D biogeochemical models (Duarte *et al.* 2003; Maar *et al.* 2009; Guyondet *et al.* 2010; Dabrowski *et al.* 2013). Some authors have coupled DEB models with biogeochemical models to provide feedback from mussel farms to phytoplankton and detritus (Maar *et al.* 2009 and references therein). Also, the same coupling has proven to be an effective means for

Table 6.3 Dynamic energy budget (DEB) parameters of bivalve species at reference temperature 20 °C.

Species	z	δ_M	κ_X	p_{Am}	v	κ	κ_R	$[p_M]$	$[E_G]$
Mya arenaria	3.06	0.27	0.80	52.50	0.01	0.896	0.95	15.37	3069
Brachidontes pharaonis	0.04	0.31	0.75	16.40	0.02	0.998	0.95	14.24	2478
Mytilaster minimus	0.02	0.43	0.80	0.61	0.02	0.999	0.95	39.61	2420
Mytilus californianus	1.72	0.24	0.72	26.42	0.01	0.989	0.92	15.15	3140
Mytilus edulis	4.69	0.29	0.53	79.70	0.07	0.809	0.95	13.74	4783
Perna viridis	0.88	0.16	0.70	26.79	0.01	0.990	0.95	30.19	3123
Crassostrea gigas	1.40	0.12	0.33	61.04	0.01	0.290	0.95	12.69	2674
Pecten maximus	2.01	0.38	0.80	72.08	0.02	0.812	0.95	29.05	2976
Cerastoderma edule	0.67	0.35	0.80	11.02	0.02	0.980	0.95	16.13	2527
Ensis leei (directus)	0.55	0.16	0.80	39.27	0.01	0.991	0.95	70.51	2491
Macoma balthica	0.30	0.29	0.80	21.36	0.01	0.990	0.95	69.48	2279
Ruditapes philippinarum	0.09	0.35	0.70	2.57	0.01	0.987	0.95	28.24	3050

Source: Based on Montalto *et al.* (2014).

predicting spatial and temporal patterns in reproduction and growth under changing environmental conditions (Kearney *et al.* 2010, 2013; Sarà *et al.* 2011). Others have coupled the DEB model with an ecosystem model that was developed to estimate the potential production of shellfish culture and its effect on the pelagic ecosystem in sheltered coastal waters (Ren *et al.* 2010 and references therein). Methods using the relationship between a biological response (e.g. growth or abundance) and explanatory environmental variables are being used to fill gaps in information. This has led to an increased use of predictive models that forecast the distribution and abundance of species in aquatic environments (Bergström *et al.* 2013, 2015; Larsen & Riisgård 2016). Larsen & Riisgård (2016) developed a predictive model that can predict the growth time of mussels on farm ropes, from newly settled to mini-mussels, under specified chl *a* and temperature conditions and when intraspecific competition does not occur (i.e. on optimally thinned farm ropes). Ultimately, information from the preceding model-based studies is used to evaluate the suitability of coastal ecosystems for bivalve culture.

Growth Modulators

There are many factors that influence growth in bivalves. Of these, food supply is considered to be the most important, since without this, sustained growth is impossible (Seed & Suchanek 1992). But food supply to bivalves is influenced by factors such as temperature, aerial exposure, water depth and population density. Also, some modulators of growth such as temperature and salinity interact, often synergistically. It is for these reasons that it is usually very difficult to quantify the precise influence of a single environmental factor on growth in natural populations of bivalves (see Bergström *et al.* 2016). In addition, endogenous influences that are inherent to the organism (e.g. genotype and physiological status) interact in a complex way with environmental factors. Although the bulk of published information deals with the effect of environmental factors on absolute growth, where possible their effect on SFG will also be dealt with in the following sections.

Exogenous Growth Modulators

Food

In the wild, bivalves derive nutrients from a variety of particulate organic material (POM), including phytoplankton, resuspended benthic microalgae, detritus from a mixture of sources, bacteria and micro-heterotrophs (Pernet *et al.* 2012; details in Chapter 4). The term 'seston' includes all of the preceding, as well as inorganic material. Results from several studies have highlighted the importance of seasonal development of phytoplankton as an available food source during the critical periods of gamete production and growth for both suspension- and deposit-feeding bivalves (Hilbish & Zimmerman 1988; Boon *et al.* 1998; Wong & Cheung 2003; Kang *et al.* 2006).

Higher abundances and growths of filter feeders have been reported at sites with consistently higher levels of upwelling and phytoplankton (Menge *et al.* 1997). Very often, however, enhanced productivity from upwelling develops well offshore, rarely becoming available to coastal benthic organisms (Blanchette *et al.* 2007 and references therein).

Controlled upwelling of nutrient-rich deep water in oligotrophic waters (i.e. low-nutrient coastal waters) has been proposed as a means of increasing phytoplankton. This was tested as part of a large-scale upwelling experiment in an oligotrophic fjord on the west coast of Norway (Strohmeier *et al.* 2015). The mean chlorophyll *a* concentration in the upwelling area was consistently higher than at the control site (mean ± SD: 3.3 ± 1.9 and 1.5 ± 0.6 mg chl *a* m^{-3}, respectively) during the four months of controlled upwelling. After two months with upwelling, the dry flesh weight of one- and two-year-old mussels (*M. edulis*) was 95 and 24% higher, respectively, than that of the mussels at the control site (Figure 6.11). Reproductive output was also higher at the upwelling site. This study illustrates how controlled upwelling is a simple and effective method of increasing phytoplankton concentrations, leading to an increase in mussel tissue growth and an enhancement of human food production.

But phytoplankton quality is often more important than overall biomass, since bivalves preferentially select phytoplankton for ingestion based on size, morphology and nutritional content (Safi & Gibbs 2003; Pernet *et al.* 2012; Bracken *et al.* 2012; see Chapter 4). There is

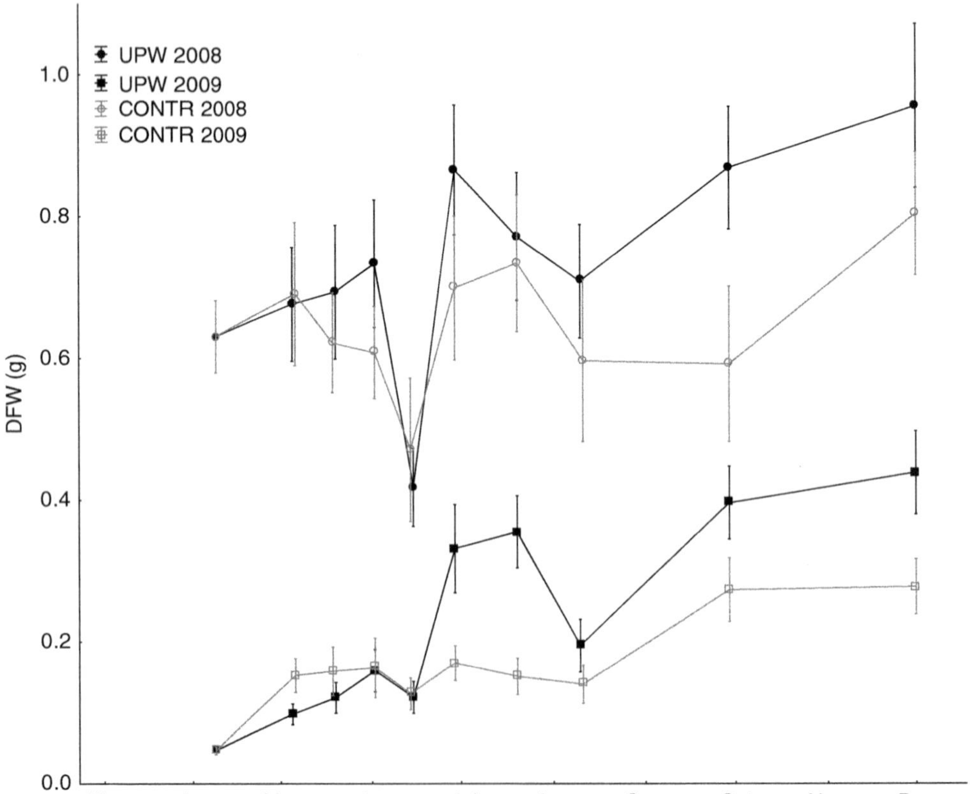

Figure 6.11 Temporal changes from March 2010 to December 2010 in the mean dry flesh weight (DFW) of two *Mytilus edulis* cohorts (2008 and 2009) held at 7 m depth at the upwelling (UPW) and control (CONTR) sites in Lysefjord, on the south coast of Norway. Error bars indicate ±95% confidence limits. *Source:* From Strohmeier *et al.* (2015). CC BY 4.0.

growing evidence, however, to show that growth may in fact be phytoplankton-limited, owing to the enormous filtration capacity of bivalves (Frechette *et al.* 1989; Page & Ricard 1900; Dolmer 2000; Allen & Williams 2003; Petersen *et al.* 2008, 2013; Nielsen *et al.* 2016). Consequently, in order to meet their energy requirements, bivalve populations must exploit nonphytoplanktonic carbon in the form of seston. A study on *M. edulis* reported an increase in growth at low or moderate seston concentrations of 1–25 mg l^{-1}, with a general decrease at high seston concentrations of 40–50 mg l^{-1} (see Fréchette & Grant 1991 for references). Similar results have been obtained from SFG experiments. When SFG in mussels (*Aulacomya ater*) fed on kelp detritus was compared with that from mussels fed on algae (Griffiths & King 1979; Stuart 1982), absorption was found to differ markedly. Moreover, mussels continued to maintain positive and increasing SFG at high ration levels of detrital material (Figure 6.12). This was even more obvious when different size classes of mussels were compared. On both diets, small mussels exhibited faster relative growth rates, largely because they were metabolically more efficient, consistently filtering larger volumes per unit of oxygen consumed than larger mussels (Griffith & Griffiths 1987). Response to variation in the concentration of suspended particles was also species-specific.

Another food source for suspension-feeding bivalves is through nutrient enrichment of coastal waters, which increases algal productivity, stimulating accumulation of organic

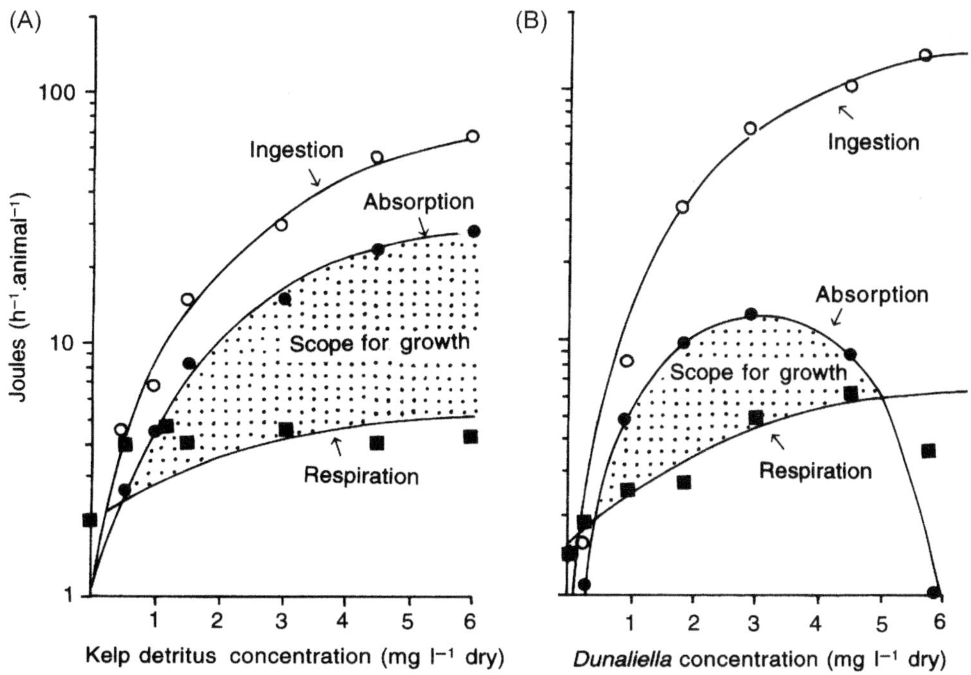

Figure 6.12 Ingested ration, absorbed ration, respiration rate and resultant scope for growth (SFG) in the mussel *Aulacomya atra* (50 mm shell length) as a function of increasing concentration of (A) kelp detritus and (B) green alga, *Dunaliella primolecta. Source:* From Stuart (1982). Reproduced with permission from Elsevier.

matter from detritus of algae and increasing the nitrogen (N) content of particles in the water and sediment. Increased primary production usually means increased food quantity and quality, with corresponding increases in growth and abundance in nutrient-enriched estuaries (Carmichael *et al.* 2012 and references therein). Indeed, mussel mitigation farming has been proposed as an eco-engineering approach for removing nutrients from eutrophic marine environments and improving water quality (discussed in Nielsen *et al.* 2016).

For additional information on the role of food in mussel growth, see Lesser *et al.* (1994), Sarà *et al.* (1998), Celik *et al.* (2009), Karayucel *et al.* (2010), Riisgård *et al.* (2013), Soon & Ransangan (2014) and Bergström & Lindegarth (2016). Diets and their impact on growth of larval and juvenile mussels are dealt with in Chapter 10.

The energetic basis of inter-individual differences in growth rate has been analysed in bivalves, mostly oysters and clams. A recent study tested the hypothesis that the physiological components acting on size differentiation of fast- versus slow-growing juvenile *M. galloprovincialis* differ as a function of nutritional conditions during the selective phase (Tamayo *et al.* 2016). For this purpose, juvenile mussels were collected from a natural population and maintained in the laboratory until size differentiation occurred under two different experimental conditions: low and intermittent phytoplankton supply involving restrictive feeding conditions, and high and continuous phytoplankton ration involving optimal feeding conditions. In the two experiments, the individuals with the lowest and highest growth rates were selected and the physiological components of their energy balances were determined at different food rations. Analysis of physiological traits indicated that under optimal food supply conditions, the capacity to ingest and absorb food plus the associated costs of growth were the main factors underlying growth rate differences. Under restrictive food conditions, the set of physiological differences changed with higher rates of absorption coupled with reduced rates of metabolism, accounting for the faster growth. The authors concluded that size differentiation of juvenile mussels in the natural environment results from a balance between opposing forces, including: (1) periods of good trophic conditions characterised by high POM concentrations and long immersion times, which would maximise the growth rates of individuals well equipped to acquire food and process it at low costs, in spite of their relatively high resting metabolic rates; and (2) periods of restricted trophic conditions of low POM, low organic content of food and/or prolonged emersion times, which would have a more adverse impact on those individuals unable to downregulate their standard metabolic rate. These results will no doubt contribute toward the ongoing controversy concerning the physiological basis of individual differences in the growth rate of bivalves (see later).

Temperature

Water temperature varies with latitude, and there has been a general consensus in the literature that bivalves from low latitudes grow more rapidly at ambient temperature than do conspecifics from higher latitudes. Growth is also seasonal, with higher growth rates during the warmer months of the year (Stirling & Okumus 1995; Peharda *et al.* 2007; Bagur *et al.* 2013). Phillips (2005) examined growth rates in *M. californianus* in both spring and summer at seven locations on the southern coast of California, United States (Figure 6.13). South of Point Conception, upwelling is typically weak, while north of there, it is often

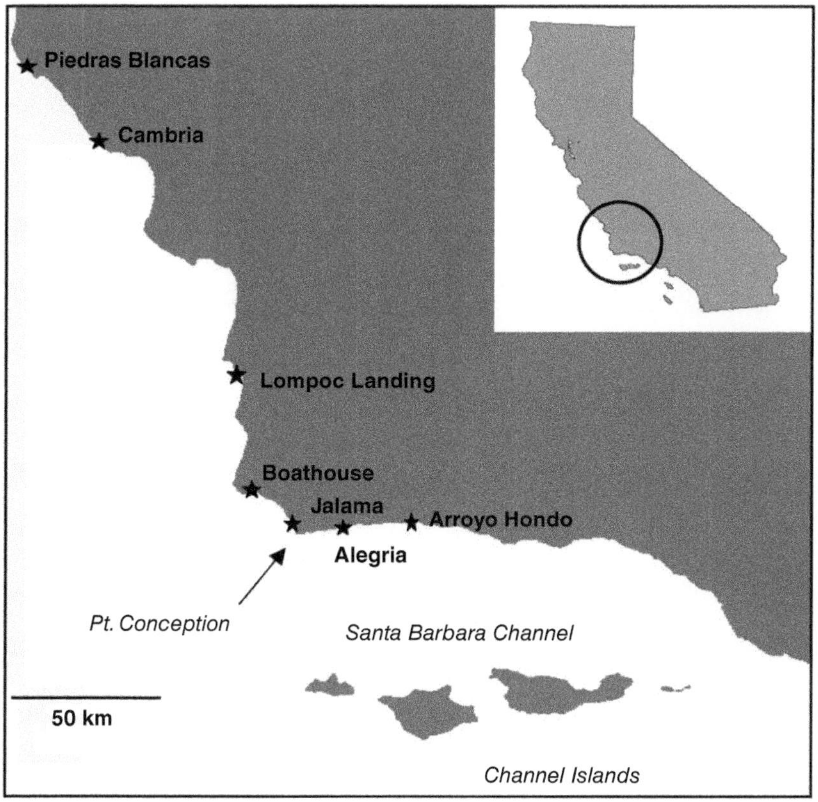

Figure 6.13 Sites in southern and central California, United States, where mussels, *Mytilus californianus*, were outplanted to bare rock in the intertidal zone for spring and summer growth studies. *Source:* From Phillips (2005). CC BY 4.0.

strong and frequent. Unexpectedly, the results showed that growth and body condition were three to six times higher at the sites furthest south of Point Conception and decreased dramatically moving north around the point (Figure 6.14). Spatial variation in food availability and food quality did not explain these patterns. Tidal height of the organisms also failed to account for the persistent spatial variation in growth rates. However, water temperatures (warmer in the south) did vary in a pattern consistent with growth rate. The author suggested that factors other than – or perhaps in conjunction with – plankton productivity were responsible for the large-scale differences in growth rates exhibited by these mussels. See Smith *et al.* (2009) for a related study on the same species.

The closely related species *M. edulis*, *M. trossulus* and *M. galloprovincialis* have distinct biogeographical ranges that are correlated with SST. Fly & Hilbish (2013) determined the SFG of these three species over a range of temperatures (5–30 °C) to determine if energetics could predict their distributions. As already mentioned, SFG represents energy available for growth and/or reproduction above that necessary for maintenance requirements. Results showed, as expected, reasonable correlation with the temperature regimes of habitats presently occupied, with *M. trossulus*, a boreal species, exhibiting positive SFG up

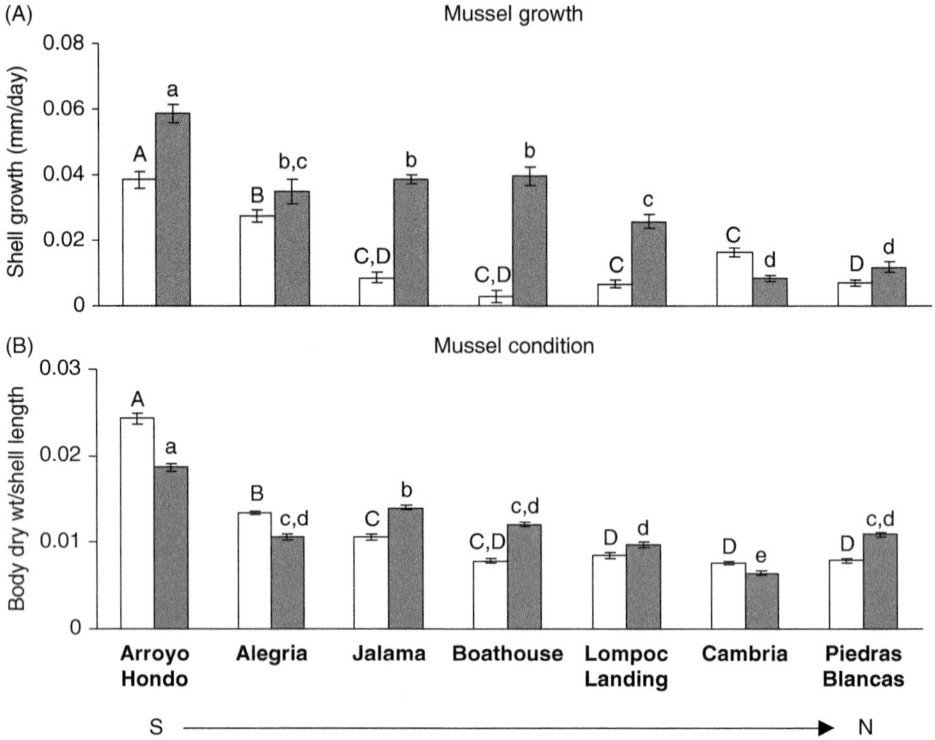

Figure 6.14 *Mytilus californianus* (A) growth and (B) condition mean ±1 SE among sites. Different letters above bars indicate significantly different means (post hoc Tukey tests; $p < 0.05$). Open bars, capital letters: spring; filled bars, lowercase letters: summer. *Source:* From Phillips (2005). CC BY 4.0.

to 17 °C, *M. edulis*, a cold-temperate species, up to 23 °C, and *M. galloprovincialis*, a warm-water species, up to 30 °C (Figure 6.15). In the spring, temperatures of peak performance (T_{opt}) for both *M. trossulus* and *M. edulis* occurred at ~15 °C, and both species exhibited negative SFG at 25 °C, which was the T_{opt} for *M. galloprovincialis* (Figure 16.15a). Overall, SFG was lower in summer than during the other two seasons. *M. trossulus* only showed positive SFG at 10 °C, while *M. edulis* and *M. galloprovincialis* maintained positive SFG at all temperatures above 5 °C, with temperatures of peak performance at 20 and 25 °C, respectively (Figure 6.15b). In the autumn (fall), all three species showed the same relationship between temperature and SFG. The T_{opt} for all species occurred at 10 °C (Figure 6.15c), and just as in summer, *M. trossulus* SFG was negative at 15 °C and higher. T_{opt} for *M. galloprovincialis* in spring and summer occurred at 25 °C, but in the autumn its SFG was very low at this temperature, while that for *M. edulis* was negative (Figure 6.15c). The warm end of each species' range correlated positively with the seawater temperature at which that species' SFG became negative (i.e. when metabolic expenditure exceeds energy acquisition) in summer and autumn. Energetics at cold temperatures did not predict the cold end of the species' ranges as there was no clear SFG advantage to explain the dominance of *M. trossulus* in cold habitats. With rising SST due to climate change, the results of

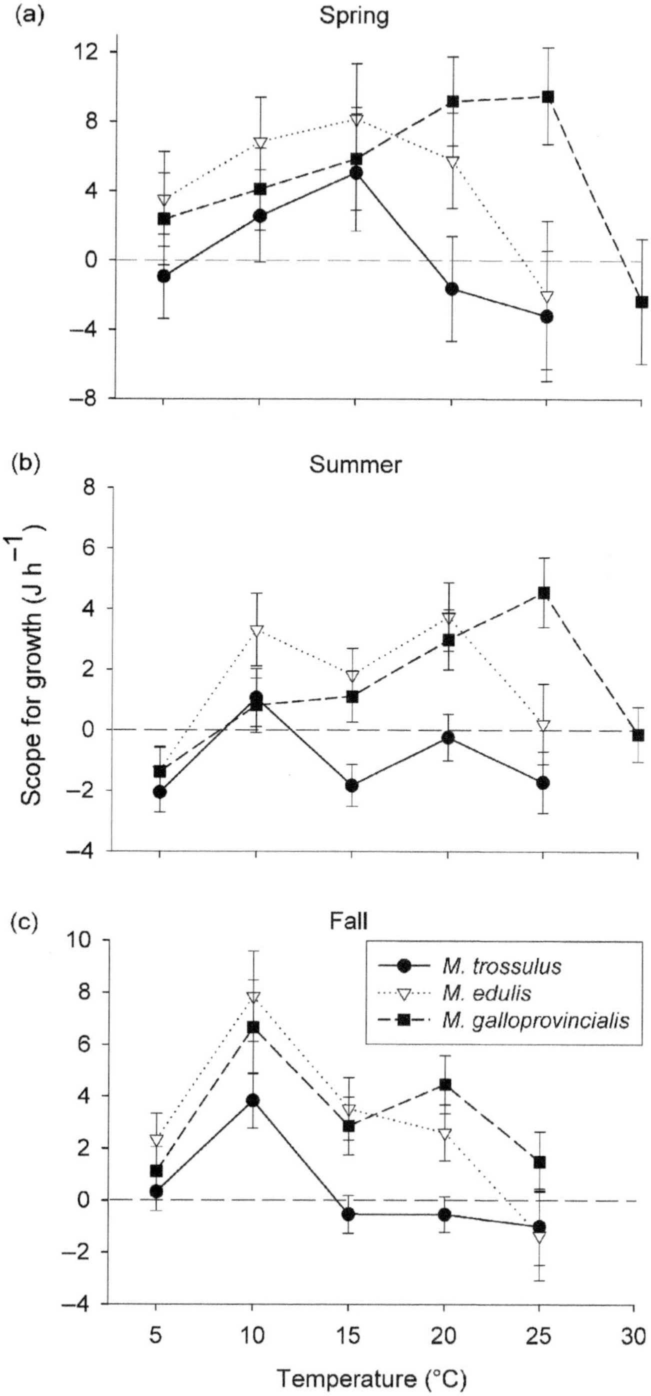

Figure 6.15 Scope for growth (SFG) (mean ± SE) of *Mytilus trossulus* (black circle, solid line), *M. edulis* (white triangle, dotted line) and *M. galloprovincialis* (black square, dashed line) during (a) spring, (b) summer and (c) autumn (fall). SFG values adjusted by ANCOVA to mean dry weights of 0.247 g in the spring, 0.191 g in the summer and 0.467 g in the autumn. Dashed grey line indicates the zero SFG line. *Source:* From Fly & Hilbish (2013). Reproduced with permission from Springer Nature.

this study could provide a basis for developing mechanistic models to predict future distribution and productivity changes in mussel populations (Fly & Hilbish 2013).

Several studies have evaluated the combined effects of food, temperature and other factors on growth in mussels. For example, Menge *et al.* (2008) examined growth responses in *M. californianus* to intertidal sea and air temperatures, phytoplankton, sea level and tidal height. Together, water temperature (32%) and food (12.5%) explained 45% of the variance in mussel growth. Pechenik *et al.* (1990) examined the combined influence of food concentration and temperature on rates of larval growth was examined in *M. edulis*. Temperature of rearing was either 12 or 16 °C, while algal concentrations ranged between 0.5×10^4 and 30×10^4 cells ml^{-1}. Larvae grew as quickly as 8 μm day^{-1}, with the highest rates recorded at 16°C and 15–30 × 10^4 cells ml^{-1}. Others have reported comparable mean growth rates (references in Pechenik *et al.* 1990).

Several studies (see later) have examined the combined effects of temperature and salinity (Resgalla *et al.* 2007; Sarà *et al.* 2008), heavy metals (Mubiana & Blust 2007) and pCO_2 (Hiebenthal *et al.* 2012; Kroeker *et al.* 2014; Wang *et al.* 2015) on mussel growth.

Salinity

Bivalves respond to changes in external salinity by closing the shell valves and adjusting the intracellular concentrations of ions, amino acids and other small molecules to maintain a relatively constant cell volume. Initially, rates of feeding and respiration are depressed, but they gradually recover as osmotic equilibration is reached. The period required for complete acclimation depends on the extent of the initial salinity change (see Chapter 7).

Bivalves grow well in brackish estuaries, but this is probably related to increased food levels in this habitat rather than to reduced salinity per se (Seed & Suchanek 1992). Mussels are able to adjust growth at changing salinities. To illustrate, Riisgård *et al.* (2012) carried out three series of laboratory experiments: in experiment 1, groups of mussels (*M. edulis*) were exposed to 10, 15, 25 and 30 psu; in experiment 2, two groups of mussels were exposed to 10 and 30 psu, respectively, for 15 days, after which they were exposed to the reversed salinities for another 15 days; in experiment 3, two groups of mussels were initially exposed to 15 and 25 psu for 22 days, after which they were exposed to the reversed salinities for another 17 days. Overall, mussels showed a reduced growth rate with falling salinity, reflected as an increase of doubling time of body dry weight (Figure 6.16). Mussels that were initially exposed to 10 psu, where growth was low, exhibited fast growth when subsequently exposed to 30 psu; this trend was reversed when 30 psu mussels were exposed to 10 psu, illustrating that mussels were able to adjust growth at changing salinities. However, in the northern Baltic Sea, where salinity ranges from 6.5 to 3.0 psu, mussels are at the limit of their salinity tolerance, and consequently as salinity declines there is a marked decline in mean mussel size (Westerbom *et al.* 2002; Riisgard *et al.* 2014). Suboptimal salinity (and temperature) may constitute a stressor that allocates energy away from shell production to cellular processes such as osmoregulation or to the repair of cellular damage. Riisgard *et al.* (2014) suggest that reduced ability to produce shell material at extremely low salinity may help to explain dwarfism of mussels in the Baltic Sea. In a climate change scenario, it is predicted that Baltic Sea temperatures will increase by 2–4 °C and salinity will decrease by an unknown amount due to increased precipitation and corresponding runoff (HELCOM 2013; Andersson *et al.* 2015). To better understand

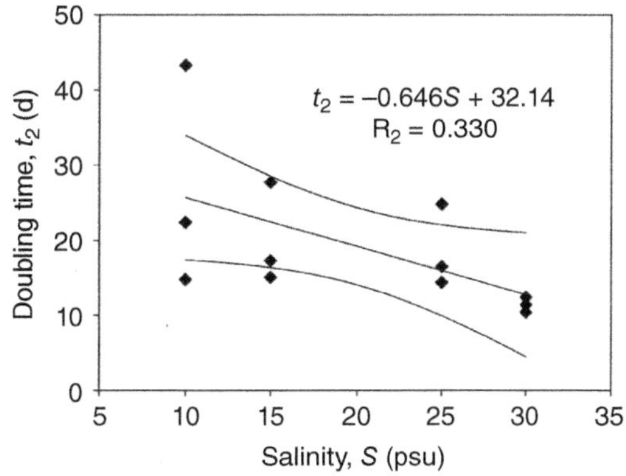

Figure 6.16 Doubling time of body dry weight for the mussel *Mytilus edulis* grown at different salinities over 11–22 days on *Rhodomonas salina* cells. The regression line, its equation and the 95% confidence interval are indicated. *Source:* From Riisgård *et al.* (2012).

the consequences of climate change-related stress, Hiebenthal *et al.* (2012) tested the isolated and interactive active effects of salinity and temperature in various combinations on a selection of response variables, including growth, in *M. edulis* and the scallop *Arctica islandica* over a range of temperatures and salinities (4, 10, 16, 20 and 25 °C for *M. edulis* and 4, 10 and 16 °C for *A. islandica*; and 15, 25 and 35 psu for both species). These treatment levels fall within the natural range of temperature and salinity in the distributional range of the two species, while the sampled populations of both species only rarely, if ever, experience salinities >25 psu and temperatures >23 °C. Interactively, temperature and salinity significantly ($p < 0.001$) affected shell increment in *M. edulis*; that is, a reduction of shell increment with increasing temperatures (especially at 25 °C), and at low (15 psu) and high (35 psu) salinity (Figure 6.17). Both factors affected shell increment of *A. islandica* interactively ($p < 0.03$); at 4 °C, the shell increment rate was highest at 35 psu, but at 10 and 16 °C it was already high at 25 psu (not illustrated). Shell increment of *M. edulis* was less affected by hyposalinity than that of *A. islandica*. In both species, condition index (see Chapter 5) decreased and lipofuscin accumulation – a biomarker for oxidative cell damage – increased with increasing temperature. The predicted scenario of warming and desalination in the Baltic Sea could impose severe stress on these and other ecologically relevant bivalve species if evolutionary adaptation does not happen at a similar rate (Hiebenthal *et al.* 2012).

The Indo-Pacific mussel, *Brachidontes pharaonis*, a species tolerant to high salinity and elevated temperatures, has invaded the eastern Mediterranean Sea, where it has formed dense mussel mats in the past few decades and is outcompeting indigenous bivalves. In recent years, there has been evidence that the species is spreading westward, colonising not only hypersalinity habitats but also wider zones of marine intertidal environments with more oceanic features. Sarà *et al.* (2008) tested the physiological tolerance of this invasive mussel to a broad range of salinities (15, 25, 37, 45, 50, 55 and 60 psu) and three temperatures (11, 15 and 20 °C), which were lower than that of the native habitat. Such information should prove useful when assessing the species' colonising process and its possible geographic spread. SFG showed negative values at the lowest salinity at all three temperatures, but there was a clear linear increase between 15 and 50 psu ($p < 0.001$), reaching maximum values at 50 psu, with marked reductions at the highest salinities, 55 and 60 psu ($P < 0.05$) (Figure 6.18). The results demonstrate that *B. pharaonis* has the capacity to maintain positive SFG and tolerate the wide range of

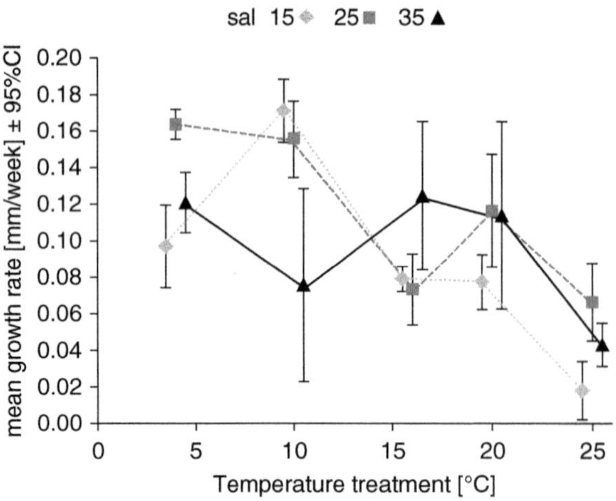

Figure 6.17 Shell increment rates (mm wk^{-1}) of all *Mytilus edulis* individuals in the experimental units at different temperatures (4,10, 16, 20 and 25 °C) and salinities (15, 25 and 35 psu). Data: mean ± 95% confidence intervals; each treatment was replicated fourfold. *Source:* From Hiebenthal *et al.* (2012). CC BY 4.0.

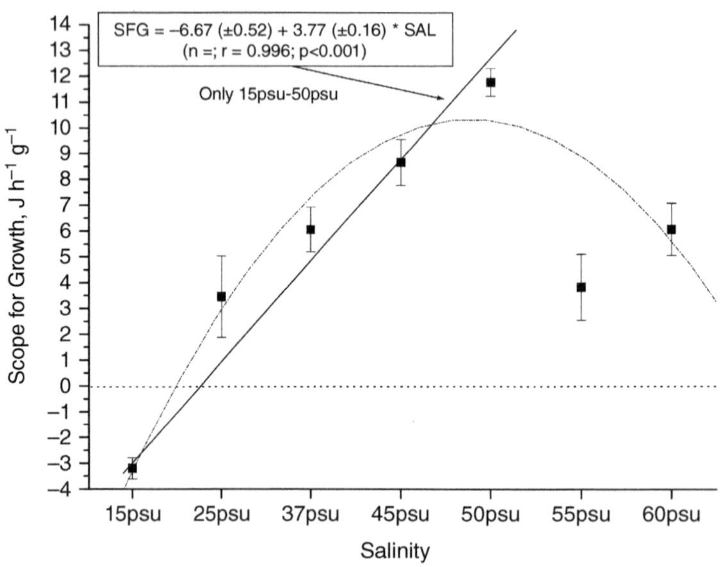

Figure 6.18 Effect of salinity (15, 25, 37, 45, 50, 55 and 60 psu) on the scope for growth (SFG; J $h^{-1}g^{-1}$) of *Brachidontes pharaonis*. *Source:* From Sarà *et al.* (2008). Reproduced with permission from Elsevier.

temperature/salinity conditions present throughout the Mediterranean. Less tolerant indigenous species such as *M. galloprovincialis* and *Mytilaster minimus* may be forced to abandon habitat patches in favour of *B. pharaonis* or other tolerant invasive species (Sarà *et al.* 2008).

Tidal Level and Wave Action

The intertidal zone is one of the most physiologically harsh environments on Earth. Stress increases along a vertical gradient, with higher physical stress occurring in the higher intertidal zone, due to both reduced feeding time and longer exposure to aerial conditions.

Therefore, it is to be expected that animals in the high intertidal zone exhibit markedly reduced growth rates compared to those that are permanently covered by the tide.

In the black mussel, *Choromytilus meridionalis*, there is a progressive reduction in growth rates with increasing tidal exposure resulting from reduction in the time available for feeding and the priority given by high-shore individuals to reproductive output at the expense of somatic growth (Griffiths & Hockey 1987 and references therein; Figure 6.19). This species could be expected to survive up to ~50% of exposure to air, except that settlements highest on the shore are killed during hot-weather periods and permanent populations seldom persist where they are exposed for more than 30% of the tidal cycle. Ribbed mussels, *Geukensia demissa*, at a salt marsh site in Jamaica Bay, New York also allocate a much smaller proportion of total production to growth, exceeding the reproductive effort of lower-shore mussels (Franz 1997). Species differ in their ability to tolerate aerial exposure. For example, van Erkom Shurink & Griffiths (1993) found that *Mytilus galloprovincialis* and *Perna perna* continue to grow at about 80% of the submerged rate even at 50% exposure level, compared to 66% for *Choromytilus meridionalis* and only 54% for *Aulacomya atra*. But, for most bivalve species, a figure of around 50% exposure represents the point of zero growth, when the energy required for metabolism during aerial exposure exceeds that available during the feeding period (Seed & Suchanek 1992).

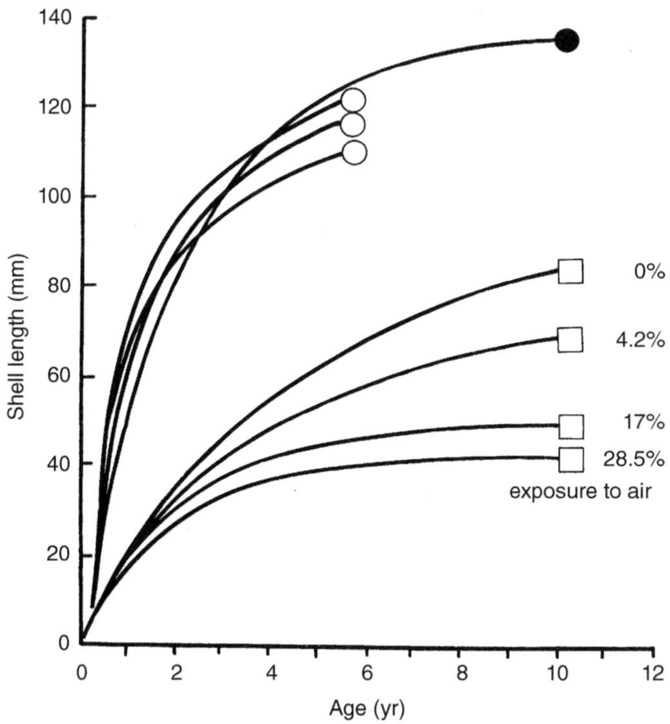

Figure 6.19 Growth rates of South African *Choromytilus meridionalis* at different sites and tidal levels. Table Bay, offshore (solid circle); Saldanha Bay, raft culture (open circles); and False Bay (open squares). *Source:* From Griffiths & Hockey (1987). Reproduced with permission from Taylor & Francis Ltd.

In reality, this figure is an overestimate, since other factors such as tolerance to extremes of temperature and desiccation may become limiting well before SFG declines to zero (van Erkom Shurink & Griffiths 1993).

The effect of strong wave action on bivalve growth is often site- or species-specific; the reasons for this are presently unclear. In some regions, such as western Europe, Chile and northern Russia, mussels are smaller on shores with strong wave action (Seed 1969; Alvaredo & Castilla 1996; Sukhotin *et al.* 2006, 2007; Figure 6.20), while in South Africa, *Perna perna* grows almost twice as fast on exposed as on sheltered shores, probably due to greater food availability at sites with greater water flow (McQuaid & Lindsey 2000, 2007). The same finding was reported for *M. galloprovincialis* in South Africa, except that growth declined at sites experiencing extreme wave action (Steffani & Branch 2003; Hammond & Griffiths 2004). Connor & Robles (2015) measured variation in wave force, temperature and growth rates within two mussel (*M. californianus*) beds separated by15° latitude on the West Coast of the United States. Both beds showed the same general trends: growth rates and terminal size increased toward the wave-exposed and low-shore points of headlands, where conditions for growth were optimal (i.e. cooler temperatures and longer feeding intervals). To illustrate, young adults from the two low and wave-exposed microhabitats grew six to ninefold higher than those from the high-shore wave-sheltered habitats.

Stock Density

Most of the information on density as a modulator of growth has come from studies on bivalves grown in suspended culture. Indeed, carrying capacity is often defined as the stocking density at which production is maximised without negatively affecting growth rates (Duarte *et al.* 2003, 2008; see Chapter 10). Cubillo *et al.* (2012) investigated the effect of mussel density on growth using the techniques commonly employed in the culture of

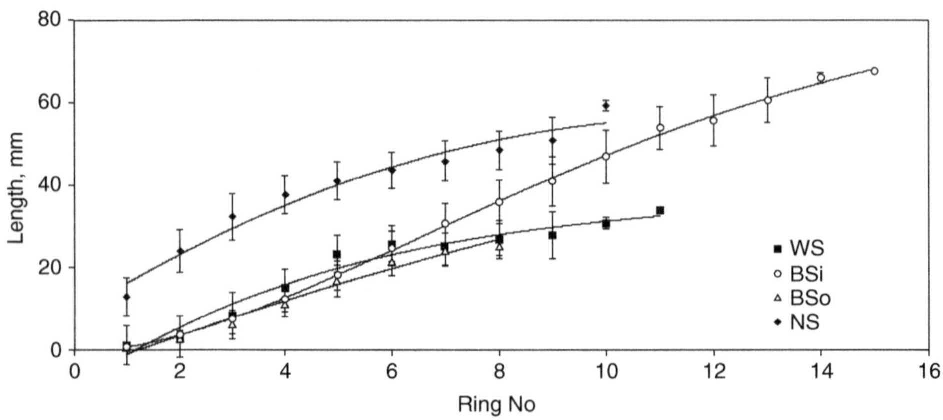

Figure 6.20 Von Bertalanffy growth curves in *Mytilus edulis* populations. NS, North Sea, $L_t = 62.8 \cdot (1- exp^{-0.197(t+0.34)})$, n = 215; BSi, Barents Sea, innermost (sheltered) part of the bay, $L_t = 412.0 \cdot (1-exp^{-0.014(t-1.380)})$, n = 194; BSo, Barents Sea, outer (open) part of the bay, $L_t = 126.2 \cdot (1-exp^{-0.035(t-1.206)})$, n = 30; WS, White Sea, $L_t = 39.3 \cdot (1 - exp^{-0.181(t-1.160)})$, n = 238. Data are mean ± SE. *Source:* From Sukhotin *et al.* (2006). Reproduced with permission from *Journal of Shellfish Research*.

M. galloprovincialis in the Galician Rías in northwestern Spain. Ropes cultured at different densities (220, 370, 500, 700, 800 and 1150 ind/m) were hung from a commercial raft, and growth indicators were monitored monthly during the second phase of cultivation (i.e. from thinning out to harvest). Mussels cultured at lower densities exhibited higher growth rates, and consequently reached greater weight and length values at the end of the experimental period, than those cultured at higher densities. The results suggest that adjusting mussel density to the characteristics of the growing area (e.g. food availability) could yield an increase of 10–15% in biomass production and at least 20% in economic benefits. They also show an increased effect of intraspecific competition with mussel size. Higher densities in a management programme could be supported without affecting growth performance, provided the final product was limited to a smaller shell length (Cubillo *et al.* 2012). Similar results have been reported for rope-grown *M. edulis* (Rodhouse *et al.* 1984; Mallet & Carver 1991; Fréchette *et al.* 1996; Alunno-Bruscia *et al.* 2000), suggesting mussels cultivated at high densities present lower growth due to low food availability caused by mussel crowding and not by the physical interference of density itself. However, Lauzon-Guay *et al.* (2005a) found no influence of density on shell morphology or tissue/shell ratio in cultivated *M. edulis* mussels, which agrees with results reported for *P. perna* in suspended culture (Bordon *et al.* 2014). Lauzon-Guay *et al.* (2005b) warn that increasing density would likely result in increased mortality and loss of seed mussels. Few studies take population size structure within the groups assayed into account, and this could be a strong determinant of the outcome of stocking experiments (discussed by Fréchette *et al.* 2005). Food depletion is the factor most often cited as the cause of decreased growth at high densities.

Wild populations also show a negative relationship between growth rate and density. Seed (1969) found that *M. edulis* spat that recruited into a population of one-year-old mussels grew at less than half the rate of that recruiting on to an adjacent bare rock surface. For *Geukensia demissa*, which lives embedded in and among salt marsh sediments, food supplies may be limiting, especially for young mussels, and growth is density-dependent, since field manipulations show that increases in density lead to decreases in growth rates (Stiven & Gardner 1992 and references therein).

The effect of density on growth appears to be species-specific. For example, in mussels, growth in *Choromytilus meridionalis* is more sensitive to density than is the case for *Aulacomya atra* or *M. galloprovincialis*. This appears to correlate with natural packing densities in the wild: *C. meridionalis* usually occurs as a monolayer, while both *A. atra* and *M. galloprovincialis* form dense multilayered beds (van Erkom Schurink & Griffiths 1993).

Carbon dioxide and Oxygen Concentrations

Human-induced rise in atmospheric CO_2 concentration has been linked to changes in seawater carbonate chemistry and a decrease in ocean pH, a process known as ocean acidification (Ventura *et al.* 2016; see Chapter 3). OA acts together with other global changes (e.g. warming and progressively lower oxygen levels; IPCC 2014). Increased acidification of seawater, due to CO_2, leads to decreased calcification in bivalves, as well as in other marine species with calcified parts (Michaelidis *et al.* 2005). To my knowledge, all studies on the potential effects of OA on bivalves, including mussels, are laboratory-based. One such study evaluated the impact of medium-term (70 days) exposure to elevated pCO_2 levels (380, 750 and 1200 ppm) on physiological processes of juvenile *M. chilensis*

(Navarro *et al.* 2013). SFG was reduced by 13% at 750 and 28% at 1200 ppm CO_2 compared to the control treatment, 380 ppm (Figure 6.21). This could represent a significant loss in annual production for commercial operations. A reduction in growth to 55% was also reported for juvenile and adult *M. edulis* maintained under long-term moderate CO_2 levels (Michaelidis *et al.* 2005). Subsequently, Navarro *et al.* (2016) examined the combined effect of temperature (12 and 16 °C) and elevated pCO_2 levels (380, 750 and 1000 μatm) on juvenile *M. chilensis*. They found, as in their previous study, that SFG was significantly lower at the highest pCO_2 concentration compared with the control, and mussels exposed to 700 μatm did not show a significantly different SFG from the other two treatments. SFG was significantly higher at 16 °C than at 12 °C, which may be because these temperatures are within the thermal tolerance of *M. chilensis* in southern Chile. When strong pCO_2 stress was coupled to food limitation, each significantly decreased shell length growth, and both significantly influenced the magnitude of inner shell surface dissolution in *M. edulis* (Melzner *et al.* 2011). In contrast, Range *et al.* (2012) found that even under extreme levels of CO_2-induced acidification, juvenile *M. galloprovincialis* could continue to calcify and grow. Similar results have been reported for larval (Ventura *et al.* 2016) and juvenile (Gazeau *et al.* 2010) *M. edulis*. The complex responses of mussels, in terms of growth rate, suggest a large degree of inter- and intraspecific variability in their sensitivity to OA. It is evident that further work is needed in order to determine the true extent of the threat posed by OA to mussel populations (Ventura *et al.* 2016).

Several studies have examined the combined effects of warming and OA on mussel growth. In one, adult mussels (*M. galloprovincialis*) were exposed to low and high pCO_2 and 12, 14, 16, 18, 20 and 24°C for one month (Kroeker *et al.* 2014). Although high pCO_2 significantly reduced mussel growth at 14 °C, this effect gradually lessened with increasing temperature, illustrating how warming can moderate the effects of OA in this species. Similar results have also been reported for adult *M. galloprovincialis* (Gazeau *et al.* 2014) and juveniles of *M. edulis* (Hiebenthal *et al.* 2013) and *M. coruscus* (Wang *et al.* 2015).

Another serious consequence of global warming is the decrease in dissolved O_2 in the world's oceans, a problem that is exacerbated through nutrient input from wastewater and agriculture in coastal and inland waters (details in Chapter 3). Short-term (seven days) exposure of adult *M. edulis* to environmental hypoxia (2.11 mgO_2L^{-1}; control 6.96 mgO_2L^{-1}) caused a significant reduction in SFG but not body condition. This was attributed primarily

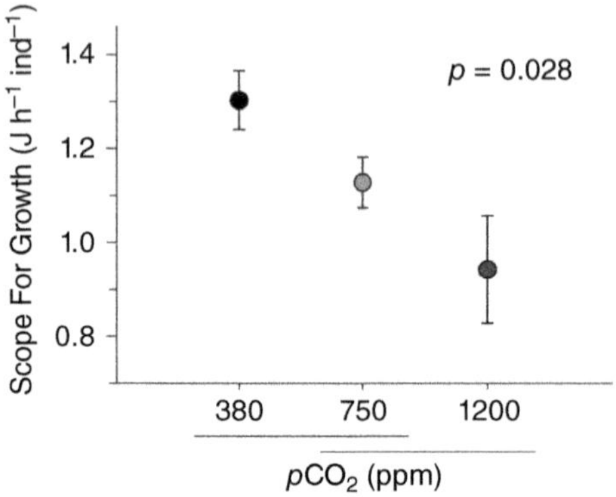

Figure 6.21 Scope for growth (SFG) measured on mussels (*Mytilus chilensis*) exposed over 70 days to 380, 750 and 1200 ppm pCO_2 levels. Values represent the integrated means ± SE of 12 experiments; *p* values are from ANOVA test comparing different treatments. Lines below the *x* axis link mean not significantly different from one another ($p > 0.05$). *Source:* From Navarro *et al.* (2013). Reproduced with permission from Elsevier.

to the reduction in energy acquisition as a result of reduced clearance and ingestion rate (Sanders *et al.* 2014). A significant hypoxia-related reduction in SFG has also been documented for *Perna viridis* (Wang *et al.* 2011). To date, combined pCO_2 and pO_2 stressor experiments on marine bivalves are largely missing (Melzner *et al.* 2013).

Other Factors

Groundwater discharge is today recognised as an important pathway for freshwater, nutrients and other dissolved chemical substances to coastal systems. While its effect in supporting primary production in coastal ecosystems is increasingly recognised, its impact on growth of animals at higher trophic levels (primary and secondary consumers) is less well documented. Recently, Andrisoa *et al.* (2019) investigated the impact of groundwater discharge on the growth of the Mediterranean mussel, *M. galloprovincialis* in a coastal lagoon. Growth rates and condition index (tissue weight/shell weight) of mussels growing at groundwater-exposed sites and at a control site in Salses-Leucate lagoon, France were measured. Mussels from groundwater-influenced sites have higher growth rate and condition index compared to those from a control site. Importantly, growth rates from groundwater-influenced sites are among the highest rates reported for the Mediterranean region (41 ± 9 μm d^{-1}). The higher growth rates at groundwater-influenced sites are likely a consequence of both the higher winter temperatures of lagoon water as a result of groundwater discharging with relatively constant temperatures and the groundwater-driven nutrient supply, which increase the food availability to support mussel growth. Overall, the study demonstrates that groundwater discharge to Mediterranean lagoons provides favourable environmental conditions for fast growth of mussels of high commercial quality.

Water depth is another factor that influences mussel growth, and it is of particular relevance in culture operations. Fuentes *et al.* (2000) examined growth rate of rope-grown mussels (*M. galloprovincialis*) in the Ría de Arosa (NW Spain), a region that is one of the largest producers of cultured mussels in the world (see Chapter 10). Their results indicated that depth of cultivation was the major factor affecting growth; mussels cultivated in the upper part of the water column (2.5 m depth) were significantly longer and heavier at harvest than those from deeper waters (7.5 m). Difference in growth rate was unlikely to be due to depth per se, but rather was a result of differences in temperature at the two depths (i.e. temperature at 2.5 m was always higher than at 7.5 m throughout the growing season). In contrast, cultured mussels (*P. viridis*) grown simultaneously at 5 and 17 m for up to 96 days showed no significant differences in growth rate between the two depths, despite the fact that chlorophyll *a* is at a maximum in deeper waters (Ogilvie *et al.* 2004).

High turbidity (Secchi-disc reading of less than 25 cm) has been shown to negatively affect mussel culture because it can result in failure of the filtering activities of *P. viridis* (Soon & Ransangan 2014 and references therein; but see Hawkins *et al.* 1996). Moreover, the low light penetration due to high turbidity also results in low primary productivity, which causes insufficient supply of food and leads to failure of mussel culture. Current speed is another important factor determining the amount of food available for mussel culture. A poorer water movement is often unfavourable to mussel growth, while too strong water currents can result in high turbidity and causes settlement difficulty for young mussels. The optimal current speed for growth in this species was proposed to be in the range of 0.1–0.3 m s^{-1}.

Space is usually the main limiting resource for sessile organisms on rocky-shore intertidal substrates. Competitive interactions between individuals from a single or different species can regulate population density and individual growth. Intraspecific competition is much stronger than interspecific competition, which is generally attributed to the fact that individuals of the same species tend to use spaces in a more similar way than individuals of different species (Arribas *et al.* 2016 and references therein). Okamura (1986), investigating the role of intraspecific competition in *M. edulis*, showed that isolated individuals and individuals associated in relatively small groups (6–9 mussels/group) grew more than those associated with relatively large groups (21–28 individuals). Also, mussels located in the centre of groups exhibited reduced growth relative to those on the edges. See also Newell (1990), Brichette *et al.* (2001), Fréchette *et al.* (2005) and Rius & McQuaid (2006, 2009). Shinen & Morgan (2009) examined the role of interspecific competition between the invasive species *M. galloprovincialis* and the two native mussels *M. trossulus* and *M. californianus* on the northern Californian coast, United States. In both field and laboratory manipulations, the performance of *M. galloprovincialis* was always better in terms of growth and survival than that of the other two species. It appears that the invader restricted movement and smothered and interfered with filter feeding of the native species. Rather than limiting invasion, competition gave *M. galloprovincialis* an advantage over the native mussels, contributing to the displacement of *M. trossulus* along much of its historic range (see Chapter 3).

Epibionts – organisms that live on the surface of another living organism – have been shown to influence mussel growth. For example, Dittmann & Robles (1991) observed decreasing length increment in mussels (*M. californianus*) overgrown by algae, while experimental removal of epiphytes resulted in increased growth rates. In contrast, the presence of algal epiphytes had no effect on mussels (*Mytilus* spp.) growth either on wave-exposed or sheltered shores on Atlantic coasts in Ireland (O'Connor 2010). However, Buschbaum & Saier (2001) found that suspension-feeding barnacle (*Semibalanus balanoides* and *Balanus crenatus*) epigrowth significantly reduced growth in *M. edulis*. Surprisingly, food competition was not responsible, since mussels with experimentally sealed barnacles showed no difference in growth compared to those overgrown with living barnacles. The authors suggested that the structure of barnacle clusters may change hydrodynamic conditions, leading to differences in microcurrents. Similar results were reported by Thieltges (2005), who found that mussels with the attached pest *Crepidula fornicata*, a limpet unintentionally introduced into Europe in the 1870s, exhibited a reduction in shell growth three to five times that of unfouled mussels, but similar to that with artificial limpets.

Bivalves, in particular mussels, have been widely used to assess the effects of specific toxicants on mortality, shell and tissue growth and SFG. Indeed, the latter is one of the most sensitive measures of pollution-induced stress, and has proved to be a powerful and cost-effective method of assessing environmental pollution (Widdows & Staff 2006). Using this approach in the mussel *M. edulis*, Widdows *et al.* (1997) found significant negative correlations between SFG and tissue concentrations of petroleum hydrocarbons, polychlorinated biphenyls (PCBs), dichlorodiphenyltrichloroethane (DDT) and hexachlorocyclohexanes (HCHs), but no association between SFG and tissue levels of metals (Cd, Co, Cu, Fe, Hg, Mn, Ni, Pb and Zn). Similar studies have been carried out on this (Widdows & Johnson 1988;

Widdows & Donkin 1992; Widdows *et al.* 1995; Mubiana & Blust 2007; Booth *et al.* 2007) and other (Soto *et al.* 2000; Tsangaris *et al.* 2007; Albentosa *et al.* 2012) mussel species. Culbertson *et al.* (2008) studied the effect of exposure on the ribbed mussel, *Geukensia demissa* to residual petroleum from a major oil spill in 1969, which still remains buried in the sediments of this species' habitat. Examination of short-term exposure was done by transplanting mussels from Great Sippewissett Marsh, a control site, into Wild Harbor, the site of the oil spill, both situated on Cape Cod, Massachusetts, United States. The authors also examined the effects of long-term exposure with transplantation of mussels from Wild Harbor into Great Sippewissett. Both the short- and long-term exposure transplants exhibited slower growth rates, shorter mean shell lengths, lower condition indices and decreased filtration rates. See also Peteiro *et al.* (2006) on the effects of the *Prestige* oil spill on cultured mussel (*M. galloprovincialis*) growth in NW Spain. The effects of environmental contaminants on marine mussels are discussed in greater detail in Chapter 8.

Other factors that have been shown to decrease mussel growth are disturbance (Calcagno *et al.* 2012), substrate type (Frandsen & Dolmer 2002), chlorinated water (Thompson *et al.* 2000), fouling (de Sá *et al.* 2007) and disease and parasites (see Chapter 11)

Endogenous Growth Modulators

Genotype

It is generally accepted that variation of growth in bivalves has both environmental and genetic components. This is because highly variable rates of growth are routinely observed in bivalve larvae reared under identical, controlled laboratory conditions. While considerable progress has been made in understanding the role of exogeneous factors in growth modulation (see previous section), it has been more difficult to understand the role that endogenous mechanisms play in the process, although allozyme[2] analysis, breeding studies and chromosome manipulations have gone some way in helping to elucidate the mechanistic basis of genotype-dependent growth (see Chapter 9).

Genetic variability, or the heterozygosity of a population, can be calculated by first obtaining the frequency of heterozygotes at each gene locus and then averaging these over all analysed loci to get a mean heterozygosity value (see Chapter 9 for details). Positive correlations between multiple locus heterozygosity (MLH), as detected by protein electrophoresis, and fitness-related traits have been reported in a variety of marine molluscs. Growth rate, as well as viability, fecundity, fluctuating asymmetry and metabolic parameters such as decreased O_2 consumption and more efficient protein synthesis, have been found to be positively associated with heterozygosity in many wild populations of juvenile and adult bivalves, including mussels (references in Gosling 1992, Britten 1996 and Myrand *et al.* 2009). The positive association is believed to be due to a higher mean efficiency of basal metabolism in more heterozygous individuals (Goff 2011 and references therein). But the magnitude of the correlation appears to depend on factors such as age, reproductive state, environmental conditions, background genetic effects and number and type of loci analysed (Saavedra & Guerra 1996; Myrand *et al.* 2002 and references therein). Even when the same species and set of loci are used, conflicting results have been reported, with some showing a positive correlation and others no correlation at all (Gosling 1992). It is important to note that where a heterozygosity/growth correlation has been observed, heterozygosity typically explains only a small percentage (2–4%) of the variation in size among individuals, although a higher figure of 27% has been reported by Bayne & Hawkins (1997). There are conflicting views regarding the genetic interpretation of the heterozygosity/growth correlation. One is that allozymes are merely neutral indicators of selection acting against deleterious recessive mutations at linked loci with fitness effects (associative overdominance hypothesis). Another is that selection acts directly on allozyme genotypes and the heterozygosity/growth correlation is the result of functional superiority of allozyme heterozygotes (overdominance hypothesis) (see David *et al.* 1997; Launey & Hedgecock 2001). To my knowledge, there is no evidence of a direct connection between an electrophoretic locus and bivalve growth.

A more direct approach to the study of growth heterosis – a term that has replaced the older 'heterozygosis' – is to use established inbred lines that can be crossed to produce larvae with reproducibly different growth phenotypes. Alternatively, experimentally produced full- or half-sib lines, or indeed populations, can be used. One such study (Pace *et al.* 2006) focused on the physiological bases that underlie endogenous variation in growth potential of larvae, albeit using the oyster *Crassostrea gigas*, the best studied bivalve group on this topic. Large numbers of larvae with different growth rates, produced from experimental crosses of ripe adults from pedigreed families, were reared under similar environmental conditions of food and temperature. A total of 35 larval families were used, in which growth varied over a fivefold range from 3.4 (±0.5, S.E.) to 17.6 (±0.6) μm day^{-1}. The suite of integrated measurements applied to study growth variation included size, biochemical composition, rate of particulate and dissolved nutrient acquisition, absorption efficiency, respiration rate and enzyme activity. The authors showed that a complex set of physiological processes regulated differences in the genetically determined growth rates of larvae. About 50% of the energy required to account for genetically determined growth variation came from variation in size-specific feeding capacity, and the authors proposed that different protein depositional efficiencies could account for this (Figure 6.22). Differences in absorption rates were not significant for slow- and fast-growing larvae; nor were differences in size-specific respiration rates. Bayne *et al.* (1999) reported similar results when they compared various physiological parameters in a selected fast-growing line of adult oysters (*Saccostrea commercialis*) with a control (unselected) line. Fast growth was associated with faster rates of feeding, reduced metabolic rates and lower metabolic costs of growth. The reduced costs of growth were due in part to increased efficiencies of protein turnover in fast growers (Bayne 2004). Control oysters, however, deposited more lipid than the selected oysters, although the unit costs of lipid deposition did not differ between selected oysters and controls.

An additional study, this time on mussels (*M. galloprovincialis*), was carried out on populations of mussel seed gathered from the intertidal and subtidal regions of a Galician embayment in NW Spain (Fernández-Reiriz *et al.* 2016). The seed was maintained in the laboratory for five months to select fast (F)- and slow (S)-growing mussels, ultimately in order to compare the physiological basis for inter-individual growth variations of mussels from both seed origins. Faster growth was linked to greater energy intake (20% higher clearance and ingestion rate) and higher food absorption rate coupled with low metabolic costs. The enhanced energy absorption (around 65% higher) resulted in three times higher

Figure 6.22 Relationship of biochemical content as a function of size for larvae of *Crassostrea gigas*. Open squares, slow-growing inbred larvae; solid circles, fast-growing hybrid larvae. (A) Increase in protein content as a function of shell length (N = 3 for each data point ±1 SEM). (B) Increase in free amino acid pool content (taurine) as a function of shell length (each data point is a single measurement by high-performance liquid chromatography, HPLC). Taurine is the major constituent of the free amino acid pool of *C. gigas* larvae. Separate regressions of protein and taurine content in slow- and fast-growing larvae are not significantly different. The regression line shown is for the pooled data for all measurements for protein (A) and taurine (B) content. Regression equation for protein content (A): log (protein content) = 2.57 log (shell length) −3.53, $r^2 = 0.77$. Regression equation for taurine content (B): log (taurine content) = 2.91 log (shell length) −4.48, $r^2 = 0.93$. *Source:* From Pace *et al.* (2006). Reproduced with permission from Elsevier.

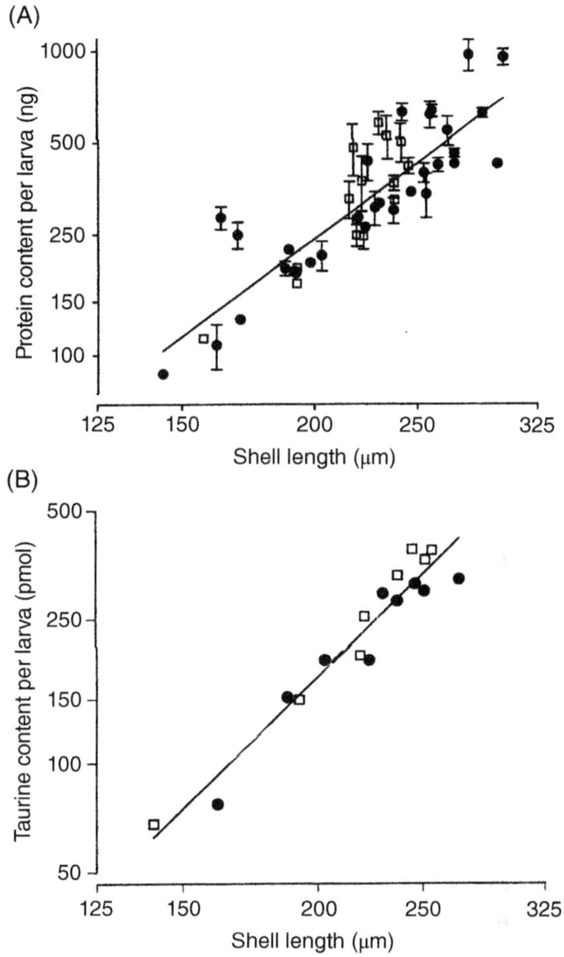

SFG in F mussels (20.5 ± 4.9 J h^{-1}) than in S ones (7.3 ± 1.1 J h^{-1}), irrespective of seed origin. Results from these three studies indicate a wide plasticity of physiological rates and efficiencies and show how, through selection, interactions between physiological traits can serve to enhance growth (Bayne 2000).

Quantitative trait loci (QTLs) are unidentified genes that affect performance characteristics such as growth rate and disease resistance of importance to breeders. They are mapped by identifying which molecular markers correlate with an observed trait. (see Chapter 9). When QTLs have been identified, the QTL-linked marker information can be incorporated into breeding programmes. The majority of studies on QTL detection have been on cultured fish, with far fewer on cultured bivalves – although numbers are increasing. Not surprisingly, almost all deal with growth-related and disease-resistance QTL detection in oyster species. To date, more than 80 growth-related QTLs have been detected in oyster species (Li & He 2014; reviewed in Gosling 2015; Jones *et al.* 2014; Wang *et al.* 2016). But there has not been a single published study on QTLs for growth-related or any other trait in marine mussels.

Chromosome set manipulation can produce phenotypic changes that are commercially important in bivalve aquaculture. Triploids have three sets of homologous chromosomes in their somatic cells instead of the usual two pairs found in diploids. Because triploids (3n) have a considerably reduced reproductive potential compared to diploids (2n), they should exhibit faster growth rates once maturity has been reached. This is indeed the case for many bivalve species (reviewed by Piferrer *et al.* 2009; Gosling 2015). However, while triploid mussels (*Mytilus*) grow faster than diploids in field tests (Davis *et al.* 1997; Brake *et al.* 2004), a hatchery-based study by Kamermans *et al.* (2013) did not confirm this. Overall, the main advantage of triploid induction in the hatchery might be the production of nonmaturing mussels for year-round marketing.

Size

In bivalves, growth is rapid in the first year(s) of life but progressively slows with increasing age (Figure 6.2). For animals, the surface area available for oxygen diffusion limits general metabolism. Metabolism is proportional to a constant power of the body weight, as described by the allometric equation:

$$Y = aX^b$$

where Y is the metabolic rate (measured as oxygen consumption), X the body mass, a the intercept and b the allometric scaling exponent. For quite some time it has been commonly accepted that the b coefficient in most groups of organisms is ~0.75 (Riisgård 1998). And for aerobic expenditure in *Mytilus*, the b coefficient does approximate 0.75 when organisms covering a large spread in body mass are compared (Hawkins & Bayne 1992 and references therein). However, when respiration rate, as a function of size, was studied throughout the ontogeny of *M. edulis* from early larval to adult stage, the b exponent was ~0.9 throughout the early pelagic larval stage up to the juvenile stage and decreased to about 0.7 in the adult stage. From this, it is clear that as size increases, the metabolic rate decreases. In addition, ingestion rate also declines with increasing size, but at a faster rate than metabolic rate. Both these factors contribute to a decline in growth efficiency (growth per unit of absorbed ration) with size (Figure 6.23). Also, there is evidence that metabolic efficiency per se decreases with increasing mussel size from 55 to 15% in *M. edulis* of 0.25 to 2.0g dry flesh weight. But it is probably the increasing proportional allocation of available energy into gamete production that ultimately puts the brake on growth (Griffiths & Griffiths 1987; Hawkins & Bayne 1992). Because of various exogenous and endogenous influences, the proportional allocation of energy to somatic growth or reproduction varies widely between species, and even between populations of the same species (Arranz *et al.* 2016 and references therein; see Figure 5.8)

Growth Factors

Growth factors are widespread, locally acting, signalling polypeptides that are involved in the regulation of cell proliferation and differentiation during the development of organs and tissues in embryos and adults (Lelong *et al.* 2000). The majority of growth factors were initially identified in vertebrates. To investigate the possibility that known growth factors may have counterparts in molluscs, studies on the effects of some of these factors on tissue cells have been carried out. Various growth factors have now been identified in bivalves. mGDF (molluscan growth and differentiation factor), which functions as a broadly

Figure 6.23 Net growth efficiency (net energy balance/net energy absorbed) in relation to absorbed energy and body size (mg total dry tissue) in the mussel *Mytilus edulis*. *Source:* From Hawkins & Bayne (1992) based on data in Thompson & Bayne (1974). Reproduced with permission from Elsevier.

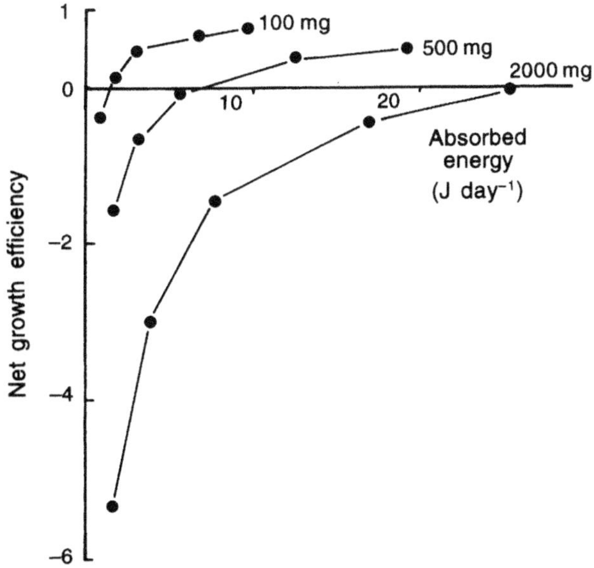

multifunctional growth and differentiation factor in molluscs, like the transforming growth factor-β (TGF-β) superfamily of proteins in vertebrates (Favrel *et al.* 1998), has been identified in *Crassostrea gigas* (Lelong *et al.* 2000) and, more recently, in a homologue of vertebrate TGF-β, *Crassostrea gigas* TGF-β (cg-TGF-β), which seems to act as a cell-proliferative regulatory factor (Lelong *et al.* 2007). A novel TGF-β, named oyster-gonadal-TGF-like (og-TGFβ-like), a derived member of the TGF-β superfamily, appears to play an essential role in *C. gigas* germ cell development by functioning as an activator of germ cell proliferation in both male and female oysters and of vitellogenesis in females (Fleury *et al.* 2008; Corporeau *et al.* 2011; Huvet *et al.* 2012). Another signalling system, the BMP4-Chordin system, also included in the TGF-β signalling pathway, is employed in dorso-ventral determination in both vertebrate and many invertebrate species, including *C. gigas* (Liu *et al.* 2014, 2017). Finally, myostatin (MSTN), another member of the TGF-β superfamily, significantly affects shell length by altering MSTN protein structure in the clam *Sinonovacula constricta* (Niu *et al.* 2015).

Insulin-related peptides have also been identified and appear to stimulate protein synthesis in mantle-edge cells in *M. edulis* (Kellner-Cousin *et al.* 1994), *C. gigas* (Gricourt *et al.* 2003) and both mantle-edge and digestive-gland cells in the scallop *Pecten maximus* (Giard *et al.* 1998). In *C. gigas*, the growth-promoting effect is mediated by specific receptors that were initially characterised and expressed in mantle and mantle-edge cells. In the same species, insulin-like substances are involved in gonadal tubule rebuilding and the development of germinal cells and maturation of gametes (Gricourt *et al.* 2006). Hamano *et al.* (2005) conducted cDNA cloning and expression analysis of the oyster (*C. gigas*) insulin-related peptide gene. The gene showed seasonal variation in expression, with the period of highest expression at the onset of body growth and gamete development (see also Shi *et al.* 2013). As has been seen, the majority of studies have been carried out on the commercially important Pacific oyster, *C. gigas*. However, it is highly likely that the growth factors described play similar or related roles in mussels and other marine bivalve groups.

Notes

1 The study of periodicities stored in accreted hard parts (e.g. shells and skeletons) of various extant and extinct organisms.
2 Allozymes are variant forms of an enzyme that are coded by different alleles at the same gene locus.

References

Abades, S., Richardson, C.A. & Otaíza, R. (2000) Dynamical assessment of the accretionary record in the shell of the mussel *Semimytilus algosus* from a rocky shore in Chile. *Biosystems*, **57**, 163–72.

Adams, D.C., Rohlf, F.J. & Slice, D.E. (2004) Geometric morphometrics: ten years of progress following the 'revolution'. *Italian Journal of Zoology*, **71**, 5–16.

Aguirre, M.L., Perez, S.I. & Negro Sirch, Y. (2006) Morphological variability of *Brachidontes* Swainson (Bivalvia, Mytilidae) in the marine Quaternary of Argentina (SW Atlantic). *Palaeogeography, Palaeoclimatology, Palaeoecology*, **239**, 100–125.

Albentosa, M., Viñas, L., Besada, V., Franco, A. & González-Quijano, A. (2012) First measurements of the scope for growth (SFG) in mussels from a large scale survey in the North-Atlantic Spanish coast. *Science of the Total Environment*, **435–436**, 430–445.

Alfaro, A.C., Webb, S.C. & Barnaby, C. (2008) Variability of growth, health, and population turnover within mussel beds of *Perna canaliculus* in northern New Zealand. *Marine Biology Research*, **4**, 376–383.

Allen, B.J. & Williams, S.L. (2003) Native eelgrass *Zostera marina* controls growth and reproduction of an invasive mussel through food limitation. *Marine Ecology Progress Series*, **254**, 57–67.

Alunno-Bruscia, M., Petraitis, P.S., Bourget, E. & Fréchette, M. (2000) Body size-density relationships for *Mytilus edulis* in an experimental food-regulated situation. *Oikos*, **90**, 28–42.

Alunno-Bruscia, M., Bourget, E. & Fréchette, M. (2001) Shell allometry and length-mass-density relationship for *Mytilus edulis* in an experimental food-regulated situation. *Marine Ecology Progress Series*, **219**, 177–188.

Alvarado, J.L. & Castilla, J.C. (1996) Tridimensional matrices of mussels *Perumytilus purpuratus* on intertidal platforms with varying wave forces in central Chile. *Marine Ecology Progress Series*, **133**, 135–141.

Andersson, A., Meier, H.E.M., Ripszam, M., Rowe, O., Wikner, J., Haglund, P. *et al.* (2015) Projected future climate change and Baltic Sea ecosystem management. *Ambio*, **44**, S345–S356.

Andrisoa, A., Lartaud, F., Rodellas V., Neveu, I. & Stieglitz, T.C. (2019) Enhanced growth rates of the Mediterranean mussel in a coastal lagoon driven by groundwater inflow. *Frontiers in Marine Science*, **6**, art. 753.

Arranz, K., Labarta, U., Fernández-Reiriz, M.J. & Navarro, E. (2016) Allometric size-scaling of biometric growth parameters and metabolic and excretion rates. A comparative study of intertidal and subtidal populations of mussels (*Mytilus galloprovincialis*). *Hydrobiologia*, **772**, 261–275.

Arribas, L.P., Bagur, M., Soria, S.A., Gutiérrez, J.L. & Palomo, M.G. (2016) Competition between mussels at the rocky intertidal zone of La Lobería, Río Negro, Argentina. *Revista del Museo Argentino de Ciencias Naturales*, **18**, 1–7.

Asmus, H. (1987) Secondary production of an intertidal mussel bed community related to its storage and turnover compartments. *Marine Ecology Progress Series*, **39**, 251–266.

Bagur, M., Richardson, C.A., Gutiérrez, J.L., Arribas, L.P., Socorro Doldan, M.S. & Palomo, M.G. (2013) Age, growth and mortality in four populations of the boring bivalve *Lithophaga patagonica* from Argentina. *Journal of Sea Research*, **81**, 49–56.

Barillé, L., Lerouxel, A., Dutertre, M., Haure, J., Barillé, A.-L., Pouvreau, S. *et al.* (2011) Growth of the Pacific oyster (*Crassostrea gigas*) in a high-turbidity environment: comparison of model simulations based on scope for growth and dynamic energy budgets. *Journal of Sea Research*, **66**, 392–402.

Bayne, B.L. (1998) The physiology of suspension feeding by bivalve mollusks: an introduction to the Plymouth 'TROPHEE' workshop. *Journal of Experimental Marine Biology and Ecology*, **219**, 1–19.

Bayne, B.L. (2000) Relations between variable rates of growth, metabolic costs and growth efficiencies in individual Sydney rock oysters (*Saccostrea commercialis*). *Journal of Experimental Marine Biology and Ecology*, **251**, 185–203.

Bayne, B.L. (2004) Phenotypic flexibility and physiological tradeoffs in the feeding and growth of marine bivalve molluscs. *Integrative Comparative Biology*, **44**, 425–432.

Bayne, B.L. & Worrall. C.M. (1980) Growth and production of mussels *Mytilus edulis* from two populations. *Marine Ecology Progress Series*, **3**, 317–328.

Bayne, B.L. & Newell, R.C. (1983) Physiological energetics of marine molluscs. In: *The Mollusca, Vol. 4, Physiology, Part 1* (eds A.S.M. Saleuddin & K.M. Wilbur), pp. 407–515. Academic Press, New York.

Bayne, B.L. & Hawkins, A.J.S. (1997) Protein metabolism, the costs of growth, and genomic heterozygosity: experiments with the mussel *Mytilus galloprovincialis* Lmk. *Physiological Zoology*, **70**, 391–402.

Bayne, B.L., Svensson, S. & Nell, J.A. (1999) The physiological basis for faster growth in the Sydney rock oyster *Saccostrea commercialis*. *Biological Bulletin*, **197**, 377–387.

Beadman, H.A., Willows, R.I. & Kaiser, M.J. (2002) Potential applications of mussel modeling. *Helgoland Marine Research*, **56**, 76–85.

Bergström, P. & Lindegarth, M. (2016) Environmental influence on mussel (*Mytilus edulis*) growth–a quantile regression approach. *Estuarine, Coastal and Shelf Science*, **171**, 123–132.

Bergström, P., Lindegarth, S. & Lindegarth, M. (2013) Temporal consistency of spatial pattern in growth of the mussel, *Mytilus edulis*: implications for predictive modelling. *Estuarine, Coastal and Shelf Science*, **131**, 93–102.

Bergström, P., Lindegarth, S. & Lindegarth, M. (2015) Modeling and predicting the growth of the mussel, *Mytilus edulis*: implications for planning of aquaculture and eutrophication mitigation. *Ecology and Evolution*, **5**, 5920–5933.

Bernhard, J.M., Phalen, W.G., McIntyre-Wressnig, A., Mezzo, F., Wit, J.C., Jeglinski, M. *et al.* (2015) Technical note: towards resolving in situ, centimeter-scale location and timing of biomineralization in calcareous meiobenthos – the calcein–osmotic pump method. *Biogeosciences*, **12**, 5515–5522.

Blanchette, C.A., Helmuth, B. & Gaines, S.D. (2007) Spatial patterns of growth in the mussel, *Mytilus californianus*, across a major oceanographic and biogeographic boundary at Point Conception, California, USA. *Journal of Experimental Marine Biology and Ecology*, **340**, 126–148.

Booth, A.M., Sutton, P. A., Lewis, C.A., Lewis, A.C., Scarlett, A., Chau, W. *et al.* (2007) Unresolved complex mixtures of aromatic hydrocarbons: thousands of overlooked persistent, bioaccumulative, and toxic contaminants in mussels. *Environmental Science and Technolology*, **41**, 457–464.

Bonel, N., Solari, L.C. & Lorda. J. (2013) Differences in density, shell allometry and growth between two populations of *Limnoperna fortunei* (Mytilidae) from the Río de aa Plata basin, Argentina. *Malacologia*, **56**, 43–58.

Bookstein, F.L. (1991) *Morphometric Tools for Landmark Data: Geometry and Biology*. Cambridge University Press, Cambridge.

Boon, A.R., Duineveld, C.A., Berghuis, E.M. & van der Weele, J.A. (1998) Relationships between benthic activity and the annual phytopigment cycle in near-bottom water and sediments in the southern North Sea. *Estuarinei, Coastal and Shelf Science*, **46**,1–13.

Bordon, I.C.A.C., Marques, H.L.A., Alves, J.L., Rossi, A. & Pecora, I.L. (2014) Influence of densities on the growth of brown mussel *Perna perna* L. cultivated in suspended socks at Caraguatatuba, southeastern Brazil. *Aquaculture international*, **22**, 541–549.

Borrero, F.J. & Hilbish, T.J. (1988) Temporal variation in shell and soft tissue growth of the mussel *Geukensia demissa*. *Marine Ecology Progress Series*, **42**, 9–15.

Bownes, S.J. & McQuaid, S.D. (2010) Mechanisms of habitat segregation between an invasive (*Mytilus galloprovincialis*) and an indigenous (*Perna perna*) mussel: adult growth and mortality. *Marine Biology*, **157**, 1799–1810.

Bracken, M.E.S., Menge, B.A., Foley, M.M., Sorte, C.J.B., Lubchenco, J. & Schiel, D.R. (2012) Mussel selectivity for high-quality food drives carbon inputs into open-coast intertidal ecosystems. *Marine Ecology Progress Series*, **459**, 53–62.

Brake, J., Davidson, J. & Davis, J. (2004) Field observations on growth, gametogenesis, and sex ratio of triploid and diploid *Mytilus edulis*. *Aquaculture*, **236**, 179–191.

Brichette, I., Reyero, M.I. & García, C. (2001) A genetic analysis of intraspecific competition for growth in mussel cultures. *Aquaculture*, **192**, 155–169.

Britten, H.B. (1996) Meta-analyses of the association between multilocus heterozygosity and fitness. *Evolution*, **50**, 2158–2164.

Buschbaum, C. & Saier B. (2001) Growth of the mussel *Mytilus edulis* L. in the Wadden Sea affected by tidal emergence and barnacle epibionts. *Journal of Shellfish Research*, **45**, 27–36.

Calcagno, J.A., Curelovich, J.N., Fernandez, V.M., Thatje, S. & Jovrich, G.A. (2012) Effects of physical disturbance on a sub-Antarctic middle intertidal bivalve assemblage. *Marine Biology Research*, **8**, 937–953.

Cardoso, J.F.M.F., Witte, I.J.J. & van der Veer, H.W. (2006) Intra- and interspecies comparison of energy flow in bivalve species in Dutch coastal waters by means of the Dynamic Energy Budget (DEB) theory. *Journal of Sea Research*, **56**, 182–197.

Carmichael, R.H., Shriver, A.C. & Valiela, I. (2012) Bivalve response to estuarine eutrophication: the balance between enhanced food supply and habitat alterations. *Journal of Shellfish Research*, **31**, 1–11.

Celik, M.Y., Karayücel, S. & Karayücel, I. (2009) Effects of environmental factors on growth and mortality of raft cultivated mussel (*Mytilus galloprovincialis* L.) cultivated in lantern nets in Black Sea. *Aquaculture, Aquarium, Conservation and Legislation International Journal of the Bioflux Society*, **2**, 97–108.

Cerrato, R.M. (1980) Demographic analysis of bivalve populations. In: *Skeletal Growth of Aquatic Organisms* (eds D.C. Rhoads & R.A. Lutz), pp. 417–468. Plenum Press, New York.

Chute, A.S., Wainright, S.C. & Hart, D.R. (2012) Timing of shell ring formation and patterns of shell growth in the sea scallop *Placopecten magellanicus* based on stable oxygen isotopes. *Journal of Shellfish Research*, **31**, 649–662.

Connor, K.M & Robles, C.D. (2015) Within-site variation of growth rates and terminal sizes in *Mytilus californianus* along wave exposure and tidal gradients. *Biological Bulletin*, **228**, 39–51.

Conover, R.J. (1966) Assimilation of organic matter by zooplankton. *Limnology and Oceanography*, **11**, 338–354.

Corporeau, C., Groisillier, A., Jeudy, A., Barbeyron, T., Fleury, E., Fabioux, C. *et al.* (2011) A functional study of transforming growth factor-beta from the gonad of Pacific oyster *Crassostrea gigas*. *Marine Biotechnology*, **13**, 971–980.

Costa, C., Aguzzi, J., Menesatti, P., Antonucci, F., Rimatori, V. & Mattoccia, M. (2008) Shape analysis of different populations of clams in relation to their geographical structure. *Journal of Zoology*, **276**, 71–80.

Cubillo, A.M., Peteiro, L.G., Fernández-Reiriz, M.J. & Labarta, U. (2012) Influence of stocking density on growth of mussels (*Mytilus galloprovincialis*) in suspended culture. *Aquaculture*, **342–343**, 103–111.

Culbertson, J.B., Valiela, I., Olsen, Y.S. & Reddy, C.M. (2008) Effect of field exposure to 38-year-old residual petroleum hydrocarbons on growth, condition index, and filtration rate of the ribbed mussel, *Geukensia demissa*. *Environmental Pollution*, **154**, 312–319.

Cusack, M., Parkinson, D., Freer, A., Pérez-Huerta, A., Fallick, A.E. & Curry, G.B. (2008) Oxygen isotope composition in *Modiolus modiolus* aragonite in the context of biological and crystallographic control. *Mineralogical Magazine*, **72**, 569–577.

Dabrowski, T., Lyons, K., Curé, M., Berry, A. & Nolan, G. (2013) Numerical modelling of spatio-temporal variability of growth of *Mytilus edulis* (L.) and influence of its cultivation on ecosystem functioning. *Journal of Sea Research*, **76**, 5–21.

Dame, R.F. (2012) *Ecology of Marine Bivalves: An Ecosystem Approach*, 2nd edn. CRC Press, Boca Raton, FL.

David, P., Delay, B. & Jarne, P. (1997) Heterozygosity and growth in the marine bivalve *Spisula ovalis*: testing alternative hypotheses. *Genetical Research Cambridge*, **70**, 215–223.

Davis, J.P., 1997. Optimizing triploid production techniques and comparative field performance of Mediterranean mussels (*Mytilus galloprovincialis*) in Puget Sound. *Journal of Shellfish Research*, **20**, 311–312.

de Sá, F.S., Nalesso, R.C. & Paresque, K. (2007) Fouling organisms on *Perna perna* mussels: is it worth removing them? *Brazilian Journal of Oceanography*, **55**,155–161.

Devescovi, M. (2009) Biometric differences between date mussels *Lithophaga lithophaga* colonizing artificial and natural structures. *Acta Adriatica*, **50**, 129–138.

Dittmann, D. & Robles, C. (1991) Effect of algal epiphytes on the mussel *Mytilus californianus*. *Ecology*, **72**, 286–296.

Dolmer, P. (2000) Feeding activity of mussels *Mytilus edulis* related to near-bed currents and phytoplankton biomass. *Journal of Sea Research*, **44**, 221–231.
Dowd, M. (1997) On predicting the growth of cultured bivalves. *Ecological Modelling*, **104**, 113–131.
Duarte, P., Meneses, R., Hawkins, A.J.S., Zhu, M., Fang, J. & Grant, J. (2003) Mathematical modelling to assess the carrying capacity for multi-species culture within coastal waters. *Ecological Modelling*, **168**, 109–143.
Duarte, P., Labarta, U. & Fernández-Reiriz, M.J. (2008) Modelling local food depletion effects in mussel rafts of Galician Rias. *Aquaculture*, **274**, 300–312.
Duarte, P., Fernández-Reiriz, M.J., Filgueira, R. & Labarta, U. (2010) Modelling mussel growth in ecosystems with low suspended matter loads. *Journal of Sea Research*, **64**, 273–286.
Duarte, P., Fernández-Reiriz, M.J. & Labarta, U. (2012) Modelling mussel growth in ecosystems with low suspended matter loads using a Dynamic Energy Budget approach. *Journal of Sea Research*, **67**, 44–57.
Elliot, M., deMenocal, P.B., Linsley, B.K. & Howe, S.S. (2003) Environmental controls on the stable isotopic composition of *Mercenaria mercenaria*: potential application to paleoenvironmental studies. *Geochemistry Geophysics Geosystems*, **4**, art. 7.
van Erkom Schurink, C. & Griffiths, C.L. (1993) Factors affecting relative rates of growth in 4 South-African mussel species. *Aquaculture*, **109**, 257–273.
Favrel, P., Lelong, C. & Mathieu, M. (1998) Control of growth and differentiation in bivalve mollusc larvae. *Molecular characterization of a new factor from the oyster Crassostrea gigas. Annals of the New York Academy of Sciences*, **839**, 316–320.
Fernández-Reiriz, M. J., Irisarri, J & Labarta, U. (2016) Flexibility of physiological traits underlying inter-individual growth differences in intertidal and subtidal mussels *Mytilus galloprovincialis*. *PLoS ONE*, **11**, e0148245.
Ferson, S., Rohlf, F.J. & Koehn, R.K. (1985) Measuring shape variation of two-dimensional outlines. *Systematic Zoology*, **34**, 59–68.
Filgueira, R. & Grant, J. (2009) A box model for ecosystem-level management of mussel culture carrying capacity in a coastal bay. *Ecosystems*, **12**, 1222–1233.
Filgueira, R., Rosland, R. & Grant, J. (2011) A comparison of scope for growth (SFG) and dynamic energy budget (DEB) models applied to the blue mussel (*Mytilus edulis*). *Journal of Sea Research*, **66**, 403–410.
Fitzer, S.C., Chung, P., Maccherozzi, F., Dhesi, S.S., Kamenos, N.A., Phoenix, V.R. *et al.* (2016) Biomineral shell formation under ocean acidification: a shift from order to chaos. *Scientific Reports*, **6**, 21076.
Fleury, E., Fabioux, C., Lelong, C., Favrel, P. & Huvet, A. (2008) Characterization of a gonad-specific transforming growth factor-β superfamily member differentially expressed during the reproductive cycle of the oyster *Crassostrea gigas*, *Gene*, **410**, 187–196.
Fly, E.K. & Hilbish, T.J. (2013) Physiological energetics and biogeographic range limits of three congeneric mussel species. *Oecologia*, **172**, 35–46.
Ford, H.L., Schellenberg, S.A., Becker, B.J., Deutschman, D.L. Dyck, K.A. & Koch, P.L. (2010) Evaluating the skeletal chemistry of *Mytilus californianus* as a temperature proxy: effects of microenvironment and ontogeny. *Paleoceanography*, **25**, PA1203.
Fournier, D.A., Sibert, J.R., Majkowski, J. & Hampton, J. (1990) MULTIFAN a likelihood-based method for estimating growth parameters and age composition from multiple length

frequency data sets illustrated using data for southern bluefin tuna (*Thunnus maccoyii*). *Canadian Journal of Fisheries and Aquatic Sciences*, **47**, 301–317.

Frandsen, R.P. & Dolmer, P. (2002) Effects of substrate type on growth and mortality of blue mussels (*Mytilus edulis*) exposed to the predator *Carcinus maenas*. *Marine Biology*, **141**, 253–262.

Franz, D.R. (1997) Resource allocation in the intertidal salt-marsh mussel *Geukensia demissa* in relation to shore level. *Estuaries*, **20**, 134–148.

Fréchette, M. & Grant, J. (1991) An in situ estimation of the effect of wind-driven resuspension on the growth of the mussel *Mytilus edulis* L. *Journal of Experimental Marine Biology and Ecology*, **148**, 201–213.

Fréchette, M., Butman, C. & Geyer, W.R. (1989) The importace of boundary-layer flows in supplying phytoplankton to the benthic suspension feeder, *Mytilus edulis* L. *Limnology and Oceanography*, **34**, 19–36.

Fréchette, M., Bergeron, P. & Gagnon, P. (1996) On the use of self-thinning relationships in stocking experiments. *Aquaculture*, **145**, 91–112.

Fréchette, M., Alunno-Bruscia, M., Dumais, J.-F., Sirois, R. & Daigle, G. (2005) Incompleteness and statistical uncertainty in competition/stocking experiments. *Aquaculture*, **246**, 209–225.

Freitas, P.S., Clarke, L.J., Kennedy, H. & Richardson, C.A. (2008) Inter- and intra-specimen variability masks reliable temperature control on shell Mg/Ca ratios in laboratory- and field-cultured *Mytilus edulis* and *Pecten maximus* (Bivalvia). *Biogeosciences*, **5**, 1245–1258.

Fuentes, J., Gregorio, V., Giráldez, R. & Molares, J. (2000) Within-raft variability of the growth rate of mussels, *Mytilus galloprovincialis*, cultivated in the Ría de Arousa (NW Spain). *Aquaculture*, **189**, 39–52.

Fujikura, K., Okoshi, K. & Naganuma, T. (2003) Strontium as a marker for estimation of microscopic growth rates in a bivalve. *Marine Ecology Progress Series*, **257**, 295–301.

García-Berthou, E., Carmona-Catot, G., Merciai, R. & Ogle, D.H. (2012) A technical note on seasonal growth models. *Reviews in Fish Biology and Fisheries*, **22**, 635–640.

Gardner, J.P.A. (1996) The *Mytilus edulis* species complex in southwest England: effects of hybridization and introgression upon interlocus associations and morphometric variation. *Marine Biology*, **125**, 385–399.

Gardner, J.P.A. & Thompson, R.J. (2009) Influence of genotype and geography on shell shape and morphometric trait variation among North Atlantic blue mussel (*Mytilus* spp.) populations. *Biological Journal of the Linnean Society*, **96**, 875–897.

Gayanilo, F.C., Sparre, P. & Pauly, D. (2005) FAO-ICLARM Stock Assessment Tools II (FiSAT II). User's guide. FAO Computerized Information Series (Fisheries). No. 8, revised version. FAO, Rome.

Gazeau, F., Gattuso, J.-P., Dawber, C., Pronker, A.E., Peene, F., Peene, J. *et al.* (2010) Effect of ocean acidification on the early life stages of the blue mussel *Mytilus edulis*. *Biogeosciences*, **7**, 2051–2060.

Gazeau, F., Alliouane, S., Bock, C., Bramanti, L., López Correa, M., Gentile, M. *et al.* (2014) Impact of ocean acidification and warming on the Mediterranean mussel (*Mytilus galloprovincialis*). *Frontiers in Marine Science*, **1**, art. 62.

Geest, M. van der, Gils, J.A. van, Meer, J. van der, Olff, H. & Piersma, T. (2011) Suitability of calcein as an *in situ* growth marker in burrowing bivalves. *Journal of Experimental Marine Biology and Ecology*, **399**, 1–7.

Giard, W., Lebel, J-M., Boucaud-Camou, E. & Favrel, P. (1998) Effects of vertebrate growth factors on digestive gland cells from the mollusk *Pecten maximus* L.: an *in vitro* study. *Journal of Comparative Physiology B*, **168**, 81–86.

Goff, S.A. (2011) A unifying theory for general multigenic heterosis: energy efficiency, protein metabolism, and implications for molecular breeding. *New Phytologist*, **189**, 923–937.

Gompertz, B. (1825) On the nature of the function expressive of human mortality, and on a new mode to determining the value of life contingencies. *Philosophical Transactions of the Royal Society London*, **115**, 513–585.

Gosling, E.M. (1992) Systematics and geographic distribution of *Mytilus*. In: *The Mussel Mytilus: Ecology, Physiology, Genetics and Culture* (ed E.M. Gosling), pp. 1–20. Elsevier Science Publishers B.V., Amsterdam.

Gosling, E.M. (2015) *Marine Bivalve Molluscs*, 2nd edn. Wiley Blackwell, Hoboken, NJ.

Gosselin, M., Lazareth, C.E. & Ortlieb, L. (2013) Sclerochronological studies in the Humboldt current system, a highly variable ecosystem. *Journal of Shellfish Research*, **32**, 867–882.

Grant, J., Dowd, M., Thompson, K.R., Emerson C. & Hatcher, A. (1993) Perspectives on field studies and related biological models of bivalve growth and carrying capacity. In: *Bivalve Filter Feeders and Marine Ecosystem Processes* (ed R. Dame), pp. 371–420. Springer Verlag, New York.

Grant J., Bacher, C., Cranford, P.J., Guyondet, T. & Carreau, M. (2008) A spatially explicit ecosystem model of seston depletion in dense mussel culture. *Journal of Marine Systems*, **73**, 155–168.

Gray, A.P., Seed, R. & Richardson, C.A. (1997) Reproduction and growth of *Mytilus edulis chilensis* from the Falkland Islands. *Scientia Marina*, **61**, 39–48.

Gricourt, L., Bonnec, G., Boujard, D., Mathieu, M. & Kellner, K. (2003) Insulin-like system and growth regulation in the Pacific oyster *Crassostrea gigas*: hrIGF-1 effect on protein synthesis of mantle edge cells and expression of an homologous insulin receptor-related receptor. *General and Comparative Endocrinology*, **134**, 44–56.

Gricourt, L., Mathieu, M. & Kellner, K. (2006) An insulin-like system involved in the control of Pacific oyster *Crassostrea gigas* reproduction: hrIGF-1 effect on germinal cell proliferation and maturation associated with expression of an homologous insulin receptor-related receptor. *Aquaculture*, **251**, 85–98.

Griffiths, C.L. (1981) Production and energy flow in relation to age and shore level in the bivalve *Choromytilus meridionalis* (Kr.). *Estuarine, Coastal and Shelf Science*, **13**, 477–493.

Griffiths, C.L. & King, J.A. (1979) Some relationships between size, food availability and energy balance in the ribbed mussel *Aulacomya ater*. *Marine Biology*, **51**, 141–149.

Griffiths, C.L. & Griffiths, R.J. (1987) *Bivalvia. In: Animal Energetics, Vol. 2, Bivalvia through Reptilia (eds T.J. Pandian & F.J. Vernberg)*, pp. 1–87. Academic Press, Cambridge, MA.

Griffiths, C.L. & Hockey, P.A.R. (1987) A model describing the interactive roles of predation, competition and tidal elevation in structuring mussel populations. *South African Journal of Marine Science*, **5**, 547–556.

Guyondet, T., Roy, S., Koutitonsky, V.G., Grant, J. & Tita, G. (2010) Integrating multiple spatial scales in the carrying capacity assessment of a coastal ecosystem for bivalve aquaculture. *Journal of Sea Research*, **64**, 341–359.

Hahn, S., Rodolfo-Metalpa, R., Griesshaber, E., Schmahl, W.W., Buhl, D., Hall-Spencer, J.M. *et al.* (2012) Marine bivalve shell geochemistry and ultrastructure from modern low pH environments: environmental effect versus experimental bias. *Biogeosciences*, **9**, 1897–1914.

Hamano, K., Awaji, M. & Usuki, H. (2005) cDNA structure of an insulin-related peptide in the Pacific oyster and seasonal changes in the gene expression. *Journal of Endocrinology*, **187**, 55–67.

Hammond, W. & Griffiths, C.L. (2004) Influence of wave exposure on South African mussel beds and their associated infaunal communities. *Marine Biology*, **144**, 547–552.

Handå, A., Alver, M., Edvardsen, C.V., Halstensen, S., Olsen, A.J., Øie, G. *et al.* (2011) Growth of farmed blue mussels (*Mytilus edulis* L.) in a Norwegian coastal area: comparison of food proxies by DEB modeling. *Journal of Sea Research*, **66**, 297–307.

Hawkins, A.J.S. & Bayne, B.L. (1992) Physiological interrelations and the regulation of production. In: *The Mussel Mytilus: Ecology, Physiology, Genetics and Culture* (ed E.M. Gosling), pp. 171–222. Elsevier Science Publishers B.V., Amsterdam.

Hawkins, A.J.S., Smith, R.F.M., Bayne, B.L. & Heral, M. (1996) Novel observations underlying the fast growth of suspension-feeding shellfish in turbid environments: *Mytilus edulis*. *Marine Ecology Progress Series*, **131**, 179–190.

HELCOM (2013) *Climate change in the Baltic Sea area: HELCOM thematic assessment in 2013.* Baltic Sea Environment Proceedings, No. 137.

Herrmann, M., Lepore, M.L., Laudien, J., Arntz, W.E. & Penchaszadeh, P.E. (2009) Growth estimations of the Argentinean wedge clam *Donax hanleyanus*: a comparison between length-frequency distribution and size-increment analysis. *Journal of Experimental Marine Biology and Ecology*, **379**, 8–15.

Hiebenthal, C., Philipp, E.E.R., Eisenhauer, A. & Wahl, M. (2012) Interactive effects of temperature and salinity on shell formation and general condition in Baltic Sea *Mytilus edulis* and *Arctica islandica*. *Aquatic Biology*, **14**, 289–298.

Hiebenthal, C., Philipp, E.E.R., Eisenhauer, A. & Wahl, M. (2013) Effects of seawater pCO_2 and temperature on shell growth, shell stability, condition and cellular stress of Western Baltic Sea *Mytilus edulis* (L.) and *Arctica islandica* (L.). *Marine Biology*, **160**, 2073–2087.

Hilbish T.J. & Zimmerman, K.M. (1988) Genetic and nutritional control of the gametogenic cycle in *Mytilus edulis*. *Marine Biology*, **98**, 223–228.

Huvet, A., Fleury, E., Corporeau, C., Quillien, V., Daniel, J.Y., Riviere, G. *et al.* (2012) in vivo RNA interference of a gonad-specific transforming growth factor-β in the Pacific oyster *Crassostrea gigas*. *Marine Biotechnology*, **14**, 402–410.

Innes, D.J. & Bates, A. (1999) Morphological variation of *Mytilus edulis* and *Mytilus trossulus* in eastern Newfoundland. *Marine Biology*, **133**, 691–399.

IPCC (2014) Climate Change 2014: Synthesis Report. Contribution of Working Groups I, II and III to the Fifth Assessment Report of the Intergovernmental Panel on Climate Change (core writing team R.K. Pachauri & L.A. Meyer). IPCC, Geneva.

Jones, D.B., Jerry, D.R., Khatkar, M.S., Moser, G., Raadsma, H.W., Taylor, J.J. *et al.* (2014) Determining genetic contributions to host oyster shell growth: quantitative trait loci and genetic association analysis for the silver-lipped pearl oyster, *Pinctada maxima*. *Aquaculture*, **434**, 367–375.

Jusup, M., Sousa, T., Domingos, T., Labinac, V., Marn, N., Wang, Z. *et al.* (2017) Physics of metabolic organization. *Physics of Life Reviews*, **20**, 1–39.

Kaehler, S. & McQuaid, C.D. (1999) Use of the fluorochrome calcein as an in situ growth marker in the brown mussel *Perna perna*. *Marine Biology*, **133**, 455–460.

Kamermans, P., Galley, T., Boudry, P., Fuentes, J., McCombie, H. & Batista, F.M. (2013) Blue mussel hatchery technology in Europe. In: *Advances in Aquaculture Hatchery Technology* (eds G. Allan & G. Burnell), pp. 339–373. Woodhead Publishing, Cambridge.

Kang, C.K., Lee, Y.W., Choy, E.J., Shin, J.K., Seo, I.S. & Hong, J.S. (2006) Microphytobenthos seasonality determines growth and reproduction in intertidal bivalves. *Marine Ecology Progress Series*, **315**, 113–127.

Karakousis, Y. & Skibinski, D.O.F. (1992) An analysis of allozyme, mitochondrial DNA and morphological variation in mussel (*Mytilus galloprovincialis*) populations from Greece. *Experientia*, **48**, 878–881.

Karayücel, S., Çelik, M.Y., Karayücel, I. & Gökhan, Erik. (2010) Growth and production of raft cultivated Mediterranean mussel (*Mytilus galloprovincialis* Lamarck, 1819) in Sinop, Black Sea. *Turkish Journal of Fisheries and Aquatic Sciences*, **10**, 9–17.

Katsanevakis, S., Thessalou-Legaki, M., Karlou-Riga, C., Lefkaditou, E., Evagelos, D. & Verriopoulos, G. (2007) Information-theory approach to allometric growth of marine organisms. *Marine Biology*, **151**, 949–959.

Kautsky, N. (1982) Growth and size structure in a Baltic *Mytilus edulis* population. *Marine Biology*, **68**, 117–133.

Kearney, M., Simpson, S.J., Raubenheimer, D. & Helmuth, B. (2010) Modelling the ecological niche from functional traits. *Philosophical Transactions of the Royal Society B*, **365**, 3469–3483.

Kearney, M.R., Simpson, S.J., Raubenheimer, D. & Kooijman, S.A.L.M. (2013) Balancing heat, water and nutrients under environmental change: a thermodynamic niche framework. *Functional Ecology*, **27**, 950–966.

Kellner-Cousin, K., Mialhe, E. & Mathieu, M. (1994) Identification of insulin-like peptides in cerebral ganglion neurosecretory cells of the mussel *Mytilus edulis*. *Tissue and Cell*, **26**, 891–899.

Kenchington, E.L. & Full, W.E. (1994) Fourier analysis of sea scallop (*Placopecten magellanicus*) shells determining population structure. *Canadian Journal of Fisheries and Aquatic Sciences*, **51**, 348–356.

Klein, R.T., Lohmann, K.C. & Thayer, C.W. (1996) Sr/Ca and $^{13}C/^{12}C$ ratios in skeletal calcite of *Mytilus trossulus*: covariation with metabolic rate, salinity, and carbon isotopic composition of seawater. *Geochimica et Cosmochimica Acta*, **60**, 4207–4221.

King, M. (2007) *Fisheries Biology, Assessment and Management*, 2nd edn. Blackwell Publishing, Oxford.

Kooijman, S.A.L.M. (2010) *Dynamic Energy Budget Theory for Metabolic Organisation*, 3rd edn. Cambridge University Press, Cambridge.

Kooijman, S.A.L.M. (2012) Energy bugets. In: *Sourcebook in Theoretical Ecology* (eds A. Hastings & L. Gross), pp 249–258. University of California Press, Berkeley, CA.

Kooijman, S.A.L.M. (2013) Waste to hurry: dynamic energy budgets explain the need of wasting to fully exploit blooming resources. *Oikos*, **122**, 348–357.

Krapivka, S., Toro, J.E., Alcapán A.C., Astorga, M., Presa, P., Pérez, M. *et al.* (2007) Shell-shape variation along the latitudinal range of the Chilean blue mussel *Mytilus chilensis* (Hupe 1854). *Aquaculture Research*, **38**, 1770–1777.

Kroeker, K.J., Gaylord, B., Hill, T.M., Hosfelt, J.D., Miller, S.H. & Sanford, E. (2014) The role of temperature in determining species' vulnerability to ocean acidification: a case study using *Mytilus galloprovincialis*. *PLoS ONE*, **9**, e100353.

Kuenzler, E.J. (1961) Structure and energy flow of a mussel population in a Georgia salt marsh. *Limnology and Oceanography*, **6**, 191–204.

Larsen, P.S. & Riisgård. H.U. (2016) Growth-prediction model for blue mussels (*Mytilus edulis*) on future optimally thinned farm-ropes in Great Belt (Denmark). *Journal of Marine Science and Engineering*, **4**, 42.

Larsen, P.S., Filgueira, R. & Riisgård, H.U. (2014) Somatic growth of mussels *Mytilus edulis* in field studies compared to predictions using BEG, DEB, and SFG models. *Journal of Sea Research*, **88**, 100–108.

Lartaud, F., Emmanuel, L., de Rafelis, M., Pouvreau, S. & Renard, M. (2010) Influence of food supply on the $\delta^{13}C$ signature of mollusc shells:implications for palaeoenvironmental reconstitutions. *Geo-Marine Letters*, **30**, 23–34.

Launey, S. & Hedgecock, D. (2001) High genetic load in the Pacific oyster *Crassostrea gigas*. *Genetics*, **159**, 255–265.

Lauzon-Guay, J.S., Hamilton, D.J. & Barbeau, M.A. (2005a) Effect of mussel density and size on the morphology of blue mussels (*Mytilus edulis*) grown in suspended culture in Prince Edward Island, Canada. *Aquaculture*, **249**, 265–274.

Lauzon-Guay, J.S., Barbeau, M.A. & Hamilton, D.J. (2005b) Effects of seed size and density on growth, tissue-to shell ratio and survival of cultivated mussels (*Mytilus edulis*) in Prince Edward Island, Canada. *Aquaculture*, **250**, 652–665.

Lelong, C., Mathieu, M. & Favrel, P. (2000) Structure and expression of mGDF, a new member of the transforming growth factor-β superfamily in the bivalve mollusc *Crassostrea gigas*. *European Journal of Biochemistry*, **267**, 3986–3993.

Lelong, C., Badariotti, F., Le Quéré, H., Rodet, F., Dubos, M.P. & Favrel, p. (2007) Cg-TGF-β, a TGF-β/activin 1homologue in the Pacific oyster *Crassostrea gigas*, is involved in immunity against Gram-negative microbial infection. *Developmental and Comparative Immunology*, **31**, 30–38.

Lesser, M.P., Witman, J.D. & Sebens, K.P. (1994) Effects of flow and seston availability on scope for growth of benthic suspension-feeding invertebrates from the Gulf of Maine. *Biological Bulletin*, **187**, 319–335.

Li Y. & He, M. (2014) Genetic mapping and QTL analysis of growth-related traits in *Pinctada fucata* using restriction-site associated DNA sequencing. *PLoS ONE*, **9**, e111707.

Liu, G., Huan, P. & Liu, B. (2014) Cloning and expression patterns of two Smad genes during embryonic development and shell formation of the Pacific oyster *Crassostrea gigas*. *Chinese Journal of Oceanology and Limnology*, **32**, 1224–1231.

Liu, G., Huan, P. & Liu, B. (2017) A SoxC gene related to larval shell development and co-expression analysis of different shell formation genes in early larvae of oyster. *Development Genes and Evolution*, **227**, 181–188.

Lucas, T., Palmer, P.J., Wang, S., Scoones, R. & O'Brien, E. (2008) Marking the shell of the saucer scallop *Amusium balloti* for sea ranching using oxytetracycline, calcein and alizarin red S. *Journal of Shellfish Research*, **27**, 1183–1188.

Lutz, R.A. (1976) Annual growth patterns in the inner shell layer of *Mytilus edulis* L. *Journal of the Marine Biological Association of the UK.*, **56**, 723–731.

Margosian, A., Tan, F.C., Cai, D. & Mann, K.H, (1987) Seawater temperature records from stable isotopic profiles in the shell of *Modiolus modiolus*. *Estuarine, Coastal and Shelf Science*, **25**, 81–89.

Maar, M., Bolding, K., Petersen, J.K., Hansen, J.L.S. & Timmermann, K. (2009) Local effects of blue mussels around turbine foundations in an ecosystem model of Nysted off-shore wind farm, Denmark. *Journal of Sea Research*, **62**, 159–174.

Maar, M., Ysebaert, T. & Herman, P.M.J. (2013) Near-bed gradients in particles and nutrients above a mussel bed in the Limfjorden: influence of physical mixing and mussel filtration. *Marine Ecology Progress Series*, **490**, 137–146.

Maar M., Saurel, C., Landes, A., Dolmer, P. & Petersen, J.K. (2015) Growth potential of blue mussels (*M. edulis*) exposed to different salinities evaluated by a Dynamic Energy Budget model. *Journal of Marine Systems*, **148**, 48–55.

Mallet, A.L. & Carver, C.E. (1991) An assessment of strategies for growing mussels in suspended culture. *Journal of Shellfish Research*, **10**, 471–477.

Márquez, F. & Van der Molen, S. (2011) Intraspecific shell-shape variation in the razor clam *Ensis macha* along the Patagonian coast. *Journal of Molluscan Studies*, **77**, 123–128.

Márquez, F., Amoroso, R., Gowland Sainz, M.F. & Van der Molen, S. (2010) Shell morphology changes in the scallop *Aequipecten tehuelchus* during its life span: a geometric morphometric approach. *Aquatic Biology*, **11**, 149–155.

Márquez, F., Adami, M.L., Trovan, B., Nieto-Vilela, R.A. & González-José, R. (2018) Allometric differences on the shell shape of two scorched mussel species along the Atlantic South American Coast. *Evolutionary Ecology*, **32**, 43–56.

Matzelle, A., Montalto, V., Sarà, G., Zippay, M. & Helmuth, B. (2014) Dynamic Energy Budget model parameter estimation for the bivalve *Mytilus californianus*: Application of the covariation method. *Journal of Sea Research*, **94**, 105–110.

McConnaughey, T.A. & Gillikin, D.P. (2008) Carbon isotopes in mollusk shell carbonates. *Geo-Marine Letters*, **28**, 287–299.

McDonald, J.H., Seed, R., Koehn, R.K. (1991) Allozymes and morphometric characters of three species of *Mytilus* in the Northern and Southern Hemispheres. *Marine Biology*, **111**, 323–333.

McQuaid, C.D. & Lindsay, T.L. (2000) Effect of wave exposure on growth and mortality rates of the mussel *Perna perna*: bottom-up regulation of intertidal populations. *Marine Ecology Progress Series*, **206**, 147–154.

McQuaid, C.D. & Lindsay, T.L. (2007) Wave exposure effects on population structure and recruitment in the mussel *Perna perna* suggest regulation primarily through availability of recruits and food, not space. *Marine Biology*, **151**, 2123–2131.

Melzner, F., Stange, P., Trübenbach, K., Thomsen, J., Casties, I., Panknin, U. *et al.* (2011) Food supply and seawater pCO_2 impact calcification and internal shell dissolution in the blue mussel *Mytilus edulis*. *PLoS ONE*, **6**, e24223.

Melzner, F., Thomsen, J., Koeve, W., Oschlies, A., Gutowska, M.A., Bange, H.W. *et al.* (2013) Future ocean acidification will be amplified by hypoxia in coastal habitats. *Marine Biology*, **160**, 1875–1888.

Menge, B., Daley, B.A., Wheeler, P.A., Dahlhoff, E., Sanford, E. & Strub, P.T. (1997) Benthic–pelagic links and rocky intertidal communities: bottom-up effects on top-down control? *Proceedings of the National Academy of Sciences*, **94**, 14530–14535.

Menge, B.A., Chan, F. & Lubchenco, J. (2008) Response of a rocky intertidal ecosystem engineer and community dominant to climate change. *Ecology Letters*, **11**, 151–162.

Michaelidis, B., Ouzounis, C., Paleras, A. & Pörtner, H.O. (2005) Effects of long-term moderate hypercapnia on acid–base balance and growth rate in marine mussels *Mytilus galloprovincialis*. *Marine Ecology Progress Series*, **293**, 109–118.

Millstein, J. & O'Clair, C.E. (2001) Comparison of age-length and growth-increment general growth models of the Schnute type in the Pacific Blue Mussel, *Mytilus trossulus* Gould. *Journal of Experimental Marine Biology and Ecology*, **262**, 155–176.

Moneva, C.S.O., Baquiano, P.M.L., Blasco, J.O., Borlaza, K.M.E., Burias, D.M.E., Flores, K.A. *et al.* (2014) Comparative morphological descriptions of interior shell patterns of the Venerid bivalves: *Meretrix lyrata*, *Mercenaria mercenaria* and *Venerupis philippinarum* using landmark-based geometric morphometric analysis. *AACL Bioflux*, **7**, 386–395.

Montalto, V., Palmeri, V., Rinaldi, A., Kooijman, S.A.L.M. & Sarà, G. (2014) Dynamic Energy Budget parameterisation of *Brachidontes pharaonis*, a Lessepsian bivalve in the Mediterranean Sea. *Journal of Sea Research*, **94**, 47–51.

Moschino, V., Bressan, M., Cavaleri, L. & Da Ros, L. (2015) Shell-shape and morphometric variability in *Mytilus galloprovincialis* from micro-tidal environments: responses to different hydrodynamic drivers. *Marine Ecology*, **36**, 1440–1453.

Mubiana, V.K. & Blust, R. (2007) Effects of temperature on scope for growth and accumulation of Cd, Co, Cu and Pb by the marine bivalve *Mytilus edulis*. *Marine Environmental Research*, **63**, 219–235.

Muller, E.B. & Nisbet, R.M. (2000) Survival and production in variable resource environments. *Bulletin of Mathematical Biology*, **62**, 1163–1189.

Muller, E.B., Hanna, S.K., Lenihan, H.S., Miller, R.J. & Nisbet, R.M. (2014) Impact of engineered zinc oxide nanoparticles on the energy budgets of *Mytilus galloprovincialis*. *Journal of Sea Research*, **94**, 29–36.

Myrand, B., Tremblay, R. & Sévigny, J.M. (2002) Selection against blue mussels (*Mytilus edulis* L.) homozygotes under various stressful conditions. *Journal of Heredity*, **93**, 238–248.

Myrand, B., Tremblay, R. & Sévigny, J.M. (2009) Impact of suspension culture using mesh sleeves on genetic characteristics of *Mytilus edulis* L. in Canada. *Aquaculture*, **291**, 147–153.

Navarro, J.M., Torres, R., Acuña, K., Duarte, C., Manriquez, P.H., Lardies, M. *et al.* (2013) Impact of medium-term exposure to elevated pCO_2 levels on the physiological energetics of the mussel *Mytilus chilensis*. *Chemosphere*, **90**, 1242–1248.

Navarro, J.M., Duarte, C., Manríquez, P.H., Lardies, M.A., Torres, R., Acuña, K. *et al.* (2016) Ocean warming and elevated carbon dioxide: multiple stressor impacts on juvenile mussels from southern Chile. *ICES Journal of Marine Science*, **73**, 764–771.

Newell, C.R. (1990) The effects of mussel (*Mytilus edulis*, Linnaeus, 1758) position in seeded bottom patches on growth at subtidal lease sites in Maine. *Journal of Shellfish Research*, **9**, 113–118.

Nielsen, P., Cranford, P.J., Maar, M. & Petersen, J.K. (2016) Magnitude, spatial scale and optimization of ecosystem services from a nutrient extraction mussel farm in the eutrophic Skive Fjord, Denmark. *Aquaculture Environment Interactions*, **8**, 311–329.

Niu, D., Wang, L., Bai, Z., Xie, S., Zhao, H. & Li, J. (2015) Identification and expression characterization of the myostatin (MSTN) gene and association analysis with growth traits in the razor clam *Sinonovacula constricta*. *Gene*, **555**, 297–304.

Ocaña, F.A (2015) Growth and production of *Donax striatus* (Bivalvia: Donacidae) from Las Balsas beach, Gibara, Cuba. *Revista de Biología Tropical.* **63**, 639–646.

O'Connor, N.E. (2010) Shore exposure affects mussel population structure and mediates the effect of epibiotic algae on mussel survival in SW Ireland. *Estuarine, Coastal and Shelf Science*, **87**, 83–91.

Ogilvie, S.C., Fox, S.P., Ross, A.H., James, M.R. & Schiel, D.R. (2004) Growth of cultured mussels (*Perna canaliculus* Gmelin 1791) at a deep-water chlorophyll maximum layer. *Aquaculture Research*, **35**, 1253–1260.

Okamura, B. (1986) Group living and the effects of spatial position in aggregations of *Mytilus edulis*. *Oecologia*, **69**, 341–347.

Pace, A.D., Marsh, A.G., Leong, P.K., Green, A.J., Hedgecock, D. & Manahan, D.T. (2006) Physiological bases of genetically determined variation in growth of marine invertebrate larvae: a study of growth heterosis in the bivalve *Crassostrea gigas*. *Journal of Experimental Marine Biology and Ecology*, **335**, 188–209.

Page, H.M. & Ricard, Y.O. (1990) Food availability as a limiting factor to mussel *Mytilus edulis* growth in California coastal waters. *Fishery Bulletin*, **88**, 677–686.

Pauly, D. (1987) A review of the ELEFAN system for analysis of length-frequency data in fish and aquatic invertebrates. In: *Length Based Methods in Fisheries Research* (eds D. Pauly & G.R. Morgan), pp. 7–34. International Center for Living Aquatic Resources Management (ICLARM), Manila.

Pauly, D. & Arreguín-Sánchez, F. (1995) Improving Shepherd's length composition analysis (SLCA) method for growth parameter estimations. *Naga, the ICLARM Quarterly*, **18**, 31–33.

Pechenik, J.A., Eyster, L.S., Widdows, J. & Bayne, B.L. (1990) The influence of food concentration and temperature on growth and morphological differentiation of blue mussel *Mytilus edulis* L. larvae. *Journal of Experimental Marine Biology and Ecology*, **136**, 41–64.

Peharda, M., Richardson, C.A., Mladineo, I., Šestanović, S., Popović, Z., Bolotin, J. *et al.* (2007) Age, growth and population structure of *Modiolus barbatus* from the Adriatic. *Marine Biology*, **151**, 629–638.

Peharda, M., Puljas, S., Chauvaud, L., Schöne, B.R., Ezgeta-Balić, D. & Thébault, J. (2015) Growth and longevity of *Lithophaga lithophaga*: what can we learn from shell structure and stable isotope composition? *Marine Biology*, **162**, 1531–1540.

Pernet, F., Maleta, N., Pastoureaud, A., Vaquer, A., Quéré, C. & Dubroca, L. (2012) Marine diatoms sustain growth of bivalves in a Mediterranean lagoon. *Journal of Sea Research*, **68**, 20–32.

Peteiro, L.G., Babarro, J.M.F., Labarta, U. & Fernández-Reiriz, M.J. (2006) Growth of *Mytilus galloprovincialis* after the Prestige oil spill. *ICES Journal of Marine Science*, **63**, 1005–1013.

Petersen, J.K., Nielsen, T.G., van Duren, L. & Maar, M. (2008) Depletion of plankton in a raft culture of *Mytilus galloprovincialis* in Ria de Vigo, NW Spain. I. Phytoplankton. *Aquatic Biology*, **4**, 113—125.

Petersen, J.K., Maar, M., Ysebaert, T. & Herman, P.M.J. (2013) Near-bed gradients in particles and nutrients above a mussel bed in the Limfjorden: influence of physical mixing and mussel filtration. *Marine Ecology Progress Series*, **490**, 137–146.

Pfister, C.A., McCoy, S.J., Wootton, J.T., Martin, P.A., Colman, A.S. & Archer, D. (2011) Rapid environmental change over the past decade revealed by isotopic analysis of the California mussel in the Northeast Pacific. *PLoS ONE*, **6**, e25766.

Phillips, N.E. (2005) Growth of filter-feeding benthic invertebrates from a region with variable upwelling intensity. *Marine Ecology Progress Series*, **295**, 79–89.

Piferrer, F., Beaumont, A., Falguière, J-C., Flajšhans, M. Haffray, P. & Colombo, L. (2009) Polyploid fish and shellfish: production, biology and applications to aquaculture for performance improvement and genetic containment. *Aquaculture*, **293**, 125–156.

Ponnurangam, A., Bau, M., Brenner, M. & Koschinsky, A. (2016) Mussel shells of *Mytilus edulis* as bioarchives of the distribution of rare earth elements and yttrium in seawater and the potential impact of pH and temperature on their partitioning behavior. *Biogeosciences*, **13**, 751–760.

Rajaei, M., Poorbagher, H., Farahmand, H., Mortazavi, M.S. & Eagder, S. (2014) Interpopulation differences in shell forms of the pearl oyster, *Pinctada imbricata radiata* (Bivalvia: Pterioida), in the northern Persian Gulf inferred from principal component analysis and elliptic Fourier analysis. *Turkish Journal of Zoology*, **38**, 42–48.

Ramón, M., Fernández, M. & Galimany, E. (2007) Development of mussel (*Mytilus galloprovincialis*) seed from two different origins in a semi-enclosed Mediterranean Bay (NE Spain). *Aquaculture*, **264**, 148–159.

Range, P., Piló, D., Ben-Hamadou, R., Chícharo, M.A., Matias, D. Joaquim, S. *et al.* (2012) Seawater acidification by CO2 in a coastal lagoon environment: effects on life history traits of juvenile mussels *Mytilus galloprovincialis*. *Journal of Experimental Marine Biology and Ecology*, **424–425**, 89–98.

Ren, J.S. (2009) Effect of food quality on energy uptake. *Journal of Sea Research*, **62**, 72–74.

Ren, J.S. & Ross, A.H. (2005) Environmental influence on mussel growth: a dynamic energy budget model and its application to the greenshell mussel *Perna canaliculus*. *Ecological Modelling*, **189**, 347–362.

Ren, J.S., Ross, A.H., Hadfield, M.G. & Hayden, B.J. (2010) An ecosystem model for estimating potential shellfish culture production in sheltered coastal waters. *Ecological Modelling*, **221**, 527–539.

Resgalla, C., de Souza Brasil, E. & Salomão, L.S. (2007) The effect of temperature and salinity on the physiological rates of the mussel *Perna perna* (Linnaeus 1758). *Brazilian Archives of Biology and Technology*, **50**, 543–556.

Reynolds, D.J., Scourse, J.D., Halloran, P.R., Nederbragt, A.J., Wanamaker, A.D., Butler, P.G. *et al.* (2016) Annually resolved North Atlantic marine climate over the last millennium. *Nature Communications*, **7**, 13502.

Richardson, C.A. (1989) An analysis of the microgrowth bands in the shell of the common mussel *Mytilus edulis*. *Journal of the Marine Biological Association of the UK*, **69**, 477–491.

Richardson, C.A. (2001) Molluscs as archives of environmental change. *Oceanography Marine Biology Annual Review*, **39**, 103–164.

Richardson, C.A, Seed, R. & Naylor, E. (1990) Use of internal growth bands for measuring individual and population growth rates in *Mytilus edulis* from offshore production platforms. *Marine Ecology Progress Series*, **66**, 259–265.

Ricker, W.E. (1975) Computation and interpretation of biological statistics of fish populations, *Bulletin of the Fisheries Research Board of Canada*, **191**, 1–382.

Riisgård, H.U. (1998) No foundation of a '3/4 power scaling law' for respiration in biology. *Ecology Letters*, **1**, 71–73.

Riisgård, H.U., Bottiger, L. & Pleissner, D. (2012) Effect of salinity on growth of mussels, *Mytilus edulis*, with special reference to Great Belt (Denmark). *Open Journal of Marine Science*, **2**, 167–176.

Riisgård, H-U., Pleissner, D., Lundgreen, K. & Larsen, P.S. (2013) Growth of mussels *Mytilus edulis* at algal (*Rhodomonas salina*) concentrations below and above saturation levels for reduced filtration rate. *Marine Biology Research*, **9**, 1005–1017.

Riisgård, H.U., Larsen, P.S., Turja, R. & Lundgreen, K. (2014) Dwarfism of blue mussels in the low saline Baltic Sea–growth to the lower salinity limit. *Marine Ecology Progress Series*, **517**, 181–192.

Rius, M. & McQuaid, C.D. (2006) Wave action and competitive interaction between the invasive mussel *Mytilus galloprovincialis* and the indigenous *Perna perna* in South Africa. *Marine Biology*, **150**, 69–78.

Rius, M. & McQuaid, C.D. (2009) Facilitation and competition between invasive and indigenous mussels over a gradient of physical stress. *Basic and Applied Ecology*, **10**, 607–613.

Rodhouse, P.G., Roden, C.M., Hensey, M.P., McMahon, T., Ottway, B. & Ryan, T.H. (1984) Food resource, gametogenesis and growth of *Mytilus edulis* on the shore and in suspended culture: Killary Harbour, Ireland. *Journal of the Marine Biological Association of the UK*, **64**, 513–529.

Rodhouse, P.G., Roden, C.M., Hensey, M.P. & Ryan, T.H. (1985) Production of mussels, *Mytilis edulis*, in a suspended culture and estimates of carbon and nitrogen flow: Killary Harbour, Ireland. *Journal of the Marine Biological Association of the UK*, **65**, 55–68.

Rosland, R., Strand, Ø., Alunno-Bruscia, M., Bacher, C. & Strohmeier, T. (2009) Applying Dynamic Energy Budget (DEB) theory to simulate growth and bio-energetics of blue mussels under low seston conditions. *Journal of Sea Research*, **62**, 49–61.

Rosland, R., Bacher, C., Strand, O., Aure, J. & Strohmeier, T. (2011) Modelling growth variability in longline mussel farms as a function of stocking density and farm design. *Journal of Sea Research*, **66**, 318–330.

Rufino, M. M., Gaspar, M.B., Pereira, A.M. & Vasconcelos, P. (2006) Use of shape to distinguish *Chamelea gallina* and *Chamelea striatula* (Bivalvia: Veneridae): linear and geometric morphometric methods. *Journal of Morphology*, **267**, 1433–1440.

Saavedra, C. & Guerra, A. (1996) Allozyme heterozygosity, founder effect and fitness traits in a cultivated population of the European oyster, *Ostrea edulis*. *Aquaculture*, **139**, 203–224.

Safi, K.A. & Gibbs, M.M. (2003) Importance of different size classes of phytoplankton in Beatrix Bay, Marlborough Sounds, New Zealand, and the potential implications for the aquaculture of the mussel, *Perna canaliculus*. *New Zealand Journal of Marine and Freshwater Research*, **37**, 267–272.

Sanders, T., Widdicombe, S., Calder-Potts, R. & Spicer, J.I. (2014) Environmental hypoxia but not minor shell damage affects scope for growth and body condition in the blue mussel *Mytilus edulis* (L.). *Marine Environmental Research*, **95**, 74–80.

Sansom, B.J., Hornbach, D.J., Hove, M.C. & Kilgore, J.S. (2013) Effects of flow restoration on mussel growth in a Wild and Scenic North American River. *Aquatic Biosystems*, **9**, 6.

Sarà, G., Manganaro, A., Cortese, G., Pusceddu, A. & Mazzola, A. (1998) The relationship between food availability and growth in *Mytilus galloprovincialis* in the open sea (southern Mediterranean). *Aquaculture*, **167**, 1–15.

Sarà, G., Romano, C., Widdows, J. & Staff, F.J. (2008) Effect of salinity and temperature on feeding physiology and scope for growth of an invasive species (*Brachidontes pharaonis* – Mollusca: Bivalvia) within the Mediterranean Sea. *Journal of Experimental Marine Biology and Ecology*, **363**, 130–136.

Sarà, G., Kearney, M. & Helmuth, B. (2011) Combining heat-transfer and energy budget models to predict thermal stress in Mediterranean intertidal mussels. *Chemical Ecology*, **27**, 135–145.

Scalici, M., Traversetti, L., Spani, F., Bravi, R., Malafoglia, V., Persichini, T. *et al.* (2016) Using 3D virtual surfaces to investigate molluscan shell shape. *Aquatic Living Resources*, **29**, 207–211.

Schnute, J.T. & Richards, L.J. (1990) A unified approach to the analysis of fish growth, maturity, and survivorship data. *Canadian Journal of Fisheries and Aquatic Sciences*, **47**, 24–40.

Schöne, B.R. & Giere, O. (2005) Growth increments and stable isotope variation in shells of the deep-sea hydrothermal vent bivalve mollusk *Bathymodiolus brevior* from the North Fiji Basin, Pacific Ocean. *Deep-Sea Research Part I*, **52**, 1896–1910.

Sebens, K., Sarà, G. & Nishizaki, M. (2016) Energetics, particle capture, and growth dynamics of benthic suspension feeders. In: *Marine Animal Forests* (eds S. Rossi, L. Bramanti, A. Gori & C.O. Saco del Valle), pp. 813–854. Springer International Publishing, Switzerland.

Seed, R. (1968) Factors influencing shell shape in *Mytilus edulis*. *Journal of the Marine Biological Association of the UK*, **48**, 561–584.

Seed, R. (1969) The ecology of *Mytilus edulis* L. (Lamellibranchiata) on exposed rocky shores. 2. Growth and mortality. *Oecologia*, **3**, 317–50.

Seed, R. (1980) Shell growth and form in the Bivalvia. In: *Skeletal Growth of Aquatic Organisms* (eds D.C. Rhoads & R.A. Lutz), pp. 23–67. Plenum Press, New York.

Seed, R. & Suchanek, T.H. (1992) Population and community ecology of *Mytilus*. In: *The Mussel Mytilus: Ecology, Physiology, Genetics and Culture* (ed E.M. Gosling), pp. 87–169. Elsevier Science Publishers B.V., Amsterdam.

Seed, R., Richardson, C.A. & Smith, K. (2000) Marine mussels, their evolutionary success, ecological significance and use as chronometers of environmental change. In: *The Evolutionary Biology of the Bivalvia* (eds E.M. Harper, J.D. Taylor & J.A. Crame), pp. 465–478. The Geological Society of London, London.

Serb, J.M., Alejandrino, A., Otárola-Castillo, E. & Adams, D.C. (2011) Morphological convergence of shell shape in distantly related scallop species (Mollusca: Pectinidae). *Zoological Journal of the Linnean Society*, **163**, 571–584.

Shepherd, J.G. (1987) A weakly parametric method for estimating growth parameters from length composition data. In: *Length Based Methods in Fisheries Research* (eds D. Pauly & G.R. Morgan), pp. 113–119. International Center for Living Aquatic Resources Management (ICLARM), Manila.

Shi, Y., Guan, Y. & He, M. (2013) Molecular identification of insulin-related peptide receptor and its potential role in regulating development in *Pinctada fucata*. *Aquaculture*, **408–409**, 118–127.

Shinen, J.S. & Morgan, S.G. (2009) Mechanisms of invasion resistance: competition among intertidal mussels promotes establishment of invasive species and displacement of native species. *Marine Ecology Progress Series*, **383**, 187–197.

Smith, J.R., Fong, P. & Ambro, R.F. (2009) Spatial patterns in recruitment and growth of the mussel *Mytilus californianus* (Conrad) in southern and northern California, USA, two regions with differing oceanographic conditions. *Journal of Shellfish Research*, **61**, 165–173.

Soon, T.K. & Ransangan, J. (2014) A review of feeding behavior, growth, reproduction and aquaculture site selection for green-lipped mussel, *Perna viridis*. *Advances in Bioscience and Biotechnology*, **5**, 462469.

Soto, M., Ireland, M.P. & Marigómeza I. (2000) Changes in mussel biometry on exposure to metals: implications in estimation of metal bioavailability in 'Mussel-Watch' programmes. *Science of the Total Environment*, **247**, 175–187.

Sousa, T., Domingos, T. & Kooijman, S.A.L.M. (2008) From empirical patterns to theory: a formal metabolic theory of life. *Philosophical Transactions of the Royal Society B*, **363**, 2453–2464.

Sousa, T., Domingos, T., Poggiale, J-C. & Kooijman, S.A.L.M. (2010) Dynamic energy budget theory restores coherence in biology. *Philosophical Transactions of the Royal Society B*, **365**, 3413–3428.

Sparre, P. & Venema, S.C. (1992) Introduction to tropical fish stock assessment. *FAO Fishery Technical Paper*, **306**.

Steffani, C.N. & Branch, G.M. (2003) Growth rate, condition, and shell shape of *Mytilus galloprovincialis*: responses to wave exposure *Marine Ecology Progress Series*, **246**, 197–209.

Stirling, H.P. & Okumus, I. (1995) Growth and production of mussels (*Mytilus edulis* L.) suspended at salmon cages and shellfish farms in two Scottish sea lochs. *Aquaculture*, **134**, 193–210.

Stiven, A.E. & Gardner, S.A. (1992) Population processes in the ribbed mussel *Geukensia demissa* (Dillwyn) in a North Carolina salt marsh tidal gradient: spatial pattern, predation, growth and mortality. *Journal of Experimental Marine Biology and Ecology*, **160**, 81–102.

Strohmeier, T., Strand, O., Alunno-Bruscia, M., Duinker, A., Rosland, R., Aure, J. *et al.* (2015) Response of *Mytilus edulis* to enhanced phytoplankton availability by controlled upwelling in an oligotrophic fjord. *Marine Ecology Progress Series*, **518**, 139–152.

Stuart, V. (1982) Absorbed ration, respiratory cost and resultant scope for growth in the mussel *Aulacomya ater* (Molina) fed on a diet of kelp detritus of different ages. *Marine Biology Letters*, **3**, 289–306.

Sukhotin, A.A., Abele, D. & Pörtner, H.-O. (2006) Ageing and metabolism of *Mytilus edulis*: populations from various climate regimes. *Journal of Shellfish Research*, **25**, 893–899.

Sukhotin, A.A., Strelkov, P.P., Maximovich, N.V. & Hummel, H. (2007) Growth and longevity of *Mytilus edulis* (L.) from northeast Europe. *Marine Biology Research*, **3**, 155–167.

Tamayo, D., Azpeitia, K., Markaide, P., Navarro, E. & Ibarrola, I. (2016) Food regime modulates physiological processes underlying size differentiation in juvenile intertidal mussels *Mytilus galloprovincialis*. *Marine Biology*, **163**, art. 131.

Thieltges, D.W. (2005) Impact of an invader: epizootic American slipper limpet *Crepidula fornicata* reduces survival and growth in European mussels. *Marine Ecology Progress Series*, **286**, 13–19.

Thomas, Y., Mazuri, J., Alunno-Bruscia, M., Bacher, C., Bouget, J-F., Gohin, F. *et al.* (2011) Modelling spatio-temporal variability of *Mytilus edulis* (L.) growth by forcing a dynamic energy budget model with satellite-derived environmental data. *Journal of Sea Research*, **66**, 308–317.

Thompson, I.S., Richardson, C.A., Seed, R. & Walker, G. (2000) Quantification of mussel (*Mytilus edulis*) growth from power station cooling water in response to chlorination procedures. *Biofouling*, **16**, 1–15.

Thompson, R.J. & Bayne, B.L. (1974) Some relationships between growth, metabolism and food in the mussel *Mytilus edulis*. *Marine Biology Letters*, **27**, 317–326.

Troost, T.A., Wijsman, J.W.M., Saraiva, S. & Freitas, V. (2010) Modelling shellfish growth with dynamic energy budget models: an application for cockles and mussels in the Oosterschelde (southwest Netherlands). *Philosophical Transactions of the Royal Society B*, **365**, 3567–3577.

Tsangaris, C., Papathanasiou, E. & Cotou, E. (2007) Assessment of the impact of heavy metal pollution from a ferro-nickel smelting plant using biomarkers. *Ecotoxicology and Environmental Safety*, **66**, 232–243.

Tumanda Jr., M.I., Yap, H.T., McManus, L.T., Ingles, J.A. & Lopez, M.G. (1997) Growth, mortality and recruitment pattern of the brown mussel, *Modiolus metcalfei* (Bivalvia: Mytilacea), in Panguil Bay, Southern Philippines. *Aquaculture*, **154**, 233–245.

Vadopalas B., Weidman, C. & Cronin, E.K. (2011) Validation of age estimation in geoduck clams using the bomb radiocarbon signal. *Journal of Shellfish Research*, **30**, 303–307.

Valladares, A., Manríquez, G. & Suárez-Isla, B.A. (2010) Shell shape variation in populations of *Mytilus chilensis* (Hupé 1854) from southern Chile: a geometric morphometric approach. *Marine Biology*, **157**, 2731–2738.

Van der Molen, S., Márquez, F., Idaszkin, Y. & Adami, M. (2013) Use of shell-shape to discriminate between *Brachidontes rodriguezii* and *Brachidontes purpuratus* species (Mytilidae) in the transition zone of their distributions (south-western Atlantic). *Journal of the Marine Biological Association of the UK*, **93**, 803–808.

Ventura, A., Schulz, S. & Dupont, S. (2016) Maintained larval growth in mussel larvae exposed to acidified undersaturated seawater. *Scientific Reports*, **6**, 23728.

Versteegh, E.A.A., Blicher, M.E., Mortensen, J., Rysgaard, S., Als, T.D. & Wanamaker, A.D. (2012) Oxygen isotope ratios in the shell of *Mytilus edulis*: archives of glacier meltwater in Greenland. *Biogeosciences*, **9**, 5231–5241.

von Bertalanffy, L. (1938) A quantitative theory of organic growth. *Human Biology*, **10**, 181–213.

Walford, L.A. (1946) A new graphic method of describing the growth of animals. *Biological Bulletin*, **90**, 141–147.

Wanamaker, A.D., Kreutz, K.J., Borns, H.W., Introne, D.S., Feindel, S., Funder, S. *et al.* (2007) Experimental determination of salinity, temperature, growth, and metabolic effects on shell isotope chemistry of *Mytilus edulis* collected from Maine and Greenland. *Paleoceanography*, **22**, PA2217.

Wanamaker, A.D., Kreutz, K.J., Wilson, T., Borns, H.W., Introne, D.S. & Feindel, S. (2008) Experimentally determined Mg/Ca and Sr/Ca ratios in juvenile bivalve calcite for *Mytilus edulis*: implications for paleotemperature reconstructions. *Geo-Marine Letters*, **28**, 359–368.

Wanamaker, A.D., Baker, A., Butler, P.G., Richardson, C.A., Scours, J.D., Ridgway, I. *et al.* (2009) A novel method for imaging internal growth patterns in marine mollusks: a fluorescence case study on the aragonitic shell of the marine bivalve *Arctica islandica* (Linnaeus). *Limnology and Oceanography: Methods*, **7**, 673–681.

Wang, J., Li, L. & Zhang G. (2016) A high-density SNP genetic linkage map and QTL analysis of growth-related traits in a hybrid family of oysters (*Crassostrea gigas* × *Crassostrea angulata*) using genotyping-by-sequencing. *Genes, Genomes, Genetics*, **6**, 1417–1426.

Wang, Y., Li, L., Hu, M. & Lu, W. (2015) Physiological energetics of the thick shell mussel *Mytilus coruscus* exposed to seawater acidification and thermal stress. *Science of the Total Environment*, **514**, 261–272.

Wang, Y.J., Hu, M.H., Wong, W.H., Shin, P.K.S. & Cheung, S.G. (2011) The combined effects of oxygen availability and salinity on physiological responses and scope for growth in the green-lipped mussel *Perna viridis*. *Marine Pollution Bulletin*, **63**, 255–261.

Warren, C.E. & Davis, G.E. (1967) Laboratory studies on the feeding, bioenergetics and growth of fish. In: *The Biological Basis of Freshwater Fish Production* (ed S.D. Gerking), pp. 175–214. Blackwell Publishing, Oxford.

Westerbom, W., Kilpi, M. & Mustonen, O. (2002) Blue mussels, *Mytilus edulis*, at the edge of the range: population structure, growth and biomass along a salinity gradient in the north-eastern Baltic Sea. *Marine Biology*, **140**, 991–999.

Widdows, J. (1993) Marine and estuarine invertebrate toxicity tests. In: *Handbook of Ecotoxicology*, volume 1. (ed P. Calow), pp. 145–166. Blackwell Publishing, Oxford.

Widdows, J. & Johnson, D. (1988) Physiological energetics of *Mytilus edulis*: scope for growth. *Marine Ecology Progress Series*, **46**, 113–121.

Widdows, J. & Donkin, P. (1992) Mussels and environmental contaminants: bioaccumulation and physiological aspects. In: *The Mussel Mytilus: Ecology, Physiology, Genetics and Culture* (ed E.M. Gosling), pp. 383–424. Elsevier Science Publishers B.V., Amsterdam.

Widdows, J. & Staff, F. (2006) Biological effects of contaminants: measurement of scope for growth in mussels. *ICES Techniques in Marine Environmental Science*, **40**, 1–30.

Widdows, J., Donkin, P., Brinsley, M.D., Evans, S.V., Salkeld, P.N., Franklin, A. *et al.* (1995) Scope for growth and contaminant levels in North Sea mussels *Mytilus edulis*. *Marine Ecology Progress Series*, **127**, 131–148.

Widdows, J., Nasci, C. & Fossato, V.U. (1997) Effects of pollution on the scope for growth of mussels (*Mytilus galloprovincialis*) from the Venice Lagoon, Italy. *Marine Environmental Research*, **43**, 69–79.

Willmore, K.E., Young, N.M. & Richtsmeier, J.T. (2007) Phenotypic variability: its components, measurement, and underlying developmental processes. *Evolutionary Biology*, **34**, 99–120.

Wong, W.H. & Cheung, S.G. (2003) Seasonal variation in the feeding physiology and scope for growth of green mussels, *Perna viridis* in estuarine MaWan, Hong Kong. *Journal of the Marine Biological Association of the UK*, **83**, 543–552.

Yonge, C.M. & Campbell, J.I. (1968) On the heteromyarian conditions in the Bivalvia with special reference to *Dreissena polymorpha* and certain Mytilacea. *Transactions of the Royal Society Edinburgh*, **68**, 21–43.

7

Physiology of the Circulatory, Respiratory and Excretory Systems

Cardiovascular System

Heart and Haemolymph

A detailed description of the heart and associated haemolymph ('blood') vessels is presented in Chapter 2. Briefly, haemolymph flows from the gills into the heart, which contracts to drive the fluid into a single vessel, the anterior aorta, which divides into many arteries. The most important of these are the pallial arteries, which supply the mantle, and the visceral arteries (gastro-intestinal, hepatic and terminal), which supply the stomach, intestine, shell muscles and foot (see Figure 2.13). The arteries break up into a network of vessels in all tissues, and these then join to form veins, which empty into extensive spaces called sinuses. The haemolymph system is therefore an open circulatory system, with haemolymph in the sinuses bathing the tissues directly. As well as having a respiratory function, the sinuses also serve as a fluid skeleton, giving temporary rigidity to various parts of the body (e.g. labial palps, foot and mantle edges). A consequence of an open circulatory system is that the animal is continually exposed to fluctuations in environmental factors, such as temperature, salinity and contaminants. From the sinuses, the haemolymph flows to the kidneys for purification, and from there, in mussels, a branch enters the gill circulation for oxygenation and returns to the heart via the kidney. In all bivalves, the mantle has an extensive system of haemolymph sinuses, and thus may function as an additional respiratory site to the gills.

Haemolymph Plasma and Haemocytes

Haemolymph plays a number of important roles in bivalve physiology. These include gas exchange, osmoregulation, nutrient distribution, waste elimination and internal defence (see later and Chapter 11). Also, proteomics[1] and metabolomics[2] of haemolymph may be used to follow changes of mussel physiological status in response to external stimuli such as exposure to xenobiotics or environmental challenges such as temperature (Rocher *et al.* 2015; Digilio *et al.* 2016). The haemolymph contains cells called haemocytes that float in a colourless plasma. Most bivalves lack circulating respiratory pigments, although

Marine Mussels: Ecology, Physiology, Genetics and Culture, First Edition. Elizabeth Gosling.

Table 7.1 Organic constituents over an annual cycle in the haemolymph of the mussel *Mytilus edulis*.

	Concentration (mg 100 ml^{-1})
Protein	115–282
Non-protein-nitrogen	3.5–23.4
Ammonia	0.31–1.79
Sugars	9.8–35.7
Lipid	20.4–84.3

Source: Based on Bayne (1973).

haemocyanin is found in some protobranchs (Morse *et al.* 1986) and haemoglobin has been reported in several pteriomorph (clam) families (Alyakrinskaya 2003 and references therein), but not in mussels. The osmotic concentration of haemolymph is determined chiefly by inorganic ions (Na^+, K^+, Ca^{++}, Mg^{++}, Cl^-, SO_4^{2-}) and is equal to or marginally greater than the osmotic concentration of seawater (Bayne *et al.* 1976a). The metabolic costs of maintaining a steep plasma–seawater osmotic gradient would be prohibitive because of the vast surface areas exposed to the external environment. However, the K^+ concentration of bivalve plasma is greater than that of seawater by a factor of 1–2 (Burton 1983), which probably reflects the normal intracellular–extracellular K^+ gradient. Haemolymph plasma also contains numerous dissolved organic molecules (Table 7.1), and concentrations fluctuate according to season and locality, food ration, gametogenesis, parasitism (Santarém *et al.* 1994; Robledo *et al.* 1995) and short-term osmoregulation (see later). Recently, Campos *et al.* (2015), using high-throughput methodology, have described the haemolymph proteome from *Mytilus edulis*. Close to 600 hemolymph proteins were identified, with functions encompassing primary cellular metabolic processes: energy production and metabolism of biomolecules, as well as processes related to oxidative stress defence, xenobiotic detoxification, drug metabolism and immune response.

Haemocytes are not confined to the haemolymph system but move freely out of the sinuses (haemolymph spaces) into surrounding connective tissue, mantle cavity and gut lumen. As seen in Chapter 11, haemocytes represent the most important internal defence mechanism in marine bivalves. They also have a digestive function, ingesting particles of nutritional value that are too large to enter the cells of the digestive gland (see Chapter 4).

Mussel haemocytes have been classified/characterised using a variety of methods, including density gradient centrifugation, morphological analysis via light and transmission electron microscopy (TEM), lectin binding assays (Tame *et al.* 2015 and references therein) and the use of monoclonal antibodies that react with haemocyte surface proteins (Sekine *et al.* 2016). Several different haemocyte classification schemes have been published for mussels (Noël *et al.* 1994; Cajaraville *et al.* 1995; Dyrynda *et al.* 1997; Pipe *et al.* 1993; Bettencourt *et al.* 2009; Tame *et al.* 2015), which generally report three different types: agranulocytes (AGs) and two types of granular haemocytes, basophilic granulocytes (BGs) and eosinophilic granulocytes (EGs). Tame *et al.* (2015) characterised these haemocyte types in three *Bathymodiolus* species. EGs were the largest (diameter 8.4–10.0 μm) and had cytoplasm with numerous granules (diameter 0.8–1.2 μm), while BGs were of medium size (6.7–8.0 μm) and contained small cytoplasmic granules (0.3–0.4 μm) and AGs were the smallest (4.8–6.0 μm), with a cytoplasm that lacked granules (Figure 7.1). The total haemocyte population densities within the hemolymph of all three *Bathymodiolus* mussel species were similar ($8.4–13.3 \times 10^5$ cells ml^{-1}), and the percentages of circulating AGs, BGs and EGs in the haemolymph of these mussels were 44.7–48.5, 14.3–17.6 and 34.3–41.0%, respectively. To analyse functional differences between types, the phagocytic activity and post-phagocytic phagosome–lysosome fusion events (see Chapter 11) were analysed in each. While the AGs exhibited no phagocytic activity, both types of granulocyte were phagocytic. Of the three haemocyte types, EGs exhibited the highest level of phagocytic activity, as well as rapid phagosome–lysosome fusion. The BGs showed lower phagocytic activity and lower rates of phagosome-lysosome fusion than EGs, which suggests that the two types of granulocyte play distinct roles in the defence system. Granulocytes contain high levels of intracellular enzymes, which enable them to kill pathogens after phagocytosis. Also, this cell type plays a prominent role in encapsulation, a process used by molluscs to sequester pathogens that are too

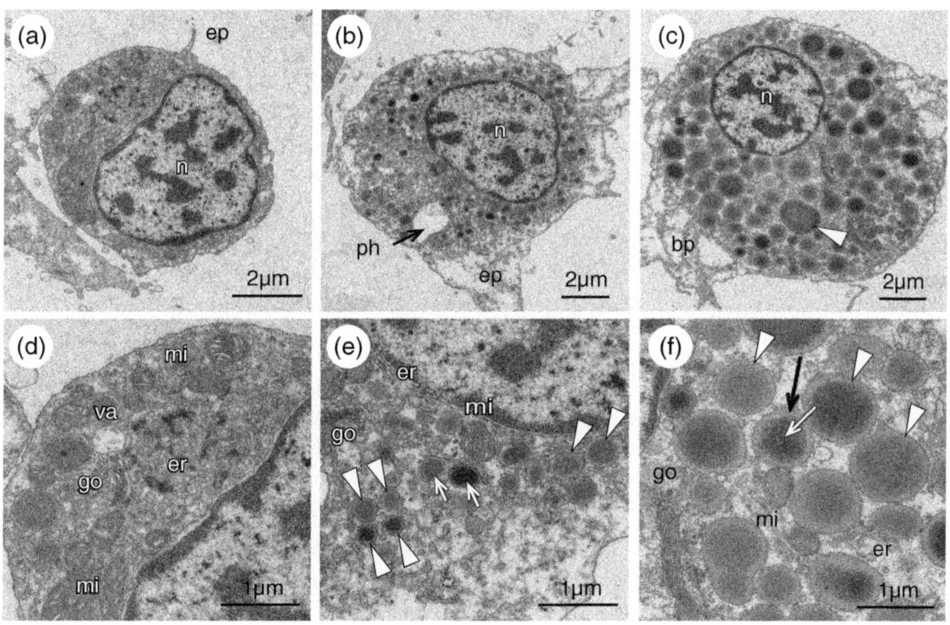

Figure 7.1 Transmission electron micrographs (TEMs) of the haemocytes of *Bathymodiolus japonicus*. (a,d) Agranulocytes (AGs). (b,e) Basophilic granulocytes (BGs), which contain small electronlucent to electron-dense granules (arrowheads) with electron-dense cores (arrows). (c,f) Eosinophilic granulocytes (EGs), which contain large electron-dense granules (arrowhead) with electron-dense cores (white arrow) and more electron-lucent peripheries (black arrow). bp, broad pseudopodia; ep, elongated pseudopodia; er, endoplasmic reticulum; go, Golgi apparatus; mi, mitochondria; n, nucleus; ph, phagosome-like vacuole; va, vacuole. *Source:* From Tame *et al.* (2015). Reproduced with permission of Elsevier.

big to be phagocytosed (see Chapter 11). While AGs play no role in phagocytosis, they have been shown to have a central role in haemocyte aggregation processes associated with wound healing (Wootton & Pipe 2003).

The analysis of DNA damage in mussel haemocytes is a valuable tool for assessing the impact of contaminant mixtures (Brooks *et al.* 2009), but sampling issues related to storage, handling and transportation have often limited its application in large-scale monitoring programmes. Kwok *et al.* (2013) have developed a suitable protocol for cryopreservation of mussel (*Mytilus* spp.) haemocytes, allowing material collected in the field to be analysed later under controlled laboratory conditions.

Heart Rate

Heart beat rate, or cardiac activity, has been shown to serve as an effective index for whole-organism physiology in a range of intertidal invertebrates, including molluscs (Burnett *et al.* 2013). Marine mussels have been particularly well studied because of their worldwide distribution and because, being sessile, they must withstand diurnal and seasonal fluctuations in their environments (see Chapter 3). Responses to changing conditions, such as immersion and aerial exposure on a tidal cycle, include changes in heart beat rate, metabolic rate, oxygen consumption and valve opening/gape (Connor & Gracey 2012 and references therein).

Several methods have been used to study heart rate (HR) in mussels. Investigators have drilled windows in the valves of bivalves to view the heart directly, a method that was later modified by covering the windows with glass or cellophane. Subsequently, heat beat frequency and amplitude were measured using platinum electrodes implanted alongside the pericardium (see Figure 2.13), activated through an impedance pneumograph linked to a multipen recorder (details in Widdows 1973a). This method, which has been modified in more recent times (Nicholson 2002; Braby & Somero 2006), creates minimal disturbance to the animal and allows HR to be monitored from a distance. Haefner *et al.* (1996) used a noninvasive ultrasound technique to monitor HR in the mussel *Mytilus edulis*. This involved positioning an ultrasound probe parallel to the long axis of a submerged mussel. Dual video monitors provided a rectilinear image scan of the heart, as well as its frequency of movement. Another noninvasive technique, introduced by Depledge *et al.* (1996) and modified by Burnett *et al.* (2013), uses infrared (IR) sensors glued externally to the mid-dorsal posterior hinge area, which corresponds to the position of the heart. Changes in the shape or volume of the circulatory structures during a heart contraction, or heart beat, cause a change in the amount of IR light reflected from the animal's internal anatomy back to the IR detector. These changes in reflected IR light, transduced to changes in electrical current, are then electronically amplified, filtered and processed by a laptop computer using specific software such as PicoScope 6.0 (Bakhmet & Zdorovenov 2010; Sarà & de Pirro 2011; Burnett *et al.* 2013; Saggese *et al.* 2016; Tagliarolo & McQuaid 2016). Known as an infrared photoplethysmogram (IR-PPG), this is a noninvasive, economical, robust method that can be used in field experiments (Seo *et al.* 2016). HR has been used successfully in bivalves as a proxy for whole-animal stress in response to temperature, salinity, hypoxia, food level, contaminants and predation.

Factors Affecting HR

Nicholson (2002) found a significant positive relationship between HR and temperature (15–31 °C) in the subtropical mussel, *Perna viridis* (Figure 7.2). When the mussel was exposed to higher temperatures of 31.05 and 34.3°C, mortality was 30 and 70%, respectively. These higher temperatures represented the upper limits of temperature tolerance for this species. The upper thermal limit varies dependent on the species but is coincident with the temperature at which other physiological processes are disrupted. For example, in *M. edulis*, this temperature is 25–27 °C, whereas for the warm-water mussel *P. perna*, it is as high as 32 °C (Bayne *et al.* 1976a). Braby & Somero (2006) examined the effect of temperature on HR in three species of closely related mussels, *M. edulis*, *M. galloprovincialis* and *M. trossulus*, with distinct global distribution patterns (see Chapter 3). HR increased with rising temperature until a high critical temperature (H_{crit}) was reached, at which point activity fell sharply. H_{crit} increased with increasing acclimation temperature and differed among species in a pattern that may reflect their evolutionary adaptation temperatures. HRs of two genetic lineages (referred to as east and west) of *P. perna* (Zardi *et al.* 2015) were recorded over a full tidal cycle in situ at two sites, located ~370 km apart, on the South African coast (Tagliarolo & McQuaid 2016), in order to evaluate the cardiac responses of the two genetic lineages present in the region to temperature and the immersion/emersion

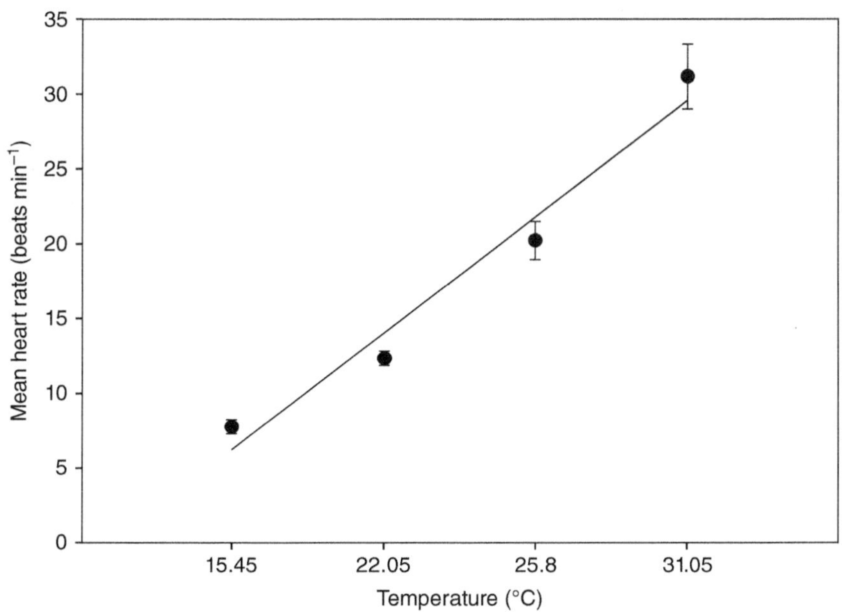

Figure 7.2 Relationship between mean (±SE, N = 5–7) heart rate (HR) and temperature in immersed *Perna viridis*. Line represents regression equation, where $y = 7.812 + 3.072\,x$; $r = 0.99$; $P < 0.0001$. The upper tolerance limits for cardiac activity were between 30 and 35 °C. When seawater temperature was allowed to return to ambient conditions prior to thermal manipulation, cardiac activity returned to initial rates, indicating that the elevated HR shown by mussels exposed to the higher temperatures was transient. *Source:* From Nicholson (2002). Reproduced with permission from Elsevier.

cycle. Robomussel temperature loggers were used to monitor thermal conditions at each site over a one-year period. These biomimetic devices have been shown to be an effective tool for recording the body temperature of mussels in their natural environment (Helmuth *et al.* 2016; see Chapter 3). Comparison with live animals showed that robomussels provided a good estimate of mussel body temperature. At each site, over a 24-hour period, body temperature was estimated by deploying two live mussels fitted with a thermocouple, and recorded body temperatures were compared with values obtained from two robomussels that were placed for the same period next to natural mussel beds in sun- and wave-exposed areas of the shore. In the laboratory, the body temperatures of two live mussels from each site were recorded and compared with values recorded from six robomussels (details in Tagliarolo & McQuaid 2016). HR was greatly affected by the tidal regime and temperature at the two sites (Figure 7.3). Minimum mean HR values occurred during emersion for both the east and the west lineages (18.3 and 27.7 beats min^{-1}, respectively). The highest mean HR was recorded underwater during the daytime for both lineages (39.6 beats min^{-1} for the east lineage and 79.4 beats min^{-1} for the west lineage) (Figure 7.3). Unexpectedly, when field HR data were compared with data measured in the laboratory, the latter seriously underestimated HR activity (by as much as 75%), especially during immersion. Indeed, mussels in the field tolerated high temperatures that would be regarded as lethal in laboratory experiments, which suggests that the interaction of abiotic conditions in the field does not necessarily raise vulnerability to high temperatures.

Salinity also affects heart activity by interfering with the osmotic balance of heart tissue. Changes in HR of *Brachidontes pharaonis* (an invasive species) and *Mytilaster minimus* (a native Mediterranean species) in response to changes in salinity values from brackish (20 psu) to extreme hyper-salinity (75 psu) were investigated both in situ and in the laboratory (Sarà & de Pirro 2011). The two species displayed different responses to varying salinity, with clear differences in HR observed between groups investigated at different salinities and between groups originating from different environments. The native species, adapted to narrow salinity changes as found in the Mediterranean Sea, showed signs of stress at salinities slightly above 37 psu (in the range 37–45 psu). On the other hand, the invasive species only displayed clear signs of stress at salinities over 45 psu, and exhibited heart activity even under hyper-saline conditions (i.e. 75 psu), indicating that the invasive species has the potential to invade most of the transitional environments across the entire Mediterranean basin. See Nicholson (2002), Bakhmet *et al.* (2005) and Braby & Somero (2006) for additional studies on the effects of salinity (plus temperature in some cases) on HR in *Mytilus* and *Perna* species.

Reduced oxygen tensions, caused by emersion or seawater hypoxia, elicit a rapid and significant decrease in HR, referred to as bradycardia – a common response in bivalves to lower ambient oxygen tensions (Nicholson 2002). During prolonged hypoxia, bivalves may close their valves and shift to anaerobic metabolism, leading to internal hypoxia and bradycardia or cardiac arrest (see later). When mussels (*P. viridis*) were held in hypoxic (oxygen tensions 1–5%) or normoxic (oxygen tensions 10–21%) seawater, they were allowed to equilibrate to the oxygen regime for 15 minutes, after which HR was monitored hourly for a total exposure time of five hours. Mussels exposed to hypoxic seawater showed an immediate decrease in HR and a slight overshoot when the water was reoxygenated, although this increase was transient. Mussels held in normoxic seawater showed little variation in

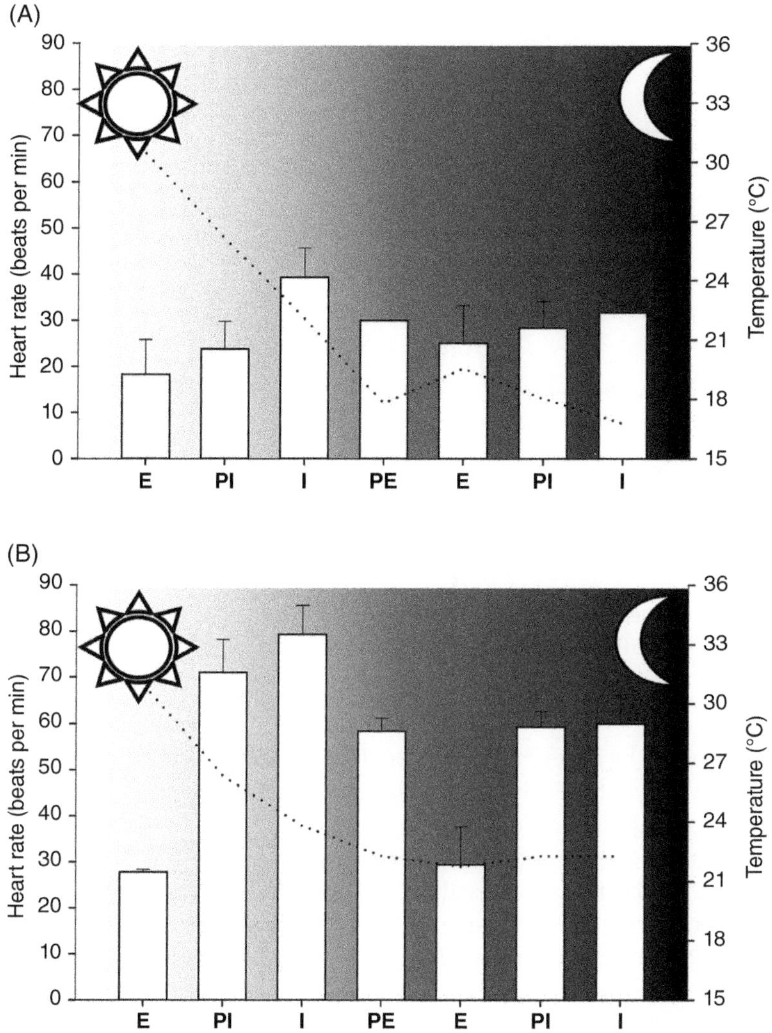

Figure 7.3 Heart rate (HR) variability during a complete tidal cycle measured for *Perna perna*. (A) East site (average ± SD, N = 2). (B) West site (average ± SD, N = 4). Dotted line shows the variation in the average temperature measured during the recordings. Average water and air temperatures during the 24-hour experimental period at the west site were 23.2 and 24.6 °C, respectively, while those at the east site were 17.3 and 20.2 °C, respectively. Sun and moon symbols indicate day and night periods. E, emersion; PE, partially emersed; I, immersion; PI, partially immersed. *Source:* From Tagliarolo & McQuaid (2016). CC BY 4.0.

HR over the monitoring period. At the end of hypoxic exposure, mussels had significantly lower HRs compared to those held in normoxic seawater ($P < 0.0001$; Figure 7.4).

The HR of *M. edulis* increased significantly when it was in the presence of effluent from the dogwhelk, *Nucella lapillus*, probably due to an olfactory-mediated perception of predation threat (Rovero *et al.* 1999). When attacked through a drilled hole, the HR increased

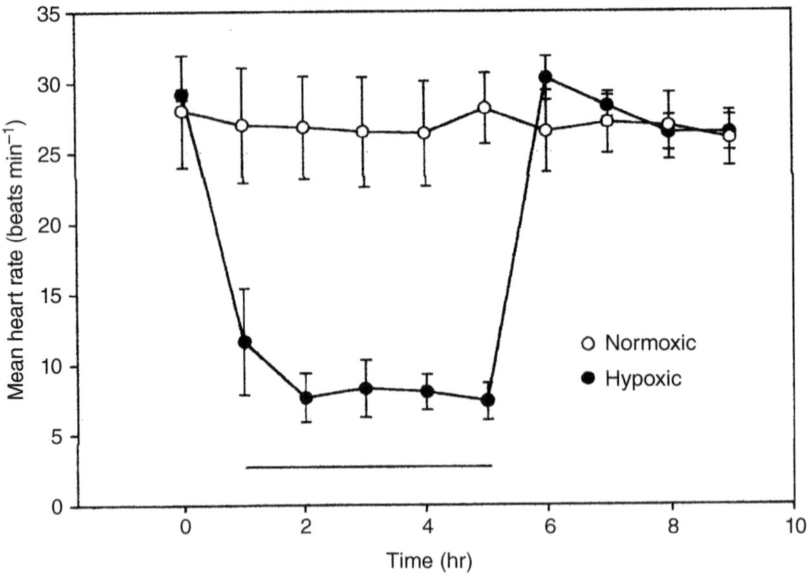

Figure 7.4 Mean (± SE, N = 2–7) heart rate (HR) of *Perna viridis* exposed to hypoxic and normoxic conditions (bar denotes period of hypoxia; $P < 0.0001$). *Source:* From Nicholson (2002). Reproduced with permission from Elsevier.

significantly, but when attacked by penetration between the valves, HR decreased throughout the attack, supporting the hypothesis that dogwhelks that penetrate between the valves induce muscular paralysis by injection of toxins.

Cardiac activity can be used as an indication of the stress induced in marine mussels when exposed to toxicants. Hence, the widespread use of HR as a physiological biomarker[3] in pollution monitoring programmes (see Chapter 8). A wide range of contaminants has been reported to inhibit HR (bradycardia) in mytilids (reviewed in Nicholson & Lam 2005). Neuronal toxicity has been proposed as a mechanism for bradycardia in mytilids during pollutant exposure, although many studies appear to have been conducted using unrealistically elevated contaminant concentrations or not to have fully accounted for avoidance behaviour including shell closure, which also inhibits HR (Nicholson 2003a). When *P. viridis* was exposed to copper (48 hr continuous exposure of 100–400 μg l^{-1} Cu; Nicholson 2003a) and tributyltin (TBT; Nicholson 2003b), reflecting conditions present in estuaries, HR tended to be elevated over a range of concentrations, suggesting that tachycardia (fast resting HR) is the typical response of mytilids in polluted waters (Figure 7.5; Nicholson & Lam 2005). Injection of cardio-inhibitory dopamine into copper-exposed tachycardic individuals reversed the elevated HRs with respect to controls. Likewise, *P. viridis* transplanted to various locations in Hong Kong coastal waters showed tachycardic responses that were considered to be indicative of enhanced metabolism during the detoxication of contaminants (Nicholson 1999).

In summary, various parameters of HR are regulated by neurons, and bradycardia is a common neuronal response in bivalves exposed to high ambient metal concentrations,

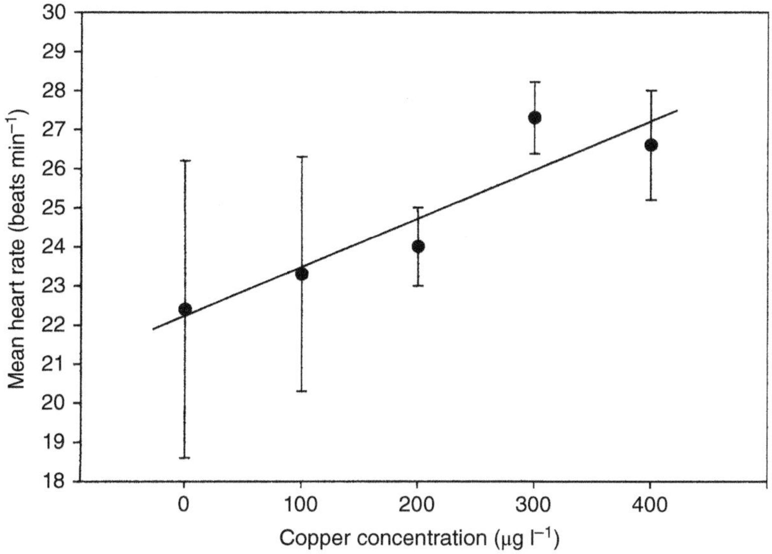

Figure 7.5 Relationship between mean (± SE, N = 4–6) heart rate (HR) of *Perna viridis* and copper concentration. HRs ranged from 22.4 ± 3.85 to 27.3 ± 0.92 beats min^{-1} (mean ± SE, N = 5) in the control and 300 µg 1^{-1} copper groups, respectively. Line represents regression equation, where $y = 22.24 + 0.0124x$; $r^2 = 0.85$; $P = 0.02$. *Source:* From Nicholson (2003a). Reproduced with permission from Elsevier.

although lower levels induce tachycardia. Mild tachycardia evident in *P. viridis* (Figure 7.5) was likely the consequence of serotonergic-mediated elevation of HR required for metal detoxication. Additional information on the use of mussel HR as a biomarker in pollution studies can be found in Brown *et al.* (2004), Halldórsson *et al.* (2008), Rajkumar (2009), Hagger *et al.* (2010), Turja *et al.* (2014), Martinović *et al.* (2015, 2016), Saggese *et al.* (2016) and Nikolic *et al.* (2019).

Other factors, such as body size, food, oxygen availability and aerial exposure, also affect HR. Since these same factors can also affect oxygen consumption, they will be discussed in more detail in the next section.

Respiratory System

In bivalves, the gill filaments have a respiratory as well as a feeding role. Their large surface area and rich haemolymph supply make them well suited for gas exchange. Deoxygenated haemolymph is carried from the kidneys to the gills by way of the afferent gill vein, which gives off a small branch to each filament, descending on one side and ascending on the other. As water passes through the gills, gas exchange takes place across the thin walls of the filaments. The oxygenated haemolymph from each filament is collected into the efferent gill vein that goes either to the kidney or directly to the heart (see Figure 2.13). It is likely that gas exchange also occurs over the general mantle surface. The efficiency with which this occurs, however, is low, usually in the range 1–13%, although this can rise up to 30% or more by an increase in the HR (Bayne *et al.* 1976b).

Widdows & Staff (2006) have developed a well-cited protocol for measuring oxygen consumption rate (OCR), otherwise known as VO_2, respiration or metabolic rate, in mussels. OCRs of individual mussel are measured in closed glass respirometers (500 ml) held in a temperature-controlled water bath mounted on a multi-point magnetic stirrer. Air-saturated seawater is added to each respirometer and stirred with a magnetic stirrer bar beneath a perforated glass plate (e.g. small Petri dish with many 1 cm holes) supporting a mussel. A calibrated oxygen electrode connected to an oxygen meter measures the rate of decline in oxygen partial pressure (pO_2) in each chamber. A number of respirometers, including a control without a mussel, are usually run simultaneously, and each oxygen meter is coupled to a multi-channel chart recorder or a computer. Twenty minutes are allowed for the mussels to open and to resume pumping; oxygen uptake is then measured for the next hour. The rate of oxygen consumption should not be measured below a partial pressure of ~100 mm Hg (13 kPa) because the rate then becomes dependent on the external pO_2. OCR is calculated as follows:

$$\text{Rate of O}_2\text{ uptake}\left(\mu\text{moles O}_2\text{ g}^{-1}\text{ h}^{-1}\right)=\left[C\left(t_0\right)\quad C\left(t_1\right)\right]\times\left(V_r\right)\times 60/\left(t_1-t_0\right)$$

where t_0 and t_1 are the start and finish times (min) of the measurement period, $C(t)$ is concentration of oxygen in the water (μmoles $O_2\ l^{-1}$) at time t and Vr is volume of respirometer minus the animal. Individual OCRs are standardised to 1 g dry tissue weight. Modifications to this method can be found in the following investigations on mytilids: Arifin *et al.* (2001), Michaelidis *et al.* (2005a), Wang *et al.* (2011), Barrento *et al.* (2013), Thyrring *et al.* (2015), Landes *et al.* (2015), Zittier *et al.* (2015), Arranz *et al.* (2016), Olabarria *et al.* (2016) and Saggese *et al.* (2016). To gain a full understanding of oxygen consumption in a given species, measurements should be made at different times of the year and should include animals of different sizes and at different reproductive stages.

Factors Affecting Oxygen Consumption

Size

The relationship between body size (usually expressed as weight) and OCR is generally described by the allometric equation, $y = axb$, where y is the metabolic rate as measured by OCR, x is the body weight and the intercept a and slope b are fitted parameters (see Chapter 6 for details). The intercept is a measure of the OCR of an individual of unit weight (or length), and its value changes depending on environmental conditions and the species being tested. The slope is a measure of the rate of increase in OCR with size. Based on experimental data, the b value is often close to 0.75 when organisms covering a large range in body mass are compared. However, research on respiration rate as a function of size throughout the ontogeny of *M. edulis*, from early larval to the adult stage, has shown that the b exponent is about 0.9 throughout the early pelagic larval stage and during the juvenile stage up to a size of 1–10 mg tissue dry mass, above which it decreases to about 0.7 in the adult stage (Riisgård 1998 and references therein). Riisgård (1998) concludes that the '3/4 power scaling law is a formalism and not a biological law because respiration cannot be separated from growth, a fact which becomes obviously clear the smaller the organisms are' (discussed in Glazier 2006, 2010, 2014, 2015; Harte *et al.* 2008).

Food

Three different rates of metabolism have been recognised in *M. edulis* and other bivalves (Bernard 1983; Resgalla *et al.* 2006; references in Tang & Riisgård 2016). Standard metabolic rate (SMR) is defined as the minimum amount of energy required to maintain the vital functions of an inactive bivalve. Filtration rate and OCR are low, and there may be reduced sensitivity to temperature. Routine metabolic rate (RMR) corresponds to the energy required for ventilation, metabolism of ingested ration and gonadal and somatic growth, as well as the basic vital functions (SMR). In other words, RMR is the level of activity of a static animal dealing with an optimal concentration of suspended particles. Active metabolic rate (AMR) is the temporarily high level of activity stimulated by temperature increase or presence of food. Maximum AMR is elicited by a combination of warmer water and food, while the size of the response is dependent on whether it occurs during SMR or RMR. There can be as much as a fivefold difference between SMR and AMR in mussels (Griffiths & Griffiths 1987). The difference between RMR and SMR represents the energy, above basal requirements, that is available for the various activities of feeding, digestion and biosynthesis involved in tissue growth (Tamayo *et al.* 2013).

The effect of ration and, in particular, starvation on the metabolic rate of *M. edulis* appears to vary depending upon the season and the physiological condition of the animal (Widdows 1978). In winter (ambient temperature 5 °C), when mussels in SW England do not contain mature gametes and there is less food available in the natural environment, the rate of oxygen consumption of starved animals is reduced to a 'standard rate', as shown in several studies (Widdows & Bayne 1971; Bayne 1973; Widdows 1973a,b; Bayne *et al.* 1973; Thompson & Bayne 1972, 1974). However, at all other times of the year there is no significant reduction in the rate of oxygen uptake following five weeks of starvation. During this period (April–September), the presence of mature gametes in the mantle may inhibit the reduction of activity and metabolism of *M. edulis* to a standard level. Gabbott & Bayne (1973) recorded a similar response in an earlier work and suggested that there may be a mechanism that prevents a reduction in metabolic rate as long as reserves are available in the body tissues to meet the energy demand of gametogenesis. Thompson & Bayne (1974) have shown that at low ration levels, only slightly in excess of maintenance, oxygen consumption in *M. edulis* of different dry flesh weights increases rapidly with any increase in the concentration of food particles, reaching an asymptotic value, which is then maintained over a considerable range of additional ration increase.

The concentration of naturally occurring marine phytoplankton biomass is typically between ~1 and 5 μg chl *a* l^{-1} (Riisgård *et al.* 2013), but because mussels often live in dense aggregations, the actual phytoplankton biomass to which they are exposed can often be very much reduced. Also, starvation often occurs in winter, when there is no primary production. During such periods, *M. edulis* reduces its valve gape, thereby strongly reducing filtration rate, leading to a reduced oxygen uptake. Tang & Riisgård (2016) tested the hypothesis that, during starvation, *M. edulis* regulates the opening degree of its valves in such a way that the oxygen concentration in the mantle cavity is reduced in order to minimise respiration and at the same time prevent anaerobic metabolism, which is energetically expensive (see Riisgård & Larsen 2015). They measured the oxygen concentration changes in the mantle cavity of both starved and fed mussels and found that when there were no algal cells in the ambient water, mussels gradually closed their valves, causing a decline in the filtration rate, along with a simultaneous decrease in the oxygen concentration in the mantle cavity and a subsequent remarkable decrease in respiration rate. Addition of algal cells stimulated the mussels to open their valves, resulting in a rapid increase in oxygen concentration. Tang & Riisgård (2016) interpreted the low OCR of *M. edulis* in phytoplankton-depleted water as an efficient physiologically regulated mechanism that allows mussels to save energy during a starvation period. Hence, the different respiration rates observed in their study are simply a consequence of the varying valve-opening degree, rather than a specific type of metabolic rate (see earlier).

The relationship between OCR and food supply is complex, involving not only energy expenditure associated with the mechanical process of filtration but also the physiological costs of feeding, which include energy utilised in digestion and assimilation of food and energy lost during deamination and other catabolic processes that accompany digestion (Warren & Davis 1967). In *M. californianus*, Bayne *et al.* (1976c) estimated that the mechanical cost of feeding is 18% of ingested ration, while the physiological cost is about 6%. For *M. edulis*, the equivalent values are 24 and 4%, respectively (Bayne & Scullard 1977a). However, in *M. chilensis*, while the physiological costs of feeding are similar to those observed for *M. californianus* and *M. edulis*, the mechanical costs are about 5%, reflecting the tendency of this species to reduce its filtration rate in response to increased ration (Navarro & Winter 1982). In a later study, an even lower value of ~1% of mechanical costs relative to total metabolic expenditure was reported for *M. edulis* (Riisgard & Larsen 2001).

Temperature

Temperature is one of the major factors influencing OCR in marine bivalves. Respiration rates of bivalves generally increase with temperature until a threshold is reached. Beyond this, physiological rates can drop drastically and anaerobic metabolic end products begin to accumulate. The threshold is often referred to as the Arrhenius break temperature (ABT), the temperature at which a sharp discontinuity occurs in an Arrhenius plot (log rate versus reciprocal of absolute temperature; Somero 2002), and indicates a general collapse of metabolic functioning (Sarà *et al.* 2014). Breakpoint temperatures tend to correlate with the maximum habitat temperatures of marine invertebrates (Somero 2002). In addition to a high temperature threshold, organisms also exhibit a low temperature threshold for aerobic respiration (Jansen *et al.* 2009 and references therein). In response to declining oxygen availability, organisms have been shown to exhibit oxyregulation by enhancing ventilation and heart beat rate.

Acute Temperature Response

Changes in temperature elicit two types of response. If the animal is subject to a sudden change in temperature, it responds by a change in OCR, which may initially result in overshoot, but is followed by a period of stabilisation that takes place over a period of minutes or hours (acute or short-term response). If the temperature change persists over days or weeks, there is a gradual adjustment of the OCR to a level comparable with that preceding the temperature change (acclimation) (Bayne & Newell 1983). However, these responses are by no means universal among bivalves.

The effect of change in temperature on OCR is often expressed in terms of temperature coefficient or Q_{10} values:

$$Q_{10} = \left[\frac{V_2}{V_1}\right]^{\frac{10}{t_2 - t_1}}$$

where V_1 and V_2 are the rates of oxygen consumption at temperatures t_1 and t_2, respectively. An organism's Q_{10} reflects the rate of change for biological, chemical or physiological processes as a consequence of a 10 °C increase in temperature. The values obtained are dependent on the thermal history of the animal, body size, activity, season and reproductive condition. Temperature effects on OCR in terms of Q_{10} values can provide an understanding of an organism's tolerance to the environment; examples of some such studies using mussels are reviewed in Zippay & Helmuth (2012). When *M. californianus* was exposed to nonstressful temperatures (13–22 °C), oxygen consumption in the field was found to be relatively temperature-independent, with a Q_{10} of 1.20 (Bayne *et al.* 1976c). Similarly, Widdows (1976) reported a Q_{10} for *M. edulis* of 1.5–2.1 after using a cyclic temperature regime between 11 and 19 °C for 30 days. Respiration rate in *M. galloprovincialis* increased with temperature, with Q_{10} values >2 and good acclimation between 6 and 19 °C (Barbariol & Razouls 2000). Generally, for most reactions under 'nonstressful' conditions, Q_{10} values range between 1 and 3, while some energy-intensive reactions show Q_{10} values between 4 and 6. Low Q_{10} values illustrate temperature independence where no significant acclimation to a temperature change is evident (Zippay & Helmuth 2012).

Temperature effects on OCR are usually depicted in the form of rate/temperature or R/T curves (Figure 7.6). Hicks & McMahon (2002a) investigated the effects of temperature acclimation on OCR of *P. perna* at full air O_2 saturation and during progressive hypoxia over the range of acclimation and test temperatures spanning its natural ambient range of 10–30 °C in the Gulf of Mexico. In all three acclimation groups, 15, 20 and 25 °C, mean OCR increased from minimum values at test temperatures of 10 °C to maximum values at 25 °C. Q_{10} values for mean OCRs indicate that OCR increased acutely between the test temperatures of 10 and 15 °C ($P < 0.05$) in all three acclimation groups ($Q_{10} = 5.21$, 3.55 and 3.9 in the 15, 20 and 25 °C acclimation groups, respectively; Figure 7.6). In contrast, OCRs between 15 and 20 °C were not significantly different, resulting in reduced Q_{10} values of 1.45, 1.31 and 1.45 for this temperature range in the 15, 20 and 25 °C acclimation groups, respectively. There was a significant increase in OCR between 20 and 25 °C, with Q_{10} values in all acclimation groups ranging from 2.24 to 3.24. There was no difference in OCR between 25 and 30 °C in any group; hence, Q_{10} values were depressed in this temperature range, falling to 0.76, 1.43 and 1.64 in the 15, 20 and 25 °C acclimated groups, respectively. This result suggests metabolic suppression[4] as the 30–31 °C incipient upper lethal temperature of *P. perna* is approached (Hicks & McMahon 2002b). Metabolic suppression at 10 °C, a temperature close to the lower thermal limit of 7.5 °C of *P. perna* (Hicks & McMahon 2002b), reflects the subtropical nature of this species. In *M. edulis*, OCR is not suppressed at temperatures as low as 5 °C (Widdows 1973b). The thermal insensitivity of OCR in *P. perna* between 15 and 20 °C (Q_{10} values 1.31–1.48) suggests regulation of metabolic rate within

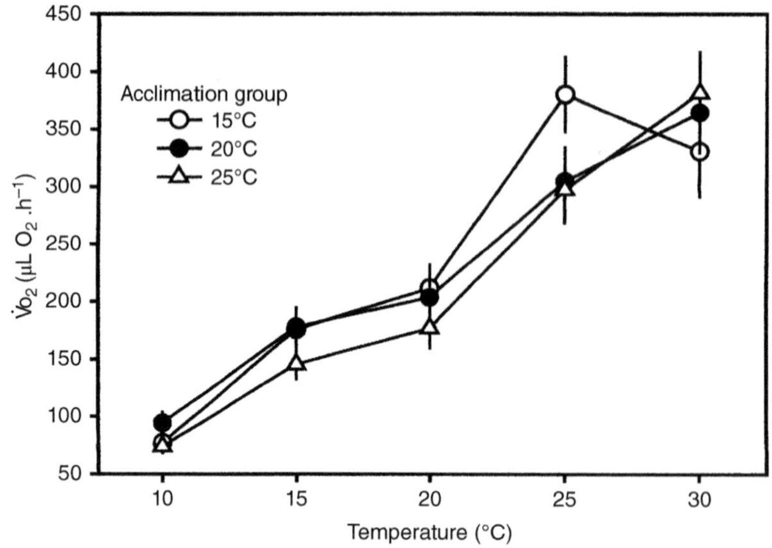

Figure 7.6 Mean oxygen consumption rates (OCR; μl O_2 h $^{-1}$) adjusted to a standardised individual dry tissue weight of 0.4 g, in response to acute temperature change among specimens of *Perna perna* from the Gulf of Mexico acclimated to 15, 20 or 25 °C. Error bars represent one standard error of the mean (N = 101). *Source:* From Hicks & McMahon (2002a). Reproduced with permission from Elsevier.

this temperature range. But, between 20 and 25 °C, the increase in Q_{10} values indicates a lack of acclimatory metabolic temperature compensation.

Many intertidal molluscs, including mussels, have developed a capacity for acute thermal metabolic regulation within a portion of their normal ambient temperature range, as illustrated earlier for *P. perna*. For example, Bayne (1967) recorded acute thermal regulation of OCR in Brazilian specimens of *P. perna* over 15–30 °C (Q_{10} range = 1.20–1.49), and van Erkom Schurink & Griffiths (1992) over the range 10–15 °C (Q_{10} = 1.56) for the same species in South Africa, while Berry and Schleyer (1983) in contrast recorded little capacity for acute thermal metabolic regulation in *P. perna* from the same region over 15–25 °C (Q_{10} = 1.74–2.04). Acute temperature regulation of OCR has also been recorded in a number of temperate intertidal mytilid species, including *Geukensia demissa* over 15–20 °C (Read 1962), *M. edulis* over 5–20 °C (Widdows 1973b) and *M. californianus* over 10–20 °C (Bayne *et al.* 1976c).

The per cent O_2 regulation value, which represents the organism's capacity of regulation of oxygen uptake, can be estimated directly from VO_2/pO_2 response curves that allow direct comparisons of regulatory ability across species. For example, in *P. perna* (Hicks & McMahon 2002a), regulatory ability is temperature-dependent, with regulatory capacity increasing from poor (69.5%) at 10 °C to good (81.7%) at 30 °C (Table 7.2). Table 7.2 shows that mytilids as a group are all moderate regulators of OCR, irrespective of their particular habitat, so that estuarine species such as *P. viridis*, *M. edulis* and *G. demissa*, which are more likely to experience periodic hypoxia, have per cent O_2 regulatory values of 81.6, 62.8–78.1 and 64.3%, respectively, falling within the same range as that of open-water, rocky-shore intertidal species, which are much less likely to experience hypoxia (e.g. *P. perna*, *M. californianus* and *Choromytilus meridionalis*; range = 64.0–81.7%).

Table 7.2 Per cent O_2 regulation values estimated from published VO_2/PO_2 response curves of mussel species. M/e, marine/epibenthic; E/e, estuarine/epibenthic.

Species	Per cent O_2 regulation	Degree of regulation	Temperature (°C)	Habitat	Reference
Perna perna	69.5	Poor	10	M/e	Hicks *et al.* (2002a)
	75.9	Moderate	15		
	77.7	Moderate	20		
	78.3	Moderate	25		
	81.7	Good	30		
Perna viridis	81.6	Good	25	E/e	Davenport (1983)
Mytilus edulis	62.8	Poor	15	E/e	Bayne (1971a)
	78.1	Moderate	15		Bayne (1971b)
Mytilus californianus	64.0	Poor	10	M/e	Moon & Prichard (1970)
	71.6	Moderate	13		Bayne *et al.* (1976c)
Geukensia demissa	64.3	Poor	20–23	E/e	Booth & Mangum (1978)
Choromytilus meridionalis	74.4	Moderate	20	M/e	Marshall & McQuaid (1993)

Source: Adapted from Hicks *et al.* (2002a).

Acclimation to Temperature Change

Thermal acclimation was first demonstrated in bivalve molluscs, including mussels, in the 1930s. When *M. edulis* was investigated in populations from different latitudes, those from high latitudes had higher OCRs than those from lower ones when measured at the same temperature. Subsequently, similar findings were reported for *M. californianus* (Bayne *et al.* 1976b and references therein). Mussels also experience quite different temperature regimes depending on the season, and in the case of intertidal species can experience daily fluctuations in temperature associated with the tidal cycle. Some species respond to temperature changes lasting days or weeks by adjusting OCR to a level comparable with that which occurred before the temperature changed. During a 30-day experiment, one group of mussels (*M. edulis*) was held at the ambient (field) temperature in November (10 °C), another group at 5 °C above ambient and another at 5 °C below it (Widdows & Bayne 1971). At seven-day intervals, oxygen consumption was measured, along with other physiological parameters. Acclimation was attained within 14 days of the start of experiment. The compensatory changes in oxygen consumption during cold and warm temperature acclimation are shown in Figure 7.7. The initial response to change was an increase in OCR with a rise in temperature or a decline in OCR with a decrease in temperature. During the 14-day period of acclimation, OCR declined in mussels held at higher temperatures and increased in those at lower temperatures. The slight increase in oxygen consumption of animals maintained at the 'ambient temperature' of 10 °C suggests that 10 °C was in fact a few degrees below the field temperature at that time of year. In *M. edulis*, compensation is achieved by shifting the rate-temperature curve to the right following warm acclimation. This means that the animal tolerates higher temperatures but still maintains oxygen levels

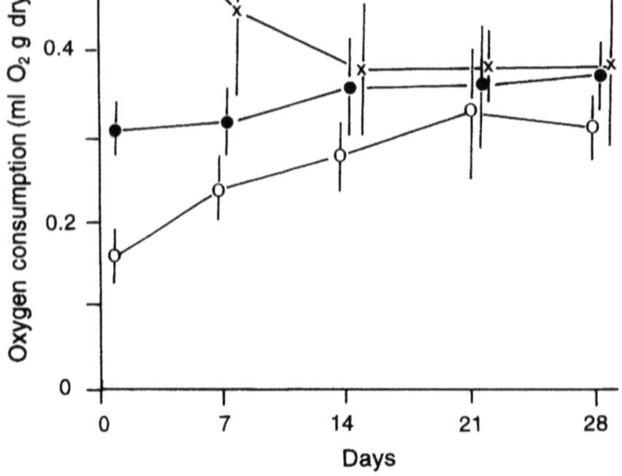

Figure 7.7 Oxygen consumption in the mussel *Mytilus edulis* acclimated to 5 °C (empty circles) and 15 °C (crosses), compared with mussels at 10 °C (the ambient temperature in November; filled circles). *Source:* From Widdows & Bayne (1971). Reproduced with permission from Cambridge University Press.

relatively constant. This acclimation process ensures that respiration, and also feeding and excretion, is maintained at approximately constant levels, but this is only possible over a temperature range that is species-dependent.

In a large-scale study, Jansen *et al.* (2007a) investigated seasonal variation in metabolic adjustments in 10 populations of *Mytilus* spp. distributed from 39 to 58 °N. Field temperature profiles were obtained between April 2004 and May 2005 using temperature loggers positioned in the vicinity of mussels at selected sites. Logger output was compared to sea surface temperature (SST) profiles obtained by satellite. Mean habitat temperature of shallow-water and intertidal habitats from logger data showed a constant relationship with SST. Using this, the authors estimated the acclimatisation temperature of each mussel population for each sampling occasion. This acclimatisation temperature was defined as the mean water temperature for the period of 30 days before sampling, assuming that this was a true indication of the temperature to which the mussels should have been well adjusted. Respiration rates of three to six mussels from each site were measured at 3, 10, 17, 24 and 31 °C until oxygen tension had decreased by 20–30%. Respiration rates and Q_{10} values were based on the rates that corresponded to the seven-degree temperature interval (e.g. 10–17 °C) nearest to the acclimatisation temperature. Respiration rates at experimental temperatures usually declined at between 24 and 31 °C (Figure 7.8). Some cold-acclimatised populations (e.g. sites 5, 6 and 9) sampled in January showed maximum OCRs at between 17 and 24 °C. Peaks in OCRs, between 1.5 and 2.5 (mg O_2 g^{-1} h^{-1}) at 24 °C, were frequently observed in springtime (April–May). Respiration rates never exceeded 1.5 (mg O_2 g^{-1} h^{-1}) in mussels from sites 5 and 6. Respiration rates increased with acclimatisation temperature (Figure 7.9A). Maximum acclimatised respiration rates occurred at between 8 and 14 °C in the Baltic Sea group (◇), between 15 and 19 °C in the North Sea group (Δ) and between 20 and 24 °C in the Mediterranean Sea and Bay of Biscay groups (□) (Figure 7.9B). The rate deviations found for the Mediterranean Sea and Bay of Biscay groups differed significantly from those of the Baltic Sea group. However, several data points for the North Sea group were found to fit the reaction norm of the Mediterranean Sea and Bay of Biscay groups between 15 and 20 °C. Hence, the rate deviations of these two groups could not be distinguished ($P > 0.05$).

Investigating seasonal sensitivity to temperature of the metabolic rate, Jansen *et al.* (2007a) found that a broad range of Q_{10} values occurred under relatively cold conditions. As habitat temperatures increased, the range of Q_{10} values narrowed, reaching a bottleneck in southern marginal populations during summer. Group-specific clines and limits on Q_{10} values were observed at temperatures corresponding to the maximum climatic conditions these geographic populations presently experience, indicating differential thermal adaptation among the three groups. Seasonal variation in thermal sensitivity of the metabolic rate showed that Q_{10} values greater than 4 were only found in January and April, when values decreased with latitude. During the following seasons, July and October, mean Q_{10} values decreased further and latitudinal variation disappeared.

Adaptation and Acclimation to Low Temperatures

The continued warming of the Arctic is facilitating the northward expansion of temperate-boreal species into the region, while the southern distributional limit of Arctic species is

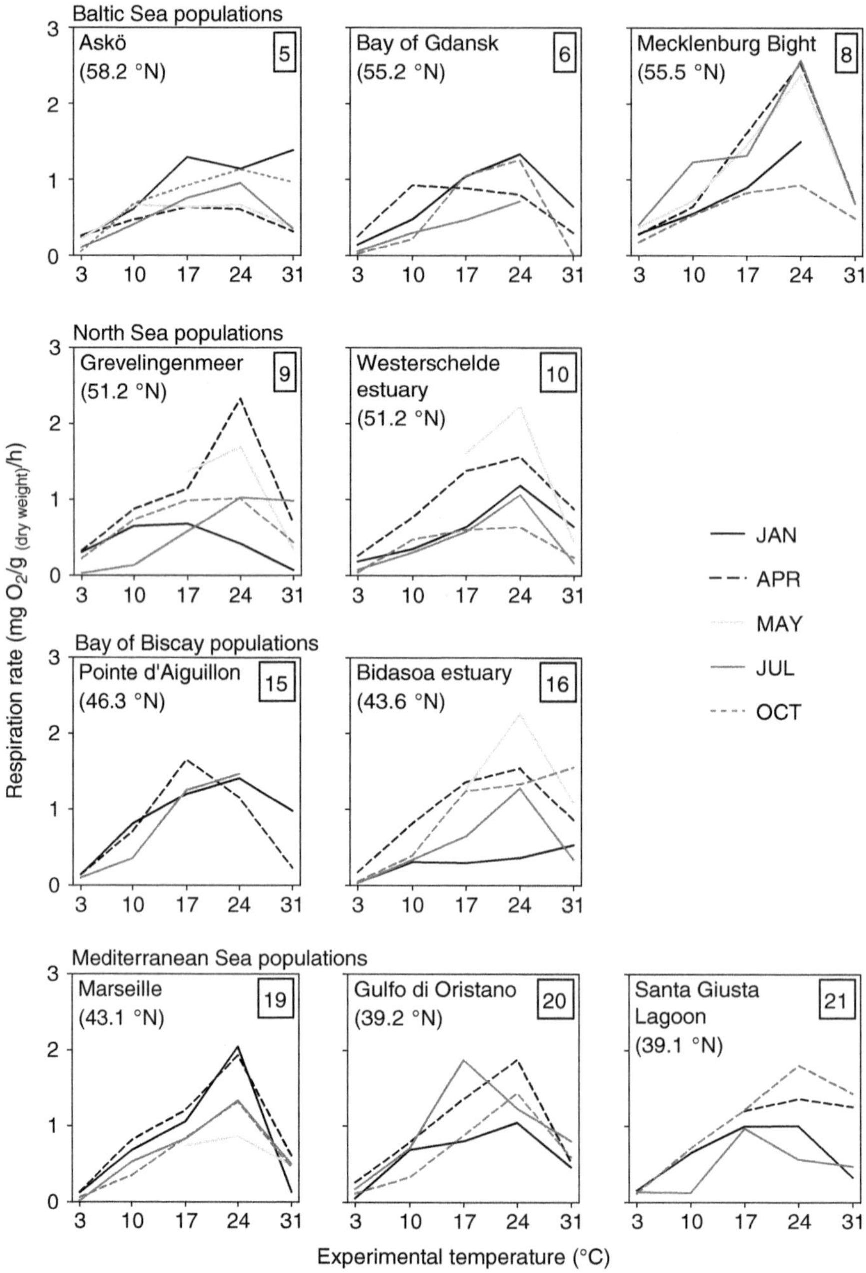

Figure 7.8 Seasonal variation in the respiratory response to an experimental temperature range for 10 European *Mytilus* spp. populations. Site names and degrees north are given in the graphs. Numbers in square boxes refer to the site number. *Source:* From Jansen *et al.* (2007a). Reproduced with permission from Springer Nature.

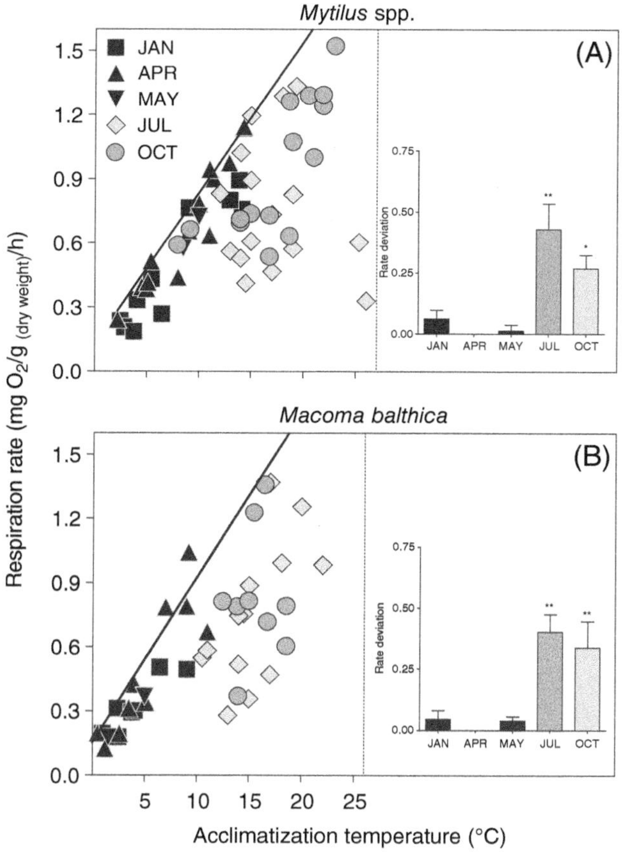

Figure 7.9 Respiration rates as a function of acclimatisation temperature. The black lines are the upper edge slopes, which represent the metabolic scope (positive energy balance) of *Mytilus* spp. These graphs include measurements from all 16 research sites. (A) Different symbols represent different sampling seasons. (B) Respiratory response to temperature in different groups. The grey diamonds, white triangles and black circles represent the Baltic Sea group, the North Sea group and the Mediterranean Sea and Bay of Biscay group, respectively. Included are the rate deviations calculated per season and per group. BS, Baltic Sea group; NS, North Sea group; MS, Mediterranean Sea group; BC, Bay of Biscay group. $^{*}P < 0.05$ and $^{**}P < 0.01$ refer to the statistical differences for a given season compared to January in (A) and for a given group compared to BS in (B). *Source:* From Jansen *et al.* (2007a). Reproduced with permission from Springer Nature.

retracting (see Chapter 3). For example, since the 1950s, the intertidal distribution of blue mussels in the western Atlantic has shifted in a poleward direction, with the southern edge contracted by an estimated 350 km due to high-temperature mortality (Jones *et al.* 2010). Although there is abundant information on the physiological constraints and adaptations related to high temperatures experienced by mussels (see Chapter 3), there is a paucity of information pertaining to low-temperature exposure. At high latitudes, cold water slows down metabolic rate so that benthic invertebrates are characterised by slow growth, low production-to-biomass ratios, prolonged longevity and low activity (Heilmeyer & Brey 2003). It has been postulated that metabolic rates should be elevated to compensate for the physiological constraints imposed by living at temperatures near or below 0 °C. This has been referred to as 'metabolic cold adaptation' (MCA; see Peck 2002 for review). However, data for the past two decades show that a low standard or RMR is the norm for polar ectotherms, including bivalves (Peck & Conway 2000 and references therein; Heilmeyer & Brey 2003). Reduced metabolism may be an important strategy of energy conservation for bivalves in the Antarctic, where food can be in short supply for up to nine months of the year (Ahn & Shim 1998).

Sukhotin *et al.* (2006) examined the impact of age and temperature on metabolic key parameters, one of which was respiration rate, from geographically separated populations of *M. edulis* from the relatively warm (5–17 °C) temperate North Sea (NS), the cold (0–14 °C) sub-Arctic White Sea (WS) and the cold but thermally more stable (3–9 °C) Barents Sea (BS). Mussels were collected in October (BS and WS) and November (NS) 2002, when seawater temperatures were +5.5 and +9 °C, respectively, and in all individuals age was determined from external shell rings, a method that had been previously verified by comparing the number of internal rings and the seasonal growth of individually marked mussels (references in Sukhotin *et al.* 2006). Mussels were maintained in aquaria at 5.7 °C and 30 psu salinity until January 2003, when OCRs were analysed. In all populations, a similar pattern was observed: low respiration rates in the youngest mussels, followed by a rapid increase and then a gradual decline to nearly constant and similar values in the oldest animals. In other words, cold adaptation – specifically, latitudinal differences in metabolic rates according to different temperatures – was manifested in the most metabolically active specimens, possibly because of active somatic growth and substantial reproductive effort, which are costly in terms of assimilated energy. By contrast, old and very young, sexually immature mussels had lower and uniform respiration rates in all populations, so that population differences were abolished in the old animals. When combined, both findings support the view that in the northern *Mytilus* populations, lifetime energy expenses peak early to support reproductive output in young animals, while metabolic activity declines rapidly thereafter. This may be an adaptation to the harsh environmental conditions and may also reflect an age-related decrease of energy demands for somatic growth and reproduction, until a baseline level of energy turnover (as required for maintenance) is reached that allows no further reduction in the old animal (Sukhotin *et al.* 2006).

To compensate for the effects of low temperature, eurythermal bivalves such as mussels modify physiological processes to maintain body function. One such adaptation is adjustment in mitochondrial density and/or capacity to regulate ATP production (Sukhotin *et al.* 2006; Thyrring *et al.* 2015, 2017 and references therein). Thyrring *et al.* (2015) compared *M. edulis* populations from three different climate zones, represented by Denmark (56 °N, temperate), South Greenland (64 °N, subarctic) and North Greenland (77 °N, High Arctic), and found that mussels in North Greenland displayed MCA and lower thermal tolerance compared to those in South Greenland and Denmark, providing one of the first insights into the performance, adaptability and capability of a mussel species that has expanded to and inhabited the High Arctic region. Subsequently, Mathiesen *et al.* (2017) identified NW Greenland mussels as *Mytilus trossulus*, which provided Thyrring *et al.* (2017) with the opportunity to test the hypothesis that latitudinal variation in whole-animal metabolic rate of two congeners, *M. edulis* collected at 56 °N and *M. trossulus* collected at 77 °N, is related to differences in mitochondrial respiration.[5] They also investigated the changes in mitochondrial respiration following long-term cold water acclimation of

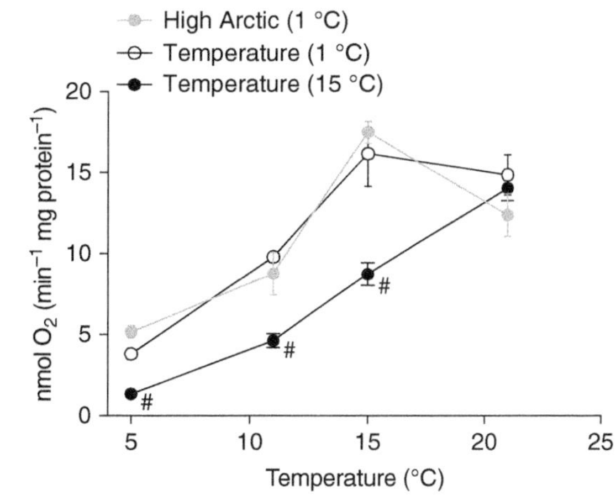

Figure 7.10 Mitochondrial respiration of cold-adapted *Mytilus trossulus* acclimated at 1 °C (grey circles) and *Mytilus edulis* collected in the temperate region (Denmark) and acclimated at either 15 °C (filled black circles) or 1 °C (open circles). Values are presented as mean ± standard error, N = 5. Significant differences in temperature-specific respiration are indicated by #. *Source:* From Thyrring *et al.* (2017). Reproduced with permission from Springer Nature.

M. edulis. Mitochondrial respiration was assessed after five months of acclimation to 1 and 15 °C. At similar experimental temperatures, mitochondrial respiration in mussels acclimated to 1 °C was significantly higher than in those acclimated to 15 °C (Figure 7.10). After five months of acclimation, 1 °C-acclimated *M. edulis* had similar mitochondrial respiration to 1 °C-acclimated *M. trossulus* despite their different geographical origins. Therefore, the data indicate that mitochondrial capacity does not support latitudinal observed differences in whole-animal metabolism between *M. edulis* and *M. trossulus*. See Gosling (2015) for a review on acclimation to low temperatures in other polar bivalves beside mussels.

Temperature Tolerances

Studies have shown that oxygen supply is an important factor in setting upper lethal limits, from observations of a transfer to anaerobic metabolism and the accumulation of anaerobic end products such as acetate, succinate and octopine in tissues in both polar and temperate marine invertebrates. This transfer to anaerobic metabolism occurs close to the experimental upper lethal limit and has been called the 'critical temperature', the physiological limit at which oxygen supply and energy balance can no longer be maintained, and indicates the maximum temperature for long-term survival in terms of weeks or months. Extended exposure to temperatures above high critical temperature (critical thermal maximum, CT_{max}) or below low critical temperature (CT_{min}) finally leads to death of the animal unless thermal acclimation (i.e. a shift of CT values) occurs. Tolerance windows in marine species, estimated by lethal and CT limits, impact species' distribution patterns on a global and local scale (see Chapter 3). Ectotherm species in Antarctic and tropical regions have narrow thermal tolerance windows compared to temperate species, which show broad thermal tolerance windows that reflect the higher variation in their environmental temperatures (reviewed in Peck *et al.* 2014). In Antarctic waters, temperature variation is about 1–2 °C – similar to the 1–3 °C in tropical waters, but very different from the 10–15 °C reported for temperate ones (references in Peck *et al.* 2010). This relationship between climate variation and thermal tolerance windows has been termed the 'climate variability hypothesis'. To test this hypothesis, Compton *et al.* (2007) compared the upper and lower lethal thermal tolerance limits of numerous bivalve species, including mussels, from a tropical and a temperate location, and found that species from the tropical site had higher upper and lower lethal thermal limits than species from the temperate site. However, temperate species were able to survive freezing temperatures, and this resulted in thermal tolerance windows that were on average 7 °C higher than for tropical species. Also, when the authors reviewed previous studies on a global scale, they found that upper lethal thermal limits of tropical species were closer to maximum habitat temperatures than those of temperate species, suggesting that temperate species are better adapted to temperature variation – findings that are consistent with the climate variability hypothesis. The following examples concentrate on studies pertaining to mussels from temperate and tropical regions, as these regions are the most common habitats of mussel species. The majority of published bivalve thermal limits are acute lethal temperatures at which instantaneous death occurs on exposure to temperature increasing at a specific rate, while chronic testing involves holding specimens at constant lethal temperatures for periods long enough to induce mortality (Hicks & McMahon 2002b).

To illustrate, Segnini de Bravo *et al.* (1998) investigated upper and lower tolerance limits in *P. viridis* and *P. perna* collected from natural mussel beds on the Caribbean coast of Venezuela. The former is an introduced tropical mussel from the Indo-Pacific Ocean (temperatures range 26–28 °C) that appears to be out-competing the native subtropical species, *P. perna*, which has a typical temperature range of 15–25 °C (see Hicks & McMahon 2002b; Laurenço *et al.* 2012). Mussels were acclimatised for 15 days under laboratory conditions at 25 °C, the temperature at the collection site. To determine upper temperature tolerance limits, 600 unfed mussels (5–100 mm shell length) of each species were placed in 95 l-capacity aquaria, and water temperature was increased by 1 °C every day starting from 25 °C, the acclimation temperature. For the lower temperature tolerance limits, mussels were placed in a container (Frigid Unit Inc.) specially designed for that purpose, and temperature was decreased by approximately 1 °C per day. To determine CT_{max} and CT_{min}, two mussels (30–46 mm shell length) of each species were selected at random and placed in a 5 l container with 2 l of well-aerated seawater. In the case of CT_{max} determination, temperatures in each aquarium were increased by 0.2, 0.5 or 0.8 °C min^{-1}. Five replicates of two mussels were used for each heating rate. In the case of CT_{min} determination, temperature was lowered by 0.5 °C min^{-1} in a cold bath (1 l) obtained by regularly adding ice cubes. The death criterion for CT estimates was the mussel opening but not closing its valves. A mortality criterion was defined taking into account that stressed mussels have their valves open and when a metal bar is introduced into them they close them quickly. The results indicated that *P. viridis* had wider thermal tolerance limits than *P. perna*. Lower and upper lethal temperatures were 6 and 37.5 °C for *P. viridis* and 3 and 34.5 °C for *P. perna*. The CT_{max} values indicated that at the rate of 0.2 °C min^{-1}, *P. perna* died at 38.3 °C and *P. viridis* at 40.8 °C. Equivalent values at 0.5 °C min^{-1} were 39.6 and 41.8 °C for *P. perna* and *P. viridis*, respectively, while at 0.8 °C min^{-1} the former died at 40.5 °C and the latter at 43.1 °C. CT_{min} values for *P. perna* and *P. viridis* were 2.4 and 5.4°C, respectively. CT_{max} values for both species were always higher than upper tolerance limits, and the faster the rate of increasing temperature, the higher the difference. Wider temperature tolerance limits (and salinity tolerances; results not shown; Segnini de Bravo *et al.* 1998), together with a fast growth rate and a lack of competitors, have led to populations of *P. viridis* rapidly outcompeting *P. perna* at introduced locations (Dias *et al.* 2013).

Hicks & McMahon (2002b) found that lower and upper thermal limits in *P. perna* from the Gulf of Mexico were 7.5 and 30 °C, respectively – limits that vary somewhat, especially the upper one, from those reported by Segnini de Bravo *et al.* (1998). Thermal tolerance of animals is a plastic, environmentally induced trait that responds to planned experimental sources of variance including thermal history, acclimation, ontogenetic stage and seasonal and daily biological cycles (Ribeiro *et al.* 2012). Another source is lack of standardisation of experimental protocols, although in more recent studies improved procedures are being used to estimate lethal and critical temperature tolerances in mussels (Hicks & McMahon 2002b; Jansen *et al.* 2007b; Logan *et al.* 2012; Yuan *et al.* 2016). Critical thermal limits, which define the breadth of temperatures over which an organisms can tolerate at least short exposure, are normally indexed by one of two metrics: upper and lower LT_{50}, or CT_{max} and CT_{min}.

Salinity

The open ocean typically has surface salinities between 33 and 37 psu, with an average of 35 psu. In contrast, estuaries and bays are subject to pronounced salinity fluctuations because of evaporation, rainfall and inflow from rivers. The salinity range tolerated is, however, species-dependent. Many mussels, in particular *Mytilus* spp., are euryhaline; that is, they can tolerate an extremely wide range of salinity (4–40 psu) in their natural environment. In the northern Baltic, *M. trossulus* is living at the margin of its salinity tolerance (4.5 psu), and although dwarfed by the low-salinity conditions, the species is very abundant in this area (Westerbom *et al.* 2002). *Geukensia demissa* can survive in salinities ranging from 6 to 70 psu, and *Brachidontes exustus* also exhibits a broad salinity tolerance range of 10 to 41 psu (Yuan *et al.* 2010). In contrast, the lowest salinities tolerated by *M. galloprovincialis* and *M. californianus* are 12 and 17 psu, respectively (references in Bayne *et al.* 1976b). *P. viridis*, with its extraordinary invasive ability, has a wide salinity (and thermal) tolerance compared to its congener *P. perna* (see earlier). The former can survive in salinities as low as 20 psu, but its optimal range is 27–33 psu (Rajagopal *et al.* 2006). The mussel *Mytilaster minimus*, adapted to narrow salinity changes as found in the Mediterranean Sea, shows signs of stress at salinities slightly above 37 psu (in the range 37–45 psu), while the invasive species *Brachidontes pharaonis* only displays clear signs of stress at salinities over 45 psu, indicating that this species has the potential to invade most of the transitional environments across the entire Mediterranean basin (Sarà & de Pirro 2011). *Mytella charruana* is a newly introduced species to the SE Atlantic coast of the United States. Yuan *et al.* (2010) found that it exhibits a very wide range of salinity tolerances, with limits for survival ranging from 2 to 40 psu for long-term exposures, and from 0 to 45 psu for short-term fluctuations, indicating that these mussels could invade a wide variety of saline environments with significant fresh- or saltwater input.

As *M. edulis* has the capacity to adapt to a broad range of salinity, Landes *et al.* (2015) hypothesised that OCRs in locally adapted mussels do not differ among locations, but display similar values within a certain range. They collected mussels from six locations in Denmark, covering a salinity range of 10–30 psu, and randomly sorted them into three groups (N = 20) of large (~25 mm) and small (~19 mm) mussels, which they then transferred to net bags that they deployed in the water column near the original collection site. Salinity, temperature and chlorphyll *a* were monitored over the deployment period of

29–50 days. Salinities at all locations were relatively constant during the experimental period, except at site 2 where the spatial proximity to a sluice induced heavy temporal fluctuations. Weight-standardised OCRs were similar in mussels from sites 1, 3, 4, 5 and 6 ($P \leq 0.83$; Figure 7.11), but measurements at site 2 yielded a significantly ($P \leq 0.038$) lower average value than all others, except site 4 ($P = 0.193$). There was no evidence that OCRs vary significantly among mussels from different salinity regimes. On the contrary, there is reason to believe that oxygen consumption differs more with individual physiological condition and season than with salinity differences between localities (Landes *et al.* 2015 and references therein).

When investigating the effect of salinity on OCR, it is essential to carefully consider the time course of metabolic adaptations to change in salinity (Bayne *et al.* 1976b). When the respiration rates of mussels from three different salinities (5–6 psu, brackish water; 15 psu, intermediate water; 30 psu, fully marine water) were measured in the field, they were found to be similar (Remane & Schlieper 1971). But if mussels are exposed to a sudden change in salinity, the immediate response is to close the shell valves. *Geukensia demissa* may remain shut for up to 10 days, depending on the extent of the salinity change (Pierce 1971). During this time, the valves open periodically for short periods. Eventually, ventilation is resumed or else death occurs. During this initial period of 'shock', respiration rate is depressed. However, should the new salinity regime persist, there is gradual acclimation, which may take as long as four to seven weeks depending on the magnitude of the salinity change, the ambient temperature and the size of the animal (Bayne & Newell 1983). The length of the acclimation period far exceeds the two weeks normally required for temperature acclimation, probably a reflection of the time needed for intracellular osmotic

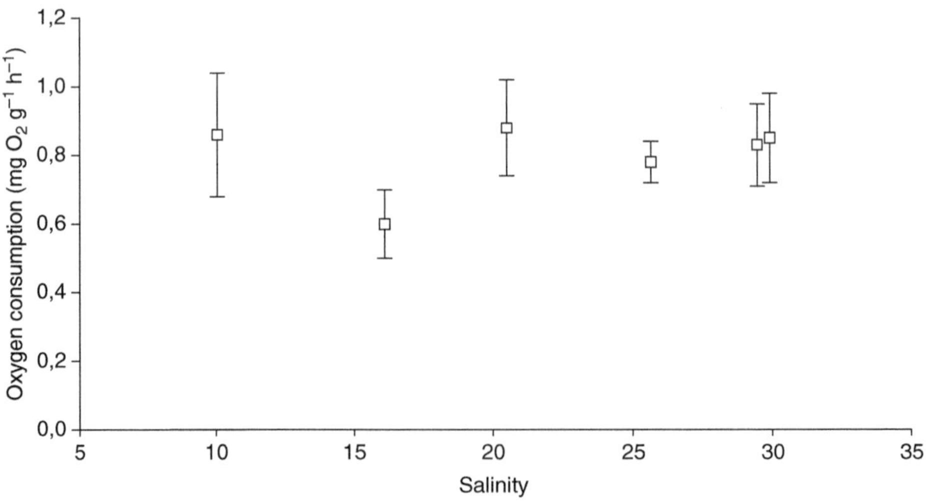

Figure 7.11 Oxygen consumption of small mussels. Rates are standardised per gram dry weight and normalised according to Bayne & Newell (1983). Data are mean ± SD of six randomly picked and individually measured mussels per station/cohort. *Source:* From Landes *et al.* (2015). Reproduced with permission from the *Journal of Shellfish Research*.

regulation (see later). However, a higher acclimation capacity for salinity than for temperature, for both chronic and acute conditions, has been reported in the mussel *Perna perna* (Resgalla *et al.* 2007).

Several studies have investigated the combined effects of salinity and an environmental parameter on OCR. Sarà *et al.* (2008) examined the combined effect of salinity and temperature on the OCR of the Indo-Pacific invasive mussel, *Brachidontes pharaonis*, which is outcompeting indigenous mussel species such as *Mytilaster minimus* and *Mytilus galloprovincialis* in the Mediterranean Sea. In experiment I, mussels of similar shell length (26.4 ± 0.02 mm) were divided randomly into groups of 20 individuals held at temperatures of 11, 15 and 20 °C and salinities of 45, 37, 25 and 15 psu. The acclimation period was seven days at each temperature, with a gradual stepwise reduction in salinity with animals held for a day at 45, 37, 25 and 15 psu. The control group was maintained at 45 psu and 20 °C – conditions similar to those at the sampling site. In experiment II, mussels were divided into three groups and held at a single temperature of 20 °C and in hyper-saline conditions of 50, 55 and 60 psu, providing a gradient of seven salinities when combined with the original four.

In experiment I, OCRs showed a general temperature-dependent relationship ($P < 0.001$) with the highest OCR at 20 °C (Figure 7.12). This value is relatively low in comparison to *M. edulis* (Widdows 1973b), which suggests that if temperatures higher than 20 °C had also been tested, *B. pharaonis* would have had sufficient capacity to increase its metabolic rate, taking into consideration that the species can survive in water temperatures over 30–34 °C. At each temperature, there was no significant difference in OCR at salinities ranging from

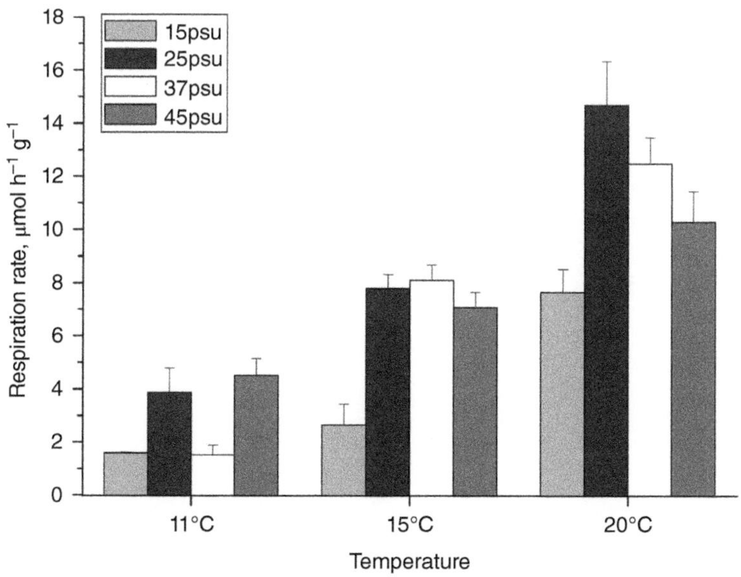

Figure 7.12 Combined effects of salinity (15, 25, 37 and 45 psu) and temperature (11, 15 and 20 °C) on the respiration rate (µmol O_2 h^{-1} g^{-1}) of *Brachidontes pharaonis*. *Source:* From Sarà *et al.* (2008). Reproduced with permission from Elsevier.

25 to 45 psu. However, mussels at 15 psu had a significantly lower OCR than those at 25, 37 and 45 psu at all experimental temperatures. In experiment II, the lowest OCR value was at 15 psu ($P < 0.05$), with higher and relatively constant values between 25 and 50 psu and maximum values (~2-fold higher) at 55 psu (Figure 7.13). An acclimation period of just seven days (see earlier) was probably insufficiently long to allow adaptation of OCR to the higher salinities (50–60 psu).

Wang *et al.* (2011) examined the combined effect of dissolved oxygen (DO; see later) and salinity in *P. perna*, the ranges of which were selected to cover those experienced by the species in Hong Kong waters. The study involved a 3 × 4 factorial design, with three levels of DO (1.5, 3.0 and 6.0 mg l^{-1}) and four levels of salinity (15, 20, 25 and 30 psu). Each treatment consisted of three replicates, with 25 mussels per replicate, which were maintained in an aquarium for four weeks, after a period of one week's acclimation at 21 °C and 30 psu. Measurement of OCR, as well as clearance rate, absorption efficiency, ingestion rate and excretion rate, was conducted weekly. OCR reduced significantly ($P < 0.001$), with decreasing DO and salinity, with the lowest value at 1.5 mg O_2 l^{-1} ×15 and the highest at 6.0 mg O_2 l^{-1} ×30 (Figure 7.14). Interactions between time and salinity, as well as between salinity and DO, were significant. Very low mortality was observed after four weeks, which indicates that *P. viridis* is very tolerant of both hypoxic and low-salinity stress. Since *P. viridis* reduced its OCR gradually with decreasing DO, with a more abrupt decrease when DO was 1.5 mg O_2 l^{-1}, the species is considered a limited oxyregulator,; that is, it has a limited ability to regulate its OCR under low DO (see later).

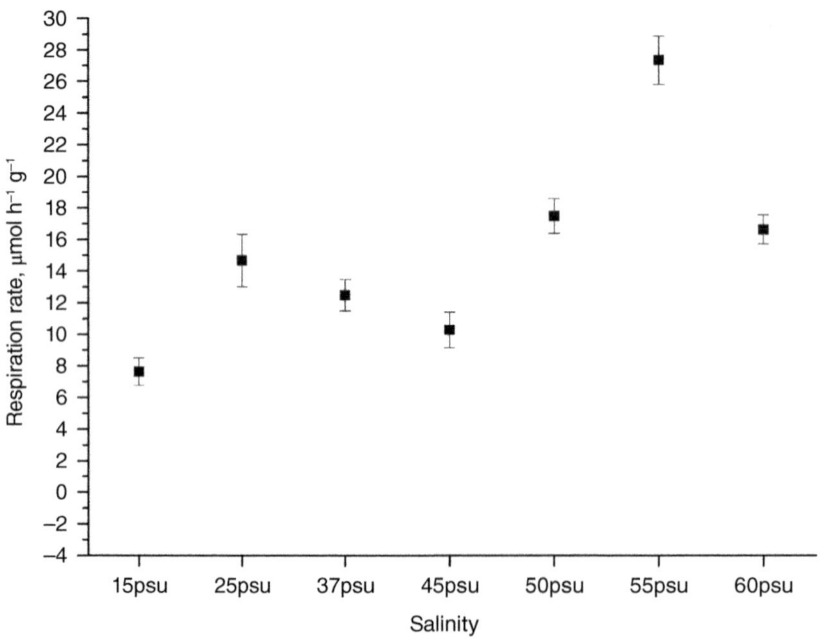

Figure 7.13 Effect of salinity (15, 25, 37, 45, 50, 55 and 60 psu) on the respiration rate (μmol O_2 h^{-1} g^{-1}) of *Brachidontes pharaonis*. *Source:* From Sarà *et al.* (2008). Reproduced with permission from Elsevier.

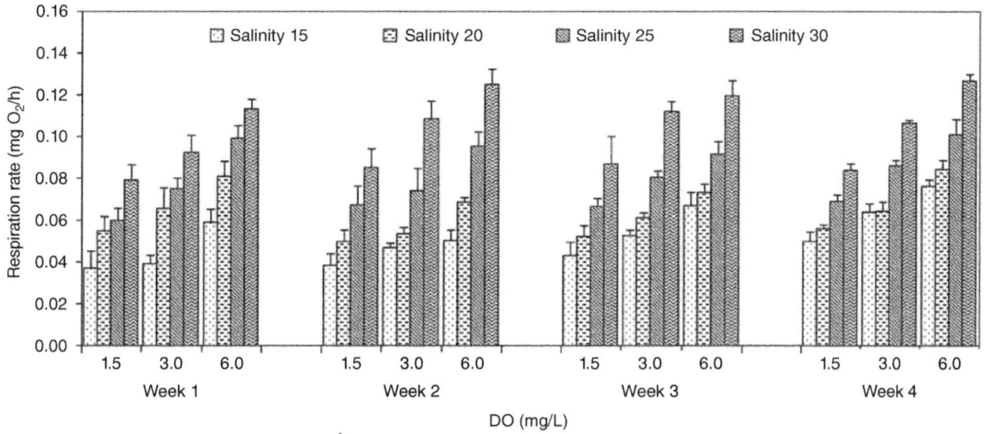

Figure 7.14 Respiration rate of *Perna viridis* exposed to three different combinations of dissolved oxygen (DO) concentration (1.5, 3.0 and 6.0 mg l^{-1}) and four levels of salinity (15, 20, 25 and 30 psu) over a period of four weeks. *Source:* From Wang *et al.* (2011). Reproduced with permission from Elsevier.

Oxygen Availability

Hypoxia, or oxygen depletion, is an environmental phenomenon whereby the concentration of DO in the water column decreases to a level that can no longer support living aquatic organisms (see Chapter 3). The main factors that lead to the development of hypoxia, and sometimes to anoxia (depletion of DO), are water column stratification (which isolates the bottom water from exchange with oxygen-rich surface water), decomposition of organic matter in the bottom water (which reduces oxygen levels), nutrient-enhanced primary production (or eutrophication) and pollution of the water column (Diaz & Rosenberg 2008). Over the past few decades, there has been an exponential increase (>5.5% per year) in the number of coastal sites reporting hypoxic conditions. This growth is expected to increase in the coming decades due to the combined effects of the continued spread of coastal eutrophication and global warming (Vaquer-Sunyer & Duarte 2008; Rabalais *et al.* 2009; Conley *et al.* 2011). Managing nutrients can create positive feedbacks allowing oxygen recovery to occur, but in the absence of nutrient reductions, recovery from hypoxia in coastal marine ecosystems is unlikely (Caballero-Alfonso *et al.* 2015).

Among bivalves, low oxygen concentration is a normal occurrence for species that live deep within soft sediments or for intertidal species during shell closure on the low tide or in crowded tide pools. However, bivalves also frequently experience periods of hypoxia when DO concentrations drop to ≤2 mg l^{-1} (Rabalais *et al.* 2010). There are two responses that bivalves utilise with respect to declining oxygen concentrations (Meltzner *et al.* 2013). The first is oxyregulation, whereby mussels, and bivalves in general, maintain rates of oxygen consumption with declining water pO_2, typically by increasing diffusion gradients across their gill epithelium. This can be achieved by increasing ventilation rate – an energy-intensive process – or by increasing the efficiency with which oxygen is removed from the water, or through a combination of the two strategies (Bayne & Newell 1983). For example, in

M. edulis, extraction efficiency can more than treble as pO_2 falls from 160 to 20 mm Hg (Bayne 1971a). Species that regulate can only maintain a constant rate of oxygen consumption down to some minimum 'critical' oxygen tension ($pO_{2,crit}$), which is species-dependent. Once $pO_{2,crit}$ is reached, oxygen consumption and demand are downregulated. Eventually, anaerobic metabolism (see later) is needed to support the organism's energetic requirements. In some species (e.g. *Geukensia demissa*), the rate of oxygen consumption is directly dependent on the ambient O_2 concentration, which therefore declines as the O_2 concentration surrounding the gill surface is reduced. Such species, called oxyconformers, also have to rely on Meltzner *et al.*'s (2013) second strategy, anaerobic energy generation, once $pO_{2,crit}$ is reached. Onset of anaerobic metabolism marks time-limited survival. The depletion of fermentable substrate (e.g. glycogen reserves) eventually results in catastrophic drops in cellular adenosine triphosphate (ATP), which induce a cascade of events leading to depolarisation of membranes, Ca^{2+} entry into the cytoplasm and cell death (see Boutilier 2001 for review). Some species (e.g. the ocean quahog, *Arctica islandica*) survive hypoxic conditions by entering a metabolic depressed state, down to 10% (or under longer anoxic conditions, even down to 1%) of their normoxic metabolic activity. Onset of metabolic rate depression (MRD), an energy-saving mechanism, occurs below environmental oxygen levels of 5 kPa (37.5 mm Hg), while the species remains fully aerobic down to 2 kPa (Strahl *et al.* 2011). During anaerobiosis, acidic end products of glycolysis (e.g. acetate, alanine, opines, propionate and succinate) accumulate, causing a decrease in haemolymph pH and an increase in CO_2 concentration due to fermentation of glycogen. In addition, increases in the expression levels and enzymatic activities of a number of glycolytic enzymes are also observed (Kawabe *et al.* 2010, Strahl *et al.* 2011, Sussarellu *et al.* 2010, 2012). Another hypoxia-tolerant species, the oyster *Crassostrea gigas*, has optimised fermentation pathways to increase ATP yield and produce less acidic alternative end products (Michaelidis *et al.* 2005b). On a return to normoxic conditions, concentrations of reactive oxygen species (ROS), which cause oxidative damage to cells and subcellular structures (reviewed in Lesser 2006), decrease greatly during anoxia but increase upon reoxygenation. It is postulated that MRD is related to an attempt to reduce the accumulation of ROS. In mussels, the extent of biological damage caused by ROS is ameliorated by specific antioxidant enzymes, such as catalase, superoxide dismutase and glutathione peroxidase (Cancio & Cajaraville 1997, 1999; Almeida *et al.* 2005, 2006; Place *et al.* 2012; Rivera-Ingraham *et al.* 2013). It has often been suggested that the degree of hypoxia tolerance relates to a species-specific capacity to activate antioxidant defences in preparation for the reoxygenation injury (reviewed in Welker *et al.* 2013).

Individuals within a species may show different responses to hypoxia, depending on body size (Wang & Widdows 1991; Hole *et al.* 1995; Hicks & McMahon 2005), temperature (Hicks & McMahon 2005; Jansen *et al.* 2009; Pörtner 2010), hyposalinity (Wang *et al.* 2012, 2013), aerial exposure (Altieri 2006), pCO_2 (Pörtner 2010; Melzner *et al.* 2013; Sui *et al.* 2017) and food availability (Huhn *et al.* 2016). To illustrate, Hicks & McMahon (2005) investigated the effects of temperature (15, 20 and 25 °C), pO_2 (0, 1, 2, 4 and 6 kPa, corresponding to 0, 7.5, 15, 30 and 45 mm Hg, respectively, where 1 kPa = 7.5006 mm Hg) and individual size (12–79 mm shell length, SL) on survivorship of the non-ingenous mussel, *P. perna*, collected from the Texas Gulf of Mexico coast in the United States. Hypoxia tolerance was temperature-dependent, survivorship being significantly extended at lower temperatures under all tested lethal pO_2 (Figure 7.15). Test temperature/pO_2

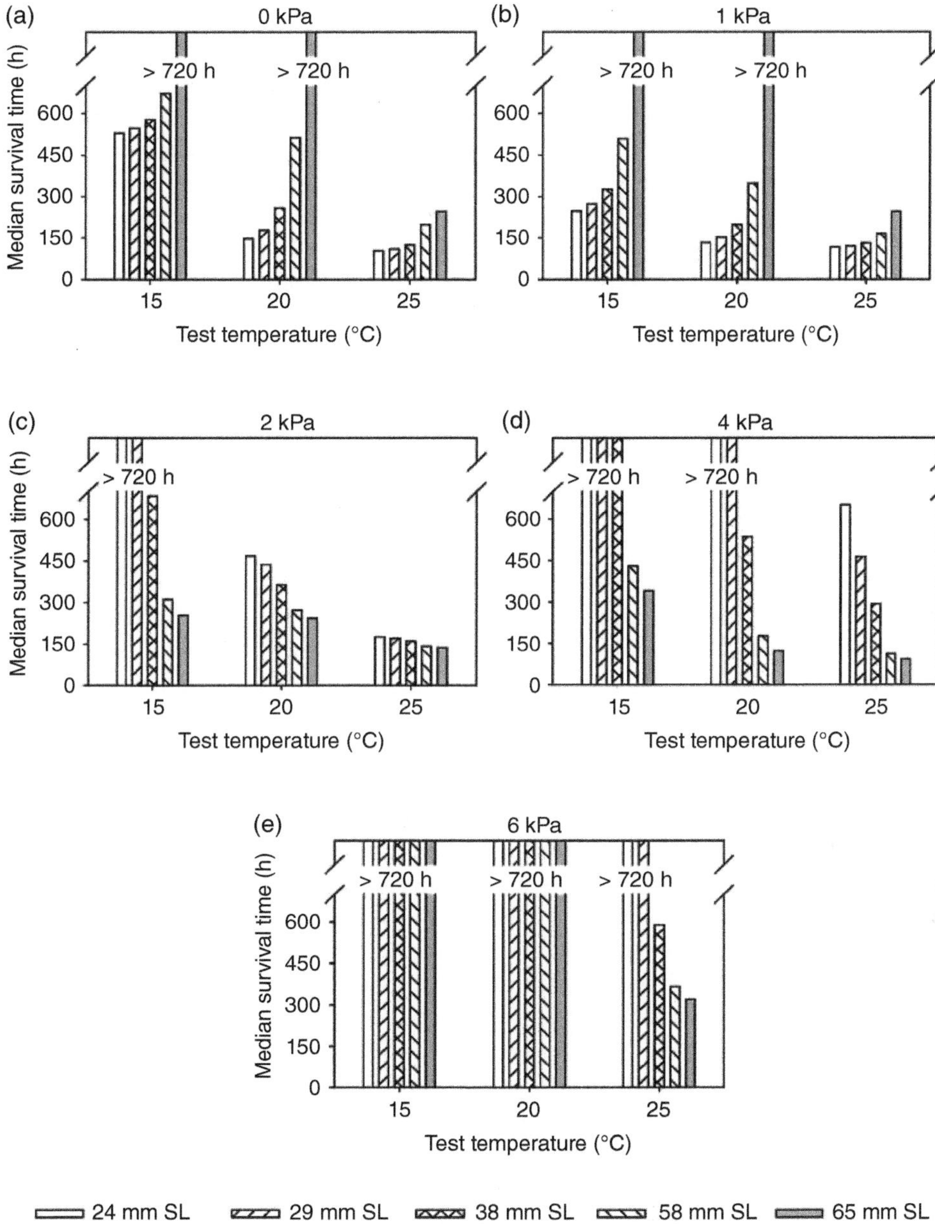

Figure 7.15 Median survival times (hr) for Texas Gulf of Mexico coast *Perna perna* on chronic exposure to P_{O2} of (a) 0, (b) 1, (c) 2, (d) 4 and (e) 6 kPa at test temperatures of 15, 20 and 25 °C, adjusted to the 10th (24 mm shell length, SL, open histograms), 25th (29 mm SL, left-hatched histograms), 50th (38 mm SL, cross-hatched histograms), 75th (58 mm SL, right-hatched histograms) and 90th (65 mm SL, shaded histograms) percentiles of the individual SL distribution. *Source:* From Hicks & McMahon (2005). Reproduced with permission of Oxford University Press.

combinations of 15 °C and 6 kPa and 20 °C and 6 kPa appeared nonlethal to this species, with 96% of individuals in these treatments surviving the 30-day (720-hour) exposure period and >50% mortality occurring at 25 °C at all tested levels of hypoxia. Median survival times were comparable with those cited for other intertidal and estuarine infaunal bivalve species (Hicks & McMahon 2005 and references therein). Survivorship was size-dependent, with median survival times increasing with increasing SL in anoxia and $pO_2 = 1$ kPa, but at 2, 4 and 6 kPa, smaller individuals survived longer than larger individuals. The larger surface area to volume ratios and reduced diffusive distances from epithelial surfaces to internal tissues of smaller individuals may support greater O_2 uptake and transport rates, accounting for their increased survival at pO_2 between 1 and 6 kPa (Hicks & McMahon 2005). The following are additional studies on hypoxia tolerance of mussels, considered either alone or in combination with an environmental factor: Altieri *et al.* (2006), *Mytilus edulis*; Erkes-Medrano *et al.* (2013), *M. californianus*; Artigaud *et al.* (2014), *Mytilus* spp.; Huhn *et al.* (2016), *P. viridis*; Noguerira *et al.* (2017), *P. perna*; and Sui *et al.* (2017), *M. coruscus*.

Thresholds of hypoxia for benthic organisms range from 0.28 to 4 mg O_2 l^{-1}, although, as already mentioned, the majority of reports refer to a value of 2 mg O_2 l^{-1} or lower. Vaquer-Sunyer & Duarte (2008) carried out a broad comparative analysis across a range of contrasting marine benthic organisms, including bivalves, and showed that hypoxia thresholds varied greatly between organisms and that the conventional definition of 2 mg O_2 l^{-1} to designate waters as hypoxic (see earlier) was below the empirical sublethal and lethal O_2 thresholds for half of the 206 species tested. Median lethal concentration ($LC_{50} \pm$ SE) for all organisms tested was 1.60 ± 0.12 mg O_2 l^{-1}, and the coefficient of variation across experiments was 78%, indicative of considerable variability in these thresholds across taxa. They found that 90% of their experiments showed LC_{50} values below 4.59 mg O_2 l^{-1}. Bivalves had a value of 1.42 ± 0.14, which was lower than that for fish species (1.54 ± 0.07), and lower again than that for crustaceans (2.45 ± 0.14). The value for bivalves is actually lower (i.e. 1.312 ± 0.43) than that calculated by Vaquer-Sunyer & Duarte (2008), because they included gastropod data in their estimate. The authors found that median lethal O_2 thresholds differed significantly with the extent of mobility of the organisms tested; that is, motile groups such as fish and crustaceans had higher threshold values than more sedentary groups. The median sublethal concentration ($SLC_{50} \pm$ SE) was 2.24 ± 0.21 mg O_2 l^{-1}, and the coefficient of variation was 76% across experiments. Once again, fish (4.41 ± 0.39) and crustaceans (3.21 ± 0.28) showed the highest median sublethal O_2 thresholds, with the bivalve value (albeit based on few experiments) considerably lower (0.75 ± 0.18).

Aerial Exposure

Two important consequences of aerial exposure at low tide are the effects of the environment on body temperature and water loss rates (Helmuth 1998). During tidal exposure to air, bivalves either close their valves and switch to anaerobic metabolism or exhibit alternate closure and opening of the shell (gaping), thereby allowing the maintenance of aerobic respiration. Shell closure results in reduced evaporative water loss at the cost of inefficient use of stored energy. While gaping is more compatible with optimal functioning of the metabolic machinery, valve movements push water out of the mantle cavity, thereby increasing levels of water loss and the risk of desiccation (Nicastro *et al.* 2010 and references therein). Despite the increase in water loss, gaping does not appear to have a significant influence on body temperature through evaporative cooling for the mussel species examined to date (Fitzhenry *et al.* 2004; Jost & Helmuth 2007), and it seems most likely that gaping occurs as a means of increasing aerobic respiration rather than evaporative cooling. However, Nicastro *et al.* (2012) investigated the group effects of gaping behaviour by comparing gaping (*P. perna*) and nongaping (*M. galloprovincialis*) species as isolated individuals and in aggregations. Gaping behaviour had no effect on body temperatures of isolated individuals, but it led to increased humidity and decreased temperatures in dense groups (beds). Also, in aggregations, gaping in *P. perna* resulted in cooler body temperatures than in *M. galloprovincialis*. As mussels normally occur in dense aggregations, the results of this study revealed an additional ecological role of gaping behaviour that is manifested in mussel aggregations, but not in solitary animals.

Other Factors

As Bayne *et al.* (1976b) so rightly say, in the light of the many varied effects of environmental change on OCR (as seen earlier), it is not surprising to find a regular seasonal pattern in OCR by some species. Widdows (1978) investigated the effects of size, food concentration and season on *M. edulis* OCR and found that the main factor that affected OCR was the seasonal cycle of reproductive condition. The rate of oxygen uptake by a 1 g dry weight individual was at a maximum (0.39 ml O_2 h^{-1}) in late spring and summer, when the mussels were fully mature, and declined to a minimum (0.21 ml O_2 h^{-1}) in the autumn and early winter. In the summer, gametogenesis is in the resting stage, and reserves of glycogen are high. This reserve of metabolically inert material results in a low weight-specific OCR. During the winter, gametogenesis is active and glycogen stores are being utilised resulting in an increased metabolic demand. In the spring, the large mass of developing and mature gametes, forming up to 35% of the total body weight, continues to impose a high demand for oxygen, which is only reduced after spawning (Bayne *et al.* 1976b; Widdows 1978). Similar results have been reported for other mussel and bivalve species (Bayne 1973; Bayne *et al.* 1976b; Sukhotin 1992; Ekaratne & Davenport 1993; Rueda & Smaal 2004). The form of the annual metabolic cycle varies because mussel reproductive cycles vary from population to population depending upon annual temperature cycles and latitude (see Chapter 5 and references therein). While changes in OCR appear to mirror environmental temperatures, Shumway *et al.* (1988) suggest that seasonal changes in food availability and reproductive stage have a greater effect on OCR than temperature per se (discussed in Bayne *et al.* 1976b).

Another factor that affects OCR is a high level of CO_2 in the haemolymph (hypercapnia). In the mussel *M. galloprovincialis*, this leads to a reduction in haemolymph pH. To limit the degree of acidosis, mussels increase bicarbonate levels through shell dissolution. This is accompanied by a significant reduction in OCR, indicating a lowering of metabolic rate (Michaelidis *et al.* 2005b). A close correlation is evident between the reduction in extracellular pH and the reduction in metabolic rate during hypercapnia. Such effects are relevant in the context of future scenarios of progressive accumulation of anthropogenic CO_2 in marine surface waters (see Chapter 3). However, there are indications that there is intra-population variability in the capacity of mussels – larvae, juveniles and adults – for metabolic adaptation to different CO_2 levels (Fernández Reiriz *et al.* 2012; Stapp *et al.* 2017).

Pollutants such as the heavy metals, copper, mercury, cadmium and zinc, as well as benzene, zinc sulphate, chlorine, organotins and polyaromatic hydrocarbons (PAHs), significantly depress respiration rate in mussels (Akberali *et al.* 1984, 1985; Baby & Menon 1986; Cheung & Cheung 1995; Rajagopal *et al.* 2003; Halldórsson *et al.* 2005; Jorge *et al.* 2007). Some, such as sulphide (Lee *et al.* 1996), microplastics (Van Cauwenberghe *et al.* 2015) and nanoparticles (Saggese *et al.* 2016), cause an increase in respiration rate, which is often characteristic of increased physiological stress.

Excretory System

Excretion

Two tubular nephridia (kidneys) and pericardial glands represent the major excretory organs of bivalves (see Chapter 2 for details). Waste accumulates in certain cells of the pericardial glands and is periodically discharged into the pericardial cavity; from there, the pericardial gland cells, which possess an exocytotic excretory function (Moore *et al.* 1980), eliminate the waste via the kidneys. Other cells of the pericardial glands are probably involved in filtering the haemolymph – the first stage of urine formation. The filtrate then flows through the reno-pericardial canals to the glandular part of the kidneys, where the process of secretion and reabsorption of ions occurs. The end result is a urine that has a high concentration of ammonium ion (NH_4^+) (37–100%) plus smaller amounts of urea, organic nitrogen (amino acids) and uric acid-nitrogen (Table 7.3). Ammonia is the major excretory end product of protein and amino acid catabolism in *M. edulis* (Bayne *et al.* 1976e; Widdows 1978) and most aquatic molluscs. Therefore, the rate of ammonia-nitrogen excretion may be regarded as reflecting the rate of protein catabolism. While some of the ammonia may be utilised to synthesise organic products, there is always a surplus that is excreted to the external environment (Dame 2012). Because ammonium is soluble in water, most excreted ammonium readily diffuses across the body surface, particularly the gills, and into the surrounding water. Excretion of other nitrogenous waste (see earlier), salts and byproducts of metabolism occurs through ultrafiltration and reabsorption in the tubular nephridia (Bracken 2007). Recently, Thomsen *et al.* (2016) investigated the localisation and mechanisms of ammonia excretion in two mussels, *M. californianus* and *M. edulis*. The plicate organ (PO), previously described as a solely respiratory organ, has a thin single-layered epithelium with an enlarged plicate surface area and is located on both sites of the gill central axis in the proximity of the kidney. Thomsen *et al.* (2016) found that a rhesus-like protein was abundantly expressed in this organ, as well as in gill tissue, albeit at lower concentrations. Rhesus (Rh) glycoproteins may serve as ammonia transporters in the respiratory surfaces of aquatic animals (reviewed in Wright & Wood 2009). Rh was not detectable in kidney tissue, suggesting that the kidneys are not involved in active NH_3/NH_4^+ excretion. In mussels, excretion of these products, collectively referred to as 'total ammonia' (TAM), takes place by passive diffusion across respiratory epithelia via the Rh protein. Diffusion is chiefly facilitated by cilia beating of the gill and PO boundary layers, a mechanism that is energy-efficient as it takes advantage of the water current that is already generated for filter feeding. A subsequent study on the distribution of the PO revealed a

Table 7.3 Nitrogenous compounds in the urine of some bivalves. –, component not measured.

	Excreted components as % of total measured nitrogen (N)				
Species	**NH_4-N**	**Urea-N**	**Amino-N**	**Uric acid-N**	**Reference**
Geukensia demissa	62–75	0	25–38	–	Lum & Hammen (1964)
	64	–	36	–	
Mytilus californianus	100	–	0	–	Bayne & Scullard (1977b)
Mytilus galloprovincialis					
Summer	37–58	–	42–63	–	Bayne & Scullard (1977b)
Autumn	100	–	0	–	Bayne & Scullard, (1977b)
Mytilus edulis					
Summer	71	–	29	–	Bayne & Scullard, (1977b)
Winter	97–100	–	0–3	–	Bayne & Scullard, (1977b)

diverse picture of the absence or presence of the organ within the superfamily Mytiloidea (Thomsen *et al.* 2018).

While ammonia is the dominant excretory product, significant amounts of amino-nitrogen are lost (0–63%), although amounts of lost urea and uric acid are negligible in mytilids, unlike in other bivalves (Bayne *et al.* 1976a; Dame 2012). The proportion of amino-nitrogen to total excreted nitrogen may increase above normal when mussels are subject to environmental stress. Bayne (1973) found that the proportion of amino-nitrogen increased from 55 to >75% during stress from temperature and starvation in *M. edulis*. This represents a considerable energy loss for the mussel, with a mean loss of 10.6% RMR throughout the year, with the highest value in May (17.5%) and the lowest in August (3.9%). Under temperature and nutritive stress, the loss increased to 63% (Bayne 1973).

Nitrogen excretion rates are extremely variable in bivalves. For example, rates of ammonia-nitrogen loss by *M. edulis* varied from a mean maximum 67.2 μg $d^{-1}g^{-1}$ fresh weight during winter to a mean minimum 9.6 μg $d^{-1}g^{-1}$ fresh weight in the summer, with an overall mean value of 19.2 μg $d^{-1}g^{-1}$ fresh weight (Bayne 1973). Rates of ammonium (NH_4^+) excretion in mytilids can be estimated using a protocol described by Tagliarolo *et al.* (2012). These authors collected samples of water at the beginning and end of respiration incubations and stored them in 100 ml bottles. Ammonium concentration was determined according to the phenol-hypochlorite method (Solorzano 1969). Reagents were added to the water samples immediately after sampling. After 24 hours, sample coloration was measured at 630 nm (UV-1700 Pharma Spec, UV-VIS spectrophotometer, Shimadzu) using Milli-Q water as the blank. Ammonium excretion rates were expressed as μmol NH_4^+ g ash-free dry weight (AFDW)$^{-1}$ h^{-1}. These rates can be converted to energy terms using a conversion factor of 0.025J μg^{-1} NH_4^+ (Elliott & Davison 1975). With the preceding method, one should be aware that some of the measured nitrogen might come from bacterial contamination, from faeces voided during the experiment or from 'leakage' rather than true excretion (Bayne *et al.* 1976a; Griffiths & Griffiths 1987). This can be avoided by using high-performance liquid chromatography (HPLC), which directly measures the net fluxes of metabolites during a single passage of natural seawater over the bivalve gill (Hawkins & Bayne 1992). Bartberger & Pierce (1976) did find that quantities of ammonia that could be attributed to non-mussel sources (e.g. bacteria) on empty scrubbed mussel (*Geukensia demissa*) valves were less than the standard errors of whole-animal excretion rates.

Factors Influencing Excretion Rates

It now appears that much of the reported variability in excretion rates can be attributed to feeding history and the gametogenic stage of the animals being analysed. For example, Bayne & Scullard (1977b) have shown that in *Mytilus edulis*, nitrogen excretion peaks prior to spawning, before experiencing a progressive decline to minimum rates in autumn and early winter (Figure 7.16). Just before spawning, glycogen reserves are at a minimum and mussels must catabolise protein to offset nutritional stress. In the autumn, when glycogen reserves are high but food is at a premium, ammonia excretion rates are low, which indicates that glycogen is preferentially being used to meet metabolic requirements. A similar seasonal pattern in ammonia excretion has been reported for rope-grown mussels (*M. galloprovincialis*) in NW Spain (Babarro *et al.* 2000) and for a *Geukensia demissa* salt marsh population in New England, United States (Huang *et al.* 2002). In the latter study, excretion rate gradually

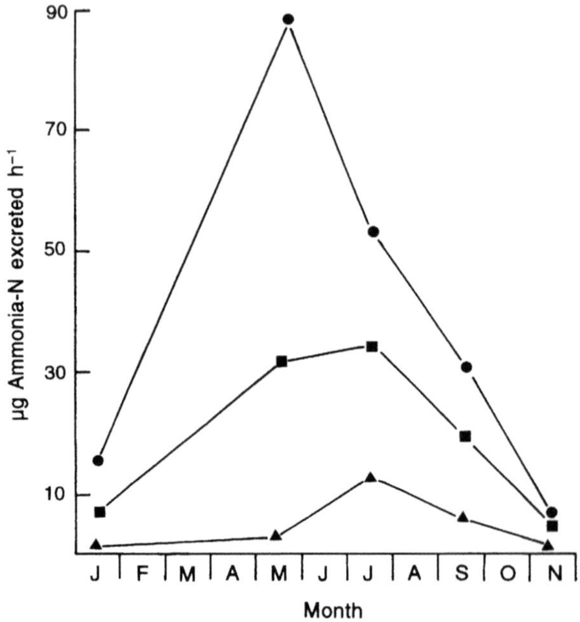

Figure 7.16 Rates of excretion of ammonia (NH_4-N) by the mussel *Mytilus edulis* of 0.2 g (black triangle), 1.0 g (filled square) and 2.0 g (filled circle) dry flesh weight during different months of the year. *Source:* From Bayne & Scullard (1977b). Reproduced with permission of Cambridge University Press.

increased from spring (13.8 μg NH_3–N h^{-1} g^{-1}) to a peak in summer (33.7 μg NH_3–N h^{-1} g^{-1}) during the spawning period for this species, before decreasing again during autumn (11.8 μg NH_3–N h^{-1} g^{-1}). Huang *et al.* (2002) observed that the seasonal changes in the rate of ammonium excretion did not interact with body size (i.e. both large and small individuals changed their rates of ammonium excretion in the same proportion seasonally – a pattern that differs from that observed in other bivalve species). Bayne & Scullard (1977b) found that the numerical values for the parameters of the allometric equation (a and b in the expression ammonia-nitrogen excretion = $a \times$ body size$b$) were extremely variable. For example, in *M. edulis*, the regression coefficient (b) for ammonium excretion ranged from 0.48 in November to 1.48 in May, and for mussels that were fed and starved in the laboratory for four months, the coefficient varied between 0.35 for starved mussels and 1.20 for fed mussels in winter, while both fed and starved mussels had a similar coefficient of 0.72 in summer (Bayne *et al.* 1976a). The utilisation of protein reserves under stress conditions has been further substantiated by the results of Bayne & Thompson (1970) for starved *M. edulis* in the laboratory, where protein was utilised more rapidly at higher (16 °C) than at lower (6 °C) temperatures. Salinity also affects excretion rate, with increased amounts of ammonia and amino acids being excreted when animals are initially exposed to reduced salinities. In *M. edulis*, both rates declined over a period of 20 days at 14.5 psu, eventually returning to values typical of mussels at full salinity. In *M. edulis*, the rate of ammonia excretion increased with a rise in temperature and, unlike with salinity, Bayne & Scullard (1977b) observed that there was no acclimation of this rate. During exposure to air, mussels (*M. californianus*) continued to produce ammonia, which accumulated in the fluid trapped in the mantle cavity (Bayne *et al.* 1976d). Upon re-immersion, bivalves typically show a distinct overshoot of ammonia (de Vooys & de Zwaan 1978). Barrento *et al.* (2013) reported such an ammonia peak when they investigated the combined effects of temperature (5, 11 and 15 °C) and aerial exposure of different durations on ammonia excretion rates in rope-grown *M. edulis*. The rate was greater at higher temperatures and after a 48-hour period out of water. Immediately after re-immersion at 15 °C, ammonia concentration was greater (2.1 ± 0.3 μmol g^{-1} h^{-1}) than that at 11 °C (0.7 ± 0.2 μmol g^{-1} h^{-1}) or 5 °C (<0.01 μmol g^{-1} h^{-1}; Figure 7.17). Following an eight-hour period out of water (harvest simulation period), ammonia excretion at 5 °C was delayed, and only after 24 hours was there an overshoot. Ammonia excretion then decreased to levels below 0.01 μmol g^{-1} h^{-1} after 72 hours. Following 48 hours of air exposure (long transport simulation period), the pattern was similar, but at 5 and 11 °C excretion was greater after 48 hours than after eight hours. There was no difference in ammonia excretion rates for mussels at 11 and 15 °C. Sadok *et al.* (1999) studied the effects of starvation and emersion on the ammonia effluxes of intertidal *M. edulis*. They did not observe an ammonia overshoot immediately after re-immersion and concluded that this could be a strategy used by intertidal mussels to conserve energy by conserving ammonia, and suggested that some ammonia utilisation occurred during exposure to air.

Michaelidis *et al.* (2005a) found that long-term exposure to moderately elevated CO_2 levels (seawater pH 7.3) caused interdependent changes in acid–base balance, oxygen consumption, nitrogen excretion and growth rate in the Mediterranean mussel *M. galloprovincialis*. The rate of ammonia excretion increased by 60% within the first five hours of exposure to pH 7.3; thereafter, the rate remained constantly elevated at 10 and 20 hours. The stimulation of amino acid catabolism in this species during exposure to hypercapnic conditions may

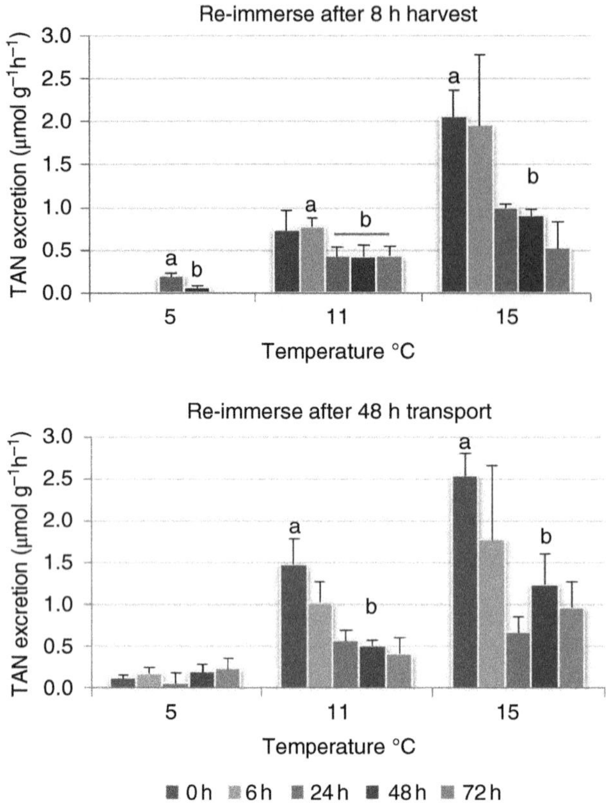

Figure 7.17 Ammonia excretion (μmol g^{-1} h^{-1}) of mussels immediately after re-immersion (0 hr) and after 6, 24, 48 and 72 hr in different water temperatures (5, 11 and 15 °C). Top: harvest simulation – mussels kept in a cooling box without water for an 8 hr period before ammonia analysis. Bottom: simulation of long transport period of 48 hr out of the water. For each temperature, different letters represent significant differences between lengths of time out of water. *Source:* From Barrento *et al.* (2013). Reproduced with permission from EDP Sciences.

damage the cellular protein pool and cause a decrease in growth performance, as seen in the same study (Michaelidis *et al.* 2005a), involving both metabolic depression and reduced shell growth. Exposure to an elevated seawater pH of 8.5 transiently reduced ammonia excretion rates in *M. edulis* and *M. californianus*, but rates returned to control values following 24 hours' acclimation (Thomsen *et al.* 2016).

The origin of the test species may also influence ammonia excretion. For example, when mussel seed from two origins (rocky shore and collector ropes) was cultivated on a raft in NW Spain, the rocky-shore mussels showed significantly higher ammonia excretion rates than the collector-rope ones over the experimental period of seven months, which probably reflects the slow adaptive process of osmotic adjustment in these mussels to the littoral habitat (Babarro *et al.* 2000). Other factors such as pollutants (Jorge *et al.* 2007) and pesticides (Führer *et al.* 2012), which induce shell closure and a switch from aerobic to anaerobic metabolism, alter various physiological parameters, including promoting a reduction in ammonia excretion.

Oxygen–Nitrogen (O:N) Ratio

This is the atomic ratio of oxygen consumed to nitrogen excreted [(mg O_2) ÷ 16] ÷ [(mg NH_4) ÷ 14], which provides an index of the balance in an animal's tissues between rates of catabolism of protein, lipid and carbohydrate substrates. If amino acids resulting from protein catabolism are deaminated (i.e. the amino group NH_2 is removed) and the resulting nitrogen is excreted while the carbon skeletons of the amino acids are completely oxidised, the theoretical minimum for the O:N ratio is approximately 7, signifying a catabolism based entirely upon protein. A low value of ~10 indicates considerable protein catabolism (e.g. during starvation or low food rations), while higher values signify that greater proportions of carbohydrate and/or lipid are being metabolised (Bayne & Newell 1983). The O:N ratio can therefore be considered as an index of stress (Bayne *et al.* 1976e and references therein). Bayne (1975) defined stress as a 'measurable alteration of a physiological, behavioural, biochemical or cytological steady-state that is induced by an environmental change and which renders the individual, or the population or the community more vulnerable to further environmental change'. The O:N ratio has been examined in relation to several environmental factors (e.g. ration, temperature, salinity and pollutants). Bayne (1973) recorded changes in O:N ratios in mussels (*M. edulis*) and found that for most of the year, excluding the summer (May–July; Figure 7.18), acclimated values for the O:N ratio were stable around a value of 100, indicating that protein, carbohydrate and lipid were catabolised in constant proportion. Theoretical calculations show that complete catabolism of protein as the sole energy substrate would give an O:N ratio of 9.33. The value of 100 for the O:N ratio indicates that 90% or more of the energy metabolism is derived from carbohydrate and lipid (Bayne 1973). During the

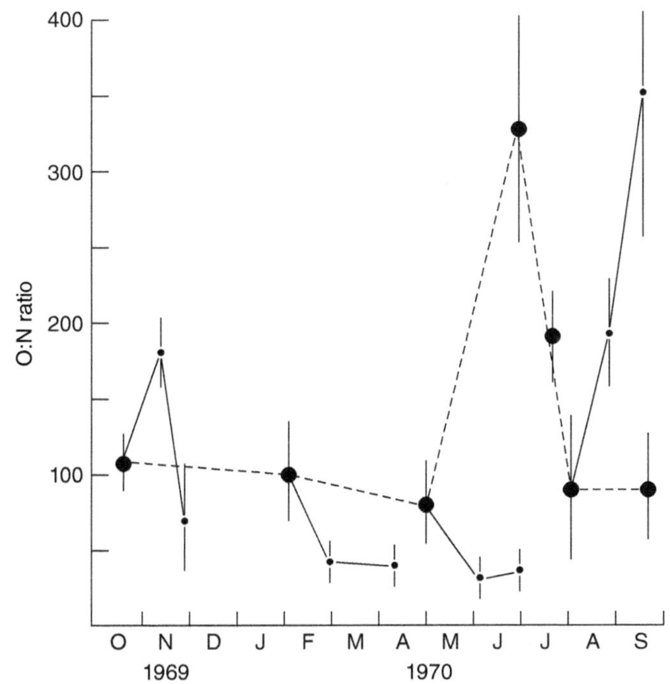

Figure 7.18 Ratio of oxygen uptake: ammonia-N excretion, by atomic equivalents (the O:N ratio) in *Mytilus edulis*. Seasonal data (large filled circles) for animals acclimated to 15 °C, while experimental data (small filled circles) indicate animals kept at 15 °C in routine stress experiments. The vertical bars represent the range of determinations for four animals. *Source:* From Bayne (1973). Reproduced with permission from Cambridge University Press.

summer, there is a build-up of carbohydrate reserves in the body tissues. Consequently, the O:N ratio increases to a value of >300, due to a decreased use of protein as an energy substrate. In the winter and late spring, carbohydrate (glycogen) reserves are low and stress results in a decline in the O:N ratio to new steady-state values of ~100. The decline in the O:N ratio is associated with an increase in ammonia-N excretion due to an increased use of protein as an energy substrate. The results suggest that O:N ratio is a sensitive index of nutritive stress (see also Widdows 1978; Bayne & Scullard 1977b). Bayne & Scullard (1977b) found that when mussels (*M. edulis*) were maintained in the laboratory at a ration level above the maintenance requirement (i.e. at a ration greater than that for which there is neither a loss nor a gain in weight), the O:N ratio was higher in smaller than in larger individuals. During starvation, however, mussels of all sizes showed a reduction in O:N ratio, which was particularly marked in smaller individuals, which probably had a smaller glycogen reserve to call on during nutritive stress.

Bayne (1975) demonstrated short-term changes in O:N ratio in *M. edulis* following an increase in temperature (ambient + 7 °C). There was a transient increase in the O:N ratio from 100 to almost 200 during the first few days of the experiment (Figure 7.19), signifying a period of rapid utilisation of carbohydrate during the early stages of acclimation. This was followed by a decrease to ~60 at day 10, then by stabilisation to a lower, steady state

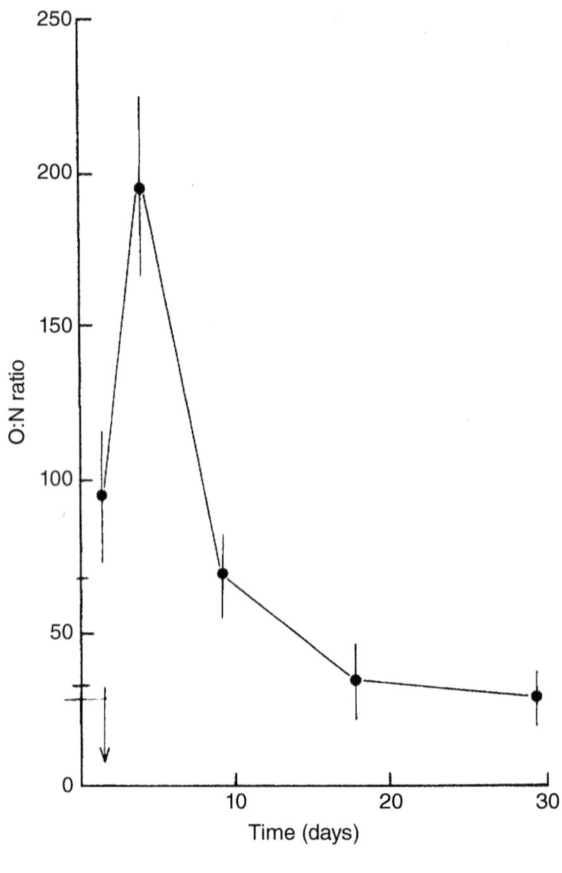

Figure 7.19 Atomic ratio of oxygen consumed to ammonia-N excreted (O:N ratio) in *Mytilus edulis* subjected to an increase in temperature (ambient + 7 °C) from the time indicated by the arrow. Means ± 1 SE for five individuals. From Bayne (1975). Reproduced with permission from Cambridge University Press.

over the remaining 20 days. It should be acknowledged that a low O:N ratio is not always indicative of stress (Huang *et al.* 2002).

The O:N ratio can also be used as a biomarker for assessing the deleterious effects of contaminants. Hu *et al.* (2017) examined the effect of nanoparticles in combination with CO_2-induced pH on O:N ratios in *M. coruscus*, and found that the ratios were significantly influenced by both factors, with the lowest ratios (<7) observed for both the highest nanoparticle concentration (10 mg l^{-1}) and the lowest pH (7.3), signifying a catabolism based entirely upon protein. Fuhrer *et al.* (2012) investigated the effect of the organophosphate pesticide Lorsban 4E on the O:N ratio in *Aulacomya atra*. They found a gradual decrease in ratio with increasing concentrations of the pesticide (from 40 in the control to a value close to 15 at a concentration of 1.6 μg l^{-1}), indicating that individuals exposed to high concentrations of this pesticide would use only protein as an essential source of amino acids.

Osmoregulation

Bivalves exhibit a wide variation in salinity tolerance. Euryhaline species that tolerate a wide range of salinities include *Mytilus* spp. (4–40 psu), *Geukensia demissa* (6–70), *Brachidontes exustus* (10–41), *P. viridis* (18–35) and *Xenostrobus securis* (1–31), whereas stenohaline species such as *Modiolus modiolus* are confined to a much narrower range. Despite such differences, all bivalves are osmoconformers that maintain an internal environment that is osmotic to their external environment. This means that the osmotic pressure, or osmolarity, of the organism's cells is equal to that of their surrounding environment, although the cellular ionic composition may be different from that of seawater. Osmoregulators, on the other hand, tightly regulate their body osmolarity, maintaining constant internal conditions despite changes in their external environment. Bayne *et al.* (1976a) describe three different mechanisms that control body fluid composition in bivalves: control of ionic concentration (already described), extracellular anisosmotic regulation and intracellular isosmotic regulation; all three are interrelated and are elements of a fourth process, cell volume control. Before discussing the last two physiological mechanisms, the first and main response to a stressful salinity change is valve closure. This behavioural response isolates the animal from the external environment and can allow for the internal osmotic concentration to remain high when external salinity is low, possibly preventing a rapid decrease in internal osmotic concentration leading to osmotic shock. But valve closure is a short-term response, and should salinity change persist, the animal may be compelled to regulate cell volume through its demand for oxygen and food. Hence, there is a trade-off between the need to interact with the environment and the need to be isolated from it. McFarland *et al.* (2013) examined the valve opening/closing behaviour of the euryhaline mussel *P. viridis* when subjected to acute changes in salinity. All mussels remained open at salinities of 25 and 35 psu, but at 10 and 15 psu they increasingly closed their valves over time, and within 120 hours of exposure all were either closed or dead (Figure 7.20). Davenport (1979) found that the osmolality of the retained mantle fluid of closed mussels (*M. edulis*) was higher in individuals from areas characterised by severe salinity changes than in those from places where environmental changes were less extreme.

Extracellular anisosmotic control is osmotic regulation between body fluids (in haemolymph and pericardium) and the external environment, while isosmotic

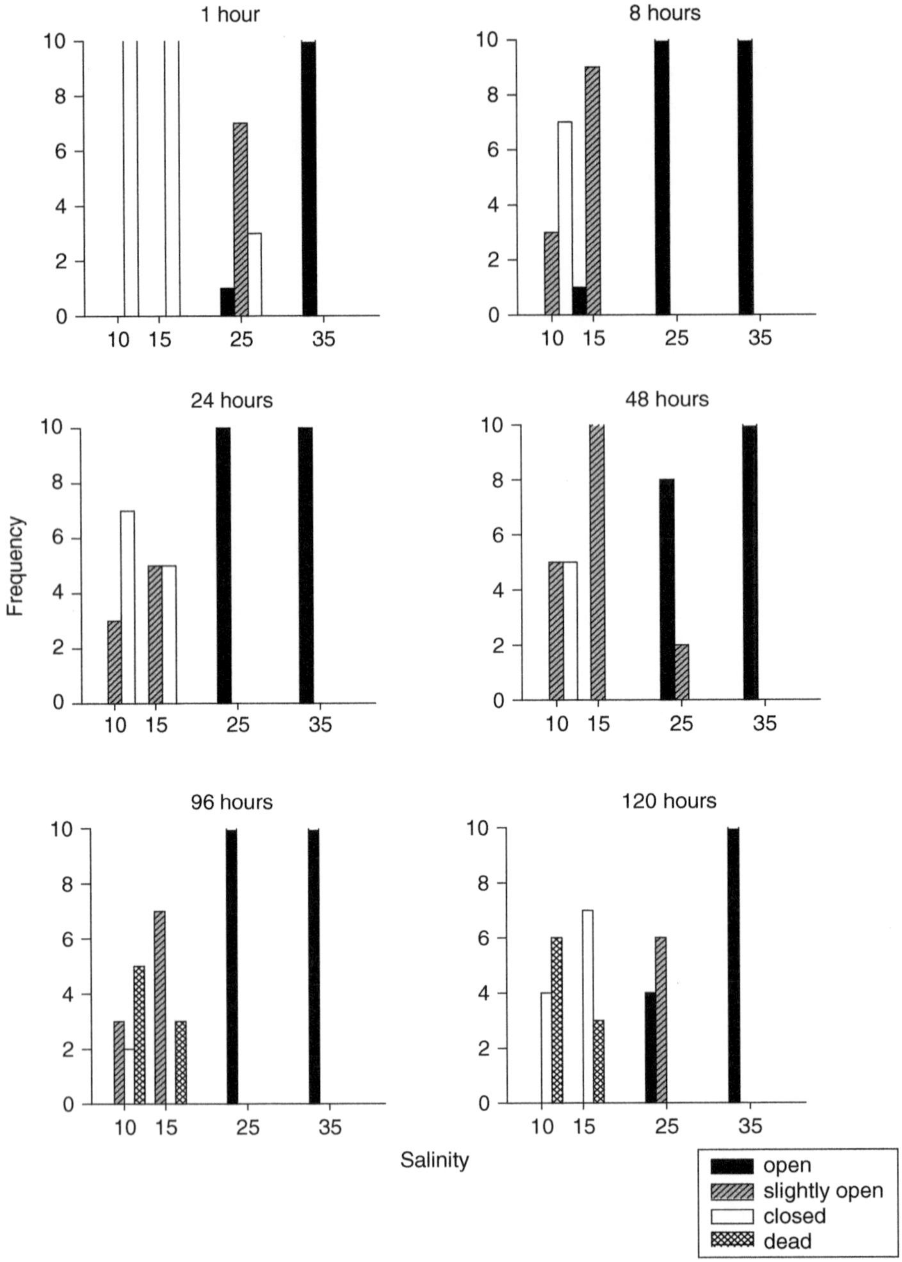

Figure 7.20 Valve opening/closing behaviour over time of the euryhaline mussel *Perna perna* when subjected to acute changes in salinity. At all time periods, valve closure response at salinities of 10 and 15 psu were significantly different from the response at 25 and 35 psu ($P < 0.0005$). *Source:* From McFarland *et al.* (2013). CC BY 4.0.

regulation is osmotic regulation between tissue and body fluids. Although the cells and haemolymph of most marine invertebrates are in osmotic equilibrium, the osmotic constituents are different. The main osmotic constituents of the cells are potassium and bicarbonate ions, phosphate compounds, amino acids and a variety of other low-molecular-weight compounds, while those of the haemolymph are mainly inorganic ions (Shumway *et al.* 1977 and references therein). To date, the general opinion is that mussels are incapable of extracellular osmotic control (Bayne *et al.* 1976a; Shumway *et al.* 1977; Rivera-Ingraham & Lignot 2017) and therefore rely solely on isosmotic intracellular regulation of volume.

Isosmotic intracellular regulation, as defined by Florkin (1962), is a cellular mechanism that, upon a change in surrounding salinity, leads to the adjustment of the intracellular osmotic pressure to meet that of the environment, thus minimising variations in cell hydration. This involves increasing or decreasing the concentrations of osmotically active solutes (e.g. ninhydrin-positive substances, NPS; i.e. substances containing an amino group), K^+ and free amino acids (FAA) to achieve cell volume regulation, as well as modifying membrane-bound transporters. Osmosensing[6] is achieved through a wide variety of internal mechanisms, such as Ca^{2+} gradients, transient receptor potential ion channels and cell volume sensors, and is often controlled by specific hormones (Rivera-Ingraham & Lignot 2017 and references therein).

The key role of FAAs in isosmotic intracellular regulation has been well documented (Bayne *et al.* 1976a). In response to a decrease in salinity, cell swelling is controlled by the release of FAAs from the intracellular pool to the haemolymph. Such losses represent a major component cost of cell volume regulation, and Hawkins & Hilbish (1992) have suggested that this may help to explain stress and even mortality resulting from small but frequent fluctuations in salinity. Cell shrinkage, on the other hand, is controlled by cytoplasmic increase in FAAs. The principal source of FFAs is from breakdown of endogenous body protein, with minor contributions from dietary sources, de novo synthesis and direct uptake of dissolved amino acids from the medium (Hawkins & Hilbish 1992). Bivalves are thus able to maintain their cells isoosmotic with their extra-cellular fluids while at the same time regulating their volume, although the capacity for volume regulation is species-dependent. Euryhaline species exhibit near-perfect volume regulation in comparison to stenohaline species (Gainey 1994). The prevalence of these nitrogenous solutes over inorganic ions for such regulation was established on experimental evidence, which showed that they preserved a solution microenvironment compatible with better enzyme functioning than when inorganic ions were used as osmotic effectors during high or varying external salinity conditions (Yancey *et al.*1982; Somero & Bowlus 1983).

The amino acids alanine, arginine, aspartic and glutamic acids, glycine, taurine and betaine are particularly important in the process of cell volume regulation (Shumway *et al.* 1977; May *et al.* 2017; Table 7.4). The quaternary base glycine-betaine also plays a role in osmoregulation (de Vooys & Geenevasen 2002; Kube *et al.* 2006). Amino acids are generally separated and quantified from lyophilised bivalve tissue using HPLC (see Kube

Table 7.4 Concentration of amino acids in the adductor muscle of *Mytilus edulis* in 100% and 50% seawater.

	Concentration (μmoles per g wet weight)	
Amino acid	**100% seawater**	**50% seawater**
Alanine	18.40	13.0
Arginine	11.94	8.38
Aspartic acid	9.77	1.73
Glutamic acid	9.79	10.61
Glycine	61.87	18.40
Histidine	1.80	1.22
Isoleucine	0.31	0.39
Leucine	0.42	0.60
Lysine	2.81	1.71
Phenylalanine	0.51	0.28
Proline	6.29	2.80
Serine	5.99	3.05
Threonine	4.37	1.93
Tyrosine	0.77	0.48
Valine	0.94	0.50
Taurine	58.09	44.51
Betaine	62.99	40.66
Water content	69.80	80.20
Osmotic pressure (mOsm per kg H_2O) due to:		
Amino acids	216.54	90.19
Taurine	92.47	61.66
Betaine	100.27	56.23
Osmotic pressure of the medium (mOsm)	1180	573

Source: Adapted from Bayne *et al.* (1976a).

et al. 2006). Species, or even populations within a species, may differ in the size and content of their FAA pool. In response to an increase in salinity, alanine, proline and glycine accounted for most of the increase in the amino acid pool in the euryhaline mussel *Geukensia demissa* (Deaton 2001). In the brackish waters of the southern Baltic Sea, the main pool of FAA in the clam *Macoma balthica* was composed of alanine, glutamate, arginine, glycine and ornithine, which represented approximately 80% of the FAA total (Sokolowski *et al.* 2003). The composition differed substantially to that found in clams from fully saline environments, which can be attributed to the lower salinity in the southern

Baltic. Also, some amino acids play a more predominant role than others, depending on salinity. For example, in the Baltic, glycine seems to be most important in regulating intracellular osmolarity in *M. balthica*, a function performed primarily by taurine in Atlantic and North Sea populations (Sokolowski *et al.* 2003). Another study that examined differences in the concentration of FAA in populations of *M. balthica* and *Mytilus* spp. exposed to high versus low intra-annual salinity fluctuations along their European distribution found that seasonal FAA variations were more pronounced in *M. balthica*, and that in both species at different locations within the Baltic the highest FAA concentrations were found in autumn and winter, with low concentrations measured in summer (Kube *et al.* 2007). Seasonal patterns were less pronounced in both taxa at locations with constant salinity conditions. In contrast to Baltic Sea populations, Atlantic and Mediterranean bivalves showed high FAA concentrations in summer and low values in winter, regardless of seasonal salinity fluctuations. The authors concluded that seasonal patterns of the main FAA pool components (alanine, glycine and taurine) within and between populations should be considered in the context of seasonal variations of other environmental factors beside salinity and physiological state (glycogen content, reproductive stage) of the test species. Babarro *et al.* (2011) investigated whether the pool of FAA remained constant in raft-cultured *M. galloprovincialis* throughout the growing season. It did indeed remain fairly constant despite significant changes in some of the main components of the pool. Specifically, in both field and laboratory experiments, there was an up to 40% drop in taurine content with increase in soft tissues. This was counterbalanced by an increase in alanine, whereas glycine showed high variability and inconsistency. The main function of taurine, which is the most abundant FAA in mussels, is isosmotic intracellular regulation in soft tissues. Therefore, it might be expected that taurine scales up with the amount of tissues of the individual. However, for reasons unknown, an inverse relationship between relative taurine amounts and bivalve body size has been reported (Allen & Garret 1972; Babarro & Fernández Reiriz 2006).

Nuclear magnetic resonance (NMR) spectroscopy is a commonly used tool for studying the metabolic profiles of bivalve molluscs. May *et al.* (2017) used NMR spectroscopy to examine the baseline composition of the metabolome, with attention to the composition of the FAA pools in gill, mantle and muscle tissues from larval (veliger and pediveliger stages) and juvenile *M. edulis* reared under ambient conditions and low salinity exposure. In juveniles, the FAA pools were dominated by taurine and betaine, and to a lesser extent, glycine and alanine, while in larvae, glycine was just as abundant as taurine and betaine. The compositions of the FAA pools of veliger and pediveliger larvae were similar, although concentrations of the six FAAs analysed, especially taurine and betaine, tended to be higher in the pediveligers than in the veligers and juvenile tissues, but these differences were not statistically significant. Tissue-specific variation in the concentration of individual FAAs was also observed. This was the first metabolomics study to include multiple tissue types and developmental stages in a mussel species (*M. edulis*) in evaluating organismal responses to salinity stress. FAA profiles have also been used as a convenient indicator of stress in contaminated environments (see Babarro *et al.* 2006 and references therein), although their use has inherent risks of misinterpretation in view of the large temporal fluctuations in FAA profiles discussed earlier.

Notes

1 The large-scale study of proteins.
2 The study of the metabolome, the complete set of small-molecule chemicals found within a cell, cellular organelle, organ, tissue, tissue extract, biofluid or entire organism.
3 An anthropogenically induced variation in biochemical, physiological or ecological components or processes, structures or functions that is measurable in a biological sample or system (National Research Council 1987).
4 An adaptive strategy for survival under transient stressful conditions such as hypoxia, anoxia, thermal and osmotic stress.
5 The set of metabolic reactions and processes requiring oxygen that take place in mitochondria in order to convert the energy stored in macronutrients to adenosine triphosphate (ATP), the universal energy donor in the cell.
6 The cell turgor in response to changes in the osmolarity of the external environment.

References

Ahn, I.Y. & Shim, J.H. (1998) Summer metabolism of the Antarctic clam, *Laternula elliptica* (King and Broderip) in Maxwell Bay, King George Island and its implications. *Journal of Experimental Marine Biology and Ecology*, **224**, 253–264.

Akberali, H.B., Earnshaw, M.J. & Marriott, K.R.M. (1984) The action of heavy metals on the gametes of the marine mussel *Mytilus edulis* (L.). I. Copper-induced uncoupling of respiration in the unfertilized egg. *Comparative Biochemistry and Physiology Part-C*, **77**, 289–294.

Akberali, H.B., Earnshaw, M.J. & Marriott, K.R.M. (1985) The action of heavy metals on the gametes of the marine mussel *Mytilus edulis* (L.). II. Uptake of copper and zinc and their effects on respiration in the sperm and unfertilized egg. *Marine Environmental Research*, **16**, 37–59.

Allen, J.A. & Garret, M.R. (1972) Studies on taurine in the euryhaline bivalve *Mya arenaria*. *Comparative Biochemistry and Physiology Part-A*, **41**, 307–317.

Almeida, E.A. & Bainy, A.C.D. (2006) Effects of aerial exposure on antioxidant defenses in the brown mussel *Perna perna*. *Brazilian Archives of Biology and Technology*, **549**, 225–229.

Almeida, E. A., Bainy, A.C.D., Dafré, A. L., Gomes, O. F., Medeiros, M.H.G. & Di Mascio, P. (2005) Oxidative stress in digestive gland and gill of the brown mussel (*Perna perna*) exposed to air and re-submersed. *Journal of Experimental Marine Biology and Ecology*, **318**, 21–30.

Altieri, A.H. (2006) Inducible variation in hypoxia tolerance across the intertidal-subtidal distribution of the blue mussel *Mytilus edulis*. *Marine Ecology Progress Series*, **325**, 295–300.

Alyakrinskaya, I.O. (2003) Tissue hemoglobins in Bivalvia (Mollusca). *Biology Bulletin*, **30**, 617–626.

Arifin, Z. & Bendell-Young, L.I. (2001) Cost of selective feeding by the blue mussel (*Mytilus trossulus*) as measured by respiration and ammonia excretion rates. *Journal of Experimental Marine Biology and Ecology*, **260**, 259–269.

Arranz, K., Labarta, U., Fernández-Reiriz, M.J. & Navarro, E. (2016) Allometric size-scaling of biometric growth parameters and metabolic and excretion rates. A comparative study of intertidal and subtidal populations of mussels (*Mytilus galloprovincialis*). *Hydrobiologia*, **772**, 261–275.

Artigaud, S., Lacroix, C., Pichereau, V. & Flye-Sainte-Marie, J. (2014) Respiratory response to combined heat and hypoxia in the marine bivalves *Pecten maximus* and *Mytilus* spp. *Comparative Biochemistry and Physiology Part-A*, **175**, 135–140.

Babarro, J.M.F. & Fernández Reiriz, M.J. (2006) Variability of taurine concentrations in *Mytilus galloprovincialis* as a function of body size and specific tissue. *Comparative Biochemistry and Physiology Part-B*, **145**, 94–100.

Babarro, J.M.F., Fernández Reiriz, M.J., Garrido, J.L. & Labarta, U. (2000) Metabolism of the mussel *Mytilus galloprovincialis* from two origins in the Ría de Arousa (north-west Spain). *Journal of the Marine Biological Association of the UK*, **80**, 865–872.

Babarro, J.M.F., Fernández–Reiriz, M.J. & Labarta, U. (2006) Free amino acid composition in juveniles of *Mytilus galloprovincialis*: spatial variability after Prestige oil spill. *Comparative Biochemistry and Physiology Part-A*, **145**, 204–213.

Babarro, J.M.F., Fernández–Reiriz, M.J., Labarta, U. & Garrido, J.L. (2011) Variability of the total free amino acid (TFAA) pool in *Mytilus galloprovincialis* cultured on a raft system. Effect of body size. *Aquaculture Nutrition*, **17**, e448–e458.

Baby, K.V. & Menon, N.R. (1986) Oxygen uptake in the brown mussel, *Perna indica* (Kuriakose & Nair) under sublethal stress of Hg, Cd and Zn. *Indian Journal of Marine Science*, **15**, 127–128.

Bakhmet, N. & Zdorovenov, P.E. (2010) Variability in cardiac activity of the bivalves *Mytilus edulis* and *Modiolus modiolus*. *Russian Journal of Marine Biology*, **36**, 223–225.

Bakhmet, N., Berger, V.J. & Halaman, V.V. (2005) Heart rate of the blue mussel *Mytilus edulis* (Bivalvia) under salinity change. *Russian Journal of Marine Biology*, **31**, 314–317.

Barbariol, V. & Razouls, S. (2000) Experimental studies on the respiratory metabolism of *Mytilus galloprovincialis* (Mollusca Bivalvia) from the Mediterranean Sea (Gulf of Lion) *Vie et Milieu–Life and Environment*, **50**, 87–92.

Bartberger, C.A. & Pierce, S.K. Jr (1976) Relationship between ammonia excretion rates and hemolymph nitrogenous compounds of a euryhaline bivalve during low salinity acclimation. *Biological Bulletin*, **150**, 1–14.

Barrento, S., Lupatsch, I., Keay, A. & Christophersen, G. (2013) Metabolic rate of blue mussels (*Mytilus edulis*) under varying post-harvest holding conditions. *Aquatic Living Resources*, **26**, 241–247.

Bayne, B.L. (1967) The respiratory response of *Mytilus perna* L. (Mollusca: Lamellibranchia) to reduced environmental oxygen. *Physiological Zoology*, **40**, 307–313.

Bayne, B.L. (1971a) Ventilation, the heart beat and oxygen uptake by *Mytilus edulis* L. in declining oxygen tension. *Comparative Biochemistry and Physiology Part-A*, **40**, 1065–1085.

Bayne, B.L. (1971b) Oxygen consumption by three species of lamellibranch mollusc in declining oxygen tension. *Comparative Biochemistry and Physiology Part-A*, **40**, 955–970.

Bayne, B.L. (1973) Physiological changes in *Mytilus edulis* L. induced by temperature and nutritive stress. *Journal of the Marine Biological Association of the UK*, **53**, 39–58.

Bayne, B.L. (1975) Aspects of physiological condition in *Mytilus edulis* (L.) with special reference to the effects of oxygen tension and salinity. In: *Proceedings of the Ninth European Marine Biology Symposium* (ed. H. Barnes), pp. 213–238. Aberdeen University Press, Aberdeen.

Bayne, B.L. & Thompson, R.J. (1970) Some physiological consequences of keeping *Mytilus edulis* in the laboratory. *Helgoländer Wissenschaftliche Meeresuntersuchungen*, **20**, 526–552.

Bayne, B.L. & Scullard, C. (1977a) An apparent specific dynamic action in *Mytilus edulis* L. *Journal of the Marine Biological Association of the UK*, **57**, 371–378.

Bayne, B.L. & Scullard, C. (1977b) Rates of nitrogen excretion by species of *Mytilus* (Bivalvia; Mollusca). *Journal of the Marine Biological Association of the UK*, **57**, 355–369.

Bayne, B.L. & Newell, R.C. (1983) Physiological energetics of marine molluscs. In: *The Mollusca, Vol. 4, Physiology*, Part 1 (eds A.S.M. Saleuddin & K.M. Wilbur), pp. 407–515. Academic Press, New York.

Bayne, B.L., Thompson, R.J. & Widdows, J. (1973) Some effects of temperature and food on the rate of oxygen consumption by *Mytilus edulis* L. In: *Effects of Temperature on Ectothermal Organisms* (ed. W. Wieser), pp. 181–193. Springer-Verlag, Berlin.

Bayne, B.L., Widdows, J. & Thompson, R.J. (1976a) Physiology II. In: *Marine Mussels: Their Ecology and Physiology* (ed. B.L. Bayne), pp. 207–260. Cambridge University Press, Cambridge.

Bayne, B.L., Widdows, J. & Thompson, R.J. (1976b) Physiology I. In: *Marine Mussels: Their Ecology and Physiology* (ed. B.L. Bayne), pp. 121–206. Cambridge University Press, Cambridge.

Bayne, B.L., Bayne, C.J., Carefoot, T.C. & Thompson, R.J. (1976c) The physiological ecology of *Mytilus californianus* Conrad. 1. Metabolism and energy balance. *Oecologia*, **22**, 211–228.

Bayne, B.L., Bayne, C.J., Carefoot, T.C. & Thompson, R.J. (1976d) The physiological ecology of *Mytilus californianus* Conrad. 2. Adaptations to low oxygen tension and air exposure. *Oecologia*, **22**, 229–250.

Bayne, B. L., Widdows, J. & Thompson, R. J. (1976e) Physiological integrations. In: *Marine Mussels: Their Ecology and Physiology* (ed. B.L. Bayne), pp. 261–291. Cambridge University Press, Cambridge.

Bernard, F.R. (1983) *Physiology and the Mariculture of Some Northeastern Pacific Bivalve Molluscs.* Canadian Science Publishing, Ottawa, ON.

Berry, P.F. & Schleyer, M.H. (1983) The brown mussel *Perna perna* on the Natal coast, South Africa: utilization of available food and energy budget. *Marine Ecology Progress Series*, **13**, 201–210.

Bettencourt, R., Dando, P., Collins, P., Costa, V., Allam, B. & Serrão Santos, R. (2009) Innate immunity in the deepsea hydrothermal vent mussel *Bathymodiolus azoricus. Biochemistry and Physiology Part-A*, **152**, 278–289.

Booth, C.E. & Mangum, C.P. (1978) Oxygen uptake and transport in the lamellibranch mollusc *Modiolus demissus. Physiological Zoology*, **51**, 17–32.

Boutilier, R.G. (2001) Mechanisms of cell survival in hypoxia and hypothermia. *Journal of Experimental Biology*, **204**, 3171–3181.

Braby, C.E. & Somero, G.N. (2006) Following the heart: temperature and salinity effects on heart rate in native and invasive species of the blue mussels (genus *Mytilus*). *Journal of Experimental Biology*, **209**, 2554–2566.

Bracken, M.E.S. (2007) Excretion. In: *Encyclopedia of Tidepools and Rocky Shores* (eds M.W. Denny & S.D. Gaines), pp. 215–217. University of California Press, Berkeley, CA.

Brooks, S., Lyons, B.P., Goodsir, F., Bignell, J. & Thain, J.E. (2009) Biomarker responses in mussels, an integrated approach to biological effects measurements. *Journal of Toxicology and Environmental Health A*, **72**, 196–208.

Brown, R.J., Galloway, T.S., Lowe, D., Browne, M.A., Dissanayake, A., Jones, M.B. *et al.* (2004) Differential sensitivity of three marine invertebrates to copper assessed using multiple biomarkers. *Aquatic Toxicology*, **66**, 267–278.

Burnett, N.P., Seabra, R., de Pirro, M., Wethey, D.S., Woodin, S.A., Helmuth, B. *et al.* (2013) An improved noninvasive method for measuring heartbeat of intertidal animals. *Limnology and Oceanography: Methods*, **11**, 91–100.

Burton, R.F. (1983) Ionic regulation and water balance. In: *The Mollusca*, Vol. 5. Physiology, Part 2. (eds A.S.M. Saleuddin & K.M. Wilbur), pp. 291–352. Academic Press, London.

Caballero-Alfonso, A.M., Carstensen, J. & Conley, D.J. (2015) Biogeochemical and environmental drivers of coastal hypoxia. *Journal of Marine Systems*, **141**, 190–199.

Cajaraville, M.P. & Pal, S.G. (1995) Morphofunctional study of the haemocytes of the bivalve mollusk *Mytilus galloprovincialis* with emphasis on the endolysosomal compartment. *Cell Structure and Function*, **20**, 355–367.

Campos, A., Apraiz, I., da Fonseca, R.R. & Cristobal, S. (2015) Shotgun analysis of the marine mussel *Mytilus edulis* hemolymph proteome and mapping the innate immunity elements. *Proteomics*, **15**, 4021–4029.

Cancio, I. & Cajaraville, M. (1997) Histochemistry of oxidases in several tissues of bivalve molluscs. *Cell Biology International*, **21**, 574–584.

Cancio, I. & Cajaraville, M. (1999) Seasonal variation of xanthine oxidoreductase activity in the digestive gland cells of the mussel *Mytilus galloprovincialis*: a biochemical, histochemical and immunochemical study. *Biology of the Cell*, **91**, 605–615.

Cheung, S.G. & Cheung, R. (1995) Effects of heavy metals on oxygen consumption and ammonia excretion in green-lipped mussels (*Perna viridis*). *Marine Pollution Bulletin*, **31**, 381–386.

Compton, T.J., Rijkenberg, M.J.A., Drent, J. & Piersma, T. (2007) Thermal tolerance ranges and climate variability: a comparison between bivalves from differing climates. *Journal of Experimental Marine Biology and Ecology*, **352**, 200–211.

Conley, D.J., Carstensen, J., Aigars, J., Axe, P., Bonsdorff, E., Eremina, T. *et al.* (2011) Hypoxia is increasing in the coastal zone of the Baltic Sea. *Environmental Science and Technology*, **45**, 6777–6783.

Connor, K.M. & Gracey, A.Y. (2012) High-resolution analysis of metabolic cycles in the intertidal mussel *Mytilus californianus*. *American Journal of Physiology. Regulatory, Integrative and Comparative Physiology*, **302**, R103–R111.

Dame, R.F. (2012) *Ecology of Marine Bivalves: An Ecosystem Approach*, 2nd edn. CRC Press, Boca Raton, FL.

Davenport, J. (1979) Is *Mytilus edulis* a short-term regulator? *Comparative Biochemistry and Physiology Part-A*, **64**, 91–95.

Davenport, J. (1983) A comparison of some aspects of the behaviour and physiology of the Indian mussel *Perna (= Mytilus) viridis* and the common mussel *Mytilus edulis* L. *Journal of Molluscan Studies*, **49**, 21–26.

Deaton, LE. (2001) Hyperosmotic volume regulation in the gills of the ribbed mussel, *Geukensia demissa*: rapid accumulation of betaine and alanine. *Journal of Experimental Marine Biology and Ecology*, **260**, 185–197.

Depledge, M.H., Lundebye, A.K., Curtis, T., Aagaard, A. & Andersen, B.B. (1996) Automated interpulse-duration assessment (AIDA): a new technique for detecting disturbances in cardiac activity in selected macroinvertebrates. *Marine Biology*, **126**, 313–319.

de Vooys, C.G.N. & de Zwaan A. (1978) The rate of oxygen consumption and ammonia excretion by *M. edulis* after various periods of exposure to air. *Comparative Biochemistry and Physiology*, **60**, 343–347.

de Vooys, C.G.N. & Geenevasen, J.A.J. (2002) Biosynthesis and role in osmoregulation of glycine-betaine in the Mediterranean mussel *Mytilus galloprovincialis* Lmk. *Comparative Biochemistry and Physiology Part-B*, **132**, 409–414.

Dias, P.J., Fotedar, S., Gardner, J.P.A. & Snow, M. (2013) Development of sensitive and specific molecular tools for the efficient detection and discrimination of potentially invasive mussel species of the genus *Perna*. *Management of Biological Invasions*, **4**, 155–165.

Díaz, R.J. & Rosenberg, R. (2008) Spreading dead zones and consequences for marine ecosystems. *Science*, **321**, 926–992.

Digilio, G., Sforzini, S., Cassino, C., Robotti, E., Oliveri, C., Marengo, E. *et al.* (2016) Haemolymph from *Mytilus galloprovincialis*: response to copper and temperature challenges studied by ^{1}H-NMR metabolomics. *Comparative Biochemistry and Physiology Part-C*, **183–184**, 61–71.

Dyrynda, E.A., Pipe, R.K. & Ratcliffe, N.A. (1997) Sub-populations of haemocytes in the adult and developing marine mussel, *Mytilus edulis*, identified by use of monoclonal antibodies. *Cell and Tissue Research*, **289**, 527–536.

Eerkes-Medrano, D., Menge, B.A., Sislak, C. & Langdon, C.J. (2013) Contrasting effects of hypoxic conditions on survivorship of planktonic larvae of rocky intertidal invertebrates. *Marine Ecology Progress Series*, **478**, 139–151.

Ekaratne, S.U.K. & Davenport, J. (1993) The relationships between the gametogenetic status of triploids or diploids of Manila clams, *Tapes philippinarum*, and their oxygen uptake and gill particle transport. *Aquaculture*, **117**, 335–349.

Elliott, J.M. & Davison, W. (1975) Energy equivalents of oxygen consumption in animal energetics. *Oecologia*, **19**, 195–201.

van Erkom Schurink, C. & Griffiths, C.L. (1992) Physiological energetics of four South African mussel species in relation to body size and temperature. *Comparative Biochemistry and Physiology Part-A*, **101**, 779– 789.

Fernández Reiriz, M.J., Range, P., Álvarez-Salgado, X.A., Espinosa, J. & Labarta, U. (2012) Tolerance of juvenile *Mytilus galloprovincialis* to experimental seawater acidification. *Marine Ecology Progress Series*, **454**, 65–74.

Fitzhenry, T., Halpin, P.M. & Helmuth, B. (2004) Testing the effects of wave exposure, site, and behavior on intertidal mussel body temperatures: applications and limits of temperature logger design. *Marine Biology*, **145**, 339–349.

Florkin, M. (1962) La régulation isosmotique intracellulaire chez les invertébrés marins euryhalins. *Bulletin de la Classe des Sciences de l'Académie Royale de Belgique*, **48**, 687–694.

Führer, E., Rudolph, A., Espinoza, C., Díaz, R., Gajardo, M. & Camaño, N. (2012) Integrated use of biomarkers (O:N ratio and acetylcholinesterase inhibition) on *Aulacomya ater* (Molina, 1782) (Bivalvia:Mytilidae) as a criteria for effects of organophosphate pesticide exposition. *Journal of Toxicology*, **2012**, art. 951568.

Gabbott, P.A. & Bayne, B.L. (1973) Biochemical effects of temperature and nutritive stress on *Mytilus edulis* L. *Journal of the Marine Biological Association of the UK*, **53**, 269–286.

Gainey, L.F. (1994) Volume regulation in 3 species of marine mussels. *Journal of Experimental Marine Biology and Ecology*, **181**, 201–211.

Glazier, D.S. (2006) The 3/4-power law is not universal: evolution of isometric, ontogenetic metabolic scaling in pelagic animals. *BioScience*, **56**, 325–332.

Glazier, D.S. (2010) A unifying explanation for diverse metabolic scaling in animals and plants. *Biological Reviews*, **85**, 111–138.

Glazier, D.S. (2014) Metabolic scaling in complex living systems. *Systems*, **2**, 451–540.

Glazier, D.S. (2015) Body-mass scaling of metabolic rate: what are the relative roles of cellular versus systemic effects? *Biology*, **4**, 187–199.

Gosling, E.M. (2015) *Marine Bivalve Molluscs*, 2nd edn. Wiley Blackwell, New York.

Griffiths, C.L. & Griffiths, R.J. (1987) Bivalvia. In: *Animal Energetics, Vol. 2, Bivalvia through Reptilia* (eds T.J. Pandian & F.J. Vernberg), pp. 1–87. Academic Press, Cambridge, MA.

Haefner, P.A., Sheppard, B., Barto, J., McNeil, E. & Cappellino, V. (1996) Application of ultrasound technology to molluscan physiology: noninvasive monitoring of cardiac rate in the blue mussel, *Mytilus edulis* Linnaeus, 1758. *Journal of Shellfish Research*, **15**, 685–689.

Hagger, J.A., Lowe, D., Dissanayake, A., Jones, M.B. & Galloway, T.S. (2010) The influence of seasonality on biomarker responses in *Mytilus edulis*. *Ecotoxicology*, **19**, 953–962.

Halldórsson, H.P., Svavarsson, J. & Granmo, A. (2005) The effect of pollution on scope for growth of the mussel (*Mytilus edulis* L.) in Iceland. *Marine Environmental Research*, **59**, 47–64.

Halldórsson, H.P., De Pirro, M., Romano, C., Svavarsson, J. & Sarà, G. (2008) Immediate biomarker responses to benzo[a]pyrene in polluted and unpolluted populations of the blue mussel (*Mytilus edulis* L.) at high-latitudes. *Environment International*, **34**, 483–489.

Harte, J., Zillio, T., Conlisk, E. & Smith, A.B. (2008) Maximum entropy and the state-variable approach to macroecology. *Ecology*, **89**, 2700–2711.

Hawkins, A.J.S. & Bayne, B.L. (1992) Physiological interrelations and the regulation of production. In: *The Mussel Mytilus: Ecology, Physiology, Genetics and Culture* (ed. E.M. Gosling), pp. 171–222. Elsevier Science Publishers B.V., Amsterdam.

Hawkins, A.J.S. & Hilbish, T.J. (1992) The costs of cell volume regulation: protein metabolism during hyperosmotic adjustment. *Journal of the Marine Biological Association of the UK*, **72**, 569–578.

Heilmayer, O. & Brey, T. (2003) Saving by freezing? Metabolic rates of *Adamussium colbecki* in a latitudinal context. *Marine Biology*, **143**, 477–484.

Helmuth, B. (1998) Intertidal mussel microclimates: predicting the body temperature of a sessile invertebrate. *Ecological Monographs*, **68**, 51–74.

Helmuth, B., Choi, F., Matzelle, A., Torossian, J.L., Morello, S.L., Mislan, K.A.S. *et al.* (2016) Long-term, high frequency *in situ* measurements of intertidal mussel bed temperatures using biomimetic sensors. *Scientific Data*, **3**, 160087.

Hicks, D.W. & McMahon, R.F. (2002a) Respiratory responses to temperature and hypoxia in the nonindigenous brown mussel, *Perna perna* (Bivalvia: Mytilidae), from the Gulf of Mexico. *Journal of Experimental Marine Biology and Ecology*, **277**, 61–78.

Hicks, D.W. & McMahon, R.F. (2002b) Temperature acclimation of upper and lower thermal limits and freeze resistance in the nonindigenous brown mussel, *Perna perna* (L.), from the Gulf of Mexico. *Marine Biology*, **140**, 1167–1179.

Hicks, D.W & McMahon, R.F. (2005) Effects of temperature on chronic hypoxia tolerance in the non-indigenous brown mussel, *Perna perna* (Bivalvia: Mytilidae) from the Texas Gulf of Mexico. *Journal of Molluscan Studies*, **71**, 401–408.

Hole, L.M., Moore, M.N. & Bellamy, D. (1995) Age-related cellular and physiological reactions to hypoxia and hyperthermia in marine mussels. *Marine Ecology Progess Series*, **122**, 173–178.

Hu, M., Lin, D., Shang, Y., Hu, Y., Lu, W. Huang, X. *et al.* (2017) CO_2-induced pH reduction increases physiological toxicity of nano-TiO_2 in the mussel *Mytilus coruscus*. *Scientific Reports*, **7**, 40015.

Huang, S.-C. & Newell, R.I.E. (2002) Seasonal variations in the rates of aquatic and aerial respiration and ammonium excretion of the ribbed mussel, *Geukensia demissa* (Dillwyn). *Journal of Experimental Marine Biology and Ecology*, **270**, 241–255.

Huhn, M., Zamani, N.P., von Juterzenka, K. & Lenz, M. (2016) Food availability in an anthropogenically impacted habitat determines tolerance to hypoxia in the Asian green mussel *Perna viridis*. *Marine Biology*, **163**, 15.

Jansen, J.M., Pronker, A., Kube, S., Sokolowski, A., Sola, J., Marquiegui, M. *et al.* (2007a) Geographic and seasonal patterns and limits on the adaptive response to temperature of European *Mytilus* spp. and *Macoma balthica* populations. *Oecologia*, **154**, 23–34.

Jansen, J.M., Wendelaar Bonga, S. & Hummel, H. (2007b) Differential cold-shock resistance among acclimated European mussel populations. *Marine and Freshwater Behaviour and Physiology*, **40**, 233–245.

Jansen, J.M., Hummel, H. & Wendelaar Bonga, S. (2009) The respiratory capacity of marine mussels (*Mytilus galloprovincialis*) in relation to the high temperature threshold. *Comparative Biochemistry and Physiology Part-C*, **153**, 399–402.

Jones, S.J., Lima, F.P. & Wethey, D.S. (2010) Rising environmental temperatures and biogeography: poleward range contraction of the blue mussel, *Mytilus edulis* L., in the western Atlantic. *Journal of Biogeography*, **37**, 2243–2259.

Jorge, R., Lemos, D. & Moreira, G.S. (2007) Effect of zinc and benzene on respiration and excretion of mussel larvae (*Perna perna*) (Linnaeus, 1758) (Mollusca; Bivalvia). *Brazilian Journal of Brazil*, **67**, 111–115.

Jost, J. & Helmuth, B. (2007) Morphological and ecological determinants of body temperature of *Geukensia demissa*, the Atlantic ribbed mussel, and their effects on mussel mortality. *Biological Bulletin*, **213**, 141–151.

Kawabe, S., Takada, M., Shibuya, R. & Yokoyama, Y. (2010) Biochemical changes in oyster tissues and hemolymph during long-term air exposure. *Fisheries Science*, **76**, 841–855.

Kube, S., Gerber, A., Jansen, J.M. & Schiedek, D. (2006) Patterns of organic osmolytes in two marine bivalves, *Macoma balthica*, and *Mytilus* spp., along their European distribution. *Marine Biology*, **149**, 1387–1396.

Kube, S., Sokolowski, A., Jansen, J.M. & Schiedek, D. (2007) Seasonal variability of free amino acids in two marine bivalves, *Macoma balthica* and *Mytilus* spp., in relation to environmental and physiological factors. *Comparative Biochemistry and Physiology Part-A*, **147**, 1015–1027.

Kwok, A., Lyons, B.P., Hodges, N.J. & Bean, T.B. (2013) Cryopreservation and storage of mussel (*Mytilus* spp.) haemocytes for latent analysis by the Comet assay. *Mutation Research*, **750**, 86–91.

Landes, A., Dolmer, P., Poulsen, L.K., Petersen, J.K. & Vismann, B. (2015) Growth and respiration in blue mussels (*Mytilus* spp.) from different salinity regimes. *Journal of Shellfish Research*, **34**, 373–382.

Lee, R.W., Kraus, D.W. & Doeller, J.E. (1996) Sulfide-stimulation of oxygen consumption rate and cytochrome reduction in gills of the estuarine mussel *Geukensia demissa*. *Biological Bulletin*, **191**, 421–430.

Lesser, M.P. (2006) Oxidative stress in marine environments: biochemistry and physiological ecology. *Annual Review of Physiology*, **68**, 253–278.

Logan, C.A., Kost, L.E. & Somero, G.N. (2012) Latitudinal differences in *Mytilus californianus* thermal physiology. *Marine Ecology Progress Series*, **450**, 93–105.

Lourenço, C.R., Nicastro, K.R., Serrão, E.A. & Zardi, G.I. (2012) First record of the brown mussel (*Perna perna*) from the European Atlantic coast. *Marine Biodiversity Records*, **5**, e39.

Lum, S.C. & Hammen, C.S. (1964) Ammonia excretion of Lingula. *Comparative Biochemistry and Physiology*, **12**, 185–190.

Marshall, D.J. & McQuaid, C.D. (1993) Differential physiological and behavioural responses of the intertidal mussels, *Choromytilus meridionalis* (Kr.) and *Perna perna* L., to exposure to hypoxia and air: a basis for spatial separation. *Journal of Experimental Marine Biology and Ecology*, **171**, 225– 237.

Martinović, R., Kolarević, S., Kračun-Kolarević, M., Kostić, J., Marković, S., Gačić, Z. *et al.* (2015) Genotoxic potential and heart rate disorders in the Mediterranean mussel *Mytilus galloprovincialis* exposed to Superdispersant-25 and dispersed diesel oil. *Marine Environmental Research*, **108**, 83–90.

Martinović, R., Kolarević, S., Kračun-Kolarević, M., Kostić, J., Jokanović, S., Gačić, Z. *et al.* (2016) Comparative assessment of cardiac activity and DNA damage in haemocytes of the Mediterranean mussel *Mytilus galloprovincialis* in exposure to tributyltin chloride. *Environmental Toxicology and Pharmacology*, **47**, 165–174.

Mathiesen, S.S., Thyrring, J., Hemmer-Hansen, J., Berge, J., Sukhotin, A., Leopold, P. *et al.* (2017) Genetic diversity and connectivity within *Mytilus* spp. in the subarctic and Arctic. *Evolutionary Applications*, **10**, 39–55.

May, M.A., Bishop, K.D. & Rawson, P.D. (2017) NMR profiling of metabolites in larval and juvenile blue mussels (*Mytilus edulis*) under ambient and low salinity conditions. *Metabolites*, **7**, 33.

McFarland, K., Donaghy, L. & Volety, A.K. (2013) Effect of acute salinity changes on hemolymph osmolality and clearance rate of the non-native mussel, *Perna viridis*, and the native oyster, *Crassostrea* in Southwest Florida. *Aquatic Invasions*, **8**, 299–310.

Melzner, F., Thomsen, J., Koeve, W., Oschlies, A., Gutowska, M.A., Bange, H.W. *et al.* (2013) Future ocean acidification will be amplified by hypoxia in coastal habitats. *Marine Biology*, **160**, 1875–1888.

Michaelidis, B., Ouzounis, C., Paleras, A. & Pörtner, H.O. (2005a) Effects of long-term moderate hypercapnia on acid–base balance and growth rate in marine mussels *Mytilus galloprovincialis*. *Marine Ecology Progress Series*, **293**, 109–118.

Michaelidis, B., Haas, D. & Grieshaber, M.K. (2005b) Extracellular and intracellular acid–base status with regard to the energy metabolism in the oyster *Crassostrea gigas* during exposure to air. *Physiological and Biochemical Zoology*, **78**, 373–383.

Moon, T.W. & Pritchard, A.W. (1970) Metabolic adaptations in vertically separated populations of *Mytilus californianus*. *Journal of Experimental Marine Biology and Ecology*, **5**, 35–46.

Moore, M.N., Bubel, A. & Lowe, D.M. (1980) Cytology and cytochemistry of the pericardial gland cells of *Mytilus edulis* and their lysosomal responses to injected horseradish peroxidase and anthracene. *Journal of the Marine Biological Association of the UK*, **60**, 135–149.

Morse, M.P., Meyhöfer, E., Otto, J.J. & Kuzirian, A.M. (1986) Hemocyanin respiratory pigment in bivalve mollusks. *Science*, **231**, 1302–1304.

National Research Council (1987) Committee on Biological Markers. *Environmental Health Perspectives*, **74**, 3–9.

Navarro, J.M. & Winter, J.E. (1982) Ingestion rate, assimilation efficiency and energy balance in *Mytilus chilensis* in relation to body size and different algal concentrations. *Marine Biology*, **67**, 255–266.

Nicastro, K.R., Zardi, G.I., McQuaid, C.D., Stephens, L., Radloff, S. & Blatch, G.L. (2010) The role of gaping behaviour in habitat partitioning between coexisting intertidal mussels. *BMC Ecology*, **10**, 1–11.

Nicastro, K.R., Zardi, G.I., McQuaid, Pearson, G.A. & Serrão, E.A. (2012) Love thy neighbour: group properties of gaping behaviour in mussel aggregations. *PLoS ONE*, **7**, e47382.

Nicholson, S. (1999) Cytological and physiological biomarker responses from green mussels, *Perna viridis* (L.) transplanted to contaminated sites in Hong Kong coastal waters. *Marine Pollution Bulletin*, **39**, 261–268.

Nicholson, S. (2002) Ecophysiological aspects of cardiac activity in the subtropical mussel *Perna viridis* (L.) (Bivalvia: Mytilidae). *Journal of Experimental Marine Biology and Ecology*, **267**, 207–222.

Nicholson, S. (2003a) Cardiac and branchial physiology associated with copper accumulation and detoxication in the mytilid mussel *Perna viridis* (L.). *Journal of Experimental Marine Biology and Ecology*, **295**, 157–171.

Nicholson S. (2003b) Tachycardia in the mussel, *Perna viridis* (L.) on exposure to tributyltin antifouling paint. *Australasian Journal of Ecotoxicology*, **9**, 137–140.

Nicholson, S. & Lam, P.K.S. (2005) Pollution monitoring in Southeast Asia using biomarkers in the mytilid mussel *Perna viridis* (Mytilidae: Bivalvia). *Environment International*, **31**, 121–132.

Nikolic, M., Kuznetsova, T., Kholodkevich, S., Gvozdenovic, S., Mandic, M., Joksimovic, D. *et al.* (2019) Cardiac activity in the Mediterranean mussel (*Mytilus galloprovincialis* Lamarck, 1819) as a biomarker for assessing sea water quality in Boka Kotorska Bay, South Adriatic Sea. *Mediterranean Marine Science*, **20**, 680687.

Noël, D., Pipe, R., Elston, R., Bachère, E. & Mialhe, E. (1994) Antigenic characterization of hemocyte subpopulations in the mussel *Mytilus edulis* by means of monoclonal antibodies. *Marine Biology*, **119**, 549–556.

Nogueira, L., Mello, D. F., Trevisan, R., Garcia, D., da Silva Acosta, D., Dafre, A.L. *et al.* (2017) Hypoxia effects on oxidative stress and immunocompetence biomarkers in the mussel *Perna perna* (Mytilidae, Bivalvia). *Marine Environmental Research*, **126**, 109–115.

Olabarria, C., Gestoso, I., Lima, F.P., Vázquez, E., Comeau, L.A., Gomes, F. *et al.* (2016) Response of two mytilids to a heatwave: the complex interplay of physiology, behaviour and ecological interactions. *PLoS ONE*, **11**, e0164330.

Peck, L S. (2002) Ecophysiology of Antarctic marine ectotherms: limits to life. *Polar Biology*, **25**, 31–40.

Peck, L.S. & Conway, L.Z. (2000) The myth of metabolic cold adaptation: oxygen consumption in stenothermal Antarctic bivalves. *Geological Society Special Publication*, **177**, 441–450.

Peck, L.S., Morley, S.A. & Clark, M.S. (2010) Poor acclimation capacities in Antarctic marine ectotherms. *Marine Biology*, **157**, 2051–2059.

Peck, L.S., Morley, S.A., Richards, J. & Clark, M.S. (2014) Acclimation and thermal tolerance in Antarctic marine ectotherms. *Journal of Experimental Biology*, **217**, 16–22.

Pierce, S.K. (1971) Volume regulation and valve movements by marine mussels. *Comparative Biochemistry and Physiology Part-A*, **39**, 103–117.

Pipe, R.K., Porte, C. & Livingstone, D.R. (1993) Antioxidant enzymes associated with the blood cells and haemolymph of the mussel *Mytilus edulis*. *Fish and Shellfish Immunology*, **3**, 221–233.

Place, S.P., Menge, B.A. & Hofmann, G.E. (2012) Transcriptome profiles link environmental variation and physiological response of *Mytilus californianus* between Pacific tides. *Functional Ecology*, **26**, 144–155.

Pörtner, H.-O. (2010) Oxygen- and capacity-limitation of thermal tolerance: a matrix for integrating climate-related stressor effects in marine ecosystems. *Journal of Experimental Biology*, **213**, 881–893.

Rabalais, N.N., Turner, R.E., Díaz, R. J. & Justić, D. (2009) Global change and eutrophication of coastal waters. *ICES Journal of Marine Science*, **66**, 1528–1537.

Rabalais, N.N., Díaz, R.J., Levin, L.A., Turner, R.E., Gilbert, D. & Zhang, J. (2010) Dynamics and distribution of natural and human-caused hypoxia. *Biogeosciences*, **7**, 585–619.

Rajagopal, S., Venugopalan, V.P., van der Velde, G. & Jenner, H.A. (2003) Response of fouling brown mussel, *Perna perna* (L.), to chlorine. *Archives of Environmental Contamination and Toxicology*, **44**, 369–376.

Rajagopal, S., Venugopalan, V.P., van der Velde G. & Jenner, H.A. (2006) Greening of the coasts: a review of the *Perna viridis* success story. *Aquatic Ecology*, **40**, 273–297.

Rajkumar, A. (2009) *Physiological and cytological biomarker studies using Perna perna for marine pollution monitoring*. MSc thesis, University of KwaZulu-Natal Durban, South Africa.

Read, K.R.H. (1962) Respiration of the bivalve molluscs *Mytilus edulis* L. and *Brachidontes demissus plicatulus* Lamarck as a function of size and temperature. *Comparative Biochemistry and Physiology*, **7**, 89–101.

Remane, A. & Schlieper, C. (1971) *Biology of Brackish Water*. Wiley-Interscience, New York.

Resgalla, Jr. C., Brasil, E.S. & Salomão, L.C. (2006) Physiological rates in different classes of sizes of *Perna perna* (Linnaeus, 1758) submitted to experimental laboratory conditions. *Brazilian Journal of Biology*, **66**, 325–336.

Resgalla, Jr. C., Brasil, E.S. & Salomão, L.C. (2007) The effect of temperature and salinity on the physiological rates of the mussel *Perna perna* (Linnaeus 1758). *Brazilian Archives of Biology and Technology*, **50**, 543–556.

Rivera-Ingraham, G.A., Rocchetta, I., Meyer, S. & Abele, D. (2013) Oxygen radical formation in anoxic transgression and anoxia reoxygenation: foe or phantom? Experiments with a hypoxia tolerant bivalve. *Marine Environmental Research*, **92**, 110–119.

Rivera-Ingraham, G.A. & Lignot, J.H. (2017) Osmoregulation, bioenergetics and oxidative stress in coastal marine invertebrates: raising the questions for future research. *Journal of Experimental Biology*, **220**, 1749–1760.

Ribeiro, P.L., Camacho, A. & Navas, C.A. (2012) Considerations for assessing maximum critical temperatures in small ectothermic animals: insights from leaf-cutting ants. *PLoS ONE*, **7**, e32083.

Riisgård, H.U. (1998) No foundation of a '3/4 power scaling law' for respiration in biology. *Ecology Letters*, **1**, 71–73.

Riisgård, H.A. & Larsen, P.S. (2001) Ciliary filter feeding and bio-fluid mechanics-present understanding and unsolved problems. *Limnology and Oceanography*, **46**, 882–891.

Riisgård, H.U. & Larsen, P.S. (2015) Physiologically regulated valve-closure makes mussels long-term starvation survivors: test of hypothesis. *Journal of Molluscan Studies*, **81**, 303–307.

Riisgård, H.U., Pleissner, D., Lundgreen, K. & Larsen, P.S. (2013) Growth of mussels *Mytilus edulis* at algal (*Rhodomonas salina*) concentrations below and above saturation levels for reduced filtration rate. *Marine Biology Research*, **9**, 1005–1017.

Robledo, J.A.F., Santarém, M.M., González, P. & Figueras, A. (1995) Seasonal variations in the biochemical composition of the serum of *Mytilus galloprovincialis* Lmk. and its relationship to the reproductive cycle and parasitic load. *Aquaculture*, **133**, 311–322.

Rocher, B., Bultelle, F., Chan, P., Le Foll, F., Letendre, J., Monsinjon, T. *et al.* (2015) 2-DE mapping of the blue mussel gill proteome: the usual suspects revisited. *Proteomes*, **3**, 3–41.

Rovero, F., Hughes, R.N. & Chelazzi, G. (1999) Cardiac and behavioural responses to risk of predation by dogwhelks. *Animal Behaviour*, **58**, 707–714.

Rueda, J.L. & Smaal, A.C. (2004) Variation of the physiological energetics of the bivalve *Spisula subtruncata* (da Costa, 1778) within an annual cycle. *Journal of Experimental Marine Biology and Ecology*, **301**, 141–157.

Sadok S., Uglow R.F. & Haswell S.J. (1999) Some aspects of nitrogen metabolism in *Mytilus edulis*: effects of aerial exposure. *Marine Biology*, **135**, 297–305.

Saggese, I., Sarà, G. & Dondero, F. (2016) Silver nanoparticles affect functional bioenergetic traits in the invasive Red Sea mussel *Brachidontes pharaonis*. *BioMed Research International*, **2016**, art. 1872351.

Santarém, M.M., Robledo, J.A.F. & Figueras, A. (1994) Seasonal changes in hemocytes and serum defense factors in the blue mussel *Mytilus galloprovincialis*. *Diseases of Aquatic Organisms*, **18**, 217–222.

Sarà, G. & de Pirro, M. (2011) Heart beat rate adaptations to varying salinity of two intertidal Mediterranean bivalves: the invasive *Brachidontes pharaonis* and the native *Mytilaster minimus*. *Italian Journal of Zoology*, **78**, 193–197.

Sarà, G., Romano, C., Widdows, J. & Staff, F.J. (2008) Effect of salinity and temperature on feeding physiology and scope for growth of an invasive species (*Brachidontes pharaonis* – Mollusca: Bivalvia) within the Mediterranean Sea. *Journal of Experimental Marine Biology and Ecology*, **363**, 130–136.

Sarà, G., Milanese, M., Prusina, I., Sarà, A., Angel, D.L., Glamuzina, B. *et al.* (2014) The impact of climate change on Mediterranean intertidal communities: losses in coastal ecosystem integrity and services. *Regional Environmental Change*, **14**, S5–S17.

Segnini de Bravo, M.I., Chung, K.S. & Pérez, J.E. (1998) Salinity and temperature tolerances of the green and brown mussels, *Perna viridis* and *Perna perna* (Bivalvia: Mytilidae). *Revista Biología Tropical*, **5**, 121–125.

Seo, E., Sazi, T., Togawa M., Nagata, O., Murakami, M., Kojima, S. *et al.* (2016) A portable infrared photoplethysmograph: heartbeat of *Mytilus galloprovincialis* analyzed by MRI and application to *Bathymodiolus septemdierum*. *Biology Open*, **5**, 1752–1757.

Sekine, D., Ohishi, K., Nakamura, Y., Kusaka, C., Tame, A. Inoue, K. *et al.* (2016) Monoclonal antibodies to hemocytes of the deep-sea symbiotic mussel, *Bathymodiolus japonicas*. *JAMSTEC Report of Research and Development*, **23**, 27–33.

Shumway, S.E., Gabbott, P.A. & Youngson, A. (1977) The effect of fluctuating salinity on the concentrations of free amino acids and ninhydrin-positive substances in the adductor muscle of eight species of bivalve molluscs. *Journal of Experimental Marine Biology and Ecology*, **29**, 131–150.

Shumway, S.E., Barter, J. & Stahlnecker, J. (1988) Seasonal changes in oxygen consumption of the giant scallop, *Placopecten magellanicus* (Gmelin). *Journal of Shellfish Research*, **7**, 77–82.

Sokolowski, A., Wolowicza, M. & Hummel, H. (2003) Free amino acids in the clam *Macoma balthica* L. (Bivalvia, Mollusca) from brackish waters of the southern Baltic Sea. *Comparative Biochemistry and Physiology Part-A*, **134**, 579–592.

Solorzano, L. (1969) Determination of ammonia in natural waters by the phenolhypochlorite method. *Limnology and Oceanography*, **14**, 799—801.

Somero, G.N. (2002) Thermal physiology and vertical zonation of intertidal animals: optima, limits, and costs of living. *Integrative and Comparative Biology*, **42**, 780–789.

Somero, G.N. & R.D. Bowlus (1983) Osmolytes and metabolic end products of molluscs: the design of compatible solute systems. In: *The Mollusca. Environmental Biochemistry and Physiology*, Vol 2 (eds P.W. Hochakcha & K.M. Wilbur), pp. 77–98. Academic Press, New York.

Strahl, J., Dringen, R., Schmidt, M.M., Hardenberg, S. & Abele, D. (2011) Metabolic and physiological responses in tissues of the long-lived bivalve *Arctica islandica* to oxygen deficiency. *Comparative Biochemistry and Physiology Part -A*, **158**, 513–519.

Stapp, L.S., Thomsen, J., Schade, H., Bock, C., Melzner, F., Pörtner, H.O. *et al.* (2017) Intra-population variability of ocean acidification impacts on the physiology of Baltic blue mussels (*Mytilus edulis*): integrating tissue and organism response. *Journal of Comparative Physiology B*, **187**, 529–543.

Sui, Y., Liu, Y., Zhao, X., Dupon, S., Hu, M., Wu, F. *et al.* (2017) Defense responses to short-term hypoxia and seawater acidification in the thick shell mussel *Mytilus coruscus*. *Frontiers in Physiology*, **8**, art. 145.

Sukhotin, A.A. (1992) Respiration and energetics in mussels (*Mytilus edulis* L.) cultured in the White Sea. *Aquaculture*, **101**, 41–57.

Sukhotin, A.A., Abele, D. & Pörtner, H.O. (2006) Ageing and metabolism of *Mytilus edulis*: populations from various climate regimes. *Journal of Shellfish Research*, **25**, 893–899.

Sussarellu, R., Fabioux, C., Le Moullac, G., Fleury, E. & Moraga, D. (2010) Transcriptomic response of the Pacific oyster *Crassostrea gigas* to hypoxia. *Marine Genomics*, **3**, 133–143.

Sussarellu, R., Fabioux, C., Camacho Sanchez, M., Le Goïc, N. Lambert, C., Soudant, P. *et al.* (2012) Molecular and cellular response to short-term oxygen variations in the Pacific oyster *Crassostrea gigas*. *Journal of Experimental Marine Biology and Ecology*, **412**, 87–95.

Tagliarolo, M. & McQuaid, C.D. (2016) Field measurements indicate unexpected, serious underestimation of mussel heart rates and thermal tolerance by laboratory studies. *PLoS ONE*, **11**, e0146341.

Tagliarolo, M., Clavier, J., Chauvaud, L., Koken, M. & Grall, J. (2012) Metabolism in blue mussel: intertidal and subtidal beds compared. *Marine Ecology Progress Series*, **17**, 167–180.

Tamayo, D., Ibarrola, I. & Navarro, E. (2013) Thermal dependence of clearance and metabolic rates in slow- and fastgrowing spats of Manila clam *Ruditapes philippinarum*. *Journal of Comparative Physiology B*, **183**, 893–904.

Tame, A., Yoshida, T., Ohishi, O. & Maruyama, T. (2015) Phagocytic activities of hemocytes from the deep-sea symbiotic mussels *Bathymodiolus japonicus*, *B. platifrons*, and *B. septemdierum*. *Fish and Shellfish Immunology*, **45**, 146–156.

Tang, B.J. & Riisgård, H.U. (2016) Physiological regulation of valve-opening degree enables mussels *Mytilus edulis* to overcome starvation periods by reducing the oxygen uptake. *Open Journal of Marine Science*, **6**, 341–352.

Thompson, R.J. & Bayne, B.L. (1972) Active metabolism associated with feeding in the mussel *Mytilus edulis* L. *Journal of Experimental Marine Biology and Ecology*, **8**, 191–212.

Thompson, R.J. & Bayne, B.L. (1974) Some relationships between growth, metabolism and food in the mussel, *Mytilus edulis*. *Marine Biology*, **27**, 317–326.

Thomsen, J., Himmerkus, N., Holland, N., Sartoris, F.J., Bleich, M. & Tresguerres, M. (2016) Ammonia excretion in mytilid mussels is facilitated by ciliary beating. *Journal of Experimental Biology*, **219**, 2300–2310.

Thomsen, J., Morton, B., Ossenbrügger, H., Crooks, J.A., Valentich-Scott, P. & Haynert, K. (2018) The mytilid plicate organ: revisiting a neglected organ. *Journal of Molluscan Studies*, **84**, 486–489.

Thyrring, J., Rysgaard, S., Blicher, M.E. & Sejr, M.K. (2015) Metabolic cold adaptation and aerobic performance of blue mussels (*Mytilus edulis*) along a temperature gradient into the High Arctic region. *Marine Biology*, **162**, 235–243.

Thyrring, J., Bundgaard, A. & Sejr, M.K. (2017) Seasonal acclimation and latitudinal adaptation are of the same magnitude in *Mytilus edulis* and *Mytilus trossulus* mitochondrial respiration. *Polar Biology*, **40**, 1885–1891.

Turja, R., Höher, N., Snoeijs, P., Baršienė, J., Butrimavičienė, L., Kuznetsova, T. *et al.* (2014) A multibiomarker approach to the assessment of pollution impacts in two Baltic Sea coastal areas in Sweden using caged mussels (*Mytilus trossulus*). *Science of the Total Environment*, **473**–474, 398–409.

Van Cauwenberghe, L., Claessens, M., Vandegehuchte, M.B. & Janssen, C.R. (2015) Microplastics are taken up by mussels (*Mytilus edulis*) and lugworms (*Arenicola marina*) living in natural habitats. *Environmental Pollution*, **199**, 10–17.

Vaquer–Sunyer, R. & Duarte, C.M. (2008) Thresholds of hypoxia for marine biodiversity. *Proceedings of the National Academy of Sciences*, **105**, 15452–15457.

Wang, W.X. & Widdows, J. (1991) Physiological responses of mussel larvae *Mytilus edulis* to environmental hypoxia and anoxia. *Marine Ecology Progress Series*, **70**, 223–236.

Wang, Y., Hu, M., Wong, W.H., Shin, P.K.S. & Cheung, S.G. (2011) The combined effects of oxygen availability and salinity on physiological responses and scope for growth in the green-lipped mussel *Perna viridis*. *Marine Pollution Bulletin*, **63**, 255–261.

Wang, Y., Hu, M., Cheung, S.G., Shin, P.K.S., Lu, W. & Li, J. (2012) Immune parameter changes of hemocytes in green-lipped mussel *Perna viridis* exposure to hypoxia and hyposalinity. *Aquaculture*, **356**–357, 22–29.

Wang, Y., Hu, M., Cheung, S.G., Shin, P.K.S., Lu, W. & Li, J. (2013) Antipredatory responses of *Perna viridis* (Linnaeus,1758) under acute hypoxia and low salinity. *Journal of Molluscan Studies*, **79**, 42–50.

Warren, C.E. & Davies, G. E. (1967) Laboratory studies on the feeding, bioenergetics, and growth of fish. In: *The Biological Basis of Freshwater Fish Production* (ed. S.D. Gerking), pp. 175–214. Blackwell Scientific, Oxford.

Welker, A.F., Moreira, D.C., Campos, E.G. & Hermes-Lima, M. (2013) Role of redox metabolism for adaptation of aquatic animals to drastic changes in oxygen availability. *Comparative Biochemistry and Physiology Part-A*, **165**, 384–404.

Westerbom, M., Kilpi, M. & Mustonen, O. (2002) Blue mussels, *Mytilus edulis*, at the edge of the range: population structure, growth and biomass along a salinity gradient in the north-eastern Baltic Sea. *Marine Biology*, **140**, 991–999.

Widdows, J. (1973a) Effect of temperature and food on the heart beat, ventilation rate and oxygen uptake of *Mytilus edulis*. *Marine Biology*, **20**, 269–276.

Widdows, J. (1973b) The effects of temperature on the metabolism and activity of *Mytilus edulis*. *Netherlands Journal of Sea Research*, **7**, 387–398.

Widdows, J. (1976) Physiological adaptation of *Mytilus edulis* to cyclic temperatures. *Journal of Comparative Physiology*, **105**, 115–128.

Widdows, J. (1978) Combined effects of body size, food concentration and season on the physiology of *Mytilus edulis*. *Journal of the Marine Biological Association of the UK*, **58**, 109–124.

Widdows, J. & Bayne, B.L. (1971) Temperature acclimation of *Mytilus edulis* with reference to its energy budget. *Journal of the Marine Biological Association of the UK*, **51**, 827–843.

Widdows, J. & Staff, F. (2006) Biological effects of contaminants: measurement of scope for growth in mussels. *ICES Techniques in Marine Environmental Science*, **40**, 1–30.

Wootton1, E.C. & Pipe, R.K. (2003) Structural and functional characterisation of the blood cells of the bivalve mollusc, *Scrobicularia plana*. *Fish & Shellfish Immunology*, **15**, 249–262.

Wright, P.A. & Wood, C.M. (2009) A new paradigm for ammonia excretion in aquatic animals: role of Rhesus (Rh) glycoproteins. *Journal of Experimental Biology*, **212**, 2303–2312.

Yancey, P.H., Hand, S.C., Bowlus, R.D. & Somoro, G.N. (1982) Living with water stress: evolution of osmolyte systems. *Science*, **217**, 1214–1222.

Yuan W., Walters, L.J., Schneider, K.Y.R. & Hoffman, E.A. (2010) Exploring the survival threshold: a study of salinity tolerance of the nonnative mussel *Mytella charruana*. *Journal of Shellfish Research*, **29**, 415–422.

Yuan, W.S., Walters, L.J., Brodsky, S.A., Schneider, K.R. & Hoffman, E.A. (2016) Synergistic effects of salinity and temperature on the survival of two nonnative bivalve molluscs, *Perna viridis* (Linnaeus 1758) and *Mytella charruana* (d'Orbigny 1846). *Journal of Marine Biology*, **2016**, art. 9261309.

Zardi, G.I., Nicastro, K.R., McQuaid, C.D., Castilho, R., Costa, J., Serrão, E.A. *et al.* (2015) Intraspecific genetic lineages of a marine mussel show behavioural divergence and spatial segregation over a tropical/subtropical biogeographic transition. *BMC Evolutionary Biology*, **15**, 100.

Zippay, M.L. & Helmuth, B. (2012) Effects of temperature change on mussel, *Mytilus*. *Integrative Zoology*, **7**, 312–327.

Zittier, Z.M.C., Bock, C., Lannig, G. & Pörtner, H.O. (2015) Impact of ocean acidification on thermal tolerance and acid–base regulation of *Mytilus edulis* (L.) from the North Sea. *Journal of Experimental Marine Biology and Ecology*, **473**, 16–25.

8

Mussels and Marine Environmental Contaminants

Introduction

Oceanic ecosystems are today suffering from both past and novel, rapidly diversifying modern human activities (Vallaeys *et al.* 2017 and references therein). This common reservoir suffers from environmental pressure exerted by humans on the marine ecosystems itself, such as offshore petroleum production, sea transport and fishing, but also, more recently, exploitation of deep sea metallic nodules, marine aquaculture, installation of offshore windmills fields and, indirectly, the exploitation of nearby terrestrial ecosystems by tourism, agriculture and industry. Even the exponential increase in marine aquaculture impacts marine ecosystems through the release of antimicrobials, food supplements, nutrients and pharmaceuticals from marine farms. While most contaminants are concentrated in coastal zones, mainly affecting pelagic and benthic food webs of the continental shelf, long-range transport of contaminants across large distances has been described in the literature (Vorkamp & Rigét 2014). Among all the diversity of human activities and sources of pollution, there are four main ways that pollutants enter the marine environment: direct discharge of effluents and solid wastes into the seas and oceans; land runoff into the coastal zone, mainly into rivers; ship pollution; and atmospheric fallout of pollutants transferred by the air mass on to the sea surface. It is noteworthy, however, that natural oceanographic and geological factors, including geothermal activity, can sometimes be responsible for elevated levels of some contaminants (e.g. heavy metals).

Contamination caused by hazardous substances is a major environmental concern in European waters and consequently is addressed by a number of EU legislative measures and policies. The Water Framework Directive (WFD; European Commission 2000) provides for measures against chemical pollution of surface waters. It has two components: the selection and regulation of substances of EU-wide concern (priority substances, PS) as a means to assess the chemical status of water bodies up to 12 nautical miles from the straightened coastline; and the selection by Member States of substances of national or local concern. The Marine Strategy Framework Directive (MSFD 2008/56/EC) aims to provide an integrative marine environment status assessment and considers both coastal and offshore environment, thus overlapping with WFD for some parts of the marine environment. Contaminants are defined in European legislation as:

Marine Mussels: Ecology, Physiology, Genetics and Culture, First Edition. Elizabeth Gosling.

> substances (i.e. chemical elements and compounds) or groups of substances that are toxic, persistent and liable to bio-accumulate and other substances or groups of substances which give rise to an equivalent level of concern. (European Commission 2000)

Contaminants affecting marine ecosystems include a wide range of synthetic organic chemicals, including some industrial solvents, persistent toxins (e.g. PCBs, DDT, pesticides, furans, phenols, dioxins and radioactive waste), diverse heavy metals that are toxic or poisonous at low concentration (e.g. mercury, nickel, arsenic and cadmium), plastic debris, endocrine disrupters, genotoxicants, pharmaceuticals and personal care products (PCPs).

In 2001, a list of PS was adopted, and in 2008 the environmental quality standards (EQSs) for those substances and eight other pollutants, already regulated at EU level, were set out in the EQS Directive, later amended by Directive 2013/39/EU (European Commission 2013). The list identified 33 substances or groups of substances shown to be of major concern for European waters. Within this list, 11 substances were identified as priority hazardous substances and therefore subject to cessation or phasing out of discharges, emissions and losses within an appropriate timetable not exceeding 20 years. A further 14 substances were identified as being subject to later review. Their status was addressed in the Commission proposal on EQSs, which became Directive 2008/105/EC on quality standards (European Commission 2008). Subsequently, Tornero & Hanke (2016) compiled a list of contaminants potentially released into the sea from sea-based sources. This list contains 276 substances, each with its CAS identification number[1] and major sea-based sources: 19 metals/metalloids, 10 organometallic compounds, 24 inorganic compounds, 204 organic compounds and 19 radionuclides, with specific information on concentrations and toxicity data for each. Offshore oil and gas operations contributed the highest percentage of substances (39%) to this list, followed by shipping (32%), dredge/dumping of dredged material (28%) and mariculture activities (24%). Although most substances were linked to only one sea-based activity, there were cases in which they were associated with more than one, thus increasing their potential risks. The list also gives information on the relevant international regulations/legislations and Regional Sea Conventions (RSCs), such as the Barcelona Convention for protection of the Mediterranean Sea, the Black Sea Commission, the Baltic Marine Environment Protection Commission (HELCOM) and the Commission for the protection of the marine environment of the North-East Atlantic (OSPAR), which, through monitoring programmes, aim to protect the marine environment and bring together Member States and neighbouring countries.

Reporting updates for the MSFD Articles 8, 9 and 10 (assessment, determination of good environmental status and establishment of environmental targets, respectively, for public consultation) were made through web forms, to aid completion of drop-down lists by Member States. Tornero & Hanke (2018a) outlined a process to prepare a list of contaminants for use in 2018 MSFD Article 8 reporting. It was planned to use, wherever possible, drop-down lists to facilitate data entry, thereby helping to ensure consistency in the data entered. This work is part of a process to support regulators in order to select relevant chemical contaminants to be assessed in their jurisdictional area, thus aiming at EU national authorities, but also at Regional Sea Conventions (RSC) in the shared marine basins. A subsequent list of more than 2700 substances or groups of substance came from main lists of chemicals compiled by relevant global conventions (see earlier), European legislation, European regional seas and dedicated research work (Tornero & Hanke 2018b). Although not all of these contaminants are of concern for the marine environment or for a particular MSFD region/sub-region/Member State, this general overview should help in understanding the different options and support further developments for the monitoring and assessment of chemical pollution in European marine waters. Discussion and collaboration between the different stakeholders and expert groups will be necessary to find an efficient way toward an equally high level of protection for healthy and productive oceans. This compilation also assists in using a harmonised nomenclature for the unambiguous identification of the substances of concern.

Mussels (*Mytilus* spp.) are widely used for monitoring of coastal water pollution (mussel watch programmes). This chapter will examine the suitability of mussels as 'sentinel' or 'bioindicator' organisms in environmental monitoring; sources, properties, uptake, bioaccumulation patterns and analytical methods of key pollutant classes; use of *Mytilus* spp. in mussel watch programmes; EQSs and background concentrations of key contaminants; biomarkers of pollution and biological response in mussels occurring at molecular, cellular or physiological levels as a consequence of exposure to or toxic effects of environmental chemicals; harmonisation of monitoring procedures; and the use of deployed mussels (transplant caging) and artificial mussels (AMs) in pollution monitoring.

The advantages of using *Mytilus* spp. as environmental sentinels[2] arise from their biological and ecological characteristics, which make them virtually ideal for pollution monitoring (Widdows & Donkin 1992; Dailianis 2010; Beyer *et al.* 2017; Yancheva *et al.* 2018). Mussels are dominant in coastal and estuarine communities and have a wide geographical distribution in temperate regions of the both Northern and the Southern Hemispheres. Because they are sedentary and long-lived, they provide location-specific information, and because they form mussel beds in shallow waters, they can easily be collected. They are relatively tolerant, but not insensitive, to a wide range of environmental conditions (see Chapters 3 and 7), including moderately high levels of a wide range of contaminants. As suspension feeders, they filter large volumes of seawater and concentrate many chemicals in their tissues by factors of 10 to 10^5 relative to the concentration in seawater. Furthermore, they are easy to maintain in culture and provide sufficient tissue to be suitable for ecotoxicological laboratory exposure studies as well as in situ analysis. Mussel populations are relatively stable and can be large enough for repeated sampling, thereby providing data on both short- and long-term changes in contaminant levels. They can be easily transferred and maintained in cages in different areas, allowing data to be obtained on the impact of environmental pollution in low- versus high-pollution areas. Finally, since mussels are commercially important worldwide, their role in contaminant accumulation is important from a public health perspective (see Chapter 12).

Contaminants in the Marine Environment

This section describes the major chemical substances that are released into the marine environment due to anthropogenic activities. The main sources of contamination and physico-chemical properties of these substances are described, as are the common analytical methods used for their detection and quantification in the marine environment. Tables 8.1A–G provides the ranges of concentration for chemical contaminants recorded in mussels sampled from different coastal regions of the world.

Table 8.1A Concentrations of BDEs (Brominated diphenyl ethers) congeners in mussels from European marine environments.

Region	Location	Species	BDE-47	BDE-49[a]	BDE-99	BDE-100	Reference
France	English Channel	*Mytilus edulis*	1670	1460	7 40	3 00	Johansson *et al.* (2006)
			9960	4540	4340	2500	
			5040	2340	2460	1330	
	Atlantic Ocean	*M. edulis*	130	<20	60	30	
			900	120	160	100	
			290	30	80	70	
	Mediterranean Sea	*M. galloprovincialis*	540	<20	40	20	
			810	60	380	160	
			590	20	190	90	
Italy	Central Adriatic Sea	*M. galloprovincialis*	69	<20	27	17	Piersanti *et al.* (2015)
			819	209	510	271	
			335	74	136	82	
Italy	NW Adriatic Sea	*M. galloprovincialis*	305	ns	535	40	Pizzini *et al.* (2015)
			12600		4830	3270	
			2940		1510	580	
Spain	Galician coast	*M. galloprovincialis*	4400	ns	1500	1500	Rodil *et al.* (2019)[b]
			510		310	130	

[a] Tentatively identified as BDE-49 in Johanesson *et al.* (2006); ns: not searched; Concentrations are in pg g^{-1}dw. Min, max and median values are presented respectively for each BDE, except in Rodil *et al.* (2019) where only maximum and average values are presented. [b]published values are converted in pg g^{-1}dw assuming a water content of 80 %.

Table 8.1B Concentration (pg g^{-1} ww.) of polychlorinated dibenzo-p-dioxins (PCDDs) and dioxin-like PCBs (non-ortho) in samples of mussels (*Mytilus galloprovincialis*) collected in 2014 from Galician Rías (NW Spain). [a]wild mussels; [b]cultured mussels. *Source:* From Carro *et al.* (2018). Reproduced with permission from Elsevier.

Location	2,3,7,8-TCDF	2,3,4,7,8-PeCDF	2,3,4,6,7,8-HxCDF	2,3,7,8-TCDD	1,2,3,7,8-PeCDD	OCDD	PCB77	PCB 126	PCB169
Ría de Ferrol[a]	0.237	0.265	0.015	0.016	0.032	0.0003	0.00440	1.0022	0.1565
Ría de Coruña[a]	0.048	0.042	0.002	0.016	0.032	0.0002	0.00088	0.0160	0.0111
Ría de Pontevedra[a]	0.024	0.008	0.002	0.016	0.032	0.0001	0.00091	0.3874	0.0111
Sample I from Ría de Vigo[b]	0.003	0.003	0.002	0.016	0.032	0.0001	0.00232	0.7542	0.0111
Sample II from Ría de Vigo[b]	0.031	0.048	0.008	0.016	0.032	0.0002	0.00441	0.6336	0.0832

Table 8.1C Levels of PAHs (ng g^{-1} dw) in *Mytilus galloprovincialis* tissue collected from locations in the Bizerte lagoon, Tunisia. *Source:* Data from Barhoumi *et al.* (2016).

	Site 1		Site 2		Site 3	
	Mean±S.D.	Range	Mean±S.D.	Range	Mean±S.D.	Range
Nap	31.5±9.6	21.1–39.8	35.5±6.6	28.9–42.1	66.9±2.3	64.6–69.2
Ace	42.4±23.6	27.1–69.6	<0.02		24.1±8.8	14.0–29.8
Fl	<0.02		<0.02		<0.02	
Phe	55.5±13.1	41.5–67.5	18.6±2.3	16.2–20.8	41.5±1.1	40.5–42.6
Ant	14.11±	0.5–2.7	0.2±0.1	0.1–0.3	0.3±0.2	0.1–0.5
Flu	45.4±6.1	38.4–49.8	26.6±11.9	14.8–38.5	11.0±5.9	4.8–16.5
Pyr	18.3±4.4	15.4–23.3	17.2±0.6	16.5–17.6	4.0±1.6	2.4–5.7
BaA	26.8±9.9	15.5–33.8	<0.11		<0.11	
Ch	<0.12		3.5±0.5	3.0–4.0	25.4±0.9	24.8–26.5
BbF	17.8±2.3	15.1–19.3	1.7.01±0.1	1.6–1.9	1.7±0.7	0.9–2.1
BkF	4.2±0.6	3.6–4.7	3.4±2.5	0.95.9	0.7±0.4	0.5–1.1
BaP	<0.07		<0.07		<0.07	
DahA	<0.06		<0.06		<0.06	
BghiP	33.7±10.9	25.9–46.1	<0.16		33.8±2.1	31.8–35.9
InP	7.1±1.4	6.3–8.7	0.7±0.1	0.6–0.8	0.3±0.1	0.3–0.4

Abbreviations: Nap: naphthalene; Ace: acenaphthene; Fl: fluorene; Phe: phenanthrene; Ant: anthracene; Flu: fluoranthene; Pyr: pyrene: BaA: benzo[a]anthracene; Ch: chrysene(Ch); BbF: benzo[b]fluoranthene; BkF benzo[k] fluoranthene; BaP: benzo[a]pyrene; DahA: dibenzo[a,h]anthracene; BghiP: benzo[g,*h*,*i*]perylene and InP: indeno[1,2,3-cd] pyrene.

Table 8.1D (a) Concentrations of polychlorinated biphenyls (PCBs) (ng g^{-1} dw) in *Perna viridis* from eight locations in coastal waters of Singapore. BLD are concentrations that were below the limit of detection. *Source:* From Bayen *et al.* (2003). Reproduced with permission from John Wiley and Sons.

(a)	Sungei Buloh	Lim Chu Kang	Tuas	Jurong	Changi	Pulau Ubin	Punggol	Sembawang
PCB 49	0.35	0.53	BLD	0.87	0.64	0.88	1.1	4.6
PCB 52	0.59	0.86	BLD	1.2	0.85	1.2	1.2	4.6
PCB 90/101	1.4	1.5	BLD	3.6	1.4	1.5	1.6	8.4
PCB 110	0.85	0.99	0.35	3.3	1.1	1.1	1.3	5.8
PCB 118	0.56	0.57	0.22	2.0	0.96	0.91	0.93	4.4
PCB 138	2.3	1.9	1.1	6.1	1.8	1.7	2.0	9.2
PCB 149	2.1	1.9	0.93	4.7	1.2	1.2	1.5	7.4
PCB 153	2.5	1.9	1.1	5.7	1.8	1.8	2.1	9.5

Table 8.1D (b) Mean± S.D. concentrations of polychlorinated biphenyls (PCBs) (ng g^{-1} lipid weight) in *Brachidontes rodriguezii* from five locations at Mar del Plata on the east coast of Argentina. *Source:* From Laitano *et al.* (2016b). Reproduced with permission from Elsevier.

La Tapera	Las Chacras	Seco	Chapadmalal	Barrera (reference site)
847 ± 240	652 ± 43	1002 ± 580	840 ± 170	929 ± 640
121 ± 57	827 ± 980	192 ± 32	80 ± 24	50 ±5 6
82 ± 49	407± 380	199 ± 37	27 ± 41	133 ± 21
29 ± 10	840 ± 110	871 ± 170	17 ± 17	33 ± 32
282 ± 34	460 ± 280	322 ± 85	146 ± 100	184 ± 87
207 ± 140	117 ± 170	190 ± 270	104 ± 120	238 ± 13
606 ± 430	722 ± 180	856 ± 240	549 ± 80	672 ± 100
16.4 ± 9	408 ± 570	340 ± 620	14 ± 12	320 ± 550
119 ± 81	142 ± 30	76 ± 91	83 ± 57	141 ± 6
144 ± 105	9 ± 13	225 ± 280	43 ± 85	194 ± 12

Heavy Metals

Heavy metals are natural components of the Earth's crust, occurring in the continental (soil, water) and marine compartment (water and sediments) and in the atmosphere and the biosphere. Other phenomena, such as volcanic activity, biomass fires and thermal sources, contribute to metal releases in the environment (Azizi *et al.* 2018 and references therein). Heavy metals refer to all metal with high atomic mass, including lead (Pb), chromium (Cr), cadmium (Cd), arsenic (As) and mercury (Hg). The main inputs whereby metals are introduced into seawater are river runoff, atmospheric transport and hydrothermal venting (Álvarez-Muñoz *et al.* 2016). Some metals used as antifouling chemicals (mainly copper) and corrosion anodes (mainly zinc) are deliberately placed in the marine environment, through their use on ships' hulls or marine installations, causing hot spots of metal concentrations in and around harbours. Metals with ecotoxicological relevance (Álvarez-Muñoz *et al.* 2016) are: silver (Ag), arsenic (As), gold (Au), cadmium (Cd), cobalt (Co), chromium (Cr), copper (Cu), iron (Fe), gallium (Ga), mercury (Hg), iridium (Ir), manganese (Mn), molybdenum (Mo), nickel (Ni), lead (Pb), platinum (Pt), selenium (Se), tin (Sn), titanium (Ti), vanadium (V) and zinc (Zn). Unlike other pollutants, metals are not biodegradable. They can be adsorbed, and ultimately settle down on the sediment. Colloid particles in the water column (e.g. clay, iron and manganese oxoids or hydroxides, calcium carbonate and organic matter) act as reactive surfaces for the binding of metals (Álvarez-Muñoz *et al.* 2016). The physico-chemical form of a metal (its speciation) determines its mobility and consequently its bioavailability.[3] In general, the free or ionic form, the most bioavailable, is the most toxic (Azizi *et al.* 2018).

Analytical techniques have played an important role in solving marine pollution problems and have been extensively applied to measure heavy metal levels in seawater, marine sediments and biota (Ansari *et al.* 2004). Seawater samples for heavy metal analysis are collected in 1 l plastic containers, which are initially washed with detergent and rinsed

Table 8.1E (a) Average concentrations (ng g^{-1}ww) of organochlorine pesticides in semi-permeable membrane devices (SPMDs) i.e. passive samplers (ng/g lipid; mean of two replicates) and mussels *Perna viridis* (ng/g dry weight (DW) and ng g^{-1} lipid weight (LW) at five locations in Hong Kong coastal waters. nd: not detected. *Source:* From Richardson *et al.* (2001). Reproduced with permission from Elsevier.

Locations	Heptachlor			Aldrin			p,p'-DDT			Dieldrin			Endrin			Endosulfan		
	SPMDs	DW	LW	SPMDs	DW	LW	SPMDs	DW	LW	SPMDs	DW	LW	SPMDs	DW	LW	SPMDs	DW	LW
Kat O	38.5	4.6	46.8	15.8	6.6	67.1	9.4	5.9	60.0	nd	3.8	38.6	8.5	3.8	38.5	10.1	nd	nd
Sai Wan Ho	15.8	nd	nd	38.9	10.0	101.7	17.8	7.9	80.3	19.5	2.9	29.5	33.3	2.2	22.4	25.3	nd	nd
Tsim Sha Tsui	27.5	nd	nd	39.0	5.7	58.0	15.7	12.9	131.2	nd	3.8	38.6	nd	4.0	40.7	19.9	nd	nd
Tolo Harbour	nd	10.9	110.9	44.1	1.9	19.4	29.4	8.0	81.4	nd	2.0	20.3	44.5	3.7	37.6	nd	nd	nd
Kwun Tong	49.3			nd			nd			nd			45.5			nd		

Table 8.1E (b) Average concentrations (ng g^{-1} ww) of organochlorine pesticides in *Mytilus galloprovincialis* from the mid-Black Sea coast of Turkey. Italicized data are detection limits. *Source:* From Ozkoc *et al.* (2007). Reproduced with permission from Springer Nature.

Sampling sites	Heptachlor	Aldrin	p,p'-DDT	Dieldrin	Endrin	Endosulfan
SM3	*<0.05*	0.879	*<0.18*	*<0.12*	7.782	*<0.10*
SM5	*<0.05*	*<0.12*	*<0.18*	*<0.12*	0.205	0.800
SM6	2.419	*<0.12*	*<0.18*	*<0.12*	*<0.15*	*<0.10*
SM8	0.388	*<0.12*	*<0.18*	*<0.12*	*<0.15*	*<0.10*

Table 8.1F Concentrations (ng /g^{-1}dw) of pharmaceuticals in marine mussels from a range of countries worldwide. *M.*spp: *Mytilus* species; *G.d*: *Geukensia demissa*; *Me*: *Mytilus edulis*; *Mg*: *Mytilus galloprovincialis*.

Pharmaceutical compound	Therapeutic use	Concentration	Country	Reference
Azithromycin Erythromycin	Antibiotic	<20 ng/g (*G.d*) 2.0 ng/g (*M.*spp))	Portugal, Spain, Italy, The Netherlands & Norway; California, USA	Álvarez-Muñoz *et al.* 2015a; Dodder *et al.* (2014)
Diflofenac Ketoprofen	Analgesics	29 ng/g (*M.*spp.) 5 ng/g (*Me*)	Ireland Belgium, western Europe	McEneff *et al.* (2014) Wille *et al.* (2011)
Carbamazepine	Antiepileptic	11 ng/g (*Me*) 6 ng/g (*M.*spp)	Belgium, western Europe Ireland	Wille *et al.* (2011) McEneff *et al.* (2014)
Fluoxetine Venlfaxine	Antidepressant	8.4–192.9 ng/g (*Mg*) 2.7 ng/g (*Mg*)	Cesenatico, Italy Ebro Delta, Spain	Franzellitti *et al.* (2014) Álvarez-Muñoz *et al.* 2015a
Diphenhy-dramamine	Antihistamine	0.03 ng/g (*G.d*)	San Francisco, USA	Klosterhaus *et al.* (2013)
Propranolol Atenolol	ß-blocker	52 ng/g (*Me*) 0.3 ng/g (*G.d*)	Belgium, western Europe San Francisco, USA	Wille *et al.* (2011) Klosterhaus *et al.* (2013)
Gemfibrozil	Lipid regulator	18.0 ng/g (*M.*spp)	Ireland	McEneff *et al.* (2014)
Cocaine Amphetamine	Stimulant	0.3 ng/g (*G.d*) 4.2 ng/g (*G.d*)	San Francisco, USA San Francisco, USA	Klosterhaus *et al.* (2013) Klosterhaus *et al.* (2013)

Table 8.1G Concentrations (mg/kg dw) of heavy metals in the soft tissue of marine mussel species from various regions worldwide. From Krishnakumar *et al.* (2018). CC BY 3.0.

Country	Species	As	Cd	Cr	Cu	Fe	Ni	Pb	Zn	Reference
W. Australia†	*X.* sp.	0.46–0.75	0.21–0.27	0.05–0.17	1.7–2.2		0.22–85	0.08–0.52	6–9.6	Nice & Fisher (2011)
S. Island, NZ	*P.c*	5.35–27.48	0.14–1.67		2.19–18.25		0.08–5.83	0.13–1.53	45.31–147.18	Chandurvelan *et al.* (2015)
S. China Sea	*B.f*	5.89–10.03	0.78–4.35	0.81–1.72	5.53–42.31	14.28–56.07	0.42–1.25	4.41–4.8	33.76–79.04	Wang *et al.* (2017)
East coast of China	*P.v*	12.64–20.95	0.48–5.31	1.51–10.93	1.45–28.55	96.62–1002	1.3–4.78	0.44–2.93	66.05–231	Fung *et al.* (2004)
East Adriatic Sea, Croatia	*M.g*	4–30		1–2.9	3.7–11.1	53.4–719	0.8–5	2–7	59.1–273	Orescanin *et al.* (2006)
Spain Cantabrian Coast	*M.g*	14.6–31.5	0.4–2.3	2.6–5.7	9.1–34.8		1.5–15.4	1.1–13.3	202.7–300.8	Bartolome′ *et al.* (2010)
Trinidad†	*P.v*		0.01–0.61	0.06–0.2	1.02–1.98		0.3–0.75		11.3–40.37	Astudillo *et al.* (2002)
Venezuela†	*P.v*		0.02–0.05	0.12–0.16	1.42–3.43		0.22–1.3		8.75–16.38	Astudillo *et al.* (2002)
Italy Tyrrhenian Coast	*M.g*		0.33–0.49	0.46–1.31	5.51–11.5			1.67–2.49	123–180	Conti & Cecchetti (2003)
Turkey, E. Aegean Sea	*M.g*		0.24–0.49	0.32–7.27	2.44–5.49			0.84–2.41	75.9–201	Bilgin & Uluturhan-Suzer (2017)

† concentrations in mg/kg ww; *X.sp*: *Xenostrobus* species; *P.c*: *Perna canaliculus*; *B.f*: *Bathymodiolus platifrons*; *P.v*: *Perna viridis*; *M.g*: *Mytilus galloprovincialis*.

Table 8.2 List of acronyms of analytical techniques used in contaminant analysis.

ASV Anodic stripping voltammetry	LC–Fl Liquid chromatography with fluorescence detection
CE Capillary electrophoresis	LC-MS Liquid chromatography mass spectrometry
CSV Cationic stripping voltammetry	LLE Liquid-liquid extraction
DLS Dynamic light scattering	MSPD Microwave solid-phase dispersion
EM Electron microscopy	MAE Microwave-assisted extraction
FAAS Flame atomic absorption spectroscopy	MASE Microwave-assisted solvent extraction
FFF Field-flow fractionation	MALS Multi angle light scattering
FIA Flow injection analysis	NTA Nanoparticle tracking analysis
FLD Fluorescence detector	PLE Pressurized liquid extraction
FT-IR Fourier transform infrared spectroscopy	py-GC-MS Pyrolysis gas chromatography mass spectrometry
GC-ECD Gas chromatography electron capture detector	QuEChERS Quick, easy, cheap, effective, rugged and safe
GC-MS Gas chromatography mass spectrometry	RM Raman microspectroscopy
GPC Gel permeation chromatography	SEM Scanning electron microscopy
GF-AAS Graphite furnace atomic absorption spectrometry	SPM Scanning probe microscopy
GCB Graphitized carbon black	SIA Sequential injection analysis
HRMS High resolution mass spectrometry	SEC Size exclusion chromatography
HSSE High speed solvent extraction	SPE Solid phase extraction
HPLC-MS High-performance liquid chromatography mass spectrometry	SPME Solid-phase microextraction
HDC Hydrodynamic chromatography	SAX Strong anionic exchange
ICP Inductively coupled plasma	SCX Strong cationic exchange
ICP-AE Inductively coupled plasma atomic emission	SFE Supercritical fluid extraction ; SWE Subcritical water extraction
ICP-MS ICP-mass spectrometry	TED-GC-MS Thermal extraction desorption GC-MS
ICP-OES Inductively coupled plasma optical emission spectrometry	USE Ultrasonic assisted extraction
LD Laser diffraction	XPS X-ray photoelectron spectroscopy

with distilled water, before being soaked in 10% nitric acid and rinsed at least three times with purified water. The samples are preserved by adding concentrated HNO_3 to each container and adjusting the pH to 2.0. They are stored at ~4 °C for later analysis. Before analysis, samples are dried at 105 °C for 24 hours to ensure they have no water content. As samples may contain particulate or organic materials, pre-treatment in the form of digestion is required before analysis (Fatoki *et al.* 2012; Dabwan & Taufiq 2016). European Commission Regulation (EU) No 836/2011 lays down the methods of sampling and analysis for the official control of the levels of lead, cadmium, mercury, inorganic tin, 3-MCPD and benzo(a)pyrene in foodstuffs (European Commission 2011a).

A variety of analytical techniques used for metal detection in seawater samples are reviewed by Álvarez-Muñoz *et al.* (2016). See Table 8.2 for a list of acronyms. Metal detection methods include graphite furnace atomic absorption spectrometry (GF-AAS), inductively coupled plasma atomic emission (ICP-AE), inductively coupled plasma mass spectrometry (ICP-MS), flowing stream systems such as flow injection analysis (FIA) or sequential injection analysis (SIA) with sensitive detection techniques (spectrophotometer, fluorometer) and voltammetery (anodic or cationic stripping voltammetery, ASV or CSV). ICP-AE and ICP-MS are the most used techniques because of their multielemental characteristic, but their direct use in seawater analysis is only possible with very diluted samples.

El-Moselhy *et al.* (2016) have used the following protocol for treatment of sediment samples prior to heavy metal analysis. Samples were collected in plastic bags, air-dried at room temperature and stored before analysis. Samples were sieved and 0.5 g of the fine fraction (<0.063 mm) was digested. Concentrated nitric, perchloric and hydrofluoric acids were added in the ratio of 3:2:1 to the samples in Teflon vessels and left overnight. Samples were then heated to 100 °C for about two hours, cooled, filtered and diluted to 25 ml with deionised water. Metals were determined using flame atomic absorption spectroscopy (FAAS). Additional methods used in sediment analysis include GF-AAS, inductively coupled plasma optical emission spectrometry (ICP-OES) and ICP-MS (see details in Duarte *et al.* 2011; Ponnusamy *et al.* 2014; HELCOM 2018).

According to Dokmeci *et al.* (2017), prior to metal analysis in mussels, soft tissue from a number of individuals is pooled as a composite sample, homogenised and freeze-dried. To avoid contamination, all parts of the homogeniser that come into contact with a sample are covered with Teflon adaptors. The homogenates are then lyophilised and stored in polyethylene bottles. An aliquot of the homogenised tissue is used to calculate dry weight using oven drying at 105 °C until a constant weight is attained. Subsequently, a 0.25 g lyophilised sample is mineralised with 9 ml of concentrated HNO_3 (70% v/v) and 1 ml of H_2O_2 (30% v/v) in a closed Teflon vessel using a microwave digestion system (heating to 175 °C for 5.5 min and holding at 175 °C for 4.5 min). After mineralisation, digests are cooled and the resulting solutions are diluted to a known volume (50 ml) with purified water and stored in polyethylene bottles until analysis. Heavy metal instrumental analysis of mussel samples can include FAAS (El-Moselhy *et al.* 2016), ICP-MS (Fatoki *et al.* 2012; Mok *et al.* 2014) and ICP-AE (Gonzalez-Rey *et al.* 2011; Ponnusamy *et al.* 2014).

Persistent Organic Contaminants

Persistent organic pollutants (POPs) are a group of very toxic organic chemical compounds which bioaccumulate in animals and humans. They are primarily the products and byproducts of human industrial processes. They are persistent in the environment and travel vast distances via air and water. POPs include industrial chemicals like polychlorinated biphenyls (PCBs), polycyclic aromatic hydrocarbons (PAHs), pesticides like dichloro-diphenyl-trichloroethane (DDT), endrin, dieldrin, aldrin, chlordane, toxaphene, heptachlor, mirex and hexachlorobenzene (HCB) and unwanted wastes like dioxins and furans. The Stockholm Convention on Persistent Organic Pollutants, an international environmental treaty that aims to eliminate or restrict the production and use of almost 30 POPs, was signed and took effect in May 2004. As of April 2017, 180 states plus the European Union are party to the Convention; nonratifying states include the United States, Israel, Malaysia and Italy. Despite the ban, POPs continue to exist in the environment, albeit at low concentrations (see Chapter 12).

Sample preparation for the determination and quantification of organic compounds is based on three main steps: extraction, clean-up and preconcentration. The extraction step is a procedure performed to separate the analytes[4] from the sample matrix. In the clean-up step, the possible interferences in the analysis, which make it difficult to identify and quantify the compounds of interest, are removed. Some contaminants such as endocrine disrupters (EDs) in environmental samples are usually present at concentration levels below the limit of detection (LOD) of most analytical methods. Consequently, preconcentration is a crucial step to improve detectability and to meet the LOD and limit of quantification (LOQ) of analytical methods. Sample preparation techniques are specific for each sample type. For instance, solid phase extraction (SPE) has been the preferred technique in studies of EDs in liquid samples. The SPE technique applied to aqueous samples has the advantage of aggregating the three steps (extraction, clean-up and preconcentration) in a single stage.

Polycyclic Aromatic Hydrocarbons

These are one of the most important classes of environmental pollutants. PAHs are complex chemicals that are formed and released during incomplete combustion or pyrolysis (burning) of organic matter such as waste or food, during industrial processes and through other human activities (e.g. oil spillages). They are also formed in natural processes such as forest fires and volcanoes (FSAI 2015). Their molecular structure contains two or more fused aromatic rings made up of carbon and hydrogen atoms. Benzo[a]pyrene is one of the best characterised members of the PAH family. The general characteristic common to PAHs are high melting and boiling points, low vapour pressure and very low water solubility, which tend to decrease with increasing molecular mass. PAHs are soluble in many organic solvents and are therefore highly lipophilic (soluble in fat). Many of them adsorb to suspended particulate matter, which then settles through the water column and accumulates in aquatic sediments and biota. In the air, heavy PAHs (more than four rings) tend to adsorb to particulate matter, while lighter PAHs tend to remain gaseous until they are removed through precipitation (Skupinska *et al.* 2004). European Commission Regulation (EU) 835/2011 defines maximum levels for PAHs in foodstuffs (European Commission 2011b).

Due to their high hydrophobicity in seawater, PAHs are mostly accumulated in organisms, particulate matter and sediments (Álvarez-Muñoz *et al.* 2016). Hence, they are usually found in the water phase at low concentrations, necessitating an extraction and enrichment step such as SPE, liquid-liquid extraction (LLE) or solid phase micro extraction (SPME) and a selective chromatographic/detection step such as gas chromatography mass spectrometry (GC-MS), liquid chromatography mass spectrometry (LC-MS) or liquid chromatography with fluorescence detection (LC-Fl) in every analytical procedure. Depending on the analytes chosen, water body studied and expected pollutant concentration, clean-up may be necessary. Extraction protocols for sediments and biota use nonpolar solvents and different techniques, including supercritical fluid extraction (SFE), pressurised liquid extraction (PLE), subcritical water extraction (SWE), microwave assisted extraction (MAE) and ultrasonic assisted extraction (USE) (Álvarez-Muñoz *et al.* 2016 and references therein). GC, rather than LC, is often the technique of choice for separation and identification of PAHs, largely because

GC has greater selectivity, resolution and sensitivity than LC (Poster *et al.* 2006). Additional information on PAH analysis can be found in Li *et al.* (2015) for seawater, Palma-Fleming *et al.* (2012) and HELCOM (2016) for sediment and Storelli & Marcotrigiano (2001), Galgani *et al.* (2011), El Nemr *et al.* (2004) and Booth *et al.* (2007) for mussel tissue.

Surfactants

Surfactants are the basic cleaning agent in consumer and industrial cleaning products such as detergents for washing clothes, dishes, carpets, floors and walls, as well as shampoos, lotions, toothpastes and cosmetics. Other applications include fabric softeners, lubricating ingredients, emulsifiers for spreading agricultural products and industrial applications such as latex systems, plastics and composites. Surfactants are molecules that lower the surface tension function of a liquid or the interfacial tension between two liquids. They are able to act in this way because they contain both a hydrophilic (water attracting) group, such as an acid anion (carbon dioxide CO_2^- or sulphur trioxide SO_3^-) and a hydrophobic (water-repelling) group, such as an alkyl chain (e.g. CH_{3-}) that has a carbon and hydrogen atom present in its structure. Thus, a surfactant contains both a water-insoluble (or oil-soluble) component and a water-soluble component. There are four types of surfactants, based upon the composition of the polarity of the head group: nonionic, anionic, cationic and zwitterionic. A nonionic surfactant has no charge groups in its head, while the head of an anionic surfactant carries a net negative charge in aqueous solutions due to the presence of a sulphonate, sulphate, carboxylate or phosphate group. The largest volume of anionic surfactants is used in consumer products such as laundry detergents, cleaning and dishwashing agents and PCPs. Cationic surfactants have a positive charge on their hydrophilic end. The positive charge makes them useful in antistatic products, like fabric softeners, cosmetics and medicines. They can also serve as antimicrobial agents, so they are often used in disinfectants. If a surfactant contains a head with two oppositely charged groups, it is termed zwitterionic. These surfactants are often used in PCPs such as shampoos and cosmetics.

After use, an enormous number of surfactants (and/or their degradation products) enter wastewater treatment plants (WWTPs), where removal has been shown to be highly efficient (average rate 95–99%) (Jackson *et al.* 2016 and references therein). The marine environment, therefore, receives almost continuous surfactant input, either directly (from treated and untreated wastewaters discharges) or indirectly (via contaminated rivers), and hence is regarded as the final site of deposition for surfactants. Once within the marine environment, surfactants may be removed via volatilisation (albeit only to a limited extent), adsorption to particles, abiotic or microbial degradation or uptake by marine organisms. Degradation and sorption are the main mechanisms involved. Moreover, some surfactants can be transformed to more toxic degradation products of compounds from the group of alkylphenol ethoxylates, used in production of cleaners, degreasers, paints, textiles and agricultural chemicals (Olkowska *et al.* 2014). For details on the degradation process in anionic and nonionic surfactants, see Lara-Martín *et al.* (2010), Corada-Fernández *et al.* (2013), Okbah *et al.* (2013) and Traverso-Soto *et al.* (2015). European Commission Regulation (EU) No 259/2012 of the European Parliament and of the Council of 14 March 2012 amending Regulation (EC) No 648/2004 details the use of phosphates and other phosphorus compounds in consumer laundry detergents and consumer automatic dishwasher detergents (European Commission 2012).

Enormous volumes of surfactants are used every day. Invariably, they enter the environment, where they and/or their degradation products may cause damage depending on their concentrations. Therefore, it is necessary to use reliable analytical methods for determination of levels of surfactants in environmental matrices. This can be a complex process, as surfactants are often sold as commercial mixtures, which can comprise hundreds of different compounds with different physico-chemical properties (Traverso-Soto *et al.* 2012; Álvarez-Muñoz *et al.* 2016 and references therein). Water samples are immediately preserved upon collection by the addition of a biocide, usually formaldehyde, to minimise biodegradation (Lara-Martín *et al.* 2006) and kept in the freezer for a period of 48 hours. Solid samples (sewage sludges, soils or sediments) are also kept at low temperature once collected, to avoid any degradation of the analytes during transport to the laboratory. Subsequently, the samples are usually dried in a heater, kept at room temperature or frozen at −20 °C and later freeze-dried. SPE is the most commonly used technique for the extraction and purification of surfactants from liquid and solid samples. It can be used in conjunction with strong anionic exchange (SAX) for anionic surfactants, while for nonionic surfactants graphitised carbon black (GCB) and silica (C2–C10) cartridges combined with strong cationic exchange (SCX) and SAX are also used. Most use C18 and GBC cartridges, because they are suitable for simultaneous extraction and isolation of a wide range of surfactants in a single stage. Regarding the separation, identification and quantification of surfactants, high-performance liquid chromatography mass spectrometry (HPLC-MS) and, to a lesser extent due to the nonvolatility of most analytes, GC-MS are the main tools currently employed as they allow for determination of every single homologue, ethoxymer and/or isomer from surfactant mixtures in different environmental matrices (solids, water and biota), although other detectors like ultraviolet (UV) and fluorescence detection (FLD) have also been used.

Polychlorinated Biphenyls

PCBs are a group of artificial organic chemicals consisting of carbon, hydrogen and chlorine atoms. They were commercially produced worldwide on a large scale between the 1930s and 1980s. They have a number of chlorine atoms (between 1 and 10) attached to a biphenyl structure with a common chemical formula $C_{12}\ H_{10-x}Cl_{x.}$ There are 209 possible chlorine-substituted biphenyl congeners, but PCBs in samples analysed for environmental and regulatory purposes usually consist of complex mixtures of about 130 congeners originating from technical mixtures (Jaikanlaya *et al.* 2009). The number of chlorine atoms and their location in a PCB molecule determine many of the PCB's physical and chemical properties. PCBs have a range of toxicities and vary in consistency from thin, light-coloured liquids to yellow or black waxy solids. Due to their nonflammability, chemical stability, high boiling point and electrical insulating properties, they may be used in hundreds of industrial and commercial applications, including electrical, heat transfer and hydraulic equipment; plasticisers in paints, plastics and rubber products; pigments, dyes and carbonless copy paper; adhesives, pesticides and fire retardants; and other industrial applications (Álvarez-Muñoz *et al.* 2016; EPA 2020). Although the manufacture of PCBs is forbidden under the Stockholm Convention on Persistent Organic Pollutants, they are still released to the environment by disposal of large-scale electrical equipment and waste. European Commission Directive 96/59/EC governs the disposal of PCBs and polychlorinated terphenyls (PCTs) (European Commission 1996).

Analysis of PCBs in the marine environment focuses on the analysis of sediments and biota because, due to their high hydrophobicity, PCBs are not present in seawater. Analysis involves an extraction of lyophilised matrices, followed by a strong clean-up of the matrix (Álvarez-Muñoz *et al.* 2016). At present, PLE is the most widely used extraction technique. PLE and MAE have the shared advantage over SFE in that they are matrix-independent (Webster *et al.* 2013). The methods discussed here are suitable for extracting PCBs from both sediment and biota. The PCB content of environmental samples is commonly monitored using gas chromatography with electron capture detection (GC-ECD) or GC-MS. Dioxin-like polychlorinated biphenyls (DL-PCBs) should be analysed by GC-MS, preferably using high-resolution MS, although low-resolution MS may be a suitable alternative (Vorkamp *et al.* 2012; Webster *et al.* 2013). Additional information on the analysis of sediment samples can be found in Solaun *et al.* (2015) and Bayan *et al.* (2003), while details on the analysis of biota samples can be found in Mari *et al.* (2008), Jaikanlaya *et al.* (2009), Solaun *et al.* (2015), Georgieva *et al.* (2016) and Kampire *et al.* (2016).

Pesticides

Pesticides are chemical substances used to kill, reduce or repel insects, weeds, rodents, fungi or other organisms that are linked with a high incidence of severe or irreversible adverse effects on human health or the environment. They are often classified by target pest (e.g. fungicides, herbicides, insecticides, nematicides, rodenticides) or by chemical identity (e.g. organophosphates, organochlorines, carbamates) (Saravi & Shokrzadeh 2011; Álvarez-Muñoz *et al.* 2016). They are used especially in agriculture and around areas where humans live. Some are harmful to humans, either through direct contact or as residue on food, and some are harmful to the environment as a result of their high toxicity – such as dichloro-diphenyl-trichloroethane, commonly known as DDT, which is now banned in many countries. In March 2019, there were more than 300 chemicals on the Pesticide Action Network (PAN) International List of Highly Hazardous Pesticides (Pesticide Action Networks UK 2019). Pesticide formulation includes the main molecule of the pesticide, called the active ingredient, and other ingredients, which may be inert or may enhance the pesticide properties (Álvarez-Muñoz *et al.* 2016). They are mainly introduced into rivers via run-off, and then may enter marine areas, particularly estuarine and coastal zones. Leaching of water through the soil can also carry pesticides down to the groundwater. The more water-soluble a pesticide is, the greater the potential for runoff and leaching. But if a pesticide is not water-soluble, it may accumulate in the soil or biota (Álvarez-Muñoz *et al.* 2016). European Commission Regulation 396/2005 governs maximum residue levels of pesticides in or on food and feed of plant and animal origin (European Commission 2005).

Analysis of pesticides is very demanding, since the group contains many different chemical classes with a wide range of physico-chemical properties, present at low concentrations in complex matrices. Sediment and biota samples are stored frozen and freeze-dried, and are homogenised prior to analysis (Okay *et al.* 2011). SPE is usually the preferred technique for water sample preparation, although LLE and passive sampling techniques have also been used. Several SPE sorbents[5] (e.g. GCB, Oasis HLB, Strata-X and Strata C18) are used to extract pesticides from water samples (reviewed in Álvarez-Muñoz *et al.* 2016). For the extraction of pesticides from mussel tissue or sediments, the quick, easy, cheap, effective, rugged and safe (QuEChERS) method is the most often used, because it exhibits better analytical performance compared with other common approaches (Bruzzoniti *et al.* 2014; Rejczak & Tuzimski 2015). SPE and QuEChERS in combination with LC-MS are the most applied techniques for the analysis of pesticide samples. LC is preferred over GC because pesticides are currently quite polar, thermally labile or not easily vaporised, and therefore detection by GC is not as good as LC (Álvarez-Muñoz *et al.* 2016). Further information on analysis of sediment samples can be found in Ozkoc *et al.* (2007), Okay *et al.* (2011), Hladik & McWayne (2012) and Choi *et al.* (2016), while information on analysis in biota samples can be found in Ferante *et al.* (2007), Kasiotis (2009), Scarpato *et al.* (2010), Choi *et al.* (2016), Georgieva *et al.* (2016) and Suryono *et al.* (2019a,b).

Dioxins

Dioxins, furans and DL-PCBs are the abbreviated names for a family of chemicals that have similar toxicity and shared chemical characteristics. Dioxins and furans are chemically classified as halogenated aromatic hydrocarbons. The most widely studied compound is 2,3,7,8-tetrachloro-*p*-dibenzodioxin TCDD ($C_{12}H_4Cl_4O_2$), which is often referred to simply as 'dioxin' and is the reference for a number of compounds (PCDDs/PCDFs and DL-PCBs) because their molecules are similar structurally and have dioxin-like toxicity. A molecule of dibenzo-*p*-dioxin (DD) and dibenzofuran (DF) is a triple-ring structure consisting of two benzene rings interconnected by a ring containing an oxygen atom. There are 75 compounds in the dioxin family and 135 in the furan family. In general, the compounds have low water solubility, low vapour pressure, are very stable and tend to bioaccumulate in tissue rich in lipids and bioconcentrate through food chains due to their high fat solubility (NPI 2019). Also, because of their high chlorine content, they persist longer in the environment and show greater bioaccumulation, leading to biomagnification[6] in marine food chains. Diоxins are unwanted byproducts of a wide range of manufacturing processes, including smelting, chlorine bleaching of paper pulp and the manufacture of some herbicides and pesticides. In terms of dioxin release into the environment, uncontrolled waste incinerators (solid waste and hospital waste) are often the worst culprits, due to incomplete burning. Technology is available that allows for controlled waste incineration with low dioxin emissions (WHO 2016). European Commission Regulation 2017/644 governs methods of sampling and analysis for the control of levels of dioxins, DL-PCBs and non-DL-PCBs in certain foodstuffs (European Commission 2017a).

Like PCBs, dioxins, because of their high hydrophobicity, are analysed solely from sediment and biota samples. Mussel samples should be transported at temperatures between 5 and 15 °C, similar to those of the area of origin, in a clean container. It is important to remove any sediment particles from their intestinal system, by depuration in a glass aquarium with filtered water, preferably from the sampling location, for approximately 24 hours. After dissection, all samples should be stored in the dark below −20 °C, freeze-dried and homogenised prior to analysis. Lipid content of biota samples should be determined to allow normalisation of dioxin concentrations to lipids (Vorkamp *et al.* 2012 and references therein). Sediments containing a large proportion of fine material are preferable for organic contaminant monitoring. Sediment samples consisting of coarse sandy material may require sieving. Because sediments may often display a rather heterogeneous distribution of contaminants and organic carbon content, pooling of a number of subsamples from a specific area per sampling location is recommended. Total organic carbon (TOC) should be determined for characterisation of sediments and for use as a possible normalisation parameter. Extractable lipid determinations may be used as well, particularly for small sample sizes and high-lipid contents.

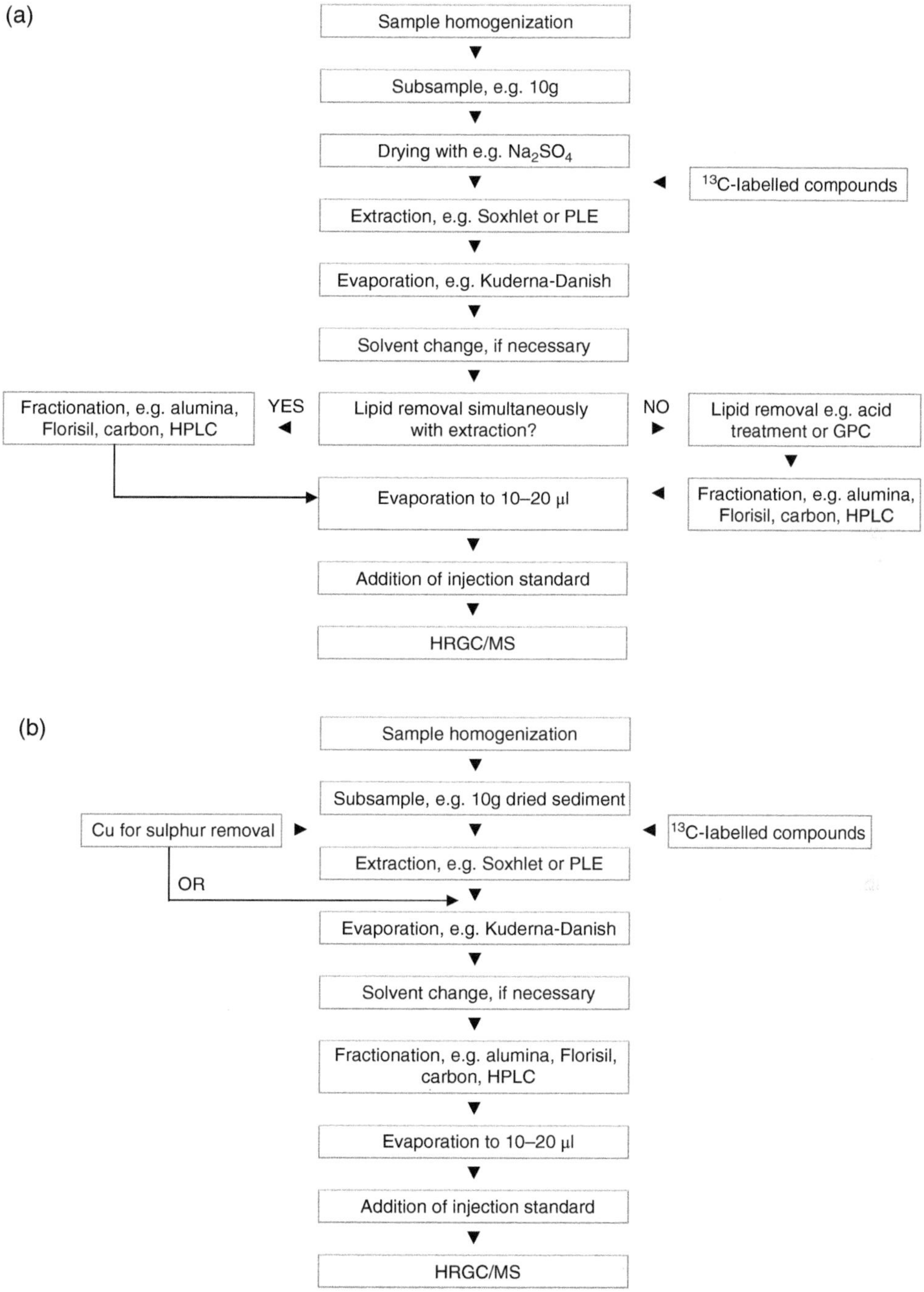

Figure 8.1 (a) Example of an analytical method for the determination of polychlorinated dibenzo-*p*-dioxins, polychlorinated dibenzo-*p*-furans (PCDD/F) and dioxin-like-polychlorinated bipenyls (DL-PCBs) in biota. (b) Example of an analytical method for the determination of PCDD/Fs and DL-PCBs in sediment. *Source:* From Vorkamp *et al.* (2012).

For both biota and sediment samples, dry weight should be determined for normalisation purposes, for instance by drying at 105 °C until constant weight (Vorkamp *et al.* 2012). For instrumental analysis of dioxins, GC coupled to high resolution mass spectrometry (HRMS) is the preferred analytical method (Álvarez-Muñoz *et al.* 2016). Examples of suitable methods for the analysis of dioxins and DL-PCBs in biota and sediment samples are illustrated in Figures 8.1a and 8.1b, respectively. For additional information on sediment treatment and analysis, see Danis *et al.* (2006) and Fernandes *et al.* (2009), while for information on biota samples, see Malvarn *et al.* (2005), Munschy *et al.* (2008), Parera *et al.* (2013), Carro *et al.* (2018) and Fleidner *et al.* (2018).

Emerging Organic Contaminants

Emerging organic contaminants (ECs) are chemicals (synthetic or naturally occurring) that have recently been detected in the environment, although insufficient information is available to determine their potential environmental or public health risk (Naidu *et al.* 2016). The lack of information can be attributed to limited financial resources and the dearth of analytical techniques for detecting the effects of ECs on ecosystems and human health, either on their own or as mixture. The most prevalent ECs include perfluorinated compounds (PFCs), pesticides, water disinfection byproducts, gasoline additives, food additives, manufactured nanomaterials, brominated flame retardants (BFRs), surfactants, human and veterinary pharmaceuticals, UV filters, PCPs and lifestyle compounds such as caffeine and nicotine (Lapworth *et al.* 2012; Lei *et al.* 2015; Vandermeersch *et al.* 2015; Naidu *et al.* 2016). Wastewater sources and their artificial recharge/infiltration to the subsurface are considered one of the most important point sources of ECs in the aquatic environment. Also, landfill sites are major point sources because ECs may leach and reach the groundwater or surface water (reviewed in Lapworth *et al.* 2012).

Brominated Flame Retardants

BFRs are a diverse group of chemicals added to a wide range of consumer products, including plastic polymers, textiles, building materials, car parts and electric and electronic equipment, to prevent or delay the propagation of fire (Abdallah 2014). Some BFRs are additives mixed into polymers and are not chemically bound to the plastics or textiles, and therefore may separate or leach from the surface of their product applications into the environment. Additives include polybrominated biphenyls (PBBs), polybrominated diphenyl ethers (PBDEs) and hexabromocyclododecane (HBCD). Others, such as tetrabromobisphenol A (TBBPA), are reactive and are chemically bonded into plastics (de Wit 2002). Polymeric BFRs include compounds like brominated polystyrene (BPS), a white granule powder with a bromine content of ≥68% that is incorporated in the backbone of the polymer, resulting in a more stable chemical structure with a very high molecular weight and a low volatility, bioavailability and toxicity. Generally, volatility decreases with rising bromine content, which means that highly brominated compounds are nonvolatile, while low-brominated compounds will be more mobile in water and have a greater inclination tendency to evaporate from surface waters (Álvarez-Muñoz *et al.* 2016).

Flame retardants (FRs) can find their way into the environment as wastewaters of industrial facilities that produce them or of manufacturing facilities that incorporate them into products, through volatilisation and leaching from products during manufacture or usage, upon breakdown of foam products, by disposal of products such as electronic equipment, through leaching from landfills, through combustion and recycling of waste products or via adsorption on to dust particles (Segev *et al.* 2009 and references therein). In addition, additive FRs are believed to be more easily released into the environment than are reactive FRs. Once an FR arrives into the environment, it can be attached to a particle for transport in water or delivery to the sediment or end up on an airborne dust particle and travel distances far from the production and/or emission site. Thus, traces of FRs are found in terrestrial, freshwater and marine ecosystems (i.e. in air, water, soil and sediments) at various locations far from where they were produced and/or used. European Commission Directive 76/769/EEC sets restrictions on the marketing and use of certain dangerous substances and preparations (e.g. pentabromodiphenyl ether, octabromodiphenyl ether) (European Commission 2003).

Because of the hydrophobicity of PBBs, PBDEs, TBBPA and HBCD and the low concentrations of these compounds expected in water, volumes as large as 1000 ml are typically required to allow their detection (Álvarez-Muñoz *et al.* 2016). This has led to the use of LLE with organic solvents for the extraction of BFRs from water samples (Dirtu *et al.* 2013). However, this technique requires large volumes of solvent, so solid SPE is often used instead of LLE. A wide range of extraction methods have been used for BFRs (see Álvarez-Muñoz *et al.* 2016). For the analysis of sediments or biota, water depletion by mixing of the sample with sodium sulphate or by freeze-drying is generally carried out before extraction, using methods such as Soxlet PLE, SPME and microwave solid phase dispersion (MSPD) (Álvarez-Muñoz *et al.* 2016 and references therein). Because of the relatively low selectivity of extraction techniques and complexity of BFR samples, a purification step is compulsory. For sediment samples, the clean-up process involves the removal of sulphur, whereas for biota, it involves the removal of fat. The best way to do this is fractionation with gel chromatography and adsorption chromatography using silica gel, alumina or a florisil column. The most commonly used instrumental techniques are GC-MS and GC-electron capture detector (ECD), although LC-MS has also been used (Álvarez-Muñoz *et al.* 2016). See also a review in Abdallah (2014) that highlights recent advances and methodological improvements in instrumental techniques applied for the analysis of different BFRs. Further information on analysis of mussel samples can be found in Johansson *et al.* (2006), Parera *et al.* (2013), Rodil *et al.* (2019), Sun *et al.* (2020) and Tavoloni *et al.* (2020).

Endocrine-Disrupting Compounds

Endocrine-disrupting compounds (EDCs) are a large class of substances of natural and/or anthropogenic origin that interfere with the endocrine (hormone) system of organisms, possibly resulting in sterility and species extinctions as a consequence of reproductive and developmental disorders. Consequently, EDCs are of great environmental concern (Filipkowska & Lubecki 2016). In the marine environment, a wide range of heterogeneous substances with different structures and physico-chemical properties can disrupt the endocrine system, including dioxin and dioxin-like compounds, 4-nonlyphenols (4-NPs), PCBs, bisphenol A (BPA), PBBs, PBDEs, DDT, plasticisers (phthalates), plastics (BPA), fungicides and pharmaceutical agents (diethylstilbestrol, DES) (Diamanti-Kandarakis *et al.* 2009; Álvarez-Muñoz *et al.* 2015a; Filipkowska & Lubecki 2016). Bisphenols (from plastics and resins), alkylphenols (found in industrial surfactants), oestrogens (reproductive hormones) and perfluorinated chemicals (used in the production of Teflon) are prevalent over other EDCs in the aquatic environment (Vandermeersch *et al.* 2015). Synthetic chemicals suspected as endocrine disruptors may reach humans and animals in a variety of ways. Some, such as pesticides, are released intentionally. Others are byproducts of industrial processes and waste disposal, incineration, landfill, industrial and municipal effluents, agricultural runoff and atmospheric transport (CCOHS 2020). The presence of EDCs in bodies of water is mainly due to the discharge of wastewater, impacting the quality of surface water and groundwater with compounds that are not entirely removed by conventional treatment processes – although the European Union, Germany, England, the United States, Australia, Canada and Japan have all installed multi-stage treatment systems that effectively reduce the concentration of EDCs in drinking water (Cortés Muñoz *et al.* 2013 and references therein). European Commission Delegated Regulation (EU) 2017/2100 sets out the scientific criteria for the determination of endocrine-disrupting properties (European Commission 2017b).

As already seen, the molecular heterogeneity of EDCs makes the development of analytical methods a difficult process. Usually, the addition of a surrogate internal standard (IS)[7] to water, sediment and biota samples is the first step in the analytical protocol. For water samples, the most used extraction techniques are SPE with C18-, $-NH_2$- or CN-modified silica (Petrovic *et al.* 2002). Eluates are evaporated near to dryness under a gentle stream of nitrogen and reconstituted in adequate solvent for chromatographic analysis with LC or GC. For solid samples (e.g. sediment or biota), a pretreatment consisting of freeze-drying is often needed before the extraction process. Sediments are extracted using solid-liquid extraction (SLE) by sonication in an ultrasonic bath, or else using PLE. The extract is then cleaned up with solid-liquid adsorption chromatography in open columns, or by SPE (Petrovic *et al.* 2002). Freeze-dried biota samples are extracted using techniques such as PLE, sonication, high-speed solvent extraction (HSSE) or MAE. As these are time-consuming processes, the faster QuEChERS technique has been successfully applied to biota samples (Jakimska *et al.* 2013). The most used purification techniques are SPE (as in sediment samples) using florisil adsorbent, C18 cartridges and gel permeation chromatography (GPC). Both water and biota samples are treated in a similar way, and instrumental analysis is based on LC-MS/MS or GC-MS, depending on the type of EDCs being analysed (Álvarez-Muñoz *et al.* 2016). Further details on analysis of EDCs in sediment and biota samples can be found in Filipkowska & Lubecki (2016).

Pharmaceuticals and personal care products

Evidence indicates that the use and consumption of pharmaceuticals per capita in the European Union has doubled or nearly tripled in a time span of only 14 years (2000–2014) and suggests an even bigger absolute consumption amount due to continuous global population increase (OECD 2018). This critical piece of information highlights the need for further investigation of these substances and their possible effects on the environment. More than 100 pharmaceutical compounds, covering different therapeutic classes, have been accounted for in drinking water, wastewater, groundwater, marine organisms, sewage and surface water around the world. Based on their therapeutic uses, pharmaceuticals are classified into the following groups (Ojemaye & Petrik 2018 and references therein): antidiabetics such as alpha-glucosidase inhibitor, β-blockers (atenolol, metoprolol), antibiotics (trimethoprim), lipid regulators (gemfibrozil), anti-epileptics (acetazolamide), tranquilisers (diazepam), antimicrobials (penicillins), antiulcer and antihistamine drugs (cimetidine and famotidine), anti-inflammatories and analgesics (ibuprofen, paracetamol, diclofenac), antidepressants (benzodiazinepines), anticancer drugs (cyclophosphamide, ifosfamide), antipyretics and stimulants (dexamphetamine, methylphenidate, modafinil) and oestrogens and hormonal compounds (estriol, estradiol, estrone). When pharmaceuticals are administered, they can be transformed by the organism into more soluble molecules. However, metabolism is often incomplete and excretion rates range from 0 to 100% (Álvarez-Muñoz *et al.* 2016). Pharmaceuticals used in human and/or veterinary medicine and their metabolites are spread into the environment in different ways, including through sewage treatment plant (STP) effluent, by heavy rain on agricultural land (which provokes surface water run-off) and, occasionally, through untreated sewage from domestic wastes and flooding. Some of the pharmaceuticals reach surface waters (rivers, lakes and estuaries) and eventually groundwaters after resisting the intended biological degradation.

However, in surface waters, they may be degraded through different processes, such as photolysis, the efficiency of which depends on factors such as intensity of solar irradiation and therefore on latitude and season of the year. For drugs that have low volatility and high solubility, distribution is mainly through aqueous transport or even food-chain dispersion (details in Santos *et al.* 2010; Sui *et al.* 2015; UNESCO & HELCOM 2017; Yang *et al.* 2017; Ojemaye & Petrik 2018). Other important sources are waste disposal, aquaculture, animal husbandry and horticulture. These provide a constant input of pharmaceuticals into the environment, and although these compounds have high transformation/removal rates, they are compensated by their continuous introduction into aquatic systems (Álvarez-Muñoz *et al.* 2016). European Commission (2004) Directive 2004/27/EC relates to medicinal products for human use (European Commission 2004).

While substantial research has been conducted on the occurrence and effects of human-use pharmaceuticals in the aquatic environment, comparatively little has been conducted on PCPs, despite the fact that they are found more frequently and at higher concentrations. PCPs are continually released into the aquatic environment and are biologically active and persistent (Brausch & Rand 2011). They represent a diverse group of compounds, including disinfectants such as triclosan (TCS) and triclocarban (TCC), which are biphenyl ethers used as antimicrobials in soaps, deodorants, skin creams, toothpaste and plastics; fragrances (musks, perfume ingredient); insect repellents (DEET); preservatives such as parabens in pharmaceuticals and PCPs; and UV filters, such as methylbenzylidene camphor in sunscreens (Brausch & Rand 2011). PCPs are products intended for external use on the human body, and consequently are not subjected to metabolic alterations. Therefore, large quantities enter wastewater unaltered via washing, bathing and showering. Because of elevated lipophilicity, most ingredients in PCPs are taken up by absorption or adsorption by sludge and sediments (Ternes *et al.* 2004). Many of these compounds are used in large quantities, and studies have indicated many are environmentally persistent, bioactive and have the potential for bioaccumulation (Brausch & Rand 2011). Wastewater is considered to be one of the most important sources of PCPs in the aquatic environment. Generally, household PCPs can enter the sewer system after excretion from the body or washing-off by tapwater via sinks or toilets (Figure 8.2). Only limited removal has been identified during primary treatment at STPs due to the limited adsorption to sludge. Once the PCP-containing effluents are discharged, the receiving surface waters are contaminated. PCPs may stay in the surface water body and be subjected to low natural attenuation. Managed aquifer recharge (MAR) invovles the use of the surface system and of reclaimed water to

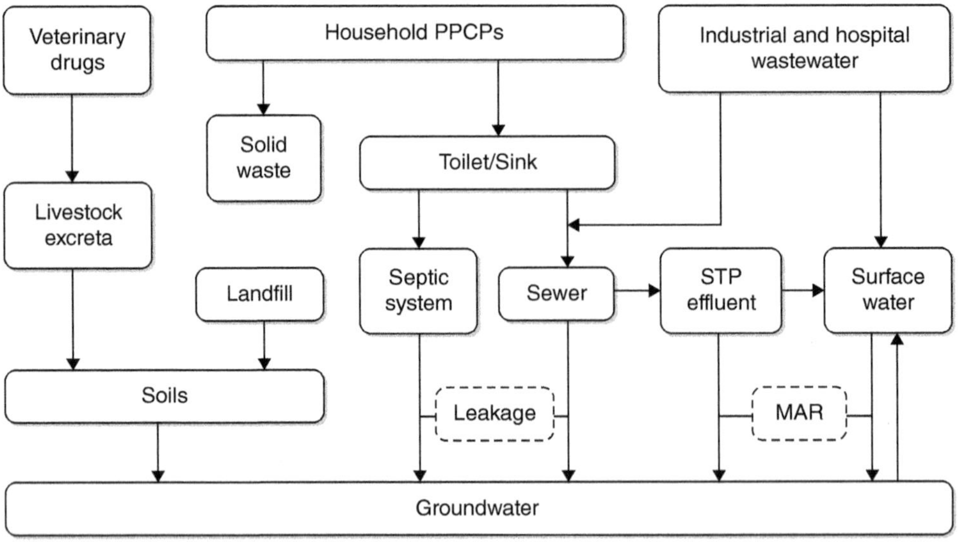

Figure 8.2 Sources and pathways for personal care products (PCPs) entering groundwater. MAR, managed aquifer recharge; STP, sewage treatment plant. From Sui *et al.* (2015). Reproduced with permission from Elsevier.

artificially recharge an aquifer (a body of rock and/or sediment that holds groundwater). It is also an important potential source of PCPs into groundwater, particularly when the residence times are short and where wastewater treatment is poorly regulated (Sui *et al.* 2015). European Commission Regulation EU 2017/1410 amending Annexes II and III to Regulation (EC) No 1223/2009 regulates cosmetic products (European Commission 2017c).

Multiresidue analytical methods are the most commonly applied methodology used in the analysis of pharmaceuticals and PCPs in the marine environment. This means that pharmaceutical classes (see earlier) are simultaneously analysed (Álvarez-Muñoz *et al.* 2016). For seawater samples, SPE followed by LC in combination with MS has been the method most often used in pharmaceutical analysis. SPE is particularly well adapted to multiresidue analysis including compounds from a wide range of polarity or characterised by various physicochemical properties (Pichon 2000). For sediment and biota samples, the target analytes should first be extracted from the matrix, using different extraction techniques. Organic solvent extraction combined with either shaking, vortexing or ultrasonication is the technique most frequently used due to its ease of execution. PLE and MASE have also been used in sediments and biota samples (Álvarez-Muñoz *et al.* 2015a,b). QuEChERS has been used as an extraction and purification technique for biota samples, although SPE on Oasis HLB is the more commonly selected clean-up technique due to its superiority (references in Álvarez-Muñoz *et al.* 2016). For sediment samples, LC in combination with MS is the preferred instrumental technique. This can be used for biota samples, although other kinds of detectors like FLD, UV and diode array detectors (DODs) have also been used, as well as enzyme-linked immunosorbent assay (ELISA) and time-resolved fluoroimmunoassay (TR-FIA). Additional details on the preparation and analysis of biota samples can be found in Álvarez-Muñoz *et al.* (2018), Martinez Bueno *et al.* (2013, 2014) and McEneff *et al.* (2013, 2014).

Microplastic Particles

Microplastic particles (MPs) are plastic particles smaller than 5 mm diameter, and are ubiquitous in various environmental compartments: the water column, sediment and biota. They enter the environment from both primary and secondary sources. Primary sources include particulates, which are produced either for direct use, such as for industrial abrasives, exfoliants and cosmetics, or as precursors for the production of consumer products. Secondary sources relate to the formation of MPs in the environment due to the fragmentation of larger plastic material as a consequence of photolytic, mechanical, thermos-oxidative, microbial and hydrolysis degradation (Browne *et al.* 2007; Cole *et al.* 2011; Ivar du Sol & Costa 2014; Vandermeersch *et al.* 2015). One source of MPs is plastic fibres (<5 mm in length), which are released via shedding from plastic-based textiles and garments either during production, use or predominantly via wastewater expelled from domestic and industrial washing machines. Another source is from industrial and domestic products, including toilet, hand, body and facial cleansers, that contain tiny polyethylene and polystyrene particles <1 mm in diameter. Also, larger particles of acrylic, melamine and polyester, ranging from 0.25 to 1.7 mm in diameter, are used to clean machinery and boat hulls in dockyards. MPs are further used in a range of medical applications, including drug delivery systems (Browne 2015). Small particles from all these sources are likely to be transported with wastewater and through sewage treatment works, and subsequently to enter aquatic habitats (Browne *et al.* 2007). WWTPs are the main sources of MPs in aquatic environments. MPs from cosmetics and other plastic waste end up at municipal WWTPs (da Costa *et al.* 2016). The content of microplastics in WWTP influent varies in the range of 10^4–10^5 microplastics m^{-3}, and their incomplete removal (70–100%) during the treatment process can result in pollution in the receiving water. In a study of effluent from 12 WWTPs in Lower Saxony, Germany, Mintenig *et al.* (2017) detected MPs in quantities ranging from 0 to 5×10^1 m^{-3} of particle size >500 μm and 1×10^1 to 9×10^3 m^{-3} of particle size <500 μm.

Microplastics differ from most other pollutants as they are not metabolised, and they cover a range of different polymers with highly different characteristics, some being bioavailable to mussels like the positively buoyant polyethylene, while the denser polymers (e.g. polyvinylchloride, PVC) are considered not to be bioavailable (Beyer *et al.* 2017). Because of their hydrophobic nature, plastics can adsorb other dangerous organic and inorganic contaminants such as EDCs, pharmaceuticals, PCBs, PAHs, PBDEs, PCPs and heavy metals from the surrounding water (Mrowiec 2017). There is a need to investigate the interaction of MPs with other contaminants, their chemical burden, absorption/desorption kinetics and the transfer of chemicals from plastic to biota (Álvarez-Muñoz *et al.* 2016), a process that is of significant concern (Cole *et al.* 2011).

Microplastics consist of different shapes, as seen for polymers found in wild and farmed mussels, although several studies have also seen a predominance of synthetic fibres in the environment (Li *et al.* 2016; Figure 8.3). Bivalves are of particular interest because their extensive filter-feeding activity exposes them directly to microplastics present in the water column. Several in vivo studies have suggested that microplastics may translocate from the gut cavity to the blood stream, enter cells and cause significant effects on the tissue and on cellular levels in mussels (Browne *et al.* 2007, 2008; Li *et al.* 2015 and references therein).

Microplastic contamination has been reported on a global scale, from the poles to the equator (Thompson 2015 and references therein), contaminating the water surface of the

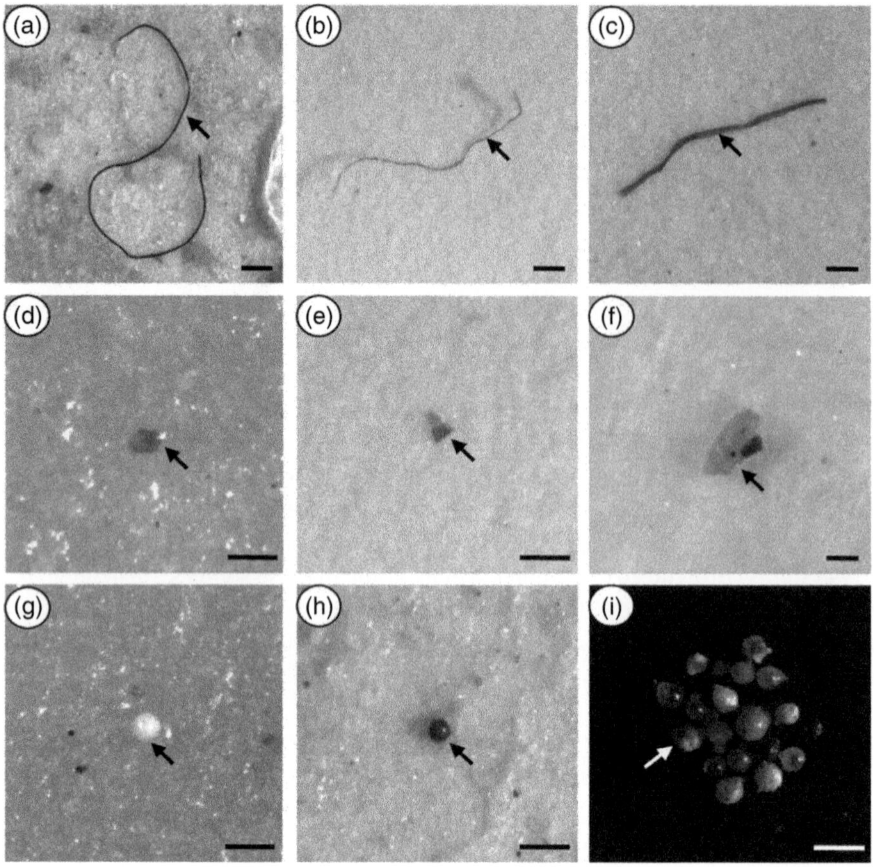

Figure 8.3 Photographs of different types of microplastics in bivalves from a fishery market in China. Arrows indicate (a–c) fibres, (d–f) fragments and (g–i) pellets. Scale bar = 100 μm. *Source:* From Li *et al.* (2015). Reproduced with permission from Elsevier.

open ocean, estuaries, rivers and lakes, together with marine and freshwater shorelines and subtidal sediments down to the deep sea. MPs have also been reported in considerable concentrations in Arctic Sea ice. Suspended in the water column, MPs can become trapped by ocean currents and accumulate in central ocean regions. Recently, Kane *et al.* (2020) have shown that the spatial distribution and ultimate fate of microplastics are strongly controlled by near-bed thermohaline currents (bottom currents). Ocean gyres and convergent zones are noteworthy areas of debris accumulation, as the rotational pattern of currents causes high concentrations of plastics to be captured and moved toward the centre of the region (Lusher 2015). As gyres are present in all of the world's oceans, microplastic accumulation can occur on a global scale, and it has been documented since the 1980s. Distribution is further influenced by wind mixing, affecting the vertical movement of plastic. Estimates suggest that there could be between 7000 and 35 000 tonnes of plastic floating in the open ocean (Cózar *et al.* 2014), although one study estimated that more than 250 000 tonnes – over five trillion pieces of plastic – are currently floating in the oceans (Eriksen *et al.* 2014).

Current microplastic research suffers from insufficient reliable data on concentrations of microplastics in the marine environment and on the composition of involved polymers because standard operation protocols for microplastic sampling and detection are not available (Hildalgo-Ruz *et al.* 2012). In a comprehensive review, Löder & Gerdts (2015) appraised the methodology currently used for assessing the concentration of MPs in water and sediments (beaches and subtidal habitats) with a focus on the most convenient techniques and approaches. Microplastics in sediments or beaches are currently more frequently analysed than MPs in the water column. The authors examined the laboratory methods used to extract, size fractionate and purify MPs from samples. Identification of MPs can be executed by visual examination (with the naked eye or a microscope), but in the case of visual examination without a microscope, attribution of a particle to MPs is usually determined by features such as the source, type, shape, degradation stage, colour, elasticity or hardness of the particles (Hidalgo-Ruz *et al.* 2012). To identify MPs, analytical techniques such as pyrolysis GC-MS, Raman spectroscopy, SEM and some of its variations and Fourier transform infrared spectrometry (FT-IR) are used (details in Löder & Gerdts 2015; Araujo *et al.* 2018; Zobkov & Esiukova 2018; Renner *et al.* 2019).

Microplastics can be consumed by a diverse array of marine organisms, including protists, zooplankton, annelids, echinoderms, cnidaria, amphipods, decapods, isopods, bivalves, cephalopods, barnacles, fish, turtles, birds and cetaceans. In a recent review of the sampling, isolation and identification of microplastics ingested by fish and invertebrates, Lusher *et al.* (2017) highlight the risk of microplastic losses and accumulation during biotic sampling and storage, and suggested protocols for mitigating contamination in the field and laboratory. They evaluated a suite of methods, including dissection, digestion and density separation for extracting microplastics ingested by biota. The most efficient digestive agent is nitric acid (HNO_3), as compared to other oxidising agents such as hydrochloric acid (HCl) and hydrogen peroxide (H_2O_2). Potassium (KOH) is a rapid digestive agent, achieving a 100% recovery rate of MPs in mussel, crab and fish species (Dehaut *et al.* 2016). In contrast to chemical digestion techniques, enzymatic digestion using proteases is a biologically specific way of hydrolysising proteins and breaking down tissues, ensuring no loss, degradation or surface change in the plastics present (Cole *et al.* 2014). Löder & Gerdts (2015) examined the applicability of visual identification and chemical analyses in categorising microplastics, the majority of which are similar to those used for water and sediment samples. For additional information on methods of microplastic detection in seawater samples, see Cole *et al.* (2014); in sediments, see Claessens *et al.* (2013) and Mathalon & Hill (2014); in effluent, see Browne *et al.* (2011); in mussels, see Claessens *et al.* (2013), De Witte *et al.* (2014), Van Cauwenberghe & Janssen (2014), Dehaut *et al.* (2016) and Santana *et al.* (2016); and in combined sample types, see EFSA CONTAM Panel (2016) and Karlsson *et al.* (2017).

Currently, there is no EU legislation for microplastics and nanoplastics (see later) as contaminants in food (EFSA CONTAM Panel 2016). In September 2018, the European Parliament called on the European Commission to introduce an EU-wide ban on intentionally added microplastics in products such as cosmetics and detergents by 2020, and to take measures to minimise the release of microplastics from textiles, tyres, paint and cigarette butts. The following countries have already taken or will be taking action by 2021 toward a ban on the manufacture, sale or import of rinse-off PCPs: Canada, France, Ireland, Italy, India, New Zealand, South Korea, Sweden, Taiwan, Thailand, The Netherlands and the United Kingdom, as well as nine US states. In January 2019, the European Chemicals Agency (ECHA) published a restriction dossier under the REACH Regulation,[8] targeting intentionally added microplastics in products in order to prevent their ending up in the environment. This dossier reviews the available scientific information on the hazards of microplastics, identifies their uses and emissions and tries to assess their risks. It was developed at the request of the European Commission in the context of the Plastics Strategy and covers a variety of sectors, from the cosmetic industry to agriculture and construction. The ECHA Committees will formulate their opinions and send them to the European Commission. It is then up to the European Commission to propose to amend the REACH Regulation, if the restriction meets the legal requirements.

Nanoparticles

Since a large amount of attention is being given to microplastics released in the environment, several researchers have now begun to consider the fragmentation of plastics down to a very small scale (i.e. below 1 mm). The term 'nanoplastics' is still under debate, and some authors have set the upper size limit at 1000 nm while others have set it at 100 nm (Gigault *et al.* 2017 and references therein). A clear definition of what constitutes a nanoplastic has not yet been provided. This is because the term does not consist of a uniform material composition. In their review of nanoplastics in the aquatic environment, Koelmans *et al.* (2015) suggest using the definition used for nonpolymer nanomaterials, implying that a plastic particle is nano-sized if it is <100 nm in at least one of its dimensions. Nanoplastics are polymer-based particles of polyethylene terephthalate (PET) or polystyrene in the nanometre size range. The European Commission (2011c) defines a nanomaterial as a natural, incidental or manufactured material containing particles in an unbound state or as an aggregate or an agglomerate and where, for 50% or more of the particles in the number size distribution, one or more external dimensions is in the size range 1–100 nm.

Nanoparticles (NPs) released to the environment come from different natural and intentional/unintentional anthropogenic activities. NPs are those that are produced in nature, such as via ocean spray, soil erosion, forest fires, dust storms and volcanic ash – all natural activities that release NPs into the environment (Turan *et al.* 2019). The intentional release of NPs is mostly related to the use of engineered nanoparticles (ENPs) for drug production, PCPs, groundwater remediation, biomedical imaging and other applications, whereas the unintentional release is related to activities such as burning fossil fuels, vehicle exhaust, mining and building demolition. ENPs exhibit a specific size ranging from 1–100 nm. They can be classified into different groups based on their properties, shape or size, including the fullerenes (graphene sheets rolled into tubes or spheres), carbon nanotubes (antennas for radios and other electromagnetic devices), metal NPs, ceramic NPs and polymeric NPs. NPs possess unique physical and chemical properties due to their high surface area and nanoscale size (Khan *et al.* 2019). Depending on their composition, size and morphology, some ENPs may be virtually nontoxic, especially if they are contained in coatings, microelectronics or other nanoengineered materials (Sadik 2013). NPs can enter the environment through release during production of raw material and nano-enabled products or during use and after disposal of NP-containing products. Emissions can be either directly into the environment or indirectly via WWTPs or landfills. Indirect emissions likely occur via the effluent of WWTPs, application of biosolids to soil or leachates from landfills (Bundschuh *et al.* 2018). Released NPs are subject to several transformation processes, depending on their intrinsic properties and on the water properties. Physical, chemical and biological transformations are the main processes that will define the behaviour of NPs in water. Physical processes cover homo/heteroaggregation, sedimentation and deposition. Aggregation is an irreversible process that binds particles together through electrostatic forces or strong chemical bonds, leading to a decrease in surface area. There are two different forms: homoaggregation, which occurs between the same kinds of NPs, and heteroaggregation, which takes place between NPs of different types and between NPs and other surrounding particles in the environment. Chemical processes encompass photochemical reactions, dissolution and redox reactions such as oxidation and sulphidation. Dissolution and sulphidation processes have a great impact on NP surface properties, persistence and toxicity. Microbial-mediated biodegradation is the main examples of a biological process. Biodegradation involves the decomposition of an organic substance. This implies that this process is not relevant for inorganic ENPs, although carbon-based ENPs have been observed to undergo biological degradation (details in Hartman *et al.* 2014; Bundschuh *et al.* 2018; Turan *et al.* 2019). Figure 8.4 illustrates the major processes that NPs may undergo in the environment. These transformations differ according to the nature of the NPs and the environmental conditions, such as water chemistry, pH, ionic strength and the presence of natural organic matter (NOM). The interaction of released NPs with NOM will change their surface properties by forming a different natural coating, which affects their fate and behaviour in water.

Among bivalve species, *Mytilus* spp. to date represent the most used model for predicting the impact of ENPs on the health of the marine environment (Rocha *et al.* 2015a; Canesi & Corsi 2016). As filter feeders, mussels remove ENPs from the water column independently of their form (individual particles, homo- and/or heteroaggregates). Their bioavailability

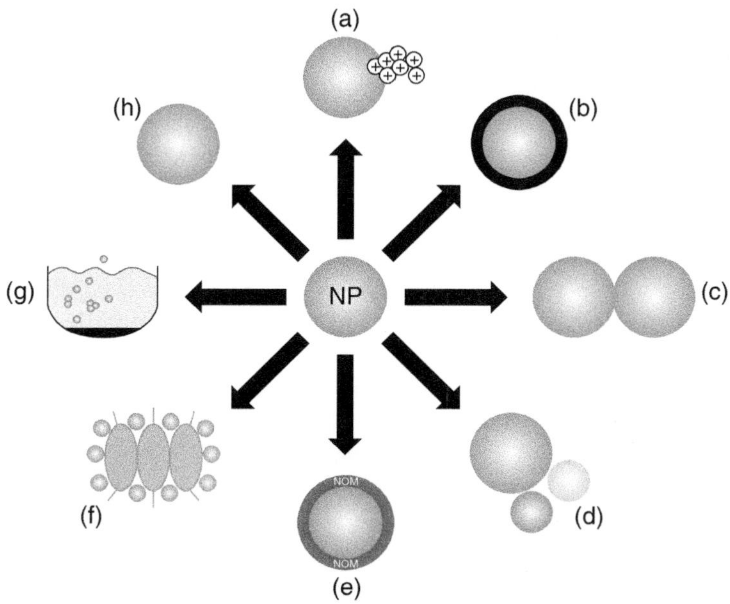

Figure 8.4 Interactions and fate of NPs in the environment: (a) dissolution, (b) sulphidation, (c) homoaggregation, (d) heteroaggregation, (e) coating with natural organic material (NOM), (f) adsorption on biological surfaces, (g) sedimentation/deposition and (h) persistence. *Source:* From Bundschuh *et al.* (2018). CC BY 4.0.

and uptake are dependent on their particular features (e.g. size, shape, surface charge), the type and composition of the dispersing media, the presence of normal organic matter, their behaviour and their fate (Corsi *et al.* 2014). Effects of NPs on mussels will be discussed in more detail later.

Schwaferts *et al.* (2019) recently published a critical review of the techniques that have been applied in sub-microplastic and nanoplastic studies. They include methods for analyte extraction and preconcentration from various environmental matrices; methods for the separation of NPs into specific size fractions; light scattering techniques and various types of microscopy to characterise particle fractions; and chemical identification of NPs to validate the obtained data. For these methods, they discuss the prospects of and limitations on developing analytical protocols for specific sampling scenarios. They have produced a roadmap to illustrate the complete analytical process from sample treatment to particle characterisation (Figure 8.5). The following section is greatly reliant on the information they provide.

The analytical process begins with the clarification of the question that is to be answered (e.g. plastic particle number, size and particle size distribution, or mass of plastic particles per mass or volume of sample), which, in turn, depends on the sample to be analysed. Samples can be from drinking water, food, seawater, sediments, biota tissue or effluent from WWTPs. Methods for the digestion of organic matrices can be an acid treatment, usually 65% nitric acid, sometimes together with 30% hydrogen peroxide and alkaline treatment with sodium hydroxide, or a mild enzymatic protocol with proteinase K used for tissue decomposition. Because the mass of sub-microplastic and nanoplastic particles is probably very low in most types of samples, a preconcentration step is unavoidable. Methods that are used for this process are membrane filtration, ultrafiltration, dialysis, ultracentrifugation and evaporation of solvent. Field-flow fractionation (FFF) is an important analytical method used for the separation and characterisation of particles in nm and μm-size ranges. The technique applies a perpendicular force on particles in flow. Depending on their diffusivity, which is determined by characteristics like density or shape, particles are retained in the flow channel for different durations, which causes a separation of the particles in the sample. Chromatic techniques, such as HPLC, size exclusion chromatography (SEC) and hydrodynamic chromatography (HDC), with their widespread and easy application, offer a notable alternative to FFF. Another separation technique is capillary electrophoresis (CE), which uses the mobility of charged particles in an electric field to achieve a spatial separation.

Dynamic light scattering (DLS) is the most widely used of many methods that apply the scattering of laser light on particles to obtain information on physical properties like size and particle size distribution. DLS measures particle sizes in the range from 1 nm to 3 μm based on the fluctuation of intensity of a laser beam that passes the suspension but does not provide any chemical information and cannot distinguish particles of similar form but different composition. A related method, multi angle light scattering (MALS), records scattered laser light at different angles to obtain information on the size of the particle. Laser diffraction (LD) is another laser scattering-based technique, capable of sizing solid particles in liquid media across a very large size range from 10 nm to 10 mm, while nanoparticle tracking analysis (NTA) records scattered laser light with a microscope and a digital camera. NTA suffers less perturbation from large particles, which makes it an alternative to

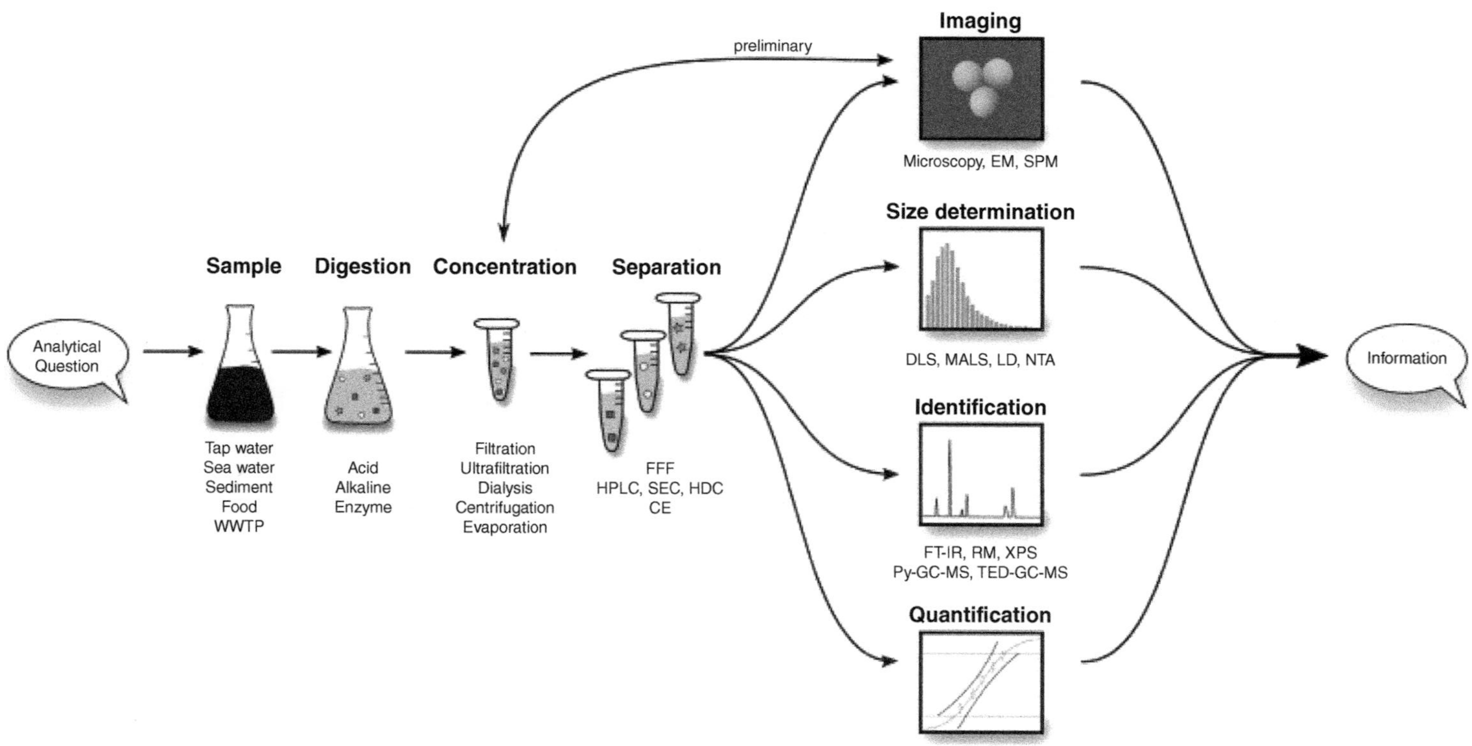

Figure 8.5 Roadmap for the analytical process of sub-microplastic and nanoplastic particles, starting from the sample in its matrix and proceeding through the single steps (digestion, concentration and separation) that lead to a full analysis. *Source:* From Schwaferts *et al.* (2019). Reproduced with permission from Elsevier.

DLS for the determination of particle size distribution of sub-microplastic and nanoplastic particles.

Microscopy is the most workable imaging method because it offers direct access to particle geometry and surface characteristics. Fluorescence microscopy in combination with fluorescently marked particles is frequently used to analyse the behaviour of sub-microparticles and NPs in organisms in order to track the translocation of particles in the tissue. Other microscopy techniques are electron microscopy (EM), scanning electron microscopy (SEM) and scanning probe microscopy (SPM), which forms images of surfaces using a physical probe that scans the specimen.

Chemical identification is of great importance for the analysis of sub-microplastic and nanoplastic particles in environmental samples, because it can provide an additional chemical characterisation of the particle (e.g. information on additive presence and/or ageing). Chemical information on NPs can be measured by FT-IR. Since most molecules can absorb infrared light, their absorption will create molecular fingerprints of the sample and show how the sample absorbs light at each wavelength. This technique is very useful for identifying the side chains, functional groups and crosslinks of the NPs, due to the fact that all of them have characteristic vibrational frequencies in the infrared range. As well as FT-IR, there is Raman microscopy (RM), which employs the inelastic scattering[9] of laser light, giving a vibrational fingerprint spectrum that is complementary to the FT-IR spectrum, and enabling the unambiguous identification of the plastic particles. X-ray photoelectron spectroscopy (XPS) is a surface-sensitive analytical technique that analyses the energy of the photoelectrons emitted when a surface is bombarded with X-rays in a vacuum. The energy of these electrons is characteristic of the atom being bombarded and thus allows identification of elements. Mass spectrometric polymer identification analysis gives information on the mass fraction of a polymer rather than particle count. There are two different methods: pyrolysis gas chromatography mass spectrometry (py-GC-MS) and thermal extraction desorption gas chromatography mass spectrometry (TED-GC-MS). The former is a method of chemical analysis in which the sample is heated to decomposition to produce pyrolysis fragments, which are subsequently thermally desorbed into GC-MS to enable the identification of the polymer. The latter method allows the analysis of plastic in environmental samples without removal of (in)organic matrix.[10]

Additional information on the effects of industrial pollutants (e.g. heavy metals, PAHs, PCBs, pesticides, radionuclides, microplastics and nanoplastics) from a public health perspective is provided in Chapter 12.

Contaminant Levels in Seawater, Sediments and Biota

Bioaccumulation

Bioaccumulation is generally defined as the gradual accumulation of substances, such as contaminants, in an organism. It is a highly dynamic process affected by the flows into (influx) and out of (efflux) the organism. It is crucial to study both these flow types in order to understand the bioaccumulation process. For aquatic organisms, there are several sources of contaminant uptake (e.g. water, suspended particles and sediments) (Wang &

Fisher 1999) and elimination (efflux) due to metabolic processes such as excretion and reproduction, so that concentration (bioaccumulation) of a contaminant is determined by the balance between influx and efflux. Growth is another important factor, since it can act to dilute contaminant concentrations in organisms. In mussels, there are three major mechanisms for the uptake of environmental contaminants (Beyer *et al.* 2017): uptake by passive diffusion from the dissolved phase over external surfaces (generally gills, but also mantle and gut wall); active uptake by transmembrane ion pump transport (gills, gut wall); and active uptake by endocytosis of contaminated particles (predominantly gut wall, but also gill surface). Bioaccumulation directly links environmental chemistry/processes and organism physiology/biochemistry, and thereby can be seen as an interface between environmental chemistry and biology (Wang 2016).

Absorption and Absorption Efficiency

Aqueous intake is critical in the accumulation of organic contaminants by marine bivalves. One of the important parameters in quantifying the trophic transfer of organic contaminants is the absorption efficiency (AE) of the ingested food sources.

For filter-feeding bivalves, absorption can be quantified using the equation:

$$I = k_u \times C_w = \alpha \times FR \times C_w$$

where I is the influx rate (g/h), k_u is the uptake rate constant from the dissolved phase (L/g/h), C_w is the contaminant concentration in the dissolved phase (μg/L), FR is the filtration (or clearance) rate of the bivalve (L/g/h) and α is the AE (%). AE defines the fraction of contaminants that is absorbed across the gut lining of an organism after ingestion. Different methods, including the ratio method, mass balance method and selectivity index method, have been employed to quantify the AE of organic contaminants (Penry 1998).

To illustrate, Wang (2001) compared the uptake rate of four trace metals and AE in the mussels *Perna viridis* and *Septifer virgatus* and the clam *Ruditapes philippinarum*. The bivalves were first measured for their clearance rates and then their metal influx rates, after which the metal α and the uptake rate constant (k_u) were computed by the following equations:

$$\alpha = I / \left(CR \times C_w\right)$$

and:

$$K_u = I / C_w$$

The *ku* values were comparable between the two mussel species but were 1.8- to 3.3-fold lower in the clam species. Interspecific difference in metal k_u was strongly related to clearance rate, but intraspecific difference in k_u was not affected. Among the four metals, the highest k_u and α were found for Zn, followed by Cd, Cr and Se. Interspecific difference in metal α was smaller than in metal k_u and was independent of the clearance rate, whereas intraspecific difference in metal α correlated with individual variations of the clearance rate.

Assimilation and Assimilation Efficiency

In contrast to absorption from the dissolved phase, assimilation refers to uptake from a dietary source (Wang 2016). When food is ingested by mussels, digestion immediately occurs in the stomach. Undigested material is subsequently ejected as faeces, while the remaining digested material is absorbed across the intestinal linings (details in Chapter 4) or lost from the mussel through elimination, respiration or excretion, with the remainder incorporated into tissues, in a process referred to as assimilation. The following equations summarise this energetically:

$$IR = AR + Faeces$$

$$AR = AE + Excretion + Respiration$$

where *IR* is the total ingestion (feeding) rate, *AR* is the absorption rate and *AE* is the assimilation rate. The assimilation efficiency is then calculated as the assimilation rate divided by the total ingestion (Wang 2016).

Wang & Zhang (2013) quantified the aqueous uptake and dietary assimilation of two EDCs (dioxin and phathalic acid) in the green mussel, *Perna viridis*. Over a short exposure period, dioxin rapidly sorbed on to phytoplankton, and its accumulation was much higher than that of phthalate. The dietary assimilation of the two contaminants was moderate – 10–64% for dioxin and 20–47% for phthalate – and depended greatly on food type and concentration. Gut passage time (see Chapter 4) was partially responsible for the variable dietary assimilation. A longer retention in the mussel gut may result in a more efficient assimilation, as suggested by Wang & Fisher (1999). Similarly, Wang & Chow (2002) demonstrated that the differences in benzo[a]pyrene assimilation among different food types are largely attributable to differences in the gut passage of mussels (*Perna viridis*).

Bioconcentration Factor

Bioaccumulation potential can be measured using two factors: bioconcentration factor (BCF) and bioaccumulation factor (BAF). Bioconcentration is the result of direct uptake of a chemical by an organism from water only. Experimentally, the result of such a process is reported as the BCF. Consequently, the BCF is defined as the ratio of steady state[11] concentration of the chemical in aquatic organisms (C) and the corresponding freely dissolved chemical concentration in the surrounding water (C_W):

$$BCF = C\frac{\left[ng\,kg^{-1}\right]}{C_w\left[\,ng\,L^{-1}\right]}$$

The BCF gives quantitative information on the ability of a contaminant to be taken up by organisms from the water. It is not a constant but a variable that depends on different environmental and biological conditions (Wang 2016). BAF is calculated as overall bioaccumulation in the organisms divided by the concentration in the environment, in both dissolved and particulate phases (Świacka *et al.* 2019). Like BCF, BAF is not a constant but is also dependant on the ambient concentration (Wang *et al.* 2016).

Antibiotic bioaccumulation in aquatic organisms is concerning not just from an ecological point of view, but also from a human one, especially when antibiotics accumulate in edible organisms like marine mussels. In a recent study, *M. galloprovincialis* was exposed to sulphamethoxazole antibiotic (SMX) at 10 μg/l for 96 hr, followed by 24 hr depuration (Serra-Compte *et al.* 2019). The experiment was carried out under both summer and winter conditions. SMX concentration in mussels increased to 13.2 ± 0.7 ng/g dry weight after 96 hr of exposure. After 6 hr of depuration, the concentration of SMX in mussels reached its highest value, 16.8 ± 1.1 ng/g dw, whereas at the end of the 24 hr depuration phase, concentration decreased to 4.1 ± 1.1 ng/g (69% of elimination when comparing the concentration at the end of the exposure phase with that at the end of the depuration phase). The highest BCF of 1.5 ± 0.1 l/kg was achieved after 96 hr of exposure. Sulphamethoxazole uptake by mussels followed a logarithmic pattern, with an $R^2 = 0.93$ (Figure 8.6). A stabilisation of the BCF was reached after 24 hr of exposure, which suggests that the steady state was achieved after this period. Although a slight increase in the BCF was observed after 96 hr of exposure, it was not statistically different from the BCF steady state. The low bioconcentration capacity of SMX in mussel soft tissue may be related to its physico-chemical properties, in particular the low octanol–water partition coefficient (K_{ow}); log $K_{ow} = 0.89$ for SMX. Log K_{ow} is a very important parameter for predicting the distribution of a substance in various environmental compartments (water, soil, air, biota, etc.). Substances with high log K_{ow} values (>4.5) tend to adsorb more readily to organic matter in soils or sediments because of their low affinity for water. Hence, the n-octanol–water partition coefficient is often used as a screening test for bioaccumulation. A substance is considered as not bioaccumulative if it has a BCF less than 1000, as bioaccumulative if it has a BCF of 1000–5000 and as very bioaccumulative if it has a BCF greater than 5000 (ChemSafetyPro 2019).

Mytilidae are the most widely used test organisms for bioaccumulation of pharmaceuticals. In a recent study, the bioconcentration and metabolism of the pharmaceutical diclofenac were evaluated in *Mytilus trossulus* under laboratory conditions (Świacka *et al.* 2019). BCF was used to determine the accumulation. BCF (l/kg, dw, ww) was calculated as concentration in tissues (mg/kg)/concentration in water (mg/l). Mussels were

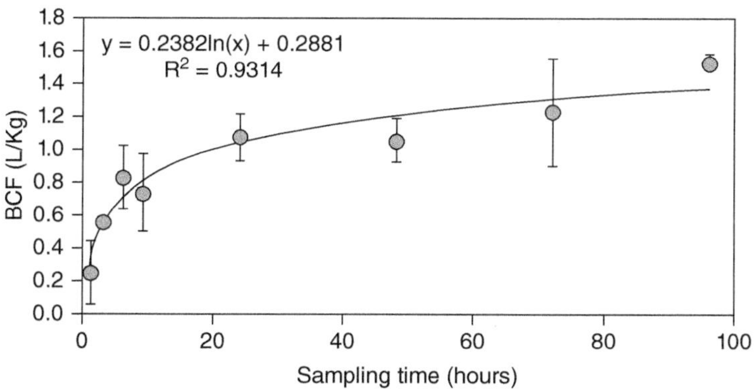

Figure 8.6 Sulphamethoxazole bioconcentration factors (BCFs; l kg^{-1} dry weight) determined in mussel soft tissue, showing mean BCF of summer and winter trials during 96 hr exposure. *Source:* From Serra-Compte *et al.* (2019). Reproduced with permission from Elsevier.

exposed to diclofenac at a concentration of 133.33 μg/l for five days, followed by a five-day depuration phase. The highest concentration of diclofenac (7.79 μg/g dw) in tissues was determined on day 3. Subsequently, the concentration of diclofenac in tissues decreased rapidly to 0.86 μg/g dw on day 5. After five days of depuration, the concentration of diclofenac was 0.21 μg/g dw. Also, two diclofenac metabolites (4-OH and 5OH) were found in mussel tissues and in water – a finding also reported in *M. galloprovincialis* after diclofenac exposure (Bonnefille *et al.* 2017).

In a related study, mussels (*M. trossulus*) were exposed to diclofenac for eight days at concentrations of 1 and 10 000 μg/L (Ericson *et al.* 2010). BCF for a nominal concentration of 1 μg/l was 180 l/kg ww, while that for a concentration of 10 000 μg/l was only 10 l/kg ww. The concentrations detected in mussel tissues were 0.18 ± 0.02 and 82 ± 6 μg/g ww, respectively. The higher concentration of the tested pharmaceutical was possibly a stress factor causing the isolation of the mussels from external conditions by shell closure and a switch to anaerobic metabolism, as a result of which the pharmaceutical was not adsorbed by them. Therefore, the BCF was lower in the group of mussels exposed to higher diclofenac concentration. Whether mussels in both studies would switch to anaerobic metabolism as a result of exposure to diclofenac has yet to be tested. See Geyer *et al.* (2000), Rosen & Lotufo (2007), Klosterhaus *et al.* (2013) and Boillot *et al.* (2015) on additional BCF studies in marine mussels.

Kinetic Modeling in Contaminant Bioaccumulation

Kinetic models have been successfully applied to the study of contaminant bioaccumulation in marine bivalves when measurements are limited to a few important physiological parameters, including influx rate from dissolved phase, assimilation efficiency from ingested food particles and efflux rates (Wang *et al.* 1996). There are several biokinetic models, from a simple one-compartmental model to multi-compartmental ones. The one-compartmental model can be simulated by the following equation (Wang 2016):

$$\frac{dC}{dt} = k_u \times C_w - k_e \times C$$

where k_u is the uptake rate constant from water (l/g/h), C_W is the contaminant concentration in the water (μg/l), k_e is the efflux rate constant and C is the accumulated concentration (μg/g). After integration, C at time t can be described by the following equation:

$$C_t = ku \times Cw / k_e \times \left[1 - e^{-K_e t}\right]$$

Under steady state conditions, $dC/dt = 0$ and the steady state concentration C_{SS} can be calculated as:

$$C_{SS} = k_u \times C_w / k_e$$

When parameters k_u, k_e and C_W are known, it is possible to predict the concentration of contaminants under steady state conditions. Since BCF = C_{SS}/C_W, BCF can also be calculated using the two kinetic parameters:

$$BCF = k_u / k_e$$

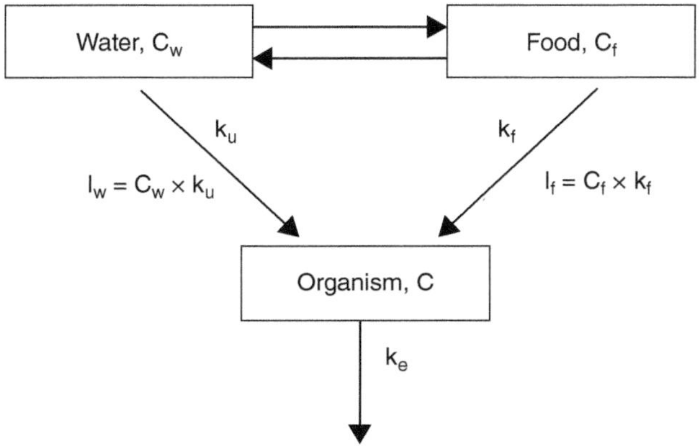

Figure 8.7 Kinetic model simultaneously considering uptake from waterborne and dietborne phases, with different kinetic parameters used to quantify bioaccumulation. *Source:* From Wang (2016). Reproduced with permission from Elsevier.

For many aquatic animals, both dissolved and food sources contribute to contaminant uptake. The simple kinetic model can incorporate the food exposure source (Figure 8.7):

$$\frac{DC}{dt} = \left[k_u \times C_w + k_f \times C_f\right] - \left(k_e \times C\right)$$

where k_f is the uptake rate constant from the food source and C_f is the contaminant concentration in the food. This model can be solved to predict the accumulated concentration at specific time t:

$$C_t = \left[k_u \times C_w + k_f \times C_f\right] / k_e \times \left[1 - e^{-K_e t}\right]$$

Under steady state conditions, $dC/dt = 0$ and the accumulated concentration can be calculated as:

$$C_{ss} = [\, k_u \times C_w + k_f \times C_f \,] / k_e$$

Thomann (1981) further developed kinetic modelling by introducing bioenergetic concepts into the equations:

$$k_u = \alpha \times FR$$

$$k_f = AE \times IR$$

where α is the AE of contaminants from the water, FR is the filtration rate of the organisms, AE is the dietary assimilation efficiency and IR is the ingestion rate of the organisms. The kinetic equation has sometimes been termed a 'bioenergetics-based model' or 'biokinetic/biodynamic model' (Wang *et al.* 1997; Wang 2016). The C_{ss} can be calculated as:

$$C_{ss} = \left[\, \alpha \times FR \times C_w + AE \times IR \times C_f \right] / \left(k_e + g\right)$$

This model, which considers the growth rate constant (g) of the organisms, treats an organism as a single compartment without going into the kinetic processes (e.g. transportation, redistribution, storage and sequestration) inside the organisms. Water is the only source of uptake for marine phytoplankton, in which the growth constant g is much higher than the efflux constant k_e. The concentration of a contaminant in phytoplankton can thus be calculated as:

$$C_{ss} = \frac{(k_u \times C_w)}{g} = I_w / g$$

where I_w is the influx from the water source. This equation is of importance when studying the uptake of contaminants in phytoplankton, which can be directly calculated when C_{ss} and g are known. If trophic transfer[12] is the only source of bioaccumulation, the penultimate equation can be simplified as:

$$C_{ss} = (k_f \times C_f) / k_e$$

The trophic transfer factor (TTF) is the ratio of a substance concentration in an organism's tissue to its concentration in the organism's food item (DeForest *et al.* 2007). It is a general term that may include biomagnification, the process whereby xenobiotic[13] substances are transferred from food to an organism, resulting in higher concentrations as compared with the source (Gray 2002). The TTF or biomagnification factor (BMF) can be calculated as:

$$TTF = BMF = \frac{C_{ss}}{C_f} = k_f / k_e$$

The concept of a TTF or BMF is very similar to that of a BCF. A simple method to quantify TTF, or to examine whether a contaminant has the potential of being biomagnified in the food chain (i.e. whether concentrations of a substance increases through three or more tropic levels via food uptake alone), involves comparing the two kinetic parameters, k_f and k_e. A TTF > 1 necessitates that k_f is greater than k_e (Wang 2016).

Factors Affecting Bioconcentration in Marine Mussels

The degree to which organic and inorganic contaminants are accumulated by marine mussels is determined by abiotic factors such as the physico-chemical properties of the contaminants and their speciation (changes in chemical form), and biotic factors such as pumping activity, growth, biochemical composition, reproductive condition and metabolism/elimination (Widdows & Donkin 1992). Both of these factor types are affected by environmental variables such as temperature, salinity and pH.

Bioconcentration is the process by which a chemical concentration in an aquatic organism exceeds that in water as a result of exposure to a waterborne chemical (see earlier). It invovles the accumulation of a substance due to passage from one phase into another phase. The BCF is the ratio of equilibrium concentrations of specific chemical between two phases. This situation is symbolically represented as:

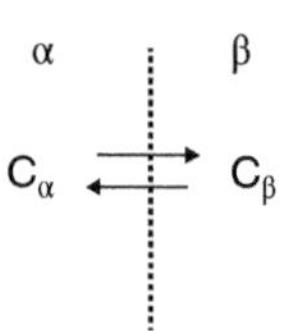

where $C\alpha$ is the concentration in the α phase and $C\beta$ is the concentration in the β phase. The BCF for this system is simply BCF = $C\beta/C\alpha$. BCF can only be measured under controlled laboratory conditions in which dietary intake of the chemical is intentionally omitted. Bioconcentration is dependent on the physico-chemical properties of the tested chemical, such as water solubility and lipophilicity, measured as octanol water partition coefficient (log K_{ow}). The higher the K_{ow} value of a chemical, the higher the bioconcentration potential in a specific aquatic organism, provided that the chemical is not metabolised (Geyer *et al.* 2000). However, there are many other external and internal factors that can influence the BCF value. Some of these are abiotic factors such as water temperature, oxygen content, total organic carbon (TOC) and salinity. Bioconcentration potential is dependent on the age, sex, strains and species of the aquatic organisms. Other important factors which may affect the bioconcentration potential of chemicals in aquatic organisms include toxic effects, bioavailability, concentration of the chemicals in water, pH of the water and, especially, the lipid content of the organisms (Geyer *et al.* 2000). For aquatic organisms, in general, the greater the lipid content, the greater the bioconcentration potential of a chemical. Because under normal conditions the lipid content of aquatic organisms increases with body weight and/or age, the concentration of a chemical and/or the BCF on a wet weight basis (BCFw) under steady state conditions is higher in organisms with higher body weight and/or age. To a great extent, lipid content of an aquatic animal reflects the lipid content of the diet, and changes in lipid consumption will affect the fatty acid composition and lipid content of the animal. Because hydrophobic chemicals must initially pass through the gill, lipids of the gill may be more important in controlling uptake than the lipid content of the entire animal (Barron 1990). During spawning, however, aquatic organisms lose a large amount of lipids. Therefore, during this time, the concentration of chemicals and/or the BCFw decreases in these organisms. In order for an organic chemical to reach lipid storage sites within an aquatic animal, it must be delivered to the absorbing epithelium (e.g. the gill in mussels), move from water through diffusion barriers (e.g. mucus, membranes) into the haemolymph and be transported by the blood to lipoidal tissues. In general, hydrophobic organics will diffuse through cells rather than moving intercellularly or through intracellular channels.

Transport of chemicals into and through biological membranes requires that the compound in the surrounding water be available in a dissolved form. Environmental factors that can reduce the chemical amount in true solution will reduce the uptake rate and bioconcentration and/or bioaccumulation potential. The most important processes that may influence and reduce the bioavailability of hydrophobic chemicals are binding to particulates and dissolved organic matter (DOM; see Chapter 4) and adsorption to humic acids, sediments and other suspended macromolecules. It is also important that formation of colloidal suspensions – especially of very hydrophobic chemicals – can reduce the effective water exposure concentration and its bioavailability (Geyer *et al.* 2000). DOM can form complexes with metals in the water, thereby effectively reducing the bioavailable fraction of metals and reducing their uptake, although DOM–metal complexes may be directly available for filter-feeding bivalves (Wang 2016). However, while Sánchez-Marín *et al.* (2011) found that lead (Pb) complexation by DOM decreased Pb bioavailability in *Mytilus edulis*, other effects of DOM, probably caused by its adsorption on cell membranes, may enhance Pb bioavailability. Which effect will prevail in natural conditions will depend on the physico-chemical characteristics of the natural DOM and the structural and functional organisation of the epithelial interface for different biological species and exposure conditions.

Environmental factors such as salinity, temperature, competing ions (e.g. H^+, Ca^{2+}, Mg^2) DOM and dissolved oxygen considerably affect the bioconcentration of contaminants. Among these, salinity is probably the most studied, especially in estuarine and coastal waters (Wang 2016). Metals are considered as important toxic pollutants, and there is an extensive literature concerning their accumulation in ecosystems (Azizi *et al.* 2018).

Salinity

Metals become more toxic as salt content of the water decreases. Salinity also modifies metal speciation, different metal species inducing different toxic effects (Azizi *et al.* 2016 and references therein). Jackim *et al.* (1977) found that a decrease in salinity from 30 to 20 psu increased cadmium uptake by *M. edulis* by 24–400% at 10 and 20 °C. These observations were confirmed by George *et al.* (1978), who noted that mussels accumulate more Cd in dilute than in 100% seawater. They concluded that this increase is due to the effect of osmolarity of the surrounding medium. Trace metal (Cd, Cr, Se and Zn) uptake from the dissolved phase, assimilation efficiency from the dietary phase and metal body burden, as well as the clearance rate, apparent water permeability and gill surface area, were measured in the green mussel, *Perna viridis*, collected from two contrasting salinity sites in Hong Kong coastal waters, following acclimation in the laboratory at different salinities (Blackmore & Wang 2003). Mussels collected from the high-salinity site accumulated metals 1.2–6.4 faster than those from the low-salinity site when they were acclimated at intermediate and high experimental salinities (>17 psu). Metal assimilation efficiencies were unaffected by salinity variation, but gut passage times were significantly longer at low salinity (10 psu), indicating some effect of lowered salinity on mussel gut physiology. Gill surface area, which was similar in both populations, and filtration rate, which was generally similar or higher in those groups with low metal uptake, could not explain the interpopulation difference in metal accumulation. Alternatively, an apparent water permeability of the high-salinity population was on average 1.6 times greater compared with that of the low-salinity mussels, which may partially account for the difference in metal uptake in the two populations. Thus, the effect of salinity on metal uptake is dependent on metal biogeochemistry as well as a range of physiological responses. There was a clear effect of acclimation to lowered salinity on metal uptake, which may have important implications for metal accumulation modelling, biomonitoring and toxicity studies.

Temperature

The role of temperature on metal uptake and accumulation in marine mussels has not been well studied, which is surprising since mussels are important biomonitoring organisms. However, uptake and accumulation of Cu, Co, Cd and Pb were investigated (Mubiana & Blust 2007) in *M. edulis* at different temperatures (6–26 °C). Results from exposure of isolated gills showed a positive relationship between temperature and metal uptake. However, in whole-organism experiments, only the accumulations of nonessential metals (Cd, Pb) showed a similar trend, while the two essential metals Co and Cu were independent of and inversely related to temperature, respectively. With the exception of Cu, the elimination process appeared to be independent of temperature. The study also showed that neither changes in the scope for growth (SFG) of mussels nor chemical speciation could fully account for the observed temperature effects. Overall, the results suggested that fundamentally (i.e. at epithelial membranes), temperature effects on uptake are largely due to changes in solution chemistry and physical kinetics, which favour higher uptake at high temperature. At the whole-organism level, complex physiological responses appeared to mask the relationship, particularly for biologically essential metals like copper.

The effect of temperature on uptake and retention of dissolved metal by invertebrates has been investigated mainly from temperate zones (Hutchins *et al.* 1996, 1998), but Baines *et al.* (2006) used radioisotopes to determine the uptake and excretion of Ag, Am, Cd, Co and Zn by *M. edulis* collected from both Arctic and temperate waters at 2 and 12 °C – near the extremes experienced in Arctic environments. For both mussel populations, uptake rates (k_u) of Cd, Co and Zn were 50–60% slower at 2 than at 12 °C, mainly attributable to differences in mussel filtration rates at these temperatures, whereas uptake of Ag and Am was not affected by temperature. For both populations, the first-order loss constant for all metals from the slowest exchanging pool (k_e) was 20–50% smaller at 2 than at 12 °C. Also, the loss of Cd and Zn was significantly slower in temperate than in Arctic mussels. Because temperature had similar effects on uptake and efflux, the BCF for all metals from water ($BCF_W = k_u/ k_e$) did not vary significantly between temperatures, but BCF_W values for Cd and Zn in temperate mussels were 15- and 2.6-fold higher, respectively, compared with Arctic mussels. Hence, while temperature did not exert a major influence on net bioaccumulation of aqueous metal by *M. edulis*, Arctic mussels may accumulate dissolved Cd and Zn to a lesser degree than temperate mussels.

ECs such as FRs and PFCs have not been recognised as pollutants or toxicological hazards, but they are becoming more frequent in the marine environment. Moreover, the effects of environmental conditions (e.g. temperature and pH) on the bioaccumulation and elimination mechanisms of these emerging contaminants in marine biota have been poorly studied up to recent times. In this context, Maulvault *et al.* (2018) assessed, for the first time, the effect of warmer seawater temperatures ($\Delta = +4$ °C) and lower pH levels ($\Delta = -0.4$ pH units), acting alone or in combination, on the bioaccumulation and elimination of emerging FRs (dechloranes) and TBBPA, a BFR, inorganic arsenic (iAs) and PFCs (organic substances in which all of the hydrogens of the hydrocarbon backbones are substituted with fluorine atoms) in two estuarine bivalves, *M. galloprovincialis* and the clam *Ruditapes philippinarum*. Overall, the results showed that warming alone or in combination with acidification promoted the bioaccumulation of some compounds (i.e. dechloranes) but facilitated the elimination of others (i.e. TBBPA). Similarly, lower pH also resulted in higher levels of dechloranes, as well as enhanced iAs and PFC elimination. Data suggest that when both abiotic stressors were combined, bivalves' capacity to accumulate contaminants was time-dependent, in view of a significant increase observed with dechloranes and TBBPA during the last 10 days of exposure when compared to reference conditions. Such changes in contaminants' bioaccumulation/elimination patterns also suggest a potential increase of human health risks from some compounds if the climate continues changing as forecasted. For a complementary study, see Serra-Compte *et al.* (2018).

Other abiotic factors that influence bioaccumulation are co-occurrence of metals and pre-exposure to metals (Azizi *et al.* 2018 and references therein). Biotic factors include age and body size, reproductive cycle and sex, location within the intertidal zone and time of year of sample collection (reviewed in Azizi *et al.* 2018).

Age and Body Size

The age of the organism is an important factor affecting the metal accumulation capability of sentinel species like mussels. The youngest stages, such as embryos, accumulate a lot, while mature individuals accumulate relatively little. This could be due to the higher surface to volume ratio of embryos compared to mature mussels (Ansari *et al.* 2004). Also, Yap *et al.* (2009) found that large and aged mussels tend to pump less water through their bodies per unit of body weight, and hence have a lower uptake of metals compared with smaller individuals. Therefore, the factor of body size should be taken into consideration in designing experiments in monitoring programmes. Boyden (1974) reported relationships between organism size, metal concentration and total metal content for several species of bivalves, and found that Mn, Fe, Ni, Cu, Zn and Pb were more concentrated in smaller *M. edulis*, while Cd concentrations were size-independent. Similarly, Jones & Walker (1979) found that concentrations of Fe, Mn, Zn and Cd in the whole soft body parts of freshwater mussels decreased with increasing body weight. See also Riget *et al.* (1996) and Mubiana *et al.* (2006).

Reproductive Cycle and Sex

Effects of the reproductive cycle have often been indicated as an explanation for the variations in metal content in mussels. The fact that during spawning up to 40% of soft tissue weight is lost shows the importance of gametogenesis in the physiology of the genus *Mytilus* (Cossa 1989). Body weight changes seasonally according to the relative rate of somatic or gonadal growth at different times of the year (highest values in summer and smallest in winter). Body weight decreases sharply to a minimum in spring–summer, coincident with spawning, and increase rapidly in summer as mussels build up reserves to use in the following autumn and winter for gonad development. Therefore, the unpredictability of the extent of gonad development also makes it more difficult to use adult animals as indicators in pollution monitoring programmes. For this reason, using immature specimens to reduce the influence of metabolism on the metal content gives a more precise indication of pollution levels. Regarding the sex of mussels, Watling & Watling (1976) separated the soft tissues of black mussels, *Choromytilus meridionalis*, and found that Mn, Cu and Zn were more concentrated in females but that Pb and Bi were more concentrated in males, with no significant differences in the concentrations of Cd, Fe, Co, Ag, Ni and Cr. However, in a subsequent study, Orren *et al.* (1980) analysed soft tissues of the same species on two occasions and found that Mn, Fe, Cu and Zn were more concentrated in females only before a major spawning event; no significant differences were found after spawning had occurred.

Location within the Intertidal Zone

This is a factor with great potential to cause high variability in the contents of heavy metals in individuals within a mussel bed (Azzi *et al.* 2018). Despite this, there are few studies on the influence of tidal exposure in metal uptake in *Mytilus* spp. De Wolf (1975) observed that concentrations of Hg in *M. edulis* and *M. galloprovincialis* were in general higher for individuals in the intertidal zone than for those continually immersed at the same site. Bourget & Cossa (1976) found no clear difference in Hg concentration in *M. edulis* in the intertidal zone. Cultured mussels (*Perna canaliculus*) grown on vertically suspended ropes to a depth of 9 m from two widely separated locations in New Zealand were analysed for variation in heavy metal (Zn, Cd, Pb and Fe) content at 1 m intervals (Nielsen 1974). At one location, Cd, Pb and Fe increased with depth, while Zn decreased with depth. At the other, concentrations of the four metals all remained essentially constant with depth. The differences in vertical concentration gradients between the two locations may be a result of differences in mixing of the water column, which could cause variations in the type of food organisms or in the ratio of particulate versus dissolved metal levels with depth. Either or both of these conditions could lead to differences in the bioavailability of these metals.

Time of Year of Sample Collection

There are a number of studies on the structure of seasonal variations in metal concentrations. Pentreath (1973) reported seasonal differences in total concentrations and tissue distributions of nonradioactive Zn, Fe and Mn in *M. edulis*. Bryan (1976) found that the concentrations of Zn, Pb, Cu, Co, Fe, Mn and Ni in mussel tissues were greatest in autumn and winter, and suggested that metal concentrations were inversely related to phytoplankton productivity. Rouane-Hacene *et al.* (2015) found enrichment in metals (Zn, Cu, Cd) in *M. galloprovincialis* tissues collected in winter, which may have resulted from inputs from urban and industrial activities. Strong Pb accumulation in summer could be related to an increase in urban populations during this period, leading to higher urban wastewater discharges and atmospheric Pb levels due to more intense traffic. In the latter case, contamination could be both direct from atmospheric deposition and indirect by the leaching of roads by rainwaters. Seasonal changes in the various metal concentrations in mussel tissues may have resulted from a combination of factors directly correlated with the weight of the animals (cycle of reproduction, metabolism, development and age, food availability, temperature) and with other more independent factors, such as salinity, modification of biogeochemical cycle and bioavailability of metals related to the sites sampled in the study. The physico-chemical characteristics of water, which vary according to seasons, modulate the bioavailability of pollutants and therefore influence both their bioaccumulation and the biological responses of organisms to these pollutants. Giarratano & Amin (2010) reported that heavy metals (Cu, Zn, Fe, Cd) in *M. chilensis* showed a seasonal trend of bioaccumulation, being higher in winter than in summer. This is probably due to slower growth at cooler temperatures, when the availability of food is lower. Reserves are used for metabolism, leading to a decreasing tissue mass and thereby increasing the metal content ratio.

Biomonitoring

Over the last few decades, various biomonitoring strategies have been developed to monitor and evaluate the adverse impact of contaminants on marine and estuarine ecosystems. Monitoring programmes (see later) use a large number of bioindicators, also known as sentinel organisms, to detect temporal and spatial variation of chemical pollutants, ultimately contributing to acquisition of knowledge on trends in marine contamination (Azizi *et al.* 2018).

Bioindicators

Gerhardt *et al.* (2002) defined a bioindicator as 'a species or group of species that readily reflects the abiotic or biotic state of an environment, represents the impact of environmental change on a habitat, community or ecosystem, or is indicative of the diversity of a subset of taxa or the whole diversity within an area'. The widespread development and application of bioindicators has occurred primarily since the 1960s. Over the years, the repertoire of bioindicators used for environmental monitoring in marine areas has come to encompass major taxonomic groups such as bacteria, phytoplankton, algae, sponges, Cnidaria, Annelida, Bivalvia, Gastropoda, Amphipoda, fish and even birds (reviewed in Parmer *et al.* 2016; Vallaeys *et al.* 2017). Bioindicators are useful when an indicated environmental factor cannot be measured, such as in situations where environmental factors in the past are reconstructed (e.g. climatic change, palaeo-biomonitoring), where the indicated factor is difficult to measure (e.g. pesticides and their residues, complex toxic effluents containing several interacting chemicals) or where the environmental factor is easy to measure but difficult to interpret (e.g. whether observed changes have ecological significance) (Gerhardt *et al.* 2002). Rees *et al.* (2008) have suggested that the ideal indicator should be:

1) capable of conveying information that is responsive and meaningful to decision making (directly tied to management questions and linked to thresholds for appropriate action relative to designated ecosystem goals);
2) linked to a conceptual stressor–response framework (with the ability to communicate potential cause–effect relationships);
3) capable of measuring change or its absence with confidence (robust to influences of confounding environmental factors);
4) highly sensitive and anticipatory (early warning of potential problems);
5) applicable over a variety of spatial scales and conditions (to support global as well as local comparisons);
6) desirable operationally (easy to measure, reproducible with minimum measurement error, cost-effective);
7) integrative (serves multiple indicator purposes);
8) nondestructive (measurement does not cause ecosystem damage);
9) easy to understand and communicate (nonspecialists need to act on findings);
10) scientifically and legally defensible (robust to peer review and wider challenge).

As already mentioned, the use of *Mytilus* spp. as marine environmental bioindicators arises from their biological and ecological characteristics, which make them virtually ideal for pollution monitoring – although Ward *et al.* (2019) have recently demonstrated that bivalves (*M. edulis* and *Crassostrea virginica*) are not robust bioindicators of microplastic pollution.

A large number of monitoring studies have been carried out using several marine mussel species as bioindicators for multiple contaminants, or looking at the effects of contaminants on mussels under laboratory conditions (Table 8.3). Because of their worldwide distribution, it is practical to analyse wild populations, but the adoption of an active biomonitoring alternative using cage experiments with transplantation has gained popularity in ecotoxicology research and monitoring. The advantages compared to the analysis

Table 8.3 Studies (2011–2019) on marine mussels as bioindicators of contaminants, and their effects on mussels. *Source:* Adapted from Strehse & Maser (2020). C, W and P refer to caged, wild and transplanted mussels, respectively. Reproduced with permission from Elsevier.

Species	Analysed contaminants	Effects	Reference
Mytilus galloprovincialis C	Metals	CI	Benedicto *et al.* (2011)
M. *galloprovincialis* W	Metals		Jović *et al.* (2011)
Perna viridis W	Metals		Putri *et al.* (2012)
Mytilus trossulus C	Metals, PAHs, PCBs, OCPs, PBDEs	CI, GST, LMS	Turja *et al.* (2014)
M. *galloprovincialis* C	Metals, PAHs, PCBs, OCPs, FRs	MT, CAT, GSH	Regoli *et al.* (2014)
M. *galloprovincialis* W	OCPs		Carro *et al.* (2014)
P. viridis W	Metals		Shue *et al.* (2014)
P. perna T	Metals		Greenfields *et al.* (2014)
Brachidontes rodriguezzii W	PAHs		Oliva *et al.* (2015)
M.edulis W	Munitions		Ballentine *et al.* (2015)
M. galloprovincialis W	PhACs, EDCs		Álvarez-Muñoz *et al.* (2015a)
M. galloprovincialis W	Polonium-210, Selenium		Kristan *et al.* (2015)
P. canaliculus W	Metals	CAT, MTLP	Chandurvelan *et al.* (2015)
M. galloprovincialis C	Metals, HCs, PAHs	Alteration of free fatty acids (FFAs)	Albergamo *et al.* (2016)
Mytilus spp.T	Metals	CI, GSI	Helmholz *et al.* (2016)
Brachidontes rodriguezii W	PCBs, PAHs	GST, Pe, PO	Laitano & Fernández-Gimenez (2016)
M. galloprovincialis W	PCBs		Carro *et al.* (2016)
M. galloprovincialis W	Metals, PAHs	DNA damage	Kacar *et al.* (2016)
P. perna W	Microplastics		Santana *et al.* (2016)
P. perna W	Metals	GSH, MT	Lavradus *et al.* (2016a,b)
M. galloprovincialis W		CAT, GST, LMS	Balbi *et al.* (2017)
M. galloprovincialis W	Metals, PCBs, PAHs	CI, CAT, GST	Benali *et al.* (2017)
M. galloprovincialis W	Metals		Bilgin & Uluturhan-Suzer(2017)
M.edulis and *M. galloprovincialis* W	Diclofenac (anti-inflammatory drug)		Cunha *et al.* (2017)
M. galloprovincialis W	Munitions		Strehse *et al.* (2017)

(*Continued*)

Table 8.3 (Continued)

Species	Analysed contaminants	Effects	Reference
M. galloprovincialis W	Metals, FRs, PFCs		Maulvault *et al.* (2018)
Aulacomya atra T	Metals	ROS, SOD,TOSC, GST, GSH	Ruiz *et al.* (2018)

CAT, catalase; CI, condition index; EDCs, endocrine disrupting compounds; FRs, flame retardants; GSH, glutathione; GSI, gonadosomatic index; GST, glutathione s-transferase; HCs, hydrocarbons; LMS, lysosomal membrane stability; MT, metallothionein; MTLP, metallothionein-like protein; OCPs, organochlorine pesticides; PAHs, polycyclic aromatic hydrocarbons; PBDEs, polybrominated diethyl ethers; PCBs, polychlorinated biphenyls; Pe, peroxidase; PFCs, perfluorinated compounds; PhAcs, pharmaceutical active compounds; PO, phenyloxidase; ROS, Reactive oxygen species; SOD, superoxide dismutase; TOSC, total oxyradical scavenging capacity.

of wild populations include the fact that caged mussels can be strategically deployed, physical and chemical gradients can be accounted, as well as site-specific differences, temporal and spatial variability and short- and long-term trends can be elucidated and suspected sources of contamination can be studied where resident bivalves would normally not be found (Strehse & Maser 2020).

To illustrate, *M. galloprovincialis* were collected from a reference site and transplanted to 15 stations in coastal areas around Ulsan and Onsan Bays, an extensively metal-polluted area, to assess metal contamination in the coastal oceans of Korea (Kim & Choi 2017). During the 21-day biomonitoring period, transplanted mussels in mesh cages, seawater and particulate materials were collected for metal analysis. The transplanted mussels showed metal enrichment compared to initial concentrations and spatial gradients consistent with dissolved and/or particulate metal concentrations in seawaters. Stations were clustered into three groups: one in Onsan Bay, which showed high Ag, Cd, Cu, Hg, Pb, Sb and Zn enrichment, presumably arising from nonferrous metal refineries and chemical industries in this area; one near the mouth of the Oehwang River, which was enriched in Co from petrochemical industries; and one at a site intermediate between the first two groups – an isolated station with independent metal sources located in Jangsaengpo Harbour, where a number of ship repair and building companies operate. Metal accumulation rates (% day^{-1}) in mussels were estimated to be between 8 (Cr) and 281% (Pb), based on accumulated metal concentrations over the 21 days. This study showed that transplantation is a useful monitoring method because it reflects the present status of seawaters. Furthermore, physiological factors such as size, growth rate and reproduction can be standardised, and bioavailable and time-integrated metal concentrations can be obtained. For similar studies, see Capello *et al.* (2013), Ng *et al.* (2013), Greenfield *et al.* (2014), de los Ríos *et al.* (2016), Helmholz *et al.* (2016), Ruiz *et al.* (2018) and Lehtonen *et al.* (2019).

Contaminant bioaccumulation was studied in blue mussels (*Mytilus* spp.) using the harbour waters of Kristiansand, Norway as a case study (Schøyen *et al.* 2017). A suite of trace metals, PAHs and PCBs were analysed in both caged and native mussels, as well as in passive samplers[14] (diffusive gradients in thin films (DGT) devices and silicone rubbers) placed alongside the mussels to allow estimation of contaminant concentrations in water and uptake rates and BAFs in mussels during a six-month deployment period. Estimated logBAFs were in the ranges 2.3–5.5, 3.8–5.2 and 3.2–4.4 for metals, PCBs and PAHs, respectively. Contaminant levels in caged mussels increased rapidly to stable levels for trace metals, whereas for the hydrophobic organic contaminants the increase was steady but slow, and for many compounds it did not reach the levels observed in native mussels. Schøyen *et al.* (2017) discuss some key issues related to mussel caging design, such as mussel deployment time and confounding influence from seasonal fluctuations.

Despite the widespread use of living organisms as indicators of metal bioaccumulation, they have some disadvantages. The kinetics of metal uptake and elimination by these organisms are sometimes not well understood and are affected by factors such as temperature, life stage (e.g. larval versus adult), size, depth and reproduction. As different indicator species have different natural distribution patterns, they are not always present in all ecosystems that need to be monitored, preventing comparison between different sites. To overcome these disadvantages, passive sampling devices such as the artificial mussel (AM) have been utilised. The advantages of using AMs to monitor metals are the provision of time-integrated estimates of metal concentrations within aquatic ecosystems, the possibility of monitoring water bodies with unfavourable conditions for living indicator species and the avoidance of organism death (Labuschagne *et al.* 2020 and references therein). An AM device consists of a nonpermeable Perspex tubing (length 6 mm, diameter 2.5 mm) containing 200 mg of the polymer-ligand Chelex-100 resin (50–100 mesh) beads suspended in artificial seawater (Figure 8.8). Both ends of the plastic tube are capped by a layer of polyacrylamide gel (thickness 1 cm) followed by a perforated plastic cap, to protect the gel from possible mechanical damages (Wu *et al.* 2007). The gel allows slow passage of metal ions through its pores before chelating with the resin (Shen *et al.* 2020). Researchers place AMs in baskets and lower them into water. The water enters through openings at the end of the tube. A gel inside traps trace metals, which accumulate in it in much the same way they would inside a living mussel. AMs are not affected by biotic or abiotic factors, and thus can be deployed in both marine and freshwater ecosystems. As part of the Global AM Watch Programme, AMs have been applied in several countries, including Australia, Bangladesh, Brazil, China, Iceland, the Philippines, Portugal, Russia, Scotland, South Korea, South Africa and the United States (Hossain *et al.* 2015).

A comprehensive study was carried out to assess metal contamination in five cities spanning temperate to tropical environments with different hydrographical conditions along the coast of China (Degger *et al.* 2016). At each city, AMs were deployed together with a native species (*P. viridis*, *M. edulis* or, most likely, *M. galloprovincialis*; see Chapter 9) at a control site and a polluted site. High levels of Cr, Cu and Hg were found in Qingdao, high levels of Cd, Hg and Pb in Shanghai and a high level of Zn in Dalian. Levels of Cu contamination in all five cities were consistently much higher than those reported in similar studies in other countries (e.g. Australia, Portugal, Scotland, Iceland, Korea, South Africa and Bangladesh). Levels of individual metal species in AMs showed a highly significant correlation with those in native mussels, except for Zn in *M. galloprovincialis* and Cd in *P. viridis*, while no significant difference was found between the regression relationships of metals in AMs and in each of the two native mussels. The results demonstrate that AMs can provide a reliable time-integrated estimate of metal concentration in contrasting environments over large biogeographic areas with different hydrographic conditions, and can overcome the shortcomings of metal monitoring in water and sediment and the use of bioindicators.

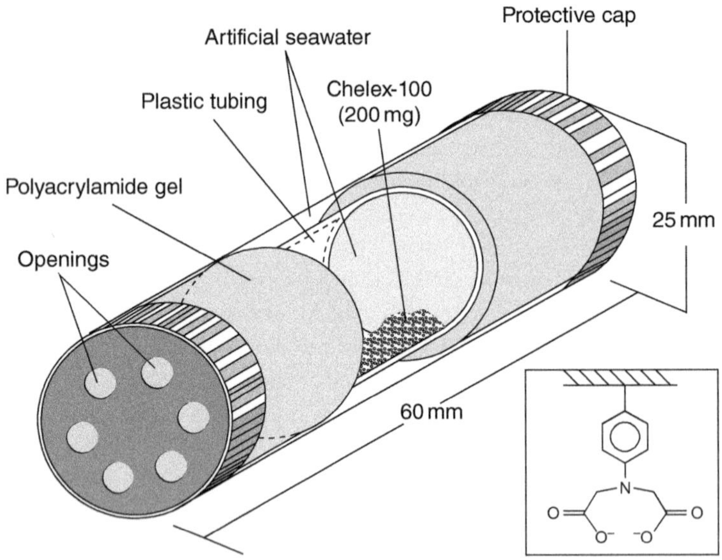

Figure 8.8 Schematic diagram showing the design of an artificial mussel (AM). The substrate chelex-100 resin (inset) has a high affinity for metal ions. See text for details. *Source:* From Wu *et al.* (2007). Reproduced with permission from Elsevier.

Additional noteworthy advantages of AMs include the following: uptake and release of metals by AMs generally resembles that by live mussels under both laboratory and field conditions; the devices are simple, reliable, easy to deploy and cost-effective; and the devices are less affected by physical factors and hydrographic conditions than real mussels, are not limited by geographical location and are not affected by pollution levels. Hence, AMs can provide worldwide comparisons of metal concentrations, disregarding differences in pollution levels and physical and hydrographic conditions (Leung *et al.* 2008). Comparable studies to that of Degger *et al.* (2016) are Wu *et al.* (2007), Gonzalez-Rey *et al.* (2011), Ra *et al.* (2014), Genç *et al.* (2018) and Shen *et al.* (2020).

Mussel Monitoring Programmes

The Mussel Watch Program

Mussel Watch is the longest-running continuous chemical contaminant monitoring programme in US coastal and Great Lakes waters, created in response to concerns over the environmental quality of the nation's coastal and estuarine ecosystems (NOAA 2019). The first known mussel surveys were performed during the late 1960s and early 1970s to monitor spatial and temporal trends of contaminants in coastal and estuarine ecosystems (Goldberg 1975). From 1986, Mussel Watch analysed chemical and biological contaminant trends in sediment and bivalve tissue collected at over 300 coastal monitoring sites. In 2008, the activity had extended to include the annual monitoring of approximately 140 prioritised pollutants in *Mytilus* spp., oyster (*Crassostrea virginica*) and zebra mussel (*Dreissena* sp., in freshwater) sentinels. Parameters monitored include over 150 organic and inorganic contaminants in sediment and bivalve tissue. Some of the chemicals regularly quantified are PAHs, PCBs, chlorinated pesticides including DDT and its metabolites, tributyltin (TBT) and its metabolites and major and trace elements. In addition, new contaminants of emerging concern (CECs) such as flame-retardant polybrominated biphenyl ethers (PBDEs) and pharmaceuticals are being analysed. These have been chosen based on their bioavailability, possible uptake and storage in animal tissues, toxicity to aquatic life and potential harm to humans. Mussel Watch activities in North America have motivated the initiation of similar systematic programmes (see later) in many other coastal countries throughout the world (Beyer *et al.* 2017).

The extensive influence over the past 40 years of Professor Edward D. Goldberg and his call in 1975 for a 'Mussel Watch' concept has been chronicled by Farrington *et al.* (2016), who highlight the concept's utility and its limitations. Between 1985 and 1986, the Mussel Watch component of the NOAA's National Status and Trends (NS&T) Program (www.coastalscience.noaa.gov) was launched. NS&T comprises three US programmes: Benthic Surveillance, Mussel Watch and Bioeffects, designed to describe the current status of – and detect changes in – the environmental quality of the nation's estuarine and coastal waters through environmental monitoring, assessment and related research. In addition, NS&T has also completed special studies designed to help assess the environmental impacts of various events. Benthic Surveillance was discontinued in 1993, but the other programmes are still active.

Mussel Watch activities in North America have motivated the initiation of similar systematic programmes in many other coastal countries throughout the world, such as the MED POL Biomonitoring Programme and the Joint Assessment and Monitoring Programme (JAMP) of the OSPAR Convention, often organised or supervised by the UN-supported International Mussel Watch (Beyer *et al.* 2017 and references therein). The Asia-Pacific Mussel Watch (APMW) program, started in 1994, is under the umbrella of the International Mussel Watch – Asia Pacific Phase, a project that mainly involves monitoring of coastal waters of countries such as Cambodia, China, Hong Kong, India, Indonesia, Japan, Korea, Malaysia, the Philippines, Far East Russia, Singapore and Vietnam (Monirith *et al.* 2003).

Mussel watch programmes generate spatial and temporal trend data for the locations and regions monitored, showing for instance whether there are significant site differences or whether pollution levels are stable, increasing or decreasing (Beyer *et al.* 2017). Further integration of data from several programmes may generate trend pictures that are representative of a broader regional or even a global scale. Some mussel monitoring studies describe contaminant concentration data and trends from remote and pristine reference locations. Such baseline data are valuable, as the information is helpful in other mussel monitoring programmes in discriminating between noncontaminated and low-contaminated locations in order to assess inputs from long-range pollution transport and evaluate EQSs (see earlier).

According to Beyer *et al.* (2017), no global integrated standard exists concerning how to perform mussel monitoring studies, although several key agencies such as UNESCO (Jernelov 1996), NOAA (Lanksbury & West 2012), ICES (Davies & Vethaak 2012), OSPARCOM (1999–2002) and the European Commission (2014) have published comprehensive instructions. Despite this, there is agreement that good comparability of monitoring data is important, for a variety of reasons (e.g. data sharing and comparisons, international monitoring collaboration, quality assurance, quality standards, quality improvement and training and competence development).

Chemical contaminant analysis in mussel watch programmes is normally performed using the whole soft tissue of the animal (i.e. the mussel without the shell), and most often using composite samples (i.e. samples in which a certain number of individual mussels within a certain size range prior to analysis are mixed into a single pool). The use of pooled mussel samples is rational for several practical reasons (especially reduction of analysis cost), although studies of toxicant distribution patterns show that different biological compartments (gills, mantle, plasma, digestive gland, gonads, muscle and other viscera) may contain variable concentrations of contaminants (Rocha *et al.* 2015b; Ricciardi *et al.* 2016; Kolandhasamy *et al.* 2018; Yap 2018) and a seemingly homogenous group of mussels collected from a field population may contain individuals that are different with respect to relevant biological factors such as taxonomy, gender, dietary condition and spawning status. Furthermore, as different mussel watch programmes may vary considerably in content, duration and other design-oriented factors, there is a question as to whether different mussel watch programmes are comparable with one another. To test this, Cantillo (1998) compared heavy metal concentration data from the World Mussel Watch (WMW) database with results from two long-term mussel watch programmes, one from France, the other the NS&T Program in the United States. They found that there was good agreement overall for medians among all three data sets, but the upper ends of the WMW data were inclined to be higher compared to their US and French counterparts. This difference probably reflects the fact that the latter two programmes emphasise collection of bivalves at representative sites rather than within small areas of extreme contamination, such as near waste discharges. This illustrates that technical differences at all levels of the monitoring activity (field work, sample procession, sample analyses, data analysis) may add to variance and hence hamper the comparability of mussel monitoring studies. To achieve better comparability, an integration of monitoring procedures and/or the use of standard materials should be mandatory. In the following section, several key monitoring programmes will be described.

Mussel Watch Monitoring on the Californian Coast

In 1977, California initiated its state-wide Mussel Watch Program to provide an indication of spatial and inter-annual trends in toxic contaminants, principally heavy metals, synthetic organic compounds and pesticides (Melwani *et al.* 2014 and references therein). The California State Mussel Watch (CSMW) conducted annual monitoring of contaminants in resident and transplanted mussels until 2003 and yielded a wealth of useful information, including evidence of rapid improvements in water quality resulting from prohibition of PCBs, several organochlorine pesticides, TBT hull coatings and reductions in metals due to air pollution regulation and wastewater treatment. However, the CSMW was mostly discontinued due to a lack of funding, although 20 sites have continued to be monitored with limited funding since 2003.

The most extensive Mussel Watch Program in California has been the National Mussel Watch (NMW), part of NOAA's National Status and Trends Monitoring Program. The programme began in 1986 and is one of the longest-running continuous coastal monitoring programmes in the world. It initially performed chemical analyses of resident oysters and mussels that spanned 145 stations. This number has steadily grown to 300 currently monitored stations throughout the United States. In California, 40 stations were sampled between 1986 and 2005, with consistent sampling performed at 35 sites. Sampling was conducted annually in the early years of the programme, but was reduced to a biennial frequency in 1994. The majority of NMW sites were in different locations than those of CSMW. The objective of combining CSMW and NMW data in California was to describe spatial and temporal trends in mussel (*M. californianus*) bioaccumulation. Long-term trends were examined in 10 contaminants at more than 50 stations.

Along the California coastline, 24 CSMW stations have been consistently monitored for approximately 120 individual contaminants in *M. californianus*. Stations sampled for 10 years or more were included in the Melwani *et al.* (2014) study; 10 were located north of Point Conception (Northern California) and 14 in Southern California. Resident mussels were sampled at about one-third of the stations, and transplanted mussels collected from relatively clean locations were deployed and analysed at the others, with 45 resident or transplanted mussels analysed per site. Quality assurance (QA) has been conducted by CSMW since the programme began in 1977, including standard reference materials and method blanks.[15] Since 1996, CSMW has participated in NOAA's inter-calibration programme, with excellent agreement achieved each year for the metals and organics included in Melwani *et al.* (2014).

The Regional Monitoring Program (RMP) for Water Quality in the San Francisco Estuary was initiated in 1993 and has extended the CSMW time series in San Francisco Bay. It began by monitoring 11 stations using transplanted mussels (*M. californianus*), oysters (*Crassostrea gigas*) and clams (*Corbicula fluminea*) to cover the range of salinities found in the Estuary. In each sampling year, approximately 300 individual contaminants were assessed. From 1994 to 2003, the number of stations fluctuated between 11 and 15, based on design modifications. The frequency of sampling was reduced from annual to biennial in 2006 (Melwani *et al.* 2008). Of the original 11 bivalve sampling stations in the RMP, six are historic CSMW sites where the long-term time series initiated by CSMW are still being extended. Three of these stations also coincide with long-term NMW monitoring stations in San Francisco Bay. Although detection of trends may be obscured to some extent by the use of different analytical laboratories and methods, the six CSMW/RMP stations represent the best data set available on bioaccumulation trends in San Francisco Bay over the last three decades.

Melwani *et al.* (2014) analysed data from 24 stations monitored by the CSMW from 1977 to 2010, six stations that were originally part of the CSMW (1980–1992) until the RMP took them over (1993–2010) and 35 stations monitored by NMW (1986–2009). Data for selected organic compounds (total chlordanes, total DDTs, dieldrin and total PCBs) and trace metals (lead, copper, silver and zinc), measured by both CSMW and NMW, were examined for temporal trends. Analysis of RMP data focused on DDTs and PCBs only, which have been contaminants of concern in San Francisco Bay. In addition, sufficient NMW data were collected on total PAHs and total butyltins for evaluation of state-wide trends. Trends in the 10 contaminants from the three major mussel watch monitoring programmes in California are summarised in Table 8.4. Chlorinated organic contaminants (DDTs, Chlordanes,

Table 8.4 Summary of trend analysis in California State Mussel Watch (CSMW) and National Mussel Watch (NMW) data, showing number of stations (N) and proportions of stations (%) with statistically significant trends. PAHs and butyltins (BTs) were not measured at CSMW stations. *Source:* From Melwani *et al.* (2014). Reproduced with permission from Elsevier.

Contaminant	CSMW[a]				NMW[b]			
	Increasing Trends		Decreasing Trends		Increasing Trends		Decreasing Trends	
	N	%	N	%	N	%	N	%
DDTs	0	0	17	90	0	0	20	57
Chlordanes	0	0	14	64	0	0	16	64
Dieldrin	0	0	9	41	0	0	7	20
PCBs	0	0	11	61	0	0	7	20
PAHs	–	–	–	–	3	9	3	9
BTs	–	–	–	–	0	0	11	31
Lead	0	0	11	52	0	0	8	23
Silver	0	0	11	52	0	0	10	29
Copper	8	38	1	<1	4	11	6	17
Zinc	1	<1	4	19	0	0	10	29

[a] Twenty-two stations assessed for Chlordanes and Dieldrin; N = 19 for DDTs; and N = 18 for PCBs. Twenty-one stations assessed for all metals.
[b] Thirty-five stations assessed for all contaminants, except for Chlordnes (N = 25).

Dieldrin and PCBs) and butyltins declined the most rapidly, with tissue concentrations in 2010 that were up to 75% lower than those during the 1980s. Silver and lead declined at about half of the stations state-wide, but generally exhibited slower rates of decline relative to the organic compounds. In contrast, copper increased at many stations, and PAHs showed little evidence for declines. Mussels from San Francisco Bay and the Southern California Bight were historically the most contaminated and had the steepest declines. The declines in these pollutants are likely the result of a combination of chemical bans, increased regulations, increased source control and improved wastewater treatment. For example, DDT was banned in 1972, and since the 1980s 90% of CSMW stations, all six of the RMP stations and 20 of 35 NMW stations have reported significant reductions of DDTs in mussels (Table 8.4). Statistical declines were more frequently detected at CSMW stations than at NMW stations, most likely due to the earlier initiation of monitoring of the CSMW and to the fact that many of those stations were in historically the most contaminated areas. Overall, the contaminant trends noted from CSMW data are consistent with those compiled by Kimbrough *et al.* (2008) and O'Connor & Lauenstein (2006). The existence of historical time series for legacy contaminants[16] of concern in bivalves has proven to be a valuable data resource in California (Maruya *et al.* 2014a,b).

To expand the utility of the Mussel Watch Program, local, regional and state agencies in California partnered with NOAA to design a pilot study that targeted CECs. Native mussels (*Mytilus* spp.) from 68 stations, stratified by land use and discharge scenario, were collected in 2009–2010 and analysed for 167 individual pharmaceuticals, industrial and commercial chemicals and current use pesticides. Passive sampling devices (PSDs) and caged *Mytilus* were co-deployed to expand the list of CECs, and to assess the ability of PSDs to mimic bioaccumulation by *Mytilus* spp. A performance-based quality assurance/quality control (QA/QC) approach was developed to ensure a high degree of data quality, consistency and comparability. Data management and analysis were streamlined and standardised using automated software tools. This pioneering study will help shape future monitoring efforts in California's coastal ecosystems, while serving as a model for monitoring CECs within the region and across the nation. The pilot study has revealed that 98% of targeted CECs were infrequently detectable at concentrations ≤1 ng/g. Selected chemicals found in commercial and consumer products were more frequently detected at mean concentrations up to 470 ng/g dry wt. The number of CECs detected and their concentrations were greatest for stations categorised as urban or influenced by stormwater discharge. Exposure to a broader suite of CECs was also characterised by PSDs, with estimated water of hydrophobic compounds correlated with *Mytilus* spp. concentrations. The results underscore the need for focused CEC monitoring in coastal ecosystems and suggest that PSDs are complementary to bivalves in assessing water quality. Moreover, the partnership established among participating agencies led to increased spatial coverage, an expanded list of analytes and a more efficient use of available resources (details in Alvarez *et al.* 2014; Dodder *et al.* 2014;Edwards *et al.* 2014; Maruya *et al.* 2014a,b).

Mussel Watch Monitoring in Coastal Waters of the Mediterranean Sea

Several European countries have implemented similar national mussel watch programmes to those described for the United States. For example, the French oceanographic institution L'Institut Français de Recherche pour l'Exploitation de la Mer (IFREMER) developed the Réseau National d'Observation (RNO) in 1974 to assess the levels of metals, organochlorines and PAHs along the French coastlines. Biota (*M. edulis* and *M. galloprovincialis* and the oyster *C. gigas*) were collected twice a year in about 100 sampling sites using passive biomonitoring. In a second phase, the Réseau Intégrateur Biologique (RINBIO) was implemented in 1996 to monitor the concentrations of chemicals and radio chemicals in bivalves caged for several weeks prior to collection (active biomonitoring) (Andral *et al.* 2004; see also Andral *et al.* 2011). Despite the presence of several ongoing national programmes, no large-scale mussel watch network was coordinated at the Mediterranean level until 2002, when the Commission Internationale pour l'Exploration Scientifique de la Méditerranée (CIESM) developed and implemented a regional Mediterranean Mussel Watch (MMW) programme, using *M. galloprovincialis* as a bioindicator species that is widespread in the region and for which background knowledge on biology and physiology is available. In sectors where wild populations are rare or absent, samples can be obtained from transplanted mussels. The main objectives of the CIESM were to develop an international trans-Mediterranean infrastructure of cooperating scientists and laboratories aiming at monitoring levels of selected contaminants in coastal waters and to document reliable baseline levels of radionuclides in the Mediterranean and Black Sea coastal waters. Some 20 marine environmental laboratories from 18 Mediterranean countries were involved in this international, dynamic network. The 23 member countries of the CIESM were Algeria, Croatia,

Cyprus, Egypt, France, Germany, Greece, Israel, Italy, Lebanon, Malta, Monaco, Morocco, Portugal, Romania, Russian Federation, Slovenia, Spain, Switzerland, Syria, Tunisia, Turkey and Ukraine. CIESM is part of the EU's MARCOM programme. It worked with other international agencies such as the General Fisheries Commission for the Mediterranean and with national projects, some funded by the Mediterranean Environmental Technical Assistance Program (METAP), organised by the World Bank in cooperation with the United Nations Development Programme, the European Investment Bank and the European Commission.

For Phase I of the CIESM MMW Network (2004–2006), ^{137}caesium (Cs) levels were quantified in mussels from 60 coastal stations, and the first distribution map of this artificial radionuclide was produced by 2009 at the scale of the entire Mediterranean Sea and Black Sea. ^{137}Cs is a radioactive isotope of caesium which is formed as one of the more common products of the nuclear fission of uranium-235 and other fissionable isotopes in nuclear reactors and nuclear weapons. In 2009, ^{137}Cs levels were found to be very low (usually <1 Bq kg^{-1} ww), although concentrations in the Black Sea and North Aegean Sea were up to two orders of magnitude higher than those in the western Mediterranean Basin. Such effects, far from representing a threat to human populations or the environment, reflect a persistent signature of the Chernobyl fallout in this area (Thébault *et al.* 2008). In European countries of the Mediterranean basin, the MMW programme is considered a good basis for the mandatory radionuclides monitoring provided for by the European Marine Strategy Framework Directive.

After a successful Phase I, the objectives of MMW Phase II were to include the extension of monitoring activities to other radionuclides, selected heavy metals and emerging contaminants and enlargement of the network via the association of six additional Mediterranean countries and 36 additional monitoring stations in order to guarantee the monitoring coverage of the entire basin. The results obtained from Phase II would reliably map the current distribution of contaminants in wild and transplanted mussels across the entire Mediterranean Basin, the objective being that this unique data set would provide critical information on the particular issue of seafood safety for human health, fishing and aquaculture. Measurements of chemicals in the marine environment are often subject to large errors in terms of both precision and accuracy. Therefore, one fundamental effort would be to enable a constructive use of the data collected in the present programme by reducing these errors to such a degree that the data from the different institutions involved (1) would be comparable among themselves and (2) would realistically reflect the status/changes of Mediterranean coastal waters. For this purpose, the MMW Phase II would be subject to a strict Analytical Quality Control Service (AQCS), which would include (1) an inter-comparison exercise involving all the laboratories of the network and (2) the routine analysis of selected reference materials (NIST-SRM 2976 for the mussel and IAEA-433 for the sediment) by each participant laboratory. All data produced by participating countries were to be synthesised in a CIESM database, where they would be fully available online to all concerned parties. However, despite repeated efforts from the coordination team and many participants, the programme did not find the necessary institutional and financial support to go beyond the initial involvement of CIESM (Thébault 2013).

Consequently, on the basis of data acquired on the French Mediterranean coast by IFREMER and the Rhône Méditerranée & Corsica water board since 1996, via the RINBIO network (IFREMER 2021), the MYTILOS project drew up a chemical contamination map on a Western Mediterranean scale, using a standardised methodology. Between 2004 and 2006, various oceanographic campaigns were conducted on the continental coasts of Spain, Italy and France, the Balearic Islands, Sicily, Sardinia and Corsica and along the Maghreb coast, taking in Algeria, Libya, Morocco and Tunisia. The areas of interest were selected along the entire coastal development of the Western Mediterranean Sea – 122 sites in total. The methodology consisted of mussel transplantation (*M. galloprovincialis*) from nonimpacted areas to selected coastal areas, characterised by potential impact from the continent due to contaminating sources. The time of mussel caging exposure was 12 weeks. Around 120 days were spent at sea deploying and retrieving 124 mussel (*M. galloprovincialis*) cages. A large proportion of research was conducted in the framework of the INTERREG III B/MEDOCC project, steered by IFREMER and backed by Toulon Var Technologies (www.tvt.fr) in cooperation with the ICRAM (Italy), IEO (Spain), PSTS (Sicily), IMEDEA (Balearic Islands), CSIC (Catalonia), Agence Catalane de l'Eau, INSTM (Tunisia), ISMAL (Algeria), INRH and University of Agadir (Morocco). MYTILOS is also backed by the PNUE/PAM – MEDPOL and Agence de l'Eau Rhône Méditerranée & Corse (Rhône Méditerranée & Corsica water board). Analyses have focused on several contaminant types: heavy metals, organochlorinated compounds (DDTs and metabolites, PCBs) and 16 PAHs. Complementary measurements were performed on a small number of cages to assess levels of dioxins and furanes, PBDEs and nonyl and octyl phenols. The most highly impacted areas were mainly urban and industrial centres and the mouths of major rivers. Toward the open sea, concentrations of organic compounds were shown to decrease faster than metal concentrations. Measured concentrations in mid-sea (i.e. around 1 km from the coast) were similar to those in natural shellfish populations living along the coast. For details on the results of surveys on these contaminants, see Scarpato *et al.* (2010), Andral *et al.* (2011), Benedicto *et al.* (2011) and Galgani *et al.* (2011).

Another mussel watch programme, the MYTIAD project, conducted in 2008, focused on the assessment of chemical contamination (metals, PAHs, PCBs and organochlorine pesticides) in Adriatic (Albania, Montenegro, Croatia, Bosnia, Slovenia, Italy) coastal waters using artificial stations of mussels. Its purpose was to harmonise and standardise strategies and methodologies for assessing the contamination status of the Adriatic Sea in the EU WFD, the MSFD and the United Nations Environment Programme (UNEP)/Mediterranean Action Plan (MAP) Integrated Monitoring and Assessment Programme. Bajt *et al.* (2019) found that data on metals pointed to some hot spots along the eastern and western Adriatic coasts, with the highest values of total PAH concentration detected in Split, Trieste and Taranto (Ionian Sea). PCB and endrin contamination were higher in the Adriatic Sea than in the western Mediterranean. This was the first comprehensive overview of contamination in the Adriatic Sea with critical comparisons to related studies over the Mediterranean. Based on the same methodology and sampling strategy used by previous related studies (projects MYTILOS, see earlier, and MYTIOR and MYTITURK, see later), the MYTIAD data set will support a better definition of baselines, targets and thresholds and further management of chemical contamination.

Within the framework of the MYTIOR project carried out in 2009, heavy metal, PCB, PAH and pesticide contamination was assessed in transplanted mussels in 16 different stations in the southern Ionian Sea, Libya. These stations were located miles offshore from industrial/urban sources, but in open sea results relate to background contamination rather than to a specific coastal source of pollutants. Overall, the results indicated a low background contamination, indicating that the Libyan coastal environment remains clean at distances from industrial/urban possible sources of contamination (Galgani *et al.* 2014).

Within the framework of the MYTITURK 2011 project, heavy metals and organic compounds were assessed in mussels transplanted to eight different bays on the Eastern Aegean coast (Kucuksezgin *et al.* 2013). In the study, seventeen PAHS, nine organochlorines (OCPs) and seven heavy metals (Cd, Cr, Cu, Hg, Pb, Zn and Mn) were evaluated in transplanted mussels (*M. galloprovincialis*). PAH levels indicated contamination was classified as low along the Aegean coast. According to world health authorities, the concentration of heavy metals in mussels for the study area can generally be considered not to be at levels posing a health risk, except in the case of Zn. Concentrations of OCPs (e.g. Aroclor1254 and 1260) were higher in transplanted mussels from Canakkale Strait Outlet due to the outflow originating from the Marmara Sea, while OCPs such as heptachlor, aldrin, dieldrin and endrin were homogeneous – although hexachlorobenzene (HCB) and lindane concentrations were found at undetectable levels along the coast. However, the DDE/DDT ratio in the mussels from Gulluk and Gokova Bays indicated recent DDT usage in these areas.

Asia-Pacific Mussel Watch

Since 1994, monitoring studies in connection with the Asia-Pacific Mussel Watch (APMW) programme, under International Mussel Watch, have focused on monitoring marine pollution in Asian coastal waters using mussels and oysters as bioindicators (Tanabe 2000). APMW is a collaborative project of scientists from various Asian countries, including Cambodia, China, India, Indonesia, Korea, Malaysia, the Philippines, Thailand, Vietnam and Japan. Earlier studies were conducted in Thailand, the Philippines and India from 1994 to 1997 (Kan-atireklap *et al.* 1997, 1998; Prudente *et al.* 1999). These detected butyltin (BT) residues and organochlorine pesticides and PCBs in green mussels, *P. viridis*, from almost all of the sampling locations investigated, suggesting widespread contamination by BTs along the coastal waters of Asian developing countries. Aquatic pollution by TBT coming from antifouling paints had been of great concern in many countries due to its reported effects on non-target marine organisms, including shell malformation in oysters, mortality in mussel larvae and imposex in gastropods. As part of the APMW project, Sudaryanto *et al.* (2002) conducted a survey from 1997 to 1999 to elucidate the status of contamination by BTs in coastal waters of Asian developing countries that were not included in earlier reports. Concentrations of BTs, including TBT and its related compounds di- (DBT) and monobutyltin (MBT), which are breakdown products of TBT, may originate from other sources, such as plastics, and were determined in the whole soft tissue of *P. viridis*, which was collected along the coastal waters of Cambodia, China (Hong Kong and southern China), India, Indonesia, Malaysia, the Philippines and Vietnam. In addition, total tin (ΣSn) concentrations were also determined in mussel samples. Sampling locations ($N = 82$) included urban population centres, commercial and fishery harbours, marinas, recreational boating and marine aquaculture areas and pristine remote locations. The relationship between BTs and ΣSn concentrations was also examined to verify the anthropogenic input of tin compounds. The results of this study and the previous ones were compared with those reported in various regions of the world to delineate the magnitude of contamination in Asian developing countries. The data obtained were also evaluated in terms of possible impacts on aquatic organisms and risks to human health. BT compounds were detected in mussels collected from all sampling locations investigated. Among butyltin derivatives, TBT was the predominant compound, indicating its ongoing usage and recent exposures in Asian coastal waters. Higher concentrations of BTs were found in mussels collected at locations with intensive maritime activities, implying that the usage of TBT as a biocide in antifouling paints was a major source of BTs. In addition, relatively high concentrations of BTs were observed in mussels from aquaculture areas in Hong Kong and Malaysia, as had also been reported in previous studies in Thailand (Kan-atireklap *et al.* 1997), India (Kan-atireklap *et al.* 1998) and the Philippines (Prudente *et al.* 1999). Also, the concentrations of TBT in some mussels from polluted areas exceeded the threshold for toxic effects on organisms and estimated tolerable average residue levels in seafoods for human consumption. A significant correlation was observed between the concentrations of ΣBTs and ΣSn in mussels, and ΣBTs made up 100% of ΣSn in mussels taken from locations with intensive maritime/human activities, which suggests that anthropogenic BTs represent the major source of tin accumulation in mussels. This was the first in-depth report on BT pollution monitoring in developing countries in the Asia-Pacific region. Relatively high TBT concentrations in Korean coastal locations (Shim *et al.* 2005) support the results of Sudaryanto *et al.* (2002).

Environmental pollution by toxic chemicals is a global problem – particularly that of organochlorine compounds (OCs), which are of great concern due to their bioaccumulative nature and toxic biological effects on wildlife and humans (Tanabe *et al.* 2000). Even though most developed countries banned or restricted the production and usage of many of these OCs during the 1970s and 1980s, they are still being used in some developing nations for agricultural and public health purposes, with some being manufactured in Asian developing regions. Because of this, Monirith *et al.* (2003) carried out a study to assess the levels of toxic contaminants in Asian coastal waters. Contamination of persistent OCs such as PCBs, DDT and its metabolites (DDTs), hexachlorocyclohexane (HCH) and its isomers, chlordane compounds (CHLs) and hexachlorobenzene (HCB) was analysed in three species of mussels, *P. viridis*, *M. galloprovincialis* and *Crenomytilus grayanus*, collected from various locations (N = 103) in Cambodia, China, Hong Kong, India, Indonesia, Japan, Korea, Malaysia, the Philippines, Far East Russia, Singapore and Vietnam in 1994, 1997, 1998, 1999 and 2001, the objectives being to elucidate contamination status, distribution and possible pollution sources and to assess the risks on aquatic organisms and humans.

OCs were detected in all mussels collected from all the sampling sites investigated. The highest residue levels of DDTs were found in mussels from China (830–54 000 ng/g lipid wt), Hong Kong (640–61 000 ng/g lw) and Vietnam (220–34 000 ng/g lw), clearly indicating the release of DDT to the environment in China, even though usage of DDTs has been officially banned since 1983. The results also indicated the recent usage of DDTs in Vietnam. In general, lower DDTs levels were observed in mussels from the Phillipines, Cambodia, Indonesia and Malaysia, which indicates less usage of DDTs in these countries.

Extremely high levels of HCH pesticides were observed in mussels from India (20–430 ng/g lw), with relatively high concentrations found in China (2.1–110 ng/g lw) and Russia (34–57 ng/g lw). India is the largest user of technical HCH in the world. The usage of HCH has been banned in Indian agriculture since 1983, but its usage is allowed for public health purposes and for certain food crops (Li *et al.* 1998). In the former Soviet Union, it was permitted to use residual stocks of pesticides (Zhulidov *et al.* 2000). It is quite possible that the active usage of these stockpiled amounts of HCH for agriculture, forestry and municipal purposes could have contributed to HCH contamination in mussels. Likewise, the higher concentrations of HCHs in mussels from India, China and Russia may have been partly caused by continued usage of HCHs in these countries.

The highest concentrations of PCBs were found in mussels from Japan (up to 12 000 ng/g lw) and Russia (up to 3700 ng/g lw). These results suggest that emission sources of PCBs still exist in Japan, even though their production was prohibited there in 1972. They also suggest the presence of local sources of PCBs in Russia. Low concentrations of PCBs were found in mussels from Cambodia, Indonesia, Malaysia, Vietnam, India and China, indicating fewer local sources. In general, lower concentrations of PCBs were detected in mussels from Asian developing countries than in those from developed countries, indicating that PCB contamination in mussels is strongly related to industrial and human activities.

The highest levels of CHL pesticides were observed in mussels from Japan (150–1800 ng/g lw), China (40–870 ng/g lw), Hong Kong (18–750 ng/g lw), Malaysia (17–630 ng/g lw) and Singapore (520 ng/g lw). China, Malaysia and Singapore showed higher concentrations of CHLs at sites in proximity to fishing harbours and urban and densely populated areas. In Japan, CHLs were used largely for termite control until 1986, but they still seem to be discharging into the marine environment. Despite lower concentrations of CHLs in mussels from the Philippines and India, relatively higher concentrations were found in certain locations, again in proximity to fishing harbours in urban and industrialised areas.

Finally, concentrations of HCB were generally uniform in mussels from the Asia-Pacific region, while relatively high concentrations were found in samples collected near industrialised cities in China, Bombay in India and Tokyo Bay in Japan. HCB is not only used as a fungicide, but is also generated as a byproduct during the production and usage of several agrochemical and industrial chemicals.

The results of this study highlighted the status of contamination by OCs in the Asian-Pacific region, suggesting serious contamination by DDTs in China, Hong Kong and Vietnamese coastal waters, by HCHs in the Indian and Chinese marine environment and by PCBs and CHLs in Philippine, Malaysian and Singapore marine waters. OC concentrations in mussels from some of the Asian countries surveyed were compared with values in bivalves reported from various global locations. Generally, concentrations of PCBs seemed to be lower in Asian developing countries than in developed nations. However, DDTs and HCHs were higher in Asian developing countries than in developed nations, clearly suggesting that these pesticides are still being used in some of them.

The occurrence of OC residues in mussels is also of great concern for human health, as mussels are valuable mariculture organisms and a commercially important seafood, particularly in developing countries. Based on their results, Monirith *et al.* (2003) suggest that continuous monitoring of OC residues – including studies on ecotoxicological risk assessment to elucidate the trend of contamination and toxic impact on humans and wildlife in Asian developing countries – is essential.

As part of the APMW programme, Isobe *et al.* (2007) reported spatial distribution of PAHs and phenolic compounds in mussels (*P. viridis*) from South East Asian countries. Also, Ramu *et al.* (2007a) analysed PBDEs, HBCDs, PCBs and organochlorine pesticides in mussels (probably *M. corusus*) from Korea during 2003 and 2005, while Ramu *et al.* (2007b) analysed PBDEs and organochlorines in mussels (*P. viridis* and *M. coruscus*) from locations in Cambodia, China, Hong Kong, India, Indonesia, Japan, Malaysia, the Philippines and Vietnam from 2003 to 2005. The level and extent of organic contaminants along the Korean coast were estimated through the APMW programme (Choi *et al.* 2010). Mussels were collected at 20 sites along the Korean coast in 2001 and at 25 sites from 2002 to 2007. Mussel tissues were analysed for PCBs, organochlorine presticides, PAHs, BTs and PCDD/Fs. However, until the study of Nakata *et al.* (2012), large-scale monitoring for emerging pollutants, synthetic musks and benzotriazole UV stabilisers (BUVs) in Asian-Pacific coastal waters was not available.

Biological Markers of Pollution

Contrary to chemical monitoring of organisms, which principally evaluates the presence of pollutants in tissues by chemical analysis, biomonitoring methods evaluate not only the presence but, more relevantly, the response of the organisms to these pollutants through the assessment of biological markers (biomarkers). Biomarkers are cellular, biochemical, molecular or physiological changes measured in cells, body fluids, tissue or organs within an organism that can be related to exposure to or toxic effects of environmental chemicals (Lam & Gray 2003). They can be used to assess the nature and extent of the exposure, to identify alterations occurring within the organism or to assess the underlying susceptibility of the organism (Figure 8.9). They can help to increase the understanding of the processes by which a chemical is absorbed and transformed within an organism to determine alterations at the cellular and molecular levels leading to a toxic effect (Lionetto *et al.* 2019).

Biomarkers were originally developed in the medical and veterinary sciences, but since the 2000s there has been an increasing emphasis on the use of invertebrate and particularly bivalve biomarkers to assess marine pollution (Nicholson & Lam 2005). Indeed, the biomarker approach has now attracted the attention of international regulatory agencies as a new and potentially powerful tool for detecting exposure to and the effect of environmental contamination (Hamza-Chaffai 2014). Detection in mussels was recognised in the 1990s (Moore 1991), and since then, according to Beyer *et al.* (2017), the number of *Mytilus* spp. biomarker studies reported in environmental science journals has steadily increased, reaching more than 1700, or approximately 10% of all reported biomarker studies.

The International Programme on Chemical Safety has identified three classes of biomarker: biomarkers of exposure of the organism to the toxic substance, biomarkers of response of the organism to that exposure and biomarkers of susceptibility of the organism to the chemical (WHO 2005). Biomarkers of exposure are the product of integration between a xenobiotic and some target molecule or cell that is measured within a compartment of an organism (Hamza-Chaffai 2014; Hedayati *et al.* 2015). Biomarkers of exposure provide an indication of the occurrence and the extent of exposure of the organism to various compounds. They can give information on the route, pathway and sometimes even the source of exposure. These markers cause early reversible cellular changes in the organism, often

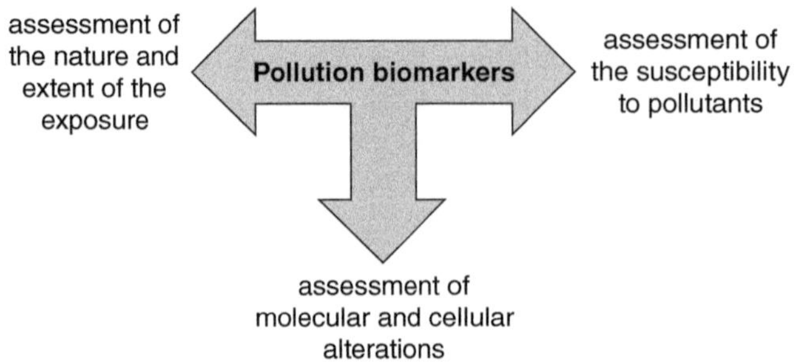

Figure 8.9 Biomarker usefulness in monitoring and assessment. *Source:* From Lionetto *et al.* (2019). CC BY 4.0.

based on the activation of detoxification mechanisms. Living organisms can show many kinds of toxic or adverse response to a chemical exposure, ranging from biochemical or physiological to pathological. Consequently, there are many biomarkers of response that can be measured. Biomarkers of effect are defined as measurable biochemical, physiological, behavioural or other alterations within an organism that, according to their magnitude, can be recognised as established or potential health impairment or disease. Biomarkers of susceptibility indicate an inherent or acquired ability of an organism to respond to specific pollutant exposure (Lionetto *et al.* 2019). Such markers cover a range of types, from deficiency in metabolic enzymes to variation in repair systems. These would typically be measured in individual members of a population. The specificity of biomarkers to pollutants ranges from highly specific, such as metallothionein induction by metals (Cu, Hg, Zn, Cd) or aminolevulinic acid dehydratase (ALAD) inhibition by lead, to unspecific, such as DNA damage or immune system impairment. However, when different specific biomarkers are used together, a complementation can occur that results in an overall higher degree of specificity (Lionetto *et al.* 2019). Some biomarkers are highly specific for individual chemicals, including inhibition of acetylcholinesterase (AChe) by pesticides such as organophosphates or carbamates and induction of metallothioneins by exposure to some metals. Others have wider applications and tend to respond to broader classes of chemicals, such as stress on stress (SoS)-related biomarkers (Hellou & Law 2003), the formation of DNA adduct and other DNA alterations (Siu *et al.* 2003; Magni *et al.* 2006) and lysosomal alteration in molluscan digestive gland cells (Cajaraville *et al.* 2000; Domouhtsidou *et al.* 2004). Overall, the most assessed biomarkers in environmental biomonitoring encompass lysosomal endpoints, oxidative stress, specific responses to metals, neurotoxic pollutants and genotoxic substances.

Biomarkers in Perna Viridis *and Other Marine Mussel Species*

The major focus of active biomonitoring with marine mussels has been in temperate regions using *Mytilus* spp. Fewer studies have been carried out with tropical and subtropical species. In recent years, the mussel *P. viridis* has been used as a biomonitor throughout the Indo-Pacific, and it is now considered a subtropical equivalent of *Mytilus* spp. Moreover, there has been a considerable research effort to develop biomarkers in this species for pollution monitoring in subtropical regions (reviewed in Nicholson & Lam 2005).

Enzyme Reactions

While conventional chemical monitoring provides information on the types and levels of potential pollutants, the interactive nature and the potential effects of chemical pollutants to organisms in the ambient environment are not reflected. Contaminants such as heavy metals and organic pollutants exert their effects on biological systems via radicals (see Chapter 11) or high-energy molecules, and such effects are ultimately reflected as oxidative stress on the organisms. These biological responses or biomarkers have been employed to assess the effect of environmental pollutants on organisms (Lau & Wong 2003). Biomarkers at the molecular and biochemical levels are often regarded as early-warning indicators for long-term ecological effects due to their rapid response to stress and their potentially high toxicological relevance.

In recent years, there has been a rapid development of enzyme biomarkers, primarily due to advances in biochemistry and improvements in analytical instruments. The suitability of various antioxidant parameters – glutathione S transferase (GST), superoxide dismutase (SOD), catalase (CAT), glutathione peroxidase (GPx), glutathione reductase (GR), NADPH DT-diaphorase (DT-d), glutathione (GSH) and lipid peroxidation (LPO)[17] – for use as biomarkers has been examined in a variety of organisms. Results utilising *P. viridis* indicated that the majority of the antioxidant parameters, except gill SOD, DT-d and LPO, were induced by increasing tissue concentrations of PAHs (Cheung *et al.* 2001). Among the various antioxidant enzyme systems, the oxyradical scavenger GSH best correlated with tissue PAH concentrations. For organochlorines, only one of the enzyme antioxidants, GPx, showed a significant response to tissue PCBs. The same study also demonstrated a correlation between GST/GSH and chlorinated hydrocarbons. Oxidative stress was also correlated with chlorinated pesticide concentration in *P. viridis* tissues, but there was no significant correlation between tissue concentrations of chlorinated hydrocarbons and other enzymatic antioxidants investigated (Cheung *et al.* 2002). In a laboratory study (Cheung *et al.* 2004), *P. viridis* was exposed to two concentrations of benzo[a]pyrene (0.3 and 3.0 $\mu g\, l^{-1}$) and two concentrations of the PCB Aroclor 1254 (0.5 and 5.0 $\mu g\, l^{-1}$). In addition, a mixture of the contaminants was used (0.3$\mu g\, l^{-1}$ B[*a*]P; +0.5. $\mu g\, l^{-1}$ Alocar 1254; 3.0 $\mu g\, l^{-1}$ B[*a*]P + 5.0 $\mu g\, l^{-1}$ Aroclor 1254). A suite of antioxidant parameters similar to those used in Cheung *et al.* (2001) were monitored in the gill and hepatopancreas of *P. viridis* over 18 days. CAT and GSH in gill tissue were positively correlated with concentration of Aroclor 1254. Activity of hepatic GST and SOD was significantly related to body burden of Aroclor 1254. LPO, GR and GPx in gill and hepatopancreas and hepatic GST were positively correlated with B[*a*]P concentration. Controlled laboratory experiments such as this are useful in ascertaining biomarkers suitable for use with complex contaminant mixtures in the marine environment.

Specific Proteins

When heavy metals enter cells, they are primarily complexed by molecules such as amino acids, glutathione and especially metal-binding detoxication proteins called metallothioneins (Livingstone & Pipe 1992). MTs are nonenzymatic proteins with a low molecular weight and a high cysteine content. The thiol groups (–SH) of cysteine residues enable MTs to bind particular heavy metals. Given the metal-binding capacity of MTs, it is generally considered that these proteins play a role in the homeostatic control of essential metals

(Cu, Zn), as they can act as essential metal stores ready to fulfil enzymatic and other metabolic demands (Amiard *et al.* 2006 and references therein). In addition, metallothionein concentrations in molluscs may also vary due to influence by nonmetallic pollutants. MTs are synthesised in the digestive gland and gill tissue of *M. galloprovincialis* (Ivanković *et al.* 2005 and Serafim *et al.* 2002, respectively). Hence, the idea arose that MTs might be used as biomarkers for environmental pollution by means of measuring their concentrations in bivalves from contaminated habitats.

Viarengo *et al.* (2007) have suggested that MT induction in fish should be considered as a general stress response, as fish are particularly sensitive to heavy metals (Authman *et al.* 2015). Conversely, in mussels, and possibly in other molluscs, MTs seem to represent a more selective biological response to heavy metals. Also, Viarengo *et al.* (2007) advised that MTs should always be used in association with other biomarkers of exposure, such as mixed function oxidase (MFO) activity or acetylcholinesterase (AChE) activity, and biomarkers of stress, such as destabilisation of lysosomal membranes, lipofuscin accumulation and DNA damage (see later). Finally, the validity of using MT induction as a biomarker is questionable as a number of studies have shown an absence of increases in MT concentration in tissues of different bivalves in response to exposure to several metals that are generally known to be potent inducers (Le *et al.* 2016).

Acetylcholinesterase Inhibition

The enzyme acetylcholinesterase (AChE) is involved in cholinergic neurotransmission and is frequently used as a biomarker of neurotoxicity in aquatic invertebrates due to its specific inhibition by organophosphate pesticides (Campillo *et al.* 2013). Other groups of chemicals present in the marine environment (e.g. heavy metals, detergents and hydrocarbons) also affect its activity. AChE activity has been identified and biochemically characterised in many aquatic invertebrate species. In particular, bivalve molluscs, including *Mytilus* spp., have high levels of AChE activity in the gills, digestive gland and haemolymph. Studies indicate that the AChE response in mussel gills is proportional to contamination levels of a number of pesticides in the aquatic environment (Lionetto *et al.* 2011 and references therein). Mora *et al.* (1999) reported a 50% inhibition in *Mytilus galloprovincialis* gill AChE activity, which corresponded to an experimental exposure of the mussels to 0.1 mg l^{-1} carbamate (pesticide) or 0.2 mg l^{-1} methyl-parathion (organophosphate) for 24 hours. Concentration-dependent inhibition of AChE after in vitro and in vivo exposure to both organophosphate and carbamate pesticides was also demonstrable in haemolymph (Galloway *et al.* 2002), suggesting that the activity of haemolymph AChE might provide a rapid, accessible means of monitoring pesticide exposure.

In a recent study, Perić & Burić (2019) investigated the combined effect of the essential metal copper (Cu) and the organophosphorous pesticide chlorpyrifos (Chp) in mussels (*M. galloprovincialis*) after short-term exposure to their sublethal concentrations. Mussels were exposed for four days to 5 and 15 mg l^{-1} Cu and 0.05 and 5 mg l^{-1} Chp, and to their binary mixtures. The investigated biomarkers were acetylcholinesterase activity (AChE), glutathione S-transferase activity (GST), metallothioneins content (MTs) and lipid peroxide levels (LPO). The effect of short-term and environmentally realistic exposure to Cu and Chp mixtures was not entirely straightforward, varying in relation to concentrations of compounds. The activity of GST was the most sensitive biomarker, revealing an oxidative challenge imposed by both the single compounds and their mixtures. A clear biomarker response pattern attributable to mixtures could not be discerned, further highlighting the challenges of stress response evaluation in the natural environment.

Environmental factors such as temperature and salinity seem to influence acetylcholinesterase activity in aquatic invertebrates, particularly in estuarine and brackish water environments characterised by a high variability of abiotic factors (temperature, salinity, trophic conditions and dissolved oxygen content). In *Mytilus* spp., AChE activity showed significant seasonal differences, with maximum activities during the summer period and minimum activities in winter (Pfeifer *et al.* 2005). These results clearly demonstrate the need to consider the potential influences of temperature and salinity on AChE activity when applying this biomarker to monitor exposure to and effect of neurotoxic substances in estuarine- and brackish-water *Mytilus* spp.

Cellular Stress Biomarkers: Lysosome Membrane Stability and Permeability

Lysosomes are ubiquitous cytoplasmic organelles that contain enzymes which degrade macromolecules to low-molecular-weight products. They accumulate many toxic metals and organic chemical contaminants that provoke loss of membrane integrity, increase in lysosomal size and accumulation of lipid and the pigment lipofuscin, granules composed of lipid-containing residues of lysosomal digestion. These responses have been used routinely to test for the effects of toxic contaminants in both experimental investigations and environmental impact assessments (Moore *et al.* 2004; Izagirre *et al.* 2009 and references therein). Lysosome enzymes (acid hydrolases) are capable of degrading cellular constituents and, in order to prevent free access to cytosolic components, are predominantly sequestered in an inactive form within the comparatively thick lysosomal membrane (Nicholson & Lam 2005). Many environmental stressors, including pollutants, are known to destabilise lysosomal membranes, and membrane damage is often proportional to the magnitude of stress (Moore 1985). Lysosomes play a major role in the cellular homeostasis of metals through storage of metallothioneins and are responsible for the sequestration of organic contaminants (Moore *et al.* 1984). Although the ability to sequester contaminants from sensitive intracellular sites is an essential protective mechanism, it also renders the lysosomal membrane particularly susceptible to elevated toxicant concentrations as excess intra-lysosomal storage can increase membrane permeability, leading to efflux of hydrolytic enzymes and enhanced autophagic activity, thereby inducing a diseased state (Nicholson & Lam 2005 and references therein).

Lysosome membrane permeability is generally quantitative to the degree of contaminant-induced stress (Moore1985) and is responsive to a range of pollutants. The assessment of membrane permeability is considered a useful nonspecific marker of pollution, and there has been much focus on developing an assay utilising lysosomal membrane responses in marine mussels. A technique has been developed by Lowe *et al.* (1995) to determine lysosomal membrane permeability in mussel cells. This involves incubating haemocyte suspensions with the dye neutral red and measuring the time required for it to leak from the lysosome to the cytosol (the fluid in which organelles of the cell reside). This method is usually referred to as neutral red retention time (NRR or NRRT). The dye is only retained in healthy cell lysosomes, and intra-lysosomal retention can be used as an indicator of membrane permeability. Viarengo *et al.* (2007) suggest that the assay should be considered an SoS test, in that neutral red is itself toxic for the cells, and therefore further damages lysosomal membranes that are potentially already damaged.

The assay was used in Hong Kong to identify a putative pollution gradient in Tolo Harbour (Cheung *et al.* 1998) and proved to be a useful general marker of stress, although a direct relationship between metal body burden and lysosome labilisation time was not evident, further suggesting that the response is best suited as a nonspecific marker of pollution. It was also used to determine spatial pollution gradients in various locations in Hong Kong coastal waters using transplanted *P. viridis*, and proved to be a useful easily measured rapid marker of metal pollution, although it was suggested that it should be used in conjunction with other markers (Nicholson 1999). In a subsequent study, a suite of cytological and physiological responses to acute (48–72 hr) copper exposures of 50–200 $\mu g\ l^{-1}$ were assessed in order to determine the suitability of *P. viridis* for marine pollution monitoring (Nicholson 2003a). Copper elicited significant destabilisation of the haemocyte lysosomal membranes and impaired phagocytosis. Survival during emersion following exposure to copper was not related to the experimental copper exposures, suggesting that higher metal concentrations may be required to interfere with the anaerobic enzymes responsible for suppression of metabolism. Based on this study, cytological biomarkers evaluated in the haemocytes extracted from *P. viridis* should prove an effective nondestructive means of assessing metal pollution throughout the mussel's subtropical range. As a good indicator of contaminant-induced lysosomal membrane damage, lysosomal membrane stability (LMS) assays have been used in a number of field studies, with successful results using mussels (*M. edulis*, Lowe *et al.* 1995; Fernley *et al.* 2000; Einsporn *et al.* 2005; Schiedek *et al.* 2006; *M. galloprovincialis*, Castro *et al.* 2004) and other invertebrate species, including oysters (Hauton *et al.* 2001; Ringwood *et al.* 2002), scallops (Hauton *et al.* 2001), limpets (Brown *et al.* 2004) and crabs (Wedderburn *et al.* 1998).

Peroxisome Proliferation

Peroxisome proliferation is a specific marker of pollution by organic chemicals such as PAHs and PCBs. Peroxisomes are universal cytoplasmic organelles involved in the metabolism of lipids and reactive oxygen free radicals (Cajaraville *et al.* 2000 and references therein). Peroxisome proliferation is defined as an increase in peroxisomal volume and numbers, usually but not always accompanied by induction of peroxisomal enzymes, particularly those of the fatty acid β-oxidation pathway. Peroxisomes of marine organisms, including bivalve molluscs and fish, are able to proliferate under experimental exposure to organic xenobiotics. In terms of field studies, peroxisome proliferation has been used to assess environmental pollution in different European coastal systems as well as the Pacific coast of the United States.

Ortiz-Zarragoitia & Cajaraville (2006) investigated the effects of different endocrine disruptors in mussels (*Mytilus* spp.) using peroxisome proliferation as a biomarker of exposure to organic pollutants and levels of vitellogenin (Vtg)-like proteins, measured as alkali-labile phosphate (ALP) levels, as a biomarker of endocrine disruption in combination with gonad histology. In one experiment, mussels were exposed for three weeks to North Sea crude oil (NSO 0.5 ppm) and a mixture of 0.5 ppm NSO, 0.1 ppm alkylphenol mix and 0.1 ppm extra PAHs (MIX). In a second experiment, mussels were exposed for three weeks to the plasticiser diallylphthalate (DAP 50 ppb), the epoxy resin bisphenol-A

(BPA 50 ppb) and the organobromine compound and aromatic ether tetrabromodiphenylether (TBDE 5 ppb). The results confirmed previous results indicating that oil derivatives and phthalate esters are peroxisome proliferators in mussels. Furthermore, the authors reported that polybrominated FRs such as TBDE act as peroxisome proliferators in mussels, whereas BPA exerts no such effect. Exposure to NSO and a mixture of oil and alkylphenols caused significant changes in gonad development, marked oocyte atresia and alterations in Vtg-like protein levels. No such effects were observed in mussels exposed to DAP, TBDE or BPA, but widespread gamete resorption was found in mussels exposed to BPA. These results confirm the usefulness of peroxisome proliferation as a biomarker of exposure to organic contaminants in mussels and indicate that changes in Vtg-like proteins could be used as potential indicators of pollutant effects on mussel reproduction. However, the use of vitellogenin as a reliable biomarker of endocrine disruption in marine mussels has recently been questioned (Fernández-González *et al.* 2020).

Morphological Damage

It is possible to identify structural and functional alterations resulting from exposure to environmental contaminants in tissues and cell types associated with feeding, digestion and reproduction (Livingstone & Pipe 1992). Gill cilia can be damaged due to mechanical injury associated with exposure to both metals (Viarengo *et al.* 1994) and suspended solids (Shin *et al.* 2002). Studies in *P. viridis* exposed to copper indicated that impaired clearance rates were likely caused by structural damage of the cilia (Nicholson 2003b). It is likely that damage to the cilia will interfere with feeding and ultimately growth. As the assessment of damage to gill filaments is relatively simple, this general morphological marker could prove useful in toxicant monitoring (Nicholson & Lam 2005).

Digestive diverticula, which contain numerous lysosomes, are the major organs of cellular digestion in bivalves. Since lysosomes have a significant capacity for sequestration of both inorganic and organic contaminants, digestive diverticula are the most important organ for toxicant accumulation. Histological examination of the digestive cells of *M. edulis* individuals from polluted locations has revealed the presence of structural abnormalities, including a high percentage of tubule breakdown and dilation related to lysosome dysfunction (Nicholson & Lam 2005 and references therein). Morphological alterations have also been assessed in *P. viridis* digestive cell tubules, and the percentage of dilation, breakdown and cilia loss has been shown to be elevated following exposure to copper and mercury. It is probable that these degenerative alterations are related to self-digestion processes as a consequence of lysososomal membrane destabilisation (Krishnakumar *et al.* 1990; see Livingstone & Pipe 1992 and Cajaraville *et al.* 1992 for details).

Physiological Markers

In mussels from contaminated sites, part of their metabolic energy is spent on detoxification processes, thus depleting the ATP needed for other physiological functions. The reduction of survival in air, or SoS biomarker, is a simple and low-cost whole-organism response that can show pollutant-induced alterations in the organism's physiology that render it more sensitive to further environmental changes (Thain *et al.* 2019).

Various physiological responses have been utilised as biomarkers, including such basic physiological functions as reproduction, respiration and changes in growth rate and feeding. Indeed, growth has been widely used in marine bivalves and gastropods to provide a measure of environmental quality (Lam 2009 and references therein). Scope for growth (SFG; see Chapter 6) in particular has been used to measure the amount of energy available to an organism for various functions, including growth and reproduction. In a contaminated environment, an organism will in theory use energy to deal with the toxic stress and therefore have less available for growth. Such physiological responses are used to provide integrated measures of an organism's well-being, based on a range of different functional attributes, and thus are arguably more ecologically relevant than molecular or biochemical biomarkers (Lam 2009).

The following physiological biomarkers have been utilised in marine mussels: heart rate (HR), clearance rate, condition indices and SFG (Nicholson & Lam 2005). HR in mytilid mussels is a relatively constant rate function when the shell remains open, individuals are continually immersed and hydrological conditions are stable (Nicholson 2002). A wide range of contaminants, including metals and aromatic and chlorinated pesticides (DDT, dieldrin, endrin and endosulphan), have been reported to elicit a rapid decrease in bivalve HR, referred to as bradycardia (Nicholson & Lam 2005 and references therein; see Chapter 7). Neuronal toxicity has been proposed as a mechanism for bradycardia in mytilids during pollutant exposure, although many studies appear to have been conducted using unrealistically elevated contaminant concentrations, or have not fully accounted for avoidance behaviour including shell closure, which also inhibits HR (Nicholson 2003b).

When *P. viridis* was exposed to copper (48 hr continuous exposure of 100–400 μg l^{-1} Cu; Nicholson 2003b) and TBT (Nicholson 2003c), reflecting conditions present in estuaries, HR tended to be elevated over a range of concentrations, suggesting that tachycardia (rapid resting HR) is the typical response of mytilids in polluted waters (Nicholson & Lam 2005). Injection of cardio-inhibitory dopamine into copper-exposed tachycardic individuals reversed the elevated HR with respect to controls. Similarly, *P. viridis* transplanted to various locations in Hong Kong coastal waters showed tachycardic responses that were considered to be indicative of enhanced metabolism during the detoxication of contaminants (Nicholson 1999). It appears that HR is a useful nonspecific physiological biomarker of general metabolism, although it is usually only significantly altered at elevated contaminant concentrations, and results may be confounded if shell closure is not adequately considered during its application. Additional information on the use of mussel HR as a biomarker in pollution studies can be found in Brown *et al.* (2004), Halldórsson *et al.* (2008), Rajkumar (2009), Hagger *et al.* (2010), Turja *et al.* (2014), Martinović *et al.* (2015, 2016), Saggese *et al.* (2016) and Nikolic *et al.* (2019); see also Chapter 7.

Among the physiological responses of bivalves, clearance rate is one of the most important, since it is not only related to the acquisition of the energy necessary for vital functions, but is also sensitive to environmental changes such as organic pollutants. Contaminants have deleterious effects on clearance rate, and at higher ambient concentrations mussels elicit avoidance behaviour through shell closure to prevent tissue exposure, while lower toxicant levels inhibit ciliary beating, thereby reducing food intake. Reduction in clearance rate was observed in the mussel *Choromytilus chorus* in relation to the degree of environmental pollution (Toro *et al.* 2003a). Sampling sites included the polluted San Vicente Bay, the averagely polluted Corral Bay and the weakly polluted Yaldad Bay, all in southern Chile. Pollution levels were estimated by quantitative analysis of PAHs in mussel tissue from all three sites. Mussels from San Vicente showed the lowest clearance rates and highest PAH levels in spring and summer, while those from Corral and Yaldad showed generally low levels of PAHs. Mussels from Yaldad showed the highest clearance rates in comparison to Corral and San Vicente populations. The significant negative relationship ($P < 0.001$) demonstrated that tissue pollution of mussels and their clearance rates support the usefulness of this nondestructive biomarker in evaluating the impact of environmental pollution on marine mussels. See Shuhong *et al.* (2005), Solé *et al.* (2010) and Brooks *et al.* (2009) for comparable studies.

The condition index (CI) in bivalves is mainly used for two purposes: as a measure of the quality of meat for the marketplace and as an ecophysiological measure of health status (Pampanin *et al.* 2005). Transplanted *P. viridis* has been used to ascertain pollution at several localities in Hong Kong (Nicholson 1999). Although it was recognised that results could be indicative of differences in food availability, CIs proved useful as a simple measure of mussel condition and generally reflected the contaminant concentrations present at each site. In the search for good indicators of water quality in the Venice urban area, two physiological indices for mussels, survival in air and CI, were evaluated (Pampanin *et al.* 2005). Mussels (*M. galloprovincialis*) were collected from three sites located in the canals of the city and a clean reference site. At the same time, two transplantation experiments were performed, deploying farmed mussels for 1, 2, 4 and 12 weeks in the historic centre of Venice and at a reference site. Both native mussels and those transplanted to the urban area showed reduced survivability in air and decreased CI values, indicating a less healthy status in animals from the canals of the city. Survival-in-air data demonstrated a low capability of mussels from the urban area to deal with an additional stress. Moreover, since CI is strictly correlated with the reproductive cycle, a decrease indicates not only a low capability to grow, but probably also a reduction in the production of gonadic tissue, impairing reproductive capability. These two physiological indices clearly show their potential for inclusion within biomonitoring programmes as a fundamental part of a battery of biomarkers.

An organism can only allocate energy to growth or reproduction if the energy gain from absorbed ration exceeds total metabolic losses. This energy, surplus to metabolic demands, is referred to as 'scope for growth' or SFG (details in Chapter 6). SFG has been employed in a number of pollution monitoring investigations. For example, Toro *et al.* (2003b) used SFG as a physiological index to evaluate stress produced by pollutants on the mussel *Choromytilus chorus* at three locations on the Chilean coast. Individuals from a highly polluted site showed negative SFG values in spring (-4.6 J/h g^{-1}) and summer (-3.5 J/h g^{-1}), possibly indicating severe stress related to the accumulation of organochlorines and PAHs in mussel tissues. Those from a site with an intermediate level of pollution had an SFG of 15.5 J/h g^{-1} in spring and 6.5 J/h g^{-1} in summer, while those from a low-pollution site had values of 6.2 J/h g^{-1} in spring and 25.7 J/h g^{-1} in summer. There was a significant negative correlation between the SFG of the three different populations and the concentrations of organochlorines and PAHs in their tissues. Mussels showing negative SFG values are under constant stress, making them more susceptible to diseases, predation and parasites. If these conditions persist, populations in highly polluted sites may disappear in the long term (Toro *et al.* 2003b). SFG is the only truly informative biomarker in understanding the effects of pollutants at the organism level, while also providing an indication of potential effects at the population level (Viarengo *et al.* 2007).

Biomarkers of Genotoxicity

The continuous discharge of genotoxic compounds into the environment is of major concern. These compounds can lead to several alterations when they interact with DNA, including point mutations, chromosomal rearrangements, DNA adducts (segments of DNA bound to a cancer-causing chemical), DNA strand breaks and increased numbers of micronuclei (Lionetto *et al.* 2019). They are particularly harmful in that they can induce changes that are passed on to future generations and have an impact on populations long after the original exposure (Nicholson & Lam 2005).

A large number of methods have been applied to evaluate genotoxic damage in different aquatic and terrestrial species. The comet assay, as a marker for detecting DNA alterations, and the micronucleus test, as a marker of chromosomal damage, are the most widely applied and validated in field studies, and are frequently employed in biomonitoring programmes (Bolognesi & Cirillo 2014). In the comet assay, individual cells are directly embedded in agarose, and the nuclear DNA is then electrophorised through the gel, in which the cleaved DNA fragments migrate away from the residual chromatin nucleosomal core structure (see Mariño-Ramírez *et al.* 2005 for details). DNA stained with a fluorescent dye shows a 'comet', with the distance of DNA migration from the core (i.e. the comet tail length) reflecting the level of double-strand DNA breakage (Siu *et al.* 2004). The sensitivity and specificity of the assay are greatly enhanced if the nucleoids[18] are incubated with bacterial repair endonucleases that recognise specific kinds of damage in the DNA and convert lesions to DNA breaks, increasing the amount of DNA in the comet tail (Collins 2004). The technique has been successfully employed with DNA extracted from fish and mollusc cells (Viarengo *et al.* 2007 and references therein). The micronucleus test, meanwhile, is a well-known cytogenetic technique commonly used for the assessment of genotoxic effects caused by environmental stressors. Micronuclei (MN) are extranuclear bodies that contain damaged chromosome fragments and/or whole chromosomes that were not integrated in the daughter nuclei during mitosis and remain in the cytoplasm after cell division. MN can be induced by defects in the cell repair machinery, accumulation of DNA damage and chromosomal aberrations. A variety of genotoxic agents may induce MN formation, leading to cell death, genomic instability or cancer development (Luzhna *et al.* 2013). Haemolymph is regarded as the most suitable tissue for biomonitoring, because its physiological role in the transport of toxic materials and in several defence mechanisms means that it is directly exposed to environmental contaminants (Villela *et al.* 2006).

P. viridis were exposed to waterborne benzo[a]pyrene (B[a]P) at nominal concentrations of 0, 0.3, 3.0 and 30g l^{-1} for up to 12 days and both the relative levels of DNA strand breaks (assessed using an alkaline comet assay) and the proportion of MN formation were monitored in mussel haemocytes at days 0, 1, 3, 6 and 12. The results of the comet assay indicated that an increase in the proportion of strand breaks occurred generally with increasing B[a]P concentration, but a significant decrease in the levels of DNA damage was observed after exposure for 12 days at all concentrations tested, suggesting that the patterns of changes in the levels of DNA strand breakage can be explained by the threshold-dependent DNA repair theory; that is, a DNA repair system may be activated after the mussel tissue has accumulated sufficient toxicant above a threshold level (Siu *et al.* 2003). Monitoring the frequency of MN development in mussel haemocytes indicated both dose-and time-response relationships within the exposure period. Moreover, the levels of DNA strand breakage correlated well with the levels of MN induction, suggesting a possible cause-and-effect relationship between the two damage types, which further suggests that DNA strand breakage and MN formation in mussel haemocytes can be used as convenient biomarkers of exposure to genotoxicants in the marine environment.

Heavy metals (Zn, Cu, Cd, Pb) and PCBs were selected as representative xenobiotics to verify the ability of an in vitro model to assess genotoxic effects in cells of *M. galloprovincialis* (Zhang *et al.* 2017). DNA damage and chromosome aberration were assessed in freshly isolated cells from haemolymph, gill and digestive gland by comet and MN assays. Of the three cell types, gill cells were more sensitive to Zn exposure, indicating tissue-specific genotoxicity. Significantly higher DNA aberrations were induced by Cu in haemocytes compared to Cd and Pb, indicating chemical-specific genotoxicity. An additive effect was detected after a combined heavy metal and PCB exposure, suggesting the interaction of selected xenobiotics. The authors emphasised that their study was the first attempt to investigate the complex effects of organic and inorganic contaminants using freshly isolated cells from marine mussels.

Copper oxide nanoparticles (CuO-NPs) are among the most widely used metal oxide NPs, meaning they have a relatively high chance of being released into the marine environment. As the applications of these particles have increased in recent years, their potential impact on the health of marine biota has also increased, but to date their toxicological effects in the marine environment are relatively unknown. Consequently, the DNA-damaging potential of CuO-NPs in *M. trossulus* was evaluated and compared to that of dissolved copper exposures (Chelomin *et al.* 2017). Genotoxicity was assessed using the comet assay in mussel gill and digestive gland cells. Results showed that copper in both forms (CuO-NPs and dissolved copper) was accumulated to different extents in mussel tissues. The mussel exposed to the dissolved copper attained higher concentrations of copper in the gills than in the digestive gland, while CuO-NPs induced much higher copper accumulation in the digestive gland than in the gills. However, CuO-NP exposure increased DNA damage significantly in gill cells only – damage that could ultimately lead to the initiation of mutations and malignant transformation in cells.

An Integrative Biomarker Approach

Due to the variation in the mechanisms involved in organisms when they are exposed to a pollutant, results interpretation from a combination of biomarkers (i.e. a multi-biomarker approach) has an advantage over interpretation of single biomarker results (D'Agata *et al.* 2014; Brenner *et al.* 2014; Gonzalez & Bebiano 2014; Franzellitti *et al.* 2015; Maisano *et al.* 2017; Larsson *et al.* 2018). However, the use of a multi-biomarker approach will often produce complex response data that can be hard to interpret, not to mention integrate into environmental policy frameworks. To circumvent this problem, several methods for simplifying complex data have been developed, among which is the integrated biomarker response (IBR) index. The IBR combines different biomarkers into single values, which can be used to describe the toxically-induced stress level of populations in different areas (Beliaeff & Burgeot 2002; Broeg *et al.* 2006; Brooks *et al.* 2012; Marigómez *et al.* 2013a,b; Devin *et al.* 2014; Turja *et al.* 2014; Bebianno *et al.* 2015). It has met with a certain amount of distrust because of frequent misuse and a bias in its calculation. However, since Devin *et al.* (2014) proposed a new calculation method based on a more simple formula and a permutation procedure, it has become possible to avoid the prior misuse and bias. As highlighted by Broeg & Lehtonen (2006), the IBR method represents a simplification of very complex exposure situations in the field, and the response index should not be taken literally but rather as a tool to direct further actions. Most crucially, it must be kept in mind that in all cases indices like this are oversimplifications of very complex exposure situations prevailing in the field, including combined effects of pollutants and abiotic factors, and of the multiple physiological responses in the target organisms.

To assess how mussels and other sentinel species in such exposed ecosystems respond to complex stresses, the International Council for the Exploration of the Sea (ICES) suggested the integrated ecosystem assessment (IEA) approach with an integrative mussel monitoring strategy as a key component, involving the monitoring of biological responses in mussels at different levels of biological complexity, including subcellular responses, tissue responses and whole-organism responses, combined with contaminant concentrations measured in the whole soft tissue of mussels (Davies & Vethaak 2012). IEA has been developed in recent years as a holistic integrated approach to biological effects monitoring. But what are biological effects? They may be defined as the response of an organism, a population or a community to changes in its environment, whether anthropogenic or natural. The usefulness of any biological effect method will obviously depend on how well it is able to separate anthropogenic stressors from the influence of variability in natural processes. The output should be a measure of the well-being or health of some ecosystem component (Thain *et al.* 2008). The most widely used biological effect tools are measures of the biochemical and/or physiological state of selected organisms. Biological effects measurements assist the assessment of the significance of measured concentrations of contaminants in biota or sediments, and can include an assessment of the impact of concurrent exposure to multiple contaminants. Biological effect techniques range from responses measured at the subcellular level (e.g. metallothionein, lysosomal stability, AChE) to whole-organism responses (e.g. SGF, SoS). Those that are to be included in an integrated programme to assess the impact of contaminants on the marine environment require the following characteristics:

- the ability to separate contaminant-related effects from influences caused by other factors (e.g. natural variability, food availability);
- sensitivity to contaminants (i.e. providing 'early warning');
- a suite of methods that covers a range of mechanisms of toxic action (e.g. oestrogenicity/androgenicity, neurotoxicity, carcinogenicity, genotoxicity mutagenicity);
- the inclusion of at least one method that measures the 'general health' of the organism (Davies & Vethaak 2012).

When biological effects measurements are carried out in combination with chemical measurements (or additional effects measurements), it allows the identification (in some cases) of the substance/group of substances contributing to the observed effects. By bringing together these monitoring (and assessment) disciplines, which are usually conducted separately, an integrated assessment can improve the ability to describe the reasons for the appearance of areas with decreased or poor environmental status. The economic benefit of such an integrated approach comes from the fact that the samples and data are gathered during the same surveys and that the data can be directly compared/used with holistic assessment tools to provide truly integrated results with respect to contaminant concentrations and their effects assessments (Vethaak *et al.* 2017).

In addition, Vethaak *et al.* (2017) provide the scientific basis for a framework for integrated chemical and biological effects monitoring and assessment in the marine environment. The framework comprises a core set of biological effect techniques developed by ICES and included or recommended in the OSPAR (www.ospar.org) monitoring programmes that can be used in an integrated manner together with chemical contaminant measurements in biota, sediments and water across OSPAR maritime areas (OSPAR Commission 2013; Figure 8.10). Furthermore, it incorporates an assessment framework that integrates contaminant and biological effects monitoring data and allows assessments to be made across matrices, sites and regions. The assessment framework is simple and transparent and allows for multiple levels of aggregation for different assessment requirements. The presented integrated framework and methodology can generally be applied to other marine regions, including the Baltic and Mediterranean seas. However, this may require some development and application of region-specific methods in key species and associated assessment criteria.

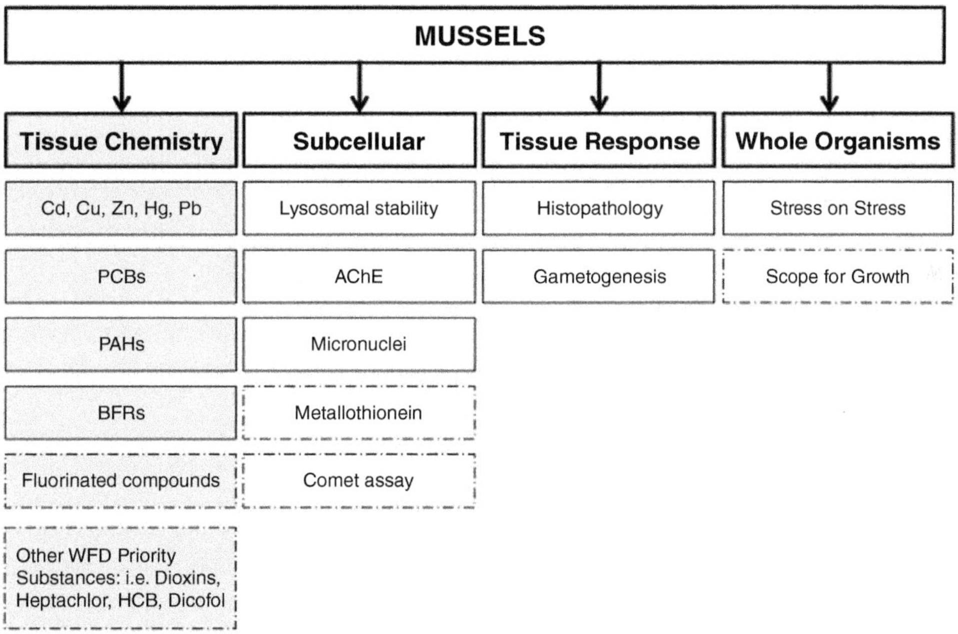

Figure 8.10 Determinands and measurements included in the mussel component of the ICES/OSPAR integrated monitoring framework. Solid lines, core methods; broken lines, additional methods. PCBs, polychlorinated biphenyls; PAH, polycyclic aromatic hydrocarbon; BFRs, brominated flame retardants; AChE, acetylcholinesterase; WFD, Water Framework Directive. WFD priority substances are required in biota under Directive 2013/39/EU. Supportive factors for biota are not shown (details can be found in OSPAR Commission 2013). *Source:* From Veethaak *et al.* (2017). Reproduced with permission from Elsevier.

Essential components of the integrated monitoring and assessment framework are:

- defined combinations of chemical and biological effects measurements;
- carefully managed sampling programmes;
- measurement methods and determinands that are understood and well supported by background documents, technical annexes, standard protocols and quality control;
- a coherent set of assessment criteria that represent similar levels of environmental risk across determinands;
- data integration methods that enable combination of data over a range of geographical scales and which can accommodate limited or incomplete data sets (Vethaak *et al.* 2019).

The ICES/OSPAR framework links chemical contaminants with the health of the ecosystem and can provide a suitable approach for the assessment of Good Environmental Status (GES) for Descriptor 8 of the MSFD (see the beginning of this chapter for information on EU legislative measures and policies). In order to give some stability to assessments, it is important that future revisions of techniques and assessment criteria are harmonised with the MSFD cycle. Currently, the background documents and assessment criteria are available for all biological effect techniques relevant to the ecosystem components for integrated monitoring of contaminants and their effects, apart from benthic fauna and passive samplers. These are important elements of the integrated scheme, and work to prepare background documents and assessment criteria needs to be undertaken as soon as possible. However, it should be noted that our knowledge regarding integrated monitoring and assessment will continue to evolve and new emerging contaminants and techniques should be added or should replace old ones. The ICES/OSPAR framework has been validated through the international ICON project and a set of case studies.

In 2017, an international workshop on marine integrated contaminant monitoring (ICON) was initiated to test the preceding framework in practice on a Europe-wide scale. The outcomes of the ICON project are reported in *Marine Environmental Research*, Volume 124 in a series of method-orientated papers addressing chemical analyses, PAH metabolites, oxidative stress, biotransformation, LMS, genotoxicity, disease and sediment assessment, as well as papers assessing specific areas. ICON was initially planned to evaluate effects of contaminants in the North Sea, with Iceland as a reference area, but was later extended to the Baltic, France and Spanish Mediterranean waters. Hylland *et al.* (2017a) provided a background and introduction to the ICON project, reviewing how the effects of contaminants on marine organisms (mussel and fish species) can be monitored and describing strategies that have been employed to monitor and assess such effects. Through the ICON project, they demonstrated the use of an integrating framework and gleaned more knowledge than had previously been gained in any single field campaign about the impacts contaminants may have in European marine areas. In addition, Hylland *et al.* (2017b) reported a full assessment of results from ICON, encompassing different matrices (sediment, fish, mussels, gastropods), areas (Iceland, North Sea, Baltic, Wadden Sea, Seine estuary (France), western Mediterranean) and end points (chemical analyses, biological effects).

ICON has demonstrated the use of a framework for integrated contaminant assessment on European coastal and offshore areas. The assessment showed that chemical contamination did not always correspond with biological effects, indicating that both are required. The framework can be used to develop assessments for EU directives. If a 95% target were to be used as a regional indicator of MSFD GES, Iceland and the offshore North Sea would achieve that target using the ICON data set, but inshore North Sea, Baltic and Spanish Mediterranean regions would fail.

In another ICON study, Martínez-Gömez *et al.* (2017) investigated whether general stress biomarkers in mussels can be applied as common first-tier biomarkers in regional biomonitoring programmes in the North Sea (including Iceland) and western Mediterranean Sea. SoS and LMS biomarkers were analysed in resident mussels (*Mytilus* spp.) from eight coastal sites and in transplanted mussels (*M. galloprovincialis*) from two Spanish Mediterranean coastal sites. The assessment of results, as input to pollution monitoring strategies, was performed jointly for LMS and SoS data from the two regions. The contaminant body burden of the mussels was compared with biomarker results. The results demonstrated that these two general and nonexpensive stress biomarkers in mussel can be applied throughout European waters, providing a cost-effective and harmonised approach to screen contaminant-related biological effects within the framework of wide-scale pollution biomonitoring programmes, such as that proposed by the European Union, the MSFD.

In another study carried out in the framework of the ICON project, Robinson *et al.* (2017) describe and assess the concentrations of trace metals, PAHs and PCBs in sediments, mussels (*Mytilus* spp.) and fish collected from estuarine, coastal and offshore waters from Iceland to the Mediterranean Sea. For organic contaminants, concentrations progressively increased from Iceland to the offshore North Sea to the coastal seas, and were highest in estuaries. Metals had a more complex distribution, reflecting local anthropogenic inputs, natural sources and hydrological conditions. Use of internationally recognised assessment criteria indicated that at no site were concentrations of all contaminants at background levels, and that concentrations of some contaminants were of significant concern in all areas except the central North Sea.

Between 2014 and 2016, an interdisciplinary team of researchers, including physical oceanographers, biologists, economists and anthropologists, developed a working example of an IEA for three ecologically distinct regions of the Northwest Atlantic – Georges Bank, the Gulf of Maine and the Grand Banks – as part of the ICES Working Group on the Northwest Atlantic Regional Sea (WGNARS). IEAs were subsequently envisioned as the cornerstone integrated science product for the US National Oceanic and Atmospheric Administration (NOAA), the agency most responsible for research, management and conservation of oceans in US waters. This led to the development of an NOAA IEA Program that would oversee regional implementation of the national IEA framework. Recently, De Piper *et al.* (2017) reviewed the transdisciplinary and collaborative process by which the IEA was developed, with a particular focus on the decision points arising from the IEA construct itself. The aim was to identify key issues faced in developing any IEA, practical decisions made to address these issues within the working group and lessons learned from the process. See also Foley *et al.* (2013), Fletcher *et al.* (2014), Levin *et al.* (2014), Samhouri *et al.* (2014) and Harvey *et al.* (2017).

Notes

1 A unique numerical identifier assigned by the Chemical Abstracts Service to every chemical substance.
2 Sentinel species are defined as biological monitors that accumulate a pollutant in their tissues without significant adverse effects and can be used to measure in a sensitive manner the amount of that pollutant that is biologically available (Beeby 2001).
3 The potential for uptake of a substance by a living organism. Usually expressed as the fraction that can be taken up by the organism in relation to the total amount of the substance available (Nikinmaa 2014).
4 An analyte is a substance whose chemical constituents are being identified and measured.
5 Materials that can soak up oil from water by either absorption or adsorption.
6 The process whereby contaminants are transferred from food to an organism, resulting in higher concentrations compared with the source (Grey 2002).
7 Surrogate standards are analytes added to a sample at a known concentration in order to determine extraction efficiency. The internal standard is an analyte chemically similar to those being evaluated. It is typically added in GC-MS methods to correct analyte concentrations during analysis.
8 The REACH Regulation Registration, Evaluation, Authorisation and Restriction of Chemicals is an EU regulation from December 2006 addressing the production and use of chemical substances and their potential impacts on both human health and the environment.
9 A fundamental scattering process in which the kinetic energy of an incident particle is not conserved (in contrast to elastic scattering). In an inelastic scattering process, some of the energy of the incident particle is lost or increased.
10 Matrix refers to the components of a sample other than the analyte of interest.
11 A system or process is in a steady state if the variables which define its behaviour are unchanging in time.
12 Marine animals eat microplastics, for example, when they mistake them for food. When they themselves are eaten, the microplastics are unintentionally ingested by the predator. This process is referred to as the trophic transfer of microplastics.
13 A chemical substance found within an organism that is not naturally produced or expected to be present within the organism.
14 Passive sampling is an environmental monitoring technique involving the use of a collecting medium, whether an artificial device or biological organism, to accumulate chemical pollutants in the environment over time (see Challis *et al.* 2016 for details).
15 Analyte-free matrices such as DI Water for liquids or cleaned sand for solids and/or soils that are processed in exactly the same manner as samples. The main function of a method blank (MB) is to document contamination resulting from the analytical process.
16 Chemicals often used or produced by industry that remain in the environment long after they were first introduced. Often, they were not recognised as harmful when they were being used.
17 Lipid peroxidation is the metabolic process in which reactive oxygen species (ROS) cause the oxidative deterioration of lipids, which may significantly affect cell membrane structure and function (Vasilaki & McMillan 2011).
18 Nucleoids are essential for controlling the activity of the cell and reproduction. They are where transcription and replication of DNA take place.

References

Abdallah, M.A.-E. (2014) Advances in instrumental analysis of brominated flame retardants: current status and future perspectives. *International Scholarly Research Notices*, **2014**, art. 651834.

Albergamo, A., Rigano, F., Purcaro, G., Mauceri, A., Fasulo, S. & Mondello, L. (2016) Free fatty acid profiling of marine sentinels by nanoLC-EI-MS for the assessment of environmental pollution effects. *Science of the Total Environment*, **571**, 955–962.

Alvarez, D., Maruya, K., Dodder, N.G., Lao, W., Furlong, E. & Smalling, K. (2014) Occurrence of contaminants of emerging concern along the California coast (2009–10) using passive sampling devices. *Marine Pollution Bulletin*, **81**, 347–354.

Álvarez-Muñoz, D., Rodríguez-Mozaz, S., Maulvault, A.L., Tediosi, A., Fernández-Tejedor, M. Van den Heuvel, F. *et al.* (2015a) Occurrence of pharmaceuticals and endocrine disrupting compounds in macroalgaes, bivalves, and fish from coastal areas in Europe. *Environmental Research*, **143**, 56–64.

Alvarez-Muñoz, D., Huerta, B., Fernandez-Tejedor, M., Rodríguez-Mozaz, S., Barceló, D. (2015b) Multi-residue method for the analysis of pharmaceuticals and some of their metabolites in bivalves. *Talanta*, **136**, 174–182.

Álvarez-Muñoz, D., Llorca, M., Blasco, J. & Barceló (2016) Contaminants in the marine environment. In: *Marine Ecotoxicology* (eds J. Blasco, P.M. Chapman, O. Campana & M. Hampel), pp. 1–34. Academic Press, New York.

Álvarez-Muñoz, D., Rodríguez-Mozaz, S., Jacobs, S., Serra-Compte, A., Cáceres, N., Sioen, I. *et al.* (2018) Pharmaceuticals and endocrine disruptors in raw and cooked seafood from European market: concentrations and human exposure levels. *Environment International*, **119**, 570–581.

Amiard, J.C., Amiard-Triquet, C., Barka, S., Pellerin, J. & Rainbow, P.S. (2006) Metallothioneins in aquatic invertebrates: their role in metal detoxification and their use as biomarkers. *Aquatic Toxicology*, **76**, 160–202.

Andral, B., Stanisière, J.-Y., Damier, E., Thébault, H., Galgani, F. & Boissery, P. (2004) Monitoring chemical contamination levels in the Mediterranean based on the use of mussel caging. *Marine Pollution Bulletin*, **49**, 704–712.

Andral, B., Galgani, G., Tomasino, C., Bouchoucha, M., Blottiere, C., Scarpato, A. *et al.* (2011) Chemical contamination baseline in the Western Basin of the Mediterranean Sea based on transplanted mussels. *Archives of Environmental Contamination and Toxicology*, **6**, 261–271.

Ansari, T.M., Marr, L.L. & Tariq, N. (2004) Heavy metals in marine pollution perspective: a mini review. *Journal of Applied Sciences*, **4**, 1–20.

Araujo, C.F., Nolasco, M.M., Ribeiro, A.M.P. & Ribeiro-Claro, P.J.A. (2018) Identification of microplastics using Raman spectroscopy: latest developments and future prospects. *Water Research*, **142**, 426–440.

Astudillo, L.R.D., Yen, I.C., Agard, J., Bekele, I. & Hubbard, R. (2002) Heavy metals in green mussel (*Perna viridis*) and oysters (*Crassostrea* sp.) from Trinidad and Venezuela. *Archives of Environmental Contamination and Toxicology*, **42**, 410–415.

Authman, M.M.N., Zaki, M.S., Khallaf, E.A. & Abbas, H.H. (2015) Use of fish as bio-indicator of the effects of heavy metals pollution. *Journal of Aquaculture Research and Development*, **6**, 328.

Azizi, G., Akodad, M., Baghour, M., Layachi, M. & Moumen, A. (2018) The use of *Mytilus* spp. mussels as bioindicators of heavy metal pollution in the coastal environment. A review. *Journal of Materials and Environmental Sciences*, **9**, 1170–1181.

Baines, S.B., Fisher, N.S. & Kinney, E.L. (2006) Effects of temperature on uptake of aqueous metals by blue mussels *Mytilus edulis* from Arctic and temperate waters. *Marine Ecology Progress Series*, **308**, 117–128.

Bajt, O., Ramšak, A., Milun, V., Andral, B., Romanelli, G., Scarpato, A. *et al.* (2019) Assessing chemical contamination in the coastal waters of the Adriatic Sea using active mussel biomonitoring with *Mytilus galloprovincialis. Marine Pollution Bulletin*, **141**, 283–298.

Balbi, T., Fabbri, R., Montagna, M., Camisassi, G. & Canesi, L. (2017) Seasonal variability of different biomarkers in mussels (*Mytilus galloprovincialis*) farmed at different sites of the Gulf of La Spezia, Ligurian Sea, Italy. *Marine Pollution Buletin*, **116**, 348–356.

Ballentine, M., Tobias, C., Vlahos, P., Smith, R. & Cooper, C. (2015) Bioconcentration of TNT and RDX in coastal marine biota. *Archives of Environmental Contamination and Toxicology*, **68**, 718–728.

Barhoumi, B., El Megdiche,Y., Clérandeau, C., Ameur, W.B., Mekni, S., Bouabdallah, S. *et al.* (2016) Occurrence of polycyclicaromatic hydrocarbons(PAHs) in mussel (*Mytilus galloprovincialis*) and eel (*Anguilla anguilla*) from Bizerte lagoon,Tunisia, and associated human health risk assessment. *Continental Shelf Research*,**124**, 104–116.

Barron, M.G. (1990) Bioconcentration. Will water-borne organic chemicals accumulate in aquatic animals? *Environmental Science & Technology*, **24**, 1612–1618.

Bayen, S., Thomas, G. O., Lee, H.K. & Obbard, J.P. (2003) Occurrence of polychlorinated biphenyls and polybrominated diphenyl ethers in green mussels (*Perna viridis*) from Singapore, Southeast Asia. *Environmental Toxicology and Chemistry*, **22**, 2432–2437.

Bebianno, M.J., Pereira, C.G., Rey, F., Cravo, A., Duarte, D. & D'Errico, G. *et al.* (2015) Integrated approach to assess ecosystem health in harbor areas. *Science of the Total Environment*, **514**, 92–107.

Beeby, A. (2001) What do sentinels stand for? *Environmental Pollution*, **112**, 285–298.

Beliaeff, B. & Burgeot, T. (2002) Integrated biomarker response: a useful tool for ecological risk assessment. *Environmental Toxicology and Chemistry*, **21**, 1316–1322.

Benali, I., Boutiba, Z., Grandjean, D., de Alencastro, L.P., Rouane-Hacene, O. & Chvre, N. (2017) Spatial distribution and biological effects of trace metals (Cu, Zn, Pb, Cd) and organic micropollutants (PCBs, PAHs) in mussels *Mytilus galloprovincialis* along the Algerian west coast. *Marine Pollution Bulletin*, **115**, 539–550.

Benedicto, J., Andral, B., Martínez-Gómez, C., Guitart, C., Deudero, S., Cento, A. *et al.* (2011) A large scale survey of trace metal levels in coastal waters of the Western Mediterranean basin using caged mussels (*Mytilus galloprovincialis*). *Journal of Environmental Monitoring*, **13**, 1495–1505.

Beyer, J., Green, N.W., Brooks, S., Allan, I.J., Ruus, A., Gomes, T. *et al.* (2017) Blue mussels (*Mytilus edulis* spp.) as sentinel organisms in coasta pollution monitoring: a review. *Marine Environmental Research*, **130**, 338–365.

Bilgin, M. & Uluturhan-Suzer, E. (2017) Assessment of trace metal concentrations and human health risk in clam (*Tapes decussatus*) and mussel (*Mytilus galloprovincialis*) from the Homa Lagoon (Eastern Aegean Sea). *Environmental Science and Pollution Research*, **24**, 4174–4184.

Blackmore, G. & Wang, W.X. (2003) Inter-population differences in Cd, Cr, Se, and Zn accumulation by the green mussel *Perna viridis* acclimated at different salinities. *Aquatic Toxicology*, **62**, 205–218.

Boillot, C., Martinez Bueno, M.J., Munaron, D., Le Dreau, M., Mathieu, O., David, A. *et al.* (2015) in vivo exposure of marine mussels to carbamazepine and 10-hydroxy-10,11-dihydro-carbamazepine: bioconcentration and metabolization. *Science of the Total Environment*, **532**, 564–570.

Bolognesi, C. & Cirillo, S. (2014) Genotoxicity biomarkers in aquatic bioindicators. *Current Zoology*, **60**, 273–284.

Bonnefille, B., Arpin-Pont, L., Gomez, E., Fenet, H. & Courant, F. (2017) Metabolic profiling identification of metabolites formed in Mediterranean mussels (*Mytilus galloprovincialis*) after diclofenac exposure. *Science of the Total Environment*, **583**, 257–268.

Booth, A.M., Sutton, P.A., Lewis, C.A., Lewis, A.C., Scarlett, A., Chau, W. *et al.* (2007) Unresolved complex mixtures of aromatic hydrocarbons: thousands of overlooked persistent, bioaccumulative, and toxic contaminants in mussels. *Environmental Science and Technology*, **41**, 457–464.

Bourget, E. & Cossa, D. (1976) Mercury content of mussels from the St. Lawrence Estuary and northwestern Gulf of St. Lawrence, Canada. *Marine Pollution Bulletin*, **7**, 237–239.

Boyden, C. (1974) Trace element content and body size in molluscs. *Nature*, **251**, 311–314.

Brausch, J.M. & Rand, G.M. (2011) A review of personal care products in the aquatic environment: environmental concentrations and toxicity. *Chemosphere*, **82**, 1518–1532.

Brenner, M., Fraser, D., Van Nieuwenhove, K., O'Beirn, F.X., Buck, B.H., Mazurie, J. *et al.* (2014) Bivalve aquaculture transfers in Atlantic Europe. Part B: environmental impacts of transfer activities. *Ocean and Coastal Management*, **89**, 139–146.

Broeg, K. & Lehtonen, K.K. (2006) Indices for the assessment of environmental pollution of the Baltic Sea coasts: integrated assessment of a multi-biomarker approach. *Marine Pollution Bulletin*, **53**, 508–522.

Brooks, S., Lyons, B., Goodsir, F., Bignell, J. &Thain, J. (2009) Biomarker responses in mussels, an integrated approach to biological effects measurements. *Journal of Toxicology and Environmental Health, Part A: Molecular and Integrative Physiology*, **72**, 196–208.

Brooks, S., Harman, C., Soto, M., Cancio, I., Glette, T. & Marigómez, I. (2012) Integrated coastal monitoring of a gas processing plant using native and caged mussels. *Science of the Total Environment*, **426**, 375–386.

Brown, R.J., Galloway, T.S., Lowe, D., Browne, M.A., Dissanayake, A., Jones, M.B. *et al.* (2004) Differential sensitivity in three marine invertebrates to copper assessed using multiple biomarkers. *Aquatic Toxicology*, **66**, 267–278.

Browne, M.A. (2015) Sources and pathways of microplastics to habitats. In: *Marine Anthropogenic Litter* (eds M. Bergmann, L. Gutow & M. Klages), pp. 229–244. Springer Open, Philadelphia, PA.

Browne, M.A., Galloway, T. & Thompson, R. (2007) Microplastic-an emerging contaminant of potential concern? *Integrated Environmental Assessment and Management*, **3**, 559–566.

Browne, M.A., Dissanayake, A., Galloway, T., Lowe, D. M. & Thompson, R.C. (2008) Ingested microscopic plastic translocates to the circulatory system of the mussel, *Mytilus edulis* (L.). *Environmental Science and Technology*, **42**, 5026–5031.

Browne, M.A., Crump, P., Niven, S.J., Teuten, E., Tonkin, A., Galloway,T. *et al.* (2011) Accumulation of microplastic on shorelines worldwide: sources and sinks. *Environmental Science and Technology*, **45**, 9175–9179.

Bruzzoniti, M.C., Checchini, L., De Carlo, R.M., Orlandini, S., Rivoira, L. & Del Bubba, M. (2014) QuEChERS sample preparation for the determination of pesticides and other organic residues in environmental matrices: a critical review. *Analytical and Bioanalytical Chemistry*, **406**, 4089–4116.

Bryan, G.W. (1976) Heavy metal contamination in the sea. In: *Marine Pollution* (ed. R. Johnston), pp. 214–228. Academic Press, New York.

Bundschuh, M., Filser, J., Lüderwald, S., McKee, M.S., Metreveli, G., Schaumann, G.E. *et al.* (2018) Nanoparticles in the environment: where do we come from, where do we go to? *Environmental Sciences Europe,* **30**, *art.* 6.

Cajaraville, M.P., Marigomez, J.A., Diez, G. & Angulo, E. (1992) Comparative effects of the water accommodated fraction of three oils on mussels 2. Quantitative alterations in the structure of the digestive tubules. *Comparative Biochemistry and Physiology Part C: Comparative Pharmacology and Toxicology*, **102**, 113–123.

Cajaraville, M.P., Bebianno, M.J., Blasco, J., Porte, C., Sarasquete, C. & Viarengo, A. (2000) The use of biomarkers to assess the impact of pollution in coastal environments of the Iberian Peninsula: a practical approach. *Science of the Total Environment*, **247**, 295–311.

Campillo, J.A., Albentosa, M., Valdés, N.J., Moreno-González, R. & León, V.M. (2013) Impact assessment of agricultural inputs into a Mediterranean coastal lagoon (Mar Menor, SE Spain) on transplanted clams (*Ruditapes decussatus*) by biochemical and physiological responses. *Aquatic Toxicology*, **142–143**, 365–379.

Canesi, L. & Corsi, I. (2016) Effects of nanomaterials on marine invertebrates. *Science of the Total Environment*, **565**, 933–940.

Cantillo, A.Y. (1998) Comparison of results of mussel watch programs of the United States and France with worldwide mussel watch studies. *Marine Pollution Bulletin*, **36**, 712–717.

Cappello, T., Maisano, M., D'Agata, A., Natalotto, A., Mauceri, A. & Fasulo, S. (2013) Effects of environmental pollution in caged mussels (*Mytilus galloprovincialis*). *Marine Environmental Research*, **91**, 52–60.

Carro, N., Cobas, J., García, I., Ignacio, M. & Mouteira, A. (2014) Distribution and trend of organochlorine pesticides in Galicia coast using mussels as bioindicator organisms. Possible relationship to biological parameters. *Chemosphere*, 1**02**, 37–47.

Carro, N., García, I., Ignacio, M. & Mouteira, A. (2016) Levels of PCBs in oysters coming from Galicia coast: comparison to mussels from the same region. *Bulletin of Environmental Contamination and Toxicology*, **96**, 608–615.

Carro, N., García, I. Ignacio, M. & Mouteira, A. (2018) Polychlorinated dibenzo-P-dioxins and dibenzofurans (PCDD/Fs) and dioxin-like polychlorinated biphenyls (dl-PCBS) in bivalve mollusc from Galician Rías (NW SPAIN). *Chemosphere*, **197**, 782–792.

Castro, M., Santos, M.M., Monteiro, N.M. & Vieira, N. (2004) Measuring lysosomal stability as an effective tool for marine coastal environmental monitoring. *Marine Environmental Research*, **58**, 741–745.

CCOHS (2020) Chemicals and materials: endocrine disruptors. Canadian Centre for Occupational Health and Safety. Available at: https://www.ccohs.ca/oshanswers/chemicals/endocrine.html (accessed 28 May 2021).

Challis, J.K., Hanson, M.L. & Wong, C.S. (2016) Development and calibration of an organic-diffusive gradients in thin films aquatic passive sampler for a diverse suite of polar organic contaminants. *Analytical Chemistry*, **88**, 10583–10591.

Chandurvelan, R., Marsden, I.D., Glover, C.N. & Gaw, S. (2015) Assessment of a mussel as a metal bioindicator of coastal contamination: relationships between metal bioaccumulation and multiple biomarker responses. *Science of the Total Environment*, **511**, 663–675.

Chelomin, V.P., Slobodskova, V.V., Zakhartsev, M. & Kukla, S. (2017) Genotoxic potential of copper oxide nanoparticles in the bivalve mollusk *Mytilus trossulus*. *Journal of Ocean University of China*, **16**, 339–345.

ChemSafetyPro (2019) Bioaccumulation. Available at: https://www.chemsafetypro.com/Topics/CRA/Bioconcentration_Factor_BCF.html (accessed 28 May 2021).

Cheung, V.V., Wedderburn, R.J. & Depledge, M.H. (1998) Molluscan lysosomal responses as a diagnostic tool for the detection of a pollution gradient in Tolo Harbour, Hong Kong. *Marine Environmental Research*, **46**, 237–241.

Cheung, C.C.C., Zheng, G.J., Li, A.M.Y., Richardson, B.J. & Lam, P.K.S. (2001) Relationships between tissue concentrations of polycyclic aromatic hydrocarbons and antioxidative responses of marine mussels, *Perna viridis*. *Aquatic Toxicology*, **52**, 189–203.

Cheung, C.C.C., Zheng, G.J., Lam, P.K.S. & Richardson, B.J. (2002) Relationships between tissue concentrations of chlorinated hydrocarbons (polychlorinated biphenyls and chlorinated pesticides) and antioxidative responses of marine mussels, *Perna viridis*. *Marine Pollution Bulletin*, **45**, 181–191.

Cheung, C.C.C., Siu, W.H.L., Richardson, B.J., De Luca-Abbott, S.B. & Lam, P.KS. (2004) Antioxidant responses to benzo[*a*]pyrene and Aroclor 1254 exposure in the green-lipped mussel, *Perna viridis*. *Environmental Pollution*, **128**, 393–403.

Choi, H.G., Moon, H.B., Choi, M., Yu, J. & Kim, S.S. (2010) Mussel watch program for organic contaminants along the Korean coast, 2001–2007. *Environmental Monitoring and Assessment*, **169**, 473–485.

Choi, J.Y., Yang, D.B., Hong, G.H., Kim, K. & Shin, K.H. (2016) Ecological and human health risk from polychlorinated biphenyls and organochlorine pesticides in bivalves of Cheons Bay, Korea. *Environmental Engineering Research*, **21**, 373–383.

Claessens, M., Van Cauwenberghe, L., Vandegehuchte, M.B. & Janssen, C.R. (2013) New techniques for the detection of microplastics in sediments and field collected organisms. *Marine Pollution Bulletin*, **70**, 227–233.

Cole, M., Lindeque, P., Halsband, C. & Galloway, T.S. (2011) Microplastics as contaminants in the marine environment: a review. *Marine Pollution Bulletin*, **62**, 2588–2597.

Cole, M., Webb, H., Lindeque, P.K. Fileman, E.S., Halsband, C. & Galloway, T.S (2014) *Isolation of microplastics in biota-rich seawater samples and marine organisms. Scientific Reports*, **4**, 4528.

Collins, A.R. (2004) The comet assay for DNA damage and repair: principles, applications, and limitations. *Molecular Biotechnology*, **26**, 249–261.

Conti, M.E. & Cecchetti, G. (2003) A biomonitoring study: trace metals in algae and molluscs from Tyrrhenian coastal areas. *Environmental Research*, **93**, 99–112.

Corada-Fernández, C., Lara-Martín, P.A., Candela, L. & González-Mazo, E. (2013) Vertical distribution profiles and diagenetic fate of synthetic surfactants in marine and freshwater sediments. *Science of the Total Environment*, **461–462**, 568–575.

Corsi, I., Cherr, G.N., Lenihan, H.S., Labille, J., Hassellov, M., Canesi, L. *et al.* (2014) Common strategies and technologies for the ecosafety assessment and design of nanomaterials entering the marine environment. *ACS Nano*, **8**, 9694–9709.

Cortés Muñoz, J.E., Calderón Mólgora, C.G., Martín Domínguez, A., Espino de la O, E.E., Gelover Santiago, S.L., Hernández Martínez, C.L. *et al.* (2013) Endocrine disruptors in water sources: human health risks and EDs removal from water through nanofiltration. In: *International Perspectives on Water Quality Management and Pollutant Control* (ed. N.W.T. Quinn), pp 25–59. InTechOpen, London.

Cossa, D. (1989) A review of the use of *Mytilus* spp. as quantitative indicators of cadmium and mercury contamination in coastal waters. *Oceanologica Acta*, **12**, 417–432.

Cózar, A., Echevarría, F., González-Gordillo, J. I., Irigoien, X., Úbeda, B., Hernández-León, S., *et al.* (2014) Plastic debris in the open ocean. *Proceedings of the National Academy of Sciences*, **111**, 10239–10244.

Cunha, S.C., Pena, A. & Fernandes, J.O. (2017) Mussels as bioindicators of diclofenac contamination in coastal environments. *Environmental Pollution*, **225**, 354–360.

Dabwan, A.H.A. & Tauiq, M. (2016) Bivalves as bio-indicators for heavy metals detection in Kuala Kemaman, Terengganu, Malaysia. *Indian Journal of Science and Technology*, **9**, doi 10.17485.

Da Costa, J.P., Santos, P.S., Duarte, A.C. & Rocha-Santos, T. (2016) (Nano)plastics in the environment– sources, fate and effects. *Science of the Total Environment*, **566–567**, 15–26.

D'Agata, A., Cappello, T., Maisano, M., Parrino, V., Giannetto, A., Brundo, M.V. *et al.* (2014) Cellular biomarkers in the mussel *Mytilus galloprovincialis* (Bivalvia: Mytilidae) from Lake Faro (Sicily, Italy). *Italian Journal of Zoology*, **81**, 43–54.

Dailianis, S. (2010) Environmental impact of anthropogenic activities: the use of mussels as a reliable tool for monitoring marine pollution. In: *Mussels: Anatomy, Habitat and Environmental Impact* (ed. L.E. McGevin), pp. 43–72. Nova Science Publishers, New York.

Danis, B., Debacker, V., Mirandab, T. & Dubois, P. (2006) *Ecotoxicology and Environmental Safety*, **65**, 188–200.

Davies, I.M. & Vethaak, A.D. (2012) Integrated marine environmental monitoring of chemicals and their effects. ICES Cooperative Research Report, No. 315.

DeForest, D.K., Brix, K.V. & Adams, W.J. (2007) Assessing metal bioaccumulation in aquatic environments: the inverse relationship between bioaccumulation factors, trophic transfer factors and exposure concentration. *Aquatic Toxicology*, **84**, 236–246.

Degger, N., Chiu, J.M.Y., Po, B.H.K., Tse, A.C.K., Zheng, G.J., Zhao, D.-M. *et al.* (2016) Heavy metal contamination along the China coastline: a comprehensive study using Artificial Mussels and native mussels. *Journal of Environmental Management*, **180**, 238–246.

Dehaut, A., Cassone, A-L., Frère, L., Hermabessiere, L., Himber, C., Rinnert, E. *et al.* (2016) Microplastics in seafood: benchmark protocol for their extraction and characterization. *Environmental Pollution*, **215**, 223–233.

DePiper, G.S., Gaichas, S.K., Lucey, S.M., Pinto da Silva, P., Anderson, M.R., Breeze, H. *et al.* (2017) Operationalizing integrated ecosystem assessments within a multidisciplinary team: lessons learned from a worked example. *ICES Journal of Marine Science*, **74**, 2076–2086.

Devin, S., Burgeot, T., Giamberini, L., Minguez, L. & Pain-Devin, S. (2014) The integrated biomarker response revisited: optimization to avoid misuse. *Environmental Science and Pollution Research*, **21**, 2448–2454.

de Wit, C.A. (2002) An overview of brominated flame retardants in the environment. *Chemosphere*, **46**, 583–624.

De Witte, B., Devriese, L., Bekaert, K., Hoffman, S., Vandermeersch, G., Cooreman, K. *et al.* (2014) Quality assessment of the blue mussel (*Mytilus edulis*): comparison between commercial and wild types. *Marine Pollution Bulletin*, **85**, 146–155.

De Wolf, P. (1975) Mercury content of mussels from west European coasts. *Marine Pollution Bulletin*, **6**, 61–63.

Diamanti-Kandarakis, E., Bourguignon, J-P., Giudice, L.C., Hauser, R., Prins, G.S. Soto, A.M. *et al.* (2009) Endocrine-disrupting chemicals: an endocrine society scientific statement. *Endocrine Reviews*, **30**, 293–342.

Dirtu, A.C., Abdallah, M. & Covaci, A. (2013) Advances in the sample preparation of brominated flame retardants and other brominated compounds. *Trends in Analytical Chemistry*, **43**, 189–203.

Dodder, N.G., Maruya, K.A., Ferguson, P.L., Grace, R., Klosterhaus, S., La Guardia, M.J. *et al.* (2014) Occurrence of contaminants of emerging concern in mussels (*Mytilus* spp.) along the California coast and the influence of land use, storm water discharge, and treated wastewater effluent. *Marine Pollution Bulletin*, **81**, 340–346.

Dokmeci, A.H. (2017) Assessment of heavy metals in wild mussels *Mytilus galloprovincialis* from the Marmara Sea coast of Tekirdag (Turkey). 15th International Conference on Environmental Science and Technology, Rhodes, Greece, 31 August–2 September, 2017. CEST2017_00569.

Domouhtsidou, G.P., Dailianis, S., Kaloyianni, M. & Dimitriadis, V.K. (2004) Lysosomal membrane stability and metallothionein content in *Mytilus galloprovincialis* (L.), as biomarkers combination with trace metal concentrations. *Marine Pollution Bulletin*, **48**, 572–586.

Duarte, C.A., Giarratano, E., Amin, O.A. & Comoglio, L.I. (2011) Heavy metal concentrations and biomarkers of oxidative stress in native mussels (*Mytilus edulis chilensis*) from Beagle Channel coast (Tierra del Fuego, Argentina). *Marine Pollution Bulletin*, **62**, 1895–1904.

Edwards, M., Jacob, A.P., Kimbrough, K.L., Davenport, E. & Johnson, W.E. (2014) Assessment of contaminant concentrations in California mussels (*Mytilus* spp): relationship to land use and outfalls. *Marine Pollution Bulletin*, **81**, 325–333.

EFSA CONTAM Panel (2016) Statement on the presence of microplastics and nanoplastics in food, with particular focus on seafood. *EFSA Journal*, **14**, e04501.

Einsporn, S., Broeg, K. & Koehler, A. (2005) The Elbe flood 2002–toxic effects of transported contaminants in flatfish and mussels of the Wadden Sea. *Marine Pollution Bulletin*, **50**, 423–429.

El-Moselhy, K.M., Saad, E.M., El-Shaarway, R.F., Mohamadein, L.I. & Mahmoud, S.A. (2016) Assessment of heavy metals pollution using sediments and bivalve *Brachidontes variabilis* as bioindicator in the Gulf of Suez, Egypt. *International Journal of Marine Science*, **6**, 1–13.

El Nemr, A., El-Sikaily, A., Khalad, A., Said, T.O. & Abd-Alla, A.M.A. (2004) Determination of hydrocarbons in mussels from the Egyptian Red Sea coast. *Environmental Monitoring and Assessment*, **96**, 251–261.

EPA (2020) Learn about polychlorinated biphenyls (PCBs). US Environmental Protection Agency. Available at: https://www.epa.gov/pcbs/learn-about-polychlorinated-biphenyls-pcbs (accessed 28 May 2021).

Ericson, H., Thorsén, G. & Kumblad, L. (2010 Physiological effects of diclofenac, ibuprofen and propranolol on Baltic Sea blue mussels. *Aquatic Toxicology*, **99**, 223–231.

Eriksen, M., Lebreton, L.C.M., Carson, H.S., Thiel, M., Moore, C. J., Borerro, J. C. *et al.* (2014) Plastic pollution in the world's oceans: more than 5 trillion plastic pieces weighing over 250 000 tons afloat at sea. *PLoS ONE*, **9**, e111913.

European Commission (1996) Directive 96/59/EC of the European Parliament and of the Council of 16 September 1996 on the disposal of polychlorinated biphenyls and polychlorinated terphenyls (PCB/PCT). *Official Journal of the European Union*, **L243**, 31–35. Available at: http://data.europa.eu/eli/dir/1996/59/oj (accessed 28 May 2021).

European Commission (2000) Directive 2000/60/EC of the European Parliament and of the Council of 23 October 2000 establishing a framework for Community action in the field of water policy (Water Framework Directive). *Official Journal of the European Union*, **L32**, 1–73. Available at: http://data.europa.eu/eli/dir/2000/60/oj (accessed 28 May 2021).

European Commission (2003) Directive 2003/11/EC of the European Parliament and of the Council of 6 February 2003 amending for the 24th time Council Directive 76/769/EEC relating to restrictions on the marketing and use of certain dangerous substances and preparations (pentabromodiphenyl ether, octabromodiphenyl ether). *Official Journal of the European Union*, **L042**, 45–46. Available at: http://data.europa.eu/eli/dir/2003/11/oj (accessed 28 May 2021).

European Commission (2004) Directive 2004/27/EC of the European Parliament and of the Council of 31 March 2004 amending Directive 2001/83/EC on the Community code relating to medicinal products for human use. (Text with EEA relevance) *Official Journal of the European Union*, **L136**, 34–57. Available at: http://data.europa.eu/eli/dir/2004/27/oj (accessed 28 May 2021).

European Commission (2005) Regulation (EC) 396/2005 of the European Parliament and of the Council of 23 February 2005 on maximum residue levels of pesticides in or on food and feed of plant and animal origin and amending Council Directive 91/414/EEC (Text with EEA relevance). *Official Journal of the European Union*, **L70**, 4–8. Available at: http://data.europa.eu/eli/reg/2005/396/oj (accessed 28 May 2021).

European Commission (2008) Directive 2008/105/EC of the European Parliament and of the Council of 16 December 2008 on environmental quality standards in the field of water policy, amending and subsequently repealing Directive 2000/60/EC of the European Parliament and of the Council. *Official Journal of the European Union*, **L348**, 84–97. Available from http://data.europa.eu/eli/dir/2008/105/oj (accessed 28 May 2021).

European Commission (2011a) Commission Regulation (EU) No 836/2011 of 19 August 2011 amending Regulation (EC) No 333/2007 laying down the methods of sampling and analysis for the official control of the levels of lead, cadmium, mercury, inorganic tin, 3-MCPD and benzo(a) pyrene in foodstuffs. (Text with EEA relevance). *Official Journal of the European Union*, **L215**, 9–16. Available at: http://data.europa.eu/eli/reg/2011/836/oj (accessed 28 May 2021).

European Commission (2011b) Regulation (EU) No 835/2011 of 19 August 2011 amending Regulation (EC) 1881/2006 as regards maximum levels for polycyclic aromatic hydrocarbons in foodstuffs. *Official Journal of the European Union*, **L215**, 4–8. Available at: http://data.europa.eu/eli/reg/2011/835/oj (accessed 28 May 2021).

European Commission (2011c) Commission Recommendation of 18 October 2011 on the definition of nanomaterial 2011/696/EU. (Text with EEA relevance). *Official Journal of the*

European Union, **L278**, 38–40. Available at: https://eur-lex.europa.eu/LexUriServ/LexUriServ.do?uri=OJ:L:2011:275:0038:0040:en:PDF (accessed 28 May 2021).

European Commission (2012) Regulation (EU) No 259/2012 of the European Parliament and of the Council of 14 March 2012 amending Regulation (EC) No 648/2004 as regards the use of phosphates and other phosphorus compounds in consumer laundry detergents and consumer automatic dishwasher detergents. (Text with EEA elevance.) *Official Journal of the European Union*, **L94**, 16–21. Available at: http://data.europa.eu/eli/reg/2012/259/oj (accessed 28 May 2021).

European Commission (2013) Directive 2013/39/EU of the European Parliament and of the Council of 12 August 2013 amending Directives 2000/60/EC and 2008/105/EC as regards priority substances in the field of water policy. *Official Journal of the European Union*, **L226**, 1–17. Available at: http://data.europa.eu/eli/dir/2013/39/oj (accessed 28 May 2021).

European Commission (2014) Common Implementation Strategy for the Water Framework Directive (2000/60/EC): Guidance Document No. 32 on Biota Monitoring (The Implementation of EQS_{Biota}) under the Water Framework Directive. European Commission. Available at: https://circabc.europa.eu/sd/a/62343f10-5759-4e7c-ae2b-12677aa57605/Guidance%20No%2032%20-%20Biota%20Monitoring.pdf (accessed 28 May 2021).

European Commission (2017a) Commission Regulation (EU) 2017/644 of 5 April 2017 laying down methods of sampling and analysis for the control of levels of dioxins, dioxin-like PCBs and non-dioxin-like PCBs in certain foodstuffs and repealing Regulation (EU) No 589/2014 (Text with EEA relevance). *Official Journal of the European Union*, **L92**, 9–34. Available at: data.europa.eu/eli/reg/2017/644/oj (accessed 28 May 2021).

European Commission (2017b) Delegated Regulation (EU) 2017/2100 of 4 September 2017 setting out scientific criteria for the determination of endocrine-disrupting properties pursuant to Regulation (EU) No 528/2012 of the European Parliament and Council. (Text with EEA relevance.) *Official Journal of the European Union*, **L301**, 1–5. Available at: http://data.europa.eu/eli/reg_del/2017/2100/oj (accessed 28 May 2021).

European Commision (2017c) Regulation EU 2017/1410 of 2 August 2017 amending Annexes II and III to Regulation (EC) No 1223/2009 of the European Parliament and of the Council on cosmetic products (Text with EEA relevance) *Official Journal of the European Union*, **L202**, 1–3. Available at: http://data.europa.eu/eli/reg/2017/1410/oj (accessed 28 May 2021).

Farrington, J.W., Tripp, B.W., Tanabe, S., Subramanian, A., Sericano, J.L., Wade, T.L. *et al.* (2016) E.D. Goldberg's proposal of 'The Mussel Watch': reflections after 40 years. *Marine Pollution Bulletin*, **110**, 501–510.

Fatoki, O.S., Okoro, H.K., Adekola, F.A., Ximba, B. J. & Snyman, R.J. (2012) *Bioaccumulation of metals in black mussels (Mytilus galloprovincialis) in Cape Town Harbour, South Africa. The Environmentalist*, **32**, 48–57.

Fernandes, A.R., Mortimer, D., Gem, M., Dicks, P., Smith, F. White, S. *et al.* (2009) Brominated dioxins (PBDD/Fs) and PBDEs in marine shellfish in the UK. *Food Additives and Contaminants*, **26**, 918–927.

Fernández-González, L.E., Diz, A.P., Grueiro, N.G., Muniategui-Lorenzo, S., Beiras, R. & Sánchez-Marín, P. (2020) No evidence that vitellogenin protein expression is induced in marine mussels after exposure to an estrogenic chemical. *Science of the Total Environment*, **721**, 137638.

Fernley, P.W., Moore, M.N., Lowe, D.M., Donkin, P. & Evans, S. (2000) Impact of the Sea Empress oil spill on lysosomal stability in mussel blood cells. *Marine Environmental Research*, **50**, 451–455.

Ferrante, M.C., Cirillo, T., Naso, B., Clausi, M.T., Lucisano, A. & Cocchieri, R.A. (2007) Polychlorinated biphenyls and organochlorine pesticides from seafood from the Gulf of Naples (Italy). *Journal of Food Protection*, **70**, 706–715.

Filipkowska, A. & Lubecki, L. (2016) Endocrine disruptors in blue mussels and sediments from the Gulf of Gdańsk (Southern Baltic). *Environmental Science and Pollution Research*, **23**, 13864–13876.

Fliedner, A., Rüdel, H., Knopf, B., Lohmann, N., Paulus. M., Jud, M. *et al.* (2018) Assessment of seafood contamination under the marine strategy framework directive: contributions of the German environmental specimen bank. *Environmental Science and Pollution Research*, **25**, 26939–26956.

Fletcher, P.J., Kelble, C.R., William K. Nuttle, W.K. & Kiker, G.A. (2014) *Using the integrated ecosystem assessment framework to build consensus and transfer information to managers. Ecological Indicators*, **11**, 11–25.

Foley, M. M., Armsby, M. H., Prahler, E. E., Caldwell, M. R., Erickson, A. L., Kittinger, J. N. *et al.* (2013) Improving ocean management through the use of ecological principles and integrated ecosystem assessments. *BioScience*, **63**, 619–631.

Franzellitti, S., Buratti, S., Capolupo, M., Du, B., Haddad, S.P., Chambliss, C.K. *et al.* (2014) An exploratory investigation of various modes of action and potential adverse outcomes of fluoxetine in marine mussels. *Aquatic Toxicology*, **151**, 14–26.

Franzellitti, S., Buratti, S., Du, B., Haddad, S.P., Chambliss, C.K., Brooks, B.W. *et al.* (2015) A multibiomarker approach to explore interactive effects of propranolol and fluoxetine in marine mussels. *Environmental Pollution*, 2**05**, 60–69.

FSAI (2015) Polycyclic aromatic hydrocarbons (PAHs) in food. Food Safety Authority Ireland. Toxicology Fact Sheet Series, No. 2.

Fung, C.N., Lam, J.C.W., Zheng, G.J., Connell, D.W., Monirith, I., Tanabe, S. *et al.* (2004) Mussel-based monitoring of trace metal and organic contaminants along the east coast of China using *Perna viridis* and *Mytilus edulis*. *Environmental Pollution*, **127**, 203–216.

Galgani, F., Martínez-Gómez, C., Giovanardi, F., Romanelli, G., Caixach, J., Cento, A. *et al.* (2011) Assessment of polycyclic aromatic hydrocarbon concentrations in mussels (*Mytilus galloprovincialis*) from the Western basin of the Mediterranean Sea. *Environmental Monitoring an Assessment*, **172**, 301–317.

Galgani, F., Chiffoleau, J.F., Barrah, M., Drebika, U., Tomasino, C. & Andral, B. (2014) Assessment of heavy metal and organic contaminants levels long the Libyan coast using transplanted mussels (*Mytilus galloprovincialis*). *Environmental Science and Pollution Research*, **21**, 11331–11339.

Galloway, T.S., Millward, N., Browne, M.A. & Depledge, M.H. (2002) Rapid assessment of organophosphorous/carbamate exposure in the bivalve mollusk *Mytilus edulis* using combined esterase activities as biomarkers. *Aquatic Toxicology*, **61**, 169180.

Genç, T.O., Po, B.H.K., Yılmaz, F., Lau, T.C. Wu, R.S.S. & Chiu, J.M.Y. (2018) Differences in metal profiles revealed by native mussels and artificial mussels in Sarıçay Stream, Turkey: implications for pollution monitoring. *Marine and Freshwater Research*, **69**, 1372–1378.

George, S.G., Carpene, E. & Coombs, T.L. (1978) The effect of salinity on the uptake of cadmium by the common mussel, *Mytilus edulis* (L.). In: *Physiology and Behaviour of Marine*

Organisms (eds D.S. McClusky & A.J. Berry). Proceedings of the 12th European Marine Biological Symposium, pp. 189–193. Pergamon Press, Oxford.

Georgieva, S., Stancheva, M. & Makedonski, L. (2016) Investigation about the presence of organochlorine pollutants in mussels from the Black Sea, Bulgaria. *Ovidius University Annals of Chemistry*, **27**, 8–12.

Gerhardt, A. (2002) Bioindicator species and their use in biomonitoring. In: *Environmental Monitoring*, Vol. I (eds H.L. Inyang & J.L. Daniels), pp. 77–123. EOLSS Publishers, Oxford.

Geyer, H.J., Rimkus, G.G., Scheunert, I., Kaune, A., Schramm, K-W., Kettrup, A. *et al.* (2000) Bioaccumulation and occurrence of endocrine-disrupting chemicals (EDCs), persistent organic pollutants (POPs), and other organic compounds in fish and other organisms including humans. In: *The Handbook of Environmental Chemistry, Vol. 2, Part J: Bioaccumulation – New Aspects and Developments* (ed. B. Beek), pp. 1–166. Springer-Verlag, Berlin Heidelberg.

Giarratano, E. & Amin, O.A. (2010) Heavy metals monitoring in the southernmost mussel farm of the world (Beagle Channel, Argentina). *Ecotoxicology and Environmental Safety*, **73**, 1378–1384.

Gigault, J., ter Halle, A., Baudrimont, M., Pascal, P-Y., Gauffre, F., Phi, T.-L. *et al.* (2018) Current opinion: what is a nanoplastic? *Environmental Pollution*, **235**, 1030–1034.

Goldberg, E.D. (1975) The mussel watch: a first step in global marine monitoring. *Marine Pollution Bulletin*, **6**, 111–132.

Gonzalez-Rey, M., Lau, T.C., Gomes, T., Maria, V.L., Bebianno, M.J. & Wu, R. (2011) Comparison of metal accumulation between 'Artificial Mussel' and natural mussels (*Mytilus galloprovincialis*) in marine environments. *Marine Pollution Bulletin*, **63**, 149–153.

Gonzalez-Rey, M. & Bebiano, M.J. (2014) Effects of non-steroidal anti-inflammatory drug (NSAID) diclofenac exposure in mussel *Mytilus galloprovincialis*. *Aquatic Toxicology*, **148**, 221–230.

Gray, J.S (2002) Biomagnification in marine systems: the perspective of an ecologist. *Marine Pollution Bulletin*, **45** 46–52.

Greenfield, R., Brink, K., Degger, N. & Wepener, V. (2014) The usefulness of transplantation studies in monitoring of metals in the marine environment: South African experience. *Marine Pollution Bulletin*, **85**, 566–573.

Hagger, J.A., Lowe, D., Dissanayake, A., Jones, M.B. & Galloway, T.S. (2010) The influence of seasonality on biomarker responses in *Mytilus edulis*. *Ecotoxicology*, **19**, 953–962.

Halldórsson, H.P., De Pirro, M., Romano, C., Svavarsson, J. & Sarà, G. (2008) Immediate biomarker responses to benzo[a]pyrene in polluted and unpolluted populations of the blue mussel (*Mytilus edulis* L.) at high-latitudes. *Environment International*, **34**, 483–489.

Hamza-Chaffai, A. (2014) Usefulness of bioindicators and biomarkers in pollution biomonitoring. *International Journal of Biotechnology for Wellness Industries*, **3**, 19–26.

Hartmann, N.I.B., Skjolding, L.M., Hansen, S.F., Baun, A., Kjølholt, J. & Gottschalk, F. (2014) *Environmental fate and behaviour of nanomaterials: new knowledge on important transfomation processes.* Copenhagen K: Danish Environmental Protection Agency. Environmental Project, No. 1594.

Harvey, C. J., Kelble, C.R. & Schwing, F.B. (2017) Implementing 'the IEA': using integrated ecosystem assessment frameworks, programs, and applications in support of operationalizing ecosystem-based management. *ICES Journal of Marine Science*, **74**, 398–405.

Hauton, C., Hawkins, L.E. & Hutchinson, S. (2001) Response of haemocyte lysosomes to bacterial inoculation in the oysters *Ostrea edulis* L. and *Crassostrea gigas* (Thunberg) and the scallop *Pecten maximus* (L.). *Fish and Shellfish Immunology*, **11**, 143–153.

Hedayati, A. (2015) Biomarkers as new monitoring method of coastal pollutants. *Journal of Coastal Zone Management*, **18**, 1–1.

HELCOM (2016) *Guidelines on the Determination of Polycyclic Aromatic Hydrocarbons (PAH) in Sediment*. Baltic Marine Environment Protection Commission Report. Available at: https://helcom.fi/helcom-at-work/publications/ (accessed 28 May 2021).

HELCOM (2018) *Guideline for the Determination of Heavy Metals in Sediment*. Baltic Marine Environment Protection Commission Report. Available at: https://helcom.fi/helcom-at-work/publications/manuals-and-guidelines/ (accessed 28 May 2021).

Hellou J. & Law R.J. (2003) Stress on stress response of wild mussels, *Mytilus edulis* and *Mytilus trossulus*, as an indicator of ecosystem health. *Environmental Pollution*, **126**, 407–416.

Helmholz, H., Ruhnau, C., Pröfrock, D., Erbslöh H.-B. & Prange, A. (2016) Seasonal and annual variations in physiological and biochemical responses from transplanted marine bioindicator species *Mytilus* spp. during a long term field exposure experiment. *Science of the Total Environment*, **565**, 626–636.

Hidalgo-Ruz, V., Gutow, L., Thompson, R. C. & Thiel, M. (2012) Microplastics in the marine environment: a review of the methods used for identification and quantification. *Environmental Science and Technology*, **46**, 3060–3075.

Hladik, M.L. & McWayne, M.M. (2012) Methods of analysis–determination of pesticides in sediment using gas chromatography/mass spectrometry. *US Geological Survey Techniques and Methods*, **5–C3**.

Hossain, Md. M. M., Kibria, G., Nugegoda, D., Lau, T.C. & Wu, R.S.S. (2015) A training manual for assessing pollution (trace/heavy metals) in rivers, estuaries and coastal waters-using innovative 'Artificial Mussel (AM) Technology' – Bangladesh Model. Research collaboration between scientists of the IMSF, University of Chittagong, Bangladesh, RMIT University, Australia, the City University of Hong Kong, and the University of Hong Kong.

Hutchins, D.A., Stupakoff, I. & Fisher, N.S. (1996) Temperature effects on accumulation and retention of radionuclides in the sea star, *Asterias forbesi*: implications for contaminated northern waters. *Marine Biology*, **125**, 701–706.

Hutchins, D.A., Stupakoff, I., Hook, S., Luoma, S.N. & Fisher, N.S. (1998) Effects of Arctic temperatures on distribution and retention of the nuclear waste radionuclides ^{241}Am ^{57}Co, and ^{137}Cs in the bioindicator bivalve *Macoma balthica*. *Marine Environmental Research*, **45**, 17–28.

Hylland, K., Burgeot, T., Martínez-Gómez, C., Lang, T., Robinson, C.D., Svavarsson, J. *et al.* (2017a) How can we quantify impacts of contaminants in marine ecosystems? The ICON project. *Marine Environmental Research*, **124**, 2–10.

Hylland, K., Robinson, C.D., Burgeot, T., Martínez-Gómez, C., Lang, T., Svavarsson, J. *et al.* (2017b) Integrated chemical and biological assessment of contaminant impacts in selected European coastal and offshore marine areas. *Marine Environmental Research*, **124**, 130–138.

IFREMER (2021) Ifremer et l'environnement littoral. Available at: https://wwz.ifremer.fr/envlit#RINBIO (accessed 28 May 2021).

Isobe, T., Takada, H., Kanai, M., Tsutsumi, S., Isobe, K., Boonyatumanond, R. *et al.* (2007) Distribution of polycyclic aromatic hydrocarbons (PAHs) and phenolic endocrine disrupting chemicals in South and Southeast Asian mussels. *Environmental Monitoring and Assessment*, **135**, 423–440.

Ivanković, D., Pavičić, J., Erk, M., Filipović-Marijić, V. & Raspor, B. (2005) Evaluation of the *Mytilus galloprovincialis* Lam. digestive gland metallothionein as a biomarker in a long-term field study: seasonal and spatial variability. *Marine Pollution Bulletin*, **50**, 1303–1313.

Ivar du Sol, J.A. & Costa, M.F. (2014) The present and future of microplastic pollution in the marine environment. *Environmental Pollution*, **185**, 352–364.

Izagirre, U., Ruiz, P. & Marigómez, I. (2009) Time-course study of the early lysosomal responses to pollutants in mussel digestive cells using acid phosphatase as lysosomal marker enzyme. *Comparative Biochemistry and Physiology Part C: Toxicology and Pharmacology*, **149**, 587–597.

Jackim, E., Morrison, G. & Steele, R. (1977) Effects of environmental factors on radiocadmium uptake by four species of marine bivalves. *Marine Biology*, **40**, 303–308.

Jackson, M., Eadsforth, C., Schowanek, D., Delfosse, T., Riddle, A. & Budgen, N. (2016) Comprehensive review of several surfactants in marine environments: fate and ecotoxicity. *Environmental Toxicology and Chemistry*, **35**, 1077–1086.

Jaikanlaya, C., Settachan, D., Denison, M.S., Ruchirawat, M. & van den Berg, M. (2009) PCBs contamination in seafood species at the Eastern Coast of Thailand. *Chemosphere*, **76**, 239–249.

Jakimska, A., Huerta, B., Bargańska, Z., Kot-Wasik, A., Rodríguez-Mozaz, S. & Barceló, D. (2013) Development of a liquid chromatography–tandem mass spectrometryprocedure for determination of endocrine disrupting compounds in fish from Mediterranean rivers. *Journal of Chromatography A*, **1306**, 44–58.

Jernelov, A. (1996) The international mussel watch: a global assessment of environmental level of chemical contaminants. *Science of the Total Environment*, **188**(Suppl. 1), S37–S44.

Johansson, I., Héas-Moisan, K., Guiot, N., Munschy, C. & Tarczyński, J. (2006) Polybrominated diphenyl ethers (PBDEs) in mussels from selected French coastal sites: 1981–2003. *Chemosphere*, **64**, 296–305.

Jones, W.G. & Walker, K.F. (1979) Accumulation of iron, manganese, zinc and cadmium by the Australian freshwater mussel *Velesunio ambiguus* (Phillipi) and its potential as a biological monitor. *Australian Journal of Marine and Freshwater Research*, **30**, 741–751.

Jović, M., Stanković, A., Slavković-Beskoski, L., Tomić, I., Degetto, S. & Stanković, S. (2011) Mussels as a bio-indicator of the environmental quality of the coastal water of the Boka Kotorska Bay (Montenegro). *Journal of the Serbian Chemistry Society*, **76**, 933–946.

Kampire, E., Rubidge, G., Adams, J.B. & Human, L. (2016) Congener profiles of polychlorinated biphenyls and the effect on marine mussels at an outfall site, Port Elizabeth, South Africa. *Water SA*, **42**, 496–504.

Kim, C.-K. & Choi, M.S. (2017) Biomonitoring of trace metals using transplanted mussels, *Mytilus galloprovincialis*, in coastal areas around Ulsan and Onsan Bays, Korea. *Ocean Science Journal*, **52**, 31–42.

Kimbrough, K.L., Johnson, W.E., Lauenstein, G.G., Christensen, J.D. & Apeti, D.A. (2008) An assessment of two decades of contaminant monitoring in the nation's coastal zone. *NOAA Technical Memorandum NOS NCCOS*, 74.

Kolandhasamy, P., Su, L., Li, J., Qu, X., Jabeen, K., Shi, H. *et al.* (2018) Adherence of microplastics to soft tissue of mussels: a novel way to uptake microplastics beyond ingestion. *Science of the Total Environment*, **610–611**, 635–640.

Kacar, A., Pazi, I., Gonul, T. & Kucuksezgin, F. (2016) Marine pollution risk in a coastal city: use of an eco-genotoxic tool as a stress indicator in mussels from the Eastern Aegean Sea. *Environmental Science and Pollution Research*, **23**, 16067–16078.

Kan-atireklap, S., Tanabe, S., Sanguansin, J., Tabucanon, M.S. & Hungspreugs, M. (1997) Contamination by butyltin compounds and organochlorine residues in green mussel (*Perna viridis* L.) from Thailand coastal waters. *Environmental Pollution*, **97**, 79–89.

Kan-atireklap, S., NTH, Yen, Tanabe, S. & Subramanian, A.N. (1998) Butyltin compounds and organochlorine residues in green mussel (*Perna viridis* L.) from India. *Toxicology and Environmental Chemistry*, **67**, 409–424.

Kane, I.A., Clare, M.A., Miramontes, E., Wogelius, R., Rothwell, J.J., Garreau, P. *et al.* (2020) Seafloor microplastic hotspots controlled by deep-sea circulation. *Science*, 368, 1140–1145.

Karlsson, T.M., Vethaak, A.D., Almroth, B.C., Ariese, F., van Velzen, M., Hassellöv, M. *et al.* (2017) Screening for microplastics in sediment, water, marine invertebrates and fish: method development and microplastic accumulation. *Marine Pollution Bulletin*, **122**, 403–408.

Kasiotis, K.M. (2009) Organochlorine pesticides residues in mussels of Greek Island Evia. *International Journal of Chemistry*, **1**, 2–9.

Khan, I., Saeed, K. & Khan, I. (2019) Nanoparticles: properties, applications and toxicities. *Arabian Journal of Chemistry*, **12**, 908–931.

Klosterhaus, S.L., Grace, R., Hamilton, M.C. & Yee, D. (2013) Method validation and reconnaissance of pharmaceuticals, personal care products, and alkylphenols in surface waters, sediments, and mussels in an urban estuary. *Environment International*, **54**, 92–99.

Koelmans, A.A., Besseling, E. & Shim, W.J. (2015) Nanoplastics in the aquatic environment. Critical review. In: *Marine Anthropogenic Litter* (eds M. Bergmann, L. Gutow & M. Klages), pp. 325–340. Springer Open, Philadelphia, PA.

Krishnakumar, P.K., Asokan, P.K. & Pillai, V.K. (1990) Physiological and cellular esponses to copper and mercury in the green mussel *Perna viridis* (Linnaeus). *Aquatic Toxicology*, **18**, 163–174.

Krishnakumar, P.K., Qurban, M.A. & Sasikumar, G. (2018) Biomonitoring of trace metals in the coastal waters using bivalve molluscs. In: *Trace Elements–Human Health and Environment* (ed. H. Saled), pp. 153–176. InTech Open, London.

Kristan, U., Planinsek, P., Benedik, L., Falnoga, I. & Stibilj, V. (2015) Polonium-210 and selenium in tissues and tissue extracts of the mussel *Mytilus galloprovincialis* (Gulf of Trieste). *Chemosphere*, **119**, 231–241.

Kucuksezgin, F., Pazi, I., Yucel-Gier, G., Akcali, B. & Galgani, F. (2013) Monitoring of heavy metal and organic compound levels along the Eastern Aegean coast with transplanted mussels. *Chemosphere*, **93**, 1511– 1518.

Labuschagne, M., Wepener, V., Nachev, M., Zimmermann, S., Sures, B. & Smit, N.J. (2020) The application of artificial mussels in conjunction with transplanted bivalves to assess elemental exposure in a platinum mining area. *Water*, **12**, 32.

Laitano, M.V. & Fernández-Gimenez, A.V. (2016) Are mussels always the best bioindicators? Comparative study on biochemical responses of three marine invertebrate species to chronic port pollution. *Bulletin of Contamination and Toxicology*, **97**, 50–55.

Lam, P.K.S. (2009) Use of biomarkers in environmental monitoring. *Ocean and Coastal Management*, **52**, 348–354.

Lam, P.K.S. & Gray, J.S. (2003) The use of biomarkers in environmental monitoring programmes. *Marine Pollution Bulletin*, **46**, 182–186.

Lanksbury, J. & West, J. (2012) 2011/2012 Mussel Watch Phase 1 – Sampling Summary and Progress Report, Puget Sound Ecosystem Monitoring Program – Toxics in Biota. NOAA, Washington Department of Fish and Wildlife.

Lapworth, D.J., Baran, N., Stuart, M.E. & Ward. R.S. (2012) Emerging organic contaminants in groundwater: a review of sources, fate and occurrence. *Environmental Pollution*, **163**, 287–303.

Lara-Martín, P.A., Gómez-Parra, A. & González-Mazo, E. (2006) Simultaneous extraction and determination of anionic surfactants in waters and sediments. *Journal of Chromatography A*, **1114**, 205–210.

Lara-Martín, PA, Gómez-Parra, A., Sanz, J.L. & González-Mazo, E. (2010) Anaerobic degradation pathway of linear alkylbenzene sulfonates (LAS) in sulfate-reducing marine sediments. *Environmental Science and Technology*, **44**,1670–1676.

Larsson, J., Smolarz, K., Świeżak, J., Turower,M., Czerniawska, N. & Grahn, M. (2018) *Multi biomarker analysis of pollution effect on resident populations of blue mussels from the Baltic Sea. Aquatic Toxicology*, **198**, 240–256.

Lau, P.S. & Wong, H.L. (2003) Effect of size, tissue parts and location on six biochemical markers in the green-lipped mussel, *Perna viridis. Marine Pollution Bulletin*, **46**,1563–1572.

Lavradas, R.T., Rocha, R.C.C., Bordon, I.C.A.C., Saint Pierre, T.D., Godoy, J.M. & Hauser-Davis, R.A. (2016a) Differential metallothionein, reduced glutathione and metal levels in *Perna perna* mussels in two environmentally impacted tropical bays in southeastern Brazil. *Ecotoxicology and Environmental Safety*, **129**, 75–84.

Lavradas, R.T., Rocha, R.C.C., Saint Pierre, T.D., Godoy, J.M. & Hauser-Davis, R.A. (2016b) Investigation of thermostable metalloproteins in *Perna perna* mussels from differentially contaminated areas in southeastern Brazil by bioanalytical techniques. *Journal of Trace Elements in Medicine and Biology*, **34**, 70–78.

Le, T.T.Y., Zimmermann, S. & Sures, B. (2016) How does the metallothionein induction in bivalves meet the criteria for biomarkers of metal exposure? *Environmental Pollution*, **212**, 257–268.

Lei, M., Zhang, L., Lei, J., Zong, L., Li, J., Wu, Z., *et al.* (2015) Overview of emerging contaminants and associated human health effects. *BioMed Research International*, **2015**, art. 404796.

Lehtonen, K.K., d'Errico, G., Korpinen, S, Regoli, F., Ahkola, H., Kinnunen, T. *et al.* (2019) Mussel caging and the weight of evidence approach in the assessment of chemical contamination in coastal waters of Finland (Baltic Sea). *Frontiers in Marine Science,* **6**, *art.* 688.

Leung, K.M.Y., Furness, R.W., Svavarsson, J., Lau, T.C. & Wu, R.S.S. (2008) Field validation, in Scotland and Iceland, of the artificial mussel for monitoring trace metals in temperate seas. *Marine Pollution Bulletin*, **57**, 790–800.

Levin, P.S., Kelble, C. R., Shuford, R.L., Ainsworth, C., deReynier, Y., Dunsmore, R. *et al.* (2014) Guidance for implementation of integrated ecosystem assessments: a US perspective. *ICES Journal of Marine Science*, **71**, 1198–1204.

Li, J., Yang, D., Li, L., Jabeen, K. & Shi, H. (2015) Microplastics in commercial bivalves from China. *Environmental Pollution*, **207**, 190–195.

Li, J., Qu, X., Su, L., Zhang, W., Yang, D., Kolandhasamy, P. *et al.* (2016) Microplastics in mussels along the coastal waters of China. *Environmental Pollution*, **214**, 177–184.

Li, J-Y., Cui, U., Su, L., Chen, Y. & Jin, L. (2015) Polycyclic aromatic hydrocarbons in the largest deepwater port. *Environmental Science and Pollution Research*, **22**, 12355–12365.

Li, Y.F., Bidelman, T.F., Barrie, L.A. & McConnell, L.L. (1998) Global hexachlorocyclohexane use trends and their impact on the Arctic atmospheric environment. *Geophysical Research Letters*, **25**, 39–41.

Lionetto, M.G., Caricato, R., Calisi, A & Schettino, T. (2011) Acetylcholinesterase inhibition as a relevant biomarker in environmental biomonitoring: new insights and perspectives. In: *Ecotoxicology around the Globe* (ed. J.E. Visser), pp. 87–115. Nova Science Publishers, New York.

Lionetto, M.G., Caricato, R. & Giordano, M.E. (2019) Pollution biomarkers in environmental and human biomonitoring. *The Open Biomarkers Journal*, **9**, 1–9.

Livingstone, D.R. & Pipe, R.K. (1992) Mussels and environmental contaminants: molecular and cellular aspects. In: *The Mussel Mytilus*: *Ecology, Physiology, Genetics and Culture* (ed. E.M. Gosling), pp. 425–464. Elsevier Science Publishers B.V., Amsterdam.

Löder, M.G.J. & Gerdts, G. (2015) Methodology used for the detection and identification of microplastics – a critical appraisal. In: *Marine Anthropogenic Litter* (eds M. Bergmann, L. Gutow & M. Klages), pp. 201–228. Springer Open, Philadelphia, PA.

Lowe, D.M., Soverchia, C. & Moore, M.M. (1995) Lysosomal membrane responses in the blood and digestive cells of mussels experimentally exposed to fluoranthene. *Aquatic Toxicology*, **33**, 105–112.

Lusher, A. (2015) Microplastics in the marine environment: distribution, interactions and effects. In: *Marine Anthropogenic Litter* (eds M. Bergmann, L. Gutow & M. Klages), pp. 245–307. Springer Open, Philadelphia, PA.

Lusher, A.L., Welden, N.A., Sobral, P. & Cole, M. (2017) Sampling, isolating and identifying microplastics ingested by fish and invertebrates. *Analytical Methods*, **9**, 1346–1360.

Luzhna, L., Kathiria, P. & Kovalchuk, O. (2013) Micronuclei in genotoxicity assessment: from genetics to epigenetics and beyond. *Frontiers in Genetics*, **4**, art. 131.

Magni, P., De Falco, G., Falugi, C., Franzoni, M., Monteverde, M., Perrone, E. *et al.* (2006) Genotoxicity biomarkers and acetylcholinesterase activity in natural populations of *Mytilus galloprovincialis* along a pollution gradient in the Gulf of Oristano (Sardinia, western Mediterranean). *Environmental Pollution*, **142**, 65–72.

Maisano, M., Cappello, T., Natalotto, A., Vitale, V., Parrino, V., Giannetto, A. *et al.* (2017) Effects of petrochemical contamination on caged marine mussels using a multi-biomarker approach: histological changes, neurotoxicity and hypoxic stress. *Marine Environmental Research*, **128**, 114–123.

Malvärn, A., Zebühr, Y., Jensen, S., Kautsky, L, Greyerz, E., Nakano, T. *et al.* (2005) Identification of polybrominated dibenzo-p-dioxins in blue mussels (*Mytilus edulis*) from the Baltic Sea. *Environmental Science and Technology*, **39**, 8235–8242.

Mari, M., Nadal, M., Schuhmacher, M. & Domingo, J.L (2008) Monitoring PCDD/Fs, PCBs and metals in the ambient air of an industrial area of Catalonia, Spain. *Chemosphere*, **73**, 990–998.

Marigómez, I., Garmendia, L., Soto, M., Orbea, A., Izagirre, U. & Cajaraville, M.P. (2013a) Marine ecosystem health status assessment through integrative biomarker indices: a comparative study after the Prestige oil spill 'Mussel Watch'. *Ecotoxicology*, **22**, 486–505.

Marigómez, I., Zorita, I., Izagirre, U., Ortiz-Zarragoitia, M., Navarro, P., Etxebarria, N. *et al.* (2013b) Combined use of native and caged mussels to assess biological effects of pollution through the integrative biomarker approach. *Aquatic Toxicology*, **136**, 32–48.

Mariño-Ramírez, L., Kann, M.G., Shoemaker, B.A. & Landsman, D. (2005) Histone structure and nucleosome stability. *Expert Review of Proteomics*, **2**, 719–729.

Martínez Bueno, M.J., Boillot, C., Fenet, H., Chiron, S., Casellas, C. & Gómez, E. (2013) Fast and easy extraction combined with high resolution-mass spectrometry for residue analysis of two anticonvulsants and their transformation products in marine mussels. *Journal of Chromatography A*, **1305**, 27–34.

Martínez Bueno, M.J., Boillot, C., Munaron, D., Fenet, H., Casellas, C. & Gómez, E. (2014) Occurrence of venlafaxine residues and its metabolites in marine mussels at trace levels: development of analytical method and a monitoring program. *Analytical and Bioanalytical Chemistry*, **406**, 601–610.

Martínez-Gómez, C., Robinson, C.D., Burgeot, T., Gubbins, M., Halldorsson, H.P., Albentosa, M. *et al.* (2017) Biomarkers of general stress in mussels as common indicators for marine biomonitoring programmes in Europe: the ICON experience. *Marine Environmental Research*, **124**, 70–80.

Martinović, R., Kolarević, S., Kračun-Kolarević, M., Kostić, J., Marković, S., Gačić, Z. *et al.* (2015) Genotoxic potential and heart rate disorders in the Mediterranean mussel *Mytilus galloprovincialis* exposed to Superdispersant-25 and dispersed diesel oil. *Marine Environmental Research*, **108**, 83–90.

Martinović, R., Kolarević, S., Kračun-Kolarević, M., Kostić, J., Jokanović, S., Gačić, Z. *et al.* (2016) Comparative assessment of cardiac activity and DNA damage in haemocytes of the Mediterranean mussel *Mytilus galloprovincialis* in exposure to tributyltin chloride. *Environmental Toxicology and Pharmacology*, **47**, 165–174.

Maruya, K., Dodder, N., Weisberg, S., Gregorio, D., Klosterhaus, S., Alvarez, D. *et al.* (2014a) Refocusing Mussel Watch on contaminants of emerging concern (CECs): the California pilot study (2009–10). *Marine Pollution Bulletin*, **81**, 334–339.

Maruya, K.A., Dodder, N.G., Weisberg, S.B., Gregorio, D., Bishop, J.S., Klosterhaus, S. *et al.* (2014b) The Mussel Watch California pilot study on contaminants of emerging concern (CECs): synthesis and next steps. *Marine Pollution Bulletin*, **81**, 355–363.

Mathalon, A. & Hill, P. (2014) Microplastic fibers in the intertidal ecosystem surrounding Halifax Harbor, Nova Scotia. *Marine Pollution Bulletin*, **81**, 69–79.

Maulvault, A.L., Camacho, C., Barbosa, V., Alves, R., Anacleto, P. Fogaça, F. *et al.* (2018) Assessing the effects of seawater temperature and pH on the bioaccumulation of emerging chemical contaminants in marine bivalves. *Environmental Research*, **161**, 236–247.

McEneff, G., Barron, L., Kelleher, B., Paull, B. & Quinn, B. (2013) The determination of pharmaceutical residues in cooked and uncooked marine bivalves using pressurised liquid extraction, solid-phase extraction and liquid chromatography–tandem mass spectrometry. *Analytical and Bioanalytical Chemistry*, **405**, 9509–9521.

McEneff, G., Barron, L., Kelleher, B., Paull, B. & Quinn, B. (2014) A year-long study of the spatial occurrence and relative distribution of pharmaceutical residues in sewage effluent, receiving marine waters and marine bivalves. *Science of the Total Environment*, **476–477**, 317–326.

Melwani, A.R., Greenfield, B.K., Jahn, A., Oram, J.J., Sedlak, M. & Davis, J.A. (2008) Power analysis and optimization of the RMP Status and Trends Program. San Francisco Estuary Institute, Oakland, CA.

Melwani, A.R., Gregorio, D., Jin, Y., Stephenson, M., Ichikawa, G., Siegel, E. *et al.* (2014) Mussel watch update: long-term trends in selected contaminants from coastal California, 1977–2010. *Marine Pollution Bulletin*, **81**, 291–302.

Mintenig, S.M., Int-Veen, I., Löder, M.G.J., Primpke, S. & Gerdts, G. (2017) Identification of microplastic in effluents of waste water treatment plants using focal plane array-based micro-Fourier-transform infrared imaging. *Water Research*, **108**, 365–372.

Mok, J.S., Yoo, H.D., Kim, P.H., Yoon, H.D., Park, Y.C., Kim, J.H. *et al.* (2014) Bioaccumulation of heavy metals in the mussel *Mytilus galloprovincialis* in the Changseon area, Korea, and assessment of potential risk to human health. *Fisheries and Aquatic Sciences*, **17**, 313–318.

Monirith, I., Ueno, D., Takahashi, S., Nakata, H., Sudaryanto, A., Subramanian, A. *et al.* (2003) Asia-Pacific mussel watch: monitoring contamination of persistent organochlorine compounds in coastal waters of Asian countries. *Marine Pollution Bulletin*, **46**, 281–300.

Moore, M.N. (1985) Cellular responses to pollutants. *Marine Pollution Bulletin*, **16**,134–139.

Moore, M.N. (1991) Lysosomal changes in the response of molluscan hepatopancreatic cells to extracellular signals. *The Histochemistry Journal*, **23**, 495–500.

Moore, M.N., Widdows, J., Cleary, J.J., Pipe, R.K., Salkeld, P.N., Donkin, P. *et al.* (1984) Responses of the mussel *Mytilus edulis* to copper and phenanthrene: interactive effects. *Marine Environmental Research*, **14**,167– 183.

Moore, M.N., Lowe, D.M. & Koehler, A. (2004) Biological effects of contaminants: measuring lysosomal membrane stability. ICES Techniques in Marine Environmental Sciences, No. 36.

Mora, P., Miche, X. & Narbonne, J. F. (1999) Cholinesterase activity as potential biomarker in two bivalves. *Environmental Toxicology and Pharmacology*, **7**, 253–260.

Mrowiec, B. (2017) Plastic pollutants in water environment. *Environmental Protection and Natural Resources*, **28**, 51–55.

Mubiana, V.K. & Blust, R. (2007) Effects of temperature on scope for growth and accumulation of Cd, Co, Cu and Pb by the marine bivalve *Mytilus edulis*. *Marine Environmental Research*, **63**, 219–235.

Mubiana, V.K., Vercauteren, K. & Blust, R. (2006) The influence of body size, condition index and tidal exposure on the variability in metal bioaccumulation in *Mytilus edulis*. *Environmental Pollution*, **144**, 272–279.

Munschy, C., Guiot, N., Héas-Moisan, K., Tixier, C. & Tronczyjski, J. (2008) Polychlorinated dibenzo-p-dioxins and dibenzofurans (PCDD/Fs) in marine mussels from French coasts: levels, patterns and temporal trends from 1981 to 2005. *Chemosphere*, **73**, 945–953.

Nakata, H., Shinohara, R-I., Nakazawa, Y., Isobe, T., Sudaryanto, A., Subramanian, A. *et al.* (2012) Asia–Pacific mussel watch for emerging pollutants: distribution of synthetic musks and benzotriazole UV stabilizers in Asian and US coastal waters. *Marine Pollution Bulletin*, **64**, 2211–2218.

Ng, Y.J., Yap, C.K., Zakaria, M.P. & Tan, S.G. (2013) Assessment of heavy metal pollution in the Straits of Johore by using transplanted caged mussel, *Perna viridis*. *Pertanika Journal of Science and Technology*, **21**, 75–96.

Naidu, R., Arias Espana, V.A., Liu, Y. & Jit, J. (2016) Emerging contaminants in the environment: risk-based analysis for better management. *Chemosphere*, **154**, 350–357.
Nice, H.E. & Fisher, S.J. (2011) Ecotoxicological and bioaccumulation investigations of the Swan Estuary in the vicinity of Claisebrook. Water Science Technical Series, Report No. 28. Western Australia: Department of Water.
Nicholson, S. (1999) Cytological and physiological biomarker responses from green mussels, *Perna viridis* (L.) transplanted to contaminated sites in Hong Kong coastal waters. *Marine Pollution Bulletin*, **39**, 261–268.
Nicholson, S. (2002) Ecophysiological aspects of cardiac activity in the subtropical mussel *Perna viridis* (L.) (Bivalvia: Mytilidae). *Journal of Experimental Marine Biology and Ecology*, **267**, 207–222.
Nicholson, S. (2003a) Lysosomal membrane stability, phagocytosis and tolerance to emersion in the mussel *Perna viridis* (Bivalvia: Mytilidae) following exposure to acute, sublethal, copper. *Chemosphere*, **52**, 1147–1151.
Nicholson, S. (2003b) Cardiac and branchial physiology associated with copper accumulation and detoxication in the mytilid mussel *Perna viridis* (L.). *Journal of Experimental Marine Biology and Ecology*, **295**, 157–171.
Nicholson S. (2003c) Tachycardia in the mussel, *Perna viridis* (L.) on exposure to tributyltin antifouling paint. *Australasian Journal of Ecotoxicology*, **9**, 137–140.
Nicholson, S. & Lam, P.K.S. (2005) Pollution monitoring in Southeast Asia using biomarkers in the mytilid mussel *Perna viridis* (Mytilidae: Bivalvia). *Environment International*, **31**, 121–132.
Nielsen, S.A. (1974) Vertical concentration gradients of heavy metals in cultured mussels, *New Zealand. Journal of Marine and Freshwater Research*, **8**, 631–636.
Nikinmaa, M. (2014) Factors affecting the bioavailability of chemicals. In: *An Introduction to Aquatic Toxicology* (ed. M. Nikinmaa), pp. 65–72. Academic Press, New York.
Nikolic, M., Kuznetsova, T., Kholodkevich, S., Gvozdenovic, S., Mandic, M., Joksimovic, D. *et al.* (2019) Cardiac activity in the Mediterranean mussel (*Mytilus galloprovincialis* Lamarck, 1819) as a biomarker for assessing sea water quality in Boka Kotorska Bay, South Adriatic Sea. *Mediterranean Marine Science*, **20**, 680687.
NOAA (2019) National Status and Trends: Mussel Watch Program. Available at: https://catalog.data.gov/dataset/national-status-and-trends-mussel-watch-program (accessed 28 May 2021).
NPI (2019) The National Pollutant Inventory. Polychlorinated dioxins and furans. Available at: http://www.npi.gov.au/resource/polychlorinated-dioxins-and-furans (accessed 28 May 2021).
O'Connor, T.P. & Lauenstein, G.G. (2006) Trends in chemical concentrations in mussels and oysters collected along the US coast: update to 2003. *Marine Environmental Research*, **62**, 261–285.
OECD/EU (2018) *Health at a Glance: Europe 2018: State of Health in the EU Cycle*, OECD Publishing, Paris.
Ojemaye, C.Y. & Petrik, L. (2018) Pharmaceuticals in the marine environment: a review. *Environmental Reviews*, **27**, 151–165.
Okay, O.S., Karacık, B., Henkelmann, B. & Schramm, K.W. (2011) *Distribution of organochlorine pesticides in sediments and mussels from the Istanbul Strait. Environmental Monitoring and Assessment*, **176**, 51–65.

Okbah, A.M.A., Ibrahim, A.M.A. & Gamal, M.N.M (2013) Environmental monitoring of linear alkylbenzene sulfonates and physicochemical characteristics of seawater in El-Mex Bay (Alexandria, Egypt). *Environmental Monitoring and Assessment*, **185**, 3103–3115.

Oliva, A.L., Ovaert, J., Arias, A.H., Souissi, S. & Marcovecchio, J.E. (2015) Mussels as bioindicators of PAHs pollution within Argentinean coastal environments. *International Journal of Environmental Research*, **9**, 1293–1304.

Olkowska, E., Ruman, M. & Polkowska, Z. (2014) Occurrence of surface active agents in the environment. *Journal of Analytical Methods in Chemistry*, 2**014**, art. 769708.

Orescanin, V., Lovrencic, I., Mikelic, L., Barisic, D., Matasin, Z., Lulic, S. *et al.* (2006) Biomonitoring of heavy metals and arsenic on the east coast of the middle Adriatic Sea using *Mytilus galloprovincialis*. *Nuclear Instruments and Methods in Physics Research, Section B*, **245**, 495–500.

Orren, M.J., Eagle, G.A., Hennig, H.F.K.O. & Green, A. (1980) Variations in trace metal content of the mussel *Choromytilus meridionalis (*Kr.*)* with season and sex. *Marine Pollution Bulletin*, **11**, 253–257.

Ortiz-Zarragoitia, M. & Cajaraville, M.P. (2006) Biomarkers of exposure and reproduction-related effects in mussels exposed to endocrine disruptors. *Archives of Environmental Contamination and Toxicology*, **50**, 361–369.

OSPARCOM (1999–2002) CEMP Guidelines for Monitoring Contaminants in Biota. Revised 2018.

OSPAR Commission (2013) Technical annex on supporting parameters for biological effects measurements in fish and mussels. Background document and technical annexes for biological effects monitoring, Update 2013. *Monitoring and Assessment Series*, 589/2013.

Ozkoc, H.B., Bakan, G. & Ariman, S. (2007) Distribution and bioaccumulation of organochlorine pesticides along the Black Sea coast. *Environmental Geochemistry and Health*, **29**, 59–68.

Palma-Fleming, H., Quiroz, E. R., Campillay, C., Figuero, M., Varas, A., Velásque, D. *et al.* (2012) Temporal and spatial trends of total aliphatic hydrocarbons of diesel range and trace elements in sediments and mussels of the Corral Bay area, Valdivia, South Central Chile. *Journal of the Chilean Chemical Society*, **57**, 1074–1082.

Pampanin, D.M., Volpato, E., Marangon, I. & Nasci, C. (2005) Physiological measurements from native and transplanted mussel (*Mytilus galloprovincialis*) in the canals of Venice. Survival in air and condition index. *Comparative Biochemistry and Physiology, Part A: Molecular and Integrative Physiology*, **140**, 41–52.

Parera, J., Ábalos, M., Santos, F. J., Galceran, M.T. & Abad, E. (2013) Polychlorinated dibenzo-p-dioxins, dibenzofurans, biphenyls, paraffins and polybrominated diphenyl ethers in marine fish species from Ebro River Delta (Spain). *Chemosphere*, **93**, 499–505.

Parmar, T.K., Rawtani, D. & Agrawal, Y.K. (2016) Bioindicators: the natural indicator of environmental pollution. *Frontiers in Life Science*, **9**, 110–118.

Penry, D.L. (1998) Applications of efficiency measurements in bioaccumulation studies: definitions, clarifications, and a critique of methods. *Environmental Toxicology and Chemistry*, **17**, 1633–1639.

Pentreath, R.J. (1973) The accumulation from water of ^{65}Zn, ^{54}Mn, ^{58}Co and ^{59}Fe by the mussel, *Mytilus edulis*. *Journal of the Marine Biological Association UK*, **53**, 127–143.

Perić, L. & Burić, P. (2019) The effect of copper and chlorpyrifos co-exposure on biomarkers in the marine mussel *Mytilus galloprovincialis*. *Chemosphere*, **225**, 126134.

Pesticide Action Networks UK (2019) Highly hazardous pesticides (HHPs). Available at: https://www.pan-uk.org/highly-hazardous-pesticides/ (accessed 28 May 2021).

Petrovic, M., Eljarrat, E., López de Alda, M.J. & Barceló, D. (2002) Recent advances in the mass spectrometric analysis related to endocrine disrupting compounds in aquatic environmental samples. *Journal of Chromatography A*, **974**, 23–51.

Pfeifer, S., Schiedek D. & Dippner, J.W. (2005) Effect of temperature and salinity on acetylcholinesterase activity, a common pollution biomarker, in *Mytilus* sp. from the south-western Baltic Sea. *Journal of Experimental Marine Biology and Ecology*, **320**, 93103.

Pichon, V. (2000) Solid-phase extraction for multiresidue analysis of organic contaminants in water. *Journal of Chromatography A*, **885** 195–215.

Piersanti, A., Tavoloni, T., Bastari, E., Lestingi, C., Romanelli, S., Saluti, G. *et al.* (2015) Polybrominated diphenyl ethers in mussels (*Mytilus galloprovincialis*) collected from Central Adriatic Sea. *Marine Pollution Bulletin*, **101**, 417–421.

Pizzini, S., Marchiori, E., Piazza, R., Cozzi, G. & Barbante, C. (2015) Determination by HRGC/HRMS of PBDE levels in edible Mediterranean bivalves collected from north-western Adriatic coasts. *Microchemical Journal*, **121**, 184–191.

Ponnusamy, K., Sivaperumal, P., Arularasan, S.M., Munilkumar, S. & Pal, A.K. (2014) Heavy metal concentration from biologically important edible species of bivalves (*Perna viridis* and *Modiolus metcalfei*) from Vellar Estuary, South East Coast of India. *Journal of Aquaculture Research and Development*, **5**, 258.

Poster, D.J., Schantz, M.M., Sader, L.C. & Wise, S.A. (2006) Analysis of polycyclic aromatic hydrocarbons (PAHs) in environmental samples: a critical review of gas chromatographic (GC) methods. *Analytical and Bioanalytical Chemistry*, **386**, 859–881.

Prudente, M., Ichihashi, H., Kan-atireklap, S., Watanabe, I. & Tanabe, S. (1999) Butyltins, organochlorines and metal levels in green mussel, *Perna viridis* L. from the coastal waters from the Philippines. *Fisheries Science*, **65**, 441–447.

Putri, L.S.E., Prasetyo, A.D & Arifin, Z. (2012) Green mussel (*Perna viridis* L.) as bioindicator of heavy metals pollution at Kamal Estuary, Jakarta Bay, Indonesia. *Journal of Environmental Research and Development*, **6**, 389–396.

Ra, K., Kim, J.-K., Kim, K.-T., Lee, S.Y., Kim, E.-S., Lee, J.M. & Wu, R.S.S. (2014) Application of the artificial mussel for monitoring heavy metal levels in seawater of the coastal environments, Korea. *Journal of the Korean Society for Marine Environment and Energy*, **17**, 131–145.

Rajkumar, A. (2009) Physiological and cytological biomarker studies using *Perna perna* for marine pollution monitoring. MSc thesis, University of KwaZulu-Natal Durban, South Africa.

Ramu, K., Kajiwara, N., Isobe, T., Takahashi, S., Kim, E-U., Min, B-Y. *et al.* (2007a) Spatial distribution and accumulation of brominated flame retardants, polychlorinated biphenyls and organochlorine pesticides in blue mussels (*Mytilus edulis*) from coastal waters of Korea. *Environmental Pollution*, **148**, 562–569.

Ramu, K., Kajiwara, N., Sudaryanto, A., Isobe, T., Takahashi, S., Subramanian, A. *et al.* (2007b) Asian mussel watch program: contamination status of polybrominated diphenyl ethers and organochlorines in coastal waters of Asian countries. *Environmental Science and Technology*, **41**, 4580–4586.

Rees, H.L., Hyland, J.L., Hylland, K., Mercer Clarke, C.S.L., Roff, J.C. & Ware, S. (2008) Environmental indicators: utility in meeting regulatory needs. An overview. *ICES Journal of Marine Science*, **65**, 1381–1386.

Regoli, F., Pellegrini, D., Cicero, A.M., Nigro, M., Benedetti, M., Gorbi, S. *et al.* (2014) A multidisciplinary weight of evidence approach for environmental risk assessment at the Costa Concordia wreck: integrative indices from Mussel Watch. *Marine Environmental Research*, **96**, 92–104.

Rejczak, T. & Tuzimski (2015) A review of recent developments and trends in the QuEChERS sample preparation approach. *Open Chemistry*, **13**, 980–1010.

Renner, G., Nellessen, A., Schwiers, A., Wenzel, M., Schmidt, T.C. & Schram, J. (2019) Data preprocessing and evaluation used in the microplastics identification process: a critical review and practical guide. *Trends in Analytical Chemistry*, **111**, 229–238.

Ricciardi, K.L., Poynton, H.C., Duphily, B.J., Blalock, B.J. & Robinson, W.E. (2016) Bioconcentration and depuration of ^{14}C-labeled 17 α-ethinyl estradiol and 4-nonylphenol in individual organs of the marine bivalve *Mytilus edulis* L. *Environmental Toxicology and Chemistry*, **35**, 863–873.

Richardson, B.J., Zeng, G. J., Tse, E.S.C. & Lam, P.K.S. (2001) A comparison of mussels (*Perna viridis*) and semi-permeable membrane devices (SPMDs) for monitoring trace chlorinated organic contaminants in Hong Kong coastal waters. *Chemosphere*, **45**, 1201–1208.

Riget, F., Johansen, P. & Asmund, G. (1996) Influence of length on element concentrations in the blue mussels (*Mytilus edulis*). *Marine Pollution Bulletin*, **32**, 745–751.

Ringwood, A.H., Hoguet, J. & Keppler, C.J. (2002) Seasonal variation in lysosomal destabilization in oysters, *Crassostrea virginica*. *Marine Environmental Research*, **54**, 793–797.

de los Ríos, A., Pérez, L., Echavarri-Erasun, B., Serrano, T., Barberoa, M.C., Ortiz-Zarragoitia, M. *et al.* (2016) Measuring biological responses at different levels of organisation to assess the effects of diffuse contamination derived from harbour and industrial activities in estuarine areas. *Marine Pollution Bulletin*, **103**, 301–312.

Robinson, C.D., Webster, L., Martínez-Gómez, C., Burgeot, T., Gubbins, M.J., Thain, J.E., *et al.* (2017) Assessment of contaminant concentrations in sediments, fish and mussels sampled from the North Atlantic and European regional seas within the ICON project. *Marine Environmental Research*, **124**, 21–31.

Rocha, T.L., Gomes, T., Sousa, V.S., Mestre, N.C. & Bebianno, M.J. (2015a) Ecotoxicological impact of engineered nanomaterials in bivalve molluscs: an overview. *Marine Environmental Research*, **111**, 74–88.

Rocha, T.L., Gomes, T., Mestre, N.C., Cardoso, C. & Bebianno, M.J. (2015b) Tissue specific responses to cadmium-based quantum dots in the marine mussel *Mytilus galloprovincialis*. *Aquatic Toxicology*, **169**, 10–18.

Rodil, R., Villaverde-de-Sáa, E., Cobas, J., Quintana, J.B., Cela, R. & Carro, N. (2019) Legacy and emerging pollutants in marine bivalves from the Galician coast (NW Spain). *Environment International*, **129** 364–375.

Rosen, G. & Lotufo, G.R. (2007) Bioaccumulation of explosive compounds in the marine mussel, *Mytilus galloprovincialis*. *Ecotoxicology and Environmental Safety*, **68**, 237–245.

Rouane-Hacene, O., Boutiba, Z., Belhaouari, B., Guibbolini-Sabatier, M.E., Francour, P. & Risso-de Faverney, C. (2015) Seasonal assessment of biological indices, bioaccumulation and bioavailability of heavy metals in mussels *Mytilus galloprovincialis* from Algerian west coast, applied to environmental monitoring. *Oceanologia*, **57**, 362–374.

Ruiz, M.D., Iriel, A., Yusseponec, M.S., Ortiz, N., Di Salvatore, P., Fernández Cirelli, A. *et al.* (2018) Trace metals and oxidative status in soft tissues of caged mussels (*Aulacomya atra*) on the North Patagonian coastline. *Ecotoxicology and Environmental Safety*, **155**, 152–161.

Sadik, O.A. (2013) Anthropogenic nanoparticles in the environment. *Environmental Science: Processes & Impacts*, **15**, 19–20.

Saggese, I., Sarà, G. & Dondero, F. (2016) Silver nanoparticles affect functional bioenergetic traits in the invasive Red Sea mussel *Brachidontes pharaonis*. *BioMed Research International*, **2016**, art. 1872351.

Samhouri, J. F., Haupt, A. J., Levin, P. S., Link, J. S. & Shuford, R. (2014) Lessons learned from developing integrated ecosystem assessments to inform marine ecosystem-based management in the USA. *ICES Journal of Marine Science*, **71**, 1205–1215.

Sánchez-Marín, P., , Bellas, J., Mubiana, V.K., Lorenzo, J.I., Blust, R. & Beiras, R. (2011) Pb uptake by the marine mussel *Mytilus* sp. Interactions with dissolved organic matter. *Aquatic Toxicology*, **102**, 48–57.

Santana, M.F.M., Ascer, L.G., Custódio, M.R., Moreira, F.T. & Turra, A. (2016) Microplastic contamination in natural mussel beds from a Brazilian urbanized coastal region: rapid evaluation through bioassessment. *Marine Pollution Bulletin*, **106**, 183–189.

Santos, L.H.M.L.M., Araújo, A.N., Fachini, A., Pena, A., Delerue-Matos, C. & M.C.B.S.M. Montenegro (2010) Ecotoxicological aspects related to the presence of pharmaceuticals in the aquatic environment. *Journal of Hazardous Materials*, **175**, 45–95.

Saravi, S.S.S. & Shokrzadeh, M. (2011) Role of pesticides in human life in the modern age: a review. In: *Pesticides in the Modern World – Risks and Benefits* (ed. M. Stoytcheva), pp 3–12. InTechOpen, London.

Scarpato, A., Romanelli, G., Galgani, F., Andral, B., Amici, M., Giordano, P. *et al.* (2010) Western Mediterranean coastal waters–monitoring PCBs and pesticides accumulation in *Mytilus galloprovincialis* by active mussel watching: the Mytilos project. *Journal of Environmental Monitoring*, **12**, 924–935.

Schiedek, D., Broeg, K., Baršienė, J., Lehtonen, K.K., Gercken, J., Pfeifer, S. *et al.* (2006) Biomarker responses as indication of contaminant effects in blue mussel (*Mytilus edulis*) and female eelpout (*Zoarces viviparus*) from the southwestern Baltic Sea. *Marine Pollution Bulletin*, **53**, 387–405.

Schøyen, M., Allan, I., Ruus, A., Håvardstun, J., Hjermann, Dag Ø. & Beyer, J. (2017) Comparison of caged and native blue mussels (*Mytilus edulis* spp.) for environmental monitoring of PAH, PCB and trace metals. *Marine Environmental Research*, **130**, 221–232.

Schwaferts, C., Niessner, R., Elsner, M. & Ivleva, N.P. (2019) Methods for the analysis of submicrometer- and nanoplastic particles in the environment. *Trends in Analytical Chemistry*, **112**, 52–65.

Segev, O., Kushmaro, A. & Brenner, A. (2009) Environmental impact of flame retardants (persistence and biodegradability). *International Journal of Environmental Research and Public Health*, **6**, 478–491.

Serafim, M.A., Company, R.M., Bebianno, M.J. & Langston, W.J. (2002) Effect of temperature and size on metallothionein synthesis in the gill of *Mytilus galloprovincialis* exposed to cadmium. *Marine Environmental Research*, **54**, 361–365.

Serra-Compte, A., Maulvault, A.L., Camacho, C., Álvarez-Muñoz, D., Barceló, D., Rodríguez-Mozaz, S. *et al.* (2018) Effects of water warming and acidification on bioconcentration, metabolization and depuration of pharmaceuticals and endocrine disrupting compounds in marine mussels (*Mytilus galloprovincialis*) *Environmental Pollution*, **236**, 824–834.

Serra-Compte, A., Álvarez-Muñoz, D., Solé, M., Cáceres, N. Barceló, D. & Rodríguez-Mozaz, S. (2019) Comprehensive study of sulfamethoxazole effects in marine mussels: bioconcentration, enzymatic activities and metabolomics. *Environmental Research*, **173**, 12–22.

Shen, H., Kibria, G., Wu, R.S.S., Morrison, P. & Nugegoda, D. (2020) Spatial and temporal variations of trace metal body burdens of live mussels *Mytilus galloprovincialis* and field validation of the Artificial Mussels in Australian inshore marine environment. *Chemosphere*, **248**, 126004.

Shim, W.J., Hong, S.H., Kim, N.S., Yim, U.H., Li, D. & Oh, J.R. (2005) Assessment of butyl- and phenyltin pollution in the coastal environment of Korea using mussels and oysters. *Marine Pollution Bulletin*, **51**, 922–931.

Shin, P.K.S., Yau, F.N., Chow, S.H., Tai, K.K. & Cheung, S.G. (2002) Responses of the green-lipped mussel *Perna viridis* (L.) to suspended solids. *Marine Pollution Bulletin*, **45**, 157–162.

Shue, M.-F., Chen, W.-D., Bellotindos, L.M. & Lu, M.-C. (2014) Seasonal variations of heavy metals content in muscle and viscera of green-lipped mussel *Perna viridis* from Da–Peng Bay Lagoon in Taiwan. *Journal of Toxicology and Environmental Health A*, **77**, 1222–1228.

Shuhong, W., Huasheng, H. & Xinhong, W. (2005) Bioenergetic responses in green lipped mussels (*Perna viridis*) as indicators of pollution stress in Xiamen coastal waters, China. *Marine Pollution Bulletin*, **51**, 738–743.

Siu, W.H.L., Hung, C.L.H., Wong, H.L., Richardson, B.J. & Lam, P.KS. (2003) Exposure and time dependent DNA strand breakage in hepatopancreas of green-lipped mussels (*Perna viridis*) exposed to Aroclor 1254, and mixtures of B[a]P and Aroclor 1254. *Marine Pollution Bulletin*, **46**, 1285–1293.

Siu, W.H.L., Cao, J., Jack, R.W.,Wu, R.S.S., Richardson, B.J., Xu, L. *et al.* (2004) Application of the comet and micronucleus assays to the detection of B[a]P genotoxicity in haemocytes of the green-lipped mussel (*Perna viridis*). *Aquatic Toxicology*, **66**, 381–392.

Skupiska, K., Misiewicz, I. & Kasprzycka-Guttman (2004) Polycyclic aromatic hydrocarbons: physicochemical properties, environmental appearance and impact on living organisms. *Acta Poloniae Pharmaceutica*, **61**, 233–240.

Solaun, O., Rodríguez, J.G., Borja, A., Larreta, J. & Valencia, V. (2015) Relationships between polychlorinated biphenyls in molluscs, hydrological characteristics and human pressures, within Basque estuaries (northern Spain). *Chemosphere*, **118**, 130–135.

Solé, M., Shaw, J.P., Frickers, P.E. Readman, J.W. & Hutchinson, T.H. (2010) Effects on feeding rate and biomarker responses of marine mussels experimentally exposed to propranolol and acetaminophen. *Analytical and Bioanalytical Chemistry*, **396**, 649–656.

Storelli, M.M. & Marcotrigiano, G.O. (2001) Polycyclic aromatic hydrocarbons in mussels (*Mytilus galloprovincialis*) from the Ionian Sea, Italy. *Journal of Food Protection*, **64**, 405–409.

Strehse, J.S. & Maser, E. (2020) Marine bivalves as bioindicators for environmental pollutants with focus on dumped munitions in the sea: a review. *Marine Environmental Research*, **158**, 105006.

Strehse, J.S., Appel, D., Geist, C., Martin, H.-J. & Maser, E.(2017) Biomonitoring of 2,4,6-trinitrotoluene and degradation products in the marine environment with transplanted blue mussels (*M. edulis*). *Toxicology*, **390**, 117–123.

Sudaryanto, A., Takahashi, S., Monirith, I., Ismail, A., Muchtar, M., Zheng. J. *et al.* (2002) Asia-Pacific mussel watch: monitoring of butyltin contamination in coastal waters of Asian developing countries. *Environmental Toxicology and Chemistry*, **21**, 2119–2130.

Sui, Q., Cao, X., Lu, S., Zhao, W., Qiu, Z. & Yu, G. (2015) Occurrence, sources and fate of pharmaceuticals and personal care products in the groundwater: a review. *Emerging Contaminants*, **1**, 14–24.

Sun, R., Pan, C., Peng, F., Wu, U., Chen, X. & Mai, B. (2020) Alternative halogenated flame retardants (AHFRs) in green mussels from the south China Sea. *Environmental Research*, **182**, 109082.

Suryono, C.A., Sabdono, A., Subagyo, Setyati, W.A., Rochaddi, B., Suryono, E. *et al.* (2019a) The presence of organochlorine pesticide in Semarang Indonesia marine waters and their contamination on green mussel *Perna viridis* (Bivalvia: Mytilidae, Linnaeus, 1758). *Earth and Environmental Science*, **246**, 012069.

Suryono, C.A., Sandono, A., Subagyo. & Setyati.W.A. (2019b) The preliminary investigation of organophosphate pesticide residues on green mussel *Perna viridis* (Bivalvia: Mytilidae, Linnaeus, 1758) at Demak coastal waters central Java Indonesia. *Earth and Environmental Science*, **246**, 012071.

Świacka, K., Szaniawska, A. & Caban, M. (2019) Evaluation of bioconcentration and metabolism of diclofenac in mussels *Mytilus trossulus*–laboratory study. *Marine Pollution Bulletin*, **141**, 249–255.

Tanabe, S., Prudente, M.S., Kan-atireklap, S. & Subramanian, A. (2000) Mussel watch: marine pollution monitoring of butyltins and organochlorines in coastal waters of Thailand, Philippines and India. *Ocean & Coastal Management*, **43**, 819–839.

Tavoloni, T., Stramenga, A., Stecconi, T., Siracusa, M., Bacchiocchi, S. & Piersanti, A. (2020) Single sample preparation for brominated flame retardants in fish and shellfish with dual detection: GC-MS/MS (PBDEs) and LC-MS/MS (HBCDs). *Analytical and Bioanalytical Chemistry*, **412**, 397–411.

Ternes, T.A., Joss, A. & Siegrist, H. (2004) Scrutinizing pharmaceuticals and personal care products in wastewater treatment. *Environmental Science and Technology*, **38**, 392A–399A.

Thain, J.E., Vethaak, A.D. & Hylland, K. (2008) Contaminants in marine ecosystems: developing an integrated indicator framework using biological-effect techniques. *ICES Journal of Marine Science*, **65**, 1508–1514.

Thain, J., Fernández, B. & Martínez-Gómez, C. (2019) Biological effects of contaminants: stress on stress (SoS) response in mussels. ICES Techniques in Marine Environmental Sciences, No. 59.

Thébault, H., Rodriguezy Baena, A.M., Andral, B., Barisic, D., Albaladejo, B.J., Bologa, A.S. *et al.* (2008) ^{137}Cs baseline levels in the Mediterranean and Black Sea: a cross-basin survey of the CIESM Mediterranean Mussel Watch programme. *Marine Pollution Bulletin*, **57**, 801–806.

Thébault, H. (2013) The Mediterranean Mussel Watch, ten years after: success and end. *Rapport du Comité International de la Mer Méditerranée*, **40**, 246.

Thomann, R.V. (1981) Equilibrium model of fate of microcontaminants in diverse aquatic food chains. *Canadian Journal of Fisheries and Aquatic Sciences*, **38**, 280–296.

Thompson, R.C. (2015) Microplastics in the marine environment: sources, consequences and solutions. In: *Marine Anthropogenic Litter* (eds M. Bergmann, L. Gutow & M. Klages), pp. 185–200. Springer Open, Philadelphia, PA.

Tornero, V. & Hanke, G. (2016) Chemical contaminants entering the marine environment from sea-based sources: a review with a focus on European seas. *Marine Pollution Bulletin*, **112**, 17–38.

Tornero, V. & Hanke, G. (2018a) Marine chemical contaminants–support to harmonized MSFD reporting: substances considered for MSFD descriptor 8, EUR 29281, Publications Office of the European Union.

Tornero, V. & Hanke, G. (2018b) Potential chemical contaminants in the marine environment: an overview of main contaminant lists. Available at: https://publications.jrc.ec.europa.eu/repository/bitstream/JRC108964/potential_chemical_contaminants_in_the_marine.pdf (accessed 28 May 2021).

Toro, B., Navarro, J.M. & Palma-Fleming, H. (2003a) Use of clearance rate in *Choromytilus chorus* (Bivalvia: Mytilidae) as a non-destructive biomarker of aquatic pollution. *Revista Chilena de Historia Natural*, **76**, 267–274.

Toro, B., Navarro, J.M. & Palma-Fleming, H. (2003b) Relationship between bioenergetics responses and organic pollutants in the giant mussel, *Choromytilus chorus* (Mollusca: Mytilidae). *Aquatic Toxicology*, **63**, 257–269.

Traverso-Soto, J.M., González-Mazo, E. & Lara-Martín, P.A. (2012) Analysis of surfactants in environmental samples by chromatographic techniques. In: *Chromatography the Most Versatile Method of Chemical Analysis* (ed. L. Calderon), pp 187–216. InTechOpen, London.

Traverso-Soto, J.M., Lara-Martín, P.A., González-Mazo, E. & León, V.M. (2015) Distribution of anionic and nonionic surfactants in a sewage-impacted Mediterranean coastal lagoon: inputs and seasonal variations. *Science of the Total Environment*, **503–504**, 87–96.

Turan, N.B., Erkan, H.S., Engin, G.O. & Bilgili, M.S. (2019) Nanoparticles in the aquatic environment: usage, properties, transformation and toxicity–a review. *Process Safety and Environmental Protection*, **130**, 238–249.

Turja, R., Höher, N., Snoeijs, P., Baršienė, J., Butrimavičienė, L., Kuznetsova, Y. *et al.* (2014) A multibiomarker approach to the assessment of pollution impacts in two Baltic Sea coastal areas in Sweden using caged mussels (*Mytilus trossulus*). *Science of the Total Environment*, **473–474**, 398–409.

UNESCO and HELCOM (2017) Pharmaceuticals in the aquatic environment of the Baltic Sea region: a status report. UNESCO Emerging Pollutants in Water Series, No. 1. UNESCO Publishing, Paris.

Vallaeys, T., Klink, S.P., Fleouter, E., Le Moing, B., Hervé Lignot, J. & Smith, A.J. (2017) Bioindicators of marine contaminations at the frontier of environmental monitoring and environmental genomics. *Advances in Biotechnology and Microbiology*, **4**, ID555629.

Van Cauwenberghe, L. & Janssen, C.R. (2014) Microplastics in bivalves cultured for human consumption *Environmental Pollution*, **193**, 65–70.

Vandermeersch, G., Lourenço, H.M., Alvarez-Muñoz, D., Cunha, S., Diogène, J., Cano-Sancho, G. *et al.* (2015) Environmental contaminants of emerging concern in seafood-European database on contaminant levels. *Environmental Research*, **143**, 29–45.

Vasilaki, A.T. & McMillan, D.C. (2011) Lipid peroxidation. In: *Encyclopedia of Cancer* (ed. M. Schwab). Springer, Berlin, Heidelberg.

Vethaak, A.D., Davies, I.M., Thain, J.E., Gubbins, M.J., Martínez-Gómez, C., Robinson, C.D. *et al.* (2017) Integrated indicator framework and methodology for monitoring and assessment of hazardous substances and their effects in the marine environment. *Marine Environmental Research*, **124**, 11–20.

Viarengo, A., Arena. N., Canesi, L., Alia, F.A. & Orunesu, M. (1994) Structural and biochemical alterations in the gills of copper-exposed mussels. In: *Contaminants in the Environment* (eds A. Renzoni, N. Mattei & L. Lari), pp. 135–144. Lewis Publishers, Boca Raton, FL.

Viarengo, A., Lowe, D., Bolognesi, C., Fabbri, E. & Koehler, A. (2007) The use of biomarkers in biomonitoring: a 2-tier approach assessing the level of pollutant induced stress syndrome in sentinel organisms. *Comparative Biochemistry and Physiology Part C: Toxicology and Pharmacology*, **146**, 281–300.

Villela, E.V., Marques de Oliveira, I., da Silva, J. & Henriques, J.A. P. (2006) DNA damage and repair in haemolymph cells of golden mussel (*Limnoperna fortunei*) exposed to environmental contaminants. *Mutation Research*, **605**, 78–86.

Vorkamp, K. & Rigét, F.F. (2014) A review of new and current-use contaminants in the Arctic environment: evidence of long-range transport and indications of bioaccumulation. *Chemosphere*, **111**, 379–395.

Vorkamp, K., Roose, P., Bersuder, P., Webster, L., Lepom, P., Munschy, C. *et al.* 2012. Determination of polychlorinated dibenzo-*p*-dioxins, polychlorinated dibenzofurans and dioxin-like polychlorinated biphenyls in biota and sediment. *ICES Techniques in Marine Environmental Sciences*, No. 50.

Watling R.J. & Watling, H.R. (1976) Trace metals in *Choromytilus meridionalis. Marine Pollution Bulletin*, **7**, 91–94.

Wang, W.X (2001) Comparison of metal uptake rate and absorption efficiency in marine bivalves. *Environmental Toxicology and Chemistry*, **20**, 1367–1373.

Wang, W.X. (2016) Bioaccumulation and biomonitoring. In: *Marine Ecotoxicology* (eds J. Blasco, P.M. Chapman, O. Campana & M. Hampel), pp. 99–119. Academic Press, New York.

Wang, W.X. & Fisher, N.S. (1999) Delineating metal accumulation pathways for marine invertebrates. *The Science of the Total Environment*, **237–238**, 459–472.

Wang, W.X., Fisher, N.S. & Luoma, S.N. (1996) Kinetic determinations of trace element bioaccumulation in the mussel *Mytilus edulis. Marine Ecology Progress Series*, **140**, 91–113.

Wang, W.X. & Chow, A.T.S. (2002) Benzo[a]pyrene absorption and exposure pathways in the green mussel *Perna viridis. Environmental Toxicology and Chemistry*, **21**, 451–458.

Wang, W.X. & Zhang, Q (2013) Dioxin and phthalate uptake and assimilation by the green mussel *Perna viridis. Environmental Pollution*, **178**, 455–462.

Wang, W.X., Griscom, S.B. & Fisher, N.S. (1997) Bioavailability of Cr (III) and Cr (VI) to marine mussels from solute and particulate pathways. *Environmental Science and Technology*, **31**, 603–611.

Wang, X. Li, C. & Zhoud, L. (2017) Metal concentrations in the mussel *Bathymodiolus platifrons* from a cold seep in the South China Sea. *Deep-Sea Research, Part I*, **129**, 80–88.

Ward, J.E., Zhao, S., Holohan, B.A., Mladinich, K.M., Griffin, T.W., Wozniak, J. *et al.* (2019) Selective ingestion and egestion of plastic particles by the blue mussel (*Mytilus edulis*) and eastern oyster (*Crassostrea virginica*): implications for using bivalves as bioindicators of microplastic pollution. *Environmental Science and Technology*, **53**, 8776–8784.

Webster, L., Roose, P., Bersuder, B., Kotterman, M., Haarich, M. & Vorkamp,. K. (2013) Determination of polychlorinated biphenyls (PCBs) in sediment and biota. *ICES Techniques in Marine Environmental Sciences*, No. 53.

Wedderburn, J., Cheung, V., Bamber, S., Bloxham, M. & Depledge, M.H. (1998) Biomarkers of biochemical and cellular stress in *Carcinus maenas*: an in situ field study. *Marine Environmental Research*, **46**, 321–324.

WHO (2005) International Programme on Chemical Safety Harmonization Project Strategic Plan 2005-2008. World Health Organization, Geneva, Switzerland.

WHO (2016) Dioxins and their effects on human health. Available at: https://www.who.int/news-room/fact-sheets/detail/dioxins-and-their-effects-on-human-health (accessed 28 May 2021).

Widdows, J. & Donkin, P. (1992) Mussels and environmental contaminants: bioaccumulation and physiological aspects. In: *The Mussel Mytilus: Ecology, Physiology, Genetics and Culture* (ed. E.M. Gosling), pp. 383–424. Elsevier Science Publishers B.V., Amsterdam.

Wille, K., Kiebooms, J. A. L., Claessens, M., Rappé, K., Vanden Bussche, J., Noppe, H. *et al.* (2011) Development of analytical strategies using U-HPLC-MS/MS and LC-ToF-MS for the quantification of micropollutants in marine organisms. *Analytical and Bioanalytical Chemistry*, **400**, 1459–1472.

Wu, R.S.S., Lau, T.C., Fung, W. K. M., Ko, P.H. & Leung, K.M.Y. (2007) An 'artificial mussel' for monitoring heavy metals in marine environments. *Environmental Pollution*, **145**, 104–110.

Yancheva, V.S., Stoyanova, S.G., Georgieva, E.S. & Velcheva, I.G. (2018) Mussels in ecotoxicological studies – are they better indicators for water pollution than fish? *Ecologia Balkanica*, **10**, 57–84.

Yang, Y., Ok, Y.S., Kim, K.H., Kwon, E.E. & Tsang, Y.F. (2017) Occurrences and removal of pharmaceuticals and personal care products (PPCPs) in drinking water and water/sewage treatment plants: a review. *Science of the Total Environment*, **596–597**, 303–320.

Yap, C.K., Ismail, A.1. & Tan, S.G. (2009) Effect of body size on heavy metal contents and concentrations in green-lipped mussel *Perna viridis* (Linnaeus) from Malaysian coastal waters. *Journal of Science and Technology*,**17**, 61–68.

Yap, C.K. (2018) Selected organs of marine mussels as accurate biomonitors of metal bioavailability and contamination in the coastal waters: challenges. *EC Pharmacology and Toxicology*, **6**. 528–534.

Zhang, Y.F., Chen, S.Y., Qu, M.J., Adeleye, A.O. & Di, Y.N. (2017) Utilization of isolated marine mussel cells as an *in vitro* model to assess xenobiotics induced genotoxicity. *Toxicology in Vitro*, **44**, 219–229.

Zhulidov, A.V., Headley, J.V., Pavlov, D.F., Robarts, R.D., Korotova, L.G., Vinnikov, Y. *et al.* (2000) Riverine fluxes of the persistent organochlorine pesticides hexachlorcyclohexane and DDT in the Russian Federation. *Chemosphere*, **41**, 829–841.

Zobkov, M.B. & Esiukova, E.E. (2018) Microplastics in a marine environment: review of methods for sampling, processing, and analyzing microplastics in water, bottom sediments, and coastal deposits. *Oceanology*, **58**, 137–143.

9

Population Genetics, Genomics and Selective Breeding

Introduction

Since early prehistory, humans have modified useful species over many generations, and as a result – through a process of artificial selection – we now have thousands of varieties of crop plants and domesticated livestock. The early attempts at domestication were achieved without any real knowledge of genetic principles and concepts. Nevertheless, the procedures employed then still form the basis of genetic management programmes today: breeding from selected parents, culling inferior offspring, crossbreeding with other species and producing and crossing different strains within a species. The application of these procedures to aquatic organisms is still in its infancy. Up to three or four decades ago, aquaculture management procedures were largely concerned with the physical constraints inherent in rearing organisms in restricted bodies of water. With husbandry problems largely sorted for a wide variety of fish, bivalve and crustacean species, attention is now increasingly focused on the application of genetic techniques, such as selective breeding, chromosome manipulation and, more recently, gene transfer, to the aquaculture industry. In addition, since the 1980s, there has been a dramatic development of DNA-based genetic markers, which has allowed rapid progress in aquaculture investigations of genetic variability and inbreeding, parentage assignments, species and strain identification and the construction of high-resolution genetic linkage maps for aquaculture species (reviewed in Grover & Sharma 2016). Several features of the life cycle make aquatic organisms more amenable to these procedures than domesticated livestock, such as a shorter generation time, higher fecundity, external fertilisation and larval development and more plastic sex determination. In addition, the amount of variation for genetic manipulation is greater in fish and bivalves than in domesticated livestock – which are already considerably improved following a long history of artificial selection. It should be pointed out at this stage that because of their higher market value, genetic advances in aquaculture are much further ahead for fish than for bivalve species.

Marine mussels, especially those in the genus *Mytilus*, have been more extensively used in genetic studies than any other marine organisms. There are several reasons for this (Gosling 1992). The widespread nature of the genus lends itself particularly well to the study of populations – either of the same or of different species – on different geographic scales. In addition, because mussels occupy a wide range of environments within a

Marine Mussels: Ecology, Physiology, Genetics and Culture, First Edition. Elizabeth Gosling.

particular geographic area, genetic variability of populations can be compared between different habits. Also, the two distinct stages in the life cycle of mussels – the extended dispersal larval stage and the sedentary adult stage – facilitate the study of the relative contributions of natural selection and gene flow (N_m)[1] in maintaining patterns of genetic variation in natural populations. High fecundity combined with ease of culture makes marine mussels particularly well suited for laboratory-based genetic studies. The role of selection can be more easily assessed in laboratory cultures than in wild populations, due to the ability to experimentally alter temperature, salinity and food ration. Also, given the increasing interest in the use of proteomics techniques in marine biology, mussels are positioned to play an important future role in this area. The technology for transcriptomics in *Mytilus* is being developed, and several studies indicate the feasibility of pursuing successful proteomics work in mussels and of identifying proteins differentially expressed between species and between environments (Diz *et al.* 2009 and references therein).

This chapter deals with protein and DNA molecular markers and their use in quantifying genetic variability in populations, functional genomics (transcriptomics, proteomics and metabolomics) and quantitative genetics and the various types of breeding schemes used in selective breeding of mussels.

Population Genetics

Population genetics deals with genetic differences within and between populations. It is a part of evolutionary biology that addresses phenomena such as adaptation, speciation and population structure. Genetic variation enhances the capability of an organism to adapt to a changing environment. That large amounts of genetic variation are present in natural populations has been known since Darwin's time, but the ability to measure this variation only became possible in the 1950s with two developments: the elucidation of the structure of the DNA molecule, which ultimately clarified the direct relationship between genes and proteins, and the development of gel electrophoresis, an analytical method of protein separation, which permitted detection of allozymes (proteins differing in electrophoretic mobility as a result of allelic differences in the protein sequence, which ultimately result from the existence of variation in the corresponding DNA sequence). The usefulness of the method lies in the fact that the genotype of an individual with respect to the gene locus/loci coding for a particular protein can be inferred from the banding pattern on the gel: homozygotes are single-banded phenotypes, while heterozygote are either double-, triple- or quintuple-banded, depending on whether the enzyme is composed of one, two or four polypeptide chains. By counting all the different homozygotes and heterozygotes in a sample of individuals, one can estimate the number of alternative genes (alleles) coding for a particular protein, and also the frequency of those alleles. The genetic variability, or heterozygosity, of a population can be calculated by first obtaining the frequency of heterozygotes at each locus and then averaging these over all loci to get a mean heterozygosity value. The electrophoretic technique makes it possible to compare allele frequencies and levels of heterozygosity within and between different populations of a species, between different species and so on. Such surveys have shown that there is a large amount of genetic variation in most populations. The technique has a number of advantages: a large number of loci and individuals can be examined in a relatively short time and at moderate expense, and environmental effects on banding patterns are usually unimportant. However, the fact that only the products of structural and not regulatory (non-protein-coding) genes are detected means that it underestimates overall genomic variability. In addition, because protein expression is two steps away from coding DNA, a change in a DNA base will not necessarily result in an altered amino acid sequence of the protein (because of the degeneracy of the genetic code) or altered electrophoretic mobility. Also, fresh or freshly frozen tissue is needed in relatively large amounts, which invariably means killing the specimens. Advantages and disadvantages associated with allozymes are discussed in Sunnucks (2000), Okumuş & Çiftci (2003), Liu & Cordes (2004), Askari *et al.* (2013), Gosling (2015) and Casillas & Barbadilla (2017). Up to the end of the 20th century, allozymes were the marker of choice for population structure, phylogenetic analysis and the tracking of inbreeding, stock identification and parentage analysis in aquaculture studies. However, due to the availability of more powerful DNA markers, allozymes have largely – but not entirely – been replaced.

There are two types of DNA located within animal cells: nuclear DNA (nDNA), sometimes referred to as chromosomal DNA, and mitochondrial DNA (mtDNA), found inside mitochondria. The number of genes in mtDNA is only 37, encoding 13 proteins, 2 ribosomal RNAs and 22 transfer RNAs. The noncoding D-loop region is a highly polymorphic region relative to coding sequences such as the cytochrome *b* gene, presumably due to reduced functional constraints and relaxed selection pressure, making it useful for analysis of population genetics. nDNA contains many thousands of genes, most of which are noncoding – either single-copy or repetitive sequences. There are several additional features that distinguish mtDNA from nDNA: the genome is haploid (i.e. each mitochondrion generally contains only one type of mtDNA), the mode of inheritance is primarily through the maternal line so that all individuals with a particular mtDNA genotype probably belong to the same maternal clone (for exceptions, see Chapter 5; one consequence of maternal transmission is that the effective population size (N_e)[2] for mtDNA is smaller than that for nDNA, so that mtDNA variation is a more sensitive indicator of population phenomena such as bottlenecks and hybridisations), there are no repetitive sequences (so that most mutations are selectively neutral) and it evolves about 5–10 times faster. Because of its higher mutation rate, mtDNA is also more likely to show differences between populations and species (details in Okumuş & Çiftci 2003; Chauhan & Rajiv 2010; Askari *et al.* 2013; Maqsood & Ahmad 2017).

Molecular Markers and Applications

Molecular markers are used in molecular biology and biotechnology to identify a particular sequence in a pool of unknown DNA. With DNA markers, it is theoretically possible to observe and exploit genetic variation in the entire genome. Both mtDNA and nDNA can be extracted from fresh, frozen or alcohol-preserved tissue using a published method such as that of Hoarau *et al.* (2002) or a genomic DNA minikit (Geneaid). After extraction, DNA can be cut at specific sites by bacterial enzymes called restriction endonucleases. Cleavage results in a series of DNA fragments called 'restriction fragments', whose number and size are quantified by gel electrophoresis. These fragments can be detected by direct staining (usually with ethidium bromide), radio end-labelling or Southern blotting. Polymorphisms in the nucleotide sequence at a restriction enzyme recognition site result in varied numbers of DNA fragments, while polymorphisms in the length of DNA fragments result in an altered electrophoretic mobility of the fragment through a gel matrix. These are called restriction fragment-length polymorphisms (RFLPs). In practice, when individuals within or between populations are being compared, a number of different restriction enzymes are used to increase the probability of detecting RFLPs. Digestion of nDNA with most enzymes results in too many fragments, but the technique works well on the smaller mtDNA genome, which can be extracted from an individual in high copy number without cloning[3] or polymerase chain reaction (PCR).[4] This, together with the rapid rate of evolution and the maternal mode of inheritance, make RFLP analysis of this molecule one of the methods of choice for many population studies (Hawley 2016). Indeed, analysis of mitochondrial markers is mostly done by RFLP analysis or direct sequence analysis (see later). Being highly polymorphic, and given its ease of analysis, mtDNA has been very successful as a marker in aquaculture and fisheries research (Maqsood & Ahmad 2017). Most nDNA does not code for any gene product, and even within the DNA sequences that are recognised as genes there are noncoding regions, called introns. One class of noncoding DNA consists of a variable number of multiple copies of tandemly arranged simple sequence repeats (SSRs), which range in size from one to several hundred nucleotides. Repeat units from 10 to 100 nucleotides long are referred to as minisatellites, while shorter repeat units of one to four nucleotides are called microsatellites (MS). The key feature of these tandem repeats is that the number of repeated units within a microsatellite locus can vary within a population, producing variation in the length of the locus, which can be detected by amplifying the locuş using PCR followed by electrophoresis (Beaumont *et al.* 2010). Each SSR locus (tandem repeat array) is flanked by a unique sequence, and if these sequences are known, primers[5] can be synthesised that are complementary to them. Using primers, the tandem array of the locus can be amplified by PCR. The traditional process for developing genomic microsatellite markers is time-consuming and expensive due to the need to prepare genomic libraries and sequence large numbers of clones that potentially contain microsatellite regions. However, next-generation sequencing (NGS), also known as high-throughput (HTS) or parallel sequencing, provides a much more efficient and cost-effective method for the acquisition of genetic markers (including microsatellites) in organisms for which adequate databases are not currently available (reviewed in McCormack *et al.* 2013; see An & Lee 2012). The high rate of mutation in these loci, detected as length variants on gels, leads to extensive allelic variation (>20 alleles per locus), making microsatellite markers ideal for many types of application in aquaculture, such as: (1) identification of genetic variability within and between stocks; (2) monitoring of genetic changes in stocks; (3) parentage and pedigree analysis in selective breeding; (4) assignment tests; and (5) genomic mapping and detection of quantitative trait loci (QTLs; see later) (Selkoe *et al.* 2006; ICES 2014; Abdul-Muneer 2014; Vieira *et al.* 2016). Microsatellite markers are also very useful in detecting variability in situations where conventional markers such as allozymes or mtDNA markers have revealed low inter-population or species variability.

Another approach is the use of random amplified polymorphic DNA (RAPD) markers, whereby DNA polymorphisms are detected by randomly amplifying multiple regions of the genome using an identical pair of primers 8–10 base pair (bp) in length. Because the primers are short and relatively low annealing temperatures (often 36–40 °C) are used, the likelihood of amplifying multiple products is great, with each product (presumably) representing a different locus (Liu & Cordes 2004). Since much of the nuclear genome is non-coding, it is presumed that most of the amplified loci will be selectively neutral. Genetic variation and divergence within and between taxa of interest are assessed by the presence or absence of each product, which is dictated by changes in the DNA sequence at each locus. RAPD loci are inherited as Mendelian markers in a dominant fashion and scored as present/absent, making it impossible to distinguish between homozygotes and heterozygotes (Askari *et al.* 2013). Another disadvantage is that RAPDs are neither reliable nor repeatable (Beaumont *et al.* 2010). However, they have all the advantages of a PCR-based marker with the added benefit that primers are commercially available and do not require prior knowledge of the target DNA sequence or genome organisation. Other advantages include the ease with which a large number of loci and individuals can be screened simultaneously. RAPD analysis has been used to evaluate genetic diversity for species, subspecies and population/stock identification (Okumuş & Çiftci 2003).

Amplified fragment-length polymorphism (AFLP) is a technique used to detect polymorphisms in DNA when no information about the genome is known. AFLP protocols involve four main steps (Paun & Schönswetter 2012): (1) restriction of genomic DNA and ligation of adaptors of known sequences to the DNA fragments generated by complete digestion of genomic DNA; (2) pre-selective PCR amplification of a subset of the restricted fragments; (3) selective PCR amplification, further reducing the fragment number; and (4) electrophoretic separation of amplified DNA fragments. The resulting fragments are visualised by separation on a polyacrylamide gel followed by autoradiography, using fluorescence methods or via automated capillary sequencing instruments. There are many advantages to AFLP analysis when compared to other marker technologies, including RAPD, RFLP and microsatellites: it not only has higher reproducibility, resolution and sensitivity at the whole-genome level, but also has the capability to amplify between 50 and 100 fragments at one time. Also, no prior sequence information is needed for amplification. AFLP loci, like RAPDs, are dominant markers. The technique has become widely used in the identification of genetic variation in strains or closely related species, in determining slight differences within populations, in food traceability (Figure 9.1) and in linkage studies, where it is used to generate maps for QTL analysis.

Single-nucleotide polymorphisms (SNPs) are the newest addition to the molecular toolbox. They describe polymorphisms caused by point mutations that give rise to different alleles containing alternative bases at a given nucleotide position within a locus. For example, two sequenced DNA fragments from different individuals, ATGGCAT and ATGGTAT, contain a difference in a single nucleotide. In this case, we say there are two alleles, C and T. Almost all common SNPs are bi-allelic (i.e. they usually have only two variants). SNP markers, like allozymes, RFLPs, SSRs and ESTs (see later), are inherited as co-dominant markers (i.e. markers for which both alleles are expressed when co-occurring in an individual). In the oyster (*Crassostrea gigas*) genome, the average density of SNPs was estimated to be 1 per 60 bp in coding regions and 1 per 40 bp in noncoding regions (Sauvage *et al.* 2007). SNP genotyping through DNA sequencing is rapidly becoming a powerful tool that can be applied to a wide range of population studies, from individual, family and population identification to the discovery of genes and genomic regions affecting adaptive

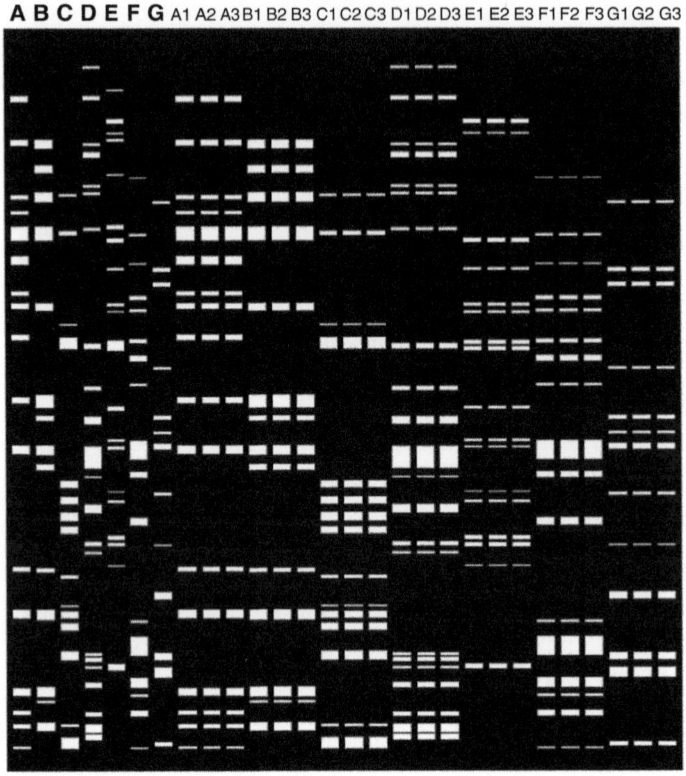

Figure 9.1 Comparison of AFLP–Genographer profiles of reference fish samples (A, *Merluccius hubbsi*; B, *Merluccius capensis*; C, *Oncorhynchus keta*; D, *Oncorhynchus mykiss*; E, *Salvelinus alpinus*; F, *Salvelinus fontinalis*; G, *Salmo trutta*) and three unknown samples of each species. *Source:* From Maldini *et al.* (2006). Reproduced with permission from Elsevier.

phenotypic variation (ICES 2011). However, because they are considerably less variable than microsatellite markers, most SNP studies have focused on gene mapping, QTL analysis and functional genomics (see later). Several approaches have been used for SNP discovery, including single-strand conformational polymorphism analysis, heteroduplex analysis and direct DNA sequencing, the latter being the most accurate and most used approach (ICES 2014). For additional information on SNPs, see Liu & Cordes (2004), ICES (2008, 2014, 2018), Beaumont *et al.* (2010), Grover & Sharma (2016), Maqsood & Ahmad (2017) and Wenne (2018).

Expressed sequence tags (ESTs) are fragments of messenger RNA (mRNA) sequences derived through single sequencing reactions performed on randomly selected clones from complementary DNA (cDNA) libraries.[6] An individual EST is a small piece (<1000 bp in length) of a cDNA sequence that is generated by sequencing either one or both ends of the expressed gene. They can be used as ‘tags’ to fish an unknown gene out of a portion of chromosomal DNA by matching base pairs. They may be submitted to a general database such as GenBank or to one specific for a particular species, such as GigasBase for the oyster *Crassostrea gigas*. Such collections can form the basis for subsequent microarray design,

SNP and microsatellite detection and the placement of novel markers on genetic maps (see later). The US National Center for Biotechnology Information (NCBI) is currently in the process of merging EST and genome survey sequence (GSS) records into the Nucleotide database.

Recently, DNA barcoding has become a standardised and broadly used molecular method for specimen identification (reviewed in Barco *et al.* 2016). Specimen identification is based on the evidence that selected DNA sequences are more variable among species than within species. Consequently, DNA sequences from identified specimens can be used as a reference to assign species names to sequences from unidentified material. The reliability of DNA barcoding is largely determined by the quality of the reference barcode libraries to which the unknown specimen is compared. The primary barcode sequence for members of the animal kingdom is a portion of the mitochondrial cytochrome *c* oxidase subunit I (COI) gene. To increase the quality of species identifications, the Barcode of Life Data Systems (www.boldsystems.org) stores barcode sequences and associates then with other taxonomically relevant data (Ratnasingham & Hebert 2007). Applications of DNA barcoding include identification of new species, safety assessment of food, identification and assessment of cryptic species, detection of alien/introduced species and identification of endangered and threatened species (Nicolè *et al.* 2012). To illustrate, in 2014 the black pygmy mussel, *Xenostrobus securis*, native to New Zealand and Australian coasts, was detected for the first time at low density inside one port in the Avilés estuary, Bay of Biscay (Pejović *et al.* 2016). It was expected that the specimens would prove to be part of a recent introduction event. Devloo-Delva *et al.* (2016) carried out a study to assess population status, detect signs of population expansion and develop a Barcode-based tool for easy detection of the species from water samples (environmental DNA approach). A total of 130 individuals were analysed for the DNA Barcode COI gene. *Xenostrobus securis* was found in two consecutive years and the population showed evidence of multiple cohorts, meaning it survived and reproduced in the Avilés estuary. High population genetic variability at the COI gene and two genetically distinct clades suggested multiple introductions. Species-specific primers were developed within the DNA barcode for PCR amplification from water samples, enabling rapid detection in initial expansion stages of the species.

Table 9.1 summarises the results from a wide variety of mussel population genetic studies, illustrating the diversity of molecular markers used.

Genetic Variation and Population Structure in Mussels

Genetic variation describes naturally occurring genetic differences among individuals of the same species. This permits flexibility and survival of a population in the face of changing environmental circumstances. Several evolutionary forces affect the amount of genetic variation, and hence the degree of differentiation, between populations. Factors such as geographic distance and physical barriers lead to reduced contact between populations. Subsequently, limited gene flow in combination with random genetic drift[7] and adaptation[8] acts to increase genetic differentiation between the fragmented populations. Molecular markers, as already described, provide a direct assessment of the pattern and distribution of genetic variation between populations, species and higher-order taxonomic groups. Also, increased computational power and mathematical models have enhanced the scope of conclusions that

Table 9.1 Utilisation of molecular markers in a variety of population genetic studies on mussels, primarily those in the *Mytilus edulis* species complex. The main objective of each study is highlighted in italics. See text for information on the different molecular markers. *Me*, *Mytilus edulis*; *Mg*, *Mytilus galloprovincialis*; *Mt*, *Mytilus trossulus*; *Mc*, *Mytilus californianus*; *Mch*, *Mytilus chilensis*; *Mcor*, *Mytilus coruscus*; *Pc*, *Perna canaliculus*. SNPs, single-nucleotide polymorphisms; AFLPs, amplified fragment-length polymorphisms; ESTs, expressed sequence tags; SSRs, simple sequence repeats; RFLPs, restriction fragment-length polymorphisms; RAPDs, random amplified polymorphic DNAs.

Study region	Marker(s)	Species	Objective	Outcome	References
Pacific coast of Chile and Canada (*Mt*), Northern Ireland (*Me*), Italy (*Mg*), New Zealand (*Mg*) and Spain (*Mg*)	SNPs	Chilean blue mussel and reference species: *Me*, *Mg* and *Mt*	Test hypothesis that native Chilean blue mussel is an endemic species, with its own unique *evolutionary history* in the Southern Hemisphere	Native Chilean blue mussel is genetically distinct from reference species and should be recognised as *Mytilus chilensis*	Larraín *et al.* (2018)
Large range of sites on the European coastline	Diagnostic nuclear markers (*Me* 15/16, EF-bis, ITS) and mtDNA marker (ND2-COIII)	*Me*, *Mg* and *Mt*	Investigate the *population structure* of *Mytilus* taxa on European coasts	Clear genetic distinction between the Black Sea and Mediterranean populations and a higher incidence of *Mt* than reported in northern Europe. Frequency of *Mg* nuclear alleles high along the Iberian Peninsula, decreasing abruptly along the French coasts, with a high frequency of *Me* alleles in the Bay of Biscay, The Netherlands, Germany, Iceland and the Barents and White seas, with little evidence of introgression between *Me* and *Mg*. *Mt* alleles observed in the Baltic Sea and Danish Straits, and reported for the first time in cold waters of Iceland, the Barents Sea and the White Sea	Kijewski *et al.* (2011)

Northern and southern Greenland	SNPs	Greenland *Mt* and reference *Mt* samples	Investigate *genetic identity* of *Mt* in Greenland and whether Greenland *Mt* populations are more closely related to those of Atlantic Canada than those of the North Pacific	The North Pacific is the most likely source for northern Greenland *Mt*, while Atlantic Canada is the most likely source for *Mt* in southern Greenland	Bach *et al.* (2019)
Mytilus taxa collected from Northern and Southern Hemisphere sites	Restriction site variation in the mitochondrial 16S rRNA gene	*Mytilus* taxa collected in both hemispheres	Assess the *phylogenetic relationship* of Southern Hemisphere *Mytilus* taxa relative to their northern counterparts	Two types of mtDNA lineages found among Southern Hemisphere mussels. >90% of the lineages probably arose from a migration event from the Northern Hemisphere during the Pleistocene era via an Atlantic route, while the remainder (<10%) suggest a more recent migration to the Southern Hemisphere (see text)	Hilbish *et al.* (2000)
North Sea and Baltic Sea	Taxa-specific DNA markers: Glu-5, Ef-bis, M7 and ITS, plus previously published SNP data	*Me*, *Mg* and *Mt* collected in a transition from the SE North Sea to the NE Baltic Sea characterised by strong gradients of environmental factors and continuous distributions of *Mytilus* populations	Assess the relative *contribution of natural environmental drivers on genetic differentiation* in *Mytilus* populations	Strong genetic differentiation in allele frequencies is correlated with environmental factors along the North Sea–inner Baltic transect. Environmental gradients explain large variability in the studied allele frequencies, with salinity contributing the most to observed genetic differences	Kijewski *et al.* (2019)

(*Continued*)

Table 9.1 (Continued)

Study region	Marker(s)	Species	Objective	Outcome	References
European Arctic	SNPs	*Mytilus edulis* complex	*Identify species, distribution and variation*	*Me* is the most abundant species found, with a clear genetic split between populations in Greenland and the eastern Atlantic. Analyses reveal the presence of *Mt* in high Arctic NW Greenland (77 °N) and of *Mg* or their hybrids in SW Greenland, Svalbard and the Pechora Sea. Also, a high degree of hybridisation and introgression is observed between species	Mathiesen *et al.* (2017)
European coasts	SNPs	*Me*, *Mg* and *Mt* diagnosed using Me15/16 marker	*Develop species-specific markers* to genotype the *Mytilus* species complex	A panel of SNP markers was developed that separate the three taxa unambiguously, and hybrids and interspecies introgression events were identified	Wilson *et al.* (2018)
Loch Etive on the west coast of Scotland	Five allozymes and nuclear marker Me15/16	*Me*, *Mg* and unidentified *Mytilus* sp. collected from mussel farms in Loch Etive	Investigate the *distribution and identification of a mussel type* that differs morphologically from *Me* and *Mg*	Mussel type identified as a mixture of *Mt* and *Mt* × *Me* hybrids. First record of *Mt* in UK waters; most likely a postglacial relict population in Loch Etive. Complete sequencing of F and M mtDNA of *Mt* genomes provides evidence that they have close affinity with those in ancestral American *Mt* rather than *Mt* from the Baltic, supporting direct invasion of Loch Etive from North America rather than from the Baltic	Beaumont *et al.* (2008), Zbawicka *et al.* (2010)

Central California, USA	Species-specific markers Glu-5' and ITS-1	*Mt* and *Mg* collected over four years from four sites within both the San Francisco Bay and Monterey Bay	*Identify coexisting native Mt and invasive Mg and their hybrids*, as well as the *influence of spatial and temporal variation in temperature and salinity on their abundance*	Proportion of *Mt* at the eight hybrid zone sites correlates negatively with average salinity and positively with maximal temperature. Proportion of *Mg* shows the opposite pattern. Proportion of hybrids is correlated neither with habitat temperature nor with salinity	Braby & Somero (2006a)
Central California, USA	SNPs	*Mt* and *Mg* in hybrid zone in central California	Examine the genome-wide *patterns of introgression between invasive Mg and native Mt* across their zone of hybridisation	*Mg* has acquired genetic material from *Mt* during its invasion of southern and central California. Although most genes entering *Mg* are likely deleterious, it is possible that some adaptive introgression is presently occurring that may facilitate its invasion potential. However, the threat posed by invasive *Mg* is more ecological than genetic	Saarman & Pogson (2015)
Baltic Sea	AFLPs	*Me* and *Mt* collected on the Swedish west coast, in the Baltic proper and in the Bothnian Sea	Examine *genetic structure and barriers to gene flow* in the *Me*/*Mt* complex at collection sites	There is distinct genetic differentiation between the three regions, with lower gene diversity in the Bothnian Sea. Oceanographic connectivity, together with salinity and to some extent species identity, provides explanations for the differentiation between the west coast and the Baltic proper and Bothnian Sea	Larsson *et al.* (2017)

(*Continued*)

Table 9.1 (Continued)

Study region	Marker(s)	Species	Objective	Outcome	References
California, USA	ESTs	*Mc*	*Use EST sequences to screen for SSRs*	The nine microsatellites developed will provide much needed powerful molecular tools for management and conservation studies in *Mc* populations and related species	Vidal *et al.* (2009)
Southern Chile	SSRs	*Mytilus* collected from six sites	Evaluate the potential of SSRs combined with allocation algorithms to *assign Mytilus to their geographical origin*, in order to verify and validate administrative traceability systems	The frequency-based algorithms show the best results, reallocating individuals to their original population at least six times more often than the three other assignment methods evaluated	Larraín *et al.* (2015)
Central–eastern Mediterranean Sea	SSRs	Cultured and wild populations of *Mg* collected from Greece, Turkey, Croatia and Italy	Investigate *genetic variation* of *Mg* mussels throughout the central–eastern Mediterranean Sea	Results indicate broad patterns of genetic homogeneity between Greek populations, as well as between the two Italian ones, and differentiation of those from Turkey and Croatia. This probably reflects specific topographic and oceanographic conditions of these regions	Giantsis *et al.* (2014)
Sampled over a wide geographic range in the Northern and Southern Hemispheres	RFLPs of sequences of adhesive foot protein and the mitochondrial cytochrome oxidase subunit I (COI) region	*Me* (Europe, Canada), *Mg* (Europe, China, New Zealand), *Mch* (Chile), *Mc* (Canada), *Mt* (Sweden, Canada) and *Mcor* (China)	*Identify* six *Mytilus* species from fresh and processed samples	An accurate, reliable and powerful tool for *Mytilus* species identification and traceability	Fernández-Tajes *et al.* (2011)

Culture sites in Europe (Spain and The Netherlands), Chile, Canada and New Zealand	RAPDs	*Me*, *Mch*, *Mg* and *Pc*	Analyse the suitability of RAPD products for searching for a diagnostic pattern that can be used in the *identification* of the four mussel species	RAPD bands on gels show that *M. galloprovincialis* can be clearly differentiated from the other species	Rego *et al.* (2002)
Manila Bay, Luzon Island, Philippines	DNA barcode mitochondrial COI gene	*Mytilus* sp.	Ascertain the *identity of a newly detected mussel species* using DNA barcoding	Results suggest that the previously identified *Mytilus* sp. are in fact *Mytella charruana*, with an average identity match of 94%. This is the first record of the invasive species in Manila Bay. As establishment is highly likely, the mussel is also likely to compete with native *Perna viridis*	Vallejo *et al.* (2017)

can be drawn from genotype data (e.g. information on effective population size, genetic bottlenecks[9] and assignment of migrants; Chauhan & Rajiv 2010 and references therein).

Intraspecific Genetic Variation

Mytilus spp.

Identifying population genetic structures of species, describing the distribution of genetic diversity and understanding the causes of these structures are important for effective management and conservation (Wennerström *et al.* 2013 and references therein). The Chilean blue mussel, *Mytilus chilensis*, is an important commercial species (see Chapter 10) distributed along approximately 1900 km of the Chilean coast from the Arauco Gulf (35 °S) in the north to Cape Horn (55 °S) in the south. Molecular analysis has confirmed that the species is genetically distinct from reference species *M. edulis*, *M. galloprovincialis* and *M. trossulus*, which supports the hypothesis of a unique evolutionary history of the native Chilean mussel on the Pacific coast of South America (Larraín *et al.* 2018). The genetic diversity and population structure of this economically important species are critical for supporting management and conservation programmes. A lack of differentiation, using RAPDs, has been reported for *M. chilensis* at seven of eight locations sampled along its range (Toro *et al.* 2004a). Genetic homogeneity was ascribed to extensive gene flow due to a long larval dispersal phase enhanced by oceanographic features such as the north-flowing West Wind Drift (WWD) and several coastal currents along this coast. The significant differentiation observed between the most southerly Punta Arenas sample and all the rest was probably due to the southerly flowing Cape Horn current, a branch of the WWD, which precludes larval transport toward northern populations.

In a later study, Araneda *et al.* (2016) used 1240 SNPs to investigated neutral and adaptive genetic variation in *M. chilensis* from six locations in southern Chile. Differentiation among locations with 891 neutral SNPs was low (*FST*[10] = 0.005). Higher differentiation was obtained with a panel of 58 putative outlier SNPs (*FST* = 0.114), indicating the potential for local adaptation. This panel identified clusters of genetically related individuals and demonstrated that much of the differentiation (~92%) could be attributed to the three major regions and environments: extreme conditions in Patagonia, inner bay influenced by aquaculture (Reloncaví Gulf; ~41–42 °S) and outer bay influenced by sea water (Chiloé Island; ~43 °S). The Patagonian samples, in the most southerly region (Isla Peel; ~51 °S), were most distinct, but additional analysis carried out excluding samples from this region also revealed adaptive divergence between inner- and outer-bay samples. The four locations within the Reloncaví Gulf area were most similar with all panels of markers, likely due to their similar environments, high gene flow from aquaculture practices and small geographical distances. This study demonstrates that mussel species with high gene flow may require analyses at different geographical scales and multiple panels of putative adaptive loci to reveal hidden population structure. A previous study (Larraín *et al.* 2015), using only nine microsatellite loci on the same species from southern Chile, achieved just 51% correct reassignment to locations of origin. The SNP marker panels developed by Araneda *et al.* (2016) increased assignment accuracy (84–100%), probably due to the use of adaptive loci and the number of markers employed.

NW Spain is one of the main producers of cultured *M. galloprovincialis* in the world. The industry is mainly based in sheltered and nutrient-rich deep sunken valleys, called rías, which provide the perfect environment for suspended mussel culture (see Chapter 10). Diz & Presa (2009) carried out a large study using six microsatellites on 27 mussel samples from the main five Galician rías (Rías Altas, consisting of Ría de Betanzos plus some other smaller estuaries, and Rías Bajas, consisting of Ría de Vigo, Ría de Pontevedra, Ría de Arousa and Ría de Muros – the former towards the north and the latter south of Cape Finisterre). Each sample consisted of 40 mature individuals of 50–70 mm shell length, which were collected by scraping on intertidal rocky shores at two to three locations evenly separated from each other (3–5 km) along the two banks of each ría. The six microsatellites analysed were moderately polymorphic with allele numbers (allelic richness) per locus and sample size ranging between 3 and 25 individuals. Diversity parameters showed very similar scores between rías. Despite the narrower geographical range of this study, allelic richness and heterozygosity were 5 and 2.7% larger, respectively, in Galician rías than in other samples from the Atlantic Iberian coast for the same set of loci (Diz & Presa 2008). This result suggests a relationship between allele diversity and the large population sizes of mussels cultivated in Galician rías. Interpopulation differentiation ($FST = 0.0122$) was half of that previously reported for Iberian Atlantic samples ($FST = 0.0240$). This weak genetic divergence suggests that no genetic structuring exists either within rías (inner versus outer and northern versus southern banks) or among the five rías compared. The greatest local differentiation was observed between the sample pools from Rías Altas versus Rías Bajas, which are separated by the oceanographic boundary at Cape Finisterre. Around this cape, the Galician seasonal upwelling is stronger due to the rise of Eastern North Atlantic Central Water (ENACW) against the Galician coast, which replaces surface water (Fraga 1981) and may act as a barrier to larval flow between Rías Altas and Rías Bajas areas. Diz & Presa (2009) suggest that transfer of seed between regions may have a homogenising effect that to some degree counteracts the major regional restriction of larval dispersal by marine currents.

Addison *et al.* (2008) tested for genetic differentiation among six populations of California sea mussels, *M. californianus*, sampled across 4000 km of their geographical range by comparing patterns of variation at four independent types of genetic markers: allozymes, single-copy nDNA markers and DNA sequences from the male and the female mitochondrial genomes. They found that all four types of marker examined were unable to detect any significant differences among populations sampled throughout the geographical range. They suggested that this striking genetic homogeneity might result from extensive gene flow through a long planktonic larval phase lasting up to 45 days and from the lack of exposure at the post-settlement stage to strong selective gradients across the range.

Perna spp.

The greenshell mussel, *Perna canaliculus*, endemic to New Zealand, is distributed from 35 to 47 °S. Several population genetics assessments of this species have identified a pronounced and ~1.3 million-year-old genetic discontinuity at ~42 °S, just south of Cook Strait, the major break between the North and South islands (Figure 9.2). Ten polymorphic microsatellite markers were employed to assay genetic variation in 14 populations of *P. canaliculus* (Wei *et al.* 2013a). Nine revealed limited evidence of genetic structure, but one outlier locus (*Pcan1-27*), which may be under selection, provided evidence of the discontinuity in central NZ. The limited multilocus evidence of genetic structure is

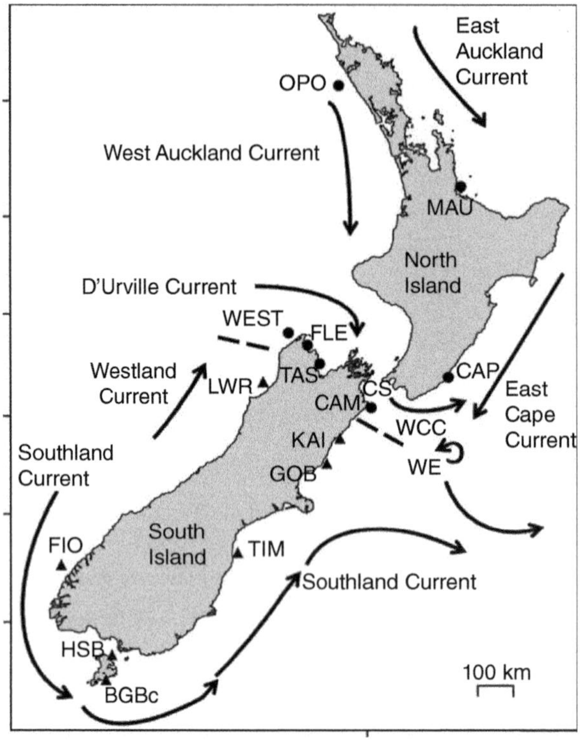

Figure 9.2 *Perna canaliculus* collection sites and hydrographic patterns around New Zealand. OPO, Opononi (N = 27); MAU, Maunganui (N = 29); CAP, Castlepoint (N = 30); WEST, Westhaven (N = 13); FLE, Fletchers Beach (N = 14); TAS, Tasman Bay (N = 20); CAM, Cape Campbell (N = 30); LWR, Little Wanganui River (N = 26); KAI, Kaikoura (N = 16); GOB, Gore Bay (N = 28); TIM, Timaru (N = 22); FIO, Fiordland (N = 19); HSB, Horseshoe Bay (N = 22); BGB, Big Glory Bay (N = 20). Black circles, northern group; filled triangles, southern group; – –, areas of coastal upwelling; c, cultured mussels; WE, Wairarapa Eddy; WCC, Wairarapa Coastal Current. *Source:* From Wei *et al.* (2013b). CC BY 4.0.

probably due to high levels of gene flow among populations of this continuously distributed species. Superimposed on top of high levels of genetic connectivity are variable but generally high rates of self-recruitment within the 14 populations and the north and south regions.

Environmental variability has long been recognised as a contributing factor to population genetic structuring, and advances in genetic marker technology provide new opportunities for assessing how environmental variation may contribute to the genetic structuring of populations, a discipline referred to as seascape genetics in the marine environment. Wei *et al.* (2013b) used the microsatellite data set from Wei *et al.* (2013a) to assess the influence of environmental variability on macrogeographic genetic variation. They obtained site-specific data for seven environmental and three geospatial variables, including annual mean solar radiation, winter solar radiation, wintertime sea surface temperature (SST), annual mean SST, mean orbital water velocity and tidal current. Seascape genetics analysis of the 10 variables revealed that SST explained far more genetic variation (among populations and individuals) than any other variable. Genetic variation across the full distributional range of *P. canaliculus* is therefore best explained by SST, which suggests that a contemporary temperature-related selective force is acting on greenshell mussels from the subtropical north to the cold temperate south. These findings thus strongly suggest that different processes (phylogeographic for the genetic break and contemporary selection for the association with SST) are influencing the genetic strucuture of *P. canaliculus* populations.

The brown mussel, *P. perna*, is found on the east, west and south coasts of Africa and the coast of South America, extending into the Caribbean Sea. Analyses of mtDNA sequences indicate a sharp phylogeographic break on the southeast coast of South Africa (corresponding to the warm temperate/subtropical transition), forming a western and an eastern lineage[11] (Zardi *et al.* 2007). The two lineages overlap in their distributions on the southeast coast over a distance of approximately 200 km. Their presence was subsequently confirmed by mitochondrial (COI) and nuclear (internal transcribed spacer, ITS) markers, supporting an Indo-Pacific origin (Cunha *et al.* 2014). Microsatellite analyses of the two lineages revealed clear genetic differentiation between western (temperate) and eastern (subtropical/tropical) populations (Zardi *et al.* 2015), confirming the divergence previously detected with mitochondrial and nuclear markers. Gene flow between regions was predominantly east-to-west, which is to be expected because the dominant ocean current between the ocean basins, the Agulhas Current, runs westerly from the Indian Ocean toward the Atlantic Ocean and is thought to play a role in limiting marine dispersal in the opposite direction (references in Zardi *et al.* 2015). Eastward gene flow could also be impeded by selection (i.e. unfavourable environmental conditions) causing outbreeding depression if lineages are locally adapted. Alternatively, poor mating success between lineages could also play a role, due either to divergent reproductive ecology (e.g. differences in the timing and frequency of spawning events) or to some level of genetic incompatibility. There were significant heterozygote deficiencies in populations where the two lineages co-occurred. Such effects can be expected when pooling discrete subpopulations with different allele frequencies that do not interbreed as a single randomly mating unit. Low levels of heterozygote deficiency were detected in Ponta do Ouro, which is the farthest population sampled from the contact zone, located on the east side of the coast. Common garden (transplant) experiments showed that gaping (periodic closure and opening of the shell) and attachment to the substratum differed significantly between the two lineages: western individuals gaped more and attached less strongly to the substratum compared to eastern ones. The authors concluded that these behavioural differences are consistent with the geographic and intertidal distributions of each lineage along sharp environmental clines, indicating their strong adaptive significance.

The Asian green mussel, *Perna viridis*, an economically important mussel (see Chapter 10), is native to the coastal marine waters of the Indo-Pacific region. Using nine microsatellite loci, Ye *et al.* (2016) examined genetic diversity and population genetic structure of samples of *P. viridis* from natural stocks at four locations in the South China Sea. Low *FST* values (0.001–0.005) and insignificant genetic divergence were observed between populations, consistent with results revealed by mitochondrial COI and 16 S rRNA sequences for *P. viridis* in the same area (Ye *et al.* 2015). Low levels of genetic differentiation for this species were also found in Malaysia, the Gulf of Thailand and Singapore (references in Ye *et al.* 2016). This may be explained by the relatively long pelagic larval stage (3–4 weeks), which provides ample opportunity for larval dispersal, thereby promoting gene flow. A stable prevailing northeastward current (the China Coastal Current) during summer (the *P. viridis* spawning season) in the South China Sea might also be a major boost to gene flow among locations. In addition, all samples showed moderate levels of genetic diversity, a large estimate of effective population size and no sign of recent bottleneck events.

Interspecific Genetic Variation

The Mytilus Edulis Species Group

Mussels in the genus *Mytilus* are a dominant component of rocky-shore communities in cooler waters of the Northern and Southern Hemispheres. Genetic analyses have revealed three closely related species that make up what is known as the *M. edulis* species group/complex: *Mytilus edulis*, *M. galloprovincialis* and *M. trossulus*. *Mytilus* taxa are difficult to identify using only shell characteristics because of adaptations of the shell to different environmental conditions, as well as hybridisation (see later) between pairs of taxa (Kijewski *et al.* 2011 and references therein). The three congener species have distinct but overlapping distributions. *M. edulis* has a wide distribution in the Northern Hemisphere, occurring in European waters from Spitsbergen (74–81 °N) to western France; in the Baltic, North and White Seas and Iceland (Śmietanka *et al.* 2004); and on the Atlantic coast of North America, from the Canadian Maritimes southward to North Carolina (Gosling 2015 and references therein). *Mytilus* spp. have been reported in Greenland from the southern tip (59 °N) to the High Arctic at 77 °N on the west coast, and reaching their northern limit at around 65 °N on the east coast (Bach *et al.* 2019 and references therein). The occurrence of *Mytilus* spp. in Greenland was believed to consist solely of *M. edulis*. However, *M. trossulus*, which is considered to be better adapted to colder temperatures and brackish environments, has recently been documented in northwest Greenland, where it co-exists and hybridises with *M. edulis* (Wenne *et al.* 2016). *M. trossulus* also inhabits cool-water regions of the Northern Hemisphere, such as the coasts of the North Pacific, eastern Canada, the Baltic Sea, Scotland, Iceland, the Barents Sea, the White Sea and Norwegian fjords (Zbawicka *et al.* 2003, 2010, 2012; Brooks & Farmen 2013; Wennerström *et al.* 2013). *M. galloprovincialis*, referred to as the Mediterranean mussel (Med), has a large natural distribution from the Black Sea to north of the British Isles (Gosling 2003 and references therein) and has been shown to be divided into two main lineages: Atlantic (Atl) and Mediterannean (Med) (Quesada *et al.* 1995a; Roux *et al.* 2014; Fraïsse *et al.* 2016). These lineages form hybrid zones (see later) in the Almeria–Oran front, a region located in the western Mediterranean Sea between Spain and Algeria and which separates salty Mediterranean surface water from inflowing Atlantic water (Quesada *et al.* 1995a,b; Sanjuan *et al.* 1994; El Ayari *et al.* 2019). Another lineage, native to New Zealand, has been identified by Westfall & Gardner (2010).

Some of the present distribution of *M. galloprovincialis* is due to multiple accidental introductions into Japan, Hong Kong, South Korea, southeastern China, South Africa, southern California (references in Gosling 1992a, 2003), southern Chile (Toro *et al.* 2005; Larraín *et al.* 2018) and Australia (Ab Rahim *et al.* 2016). It has also been reported in SW Greenland, Svalbard Island to the east of Greenland and the Pechora Sea, NW Russia (Mathiesen *et al.* 2017). In the Northern Hemisphere, where the ranges of *M. edulis*, *M. trossulus* and *M. galloprovincialis* overlap, variable amounts of hybridisation occur between species pairs (details later).

In the Northern Hemisphere, mussels in the genus *Mytilus* spp. first arrived in the Atlantic Ocean from the Pacific Ocean during the Pliocene epoch (5.3–2.58 mya) following the opening of the Bering Strait between Alaska and Siberia about 3.5 mya (Vermeij 1991). Repeated periods of glaciation throughout the Pleistocene (0.01–1.8 mya) led to re-isolation of the two ocean basins, and presumably to the allopatric divergence of *M. edulis* in the

Atlantic and *M. trossulus* in the Pacific (Riginos & Cunningham 2005). *M. trossulus* spread to the northwest Atlantic shortly after the last glacial maximum (LGM, 18 000–21 000 years ago) (Rawson & Harper 2009). Northwest and northeast Atlantic *M. trossulus* populations are closely related, suggesting a common ancestor from the Pacific before the two populations split (Väinölä & Strelkov 2011). More recently, during the Pleistocene or Holocene, *M. trossulus* colonised both sides of the Atlantic so that descendants of the species now form two distinct hybrid zones with the Atlantic-endemic *M. edulis*, one in the western Atlantic (Canadian Maritimes) and one in the eastern Atlantic (Scandinavia) (see Śmietanka *et al.* 2013). Väinölä & Hvilsom (1991) have suggested that *M. trossulus* may have been widely distributed in Europe and colonised the Baltic approximately 7500 years ago when there was an intrusion of North Sea water. *M. galloprovincialis* is believed to have originated in the Mediterranean during one of the Pleistocene ice ages when the area had restricted connection to the Atlantic Ocean. The warmer conditions that developed in the Mediterranean, together with the reduced contact between the Atlantic and the Mediterranean, probably favoured the process of vicariant speciation[12] between *M. edulis* and *M. galloprovincialis* (Barsotti & Meluzzi 1968). The present-day distribution of *M. galloprovincialis* in western Europe represents a northward expansion of its range (see earlier) since the Pleistocene. As already seen *M. edulis*, *M. galloprovincialis* and *M. trossulus* exhibit a typical antitropical distribution pattern; that is, they occur in high latitudes of the Northern and Southern Hemispheres in both the Pacific and Atlantic oceans but are absent from the tropics. Hilbish *et al.* (2000) tested various hypotheses to explain the timing and route of trans-equatorial migration by the three taxa. These yield different predictions for the phylogenetic[13] relationship of southern taxa relative to their northern counterparts. Using data from the 16S rDNA mitochondrial locus, two types of mtDNA lineage were found among populations of mussels collected from the Southern Hemisphere. Over 90% of the mtDNA lineages formed a distinct subclade,[14] which on average had 1.4% divergence from haplotypes[15] found exclusively in northern Atlantic populations of *M. galloprovincialis*. The data indicated that Southern Hemisphere mussels arose from a migration event from the Northern Hemisphere during the Pleistocene via an Atlantic route. The remainder of the Southern Hemisphere lineages (<10%) were very closely related to mtDNA haplotypes found in both *M. edulis* and *M. galloprovincialis* in the Northern Hemisphere, suggesting a second, more recent migration to the Southern Hemisphere. There was no evidence that Southern Hemisphere mussels arose from Pacific populations. A more recent phylogenetic investigation, using the COI mitochondrial gene, indicated that mussels may have diverged in the Southern Hemisphere before speciation of *M. galloprovincialis* and *M. edulis* in the Northern Hemisphere (Gérard *et al.* 2008). In addition, strong genetic structure was detected among the southern samples, revealing three genetic entities corresponding to South America and Kerguelen Island, Tasmania and New Zealand. Using the trans-Arctic interchange as a molecular clock calibration, Gérard *et al.* (2008) estimated the time of divergence of the populations from the two hemispheres to be 0.5–1.3 mya (average 0.84 mya). A robust and rapid RFLP assay has been developed to distinguish Northern Hemisphere *M. galloprovincialis* and *M. edulis* from native Southern Hemisphere *M. galloprovincialis*. This should prove useful in monitoring the introduction and possible spread of Northern Hemisphere *M. galloprovincialis* in the Southern Hemisphere (Westfall *et al.* 2010).

Perna spp.

Mussels in the genus *Perna* are currently recognised as three species: *Perna perna*, *P. viridis* and *P. canaliculus* (Dias *et al.* 2013 and references therein). *P. perna* is found throughout South Africa and on the Atlantic coasts of Namibia and Angola; in the Cape Verde Islands and from Mauritania northward; on the Indian Ocean coast of Mozambique and the west coast of Madagascar; extending through the Gulf of Aden into the Red Sea; along the Atlantic coast of South America, from Rio de la Plata, Argentina to Recife, Brazil; and on to the Caribbean shores of Venezuela (Wood *et al.* 2007). *P. viridis* has an extensive Indo-Pacific distribution from the Persian Gulf, through India and South East Asia, as far north as Japan, Korea and southern China and as far east as Papua New Guinea (Wood *et al.* 2007; McDonald 2012). *P. canaliculus* is only found in New Zealand – with one exception, when very small numbers were reported in 1996 at Port Adelaide in southern Australia; there is no evidence that the species has since become established in the area (Wiltshire *et al.* 2010). *P. perna* and *P. viridis* introductions have been recorded outside their native ranges in several regions (see Chapter 10). Rapid detection of marine pests such as *Perna* species remains fundamental to their effective containment and control. Dias *et al.* (2013) developed a potentially high-throughput real-time PCR assay capable of rapidly detecting and discriminating juvenile and adult specimens of *P. perna*, *P. canaliculus* and *P. viridis*. The adoption of this assay by the Western Australian biosecurity programme represented an important step toward developing capacity for the rapid and efficient identification and management of *Perna* species incursions in Australian waters.

Wood *et al.* (2007) constructed a molecular phylogeny based on nuclear (ITS1, ITS2) and mitochondrial (COI) DNA sequence data for the three generally recognised species (*P. viridis*, *P. perna* and *P. canaliculus*) and one putative species (*P. picta*) collected from several locations within their known geographic distributions. Their aims were to investigate current taxonomic assignments, determine the evolutionary relationships among these mussels and evaluate several biogeographical hypotheses to explain how their modern distributions became established. Phylogenetic analyses produced three well-supported clades (a group of organisms believed to comprise all the evolutionary descendants of a common ancestor), corresponding to the recognised species *P. perna* (Atlantic), *P. viridis* (Indo-West Pacific) and *P. canaliculus* (New Zealand). Putative *P. picta* from North Africa clustered within the *P. perna* clade, and therefore its specific status cannot be supported – a finding that is in agreement with Thankakkon & Patterson (2013) and the World Register of Marine Species (www.marinespecies.org). *P. perna* and *P. canaliculus* were the most closely related of the three species. Wood *et al.* (2007) proposed that the most recent common ancestor of extant *Perna* species occurred in the Indo-West Pacific or the Atlantic and dispersed between these areas before dispersing from the Atlantic to New Zealand. Dispersal between the Indo-West Pacific and the Atlantic could have been around the southern tip of South Africa or alternatively through the Tethys Sea, a sea that existed between Africa and Eurasia until the early Miocene (~18 mya). The estimated divergence time of the ancestor of *P. viridis* from the common ancestor of *P. perna* and *P. canaliculus* (9.4–32.1 mya) is consistent with the possibility that the closure of the Tethys Sea (50 mya) resulted in allopatric divergence of these lineages. It is highly probable that the ancestor of *P. canaliculus* reached New Zealand by dispersal in the WWD from South Africa or South America; alternatively, it could have evolved from *Perna tetleyi*, which was present in New

Zealand in the early Miocene (references in Wood *et al.* 2007). Another hypothesis, given that *Perna* was present on Antarctica in the late Eocene (37–40 mya), is that the common ancestor of *P. perna* and *P. canaliculus* dispersed from Antarctica to South America and to New Zealand, possibly via Australia. The divergence time of *P. perna* and *P. canaliculus* has been estimated to be 19.9–5.8 mya. An earlier dispersal event to or from Antarctica (32.1–9.4 mya) would need to be inferred for the precursor of *P. viridis*. Alternatively, it is feasible that the *Perna* lineage became extinct before or after dispersal, and the extant species may have arisen from a different ancestor. It is clear that further research is essential before the origin and likely dispersal routes of extant *Perna* species are elucidated.

Brachidontes and Bathymodiolus spp.

Rocky-shore intertidal communities along the cold- and warm-temperate coasts of the southwestern Atlantic are dominated by small mussels of the genus *Brachidontes*. In the past, *Brachidontes* spp. were separated solely on the basis of shell phenotypic characters, but high phenotypic variation has led to considerable confusion. For example, molecular studies clarified the systematics and phylogeography of *Brachidontes* in Central and North America by finding that what was generally considered a single species, *B. exustus*, is actually a complex of five species in the Gulf of Mexico and the Caribbean Sea plus two little-differentiated species on the Pacific coast of Central America (Lee & O'Foighil 2004, 2005). Based on mitochondrial and nuclear genes (COI, 28S rDNA and ITS1) from nine localities in Uruguay and Argentina, Trovant *et al.* (2013) confirmed the occurrence of three species in the southwestern Atlantic: *Brachidontes darwinianus* and *B. rodriguezii* in the warm-temperate sector and *B. purpuratus* in the cold-temperate sector (Figure 9.3). Analysis of molecular variance (AMOVA) for all three genes, pairwise *FST* values and genetic distance analyses for COI consistently indicated a high level of genetic divergence among the three species. Thus, the hypothesis of panmixia, tested with AMOVA with all sampling sites of the three species as one gene pool, was rejected. In all the phylogenetic trees, *B. purpuratus* was sister to the well-supported clade formed by *B. rodriguezii* and *B. darwinianus*. Phylogenies based on COI and 28S rDNA sequences indicate an early divergence of *B. purpuratus*. At the intraspecific level, low genetic differentiation and the absence of a fossil record for *B. purpuratus* from the earlier Quaternary Patagonian marine terraces likely result from a relatively recent (post-LGM) colonisation from populations in the southeastern Pacific. In the case of *B. rodriguezii*, by contrast, genetic intraspecific differentiation, a fossil record of phenotypically related forms going back to the Late Miocene and its phylogenetic position in the COI-based phylogeny suggest that this species may be derived from a local stock with a long history of occurrence in the warm-temperate region of the southwestern Atlantic. The origin of *B. darwinianus*, an estuarine species, has not been resolved. One plausible scenario involves a shallow sea with variable conditions of salinity, from normal to low, that covered a wide area of Argentina, Uruguay and southern Brazil during the Middle Miocene (16.0–11.6 mya), giving the conditions for speciation (Trovant *et al.* 2016 and references therein).

Symbiont-bearing bivalve molluscs are among the dominant animals found at vents worldwide, and *Bathymodiolus* mussels are a major element of this bivalve fauna. Phylogenetic analysis based on 18S ribosomal DNA sequences closely allies *Bathymodiolus* with other deep-sea mussel genera within the subfamily Bathymodiolinae (details in

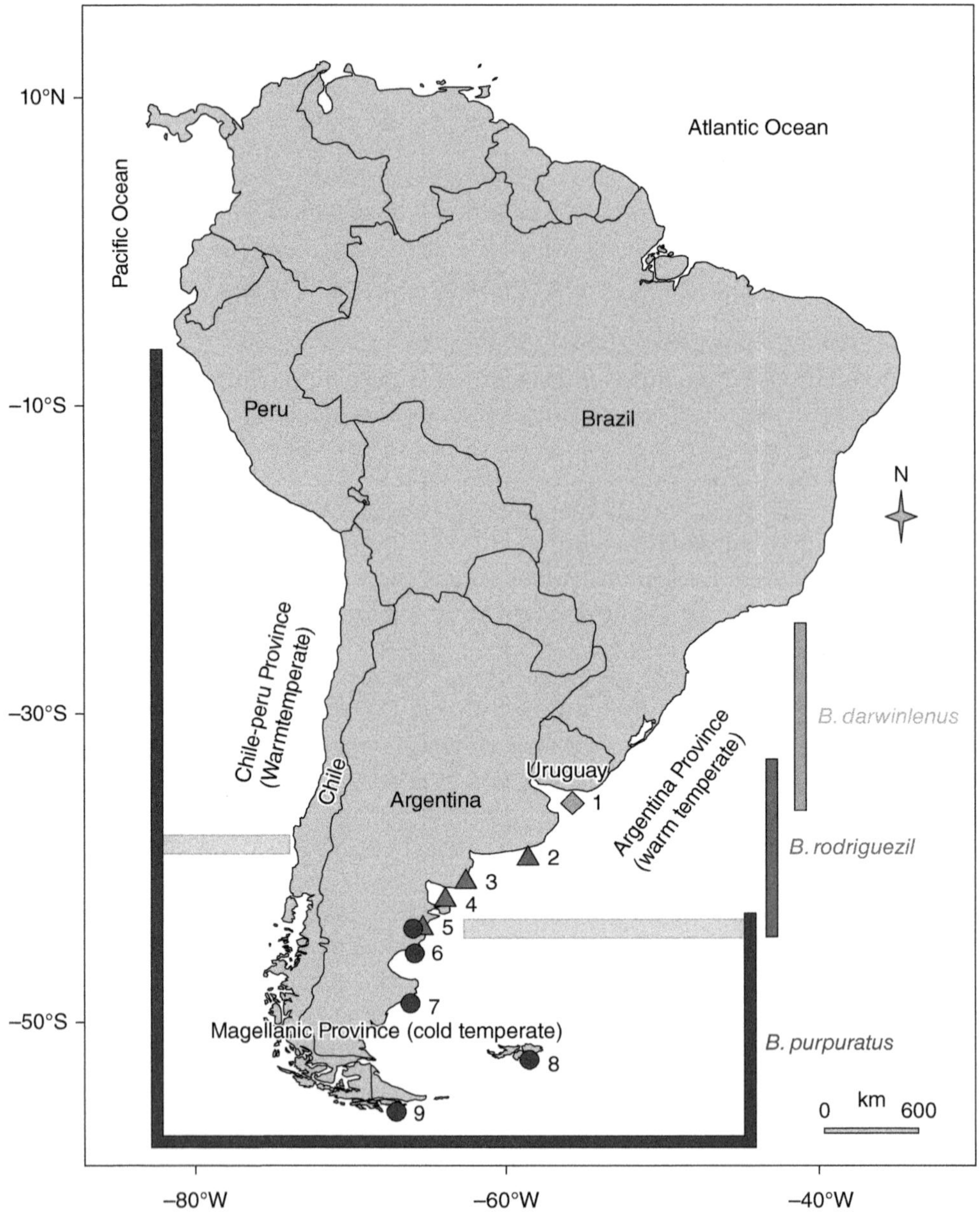

Figure 9.3 Latitudinal range of distribution of *Brachidontes* s.*l*. species present in the southwestern Atlantic (bars), and sampling sites for this study (symbols). Diamonds = *B. darwinianus*; triangles = *B. rodriguezii*; circles = *B. purpuratus*. Numbers indicate locations: 1, Punta Canario; 2, Santa Clara del Mar; 3, Bahia San Blas; 4, Bahia Rosas; 5, Puerto Madryn; 6, Caleta Carolina; 7, Puerto Deseado; 8, Surfer Bay; 9, Bahia Ensenada. Regarding *B. purpuratus*, the World Register of Marine Species does not accept the genus *Brachidontes* but advises that the species should be named as *Perumytilus purpuratus*. *Source:* From Trovant *et al*. (2013). Reproduced with permission from Cambridge University Press.

Chapter 1). Between 1985 and 2013, new *Bathymodiolus* species were being discovered at a rate of about one per year (www.marinespecies.org). New species found at a great range of depths and in a variety of habitats provide unique opportunities to test hypotheses regarding evolution, biogeography and factors affecting dispersal of deep-sea faunas (Smith *et al.* 2004). Smith *et al.* (2004) provided evidence for three new species of *Bathymodiolus* (NZ-1, NZ-2 and NZ-3) in relatively shallow water (<750 m depth) associated with the Kermadec Arc, an active volcanic island arc stretching north-northeast from New Zealand's North Island toward Tonga, an archipelago in the South Pacific Ocean. Mitochondrial COI sequences from the three putative species differed substantially from those of other known bathymodiolin species from the Pacific and Indian oceans. Sufficient samples of one of the new species (NZ-1) were collected from two seamounts to examine allozyme variation at 13 gene loci and test for genetic differentiation among populations. Seamounts are generally underwater extinct volcanoes that rise at least 1000 m from the seafloor. In contrast to the relative uniformity of mtDNA sequences in *Bathymodiolus* new species NZ-1 from the two seamounts, there was significant heterogeneity in allozyme gene frequencies between four samples (mean size = 17–31), which was surprising as the seamounts were only ~50 m apart. Their data did not allow the authors to assess whether random processes (e.g. genetic drift, founder events) or adaptive divergence was responsible for allozyme differences between seamount populations of *Bathymodiolus* new species NZ-1.

Two distinct species, *Bathymodiolus azoricus* and *B. puteoserpentis*, dominate hydrothermal vent sites along the northern part of the Mid-Atlantic Ridge (MAR; Figure 9.4). Using both the internal transcribed spacer ITS2 of ribosomal DNA and a combination of mtDNA sequences and allozymes, Faure *et al.* (2015) showed the hybridising potential of these two species in a very restricted area, Broken Spur, where their distributions meet. From the hybrid zone, the distribution of *B. azoricus* extends northward, while that of *B. puteoserpentis* extends southward over 3000 km (Faure *et al.* 2015). Mussels from the Broken Spur site comprise a mixture of hybrids and parental forms (O'Mullan *et al.* 2001; Won *et al.* 2003a). Samples of DNA sequences were obtained for seven nuclear loci and a mitochondrial locus in order to conduct a multi-locus analysis of divergence and gene flow between these two closely related and hybridising species (Faure *et al.* 2009). The analysis of populations located at both extremities of the mussel range, using a multilocus approach, indicated that (1) the two species started to diverge approximately 0.760 mya; (2) the *B. azoricus* population size was two to five time greater than those of *B. puteoserpentis* and the ancestral population, suggesting population expansions in *B. azoricus* – a finding that is in agreement with census population sizes observed for both species (dense mussel beds in the case of *B. azoricus* versus scattered individuals in the case of *B. puteoserpentis*; Comtet & Desbruyères 1998; Gebruk *et al.* 2000); and (3) gene flow between the two species occurred over the complete species range, although it was mainly asymmetrically from *B. azoricus* to *B. puteoserpentis* – there was also some evidence that gene flow was due to secondary introgression[16] after a period of allopatry (two species living in different regions).

Deep-sea vents and seeps are patchily distributed and are separated from one another by a few hundred kilometres in vent sectors and by hundreds of metres to a few kilometres in vent fields within the vent sector (Jollivet 1996). One crucial issue in deep-sea biology is how animals in these habitats can disperse to maintain their communities and avoid extinction, and how they can be isolated to accumulate genetic differentiation leading to

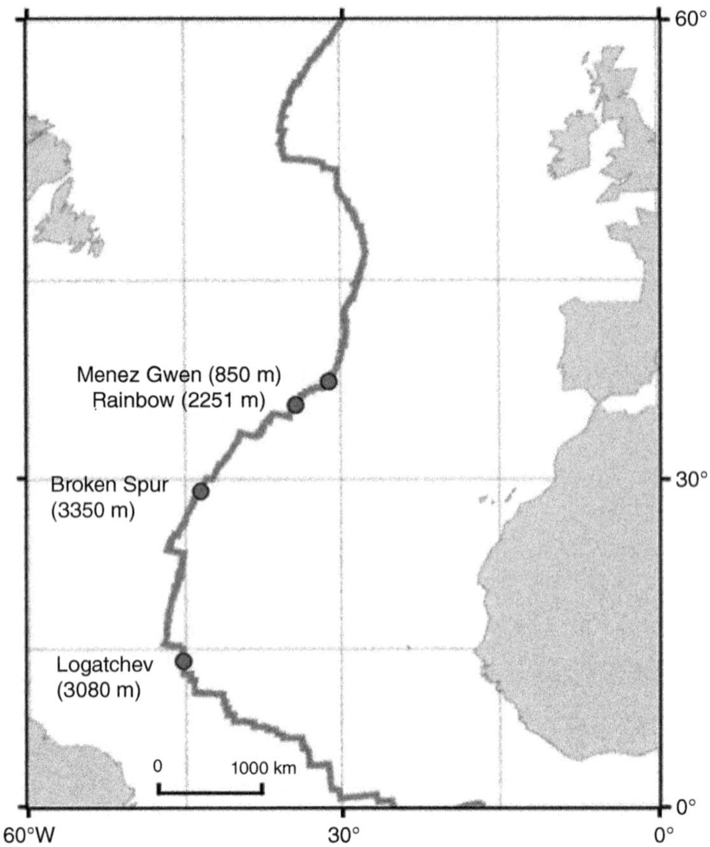

Figure 9.4 Positions of Mid-Atlantic Ridge (MAR) hydrothermal vent fields. *Source:* From Faure *et al.* (2009). CC BY 4.0.

consequent speciation. To elucidate the dispersal ability and environmental adaptability of deep-sea mussels, Miyazaki *et al.* (2013) analysed sequences of the mitochondrial ND4 gene in samples of *Bathymodiolus platifrons* and *B. japonicus* obtained from seeps of the Sagami Bay and vents of the Okinawa Trough, which are separated by approximately 1500 km. Only 3 of 20 deep-sea mussel species in the Japanese waters, including those just discussed, can inhabit both vents and seeps. The environmental differences between vents and seeps have not yet been fully elucidated, probably due to large site-to-site variations, but Miyazaki *et al.* (2013) assumed that there are indeed differences (light strength, temperature, pressure, chemical and physical conditions), since species – apart from the three in Japanese waters – can only inhabit either vents or seeps. Miyazaki *et al.* (2013) examined phylogenetic relationships, genetic divergence, gene flow and genetic population structures to compare seep and vent populations. Phylogenetic analysis showed that specimens from vents of the Okinawa Trough were not separated from, but intermingled with those from seeps of the Sagami Bay in both *B. platifrons* and *B. japonicus*, indicating no clear genetic differentiation between the vent and seep populations. There was extensive gene flow between the seep and vent populations, indicating high dispersal ability of the two species, which favours colonisation in patchy and ephemeral habitats. The results also indicated that environmental type (vent vs. seep) is not the primary factor responsible for habitat segregation in the two species.

Hybrid Zones

Hybrid zones may be formed following localised hybridisation of two formerly allopatric or parapatric[17] taxa that have come into secondary contact after a period of geographic isolation (Lowe *et al.* 2004 and references therein). Such zones provide an ideal arena for studying the forces that drive divergence of populations (Buggs 2007).

The occurrence of hybrids in nature indicates that reproductive isolating mechanisms (RIMs) are insufficient to prevent hybridisation (Barton & Hewitt 1989). Depending on the fertility and fitness of hybrids relative to the parental taxa, the potential exists for introgressive gene flow between species when hybrid individuals backcross with one or both parents. Information on the factors that limit hybridisation is essential to understanding the maintenance of hybrid zones and the continued separation of two species despite extensive gene flow. The coexistence of genetically distinct species may result from a combination of RIMs. Pre-zygotic barriers to fertilisation can occur through mating preference, habitat specialisation, asynchronous reproduction and gamete incompatibility (Palumbi 1994). Alternatively, isolation may occur post-zygotically through gametic incompatibility, zygotic inviability or hybrid sterility. Despite the potential for reproductive isolation, naturally occurring hybrids are often abundant. The width of hybrid zones can vary from tens of metres to hundreds of kilometres. The shape of the zones (clines) can be gradual or stepped. They may be ephemeral or long-lasting (Lowe *et al.* 2004).

M. Edulis and *M. Galloprovincialis* Hybrid Zones

The most intensely studied marine hybrid zones are those of mussels in the *Mytilus edulis* species complex, particularly the *M. edulis* and *M. galloprovincialis* hybrid zone in western Europe (Gosling 1992a,b; Gardner 1994). The two species are distributed in a mosaic of pure and hybrid populations in SW England and on the Atlantic coasts of France (Bierne *et al.* 2002a, 2003a and references therein) and Ireland (Gosling & Wilkins 1981; Gosling 1984; Gosling & McGrath 1990; Coghlan & Gosling 2007; Gosling *et al.* 2008; Doherty *et al.* 2009). There is no single morphological character that can be reliably used to separate mixed populations. This is particularly true at wave-exposed locations, where hybridisation and introgression between *M. edulis* and *M. galloprovincialis* generate large numbers of intermediate forms (Seed 1974). Several decades ago, allozymes were used extensively for taxonomic characterisation. About four such allozymes are well differentiated but not diagnostic between any pair of blue mussel taxa (McDonald *et al.* 1991; Gosling 1992a). Since the 1990s, PCR-based DNA markers have been developed, which diagnostically identify the three taxa *M. edulis*, *M. galloprovincialis* and *M. trossulus* and allow accurate estimation of the frequency of parental and hybrid mussel classes within hybrid zones (Inoue *et al.* 1995; Rawson *et al.* 1996; Bierne *et al.* 2002a,b, 2003a,b, 2006; Hilbish *et al.* 2002; Coghlan & Gosling 2007; Gilg *et al.* 2007; Gosling *et al.* 2008; Doherty *et al.* 2009; Brooks & Farmen 2013; El Alteri *et al.* 2019).

In SW England, relatively pure populations of *M. galloprovincialis* inhabit open coast sites on much of the northern coast of Cornwall, while pure populations of *M. edulis* occupy sheltered estuarine sites along the southern coast of England westward to Start Point in south Devon (Hilbish *et al.* 2002). Between these populations is a hybrid zone (~200 km in length) that contains mussels of mixed ancestry (Skibinski *et al.* 1983; Gardner 1994; Wilhelm & Hilbish 1998). Hybrid populations typically show a pattern of directional selection in which mussels with *M. galloprovincialis*-like genotypes are favoured over those with *M. edulis*-like genotypes and mussels with intermediate genotypes have intermediate fitness (Gardner *et al.* 1993; Wilhelm & Hilbish 1998). The transition between hybrid populations and those containing pure populations is abrupt (<5 km), which suggests that coastal circulation patterns may provide strong barriers to larval dispersal, accounting for the position and maintenance of the hybrid zone (Hilbish *et al.* 2002). The location of the zone and the genetic composition of mussels within it have not changed appreciably between the 1980s and the early 2000s (Skibinski *et al.* 1983; Gardner 1994; Hilbish *et al.* 2002; Gilg & Hilbish 2003). Hybrid populations have a high frequency (typically >0.90) of the *M. edulis* Glu-5' allele, although strong selection results in a decline of the allele with increasing size (Gardner & Skibinski 1988; Gardner *et al.* 1993; Wilhelm & Hilbish 1998; Hilbish *et al.* 2002). Using the genetics of recently settled larvae and a fine-scale model of the physical oceanography of the region, Gilg & Hilbish (2003) found that Start Point separates the *M. edulis* population from the hybrid zone and allows minimal dispersal from the zone into the population but none in the other direction. Likewise, the *M. galloprovincialis* populations along the northern coast of Cornwall regularly receive immigrating larvae from the hybrid zone, but larvae from the populations do not enter the zone. However, larvae settling at hybrid zone sites have high frequencies of alleles specific to *M. edulis*, suggesting that reproductive barriers, selection in the larval stage or gene flow from an undetermined source is effectively balancing the directional selection observed in the adults. Spawning asynchrony has been suggested as one of the factors that prevent complete introgression of the two taxa (Gardner & Skibinski 1990). However, Gilg *et al.* (2007) have shown that *M. edulis, M. gallprovincialis* and hybrids have similar reproductive cycles in the hybrid zone. Unfortunately, for over a decade now, there have been no further studies on mussels in the SW England hybrid zone, which continues to be both an intriguing and a complex scenario. In a recent study, however, Tolman *et al.* (2019) investigated the immune response of mussels from Croyde Bay, within the hybrid zone. The site is characterised by size-dependent variation in genotype frequencies, with a higher frequency of *M. galloprovincialis* alleles in large mussels, largely attributed to selective mortality in favour of the *M. galloprovincialis* genotype. To determine whether differences in immune response might contribute to this size-dependent variation in genotype frequencies, Tolman *et al.* (2019) assessed the two pure species and their hybrids in their phagocytic abilities when subject to immune challenge as a measure of immunocompetence, and measured the metabolic cost of mounting an antigen-stimulated immune response. Mussels identified as *M. galloprovincialis* had a greater immune competence response at a lower metabolic cost compared to those identified as *M. edulis*. Mussels identified as hybrids had intermediate values for both parameters, providing no evidence for heterosis but suggesting that increased susceptibility compared to *M. galloprovincialis* may be attributed to the *M.edulis* genotype. The results indicate phenotypic differences in the face of pathogenic infection, which the authors suggest may be a contributing factor to the differential mortality in favour of *M. galloprovincialis* and the size-dependent variation in genotype frequencies associated with this contact zone.

The *M. edulis*/*M. galloprovincialis* contact zone may be best understood as a succession of several hybrid zones rather than a vast single zone. For example, using several nuclear markers, three independent hybrid zones have been defined on the Atlantic coast of France: Bay of Biscay, NW France and the English Channel (Bierne *et al.* 2003a). Each zone has differentiated source populations, and the degree of hybridisation varies between zones. This mosaic structure confirms and extends the patterns observed around the British Isles by Skibinski *et al.* (1983) using allozymes and in France by Coustau *et al.* (1991) using allozymes and by Daguin *et al.* (2001) using the DNA nuclear marker actin gene locus *mac-1*. Large linkage disequilibria[18] and significant departures from Hardy–Weinberg equilibria (HWE)[19] indicate that there are restrictions to gene exchange between *M. edulis* and *M. galloprovincialis*. In NW France, habitat choice (Bierne *et al.* 2003b) contributes to the barrier, but habitat-independent mechanisms such as spawning asynchrony, assortative fertilisation and hybrid fitness depression probably also operate (Bierne *et al.* 2002a,b, 2006).

The Irish hybrid zone differs in several respects from those in SW England and NW France. For example, the zone is not situated between pure populations of *M. edulis* and *M. galloprovincialis*. Results from an Irish study (Gosling & Wilkins 1981) using partially diagnostic allozyme markers found that although pure populations of *M. edulis* occur on the Irish Sea coast, there are no pure populations of *M. galloprovincialis* or *M. edulis* outside of this region on Atlantic coasts. On Atlantic wave-exposed shores, and to a much lesser extent on wave-protected ones, mussels comprise an interbreeding mixture of *M. edulis* and *M. galloprovincialis*. In a later study (Coghlan & Gosling 2007) using the diagnostic *Me15/16* nDNA marker (Inoue *et al.* 1995; Figure 9.5) in populations on Atlantic coasts, high frequencies of hybrid genotypes (34–72%), together with nonsignificant departures from HWE, suggested the strong possibility that populations are highly introgressed. Subsequently, Gosling *et al.* (2008) analysed mussels from 20 locations on Irish coasts using the same DNA marker. Their results showed a high frequency of *M. galloprovincialis* (0.378 ± 0.198) and hybrid (0.429 ± 0.175) genotypes, and correspondingly low frequencies of the *M. edulis* (0.194 ± 0.107) genotype at both exposed and sheltered locations on Atlantic

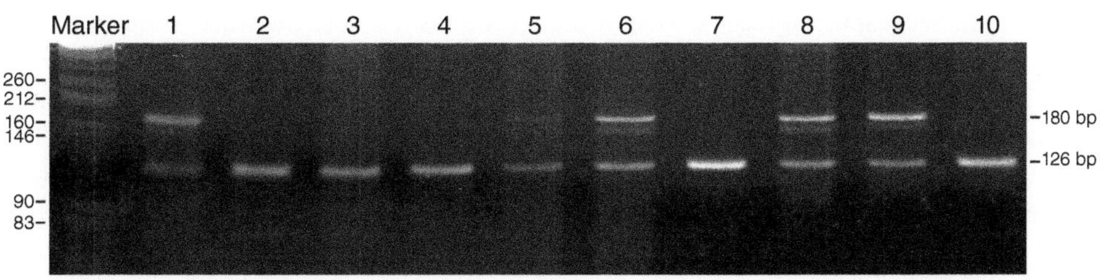

Figure 9.5 PCR amplification of the foot protein gene 1 *Me15/16* (Inoue *et al.* 1995) produces a species-specific band of about 180 bp for *M. edulis* and of 126 bp for *M. galloprovincialis*; individuals with both bands are considered hybrids. Lanes 1 to 10 = hybrid (H), Mg, Mg, Mg, H, H, Mg, H, H and Mg, respectively. *Source:* From Tagliarolo *et al.* (2012). Reproduced with permission from Inter-Research.

coasts (Figure 9.6). There was no evidence of habitat specialisation in that mussels at both wave-exposed and sheltered estuarine sites comprised a mixture of the three taxa. Also, the *M. edulis* genotype had no apparent advantage in sheltered, estuarine conditions, as had been proposed previously by several researchers (Bierne *et al.* 2003a,b and references therein). In a later study, Doherty *et al.* (2009) compared the reproductive cycles of *M. edulis*, *M. galloprovincialis* and their hybrids at Ballynahown in Galway Bay (Figure 9.6) over a 12-month period. Each month, 40 individuals (40–45 mm shell length) were collected (*M. edulis* 7–17%, *M. galloprovincialis* 28–36% and hybrids 55–57%). Results showed that there was a large overlap in the spawning cycles of the three taxa, providing ample opportunity for introgressive gene flow. Fecundity was slightly higher in *M. galloprovincialis* females compared to the other taxa (up to 30% difference, $P < 0.05$). The apparent advantage of higher fecundity was small, however, and so was probably offset by higher numbers of hybrid adults releasing gametes. The difference in fecundity between the taxa at this site was notably much less pronounced than that observed in SW England, where fecundity in *M. galloprovincialis* exceeded that of the other taxa by a factor of 2.2–2.8 (Gardner & Skibinski 1990). Several features of the Ballynahown population refute the hypothesis of depressed hybrid fitness, since the percentage of hybrid genotypes in the adult population is high, probably due to the continuous spawning by hybrids during peak spawning periods of both parental species. Hybrids are fertile with the potential to backcross to parent species, contributing again to the high frequency of the hybrid genotype. Sex ratios do not deviate from the expected 1:1. Deviations are observed when increased mortality in hybrid males, caused by nuclear–cytoplasmic incompatibility, shifts the sex ratio toward a female bias (Toro *et al.* 2002). These features highlight not just an absence of selective pressure against hybrid adults, but that ecological factors in this region of the hybrid zone actually favour hybrids. Any selection acting against the complete introgression of the parental species must be taking place either pre-zygotically through gamete incompatibility or assortative fertilisation or post-zygotically via selection against hybrids in the early life-history stages affecting larval mortality or recruitment success.

One notable feature of the Irish hybrid zone is the absence of *M. galloprovincialis* from the Irish Sea. From satellite imagery, a distinct thermal front develops during late spring at the boundary between the stratified Celtic Sea and the mixed waters of the southern Irish Sea (Brown *et al.* 2003; Hartnett *et al.* 2007; O'Boyle & Silke 2010). This breaks down with the onset of winter cooling and increased wind-induced mixing. Stratification also occurs within the Irish Sea, especially to the west of the Isle of Man, where the water is deeper and the tides weaker than to the east of the island. This front also breaks down in winter and re-forms in spring. Mussels in this area may be a self-recruiting population of *M. edulis* due to thermal front development at the northern and southern entrances to the Irish Sea in late spring, which prevents an influx of spring-spawned *Mytilus* larvae. However, because there is inter-annual variation in the timing of formation and break-up of the fronts (Xing & Davies 2001), we cannot rule out the possibility that larvae enter but due to a scarcity of suitable substrates do not recruit to sites along the southeastern or northeastern coasts, or, alternatively, that they only settle in the sublittoral region. If the latter is true, however, then we could expect to see recruitment of *M. galloprovincialis* at intertidal sites along the southeastern and northeastern coasts. This may be what is happening in Belfast Lough (Figure 9.6); or, alternatively, because Belfast is one of the busiest Irish ports,

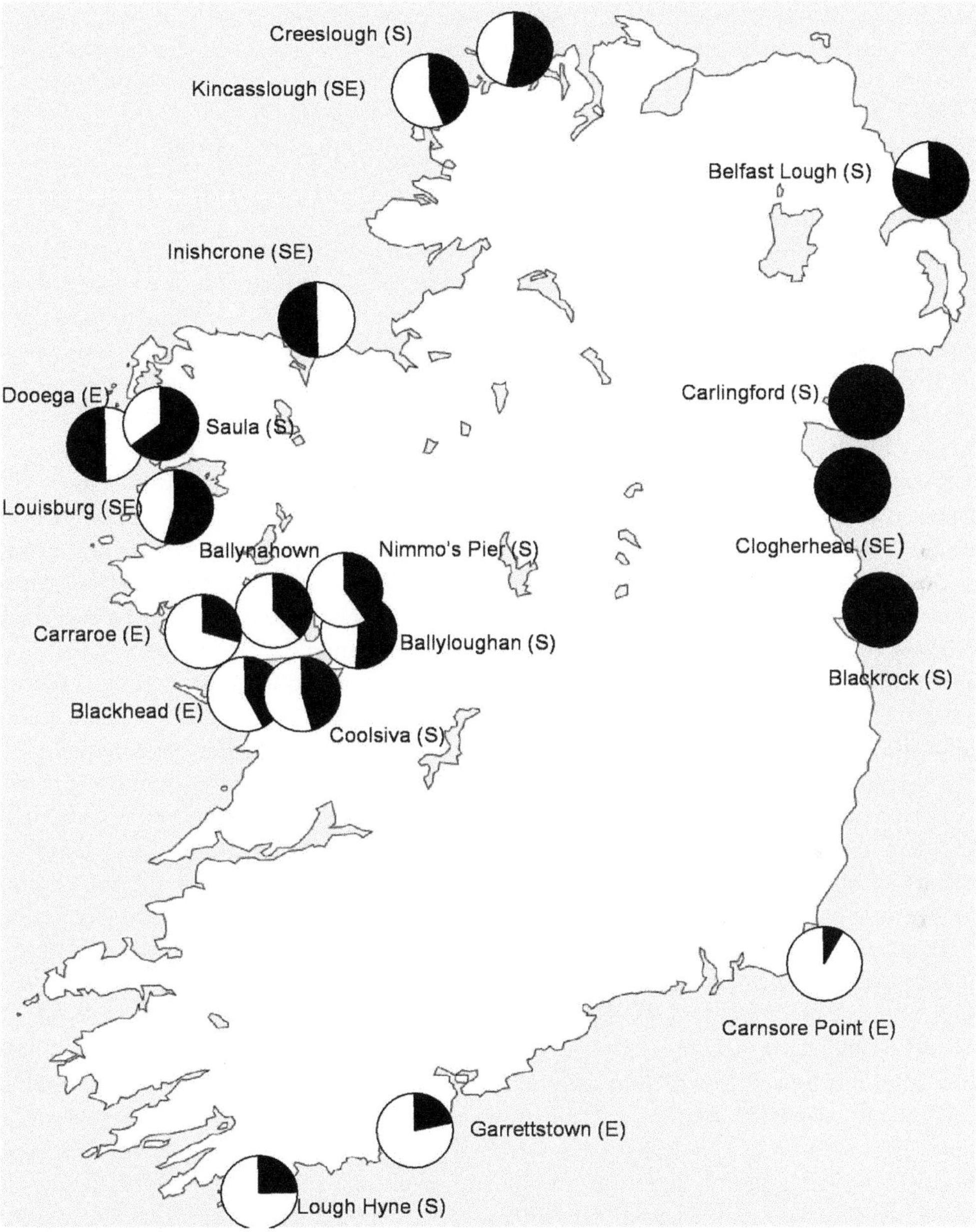

Figure 9.6 Frequency of the *Mytilus edulis* allele (black) and the *Mytilus galloprovincialis* allele (white) at Irish sampling sites. Because allele frequencies were identical for the Ballynahown wave-exposed and sheltered sites, only one frequency is presented. *Source:* From Gosling *et al.* (2008). Reproduced with permission from Cambridge University Press.

M. galloprovincialis may have been brought into the lough either as larvae in ballast water or as a fouling organism on boat hulls. The importance of fronts as biogeographical boundaries for mussel species in European and North American waters has been well documented (Quesada *et al.* 1995a; Blanchette *et al.* 2007; Hayhurst & Rawson 2009; Cole & McQuaid 2010; Trovant *et al.* 2013; Gosnell *et al.* 2014; El Ayari *et al.* 2019).

M. Edulis and *M. Trossulus* Hybrid Zones

As seen earlier, during the Pleistocene or Holocene (the current period of geological time), *M. trossulus* colonised both sides of the Atlantic, so that descendants of the species now form two distinct hybrid zones, one in the eastern Atlantic (Scandinavia; Śmietanka *et al.* 2013) and the other in the western Atlantic (Canadian Maritimes). The Baltic Sea was colonised by *Mytilus* spp. about 7000 years ago, after the most recent freshwater period and the emergence of this area as a marine (brackish) ecosystem (Zillén *et al.* 2008). The Baltic Sea has a salinity ranging from almost freshwater in the northern part (Gulf of Bothnia) through brackish (6–8 psu) in the inner region to full salinity (20–30 psu) in the northern Danish Straits (Kattegat) and rising to 33 psu in the North Sea (Zbawicka *et al.* 2014). Initially, North Sea and Baltic Sea mussels were characterised using allozymes (Bulnheim & Gosling 1988; McDonald *et al.* 1991, Väinölä & Hvilsom 1991). Allozyme surveys have identified North Sea and Skagerrak (a strait running between the southeast coast of Norway, the west coast of Sweden and the Jutland peninsula of Denmark, connecting the North Sea and the Kattegat Sea area, which leads to the Baltic Sea) populations as *M. edulis* and Baltic Sea populations as *M. trossulus*. Kattegat populations (Figure 9.7) have allele frequencies intermediate to the Baltic and North Seas, although changes in allozyme allele frequencies at six strongly differentiated loci occur coincidently and rapidly with a cline width of approximately 100 km through the islands of Denmark, the various Belt Straits and the Öresund in SW Sweden (see arrow in Figure 9.7). The genetic transition zone is located at the area of maximum salinity change (15–10 psu; Stuckas *et al.* 2017). Väinölä & Hvilsom (1991) have argued that this situation results from secondary contact between Skagerrak *M. edulis* and Baltic Sea *M. trossulus*, with limited hybridisation in the Kattegat. The structure of hybrid zones is the product of a balance between natural selection and gene flow. The nature of these interactions depends on the origin of natural selection; in some cases, selection against hybrids occurs because of incompatibilities between the parental genomes (tension zone models), while in others endogenous selection against hybrids occurs because of a mismatch between their phenotypes and the environment (exogeneous selection), which could be the case for the Öresund hybrid zone (Väinölä & Hvilsom 1991). Because the Kattegat is a region of rapidly changing salinity, primary differentiation of *M. edulis* associated with adaptation to the low-salinity conditions of the Baltic Sea is perhaps the best explanation for the observed pattern (Bulnheim & Gosling 1988; Johannesson *et al.* 1990). Between 1987 and 2005, there was a change in the position and shape of the cline (Strelkov *et al.* 2017), which shifted 25 km toward the Baltic end of the Öresund and became narrower. Reasons for such changes and their consequences are presently unresolved. While allozyme loci show differentiation across the Scandinavian hybrid zone, other genetic loci indicate extensive gene flow. For example, there has been complete asymmetric introgression of *M. edulis* female mtDNA into Baltic Sea populations, such that no (native) *M. trossulus* female (F) mtDNA are now detected (Rawson & Hilbish 1998; Quesada

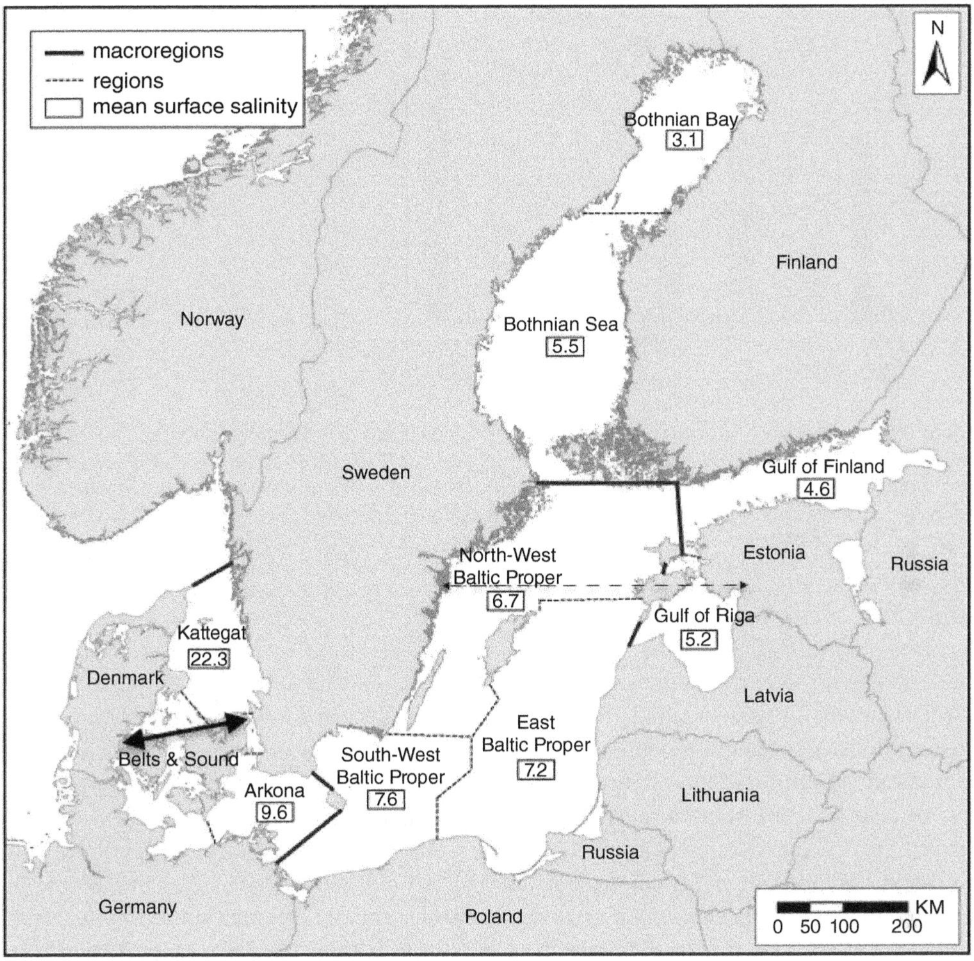

Figure 9.7 Map of macroregions and regions of the Baltic Sea (Ojaveer & Kalejs 2008). Numbers refer to salinity, with the lowest figure (3.1 psu) in Bothnian Bay and the highest (22.3 psu) in the Kattegat. The arrow indicates the location of the cline. *Source:* From Ojaveer *et al.* (2010). CC BY 4.0.

et al. 1999; see Chapter 5). Furthermore, male (M) *M. trossulus* mtDNA is also not detected (see also Kijewski *et al.* 2006). In an analysis of Baltic Sea populations using three nDNA markers (*Glu-5′*, *ITS* and *MAL-I*), Riginos *et al.* (2002) and Riginos & Cunningham (2005) found that that the majority of mussels were hybrids (96%). The distribution of Baltic genotypes follows a unimodal pattern, where complex hybrids (hybrid scores of 2–4) are most common and parental species (hybrid scores 0 and 6) are virtually absent (Figure 9.8). Although individuals with equal proportions of *M. edulis* and *M. trossulus* alleles are common (hybrid score of 3), they did not find any mussels with F1 genotypes, probably as a consequence of the negligible frequency of *M. trossulus* parental types. Consequently, based on these results, Riginos & Cunningham (2005) suggest that Baltic mussel populations are best described as a hybrid swarm.

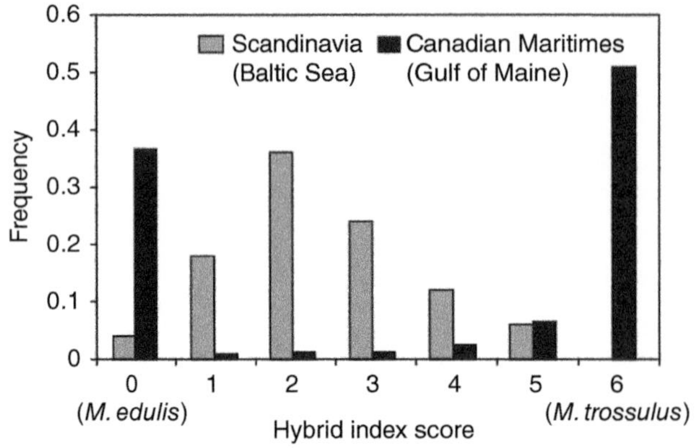

Figure 9.8 Frequency of *Mytilus* hybrid genotypes. Mussels are given a score of 0–6 based on their total number of *M. trossulus* alleles for the loci *Glu-5′*, *ITS* and *MAL-I*. A score of 0 is consistent with a pure *Mytilus edulis* genotype, whereas a score of 6 is consistent with a pure *M. trossulus* genotype. Hybrids are rare in the Gulf of Maine, but make up the majority of individuals in the Baltic Sea. Data from the Gulf of Maine are from Rawson *et al.* (2001). *Source:* From Riginos & Cunningham (2005). Reproduced with permission from John Wiley & Sons.

The strong discordance between allozyme and nDNA markers can only be explained by selection acting on some loci (Riginos *et al.* 2002). In a recent study, Kijewski *et al.* (2019) used random forest as the modelling algorithm to test if genetic differentiation of *Mytilus* populations might be related to any of the key environmental variables known to shape species distributions. Mussels were collected from 32 sites in a transition from the southeastern North Sea to the northeastern Baltic Sea. Genetic differentiation of populations was assayed using four known *Mytilus* taxa-specific nDNA markers (*Glu-5′*, *ITS*, *EFbis* and *M7*) and 26 published SNPs (Zbawicka *et al.* 2014). Strong genetic differentiation in allele frequencies was correlated with environmental factors along the North Sea–inner Baltic transect. Salinity appeared to define over 50% of the variability in the gene frequencies of the studied *Mytilus* spp. populations. Changes along this environmental gradient were not gradual, with a clear discontinuity in allele frequencies observed at salinity 12 psu. Other environmental gradients involved the water temperature and trophic status of the sea area, but these had only moderate linkage with gene frequencies.

In another hybrid zone in the western Atlantic, *M. edulis* and *M. trossulus* coexist and hybridise along the coasts of Newfoundland, Nova Scotia, New Brunswick and Québec, where the two species are sympatric from northern Newfoundland and Labrador south to the Gulf of Maine (Hayhurst & Rawson 2009 and references therein). There is currently no evidence on whether habitat affects species distributions in the Canadian Maritime hybrid zone (see Moreau *et al.* 2005). Riginos & Cunningham (2005) have described the hybrid zone as a mosaic, with *M. trossulus*, *M. edulis* and hybrid individuals found throughout the region. As shown in Figure 9.8, the frequency distribution of genotypes follows a bimodal patern in the Gulf of Maine. Individuals with parental genotypes (scores of 0 or 6) are most prevalent, followed by those with mostly *M. trossulus* or *M. edulis* alleles (scores of 1 and 5); among these backcross hybrid individuals, *M. trossulus* backcrosses are more common than *M. edulis* backcrosses. Mussels with F1 genotypes (heterozygous for both parental alleles at all nuclear loci) are rare, which suggests that that there are reproductive barriers between the species, although some interspecific matings must occur and produce fertile offspring (Reginos & Cunningham 2005). This bimodal distribution suggests that the two *Mytilus* taxa are progressing toward complete reproductive isolation (reviewed in Jiggins & Mallet 2000).

While *M. edulis* and *M. trossulus* have overlapping spawning periods in eastern Maine (Maloy *et al.* 2003), interspecific fertilisation between the two from areas of sympatry is highly inefficient as compared with intraspecific fertilisation (Rawson *et al.* 2003), implicating some barrier to cross-species fertilisation. Rawson *et al.* (2003) and Slaughter *et al.* (2008) have examined gamete incompatibility in vitro between *M. edulis* and *M. trossulus.* These studies were conducted by mixing the eggs of a single female with varying dilutions of sperm from a single male, carefully quantifying compatibility of all possible male/female combinations and comparing the results for both intraspecific combinations with those from the two reciprocal interspecific crosses. Both studies showed stronger gamete incompatibility of interspecific crosses, and also high interspecific fertilisation in some crosses involving *M. edulis* females, suggesting these are the route through which introgression occurs in the hybrid zone. Klibansky & McCartney (2014) performed gamete-crossing experiments using *M. trossulus* and *M. edulis* individuals collected from natural populations during the spring spawning season in order to detect the presence or absence of conspecific sperm precedence (CSP)[20] within the hybrid zone. They detected CSP (i.e. a reduction in heterospecific offspring from competitive fertilisations in vitro compared to those seen in noncompetitive fertilisations) in 5 of the 12 conspecific crosses. Their findings supported predictions that CSP can promote assortative fertilisation in bimodal hybrid zones, further advancing progression toward full speciation. Despite strong CSP, numerous heterospecific fertilisations remained, reinforcing the hypothesis that compatible females are a source of hybrid offspring in mixed natural spawns.

While Jiggins & Mallet (2000) suggest a necessity for some post-zygotic barrier in a bimodal hybrid zone, they do state that the pre-zygotic barriers are key. Miranda *et al.* (2010) examined pre-zygotic (gamete interactions) and post-zygotic (larval survival) components of reproductive incompatibility in laboratory crosses of *M. edulis* and *M. trossulus.* Conspecific crosses showed significantly greater rates of fertilisation than heterospecific crosses, although some females of both species showed heterospecific gamete compatibility. The proportion of fertilised eggs developing into normal larvae was not significantly different between conspecific and heterospecific crosses, but survival of normal larvae was greater for conspecific crosses. Mixed-species sperm experiments suggested that conspecific sperm preference may further limit hybridisation. The different components of reproductive incompatibility and total incompatibility varied among females of both species. Another study on pre- and post-zygotic barriers in the two taxa examined the genetic composition of early life-history stages by sampling veliger and pediveliger larvae, juveniles (<2.0 to 15.0 mm shell length) and adults (>15 mm shell length) in Trinity Bay, Newfoundland during three consecutive years to determine if differential mortality limits the relative abundance of hybrids (Toro *et al.* 2004b). The relative frequencies of the two species and the different hybrid genotypes were similar among the larvae, juveniles and small adult mussels. The double hybrid (F1) was the least common genotype observed. There was no evidence of differential mortality during the early life-history stages, and factors limiting production of hybrids appeared to operate before the late larval stage. *M. trossulus* dominated the early life-history stages, possibly due to a higher population density and a greater reproductive output compared with *M. edulis*. Toro *et al.* (2004b) concluded that the hybrid zone appears to be maintained by RIMs limiting the production of hybrids and life-history differences (e.g. differences in age-specific survival, Comesaña *et al.* 1999, Toro *et al.* 2006; and reproduction, Toro *et al.* 2002) that allow the two species to coexist. See Riginos & Cunningham (2005) for a summary of the main reasons as to why patterns of hybridisation between *M. edulis* and *M. trossulus* are so dissimilar in the Canadian Maritimes and Baltic Sea.

As seen earlier, Wenne *et al.* (2016) provided the first evidence, using 54 SNPs, for the existence of *M. trossulus* on the west coast of Greenland, where it coexists and hybridises with *M. edulis*. Identified *M. edulis* and *M. trossulus* hybrids mostly had a hybrid index of about 0.5 (a score of 0 indicates 'pure' *M. edulis*, whereas a score of 1 indicates 'pure' *M. trossulus*). There was a low number of backcrosses between 'pure' taxa and hybrids. This newly discovered hybrid zone between the two taxa is unique relative to Canadian populations. A recently discovered area of *M. edulis* and *M. trossulus* co-occurrence is in the Kandalaksha Bay of the White Sea in northwestern Russia. Using a set of four nearly diagnostic allozyme markers, Katolikova *et al.* (2016) found that in the White Sea the occurrence of *M. trossulus* is largely concentrated in harbours, mixed populations of the two taxa are always dominated by purebred individuals (with mussels classified as hybrids constituting only ~18%) and at sites of sympatry *M. trossulus* is more common on algal substrates while *M.edulis* mostly lives directly on the bottom. This segregation by substrate, which has not been documented for other *M. edulis* and *M. trossulus*, may contribute to maintaining reproductive isolation and decreasing competition between taxa. However, when the Mediterranean mussel *M. galloprovincialis* was introduced into southern California in the mid 20th century, it was able to completely displace the native *M. trossulus* throughout the southern portion of its range.

M. Trossulus and *M. Galloprovincialis* Hybrid Zones

Mytilus trossulus is native to the North Pacific, an ocean that has relatively low nearshore salinity, large diurnal tidal fluctuation, cold and variable surface water temperatures and distinct oceanic seasons. During the last 3 million years, the Mediterranean Sea, home to *M. galloprovincialis*, has had higher salinity, less tidal fluctuation, warmer water temperatures and less seasonal variation in temperature and salinity than the North Pacific (Braby & Somero 2006a). Hence, each taxon is potentially adapted for quite different environmental conditions. The range of *M. trossulus* once extended from Baja California north to Alaska and west to Japan. However, when *M. galloprovincialis* was introduced into southern California, it was able to completely displace it throughout the southern portion of its range (Geller *et al.* 1999). Fields *et al.* (2006) investigated whether biochemical adaptation to temperature might potentially play a role in invasion success. An examination of cytosolic malate dehydrogenase (cMDH), an enzyme known to exhibit distinct patterns of temperature adaptation in kinetic properties, showed that a minor change in its structure permitted *M. galloprovincialis* cMDH to function at warmer temperatures; this may thus be a part of a broad suite of molecular adaptations that has allowed this species to displace its congener throughout the warmer part of *M. trossulus*' original range in North America and Japan (see also Schneider & Helmuth 2007; Lockwood & Somero 2011a; Saarman *et al.* 2017).

In contrast to the *Mytilus* hybrid zones discussed so far, secondary contact and hybridisation between *M. galloprovincialis* and *M. trossulus* in central California (Suchanek *et al.* 1997), Japan (Inoue *et al.* 1995, 1997) and Puget Sound (Heath *et al.* 1995) has occurred in the last several hundred years as a result of anthropogenic activities.

According to Braby & Somero (2006a), the hybrid zone in California lies between Monterey and the Cape Mendocino region (latitudes 36.63–40.5 °N). In a recent study, Saarman & Pogson (2015) investigated the relative importance of isolating barriers and the demographic expansion of invasive *M. galloprovincialis* on the magnitude and the direction of introgression with the native *M. trossulus* in the central Californian hybrid zone. They sampled seven sites within the zone, one putatively pure *M. galloprovincialis* population from outside it and one *M. galloprovincialis* population from the south of France. A large number of randomly selected SNPs (N = 1337) that accurately distinguish early and advanced generation hybrids were used in the analysis. Weak levels of introgression were observed in both directions, although they were slightly more prevalent from *M. trossulus* into *M. galloprovincialis* (20.4 vs. 14.6% of SNPs, respectively). Few early and advanced backcrossed individuals were observed across the hybrid zone, most likely as a consequence of strong and pervasive selection acting against F1 hybrids and backcrossed individuals. In a study of the *M. galloprovincialis*/*M. trossulus* hybrid zone in Japan (see later), Brannock & Hilbish (2010) found that F1 hybrids exhibited arrested gametogenesis, which they estimated would reduce their fertility by an order of magnitude. In California, heterogeneous patterns of admixture across the zone of contact were consistent with the colonisation history of *M. galloprovincialis*, with more extensive introgression in northern localities farthest away from Los Angeles, the putative site of introduction. These observations reinforce the importance of dynamic spatial and demographic expansions in determining patterns of introgression between close congeners, even those with high dispersal potential and well-developed reproductive barriers. Saarman & Pogson (2015) suggest that the threat posed by *M. galloprovincialis* is more ecological than genetic, as it has displaced, and continues to displace, the native *M. trossulus* from much of central and southern California.

Hybridisation also occurs between the two taxa in coastal areas further north of the Californian coast, such as Oregon and Washington in the United States and British Columbia in Canada. As in the Californian hybrid zone, *M. galloprovincialis* has been introduced into these regions both accidentally via shipping and intentionally via aquaculture. See Sarver & Foltz (1993), Suchanek *et al.* (1997), Wonham (2004), Elliot *et al.* (2008), Shields *et al.* (2010) and Crego-Prieto *et al.* (2015) for details.

Brannock *et al.* (2009) examined the level of hybridisation and introgression between *M. galloprovincialis* and *M. trossulus* around Hokkaido, the northernmost of Japan's main islands. Using three nuclear markers (*Me15/16*, *ITS* and *MAL-1*), they found that *M. galloprovincialis* dominated the western portion of Hokkaido within the Sea of Japan, while *M. trossulus* was only common in samples from eastern Hokkaido (Figure 9.9). Two independent hybrid zones exist in Hokkaido, each with distinct genetic architectures. Along the southern coast, there is a relatively smooth genetic cline, with western populations composed almost exclusively of *M. galloprovincialis* individuals and eastern

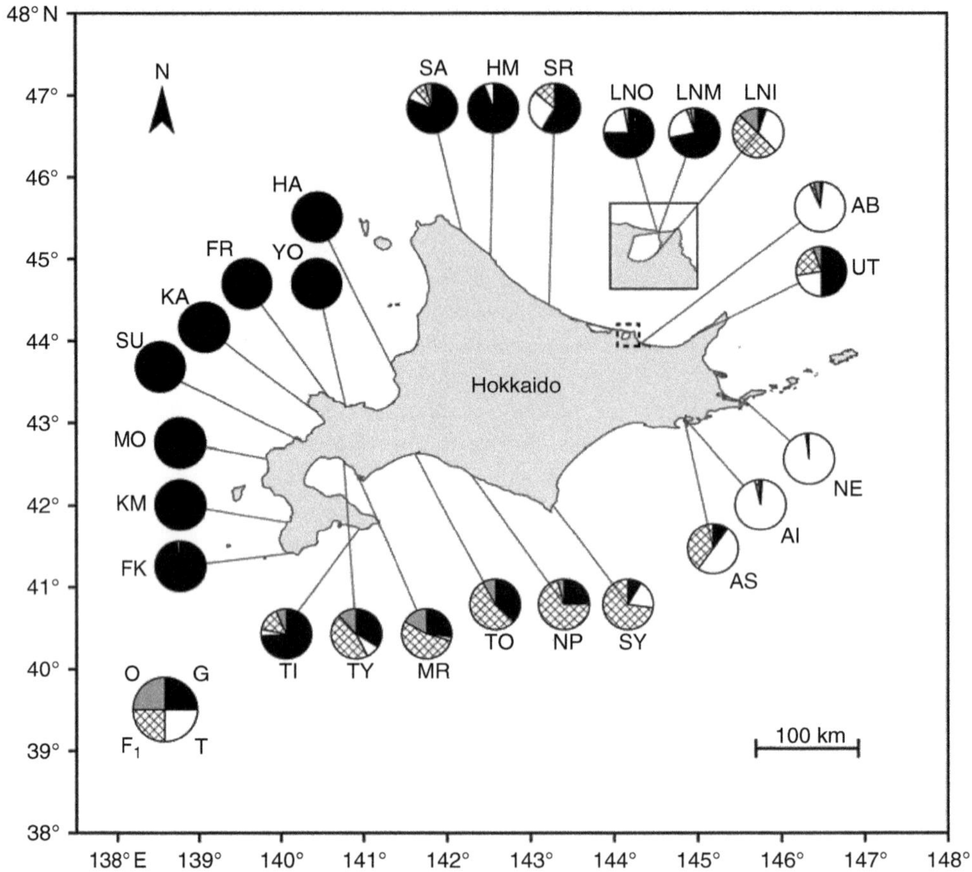

Figure 9.9 Distribution in 2006 of homozygous *Mytilus galloprovincialis* (black, G), homozygous *M. trossulus* (white, T), F1 hybrid (cross-hatched, F1) and other hybrid (grey, O) individuals around the island of Hokkaido, Japan. Data represent a multi-locus composite of genotypes from three nuclear PCR assays (*Glu-5'*, *ITS* and *MAL-I*). *Source:* Brannock *et al.* (2009). CC BY 4.0.

populations composed almost solely of *M. trossulus* individuals. Within this cline, the frequency of F1 hybrid individuals often exceeds 50%. In contrast, hybrid mussel populations in northern Hokkaido form a mosaic pattern, with radical shifts in genetic composition over small distances, especially near the eastern portion of the coast. Hybridisation levels along the northern coast of Hokkaido almost never exceed 15%, except at LNI (Figure 9.9), where >60% of the population are hybrids. These geographical distributions were similar in 2004 and 2006. Despite the high frequency of F1 genotypes in southern populations, the frequency of other hybrid genotypes is very low, suggesting either that F1 hybrids have relatively low fertility and consequently F2 and backcross hybrids are rarely formed (indicating limited introgression) or that individuals with other hybrid genotypes have poor survival.

The two hybrid zones in Japan have two distinct features (Brannock & Hilbish 2010). First, F1 hybrids occur at extraordinary levels, far exceeding expectations based upon random mating, suggesting that the incompatibilities in gamete recognition are weak or nonexistent and certainly are not limiting the formation of F1 hybrids – a conclusion consistent with previous laboratory studies that reported no difficulties in the formation of F1 hybrids from *M. galloprovincialis* and *M. trossulus* crosses (Matson *et al.* 2003). Second, despite the preponderance of F1 hybrids, there has been little introgression beyond their formation (Brannock *et al.* 2009). Hybridisation between these taxa has occurred in northern Japan for more than two decades (Inoue *et al.* 1997), which makes the low level of introgression especially striking because sufficient time has passed for hybridisation to proceed beyond the first two generations. Low introgression is a characteristic of other areas where *M. galloprovincialis* and *M. trossulus* hybridise (Brannock & Hilbish 2010 and references therein), strongly suggesting the operation of post-zygotic selection in determining the level of introgression between these species and in shaping their hybrid-zone characteristics. Brannock & Hilbish (2010) have demonstrated that hybrids between *M. galloprovincialis* and *M. trossulus* in northern Japan begin gametogenesis but rarely produce mature gametes. Disruption of gametogenesis in hybrids provides a potent mechanism for post-mating reproductive isolation that restricts introgression between these species.

The features of a hybrid zone between *B. azoricus* and *B. puteoserpentis* have already been described. Additional information on *Bathymodiolus* hybrid zones can be found in Won *et al.* (2003b), Johnson *et al.* (2013) and Breusing *et al.* (2016, 2017). There are no published reports of hybridisation between *Perna* species (*P. perna*, *P. viridis* and *P. canaliculus*), merely because on a global scale their geographic ranges do not overlap (see earlier and Wood *et al.* 2007 fig. 2).

Invasive Species: Origins, Dispersal and Population Genetics

Retracing the origins of invasive species is a first critical step in identifying potential mechanisms of introduction, implementing management strategies and forecasting the spread of invasive species (Zardi *et al.* 2018). *Mytilus galloprovincialis* is considered highly invasive due to its quick rate of spread and its ability to displace and outcompete native mussels. The species was accidentally introduced on to the South African west coast in the late 1970s, probably with shipping from a Mediterranean source population. It rapidly colonised thousands of kilometres of the west and south coastlines of southern Africa, but its spread appears to have ceased over the last decade (Assis *et al.* 2015). The arrival of the invader has had significant ecological impacts, although it appears that it has now reached at least a temporary equilibrium in its eastern limit and in its pattern of coexistence with the indigenous mussel, *Perna perna* (reviewed by Zardi *et al.* 2015). The genetic identity of *M. galloprovincialis* populations along the Namibian and South African shores was evaluated using alleles at the Me15/16 locus and the mitochondrial 16S RFLP marker to infer the most likely origin of invasion (Zardi *et al.* 2018). The 16S RFLP marker, which distinguishes between Northern and Southern Hemisphere *M. galloprovincialis*, did not show Southern Hemisphere haplotypes, indicating that the Northern Hemisphere is the sole origin of invasion. Frequencies of haplotypes at the 16S marker and alleles at the Me15/16 locus also support a European northeastern Atlantic origin throughout the species' Namibian and South African distribution, confirming results of earlier studies based on samples from a limited area on the southwestern coast of South Africa (references in Zardi *et al.* 2018).

In a related study, Pickett & David (2018) attempted to elucidate global connectivity patterns of *M. galloprovincialis* using archived cytochrome oxidase 1 (CO1) sequence data obtained from public databases. They compiled a comprehensive global CO1 dataset consisting of 209 sequences representing 14 *M. galloprovincialis* populations. The hypothesis was that *M. galloprovincialis* would exhibit marked genetic differentiation between Northern and Southern Hemisphere mussels but overall very low levels of genetic differentiation within hemispheres due to repeated introductions. There was significant genetic structuring across populations, with significant geographic patterning of haplotypes. In particular, South Korea, southern China, Turkey and Australasia were the most genetically isolated populations. However, there was no indication of a Northern and Southern Hemisphere grouping as found in previous studies that used RFLPs or 16S rRNA sequences (Hilbish *et al.* 2000; Gérard *et al.* 2008; Westfall *et al.* 2010). The authors concluded that the results were indicative of a complex dispersal pattern for *M. galloprovincialis*, likely involving a combination of natural and anthropogenic dispersal coupled with local adaptation and hybridisation events.

Perna viridis is a recent invader of the Caribbean region and Florida in the southeast of the United States, and has also been introduced into parts of coastal Australia, Japan and North and South America via ship ballast, hull fouling and experimental introduction for farming (see Chapter 10). Gilg *et al.* (2013) set out to determine whether multiple, geographically isolated non-native populations of *P. viridis* were the product of a stepping-stone expansion of a single introduction or of multiple independent introductions from the native range. Measurements of genetic variation were compared among five introduced populations (from Florida, Jamaica and Trinidad) and three populations from within the native range (south of Kochi in India, Singapore and Hong Kong). Based on 650 bp of the CO1 mitochondrial gene from 280 samples from the introduced populations and another 190 samples from the three native populations, haplotype frequencies of all introduced populations were not significantly different from each other, but virtually all populations differed from samples taken from the native range. Introduced populations had less variation than most native populations and there was no evidence for admixture. The lack of genetic variation in introduced populations suggests that the initial introduction was relatively small, while the lack of admixture suggests a single original source population. There was strong evidence of a stepping-stone pattern of introduction from the native range to Trinidad and from Trinidad to other locations in the Caribbean and the United States. The subsequent development of 16 polymorphic microsatellite markers should be useful in future genetic diversity and bioinvasion studies of *P. viridis*, supporting a science-based management approach to the future prevention and management of this species (Lukehurst *et al.* 2017).

The charru mussel, *Mytella charruana*, is native to Central and South America, with a distribution on the Pacific coast from Mexico to Ecuador and on the south Atlantic coast from Colombia to Argentina (see Chapter 10). Since 2004, it has been found along the southeastern coast of the United States from Florida to South Carolina. The main life history characters that favor survival of *M. charruana* in new environments are high salinity and temperature tolerances and the ability to change sex under stressful conditions (Yuan *et al.* 2016; see Chapter 10). Using a 722 bp fragment of the CO1 mitochondrial gene, Calazans *et al.* (2018) compared genetic variation throughout the mussel's native range in

South America to its invasive range in the southeastern United States, where samples were collected in 2006 and 2010. The aim was twofold: to determine the potential source populations (or geographic regions) for the southeastern US invasion and to discover how genetic diversity had changed between the two sampling periods. The results suggested that the population originated from multiple sources, with the Panama region as the primary one, while the genetic composition of the non-native populations was unchanged between the two sampling periods. Throughout its native range, the authors found that *M. charruana* exhibited a pattern of minimal dispersal, yet it was both persistent and spreading throughout the invasive range. It appears that although this species will not likely be transported long distances on its own, anthropogenic transportation can lead to foundation events in a variety of habitats that will likely make it difficult to eradicate.

Mussel Genomics

The genome comprises the total DNA complement of genetic material in the cell nucleus (nDNA) and mitochondria (mtDNA), and genomics is the discipline that looks at genome structure, function, evolution, mapping and editing. The entire RNA content of an organism is called its transcriptome. The genome and transcriptome are closely linked, and between them define and produce the proteome, which is the entire set of proteins expressed by a genome, cell, tissue or organism at a certain time (Beaumont *et al.* 2010; see later). Genome size can vary quite a bit between bivalve species, and even in the same species in different studies. Mugarella *et al.* (2016) estimated a genome size of 1600 million base pairs (Mb) in *M. galloprovincialis* and of 545 and 1500 Mb in the oysters *Crassostrea gigas* and *Pinctada fucata*, respectively. In eukaryotes, unlike prokaryotes, there is no relationship between genome size or number of genes and number of chromosomes (El-Shehawi & Elseehy 2017).

Genome Mapping

Genome mapping is used to identify and record the locations of genes and the distances between genes on a chromosome. This can be achieved in two different ways: through genetic (linkage) mapping or through physical mapping. Genetic mapping provides the positions of known genes or genetic markers relative to one another in terms of recombination frequency, rather than specific physical distance along a chromosome. The greater the frequency of recombination (segregation) between two genetic markers, the farther apart they are assumed to be. Conversely, the lower the frequency of recombination between the markers, the smaller the physical distance between them. The frequency of recombination is estimated from an analysis of the progeny of experimental crosses between parents of known genotype. By finding out how often various characteristics are inherited together, it is possible to estimate the distance between the genes. The standard unit of genetic distance is the centiMorgan (cM), an expression of the percentage of progeny in which a recombination event has occurred between two markers (Gibson & Muse 2009). A map of where the genes are in relationship to each other on the chromosomes, known as a linkage map, can then be drawn. Construction of a linkage map requires four major components: polymorphic markers, genotyping platforms, reference families and software for analysis of linkage between pairwise markers and among markers (Yue 2014). Currently, the most widely used DNA markers in genetic mapping are microsatellites and SNPs. MS markers are highly polymorphic and easily mapped, while SNPs are very abundant in the genome, easily identified using high-throughput technology and can be developed from any gene, which offers the chance of identifying and mapping functional polymorphisms. They are also co-dominant markers (i.e. both alleles can be identified at a locus), in contrast to AFLPs, which are dominant markers (see earlier). Genetic maps are the most common type of map available for fish and shellfish, largely because they are affordable and not too complex to prepare. The very first linkage map was constructed for the eastern oyster, *Crassostrea virginica* (Yu & Guo 2003), and shortly after that one was prepared for the Pacific oyster, *C. gigas* (Li & Guo 2004). Lallias *et al.* (2007) produced the first linkage maps for *Mytilus edulis*. Sex-specific linkage maps were built with MapMaker 3.0 software. The female framework map consisted of 121 markers ordered into 14 linkage groups,[21] spanning 862.8 cM, with an average marker spacing of 8.0 cM. The male framework map consisted of 116 AFLP markers ordered into 14 linkage groups, spanning 825.2 cM, with an average marker spacing of 8.09 cM. Since this study, no further linkage maps have been published for marine mussel species. However, maps are now available for several scallop, clam and oyster species (see Hollenbeck & Johnston 2018 for details). Linkage maps are usually carried out in the framework of quantitative genomics programmes (i.e. mid- to long-term studies directed to locate genomic regions underlying phenotypic variation for a given trait, usually of economic importance; Liu 1997; see later).

Physical mapping provides the true distance, measured in number of DNA base pairs, between two markers. The most common method used to construct physical maps is bacterial artificial chromosome (BAC) fingerprinting. The donor's DNA is cut using restriction enzymes into size-selected fragments 100–300 kb in length and these are inserted into BACs, which act as vectors to artificially carry DNA into bacterial cells (e.g. *Escherichia coli*). As the bacterial cells grow and divide, they amplify the BAC DNA, ultimately producing what is known as a BAC library. This allows areas of overlap between fragments to be identified, and thus the restriction profiles of BAC clones can be assembled in the correct order to construct a physical map (details in Beaumont *et al.* 2010). These fragments are much simpler to sequence than an entire genome. Besides physical mapping, BAC libraries are also essential for chromosome mapping, genome sequencing and positional cloning of genes and their regulatory elements (Guo *et al.* 2008). Other methods used to construct physical maps are radiation hybrid (RH) mapping, cytogenetic mapping by fluorescence in situ hybridisation (FISH) and whole-genome sequencing (see Canario *et al.* 2008 for review).

To date, complete or draft genomes of the following, mostly commercially important bivalves have been sequenced: *Mytilus galloprovincialis* (Murgarella *et al.* 2016), the horse mussel *Modiolus philippinarum* and the deep-sea vent mussel *Bathymodiolus platifrons* (Sun *et al.* 2017); the Pacific oyster *Crassostrea gigas* (Zhang *et al.* 2012), the Eastern oyster *C. virginica* (Gómez-Chiarri *et al.* 2015), the pearl oyster *Pinctada fucata* (Takeuchi *et al.* 2016) and the Sydney rock oyster *Saccostrea glomerata* (Powell *et al.* 2018); the scallop *Chlamys farreri* (Li *et al.* 2017), the Yesso scallop *Patinopecten yessoensis* (Wang *et al.* 2017) and the Peruvian scallop *Argopecten purpuratus* (Li *et al.* 2018); and the Manila clam *Ruditapes philippinarum* (Mun *et al.* 2017) and the blood clam *Scapharca (Anadara) broughtonii* (Bai *et al.* 2019). *M. galloprovincialis* was found to have a highly complex genome, in which repeated elements seemed to be widespread (~30% of the genome) – a feature that is also shared with other marine molluscs (Murgarella *et al.* 2016). Murgarella *et al.* (2016) found genes with functional annotations that would explain how mussels cope with the specific characteristics of their environment, such as those related to immunity, resistance to hypoxia, shell formation and adhesion to surfaces. They looked for genes with immune-related functions in mussels that provide defence against bacteria and other environmental challenges and found 20 for immune response, 55 for immune system process and 35 for defence response. These genes had functions related to the Toll-like receptor (TLR) signaling pathway, inflammatory response, regulation of innate immune response and defence response to fungi, viruses and bacteria (see Chapter 11). Other genes that are important for adaptation in mussels are those that confer resistance to hypoxia and oxidative stress during tidal emersion and immersion cycles. Murgarella *et al.* (2016) found 7 for resistance to hypoxia, 24 for response to oxidative stress and 55 for response to abiotic stress. They further searched for genes that provide physical protection against the environment and found 10 involved in byssus attachment to surfaces and 24 involved in shell formation.

The sequences of the two mitochondrial genomes, maternal (F) and paternal (M), have been isolated from the following marine mussels: *M. edulis* (Śmietanka *et al.* 2009, 2016), *M. galloprovincialis* (Mizi *et al.* 2005; Śmietanka *et al.* 2009; Murgarella *et al.* 2016), *M. trossulus* (Śmietanka *et al.* 2009, 2013), *M. californianus* (Cao *et al.* 2009), *M. chilensis* (Śmietanka & Burzyński 2017), *Perna viridis* (Li *et al.* 2012) and *Perumytilus purpuratus* (Vargas *et al.* 2015; Śmietanka *et al.* 2018). Gene maps of the complete F and M mitochondrial genomes of *M. galloprovincialis* are provided in Figure 9.10 (Mizi *et al.* 2005). The two genomes have diverged by about 20% in nucleotide sequence but retain identical gene content and gene arrangement, and similar nucleotide composition and codon usage bias.[22] Mizi *et al.* (2005) found that the F genome is virtually identical to that of *M. edulis*, in that the sequence is 16 744 bp – only four nucleotides longer than the *M. edulis* F genome assembled by Boore *et al.* (2004). The gene order of the *M. galloprovincialis* F genome is also identical to that of the *M. edulis* F genome, and the overall nucleotide divergence is within the range expected for conspecific mtDNA genomes.

Functional Genomics

Transcriptomics

By using the vast wealth of data produced by genome sequencing projects, functional genomics attempts to answer questions about the function of DNA at the levels of genes, RNA transcripts and protein products. In other words, functional genomics aims to bridge the gap between the blueprint (genome sequence or genotype) and the living organism (trait or phenotype) under various environmental conditions. Gene expression is the term used to describe the transcription of the information contained within the DNA into mRNA molecules, which are then translated into the proteins that perform most of the critical functions of the cell. Consequently, by analysing the entire collection of RNA sequences in

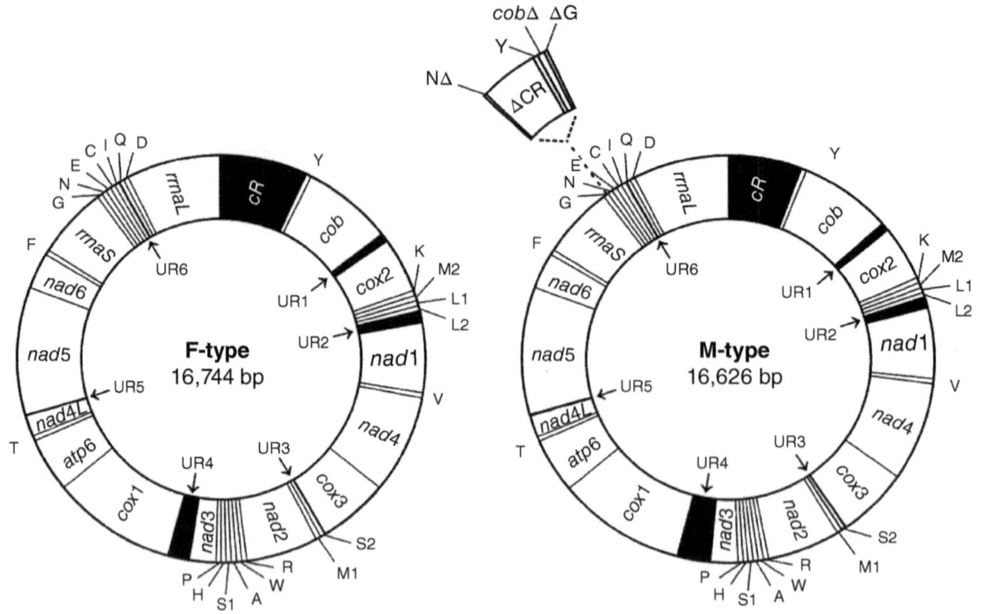

Figure 9.10 Gene maps of the mitochondrial genomes of *Mytilus galloprovincialis*. All genes are transcribed clockwise. The one-letter amino acid code is used for tRNA designation. L1, L2, M1, M2, S1 and S2 designate tRNAs recognising codons CUN, UUR, AUA, AUG, AGN and UCN, respectively. Black areas indicate noncoding regions. CR, putative CR; UR1–UR6, unassigned regions 1 to 6. Noncoding sequences of less than 10 bp are not shown. *atp6*, ATP synthase subunit 6; *cox1–3*, cytochrome c oxidase subunits I, II and III; *cob*, cytochrome b apoenzyme; *nad1–6* and *nad4L*, nicotinamide adenine dinucleotide dehydrogenase subunits 1–6 and 4L; *rrnaS* and *rrnaL*, small and large subunits of ribosomal RNA. The dotted line shows the position of a 1045 bp insertion that was found in the sequenced M genome. NΔ corresponds to the first 11 bp of the normal *trnN*; ΔCR corresponds to bases 77–902 of the CR; Y is a full copy of the *trnY*; *cob*Δ corresponds to the first 88 bp of *cob*; ΔG corresponds to the last 56 bp of *trnG*. Numbers inside the circles indicate the size of the genome. *Source:* From Mizi *et al.* (2005). Reproduced with permission from Oxford University Press.

a cell (the transcriptome; Figure 9.11), researchers can determine when and where each gene is turned on or off in the cells and tissues of an organism. A search of a transcriptome database can give us a list of all the tissues in which a gene is expressed, providing clues about its possible function (NIH 2015a). Various technologies have been developed to deduce and quantify the transcriptome, including hybridisation (e.g. microarrays) and sequence-based approaches (e.g. RNA-Seq). Hybridisation-based approaches typically involve incubating fluorescently labelled cDNA with custom-made microarrays or commercial high-density oligo microarrays. RNA-Seq refers to the sequencing of transcript cDNAs, in which abundance is derived from the number of counts from each transcript (reviewed in Wang *et al.* 2009; Lowe *et al.* 2017). The technique has been heavily influenced by the development of sequencing technologies. Next-generation sequencing (NGS), also known as high-throughput sequencing (HTS), is the catch-all term used to describe a

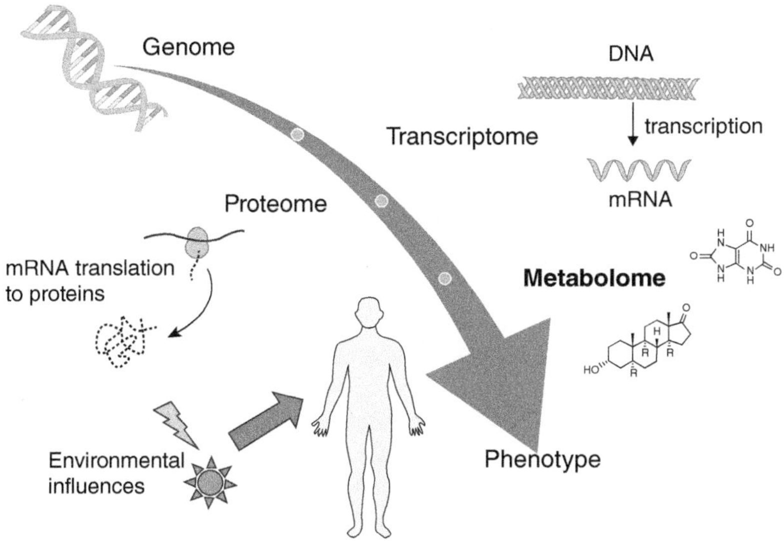

Figure 9.11 Overview of different omics sciences, such as genomics, transcriptomics and proteomics. Metabolomics represents the downstream output of the genome, but also the upstream input from the environment, and is therefore positioned at the bottom of the omics cascade. *Source:* From Steuer *et al.* (2019). CC BY 4.0.

number of different modern sequencing technologies (Reuter *et al.* 2015). RNA-Seq data sets can be uploaded via the Gene Expression Omnibus (GEO; www.ncbi.nlm.nih.gov/geo/), an international public repository that archives and freely distributes microarray, NGS and other forms of high-throughput functional genomics data submitted by the research community. Microarrays allow detection of DNA or RNA molecules by hybridising or sticking to target DNA molecules or RNA molecules on a small glass plate encased in plastic, with detection of adherent DNA or RNA molecules by various labels or dyes that allow them to be seen under a microscope. Using an array containing many DNA samples, the expression levels of hundreds or thousands of genes within a cell can be determined by measuring the amount of mRNA bound to each site, generating a profile of gene expression in the cell (NIH 2015b; Lowe *et al.* 2017).

Marine mussels, such as *Mytilus galloprovincialis* and *M. edulis*, are often employed as sentinel organisms of aquatic pollution because they are resistant to high doses of contaminants and show alterations of physiological and biochemical parameters, which can be used in the biological monitoring of seawater (Cajaraville *et al.* 2000; see Chapter 8). The first microarray for mussels (*Mytilus galloprovincialis*) was enriched for genes involved in response to a mixture of hydrocarbons (North Sea oil) and inorganic contaminants, represented by mercury (Hg) (Dondero *et al.* 2006). Subsequently, Wong *et al.* (2015) sequenced the transcriptome of gill, foot and mantle tissues in the cold-seep mussel *Bathymodiolus platifrons*, from which they generated a transcriptomic database containing 96 683 transcript sequences. They highlighted a repertoire of genes involved in the innate immune system, heavy-metal binding and sulphide detoxification. The expression patterns of these genes were reported in all tissue types. The mussel *Perna viridis* plays an important role in the biomonitoring of aquatic pollution. Byssus, secreted by the foot gland, has been shown to bind heavy metals effectively (Yap *et al.* 2016). Consequently, Zhang *et al.* (2017) used RNA-Seq to carry out a comparative transcriptomic analysis on mussel foot tissue in three different groups: one exposed to 50 μg l^{-1} of cadmium chloride ($CdCl_2$), one exposed to 100 μg l^{-1} $CdCl_2$ and one control. Mussels in the three groups were dissected at 24, 48 or 72 hours. The authors identified 9048 differentially expressed genes between Cd treatment (50 or 100 μg l^{-1}) at 48 hr and the control, suggesting an extensive transcriptome response of the mussels during Cd stimulation. Moreover, they found that the expression levels of 54 byssus protein-coding genes increased significantly after the 48 hr Cd stimulation. Sixteen of the critical byssus protein-coding genes were picked for profiling by quantitative real-time PCR. These proteins have a high content of the amino acids tyrosine, cysteine and histidine, which may form metal complexes with many heavy metal ions in byssus. In the ocean, the fragmentation of larger plastic items by mechanical, photolytic or biological processes is one of the main sources of microplastics (small plastic particles <5 mm in size; see Choy *et al.* 2019 for review; see also Chapter 8). Détrée & Gallardo-Escárate (2017) evaluated the expression of 12 genes associated with cellular stress, immune response and metabolism on four tissues (gills, digestive gland, haemolymph and mantle) of *M. galloprovincialis* after an acute (24 hr) exposure to a concentration of 1.5E + 7 microbeads of polyethylene per litre (ranging from 1 to 50 μm in size). This exposure concentration is higher than that generally found in the field, but was chosen to correspond to a pollution hot spot (Détrée & Gallardo-Escárate 2017 and references therein). Analysis showed upregulation of genes concerned with carbon metabolism, oxidative stress, immune response and apoptosis (programmed cell death) in the mantle and digestive gland, while a global downregulation of genes involved in carbon metabolism was observed in the haemolymph and gills. Altogether, the results showed disruptive and tissue-dependent effects on major biological processes following a short exposure to microplastics in *M. galloprovincialis*.

Saxitoxin (STX) is one of the main phycotoxins that contribute to paralytic shellfish poisoning (PSP; see Chapter 12). While there is evidence of changes at the physiological level in bivalves exposed to STX, Núñez *et al.* (2013) were the first to analyse transcriptional responses to STX in mussels (*M. chiensis*), with the objective of evaluating the expression of 13 candidate genes associated with thermal stress (heat-shock proteins 70 and 90, hsp70 and hsp90), response to metal contamination (ferritin, metallothionein), response to oxidative stress (superoxide dismutase and catalase) and response to pathogens (galectin, peptidoglycan recognition protein (PGRP), mytilin A, B, lipopolysaccharide-induced tumour necrosis factor (TNF), LITAF (a transcription factor involved in activation of the TNF gene), fibrinogen and ependymin; see Chapter 11). High levels of differential gene expression were observed for superoxide dismutase, catalase, ferritin and hsp genes. To a lesser extent, ependymin, fibrinogen-like, galectin and mytilin B genes were also significantly more expressed in haemocytes of mussels injected with STX than in control mussels. The transcriptional patterns of candidate genes reported in this study provide information that could be used for early detection of the effects of toxins on marine bivalves (Núñez *et al.* 2013). In a corresponding study, Détrée *et al.* (2016) used RNA-seq and RT-qPCR approaches to evaluate the transcriptomic response of *M. chilensis* to purified STX, as well as *in vivo Alexandrium catenella* (toxic dinoflagellate) exposure. Pattern recognition receptors (PRRs) involved in mussel immunity, such as TLRs (toll like receptors), TNF receptors and scavenger-like receptors (see Chapter 11) were found to be strongly upregulated at 8 and 16 hours post-STX injection. The results suggest an involvement of PRRs in the response to STX, as well as identifying potential novel STX-interacting receptors in *M. chilensis*. The most significant contribution of this study was the identification of immune receptors and pathways potentially involved in the recognition and defense against STX's toxicity and its impact on wild and cultivated mussel populations (see also Astuya *et al.* 2015). Comparable studies on the transcriptome response of *M. galloprovincialis* against other toxins are: Manfrin *et al.* (2010, 2012), Li *et al.* (2013), Suárez-Ulloa *et al.* (2013), Varotto *et al.* (2013), Gerdol *et al.* (2014) and Prego-Faraldo *et al.* (2018).

Mytilus coruscus is cultured widely on the eastern coast of China, but in recent years it has been seriously affected by pathogenic infections. To elucidate defense mechanisms against pathogenic challenge, the haemocyte transcriptomes of *M. coruscus* were analysed before and after *Vibrio alginolyticus* infection (Dong *et al.* 2017; see Chapter 11). The immune recognition system triggers a wide range of abelling pathways and the production of specific reactive effectors, leading to neutralisation of dangerous threats through phagocytosis (see Chapter 11). After bacterial challenge, 1270 and 265 genes were identified as remarkably up- or downregulated, respectively; among them, 1154 genes were associated with 122 pathways, which included classical immune-related pathways such as TLR signalling, complement cascades, the MAPK signalling pathway and apoptosis. These findings provide new insights on the innate immunity of *M. coruscus*, which could be crucial for management of disease in this commercially important species. In a related study, Rey-Campos *et al.* (2019) conducted Illumina sequencing of the transcriptome of individual mussels (*M. galloprovincialis*) before and after infection with *Vibrio splendidus* (see Chapter 11); control mussels were injected with filtered seawater. There was great variability among individual transcriptomes, and each mussel showed an exclusive repertoire of genes not shared with other individuals. The regulated genes in both control and infected mussels were also analysed and, unexpectedly, the sampling before the injection was considered a stress stimulus strong enough to trigger and modulate the response in haemocytes, promoting cell migration and proliferation. The authors found an obvious response against the injection of filtered seawater, suggesting a reaction against a tissue injury in which the myticins – the most expressed antimicrobial peptides in mussels – appeared significantly upregulated. See section on genomics of mussel immunity in Chapter 11.

As already described, *M. edulis* and *M. galloprovincialis* hybridise extensively at many localities, and both pre- and post-zygotic isolating mechanisms may be in operation. Sperm thus represent key targets in investigations of the molecular mechanisms underlying reproductive isolation. Recently, Romero *et al.* (2019) carried out a deep transcriptome sequencing (RNA-seq) of mature male gonads and a 2DE/MS (two-dimensional mass spectrometry)-based proteome analysis of sperm from both species raised in a common environment. They provided raw data from a transcriptomic analysis of mature male gonads in the two species using NGS (Illumina) technology and a preliminary list of transcripts that were functionally annotated showing species-specific differential expression, which was marked for mitochondrial genes and for those involved in spermatogenesis, sperm motility, sperm–egg interactions, the acrosome reaction, sperm capacitation, ATP reserves and ROS production. Proteins and their corresponding genes might thus be good targets in further genomic analysis of reproductive barriers between these closely related species.

As already discussed, *M. galloprovincialis* is a successful invader on the central and southern coasts of California, where it has largely displaced the native congener, *M. trossulus*. Thermal responses of several physiological traits may underlie its capacity to out-compete *M. trossulus* in warm habitats. For example, heart function can be sustained at higher temperatures in *M. galloprovincialis* than in the native species, whereas *M. trossulus* can maintain heart function down to lower temperatures (Braby & Somero 2006b). Also, induction of proteins (hsps) has been shown to be more pronounced in *M. galloprovincialis* (Hofmann & Somero 1996), suggesting a more robust ability of this species to initiate transcription of genes encoding hsps. To clarify possible differences in stress-induced gene expression between the two species, Lockwood *et al.* (2010) developed an oligonucleotide microarray with 8874 probes representing 4488 different genes that recognised mRNAs of both species. In laboratory acute heat-stress experiments, 1531 of these genes showed temperature-dependent changes in expression that were highly similar in the two species. In contrast, 96 showed species-specific responses to heat stress, functionally characterised by their involvement in oxidative stress, proteolysis, energy metabolism, ion transport, cell abelling and cytoskeletal reorganisation. The gene that showed the biggest difference between the species was that for the molecular chaperone small hsp24, which was highly induced in *M. galloprovincialis* and showed only a small change *in M. trossulus*. The authors concluded that these different responses to acute heat stress may help to explain and predict the invasive success of *M. galloprovincialis* in a warming world. While *M. galloprovincialis* has displaced *M. trossulus* from central and southern California, the native species remains dominant at more northerly sites that have high levels of freshwater input. Consequently, Lockwood & Somero (2011b) compared the transcriptional responses of these two species to acute salinity stress (i.e. a hypo-osmotic shock or sudden drop in salinity). Probably the most striking finding of this study was the relatively minor transcriptional responses to salinity treatment compared with the robust responses to heat stress in Lockwood *et al.* (2010). In fact, <2% of the genes represented on the microarray showed significant responses to salinity stress. This is in contrast to the 36% of the transcriptome that changed in response to heat stress (Lockwood *et al.* 2010). Among the genes that did change expression in response to acute salinity stress, most showed a shared response between the species. These shared-response genes were characterised as being involved in osmoregulation (adjusting the intracellular concentration of solutes to match that of the surrounding environment) and cell-cycle control. Conversely, very few genes were different between the species in their transcriptional responses. Lockwood *et al.* (2015) suggest that this lack of transcriptional response was because the mussels were not that stressed; that is, that the exposure conditions were not harsh enough to elicit a robust response. However, Tomanek *et al.* (2012) conducted a study on the same set of samples as those used by Lockwood & Somero (2011b) and found robust changes at the level of the proteome in response to salinity treatment (see later).

In bivalves, the shell secreted by mantle tissue during the biomineralisation process is the first barrier against predators and mechanical or toxic damage. Changing environmental conditions, such as ocean acidification, influence shell strength and thus protection of the soft body within. Despite this, the proteins involved in the biomineralisation and pigmentation processes in *Mytilus* spp. Remain unclear. Hence, Malachowicz & Wenne (2019) aimed to identify transcripts putatively involved in biomineralisation and melanogenesis and to perform a comparative transcriptome mantle tissue analysis among different *Mytilus* taxa. They assembled and characterised the mantle transcriptomes of five populations representing four taxa: *M. edulis* (North Sea), *M. galloprovincialis* (Mediterranean Sea and Tasmania), *M. trossulus* (Vancouver, Canada) and *M. chilensis* (Chile). They found a total of 1292 contigs[23] putatively related to biomineralisation and pigmentation. Most of these transcripts had previously been reported in mantle transcriptomes of *M. edulis* (Freer *et al.* 2014; Hüning *et al.* 2016; Yarra *et al.* 2016) and *M. galloprovincialis* (Moreira *et al.* 2015; Bjärnmark *et al.* 2016). Several biomineralisation proteins were selected for more comprehensive analysis, since they were represented in at least four of the five *Mytilus* groups: dermatopontin (DPD), glyceraldehyde-3-phosphate dehydrogenase (GAPDH) and arginine kinase (AK). DPD is an extracellular matrix protein that may participate in nacre formation, GAPDH is an enzyme involved in carbohydrate metabolism that may be involved in the shell growth process and AK is a protein that may play a role in buffering intracellular pH, which would be important in the extrapallial space (references in Malachowicz & Wenne 2019; see Chapter 2). Phylogenetic relationships based on transcriptome data within the genus *Mytilus* were investigated. The analysis showed that *M. edulis*, *M. chilensis*, *M. galloprovincialis* and *M. trossulus* created a separate clade from *M. californianus* and *M. coruscus*, which is consistent with previous studies of their mtDNA (Gaitán-Espitia *et al.* 2016).

Perumytilus purpuratus is a marine mussel with a broad distribution on the Pacific and Atlantic coasts of South America. Studies have shown two geographically and genetically differentiated subpopulations at the molecular level (Guiñez *et al.* 2016) and in sperm morphological traits (Briones *et al.* 2012). A global de novo transcriptome from gonadal tissue of mature males was sequenced (Briones *et al.* 2018) and almost 40 765 transcripts were successfully annotated using BUSCO analysis (https://busco.ezlab.org/), which provides quantitative measures for the assessment of genome assembly, gene set and transcriptome completeness; a value of 89% completeness was determined for the assembled transcriptome. Most of the transcripts were associated with molecular function (93%), biological processes (92%) or cellular components (90%). Among the total number of transcripts, 82% were associated with at least one of these categories, indicating that most of the annotated genes were involved in multiple biological functions. The biological function category was the most diverse, with transcripts that were associated with cellular, single-organism and metabolic processes. The molecular function category was the least variable, with transcripts that were mainly associated with binding and catalytic processes. In the cellular category, most transcripts were assigned to cell- and organelle-related genes and were involved in some type of cellular structural organisation. Additionally, a subset of annotations generated using the word 'sperm' resulted in a total of 1294 transcripts. Those from the cellular component category were related to specific sperm structures such as the acrosomal vesicle, sperm flagellum and sperm midpiece (see Chapter 5), whereas biological process transcripts were strongly associated with sperm processes required for fertilisation such as spermatogenesis, sperm capacitation, sperm motility and sperm ejaculation, as well as with processes where the sperm–egg interaction could be implicated, such as prevention of polyspermy, regulation of fusion of sperm to egg plasma membrane, sperm–egg recognition and binding of sperm to the oocyte zona pellucida, which protects oocytes, eggs and embryos during development and regulates interactions between ovulated eggs and free-swimming sperm during and following fertilisation.

Proteomics

The proteome is the entire complement of proteins expressed in a given organism, tissue/organ or cell at a certain point in time, and proteomics is the large-scale study of the expression, identification and interactions of those proteins. Studying the proteome is a challenging task, since, unlike the genome, which is relatively static, it changes constantly in response to tens of thousands of intra- and extracellular environmental signals. As a consequence, a particular cell type may make different sets of proteins at different times or under different conditions. Furthermore, any one protein can undergo a wide range of post-translational modifications (e.g. the addition of sugars, fats, phosphates and even other proteins; Adams 2008).

Proteomics involves the application of technologies for the identification and quantification of overall proteins present in the content of a cell, tissue or organism (reviewed in Aslam *et al.* 2017; Kumar *et al.* 2017). The initial step in all proteomic studies is the separation of a mixture of proteins. This can be carried out using high-resolution two-dimensional (2-DE) electrophoresis, in which proteins are first separated on the basis of their isoelectric points and the gel is then turned 90 degrees from its initial position to separate proteins based on differences in their molecular mass. The spots obtained in 2-DE electrophoresis are excised, and are most commonly detected using fluorescent dyes. Autoradiography is also used to visualise and quantitate radiolabelled proteins. Proteins, resolved by 2-DE or the related technique, two-dimensional differential gel electrophoresis (2-D DIGE), are characterised by MS techniques such as matrix-assisted laser desorption/ionisation-time of flight (MALDI-TOF-MS) or electrospray ionisation (ESI-MS), or various combinations of these. The data generated by MS are then matched with available databases using various bioinformatics softwares for identifying peptides and their sequences (Chang *et al.* 2018; NCBI 2021). For the separation of complex peptide mixtures, gel-independent methods like liquid chromatography coupled with MS (LC-MS/MS) systems have been developed. One such approach is known as 'shotgun proteomics', in which LC-MS/MS and various protein sequence databases are used together to study complex mixture of peptides generated through proteolytic digestion of proteins isolated from a sample.

Protein microarrays, also known as protein chips, are one of the most promising high-throughput techniques developed for the characterisation of proteins. A basic protein microarray consists of an array of immobilised protein spots filled with 'bait' proteins, which are hybridised with small molecular probes labeled with fluorescent tags. Protein microarrays are classified into three categories: analytical protein microarrays, functional protein microarrays and reverse-phase protein microarrays (Aslam *et al.* 2017; Kumar *et al.* 2017). The antibody microarray is the most representative class of analytical protein microarray. After antibody capture, proteins are detected by direct protein labeling. Functional microarrays are employed for probing of many protein activities and protein interactions. Reverse-phase microarray is used to determine the presence of modified disease-causing proteins and post-translational modifications. Finally, it should be emphasised that since proteomics is a young science, many of the technologies used in proteomics are still prototypical, and better materials, instrument designs and methodologies are expected to improve sensitivity, resolution and repeatability in the future (Kumar *et al.* 2017, p. 8).

The majority of proteomic studies are concerned with detection of human biomarkers for disease diagnosis and drug design, understanding the pathogenicity mechanisms, alteration of expression patterns in response to different signals and interpretation of functional protein pathways in different diseases. However, in bivalves, the main focus has been on the recognition of specific biomarker patterns or protein expression signatures (PESs) in response to environmental stressors. The mussel has been the bivalve of choice in proteomic studies, mainly because of its role as an ideal sentinel organism in marine monitoring programmes. Shepard *et al.* (2000) were the first to analyse the proteome of *M. edulis* exposed to copper, polychlorinated biphenyls (PCBs) and salinity stress, albeit under laboratory conditions. Using 2-DE and quantification analysis of protein abundance by gel image analysis, they got clear separation of 500–600 protein spots, and ultimately obtained a unique PES for each stressor. Since then, proteomics has been used to examine PES in mussels in response to a wide range of environmental contaminants (xenobiotics), including oil (Monsinjon *et al.* 2006; Apraiz *et al.* 2009), metals (Leung *et al.* 2011; Hu *et al.* 2014), polybrominated diphenyl ethers (Apraiz *et al.* 2006; Chenlong *et al.* 2013), phthalates (Apraiz *et al.* 2006), polyaromatic hydrocarbons (PAHs) (Knigge *et al.* 2004; Fang *et al.* 2009), PCBs (Fang *et al.* 2009), biotoxins (Nzoughet *et al.* 2009; Huang *et al.* 2015), cyanobacteria (Puerto *et al.* 2011), pharmaceuticals (Schmidt *et al.* 2014; Campos *et al.* 2016), nanomaterials (Matysiak *et al.* 2016) and other stressors (e.g. thermal, Tomanek & Zuzow 2010, Péden *et al.* 2014, Tomanek 2012; salinity, Tomanek *et al.* 2012; and oxidative,[24] McDonagh & Sheehan 2006; Prevodnik *et al.* 2007; Sheehan & McDonagh 2008; Tomanek 2015). A few examples from the preceding studies will now be described in some detail.

Along the Californian coast, *M. galloprovincialis* is replacing the native species *M. trossulus*, which may reflect an increase in the water temperature in this region. *M. galloprovincialis* is regarded as a warm climate-adapted species, with a natural distribution in the Mediterranean extending northward to the coast of France and the British Isles, whereas *M. trossulus* is a cold-water mussel, present in the Baltic Sea and along the north coast of California and Canada. By comparing the responses of their proteomes to acute heat stress, Tomanek & Zuzow (2010) sought to identify responses common to both species, as well as differences that could account for the greater heat tolerance in *M. galloprovincialis*. Mussels were acclimated to 13 °C for four weeks and exposed to acute heat stress (24, 28 and 32 °C) for 1 hour, before being returned to 13 °C to recover for 24 hours. Using 2-DE and MS, the authors identified 47 and 61 distinct proteins that changed abundance in *M. galloprovincialis* and *M. trossulus*, respectively. The identification of proteins involved in protein folding, proteolysis, energy metabolism, oxidative damage, cytoskeleton and deacetylation revealed common loci of heat stress in both mussels. *M. galloprovincialis* showed a lower sensitivity to high-temperature damage than *M. trossulus*, providing biochemical evidence for its capacity to colonise warmer waters. Compared with *M. gallorovicincialis*, the more cold-adapted *M. trossulus* showed increasing levels of several hsps – molecular chaperones that regulate protein folding – underlining their crucial role in protecting cells from acute heat stress and indicating interspecific differences in thermal tolerance.

In a related study, Tomanek *et al.* (2012) examined the proteomic responses to hyposaline stress in the same two species. Mussels were exposed to short periods (4 hr) of hyposaline stress, followed by a recovery period to mimic conditions typical for bays and coastal areas experiencing heavy freshwater input, with a quick return to full salinity with incoming tides and mixing with full-strength seawater. The differences in protein abundances in gill tissue suggested that *M. trossulus* was able to respond to a greater hyposaline challenge (24.5 psu) than *M. galloprovincialis* (29.8 psu), as shown by changing abundances of proteins involved in protein chaperoning, vesicle transport, cytoskeletal adjustments by actin-regulatory proteins, energy metabolism and oxidative stress. While levels of proteins involved in energy metabolism were lower in *M. trossulus* during recovery from hyposaline stress, *M. galloprovincialis* showed higher abundances at 29.8 psu, suggesting an energetic constraint in the invader but not the native congener. These differences, in a scenario of reduced coastal salinities, may enable *M. trossulus* to cope with greater hyposaline stress and to outcompete *M. galloprovincialis* in the southern part of the *M. trossulus* range, thereby preventing *M. galloprovincialis* from expanding northward.

As seen earlier, most studies to date have been concerned with changes in protein expression when mussels are exposed to environmental pollutants/toxins. For example, stress responses in *M. edulis* exposed to organic pollutants were measured using several physiological measures and changes in protein expression (Olsson *et al.* 2004). Mussels collected near Askö from the Baltic Sea were not pure *M. edulis* but a mixture of *M. edulis* and *M. trossulus* (Kijewski *et al.* 2011). They were exposed for six days in a flow-through system to two fractions of extracted Baltic sediments (containing primarily PAHs or PCBs) from one industrially impacted site and one off-shore site. Exposure to Aroclor1248 (a commercial PCB mixture) was included as a reference treatment. Physiological response was measured as changes in respiration, excretion, clearance rate and scope for growth. Only clearance rate and scope for growth in the Aroclor and impacted PCB site differed significantly ($P < 0.05$) from control organisms. Seven proteins, presumed to be from stress protein families (hsp60, hsp70 and hsp90), were observed on one-dimensional electrophoresis gels, although a wider sample was observed using 2-DE. In response to PCB exposure from the industrial site, all protein levels, except in three proteins, were significantly higher ($P < 0.05$) in treated than in control mussels, suggesting the use of stress inducible proteins as diagnostics in risk assessment. Overall, the protein response showed less variation among individuals compared to the physiological parameters. Could these differences be related to the genetic makeup of collected mussels?

Microalgae, which are critical food for filter-feeding bivalves, have the capacity to produce phycotoxins. These may ultimately be consumed by humans via the food chain, causing a variety of gastrointestinal and neurological illnesses (see Chapter 12). Huang *et al.* (2015) compared protein profiles in gills of *Perna viridis* after exposure to *Prorocentrum lima*, a dinoflagellate producing diarrhoetic shellfish poisoning (DSP) toxins, and identified the differential abundances of protein spots using 2-DE. After exposure to *P. lima*, the level of okadaic acid (a main component of DSP toxins) in the gills of *P. viridis* significantly increased at 6 hours, although all mussels were apparently healthy. Among the 28 identified protein spots, 12 proteins were upregulated and 16 were downregulated in *P. lima*-exposed mussels. The identified proteins were involved in metabolism, the cytoskeleton (which gives shape to the cell), signal transduction (see Chapter 11) and response to oxidative stress and detoxification. The results indicated that the presence of *P. lima* caused DSP toxin accumulation in mussel gill, and might consequently induce cytoskeletonal disorganisation, oxidative stress and dysfunction in metabolism and the ubiquitination proteasome system, which plays important roles in cellular processes such as regulation of cell-cycle progression, division, development and differentiation, apoptosis, cell trafficking and modulation of immune and inflammatory responses (Wang & Maldonadao 2006).

Proteomics has also been used to compare PES in mussels grown in different habitats (López *et al.* 2001), search for species-specific larval (López *et al.* 2005) and sperm (Zhang *et al.* 2015; Romero *et al.* 2019) proteins, identify *Mytilus* species (López *et al.* 2002) and hybrids (Diz & Skibinski 2007) and investigate the proteomes of gill (Rocher *et al.* 2015), shell (Marie *et al.* 2011; Gao *et al.* 2015; Arivalagan *et al.* 2017), foot (Atasaral-Şahin *et al.* 2015), haemolymph (Campos *et al.* 2015) and larval eggs (Diz *et al.* 2009).

Metabolomics

Metabolomics is the field of science that characterises endogenous and exogenous metabolites within a cell, tissue or biofluid of an organism in response to external stressors such as disease, contaminant exposure or nutritional imbalances (Lankadurai *et al.* 2013 and references therein). Metabolites are small molecules (<1 kD) that are chemically transformed by enzymes during metabolism and which play critical roles in various biological processes (Bunnik & Le Roch 2013). In contrast to genomics and proteomics, metabolomics links both gene and environmental interactions (Steuer *et al.* 2019), and represents not only the downstream output of the genome but also the upstream input from the environment; it is therefore positioned at the bottom of the 'omics' cascade (Figure 9.11). At present, nuclear magnetic resonance (NMR), liquid and gas chromatography coupled to mass spectrometry (LC-MS, GC-MS) and direct infusion tandem mass spectrometry (DI-MS/MS) are the most widely used techniques for the identification of large numbers of metabolites. The main advantages of NMR are its high reproducibility and ease of sample preparation. However, the sensitivity of MS-based techniques is higher than that of NMR, allowing the detection of most metabolites present in a cell. The advantages and disadvantages of these different techniques are discussed in Lin *et al.* (2006), Alonso *et al.* (2015), Mirsaedi *et al.* (2015) and Steuer *et al.* (2019).

The subdiscipline of environmental metabolomics involves the application of metabolomic techniques to analyse the interactions of organisms with their environment (reviewed in Lankadurai *et al.* 2013). Metabolomics has been used in mussels to describe toxicological effects of pollutants (Jones *et al.* 2008; Tuffnail *et al.* 2009; Wu & Wang 2010; Fasulo *et al.* 2012; Kwon *et al.* 2012; Capello *et al.* 2013, 2015; Bonnefille *et al.* 2018a), antibiotics (Serra-Compte *et al.* 2019) and pesticides (Zhang *et al.* 2017), as well as stress responses to heat shock (Dunphy *et al.* 2015, 2018), hyposalinity (Bussell *et al.* 2008; May *et al.* 2017; see Chapter 7), hypoxia (Hines *et al.* 2007; Tuffnail *et al.* 2009) and reduced seawater pH (Ellis *et al.* 2014). Metabolomic analysis has also been used in quality assessment of hatchery-reared larvae (Young *et al.* 2015) and in gender determination (Hines *et al.* 2007).

Capello *et al.* (2013) investigated biological responses to environmental pollution, mainly related to Hg and PAHs, in caged mussels (*M. galloprovincialis*) transplanted to an anthropogenic polluted and a pristine site along the Augusta coastline in Sicily, Italy. The use of transplanted mussels, originating from an aquaculture area, allows comparison of control organisms with those caged in potentially polluted sites. Also, use of caged mussels from a single population minimises confounding factors such as the age and reproductive status of the organisms, which can influence both contaminant bioaccumulation and biomarker responses. The gills were selected as the target organ as they represent the first organ to suffer from pollutant toxic effects due to their large surface, which is constantly exposed to seawater and its substances, whether harmful or not (see Chapter 2). Severe histological alterations in gill tissue were observed in mussels from the industrial area compared with the control, while gill metabolic profiles, obtained by proton (^{1}H) NMR, exhibited significant changes in amino acids, energy metabolites, osmolytes and neurotransmitters. Findings revealed that environmental pollution has adverse and compromising effects on caged *M. galloprovincialis* individuals, inducing severe morphological damage to the gills, as well as metabolic alterations. Similar studies on biological responses of mussel digestive gland to PAHs and of gills to petrochemicals are reported in Fasulo *et al.* (2012) and Capello *et al.* (2015), respectively.

Antibiotics accumulation in aquatic organisms may be of great concern from an ecological point of view, but it is also of concern from a human perspective, especially when they are accumulated in edible animals like marine mussels. In a laboratory study (Serre-Compte *et al.* 2019), carried out at summer and winter water temperatures, mussels (*M. galloprovincialis*) were exposed to the antibiotic sulphamethoxazole (SMX) at 10 $\mu g\ l^{-1}$ for 96 hours, followed by 24 hours of depuration. SMX showed a low bioconcentration factor in mussels of 1.5 l kg dw^{-1}, and 69% of the compound was eliminated from the organism in 24 hours. The low bioconcentration capacity of SMX and fast depuration time indicate that there is no risk for consumers from mussel ingestion. Metabolomic analysis, using high-performance LC-high-resolution MS, revealed alterations in amino acids levels (aspartate, phenylalanine, valine and tryptophan), thereby identifying disturbances in osmotic regulation and energy metabolism. In addition, the levels of some nucleotides (guanosine and inosine) and a carboxylic acid were also affected. However, SMX-exposed mussels did not show any significant alteration in the enzymatic activities related to xenobiotic metabolism and oxidative stress, although there was some alteration to the mussel metabolome, mainly during winter conditions.

Over the last two decades, the presence of pharmaceutically active compounds in aquatic environments has become a major concern, in particular their potential effects on nontarget organisms, including marine organisms. Diclofenac (DCF) is one of the most widely prescribed nonselective nonsteroidal anti-inflammatory drugs (NSAIDs), with the highest environmental concentration and detection rates, along with ibuprofen (He *et al.* 2017; Bonnefille *et al.* 2018a). A study by Bonnefille *et al.* (2018b) investigated the potential of a metabolomic approach to acquire information without any a priori hypothesis about DCF effects on marine mussels. For this purpose, they generated mussel profiles using LC combined with high-resolution MS. Their study showed that two main metabolic pathways were impacted by DCF exposure: tyrosine metabolism was mostly downmodulated and tryptophan metabolism was mostly upmodulated. The results indicated that modulation of the two pathways could lead to osmoregulation disruption and potential impairment of byssus formation and reproductive functions, such as sexual maturation and gamete multiplication and release.

Dunphy *et al.* (2015) were the first to undertake the first metabolomic analysis of thermal stress in the green-lipped mussel, *P. canaliculus*. Their aims were twofold: to determine acute thermal tolerances for adult *P. canaliculus* and to assess whether metabolomic signatures could be used to identify biomarkers of thermal stress in this species. Mussels were exposed to a three-hour acute temperature challenge using temperatures of 20 (ambient), 25, 29, 31, 33 and 35 °C. Thermally-induced mortality of adult *P. canaliculus* exhibited an abrupt transition from 0 to 100% mortality between 31 and 33 °C, with mortality assured at temperatures greater than 33 °C. Gill tissues were harvested from mussels for metabolomics analysis of thermal stress biomarkers using GC-MS. Discriminant function analysis (DFA) of 52 metabolites identified eight key biomarkers: *cis*-11-eicosenoic acid, palmitic acid, proline, gamma amino butyric acid (GABA), aspartic acid, fumaric acid, β-alanine and asparagine. These metabolites were entirely consistent in their ability to classify the exposure temperature that mussels experienced, indicating that their discriminatory capacity was strong, with the fatty acid *cis* -11-eicosenoic acid exerting the most impact on discriminatory capacity of a DFA model. Fatty acids like this one are well known as hormone precursors (e.g. prostaglandin), suggesting that perturbation of hormonal pathways may have occurred in thermally-stressed mussels. Alternatively, isolates of this fatty acid have been shown to have strong antioxidant capacities. The increase of this metabolite with increasing temperature may thus be a defensive response, as increased oxidative stress is known to be associated with rapid increases in temperature (Lesser 2006). There is a tight coupling between thermal stress and a transition to anaerobic metabolism (Dunphy *et al.* 2015 and references therein). Reductions in β-alanine, and to a lesser degree aspartic acid and asparagine, are suggestive of a transition to anaerobic metabolism in thermally-stressed mussels, and the buildup of proline certainly lends evidence to a shift to anaerobic metabolism with increasing temperature stress. In summary, pairing the per cent mortality with more detailed metabolomic analyses enabled the authors to elucidate the metabolic biomarkers associated with acute thermal stress in adult *P. canaliculus*. May *et al.* (2017) were the first to carry out a metabolomics study on multiple tissue types and developmental stages in a mussel species (*M. edulis*) in order to evaluate organismal responses to salinity stress. This study has already been described in the osmoregulation section of Chapter 7.

Quantitative Genetics

The closely related fields of population genetics and quantitative genetics both focus on the genetic basis of phenotypic variation among the individuals in a population. Population genetics traditionally deals with frequencies of alleles and genotypes, whereas quantitative genetics concentrates on the ways that individual variation in genotype and environment contribute to variance in phenotype (Nature Education 2014). In bivalves, the traits of most interest from a production viewpoint are growth rate, meat yield, survival, food conversion efficiency and disease resistance. These traits can vary among individuals over a range to produce a continuous distribution of phenotypes. Typically, quantitative traits are controlled by a large number of genes on different chromosomes, collectively referred to as quantitative trait loci (QTLs; see later). In a selective breeding programme, the success of selection depends upon the extent to which the variation for a particular trait in the population from which the parents were chosen is inherited.

Genetic analysis of quantitative traits involves working with means and variances, trying to partition the phenotypic variance of the trait (*VP*) into the variance due to genes (*VG*) and to the environment (*VE*):

$$V_G = V_P + V_E$$

Genetic sources of variation can themselves be divided into several components: additive variance (*VA*), dominance variance (*VD*) and epistatic variance (*VI*). Together, the values for each of these subcategories yield the total amount of genetic variation (V_G) responsible for a particular phenotypic trait:

$$V_G = V_A + V_D + V_I$$

VA results from the cumulative contribution of alleles at all the loci governing a quantitative trait. It is the most important element of genetic variation for the breeder, since the presence or absence of particular alleles is a character that is passed unchanged to the next generation, leading to cumulative gains over generations (Beaumont *et al.* 2010). *VD* results from interactions among alternative alleles at the same gene locus, while *VI* involves an interaction between alleles associated with different loci (Byers 2008). Neither of these is passed on to progeny, due to segregation of parental alleles at meiosis.

When the total amount of genetic variation responsible for a trait has been determined, the heritability of the trait can be calculated. Broad-sense heritability ($h^2 = VG/VP$) captures the proportion of phenotypic variation due to genetic values perhaps including effects due to dominance and epistasis. Narrow-sense heritability ($h^2 = VA/VP$) captures only that proportion of genetic variation that is due to additive genetic values (*VA*). It is the narrow-sense heritability that is used by breeders to adopt appropriate selection strategies and to predict rate of phenotypic change (Wray & Visscher 2008).

One method for estimating the heritability of a trait is to compare the mean phenotypic value of full sibs (individuals that share the same two parents) to the mean phenotypic value of their parents (midparent mean) in a regression analysis. In a randomly mating population, the slope of the line obtained from a regression of full-sib means on midparent means is the heritability (Figure 9.12). Another method is to compare full and half sibs (individuals that share one parent) using analysis of variance (Falconer & Mackay 1996; Wray & Visscher 2008). With the availability of high-density markers, there is also growing interest in marker-based estimates of narrow-sense heritability, using mixed models in which genetic relatedness is estimated from genetic markers (Visscher *et al.* 2006; Kruijer *et al.* 2015). Estimates of h^2 may theoretically range between zero, where V_P is entirely due to environmental effects, and one, where all the variance is due to additive genetic effects. The response to selection of a particular trait can be predicted from the heritability estimate and the phenotypic variance, and selection methods (see later) are chosen on the basis of these values. Heritability is a population parameter, and therefore depends on population-specific factors such as allele frequencies, the effects of gene variants and variation due to environmental factors. It does not necessarily predict the value of heritability in other populations (or other species).

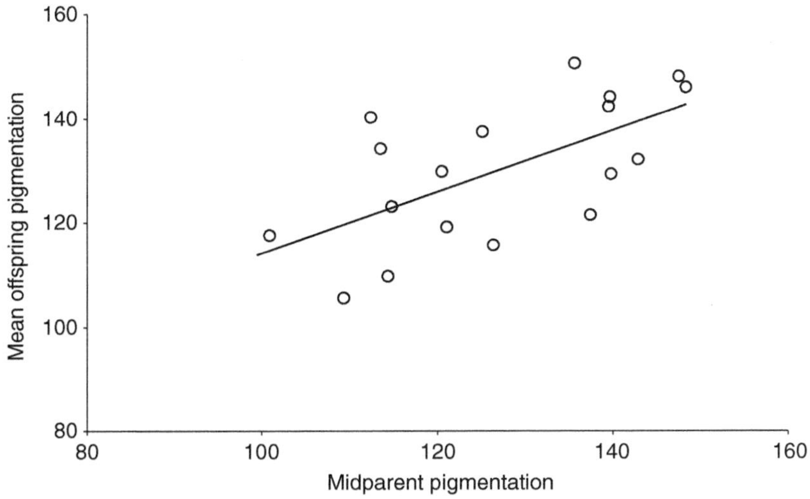

Figure 9.12 Midparent–offspring regression of total left-shell pigmentation in the Pacific oyster, *Crassostrea gigas*, based on 18 full-sib families. The narrow-sense heritability of total left-shell pigmentation is derived from the slope of the midparent–offspring regression ($b = h^2 = 0.59$, $SE = 0.19$, $P < 0.01$). *Source:* From Evans *et al.* (2009). Figure courtesy of F. Evans, Aquaculture and Fisheries CRSP, Oregon State University, Corvallis, Oregon, USA. Reproduced with permission from Elsevier.

Nevertheless, it is surprising how constant heritability estimates are across populations and species (Visscher *et. al.* 2008). Heritability estimates for a broad range of quantitative traits in marine mussels are given in Table 9.2.

Selective Breeding

The main objectives in a breeding plan are to ascertain the breeding objectives, identify the broodstock, choose a mating scheme for the reproduction of selected broodstock and identify how the selected stock can be expanded for production purposes.

Breeding Objectives

These are usually based on input from industry and the consumer, and should be clearly defined at the start of the breeding programme. Traits chosen for selection often consist of multiple economically important traits for simultaneous improvement. For each trait, the extent of phenotypic and genetic variation must be quantified by establishing its range, mean, variance and heritability (Beaumont *et al.* 2010). Frequently, the objectives differ from one species to the next, from one country to the next and even between different regions of a country (Gjedrem 1983). The Cawthron Institute, Nelson, New Zealand has established a programme for the greenshell mussel, *Perna canaliculus*. Beginning in 2013, when the programme was commercialised as part of a private growth partnership between the cultured mussel industry and government, it developed a more intensive and rigorous selection index

Table 9.2 Narrow-sense heritability estimates (with standard error in brackets) for various quantitative traits in mussel species.

Species	Trait	h^2	References
Mytilus chilensis	Larval valve length (25 days)	0.23 (0.30)	Alcapán *et al.* (2007)
	Larval valve length (40 days)	0.33 (0.003)	
	Larval valve length (300 days)	0.001 (0.076)	
	Larval valve weight (300 days)	0.030 (0.102)	
	Shell thickness	0.294 (0.194)	Guiñez *et al.* (2017)
	Shell length	0.731 (0.379)	
M. galloprovincialis	Total weight	(0.051 (0.027)	Nguyen *et al.* (2011)
	Shell shape	0.085 (0.038)	
	Meat yield	0.049 (0.028)	
Perna canaliculus	Shell height	0.51 (0.08)	Hess *et al.* (2018)
	Shell weight	0.92 (0.09)	
	Meat cover and gonad scores	0.21 (0.04)	

approach that combines four production-related traits: shell length, meat weight, shell weight and a semi-quantitative, visually-assessed meat condition score (Camara & Symonds 2014). Shell length is included as the simplest measure of growth rate. Meat weight has been identified by industry as a top priority. Genetic analyses (see later) have found that shell weight has an extremely high genetic correlation with shell breaking strength, which has also been identified as an economically important trait by industry. Meat condition is based on how much of the shell is covered by the meat in a halfshell presentation, as well as by plumpness – an important consumer attribute. Research is in progress to evaluate the potential for incorporating the seasonal time course of meat condition (i.e. the reproductive cycle) to mitigate the current 'winter window' when most mussels are spawned out and unsuitable for harvesting, as well as the potential for changing consumer attributes such as shell colour and high-value lipid content. Consistency of product within a growing unit (line) at the time of harvest has been suggested as an additional breeding objective.

Broodstock Establishment

It is inevitable that some of the natural variability of the species is lost when individuals are chosen as broodstock, since they represent but a small fraction of the population. In addition, over the following generations the offspring of these parents will be bred together, with possible inadvertent loss of variation through inbreeding. For example, Gurney-Smith *et al.* (2017) showed that allele frequencies at microsatellite loci did not conform to Hardy–Weinberg expectations, and substantially less genetic diversity and very low effective population size estimates (see later) were observed in farmed populations (*M. edulis*) compared to a wild reference population. Bentsen & Olesen (2002) used a series of replicated stochastic simulations to determine the effect of the number of breeders selected (4–100 pairs), the number of progeny tested (5–150 progeny per pair) and the magnitude of the heritability (0.1–0.4) on the rate of inbreeding and the response to selection through 15 generations of mass selection (see later). They found that in order to keep inbreeding rates low (about 1% per generation), a minimum of 50 pairs of breeders would need be selected and the number of progeny tested would have to be restricted and standardised to not less than 30–50 per pair; the gain by testing larger numbers was limited. Even if 50 pairs of breeders were used, there could still be an unequal contribution between males and females. The effective population size in a spawning broodstock is estimated using the following equation:

$$Ne = \left(4 * Nm * Nf\right) / \left(Nm + Nf\right)$$

where *Ne* is the effective number of spawners and *Nf* and *Nm* are the numbers of females and males, respectively. Two *Ne* estimators based on sibship reconstruction using multilocus genotype data from (e.g.) SNP markers have been developed in recent years: sibship assignment and analysis without parents (Ackerman *et al.* 2017). If there is a difference between the numbers of spawning males and females then *Ne* will vary depending on the extent of this difference. Also, bivalves have a large variance in reproductive success, reducing effective population numbers to a small fraction of breeding numbers (Boudry *et al.* 2002; Lallias *et al.* 2010). The uneven contribution of parents has the potential to rapidly decrease the *Ne* of the breeding population, which can lead to both loss of genetic diversity through genetic drift and rapidly increased levels of inbreeding. Determining the relatedness among the selected individual broodstock using microsatellite DNA genotyping can reduce the inbreeding risk (Camara *et al.* 2014; Nguyen *et al.* 2014; Morvezen *et al.* 2016).

Breeding Schemes

The breeding schemes most used in aquaculture are mass selection, family selection, within-family selection, marker-assisted selection (MAS) and, more recently, genomic selection. The choice of selection method will depend on a range of factors, including:

- target traits chosen for genetic improvement;
- feasibility of recording the trait on live animals;
- magnitude of heritability for the chosen trait(s);
- reproductive capacity of the species (Farias *et al.* 2017).

The terms 'mass' or 'individual selection', often used interchangeably, refer to selection solely based on the individual's phenotype. The best individuals, based on their phenotypic performance, are selected from the population for breeding, and the rest are discarded. This type of selection usually results in an increase in the mean of the trait in the offspring compared to the parental population (Beaumont *et al.* 2010). This can be repeated in each generation until the desired phenotypic change is achieved. The difference between the mean phenotypic value of the population and the mean phenotypic value of those

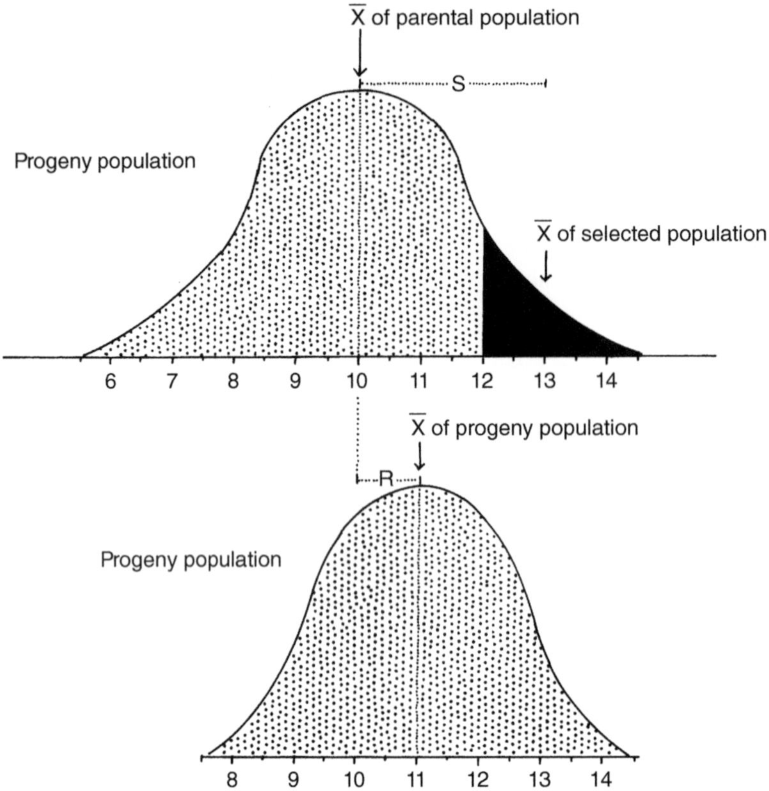

Figure 9.13 Response to individual (mass) selection. A population of fish has a mean weight of 10 g at six months of age. The adults selected for breeding (dark section of parental distribution) have a mean weight of 13 g at six months of age. The selection differential $S = 13 - 10 = 3$ g. The heritability for weight at six months of age is 0.33, so the predicted response to one generation of selection is $R = h2S = 3 \times 0.33 = 1$ g. *Source:* From Kapuschinski & Miller (2007). Reproduced with permission from Minnesota Sea Grant.

individuals selected for breeding is called the selection differential (*S*) (Figure 9.13). The response to selection (*R*) is the difference between the mean of the progeny from the selected parents and the mean of the original population. If estimated heritability is unknown, it can be estimated from the ratio of the observed selection response (*R*) to the observed selection differential (*S*) in artificial selection experiments. This relationship is summarised in the 'breeder's equation':

$$R = h^2 S$$

It is important that the heritability value for the trait under selection has been determined in the same environment and for the same population on which selection is being carried out. Otherwise, the predicted response to selection may be unreliable. Individual selection is particularly appropriate when single traits with high heritability (>0.30) are chosen, or when a population is large so that large selection differentials can be used.

When the heritability for a trait is low (<0.3), the phenotype of an individual is not an accurate estimate of the breeding value. It is therefore more efficient in terms of gain per generation to base selection on family performance rather than on individual performance. Entire families, usually groups of full or half sibs, are selected for breeding based on their mean phenotypic value. Environmental variance among family means is kept low by raising families in the same environment and by averaging over a large number of individuals in calculating the family mean. If a reasonable intensity of selection is to be achieved, and at the same time inbreeding is to be avoided, then the number of families raised and measured has to be two to four times greater than the number selected for breeding (Falconer & Mackay 1996). The efficiency of family selection is determined by the fact that environmental deviations of the individuals tend to cancel one another out in the mean value of the family. Consequently, the phenotypic mean of the family comes close to being a measure of its genotypic mean. Until recently, family selection was costly in terms of space, because each family had to be maintained separately. The use of highly variable microsatellite DNA markers (MacEvoy *et al.* 2008; Pino-Querido *et al.* 2015) and SNPs (Vera *et al.* 2010; Nguyen *et al.* 2014) allows the genetic tagging of individuals within a specific family, which means that many families may now be reared communally.

The reverse of family selection is within-family selection, where the best individuals – those that exceed the mean phenotypic value of the family to which they belong – are selected for breeding. This type of selection is useful when phenotypic differences among families are mainly due to environmental factors. If two members of every family are chosen to be parents of the next generation then the variance of family size is zero and, provided that the sexes are equal in numbers, $Ne = 2N$ (Falconer & Mackay 1996). This means that the population size Ne (the number of breeding individuals in the population) is twice the actual number, and thus the rate of inbreeding is half that under individual selection. A practical advantage of within-family selection is that fewer families need to be maintained – a big advantage in hatcheries where the cost of holding broodstock families separately can be excessively high (Beaumont *et al.* 2010). It is possible to combine family selection with within-family selection by choosing the best individuals from the best families for breeding. The advantage of combined selection is an increased response, because the additive genetic variance among individuals and within families is exploited.

The breeding value – the value of an individual in a breeding programme for a particular trait – is estimated from the animal's own records and the performance of known relatives. An animal's breeding value is expected to be twice the expected performance of its progeny. The reason for doubling the expected progeny performance is that only half of the genes from the individual are transmitted to any offspring (with the remaining half coming from the other parent). More than one trait in breeding programmes can be used to estimate the breeding value of an individual. Consequently, an individual can be given an index that is the weighted sum of its breeding value for all of the traits where performance of its offspring is known. The expected progeny performance, called the transmitting ability, is therefore half of the breeding value (Beaumont *et al.* 2010; ICBF 2013).

QTL Mapping and Marker-Assisted Selection

A QTL is a region of DNA associated with a particular phenotypic trait that displays continuous variation and polygenic inheritance. In bivalves, such traits are weight, growth rate, disease resistance, meat quality and food efficiency – traits that are important to breeders. The relative positions of QTLs on a chromosome are identified by first constructing a genetic (linkage) map for a species (see earlier) and then using it in conjunction with studies of breeding and assessment of quantitative traits to identify markers that are closely associated (linked) to QTLs of interest, thus allowing the QTLs to be positioned on the linkage map. Once QTLs for a trait are identified, individuals can be selected for breeding on the basis of marker alleles that segregate with favourable phenotypes. This is referred to as marker-assisted selection (MAS). MAS is an indirect selection method because the quantitative trait of interest is selected not on the basis of the trait itself but based on a genetic marker that is linked to it. MAS significantly increases the efficiency of breeding programmes compared to non-MAS programmes (>50% in stochastic simulations; Sonesson 2007), but this is dependent on the heritability of the trait, the proportion of genetic variance associated with marker(s) and the selection scheme used. The majority of studies on QTL detection have been on cultured fish species, with most of these being on growth-related traits, followed by disease-resistance traits (Liu & Cordes 2004; Wenne *et al.* 2007). There have been far fewer studies on cultured bivalves, although numbers are increasing. Like with fish, bivalve QTL studies (N = 10; Hollenbeck & Johnston 2018 and references therein) focus mainly on growth, shell parameters and disease-resistance traits, and almost all deal with oyster and scallop species, with no QTL studies to date on marine mussels. Therefore, a QTL study on an oyster species will be used as an example.

The European flat oyster, *Ostrea edulis*, is native to the European coasts. Since the 1970s, its European production has drastically decreased due to the emergence of two protozoan diseases: marteiliosis, caused by *Marteilia refringens*, and bonamiosis, caused by *Bonamia ostreae* (see Chapter 11). Harrang *et al.* (2015) aimed to identify QTLs in response to an infection with the parasite *B. ostreae* in two segregating families of *O. edulis*. First, they constructed a genetic linkage map for each studied family and improved the existing genetic linkage map for *O. edulis* with a set of SNP markers. This latter map now combines the best accuracy and the best estimate of the genome coverage available for an oyster species. Over a one-year period, individuals from the two families were reared in contact with wild oysters experimentally over-infected with *B. ostreae*. This infection method enabled the authors to mimic the spread of the parasite in the natural environment. They identified QTLs for two phenotypes, survival/mortality and parasite load, and found several potential QTLs that were identical between the different families, as well as new specific QTLs. Also, they detected, within the confidence interval of several QTL regions, some previously predicted candidate genes differentially expressed during an infection with *B. ostreae*, providing new candidate genomic regions that may provide valuable information and tools for MAS, accelerating the selective breeding process and removing the need for disease exposure as a selection method.

The ability to genotype many thousands of SNP loci has now made possible a different statistical approach to identifying loci associated with traits of interest. Genome-wide association studies (GWASs) identify marker–trait associations at the population rather than the family level by identifying genetic markers that are in linkage disequilibrium with QTLs. Although GWAS typically requires many more loci than linkage-based QTL mapping, it is able to provide much more precise locations of QTLs. As a result of enormous improvements in genotyping, it has begun to be applied in aquaculture, but so far for fish species only (Hollenbeck & Johnston 2018 and references therein).

Mussel Breeding Programmes

Bivalves are prime candidates for selective breeding programmes for a number of reasons: their high economic value in temperate regions of the world; increasing control over the complete life cycle of many species, particularly oysters; high levels of genetic variability, as evidenced from genetic marker studies (see earlier); and high fecundity, on the order of 10^6 eggs per female per season. Until recent times, the bulk of the research was conducted in universities and research institutes, but of late commercial concerns have begun taking charge of their own research programmes.

A selection of the ongoing bivalve breeding programmes described in the literature has been summarised by Hollenbeck & Johnston (2018), as presented in Table 9.3 (see Gosling 2015 for details). There are two large-scale, commercial selective breeding programmes for mussel species: family-based programmes for the greenshell mussel, *P. canaliculus* in New Zealand and for *M. galloprovincialis* in Victoria, Australia. The following passage, describing the former, is heavily reliant on published information in Camara & Symonds (2014).

In 2002, a breeding programme was started for *P. canaliculus* at the Cawthron Institute, Nelson, South Island, New Zealand (www.cawthron.org.nz). To develop the technology and background knowledge required to deliver genetic gains, Cawthron developed a novel high-density flow-through larval rearing system and optimised conditions within it to enable the simultaneous production of large numbers of families. Each family represented the offspring from a male and a female mussel chosen carefully based on desired characteristics. Cawthron produced its first cohort of 75 full-sib founder families derived from wild parents in 2002, and another 69 in 2003. Since then, six additional cohorts of about 50 families each have been produced, all of which consist mainly of crosses between top-performing families from earlier cohorts and a few wild families. The selection protocol for these cohorts was relatively simple: site-specific family means were calculated for each trait (shell length and meat weight), and family-level estimates of genetic merit weighted these means according to site-specific heritabilities of the traits. Beginning in 2013, when the programme was commercialised as part of a private growth partnership between the cultured mussel industry and government, Cawthron developed a more intensive and rigorous selection-index approach that combines four production-related traits: shell length, meat weight (identified by industry as a top priority), shell weight and a semi-quantitative, visually-assessed meat condition score. After two rounds of selection, shell length and meat weight show strong responses, even with the conscious decision to impose only moderate selection intensity in the early generations of the programme, as do shell weight, gonad condition and the weighted four-trait selection index, despite these traits not being selected for directly. Other traits of interest include shell shape and food conversion efficiency, and the breeding programme has recently developed a bioeconomic model to prioritise target traits and adopted the best linear unbiased prediction (BLUP) animal model for genetic evaluation in order to utilise all available phenotypic information and predict multi-trait breeding values adjusted for genetic correlations and systematic nongenetic effects. Cawthron is also devoting considerable effort to physiological-level research aimed at developing ways of measuring metabolic efficiency in order to ensure that fast growth does not lead to a reduction in farm-level carrying capacity, biochemical research to better evaluate stress responses and product quality traits such as lipid and carbohydrate content. Molecular tools to enable QTL mapping, MAS and whole-genome selection are

Table 9.3 Survey of large-scale breeding programmes for cultured bivalves. Letters in brackets refer to whether the programme is integrated with industry (I), for experimental purposes (E) or unknown (U).

Species	Location	Type of selection	Target traits	Founded	References
Mussel, *Mytilus galloprovincialis*	Australia	Family	Total weight, shell shape, meat yield	2008 (I)	Nugyen *et al.* (2011, 2012, 2014)
Mussel, *Perna canaliculus*	New Zealand	Family	Shell length, meat weight, shell weight, meat condition	1999 (I)	Camara & Symonds (2014)
Oyster, *Crassostrea gigas*	USA	Family	Shell length, shell width, wet meat weight	1996 (I)	Langdon *et al.* (2003), de Melo *et al.* (2016)
Crassostrea gigas	Australia	Family; Mass	Meat yield, growth rate, shell shape	1997 (I)	Ward *et al.* (2005), Kube *et al.* (2011)
Crassostrea gigas	New Zealand	Family	Growth rate, meat to shell ratio, shell shape, Herpes virus (OsHV-1) resilience	1999 (I)	Camara & Symonds (2014)
Crassostrea gigas	France	Mass	Survival and resistance to Herpes virus (OsHV-1) infection	2009 (E)	Dégremont *et al.* (2015)
Crassostrea gigas	China	Mass	Shell height (proxy for growth)	2007 (E)	Li *et al.* (2011), Zhong *et al.* (2016)
Oyster, *Saccostrea glomerata*	Australia	Mass	Whole weight, resistance to QX disease (*Marteilia sydneyi*) and winter mortality (*Bonamia roughleyi*)	1990 (I)	Nell *et al.* (1996, 1999); Dove *et al.* (2013)
Scallop, *Argopecten irradians*	China	Mass	Shell length	2001 (U)	Zheng *et al.* (2004, 2006)

Source: Adapted from Hollenbeck & Johnston (2018).

also being developed. Videos on mussel breeding at Cawthron are provided in Cawthron Institute (2013, 2014, 2017).

Victoria has historically dominated the blue mussel faming industry in Australia. In 2008, a joint venture project was established between Fisheries Victoria (FV) and the mussel industry group Victorian Shellfish Hatchery Pty Ltd (VSH) to develop hatchery technologies for a commercial-scale mussel seed production system at Queenscliff, Victoria. The same year, a selective breeding programme was initiated in the VSH. A genetically diverse founder population of 74 full-sib families was established by crossing parents collected from different parts of nearby Port Phillip Bay (Nguyen *et al.* 2011). These mussels were reared communally, and microsatellite markers were used to identify families. From the first-generation (G_1) population, 48 individuals of each sex were selected as broodstock based on their breeding values for total weight (TW: the weight in grams at harvest of whole closed mussel, including shell, flesh and enclosed water), meat yield (MY1: the weight of the meat divided by total weight) and shell shape (SH: the ratio of width to length), the results of which would be used to generate the second generation (G_2). Estimates for heritability of TW, SH and MY1 were low (0.051 ± 0.027, 0.085 ± 0.038 and 0.049 ± 0.028, respectively), but this reflected large environmental variation rather than limited genetic variation, suggesting a family-based breeding programme could improve these traits. Response to selection in G_1 for TW, SH and MY1 was 10, 3.0 and 2.3%, respectively (Nguyen *et al.* 2014). For G_2, a total of 77 families were generated in June 2010 (Nguyen *et al.* 2014): 43 of G_1 ♂ × G1 ♀ (SS), 5 of G_1 ♂ × wild ♀ (SW), 5 of wild ♂ × G_1 ♀ (WS) and 24 of wild ♂ × wild ♀ (WW) (Figure 9.14). Wild broodstock were included to

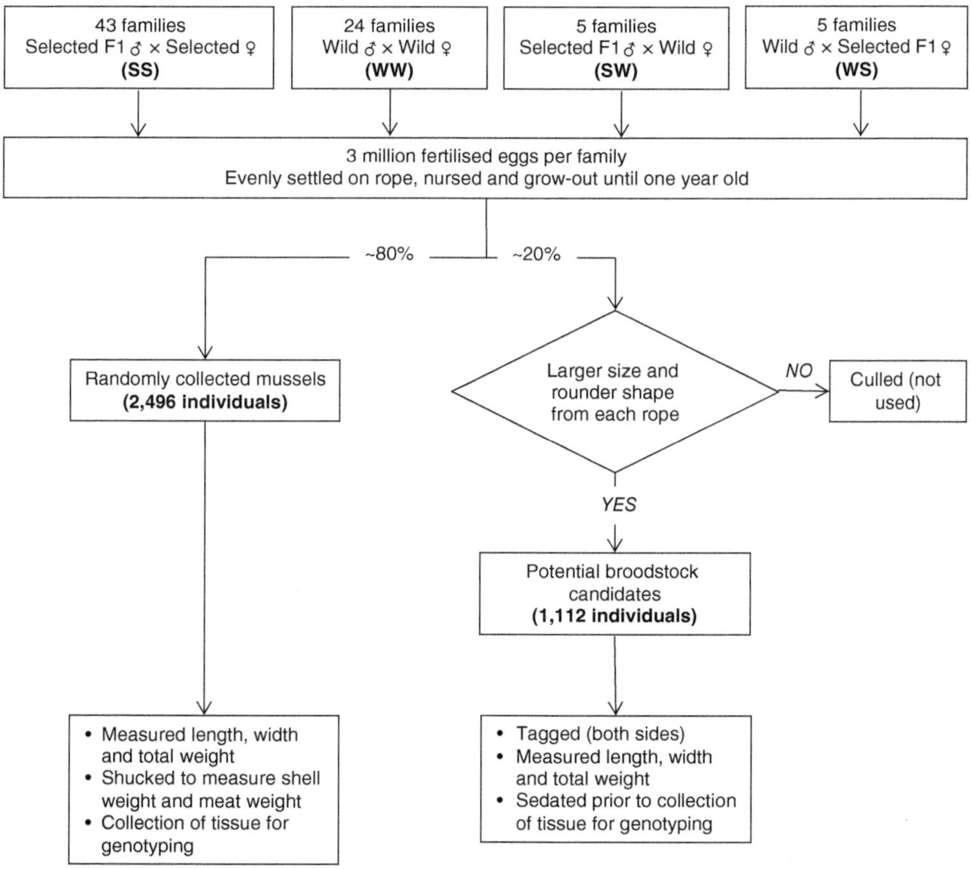

Figure 9.14 Steps involved in selection of potential broodstock and phenotypic measurements undertaken in the 2011 breeding programme for second-generation mussel (*Mytilus galloprovincialis*) at the Victorian Shellfish Hatchery Pty Ltd. (VSH), Queenscliff, Victoria, Australia. *Source:* From Nguyen *et al.* (2014). Reproduced with permission from Elsevier.

introduce new genetic material into the selected line and for comparison with families generated from selected G_1 individuals. When offspring of the 77 families reached 12 months of age, sampling was conducted to collect phenotypic data and tissue for genotyping, following the procedures in Nguyen *et al.* (2011). The same three traits, TW, SH and MY, that were investigated for G_1 were also used for G_2, and a second meat yield trait (MY2) – the ratio of meat weight to the sum of shell weight and meat weight (i.e. enclosed water removed) – was also included. Phenotypic data were collected from 3608 mussels, and tissue samples were collected for genotyping and identification of families/individuals using SNPs, which are more informative than microsatellites in parentage assignment (Nguyen *et al.* 2011). G_2 mussels ($N = 1112$) were tagged and retained as potential broodstock for the creation of the G_3 population. Estimates of heritability for the G_2 generation were higher for all traits compared to those reported in Nguyen *et al.* (2011). SH showed the greatest heritability (0.64 ± 0.10), followed by MY2 (0.46 ± 0.10), TW (0.35 ± 0.09) and MY1 (0.23 ± 0.08). This could be attributed to the increase in parentage assignment rate and in turn the number of records used to estimate genetic parameters, as well as to the presence of two generations of pedigree rather than one. Heritability of the new meat yield trait (MY2, 0.46 ± 0.10) was higher than that of MY1. MY2 is probably a better trait to include in the selection index given that it is closer to the trait consumers will experience. Selection responses were positive, with TW (10%), SH (3%) and MY1 (2.3%) values identical to those reported for G_1 (value for MY2 = 1.1%), indicating that further genetic gains can be achieved through this family-based breeding programme.

Notes

1 The transfer of genetic material from one population to another. Also called gene migration.
2 The number of individuals in a population who contribute offspring to the next generation.
3 Transfer of a DNA sequence, such as a gene, from one organism to another, followed by its replication using genetic engineering techniques.
4 A method widely used in molecular biology to make many copies of a specific DNA segment. Using PCR, copies of DNA sequences can be exponentially amplified to generate thousands to millions more.
5 Short, single strands of RNA or DNA (generally about 18–22 bases) that serve as a starting point for DNA synthesis. Primers are required for DNA replication because the enzymes that catalyse this process, DNA polymerases, can only add new nucleotides to an existing strand of DNA.
6 Combinations of cloned cDNA fragments inserted into a collection of host cells, which constitute some portion of the transcriptome of the organism and are stored as a 'library'.
7 A mechanism of evolution in which allele frequencies of a population change over generations due to chance (sampling error). Its effects are strongest in small populations.
8 A mutation or genetic change that helps an organism, such as a plant or animal, survive in its environment.
9 Sharp reductions in the size of a population due to environmental events.
10 Fixation index (F_{ST}) is a measure of population substructure. It is most useful for examining overall genetic divergence among subpopulations. F_{ST} can range from 0 to 1,

where 0 means complete sharing of genetic material and 1 means no sharing. It is frequently estimated from genetic polymorphism data, such as SNPs or microsatellites.

11 An evolutionary lineage is a temporal series of organisms, populations, cells or genes connected by a continuous line of descent from ancestor to descendent.

12 Speciation that occurs when populations of the same species become isolated from each other to an extent that prevents or interferes with gene flow. Also called allopatric speciation.

13 Phylogeny is the evolutionary development or history of a species or of a taxonomic group of organisms.

14 A clade is group of organisms consisting of a common ancestor and all its lineal descendants.

15 Sets of alleles within an organism that were inherited together from a single parent.

16 The transfer of genetic information from one species to another as a result of hybridisation.

17 Parapatric taxa evolve reproductive isolation from one another while continuing to exchange genes.

18 Linkage disequilibrium is the non-random association of alleles at linked loci in a given population.

19 HWE states that allele and genotype frequencies in a population will remain constant from generation to generation in the absence of other evolutionary influences, such as selection, mutation, gene flow, genetic drift, assortative mating, population bottleneck, founder effect or inbreeding.

20 A post-mating, pre-zygotic reproductive barrier defined as a nonrandom increase in the fertilisation success of conspecific sperm when an egg is challenged with both conspecific and heterospecific sperm simultaneously.

21 A set of linked genes at different loci on the same chromosome.

22 A codon is a sequence of three DNA or RNA nucleotides that correspond with a specific amino acid or stop signal during protein synthesis. Codon usage bias refers to the phenomenon whereby specific codons are used more often than other synonymous codons during translation of genes (Behura & Severson 2013).

23 A set of overlapping DNA segments that are related to one another by overlap of their sequences.

24 Oxidative stress is the production and accumulation of reduced oxygen intermediates such as superoxide radicals, singlet oxygen, hydrogen peroxide and hydroxyl radicals, which can damage lipids, proteins and DNA (Lesser *et al.* 2006).

References

Ab Rahim, E.S., Nguyen, T.T.T., Ingram, B., Rigino, C., Weston, K.J. & Sherman, C.D.H. (2016) Species composition and hybridisation of mussel species (Bivalvia: Mytilidae) in Australia. *Marine and Freshwater Research*, **67**, 1955–1963.

Abdul-Muneer, P.M. (2014) Application of microsatellite markers in conservation genetics and fisheries management: recent advances in population structure analysis and conservation strategies. *Genetics Research International,* **2014**, *art. 691759*. Hindawi Publishing Corporation.

Ackerman, M.W., Hand, B.K., Waples, R.K., Luikart, G., Waples, R.S., Steele, C.R. *et al.* (2017) Effective number of breeders from sibship reconstruction: empirical evaluations using hatchery steelhead. *Evolutionary Applications*, **10**, 146–160.

Adams, J. (2008) The proteome: discovering the structure and function of proteins. *Nature Education*, **1**, 6.

Addison, J.A., Ort, B.S., Mesa, K.A. & Pogson, G.H. (2008) Range-wide genetic homogeneity in the California sea mussel (*Mytilus californianus*): a comparison of allozymes, nuclear DNA markers, and mitochondrial DNA sequences. *Molecular Ecology*, **17**, 4222–4232.

Alcapán, A.C., Nespolo, R.F. & Toro, J.E. (2007) Heritability of body size in the Chilean blue mussel (*Mytilus chilensis* Hupé 1854): effects of environment and ageing. *Aquaculture Research*, **38**, 313–320.

Alonso, A., Marsal, S. & Antonio Julià, A. (2015) Analytical methods in untargeted metabolomics: state of the art in 2015. *Frontiers in Bioengineering and Biotechnology*, **3**, *art.* 23.

An, H.S. & Lee, J.W. (2012) Development of microsatellite markers for the Korean mussel, *Mytilus coruscus* (Mytilidae) using next-generation sequencing. *International Journal of Molecular Sciences*, **13**, 10583–10593.

Apraiz, I., Mi, J. & Cristobal, S. (2006) Identification of proteomic signatures of exposure to marine pollutants in mussels (*Mytilus edulis*). *Molecular Cellular Proteomics*, **5**, 1274–1285.

Apraiz, I., Cajaraville, M.P. & Cristobal, S. (2009) Peroxisomal proteomics: biomonitoring in mussels after the Prestige's oil spill. *Marine Pollution Bulletin*, **58**, 1815–1826.

Araneda, C., Larraín, M.A., Hecht, B. & Narum, S. (2016) Adaptive genetic variation distinguishes Chilean blue mussels (*Mytilus chilensis*) from different marine environments. *Ecology and Evolution*, **6**, 3632–3644.

Arivalagan, J., Yarra, T., Marie, B., Sleight, V.A., Duvernois-Berthet, E., Clark, M.S. *et al.* (2017) Insights from the shell proteome: biomineralization to adaptation. *Molecular Biology and Evolution*, **34**, 66–77.

Askari, G., Shabani, A. & Kolangi Miandare, H. (2013) Application of molecular markers in fisheries and aquaculture. *Scientific Journal of Animal Science*, **2**, 82–88.

Aslam, B., Basit, M., Nisar, M.A., Khurshid, M. & Rasool, M.H. (2017) Proteomics: technologies and their applications. *Journal of Chromatographic Science*, **55**, 182–196.

Assis, J., Zupan, M., Nicastro, K.R., Zardi, G.I., McQuaid, C.D & Serrão, E.A. (2015) Oceanographic conditions limit the spread of a marine invader along southern African shores. *PLoS One*, **10**, e0128124.

Astuya, A., Carrera, C., Ulloa, V., Aballay, A., Núñez-Acuña, G. & Hégaret, H. *et al.* (2015) Saxitoxin modulates immunological parameters and gene transcription in *Mytilus chilensis* hemocytes. *International Journal of Molecular Sciences*, **16**, 15235–15250.

Atasaral-Şahin, S., Romero, M.R., Cueto, R., González-Lavín, N., Marcos, M. & Diz, A.P. (2015) Subtle tissue and sex-dependent proteome variation in mussel (*Mytilus galloprovincialis*) populations of the Galician coast (NW Spain) raised in a common environment. *Proteomics*, **15**, 3993–4006.

Bach, L., Zbawicka, M., Strand, J. & Wenne, R. (2019) *Mytilus trossulus* in NW Greenland is genetically more similar to North Pacific than NW Atlantic populations of the species. *Marine Biodiversity*, **49**, 1053–1059.

Bai, C.M., Xin, L.S., Rosani, U., Wu, B., Wang, Q.C., Duan, X.K. *et al.* (2019) Chromosomal-level assembly of the blood clam, *Scapharca (Anadara) broughtonii*, using long sequence reads and Hi-C. *GigaScience*, **8**, 1–8.

Barco, A., Raupach, M. J., Laakmann, S., Neumann, H. & Knebelsberger, T. (2016) Identification of North Sea molluscs with DNA barcoding. *Molecular Ecology Resources*, **16**, 288–297.

Barsotti, G. & Meluzzi, C. (1968) Osservazioni su *Mytilus edulis* L. e *Mytilus galloprovincialis* Lamarck. *Conchiglie (Milan)*, **4**, 50–58.

Barton, N.H. & Hewitt, G.M. (1989) Adaptation, speciation and hybrid zones. *Nature*, **431**, 497–503.

Beaumont, A.R., Hawkins, M.P., Doig, F.L., Davies, I.M. & Snow, M. (2008) Three species of *Mytilus* and their hybrids identified in a Scottish Loch: natives, relicts and invaders? *Journal of Experimental Marine Biology and Ecology*, **367** 100–110.

Beaumont, A.R., Boudry, P. & Hoare, K. (2010) *Biotechnology and Genetics in Fisheries and Aquaculture*. 2nd edn. Wiley-Blackwell, Oxford.

Behura, S.K. & Severson, D.W. (2013) Codon usage bias: causative factors, quantification methods and genome-wide patterns: with emphasis on insect genomes. *Biological Reviews of the Cambridge Philosophical Society*, **88**, 49–61.

Bentsen, H.B. & Olesen, I. (2002) Designing aquaculture mass selection programmes to avoid high inbreeding rates. *Aquaculture*, **204**, 349–359.

Bierne, N., David, P., Langlade, A. & Bonhomme, F. (2002a) Can habitat specialisation maintain a mosaic hybrid zone in marine bivalves? *Marine Ecology Progress Series*, **245**, 157–170.

Bierne, N., David, P., Boudry, P. & Bonhomme, F. (2002b) Assortative fertilization and selection at larval stage in the mussels *Mytilus edulis* and *M. galloprovincialis*. *Evolution*, **56**, 292–298.

Bierne, N., Borsa, P., Daguin, C., Jollivet, D., Viard, F., Bonhomme, F. *et al.* (2003a) Introgression patterns in the mosaic hybrid zone between *Mytilus edulis* and *M. galloprovincialis*. *Molecular Ecology*, **12**, 447–461.

Bierne, N., Bonhomme, F. & David, P. (2003b) Habitat preference and the marine-speciation paradox. *Proceedings of the Royal Society B-Biological Science*, **270**, 1399–1406.

Bierne, N., Bonhomme, F., Boudry, P., Szulkinm M. & David, P. (2006) Fitness landscapes support the dominance theory of post-zygotic isolation in the mussels *Mytilus edulis* and *M. galloprovincialis*. *Proceedings of the Royal Society B-Biological Science*, **273**, 1253–1260.

Bjärnmark, N.A., Yarra, T., Churcher, A.M., Felix, R.C., Clark, M. & Power, D.M. (2016) Transcriptomics

provides insight into *Mytilus galloprovincialis* (Mollusca: Bivalvia) mantle function and its role in biomineralisation. *Marine Genomics*, **27**, 37–45.

Blanchette, C.A., Helmuth, B. & Gaines, S.D. (2007) Spatial patterns of growth in the mussel, *Mytilus californianus*, across a major oceanographic and biogeographic boundary at Point Conception, California, USA. *Journal of Experimental Marine Biology and Ecology*, **340**, 126–148.

Bonnefille, B., Gomez, E., Courant, F., Escande, A. & Fenet, H. (2018a) Diclofenac in the marine environment: a review of its occurrence and effects. *Marine Pollution Bulletin*, **131**, 496–506.

Bonnefille, B., Gomez, E., Alali, M., Rosain, D., Fenet, H. & Courant, F. (2018b) Metabolomics assessment of the effects of diclofenac exposure on *Mytilus galloprovincialis*: potential effects on osmoregulation and reproduction. *Science of the Total Environment*, **613–614**, 611–618.

Boore, J. L., Medina, M. & Rosenberg, L.A. (2004) Complete sequences of the highly rearranged molluscan mitochondrial genomes of the scaphopod *Graptacme eborea* and the bivalve *Mytilus edulis*. *Molecular Biology and Evolution*, **21**, 1492–1503.

Boudry, P., Collet, B., Cornette, F., Hervouet, V. & Bonhomme, F. (2002) High variance in reproductive success of the Pacific oyster (*Crassostrea gigas*, Thunberg) revealed by microsatellite-based parentage analysis of multifactorial crosses. *Aquaculture*, **204**, 283–296.

Braby, C.E & Somero, G.N. (2006a) Ecological gradients and relative abundance of native (*Mytilus trossulus*) and invasive (*Mytilus galloprovincialis*) blue mussels in the California hybrid zone. *Marine Biology*, **148**, 1249–1262.

Braby, C.E. & Somero, G.N. (2006b) Following the heart: temperature and salinity effects on heart rate in native and invasive species of the blue mussels (genus *Mytilus*). *Journal of Experimental Biology*, **209**, 2554–2566.

Brannock, P.M. & Hilbish, T.J. (2010) Hybridization results in high levels of sterility and restricted introgression between invasive and endemic marine blue mussels. *Marine Ecology Progress Series*, **406**, 161–171.

Brannock, P.M., Wethey, D.S. & Hilbish, T.J. (2009) Extensive hybridization with minimal introgression in *Mytilus galloprovincialis* and *M. trossulus* in Hokkaido, Japan. *Marine Ecology Progress Series*, **383**, 161–171.

Breusing, C., Biastoch, A., Drews, A., Metaxas, A., Jollivet, D., Vrijenhoek, R.C. *et al.* (2016) Biophysical and population genetic models predict the presence of 'phantom' stepping stones connecting Mid-Atlantic Ridge vent ecosystems. *Current Biology*, **26**, 2257–2267.

Breusing, C., Vrijenhoek, R.C. & Reusch, T.B.H. (2017) Widespread introgression in deep-sea hydrothermal vent mussels. *BMC Evolutionary Biology*, **17**,13.

Briones, C., Guiñez, R., Garrido, O., Oyarzun, P.A., Toro, J.E. & Pérez, M. (2012) Sperm polymorphism and genetic divergence in the mussel *Perumytilus purpuratus*. *Marine Biology*, **159**, 1865–1870.

Briones, C., Nuñez, J.J., Pérez, M., Espinoza-Rojas, D., Molina-Quiroz, C. & Guiñez, R. (2018) *De novo* male gonad transcriptome draft for the marine mussel *Perumytilus purpuratus* with a focus on its reproductive-related proteins. *Journal of Genomics*, **6**, 127–132.

Brooks, S.J. & Farmen, E. (2013) The distribution of the mussel *Mytilus* species along the Norwegian coast. *Journal of Shellfish Research*, **32**, 265–270.

Brown, J., Carrillo, L., Fernand, L., Horsburgh, K.J., Hill, A.E., Young, E.F. *et al.* (2003) Observations of the physical structure and seasonal jet-like circulation of the Celtic Sea and St. George's Channel of the Irish Sea. *Continental Shelf Research*, **23**, 533–561.

Buggs, R.J.A. (2007) Empirical study of hybrid zone movement. *Heredity*, **99**, 301–312.

Bulnheim, H.P. & Gosling, E.M. (1988) Population genetic structure of mussels from the Baltic Sea. *Helgolander Meeresuntersuchungen*, 42, 113–129.

Bunnik, E.M. & Le Roch, K.G. (2013) An introduction to functional genomics and systems biology. *Advances in Wound Care*, **2**, 490–498.

Bussell, J.A., Gidman, E.A., Causton, D.R., Gwynn-Jones, D., Malham, S.K., Laurence, M. *et al.* (2008) Changes in the immune response and metabolic fingerprint of the mussel, *Mytilus edulis* (Linnaeus) in response to lowered salinity and physical stress. *Journal of Experimental Marine Biology and Ecology*, **358**, 78–85.

Byers, D. (2008) Components of phenotypic variance. *Nature Education*, **1**, 161.

Cajaraville, M.P., Bebianno, M.J., Blaso, J., Porte, C., Sarasquete, C. & Viarengo, A. (2000) The use of biomarkers to assess the impact of pollution in coastal environments of the Iberian Peninsula: a practical approach. *Science of the Total Environment*, **247**, 295–311.

Calazans, C.S.H., Walters, L.J., Fernandes, F.C., Ferreira, C.E.L. & Hoffman, E.A. (2018) Genetic structure provides insights into the geographic origins and temporal change in the invasive charru mussel (Sururu) in the southeastern United States. *PLoS ONE*, **13**, e0195159.

Camara, M.D. & Symonds, J.E. (2014) Genetic improvement of New Zealand aquaculture species: programmes, progress and prospects. *New Zealand Journal of Marine and Freshwater Research*, **48**, 466–491.

Campos, A., Apraiz, I., da Fonseca, R.R. & Cristobal, S. (2015) Shotgun analysis of the marine mussel *Mytilus edulis* hemolymph proteome and mapping the innate immunity elements. *Proteomics*, **15**, 4021–4029.

Campos, A., Danielsson, G., Farinha, A.P., Kuruvilla, J., Warholm, P. & Cristobal, S. (2016) Shotgun proteomics to unravel marine mussel (*Mytilus edulis*) response to long-term exposure to low salinity and propranolol in a Baltic Sea microcosm. *Journal of Proteomics*, **137**, 97–106.

Canario, A.V.M., Bargelloni, L. Volckaert, F., Houston, R.D., Massault, C. & Guiguen, Y. (2008) Genomics toolbox for farmed fish. *Reviews in Fisheries Science*, **16S1**, 3–15.

Cao, L., Ort, B.S., Mizi, A., Pogson, G., Kenchington,E., Zouros, E. *et al.* (2009) The control region of maternally and paternally inherited mitochondrial genomes of three species of the sea mussel genus *Mytilus*. *Genetics*, **181**, 1045–1056.

Cappello, T., Mauceri, A., Corsaro, C., Maisano, M., Parrino, V., Lo Paro, G. *et al.* (2013) Impact of environmental pollution on caged mussels *Mytilus galloprovincialis* using NMR-based metabolomics. *Marine Pollution Bulletin*, **77**, 132–139.

Cappello, T., Maisano, M., Giannetto, A., Parrino, V., Mauceri, A. & Fasulo, S. (2015) Neurotoxicological effects on marine mussel *Mytilus galloprovincialis* caged at petrochemical contaminated areas (eastern Sicily, Italy): ^{1}H NMR and immunohistochemical assays. *Comparative Biochemistry and Physiology, Part-C*, **169**, 7–15.

Casillas, S. & Barbadilla, A. (2017) Molecular Population Genetics. *Genetics*, **205**, 1003–1035.

Cawthron Institute (2013) How we're breeding better Greenshell™ [video file]. Available at: https://vimeo.com/73817762 (accessed 28 May 2021).

Cawthron Institute (2014) Creating a better breed of mussel [video file]. Available at: https://vimeo.com/99091656 (accessed 28 May 2021).

Cawthron Institute (2017) Cawthron mussel breeding [video file]. Available at: https://www.youtube.com/watch?v=JPhxRC-CrBM (accessed 28 May 2021).

Chang, C., Xu, K., Guo, C., Wang, J., Yan, Q., Zhang, J. *et al.* (2018) PANDA-view: an easy-to-use tool for statistical analysis and visualization of quantitative proteomics data. *Bioinformatics*, **34**, 3594–3596.

Chauhan, T. & Rajiv, K. (2010) Molecular markers and their applications in fisheries and Aquaculture. *Advances in Bioscience and Biotechnology*, **1**, 281–291.

Chenglong, Ji., Wu, H., Wei, L., Zhaoa, J. & Yu, J. (2013) Proteomic and metabolomic analysis reveal gender-specific responses of mussel *Mytilus galloprovincialis* to 2,2',4,4'-tetrabromodip henylether (BDE 47). *Aquatic Toxicology*, **140–141**, 449–457.

Choy, C.A., Robison, B.H., Gagne, T.O., Erwin, B., Firl, E., Halden, R.U. *et al.* (2019) *Scientific Reports*, **9**, 7843.

Coghlan, B. & Gosling, E. (2007) Genetic structure of hybrid mussel populations in the west of Ireland: two hypotheses revisited. *Marine Biology*, **150**, 841–852.

Cole, V.J. & McQuaid, C.D. (2010) Bioengineers and their associated fauna respond differently to the effects of biogeography and upwelling. *Ecology*, **91**, 3549–3562.

Comesaña, A.S., Toro, J.E., Innes, D.J. & Thompson, R.J. (1999) A molecular approach to the ecology of a mussel (*Mytilis edulis–M. trossulus*) hybrid zone on the east coast of Newfoundland, Canada. *Marine Biology*, **133**, 213–221.

Comtet, T. & Desbruyères, D. (1998) Population structure and recruitment in mytilid bivalves from the Lucky Strike and Menez Gwen hydrothermal vent fields (37°17′ N and 37°50′ N on the Mid-Atlantic Ridge). *Marine Ecology Progress Series*, **163**, 165–177.

Coustau, C., Renaud, F. & Delay, B. (1991) Genetic characterization of the hybridization between *Mytilus edulis* and *M. galloprovincialis* on the Atlantic coast of France. *Marine Biology*, **111**, 87–93.

Crego-Prieto, V., Ardura, A., Juanes, F., Roca, A., Taylor, J.S. & Garcia-Vazquez, E. (2015) Aquaculture and the spread of introduced mussel genes in British Columbia. *Biological Invasions*, **17**, 2011–2026.

Cunha, R.L., Nicastro, K.R., Costa, J., McQuaid, C.D., Serrão, E.A. & Zardi, G.I. (2014) Wider sampling reveals a non-sister relationship for geographically contiguous lineages of a marine mussel. *Ecology and Evolution*, **4**, 2070–2081.

Daguin, C., Bonhomme, F. & Borsa, P. (2001) The zone of sympatry and hybridization of *Mytilus edulis* and *M. galloprovincialis*, as revealed by intron length polymorphism at locus mac-1. *Heredity*, **86**, 342–354.

Dégremont, L., Nourry, M. & Maurouard, E. (2015) Mass selection for survival and resistance to OsHV-1 infection in *Crassostrea gigas* spat in field conditions: response to selection after four generations. *Aquaculture*, **446**, 111–121.

de Melo, C. M. R., Durland, E. & Langdon, C. (2016) Improvements in desirable traits of the Pacific oyster, *Crassostrea gigas*, as a result of five generations of selection on the West Coast, USA. *Aquaculture*, **460**, 105–115.

Détrée, C., NuÂñez-Acuña, G., Roberts, S. & Gallardo-Escárate, C. (2016) Uncovering the complex transcriptome response of *Mytilus chilensis* against saxitoxin: implications of harmful algal blooms on mussel populations. *PLoS ONE*, **11**, e0165231.

Détrée, C. & Gallardo-Escárate, C. (2017) Polyethylene microbeads induce transcriptional responses with tissue-dependent patterns in the mussel *Mytilus galloprovincialis*. *Journal of Molluscan Studies*, **83**, 220–225.

Devloo-Delva, F., Miralles, L., Ardura, A., Borrell, Y.J., Pejovic, I., Tsartsianidou, V. *et al.* (2016) Detection and characterisation of the biopollutant *Xenostrobus securis* (Lamarck 1819). *Marine Pollution Bulletin*, **105**, 23–29.

Dias, J.P., Fotedar, S., Gardner, J.P.A. & Snow, M. (2013) Development of sensitive and specific molecular tools for the efficient detection and discrimination of potentially invasive mussel species of the genus *Perna*. *Management of Biological Invasions*, **4**, 155–165.

Diz, A.P. & Skibinski, D.O.F. (2007) Evolution of 2-DE protein patterns in a mussel hybrid zone. *Proteomics*, **7**, 2111–2120.

Diz, A.P. & Presa, P. (2008.) Regional patterns of microsatellite variation in *Mytilus galloprovincialis* from the Iberian Peninsula. *Marine Biology*, **154**, 277–286.

Diz, A.P. & Presa, P. (2009) The genetic diversity pattern of *Mytilus galloprovincialis* in Galician Rías (NW Iberian estuaries). *Aquaculture*, **287**, 278–285.

Diz, A.P., Dudley, E., MacDonald, B.W., Piña, B., Kenchington, E.L.R., Zouros, E. *et al.* (2009) Genetic variation underlying protein expression in eggs of the marine mussel *Mytilus edulis*. *Molecular & Cellular Proteomics*, **8**, 132–144.

Doherty S.D., Brophy, D. & Gosling, E.M (2009) Synchronous reproduction may facilitate introgression in a hybrid mussel (*Mytilus*) population. *Journal of Experimental Marine Biology and Ecology*, **378**, 1–7.

Dondero, F., Piacentini, L., Marsano, F., Rebelo, M., Vergani, L., Venier, P. *et al.* (2006) Gene transcription profiling in pollutant exposed mussels (*Mytilus* spp.) using a new low-density oligonucleotide microarray. *Gene*, **376**, 24–36.

Dong, W., Chen, Y., Lu, W., Wu, B. & Qi, P. (2017) Transcriptome analysis of *Mytilus coruscus* hemocytes in response to *Vibrio alginnolyficus* infection. *Fish and Shellfish Immunology*, **70**, 560–567.

Dove, M.C., Nell, J.A. & O'Connor, W.A. (2013) Evaluation of the progeny of the fourth-generation Sydney rock oyster *Saccostrea glomerata* (Gould, 1850) breeding lines for resistance to QX disease (*Marteilia sydneyi*) and winter mortality (*Bonamia roughleyi*). *Aquaculture Research*, **44**, 1791–1800.

Dunphy, B.J., Watts, E. & Ragg, N.L.C. (2015) Identifying thermally-stressed adult green-lipped mussels (*Perna canaliculus* Gmelin, 1791) via metabolomic profiling. *American Malacological Bulletin*, **33**, 1–9.

Dunphy, B.J., Ruggiero, K., Zamora, L. N. & Ragg, N.L.C. (2018) Metabolomic analysis of heat-hardening in adult green-lipped mussel (*Perna canaliculus*): a key role for succinic acid and the GABAergic synapse pathway. *Journal of Thermal Biology*, **74**, 37–46.

El Ayari, T., Trigui El Menif, N., Hamer, B., Cahill, A.E. & Bierne, N. (2019) The hidden side of a major marine biogeographic boundary: a wide mosaic hybrid zone at the Atlantic–Mediterranean divide reveals the complex interaction between natural and genetic barriers in mussels. *Heredity*, **122**, 770–784.

Ellis, R.P., Spicer, J.I., Byrne, J.J., Sommer U., Viant M.R., White, D.A. *et al.* (2014) ^{1}H NMR metabolomics reveals contrasting response by male and female mussels exposed to reduced seawater pH, increased temperature, and a pathogen. *Environmental Science and Technology*, **48**, 7044–7052.

Elliott, J., Holmes, K., Chambers, R., Leon, K. & Wimberger, P. (2008) Differences in morphology and habitat use among the native mussel *Mytilus trossulus*, the non-native *M. galloprovincialis*, and their hybrids in Puget Sound, Washington. *Marine Biology*, **156**, 39–53.

EI-Shehawi, AM. & Elseehy, M.M. (2017) Genome size and chromosome number relationship contradicts the principle of Darwinian evolution from common ancestor. *Journal of Phylogenetics and Evolutionary Biology*, **5**, 179.

Evans, S., Camara, M.D. & Langdon, C.J. (2009) Heritability of shell pigmentation in the Pacific oyster, *Crassostrea gigas*. *Aquaculture*, **286**, 211–216.

Falconer, D.S. & Mackay, T.F.C. (1996) *Introduction to Quantitative Genetics*, 4th edn. Longman Scientific and Technical, Harlow.

Fang, J.K., Au, D.W., Wu, R. S., Zheng, G.J., Chan, A.K., Lam, P.K. *et al.* (2009) Concentrations of polycyclic aromatic hydrocarbons and polychlorinated biphenyls in green-lipped mussel *Perna viridis* from Victoria Harbour,Hong Kong and possible human health risk. *Marine Pollution Bulletin*, **58**, 615–620.

Farias, T.F., César, J.R.D.O. & Silva, L.P.D. (2017) Methods of selection using quantitative genetics in aquaculture-a short review. *Insights in Aquaculture and Biotechnology*, **1**, 1–8.

Fasulo, S., Iacono, F., Cappello, T., Corsaro, C., Maisano, M., D'Agata, A. *et al.* (2012) Metabolomic investigation of *Mytilus galloprovincialis* (Lamarck1819) caged in aquatic environments. *Ecotoxicology and Environmental Safety*, **84**, 139–146.

Faure, B., Jollivet, D., Tanguy, A., Bonhomme, F. & Bierne, N. (2009) Speciation in the deep sea: multi-locus analysis of divergence and gene flow between two hybridizing species of hydrothermal vent mussels. *PLoS ONE*, **4**, e6485.

Faure, B., Schaeffer, S.W. & Fisher, C.R. (2015) Species distribution and population connectivity of deep-sea mussels at hydrocarbon seeps in the Gulf of Mexico. *PLoS ONE*, **10**, e0118460.

Fernández-Tajes, J., Longa, A., García-Gil, J., Chiu, Y.W., Huang, Y.S., Méndez, J. *et al.* (2011) Alternative PCR–RFLP methods for mussel *Mytilus* species identification. *European Food Research and Technology*, **233**, 791–796.

Fields, P.A., Rudomin, E.L. & Somero, G.N. (2006) Temperature sensitivities of cytosolic malate dehydrogenases from native and invasive species of marine mussels (genus *Mytilus*): sequence–function linkages and correlations with biogeographic distribution. *Journal of Experimental Biology*, **209**, 656–667.

Fraga, F. (1981) Upwelling off the Galician coast, Northwest Spain. In: *Coastal Upwelling* (ed. F.A. Richards), pp. 176–182. American Geophysical Union, Washington, DC.

Fraïsse, C., Belkhir, K., Welch, J.J. & Bierne, N. (2016) Local interspecies introgression is the main cause of extreme levels of intraspecific differentiation in mussels. *Molecular Ecology*, **25**, 269–286.

Freer, A., Bridgett, S., Jiang, J.H. & Cusack, M. (2014.) Biomineral proteins from *Mytilus edulis* mantle tissue transcriptome. *Marine Biotechnology*, **16**, 34–45.

Gaitán-Espitia, J.D., Quintero-Galvis, J.F., Mesas, A. & D'Elía, G. (2016) Mitogenomics of southern hemisphere blue mussels (Bivalvia: Pteriomorphia): insights into the evolutionary characteristics of the *Mytilus edulis* complex. *Scientific Reports*, **6**, 26853.

Gao, P., Liao, Z., Wang, X.X., Baom, L.F., Fan, M.H., Li, X.M. *et al.* (2015) Layer-by-layer proteomic analysis of *Mytilus galloprovincialis* shell. *PLoS ONE*, **10**, e0133913.

Gardner, J.P.A. (1994) The structure and dynamics of naturally occurring hybrid *Mytilus edulis* Linnaeus, 1758 and *Mytilus galloprovincialis* Lamarck, 1819 (Bivalvia, Mollusca) populations: review and interpretation. *Archiv für Hydrobiologie*, **99**, 37–71.

Gardner, J.P.A. & Skibinski, D.O.F. (1988) Historical and size dependent genetic variation in hybrid mussel populations. *Heredity*, **61**, 93–105.

Gardner, J.P.A. & Skibinski, D.O.F. (1990) Genotype-dependent fecundity and temporal variation of spawning in hybrid mussel (*Mytilus*) populations. *Marine Biology*, **105**, 153–162.

Gardner, J.P.A., Skibinski, D.O.F. & Bajdik, C.D. (1993) Shell growth and viability differences between marine mussels *Mytilus edulis* (L.), *Mytilus galloprovincialis* (Lmk.), and their hybrids from two sympatric populations in SW England. *Biological Bulletin*, **185**, 405–416.

Gebruk, A.V., Chevaldonne, P., Shank, T., Lutz, R.A. & Vrijenhoek, R.C. (2000) Deep-sea hydrothermal vent communities of the Logatchev area (14°45′ N, Mid-Atlantic Ridge): diverse biotopes and high biomass. *Journal of the Marine Biological Association of the UK*, **80**, 383–393.

Geller, J. (1999) Decline of a native mussel masked by sibling species invasion. *Conservation Biology*, **13**, 661–664.

Gérard, K., Bierne, N., Borsa, P., Chenuil, A. & Feral, J-P. (2008) Pleistocene separation of mitochondrial lineages of *Mytilus* spp. mussels from Northern and Southern hemispheres and strong genetic differentiation among southern populations. *Molecular Phylogenetics and Evolution*, **49**, 84–91.

Gerdol, M., De Moro, G., Manfrin, C., Milandri, A., Riccardi, E., Beran, A. *et al.* (2014) RNA sequencing and de novo assembly of the digestive gland transcriptome in *Mytilus galloprovincialis* fed with toxinogenic and non-toxic strains of *Alexandrium minutum*. *BMC Research Notes*, **7**, 722.

Giantsis, I.A., Mucci, N., Randi, E., Abatzopoulos, T.J. & Apostolidis, A.P. (2014) Microsatellite variation of mussels (*Mytilus galloprovincialis*) in central and eastern Mediterranean: genetic panmixia in the Aegean and the Ionian Seas. *Journal of the Marine Biological Association of the UK*, **94**, 797–809.

Gibson, G. & Muse, S.V. (2009) *A Primer of Genome Science*, 3rd edn. Sinauer Associates, Sunderland, MA.

Gilg, M.R. & Hilbish, T.J. (2003) Patterns of larval dispersal and their effects on the maintenance of a blue mussel hybrid zone in Southwest England. *Evolution*, **57**, 1061–1077.

Gilg, M.R., Kirby, S.E., Sullivan, R., Knapp, L.W. & Hilbish, T.J. (2007) Dispersal vs. retention: correspondence of species-specific reproductive cycles and settlement periods in a blue mussel hybrid zone. *Marine Ecology Progress Series*, **351**, 151–161.

Gilg, M.R., Johnson, E.G., Gobin, J., Bright, B.M. & Ortolaza, A.I (2013) Population genetics of introduced and native populations of the green mussel, *Perna viridis*: determining patterns of introduction. *Biological Invasions*, **15**, 459–472.

Gjedrem, T. (1983) Genetic variation in quantitative traits and selective breeding in fish and shellfish. *Aquaculture*, **33**, 51–72.

Gómez-Chiarri, M., Warren, W.C., Guo, X. & Proestou, D. (2015) Developing tools for the study of molluscan immunity: the sequencing of the genome of the eastern oyster, *Crassostrea virginica*. *Fish and Shellfish Immunology*, **46**, 2–4.

Gosling, E.M. (1984) The systematic status of *Mytilus galloprovincialis* in western Europe: a review. *Malacologia*, **25**, 551–568.

Gosling, E.M. (1992) *The Mussel Mytilus: Ecology, Physiology, Genetics and Culture*. Elsevier Science Publishers B.V., Amsterdam.

Gosling, E.M. (1992a) Systematics and geographic distribution of *Mytilus*. In: *The Mussel Mytilus: Ecology, Physiology, Genetics and Culture* (ed. E.M. Gosling), pp. 1–20. Elsevier Science Publishers B.V., Amsterdam.

Gosling, E.M. (1992b) Genetics of *Mytilus*. In: *The Mussel Mytilus: Ecology, Physiology, Genetics and Culture* (ed E.M. Gosling), pp. 309–382. Elsevier Science Publishers B.V., Amsterdam.

Gosling, E.M. (2003) *Bivalve Molluscs: Biology, Ecology and Culture*. Blackwell Science, Hoboken, NJ.

Gosling, E.M. (2015) *Marine Bivalve Molluscs*, 2nd edn. Wiley-Blackwell, Oxford.

Gosling, E.M. & Wilkins, N.P. (1981) Ecological genetics of the mussels *Mytilus edulis* and *Mytilus galloprovincialis* on Irish coasts. *Marine Ecology Progress Series*, **4**, 221–227.

Gosling, E.M. & McGrath, D. (1990) Genetic variability in exposed-shore mussels, *Mytilus* spp., along an environmental gradient. *Marine Biology*, **104**, 413–418.

Gosling, E.M., Doherty, S.D. & Howley, N. (2008) Genetic characterization of hybrid mussel (*Mytilus*) populations on Irish coasts. *Journal of the Marine Biological Association UK*, **88**, 341–346.

Gosnell, J.S., Macfarlan, R.J.A., Shears, N.T. & Caselle, J.E. (2014) A dynamic oceanographic front drives biogeographical structure in invertebrate settlement along Santa Cruz Island, California, USA. *Marine Ecology Progress Series*, **507**, 181–196.

Grover, A. & Sharma, P.C. (2016) Development and use of molecular markers: past and present. *Critical Reviews in Biotechnology*, **36**, 290–302.

Guiñez, R., Pita, A., Pérez, M., Briones, C., Navarrete, S.A., Toro, J. *et al.* (2016) Present-day connectivity of historical stocks of the ecosystem engineer *Perumytilus purpuratus* along 4500 km of the Chilean coast. *Fisheries Research*, **179**, 322–332.

Guiñez, R., Toro, J.E., Krapivka, S., Alcapán, A.C. & Oyarzún, P.A. (2017) Heritabilities and genetic correlation of shell thickness and shell length growth in a mussel, *Mytilus chilensis* (Bivalvia:Mytilidae). *Aquaculture Research*, **48**, 1450–1457.

Guo, X., Wang, Y., Wang, L. & Lee, J.-H. (2008) Oysters. In: *Genome Mapping and Genomics in Fishes and Aquatic Animals*, Volume 2. (eds T. D. Kocher & C. Kole), pp. 163–75. Springer-Verlag, Berlin.

Gurney-Smith, H.J., Wade, A.J. & Abbott, C.L. (2017) Species composition and genetic diversity of farmed mussels in British Columbia, Canada. *Aquaculture*, **466**, 33–40.

Harrang, E., Heurtebise, S., Faury, N., Robert, M., Arzul, I. & Lapègue, S. (2015) Can survival of European flat oysters following experimental infection with *Bonamia ostreae* be predicted using QTLs? *Aquaculture*, **448**, 521–530.

Hartnett, M., Berry, A., Tully, O. & Dabrowski, T. (2007) Investigations into the transport and pathways of scallop larvae – the use of numerical models for managing fish stocks. *Journal of Environmental Monitoring*, **9**, 403–410.

Hawley, A. (2016) RFLP and gel electrophoresis [video file]. HeritageHS. Available at: https://www.youtube.com/watch?v=VraA7Tj6AZM (accessed 28 May 2021).

Hayhurst, S. & Rawson, P.D. (2009) Species-specific variation in larval survival and patterns of distribution for the blue mussels *Mytilus edulis* and *Mytilus trossulus* in the Gulf of Maine. *Journal of Molluscan Studies*, **75**, 215–222.

He, B.S., Wang, J., Liu, J. & Hu X.M. (2017) Eco-pharmacovigilance of non-steroidal anti-inflammatory drugs: necessity and opportunities. *Chemosphere*, **181**, 178–189.

Heath, D. D., Rawson, P.D. & Hilbish, T.J. (1995) PCR-based nuclear markers identify alien blue mussel (*Mytilus* spp.) genotypes on the west coast of Canada. *Canadian Journal of Fisheries and Aquatic Sciences*, **52**, 2621–2627.

Hess, A.S., Dodds, K.G., Roberts, R.D., King, N., Symonds, J. & Clarke, S.M. (2018) Genetic parameters of Greenshell™ mussel traits under two harvest seasons. *New Zealand Journal of Animal Science and Production*, **78**, 66–70.

Hilbish, T.J., Mullinax, A., Dolven, S.I., Meyer, A., Koehn, R.K. & Rawson, P.D. (2000) Origin of the antitropical distribution pattern in marine mussels (*Mytilus* spp.): routes and timing of transequatorial migration. *Marine Biology*, **136**, 69–77.

Hilbish, T.J., Carson, E.W., Plante, J.R., Weaver, L.A. & Gilg, M.R. (2002) Distribution of *Mytilus edulis, M. galloprovincialis*, and their hybrids in open-coast populations of mussels in southwestern England. *Marine Biology*, **140**, 137–142.

Hilbish, T.J., Brannock, P.M., Jones, K.R., Smith, A.B., Bullock, B.N., Wethey, D.S. *et al.* (2010) Historical changes in the distributions of invasive and endemic marine invertebrates are contrary to global warming predictions: the effects of decadal climate oscillations. *Journal of Biogeography*, **37**, 423–431.

Hines, A., Yeung, W.H., Craft, J., Brown, M., Kennedy, J., Bignell, J. *et al.* (2007) Comparison of histological, genetic, metabolomics, and lipid-based methods for sex determination in marine mussels. *Analytical Biochemistry*, **369**, 175–186.

Hoarau, G., Rijnsdorp,A.D., Van der Veer, H.W., Stam, W.T. & Olsen, J.L. (2002) *Population structure of plaice (Pleuronectes platessa L.) in northern Europe: microsatellites revealed large-scale spatial and temporal homogeneity. Molecular Ecology*, **11**, 1165–1176.

Hofmann, G.E. & Somero, G.N. (1996) Interspecific variation in thermal denaturation of proteins in the congeneric mussels *Mytilus trossulus* and *M. galloprovincialis*: evidence from the heat-shock response and protein ubiquitination. *Marine Biology*, **126**, 65–75.

Hollenbeck, C.M. & Johnston, I.A. (2018) Genomic tools and selective breeding in molluscs. *Frontiers in Genetics*, **9**, 253.

Hu, W., Culloty, S., Darmody, G., Lynch, S., Davenport, J., Ramirez-Garcia, S. *et al.* (2014) Toxicity of copper oxide nanoparticles in the blue mussel, *Mytilus edulis*: a redox proteomic investigation. *Chemosphere*, **108**, 289–299.

Huang, L., Zou, Y., Weng, H.W., Li, H.Y., Liu, J.S. & Yang, W.D. (2015) Proteomic profile in *Perna viridis* after exposed to *Prorocentrum lima*, a dinoflagellate producing DSP toxins. *Environmental Pollution*, **196**, 350–357.

Hüning, A.K., Lange, S.M., Ramesh, K., Jacob, D.E., Jackson, D.J., Panknin, U. *et al.* (2016) A shell regeneration assay to identify biomineralization candidate genes in mytilid mussels. *Marine Genomics*, **27**, 57–67.

ICBF (2013) Glossary. Available at: https://www.icbf.com/wp/?page_id=321 (accessed 28 May 2021).

ICES (2008) Report of the Working Group on the Application of Genetics in Fisheries and Mariculture (WGAGFM). Report no. ICES CM 2008/MCC: 04, 1–4 April, Pitlochry, Scotland.

ICES (2011) *Report of the Working Group on the Application of Genetics in Fisheries and Mariculture (WGAGFM). Report no. ICES CM 2011/SSGHIE:13, 4–6 May* 2011, Bangor, Wales.

ICES (2014) Report of the Working Group on the Application of Genetics in Fisheries and Mariculture (WGAGFM). Report no. ICES CM 2014/SSGHIE:13, 7–9 May, 2014, Olhãu, Portugal.

ICES (2018) Report of the Working Group on the Application of Genetics in Fisheries and Mariculture (WGAGFM). Report no. ICES CM 2018/ASG:03, 15–17 May, 2018, Brest, France.

Inoue, K., Waite, J.H., Matsuoka, M., Odo, S. & Harayama, S. (1995) Interspecific variations in adhesive protein sequences of *Mytilus edulis, M. galloprovincialis*, and *M. trossulus. The Biological Bulletin*, **189**, 370–375.

Inoue, K., Odo, S., Noda, T., Nakao, S., Takeyama, S., Yamaha, E. *et al.* (1997) A possible hybrid zone in the *Mytilus edulis* complex in Japan revealed by PCR markers. *Marine Biology*, **128**, 91–95.

Johannesson, K. Kautsky, N. & Tedengren, M. (1990) Genotypic and phenotypic differences between Baltic and North-Sea populations of *Mytilus edulis* evaluated through reciprocal transplantations. II. Genetic variation. *Marine Ecology Progress Series*, **59**, 211–219.

Jones, O.A.H., Dondero, F., Viarengo, A. & Griffin, J.L. (2008) Metabolic profiling of *Mytilus galloprovincialis* and its potential applications for pollution assessment. *Marine Ecology Progress Series*, **369**, 169–179.

Jiggins, C.D. & Mallet, J. (2000) Bimodal hybrid zones and speciation. *Trends in Ecology and Evolution*, **15**, 250–255.

Johnson, S.B., Won, Y., Harvey, J.B.J. & Vrijenhoek, R.C. (2013) A hybrid zone between *Bathymodiolus* mussel lineages from eastern Pacific hydrothermal vents. *BMC Evolutionary Biology*, **13**, 21.

Jollivet, D. (1996) Specific and genetic diversity at deep-sea hydrothermal vents: an overview. *Biodiversity and Conservation*, **5**, 1619–1653.

Kapuscinski, A.R. & Miller, L.M. (2007) *Genetic Guidelines for Fisheries Management*, 2nd edn. Minnesota Sea Grant, University of Minnesota, Minneapolis, MN.

Katolikova, M., Khaitov, V., Väinölä, R., Gantsevich, M. & Strelkov, P. (2016) Genetic, ecological and morphological distinctness of the blue mussels *Mytilus trossulus* Gould and *M. edulis* L. in the White Sea. *PLoS ONE*, **11**, e0152963.

Kijewski, T.K., Zbawicka, M., Väinölä, R. & Wenne, R. (2006) Introgression and mitochondrial DNA heteroplasmy in the Baltic populations of mussels *Mytilus trossulus* and *M. edulis*. *Marine Biology*, **149**, 1373–1385.

Kijewski, T., Śmietanka, B., Zbawicka, M., Gosling, E., Hummel, H. & Wenne, R. (2011) Distribution of *Mytilus* taxa in European coastal areas as inferred from molecular markers. *Journal of Sea Research*, **65**, 224–234.

Kijewski, T., Zbawicka, M., Strand, J., Kautsky, H., Kotta, J., Rätsep, M. *et al.* (2019) Random forest assessment of correlation between environmental factors and genetic differentiation of populations: case of marine mussels *Mytilus*. *Oceanologia*, **61**, 131–142.

Klibansky, L.K.J. & McCartney, M.A. (2014) Conspecific sperm precedence is a reproductive barrier between free-spawning marine mussels in the northwest Atlantic *Mytilus* hybrid zone. *PLoS ONE*, **9**, e108433.

Knigge, T., Monsinjon, T. & Andersen, O.K. (2004) Surface-enhanced laser desorption/ionization-time of flight-mass spectrometry approach to biomarker discovery in blue mussels (*Mytilus edulis*) exposed to polyaromatic hydrocarbons and heavy metals under field conditions. *Proteomics*, **4**, 2722–2727.

Kruijer, W., Boer, M.P., Malosetti,M., Flood, P.J., Engel, B., Kooke, R. *et al.* (2015) Marker-based estimation of heritability in immortal populations. *Genetics*,**199**, 379–398.

Kube, P. D., Cunningham, M., Dominik, S., Parkinson, S., Henshall, J., Finn, B. *et al.* (2011) Enhancement of the Pacific Oyster Selective Breeding Program. FRDC Project 2006/227. *CSIRO Marine; Atmospheric Research*, Australian Seafood Industries P/L, Seafood CRC; Fisheries Research and Development Corporation (FRDC).

Kumar, S., Gaur, V., Khurana, S., Bose, S., Kiran, M., Kiran, M. *et al.* (2017) Proteomics tools – an update. *Clinical Oncology*, **2**, 1358.

Kwon, Y.K., Jung, Y.S., Park, J.C., Seo, J., Choi, M.S. & Hwang, G.S. (2012) Characterizing the effect of heavy metal contamination on marine mussels using metabolomics. *Marine Pollution Bulletin*, **64**, 1874–1879.

Lallias, D., Lapègue, S., Hecquet, C., Boudry, P. & Beaumont, A.R. (2007) AFLP-based genetic linkage maps of the blue mussel (*Mytilus edulis*). *Animal Genetics*, **38**, 340–349.

Lallias, D., Taris, N., Boudry, P., Bonhomme, F. & Lapegue, S. (2010) Variance in the reproductive success of flat oyster *Ostrea edulis* L. assessed by parentage analyses in natural and experimental conditions. *Genetics Research*, **92**, 175–187.

Langdon, C., Evans, F., Jacobson, D. & Blouin, M. (2003) Yields of cultured Pacific oysters *Crassostrea gigas* Thunberg improved after one generation of selection. *Aquaculture*, **220**, 227–244.

Lankadurai, B.P., Nagato, E.G. & Simpson, M.J. (2013) Environmental metabolomics: an emerging approach to study organism responses to environmental stressors. *Environmental Reviews*, **21**, 180–205.

Larraín, M.A., Díaz, N.F., Lamas, C., Uribe, C., Jilberto, F. & Araneda. C. (2015) Heterologous microsatellite-based genetic diversity in blue mussel (*Mytilus chilensis*) and differentiation among localities in southern Chile. *Latin American Journal of Aquatic Research*, **43**, 998–1010.

Larraín, M.A., Zbawicka, M., Araneda, C., Gardner, J.P. A. & Wenne, R. (2018) Native and invasive taxa on the Pacific coast of South America: impacts on aquaculture, traceability and biodiversity of blue mussels (*Mytilus* spp.). *Evolutionary Applications*, **11**, 298–311.

Larsson. J., Lind, E.E., Corell, H., Grahn, M., Smolarz, K. & Lönn, M. (2017) Regional genetic differentiation in the blue mussel from the Baltic Sea area. *Estuarine, Coastal and Shelf Science*, **195**, 98–109.

Lee T. & O'Foighil D. (2004) Hidden Floridian biodiversity: mitochondrial and nuclear gene trees reveal four cryptic species within the scorched mussel, *Brachidontes exustus*, species complex. *Molecular Ecology*, **13**, 3527–3542.

Lee T. & O'Foighil D. (2005) Placing the floridian marine genetic disjunction into regional evolutionary context using the scorched mussel, *Brachidontes exustus*, species complex. *Evolution*, **59**, 39–58.

Lesser, M.P. (2006) Oxidative stress in marine environments: biochemistry and physiological ecology. *Annual Review of Physiology*, **68**, 253–278.

Leung, P.T.Y., Wang, Y., Mak, S.S.T., Ng, W.C. & Leung, K.M.Y. (2011) Differential proteomic responses in hepatopancreas and adductor muscles of the green-lipped mussel *Perna viridis* to stresses induced by cadmium and hydrogen peroxide. *Aquatic Toxicology*, **105**, 49–61.

Li, C., Liu, X., Liu, B., Ma, B., Liu, F., Liu, G. *et al.* (2018) Draft genome of the Peruvian scallop *Argopecten purpuratus. GigaScience*, **7**, 1–6.

Li, L. & Guo, X. (2004) AFLP-based genetic linkage maps of the Pacific oyster *Crassostrea gigas* Thunberg. *Marine Biotechnology*, **6**, 26–36.

Li, Q, Wang, Q.Z., Liu, S.K. & Kong, L.F. (2011) Selection response and realized heritability for growth in three stocks of the Pacific oyster *Crassostrea gigas. Fisheries Science*, **77**, 643–648.

Li, Q., Zhao, X., Kong, L. & Yu, H. (2013) Transcriptomic response to stress in marine bivalves. *Invertebrate Survival Journal*, **10**, 84–93.

Li, X., Wu, X. & Yu, Z. (2012) Complete mitochondrial genome of the Asian green mussel *Perna viridis* (Bivalvia, Mytilidae). *Mitochondrial DNA*, **23**, 358–360.

Li, Y., Sun, X., Hu, X., Xun, X., Zhang, J., Guo, X. *et al.* (2017) Scallop genome reveals molecular adaptations to semi-sessile life and neurotoxins. *Nature Communications*, **18**, 1721.

Lin, C.Y., Viant, M.R. & TjeerJdema, R.S. (2006) Metabolomics: methodologies and applications in the environmental sciences. *Journal of Pesticide Science*, **31**, 245–251.

Liu, B.H. (1997) *Statistical Genomics: Linkage*, Mapping and QTL Analysis. CRC Press, Boca Raton, FL.

Liu, Z.J. & Cordes, J.F. (2004) DNA marker technologies and their applications in aquaculture genetics. *Aquaculture*, **238**, 1–37.

Lockwood, B.L. & Somero, G.N. (2011a) Invasive and native blue mussels (genus *Mytilus*) on the California coast: the role of physiology in a biological invasion. *Journal of Experimental Marine Biology and Ecology*, **400**, 167–174.

Lockwood, B.L. & Somero, G.N. (2011b) Transcriptomic responses to salinity stress in invasive and native blue mussels (genus *Mytilus*). *Molecular Ecology*, **20**, 517–529.

Lockwood, B.L., Sanders, J.G. & Somero, G.N. (2010) Transcriptomic responses to heat stress in invasive and native blue mussels (genus *Mytilus*): molecular correlates of invasive success. *Journal of Experimental Biology*, **213**, 3548–3558.

Lockwood, B.L., Connor, K.M. & Gracey, A.Y. (2015) The environmentally tuned transcriptomes of *Mytilus* mussels. *Journal of Experimental Biology*, **218**, 1822–1833.

López, J.L., Mosquera, E., Fuentes, J., Marina, A., Vázquez, J. & Alvarez, G. (2001) Two-dimensional gel electrophoresis of *Mytilus galloprovincialis*: differences in protein expression between intertidal and cultured mussels. *Marine Ecology Progress Series*, **224**, 149–156.

López, J.L., Marina, A., Álvarez, G. & Vázquez, J. (2002) Application of proteomics for fast identification of species-specific peptides from marine species. *Proteomics*, **2**, 1658–1665.

López, J.L., Abalde, S.L. & Fuentes, J. (2005) Proteomic approach to probe for larval proteins of the mussel *Mytilus galloprovincialis. Marine Biotechnology*, **7**, 396–404.

Lowe, A., Harris, S. & Ashton, P. (2004) *Ecological Genetics: Design, Analysis, and Application.* Blackwell Science, Oxford.

Lowe, R., Shirley, N., Bleackley, M., Dolan, S. & Shafee, T. (2017) Transcriptomics technologies. *PLoS Computational Biology*, **13**, e1005457.

Lukehurst, S.S., Dias, P.J., Huhn, M., Madduppa, H.H., Lee, S.S.C., Teo, S. *et al.* (2017) Isolation and characterization of 16 polymorphic microsatellite loci for the Asian green mussel *Perna viridis* (Mollusca, Mytilidae). *Management of Biological Invasions*, **8**, 85–88.

MacAvoy, E.S., Wood, A.R. & Gardner, J.P.A. (2008) Development and evaluation of microsatellite markers for identification of individual Greenshell™ mussels (*Perna canaliculus*) in a selective breeding programme. *Aquaculture*, **274**, 41–48.

Malachowicz, M. & Wenne, R. (2019) Mantle transcriptome sequencing of *Mytilus* spp. and identification of putative biomineralization genes. *PeerJ*, **6**, e6245.

Maldini, M., Marzano, F.N., González Fortes, G., Papa, R. & Gandolfi, G. (2006) Fish and seafood traceability based on AFLP markers: elaboration of a species database. *Aquaculture*, **261**, 487–494.

Maloy, A.P., Barber, B.J. & Rawson, P.D. (2003) Gametogenesis in a sympatric population of blue mussels, *Mytilus edulis* and *Mytilus trossulus* from Cobscook Bay (USA). *Journal of Shellfish Research*, **22**, 119–124.

Manfrin, C., Dreos, R., Battistella, S., Beran, A., Gerdol, M., Varotto, L. *et al.* (2010) Mediterranean mussel gene expression profile induced by okadaic acid exposure. *Environmental Science and Technology*, **44**, 8276–8283.

Manfrin, C., De Moro, G., Torboli, V., Venier, P., Pallavicini, A. & Gerdol, M. (2012) Physiological and molecular responses of bivalves to toxic dinoflagellates. *Invertebrate Survival Journal*, **9**, 184–199.

Maqsood, H.M. & Ahmad, S.M. (2017) Advances in molecular markers and their applications in aquaculture and fisheries. *Genaqua*, **1**, 27–41.

Marie, B., Le Roy, N., Zanella-Cléon, I., Becchi, M. & Marin, F. (2011) Molecular evolution of mollusc shell proteins: insights from proteomic analysis of the edible mussel *Mytilus. Journal of Molecular Evolution*, **72**, 531–546.

Mathiesen, S.S., Thyrring, J., Hemmer-Hansen, J., Berge, J., Sukhotin, A., Leopold, P. *et al.* (2017) Genetic diversity and connectivity within *Mytilus* spp. in the subarctic and Arctic. *Evolutionary Applications*, **10**, 39–55.

Matson, S.E., Davis, J.P. & Chew, K.K. (2003) Laboratory hybridization of the mussels, *Mytilus trossulus* and *M. galloprovincialis*: larval growth, survival and early development. *Journal of Shellfish Research*, **22**, 423–430.

Matysiak, M., Kapka-Skrzypczak, L., Brzóska, K., Gutleb, A.C. & Kruszewski, M. (2016) Proteomic approach to nanotoxicity. *Journal of Proteomics*, **137**, 35–44.

May, M.A., Bishop, K.D. & Rawson, P.D. (2017) NMR profiling of metabolites in larval and juvenile blue mussels (*Mytilus edulis*) under ambient and low salinity conditions. *Metabolites*, **7**, 33.

McCormack, J.E., Hird, S.M., Zellmer, A.J., Carstens, B.C. & Brumfield, R.T.(2013) Applications of next-generation sequencing to phylogeography and phylogenetics. *Molecular Phylogenetics and Evolution*, **66**, 526–538.

McDonagh, B. & Sheehan, D. (2006) Redox proteomics in the blue mussel *Mytilus edulis*: carbonylation is not a pre-requisite for ubiquitination in acute free radical-mediated oxidative stress. *Aquatic Toxicology*, **79**, 325–333.

McDonald, J.H., Seed, R. & Koehn, R.K. (1991) Allozymes and morphometric characters of three species of *Mytilus* in the Northern and Southern Hemispheres. *Marine Biology*, **111**, 323–333.

McDonald, J.I. (2012) Detection of the tropical mussel species *Perna viridis* in temperate Western Australia: possible association between spawning and a marine heat pulse. *Aquatic Invasions*, **7**, 483–490.

Miranda, M.B.B., Innes, D.J. & Thompson, R.J. (2010) Incomplete reproductive isolation in the blue mussel (*Mytilus edulis* and *M. trossulus*) hybrid zone in the northwest Atlantic: role of gamete interactions and larval viability. *The Biological Bulletin*, **218**, 266–281.

Mirsaeidi, M., Banoei, M.M., Winston, B.W. & Schraufnagel, D.E. (2015) Metabolomics: applications and promise in mycobacterial disease. *Annals of the American Thoracic Society*, **12**, 1278–1287.

Miyazaki, J.-I., Beppu, S., Kajio, S., Dobashi, A., Kawato, M., Fujiwara, Y. *et al.* (2013) Dispersal ability and environmental adaptability of deep-sea mussels *Bathymodiolus* (Mytilidae: Bathymodiolinae). *Open Journal of Marine Science*, **3**, 31–39.

Mizi, A., Zouros,E., Moschonas, N. & Rodakis, G.C. (2005) *The complete maternal and paternal mitochondrial genomes of the Mediterranean mussel Mytilus galloprovincialis: implications for the doubly uniparental inheritance mode of mtDNA. Molecular Biology and Evolution*, **22**, 952–967.

Monsinjon, T., Andersen, O.K., Leboulenger, F. & Knigge, T. (2006) Data processing and classification analysis of proteomic changes: a case study of oil pollution in the mussel, *Mytilus edulis*. *Proteome Science*, **4**, 17–29.

Moreau, V., Tremblay, R. & Bourget, E. (2005) Distribution of *Mytilus edulis* and *M. trossulus* on the Gaspé coast in relation to spatial scale. *Journal of Shellfish Research*, **24**, 545–551.

Moreira, R., Pereiro, P., Canchaya, C., Posada, D., Figueras, A. & Novoa, B. (2015) RNA-Seq in *Mytilus galloprovincialis*: comparative transcriptomics and expression profiles among different tissues. *BMC Genomics*, **16**, 728.

Morvezen, R., Boudry, P., Laroche, J. & Charrier, G. (2016) Stock enhancement or sea ranching? Insights from monitoring the genetic diversity, relatedness and effective population size in a seeded great scallop population (*Pecten maximus*). *Heredity*,**117**, 142–148.

Mun, S., Kim, J.Y., Markkandan, K., Shin, W., Oh, S., Woo, J. *et al.* (2017) The whole-genome and transcriptome of the Manila clam (*Ruditapes philippinarum*). *Genome Biology and Evolution*, **9**, 1487–1498.

Murgarella, M., Puiu, D., Novoa, B., Figueras, A., Posada, D. & Canchaya, C. (2016) A first insight into the genome of the filter-feeder mussel *Mytilus galloprovincialis*. *PLoS ONE*, **11**, e0151561.

Nature Education (2014) Population and quantitative genetics. Available at: https://www.nature.com/scitable/topic/population-and-quantitative-genetics-21/ (accessed 28 May 2021).

NCBI (2021) Protein database. Available at: https://www.ncbi.nlm.nih.gov/protein/ (accessed 28 May 2021).

Nell, J.A., Sheridan, A.K. & Smith, I.R. (1996) Progress in a Sydney rock oyster, *Saccostrea commercialis* (Iredale and Roughley) breeding program. *Aquaculture*, **144**, 295–302.

Nell, J.A., Smith, I.R. & Sheridan, A.K. (1999) Third generation evaluation of Sydney rock oyster *Saccostrea commercialis* (Iredale and Roughley) breeding lines. *Aquaculture*, **170**, 195–203.

Nicolè, S., Negrisolo, E., Eccher, G., Mantovani, R., Patarnello, T., Erickson, D.L. *et al.* (2012) DNA barcoding as a reliable method for the authentication of commercial seafood products. *Food Technology and Biotechnology*, **50**, 387–398.

Nguyen, T.T.T., Hayes, B.J., Guthridge, K., Ab Rahim, E. & Ingram, B.A. (2011) Use of a microsatellite-based pedigree in estimation of heritabilities for economic traits in Australian blue mussel, *Mytilus galloprovincialis*. *Journal of Animal Breeding and Genetics*, **128**, 482–490.

Nguyen, T.T.T., Hayes, B. & Ingram, B.A. (2012) Progress on selective breeding program for blue mussel in Victoria. Fisheries Victoria Research Report Series No. 60. Queenscliff, VIC.

Nguyen, T.T.T., Hayes, B.J. & Ingram, B. A. (2014) Genetic parameters and response to selection in blue mussel (*Mytilus galloprovincialis*) using a SNP-based pedigree. *Aquaculture*, **420–421**, 295–301.

NIH (2015a) Transcriptome fact sheet. Available at: https://www.genome.gov/about-genomics/fact-sheets/Transcriptome-Fact-Sheet (accessed 28 May 2021).

NIH (2015b) DNA microarray tecnology fact sheet. Available at: https://www.genome.gov/about-genomics/fact-sheets/DNA-Microarray-Technology (accessed 28 May 2021).

Núñez-Acuña, G., Aballay, A.E., Hégaret, H., Astuya, A.P. & Gallardo-Escárate, C. (2013) Transcriptional responses of *Mytilus chilensis* exposed *in vivo* to saxitoxin (STX). *Journal of Molluscan Studies*, **79**, 323–331.

Nzoughet, J.K., Hamilton, J.T., Botting, C.H., Douglas, A., Devine, L., Nelson, J. *et al.* (2009) Proteomics identification of azaspiracid toxin biomarkers in blue mussels, *Mytilus edulis*. *Molecular Cellular Proteomics*, **8**, 1811–1822.

O'Boyle, S & Silke (2010) A review of phytoplankton ecology in estuarine and coastal waters around Ireland. *Journal of Plankton Research*, **32**, 99–118.

Ojaveer, E. & Kalejs, M. (2008) On ecosystem-based regions in the Baltic Sea. *Journal of Marine Systems*, **74**, 672–685.

Ojaveer, H., Jaanus, A., MacKenzie, B.R., Martin, G., Olenin, S *et al.* (2010) Status of biodiversity in the Baltic Sea. *PLoS ONE*, 5, e12467.

Okumuş, I. & Çiftci, Y. (2003) Fish population genetics and molecular markers: II – molecular markers and their applications in fisheries and aquaculture. *Turkish Journal of Fisheries and Aquatic Sciences*, **3**, 51–79.

Olsson, B., Bradley, B.P., Gilek, M., Reimer, O., Shepard, J.L. & Tedengren, M. (2004) Physiological and proteomic responses in *Mytilus edulis* exposed to PCBs and PAHs extracted from Baltic Sea sediments. *Hydrobiologia*, **514**, 15–27.

O'Mullan, G.D., Maas, P.A.Y., Lutz, R.A. & Vrijenhoek, R.C. (2001) A hybrid zone between hydrothermal vent mussels (Bivalvia: Mytilidae) from the Mid-Atlantic Ridge. *Molecular Ecology*, **10**, 2819–2831.

Palumbi, S.R., 1994. Genetic divergence, reproductive isolation and marine speciation. *Annual Review of Ecology and Systematics*, **25**, 257–272.

Paun, O. & Schönswetter, P. (2012) Amplified Fragment Length Polymorphism (AFLP)-an invaluable fingerprinting technique for genomic, transcriptomic and epigenetic studies. *Methods in Molecular Biology*, **862**, 75–87.

Péden, R., Letendre, J., Rocher, B., Durand, F., Le Foll, F., Chan, P. *et al.* (2014) Impact of heat stress on *Mytilus edulis*: contaminants affect survival and HSP 70 proteoform expression. *Journal of Xenobiotics*, **4**, 4892.

Pejovic, I., Ardura, A., Miralles, L., Arias, A., Borrell, Y.J. & Garcia-Vazquez, E. (2016) DNA barcoding for assessment of exotic molluscs associated with maritime ports in northern Iberia. *Marine Biology Research*, **12**, 168–176.

Pickett, T. & David, A.A. (2018) Global connectivity patterns of the notoriously invasive mussel, *Mytilus galloprovincialis* Lmk using archived CO1sequence data. *BMC Research Notes*, **11**, 231.

Pino-Querido, A., Álvarez-Castro, J.M., Vera, M., Pardo, B.G., Fuentes, J. & Martínez, P. (2015) A molecular tool for parentage analysis in the Mediterranean mussel (*Mytilus galloprovincialis*). *Aquaculture Research*, **46**, 1721–1735.

Powell, D., Subramanian, S., Suwansa-ard, S., Zhao, M., O'Connor, W., Raftos, D. *et al.* (2018) The genome of the oyster *Saccostrea* offers insight into the environmental resilience of bivalves. *DNA Research*, **25**, 655–665.

Prego-Faraldo, M.V., Martínez, L. & Méndez, J. (2018) RNA-Seq analysis for assessing the early response to DSP toxins in *Mytilus galloprovincialis* digestive gland and gill. *Toxins*, **10**, 417.

Prevodnik, A., Gardeström, J., Lilja, K., Elfwing, T., McDonagh, B., Petrović, N. *et al.* (2007) Oxidative stress in response to xenobiotics in the blue mussel *Mytilus edulis* L.: evidence for variation along a natural salinity gradient of the Baltic Sea. *Aquatic Toxicology*, **82**, 63–71.

Puerto, M., Campos, A., Prieto, A., Cameána, A., de Almeida, A.M., Coelho, A.V. *et al.* (2011) Differential protein expression in two bivalve species; *Mytilus galloprovincialis* and *Corbicula fluminea*; exposed to *Cylindrospermopsis raciborskii* cells. *Aquatic Toxicology*, **101**, 109–116.

Quesada, H., Zapata, C. & Alvarez, G. (1995a) A multilocus allozyme discontinuity in the mussel *Mytilus galloprovincialis*: the interaction of ecological and life-history factors. *Marine Ecology Progress Series*, **116**, 99–115.

Quesada, H., Beynon, C.M. & Skibinski, D.O.F. (1995b) A mitochondrial DNA discontinuity in the mussel *Mytilus galloprovincialis* Lmk: Pleistocene vicariance biogeography and secondary intergradation. *Molecular Biology and Evolution*, **12**, 521–524.

Quesada, H., Wenne, R. & Skibinski, D.O.F. (1999) Interspecies transfer of female mitochondrial DNA is coupled with role-reversals and departure from neutrality in the mussel *Mytilus trossulus*. *Molecular Biology and Evolution*, **16**, 655–665.

Ratnasingham, S. & Hebert P.D.N. (2007) BOLD: the barcode of life data system (http://www.barcodinglife.org). *Molecular Ecology Notes*, **7**, 355–364.

Rawson, P.D. & Hilbish, T.J. (1998) Asymmetric introgression of mitochondrial DNA among European populations of blue mussels (*Mytilus* spp.). *Evolution*, **52**,100–108.

Rawson, P.D. & Harper, F.M. (2009) Colonization of the northwest Atlantic by the blue mussel, *Mytilus trossulus* postdates the last glacial maximum. *Marine Biology*, **156**,1857–1868.

Rawson, P.D., Joyner, K.L., Meetze, K. & Hilbish, T.J. (1996) Evidence for intragenic recombination within a novel genetic marker that distinguishes mussels in the *Mytilus edulis* species complex. *Heredity*, **77**, 599–607.

Rawson, P.D., Hayhurst, S. & Vanscoyoc, B. (2001) Species composition of blue mussel populations in the northeastern Gulf of Maine. *Journal of Shellfish Research*, **20**, 31–38.

Rawson, P.D., Slaughter, C. & Yund, P.O. (2003) Patterns of gamete Incompatibility between the blue mussels *Mytilus edulis* and *M. trossulus*. *Marine Biology*, **143**, 317–325.

Rego, I., Martínez, A., González-Tizón,A., Vieites, J., Leira, F. & Méndez, J. (2002) *PCR technique for identification of mussel species. Journal of Agricultural and Food Chemistry*, **50**, 1780–1784.

Reuter, J.A., Spacek, D. & Snyder, M.P. (2015) High-throughput sequencing technologies. *Molecular Cell*, **58**, 586–597.

Rey-Campos, M., Moreira, R., Valenzuela-Muñoz, V., Gallardo-Escárate, C., Novoa, B. & Figueras, A. (2019) High individual variability in the transcriptomic response of Mediterranean mussels to *Vibrio* reveals the involvement of myticins in tissue injury. *Scientific Reports*, **9**, 3569.

Riginos, C., Sukhdeo, K. & Cunningham, C.W. (2002) Evidence for selection at multiple allozyme loci across a mussel hybrid zone. *Molecular Biology and Evolution*, **19**, 347–351.

Riginos, C. & Cunningham, C.W. (2005) Local adaptation and species segregation in two mussel (*Mytilus edulis* x *Mytilus trossulus*) hybrid zones. *Molecular Ecology*, **14**, 381–400.

Rocher, B., Bultelle, F., Chan, P., Le Foll, F., Letendre, J., Monsinjon, T. *et al.* (2015) 2-DE mapping of the blue mussel gill proteome: the usual suspects revisited. *Proteomes*, **3**, 3–41.

Romero, M.R., Pérez-Figueroa, A., Carrera, M., Swanson, W.J., Skibinski, D.O.F. & Diz, A.P. (2019) RNA-seq coupled to proteomic analysis reveals high sperm proteome variation between two closely related marine mussel species. *Journal of Proteomics*, **192**, 169–187.

Roux, C., Fraïsse, C., Castric, V., Vekemans, X., Pogson, G.H. & Bierne, N. (2014) Can we continue to neglect genomic variation in introgression rates when inferring the history of speciation? A case study in a *Mytilus* hybrid zone. *Journal of Evolutionary Biology*, **2**, 1662–1675.

Saarman, N.P. & Pogson, G.H. (2015) Introgression between invasive and native blue mussels (genus *Mytilus*) in the central California hybrid zone. *Molecular Ecology*, **24**, 4723–4738.

Saarman, N.P., Kord, K.M., Kober, Simison, W.B. & Pogson, G.H. (2017) *Sequence-based analysis of thermal adaptation and protein energy landscapes in an invasive blue mussel (Mytilus galloprovincialis) Genome Biology and Evolution*, **9**, 2739–2751.

Sanjuan, A., Zapata, C. & Alvarez, G. (1994) *Mytilus galloprovincialis* and *M. edulis* on the coasts of the Iberian Peninsula. *Marine Ecology Progress Series*, **113**, 131–146.

Sarver, S.K. & Foltz, D.W. (1993) Genetic population structure of a species' complex of blue mussels (*Mytilus* spp.). *Marine Biology*, **117**, 105–112.

Sauvage, C., Bierne, N., Lapègue, S. & Boudry, P. (2007) Single nucleotide polymorphisms and their relationship to codon usage bias in the Pacific oyster *Crassostrea gigas*. *Gene*, **406**, 13–22.

Schmidt, W., Rainville, L.C., McEneff, G., Sheehan, D. & Quinn, B. (2014) A proteomic evaluation of the effects of the pharmaceuticals diclofenac and gemfibrozil on marine mussels (*Mytilus* spp.): evidence for chronic sublethal effects on stress-response proteins. *Drug Testing and Analysis*, **6**, 210–219.

Schneider, K. & Helmuth, B. (2007) Spatial variability in habitat temperature may drive patterns of selection between an invasive and native mussel species. *Marine Ecology Progress Series*, **339**, 157–167.

Seed R (1974) Morphological variations in *Mytilus* from the Irish coasts in relation to the occurrence and distribution of *Mytilus galloprovincialis* (Lmk). *Cahiers de Biologie Marine*, **15**, 1–25.

Selkoe, K.A. & Toonen, R.J. (2006) Microsatellites for ecologists: a practical guide to using and evaluating microsatellite markers. *Ecology Letters*, **9**, 615–629.

Serra-Compte, A., Álvarez-Muñoz, D., Solé, M., Cáceres, N., Barceló, D. & Rodríguez-Mozaz, S. (2019), Comprehensive study of sulfamethoxazole effects in marine mussels: bioconcentration, enzymatic activities and metabolomics. *Environmental Research*, **173**, 12–22.

Sheehan, D. & McDonagh, B. (2008) Oxidative stress and bivalves: a proteomic approach. *Invertebrate Survival Journal*, **5**, 110–123.

Shepard, J.L., Olsson, B., Tedengren, M. & Bradley, B.P. (2000) Protein expression signatures identified in *Mytilus edulis* exposed to PCBs, copper and salinity stress. *Marine Environmental Research*, **50**, 337–340.

Shields, J.L., Heath, J.W. & Heath, D.D. (2010) Marine landscape shapes hybrid zone in a broadcast spawning bivalve: introgression and genetic structure in Canadian west coast *Mytilus*. *Marine Ecology Progress Series*, **399**, 211–223.

Skibinski, D.O.F., Beardmore, J.A. & Cross, T.F. (1983) Aspects of the population genetics of *Mytilus* (Mytilidae; Mollusca) in the British Isles. *Biological Journal of the Linnean Society*, **19**, 137–183.

Slaughter, C., McCartney, M.A. & Yund, P.O. (2008) Comparison of gamete compatibility between two blue mussel species in sympatry and in allopatry. *The Biological Bulletin*, **214**, 57–66.

Śmietanka, B. & Burzyński, A. (2017) Complete female mitochondrial genome of *Mytilus chilensis*. *Mitochondrial DNA Part B, Resources*, **2**, 101–102.

Śmietanka, B., Zbawicka, M., Wołowicz, M. & Wenne, R. (2004) Mitochondrial DNA lineages in the European populations of mussels (*Mytilus* spp.). *Marine Biology*, **146**, 79–92.

Śmietanka, B., Burzński, A. & Wenne, R. (2009) Molecular population genetics of male and female mitochondrial genomes in European mussels *Mytilus*. *Marine Biology*, **156**, 913–925.

Śmietanka, B., Zbawicka, M., Sańko, T., Wenne, R. & Burzyński, A. (2013) Molecular population genetics of male and female mitochondrial genomes in subarctic *Mytilus trossulus*. *Marine Biology*, **160**, 1709–1721.

Śmietanka, B., Wenne, R. & Burzński, A (2016) Complete male mitochondrial genomes of European *Mytilus edulis* mussels. *Mitochondrial DNA Part A, DNA Mapping, Sequencing, and Analysis*, **27**, 1634–1635.

Śmietanka, B., Lubośny, M., Przyłucka, A., Gérard, K. & Burzyński, A. (2018) Mitogenomics of *Perumytilus purpuratus* (Bivalvia: Mytilidae) and its implications for doubly uniparental inheritance of mitochondria. *PeerJ*, **6**, e5593.

Smith, P.J., McVeagh, S.M., Won, Y. & Vrijenhoek, R.C. (2004) Genetic heterogeneity among New Zealand species of hydrothermal vent mussels (Mytilidae: Bathymodiolus). *Marine Biology*, **144**, 537–545.

Sonesson, A.K. (2007) Marker assisted selection in aquaculture breeding programmes. *Aquaculture*, **272S**, S311–S311.

Steuer, A.E., Brockbals, L. & Kraemer, T. (2019) Metabolomic strategies in biomarker research–new approach for indirect identification of drug consumption and sample manipulation in clinical and forensic toxicology? *Frontiers in Chemistry*, **7**, art. 319.

Strelkov, P., Katolikova, M. & Väinölä, R. (2017) Temporal change of the Baltic Sea–North Sea blue mussel hybrid zone over two decades. *Marine Biology*, **164**, 214.

Stuckas, H., Knöbel, L., Schade, H., Breusing, C., Hinrichsen, H.H., Bartel, M. *et al.* (2017) Combining hydrodynamic modelling with genetics: can passive larval drift shape the genetic structure of Baltic *Mytilus* populations? *Molecular Ecology*, **26**, 2765–2782.

Suárez-Ulloa, V., Fernández-Tajes, J., Aguiar-Pulido, V., Rivera-Casas, C., González-Romero, R. & Ausio, J. (2013) The CHROMEVALOA database: a resource for the evaluation of okadaic acid contamination in the marine environment based on the chromatin-associated transcriptome of the mussel *Mytilus galloprovincialis*. *Marine Drugs*, **11**, 830–841.

Suchanek, T.H., Geller, J.B., Kreiser, B.R. & Mitton, J.B. (1997) Zoogeographic distributions of the sibling species *Mytilus galloprovincialis* and *M. trossulus* (Bivalvia: Mytilidae) and their hybrids in the North Pacific. *The Biological Bulletin*, **193**, 187–194.

Sun, J., Zhang, Y., Xu, T., Zhang, Y., Mu, H., Zhang, Y. *et al.* (2017) Adaptation to deep-sea chemosynthetic environments as revealed by mussel genomes. *Nature Ecology and Evolution*, **1**, 0121.

Sunnucks, P. (2000) Efficient genetic markers for population biology. *Trends in Ecology & Evolution*, **15**, 199–203.

Tagliarolo, M., Clavier, J., Chauvaud, L., Koken, M. & Grall, J. (2012) Metabolism in blue mussel: intertidal and subtidal beds compared. *Aquatic Biology*, **17**, 167–180.

Takeuchi, T., Koyanagi, R., Gyoja, F., Kanda, M., Hisata, K., Fujie, M. *et al.* (2016) Bivalve-specific gene expansion in the pearl oyster genome: implications of adaptation to a sessile lifestyle. *Zoological Letters*, **2**, 3.

Thankakkon, S. & Patterson, E.J.K. (2013) Molecular phylogenetic analysis of two closely related marine indian mussels of genus Perna (Philipsson, 1788) based on mitochondrial (COI) and nuclear (ITS) genes. *Journal of Aquatic Biology & Fisheries*, **1**, 123–139.

Tolman, D., Wood, H.L., Skibinski, D.O.F. & Truebano, M. (2019) Differential immunity as a factor influencing mussel hybrid zone structure. *Marine Biology*, **166**, Article number 151.

Tomanek, L. (2012) Environmental proteomics of the mussel *Mytilus*: implications for tolerance to stress and change in limits of biogeographic ranges in response to climate change. *Integrative and Comparative Biology*, **52**, 648–664.

Tomanek, L. (2015) Proteomic responses to environmentally induced oxidative stress. *Journal of Experimental Biology*, **218**, 1867–1879.

Tomanek, L. & Zuzow, M.J. (2010) The proteomic response of the mussel congeners *Mytilus galloprovincialis* and *M. trossulus* to acute heat stress: implications for thermal tolerance limits and metabolic costs of thermal stress. *The Journal of Experimental Biology*, **213**, 3559–3574.

Tomanek, L., Zuzow, M.J., Hitt, L., Serafini, L. & Valenzuela, J.J. (2012) Proteomics of hyposaline stress in blue mussel congeners (genus *Mytilus*): implications for biogeographic range limits in response to climate change. *Journal of Experimental Biology*, **215**, 3905–3916.

Toro, J.E., Thompson, R.J. & Innes, D.J. (2002) Reproductive isolation and reproductive output in two sympatric mussel species (*Mytilus edulis, M. trossulus*) and their hybrids from Newfoundland. *Marine Biology*, **141**, 897–909.

Toro, J.E., Ojeda, J.A. & Vergara, A.M. (2004a) The genetic structure of *Mytilus chilensis* (Hupé 1854) populations along the Chilean coast based on RAPDs analysis. *Aquaculture Research*, **35**, 1466–71.

Toro, J.E., Innes, D.J. & Thompson, R.J. (2004b) Genetic variation among life-history stages of mussels in a *Mytilus edulis–M. trossulus* hybrid zone. *Marine Biology*, **45**, 713–725.

Toro, J.E., Ojeda, J.A., Vergara, A.M., Castro, G.C. & Alcapán, A.C. (2005) Molecular characterization of the Chilean blue mussel (*Mytilus chilensis* Hupé 1854) demonstrates evidence for the occurrence of *Mytilus galloprovincialis* in southern Chile. *Journal of Shellfish Research*, **24**, 1117–1121.

Toro, J.E., Thompson, R.J. & Innes, D.J. (2006) Fertilization success and early survival in pure and hybrid larvae of *Mytilus edulis* (Linnaeus, 1758) and *M. trossulus* (Gould, 1850) from laboratory crosses. *Aquaculture Research*, **37**, 1703–1708.

Trovant, B., Ruzzante, D.E., Basso, N.G. & Orensanz, J.M. (2013) Distinctness, phylogenetic relations and biogeography of intertidal mussels (*Brachidontes*, Mytilidae) from the south-western Atlantic. *Journal of the Marine Biological Association of the UK*, **93**, 1843–1855.

Trovant, B., Basso, N.G., Orensanz, J.M., Lessa, E.P., Dincao, F. & Ruzzante, D.E. (2016) Scorched mussels (*Brachidontes* spp., Bivalvia: Mytilidae) from the tropical and warm-temperate southwestern Atlantic: the role of the Amazon River in their speciation. *Ecology and Evolution*, **6**, 1778–1798. DIDIER

Tuffnail, W., Mills, G.A., Cary, P. & Greenwood, R. (2009) An environmental 1H NMR metabolomic study of the exposure of the marine mussel *Mytilus edulis* to atrazine, lindane, hypoxia and starvation. *Metabolomics*, **5**, 33–43.

Väinölä, R. & Hvilsom, M.M. (1991) Genetic divergence and a hybrid zone between Baltic and North Sea *Mytilus* populations (Mytilidae: Mollusca). *Biological Journal of the Linnean Society*, **43**, 127–148.

Väinölä, R. & Strelkov, P. (2011) *Mytilus trossulus* in Northern Europe. *Marine Biology*, **158**, 817–833.

Vallejo, B., Conejar-Espedido, J., Manubag, L., Artiaga, K.C.C., Damatac II, A.M., Imperial, I.C.V.J. *et al.* (2017) First record of the Charru mussel *Mytella charruana* d'Orbignyi, 1846 (Bivalvia: Mytilidae) from Manila Bay, Luzon, Philippines. *BioInvasions Records*, **6**, 49–55.

Vargas, J., Pérez, M., Toro, J. & Astorga, M.P. (2015) Presence of two mitochondrial genomes in the mytilid *Perumytilus purpuratus*: phylogenetic evidence for doubly uniparental inheritance. *Genetics and Molecular Biology*, **38**,173–181.

Varotto, L., Domeneghetti, S., Rosani, U., Manfrin., C, Cajaraville, M.P., Raccanelli, S. *et al.* (2013) DNA damage and transcriptional changes in the gills of *Mytilus galloprovincialis* exposed to nanomolar doses of combined metal salts (Cd, Cu, Hg). *PLoS ONE*, **8**, e54602.

Vera, M., Pardo, B.G., Pino-Querido, A., Álvarez-Dios, J.A., Fuentes, J. & Martínez, P. (2010) Characterization of single-nucleotide polymorphism markers in the Mediterranean mussel, *Mytilus galloprovincialis*. *Aquaculture Research*, **41**, e568–e575.

Vermeij, G.J. (1991) Anatomy of an invasion: the trans-Arctic interchange. *Paleobiology*, **17**, 281–307.

Vidal, R., Peñaloza, C., Urzúa, R. & Toro, J.E. (2009) Screening of ESTs from *Mytilus* for the detection of SSR markers in *Mytilus californianus*. *Molecular Ecology Resources*, **9**, 1409–1411.

Vieira, M.L.C., Santini, L., Diniz, A.L. & de Freitas Munhoz, C. (2016) Microsatellite markers: what they mean and why they are so useful. *Genetics and Molecular Biology*, **39**, 312–328.

Visscher, P.M., Medland, S.E., Manuel, A.R., Ferreira, M.A.R., Morley, K.I., Zhu, G. *et al.* (2006) Assumption-free estimation of heritability from genome-wide identity-by-descent sharing between full siblings. *PLoS Genetics*, **2**, e41.

Visscher, P.M., Hill, W.G. & Wray, N.R. (2008) Heritability in the genomics era–concepts and misconceptions. *Nature Reviews Genetics*, **9**, 255–266.

Ward, R.D., Thompson, P.A., Appleyard, S.A., Swan, A.A. and Kube, P.D. (2005) Sustainable genetic improvement of Pacific oysters in Tasmania and South Australia. FRDC Project 2000/206. CSIRO and Fisheries Research and Development Corporation.

Wang, J. & Maldonado, M.A. (2006) The ubiquitin-proteasome system and its role in inflammatory and autoimmune diseases. *Cellular & Molecular Immunology*, **3**, 255–261.

Wang, S., Zhang, J., Jiao, W., Li, J., Xun, X., Sun, Y. *et al.* (2017) Scallop genome provides insights into evolution of bilaterian karyotype and development. *Nature Ecology and Evolution*, **1**, 0120.

Wang, Z., Gerstein, M. & Snyder, M. (2009) RNA-Seq: a revolutionary tool for transcriptomics. *Nature Reviews Genetics*, **10**, 57–63.

Wei, K., Wood, A.R. & Gardner, J.P.A. (2013a) Population genetic variation in the New Zealand greenshell mussel: locus-dependent conflicting signals of weak structure and high gene flow balanced against pronounced structure and high self-recruitment. *Marine Biology*, **160**, 931–949.

Wei, K., Wood, A.R. & Gardner, J.P.A. (2013b) Seascape genetics of the New Zealand greenshell mussel: sea surface temperature explains macrogeographic scale genetic variation. *Marine Ecology Progress Series*, **477**, 107–121.

Wenne, R. (2018) Single nucleotide polymorphism markers with applications in aquaculture and assessment of its impact on natural populations. *Aquatic Living Resources*, **31**, art. 2.

Wenne, R., Boudry, P., Hemmer-Hansen, J., Lubienieck, K.P., Was, A. & Kause, A. (2007) What role for genomics in fisheries management and aquaculture? *Aquatic Living Resources*, **20**, 241–255.

Wenne, R., Bach, L., Zbawicka, M., Strand, J. & McDonald, J.H. (2016) A first report on coexistence and hybridization of *Mytilus trossulus* and *M. edulis* mussels in Greenland. *Polar Biology*, **39**, 343–355.

Wennerström, L., Laikre, L., Ryman, N., Utter, F.M., Ab Ghani, N.I., André, C. *et al.* (2013) Genetic biodiversity in the Baltic Sea: species-specific patterns challenge management. *Biodiversity and Conservation*, **22**, 3045–3065.

Westfall, K.M. & Gardner, J.P.A, (2010) Genetic diversity of Southern hemisphere blue mussels (Bivalvia: Mytilidae) and the identification of non-indigenous taxa. *Biological Journal of the Linnean Society*, **101**, 898–909.

Westfall, K.M., Wimberger, P.H. & Gardner, J.P.A (2010) An RFLP assay to determine if *Mytilus galloprovincialis* Lmk. (Mytilidae; Bivalvia) is of Northern or Southern hemisphere origin. *Molecular Ecology Resources*, **10**, 573–575.

Wilhelm, R. & Hilbish, T.J. (1998) Assessment of natural selection in a hybrid population of mussels: evaluation of exogenous and endogenous selection models. *Marine Biology*, **131**, 505–514.

Wilson, J, Matejusova, I., McIntosh, R.E., Carboni, S. & Bekaert, M. (2018) New diagnostic SNP molecular markers for the *Mytilus* species complex. *PLoS ONE*, **13**, e0200654.

Wiltshire, K., Rowling, K. & Deveney, M. (2010) Introduced marine species in South Australia: a review of records and distribution mapping. South Australian Research and Development Institute (Aquatic Sciences), Adelaide. SARDI Publication No. F2010/000305-1. SARDI Research Report Series No. 468, 23.

Won, Y., Hallam, S.J., O'Mullan, G.D. & Vrijenhoek. R.C. (2003a) Cytonuclear disequilibrium in a hybrid zone involving deep-sea hydrothermal vent mussels of the genus *Bathymodiolus*. *Molecular Ecology*, **12**, 3185–3190.

Won, Y., Hallam, S.J., O'Mullan, G.D., Pan, I.L., Buck, K.R. & Vrijenhoek, R.C. (2003b) Environmental acquisition of thiotrophic endosymbionts by deep-sea mussels of the genus *Bathymodiolus*. *Applied and Environmental Microbiology*, **69**, 6785–6792

Wong, Y. H., Sun, J., He, L.S., Chen, L.G., Qiu, J.W. & Qian, P.Y. (2015) High-throughput transcriptome sequencing of the cold seep mussel *Bathymodiolus platifrons*. *Scientific Reports*, **5**, 16597.

Wonham, M.J. (2004) Mini-review: distribution of the Mediterranean mussel *Mytilus galloprovincialis* (Bivalvia: Mytilidae) and hybrids in the northeast Pacific. *Journal of Shellfish Research*, **23**, 535–543.

Wood, A.R., Apte, S., MacAvoy, E.S. & Gardner, J.P.A. (2007) A molecular phylogeny of the marine mussel genus *Perna* (Bivalvia: Mytilidae) based on nuclear (ITS1&2) and mitochondrial (COI) DNA sequences. *Molecular Phylogenetics and Evolution*, **44**, 685–698.

Wray, N. & Visscher, P. (2008) Estimating trait heritability. *Nature Education*, **1**, 29.

Wu, H. & Wang, W.X. (2010) NMR-based metabolomic studies on the toxicological effects of cadmium and copper on green mussels *Perna viridis*. *Aquatic Toxicology*, **100**, 339–345.

Xing, J. & Davies, A.M. (2001) A three-dimensional baroclinic model of the Irish Sea: formation of the thermal fronts and associated circulation. *Journal of Physical Oceanography*, **31**, 94–114.

Yap, C.K., Eugene Ng, Y.J., Edward Thomas, F.B., Cheng, W.H. & Ong, G.H. (2016) The use of foot of the green-lipped mussel in *Perna viridis* as an alternative method to reduce the gender effect on the bioaccumulation of Cu and Zn in the mussel. *Annals of Limnology and Oceanography*, **1**, 22–25.

Yarra, T., Gharbi, K., Blaxter, M., Peck, L.S. & Clark, M.S. (2016) Characterization of the mantle transcriptome in bivalves: *Pecten maximus*, *Mytilus edulis* and *Crassostrea gigas*. *Marine Genomics*, **27**, 9–15.

Ye, Y., Li, J. & Wu, C. (2015) Genetic diversity and population connectivity of the Asian green mussel *Perna viridis* in South China Sea, inferred from mitochondria DNA markers. *Biochemical Systematics and Ecology*, **61**, 470–476.

Ye, Y., Senanan, W., Li, J., Cai, H.W. & Wu, C. (2016) Genetic diversity and population structure of the Asian green mussel *Perna viridis* in South China Sea based on microsatellite markers. *Biochemical Systematics and Ecology*, **68**, 135–141.

Young, T., Alfaro, A.C. & Villas-Bôas, S. (2015) Identification of candidate biomarkers for quality assessment of hatchery-reared mussel larvae via GC/MS-based metabolomics. *New Zealand Journal of Marine and Freshwater Research*, **49**, 87–95.

Yu, Z. & Guo, X. (2003) Genetic linkage map of the eastern oyster *Crassostrea virginica* Gmelin. *The Biological Bulletin*, **204**, 327–338.

Yuan, Y.S., Walters, L.J., Brodsky, S.A., Schneider, K.R. & Hoffman, E.A. (2016) Synergistic effects of salinity and temperature on the survival of two nonnative bivalve molluscs, *Perna viridis* (Linnaeus 1758) and *Mytella charruana* (d'Orbigny 1846). *Journal of Marine Biology*, **2016**, art. 9261309.

Yue, G.H. (2014) Recent advances of genome mapping and marker-assisted selection in aquaculture. *Fish and Fisheries*, **15**, 376–396.

Zardi, G., McQuaid, C., Teske, P. & Barker, N. (2007) Unexpected genetic structure of mussel populations in South Africa: indigenous *Perna perna* and invasive *Mytilus galloprovincialis*. *Marine Ecology Progress Series*, **337**,135–144.

Zardi, G.I. Nicastro, K.R., McQuaid, C.D., Castilho, R., Costa, J., Serrão, E.A. *et al.* (2015) Intraspecific genetic lineages of a marine mussel show behavioural divergence and spatial segregation over a tropical/subtropical biogeographic transition. *Evolutionary Biology*, **15**,100.

Zardi, C.D., McQuaid, C.D., Jacinto, R., Lourenço, C.R., Serrão, E.A. & Nicastro, K.R. (2018) Re-assessing the origins of the invasive mussel *Mytilus galloprovincialis* in southern Africa. *Marine and Freshwater Research*, **69**, 607–613.

Zbawicka, M., Skibinski, D.O.F. & Wenne, R. (2003) Doubly uniparental transmission of mitochondrial DNA length variants in the mussel *Mytilus trossulus*. *Marine Biology*, **142**, 455–464.

Zbawicka, M., Burzyński, A., Skibinski, D.O.F. & Wenne, R. (2010) Scottish *Mytilus trossulus* mussels retain ancestral mitochondrial DNA: complete sequences of male and female mtDNA genomes. *Gene*, **456**, 45–53.

Zbawicka, M., Drywa, A., Śmietanka, B. & Wenne, R. (2012) Identification and validation of novel SNP markers in European populations of marine *Mytilus* mussels. *Marine Biology*, **159**,1347–1362.

Zbawicka, M., Sańko, T., Strand, J. & Wenne, R. (2014) New SNP markers reveal largely concordant clinal variation across the hybrid zone between *Mytilus* spp. in the Baltic Sea. *Aquatic Biology*, **21**, 25–36.

Zhang, G., Fang, X., Guo, X., Li, L., Luo, R., Xu, F. *et al.* (2012) The oyster genome reveals stress adaptation and complexity of shell formation. *Nature*, **490**, 49–54.

Zhang, L., Sun, W., Zhang, Z., Chen, H., Jia, X. & Cai, W. (2017) Gender-specific metabolic responses in gonad of mussel *Perna viridis* to triazophos. *Marine Pollution Bulletin*, **123**, 39–46.

Zhang, X., Ruan, Z., You, X., Wang, J., Chen, J., Peng, C. *et al.* (2017) De novo assembly and comparative transcriptome analysis of the foot from Chinese green mussel (*Perna viridis*) in response to cadmium stimulation. *PLoS ONE*, **12**, e0176677.

Zhang, Y., Mu, H., Lau, S.C.K., Zhang, Z. & Qiu, J.W. (2015) Sperm proteome of *Mytilus galloprovincialis*: Insights into the evolution of fertilization proteins in marine mussels. *Proteomics*, **15**, 4175–4179.

Zheng, H., Zhang, G., Liu, X., Zhang, F. & Guo, X. (2004) Different responses to selection in two stocks of the bay scallop, *Argopecten irradians irradians* Lamarck (1819). *Journal of Experimental Marine Biology and Ecology*, **313**, 213–223.

Zheng, H., Zhang, G., Liu, X. & Guo, X. (2006) Sustained response to selection in an introduced population of the hermaphroditic bay scallop *Argopecten irradians irradians* Lamarck (1819). *Aquaculture*, **255**, 579–585.

Zhong, X., Feng, D., Yu, H., Kong, L. & Li, Q. (2016) Genetic variation and breeding signature in mass selection lines of the Pacific oyster (*Crassostrea gigas*) assessed by SNP markers. *PLoS ONE*, **11**, e0150868.

Zillén, L., Conley, D.J., Andrén, T. Andrén, E. & Björck, S. (2008) Past occurrences of hypoxia in the Baltic Sea and the role of climate variability, environmental change, and human impact. *Earth-Science Reviews*, **91**, 77–92.

10

Mussel Culture

Introduction

Bivalve culture has a long history, ranging from 350 BCE to the present day – a span of almost 3400 years. Mussel culture is said to have started in the early decades of the 13th century, when Patrick Walton, an Irish sailor who had survived the wreck of his ship on the French Atlantic coast, subsisted by catching fish using nets held by poles on the strand. He found that mussels attached themselves to the poles supporting the nets, and consequently invented the first method of mussel production, now referred to as the 'bouchot'. This technique spread widely along the French Atlantic coastline over the 19th century, while Northern European countries developed subtidal culture using on bottom culture plots (FAO 2004–2018). The rental of mussel culture plots in the early 19th century resulted from the overfishing of natural beds. This allowed a rapid growth in production that was greatly augmented about a century later by the development of hatchery techniques for seed production. However, it is only in the last five decades that the full potential of the mussel as a culture organism has been realised.

High fecundity and a mobile free-living larval phase are two characteristics that have contributed to the development of mussel culture; indeed, the natural abundance of mussel larvae has been the key for such development. The natural availability of seed sources means that there is no need to utilise hatchery production, although hatchery technology is beginning to become available (see later). Other contributory features are a rapid growth rate, tolerance to a wide range of environmental conditions and a propensity to live in dense beds in the wild, allowing mussel adaptation to the high population densities necessary for an economically viable culture system. In addition, farmed mussels are cheap to feed, and despite growing at high densities, mussels do not suffer from major diseases, in contrast to oysters and clams. Also, the ability to attach and reattach repeatedly by means of byssus threads to any firm substrate particularly equips mussels for farming (Hickman 1992).

Mussel farming extends across temperate and tropical regions in inshore environments as well as in deeper, offshore waters. In 2016, total world production of cultured mussels was 2 million metric tonnes, worth US$3.8 billion, with China producing about half of this volume. Other important producing countries were Chile and Spain, which produced about 300 000 and 216 000 t, respectively. Only about 10% of world mussel production

Marine Mussels: Ecology, Physiology, Genetics and Culture, First Edition. Elizabeth Gosling.

enters international trade. This is because the majority of bivalves are eaten in the countries where they are grown, and because they are a highly perishable product. Two main types of culture are carried out: seabed/bottom culture and suspended culture. No supplementary feed is provided, and no medicines or chemicals are administered during grow-out. The industry is still reliant on the collection of wild mussel seed for on-growing (see later).

This chapter covers the different methods used to culture mussels and the phases of the culture process: seeding, grow-out and harvesting. Topics also covered in some detail are hatchery technology, production models, legislation and management and environmental effects of culture.

Wild Seed Collection

Mussel spat or seed collection is arguably the most important aspect of mussel culture, because, without a reliable seed source, a mussel farm cannot produce consistent quantities of product on an annual basis. The basic requirement for successful seed collection in the wild is information on the reproductive cycle of the population at several locations. Knowledge of spawning, larval development, larval dispersal and the settlement process of the mussel species concerned is also essential (see Chapter 5). It should be kept in mind that several studies have suggested long-term shifts in the reproductive cycles of marine invertebrates in response to climate change (Philippart *et al.* 2012, 2014). Once larvae are released into the water column, their distribution is influenced by hydrographic factors such as water currents, prevailing wind directions and tides, as well as temperature and thermocline development and chlorophyll-*a* concentrations (Heasman & Knight 2014). With this kind of information, it is possible to forecast when and where to collect mussel spat. Accurate forecasting is particularly important when collection involves setting out poles or ropes that require prior conditioning in seawater before spat will settle. If collectors are put out too early, they may become covered with fouling organisms and hinder spat settlement. Alternatively, late placement of collectors, or failure to place them at the proper place and depth, can result in low setting success. At the best of times, forecasting is a difficult task, but knowledge of the life cycle of the species and of the local hydrographic conditions will help to improve accuracy. In temperate waters, forecasting is based primarily on plankton sampling, but in tropical regions forecasting is often not necessary because of the continuous nature of settlement.

Large-scale variability in the seasonal cycle of mussels is well documented. For example, on European coasts it has been attributed to spatial variability of water temperature, food concentration (Philippart *et al.* 2012; Arribas *et al.* 2015) and upwelling–downwelling events (Nogueira *et al.* 1997; Peteiro *et al.* 2011). In order to search for interannual variability in the settlement patterns of *Mytilus galloprovincialis*, Fuentes-Santos *et al.* (2016) conducted a five-year weekly monitoring of settlement on artificial suspended substrates. The main aim of their study was to forecast the period and intensity of larval settlement in the Ría de Ares-Betanzos, NW Spain on the basis of water temperature, food availability, wind speed and direction, continental runoff and solar irradiance. Comparison between the seasonal pattern of larval settlement and solar irradiance, as well as the well-known effect of solar irradiance on water temperature and food availability, suggest that solar irradiance indirectly influences the settlement process and support the use of this meteorological variable to predict settlement occurrence. Previous to their study, late-winter solar irradiance had been found to drive the onset of spring spawning of mussels (Aguiar *et al.* 2015). Fuentes-Santos *et al.* (2016) used two predictive models: a generalised additive model (GAM) to predict the onset and end of the settlement period, and GAM plus a functional generalised kernel additive model (FGKAM) to predict settlement intensity. They found that the fortnight mean solar irradiance was a good predictor of the probability of larval settlement a month ahead, which allowed them to predict the onset and end of the settlement period. The FGKAM model, which considers the influence of solar irradiance and continental runoff on the settlement process, provided an accurate prediction of settlement intensity a fortnight in advance. See Atalah & Forrest (2019) for a comparative study. Although at present there is a paucity of similar studies, it is anticipated that predictive models will play an increasingly important role in forecasting spat settlement.

Generally speaking, larvae settle on clean, silt-free, irregular surfaces that can range all the way from filamentous algae through stone, wood, dead shell and shells of their own species. Hence, culture operations use many different materials to collect wild seed. Mussel seed for off-bottom culture is collected on poles or stakes set into the seabed, or on ropes or mesh stockings suspended from rafts or longlines, while for on-bottom culture the seed is dredged from natural beds and transplanted to culture plots where growing conditions are more favourable (see later for details on collector structure).

Hickman (1992) produced the last comprehensive review of the mussel farming methods used in different countries. This is not surprising, since methods have hardly changed since the 1970s. Rather than present a detailed account here of the techniques used to on-grow mussels, methods will be covered later when key culture species are dealt with in detail.

Hatchery Culture of Seed and Juvenile Mussels

One of the key issues for shellfish culture is seed supply. Hatchery-produced seed was initially used during the 1970s in China to supplement wild set spat, but now there is a reliable and abundant supply in China, sufficient to sustain the current high mussel production (FAO 2004–2018; Table 10.1). Compared to the collection of natural spatfall, which is very seasonal, hatcheries guarantee the supply and maintenance of broodstock, provide the means to control maturation and spawning and supply food for the early vulnerable stages in the life cycle, thereby complementing wild seed supply (Kamermans *et al.* 2013). Also, shellfish hatcheries enable the industry to produce seed consistently and at an acceptable cost, as well as to work with hybrids and selected strains (see Chapter 9). In addition, breeding techniques can be used to produce seed for desired characteristics such as fast growth and disease resistance, particularly in the case of oysters and clams. Although the use of hatchery mussel production is not yet commonplace, the technology may also provide an option for addressing the irregular spat settlement that has affected natural populations of mussels over the last 10 years in European waters. The costs involved in setting up and running hatcheries are high, and at present very little of the seed used in culture operations actually comes from hatcheries, although future improvements in technology should make it possible to produce seed at competitive prices.

Table 10.1 Yields (live weight, metric tonnes) 2012–2016 from the main producers of cultured mussels (FAO 2018a). Countries with values <3000 t y^{-1} are excluded. E, estimated value. Method of culture: R, raft; RK rack, L, longline; B, bouchot; P, bamboo pole; BP, bamboo plots; BT, bamboo trays; OB, on bottom.

Country	Method	Species	2012	2014	2016
China	R, L	Sea mussels[a]	764 395	805 583	878 771
Chile	R, L	*M. chilensis*	244 137	238 088	300 648
Spain	**R**	***Mytilus galloprovincialis***	203 664	220 449	215 855
Thailand	B, L, R	*Perna viridis*	103 203	117 014	115 488
New Zealand	L	*Perna canaliculus*	86 447	97 438	94 037
Italy	L	*M. galloprovincialis*	63 257	63 700	63 700E
Korea	L	*M. galloprovincialis*	61 310	51 463	54 642
France	B	*Mytilus edulis*	59 781	45 386	46 500E
	L	*M. galloprovincialis*	17 358	11 807	10 400
The Netherlands	OB	*M. edulis*	40 000	54 100E	54 000E
Canada	L	*M. edulis*	28 759	24 635	23 939
United Kingdom	L, OB	*M. edulis*	26 021	20 023	10 000E
Philippines	P, BT, BP	*P. viridis*	25 660	18 775	18 775
Brazil	L	*P. perna*	21 027E	19 300E	18 200E
Greece	L	*M. galloprovincialis*	16 612	16 678	23 289
Ireland	L, OB	*M. edulis*	15 188	11 375	15 121
India	R, L, P	*P. viridis*	8703	9500E	8700E
Germany	L, OB	*M. edulis*	6933	5280	22 264
Australia	L, R	*M. galloprovincialis*	3304	3012	3428

[a] Species not specified but largely *Mytilus galloprovincialis, Perna viridis* and *Mytilus coruscus* (Tang *et al.* 2002). Values from Canada may include *M. trossulus*, while those from the United Kingdom and Ireland may include *M. galloprovincialis*. *Source:* Based on FAO (2018a) Globefish-Analysis and information on world fish trade. Available at: http://www.fao.org/in-action/globefish/market-reports/bivalves/ (accessed 28 May 2021).

Because of the unpredictability of mussel seed supply, and because recently spawned mussels cannot be sold due to insufficient meat, a number of bodies came together under the funding of an EU CRAFT project (2005–2007) to develop a reliable supply of high-quality seed in blue mussel farming. The objectives of the project, known as BLUE SEED, were to investigate how to secure a reliable supply of blue mussel seed and how to develop techniques allowing farmers to market blue mussels year-round. Producers potentially could benefit greatly from a hatchery-based technique, such as triploid induction, that produces nonmaturing mussels with a higher growth potential and which can be marketed year-round. The project involved mussel farmers/sellers from France (Grainocean), The Netherlands, Spain and Wales, AquaTT (an Irish-based network for training and technology transfer), the University of Wales at Bangor and three research institutes from Spain (CIMA), France (IFREMER) and The Netherlands (IMARES). Another

EU-funded project was REPROSEED (REsearch to improve PROduction of SEED; CORDIS 2017), which ran from 2009–2014. Its key objective was to link pure and applied approaches in order to examine complex biological processes and provide innovative technology allowing an improvement in seed production of established and emerging bivalve species in European hatcheries. See CORDIS (2011b, 2015) for the final reports of the two projects.

In 2013, there were about 40 commercial shellfish hatcheries operating along the European Atlantic coast. At that time, none of these hatcheries produced mussel seed commercially, although two Dutch companies had started hatchery production of *M. edulis* on a pilot scale (Garcia & Kamermans 2015). Since then, the number of mussel hatcheries have increased, and they are now found in Scotland (*Mytilus* spp.; Scottish Funding Council 2017), on both coasts of North America (*Mytilus* spp.), on the west (*Perna viridis*; Anil *et al.* 2017) and east (*P. viridis*; Laxmilatha *et al.* 2011) coasts of India, on the west coast of Chile (*M. chilensis*; Carrasco *et al.* 2014), in Santa Catarina state, Brazil (*P. perna*; Turini *et al.* 2014), on the west and south coasts of Australia (Shellfish Production and Technology New Zealand Ltd; SPATnz), in Nelson, New Zealand (*P. canaliculus*) and in China (Zhang 1984). The basic techniques for the production of seed in hatcheries are very similar for all bivalves, although each hatchery will invariably modify them according to local conditions and the species being cultured. Procedures include the holding, conditioning and spawning of broodstock, the rearing of larvae and juveniles and the production of large quantities of microalgae as food.

Culture of Microalgae

In most hatcheries, the incoming seawater is filtered and sterilised to remove contaminants. But this leads also to the removal of potential food sources, making it necessary to produce microalgae cultures for the larvae, spat and breeding stock. Of these, juveniles consume the largest volumes of cultured algae. Infrastructure requirements for algae production can be substantial, and operating costs can represent as much as 33% of a hatchery's production costs (Duthie 2012). The species chosen for culture chiefly depends on nutritional value and ease of culture. Cell size, cell wall structure and chemical composition are also important. The most widely used species for bivalves are the diatoms *Chaetoceros gracilis*, *Chaetoceros calcitrans*, *Thalassiosira pseudonana* and *Skeletonema costatum* and the flagellates *Tetraselmis suecica*, *Isochrysis galbana*, *Isochrysis* (T-ISO) and *Pavlova lutherii* (Helm *et al.* 2004).

Algae are grown either indoors in intensive culture under artificial light or outdoors in tanks using natural light. Indoor culture, although expensive, allows control over illumination, temperature, nutrient level and contamination by predators and competing algae. Outdoor algal systems are cheaper but make it very difficult to grow specific cultures for extended periods (Lavens & Sorgeloos 1996). Contamination with bacteria, protozoa or another species of alga is a serious problem for cultures of microalgae. The most common sources of contamination include the culture medium (seawater and nutrients), the air (from the air supply as well as the environment), the culture vessel and the starter culture. To avoid contamination, the media, containers and tubing are sterilised and cultures are frequently examined under the microscope to check their purity.

The basic methods of algal culture have changed little over the years and the various steps in the process leading to production-scale cultures are as follows. Stock cultures (≤250 ml) are maintained in specialised enriched media under controlled temperature and lighting conditions (Figure 10.1). A small volume (e.g. 20 ml) of algae from these is aseptically transferred into starter cultures (250 ml to 4 l volume) and grown quickly for 7–14 days at higher temperature and light intensity with a supply of CO_2-enriched air. Later, a small volume is used for a new starter culture, and the remainder (4–20 l volume) is used to initiate batch culture or semi-continuously (see later). This may be used as food for larvae or to start a large-scale culture (80–500 l volume; see later) maintained in tall, narrow polyethylene or fibreglass cylinders.

There are three basic types of culture method: batch, continuous and semi-continuous. Batch culture, the most common culture system, consists of a single inoculation of cells into a container of fertilised seawater followed by a growing period of several days and finally harvesting, when the algal population reaches its maximum or near-maximum density. In practice, algae are transferred to larger culture volumes prior to reaching the stationary phase of algal growth, and the larger culture volumes are then brought to a maximum density and harvested. Batch culture utilises different vessel types from 2 l flasks, through 5 and 20 l carboys and 200–500 l plastic bags, to 500 l indoor tanks and 5000–25 000 l outdoor tanks. Batch production typically delivers lower-density algae concentration compared to semi-continuous or continuous algae production systems (1500 cells μl^{-1} versus 3000–20 000 cells μl^{-1}) and has a greater risk of bacterial contamination (Duthie 2012). At the Victoria Shellfish Hatchery in Victoria, Australia, the outdoor algae system for mussel

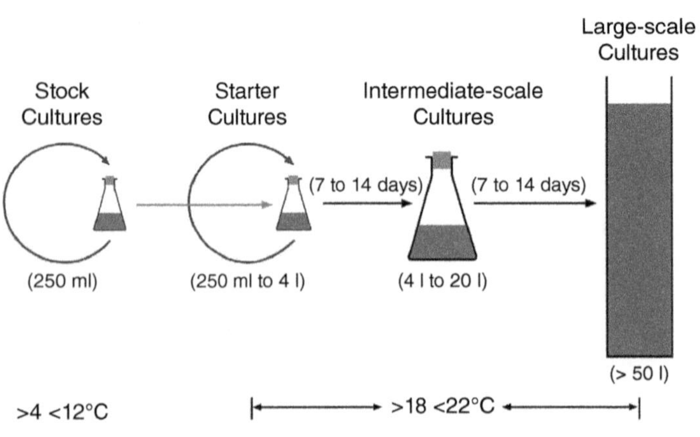

Figure 10.1 Steps in the production of algae. Stock cultures (250 ml or less) remain in isolation under light and climate control (low temperature) and are only used to inoculate starter cultures when necessary. They are not aerated, nor is carbon dioxide added. Starter cultures (250 ml to 4 l in volume) are grown quickly for 7 to 14 days at higher temperatures and light intensities with a supply of carbon dioxide-enriched air. When ready, a small portion of the volume is used to start a new starter culture and the main portion is used to begin an intermediate-scale culture. Intermediate-scale cultures (usually of between 4 l and 20 l volume) may be used as food for larvae or to start a large-scale culture. Large-scale cultures are generally of a minimum of 50 l and are frequently much greater in volume. *Source:* From Helm *et al.* (2004). Reprinted with permission from the Food and Agriculture Organization of the United Nations (FAO).

(*M. galloprovincialis*) culture is set up as a static batch culture system, consisting of 3 × 5000 l aerated polypropylene tanks, each with a halide light suspended above the water surface for lighting and heating (Figure 10.2) and a temperature range of 10–25 °C. Each outdoor algae culture tank is filled with 4000 l of either UV-sterilised or chlorinated/dechlorinated water. A modified F/2 medium, a widely used enriched seawater medium designed for growing coastal marine algae, especially diatoms, is added, and 1000 l diatom algae is introduced over a period of time. Each tank is then allowed to bloom for three days before being fed to nursery culture tanks via a magnetic drive pump and PVC feed lines over the following 24-hour period. Cultures in each tank are rotated so that one tank is being fed out while the other two are blooming. *Vibrio* is cultured from water taken from this system prior to feeding. Flagellates species are fed directly to the outdoor nursery culture tanks from the indoor algae bag culture system (Jahangard *et al.* 2010).

The continuous culture method differs from batch culture in that a supply of fertilised seawater is added to the culture at a constant rate and old media, together with some of the algae, are removed at the same rate. Therefore, the culture never runs out of nutrients. There are two types of continuous culture: turbidostat culture, in which the algal concentration is kept at a preset level by diluting the culture with fresh medium using an automatic system, and chemostat culture, in which a flow of fresh medium is introduced into the culture at a steady, predetermined rate. Chemostat culture adds a limiting vital nutrient such as nitrate at a fixed rate, thereby keeping growth rate and not cell density constant (Lavens & Sorgeloos 1996). At the Victoria Shellfish Hatchery, a continuous algal culture system is used, set up as a flow-through system. Seawater is filtered through a bank of 10 μm cartridge filters and is pasteurised at 62 °C at a flow indoor algal plastic bag culture rate of 2.1 l min^{-1} (Figure 10.3b). Water temperature in culture bags ranges between 18 and 22 °C, but can reach as high as 30 °C under abnormal culture conditions. Water is dosed at 100 l day^{-1} per bag, while F/2 medium is automatically dosed on a daily basis. Carbon dioxide-infused air is added at the base of each bag to provide the carbon source for

Figure 10.2 Outdoor algal culture tank at the Victoria Shellfish Hatchery, Victoria, Australia. *Source:* Reproduced with permission from Victoria Fisheries Authority, Australia.

(a)

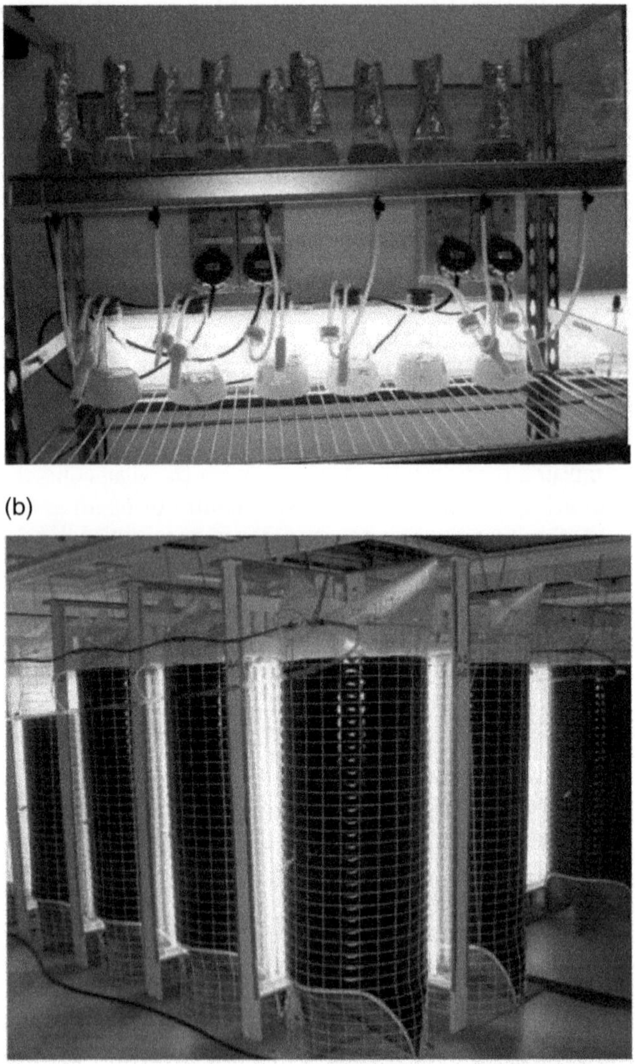

(b)

Figure 10.3 (a) Laboratory algae flask culture and (b) indoor algae bag culture system at the Victoria Shellfish Hatchery, Victoria, Australia. *Source:* From Jahangard *et al.* (2010). Reproduced with permission from Victoria Fisheries Authority, Australia.

photosynthesis and to control pH within the range 8.0–8.2. A gravity feed off a line on each bag feeds to 40 l holding flasks, keeping flagellate and diatom species separate, except in the case of the flagellate *Tetraselmis suecica*, which is kept on the diatom line. Besides *Tetraselmis suecica*, other species cultured at the Victoria Shellfish Hatchery are *Isochrysis galbana* and the diatoms *Chaetocerus muelleri* and *Thalassiosira pseudonana* Clone 3H. From the flasks, algae are distributed to either the larval culture system, the broodstock tanks or the outdoor algae tank culture system and nursery culture system (flagellates).

Bags are usually started from 5 l flasks (Figure 10.3a), but may be started from the newest bag of the species if no flasks are available. They are maintained for a period of five weeks before being replaced. The algae bag system is steam cleaned weekly, which coincides with bacterial testing for *Vibrio*. The flasks, feed lines and associated equipment are chlorinated daily. The lines feeding algae to the larval culture tanks are cleaned daily with hot water only (Jahangard *et al.* 2010). See Ragg *et al.* (2010) and Kaspar *et al.* (2014) for additional information on continuous algae culture methods for mussels (*P. canaliculus*) at the Cawthron Institute, New Zealand.

The semi-continuous technique prolongs the use of large tank cultures through partial periodic harvesting followed immediately by topping up to the original volume and supplementing with nutrients to achieve the original level of enrichment. The culture is grown up again, partially harvested and so on. Semi-continuous cultures may be kept indoors or outdoors, but usually their duration is unpredictable. Competitors, predators and/or contaminants and metabolites eventually build up, rendering the culture unsuitable for further use. Since the culture is not harvested completely, the semi-continuous method yields more algae than the batch method for a given tank size.

The relative advantages and disadvantages of the three culture systems are discussed in Lavens & Sorgeloos (1996), Helm *et al.* (2004), Andersen (2005), Duthie (2012) and Jutson *et al.* (2016).

Various methods are used to estimate the quantity of algal biomass present in cultures either by counting the number of cells or by determining volume, optical density or weight. A haemocytometer or Coulter Counter is used to measure algal cell density at the different stages in the culture process. Typical concentrations of cells in small-scale batch and semi-continuous cultures at harvest are 2000 cells μl^{-1} (*Tetraselmis suecica*) and 60 000 cells μl^{-1} (*Chaetoceros calcitrans*), respectively (Lavens & Sorgeloos 1996). A relationship between optical density and cellular concentration can be established using a spectrometer. However, because chlorophyll concentration in an algal cell varies according to the culture conditions, the relationship is affected such that a culture under low lighting conditions will be comparatively more pigmented and will eventually result in higher readings for optical density (details on methods in Jutson *et al.* 2016).

As seen earlier, microalgae are utilised as live feed for all growth stages of bivalve molluscs, but also for larval/early juvenile stages of abalone, crustaceans and some fish species and for zooplankton used in aquaculture food webs at large. It should be emphasised that the productivity of any hatchery is directly related to the quantity and quality of the food source used therein. While to date several hundred microalgal species have been tested, there are probably fewer than 20 that have become widely utilised as aquaculture feed (Guedes & Malcata 2012). Biochemical composition is a major factor determining the nutritive quality of microalgae, and it is now well established that there is a link between the presence of polyunsaturated fatty acid (PUFA), specifically eicosapentaenoic (20:5n-3, EPA) and docosahexaenoic (22:6n-3, DHA), and high nutritional value (Martínez-Fernández *et al.* 2006; Liu *et al.* 2016; González-Araya & Robert 2018). EPA and DHA are important in maintaining the functional integrity of biological membranes, while EPA and another PUFA, arachidonic acid (20:4n-6, AA), are precursors of several hormones. Therefore, it is important to take into account the overall balance between EPA, DHA and AA in conjunction with PUFA level to avoid nutritional deficiency (Pernet & Tremblay 2004;

Liu *et al.* 2016). Sterols (e.g. cholesterol) are also an essential requirement for bivalve health, as they are indispensable structural components of cell membranes (Martin-Creuzburg & von Elert 2009). As bivalves have a limited capacity to synthesise sterols, they have to acquire them from their diet, and like PUFAs, sterols vary between microalgal species (Soudant *et al.* 2000; Murphy *et al.* 2002; Tremblay *et al.* 2007). Various studies have reported beneficial effects of fatty acid (FA) content in microalgae species on conditioning in *M. galloprovincialis* broodstock (Martínez-Pita *et al.* 2016a), larval quality (*P. canaliculus*; Young *et al.* 2015), survival, growth and settlement in *M. galloprovincialis* larvae (Pettersen *et al.* 2010; Sánchez-Lazo & Martínez-Pita 2014) and growth and condition in *M. edulis* larvae (Leonardos & Lucas 2000). Also, the lipid profile of mussels themselves plays an important role in their adaptation to new conditions in their habitat, and can be used as a biochemical marker for their physiological state (Fokina *et al.* 2017). Lipid emulsions or spheres may be added as a supplement to live algae for broodstock conditioning, larvae and spat (Coutteau *et al.* 1996; Soudant *et al.* 2000; Caers *et al.* 1998; Hendricks *et al.* 2003 and references therein).

The whole area of larval nutrition is complex in that the nutritional value of a specific microalga can depend on the bivalve consuming it and the particular growth stage of that bivalve. Also, unialgal diets produce good growth and survival in some species, but others require a mixture of two or more species (see Ponis *et al.* 2006; Marshall *et al.* 2010 for review). There is evidence that the FA profile of a species changes at lag, exponential and stationary growth phases (Nalder *et al.* 2015) and under different culture conditions (Yodsuwan *et al.* 2017). To illustrate, when comparing the relative FA compositions in three *Phaeodactylum tricornutum* strains from outdoor cultures with laboratory experiments, stronger differences were observed between outdoor and laboratory-grown cultures than between strains. In one strain, a significantly higher EPA productivity was only observed under outdoor conditions, and not in laboratory experiments (Steinrucken *et al.* 2018).

Ongoing maintenance of microalga cultures is expensive. However, routine cryopreservation of strains at the Cawthron Institute, Nelson, New Zealand (www.cawthron.org.nz) makes this process more cost-effective. There is constant risk of strains becoming contaminated or lost, and cryopreservation is proving to be cleaner and safer than live maintenance. Also, the Cawthron Institute cryopreserves and stores algal strains of commercial value for clients who have identified strains that produce bioactive compounds of interest (reviewed in Blackburn & Lee-Chang 2018; Khan *et al.* 2018).

Replacement/Supplement Diets for Microalgae

As already mentioned, microalgae culture is an expensive process, representing about 40% of the costs of rearing bivalve seed to a shell length of about 5 mm in a hatchery (Helm *et al.* 2004). Costs can be even higher if special expertise, manpower and devoted facilities are included. Therefore, there is considerable interest in the development of suitable artificial diets to replace or supplement microalgae. Alternatives include dried or concentrated microalgae, microencapsulated diets, lipid emulsions, bacteria and yeast cultures, all of which have been extensively studied (reviewed by Knauer & Southgate 1999). Concentrating algal cells from mass cultures produces microalgal concentrates. McCausland *et al.* (1999) found that when concentrates of the diatoms *Skeletonema costatum* and *Chaetoceros calcitrans* were fed to juvenile oysters (*Crassostrea gigas*), they were as effective as live supplementary diets. Several methods have been used to maintain concentrate quality during storage, such as adding additives or preservatives, freezing and refrigerating. The microalgae concentrate may be stored for several days, several weeks or several months (Aji 2011), although high-speed centrifuged concentrates of *Tetraselmis* spp. and *Chaetoceros calitrans* showed very good retention of nutritional quality beyond eight weeks' storage. When using concentrates of *Pavlovi lutheri*, Tahitian *Isocrysis* and *Chaetoceros muelleri*, however, there were rapid and profound losses in nutritional quality (Heasman *et al.* 2000). The limited shelf-life and high price of the presently available concentrates have discouraged many growers from using them. Results from feeding trials on live and concentrated algal species showed that *Tetraselmis suecica* was the best suited as a preserved feed because dried and live *T. suecica* produced similar growth rates when fed to larval clams (*Ruditapes philippinarum*). Also, cold-stored *T. suecica* produced very similar growth rates to fresh *T. suecica* when fed in combination with live feed to larval oysters (*Crassostrea gigas*; Marshall *et al.* 2010 and references therein). In a more recent study, Welladsen *et al.* (2014) tested the effect of storage conditions (refrigerator and freezer) on the FA and amino acid profiles of four concentrates of microalgae: *Nannochloropsis* sp., *Dunaliella tertiolecta* and the diatoms *Entomoneis punctulata* and *Melosira dubia*. The most stable products were the first two, which showed no significant change in FA profile or protein content after two months' storage in the refrigerator. The two diatoms, however, were not as robust. *M. dubia* showed a steady decline in FA content over the course of 30 days, with a marked decrease at 60 days. And although the FA content of *E. punctulata* was stable for the first 30 days of storage, a significant decline – particularly in omega-3 content – was seen at 30–60 day in both the refrigerator and the freezer. The total protein content, however, was stable in both species throughout the course of the experiment. These findings suggest that *Nannochloropsis* sp. and *D. tertiolecta* concentrates could be used in aquaculture industries as a replacement or supplement to fresh cultures, while *E. punctulata* may be more suited to short-term storage and *M. dubia* would perform better as a fresh culture. Reed Mariculture (www.reedmariculture.com), a Californian company established in 1995, is the world's largest producer of marine microalgae concentrates. Its Shellfish Diet 1800 larviculture feeds are used by over 500 hatcheries, universities and marine ornamental operations in more than 70 countries around the world. The concentrate is a unique mix of six macroalgae species: *Isochrysis*, *Pavlova*, *Tetraselmis*, *Chaetoceros calcitrans*, *Thalassiosira weissflogii* and *Thalassiosira pseudonana* – species that have all demonstrated success with a variety of shellfish, including oysters, clams, mussels and scallops. This mixed diet provides excellent nutrition for bivalve species, from first-feeding larvae all the way up to broodstock, increasing both growth rates and survival. Other products, such as Instant Algae, have long shelf-lives so they are always available when you need them – there is no need to worry about timing phytoplankton production to coincide with the zooplankton production schedule. Instant Algae are grown in closed systems remote from the ocean using artificial seawater, so they are free from pathogens. They are used in hatcheries in 80 countries around the world for fish, shrimp and shellfish larvaculture, as well as in public aquariums.

Spray drying is a technique used to separate and dry the solid components from a solid–liquid suspension. When the supplement spray-dried *Schizochytrium*, a unicellular thraustochytrid species rich in n-3 FAs, was used as a partial substitute for live algae *Tetraselmis suecica* and *Chaetoceros* sp. in the nursery culture of juvenile oysters and clams, Boeing (1997) found that up to 40% wet weight of a mixed diet of the two algal species could be replaced with *Schizochytrium* without significantly reducing bivalve growth rate. Furthermore, when fed *T. suecica* alone, growth could be significantly improved by supplements of *Schizochytrium*. Subsequently, Davis & Campbell (1998) reported that mussels (*M. galloprovincialis*) could grow on suspensions of a *Schizochytrium*-based commercial product at rates that were about 80% those of mussels fed live microalgae alone. Similar results were also reported for *M. galloprovincialis* (Langdon & Önal 1999) and the clam *Tapes philippinarum* (Önal *et al.* 2005). In contrast, dietary incorporation of spray-dried *Schizochytrium* or the cyanobacterium *Spirulina* in juvenile geoduck (*Panopea generosa*) culture is not recommended (Arney *et al.* 2015).

Another promising supplement is microcapsules containing a mixture of animal proteins. Southgate *et al.* (1992) showed that when microcapsules were fed to oyster larvae (*Saccostrea commercialis*), they were able to support a growth rate in excess of 80% of that of algae-fed larvae. A major advantage of microcapsules is that they are made when required and, unlike microalgae, require no additional maintenance. However, possible problems include settling, clumping and bacterial degradation of the particles, leaching of nutrients and low digestibility of the cell wall material (Lavens & Sorgeloos 1996). Similar results to those of Southgate *et al.* (1992) were reported for mussel spat (*M. galloprovincialis*) fed MySpat, a microencapsulated fine powder of vegetal protein, fish oil lipids and carbohydrate (Nevejan *et al.* 2007). See also the beneficial effects of MySpat intermixed with microalgae on hatchery-reared spat of *P. canaliculus* (Gui *et al.* 2016). In a recent study, Willer & Aldridge (2017) demonstrated that a new formulation of microcapsules called BioBullets, which are rich in key nutrients for hatchery-reared spat and broodstock, could be captured by mussels (*M. edulis*) with a similar efficiency to natural foods and successfully ingested and broken down by the gut; those capsules too large for ingestion were rejected in pseudofaeces.

High larval growth rate can also be achieved when lipid spheres/emulsions are used as a supplement to live feed (Hendriks *et al.* 2003). Another algal substitute that has been tried with good results is the use of yeast-based diets. Coutteau *et al.* (1994) found that when baker's yeast mixed with algae (50:50 on a dry weight basis) was fed to clam seed (*Mercenaria mercenaria*), the growth rate was similar to that of algal-fed controls. The high lipid content of the yeast formulation probably compensated for the deficiency in calories of the algae/yeast mix. Cornmeal and cornstarch also serve as good complements to live microalgal diets (Pérez-Camacho *et al.* 1998). Adding certain species of bacteria and picoplankton – or dissolved organic material (FAs, amino acids) – to the culture water may also be beneficial, but to date there has been little research on the potential benefits to larvae (see Marshall *et al.* 2010). In a more recent study, Tanyaros & Chuseingjaw (2016) developed a single-cell detritus (SCD) product from seaweed (*Porphyra haitanensis*). A partial replacement of microalgae with SCD (25 and 50%) was unsuitable for nursing oyster larvae (*Crassostrea belcheri*) compared with the microalgae control, while a 50% substitution could be used for nursery culture of juvenile oysters. In a similar study, Rato *et al.* (2018) evaluated the microalgae replacement by dry macroalgae (*Ulva rigida*) on reproductive success and biochemical composition of the oyster *Crassostrea gigas* during broodstock conditioning. They found that the replacement of 25% of microalgae with dry *U. rigida* could lead to similar reproductive success, nutritional quality and physiological condition to broodstock oysters fed 100% live microalgae. Conversely, 100% macroalgae substitution had a negative impact in broodstock conditioning, while improved results during conditioning were obtained with a partial replacement of live microalgae (25%).

Despite the encouraging results obtained by testing a whole range of potential microalgal substitutes, industry's acceptance of these alternatives has traditionally been low – with some exceptions, as demonstrated by the company Reed Mariculture.

Hatchery Culture of Mussels

Broodstock Conditioning

Mussels may become sexually mature in their first year, although size at maturity will depend on the environmental conditions they experience at the production site. According to Bayne *et al.* (1978), an individual female mussel (*M. edulis*) can produce as many as 8×10^{10} oocytes (70 μm diameter) depending on body size. However, under hatchery conditions, females may release between 20 and 40 million eggs, but quantities of 1–8 or 2–5 million eggs in *M. edulis* are more normal according to Kamermans *et al.* (2013) and Helm *et al.* (2004), respectively.

The parents that produce larvae in a hatchery are called broodstock. The ideal broodstock of any cultured species should provide a sufficient number of ripe individuals at any time to enable year-round reproduction. It is usual for hatchery operators to maintain their own stocks for production purposes in local, sea-based grow-out. They are often the offspring of previous hatchery-reared generations, selected for desirable characteristics such as fast growth, meat weight and shell shape and appearance. Genetic improvements can potentially be achieved through selective breeding, crossbreeding, hybridisation or chromosome manipulation to produce triploid and tetraploid mussels (see Chapter 9). At the SPATnz mussel hatchery in Nelson, New Zealand, DNA fingerprinting (see Chapter 9) is used to make sure that the breeding population remains genetically diverse, so that there is a wide range of high-performing lines to choose from. At the Victoria Shellfish Hatchery in Victoria, Australia, a genetically diverse founder population was established in November 2008 by crossing parents collected from different parts of nearby Port Phillip Bay (details in Chapter 9). Measurements and genetic information from founder population mussels were used to select broodstock to produce a new strain of high-performing mussels for aquaculture. This stock was selected for faster growth, rounder shell shape and higher meat yield and is currently being evaluated in field trials by industry partners (Nguyen *et al.* 2012).

An essential factor in broodstock conditioning is to hold adults under conditions in which gametogenesis and spawning can be induced at any time of the year. This allows a greater flexibility in timing of production and increases the number of larval batches produced, thereby optimising the use of hatchery space. There are two main methods for such extension: the 'cold-and-hold' method, which extends the spawning period into the summer, and the 'heat-and-treat' method, which allows spawning before the natural period. As seen in Chapter 5, the two main factors controlling reproductive activity in bivalves are temperature and food supply. Therefore, to extend the natural spawning period, naturally ripened mussels are held at low temperatures on a minimal or irregular live algal ration. To prevent spawning, the ideal temperature for holding mussels is 6 ± 1 °C for *M. edulis* (Galley *et al.* 2010) and 9 °C for *M. galloprovincialis* (Kamermans *et al.* 2013). Using this procedure, *M. edulis* can be successfully maintained in spawning condition for four to six months. For the heat-and-treat method, a live algal diet in conjunction with a gradual rise in temperature or compressed 'winter–spring' temperature profile has been shown to be successful in enhancing the spawning condition and gamete quality of *M. edulis* (Pronker *et al.* 2008). The duration of the conditioning period is also critical; a 6–10-week period has been found to be most suitable in *M. galloprovincialis*, but the role of photoperiod is still unclear (Dominguez *et al.* 2010). The cold-and-hold method is regarded as being less expensive than heat-and-treat because it does not require a lot of algal culture. Also, provided that the mussels are well cleaned beforehand, this technique has fewer problems with growth of epibiota, a reduced need for tank cleaning and less potential for disease, since all biological processes are slowed down.

The amount of space needed for broodstock conditioning depends in part on the number of species being held and whether some or most of the conditioning will be undertaken in outdoor tanks or ponds rather than in the hatchery. In the open, heated or chilled seawater may be required at certain times of the year. In the hatchery, broodstock are usually kept in a quiet area where disturbance is minimal. Maintenance is in a flow-through system (Figure 10.4), either in trays at ambient temperature or, if out-of-season reproduction is required, in indoor tanks with heated seawater and added algal food. The water is not always filtered as it contains a diversity of food organisms that are beneficial to broodstock. Cultured microalgal species are the principal food used for conditioning, although natural phytoplankton bloomed in outdoor tanks or algal concentrates may also be employed. The recommended food ration is based on the dry weight of the adults, usually 2–4% of their mean dry meat weight at the start of the conditioning period in dry weight of algae fed per day (for details, see Helm *et al.* 2004; Pronker *et al.* 2008). Algae species that are cultured for broodstock feeding include the flagellates *Tetraselmis* spp., *Isochrysis galbana*, *Isochrysis*

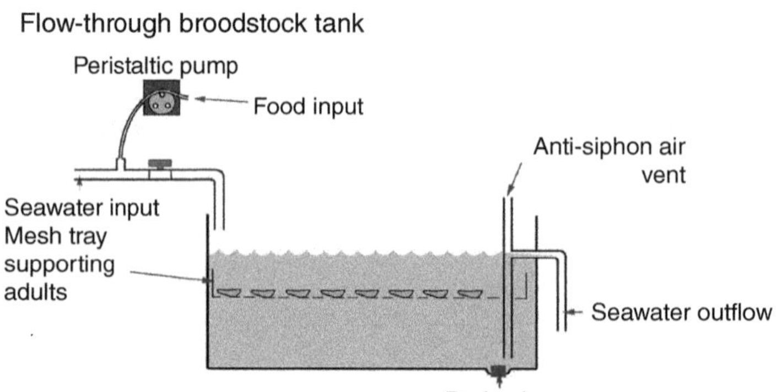

Figure 10.4 Diagrammatic representation of a flow-through broodstock tank in which adults are suspended off the bottom in a mesh tray with large apertures in the base so as not to retain faeces and detritus. From Helm *et al.* (2004). Reproduced with permission from the Food and Agriculture Organization of the United Nations (FAO).

(T-ISO) and *Pavlova lutherii* and the diatoms *Chaetoceros muelleri*, *Thalassiosira pseudonana*, *T. weisfloggii* and *Skeletonema costatum*. Feeding bivalves with a mixture of microalgae has become common practice in hatcheries, since different microalgae species vary considerable in their nutritional value and only by mixing species can optimal food conditions be obtained (Pronker *et al.* 2008 and references therein). Pronker *et al.* (2008) investigated the effects of three conditioning diets, consisting of different mixtures of microalgae at different concentrations, on the reproductive output of *M. edulis* and found that the diets had an impact on spawning success and broodstock fecundity. The best results were obtained with the diatom diet, where 84% of the conditioned animals spawned and females released 5.0×10^6 eggs on average, while those on the two flagellate diets released on average only 3.6×10^6 and 1.6×10^6 eggs, respectively. Also, microalgae lipid composition was shown to be a determinant factor in reproduction success in *M. galloprovincialis*, and egg FA was influenced by the female diet (Martínez-Pita *et al.* 2016b).

The following describes the methods used in broodstock conditioning at the Victoria Shellfish Hatchery in Victoria, Australia. The hatchery was constructed in 2008, with the objective of establishing reliable and cost-effective production of seed (*M. galloprovincialis*) for the Victorian mussel industry, a relatively new venture in southern Australia. The broodstock system comprises two 1000 l holding tanks (Figure 10.5) and one 1000 l header tank. All tanks are cleaned once a week. There is also a 400 l holding tank for final conditioning of mussels. Seawater filtered to 30 μm enters the broodstock system via the header tank at a rate of 3 l min^{-1}. Water, maintained at 9.5–10.0 °C within the broodstock system, recirculates through a temperature control unit at a rate of 1.5 l min^{-1}. Each broodstock tank receives algae from the indoor algae bag culture system at a rate of ~100 ml min^{-1} (50/50 mix of diatoms and flagellates). Broodstock are introduced to the broodstock system up to two months prior to spawning, depending on condition, but if ripe, may be brought

Figure 10.5 Broodstock holding tanks at the Victoria Shellfish Hatchery, Victoria, Australia. *Source:* From Jahangard *et al.* (2010). Reproduced with permission from Victoria Fisheries Authority, Australia.

into the hatchery from the wild the day before or on the day of spawning induction. Broodstock are held in pocket panels and stocked at a density of 130 per tank. Spawning classes are held in separate tanks to avoid accidental spawning events being triggered by other classes. Tanks were originally interconnected, but this caused problems with quarantine and unwanted spawning in some tanks. The system has since been changed to allow tanks to be operated independently of one another.

Spawning, Fertilisation and Cryopreservation

Spawning is the procedure by which conditioned mussels are induced to liberate their gametes in response to stimuli. The easiest way to assess broodstock maturity is to examine a small piece of gonad from a few sacrificed individuals under the microscope. The following description is based on the spawning procedure used by the Victoria Shellfish Hatchery. Before an expected spawning event, mussels are cleaned by scrubbing the shells to remove byssal threads and commensal organisms, and then bathed in freshwater for 10 minutes to reduce bacterial loading. Cleaned mussels are transferred to a broodstock spawning tray (Figure 10.6), which is initially filled with 1 μm filtered and UV-treated seawater at 15 °C. Water temperature is increased to 20 °C over a period of one hour to stimulate spawning. Food (algae) may also be added to the water, or mussels may be vigorously agitated in a bucket to induce spawning. Once spawning is initiated, females and males are isolated into separate buckets and allowed to continue spawning. After spawning, unfertilised eggs are rinsed on a 45 μm screen and then fertilised with washed sperm (20 μm screen) at a rate of ~5–10 sperm per egg. The sperm to egg ratio for optimal fertilisation varies widely between species, and even within a species.

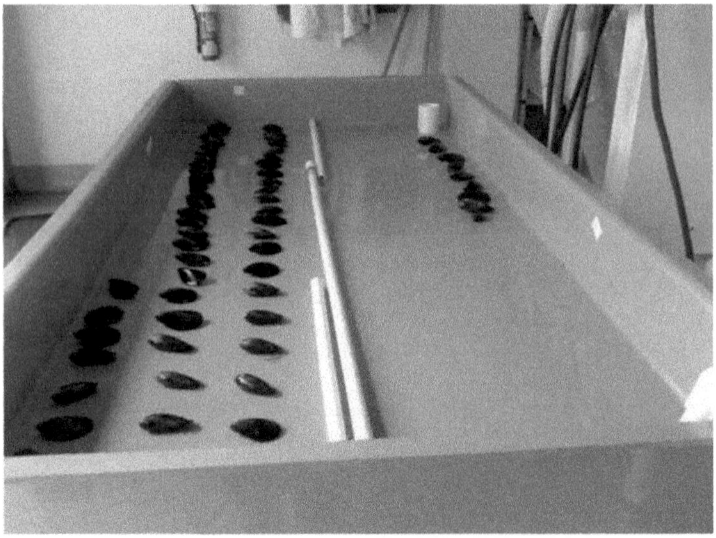

Figure 10.6 Broodstock spawning tray at the Victoria Shellfish Hatchery, Victoria, Australia. *Source:* From Jahangard *et al.* (2010). Reproduced with permission from Victoria Fisheries Authority, Australia.

For example, in the SPATnz hatchery, a ratio of 5–10 sperm per egg is used in fertilisation of *P. canaliculus* gametes, whereas a ratio of ~200 sperm per egg is used for *M. edulis* and *M. galloprovincialis* (Kamermans *et al.* 2013). Once fertilisation is evident (on microscopic examination), eggs are transferred to the incubation tanks. The day of spawning represents day 0 of each spawning run. The procedure at the SPATnz hatchery is somewhat different in that temperature shock is used to induce spawning. Once spawning is initiated, the eggs are collected from females spawned in water, while sperm is collected 'dry' from males. Eggs and sperm should be used as soon after spawning as possible. Eggs should not be older than one hour, and sperm no older than 30 minutes, although holding gametes at low temperature (e.g. in a fridge) can extend their effective life. Assessment of whether fertilised eggs are developing normally is done using a relatively low-power microscope (×20–40 magnification).

Other spawning protocols used to initiate spawning in mussels are a gradual increase in water temperature (*P. viridis*, Sreedevi *et al.* 2014), thermal cycling[1] (*P. viridis*, Anil *et al.* 2017) and intermittent exposure to air (*P. viridis*, Anil *et al.* 2017). An alternative spawning technique is to induce mussels collectively in a group, a method known as mass spawning. This involves holding the mussels together in the same vessel and exposing them to thermal cycling (Kamermans *et al.* 2013). Mussels can be removed once they begin to spawn, or else the gametes can be collected via an overflow into a submerged mesh sieve by running seawater through the spawning vessel/tank. The latter approach invariably results in the collection of developing embryos rather than separate unfertilised gametes. The use of mass spawning is not recommended if the aim is to use the gametes for the production of triploid mussels or for selective breeding experiments (see Chapter 9).

Cryopreservation is the storage of live material at temperatures so low (e.g. −196 °C) that all biological processes are suspended; thus, it is seen as preserving genetic material in perpetuity (Yang *et al.* 2012). Since the 1970s, cryopreservation of gametes has been widely used in many farmed bivalve species, particularly oysters (Martínez-Páramo *et al.* 2017 and references therein), although in the past decade more attention has been focused on mussels. For example, studies have investigated the effects of cryoprotectants on mussel sperm (*M. edulis*, Di Matteo *et al.* 2009; *P. canaliculus*, Adams *et al.* 2011 and Smith *et al.* 2012), oocytes (*M. galloprovincialis*, Liu & Li 2015 and Wang *et al.* 2014; *P. canaliculus*, Adams *et al.* 2009) and larvae (M. *galloprovincialis*, Wang *et al.* 2011 and Paredes *et al.* 2013; *P. canaliculus*, Paredes *et al.* 2012). The advantages of cryopreservation are that it provides reliable supplies of sperm without seasonal limitations and costly hatchery maintenance of adult males, and it provides a safe repository through cryobanking (reviewed in Martínez-Páramo *et al.* 2017) for improved lines with desirable traits. Cryopreserved sperm can also be used for shipment to other locations, or to cross populations that breed asynchronously. Cryopreservation of sperm is quite well defined for most bivalve species (Di Matteo *et al.* 2009). In contrast, oocytes and early embryos are more difficult to cryopreserve. The cryopreservation of trochophore larvae (*P. canaliculus*, Paredes *et al.* 2012; *M. galloprovincialis*, Paredes *et al.* 2013) could be seen as an alternative to separately cryopreserving oocytes and sperm, since previous studies with *M. edulis* (Toledo *et al.* 1989; Wang *et al.* 2011) showed that the more advanced the developmental stage, the more resistant the cells to cryoprotectant toxicity and the higher the post-thaw survival rate. Cryopreservation of gametes and larvae has yet to be applied on a commercial scale by the aquaculture industry, and will first require the standardisation of protocols across laboratories.

Embryo Development

At the Victoria Shellfish Hatchery, the egg incubation system comprises two 5000 l tanks filled with 1 μm filtered and UV-treated seawater at 16 °C on the night before or the morning of spawning. This is carried out to ensure that the water is free of pathogens and particles. The tanks are lightly aerated, and ethylenediaminetetraacetic acid (EDTA) is added at a rate of 4.5 g/tank. Without EDTA, embryos (*P. canaliculus*) show oxidative stress and ultrastructure abnormalities (Gale *et al.* 2016), although the underlying biological mechanisms associated with its use are poorly understood. Fertilised eggs are gradually added to each tank at a rate of 30 eggs ml^{-1} (150 million eggs/tank), ensuring that eggs are given time to acclimate and distribute throughout the tank. On day 1, 23 hours post-fertilisation, larval condition is checked for development to the free-swimming trochophore stage (see Figure 5.9). If that stage is reached, 5 l of *Isochrysis galbana* var Tahiti is added to the tank to provide a food for developing larvae. On day 2, 48 hours post-fertilisation, larval condition is again checked for development into the D-larval stage, and the tank is dropped through a large 45 μm screen to retain larvae, making sure to keep the larvae moist on the screen. The development period of mussel embryos is temperature-dependent: at 17.5 °C, *M. edulis* reach the D-larvae stage (Figure 10.7) within 48 hours, while at 12 °C this can take up to 72 hours (Kamermans *et al.* 2013). The harvested larvae are cleaned by rinsing with fresh filtered seawater and graded on a 60 μm screen, then distributed to the larval rearing tanks at a rate of 25 larvae ml^{-1}. Typically, <50% of eggs will develop into larvae, of which almost 100% can develop into the D-larvae stage, although this is highly dependent upon gamete quality (Kamermans *et al.* 2013). The different stages of larval development are described in Chapter 5 and illustrated in Figures 10.7.

Larval Rearing

When mussel larvae develop to the D-larval stage (Figure 10.7f,g), they can be reared to metamorphosis in conical vessels with bottom drains or in flat-bottom vessels. At the Victoria Shellfish Hatchery, the larval rearing system comprises 12 × 400 l conical fibreglass tanks set up as a flow-through system, with natural seawater entering from the bottom of the tank (Figure 10.8). Mild aeration is provided to each tank, since it improves survival in the early larval stages. Temperature is controlled by an Aquahort heat pump, and is maintained at 16–18 °C. Water flow rate ranges from five tank exchanges day^{-1} (1.4 l min^{-1}) when tanks are initially stocked to 12 tank exchanges day^{-1} (3.3 l min^{-1}) by the settlement stage. Filters (45–125 μm mesh size) are used to prevent larvae from escaping. Larvae are completely dropped every second day to grade them on appropriately sized sieves, clean the tanks and adjust stocking densities. Through a substantial culling regime, larval densities decrease from an initial 10 million D-larvae per tank (25 larvae ml^{-1}) to a final 4 million per tank (10 larvae ml^{-1}), a recovery rate of 40%. For the first week following stocking, algae are administered into each tank to maintain a residual level of 15 000 cells ml^{-1}, comprising flagellates only; in later weeks, this is increased to 25 000 cells ml^{-1},

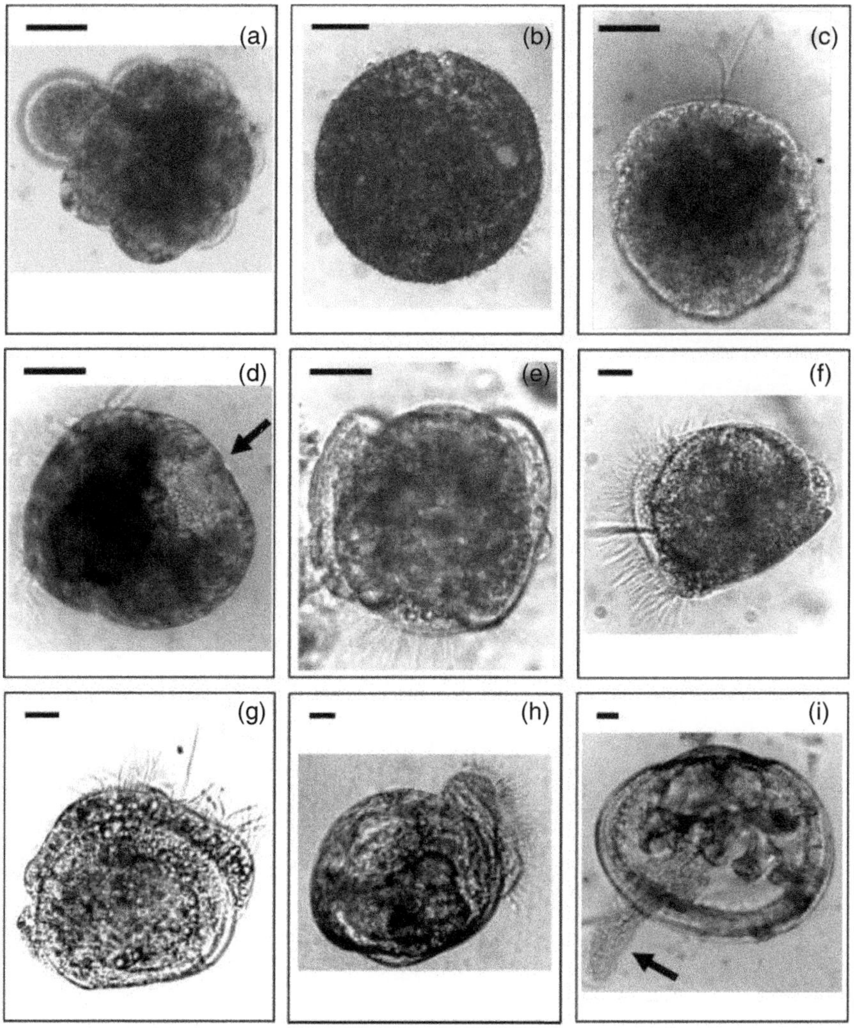

Figure 10.7 Embryonic stages in the mussel *Limnoperna fortunei*. (a) 16-cell stage. (b) Morula (ball of cells from fertilised ovum). (c) Trochophore. (d) Veliger with rudimentary valves (arrow). (e) Veliger. (f,g) Straight-hinged D-veliger. (h) Umboned veliger. (i) Juvenile (arrow points at foot). Scale bars = 25 mm. *Source:* From Cataldo *et al.* (2005). Reproduced with permission from Oxford University Press.

comprising a 50/50 mix of diatoms and flagellates. Maintaining a supply of high-quality seawater is of paramount importance to ensure healthy growing larvae in the hatchery (details in Helm *et al.* 2004; Duthie 2012; Jahangard *et al.* 2010; Kamermans *et al.* 2013; Anil *et al.* 2017).

Outcomes from the BLUE SEED project suggest that the larvae of both *M. edulis* and *M. galloprovincialis* grow successfully through their D-larval phase on a number of mixed algal diets, consisting of two or three high-nutritional-value species (Kamermans *et al.* 2013). Note that the recommended diet composition is different for the two species:

Figure 10.8 Larval rearing tanks with flow-through system at the Victoria Shellfish Hatchery, Victoria, Australia. *Source:* From Jahangard *et al.* (2010). Reproduced with permission from Victoria Fisheries Authority, Australia.

M. edulis:

- *Pavlova lutheri* + *Rhinomonas reticulata* (80:20 ratio based on cells ml^{-1});
- *Isochrysis galbana* + *Chaetoceros calcitrans* (50:50);
- *Pavlova lutheri* + *Chaetoceros calcitrans* (50:50).

M. galloprovincialis:

- *P. lutheri* + *Chaetoceros gracilis* (50:50);
- *P. lutheri* + *C. gracilis* + *Tetraselmis suecica* (50:25:25);
- *I. galbana* + *C. gracilis* + *Tetraselmis suecica* (50:25:25);
- *I. galbana* + *P. lutheri* (week 1) + *C. gracilis* (week 2) + *T. suecica* (week 3);
- *I. galbana* + *I. galbana* (T-ISO) + *P. lutherii* + *C. gracilis* (week 1) + *T. suecica* (weeks 2 and 3).

The diatom *C. calcitrans* is an important dietary component for both cultured *M. edulis* larvae and many other bivalves (*Helm et al. 2004*); it is apparently indispensable in the culture of *P. canaliculus* larvae (Ragg *et al.* 2010). However, detrimental effects are associated with high levels of the diatom, including reduced survival and metamorphosis rates and a general increase in population instability in larval rearing tanks, which indicates that appropriate feeding protocols are needed when using it.

Metamorphosis, Settlement and Nursery culture of Spat

During the veliger stage, larvae are positively phototrophic and are sensitive to pressure, keeping them in surface waters. Toward the end of the planktonic larval phase (200–260 μm shell length), they develop a pair of pigmented eyespots and an extensible ciliated foot, which rapidly becomes functional in crawling (see Chapter 5). At this stage, larvae are called pediveligers, and in contrast to veligers, are positively geotrophic and insensitive to pressure, which encourages them to descend to the bottom in preparation for settlement. This stage of development, which occurs approximately at the end of the third week of larval rearing, is defined as a process beginning with the onset of a behavioural search for a suitable substrate and ending with metamorphosis, which involves reorientation of internal structures, increased complexity of the organ systems and secretion of the adult shell (see Chapter 5). The success of metamorphosis is heavily dependent upon the production of good-quality healthy larvae with large energy reserves accumulated during the larval phase. Because the velum has been replaced by the more efficient gill system, there is a marked increase in filtering capacity and food requirements following metamorphosis (Kamermans *et al.* 2013). Almost all bivalve larvae become competent to settle in the size range 250–300 μm shell length, and once settled are referred to as spat. A wide variety of artificial compounds, as well as amino acids, ammonia (NH_3), potassium chloride (KCl) and potassium sulphate (K_2SO_4) and a variety of ions, have been shown to be effective in inducing metamorphosis and settlement in some mussel species (details in Chapter 5). If newly settled spat are placed out directly in the natural environment for on-growing, they can suffer high mortality from smothering, predation and competition. Therefore, maintaining the spat in a nursery system until they have reached a large enough size to withstand these factors often makes economic sense. Generally speaking, spat are held in the nursery until they have reached a few centimetres in size. Many hatcheries initially maintain the newly settled spat in their larval rearing containers until they have reached about 1 mm shell length. At the Victoria Shellfish Hatchery, the nursery culture system consists of 9 × 5000 l polypropylene tanks, which are aerated from a PVC grid. A total of 110 5 m ropes are hung in a V-shape across the width of each tank. Competent larvae are added at a rate of 5–10 million per tank (1–2 larvae ml^{-1}), depending on the number available.

Competency to settle depends on shell size, eyespot formation, aggregation and active pediveliger behaviour. When larvae are added to the tank, care is taken to spread them evenly; an even mix is ensured via aeration. For the first five days of settlement, the tanks are operated as static systems, with a complete drop and refill every second day, catching all unsettled larvae on a screen. If they are in good condition, they may be returned to the tanks. Algae (50/50 mix of diatoms and flagellates) are added to maintain a residual level of 100 000 cells ml^{-1}. After the initial five-day settlement period, the tanks are set up as flow-through systems (Figure 10.8), allowing two water exchanges per day (7 l min^{-1}), as well as a complete drop and refill every second day. Because the drop-and-refill process takes approximately one hour to complete, a fine seawater mist is sprayed over the ropes to keep them moist. The same algal mix and ratios are fed to maintain a residual level of 50 000 cells ml^{-1}. A second approach to spat rearing involves spreading the spat, previously detached from the settlement substrates, on 250 μm mesh size square-frame screens, which are horizontally introduced into rectangular tanks in an indoor micronursery (Kamermans *et al.* 2013). A third is the use of down-welling systems (Figure 10.9), with pediveligers placed within the confines of a nylon mesh-bottomed cylinder (120–150 μm mesh) suspended within a holding tank. A reservoir can be connected if the holding tank is not of sufficient volume. Seawater with algal food is circulated by electric pump or air-lift from the reservoir to the holding tank, passes by the spat and then returns to the reservoir (Helm *et al.* 2004; Kamermans *et al.* 2013).

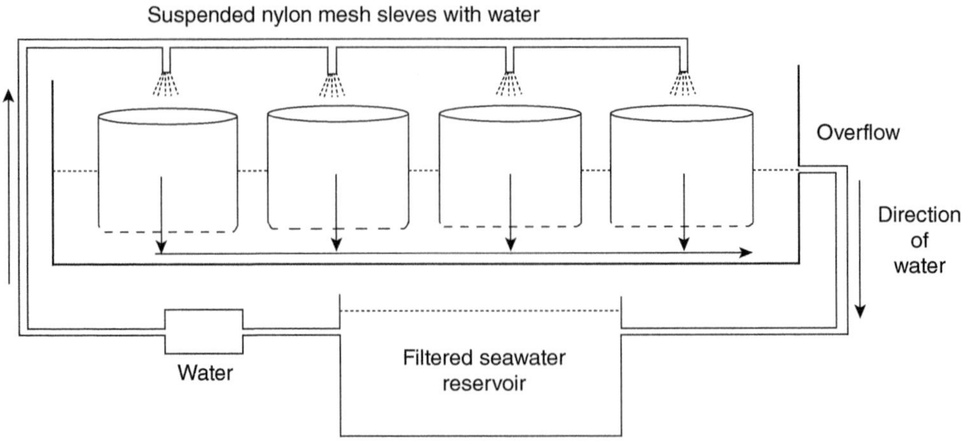

Figure 10.9 Schematic of basic down-welling system for mussel settlement. Water flows into the enclosure from above the surface at a constant rate, moving downward over the spat and out through the mesh base of the cylinder to return to the reservoir via an overflow that maintains the water level constant in the holding tank. Such a system can easily be set up in a closed recirculating system with regular water changes and addition of unicellular algae. *Source:* From Kamermans *et al.* (2013). Reproduced with permission from Elsevier.

During the nursery stage, density and average size of the spat are estimated regularly throughout the rearing process, in order to adjust culture densities, but especially at the end of the process, just before the seed is transferred to on-growing sites (Kamermans *et al.* 2013). These two variables are difficult to assess at the beginning of the rearing process because of the small size, fragility and aggregation of individuals, but become easier to estimate as the spat grow. Average spat size is estimated by measuring individuals contained in one or more random samples taken from the ropes or screens under a binocular microscope equipped with an eyepiece micrometer. Estimation of the rearing density is a more complex process and depends upon the type of substrate used to rear the spat.

Ropes are widely used for collection of mussel spat under hatchery conditions. In a study by Jahangard *et al.* (2010) at the Victoria Shellfish Hatchery, spat settlement in a single nursery tank was examined on ropes subjected to five different treatments. The highest number of spat was recorded on steam-cleaned rope, followed by pressure-washed well-cleaned rope, pressure-washed not-very-well-cleaned rope, simple cleaned rope and finally brand-new rope, although no significant differences were found between brand-new and normal cleaned rope. Spat from steamed-cleaned rope grew significantly faster ($P < 0.05$) than spat from any of the other treatments, suggesting that pressure washing and steam cleaning will improve mussel settlement in the hatchery. Location of rope in the tank and water depth also affected spat settlement on ropes. Three-dimensional visualisation (TDV) plots showed that spat recruited mostly at the middle and quarter part of the nursery tank, rather than the corner part, and that spat density decreased with increasing water depth (20, 40 and 60 cm depths tested). In a more recent and related study, Protopopescu & Beal (2015) investigated responses of cultured pediveligers (*M. edulis*) to different types of rope collectors (polyethylene and polypropylene). Competent larvae (approximately 5000 per

400 l experimental tank) were exposed to rope collectors (3 cm long) exhibiting diverse structural features, such as long loops (LL), short loops (BL), long filaments (LF), short filaments (BF) and smooth rope (SM) (Figure 10.10a). After five days, rope collectors were removed from five experimental tanks and the number of settled larvae on each 3 cm

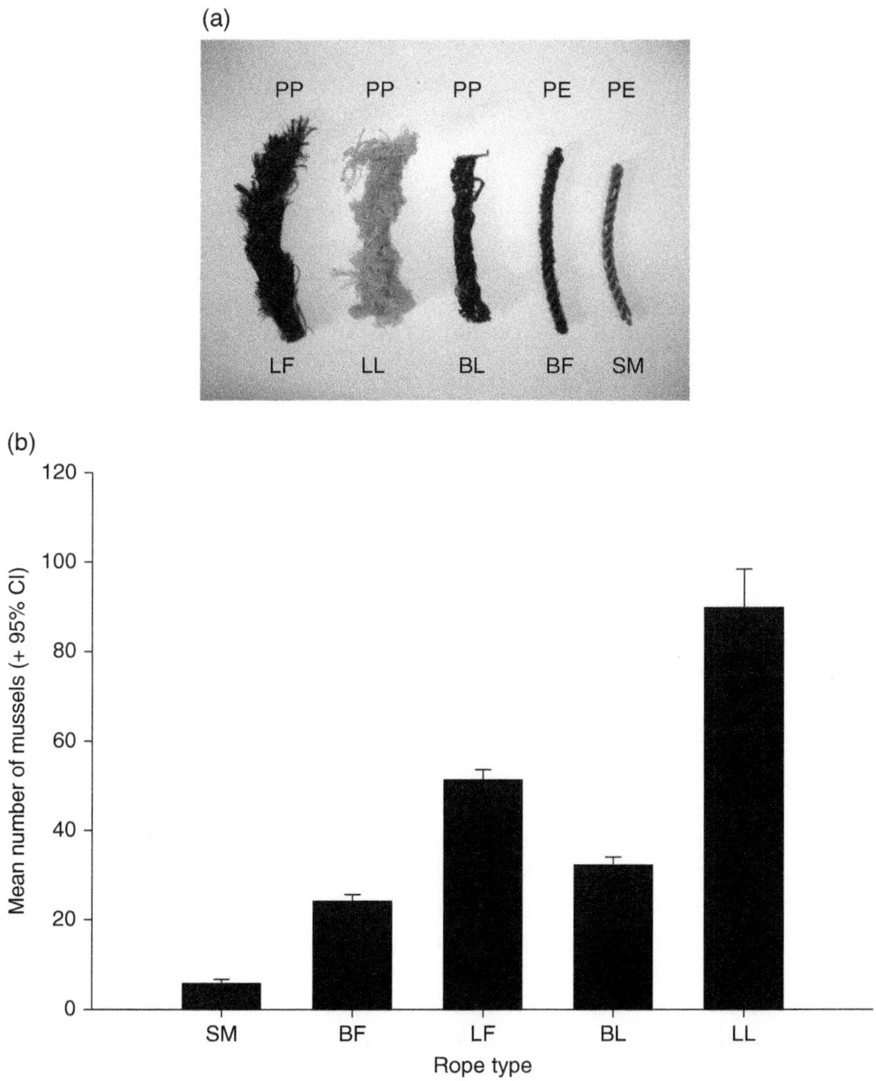

Figure 10.10 (a) Collector rope types. SM, smooth rope; BF, rope with short filaments; BL, rope with short loops; LL, rope with long loops; LF, rope with long filaments. PP, polypropylene; PE, polyethylene. (b) Mean number of mussels per rope collector type, with 95% confidence interval. SM, smooth rope collector; BF, rope collector with short filaments; LF, rope collector with long filaments; BL, rope collector with short loops; LL, rope collector with long loops. The five types of rope collectors are replicated five times in each of six tanks for a total number of 150 rope units. *Source:* From Protopopescu & Beal (2015). Reproduced with permission from Journal of Shellfish Research.

segment was counted. Rope collectors with the highest structural complexity/greatest surface area (long loops) and made of polypropylene elicited the strongest settlement response (highest densities) of mussel larvae, whereas those with the lowest complexity/least surface area (smooth) elicited the weakest response (Figure 10.10b). These findings suggest that hatcheries would be well served if they used polypropylene rope collectors with high structural complexity, since this would enhance the number of settled larvae per spawning/settlement event, thereby lowering production costs (Protopopescu & Beal 2015).

Disease in the Hatchery

The artificial elevated temperatures, high density and occasionally stressful conditions under which hatchery larvae and juveniles are reared provide an environment conducive to the spread of bacterial disease. The most important bacterial pathogens in hatchery cultures belong to the genus *Vibrio*. These bacteria infect larval, juvenile and adult molluscs, including oysters, mussels, abalones, clams and scallops (reviewed by Paillard *et al.* 2004, Beaz-Hidalgo *et al.* 2010, Travers *et al.* 2015; Dubert *et al.* 2017). Infections are usually described from larval stages, where they are associated with high mortalities in hatcheries, but only a few bacterial types are pathogenic to adults (Paillard 2004). This is because larvae are reared in static systems at elevated temperatures and high densities.

Remarkably, and in contrast to other bivalve species, no bacterial pathogens of mussels (*Mytilus* spp.) have been documented so far, and until 2009 there was no information in the published literature on bacterial pathogens of *Mytilus* larvae. Kesarcodi-Watson *et al.* (2009a) identified two strains, *Vibrio splendidus* and a *V. coralliilyticus/neptunius*-like isolate (*Vibrio* sp. DOI), as pathogens of *P. canaliculus* larvae, with the ability to cause significant mortalities >75% in vitro. This was the first report on bacterial pathogens of marine mussel larvae and, although Anguiano-Beltrán *et al.* (2004) reported on a sublethal effect caused by *V. alginolyticus* on mussel (*M. galloprovincialis*) larvae, they did not describe virulent pathogenesis (i.e. the degree of pathology caused by the organism). Histopathology by both strains indicated that the route of infection was via the digestive system. Larval problems typically begin around day 6 post-fertilisation. Characteristic signs include loss of lipids, deformed extended velum and erratic swimming. The successful outcome resulting from antibiotic treatment of larvae suggested that bacterial agents might be implicated in larval crashes in the hatchery. *Vibrio coralliilyticus* represents a serious threat for bivalve aquaculture, being one of the most important emerging pathogens responsible for bivalve larval mortalities worldwide (Balbi *et al.* 2019 and references therein).

In a related study, Kesarcodi-Watson *et al.* (2009b) demonstrated a reproducible infectious dose of the two pathogens (*Vibrio* sp. DO1 and *V. splendidus*) to *P. canaliculus* larvae under flow-through hatchery conditions using a defined protocol. Larval survival in control treatments varied between 33.8 and 89.6% (average 71.2 ± 4.4%) across all experiments, illustrating the existence of batch variability, possibly caused by genetic differences of the larvae due to variation in broodstock, potential health differences of the cohorts and differing environmental factors during experiments, such as extraneous bacteria.

Mussels (*Mytilus* spp.) are considered robust and resistant to disease (see Chapter 11). However, Eggermont *et al.* (2014) documented that mass mortality can be induced in healthy wild-caught *M. edulis* adults by stimulating mussel-associated heterotrophic bacteria. Their data indicated that wild mussels contain a reservoir of pathogenic bacteria that might enter hatchery facilities unnoticed through healthy wild-caught broodstock and then be transferred to their offspring (e.g. via the rearing water). In a subsequent study, Eggermont *et al.* (2017) demonstrated that Vibronacea isolated from wild-caught adult mussels (*M. edulis*) are capable of inducing mass mortality in mussel larvae. Approximately half of the isolates were found to belong to the *Splendidus* clade of vibrios, a group that is well known to contain bivalve pathogens, while the others belonged to the genus *Photobacterium*. The isolates caused between 17 and 98% mortality in larvae after five days of immersion challenge. These findings confirm that microbiological management is a fundamental requirement for successful larval culture (Eggremont *et al.* 2017).

In hatcheries, antibiotics (e.g. florfenicol, erythromycin, oxolinic acid and, especially, chloramphenicol) are used in bivalve hatcheries to prevent vibriosis and to improve the survival rates of larvae and juveniles (reviewed in Dubert *et al.* 2017). Because the spread of infection is so rapid, antibiotics are often ineffective, may cause larvae to stop feeding and can lead to the development of resistant strains (Dubert *et al.* 2016). An alternative to antibiotic treatment is the use of probiotics: beneficial bacteria that override pathogens by producing inhibitory substances or prevent pathogenic colonisation in the host (reviewed in Kesarcodi-Watson *et al.* 2008; Prado *et al.* 2010; Akhter *et al.* 2015; Sohn 2016).

Kesarcodi-Watson *et al.* (2009c) developed a bioassay that included test mussels to screen and select bacterial strains as probiotics for an alternative to antibiotic usage in the management of bacterial pathogens of *P. canaliculus* larvae in the hatchery. They tested 69 isolates originating from the hatchery environment of the host for probiotic activity in larval pathogen challenge bioassays. *Vibrio splendidus* and a *V. coralliilyticus/neptuniuslike* isolate (*Vibrio* sp. DO1) were the tested pathogens. Among the tested isolates, 40 (58%) afforded larval survival significantly greater than pathogen controls ($P < 0.05$). This high success rate was probably achieved through the inclusion of test larvae in the initial screening stages (screening methods reviewed in Kesarcodi-Watson *et al.* 2008).

In a subsequent study, two novel probiotic bacteria were identified and tested on *P. canaliculus* (Kesarcodi-Watson *et al.* 2010). Sequencing of the 16S rRNA gene and phylogenetic analysis identified the strains as *Alteromonas macleodii* 0444 and *Neptunomonas* sp. 0536. Both probiotics were evaluated separately during routine larval rearing and when the larvae were challenged with both a high (10^7 and 10^6 colony-forming units[2] (CFU) ml^{-1}) and a low (10^6 and 10^5 CFU ml^{-1}) pathogenic dose of *Vibrio* sp. DO1 and *V. splendidus*, respectively. In all experiments, probiotic application improved larval survival significantly when administered prior to pathogen exposure. Larvae that were exposed to the high and low dosages of pathogens averaged 14 and 36% survival, respectively, on the fourth day following pathogen exposure. The administration of probiotics prior to pathogen challenge resulted in larval survival of 50 and 66%, respectively. Noninoculated control larvae and larvae administered the probiotic alone demonstrated 67 and 79% survival, respectively. The authors emphasise that their results do not imply larval rearing success, since the larval period for *P. canaliculus* is approximately 18 days, although their results do demonstrate protection against a pathogen challenge in the early stages of larval rearing, the period when most larval mortality occurs. In a related study, Kesarcodi-Watson *et al.* (2012)

investigated whether there were additional benefits to *P. canaliculus* larvae from using the two probiotics *Alteromonas macleodii* 0444 and *Neptunomonas* sp. 0536 in a multi-strain mix, both during routine larval rearing and during pathogen challenge tests. However, in terms of survival, there were no differences between single-strain and multi-strain administration.

Probiotics strains can have species-specific protective effects for bivalve larvae. For example, Sohn *et al.* (2016) investigated the effectiveness of candidate probiotic bacteria strains *Phaeobacter inhibens* S4 (S4) and *Bacilluspumilus* RI06-95 (RI) against *Vibro coralliilyticus* challenge in larvae of four bivalve species: the clams *Mercenaria mercenaria*, the scallops *Argopecten irradians*, the mussels *M. edulis* and the razor clams *Ensis directus*. Pretreatment of larvae with 104 CFU ml^{-1} probiotics for 24 hours did not protect any of the species tested against bacterial challenge. Pretreatment of 106 CFU ml^{-1} probiotics S4 and a mixture of S4 and RI did protect scallop larvae (relative per cent survival (RPS) 69 $\pm$ 4%), but none of the other three species. Differences in disease susceptibility between bivalves may be due to differences in defence mechanisms against bacterial infection, although little is known about immune response to bacterial infection in larval stages of bivalves. Sohn *et al.* (2016) suggested that there may be differences in the mechanisms of feeding (or selective ingestion of microalgae and microbes) between species. If feeding mechanisms are shown to impact probiotic uptake and efficacy, these differences could be exploited to improve probiotic development and methods of delivery.

Grow-out of Mussel Seed

This section is heavily reliant on information in Kamermans *et al.* (2013). When mussel spat reach about 5–10 mm shell length, they are usually ready to be transferred to grow-out sites, where pre-ongrowing, thinning out and the final ongrowing phase commence. The choice of site is of paramount importance not just for good growth but also for management and economic reasons. In European mussel culture, the final grow-out sites are always natural marine environments where different types of sea-based facilities are established. In France, for example, intertidal areas on the Atlantic Coast are used for the ongrowing of mussels on bouchots (poles) placed in rows perpendicular to the coastal line (see later). In other countries, such as The Netherlands, Wales, England and parts of Ireland, mussels are traditionally seeded and ongrown on bottom plots (see later) situated in intertidal and subtidal areas of shallow coastal areas. But the most efficient environment for mussel grow-out is the water column of seston-rich estuaries and sheltered coastal areas where mussels are ongrown tied to ropes or within plastic socks suspended from rafts or longline systems. A good example of such culture is the raft system traditionally used in Galicia, NW Spain (see later). Another is longline culture, developed some years ago in New Zealand for the Greenshell mussel, which has been introduced into a wide range of range of European countries and many other regions worldwide (Table 10.1).

In some situations, such as unfavourable environmental conditions or where there is a high risk of predation, it may be more appropriate to first move small mussels to lab-scale or land-based nurseries before transferring them to ongrowing sites. Most lab-scale culture systems that test the grow-out of mussel seed in land-based facilities are variations on the same basic structure described for spat culture: down-wellers placed within holding tanks.

These seed culture systems are similar to those described for spat rearing (Figure 10.11). As the mussel seed grows, the bottom nylon net can be replaced by a larger mesh size to permit a better seawater circulation through the system. Because food requirements at this stage of mussel culture are very high, it is recommended to place all seed-containing down-wellers into large flow-through tanks (Figure 10.11), through which seawater with high seston content (only filtered by 50 μm) is continuously pumped together with a mix of several species of bag-cultured microalgae (*Tetraselmis suecica* and *Skeletonema costatum*) contained in a reservoir tank. All equipment should be cleaned frequently (every day or every other day) with freshwater to remove all retained seston material, as well as faeces and pseudofaeces.

A different approach is needed to carry out pre-ongrowing of mussel seed in large in land-based facilities. This involves transferring mussel seed, previously attached to net frames, to large outdoor raceways (long, shallow tanks most commonly made of fibreglass) in a land-based nursery. Microalgae are pumped to the raceway tanks to feed the seed. In order to promote a more homogeneous distribution of the mussel seed on the frames, it is advisable to grow them in darkness by covering the raceways with a light-reflecting material. In land-based systems, spat is regularly cleaned of fouling organisms and silt. Predation can also be a significant problem, as the young stages are particularly vulnerable to major predators such as crabs, boring gastropods and starfish (see Chapter 3). Densities are regulated to optimise growth, and samples are regularly taken for measurement of growth rate (see Chapter 6) and, as mussels mature, reproductive condition (see Chapter 5). After a month under these rearing conditions, mussel seed (4–10 mm shell length) can be separated from the frames and transferred to a final ongrowing offshore environment, as already described for European countries. Although the technique of remote settlement of bivalve spat is well known, particularly for oysters, there is no currently information on the application of the method to mussels.

Figure 10.11 Mussel seed-containing down-wellers placed in a large flow-through tank. *Source:* From Kamermans *et al.* (2013). Reproduced with permission from Elsevier.

Mussel Culture

Almost 2.0 million t of mussels were cultured in 2016, about 44% of which were produced in China, mostly for the domestic market (Table 10.1). FishStatJ Global Aquaculture Production data for China do not specify individual mussel species, but it is largely *Mytilus galloprovincialis*, *Perna viridis* and *Mytilus coruscus* (Tang *et al.* 2002) that are cultured, although there is no information on annual yields for these species. After China, global production in 2016 was highest for Chile (over 300 000 t of *Mytilus chilensis*), Spain (*M. galloprovincialis*), Thailand (*Perna viridis*), New Zealand (*Perna canaliculus*), Italy (*M. galloprovincialis*) and Korea (*M. galloprovincialis*). The Chilean mussel industry reported a record year in 2017, with production exceeding 300 000 t. Unlike in the other major producing countries, all of this production goes to exports, making Chile the world's major exporter of mussels (FAO 2018a). Spain is the main market for Chilean mussels, importing 15 000 t in 2017 – a 25% growth on 2016. These mussels are mainly used in the canning industry, and compete there heavily with the domestic mussels produced in the Galician fjords (see later). Other major importing countries of Chilean mussels are the United States, France and Italy, all three of which reported stable imports in 2017. However, the Russian Federation is becoming an important outlet for Chilean mussels, importing some 9000 t in 2017, a 51% increase from 2016. Like Chile, New Zealand produces mainly for the export market. Its main importing countries are the United States, China and Thailand. However, this situation is about to change significantly as new technologies such as hatchery-bred mussel spat are employed (see earlier). In 2018, the outlook for mussel production was positive in all other mussel-culturing countries, although demand was predicted to exceed production growth (FAO 2018a). Due to Covid-19, however, trade of mussels during the first three months of 2020 declined slightly when compared with the same period of 2019. France imported almost 4000 t less during this period, a 22% decline from 2019. Despite this, the country continued to be the main mussel-importing nation in the world. Meanwhile, Italy recorded a 7% decline. The United States was the only main player among mussel importers reporting increased imports in the first quarter of 2020. However, this is not likely to be the case during the second quarter of the year, due to lockdown orders in various US states (FAO 2020).

Culture Methods

Longline Culture in Chile

Mussel cultivation in Chile dates back to the 1940s, when the over-exploitation of mussels in the area between Valdivia and Chiloé Island on the Pacific coast of southern Chile led to experiments in collecting mussel larva for the replenishment of natural beds. By the 1960s, these efforts had resulted in Chile's first farmed mussels, and by 2003 its annual production was ~ 60 000 t (FAO 2018b). In 2016, for the first time in history, Chile surpassed Spain as a mussel producer, becoming the top exporter of *Mytilus chilensis* and the second major producer of mussels in the world (Table 10.1). In 2017, exports reached 70 000 t meat weight, an 18% increase over 2016 (FAO 2018a). Some companies have concentrated on eco-labelling to ease their way into Europe, with a number gaining Friend of the Sea certification, and one, Toralla SA, being the first Chilean company to gain the Marine Stewardship Council standard in 2014 for its mussel fishery and suspended culture. In northern Chile, the mussels *Choromytilus chorus* and *Aulacomya atra* are also cultured, but in 2016 they represented <1% of total mussel production (FAO 2018a). These two species are mainly produced for the domestic markets, the main reason being that they are not in demand by external ones. The mussel industry is clustered in fjords, channels and estuaries of the region around Chiloé Island, close to the provincial capital Puerto Montt. The main mussel culture areas in this region are Caucahue Channel and Yal Bay, as well as the Pitipalena fjord, which is far from intensive aquaculture production zones and contains important natural banks of *M. chilensis* (Blanc *et al.* 2018). Essentially, the mussel farming industry is piggy-backing on the established Chilean salmon farming sector, now the second largest farmed Atlantic salmon producer in the world (582 000 t in 2017) after Norway. Salmon production has created an excellent infrastructure base allowing mussel farmers easy access to landing facilities and the supply of equipment. The sheltered waters around Chiloé have salmon farms at intervals of 2.5 km, and increasingly mussel farms are occupying this separation space. It is an ideal polyculture situation: the salmon farms enrich the waters and the mussel farms benefit from the high levels of plankton living in the now nutrient-rich environment (BIM 2006).

In Chile, mussels are cultivated on a longline system that consists of either a single or a double polypropylene rope (headline) supported by a flotation system of buoys or high-density polyethylene (HDPE) pipes placed at regular intervals. The line(s), the length of which varies depending on location (100–200 m), are anchored at both ends to the seabed or shore (Figure 10.12). Secure anchoring is necessary to prevent the longlines shifting in heavy seas or strong currents. Most are either concrete blocks or screw anchors.

In Chile, 100% of the seed needed for the industry is obtained from natural spatfall on collectors that are hung on longline systems similar to those used for mussel growth in the fattening stage (see later). Each collector is 20–25 cm in diameter and 4–8 m in length, and there is a 10–20 cm distance between them so that on a single-line system (100 m long) there is capacity for 500–800 collectors (Figueroa & Dresdner 2016). By regulation, seed collectors can be placed in the wild from 1 October until 31 May of the following year, a period that includes the time taken to hang the collectors and market the seed (Carrasco *et al.* 2014). Collection occurs chiefly in rural areas and represents an important source of income for artisanal fishers and other small-scale aquaculture entrepreneurs (Fernández *et al.* 2018). Seed producers usually sell unit collectors (6–15 kg weight, seed 1–3 cm shell length) to mussel producers (Carrasco *et al.* 2014). They operate with sales contracts in which the price is settled before the collectors have been installed. Thus, the price does not depend strictly on the amount of seed fixed to each collector. This is because collector performance is considered exogenous by producers as it depends on environmental factors that are beyond their control (Figueroa & Dresdner 2016). Seed shortages occur periodically and are of significant concern to producers. For example, there was a 15% reduction in seed collection in 2012 when compared to 2011, and concerns were raised by industry to government (Carrasco *et al.* 2014). In Chile, there have been several studies on hatchery production of mussel seed for a number of species, including *M. chilensis*. One such study (Carrasco *et al.* 2014) found that, while technically feasible, seed production was not

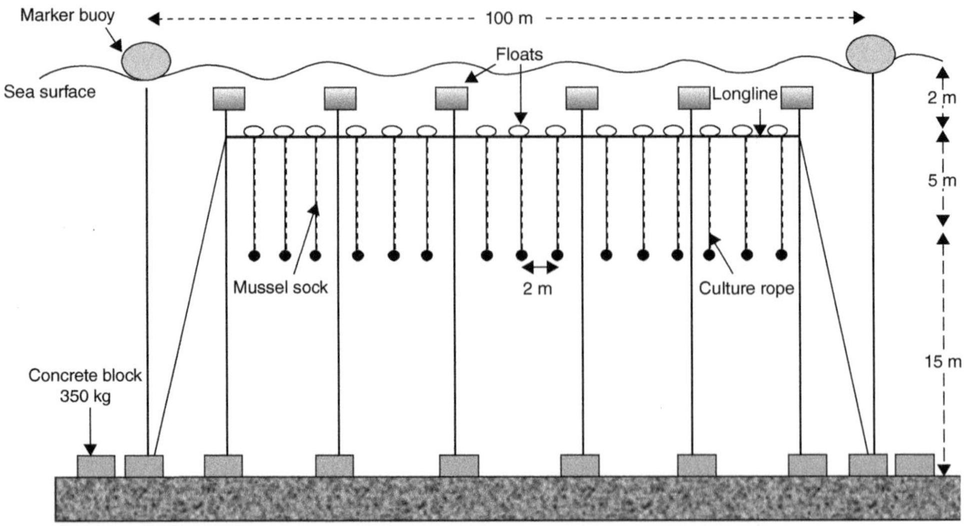

Figure 10.12 Schematic representation of the submerged longline system used for mussel culture in waters off the Atlantic coast of Morocco. Individual mussel socks are attached to the longline, which is weighted down at both ends with concrete blocks. *Source:* From Id Halla *et al.* (2017). Reproduced with permission from the International Journal of Scientific and Engineering Research.

profitable due to both the low price of Chilean mussels in national and international markets and the high cost of production, mainly associated with the production of microalgae (91% of total cost of production). Private initiatives have been scarce, and only three companies have built hatcheries that produce *M. chilensis* seed at an experimental level, and to date none of these is producing seed (details in Carrasco *et al.* 2014) except the Chinquihue Foundation, which is situated 12 km south of Puerto Montt and is the only facility producing mussel seed in Chile. It comes with a socially orientated goal: to help low-income fishermen and mussel farmers make a living wage.

Three types of legislation govern seed collection activity (Fernández *et al.* 2018): concessions, temporary permits (one season) and Management and Exploitation Areas for Benthic Resources (MEABRs). The latter are legal concessions that assign exclusive user rights to extract benthic resources in specified areas of the seabed (Gelcich *et al.* 2010). These permissions are exclusive for artisanal fisher organisations that are legally constituted. Once authorised, they can install collectors in up to 40% of the assigned area. The members of the organisation can organise in three different ways to carry out seed uptake activity in MEABRs: collectively as an organisation, in smaller groups of members of the organisation or as individual members. Mussel farms can obtain seed from their own concession or outsource to independent seed collectors. According to Fernández *et al.* (2018), the number of fishers working within the MEABR in mussel seed collection is not known because there is no official record in Chile of these producers. However, the total number of seed collector structures authorised to be installed in each MEABR known. In 2014, 86 production groups installed 805 700 collectors, representing 45% of the total collectors authorised to be installed in the MEABRs. See FAO (2005–2019) for information on aquaculture legislation in Chile.

Once collected, the seed is transported to growing (fattening) centres, where it is transplanted into a sock-like cotton tube (~17.5 cm wide), into which small mussels stripped from the collection ropes are inserted. This cotton sock, known as an independent sleeve, is then wound around individual vertical ropes known as 'droppers'. As the mussels grow, they move to the outside of the sleeve, causing the material to act in a similar way to a rope. The droppers are placed a minimum of 0.5 m apart and have at least 4 m of free space between them and the bottom. A first thinning is carried out after four months, and the entire crop is harvested within 12 to 14 (BIM 2006). Another sleeve system, known as a continuous sleeve, consists of a continuous line 6 m long laid in a sinusoidal manner and tied to the flotation system. In deeper waters, the gap between the bottom of the line and the seafloor is greater. Separation between longlines is largely dependent upon the size of the servicing vessel; sufficient clearance should be allowed to enable the vessel to work and manoeuvre safely. The density at which mussels can be cultured on longlines is about 300 m^{-1}, depending on food availability (which varies from site to site). Mussels grown on longlines can become smothered by juvenile mussels and other fouling organisms. For this reason, most farmers prefer to position their farms away from heavy spat settlement areas to avoid layers of spat attaching to larger mussels. Due to a lack of information on the Chilean situation, the following description pertains to longline mussel culture in the United Kingdom. As mussels grow on the longlines, the density on a dropper may become such that growth falls below optimal levels due to overcrowding. Some growers take action to remedy this by raising the droppers, removing the mussels, decreasing their density and then reattaching them enclosed in a new sleeve to a fresh dropper, a process called 'resocking' or 'retubing'. Other growers regard the time and effort expended in 'resocking' as not justified by the returns and allow 'natural thinning' to occur as mussels fall from the droppers under their own weight. Mussels are generally retubed at densities between 1.5 and 2 kg m^{-1} of rope. This can be done either manually or using automated machinery, the choice being largely determined by the level of production.

In Chile, once mussels reach their market size (minimum 50 mm shell length), they are harvested by machine and sent to processing plants. They are available year-round, with peak production from October through May. Processing plants are responsible for washing, declumping, debyssing, grading according to size and packing – a fully automatic, continuous process (Rivera *et al.* 2017). The final product can be sold fresh or processed depending on the market (local or international). For example, Sudmaris (www.sudmaris.com), a privately-owned processing plant on Chiloé Island, produces three products: individually quick frozen (IQF) mussels without their shell, with a glazing added to help keep their flavour and texture; IQF half-shell mussels; and whole mussels frozen in their own juices, which give all the flavour and texture of a fresh mussel, while allowing the possibility of storage for long periods of time. This method incorporates the vacuum sealing process VISTAPLUS or revitalising seawater, to retain the freshness and properties of the product before bagging and cooking processes. IQF does not enable large ice crystals to form on the meat, and since each mussel is individually frozen, they do not cohere, so that the final product is not frozen into a solid block. Sudmaris, Santa Marta produced a splendid video on *Top Quality Chilean Mussels* in 2016 that captures the full process from farm to processing plant, starting with a panoramic view of longlines in the Chiloé Island region, followed by the lifting of ropes to remove mussels and finishing with grading, packing, weighing and processing at the plant, where mussels are graded into size classes and packed for export (Sudmaris Chile SA 2016).

Besides Chile, longline systems are used in many European countries, such as Ireland, the United Kingdom, France, Germany, Greece and Italy, as well as Australia, Brazil, Canada, China, India, Korea, New Zealand and Thailand (Table 10.1). While the United States uses longlines for mussel culture, it is absent from Table 10.1 due to low annual yields.

Raft Culture in Spain

Spain is the third largest producer of cultured mussels in the world (Table 10.1). The following is a brief summary of the culture process, since a detailed description has already been published in Gosling (2015). The industry is mainly based in sheltered and nutrient-rich deep sunken valleys, called rías, which provide the perfect environment for suspended mussel culture. Production is based solely on rafts consisting of a framework of eucalyptus wood (~500 m^2) supported by cylindrical floats made of wood or steel covered with fibreglass or polyester or filled with expanded polyester (Figure 10.13a,b). The basic principle of raft culture is similar to that of longline culture in that the mussels are suspended on ropes, but these hang from the raft instead of from longlines. A raft has 200–700 hanging ropes, each a maximum of 12 m long, ~3 cm thick and made of nylon, polyethylene or esparto grass. Each raft normally holds three types of ropes: those for collecting seed, those with growing mussels and those with marketable mussels; in this way, growers maintain continuous production. Rafts are secured by mooring chains and a heavy concrete anchor and are spaced at a distance of 80–100 m from each other and located together in groups called parks. Currently, there are over 3337 mussel rafts in the Galician rías (Labarta & Fernández-Reiriz 2019). Most of these are family-owned and managed, a factor that contributes significantly to their low operational costs and ultimately to the great success of mussel culture in the region.

The culture procedure is divided into five phases: seed collection, attachment of the seed to the ropes, thinning of the ropes, on-growing and harvesting (details in The Fish Site 2010; Gosling 2015). Culture begins when farmers collect mussel seed, mainly from natural beds (60–70%, with the remainder collected from ropes hung from the rafts). The mussel seed fishery is managed on a combination of quotas and closed seasons. Mussel raft owners are entitled to harvest up to 1.5 t of mussel seed per raft from December to April (Piñeiro-Corbeira *et al.* 2018). They collect up to 1500 kg of seed per low tide, in about four hours, from the exposed rocky shores on the ocean side of the rias and islands, gathering around 4500 t of mussel seed (0.5–2.0 cm long) from these areas every culture cycle. They take the seed to the rafts, keeping it moist, and attach it to ropes, either by hand or by machine, within 24 hours after collection. To collect seed from the rafts, farmers use special nets made from old fishnets and suspend them during March and April. Within a few days, the mussels have attached themselves to the ropes. The farmers attach from one to three ropes per square metre of raft, a distribution that allows an adequate flow of water rich in food for the mussels and prevents the ropes from touching one another. The farmers install the ropes mainly from November to March. Piñeiro-Corbeira *et al.* (2018) have recently shown that harvesting young mussel seed for aquaculture is detrimental to the abundance and

(a)

(b)

Figure 10.13 (a) Galician mussel raft with others in the background. Photographed in summer. (b) Galician mussel raft showing the rope droppers on which the mussels are grown. No mussels are visible above 1.5 m depth as they must always be immersed in water. *Source:* Photos courtesy of A. Ovejero, Dpto. Geociencias Marinas y Ordenación del Territorio, Edificio Ciencias Experimentales, Universidade de Vigo, Vigo, Spain. (*See colour plate section for colour representation of this figure*).

diversity of the associated sessile assemblage not directly targeted by this activity. These differences continue until the next open season, suggesting that the closed season is too short for the recovery of the associated non-target sessile assemblage.

The next step is thinning, which prevents the mussels from falling off in rough weather and encourages rapid and uniform growth. Farmers do this mechanically, usually from June to October, when the mussels are half-grown (shell length 4–5 cm) and the average weight per rope is 48 kg. The mussels are removed from each rope, cleaned and redistributed on to two or three new ropes. The process is usually repeated before harvesting to ensure that all mussels reach a similar size at harvest time.

In the Galician region, where growth is rapid, mussels can attain market size (8–10 cm) in eight or nine months, although a high raft density can retard growth. Mussels of commercial size are available throughout the year and can be harvested at any time, but the main harvest is from October to March, when market demand is high and their condition is best. Currently, the mean annual production per raft has been estimated at 75 t, ranging between 60 and 84 depending on raft size, growing area and type of product (Labarta & Fernández-Reiriz 2019). In general, mortality of mussels is low on the rafts. Occasionally, extremely heavy rainfall may lower salinity to such an extent that mussels on the upper 0.5–1.0 m of ropes die. Predation from starfish or crabs is minimal as ropes are off the bottom, and disease and parasites do not appear to be a problem. Fouling by algae, tunicates and barnacles is more of an issue, as these organisms must be removed manually during on-growing and at harvesting before the mussels are sent for processing or depuration.

For harvesting, farmers use a crane to raise the ropes to their boat, where they separate and grade them by rubbing them over a grid of iron bars. They then wash them clear of small mussels, silt, empty shells, ascidians and other unwanted organisms. Any mussels too small for the market are wrapped on to new ropes for further growing. About 65% of harvested mussels are frozen or canned, while 35% are destined for the fresh-consumption market, but are first transported to depuration stations. Approximately 70% of the Spanish mussel production is destined for internal consumption, and the resulting 30% is exported, mainly to Italy and France (Silva *et al.* 2017).

For information on regulations governing licences, farmers' property rights and farming organisations, see Caballero Miguez *et al.* (2012), FAO (2014), Gosling (2015) and Labarta & Fernández-Reiriz (2019). Harmful algal blooms (HABs) are one of the main environmental factors affecting the cultivation of bivalves in Galicia, although, as argued by Rodríguez Rodríguez *et al.* (2011), their effect is only produced under certain circumstances. It is the responsibility of the Instituto Tecnológico para el Control del Medio Marino de Galicia (INTECMAR) to monitor the Spanish NW Atlantic coast (Dittami *et al.* 2013; Medlin *et al.* 2013).

While raft culture of mussels is a lucrative industry, it does have a number of other negative impacts apart from HABs. It causes an increase in sedimentation due to the trapping effect of the ropes and the metabolic activity of the mussels themselves (see Chapter 4). The annual input per raft to the seabed has been estimated as 69.3 t of sediment, 5219 kg of C, 620 kg N and 979 kg of Fe (references in Otero *et al.* 2009), causing significant changes in macrobenthic community structure (Ysebaert *et al.* 2009; Callier *et al.* 2009; Méndez Martínez *et al.* 2011). Spillage of chemicals used for maintenance of rafts entails a direct pollution risk to the immediate environment and the mussel culture itself, which can affect not only the quality of the product but also food safety if these chemicals are accumulated above the recommended values for human consumption (Ruiz *et al.* 2014). In addition, mussel rafts cause depletion in phytoplankton levels as high as 30% (Petersen *et al.* 2008) and depletion of different zooplankton groups of 26–77% (Maar *et al.* 2008). Other culture methods (see later) have similar negative environmental effects (see McKindsey *et al.* 2011; Grant *et al.* 2012; Toupoint *et al.* 2016).

Bouchot Culture in France

In France, *M. edulis* is cultured along the Channel and Atlantic coasts of Mont Saint-Michel, Normandy and the Côtes d'Armor region of Brittany, while *M. galloprovincialis* is cultured along the Mediterranean coast. Culture of *M. edulis* is based on the bouchot method (Gosling 2015; Toupoint *et al.* 2016; Aubin *et al.* 2018), a method that represents 65% of the country's production. Bouchot culture uses a line of poles of pine, oak or, more recently, Brazilian hardwood, 25–30 cm in diameter and 4–7 m long, with half their length embedded in the seabed (Figure 10.14). Spat settlement occurs intensively in spring, settling directly on to the wooden poles or on the horizontal coconut fibre ropes strung on them before settlement (Figure 10.15). When the seed is a few months old, it is removed from the ropes and transferred to tubular nets wound around the bouchots for on-growing. The nets eventually disintegrate, but by that time the mussel seed has spread to cover the entire post. A barrier is placed at the bottom of the pole to prevent predators such as crabs from reaching the mussels. This method of culture requires large tidal ranges, in order to supply the densely packed mussels with plankton.

Water conditions in Mont Saint-Michel Bay are not suitable for natural reproduction of *M. edulis*, so spat collection starts in March, mainly in the Charentes region, 300 km away, in the Bay of Biscay (Aubin *et al.* 2018). Seeded ropes are transported by road to Mont Saint-Michel and coiled around free bouchots, or else laid out on temporary horizontal structures built on the foreshore until new bouchots are available. A typical bouchot farm, often covering 20 ha or more, is situated perpendicular to the shoreline and consists of a single line of 110–130 poles running for about 100 m and spaced 10–30 m from the next line

Figure 10.14 Mussel (*Mytilus edulis*) culture on bouchots during low tide in the Chausey Archipelago, off the north coast of France. *Source:* Photo courtesy of P. Archambault, Départment de Biologie, Université de Laval, Canada. (*See colour plate section for colour representation of this figure*).

Figure 10.15 Spat-collecting ropes covered with *Mytilus edulis* in the Chausey Archipelago, off the north coast of France, just before the young mussels are removed from the ropes, loaded into mesh stockings and transferred to the bare wooden posts behind them. *Source:* Photo courtesy of P. Archambault, Départment de Biologie, Université de Laval, Canada.

(McKindsey *et al.* 2011). Farms are mostly family-run, each with 15 000–20 000 poles on average. Appellation d'Origine Contrôlée (AOC) restricts the percentage of bouchots that can be in use concurrently (55–65%, depending on the subarea), as well as the maximum mean annual yield (60 kg per bouchot; Aubin *et al.* 2018).

Once mussels start growing, producers stabilise the structure of the bouchot by progressively encasing each in three to seven polypropylene nets, depending on the wind and sea current exposure of their sites. To reach the culture sites, producers use tractors or small aluminium or amphibious boats (Aubin *et al.* 2018). Marketable mussels (>4 cm shell length) are harvested when they are 12–18 months old. After harvest, they are purified for at least 24 hours. They then go through a declumper, which separates them and shreds the ropes and nets, then cleaned in a brushing machine and calibrated, first mechanically, then by hand, to remove other waste (crabs, algae, broken mussels, etc.). Following this process, they are placed in bags of 'traditional' or 'ready-to-cook' mussels (byssus removed) for sale. Bouchot mussels are easily identifiable: their shell is small, they are meaty, their flesh is often yellow or orange and their taste is distinctive. In order to protect consumers, mussel growers have registered the brand 'Moules de Bouchot'. Thus, bouchot mussels' labels must have the official logo. Also, the French authorities took the unusual step in 2006 of granting the coveted AOC tag to the Mont Saint-Michel bouchot mussel. It is the first marine product to be awarded the honour, which is usually applied to fine wines and cheeses.

Bottom Culture in The Netherlands

This section has benefited greatly from information in Capelle (2017), Capelle *et al.* (2014, 2016a,b), Kamermans & Capelle (2019) and the website www.mosselen.nl. The mussel *M. edulis* is the most important aquaculture product from The Netherlands. Roughly 50 companies, with 60 vessels, produce approximately 40 000–50 000 t of mussel each year (2012–2016) on bottom-culture plots in the western part of the Wadden Sea (48%) and in the Oosterschelde estuary (52%). Large-scale bottom culture of *M. edulis* is almost the only production method in The Netherlands and Germany, and takes a substantial share in mussel production in Great Britain and Ireland.

Mussel bottom culture depends on mussel seed (0 year class mussels) dredged in autumn and spring by scraping mussels from natural subtidal seedbeds in the Wadden Sea, the Eastern Scheldt and the Dutch coastal waters of the Voordelta. Every spring and autumn, researchers from Wageningen Marine Research (commissioned by PO Mosselcultuur and the Ministry of Economic Affairs) assess the amount of mussel seed that is present at the designated beds. Next, the mussel growers apply for a permit from the Ministry of Economic Affairs granting them permission to catch some of this seed, taking into account the food needs of birds and allowing space for the development of old mussel beds. The mussel seed that is caught is next transferred to nurseries. In addition to these environmental concerns, the supply of seed shows large fluctuations. It is extremely difficult to collect sufficient amounts of mussel seed each year. As a result, Dutch mussel farmers, since 2009, have made use of seed mussel collectors (SMCs). Their introduction was the practical outcome of a covenant from 2008 between the mussel growers and the Wadden Sea Association, the Society for the Protection of Birds, the Dutch Society for Nature Conservation and the government. An SMC consists of nets or ropes suspended in the water, to which mussel larvae attach themselves and grow into mussels of 1–2 cm shell length. They are harvested from July until October, and are a reliable source of mussel spat. For example, in 2016, SMCs accounted for 32% of all seed produced in The Netherlands. According to the covenant, the mussel sector is gradually to decrease its dependence on natural mussel seed in the Wadden Sea. However, the introduction of SMCs has resulted in considerable cost increases, which will require innovations in other forms to offset, such as an increase in production efficiency. In addition, two companies in The Netherlands have started hatchery production of *M. edulis* on a pilot scale (Garcia & Kamermans *et al.* 2015).

Seed harvesting, either by fishery on natural seedbeds or from SMCs, is followed by the seeding/relay of mussels (2–3 cm shell length) on culture plots in the Wadden Sea or Oosterschelde. Commercial beds are laid at high tide by mussel vessels that flush the seed through shafts below water level. While seeding, the vessel moves in circular patterns. As a result, mussels are distributed on multiple plots in concentric patterns. There are ~7700 ha of leased culture plots, of which 4000 ha is effectively used by mussel-producing companies. Seeding of high biomasses (up to 150 t) is done as fast as possible around slack tide, to prevent mussels flushing out from the lease site via tidal currents. This results in highly concentrated mussel formations (Figure 10.16) that may be multi-layered within seeding tracks, especially where seeding tracks overlap. Seeding is the most critical step in the mussel bottom-culture cycle. Most of the seed does not survive the cycle: average survival of SMC seed or seed relayed from wild mussel beds in autumn over the culture cycle is 8 and 15%, respectively (Capelle *et al.* 2016b). The factors that contribute to losses are handling stress, damaged mussels, competition for food and space (both of which are

Figure 10.16 Mussels (*Mytilus edulis*) laid at an intertidal lease site at the Oosterschelde estuary, The Netherlands. *Source:* From Capelle *et al.* (2014). Reproduced with permission from Elsevier.

density-dependent) and predation from starfish in the subtidal area and birds and shore crabs in the intertidal zone.

During the grow-out phase, mussels are often relayed between plots within the same system or from the more dynamic Wadden Sea to the Oosterschelde, where conditions are more sheltered and the risk of storm damage (dislodgement) is lower. On the culture plots, the mussels grow from seed to half-grown size (3.0–4.5 cm) up to consumption size (>4.5 cm), when they are harvested and sold at the mussel auction in Yerseke, the only mussel auction in the world. Through the auction, mussels are transported to markets in Belgium, Germany, The Netherlands and, to a lesser extent, France. Based on both domestic production (54 000 t in 2016) and imports (23 000 t in 2012) of live mussels, The Netherlands has developed a solid processing and export-orientated industry.

Pole/Stake Culture in Thailand

Mussel culture activities have been practiced in Thailand for more than 100 years. The green mussel, *P. viridis*, which has been cultured for the past 60 years, is the most important mussel species (Keawtawee *et al.* 2018). The farms are located in intertidal areas on the Pacific and Indian Ocean coasts, and are usually close to estuaries. Production for *P. viridis*, however, has declined substantially from 291 023 t in 2002 to 115 488 t in 2016 (FAO 2018b), which can be ascribed to the rapid population growth in coastal communities, industrial development and a concomitant decline in water quality. Details on the culture methods used in Thailand can be found in Aypa (1990), Hickman (1992), Chalermwat *et al.* (2008) and Fish Consulting Group (2012, 2013).

The pole method, the traditional culture method, involves embedding bamboo poles at a depth of 4–8 m during low tide. The poles are 4–6 m in length and are staked firmly at the bottom in rows spaced 0.5–1.0 m apart. In shallow waters, the number of poles per farm is between 2000 and 8000, while in deeper waters the farms are larger and can use as many as

250 000 poles. In areas where the water current is strong, bamboo poles are kept in place by nailing long horizontal bamboo supports between the rows. Since mussels need to be submerged at all times, it is not necessary that the tips of the poles protrude above the low water level after staking. However, boundary poles should extend above the high water level. In staking, enough space between plots is allowed for the passage of the farmer's small boat during maintenance. After setting the poles/stakes, very little additional attention is needed, as this system depends completely on abundant natural spatfall. Mussels settle on the submerged poles at a rate of 2000–3000 spat m^{-1}. Growth is rapid, with a harvest size of 5–10 cm shell length attained in 6–10 months after settlement - a prerequisite to offset the constraints of the monsoon season and the rapid deterioration of the bamboo poles, both of which can lead to substantial mussel losses. Each pole yields about 8–12 kg of mussels.

Harvesting is done by pulling out the poles and bringing them ashore on a small boat. Divers are employed to pick out the larger mussels, and the small ones are left for the next harvest season. This selective harvesting results in two or more yields within the 6–8 months of the farming period. Harvested mussels are cleaned and then placed in baskets and shaken vigorously in seawater until they are clean of barnacles and dirt. Bamboo poles that are worn out are removed, while the good ones are cleaned for the next culture season. The pole/stake method is an economical and easy way of growing mussels but also has some shortcomings. Bamboo decays easily, and at times it is difficult to coincide staking operations with spatfall. The system also facilitates siltation, which makes some bays and estuaries too shallow for mussel farming.

The wigwam system, a development of the bamboo pole method, requires a central bamboo pole serving as the pivot from which eight full-length bamboo poles are made to radiate by firmly staking the butt ends into the bottom and nailing the tops to the central pole, in a wigwam fashion. The stakes are driven 1.5 m apart and 2 m away from the pivot. To further support the structure, horizontal bamboo braces are nailed to the outside frame above the low-tide mark. Spat settle on the poles and grow to harvest size in 8–10 months. Alternatively, a fixed suspended rope-web system can be used, consisting of rope strung in a zigzag fashion between two bamboo poles in shallow water. This system uses the area between the poles to grow the mussels. Another modification on the bamboo pole system is the hanging or rack method used for *P. viridis* cultivation in the Philippines and India. The racks are made by placing bamboo poles vertically and connecting them horizontally with poles. Artificial collectors are suspended from the horizontal poles over or near natural spawning grounds of the mussels. This method is suitable for estuaries and shallow bays.

Although traditional aquaculture methods are still very prevalent, the quantity and value of farmed shellfish in Thailand was increased through the introduction in 1994 of nontraditional culture systems such as rafts and longlines. Their advantage is that cultured mussels are kept totally submerged all the time, avoiding dehydration due to air exposure during low tide and thus enhancing mussel growth. Furthermore, materials used for raft farm construction, such as rope and buoys, can be reused with the next crop, unlike bamboo poles. Longline techniques also appear to have significant potential to further increase mussel production. The growth in rafts has aroused rising concern over their potential impact on the marine environment, however. In a recent study, Vichkovitten *et al.* (2017) investigated the impact of raft mussel culture on the benthic environment of mussel farms in the inner Gulf of Thailand. Not surprisingly, they found that the deleterious effects were similar to those already described for raft culture in Galicia, NW Spain.

Offshore Mussel Culture

Most of the planet's population has decided that its primary habitat in the Anthropocene[3] is on the world's coasts, hence there are fewer and fewer coastal options for the development of marine aquaculture. Farming in offshore marine waters is the only potential option for increasing seafood production, and has been a focus of international attention for more than two decades (Buck & Langan 2017 and references therein). Favourable features of open ocean waters include sufficient space for expansion, enormous carrying and assimilative capacity, reduced conflict with user groups, lower exposure to human sources of pollution, potential to reduce some of the negative environmental impacts of coastal fish farming, optimal environmental conditions for a wide variety of marine species (Holm *et al.* 2017) and reduced biofouling (Atalah *et al.* 2016).

Simple hanging culture longline systems have been adopted by a number of existing offshore mussel farms around the world. These can be surface-floating, semi-submerged or fully submerged according to local conditions. One company, Offshore Shellfish Ltd (OSL; www.offshoreshellfish.com), is developing the United Kingdom's first large-scale offshore mussel (*M. edulis*) farm, situated 5–10 km off the coast in Lyme Bay, South Devon, England. OSL is the first company in Europe to install a large-scale, rope-grown mussel farm in fully offshore conditions. In 2010, permission was granted for the farm, which will have a water depth between 20 and 28 m and an area of 15.4 square km – about the size of the entire site of Heathrow Airport, London. The farm equipment specified in the permit and lease application consists of 790 longlines. The plan is to tap into the European market, selling into main processors in Holland, which supply 100 000 t annually to France and Belgium. It is expected that the farm will create around 70 jobs and generate millions of pounds for the Lyme Bay economy. The practical procedures and requirements experienced in order to obtain the various permits and leases needed to establish this and other offshore mussel farms are described in Corbin *et al.* (2017).

One of the key charges for the OSL farm was to develop a flotation system that could cope with the rigours of severe winter storms. A unique new mussel farm flotation system, jointly developed by OSL and Fusion Marine Ltd for use in exposed sea sites, is now in successful operation and is already helping to produce high-quality, fast-growing mussels. The two companies developed a float design concept based on a vertical-axis system, as opposed to the traditional horizontal-axis barrel floats normally used to suspend mussel ropes (Figure 10.13b). The vertical-axis system consists of single headlines buoyed by vertical floats made from high-density polyethylene (HDPE). The floats also have end caps offering strong rope connection points that can be attached securely to the headlines (Figure 10.17). Compared to barrel-type surface floats, these move more gently in rough seas and can be submerged to avoid any extreme force on the ropes, thus preventing mussels from

Figure 10.17 (a) Vertical-axis flotation system developed for mussel farming operations in high-energy offshore environments. (b) Floats made from robust and durable high-density polyethylene (HDPE), which is ideal for the demanding conditions found in offshore aquaculture. (c) Machined HDPE end-cap pin securing a rope to the top of a float. *Source:* Photos courtesy of Gael Force Fusion, Argyll, Scotland, UK.

becoming dislodged. In 2018, the offshore farm was in its build-out phase, with 700 floats already supplied. It is expected to be completed in 2022, when it should be producing up to 10 000 t of mussels per year. In September 2018, OSL obtained Best Aquaculture Practices (BAP) certification. It is the first mussel farm in Europe to earn this distinction. A division of the Global Aquaculture Alliance (GAA), BAP is an international certification programme based on achievable, science-based and continuously improved performance standards for the entire aquaculture supply chain that assures healthful foods produced through environmentally and socially responsible means.

At present, other offshore mussel developments are either well established, as in Belgium, The Netherlands, Germany, Norway, France, Italy, Portugal, Morocco, the Pacific coast of the United States, China and New Zealand (Buck *et al.* 2017a; Id Halla *et al.* 2017), or operating on a pilot scale, as in Canada, Ireland, Denmark and the Atlantic coast of the United States.

High and reasonably steady wind speeds occur regularly in offshore areas, making such areas prime candidates for renewable energy production by wind energy farms (Buck *et al.* 2008). Since the 1990s, the number of offshore wind farms has increased, particularly off European coastlines (according to Wikipedia, there were 16 farms operating in Germany in 2010–2017, three in Denmark in 2002–2019, 15 in the United Kingdom in 2000–2015 and four in The Netherlands in 2008–2017). In Germany, in the early 2000s, plans for the massive expansion of wind farms in offshore areas of the North Sea triggered the idea of a combination of wind turbines with other uses.

Buck *et al.* (2004, 2008), using modified cultivation strategies, showed that the cultivation of macroalgae and shellfish was biologically and technically feasible in an offshore wind farm (Figure 10.18). The solid groundings of wind turbines can serve as attachment points for the aquaculture installations and become the key to the successful commercial cultivation of aquatic organisms. Also, culture in between the turbines is attractive, since public access is restricted, and thus the culture site is protected from outside influences. In offshore test trials in Germany, the most suitable algal and shellfish candidates

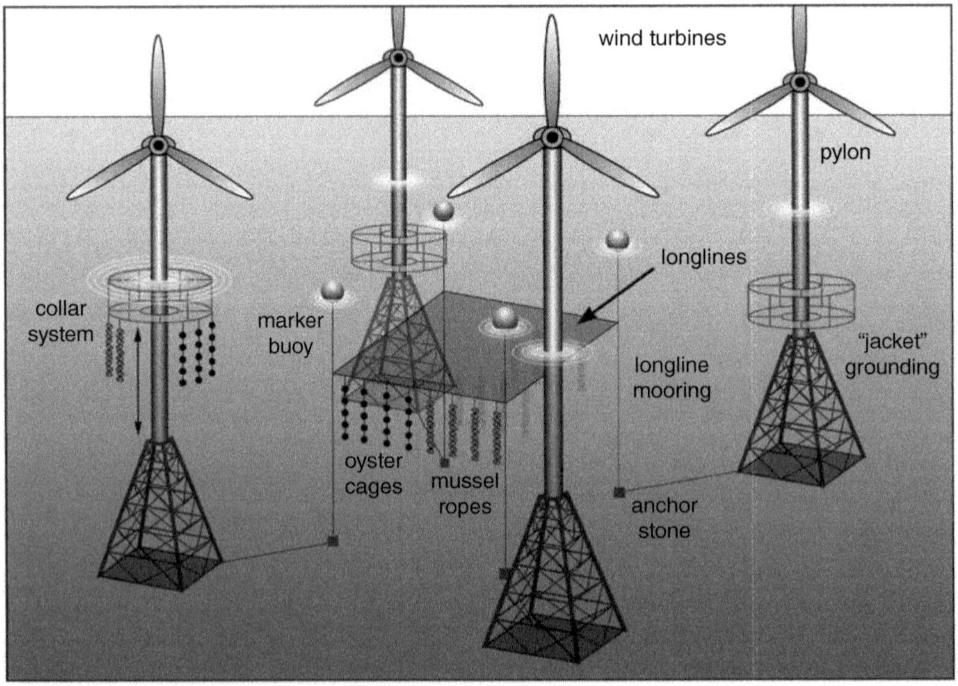

Figure 10.18 Potential multifunctional use of the fixed underwater structures of offshore wind turbines for the maintenance of aquaculture facilities. Alternative solutions include oyster cages and mussel collectors attached to longlines in the inner section of the wind farm, or offshore rings (collar systems) attached directly to the pylon. The latter system can be submersed in case of wind turbine set-up. The collar was the first modern structure in the world that enabled mass oceanic seaweed cultivation on an industrial scale. *Source:* From Buck *et al.* (2004). Reproduced with permission from Elsevier.

suggested and tested for offshore extensive aquaculture[4] were the sugar kelp (*Laminaria saccharina*), oarweed (*L. digitata*), dulse (*Palmaria palmata*), the mussel *M. edulis* and two oyster species, *Crassostrea gigas* and *Ostrea edulis*. This is not surprising, since mussels and seaweeds are cultured mainly in extensive systems throughout the world (Buck & Krause 2012 and references therein). France and Germany were the first investing countries for offshore seaweed cultivation, followed by the United Kingdom and Ireland 10 years later. Today, The Netherlands, Denmark, Spain, Portugal, Norway, many Asian nations, the United States and Canada are carrying out seaweed farming on a pilot scale. Only a few farms are in early commercial operations, including enterprises in Germany, Norway and The Netherlands, with Germany one of the leading countries in offshore seaweed science (Buck *et al.* 2017b). However, plans to connect multi-use offshore culture of mussels and macroalgae to the offshore foundations of wind turbines are still in their infancy in Germany, Belgium, Norway, The Netherlands and China (Buck *et al.* 2017a). In contrast, the Republic of Korea in 2016 started the construction of the first large-scale offshore wind farm, with the objective of ultimately using the site to combine seaweed and bivalve culture.

Mussel Culture and the Environment

Over the past 15 years, mussel production from aquaculture has increased from 877 884 t in 2001 to almost 2.0 million t in 2016 (Gosling 2015; Table 10.1), achieved through the expansion of areas of land and water under culture. Increased production translates inevitably into increased impact on the environment surrounding mussel culture activities. Suspension-feeding bivalves have a direct impact on phytoplankton communities in coastal waters. Their large grazing potential causes depletion of phytoplankton biomass above the seabed, but also increases phytoplankton productivity through recycling of inorganic nutrients from suspended particulate organic matter (POM), thereby exerting both top-down (grazing effects) and bottom-up (nutrient recycling) control on phytoplankton communities. These effects have already been dealt with in some detail in Chapter 4. Fortunately, bivalve culture does not impact on the environment in the same way that intensive fish culture operations do. Fish farms, with their high inputs of wastewater, feeds, fertiliser and chemicals, are now considered potential polluters of the aquatic environment and the cause of degradation of wetland areas. The negative effect of fish farming on the environment is often further compounded by the frequent use of chemicals, which in turn can have serious repercussions when bivalves are cultured in the vicinity of fish cages or pens.

There are many positive effects of suspension-feeding bivalve culture on shallow-water ecosystems. Bivalves help buffer estuaries and coastal ocean waters against excessive phytoplankton blooms in response to anthropogenic loading of nitrogen, counteracting the symptoms of eutrophication of coastal waters, which is a serious environmental problem in many regions (Lindahl *et al.* 2005 and references therein). The potential of mussel farming to improve coastal water quality in marine waters is well known (Petersen *et al.* 2016; Suplicy 2018). Hence, there is an increasing awareness of the mitigation potential in mussel culture in relation to eutrophication, but its implementation in European countries as well as worldwide is only in its infancy (Petersen *et al.* 2016). Mussels and oysters, in particular, remove inorganic sediments from suspension, thereby counteracting coastal water turbidity (reviewed in Gallardi *et al.* 2014). Also, biodeposition created by these bivalves through the creation of sediment anoxic microzones where denitrifying bacteria are promoted induces denitrification, which also helps to counteract eutrophication by returning nitrogen into the atmosphere as inert nitrogen gas (Gallardi *et al.* 2014 and references therein). Another way in which mussels might contribute to reduce ocean eutrophication is through their combination with marine fish farms in Integrated Multi-Trophic Aquaculture (IMTA) (Reid *et al.* 2010, 2013). This is a practice in which byproducts (wastes) from one species are recycled as inputs (fertilisers, food and energy) for another. It combines species that need supplemental feed such as fish with filter feeders (e.g. mussels; Irisarri *et al.* 2013), deposit feeders (e.g. sea urchins, sea cucumber) and seaweeds (e.g. kelps; Buck *et al.* 2017a). Species selections, combinations and proportions will be highly variable depending on the local conditions and biodiversity (reviewed in Barrington *et al.* 2009; Soto 2009; Buck *et al.* 2017a; Suplicy *et al.* 2018). The mitigation culture concept has scientifically been tested mainly on a pilot scale or through modelling for various bivalve species at both system and local scales (Petersen *et al.* 2016 and references therein), but the modelling has mainly been based on experiences from commercial mussel farming. Navarette-Mier *et al.* (2010) tested whether *M. galloprovincialis* and *Ostrea edulis* cultures located in the proximity of sea bream and sea bass floating net cages in the western Mediterranean would reduce the environmental impact caused by fish organic wastes. There was no indication that either oysters or mussels fed on fish wastes, demonstrating that polyculture of fish and bivalves does not represent an appropriate tool for reducing the environmental impact of fish aquaculture in open water. Irisarri *et al.* (2013), Cranford *et al.* (2013) and Sanz-Lazaro & Sanchez-Jerez (2017) have reported similar findings. Cranford *et al.* (2013, p. 71) point out that 'these limitations should not be interpreted as a judgment on the effectiveness of IMTA systems as a whole. It may be possible to design systems that address inefficiencies in, or impacts from, one component through the addition of other extractive species'. For example, Holdt & Edwards (2014) found that the use of mussels (*M. edulis*) and seaweeds (kelps) as biofilters of nitrogen within an IMTA system provided a possible mechanism for fish farmers to reduce the nitrogen loss from salmonid cultivation in sea cages.

Mussel farming also promotes and enhances biodiversity by creating structure and habitats for other marine species (Suplicy 2018; details in Chapter 4). The degree of biodiversity depends on the new habitat created by the method of cultivation. Murray *et al.* (2007) made a quantitative comparison of macrofaunal communities associated with natural intertidal seedbeds (*M. edulis*) and with cultivated mussels grown either on the seabed (bottom culture) or on ropes. The biomass of the macrofauna associated with mussels was significantly greater within rope-grown mussel assemblages than on mussel beds. The ropes themselves are a habitat that also attracts many epifaunal and infaunal species (Murray *et al.* 2007), although they can also provide substratum for the attachment of fouling organisms and non-indigenous species (Pochon *et al.* 2015 and references therein). Also, the mussel farming habitat can act as a floating artificial reef system for both pelagic and benthic fish species (Wang *et al.* 2015). The accumulation of shells and shell hash on the bottom and on nearby shores within mussel farm sites due to fall-off and other processes may be considerable, and adds physical structure to the benthic environment (reviewed in McKindsey *et al.* 2011). Comeau *et al.* (2015) estimated that 89% of the spat seeded on mussel lines in Prince Edward Island, Canada is lost through fall-off prior to harvesting. Such benthic accumulation should be expected to attract a variety of benthic predators. A project investigating the impact that Europe's biggest offshore mussel farm in Lyme Bay, United Kingdom (see earlier) is having on the surrounding ecosystem found that fallen mussels attract a wide range of species, such as crabs, lobsters, starfish and fish including pout, plaice and horse mackerel (The Fish Site 2017). Mussels in suspended culture have traits (i.e. high meat content and thin shells) that make them particularly appealing to a number of diving ducks (Kirk *et al.* 2007; see Chapter 3). The only effective method providing complete and long-term control of bird predation is the use of exclusion nets deployed around longlines and rafts of suspended mussel ropes (Varennes *et al.* 2013). Suspended bivalve culture structures (e.g. buoys, platforms) are used as resting places for a variety of sea birds (Figure 10.13a). Roycroft *et al.* (2004) reported a greater number of species and a greater abundance of birds in suspended mussel culture sites in Ireland than in a series of control sites, which they attributed mainly to the provision of perching areas and to the diverse communities of organisms growing on the farm-associated gear, which offer an attractive food source for a variety of species.

Mussel ropes in suspended culture create favourable habits for other invertebrates by providing refuges from predation and adverse environmental conditions, as well as a direct source of food for other invertebrates (McKindsey *et al.* 2011 and references therein). Almost 100 species, including gastropods, crustaceans, bivalves, polychaetes, ascidians, sponges, bryozoans and hydroids, have been identified on ropes (Hickman 1992). Although these organisms, collectively termed 'biofouling', enhance biodiversity of mussel culture, they also represent a costly problem for the aquaculture sector (Fitridge *et al.* 2012; Sievers *et al.* 2013; Woods *et al.* 2012; Forrest & Atalah 2017). The key impact is the direct fouling of stock, causing physical damage (e.g. around the hinge and lip, affecting feeding and susceptibility to predators), mechanical interference due to colonisation of the shell, biological competition for food and space and environmental modification due to colonisation of culture infrastructure (reviewed in Fitridge *et al.* 2012). Of particular concern is the fouling of mussel culture operations by invasive tunicates (Adams *et al.* 2011). The invasive tunicates *Ciona intestinalis* and *Styela clava* have become nuisance species and a detriment to the mussel (*M. edulis*) culture industry on Prince Edward Island and elsewhere in Atlantic Canada (Lutz-Collins *et al.* 2009 and references therein). These species have the potential to reduce water flow and the availability of food and oxygen, thereby harming mussels and sessile filter-feeding epifauna. The recent human-mediated introduction of the invasive colonial ascidian *Didemnum vexillum* (Figure 10.19) looks to threaten New Zealand's highly valued green-lipped mussel, *P. canaliculus*, industry in the Marlborough Sounds aquaculture region (Fletcher *et al.* 2013). Since 2001, when it was first detected, *D. vexillum* has been inadvertently spread within the region by anthropogenic transport mechanisms. Economic estimates of biofouling control are consistently between 5 and 10% of production costs (equivalent to US\$1.5–3.0 billion yr^{-1}), illustrating the need for effective mitigation methods. Methods to avoid, mitigate or prevent the effects of biofouling in shellfish culture fall into five broad categories: (1) avoidance of natural recruitment to

Figure 10.19 Biofouling of the green-lipped mussel, *Perna canaliculus*, by the invasive ascidian, *Didemnum vexillum*, from a farm in the Marlborough Sounds region of New Zealand. *Source:* From Fletcher *et al.* (2013). Reproduced with permission from Aquaculture Environment Interactions.

prevent settlement and growth of biofouling; (2) physical removal, ranging from scrubbing and brushing to chemical dips and sprays; (3) biocontrol using natural species; (4) coatings on shells; and (5) control and protection for equipment using antifouling coatings and organic biocides. A brief summary of these methods, based on a review by Fitridge *et al.* (2012), is provided in the following paragraph.

Temporarily removing shellfish from the depth level favoured by fouling organisms during the peak period of settlement enables fewer larvae to colonise, but this strategy is only effective for regions where fouling is predictable and seasonal. Methods of fouling removal, each with varying success, are air exposure, power washing, immersion in freshwater, exposure to brine solutions, heat treatment and spray and immersion techniques using acidic and alkaline chemicals. In mussel culture, low concentrations of acetic acid are particularly successful against soft-bodied tunicates and against algae, whether applied by immersion or spray techniques. Carman *et al.* (2016) found that brine baths killed tunicates but caused relatively high juvenile mussel (*M. edulis*) mortality, although less mussel death occurred in the short-term (10 sec) brine bath (6–17%) compared to the long-term (20 sec) brine bath (8–30%). Both long-term (24 hr) and short-term (8 hr) freshwater baths were effective against tunicates, with less mussel mortality (2%) occurring in the short-term bath. Tunicates survived short-term (5 min) freshwater sprays but not long-term

(10 min) freshwater sprays. The latter caused slightly more mussel mortality (4%). Each treatment demonstrated varying degrees of effectiveness, yet the freshwater short-term baths and sprays were able to remove tunicates while maintaining high survivorship among juvenile mussels.

Biological control, whereby predation on foulants by other marine organisms is used to manage fouling levels, is an attractive strategy as it negates the need for costly physical and chemical treatments. To my knowledge, there is only one study in which biocontrol has been tested on fouled mussel (*Mytilus* sp.) sleeves. In an effort to control overgrowth by the invasive tunicate *Ciona intestinalis*, Comeau *et al.* (2012) transferred Atlantic rock crab (*Cancer irroratus*; N = 73 000) to an estuarine mussel aquaculture site in Prince Edward Island, Canada. They found that this was not a cost-efficient method of controlling fouling tunicate populations on bivalve aquaculture sites because the effort needed to maintain the density of these extremely mobile crabs would be prohibitive.

The development of coatings technology to mitigate biofouling in shipping and on other marine infrastructure for finfish culture has instigated some development in shell coating technology within the shellfish industry. Bakker *et al.* (2011) found that the application of food-grade oil to farm buoys, ropes and floats reduced fouling by algae and tunicates in mussel culture by more than 90%. Coating of live pearl oysters with a biodegradable, wax-based, impervious, nontoxic coating was successful in treating oysters heavily infected with boring sponges, although it was unsuitable as a long-term antifoulant control, with surface integrity lasting only two or three months before degrading (de Nys & Ison 2004). Currently, the use of bivalve coatings is feasible only for certain species (e.g. oysters; Watson *et al.* 2009). Recent research has shown that mussel (*M. edulis*) periostracum extracts display some antifouling effects, while the periostracum also physically deters the settlement of algal (*Porphyra suborbiculata*) spores (Kang *et al.* 2016), suggesting that mussel periostraca or their associated biomimetic materials may become environmentally-friendly antifouling materials preventing the attachment of diverse fouling organisms.

Finally, a major concern for aquaculture is the introduction of alien (often referred to as 'exotic', 'non-native' or 'non-indigenous') species. Alien species may compete with native ones for the same resources and may carry pests, predators and diseases to which native species are more vulnerable. Introductions of exotic species can result from activities associated with bivalve aquaculture (e.g. intentional introduction of exotic species into an area for aquaculture purposes), or they may be associated with introduced bivalves or facilitated by aquaculture activities (i.e. structures or husbandry practices). Introduced species can include both 'hitchhiking' species (organisms that grow in association with or are transferred by cultured bivalves) and disease-causing organisms (reviewed in McKindsey *et al.* 2007). In addition, secondary introductions may occur from the expansion of exotic species from the initial place of establishment. Using the inventory of marine alien and cryptogenic (unknown origin) species, archived by the European Alien Species Information Network (EASIN, https://easin.jrc.ec.europa.eu/easin) and updated as of October 2013, Katsanevakis *et al.* (2013a,b) identified 1369 alien marine species in 34 countries in European Seas – a substantial increase from the 737 species previously reported in 2009. Mollusca was the most numerous phylum, followed by Arthropoda, Chordata and Annelida (Katsanevakis *et al.* 2013b; Nunes *et al.* 2014). The main pathways by which alien species have been introduced are shipping, the Suez Canal and aquaculture. Since the opening of the Suez Canal in 1869, more than 300 tropical and subtropical species of algae, invertebrates and fish coming from the Red Sea and the Indian Ocean have dispersed through the Canal and invaded the Eastern Mediterranean (Sarà *et al.* 2008). Israel is the country with the highest number of recorded first introductions in European Seas, followed by Turkey, France and Italy. However, while species introductions to Turkey and Italy were mainly due to shipping activities (58 and 67%, respectively), in France shipping represented only 39% of the introduction events, with aquaculture responsible for 51% (Nunes *et al.* 2014).

The species most widely and deliberately introduced around the world for shellfish aquaculture is the Pacific oyster, *Crassostrea gigas*, originating in Asia (McKindsey *et al.* 2007). In North America and the Australasia-Pacific regions, *C. gigas* is known to settle into dense aggregations and exclude native intertidal species. It has been documented destroying habitat and causing eutrophication of the water bodies it invades (GISD 2015a). It was repeatedly introduced into Europe for aquaculture purposes in the second half of the 20th century (Lallias *et al.* 2015; Anglès d'Auriac *et al.* 2017), accompanied by the inadvertent introduction of the seaweed *Sargassum muticum*, which has spread along the coasts of France and England, forming dense obstructive beds. In a simulated transfer of *C. gigas* in the Thau Lagoon, France, a total of 57 macroalgal taxa, including 16 not indigenous to the Thau Lagoon, were recorded, demonstrating that oyster transfers are effective as vectors in macroalgal introductions (Mineur *et al.* 2007). In the 19th century, two predatory gastropods, the oyster drill *Urosalpinx cinerea* and the slipper limpet *Crepidula fornicata*, accompanied the introduction of the oyster *Crassostrea virginica* into Europe. Phytoplanton (toxic and otherwise) may be transferred on the bivalve shell or as resting cysts in the digestive tract or faeces. To illustrate, in 1993 more than 60 species of phytoplankton were identified on oysters imported from France into Ireland (O'Mahony 1993). Disease organisms can also be introduced through transfers from one country or region to another. For example, *Bonamia*, which is responsible for the decline in flat oyster, *Ostrea edulis*, populations in Europe and elsewhere, is believed to have initially been spread by movements of oysters in the United States from California to Maine and Washington, and overseas to France and Spain (Bower 2015).

There are at least 10 alien mussel species documented in the literature, including the Mediterranean mussel, *Mytilus galloprovincialis*; the brown mussel, *Perna perna*; the green mussel, *P. viridis*; the ribbed mussel, *Geukensia demissa*; the Pacific mussel, *Semimytilus algosus*; the Asian date mussel, *Musculista senhousia*; and the charru mussel, *Mytella charruana* – the impacts of which will now be described. *M. galloprovincialis* has been recognised as being among '100 of the world's worst invasive species' (Lowe *et al.* 2000). The mussel, native to the Mediterranean coast and the Black and Adriatic seas, has established itself at widely distributed points around the globe. Ship hull fouling and transport of ballast water have been implicated in its spread (GISD 2015b). In the early 1990s, it was introduced into southern California, United States, where it has since displaced the native *M. trossulus* throughout the southern portion of its range (Lockwood & Somero 2011). In the late 1970s, it was accidentally introduced on the west coast of South Africa, where it has progressively replaced the indigenous, slower-growing mussel *Aulacomya atra* from wave-exposed shores (Branch & Steffani 2004). In the 1980s, it rapidly spread on to the South African south and east coasts, where it competes with the indigenous mussel *P. perna*. It is immune to the trematode parasites that are common in *P. perna* and that reduce both growth rates and reproductive output by castrating infected females (Calvo-Ugarteburu &

McQuaid 1998). Overall, *M. galloprovincialis* has caused significant ecological impacts, including competitive displacement of indigenous species (see earlier) and a dramatic increase in intertidal mussel biomass. These changes have also increased available habitat for many infaunal species and have resulted in enhanced food supply for intertidal predators (Robinson *et al.* 2005). However, considerable economic benefits have also resulted from this invasion, since the species forms the basis of the South African mussel culture industry. In 2009, a second marine mussel, *Semimytilus algosus*, native to Chile, invaded South Africa and rapidly spread to occupy 500 km of the west coast (Zeeman *et al.* 2018). Together with the alien *M. galloprovincialis*, *S. algosus* has become a dominant intertidal organism on wave-exposed rocky shores across this region. Both species are now much more abundant intertidally than either of the indigenous mussels *Aulacomya* and *Choromytilus meridionalis* (de Greef *et al.* 2013).

P. perna is widely distributed in tropical and subtropical regions. In 1990, the species attracted much attention because of the extension of its range to nonnative areas such as the Caribbean basin, including the western coast of the Gulf of Mexico, via ballast water release from ships from Venezuela (Hicks & Tunnel 1993; Baker *et al.* 2012). The alien is quickly becoming a nuisance of water-cooling systems for power stations. It can alter the physical structure of a habitat, like many alien species, and has been known to cause paralytic shellfish poisoning (PSP; see Chapter 12) in those that consume contaminated examples (GISD 2015c). It also has the ability to dominate certain habitats within the Florida, SE United States ecosystem and to adversely impact native species such as the oyster *Crassostrea virginica* through space competition (Baker *et al.* 2012 and references therein). Its congener, *P. viridis*, is native to the tropical Indo-Pacific region, where it is widely distributed. It is a recent invader of the Caribbean region and Florida, and has also been introduced into parts of coastal Australia, Japan and North and South America via ship ballast, hull fouling and experimental introduction for farming. It can quickly form dense colonies in a range of environmental conditions. Its impacts include accumulation on boat hulls, pilings, buoys and other artificial structures, causing blockage in intake pipes of industrial plants, clogging crab traps and clam culture bags and impeding commercial harvest. It is able to out-compete many fouling species, causing changes in community structure and trophic relationships. Finally, it has been recorded with high levels of accumulated toxins and heavy metals (GISD 2015d).

Geukensia demissa is native to intertidal salt marshes along the Atlantic coast of North America, from the Gulf of Maine to Florida and the Gulf of Mexico (Franz 2001), but in the late 1880s it was introduced into San Francisco Bay on the Pacific coast, and since then it has extended its range as far south as Venezuela. A complex mixture of ballast water, ship fouling and the movements of shellfish may have led to its transportation from central to southern California. A significant impact of *G. demissa* is its effect on vegetation, in that in California the increased mussel presence resulted in increased cordgrass, *Spartina foliosa*, production, which in turn led to a decrease in the habitat of other benthic mud-dwelling invertebrates and of many of the birds that tend to forage on tidal flats (Torchin *et al.* 2005). In addition, it caused problems for the endangered California clapper rail, *Rallus longirostris obsoletus*, a bird with whom it shares its habitat: the posterior margin of its shell, which protrudes above the surface of the mud, can clamp on to the toes and beaks of small birds, causing toe losses and sometimes death (GISD 2015e).

Musculista senhousia, native to East Asia, has successfully spread to New Zealand, Australia, the Mediterranean and the Pacific coast of the United States. It settles in aggregations and is therefore able to reach high densities. It can grow rapidly and is capable of marked habitat alteration through the construction of byssal mats on the surface of soft sediments. These mats dramatically alter the natural benthic habitat, changing both the local physical environment and the resident macroinvertebrate assemblage. Although this can result in increased species richness and abundance of some species, mussel mats reduce the densities of many common native bivalves. Also, it smothers eelgrass habitat, potentially promoting a transition from eelgrass beds (Allen & Williams 2003). The mussel may have been introduced to Australia as an accidental importation with Pacific oysters (*C. gigas*), while in the Mediterranean it has been strictly linked with shellfish farming and trading. The initial invasion of the Pacific coast of the United States is attributed to transport with oysters imported from Japan (GISD 2015f).

Mytella charruana is native to Central and South America, with a distribution on the Pacific coast from Mexico to Ecuador and on the southern Atlantic coast from Colombia to Argentina (Gillis *et al.* 2009). Since 1986, it has been reported as invasive along the southeastern coast of the United States (Gallimany *et al.* 2017 and references therein), where it is believed to have been introduced through ballast water releases in Jacksonville, Florida. In 2014, it was reported in Manila Bay, West Philippine Sea (Vallejo *et al.* 2017). Along the east coast of Florida, the mussel has been found on intertidal native oyster (*Crassostrea virginica*) reefs. In a series of field studies, Yuan *et al.* (2016a) found that survival and growth of *C. virginica* were negatively affected by *M. charruana* when the two species were grown in contact, whereas native mussels (*Geukensia demissa*) had no effect on the oysters, suggesting that feeding competition was the mechanism underlying this result. In a related study, Galimany *et al.* (2017) found that *M. charruana* had significantly higher clearance, filtration, rejection, organic ingestion and absorption rates as compared to *C. virginica*, as well as higher rejection percentage and absorption efficiency, but that it rejected the same amount of inorganic particles, suggesting that *M. charruana* can outcompete native oysters; the findings also suggest that an invasion of *M. charruana* might significantly alter plankton abundance, potentially limiting food sources available to other, less efficient native filter-feeders. Both *M. charruana* and *P. viridis*, which is also an invader of the southeastern United States (see earlier), are tolerant of many salinity–temperature combinations found in the region, and thus should be able to both persist in their present invaded range and expand into new estuaries and bays along the Atlantic coastline of Florida and west within the Gulf of Mexico (Yuan *et al.* 2016b).

To avoid indiscriminate introductions, strict criteria for selecting species are governed in the European Union by Regulation (EC) No 708/2007 – concerning use of alien and locally absent species in aquaculture; in Canada by the National Code on Introductions and Transfers of Aquatic Organisms (2017) and the Federal Fish Health Protection Regulations (FHRP) under the Fisheries Act (1985); in Australia by the Livestock (Restrictions on Entry of Aquaculture Organisms) Notice (2005) under the Livestock Act (1997, as amended in 2000); and in China by The Entry and Exit of Animal and Plant Quarantine Law (1991), Regulations for the Protection of Aquatic Wildlife (1993) and The Law on Animal Diseases (1997); in the United States, laws governing aquatic species introductions vary by state.

An Ecosystem Approach to Bivalve Culture

Global aquaculture production (including aquatic plants) in 2016 was 110.2 million t, with a first-sale value estimated at US$243.5 billion (FAO 2018c). Although world aquaculture since 2000 has not enjoyed the high annual growth rates of the 1980s and 1990s, it continues to grow faster than other major food production sector. Accompanying this has been an increasing demand for seafood from an ever-increasing population (projected to reach >9.7 billion by 2050), which aquaculture is perceived as having the greatest potential to fulfil. The most recent official statistics indicate that aquaculture provides a livelihood to 19.3 million farmers, concentrated primarily in Asia (96% of all aquaculture engagement), followed by Latin America and the Caribbean (2%) and Africa (1.6%); Europe, North America and Oceania (geographic region comprising Australasia, Melanesia, Micronesia and Polynesia) each have less than 1% of the global population engaged in the sector (FAO 2018c). The rapid increase in global aquaculture production has the potential to affect ecosystem functions and services, with adverse environmental, social and economic consequences. Therefore, there is pressure on regulators from consumers, the public, NGOs, pressure groups and scientific expert groups to come to understand the consequences of aquaculture in order to enlighten and refine regulations and codes of practice. Over the past two decades, significant progress has indeed been made in formulating legislation that, if applied, should facilitate sustainable development of the aquaculture sector. For information pertaining to aquaculture legislation in individual countries (N = 65), see FAO (2021). But regulations on their own do not provide a comprehensive framework for ensuring sustainable use of aquatic environments. This will only happen when aquaculture is treated as an integral process within the ecosystem, via an ecosystem approach to aquaculture (EAA), a strategy for the integration of the activity within the wider ecosystem such that it promotes sustainable development, equity and resilience of interlinked social-ecological systems (FAO 2010; Brugère *et al.* 2018 and references therein). In 1995, the Code of Conduct for Responsible Fisheries (CCRF) was unanimously adopted by over 170 member governments, providing principles and standards applicable to the conservation, management and development of all fisheries. The CCRF covers the capture, processing and trade of fish and fishery products, fishing operations, aquaculture, fisheries research and integration of fisheries into coastal area management (FAO 1995). It also deals with aquaculture more specifically through Article 9.1.5, which requires that 'states should establish effective procedures specific to aquaculture to undertake appropriate environmental assessment and monitoring with the aim of minimising adverse ecological changes and related economic and social consequences resulting from water extraction, land use, discharge of effluents, use of drugs and chemicals, and other aquaculture activities'. While the CCRF is voluntary, certain parts of it are based on relevant rules of international law.

The EAA is guided by three strategic principles (Soto *et al.* 2008): no degradation of ecosystem functions and services beyond their resilience capacity; improvement of human well-being and equity for stakeholders; and consideration for other relevant sectors. In order to implement the EAA strategy, relevant policy goals should be translated into operational objectives and actions, using the best available information and broad stakeholder participation (Figure 10.20). The following spatial scales have been proposed for EAA implementation: (1) at the farm level; (2) at the level of the water body and its watershed/

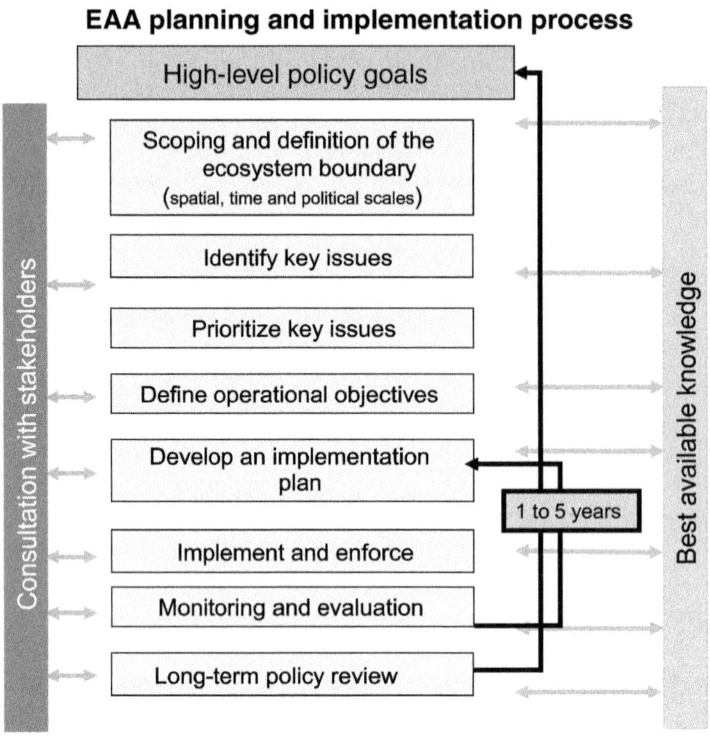

Figure 10.20 Schematic view of the EAA planning and implementation process. All stages are covered in detail in FAO (2010). EAA can start at any stage: policy formulation, monitoring and evaluation or implementation. *Source:* Figure adapted from APFIC (2009) and FAO (2010). Reproduced with permission from the Food and Agriculture Organization of the United Nations (FAO).

aquaculture zone; and (3) at the global, market-trade level. Productive (small to intensive, large-scale) and temporal scales should also be considered (Soto *et al.* 2008). At the farm level, numerous research and pilot-scale commercial initiatives have adopted an EAA, many utilising an integrated multi-trophic aquaculture (IMTA) system – a practice in which byproducts (wastes) from one species are recycled as inputs (fertilisers, food and energy) for another (see earlier). An example might be a fish/shellfish/seaweed combination (reviewed in Barrington *et al.* 2009). All stages of an EAA are carried out in accordance with current international (binding and nonbinding) and national agreements, regulations, codes of practice and guidelines that pertain to aquaculture.

A range of tools, varying from simple to complex, support the analysis and management of an EAA. To illustrate, the EU-funded Ecosystem Approach to Sustainable Aquaculture (ECASA, www.ecasa.org.uk), is evaluating the ability of quantitative ecosystem indicators[5] to discriminate between aquaculture and other anthropogenic sources of perturbation in the marine environment (CORDIS 2011a). The main outcomes of the ECASA project were presented in a virtual 'Tool-Box' containing a suite of indicators and predictive environmental models whose focus was on aiding the assessment of appropriate sites for aquaculture activities and subsequently providing a consistent framework for the application of Environmental Impact Assessments (EIAs), resulting in coherent and relevant Environmental Statements. There is also a toolbox for an Ecosystem Approach to Fisheries (FAO 2009). To date, the ECASA project has evaluated more than 50 indicators of ecosystem change, and it is currently conducting fieldwork to select the best environmental indicators in many ecological/ecosystem categories (Callier *et al.* 2008; Borja *et al.* 2009; Filgueira *et al.* 2014; Pérez-Camacho *et al.* 2014). It is also developing a range of tools, particularly predictive models, which it is testing in conjunction with indicators at a number of sites spread over a wide geographic range, each with different aquaculture impacts.

Nunes *et al.* (2011) have shown how a range of complementary tools, with different levels of complexity and scope, can be combined to provide an integrated assessment of aquaculture in a coastal system. They carried out a study in Killary Harbour, a fjord-like inlet on the west coast of Ireland, where longline mussel culture production is currently ~2000 t yr^{-1}. Tools examined included a well-tested system-scale ecological model for coastal waters, EcoWin2000 (Ferreira 1995), a local-scale carrying-capacity and environmental-effects model, FARM (Ferreira *et al.* 2009) and a management-level eutrophication-screening model, ASSETS, which is capable of qualifying system-scale trophic status (Bricker *et al.* 2003). The EcoWin2000 and FARM models use ShellSIM to simulate individual shellfish growth. Required information for EcoWin2000 includes hydrodynamic and biogeochemical boundary data (land and/or ocean) and information on species and culture practices. FARM requires input data on farm layout and dimensions, species and stocking densities, suspended food entering the farm and environmental parameters. ASSETS uses quantitative and qualitative data evaluating trophic status to provide information on how aquaculture impacts eutrophication at both the system and the farm level. All three tools were used to analyse the relationship between mussel productivity and food sources, the impacts of changes to stocking densities of mussels and the overall ecological status of Killary Harbour. Nunes *et al.* (2011) illustrated the complementary nature of the models, and how the unique capacities of each can be combined to provide integrated assessment of aquaculture in a coastal system. Their combined application revealed that: (1) the Killary Harbour system's eutrophication status can be classified as Moderate Low, with a future trend of No Change; (2) there is a strong longitudinal and vertical circulation of phytoplankton, with imports into the harbour from the ocean and exports of primary production from the harbour to the ocean; (3) the maximum mussel production of the system is 4200 t yr^{-1}, about twice the present yearly production value, although achieving this would lead to lower harvest weights and longer growth cycles; and (4) a scenario of lower stocking densities proposed for the system should lead to lower mussel productions, but could result in benefits such as higher mussel weight at harvest and/or shorter growth cycles.

While bivalve culture is potentially one of the most sustainable forms of aquaculture, requiring no artificial food input, the rapid growth of the sector has inevitably raised questions around carrying capacity and sustainability (Nunes *et al.* 2003). The concept of 'carrying capacity' is of fundamental importance within an EAA. Dame (2012 and references therein) defines bivalve carrying capacity as the total bivalve biomass supported by a given ecosystem as a function of its water residence time, primary production time and bivalve clearance time. When bivalve biomass ceases to increase, the system is said to have reached maximum carrying capacity. There are four different types of carrying capacity that could potentially limit production: physical, production, ecological and social (details in Inglis *et al.* 2000; Byron *et al.* 2013; McKindsey *et al.* 2006, 2013). The physical carrying capacity of a site (embayment, inlet, offshore area, etc.) is simply the geographic area in which conditions are suitable for the production of a given species using a given method. It is useful for quantifying the potential area available for aquaculture in the ecosystem, but offers little information toward determinations of aquaculture's limits at the water body or watershed level in the EAA. With few exceptions, carrying capacity studies have focused on determinations of production carrying capacity, which estimates maximum aquaculture production and is typically considered at the farm scale. Ecological carrying capacity is defined as the magnitude of aquaculture production that can be supported without leading to significant changes to ecological processes, species, populations or communities in the environment. Social carrying capacity is the amount of aquaculture that can be developed without adverse social impacts. The calculation of carrying capacity relies on models and tools such as Geographic Information Systems (GIS) and remote sensing (IOCCG 2009). Estimation of physical carrying capacity uses the type of data stored in a GIS (e.g. nutrient concentrations and bathymetry), while production carrying capacity uses a wide range of modelling approaches, focusing mainly on hydrodynamics, food availability and production, bivalve feeding and physiology and the influence of husbandry practices on crop production, as well as the interactions among these factors (see McKindsey *et al.* 2006 for a review of carrying capacity models). While modelling of production carrying capacity focuses on the target bivalve species and on those organisms that support its production, modelling of ecological carrying capacity should in principle consider the whole ecosystem and all culture activities from seed collection to ongrowing, harvesting and processing. In a study on carrying capacity for mussel (*P. canaliculus*) culture in South Island, New Zealand, Jiang & Gibbs (2005) define ecological carrying capacity as significant changes in modelled energy fluxes or the structure of the food web. They developed an Ecopath model (www.ecopath.org) for a marine ecosystem in which large-scale expansion of mussel aquaculture was proposed. The model estimated the mussel production capacity at 310 t km^{-2} yr^{-1}, while ecological carrying capacity was about five times lower at 65 t $km^{-2}yr^{-1}$ – the level of culture above which there would be major changes in energy fluxes with the system's food web.

Numerous models have been developed to determine carrying capacity – mainly production capacity – in bivalves. For studies pertaining to mussel culture, see Duarte *et al.* (2008; food depletion model), Callier *et al.* (2009; predictive model DEPOMOD), Filgueira & Grant (2009; ecosystem model), Weise *et al.* (2009; Shellfish-DEPOMOD), Guyondet *et al.* (2010, 2015, coupled hydrodynamic–biogeochemical model), Ibarra *et al.* (2014; ocean model ROMS and ecophysiology model SHELL-E), Filgueira *et al.* (2015; hydrodynamic model) and Srisunont & Babel (2016; dynamic energy budget (DEB) model).

In a recent review, Brugère *et al.* (2018) critically examined the use and mainstreaming of the EAA in aquaculture development to date and considered its possible evolution in the next decade. They assessed the lessons learnt from the past decade of EAA experiences, the opportunities the EAA has created and the links between the EAA and the new development agenda, finding that mainstreaming the EAA in planning processes has raised awareness of the usefulness of holistic and participatory approaches in aquaculture and helped to steer the sector toward greater sustainability. The emphasis on marine spatial planning[6] that has developed as part of the EAA implementation efforts (Aguilar-Manjarrez *et al.* 2010, 2017; Meaden *et al.* 2016) and the close links between the EAA and initiatives such as 'blue growth'[7] (FAO 2015) constitute significant opportunities for the future of the approach, although its ability to tackle increasingly complex governance issues may be limited. Despite the obstacles, Brugère *et al.* (2019, p. 13) conclude that:

> the EAA looks set to continue to be adopted, thanks in part to the priority and explicit support given to the approach by FAO member countries and requests that FAO continue its efforts to disseminate tools and guidance on the EAA. Another reason to be positive about the future of the EAA is the adoption by the European Commission of the European Union (EU) of the EAA through a number of regional aquaculture research initiatives, signalling the endorsement of the principles of the approach as a means to finding solutions to the challenges faced by aquaculture and to guide the development of the sector in Europe (e.g. EU AquaSpace and EcoAqua projects – Aguilar-Manjarrez 2016; Aguilar-Manjarrez *et al.* 2016; Haroun *et al.* 2016). The momentum should be enhanced through country-led monitoring, evaluation and communication especially at the ground level.

Notes

1 The process of cycling through two temperature extremes, typically at relatively high rates of change.
2 A CFU (colony forming unit) is used to estimate the number of viable bacteria or fungal cells in a sample.
3 The period relating to or denoting the current geological age, during which human activity is seen as having been the dominant influence on climate and the environment.
4 Extensive, as opposed to intensive, aquaculture implies a multifunctional use of ocean space by culturing several different marine organisms in a single location.
5 A subset of the physical, chemical and biological elements and processes of natural systems that are selected to represent the overall health or condition of the system.
6 Marine spatial planning provides a step-by-step approach to balance the uses and users of the marine environment with a view to providing a coordinated system that results in the development of a marine spatial plan, which defines the strategic, forward-looking planning for the regulation, zoning, management, protection and sustainability of the marine environment.
7 The Blue Growth Initiative (BGI) offers a coherent framework for the sustainable and socio-economic management of aquatic resources. Anchored in the principles set out in the benchmark Code of Conduct for Responsible Fisheries, it focuses on capture fisheries, aquaculture, ecosystem services, trade and social protection (FAO 2015).

References

Adams, C.M., Shumway, S.E., Whitlach, R.B. & Getchis, T. (2011) Biofouling in marine molluscan shellfish aquaculture: a survey assessing the business and economic implications of mitigation. *Journal of the World Aquaculture Society*, **42**, 242–252.

Adams, S.L., Tervit, H.R., McGowan, L.T., Smith, J.F., Roberts, R.D., Salinas-Flores, L. *et al.* (2009) Towards cryopreservation of Greenshell™ mussel (*Perna canaliculus*) oocytes. *Cryobiology*, **58**, 69–74.

Adams, S.L., Smith, J.F., Tervit, R., McGowan, L.T., Roberts, R.D., Janke, A.R.N. *et al.* (2011) Cryopreservation of molluscan sperm: Pacific oyster, green-lipped mussel, and Paua Abalone. In: *Cryopreservation in Aquatic Species*, 2nd edn (eds T.R. Tiersch & C.C. Green), pp. 562–573. World Aquaculture Society, Baton Rouge, LA.

Aguiar, E., Fuentes-Santos, I., Labarta, U., Álvarez-Salgado, X. A. & Fernndez-Reiriz, M. J. (2015) Empirical modelling of seston quality based on environmental factors in a mussel culture area (NW Iberian upwelling system). *Marine Ecology Progress Series*, **536**, 89–105.

Aguilar-Manjarrez, J. (2016) *Horizon 2020:* Promoting Sustainable Aquaculture. FAO Aquaculture Newsletter 54, pp. 26–27. FAO, Rome.

Aguilar-Manjarrez, J., Kapetsky, J.M. & Soto, D. (2010) *The potential of spatial planning tools to support the ecosystem approach to aquaculture. FAO/Rome. Expert Workshop. 19–21 November 2008*, Rome, Italy. FAO Fisheries and Aquaculture Proceedings No. 17. FAO, Rome.

Aguilar-Manjarrez, J., Massa, F. & Bacher, K. (2016) *Workshop on aquaculture zoning, site selection and area management: assessment of policy management issues. FAO Aquaculture Newsletter 55, pp. 28–29*. FAO, Rome.

Aguilar-Manjarrez, J., Soto, D. & Brummett, R. (2017) *Aquaculture zoning, site selection and area management under the ecosystem approach to aquaculture (Vol. 2): full document (English)*. World Bank Group, Washington, DC. Available at: http://documents.worldbank.org/curated/en/421101490644362778/full-document (accessed 28 May 2021).

Akhter, N., Wu, B., Memon, A.M. & Mohsin, M. (2015) Probiotics and prebiotics associated with aquaculture: a review. *Fish and Shellfish Immunology*, **45**, 733–741.

Aji, L.P. (2011) The use of algae concentrates, dried algae and algal substitutes to feed bivalves. *Makara Journal of Science*, **15**, 1–8.

Allen, B.J. & Williams, S.L. (2003) Native eelgrass *Zostera marina* controls growth and reproduction of an invasive mussel through food limitation. *Marine Ecology Progress Series*, **254**, 57–67.

Andersen, R.A. (2005) *Algal culturing Techniques*, 1st edn. Academic Press, London.

Anglès d'Auriac, M.B., Rinde, E., Norling, P., Lapègue, S., Staalstrøm, A., Hjermann, D.Ø. *et al.* (2017) Rapid expansion of the invasive oyster *Crassostrea gigas* at its northern distribution limit in Europe: naturally dispersed or introduced? *PLoS ONE*, **12**, e0177481.

Anguiano-Beltrán, C., Leonardo, M.L., Lizárraga-Partida. & Searcy-Bernall1, R. (2004) *Effect of Vibrio alginolyticus on larval survival of the blue mussel Mytilus galloprovincialis. Diseases of Aquatic Organisms*, **59**, 119–123.

Anil, M.K., Gomathi, P., Raheem, P.K., Rinju, M., Raju, B., Leslie, V.A. *et al.* (2017) Hatchery Technology for Mussel Seed Production. Training on Recent Advances in Mariculture. *Course Manual*. Vizhinjam Research Centre of Central Marine Fisheries Research Institute, Kerala.

APFIC (2009) APFIC/FAO Regional Consultative Workshop Practical Implementation of the Ecosystem Approach to Fisheries and Aquaculture, 18–22 May 2009, Colombo, Sri Lanka. *RAP Publication 2009/10*. FAO Regional Office for Asia and the Pacific, Bangkok.

Arney, B., Liuc, W., Forster, I.P., McKinley, R.S. & Pearce, C.M. (2015) Feasibility of dietary substitution of live microalgae with spray-dried *Schizochytrium* sp. or *Spirulina* in the hatchery culture of juveniles of the Pacific geoduck clam (*Panopea generosa*). *Aquaculture*, **444**, 117–133.

Arribas, L.P., Bagur, M., Gutiérrez, J.L. & Palomo, M.G. (2015) Matching spatial scales of variation in mussel recruitment and adult densities across southwestern Atlantic rocky shores. *Journal of Sea Research*, **95**, 16–21.

Atalah, J. & Forrest, B.M. (2019) Forecasting mussel settlement using historical data and boosted regression trees. *Aquaculture Environment Interactions*, **11**, 625–638.

Atalah, J., Fletcher, L.M., Hopkins, G.A., Heasman, K., Woods, C.M.C. & Forrest, B.M. (2016) Preliminary assessment of biofouling on offshore mussel farms. *Journal of the World Aquaculture Society*, **47**, 376–386.

Aubin, J., Fontaine, C., Callier, M. & Roque d'Orbcastel, E. (2018) Blue mussel (*Mytilus edulis*) bouchot culture in Mont-St Michel Bay: potential mitigation effects on climate change and eutrophication. *International Journal of Life Cycle Assessment*, **23**, 1030–1041.

Aypa, S.M. (1990) Mussel culture. In: *Regional Seafarming Development and Demonstration Project (RAS), Selected Papers on Mollusc Culture. UNDP/FAO(RAS/90/002)*. National Inland Fisheries Institute, Kasetsart University Campus Bangkhen, Bangkok.

Baker, P., Fajans, J.S. & Baker, R.S.M. (2012) Habitat dominance of a nonindigenous tropical bivalve, *Perna viridis* (Linnaeus, 1758), in a subtropical estuary in the Gulf of Mexico. *Journal of Molluscan Studies*, **78**, 28–33.

Bakker, J.A., Paetzold, S.C., Quijón, P.A. & Davidson, J. (2011) The use of food grade oil in the prevention of vase tunicate fouling on mussel aquaculture gear. *Management of Biological Invasions*, **2**, 15–25.

Balbi, T., Auguste, M., Cortese, K., Montagna, M., Borello, A., Pruzzo, C. *et al.* (2019) Responses of *Mytilus galloprovincialis* to challenge with the emerging marine pathogen *Vibrio coralliilyticus*. *Fish and Shellfish Immunology*, **84**, 352–360.

Barrington, K., Chopin, T. & Robinson, S. (2009) Integrated multi-trophic aquaculture (IMTA) in marine temperate waters. In: *Integrated Mariculture: A Global Review* (ed. D. Soto), pp. 7–46. FAO Fisheries and Aquaculture Technical Paper No. 529. FAO, Rome.

Bayne, B.L., Holland, D.L., Moore, M.N., Lowe, D.M. & Widdows, J. (1978) Further studies on the effects of stress in the adult on the eggs of *Mytilus edulis*. *Journal of the Marine Biological Association of the UK*, **58**, 825–841.

Beaz-Hidalgo, R., Balboa, S., Romalde, J.L. & Figueras, M.J. (2010) Diversity and pathogenecity of *Vibrio* species in cultured bivalve mollusks. *Environmental Microbiology Reports*, **2**, 34–43.

BIM (2006) Review of the Irish Rope Mussel Industry. A report jointly commissioned by Bord Iascaigh Mhara (BIM) and Enterprise Ireland (EI). Pricewaterhouse Coopers.

Blackburn, S.I. & Jye Lee-Chang, K. (2018) Microalgae: a renewable resource for food and fuels and more. In: *Blue Biotechnology: Production and Use of Marine Molecules* (eds S. La Barre & S.S. Bates), pp. 3–33. John Wiley & Sons, Chichester.

Blanc, J.M., Molinet, C., Subiabre, R. & Díaz, P.A. (2018) Cadmium determination in Chilean blue mussels *Mytilus chilensis*: implications for environmental and agronomic interest. *Marine Pollution Bulletin*, **129**, 913–917.

Boeing, P. (1997) Use of spray-dried *Schizochytrium* sp. as a partial algal replacement for juvenile bivalves. *Journal of Shellfish Research*, **16**, 284.

Borja, A., Rodríguez, J. G., Black, K., Bodoy, A., Emblow, C., Fernandes, T.F. *et al.* (2009) Assessing the suitability of a range of benthic indices in the evaluation of environmental impact of fin and shellfish aquaculture located in sites across Europe. *Aquaculture*, **293**, 231–240.

Bower, S.M. (2015) Synopsis of infectious diseases and parasites of commercially exploited shellfish: *Bonamia ostreae* of oysters. Available at: https://www.dfo-mpo.gc.ca/science/aah-saa/diseases-maladies/index-eng.html (accessed 28 May 2021).

Branch, G.M. & Steffani, C.N. (2004) Can we predict the effects of alien species? A case-history of the invasion of South Africa by *Mytilus galloprovincialis* (Lamarck). *Journal of Experimental Marine Biology and Ecology*, **300**, 189–215.

Bricker, S.B., Ferreira, J.G. & Simas, T. (2003) An integrated methodology for assessment of estuarine trophic status. *Ecological Modelling*, **169**, 39–60.

Brugère, C., Aguilar-Manjarrez, J., Beveridge, M.C.M. & Soto, D. (2019) The ecosystem approach to aquaculture 10 years on– a critical review and consideration of its future role in blue growth. *Reviews in Aquaculture*, **11**, 493–514.

Buck, B.H. & Krause, G. (2012) Integration of aquaculture and renewable energy systems. In: *Encyclopaedia of Sustainability Science and Technology*, Vol. 1. (ed R.A. Meyers), pp. 511–533. Springer Science+Business Media, Berlin.

Buck, B.H. & Langan R. (2017) *Aquaculture Perspective of Multi-Use Sites in the Open Ocean – The Untapped Potential for Marine Resources in the Anthropocene*. Springer Open, Berlin.

Buck, B.H., Krause, G. & Rosenthal, H. (2004) Extensive open ocean aquaculture development within wind farms in Germany: the prospect of offshore co-management and legal constraints. *Ocean and Coastal Management*, **47**, 95–122.

Buck, B.H., Krause, G., Michler-Cieluch, T., Brenner, M., Buchholz, C.M., Busch, J.A. *et al.* (2008) Meeting the quest for spatial efficiency: progress and prospects of extensive aquaculture within offshore wind farms. *Helgoland Marine Research*, **62**, 269–281.

Buck, B.H., Nevejan, N., Wille, M., Chambers, M.D. & Chopin, N. (2017a) Offshore and multi-use aquaculture with extractive species: seaweeds and bivalves. In: *Aquaculture Perspective of Multi-Use Sites in the Open Ocean – The Untapped Potential for Marine Resources in the Anthropocene* (eds B.H. Buck & R. Langan), pp. 23–69. Springer Open, Berlin.

Buck, B.H., Krause, G., Pogodà, B., Grote, B., Wever, L., Goseberg, N. *et al.* (2017b) The German case study: pioneer projects of aquaculture-wind farm multi-use. In: *Aquaculture Perspective of Multi-Use Sites in the Open Ocean – The Untapped Potential for Marine Resources in the Anthropocene* (eds B.H. Buck & R. Langan), pp. 253–354. Springer Open, Berlin.

Byron, C.J. & Costa-Pierce, B.A. (2013) Carrying capacity tools for use in the implementation of an ecosystems approach to aquaculture. In: *Site Selection and Carrying Capacities for Inland and Coastal Aquaculture* (eds L.G. Ross, T.C. Telfer, L. Falconer, D. Soto & J. Aguilar-Manjarrez), pp. 87–101. FAO/Institute of Aquaculture, University of Stirling, Expert Workshop, 6–8 December 2010. Stirling, the United Kingdom of Great Britain and Northern Ireland. FAO Fisheries and Aquaculture Proceedings No. 21. FAO, Rome.

Caballero Miguez, G., Garza-Gil, M.D. & Varela-Lafuente, M.M. (2012) Legal change, property rights system and institutional stability: The case of the floating raft culture in the Galician mussel sector. *Ocean and Coastal Management*, **55**, 84–93.

Caers, M., Coutteau, P., Lombeida, P. & Sorgeloos, P. (1998) The effect of lipid supplementation on growth and fatty acid composition of *Tapes philippinarium* spat. *Aquaculture*, **162**, 287–299.

Callier, M.D., McKindsey, C.W. & Desrosiers, G. (2008) Evaluation of indicators used to detect mussel farm influence on the benthos: two case studies in the Magdalen Islands, Eastern Canada. *Aquaculture*, **278**, 77–88.

Callier, M.D., Richard, M., McKindsey, C.W., Archambault, P. & Desrosiers, G. (2009) Responses of benthic macrofauna and biogeochemical fluxes to various levels of mussel biodeposition: an in situ 'benthocosm' experiment. *Marine Pollution Bulletin*, **58**, 1544–1553.

Calvo–Ugarteburu, G. & McQuaid, C.D. (1998) Parasitism and introduced species: epidemiology of trematodes in the intertidal mussels *Perna perna* and *Mytilus galloprovincialis*. *Journal of Experimental Marine Biology and Ecology*, **220**, 47–65.

Capelle, J.J. (2017) Production efficiency of mussel bottom culture. *PhD thesis*, Wageningen University, The Netherlands.

Capelle, J.J., Wijsman, J.W.M., Schellekens, T., van Stralen, M.R., Herman, M.J. & Smaal, A.C. (2014) Spatial organisation and biomass development after relaying of mussel seed. *Journal of Sea Research*, **85**, 395–403.

Capelle, J.J., Wijsman, J.W.M., van Stralenc, M.R., Herman, P.M.J. & Smaal, A.C. (2016a) Effect of seeding density on biomass production in mussel bottom culture. *Journal of Sea Research*, **110**, 8–15.

Capelle, J.J., Scheiberlich, G., Wijsman, J.W.M. & Smaal, A.C. (2016b) The role of shore crabs and mussel density in mussel losses at a commercial intertidal mussel plot after seeding. *Aquaculture International*, **24**, 1459–1472.

Carman, M.R., Lindell, S., Green-Beach, E. & Starczak, V.R. (2016) Treatments to eradicate invasive tunicate fouling from blue mussel seed and aquaculture socks. *Management of Biological Invasions*, **7**, 101–110.

Carrasco, A.V., Astorga, M., Cisterna, A., Farías, A., Espinoza, V. & Uriarte, I. (2014) Pre-feasibility study for the installation of a Chilean mussel *Mytilus chilensis* (Hupé, 1854) Seed hatchery in the Lakes Region, Chile. *Fisheries and Aquaculture Journal*, **5**, 102.

Cataldo, D., Boltovskoy, D., Hermosa, J. L. & Canzi, C. (2005) Temperature-dependent rates of larval development in *Limnoperna fortunei* (Bivalvia: Mytilidae). *Journal of Molluscan Studies*, **71**, 41–46.

Chalermwat, K., Szuster, B.W. & Flaherty, M. (2008) Shellfish aquaculture in Thailand. *Aquaculture Economics & Management*, **7**, 249–261.

Comeau, L.A., Sonier, R. & Hanson, J.M. (2012) Seasonal movements of Atlantic rock crab (*Cancer irroratus* Say) transplanted into a mussel aquaculture site. *Aquaculture Research*, **43**, 509–517.

Comeau, L.A., Filgueira, R., Guyondet, T. & Sonier, R. (2015) The impact of invasive tunicates on the demand for phytoplankton in longline mussel farms. *Aquaculture*, **441**, 95–105.

Corbin, J.S., Holmyard, J. & Lindell, S. (2017) Regulation and permitting of standalone and co-located open ocean aquaculture facilities. In: *Aquaculture Perspective of Multi-Use Sites in the Open Ocean – The Untapped Potential for Marine Resources in the Anthropocene* (eds B.H. Buck & R. Langan), pp. 187–229. Springer Open, Berlin.

CORDIS (2011a) Final Report Summary - ECASA (Ecosystem approach for sustainable aquaculture): Reporting. Available at: https://cordis.europa.eu/project/id/6540/reporting (accessed 28 May 2021).

CORDIS (2011b) Technology development for a reliable supply of high quality seed in blue mussel farming: Reporting. Available at: https://cordis.europa.eu/project/id/17729/reporting (accessed 28 May, 2021).

CORDIS (2015) REsearch to improve PROduction of SEED of established and emerging bivalve species in European hatcheries: Reporting. Available at: https://cordis.europa.eu/project/id/245119/reporting (accessed 28 May, 2021).

CORDIS (2017) REsearch to improve PROduction of SEED of established and emerging bivalve species in European hatcheries: Fact Sheet. Available at: https://cordis.europa.eu/project/id/245119 (accessed 28 May, 2021).

Coutteau, P., Hadley, N., Manzi, J. & Sorgeloos, P. (1994) Effect of algal ration and substitution of algae by manipulated yeast diets on growth of juvenile *Mercenaria mercenaria*. *Aquaculture*, **120**, 135–150.

Coutteau, P., Caers, M., Curé K. & Gajardo, G. (1996) Supplentation of lipid emulsions to algal diets in the hatchery rearing of bivalves. In: *Improvement of the Commercial Production of Marine Aquaculture Species* (eds G. Gajardo & P. Coutteau), pp. 145–154. Proceedings of a workshop on fish and larviculture. Impresora Creces, Santiago, Chile.

Cranford, P.J., Reid, G.K. & Robinson, S.M.C. (2013) Open water integrated multi-trophic aquaculture: constraints on the effectiveness of mussels as an organic extractive component. *Aquaculture Environment Interactions*, **4**, 163–173.

Dame, R.F. (2012) *Ecology of Marine Bivalves: An Ecosystem Approach*, 2nd edn. CRC Press, Boca Raton, FL.

Davis, J.P. & Campbell, C.R. (1998) The use of a *Schizochytrium* based HUFA enriched dry feed for culturing juvenile mussels – *Mytilus galloprovincialis* – and the comparative routine costs of producing live algae in a commercial bivalve hatchery. European Aquaculture Society, Special Publication No. 26, Oostende, pp. 64–65.

de Greef, K., Griffiths, C.L. & Zeeman, Z. (2013) Deja vu? a second mytilid mussel, *Semimytilus algosus*, invades South Africa's west coast. *African Journal of Marine Science*, **35**, 307–313.

Di Matteo, O., Langellotti, A.L., Masullo, P. & Sansone, G. (2009) Cryopreservation of the Mediterranean mussel (*Mytilus galloprovincialis*) spermatozoa. *Cryobiology*, **58**, 145–150.

Dittami, S.M., Pazos, Y. Laspra, M. & Medlin, L.K. (2013) Microarray testing for the presence of toxic algae monitoring programme in Galicia (NW Spain). *Environmental Science and Pollution Research*, **20**, 6778–6793.

Domínguez, L., Villalba, A. & Fuentes, J. (2010) Effects of photoperiod and the duration of conditioning on gametogenesis and spawning of the mussel *Mytilus galloprovincialis* (Lamarck). *Aquaculture Research*, **41**, 807–818.

Duarte, P., Labarta, U. & Fernández-Reiriz, M.J. (2008) Modelling local food depletion effects in mussel rafts of Galician Rias. *Aquaculture*, **274**, 300–312.

Dubert, J., Osorio, C.R., Prado, S. & Barja, J.L. (2016) Persistence of antibiotic resistant *Vibrio* spp. in shellfish hatchery environment. *Microbial Ecology*, **72**, 851–860.

Dubert, J., Barja, J.L. & Romalde, J.L. (2017) New insights into pathogenic vibrios affecting bivalves in hatcheries: present and future prospects. *Frontiers in Microbiology*, **8**, art. 762.

Duthie, I. (2012) Global perspective of bivalve hatchery processes. A report for Nuffield Australia farming scholars. Available at: https://www.nuffieldscholar.org/sites/default/files/reports/2010_AU_Ian-Duthie_Shellfish-Production-Aquaculture-Technology-Global-Perspective-Of-Bivalve-Hatchery-Processes.pdf (accessed 28 May 2021).

Eggermont, M., Tamanji, A., Nevejan, N., Bossier, P., Sorgeloos, P. & Defoirdt, T. (2014) Stimulation of heterotrophic bacteria associated with wild-caught blue mussel (*Mytilus edulis*) adults results in mass mortality. *Aquaculture*, **431**, 136–138.

Eggermont, M., Bossier, P., Pande, G.S.J., Delahaut, Y., Rayhan, A.M., Gupta, N. *et al.* (2017) Isolation of Vibrionaceae from wild blue mussel (*Mytilus edulis*) adults and their impact on blue mussel larviculture. *FEMS Microbiology Ecology*, **93**, fix039.

FAO (1995) *Code of Conduct for Responsible Fisheries.* FAO, Rome.

FAO (2004–2018) *Cultured Aquatic Species Information Programme: Mytilus edulis.* FAO, Rome.

FAO (2005–2019) National Aquaculture Legislation Overview. Chile. Available at: http://www.fao.org/fishery/legalframework/nalo_chile/en (accessed 28 May 2021).

FAO (2009) *Report of the Workshop on Toolbox for Applying the Ecosystem Approach to Fisheries. Rome, 26–29 February. FAO Fisheries and Aquaculture Report No.* 884. FAO, Rome.

FAO (2010) *Aquaculture development. 4. Ecosystem approach to aquaculture. FAO Technical Guidelines for Responsible Fisheries No. 5, Suppl.* 4. FAO, Rome.

FAO (2014) The European market for mussels. Globefish Research Programme, 115. FAO, Rome. Available at: http://www.fao.org/3/a-bb218e.pdf (accessed 28 May 2021).

FAO (2015) *Achieving Blue Growth Through Implementation of the Code of Conduct for Responsible Fisheries.* Policy Brief. FAO, Rome.

FAO (2018a) Globefish – analysis and information on world fish trade. Available at: http://www.fao.org/in-action/globefish/market-reports/bivalves/ (accessed 28 May 2021).

FAO (2018b) Fisheries and Aquaculture Statistics. Global Aquaculture Production 1950–2016 (FishStatJ). Available at: http://www.fao.org/fishery/statistics/software/fishstatj/en (accessed 28 May 2021).

FAO (2018c) *The State of World Fisheries and Aquaculture 2018 – Meeting the Sustainable Development Goals.* FAO, Rome.

FAO (2020) Globefish-information and analysis on world fish trade. Available at: http://www.fao.org/in-action/globefish/publications/details-publication/en/c/1297260/ (accessed 28 May 2021).

FAO (2021) Search aquaculture fact sheets. Available at: http://www.fao.org/fishery/nalo/search/en (accessed 28 May 2021).

Fernández, F.J., Ponce, R.D., Vásquez-Lavin, F., Figueroa, Y., Gelciche, S. & Dresdner, J. (2018) Exploring typologies of artisanal mussel seed producers in southern Chile. *Ocean and Coastal Management*, **158**, 24–31.

Ferreira, J.G. (1995) Ecowin–an object-oriented ecological model for aquatic ecosystems. *Ecological Modelling*, **79**, 21–34.

Ferreira, J.G., Sequeira, A., Hawkins, A.J.S., Newton, A., Nickell, T.D., Pastres, R. *et al.* (2009) Analysis of coastal and offshore aquaculture: application of the FARM model to multiple systems and shellfish species. *Aquaculture*, **289**, 32–41.

Figueroa, Y. & Jorge Dresdner (2016) Are mussel seed producers responsive to economic incentives? Empirical evidence from the Benthic Resource Management Areas in Chile. *Aquaculture Economics & Management*, **20**, 283–311.

Filgueira, R. & Grant, J. (2009) A box model for ecosystem-level management of mussel culture carrying capacity in a coastal bay. *Ecosystems*, **12**, 1222–1233.

Filgueira, R., Guyondet, T., Comeau, L.A. & Grant, J. (2014) Physiological indices as indicators of ecosystem status in shellfish aquaculture sites. *Ecological Indicators*, **39**, 134–143.

Filgueira, R., Guyondet, T., Bacher, C. & Comeau, L.A. (2015) Carrying capacity for mussel aquaculture in Malpeque Bay, Prince Edward Island. Canadian Science Advisory Secretariat (CSAS) Research Document 2015/001 Gulf Region.

Fish Consulting Group (2012) Harvesting of green mussel, *Perna viridis*, in Thailand. Available at: http://fishconsult.org/?p=6528 (accessed 28 May 2021).

Fish Consulting Group (2013) Culture of green mussel, *Perna viridis*, in Thailand. Available at: http://fishconsult.org/?p=9397 (accessed 28 May 2021).

Fitridge, I., Dempster, T., Guenther, J. & de Nys, R. (2012) The impact and control of biofouling in marine aquaculture: a review. *Biofouling*, **28**, 649–669.

Fletcher, L.M., Forrest, B.M. & Bell, J.J. (2013) Impacts of the invasive ascidian *Didemnum vexillum* on green-lipped mussel *Perna canaliculus* aquaculture in New Zealand. *Aquaculture Environment Interactions*, **4**, 17–30.

Fokina, N.N., Ruokolainen. T.R. & Nemova, N.N. (2017) Lipid composition modifications in the blue mussels (*Mytilus edulis* L.) from the White Sea. In: *Organismal and Molecular Malacology* (ed. S. Ray), pp. 143–159. InTechOpen, London.

Forrest, B.M. & Atalah, J. (2017) Significant impact from blue mussel *Mytilus galloprovincialis* biofouling on aquaculture production of green-lipped mussels in New Zealand. *Aquaculture Environment Interactions*, **9**, 115–126.

Franz, D. R. (2001) Recruitment, survivorship and age structure of a New York ribbed mussel population (*Geukensia demissa*) in relation to shore level – a nine-year study. *Estuaries*, **24**, 319–327.

Fuentes-Santos, I., Labarta, U., Álvarez-Salgado, X.A. & Fernández-Reiriz, M.J. (2016) Solar irradiance dictates settlement timing and intensity of marine mussels. *Scientific Reports*, **6**, 29405.

Gale, S.L., Burritt, D.J. & Adams, S.L. (2016) The role of ethylenediaminetetraacetic acid in green-lipped mussel (*Perna canaliculus*) embryo development: a biochemical and morphological characterization. *Aquaculture*, **463**, 22–27.

Galimany, E., Freeman, C.J., Lunt, J., Domingos, A., Sacks, P. & Walters, L. (2017) Feeding competition between the native oyster *Crassostrea virginica* and the invasive mussel *Mytella charruana*. *Marine Ecology Progress Series*, **564**, 57–66.

Gallardi, D. (2014) Effects of bivalve aquaculture on the environment and their possible mitigation: a review. *Fisheries and Aquaculture Journal*, **5**, 105.

Galley, T.H., Batista, F.M., Braithwaite, R., King, J. & Beaumont, A.R. (2010) Optimisation of larval culture of the mussel *Mytilus edulis* (L.) *Aquaculture International*, **18**, 315–325.

Garcia, A.B. & Kamermans, P. (2015) Optimization of blue mussel (*Mytilus edulis*) seed culture using recirculation aquaculture systems. *Aquaculture Research*, **46**, 977–986.

Gelcich, S., Hughes, T.P., Olsson, P., Folkec, C., Defeo, O. Fernández, M. *et al.* (2010) Navigating transformations in governance of Chilean marine coastal resources. *Proceeding of the National Academy of Sciences*, **107**, 16794–16799.

Gillis, N.K., Walters, L.J., Fernandes, F.C. & Hoffman, E.A. (2009) Higher genetic diversity in introduced than in native populations of the mussel *Mytella charruana*: evidence of population admixture at introduction sites. *Diversity and Distributions*, **15**, 784–795.

GISD (2015a) Species profile *Crassostrea gigas*. Available at: http://193.206.192.138/gisd/species.php?sc=797 (accessed 28 May 2021).

GISD (2015b) Species profile *Mytilus galloprovincialis*. Available at: http://www.iucngisd.org/gisd/species.php?sc=102 (accessed 28 May 2021).

GISD (2015c) Species profile *Perna perna*. Available at: http://www.iucngisd.org/gisd/species.php?sc=742 (accessed 28 May 2021).

GISD (2015d) Species profile *Perna viridis*. Available at: http://www.iucngisd.org/gisd/species.php?sc=731 (accessed 28 May 2021)

GISD (2015e) Species profile *Geukensia demissa*. Available from: http://www.iucngisd.org/gisd/species.php?sc=1153 (accessed 28 May 2021).

GISD (2015f) Species profile *Musculista senhousia*. Available from: http://www.iucngisd.org/gisd/species.php?sc=1031 (accessed 28 May 2021).

González-Araya, R. & Robert, R. (2018) Larval development and fatty acid composition of *Ostrea edulis* (L.) fed four different single diets from conditioning to pre-settlement. *Aquaculture Research*, **49**,1768–1781.

Gosling, E.M. (2015) *Marine Bivalve Molluscs*, 2nd edn. Wiley-Blackwell, New York.

Grant, C., Archambault, P., Olivier, F. & McKindsey, C.W. (2012) Influence of 'bouchot' mussel culture on the benthic environment in a dynamic intertidal system. *Aquaculture Environment Interactions*, **2**, 117–131.

Guedes, A.C. & Malcata, F.X. (2012) Nutritional value and uses of microalgae in aquaculture. In: *Aquaculture* (ed. Z. Muchlisin), pp. 59–78. InTechOpen, London.

Gui, Y., Zamora, L., Dunphy, B.J. & Jeffs, A.G. (2016) Evaluation of the formulated diet MySpat for feeding hatchery-reared spat of the green-lipped mussel, *Perna canaliculus* (Gmelin, 1791). *Aquaculture Research*, **47**, 3907–3912.

Guyondet, T., Roy, S., Koutitonsky, V.G., Grant, J. & Tita, G. (2010) Integrating multiple spatial scales in the carrying capacity assessment of a coastal ecosystem for bivalve aquaculture. *Journal of Sea Research*, **64**, 341–359.

Guyondet, T., Comeau, L.A., Bacher, C., Grant, J., Rosland, R., Sonier, R. *et al.* (2015) Climate change influences carrying capacity in a coastal embayment dedicated to shellfish aquaculture. *Estuaries and Coasts*, **38**, 1593–1618.

Haroun, R., de Courtois, V.G. & Kaushik, S. (2016) Fostering expertise in sustainable aquaculture. Impact open access and free science reports. EU FP7 EcoAqua Project.

Heasman, K.G. & Knight, B.R. (2014) Review of Greenshell mussel (GSM), *Perna canaliculus*, spat-falls in the Bay of Plenty region. Prepared for Aquaculture Direct. Cawthron Report No. 2519.

Heasman, M., Diemar, J., O'Connor, W., Sushames, T. & Foulkes, L. (2000) Development of extended shelf-life microalgae concentrate diets harvested by centrifugation for bivalve molluscs – a summary. *Aquaculture Research*, **31**, 637–659.

Helm, M.M., Bourne, N. & Lovatelli, A. (2004) Hatchery culture of bivalves: a practical manual. FAO Fisheries Technical Paper No. 471. FAO, Rome.

Hendriks, I.E., van Duren, L.A. & Herman, P.M.J. (2003) Effect of dietary polyunsaturated fatty acids on reproductive output and larval growth of bivalves. *Journal of Experimental Marine Biology and Ecology*, **296**, 199–213.

Hickman, R.W. (1992) Mussel Cultivation. In: *The Mussel Mytilus: Ecology, Physiology, Genetics and Culture* (ed. E.M. Gosling), pp. 465–510. Elsevier Science Publishers B.V., Amsterdam.

Hicks, D.W. & Tunnell, J.W. (1993) Invasion of the south Texas coast by the edible brown mussel *Perna perna* (Linneaus 1758). *The Veliger*, **36**, 92–94.

Holdt, S.L. & Edwards, M.D. (2014) Cost-effective IMTA: a comparison of the production efficiencies of mussels and seaweed. *Journal of Applied Phycology*, **26**, 933–945.

Holm, P., Buck, B.H. & Langan, R. (2017) Introduction: new approaches to sustainable offshore food production and the development of offshore platforms. In: *Aquaculture Perspective of Multi-Use Sites in the Open Ocean – The Untapped Potential for Marine Resources in the Anthropocene* (eds B.H. Buck & R. Langan), pp. 1–20. Springer Open, Berlin.

Ibarra, D.A., Fennel, K. & Cullen, J.J. (2014) Coupling 3-D Eulerian bio-physics (ROMS) with individual-based shellfish ecophysiology (SHELL-E): a hybrid model for carrying capacity and environmental impacts of bivalve aquaculture. *Ecological Modelling*, **273**, 63–78.

Id Halla, M., Hassan, N., Kassila, J., El Mustapha, A.C., Orbi, A. & Moukrim, A. (2017) Comparative production of two mussel species (*Perna perna* and *Mytilus galloprovincialis*) reared on an offshore submerged longline system in Agadir, Morocco. *International Journal of Scientific and Engineering*, **8**, 1203–1213.

Inglis, G.J., Hayden, B.J. & Ross, A.H. (2000) An overview of factors affecting the carrying capacity of coastal embayments for mussel culture. NIWA, Christchurch. Client Report CHC00/69.

IOCCG (2009) *Remote Sensing in Fisheries and Aquaculture. Forget, M.-H., Stuart, V. & Platt, T. (eds), Reports of the International Ocean-Colour Coordinating Group,No.* 8, IOCCG, Dartmouth, NS.

Irisarri, J., Fernández-Reiriz, M.J., Robinson, S.M.C., Cranford, P.J. & Labarta, U. (2013) Absorption efficiency of mussels *Mytilus edulis* and *Mytilus galloprovincialis* cultured under Integrated Multi-Trophic Aquaculture conditions in the Bay of Fundy (Canada) and Ría Ares-Betanzos (Spain). *Aquaculture*, **388–391**, 182–192.

Jahangard, S., Williams, M., Mercer, J.A., Rahim, E. & Ingram, B.A. (2010) A technical report on hatchery production of blue mussel *Mytilus galloprovincialis* at the Victoria Shellfish Hatchery (VSH), Queenscliff (2008) Fisheries Victoria Technical Report No. 112, Department of Primary Industries, Queenscliff, VIC.

Jiang, W. & Gibbs, M.T. (2005) Predicting the carrying capacity of bivalve shellfish culture using a steady, linear food web model. *Aquaculture*, **244**, 171–185.

Jutson, M.G.S., Pipe, R.K. & Tomas, C.R. (2016) The cultivation of marine phytoplankton. In: *Microalgae: Current Research and Applications* (ed. M.F. Tsaloglou), pp. 11–26. Caister Academic Press, Poole.

Kamermans, P. & Capelle, J.J. (2019) Provisioning of mussel seed and its efficient use in culture. In: *Goods and Services of Marine Bivalves* (eds AC. Smaal, J.G. Ferreira, J. Grant, J.K. Petersen & O. Strand), pp. 27–49. Springer Open, Berlin.

Kamermans, P., Galley, T., Boudry, P., Fuentes, J., McCombie, H., Batista, F.M. *et al.* (2013) Blue mussel hatchery technology in Europe. In: *Advances in Aquaculture Hatchery Technology* (eds G. Allan & G. Burnell), pp. 339–373. Woodhead Publishing, Cambridge.

Kang, J.-Y., Bangoura, I., Cho, J.-Y., Joo, J., Seong, Choi, Y. 4, Hwang, D.S. *et al.* (2016) Antifouling effects of the periostracum on algal spore settlement in the mussel *Mytilus edulis*. *Fisheries and Aquatic Sciences*, **19**, 7.

Kaspar, H.F., Keys, E.F., King, N., Smith, K.F., Kesarcodi-Watson, A. & Miller, M.R. (2014) Continuous production of *Chaetoceros calcitrans* in a system suitable for commercial hatcheries. *Aquaculture*, **420–421**, 1–9.

Katsanevakis, S., Zenetos, A., Belchior, C. & Cardoso, A.C. (2013a) Invading European seas: assessing pathways of introduction of marine aliens. *Ocean and Coastal Management*, **76**, 64–74.

Katsanevakis, S., Gatto, F., Zenetos, A. & Cardoso, A.C. (2013b) How many marine aliens in Europe? *Management of Biological Invasions*, **4**, 37–42.

Keawtawee, T., Songsangjinda, P., Sangnoi, Y. & Uppabullung, A. (2018) The current situation and environmental conditions of green mussel farming in the Gulf of Thailand. IOP Conf. Series: *Earth and Environmental Science*, **137**, 012093. IOP Publishing, Bristol.

Kesarcodi-Watson, A., Kaspar, H., Lategan, M.J. & Gibson, L. (2008) Probiotics in aquaculture: the need, principles and mechanisms of action and screening processes. *Aquaculture*, **274**, 1–14.

Kesarcodi-Watson, A., Kaspar, H., Lategan, M.J. & Gibson, L. (2009a) Two pathogens of Greenshell™ mussel larvae, *Perna canaliculus*: *Vibrio splendidus* and a *V. coralliilyticus/neptunius*-like isolate. *Journal of Fish Diseases*, **32**, 499–507.

Kesarcodi-Watson, A., Kaspar, H., Lategan, M.J. & Gibson, L. (2009b) Challenge of New Zealand Greenshell™ mussel *Perna canaliculus* larvae using two *Vibrio* pathogens: a hatchery study. *Diseases of Aquatic Organisms*, **86**, 15–20.

Kesarcodi-Watson, A., Kaspar, H., Lategan, M.J. & Gibson, L. (2009c) Screening for probiotics of Greenshell™ mussel larvae, *Perna canaliculus*, using a larval challenge bioassay. *Aquaculture*, **296**, 159–164.

Kesarcodi-Watson, A., Kaspar, H., Lategan, M.J. & Gibson, L. (2010) *Alteromonas macleodii* 0444 and *Neptunomonas* sp. 0536, two novel probiotics for hatchery-reared Greenshell™ mussel larvae, *Perna canaliculus*. *Aquaculture*, **309**, 49–55.

Kesarcodi-Watson, A., Miner, P., Nicolas, J.L. & Robert, R. (2012) Protective effect of four potential probiotics against pathogen-challenge of the larvae of three bivalves: Pacific oyster (*Crassostrea gigas*), flat oyster (*Ostrea edulis*) and scallop (*Pecten maximus*). *Aquaculture*, **344–349**, 29–34.

Khan, M.I., Shin, J.H. & Kim, J.D. (2018) The promising future of microalgae: current status, challenges, and optimization of a sustainable and renewable industry for biofuels, feed, and other products. *Microbial Cell Factories*, **17**, 36.

Kirk, M., Esler, D. & Boyd, W.S. (2007) Morphology and density of mussels on natural and aquaculture structure habitats: implications for sea duck predators. *Marine Ecology Progress Series*, **346**, 179–187.

Knauer, J. & Southgate, P.C. (1999) A review of the nutritional requirements of bivalves and the development of alternative and artificial diets for bivalve aquaculture. *Reviews in Fisheries Science*, **7**, 241–280.

Labarta, U. & Fernández-Reiriz, M.J. (2019) The Galician mussel industry: innovation and changes in the last forty years. *Ocean and Coastal Management*, **167**, 208–218.

Lallias, D., Boudry, P., Batista, F.M., Beaumont, A., King, J.W., Turner, J.R. *et al.* (2015) Invasion genetics of the Pacific oyster *Crassostrea gigas* in the British Isles inferred from microsatellite and mitochondrial markers. *Biological Invasions*, **17**, 2581–2595.

Langdon, C. & Önal, E. (1999) Replacement of living microalgae with spray-dried diets for the marine mussel *Mytilus galloprofincialis*. *Aquaculture*, **180**, 283–294.

Lavens, P. & Sorgeloos, P. (eds) (1996) *Manual on the Production and Use of Live Food for Aquaculture. Fisheries Technical Paper* No. 361. FAO, Rome.

Laxmilatha, P., Rao, G.S., Pattnaik, P., Nageswara, Rao., T., Prasad Rao, M. & Dash, B. (2011) Potential for the hatchery production of spat of the green mussel *Perna viridis* Linnaeus (1758). *Aquaculture*, **312**, 88–94.

Leonardos, N. & Lucas, I.A.N. (2000) The use of larval fatty acids as an index of growth in *Mytilus edulis* L. larvae. *Aquaculture*, **184**, 155–166.

Lindahl, O., Hart, R., Hernroth, B., Kollberg, S., Loo, L-O., Olrog, L., *et al.* (2005) Improving marine water quality by mussel farming: a profitable solution for Swedish society. *Ambio*, **34**, 131–138.

Liu, W., Pearce, C.M., McKinley, R.S. & Forster, I.P. (2016) Nutritional value of selected species of microalgae for larvae and early post-set juveniles of the Pacific geoduck clam, *Panopea generosa*. *Aquaculture*, **452**, 326–341.

Liu, Y & Li, X. (2015) Successful oocyte cryopreservation in the blue mussel *Mytilus galloprovincialis*. *Aquaculture*, **438**, 55–58.

Lockwood, B.L. & Somero, G.N. (2011) Invasive and native blue mussels (genus *Mytilus*) on the California coast: the role of physiology in a biological invasion. *Journal of Experimental Marine Biology and Ecology*, **400**, 167–174.

Lowe, S., Browne, M., Boudjelas, S. & De Poorter, M. (2000) 100 of the world's worst invasive alien species: a selection from the Global Invasive Species Database. Invasive Species Specialist Group of the Species Survival Commission (SSC) of the World Conservation Union (IUCN), Auckland, New Zealand.

Lutz-Collins, V., Ramsay, A, Quijón, P.A. & Davidson, J. (2009) Invasive tunicates fouling mussel lines: evidence of their impact on native tunicates and other epifaunal invertebrates. *Aquatic Invasions*, **4**, 213–220.

Maar, M., Nielsen, T.G. & Petersen, J.K. (2008) Depletion of plankton in a raft culture of *Mytilus galloprovincialis* in Ría de Vigo, NW Spain. II. Zooplankton. *Aquatic Biology*, **4**, 127–141.

Marshall, R., McKinley, S. & Pearce, C.M. (2010) Effects of nutrition on larval growth and survival in bivalves. *Reviews in Aquaculture*, **2**, 33–55.

Martin-Creuzburg, D. & von Elert, E. (2009) Ecological significance of sterols in aquatic food webs. In: *Lipids in Aquatic Ecosystems* (eds M. Kainz, M.T. Brett & M.T. Arts), pp. 43–64. Springer, New York.

Martínez-Fernández, E., Acosta-Salmón, H. & Southgate, P.C. (2006) The nutritional value of seven species of tropical microalgae for black-lip pearl oyster (*Pinctada margaritifera*, L.) larvae. *Aquaculture*, **257**, 491–503.

Martínez-Páramo, S., Horváth, A., Labbé, C., Zhang, T., Robles, V., Herráez, P. *et al.* (2017) Cryobanking of aquatic species. *Aquaculture*, **472**, 156–177.

Martínez-Pita, I., Sánchez-Lazo, C. & García, F.J. (2016a) Influence of microalga lipid composition on the sexual maturation of *Mytilus galloprovincialis*: a hatchery study. *Aquaculture Nutrition*, **22**, 202–216.

Martínez-Pita, I., Sánchez-Lazo, C. & Herrera, M. (2016b) A non-lethal method for establishing sexual maturation in mussels (*Mytilus galloprovincialis* (Lamarck, 1819) during broodstock conditioning in hatcheries). *Aquaculture International*, **24**,1247–1254.

McCausland, M.A., Brown, M.R., Barrett, S.M., Diemar, J.A. & Heasman, M.P. (1999) Evaluation of live microalgae and microalgal pastes as supplementary food for juvenile Pacific oysters (*Crassostrea gigas*). *Aquaculture*, **174**, 323–342.

McKindsey, C.W., Thetmeyer, H., Landry, T. & Silvert, W. (2006) Review of recent carrying capacity models for bivalve culture and recommendations for research and management. *Aquaculture*, **261**, 451–462.

McKindsey, C.W., Landry, T., O'Beirn, F.X. & Davies, I.M. (2007) Bivalve aquaculture and exotic species: a review of ecological considerations and management issues. *Journal of Shellfish Research*, **26**, 281–294.

McKindsey, C.W., Archambault, P., Callier, M.D. & Olivier, F. (2011) Influence of suspended and off-bottom mussel culture on the sea bottom and benthic habitats: a review. *Canadian Journal of Zoology*, **89**, 622–646.

McKindsey, C.W (2013) Carrying capacity for sustainable bivalve aquaculture. In: *Sustainable Food Production* (eds P. Christou, R. Savin, B. Costa-Pierce, I. Misztal & B. Whitelaw), pp. 417–448. Springer-Verlag, New York.

Meaden, G.J., Aguilar-Manjarrez, J., Corner, R.A., O'Hagan, A.M. & Cardia, F. (2016) Marine spatial planning for enhanced fisheries and aquaculture sustainability – its application in the Near East. FAO Fisheries and Aquaculture Technical Paper No. 604. FAO, Rome.

Medlin, L.K., Montresor, M., Graneli, E., Reugera, B., Raine, R., Edvardsen, B. *et al.* (2013) MIDTAL (Microarrays for the Detection of Toxic Algae). *Phytotaxa*, **127**, 201–210.

Méndez Martínez, G., Ovejero Campos, A., Gómez Vilar, E., Lastra Mier, R.E. & Pérez-Arlucea, M. (2011) Changes induced by mussel raft aquaculture in benthic environment of the Rías Baixas (Galicia, Spain). *Journal of Coastal Research,* **S164**, 786–789.

Mineur, F., Belsher, T., Johnson, M.P., Maggs, C.A. & Verlaque, M. (2007) Experimental assessment of oyster transfers as a vector for macroalgal introductions. *Biological Conserversion*, **137**, 237–247.

Murphy, K.J., Mooney, B.D., Mann, N.J., Nichols, P.D. & Sinclair, A.J. (2002) Lipid, FA, and sterol composition of New Zealand green lipped mussel (*Perna canaliculus)* and Tasmanian blue mussel (*Mytilus edulis*). *Lipids*, **37**, 587–595.

Murray, L.G., Newell, C.R. & Seed, R, (2007) Changes in the biodiversity of mussel assemblages induced by two methods of cultivation. *Journal of Shellfish Research*, **26**, 153–162.

Nalder, T.D., Miller, M.R. & Packer, M.A. (2015) Changes in lipid class content and composition of *Isochrysis* sp. (T-Iso) grown in batch culture. *Aquaculture International*, **23**, 1293–1312.

Navarrete-Mier, F., Sanz-Lazaro, C. & Marin, A. (2010) Does bivalve mollusc polyculture reduce marine fin fish farming environmental impact? *Aquaculture*, **306**, 101–107.

Nevejan, N., Davis, J., Little, K. & Kiliona, A. (2007) Use of a formulated diet for mussel spat *Mytilus galloprovincialis* (Lamarck 1819) in a commercial hatchery. *Journal of Shellfish Research*, **26**, 357–363.

Nguyen T.T.T., Hayes B. & Ingram B.A. (2012) Progress on selective breeding program for blue mussel in Victoria. *Fisheries Victoria Research Report Series No.* 60.

Nogueira, E., Pérez, F.F. & and Ríos, A.F. (1997) Seasonal patterns and long-term trends in an estuarine upwelling ecosystem (Ría de Vigo, NW Spain). *Estuarine, Coastal and Shelf Science*, **44**, 285–300.

Nunes, J.P., Ferreira, J.G., Gazeau, F., Lencart-Silva, J., Zhang, X.L., Zhu, M.Y. *et al.* (2003) A model for sustainable management of shellfish polyculture in coastal bays. *Aquaculture*, **219**, 257–277.

Nunes, J.P., Ferreira, J.G., Bricker, S.B., O'Loan, B., Dabrowski, T., Dallaghan, B. *et al.* (2011) Towards an ecosystem approach to aquaculture: assessment of sustainable shellfish cultivation at different scales of space, time and complexity. *Aquaculture*, **315**, 369–383.

Nunes, A.L., Katsanevakis, S., Zenetos, A. & Cardoso, A.C. (2014) Gateways to alien invasions in the European seas. *Aquatic Invasions*, **9**, 133–144.

de Nys, R. & Ison, O. (2004) Evaluation of antifouling products developed for the Australian pearling industry. *FRDC Project No.* 2000/254.

Önal, E., Langdon, C. & Önal, U. (2005) The evaluation of spray-dried microalgae in diets for juvenile manila clam, *Tapes philippinarum*. *Journal of Shellfish Research*, **24**, 1061–1065.

O'Mahony, J.H.T. (1993) Phytoplankton species associated with imports of the Pacific oyster *Crassostrea gigas*, from France to Ireland. ICES CM 1993/F:26.

Otero, X.L., Calvo de Anta, R.M. & Macías, F. (2009) Iron geochemistry under mussel rafts in the Galician ria system (Galicia-NW Spain). *Estuarine and Coastal Shelf Science*, **81**, 83–93.

Paillard, C. (2004) A short-review of brown ring disease, a vibriosis affecting clams, *Ruditapes philippinarum* and *Ruditapes decussates*. *Aquatic Living Resourses*, **17**, 467–475.

Paillard, C., Le Roux, F. & Borrego, J.J. (2004) Bacterial disease in marine bivalves, a review of recent studies: trends and evolution. *Aquatic Living Resources*, **17**, 477–498.

Paredes, E., Adams, S.L., Tervit, H.R., Smith, J.F., McGowan, L.T., Gale, S.L., *et al.* (2012) Cryopreservation of Greenshell™ mussel (*Perna canaliculus*) trochophore larvae. *Cryobiology*, **65**, 256–262.

Paredes, E., Bellas, J. & Adams, S.L. (2013) *Comparative cryopreservation study of trochophore larvae from two species of bivalves: Pacific oyster (Crassostrea gigas) and blue mussel (Mytilus galloprovincialis) Cryobiology*, **67**, 274–279.

Pérez-Camacho, A., Albentosa, M., Fernández-Reiriz, M.J. & Labarta, U. (1998) Effect of microalgal and inert (cornmeal and cornstarch) diets on the growth performance and biochemical composition of *Ruditapes decussatus* seed. *Aquaculture*, **160**, 89–102.

Pérez-Camacho, A., Aguiar, E., Labarta, U., Vinseiro, V., Fernández-Reiriz, M.J. & Álvarez-Salgadobal, X.A. (2014) Ecosystem-based indicators as a tool for mussel culture management strategies. *Ecological Indicators*, **45**, 538–548.

Pernet, P. & Tremblay, R. (2004) Effect of varying levels of dietary essential fatty acid during early ontogeny of the sea scallop *Placopecten magellanicus*. *Journal of Experimental Marine Biology and Ecology*, **310**, 73–86.

Peteiro, L.G., Labarta, U., Fernández-Reiriz, M.-J., Álvarez-Salgado, X.A., Álvarez-Salgado, R. & Piedracoba, S. (2011) Influence of intermittent-upwelling on *Mytilus galloprovincialis* settlement patterns in the Ría de Ares-Betanzos. *Marine Ecology Progress Series*, **443**, 111–127.

Petersen, J.K., Nielsen, T.G., van Duren, L. & Maar, M. (2008) Depletion of plankton in a raft culture of *Mytilus galloprovincialis* in Ría de Vigo, NW Spain. I. Phytoplankton. *Aquatic Biology* **4**, 113–125.

Petersen, J.K., Saurel, C., Nielsen. P. & Timmermann, K. (2016) The use of shellfish for eutrophication control. *Aquaculture International*, **24**, 857–878.

Pettersen, A.K., Turchini, G.M., Jahangard, S., Ingram, B.A. & Sherman, C.D.H. (2010) Effects of different dietary microalgae on survival, growth, settlement and fatty acid composition of blue mussel (*Mytilus galloprovincialis*) larvae. *Aquaculture*, **309**, 115–124.

Philippart, C.J.M., Amaral, A., Asmus, R., van Bleijswijk. J., Bremner, J., Buchholz, F. *et al.* (2012) Spatial synchronies in the seasonal occurrence of larvae of oysters (*Crassostrea gigas*) and mussels (*Mytilus edulis/galloprovincialis*) in European coastal waters. *Estuarine, Coastal and Shelf Science*, **108**, 52–63.

Philippart, C.J.M., van Bleijswijk, J. D., Kromkamp, J.C., Zuur, A.F. & Herman, P. M. (2014) Reproductive phenology of coastal marine bivalves in a seasonal environment. *Journal of Plankton Research*, **36**, 1512–1527.

Piñeiro-Corbeira, C., Barrientos, S., Olmedo, M., Cremades, J. & Barreiro, R. (2018) By-catch in no-fed aquaculture: exploiting mussel seed persistently and extensively disturbs the accompanying assemblage. *ICES Journal of Marine Science*, **75**, 2213–2223.

Pochon, X., Atalah, J., Wood, S.A., Hopkins, G.A., Watts, A. & Boedeker, C. (2015) *Cladophora ruchingeri* (C. Agardh) Kützing, 1845 (Cladophorales, Chlorophyta): a new biofouling pest of green-lipped mussel *Perna canaliculus* (Gmelin, 1791) farms in New Zealand. *Aquatic Invasions*, **10**, 123–133.

Ponis, E., Probert, I., Véron, B., Le Coz, J.R., Mathieu, M. & Robert R. (2006) Nutritional value of six Pavlovophyceae for *Crassostrea gigas* and *Pecten maximus* larvae. *Aquaculture*, **254**, 544–553.

Prado, S., Romalde, J.L. & Barja, J.L. (2010) Review of probiotics for use in bivalve hatcheries. *Veterinary Microbiology*, **145**, 187–197.

Pronker, A.E., Nevejan, N.M., Peene, F., Geijsen, P. & Sorgeloos, P. (2008) Hatchery broodstock conditioning of the blue mussel *Mytilus edulis* (Linnaeus 1758). Part I. Impact of different micro-algae mixtures on broodstock performance. *Aquaculture International*, **16**, 297–307.

Protopopescu, G.C. & Beal, B.F. (2015) Settlement response to various rope substrates in blue mussels (*Mytilus edulis* Linnaeus) in a hatchery setting. *Journal of Shellfish Research*, **34**, 383–391.

Ragg, N.L.C., King, N., Watts, E. & Morrish, J. (2010) Optimising the delivery of the key dietary diatom *Chaetoceros calcitrans* to intensively cultured Greenshell™ mussel larvae, *Perna canaliculus*. *Aquaculture*, **306**, 270–280.

Rato, A., Joaquim, S., Tavares, T.G., Martins, Z.E., Guedes, A.C., Pereira, L.F. *et al.* (2018) *Viability of dietary substitution of live microalgae with dry Ulva rigida in broodstock conditioning of the Pacific oyster (Crassostrea gigas). Biology Open*, **7**, bio035923.

Reid, G.K., Liutkus, M., Bennett, A., Robinson, S.M.C. & MacDonald, B. (2010) Absorption efficiency of blue mussels (*Mytilus edulis* and *M. trossulus*) feeding on Atlantic salmon (*Salmo salar*) feed and fecal particulates: implications for integrated multitrophic aquaculture. *Aquaculture*, **299**, 165–169.

Reid, G, K., Robinson, S., Chopin, T. & MacDonald, B.A. (2013) *Dietary proportion of fish culture solids required by shellfish to reduce the net organic load in open-water Integrated Multi-Trophic Aquaculture: a scoping exercise with cocultured Atlantic salmon (Salmo salar) and blue mussel (Mytilus edulis). Journal of Shellfish Research*, **32**, 509–517.

Rivera, A., Unibazoc, J., León, P., Vásquez-Lavín, F., Ponce, R., Mansur, L. *et al.* (2017) Stakeholder perceptions of enhancement opportunities in the Chilean small and medium scale mussel aquaculture industry. *Aquaculture*, **479**, 423–431.

Robinson, T.B., Griffiths, C.L., McQuaid, C.D. & Rius, M. (2005) Marine alien species of South Africa–status and impacts. *African Journal of Marine Science*, **27**, 297–306.

Rodríguez Rodríguez, G., Villasante, S. & Carme García-Negro, M.D. (2011) Are red tides affecting economically the commercialization of the Galician (NW Spain) mussel farming? *Marine Policy*, **35**, 252–257.

Roycroft, D., Kelly, T.C. & Lewis, L.J. (2004) Birds, seals and the suspension culture of mussels in Bantry Bay, a non-seaduck area in Southwest Ireland. *Estuarine, Coastal and Shelf Science*, **61**, 703–712.

Ruiz, Y., Suárez, P., Alonso, A., Longo, E. & San Juan, F. (2014) Chemicals used for maintenance of wood rafts in mussel farms: evaluation of their potential toxic risk to mussel culture. *Aquaculture Environment Interactions*, **6**, 55–66.

Sánchez-Lazo, C. & Martínez-Pita, I. (2014) Effects of different mono, bi and trispecific microalgal diets on survival, growth, development, settlement and fatty acid composition of mussel *Mytilus galloprovincialis* (Lamarck, 1819) larvae. *Aquaculture*, **426–427**, 138–147.

Sanz-Lazaro, C. & Sanchez-Jerez, P. (2017) Mussels do not directly assimilate fish farm wastes: shifting the rationale of integrated multi-trophic aquaculture to a broader scale. *Journal of Environmental Management*, **201**, 82–88.

Sarà, G., Romano, C., Widdows, J. & Staff, F.J. (2008) Effect of salinity and temperature on feeding physiology and scope for growth of an invasive species (*Brachidontes pharaonis* – MOLLUSCA: BIVALVIA) within the Mediterranean Sea. *Journal of Experimental Marine Biology and Ecology*, **363**, 130–136.

Scottish Funding Council (2017) Mussel hatchery project goes global. Available at: http://www.sfc.ac.uk/news/2017/news--mussel-hatchery-project.aspx (accessed 28 May 2021).

Sievers, M., Fitridge, I., Dempster, T. & Keough, M.J. (2013) Biofouling leads to reduced shell growth and flesh weight in the cultured mussel *Mytilus galloprovincialis*. *Biofouling: The Journal of Bioadhesion and Biofilm Research*, **29**, 97–107.

Silva, A.F., Sousa, M.C., Bernardes, C. & Dias, J.M. (2017) Will climate change endanger the current mussel production in the Rias Baixas (Galicia, Spain)? *Journal of Aquaculture and Fisheries*, **1**, 001.

Sohn, S. (2016) Evaluation of the efficacy of candidate probiotics for disease prevention in shellfish hatcheries. PhD thesis, *University of Rhode Island*, Kingston, RI.

Sohn, S., Lundgren, K.M., Tammi, K., Smolowitz, R., Nelson, D.R. Rowley, D.C. *et al.* (2016) Efficacy of probiotics in preventing vibriosis in the larviculture of different species of bivalve shellfish. *Journal of Shellfish Research*, **35**, 319–328.

Smith, J.F., Adams, S.L., Gale. S.L., McGowan, L.T., Tervit, H.R. & Roberts, R.D. (2012) *Cryopreservation of Greenshell™ mussel (Perna canaliculus) sperm. I. Establishment of freezing protocol. Aquaculture*, **334–337**, 199–204.

Soto, D., Aguilar-Manjarrez, J. & Hishamunda, N. (eds) (2008) *Building an Ecosystem Approach to Aquaculture. FAO/Universitat de les Illes Balears Expert Workshop, 7–11 May 2007, Palma de Mallorca, Spain*. FAO Fisheries and Aquaculture Proceedings No. 14. FAO, Rome.

Soto, D. (2009) *Integrated Mariculture: a Global Review*. FAO Fisheries and Aquaculture Technical Paper 529. FAO, Rome.

Soudant, P., Sanles, M. V., Quere, C., Le Coz, J.R., Marty, Y., Moal, J. *et al.* (2000) The use of lipid emulsions for sterol supplementation of spat of the Pacific oyster, *Crassostrea gigas*. *Aquaculture*, **184**, 315–326.

Sreedevi, P.R., Uthayakumar, V., Jayakumar, R. & Ramasubramanian, V. (2014) Influence of rearing water temperature on induced gonadal development and spawning behaviour of tropical green mussel, *Perna viridis*. *Asian Pacific Journal of Reproduction*, **3**, 204–209.

Srisunont, C. & Babel, S. (2016) Estimating the carrying capacity of green mussel cultivation by using net nutrient removal model. *Marine Pollution Bulletin*, **112**, 235–243.

Steinrücken, P., Prestegard, S.K., Hendrik de Vree, J., Storesund, J.E., Pree, B. Mjøs, S.A. *et al.* (2018) Comparing EPA production and fatty acid profiles of three *Phaeodactylum tricornutum* strains under western Norwegian climate conditions. *Algal Research*, **30**, 11–22.

Southgate, P.C., Lee, P.S. & Nell, J.L. (1992) Preliminary assessment of a microparticulated diet for larval culture of the Sydney rock oyster, *Saccostrea commercialis* (Iredale and Rougley). *Aquaculture*, **105**, 345–352.

Sudmaris Chile SA (2016) Top Quality Chilean Mussels. Available at: https://www.youtube.com/watch?v=LN4IXFECDfI (accessed 28 May 2021).

Suplicy, F.M. (2018) A review of the multiple benefits of mussel farming. *Reviews in Aquaculture*, **12**, 204–233.

Tang, Q.S., Fang, J.G. & Liu, H. (2002) Development of mussel aquaculture in China. *Bulletin of the Aquaculture Association of Canada*, **102**, 66–74.

Tanyaros, S. & Chuseingjaw, S. (2016) A partial substitution of microalgae with single cell detritus produced from seaweed (*Porphyra haitanensis*) for the nursery culture of tropical oyster (*Crassostrea belcheri*). *Aquaculture Research*, **47**, 2080–2088.

The Fish Site (2010) Production methods for the Mediterranean mussel. Available at: https://thefishsite.com/articles/production-methods-for-the-mediterranean-mussel (accessed 28 May 2021).

The Fish Site (2017) How mussel farms can boost biodiversity. Available at: https://thefishsite.com/articles/how-mussel-farms-can-boost-local-habitats (accessed 28 May 2021).

Toledo, J.D., Kurokura, H. & Kasahara, S. (1989) Preliminary studies on the cryopreservation of the blue mussel embryos. *Nippon Suisan Gakkaishi*, **55**, 1661.

Torchin, M.E., Hechinger, R.F., Huspeni, T.C., Whitney, K.L. & Lafferty, K.D. (2005) The introduced ribbed mussel (*Geukensia demissa*) in Estero de Punta Banda, Mexico: interactions with the native cord grass, *Spartina foliosa*. *Biological Invasions*, **7**, 607–614.

Toupoint, N., Barbier, P., Tremblay, R., Archambault, P., McKindsey, C.W., Winkler, G. *et al.* (2016) Influence of intertidal recreational fisheries and 'bouchot' mussel culture on bivalve recruitment. *Marine Environmental Research*, **117**, 1–12.

Travers, M.-A., Miller, K., Roque, A. & Friedman, C.S. (2015) Bacterial diseases in marine bivalves. *Journal of Invertebrate Pathology*, **131**, 11–31.

Tremblay, R., Cartier, S., Miner, P., Pernet, F., Quéré, C., Moal, J. *et al.* (2007) Effect of *Rhodomonas salina* addition to a standard hatchery diet during the early ontogeny of the scallop *Pecten maximus*. *Aquaculture*, **262**, 410–418.

Turini, C.S., Sühnel, S., Lagreze-Squella, F-J., Ferreira, J.F. & Rodrigues de Melo, C.M. (2014) Effects of stocking-density in flow-through system on the mussel *Perna perna* larval survival. *Acta Scientiarum Animal Sciences*, **36**, 247–252.

Vallejo, B., Conejar-Espedido, J., Manubag, L., Artiaga, K.C.C., Damatac II, A.M., Imperial, I.C.V.J. *et al.* (2017) First record of the Charru mussel *Mytella charruana* d'Orbignyi, 1846 (Bivalvia: Mytilidae) from Manila Bay, Luzon, Philippines. *BioInvasions Records*, **6**, 49–55.

Varennes, E., Hanssen, S.A., Bonardelli, J. & Guillemette, M. (2013) Sea duck predation in mussel farms: the best nets for excluding common eiders safely and efficiently. *Aquacultur Environment Interactions*, **4**, 31–39.

Vichkovitten, T., Intarachart, A. & Khaodon, K. (2017) Impact of green mussel (*Perna viridis*) raft-culture on benthic environment in Sriracha coastal water, Thailand. *GMSARN International Journal*, **11**, 116–122.

Wang, H., Li, X., Wang, M., Clarke, S., Gluis, M. & Zhang, Z. (2011) Effects of larval cryopreservation on subsequent development of the blue mussels, *Mytilus galloprovincialis* Lamarck. *Aquaculture Research*, **42**, 1816–1823.

Wang, H., Li, X., Wang, M., Clarke, S. & Gluis, M. (2014) The development of oocyte cryopreservation techniques in blue mussels *Mytilus galloprovincialis*. *Fisheries Science*, **80**,1257–1267.

Wang, Z.H., Liang, J.L. & Zhang, S.Y. (2015) Comparison of pelagic and benthic fish assemblages in mussel farming habitat. *Chinese Journal of Ecology*, **34**, 753–759.

Watson, D.I., Shumway, S.E. & Whitlatch, R.B. (2009) Biofouling and the Shellfish Industry. In: *Shellfish Safety and Quality*, 1st edn (eds S. Shumway & G. Rodrick). pp. 317–336. Woodhead Publishing, Cambridge.

Weise, A.M., Cromey, C.J., Callier, M.D., Archambault, P., Chamberlain, J. & McKindsey, C.W. (2009) Shellfish-DEPOMOD: modelling the biodeposition from suspended shellfish aquaculture and assessing benthic effects. *Aquaculture*, **288**, 239–253.

Welladsen, H., Kent, M., Mangott, A. & Li, Y. (2014) Shelf-life assessment of microalgae concentrates: Effect of cold preservation on microalgal nutrition profiles. *Aquaculture*, **430**, 241–247.

Willer, D. & Aldridge, D.C. (2017) Microencapsulated diets to improve bivalve shellfish aquaculture. *Royal Society Open Science*, **7**, 171142.

Woods, C.M.C., Floerl, O. & Hayde, B.J. (2012) Biofouling on GreenshellTM mussel (*Perna canaliculus*) farms: a preliminary assessment and potential implications for sustainable aquaculture practices. *Aquaculture International*, **20**, 537–557.

Yang, H., Hu, E., Cuevas-Uribe, R., Supan, J., Guo, X. & Tiersch, T.R. (2012) High-throughput sperm cryopreservation of eastern oyster *Crassostrea virginica*. *Aquaculture*, **344–349**, 223–230.

Yodsuwan, N., Sawayama, S. & Sirisansane, S. (2017) Effect of nitrogen concentration on growth, lipid production and fatty acid profiles of the marine diatom *Phaeodactylum tricornutum*. *Agriculture and Natural Resources*, **51**, 190–197.

Young, T., Alfaro, A.C. & Villas-Bôas, S. (2015) Identification of candidate biomarkers for quality assessment of hatchery-reared mussel larvae via GC/MS-based metabolomics. *New Zealand Journal of Marine and Freshwater Research*, **49**, 87–95.

Ysebaert, T., Hart, M. & Herman, P.M.J. (2009) Impacts of bottom and suspended cultures of mussels *Mytilus* spp. on the surrounding sedimentary environment and macrobenthic biodiversity. *Helgoland Marine Research*, **63**, 59–74.

Yuan, W.S., Hoffman, E.A. & Walters, L.J. (2016a) Effects of nonnative invertebrates on two life stages of the native eastern oyster *Crassostrea virginica*. *Biological Invasions*, **18**, 689—701.

Yuan, W.S., Walters, L.J., Brodsky, S.A., Schneider, K.R. & Hoffman, E.A. (2016b) Synergistic effects of salinity and temperature on the survival of two nonnative bivalve molluscs, *Perna viridis* (Linnaeus 1758) and *Mytella charruana* (d'Orbigny 1846). *Journal of Marine Biology*, **2016**, art. 9261309.

Zeeman, Z., Branch, G.M. & Pillay, D. (2018) Comparisons of life-history traits of the alien invasive *Semimytilus algosus* and three other mytilid mussels on the West Coast of South Africa. *Marine Ecology Progress Series*, **607**, 113–127.

Zhang, F. (1984) Mussel culture in China. *Aquaculture*, **39**, 1–10.

11

Diseases, Parasites and the Immune Response

Introduction

Because of the considerable economic importance of bivalves, there is a wealth of information available on diseases that affect them. Indeed, more than 30 years ago, we knew more about the diseases of bivalves than about the diseases of all other marine invertebrate groups put together (Kinne 1983), and that is still the case today. In contrast to oysters, there is much less known about parasites and diseases of mussels. This is probably due to their lower economic value, their shorter history of intensive culture and, most significantly, the fact that the diseases which have devastated the oyster culture industry have not been encountered by the mussel culture industry (Bower 1992). Fortunately, over the past decade, there has been an increase in published information on parasites and diseases of mussels.

This chapter describes the major disease-causing agents of marine mussels – viruses, bacteria, fungi, protistans, helminths and parasitic crustaceans – as well as the cellular and humoral defence mechanisms utilised by bivalves to combat disease. Those wishing to explore disease and parasitism in oysters, scallops and clams might be interested in the following reviews: Gosling (2015), Getchell *et al.* (2016) and Fisheries and Oceans Canada (2019).

Viruses

Viruses are ultramicroscopic infectious agents (~20–350 nm long) consisting of a nucleic acid molecule enclosed in a protein coat that can only multiply inside living cells. Once infected, the host cell begins to degenerate and is soon destroyed, liberating new virus particles in the process. These in turn infect new cells. As filter feeders, bivalves may accumulate viruses from humans and other vertebrates, thus acting as carriers of enteroviruses (e.g. hepatitis A, polio) for vertebrate hosts that feed on them (see Chapter 12). Many viruses, however, cause serious disease in the bivalves themselves and are responsible for mass mortalities in the aquaculture industry. Up to the early 1990s, only a few viruses pathogenic to marine mussels had been described. Rasmussen (1986) reported virus-like particles in mussels (*Mytilus edulis*) from Denmark. Acute cellular infiltrations were associated with particle detection and were construed as granulocytomas (inflamatory lesions).

Marine Mussels: Ecology, Physiology, Genetics and Culture, First Edition. Elizabeth Gosling.

In New Zealand, particles of about the same size were detected in farmed juvenile *Perna canaliculus* mussels suffering high (50–100%) cumulative mortalities (Jones *et al.* 1996). Extensive haemocytosis and necrosis of interstitial cells, basal cells and digestive tubule epithelial cells were observed. Identical cell pathology and virus-like particles were also observed in stunted subtidal *M. galloprovincialis* from the same area. The virus-like particles were believed to be an RNA virus, and because of their size, morphology and formation of paracrystalline arrays, they most likely belonged to the family Picornaviridae (Renault 2008), although subsequently Dayaram *et al.* (2013) suggested that the virus particles purified by Jones *et al.* (1996) could potentially have been single-strand DNA (ssDNA) viruses. In a study of parasitism in two deep-sea mussels (*Bathymodiolus* spp.), Ward *et al.* (2004) found viral-like inclusions, one of which was pathogenic, causing lysis of the digestive tubules in one species. The prevalence and intensity of infection by this pathogen were greater in hosts with shell lengths less than 100 mm. Prevalence is calculated as the percentage of mussels infected with a particular pathogen, and infection intensity as the number of a particular pathogenic species in an infected mussel. Mussels also exhibited intense infiltration of tissues and blood spaces by enlarged haemocytes. Mussels may act as intermediate hosts for some viral agents. For example, Kitamura *et al.* (2007) reported an aquatic birnavirus (ABV) in *M. galloprovincialis* that was acting as a reservoir host for infections in the Japanese flounder *Paralichthys olivaceous.*

In the past few decades, apart from the study by Kitamura *et al.* (2007), there has been no increase in the number of viral pathogens described for mussels. This is not the case for other bivalve groups. For example, the following viruses have been associated with high mortalities in the culture of larval and spat oysters (*Crassostrea*): gill necrosis virus (GNV), haemocyte infection virus (HIV), oyster velar virus disease (OVVD) and *Ostreid herpesvirus 1* (OsHV-1), which has also been reported in scallops (*Pecten maximus*) (Azul *et al.* 2001; Renault *et al.* 2008; Dang *et al.* 2009). A herpes-like virus (Renault *et al.* 2001) and a virus with strong morphological resemblance to the fish pathogen infectious pancreatic necrosis virus have been reported from *Ruditapes* and *Tellina* clams (Bower 2010a).

The main diagnostic method is still predominantly histopathology, a microscopic technique that enables the identification of cellular changes associated with infection but which does not provide conclusive identification unless accompanied by other methods such as transmission electron microscopy (TEM). But these are time-consuming and are unsuitable for epidemiological surveys. Over the past decade or so, specific and sensitive diagnostic methods that are suitable for epidemiological surveys of field samples (e.g. polymerase chain reaction (PCR), in situ hybridisation (ISH), and immunochemistry) have been developed to detect herpes-like viruses, a group associated with high mortalities in a wide range of bivalve species throughout the world. Schikorski *et al.* (2011) have developed an experimental protocol of horizontal transmission of OsHV-1 based on cohabitation between healthy and experimentally infected oysters (*Crassostrea gigas*). Through a cohabitation trial, the kinetics of OsHV-1 detection in different oyster organs and seawater samples was investigated and characterised for the first time using real-time quantitative PCR (qPCR). Despite progress in the development of diagnostic methods for virus detection, the lack of molluscan cell lines has limited the isolation of any of the previously mentioned viruses.

Larvae that are infected with a virus should be destroyed and discarded in an approved manner. There are no known control methods other than quick recognition of the problem by hatchery personnel followed by immediate disinfection of tanks and equipment to prevent tank-to-tank spread. Individuals from growing areas with a known disease (currently or historically) should not be moved to areas where the disease has not been observed. Bower (2016) provides a comprehensive account of the control measures used to combat the high mortalities associated, on a global scale, with the herpes-type virus OsHV-1 in oyster, clam and scallop species.

Eradication of viral diseases will prove difficult because of the high-density production systems used in commercial hatcheries and nurseries, further exacerbated by stock movement, one of the major risks in the spread of disease (Renault 2008). When a viral disease is diagnosed, there are at present no specific chemotherapies or vaccines available to treat it. One encouraging finding is that some species may have innate immunity against viral infection. In a study of OsHV-1 in *C. gigas*, an antiviral substance was reported in adult haemolymph, which indicates that adults could be more resistant than larvae and juveniles to infection. This substance may contribute to oyster survival despite continuing exposure to OsHV-1 (Olicard *et al.* 2005). Antiviral activity against a wider range of viruses has since been demonstrated in the haemolymph of additional bivalve species (Carriel-Gomes *et al.* 2006; Defer *et al.* 2009). Screening of mussels for antiviral activity has not been a priority in view of the paucity to date of viral diseases in mussels.

Bacteria

The aquatic environment harbours a rich bacterial flora that is easily transported in the water column, and bivalves, because of their efficient filter-feeding mechanism, normally accumulate a rich bacterial microbiota composed of various species belonging to different genera, like *Vibrio*, *Pseudomonas*, *Acinetobacter*, *Photobacterium*, *Moraxella*, *Aeromonas*, *Micrococcus* and *Bacillus* (Beaz-Hildlgo *et al.* 2010 and references therein). While the majority of marine bacteria are not harmful unless present in excessive numbers, some can be pathogenic to their host. In some cases, bivalves can act as passive carriers of human pathogens, without themselves contracting a bacterial disease (see Chapter 12). Bacterial diseases often affect bivalves differently according to their life stage. Infections are usually described from larval stages, where they are associated with high mortalities in hatcheries, but only a few bacterial types are pathogenic to adults (Paillard 2004). This is because larvae are reared in static systems at elevated temperatures and high densities.

Bacteria that evoke severe mortalities in bivalve larval culture generally belong to the genus *Vibrio*. Impacts of *Vibrio* infection range from mild to severe depending on the *Vibrio* species and the susceptibility of the host. The classical *Vibrio* infections are 'summer mortality' in juvenile oysters and 'brown ring disease' in adult clams. The former, which is caused by two bacterial strains of *V. splendidus*, infects both wild and cultured oysters (*C. gigas*) and is characterised by systemic infection of the soft tissues of larvae and juveniles that results in tissue necrosis and high but sporadic mortality rates of 60–100% (Paillard 2004). Brown ring disease, which gets its name from the brown deposit on the inside of the shell of dying clams, is caused by *V. tapetis* and affects species in the genera *Ruditapes*, *Tapes* and *Venerupis*, although only one species, *Ruditapes philippinarum*, exhibits mass mortalities. The bacteria penetrate through the mantle epithelium and into the soft tissues, causing severe damage. There is degeneration of the digestive diverticula, manifested in an inability to store reserves, mainly glygogen, and distribute them to other tissues. Growth, weight and condition indices are subsequently affected. Severely diseased clams show a reduction in clearance and respiration rates leading to a decrease in energy acquisition and, ultimately, death (Bower 2010b and references therein). Although several *Vibrio* species are associated with diseases in scallops (reviewed in Getchell *et al.* 2016), their impact appears to be less severe compared to oysters and clam species. While there is a wealth of information on bacterial diseases of oysters, clams and scallops, there are few reports on diseases of mussels, and the majority of these do not cause significant mortalities (Table 11.1). See Travers *et al.* (2015) for review of bacterial diseases in marine bivalves.

Fungi

On mussel farms in the Adriatic Sea, Italy in the early 2000s, young mussels (*M. galloprovincialis*) began losing their byssus, leading to their death and causing serious damage to the industry in the region. The pathogen responsible, which is most likely a yet unidentified fungus, provoked the progressive destruction of the foot muscles and damaged related structures such as the intra-organism part of the byssus apparatus, resulting in loss of the thread component (Franchini *et al.* 2005). It also provoked modifications in the digestive gland structure, ranging from hyperactivity to extreme cell death in the tubules. Mussel death was probably due not to the inability to attach to hard surfaces, but to the destruction of body muscles. Interestingly, Vitellaro-Zaccarello (1973) had already reported the presence of an ascomycete fungus in byssus material from *M. galloprovincialis* in the Gulfs of Trieste and Genoa, Italy. Drastic alterations of the distal portion of the byssus, accompanied by extensive hyphal intrusion and subsequent invasion by algae and protozoans, appeared to point to the fungus as being a major factor in the degradation of byssal material.

Another fungus, not definitively identified but classified as *Wallemia*, causes mycotic sloughing periostracum (MSP) in *Mytilus* in a few locations on the coasts of Newfoundland and Prince Edward Island, Canada and in other regions of Atlantic Canada, the eastern and western seaboards of the United States, British Columbia, Europe and Australia (PEIDAF 2001). Affected shells exhibit a brown discolouration associated with sloughing of the periostracum (Figure 11.1). The discolouration is most noticeable when the shell is dry, which is the usual case when serving mussels for consumption. MSP is most prevalent in areas of intense mussel cultivation. Therefore, it is no surprise that mussel farmers and processors are worried that MSP may affect marketability. The disease is diagnosed by gross observations, wet mounts for the presence of fungal hyphae beneath the remaining periostracum of affected mussels and electron microscopy of fungal hyphae associated with the periostracum of affected mussels (Bower 2004; Figure 11.1). There is no histological evidence to show that MSP affects the health of mussels, since the fungus does not penetrate through the shell and thus leaves the soft tissue unaffected.

More recently, an emergent epizootic[1] event in deep-sea mussels (*Bathymodiolus brevior*) was recorded at the Mussel Hill hydrothermal vent in the Fiji Basin, South Pacific Ocean (Van Dover *et al.* 2007). The causal agent was an ascomycete black yeast (Order Chaetothyriales) that elicits a pronounced haemocytic immune response and is associated

Table 11.1 Examples of pathogenic bacteria of mussels.

Bacteria	Disease	Species	Symptoms	Reference
Vibrio alginolyticus,	Larval vibriosis	*Mytilus galloprovincialis* larvae and hatchery-reared abalone, scallop and oyster larvae	Larval development is severely affected but does not always result in significant mortality	Anguiano-Beltrán *et al.* (2004)
V. splendidus, *V.* sp. strain DO1		*Perna canaliculus*	Mortality 23–35%	Kesarcodi-Watson *et al.* (2009)
Nocardia crassostreae	Pacific oyster nocardiosis (PON) or summer mortality	*M. galloprovincialis*, *Crassostrea gigas* and *Ostrea edulis* cultivated near infected *C.gigas*	No mortalities to date in mussels; up to 35% mortalities in oysters	Carella *et al.* (2013); Bower (2006a)
Rickettsia-like and *Chlamydia*-like organisms		*Mytilus* spp., *Mytella guyanensis* and a wide variety of other marine bivalves	Light infections of gills and digestive gland; not associated with disease	Ward *et al.* (2004); Boehs *et al.* (2010); Ceuta & Boehs (2012)
Cyanobacteria spp.	Blue-green alga shell infestation	*Perna perna*, *M. galloprovincialis*, *Aulacomya atra*, *Choromytilus meridionalis*	Shell damage; lowered reproductive output; mortalities up to 25% of total adult mortality	Kaehler (1999); Bower *et al.* (2002); Marquet *et al.* (2013); Lourenco *et al.* (2017)

Sources: Adapted from Anguiano-Beltrán, C., Lizárraga-Partida, M.L. & Searcy-Bernal, R. (2004) Effect of *Vibrio alginolyticus* on larval survival of the blue mussel *Mytilus galloprovincialis*. *Diseases of Aquatic Organisms*, **59**, 119–123; Bower, S.M. (2006) Synopsis of infectious diseases and parasites of commercially exploited shellfish: nocardiosis of oysters. Available at: http://www.dfo-mpo.gc.ca/science/aah-saa/diseases-maladies/nocardoy-eng.htm (accessed 28 May 2021).

Figure 11.1 (a) Typical manifestation of mycotic sloughing periostracum (MSP) on a dry mussel (*Mytilus edulis*) shell. (b) Electron micrograph of fungal hyphae protruding from a shell.
Source: Photos courtesy of Jeff Davidson, Atlantic Veterinary College, University of Prince Edward Island, Canada.

with tissue deterioration ranging from brown-spot discolouration of mantle tissues to black-body stages (Figure 11.2). Other tissues affected were connective tissue surrounding the gonad and digestive gland and the gill filaments. The prevalence of diseased mussels was 55% in a sample of 57 individuals, with just a single mortality, although empty shells were evident from video and digital images of the mussel beds. The fungus was not observed in associated fauna or in adjacent mussel beds. The authors concluded that this ongoing epizootic represented a disease outbreak in the early stage of the epizootic, prior to the onset of extensive mortality. To my knowledge, there have been no further reports of this pathogen either in the Fiji Basin or in mussels from other deep-sea hydrothermal vents. Several studies have shown that there is a wide diversity of yeasts at hydrothermal vents that appear to interact with endemic animal fauna, including mussels (Gadanho *et al.* 2005; Burgaud *et al.* 2009, 2010; Singh & Raghukumar 2014). These yeasts may be facultative parasites or opportunistic pathogens of endemic deep-sea animals, as shown earlier, but they might also play a role in the decomposition of abundant organic material (Burgaud *et al.* 2010).

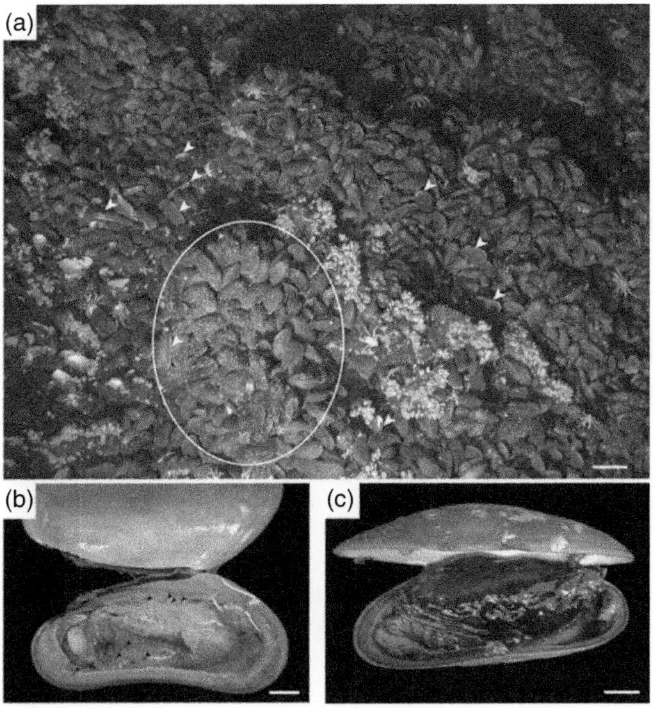

Figure 11.2 (a) Mussel beds (*Bathymodiolus brevior*) at Mussel Hill, Fiji Basin. Note the occasional paired empty mussel shells (arrowheads) and area of broken shells on the left-hand side of the photo. Ellipse surrounds the type of mussel bed targeted for sampling (i.e. a deep stack of living mussels). Scale bar = 10 cm (foreground). (b) Brown-spot stage. A few of the brown spots are highlighted with black arrowheads. (c) Black-body stage. Scale bars (b) and (c) = 1 cm. *Source:* From Van Dover *et al.* (2007). Reproduced with permission from John Wiley and Sons.

Steinhausia mytilovum in the phylum Microspora within the kingdom Fungi is a globally distributed parasite that infects the oocytes of marine mussels; hence its name, 'egg disease'. The parasite has a wide geographical range, occurring in both natural and cultured populations of *M. edulis* and *M. galloprovincialis* from the Atlantic and Pacific coasts of the United States, Europe and western Australia, and also in *Mytella guyanensis* from the Amazon Estuary, Brazil (Bower 2012). The Microsporidia constitute a group of spore-forming unicellular parasites. They were once considered protozoans or protists, but are now known to be fungi (Hibbet *et al.* 2007). The parasite infects the cytoplasm and nucleus of mussel ova, distorting the nucleus, causing egg deterioration and ultimately reducing fecundity – the extent of which varies depending on the intensity of infection (Figure 11.3). In addition, it can incite a moderate to severe diffuse-type haemocyte infiltration response, which is associated with resorption of the germinal epithelium and ova (Bower 2012). Rate of infection for other *Steinhausia* species shows a seasonal pattern that is directly related to water temperature, with lower temperatures resulting in a higher prevalence and a larger number of spores per mussel (Matos *et al.* 2005), although this has yet to be confirmed for *S. mytilovum*. Transmission is horizontal when spores released from ruptured host cells are

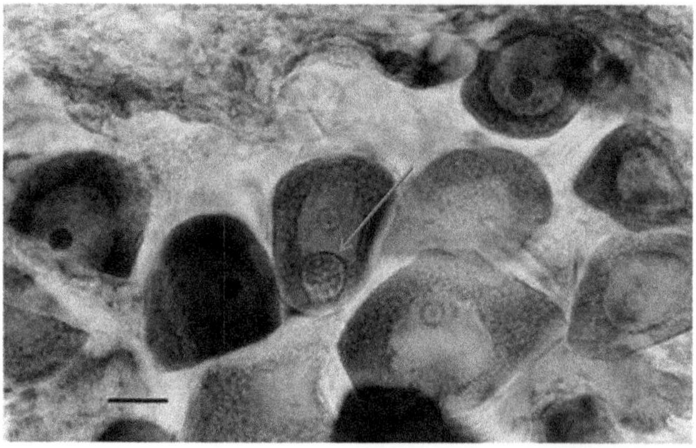

Figure 11.3 Spherical cysts (11–15 μm in diameter) of *Steinhausia mytilovum* microsporidian infection (arrow) in the cytoplasm of a *Mytilus galloprovincialis* ovocyte. Up to three cysts per ovocyte were observed in mussels collected from the Thermaikos Gulf, northern Aegean Sea, Greece. *Source:* From Rayyan & Chintiroglou (2003). Reproduced with permission from Inter-Research.

ingested by a suitable host. The parasite may also be vertically transmitted to progeny when lightly parasitised ova develop normally (Meyers & Burton 2009).

Protistans

The kingdom Protista is a diverse group of eukaryotic organisms. The majority are unicellular, with a well-defined nucleus and membrane-bound organelles contained in the cell body. Reproduction can be asexual or sexual, and locomotion is by flagella or cilia. Most protists live in aquatic environments, and they are the most common cause of disease in bivalves. Most belong to the phyla Cercozoa and Perkinsozoa (*Marteilia*, *Haplosporidium* and *Bonamia* spp.) and infect mainly oyster and clam species (reviewed in Gosling 2015).

Apicomplexa

The subphylum Apicomplexa in the phylum Myzozoa contains two groups of molluscan pathogens, the coccidians and the gregarines. The former are intracellular parasites (*Pseudoklossia* spp.) of vertebrates as well as invertebrates. They parasitise bivalves in the kidneys, reproductive tissue, connective tissues and visceral ganglia (Meyers & Burton 2009). Kidney coccidia disease has been reported from mussels, oysters, clams and scallops. In mussels, the disease appears to be confined to *Mytilus* species and has been reported from the coast of British Columbia, Canada, the east coast of the United States and NW Spain (Bower 2001a). Infected kidney epithelial cells become enlarged and kidney tubules fill with coccidia. Heavy infections may cause kidney damage, but associated mortalities appear restricted to artificial growing conditions. In British Columbia, Canada, among wild stocks of mussels, prevalence of infection is usually less than 16%, intensity of infection usually light (fewer than 50 coccidia per histological section of kidney tissue) and evidence of associated pathology (apart from infected kidney epithelial cells) not observed

(Desser *et al.* 1998). Disease diagnosis is by examination of squash preparations of kidney tissues, while all intracellular developmental stages may be observed by histological examination (details in Meyers & Burton 2009).

Species of *Cryptosporidium* are intracellular coccidian parasites that infect epithelial cells of the gastrointestinal tract and are increasingly recognised as important pathogens and a cause of enteric disease in a wide range of vertebrate hosts, including humans. There are about 16 species recognised, with *C. parvum* being the most common, responsible for clinical disease in humans and domestic animals. This species is often found in areas impacted by sewage overflow or increased urban and agricultural runoffs (Mladineo *et al.* 2009 and references therein). *Cryptosporidium* infective stages are four naked sporozoites enclosed in a spherical, thick-wall oocyst (measuring 4–6 μm and lacking the sporocyst), which enables the parasite to survive for months outside the host (Fayer *et al.* 1998). Oocysts are easily transmitted from a bivalve to coexisting individuals of the same or different species (Gómez-Couso *et al.* 2003). Detection of infective oocysts is by immunomagnetic separation and immunofluorescence assay (Li *et al.* 2006), oocyst infectivity assay (Gomez-Bautista *et al.* 2000) and molecular diagnostics (Mariné Oliveira *et al.* 2016 and references therein). As suspension feeders, bivalves act as potential vehicles of disease transmission to the consumer. Mladineo *et al.* (2009) were the first to evaluate the presence of *Cryptosporidium* spp. oocysts in cultured mussels (*M. galloprovincialis*) at four locations in the Mali Ston Bay, Adriatic Sea, Croatia over a one-year period. Prevalence was 16.8% in a sample of 184 mussels, and was correlated with the presumptive number of *E. coli* in the shellfish and seawater and abiotic factors (temperature, salinity, oxygen). A lower prevalence of 12.6% was reported in a sample (N = 56) of *Perna viridis* from fish markets located at the mouth of the Chao Phraya River, Thailand (Srisuphanunt *et al.* 2009). While the parasite does not appear to adversely effect mussel health, it is clear from the results of the preceding and similar studies (Gómez-Couso *et al.* 2006; Miller *et al.* 2005, 2006; Li *et al.* 2006; Mariné Oliveira *et al.* 2016) that mussels and other bivalves probably act as a reservoir of *Cryptosporidium* foodborne infections for humans (see Chapter 12).

Oocycsts of different gregarines species in the genus *Nematopsis*, family Porosporidae infest various tissues of different bivalve species. In the mussel *Mytella guyanensis*, for example, the most frequent tissue location is gills, followed by mantle, digestive gland and, less frequently, foot muscle (Boehs *et al.* 2010). The parasite has been reported from *M. chilensis* (Cremonte *et al.* 2015), *M. galloprovincialis* (Francisco *et al.* 2010; Özer & Güneydağ 2014), *M. edulis* (Lynch *et al.* 2014), *P. viridis* (Tuntiwaranuruk *et al.* 2004), *P. canaliculus* (Jones 1975), *Mytella guyanensis* (Azevedo & Matos 1999; Boehs *et al.* 2010; Ceuta & Boehs 2012), *Mytella falcata* (Padovan *et al.* 2003) and *M. edulis*, *M. trossulus* and *Guekensia demissa* (Bower 2001b). The life cycle involves two hosts: a marine bivalve as the intermediate host and a marine arthropod, usually a crustacean, as the final host. Trophozoites – the active, motile, feeding stage of the parasite – in the gut of a crustacean give rise to gametocysts and gamete formation. Subsequent sporulation produces gymnospores, which are released from ruptured gametocysts. Gymnospores, which represent the stage that infests bivalves, are drawn into the mantle cavity of a compatible host, where they are engulfed by phagocytes on the surface of gill or mantle epithelium. The engulfed gymnospore gives rise to sporozoites, which in *Nematopsis* spp. form resistant oocysts, each containing one sporozoite. When the final host consumes the bivalve host, sporozoites give rise to trophozoites, thus completing the gregarine life cycle (Lauckner 1983; Meyers & Burton 2009). Infection is usually associated with a benign inflammatory response without significant health effects, although some infections may result in the production of pearls or calcareous excrescencies on the inner surface of the valves of *M. edulis* (Bower 2001b). High levels of infection may cause complete destruction of the gills (Figure 11.4), degeneration of host cells and death (Francisco *et al.* 2010 and references therein). In general, prevalence of *Nematopsis* spp. tends to be high in mussels, ranging from 60 to 100% depending on species and geographic location (Jones 1975; Tuntiwaranuruk *et al.* 2004; Boehs *et al.* 2010; Ceuta & Boehs 2012).

Another pathogen in the Apicomplexa is the unknown causative agent APX (Apicomplexan parasite unknown), which affects oysters (*Ostrea chilensis*) and mussels (*Perna viridis*). Severely infected oysters may gape, and are small, shrunken, pale and transluscent due to lack of gametes in the gonads. Mortality is generally <10%, but is significantly higher (usually over 10 times higher) when co-infected with *Bonamia exitiosa*. However, in mussels, absence of mortality and disease signs suggests the agent may be undetected/under-reported. APX is known only from New Zealand. It occurs in oysters all around the country, but in mussels has been reported only on two *P. canaliculus* farms in the Marlborough Sounds (Diggles *et al.* 2002; Castinel *et al.* 2014) and on *P. canaliculus*, *M. galloprovincialis* and *Modiolus areolatus* from other New Zealand locations (Nguyan 2018).

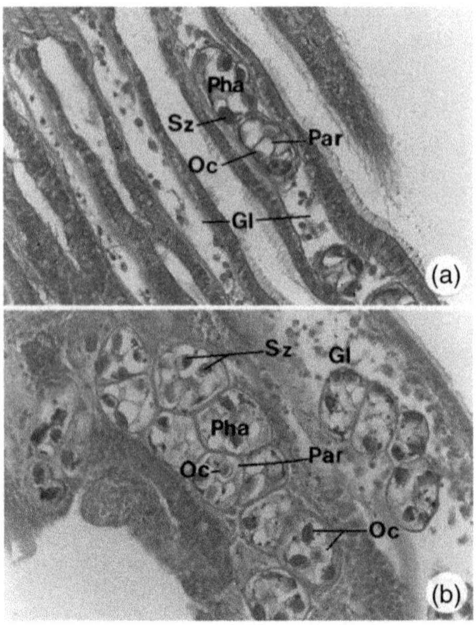

Figure 11.4 *Nematopsis* sp. infecting *Perna viridis*. Light micrographs (eosin and hematoxylin stain) of (a) moderate and (b) heavy infection, showing phagocytes (Pha) in gill lumina (Gl) and oocysts (Oc) in parasitophorous vacuole (Par); sprozoites (Sz) are clearly visible (×450). *Source:* From Tuntiwaranuruk *et al.* (2004). Reproduced with permission from Inter-Research.

Ascetosporea

The class Ascetosporea within the phylum Cercozoa comprises two important invertebrate pathogen orders, the Haplosporida, which includes the genera *Haplosporidium*, *Minchinia*, *Urosporidium* and *Bonamia*, and the Paramyxida, which contains *Marteilia*, *Paramarteilia* and *Paramyxa* (Stentiford *et al.* 2013).

Haplosporida

Haplosporidium species are obligate parasites,[2] some of which are extremely important pathogens of oysters. For example, *H. nelsoni* causes multinucleate sphere X (MSX), a disease that can cause extensive necrosis of all tissues resulting in mass mortalities in the the oyster *C. virginica* (details in Gosling 2015). *Haplosporidium* species have been reported in *M. californianus* (*H. tumefacientis*) on the Californian coast of the United States, in *M. edulis* from the coasts of Maine in the United States and Atlantic Canada (*Haplosporidian* sp.) and in *M. galloprovincialis* on the Mediterranean coast of France (*Minchinia* sp.) (Bower 2007a). A previously undescribed haplosporidian was recently detected in a single mussel (*M. edulis*) sample in northern Wales, and from DNA analysis it appears to be a *Minchinia* species (Lynch *et al.* 2014). *H. tumefacientis* causes tumefactions in *M. californianus* that can vary in size from small discrete nodules to contiguous swellings of the entire digestive gland and kidney. However, prevalence is low (~2%) and the effect on the host population is believed to be negligible. There is no evidence of infectivity to other species of *Mytilus*. Only one occurrence of *Minchinia* sp. has been reported in *M. galloprovincialis*, with just 2% of the sample appearing thin and exhibited an abnormal whitish colour (Comps & Tigé 1997). Fortunately, there is no evidence to date that *Haplosporidium* or *Minchinia* species have a significant impact on mussel health.

Bonamiosis is a haplosporidian disease caused by protists in the genus *Bonamia*. The pathogen infects the haemocytes of oysters, primarily *Ostrea* species, and causes significant and widespread mortalities (>90%) in both the Northern and the Southern Hemispheres (Culloty & Mulcahy 2007; Bower 2011c; Gosling 2015). Other species (e.g. *M. edulis*, *M. galloprovincialis*, *Ruditapes philippinarum*, *R. decussatus* and *C. gigas*) cannot be infected either naturally or experimentally, and therefore do not appear to act as vectors or intermediate hosts for the parasite (Figueras & Robledo 1994; Renault *et al.* 1995; Culloty *et al.* 1999). However, the possibility of a carrier/receiver host cannot be dismissed (Bishop *et al.* 2006; Lynch *et al.* 2010).

Paramyxida

Species in the genus *Marteilia* are paramyxean parasites that infect the digestive system of several bivalve species (Berthe *et al.* 2004). There has been some debate over the identification of *Marteilia refringens* and whether it is in fact identical to the similar *M. maurini*. Initially, the discrimination between *M. refringens* and *M. maurini* was based on histological criteria and the target host (*M. refringens* infects oysters and *M. maurini* infects mussels), but *M. refringens* has now been detected in the mussel *M. galloprovincialis*, suggesting that the same parasite may have a different morphology in different hosts. Molecular diagnosis has shown that the *Marteilia* parasites, which infect oysters and mussels, are two different strains of the same species (Lopez-Flores *et al.* 2004; Balseiro *et al.* 2007). The

World Organisation for Animal Health (Office International des Epizooties, OIE) Reference Laboratory for Infection with *Marteilia refringens* recognises two types of *Marteilia refringens*, types O and M (for oysters and mussels, respectively), as defined by Le Roux *et al.* (2001; OIE 2016). Recently, a new type, *Marteilia* sp. type C, was described in cockles (*Cerastoderma edule*) suffering mortality on the Spanish Mediterranean coast (Carrasco *et al.* 2012). However, host specificity of these genotypes is not strict: type O can be found in mussels, together with type M, and type M can be found in oysters, together with type O (Novoa *et al.* 2005). Cross-infections with type M in oysters and type O in mussels have been reported, but co-infection (both in the same host) is a more common situation, especially in areas where *M. refringens*-infected mussels and oysters cohabit. When environmental parasite loads are very high, it is probable that the predominant type infects both oysters and mussels (Carrasco *et al.* 2015). More recently, the parasite has been reported in the black pigmy mussel, *Xenostrobus securis*, a non-indigenous invasive species in the Ria de Vigo, Galicia, Spain, where it cohabits with infected *M. galloprovincialis* (Pascual *et al.* 2010).

The geographic distribution of *M. refringens* in Europe includes Albania (Pëllumb *et al.* 2006), Croatia (Zrnčić *et al.* 2001), France (Arzul *et al.* 2014), Greece (Rayyan *et al.* 2006), Italy (Serracca *et al.* 2014), Portugal (López-Sanmartín *et al.* 2015), Slovenia (Gombač *et al.* 2014), Spain (López-Sanmartín *et al.* 2015) and the United Kingdom (Murray *et al.* 2012), while in Africa it is present in Morocco (Bhaby *et al.* 2013) and Tunisia (Elgharsalli *et al.* 2013).

M. refringens is considered to be a potentially lethal pathogen. For example, in the oyster *Ostrea edulis*, infection causes a poor condition index with glycogen loss (emaciation), discolouration of the digestive gland, cessation of growth and tissue necrosis (Bower 2011a and references therein), with a 50–90% mortality rate usually reported during summer months when temperatures are >17 °C (Villalba *et al.* 1993). Mussels are usually not negatively affected by marteilioisis, although in some areas mortality attributed to this parasite is significant for the mussel farming industry. When mussels (*M. edulis*) with no previous contact with *M. refringens* or *M. maurini* in their originating countries were imported for relaying in France, they experienced mass mortality (up to 100%) when transferred to an endemic zone in which local mussel populations were still healthy (i.e. tolerant but infected; Berthe 2002; Arzul & Joly 2011).

In areas where mussels are seriously affected, the primary impact of *M. refringens maurini* infection is associated with haemocyte infiltration of the connective tissue and epithelia of the digestive gland, with extensive destruction of the gland (>10% of tubules infected; Figure 11.5) and death of the host in heavy infections (Gombač *et al.* 2014). Early stages of infection have also been detected in the stomach epithelia, the epithelium of the gill lamellae of *M. galloprovincialis* in the Mediterranean coast of Spain and the connective tissues of mantle, labial palps and gills in *M. edulis* in Normandy, France (Carrasco *et al.* 2015 and references therein). Pathologic disruption of the digestive gland tubule epithelia is most severe at the time of spore release from the sporangia. Heavy infections also cause reduction of absorption efficiency, resulting in an inhibition of gonad and storage tissue development and thus a significant loss in condition of infected mussels (Bower 2011b). Infected *M. galloprovincialis* also have lower total carbohydrate concentrations in the haemolymph (Robledo *et al.* 1995).

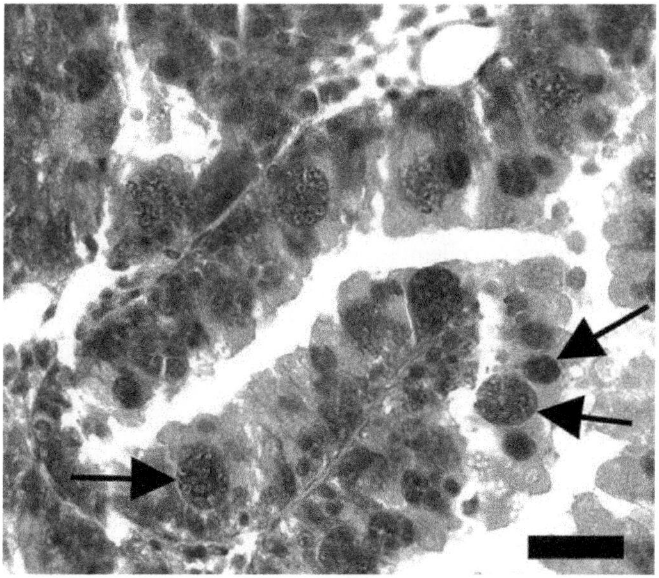

Figure 11.5 *Marteilia refringens* in the digestive tubule epithelia of the mussel *Mytilus galloprovincialis*. Arrows indicate different *M. refringens* stages of development. Scale bar = 50 μm. *Source:* From Carrasco *et al.* (2015). Reproduced with permission from Elsevier.

The complete life cycle of *M. refringens/maurini* has not yet been fully described. Comps & Joly (1980) were successful in transmitting *M. refringens* extracted from *O. edulis* to uninfected *M. galloprovincialis*. Further attempts to directly transmit the pathogen between bivalves have failed. The existence of a complex life cycle involving several hosts was postulated early on by many authors, although it remains unsolved. Developments in the DNA-based diagnosis of *M. refringens* provided new prospects for the detection of the parasite in potential hosts. However, the number of species living in the vicinity of oyster beds hindered screening. Audemard *et al.* (2001), looking for potential alternative hosts, used French shallow-water oyster ponds ('claires') where species diversity is relatively limited. They detected *M. refringens* DNA in various invertebrates, with the most promising possibility being the Calanoida copepod *Paracartia grani*, which showed gonadal tissue parasitised by *M. refringens* on ISH. Although *P. grani* could be infected with *M. refringens/maurini* from *O. edulis* and mussels (*M. edulis* and *M. galloprovincialis*), attempts to transmit *M. refringens* from *P. grani* to *O. edulis* and *M. edulis* were unsuccessful (Audemard *et al.* 2002; Carrasco *et al.* 2008). Also, infection patterns of *M. refringens/maurini* in *P. grani* were different for copepods infected via mussels and via oysters, with only early stages of infection found in the intestinal tract of those infected from mussels (*M. refringens* type M) versus a higher prevalence and intensity of infections in the intestinal tract and gonad of those infected from oysters (*M. refringens* type O) (Carrasco *et al.* 2008). Further field studies on *Marteilia* spp. life cycles carried out in the Diana lagoon, Corsica showed several additional planktonic taxa that were PCR-positive for *Marteilia* spp., including copepods, larval stages of decapods, Cladocera, Appendicularia, Chaetognatha and Polychaeta. However, parasite infection could only be confirmed in *Paracartia latisetosa*, the second copepod in this genus to be infected by *Marteilia* spp. In order to corroborate the involvement of copepods in the life cycle of these parasites, it is essential to demonstrate the transmission of *Marteilia* spp. from infected copepods to uninfected molluscs (Carrasco *et al.* 2015).

Carrasco *et al.* (2008) proposed a development cycle based on their results for *M. refringens* in the copepod *P. grani* and *O. edulis*, as shown in Figure 11.6. In the copepod, uni- and binucleated cells are found in the digestive epithelium and become nursery cells in which amoeboid daughter cells proliferate. Nursery cells migrate to the gonadal tissue through neighbouring connective tissues. In the proposed cycle, *M. refringens* continues its development in the gonad, where it forms pseudoplasmodia containing large numbers of tiny cells. It is not clear if these pseudoplasmodium-like structures are located inside or outside oocytes. Bivalves may ingest infected female copepods, although this still needs experimental verification. Infection in bivalves can develop from primary and secondary cells in the epithelium of digestive primary tubules to pansporoblasts containing sporangia in the secondary and tertiary tubules (Grizel *et al.* 1974). Sporangia can be released with faeces, and these or their spores can then be ingested by the copepod, beginning infection anew as described earlier.

Histology and cytology are common techniques used to screen for the presence of *M. refringens*. However, molecular techniques are required to detect and discriminate among species and even types (see earlier) of *Marteilia*. PCR primers that target the internal transcribed spacer 1 (ITS-1) region described by Le Roux *et al.* (2001) are recommended by the World Organisation for Animal Health (OIE 2016), as they are able to amplify only *M. refringens*. Some primers targeting the small subunit (SSU) of the rRNA gene complex are also available and allow different *Marteilia* species to be amplified (e.g. *M. refringens*

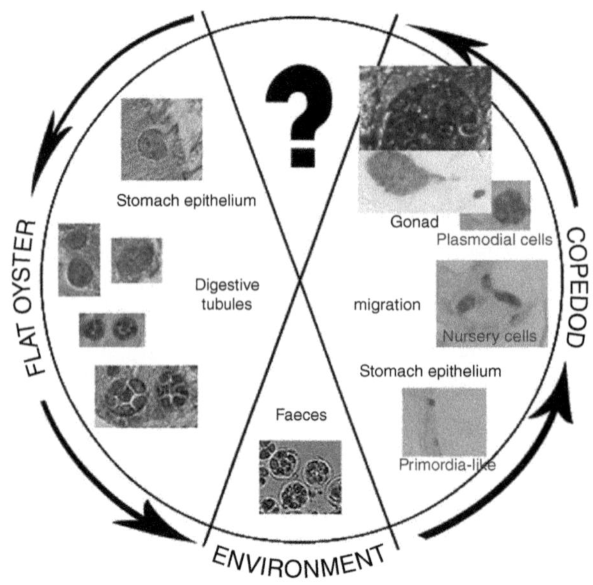

Figure 11.6 Tentative *Marteilia refrigens* life cycle and developmental stages in alternative hosts (modified from Carrasco *et al.* 2008). *Source:* From Carrasco *et al.* (2015). Reproduced with permission from John Wiley & Sons and Elsevier.

and *M. sydneyi*; Le Roux *et al.* 1999; Berthe *et al.* 2000). In addition, PCR-RFLP analysis (see Chapter 9) with HhaI enzyme on the ITS-1 region (Le Roux *et al.* 2001), which shows two different profiles, can be used to distinguish *M. refringens* type O and type M. ISH protocols have been developed (Le Roux *et al.* 1999; Berthe *et al.* 2000; see Chapter 9). Probes that target the SSU of the rRNA gene complex have been validated against histology (Le Roux *et al.* 1999; Thébault *et al.* 2005) and are recommended by the World Organisation for Animal Health (OIE 2016). See Bower (2011b) and Carrasco *et al.* (2015) for further details.

Marteilosis is listed as a notifiable disease by the OIE, and in turn the European Union, under Council Directive 2006/88/EC, list II of Annex A (www.eur-lex.europa.eu), indicates that declaration of an occurrence of marteiliosis is compulsory. As a notifiable disease, movement of infected bivalve stock is not permitted.

Chlorophyta

The genus *Coccomyxa* in the phylum Chlorophyta (green algae) family Coccomyxaceae includes both free-living marine and freshwater planktonic species but is also parasitic in starfish, scallops, mussels and clams (Vázquez *et al.* 2010 and references therein). *Coccomyxa parasitica* has been reported in *M. edulis* from the North and South Atlantic Ocean (Gray *et al.* 1999; Mortensen *et al.* 2005; Rodríguez *et al.* 2008) and the St Lawrence Estuary, Canada (Zuykov *et al.* 2014), in *M. galloprovincialis* in NW Spain (Crespo *et al.* 2009) and in *Modiolus modiolus* from the NW Pacific, Sea of Japan (Syasina *et al.* 2011). The form and size of algal cells vary depending on their maturity, with young immature cells rounded and measuring 1.2–2.0 μm in diameter, while adult mature cells have an oval or bean shape and reaching up to 5 μm in diameter (Syasina *et al.* 2012). Within the cell is a single nucleus and a chloroplast, one or two mitochondria and a dense granular cytoplasm with a lipid inclusion body, Golgi apparatus and small vesicles (Rodríquez *et al.* 2008; Figure 11.7). In mussels, reproduction of *C. parasitica* occurs by cell division into two or four daughter cells (autospores) before the parental membrane ruptures (Gray *et al.* 1999). TEM has repeatedly shown that algae are located predominantly inside the vacuoles of mussel haemocytes and are highly resistant to digestion, although disintegrating algal cells have been reported inside digestive vacuoles within haemocyte cytoplasm of *M. edulis* (Rodríquez *et al.* 2008) and *M. modiolus* (Syasina *et al.* 2011).

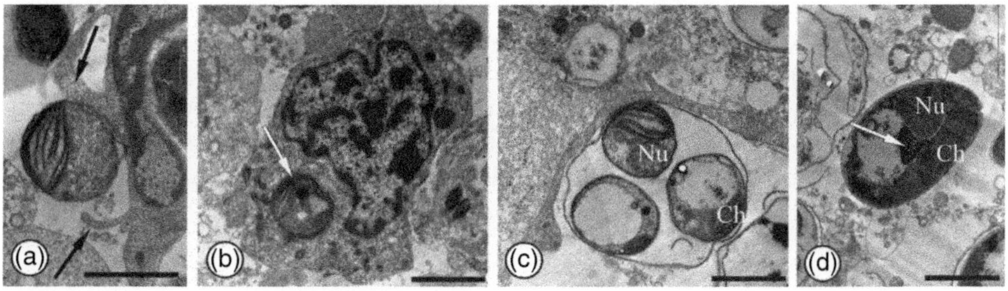

Figure 11.7 Ultrastructure of *Coccomyxa* sp. in *Modiolus modiolus* gonad. (a) Phagocytosis of microalga by haemocyte. Arrows indicate cytoplasmic extrusions. (b) Alga location (arrow) in the haemocyte cytoplasm. (c) Reproduction of alga by means of autospore formation. (d) Mature stage. Arrow indicates porous inclusion body. Ch, chloroplast; Nu, nucleus. Scale bar = 2 μm. *Source:* From Syasina *et al.* (2012). Reproduced with permission from Elsevier.

Despite this pronounced defensive reaction, pathogenic effects include a significant decrease in the condition index, deformation of the mantle and shell, high numbers of pearls in the mantle, severe haemolytic infiltration of the mantle, digestive gland, rectum, gonad and adductor muscle and necrosis of invaded tissue (Vaschenko *et al.* 2013 and references therein; Zuykov *et al.* 2014; Sokolnikova *et al.* 2016). Infection causes a mild effect on gametogenesis in *Modiolus modiolus* (probably *Modiolus kurilensis*), most likely due to the highly effective defence reaction of strong phagocytic activity of haemocytes against the algal cells. The dynamics of gonad development in healthy and infected mussels are generally similar, with spawning being partial and occurring from the beginning of June to the first half of September. However, several negative reproduction-related effects of green alga infestation occur, including general gonadal underdevelopment, which is apparent from a significant decrease in the acinus areas of the ovaries and testes and an increase in the connective tissue layer between the acini, a delay in some stages of the reproductive cycle and the production of a decreased number of spermatozoa and large vitellogenic oocytes, especially in the early spawning period. All of these results suggest a reduced reproductive output for infected mussels (Vaschenko *et al.* 2013).

A high rate of infection was observed in mussels (*M. modiolus*) collected monthly (N = ~30) from May to September 2009 from Peter the Great Bay, Sea of Japan, with values ranging from 17 to 62% (mean 43%) in different months (Syasina *et al.* 2012). For comparison, the prevalence of infected mussels from a nearby site in the same bay was on average 22% (Syasina 2011). There was no evidence for a seasonal pattern in infestation of either male or female mussels (Syasina *et al.* 2012). No significant differences were observed in size–frequency distribution for infected and noninfected mussels (*M. galloprovincialis*) sampled in the Vigo Estuary, NW Spain (Crespo *et al.* 2009). About 56% of infected mussels ranged in length from 41 to 50 mm, while those longer than 60 mm were not parasitised. Prevalence was below 8% in all size classes except the smallest (31–35 mm), where it reached 33%. The absence of infection observed in the largest size class (>60 mm) may be explained by a gradual weakening of infected mussels, leading to their premature death. On the other hand, the highest prevalence observed in the smallest size class (31–35 mm) could be indicative of the fact that infection occurs mainly in the smallest mussels. While Gray *et al.* (1999) also found no consistent seasonal pattern in infestation between September 1993 and February 1996, they did find that the highest levels of infestation (23%) occurred in individuals from the mid-mussel zone, with considerably lower levels in the upper and lower parts of the zone (<1 and 5%, respectively).

Diagnostic techniques include gross observations, histology, electron microscopy, pigment analysis of the green algae and, in recent years, ribosomal RNA sequence analysis (see Syasina *et al.* 2012). A newly developed method of isolating microalgae from bivalve tissues has facilitated the continuous pure culture of *C. parasitica*, thereby providing a description of the time course of each life stage (Sokolnikova *et al.* 2016).

Ciliophora

Cilates are protists that use their cilia for locomotion and food collection. More than 150 species have been found in the mantle cavity, on the gill surfaces or, rarely, in the digestive diverticula of marine bivalves. Most are regarded as commensal, but they can become pathogenic if their numbers become unusually high, if the physiological state of the host is compromised or if environmental stress factors shift the equilibrium (Lauckner 1983). Most belong to three classes, Kinetofragminophorea, Oligohymenophorea and Polyhymenophorea, but only a few representatives will be dealt with here. The ciliate *Ancistrum mytili*, in the class Oligohymenophorea, order Thigmotrichida, occurs, often in great numbers, on the gills of *Mytilus* spp. It appears to be ubiquitous, with reports from mussels throughout their geographic range. According to Bower (2007b and references therein), large numbers elicit no obvious host response and attachment to the gill epithelia appears to be superficial. Infection causes no visible lesions in histological sections of the gills and no significant difference in the condition index. In the eastern Maritime Provinces of Canada, *A. mytili* occurs in about 90–100% of mussels – a little higher than the value (60–90%) reported in raft-grown *M. galloprovincialis* in NW Spain (Figueras *et al.* 1991). Excessively heavy infections may have a secondary pathogenic effect under adverse growing conditions, but there has been no evidence of this to date. The ciliate has also been reported from the gonad of female mussels, where it apparently acquires nutrients from the oocytes.

Peniculistoma mytili, also in the Oligohymenophorea, but in a different order, the Pleuronematida, adheres loosely, using a patch of specialised cilia, to the foot epithelium and adjacent surfaces of *M. edulis*. Almost 100% of mussels from various locations in the North and Baltic seas were infested, some containing up to 300 individuals of this species (Fenchel 1965). It is probable that ciliate losses in the mussel exhalant current are considerable. Beers (1959) has shown that the species can survive in seawater for 48–72 hours at 14 °C, an adequate period to achieve its maintenance and dissemination in a *Mytilus* population. There is no evidence that *P. mytili* causes damage to its host (Bower 1992). When *M. edulis* is simultaneously infected with *Ancistrum mytili* and *Peniculistoma mytili*, both species occupy separate ecological niches in the host. The former ciliate feeds on smaller particles than the latter and is exclusively found on the gills, whereas the latter occurs predominantly on the foot of the mussel and is strictly host-specific to *M. edulis*. Attempts to experimentally infest *Modiolus modiolus* with this ciliate have invariably met with failure (Fenchel 1965).

Gargarius gararius and *Sphenophrya*-like ciliates in the class Kinetofragminophorea, order Rhynchodida are characterised by the presence of a suctorial 'tentacle', which they use to attach to epithelial cells of the hosts, mainly the gills or palps, and suck out nutrients (Lauckner 1983). Hosts include *Mytilus* spp. from the eastern and western United States, Canada, Europe and Scandinavia, and also oysters and clam species (Bower & McGladerry 2001c). Host specificity is most pronounced in the Rhynchodida, and it is difficult to transmit ciliates to untypical hosts (Fenchel 1965). *Gargarius gararius* is strictly host-specific to *M. edulis* and is only found in mussels carrying insignificant numbers of other ciliate species (Lauckner 1983). Large numbers can be observed with no obvious host response. Most pathology involves the effect of suctorial tubule attachment to the host cell (which in some species comprises the insertion of fibres into the host cell). No mortalities have yet been associated with infections, and prevalence of infection is usually low (2% or less; Bower 2001c and references therein). Occasionally, *Sphenophrya spp.* can produce xenomas (hypertrophied host cells containing one or many ciliates), which do not appear to cause physical harm to bivalve hosts (Meyers & Burton 2009).

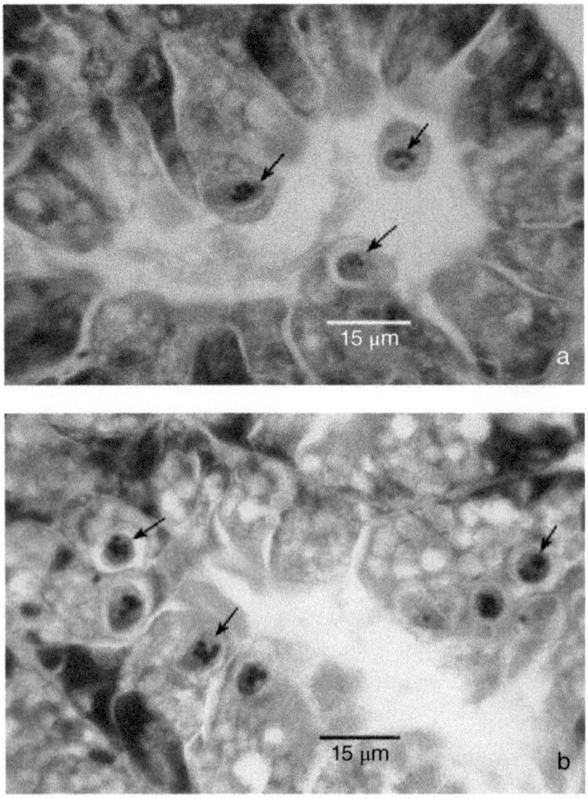

Figure 11.8 Histological sections through the digestive gland of *Mytilus trossulus* from British Columbia, Canada, with ciliates (arrows) within the cytoplasm of the epithelial cells. *Source:* From Bower (2013). Reproduced with permission of © Her Majesty the Queen in Right of Canada 2017.

Another group, referred to as intracellular ciliates or Rhynchodid-like Phyllopharyngea ciliates, infects *M. edulis*, *M. trossulus* and *M. galloprovincialis*. According to Bower (2013 and references therein), large numbers elicit no obvious host response despite disruption of the digestive tubule epithelia (Figure 11.8), and no mortalities have been associated with these infections. Prevalences up to 40% have been detected in Spain, but on both coasts of North America and on the German coast prevalence is usually less than 1%. In Slovenia, a prevalence of 21.4% was detected in *M. galloprovincialis*, but the majority of infections (94%) were classified as mild (Gombač *et al.* 2008).

Porifera

Boring sponges in the phylum Porifera have been reported globally in the shells of mussels, oysters and scallops, especially those growing on the bottom substrate. The sponge excavates into the shell, using acid to drill tunnels in the calcium material. In heavy infestations, the pest may even penetrate through to the inner shell. If perforation is extensive, the shell is weakened and can no longer offer protection, and the host will either die outright or be vulnerable to a range of predators. Ironically, shell fouling by sponges reduces predation risk for bivalves (Krug 2006). Once the shell's inhabitant is gone, the sponge will continue to grow and dissolve the shell. Apart from shell damage, infestation may interfere with adductor muscle attachment and impede feeding and cause mortality. Sponges bore to obtain shelter and do not obtain nutrients directly from their hosts.

Boring sponges belong to the class Demospongiae, family Clionaidae. There are about 350 described species in the family, most of which are in the genus *Cliona*. The genus has a global distribution, occurring from the low-tide mark down to a depth of ~100 m. For example, *Cliona celata* (Figure 11.9), a common boring sponge, is widely distributed along all western European coastlines, from Norway, Iceland and the western Baltic Sea to the Mediterranean. This species is very common on mussel and oyster beds, favouring the shells of *M. edulis*, *Ostrea edulis* and *C. gigas* (Snowden 2007). Severe penetration of the shell by *C. celata* weakens it, thereby reducing the commercial value of the mussel or oyster. Polychaete worms may inhabit sponge tunnels, also reducing market value (Bower *et al.* 1994). Growing bivalves in suspended culture is the simplest method of reducing shell damage (Bower 2001d). However, Carver *et al.* (2010) reported a 25–30% infestation rate by *Cliona celata* in off-bottom cultured *C. virginica* in New Brunswick, Canada. A comparison of various treatment strategies indicated that a six-minute brine dip (>90% NaCl saturation) every second year was completely effective at eliminating the sponge without harming the oysters.

Another boring sponge, *Alectona wallichii*, usually recorded from calcareous rocks and coral, has recently been identified in the mussel *P. indicta* from the southwest coast of India (Sunil & Thomas 2012). The low incidence (0.2%) of this sponge suggests that it is a new migrant to Indian seas. Sunil & Thomas (2015) reported eight other boring species, in the genera *Cliona*, *Pione*, *Cliothosa*, *Thoosa* and *Siphonodictyon*, in the same region, with an overall prevalence of ~18%. Sunil (2015) examined the effects of infestation on histology

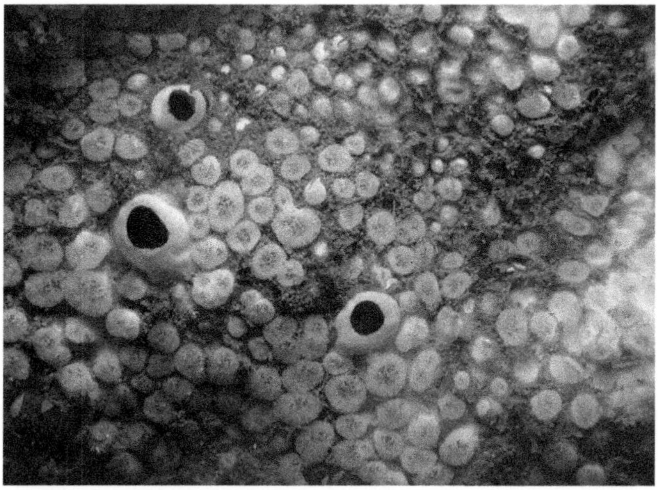

Figure 11.9 Boring sponge *Cliona celata* evenly covered by tuberculate retractable inhalant papillae (2–5 mm diameter) with larger oscula for water outflow. *Source:* Image is from World Register of Marine Species (WoRMS).

and ultra-structure of mantle tissue in *Perna indica*. TEM of the mantle tissue infested by *A. wallichii* indicated that it causes irreparable damage to the cell organelles. In sponge-infested specimens, increased cellular autophagy was recorded, along with haemocytosis, hypertrophy of the nucleus and endoplasmic reticulum (ER) and vesicle formation in the nucleolus. The most consistent change was a decrease in cell size in mantle sections, caused by loss of subcellular organelles and substances. Sponge-infested tissues also exhibited histopathological changes such as degeneration of the mantle and infiltration of haemocytes, and ultimately in severe infestations resulted in a drain in energy that affected normal growth.

In addition to rising water temperature, due to climate change, increasing amounts of atmospheric CO_2 are absorbed by the oceans, lowering the pH and leading to ocean acidification (see Chapter 3). Acidification reduces calcification rates of calcifying organisms, which in bivalves results in thinner shells (Fitzer *et al.* 2016). Duckworth & Peterson (2013) showed that while high water temperature had little effect on the growth, survival and boring rates of *Cliona celata* attached to scallop shells in seawater, lower pH slightly reduced sponge survival while greatly influencing shell boring. At pH 7.8, the sponge bored twice the number of papillary holes and removed twice as much shell weight as at pH 8.1, suggesting that lower seawater pH may increase boring rates of *C. celata* in bivalves, with potentially severe implications for wild and farmed populations.

Platyhelminths

There are about 20 000 known species in this phylum, the majority of which are parasitic. They are found in terrestrial, marine and freshwater habitats. The members, which measure 1 mm to 30 cm in length, are soft-bodied, bilaterally symmetrical and dorso-ventrally flattened. They are simple animals, composed of three cell layers (triploblastic): ectoderm, mesoderm and endoderm. They have no body cavity and generally possess an incomplete digestive tract, which opens with a mouth but ends blindly. Platyhelminths are hermaphrodites and produce a free-swimming ciliated larval stage(s) in their life cycle (Earth Life 2021). There are four classes in the phylum: Trematoda (flukes, all endoparasites), Monogenea (flukes, all ectoparasites), Cestoda (tapeworms, all endoparasites) and Catenulida (free-living flatworms). Each of these wholly parasitic classes is monophyletic, and together they form a larger monophyletic grouping, the subphylum Neodermata.

In terms of their disease-causing potential, platyhelminths are nowhere near as important as protozoan parasites. However, an enormous literature is available on this group, chiefly because they are visible to the naked eye, often have complex life cycles with several hosts and in some cases may be passed on to humans through consumption of infected shellfish. Since trematodes – in particular, the larval stages – are more important as bivalve pathogens than cestodes, most of this section will be devoted to them. For more details, see Bower (2009a) and the extensive review by Lauckner (1983) on helminth parasites.

Trematodes

Trematoda, subclass Digenea, are flattened oval or worm-like animals, usually no more than a few centimetres in length. Their most distinctive external feature is the presence of two suckers, one around the mouth, the other on the underside of the animal. The body surface comprises a tough tegument, which helps protect against digestive enzymes of host species and also serves as a surface for gas exchange and absorption. The entire body is occupied by the digestive and reproductive systems. The parasites have complex life cycles involving two or more hosts and asexual as well as sexual reproduction at different stages. Generally, there is a ciliated, free-swimming miracidium larva. After penetrating the first intermediate host, this larva forms a sporocyst. Inside the sporocyst are several individuals, called redia – another larval stage. These in turn give rise to several cercariae, which leave the host in search of a second intermediate host. These form cysts called metacercariae, which wait for the intermediate host to be eaten by the final host. When metacercariae enter the final host, they migrate to a specific organ and develop into an adult fluke, which reproduces sexually (Figure 11.10). Trematodes may use bivalves as the first host (sporocyst, redial and cercarial stages) or the secondary host (metacercarial stage), or the bivalve may be the host for all stages of the life cycle. An example of the latter is *Proctoeces maculatus*, family Fellodistomidae, which is widely distributed in temperate and tropical marine waters. The larval stages (sporocysts and cercariae) have been found in mussels such as *M. edulis*, *M. galloprovincialis*, *P. perna* and *Ischadium recurvum*, while the metacercariae have been reported in a range of molluscs, polychaetes, echinoderms and cephalopods, and in more than 60 marine fish species (references in Calvo-Ugarteburu & McQuaid 1998; Aldana *et al.* 2009; Antar & Gargouri 2016). Generally, the species that are most pathogenic to mussels and other bivalves belong to the families Fellodistomidae and Bucephalidae (see later), which utilise mussels for sporocyst and cercaria development (Bower 1992).

In infected mussels, there is an alteration in haemolymph components, a reduction in growth rate and weakness with respect to valve closure and attachment to the substrate (Bower 2009a). In heavy infections, the numerous sporocysts – as many as 28 000 per mussel – develop in the vascular tissue of the mantle. Such numbers can seriously reduce

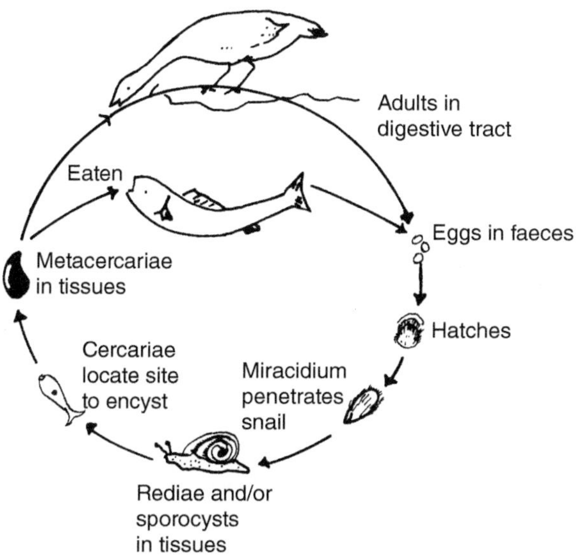

Figure 11.10 Generalised life cycle of the majority of trematode species in the marine environment, where the metacercariae encyst in (or on) mussels as well as other animals and/or plants. *Source:* From Bower (1992). Reproduced with permission from Elsevier.

glycogen content (energy reserves) of the tissues and efficiency of the circulatory system, resulting in disturbances to gametogenesis and possibly castration and death (Bower 1992). Sporocysts and cercariae can also be found in digestive gland, foot, kidney and gill tissue of some mussels (Villalba *et al.* 1997). In certain regions (e.g. western Portugal), castrated mussels do not die despite heavy infections, which indicates possible host adaptation to the parasite (Teia dos Santos & Coimbra 1995). In others, heavy infections (>60%) have been shown to limit the culture of mussels to a seasonal product (Sunila *et al.* 2004).

Prevalence of infestation varies greatly with locality and season. For example, a low prevalence (4%) was interpreted as responsible for minimal losses in natural and cultured mussel (*M. edulis*) populations in Laguna Veneta, Italy. About 10 years later, this parasite was identified as the probable cause of extensive mortality in cultured mussels (*M. edulis/galloprovincialis*) from the same location (Bower 2009a and references therein). In mussels (*M. edulis*) from New Jersey, United States, incidences of up to 46% occurred during August and September, with lows of 5–7% from December to March (Lauckner 1983). Incidence of sporocysts has been found to increase with host size, from <10% in mussels of 20–25 mm shell length to 43–60% in larger mussels, thereby eliminating a significant proportion from the breeding population due to parasitic castration (Lauckner 1983). Mussels <15 mm shell length (probably immature) are only rarely found infested, suggesting that susceptibility of *M. edulis* to *P. maculatus* is positively correlated with the onset of gonad development in young mussels. Pathogenicity has also been shown to vary with host species. For example, in wild populations, *P. perna* in S. Africa had infection rates as high as 62%, but introduced *M. galloprovincialis* was free of *P. squamatus* and other trematode species (Calvo-Ugarteburu & McQuaid 1998). This may have conferred a competitive advantage to *M. galloprovincialis*, explaining the rapid, invasive spread from its introduction on the west coast in the early 1970s on to the southern and eastern coasts of South Africa in the 1990s. There is a strong likelihood that *M. galloprovincialis* will replace *P. perna* on the southern coast as it has done with *Aulacomya* on the western one (Bownes & McQuaid 2006).

Another trematode, *Proctoeces lintoni*, requires three distinct hosts to complete its life cycle. The first, intermediate host is the mussel *Perumytilus purpuratus*, in which the parasite produces cercariae. These leave the mussel to infect the gonads of the second intermediate host, the keyhole limpet *Fissurella* spp., where they develop into the metacercarial stage. The definitive host, the clingfish (*Sicyases sanguineus*) ingests the infected limpet, and this is where the parasite reaches sexual maturity and begins egg production (Loot *et al.* 2008). In intertidal rocky habitats in central Chile, the parasite has been found to be more abundant in intermediate hosts collected in marine protected areas than in hosts collected in areas open to fishing of all three host types (Loot *et al.* 2005). The difference between protected and unprotected areas is probably due to the relatively greater density of definitive clingfish hosts in protected areas (Hechinger *et al.* 2008).

Trematodes in the family Bucephalidae are another group with complex life cycles, beginning in a bivalve, followed by a fish and then a second fish as the definitive host (Lauckner 1983). For example, in Brazil, *Bucephalus margaritae* cercariae are found in the first intermediate host *P. perna* inside the sporocysts, which have the form of orange and ramified filaments (Marchiori *et al.* 2010). The metacercariae encyst in the gills and gill cavity of the second intermediate host, the blenny (*Hypleurochilus fissicornis*). The definitive host, the southern kingcroaker (*Menticirrhus americanus*) becomes infected upon ingesting blenniids parasitised with metacercariae. Bucephalid mussel infection has been reported in *Mytilus* spp., *Mytella guyanensis*, *P. perna*, *Arcuatula senhousia*, *Semimytilus algosus*, *Brachidontes rodriguezii* and *Bathymodiolus* sp. (Lauckner 1983 and references therein; Powell *et al.* 1999; da Silva *et al.* 2002; Ceuta & Boehs 2012; Calvi Zeidan *et al.* 2012 and references therein), and also in several oyster, clam and scallop species (Calvi Zeidan *et al.* 2012; Getchell *et al.* 2016). Infected mussels have been reported from NW Europe, the east and west coasts of North America, the Gulf of Mexico (Bower & McGladdery 2009), the coasts of Brazil (Calvi Zeidan *et al.* 2012), Argentina and Chile (Lauckner 1983) and SW India (George & Nair 1974). In mussels, macroscopic evidence of infestation is the orange colour of the mantle due to sporocysts. In *Mytella guyanensis*, the most frequent parasite location is the gonad (100% of cases), although sporocysts are also found in the digestive gland and the gills, and less frequently in the foot muscle (Boehs *et al.* 2010). The gonad follicle destruction associated with this infestation is well documented (Lauckner 1983) and probably results from both mechanical damage and decomposition of tissues. According to Bower *et al.* (1994), the sporocysts reduce glycogen content of the tissues and efficiency of the circulatory system, resulting in disturbance to gametogenesis and castration. Infection also causes weakness and gaping, which can reduce product value during shipping and marketing (Bower & McGladdery 2009). Prevalence of infection is generally low (0–10%) irrespective of mussel species, region or season (Lasiak 1993; Boehs *et al.* 2010; Ceuta & Boehs 2012; da Silva *et al.* 2012; Calvi Zeidan *et al.* 2012). However, there are some exceptions in the literature. For example, in intertidal mussels from southern Chile, 20–32% of *Semimytilus algosus* examined harboured sporocysts, while <1% of *Perumytilus purpuratus* were infected and and no sporocysts were found in an adjacent population of *Choromytilus chorus* (Lasiak 1992). Prevalence was related to the size of *S. algosus* present at a particular site, and generally increased with host size. Calvo-Ugarteburu & McQuaid (1998) also found that prevalence of infection was highly dependent on the size of *Perna perna*, with over 50% of mussels larger than 90 mm being infected with *Bucephalus* sp. While George & Nair (1974) reported low prevalence of infection in *Arcuatula senhousia* (0–7%), they found that it increased from 0% for the January–May period to 7% for June–December. Finally, Powell *et al.* (1999) examined the parasite body burden in petroleum seep mussels (*Bathymodiolus* sp.) in the Gulf of Mexico. While infected mussels were not observed at some sites, high prevalence of *Bucephalus* infection (>80%) was recorded at others. Powell *et al.* (1999) suggested that higher prevalence could be due to higher susceptibility, greater exposure per time period or simply age: older animals are exposed for a longer time and thus may accumulate more parasites. Seep mussels are likely to be significantly older than shallow-water mytilids, and age alone may thus explain the higher prevalences observed (Lutz *et al.* 1985).

Throughout European coastal waters, the trematode *Renicola roscovita*, family Renicolidae, is frequently found in 100% of *M. edulis* on intertidal beds. The parasite utilises the gastropod *Littorina littorea* as the first intermediate host and a variety of bivalves as the second intermediate host (Lauckner 1983). The final host is the gull *Larus argentatus* (Lauckner 1980). On North Sea tidal flats, *R. roscovita* and *Himasthla elongata* (see later) are the two most abundant and most important larval trematodes utilising bivalves as second intermediate hosts. Several thousand encysted metacercariae can occur in individual mussels, primarily in the palps and visceral mass. To investigate the effect of the parasite on mussel growth, Thieltges (2006) carried out controlled infection of mussels in two field experiments in two habitats and at two tidal heights (low and mid intertidal). Growth was significantly lower (up to 20%) in parasitised compared to nonparasitised individuals. However, tidal height had a stronger effect on growth than parasitism, without any interaction between the two factors. The negative effect of *R. roscovita* metacercarial infection on mussel growth may result from direct tissue disruptions, interference of metacercariae with food intake ability (on the palps) or the energetic demands of the parasite (with metacercariae growing in size after initial infection) (Stier *et al.* 2015). This may be a strategy by the parasite to enhance its transmission to its final bird host (Thieltges 2006 and references therein), since mussels may thus remain within the prey size range of the parasite's final host for a longer period of time. This could have economic implications for mussel cultivation, which might be better situated at subtidal offshore sites where *R. roscovita* infection does not occur (Buck *et al.* 2005). Alternatively, Thieltges (2006) suggests that intertidal sites without the first intermediate host, *Littorina littorea*, might be suitable – albeit the species performs an important role in enhancing growth by removing epibionts such as barnacles from mussel shells (Buschbaum 2000).

When Krakau *et al.* (2006) compared prevalence of *R. roscovita* in introduced species with native bivalves from two localities in the North Sea, they found it was always lower in the introduced species. For example, at one sampling site, prevalence for the recently introduced species *Crassostrea gigas* and *Ensis americanus* was 35 and 53%, respectively, while that for the native species *M. edulis* and *Cerastoderma edule* was 98 and 85%, respectively. Similar results were obtained at the other sampling site. It is likely that, over time, pre-adaptations and adaptations by native parasites to new hosts may lead to a significant infestation. Thus, the apparent competitive advantage of alien species seen in this study may apply only to the early phase of invasion.

The trematode *Himasthla elongata*, family Himasthelidae, uses the periwinkle *Littorina littorea* as the first intermediate host, *M. edulis* and the cockle *Cerastoderma edule* as second intermediate hosts and gulls (*Larus*) as definitive hosts, in which sexual reproduction takes place. In *M. edulis*, the parasite, which encysts primarily in the foot musculature, impairs the production of byssus threads and shell cleaning behaviour, thereby favouring the attachment of fouling organisms on the shell valves, while in the cockle *C. edule* it alters burrowing ability, resulting in higher predation by gulls (Lauckner 1983). This could be seen as adaptive behaviour for the parasite, since these birds function as definitive hosts. In an intertidal population of *M. edulis* in the White Sea, Russia, Nikolaev *et al.* (2006) found that prevalence of infection reached 100% in three- to four-year-old mussels and remained at this level in older individuals. Infection intensity increased steadily with the age of the mussel, decreasing only in the oldest (nine-year-old) individuals. These patterns were associated with accumulation of cercariae in the course of the mussels' lives (see similar findings earlier). The ability of metacercariae to exist in mussels for long periods (at least two-and-a-half years) was verified in the course of this study. In a study related to that of Krakau *et al.* (2006), Thieltges *et al.* (2009) found that *C. gigas* and the American slipper limpet *Crepidula fornicata*, also an introduced species of European coastal waters, reduced the parasite load in *M. edulis* by 65–77 and 89% in single and mixed treatments, respectively. Both introduced species, which were unaffected by *H. elongata*, acted as decoys for the trematode, thereby mitigating the parasite burden of mussels. This dilution effect, which the authors showed to be relevant in the field, was density-dependent, with higher reductions at higher invader densities, thus altering the native parasite–host interactions. There are many other non-host species that interfere with the transmission of cercariae of *H. elongata* (Welsh *et al.* 2014). For example, over a three-hour period, the amphipod *Gammarus marinus* removed 100% of the cercariae; other effective diluters were the shrimp *Crangon crangon* (93%), the brown seaweed *Sargassum muticum* (87%) and the barnacle *Semibalanus balanoides* (71%). In contrast, mixed shells (*Cerastoderma edule, M. edulis, Ensis americanus* and *Littorina littorea*) had no significant effect.

The family Gymnophallidae comprises a small group of digeneans that live in the intestine, gallbladder and bursa (a lymphoid organ) of shore birds and diving ducks that feed primarily on intertidal bivalves, gastropods and benthic polychaetes (Nolsø 2002). A notable feature of the family is the flexibility of its life style: some species have bivalves as first and second intermediate hosts, some have copepods as the first and bivalves and polychaetes as the second intermediate hosts and some have bivalves as the only hosts (Mladineo & Peharda 2005). Gymnophallids infect the mussels *Modiolus barbatus* and *M. edulis* and M. *galloprovincialis*, cockles (*Cerastoderma* spp.) and a wide range of clams (Lauckner 1983). The most common site affected is connective tissue around the digestive gland, gonads and mantle tissue; sporocysts localised near gonadal tissue cause retardation in oocyte development. Gymnophallid metacercariae, situated between the mantle and the shell of *M. edulis*, are known as 'pearl trematodes' (Lauckner 1983). Pearls are formed because of an infection, probably by *Gymnophallus bursicola*. The trematode enters *M. edulis* as a cercaria and may be found between the mantle and the shell. As a protectant, the mussel encapsulates the larva with layers of nacreous shell material, and so a pearl begins to form, growing at the same rate as the thickening of the shell. Heavy infestation does not seem to harm the mussel, but it affects the potential for the development of growing mussels for the live market – the most profitable. To counteract this, operators must relay the mussels, which are grown on until a certain age (around 18 months), and sell them as 'part-grown' mussel for on-growing in another location (Wilcox 2013).

Two studies have examined whether there was any effect of several variables on intertidal mussel infection by metacercariae of trematodes in the genera *Gymnophallus*, *Himasthla* and *Renicola*, the most common trematode taxa in *M. edulis* (Lauckner 1983; Galaktionov 1996). Wilson *et al.* (2013) studied the level and incidence of infection by these trematode genera at 23 sites in North Atlantic waters. At each site, four parameters were noted: presence of bird predators of mussels, other molluscan secondary hosts, habitat and mussel population characteristics (e.g. density, biomass, size and number of individuals analysed). Density of parasites in various tissues was recorded from 10 mussels from each year class. Counts of bird species that preyed on mussels were also recorded, as well as the presence or absence of common gastropod hosts (details in Wilson *et al.* 2013). Data were analysed using PRIMER-E, a statistical package that provides a collection of specialist univariate, multivariate and graphical routines for analysing species sampling data for community ecology (Clarke & Gorley 2015). Apart from one site in the Arctic where no parasites were recorded, prevalences were up to 100% of individuals sampled, with lower rates for combined infections (Table 11.2). The analysis indicated a considerable influence of geographic location, with closely connected sites also grouped together on the basis of their parasite communities. The major influence on infestation was bird (final host) numbers, but exposure was also a strong factor in that increased exposure was associated with decreased infection rate. In a related study, Galaktionov *et al.* (2015) examined the prevalence and intensity of infection in *M. edulis* by trematode species in the same genera as previously mentioned from 19 intertidal sites from North Atlantic waters, 17 of which were common to those sampled by Wilson *et al.* (2013). The impact of a number of environmental (maximal sea surface temperature, SSTmax) and biological (mussel age and density and abundance of avian final hosts) factors on trematode infection was examined. Infection levels correlated with the interaction between local bird abundance and mussel age and density. Levels increased with mussel age, but dependence on mussel density and bird abundance varied among trematode genera. Prevalence and intensity increased with SSTmax for *Renicola* spp. but no relationship was found for *Gymnophallus* and *Himasthla* spp. It is interesting that while maximum prevalence rates were similar in the two studies (~100%), the mean prevalence rates (Table 11.2) from the Wilson *et al.* (2013) study were consistently lower than those obtained by Galaktionov *et al.* (2015), even though the 17 sites sampled were common to the two studies.

Cestodes

Cestodes are a class of parasitic flatworms commonly called tapeworms because of their extended length in the adult stage. Marine bivalves are hosts for the pre-adult stages of the tapeworm (encysted sporocysts and metacercariae), while adults are parasites in the intestine of elasmobranch fish (sharks, skates and ray), the definitive hosts. The tapeworm attaches to the intestine by scolex suckers on its head. Infection in bivalves is therefore most common in tropical and subtropical waters, where elasmobranchs constitute an important proportion of the vertebrate fauna (Lauckner 1983).

Larval cestodes of the genus *Tylocephalum* occur, very often in great numbers, in a variety of marine bivalves, especially oysters (Bower 2009b). In contrast, representatives of six genera including *Tylocephalum* have been recorded in scallops (Getchell *et al.* 2016).

Table 11.2 Prevalence (% of individuals infected) of trematode parasites, showing maximum at any one site, geometric mean rate among the sites and number of sites (out of 23) at which they were found. Values in brackets are the geometric mean % prevalence rate calculated from data in Galaktionov *et al.* (2015).

Genera recorded in an individual mussel	Max %	Mean %	No. sites
Himasthla	100	9.60 (17.29)	17
Renicola	100	18.44 (29.66)	19
Gymnophallus	100	6.05 (3.27)	15
Himasthla and *Renicola*	85.1	5.12	11
Himasthla and *Gymnophallus*	22.6	1.88	9
Renicola and *Gymnophallus*	80.6	2.72	11
Himasthla, Renicola and *Gymnophallus*	22.6	1.22	7

Source: From Wilson *et al.* (2013). Reproduced with permission from Elsevier.

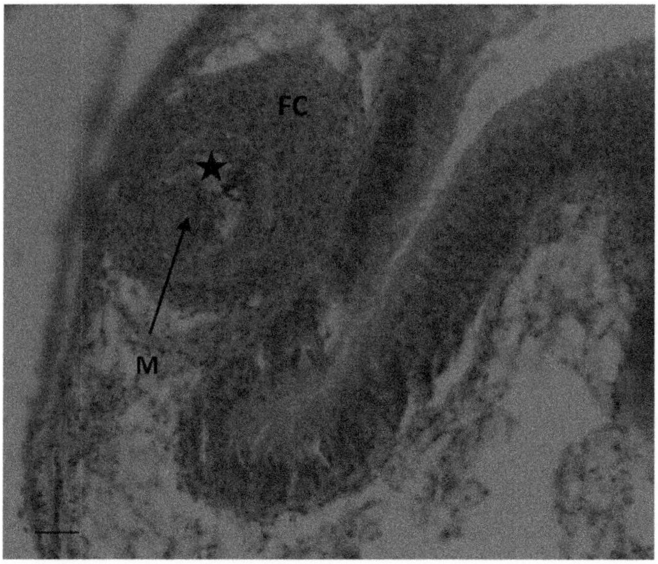

Figure 11.11 Metacestode of *Tylocephalum* sp. (★) in the mantle of the mussel, *Mytella guyanensis*, showing fibrous capsule (FC) with haemocytic response and myzorhynchus (central, muscular projection on the scolex) (M). Scale bar = 20 μm. *Source:* From Ceuta & Boehs (2012). Reproduced with permisision from *Brazilian Journal of Biology*.

Bivalves are initially infected when they ingest free-swimming larvae or cestode eggs excreted by an elasmobranch. In bivalves, these develop and encyst as metacestodes in the tissue of the digestive system. The cellular reaction to the infection is characterised by encapsulation of larval cestodes in connective tissue (Figure 11.11). The host's response is to secrete a thick fibrous capsule around each metacestode. Thus, the infection does not seem to significantly cause harm to the host bivalves (Apeti *et al.* 2014). None of the larval or post-larval tapeworms occurring in marine bivalves are known to infest humans.

There are few reports on cestode infestation in mussel species. More than three decades ago, Lauckner (1983), in his very comprehensive volume on bivalve disease, cited just one report on *Tylocephalum* infestation (11.4%) in *M. edulis* sampled from Wakayama on the east coast of Honshu Island, Japan (Sakaguchi 1973). In contrast, two oyster species examined from the same locality had infection levels of 100% (Sakaguchi 1973). Despite heavy infestation, larval cestodes cause little damage to their hosts.

In the United States, as part of the National Oceanic and Atmospheric Administration's National Status and Trends Program (NS&T), the Mussel Watch Program (MWP) is unique in being the only continuous coastal monitoring program that is national in scope (see Chapter 8). The MWP provides both national and regional assessments of patterns of occurrence and distribution of parasites and diseases in sentinel bivalves (mussels *Mytilus* spp. and oysters *C. virginica*) in US coastal waters. An array of parasite taxa (e.g. prokaryotic inclusions, cestodes, nematodes, trematodes), diseases and certain tissue conditions (e.g. Dermo, MSX, tumour, neoplasia, edema and necrosis) are quantified using direct quantitative count or estimative semi-quantitative scales. The most recent report, which was based on 1995–2009 MWP monitoring data, provides both national and regional trends and distribution of parasites and disease in mussels and oysters (Apeti *et al.* 2014). Parasite taxa richness was calculated as the total number of parasites found in a bivalve species and geographic location. Annual averages of parasite taxa richness in East Coast mussels varied between 6 and 13 (mean 9.8), while richness on the West Coast fluctuated between 10 and 16 (mean 12.2). Except for a few instances, parasite taxa richness appeared to fluctuate in a similar way in mussels from both coasts. Oysters harboured significantly more parasite taxa than mussels, due to the absence in the mussels (East and West coasts) of cestodes, nematodes (round worms) and the protistans *Perkinsus marinus* and *Haplosporidium nelsoni* – two significant disease-causing pathogens in *C. virginica*. Cestodes were likewise not detected in two small-scale studies on mussels (*Mytella guyanensis*) in Bahia, Brazil (Boehs *et al.* 2010; Calvi Zeidan *et al.* 2012), although a subsequent study on the same species in the same region reported a metacestode of a *Tylocephalum* sp. in the mantle, showing encapsulation and haemocyte infiltration (Ceuta & Boehs 2012). Also, larvae of a *Tylocephalum* sp. were observed encapsulated in connective tissues of a single brown mussel (*P. perna*) in southern Brazil (da Silva *et al.* 2012).

Turbellaria

Turbellarian flatworms, class Catenulida, are predominantly small, hermaphroditic, free-living predators or scavengers. They are found in freshwater and marine environments. Their anatomy is simple, with an anterior mouth that connects to a simple pharynx and a simple intestine that forms a ciliated sac. There are about 100 species described worldwide, but the simple anatomy makes species distinction problematic. A small proportion of turbellarians live as endosymbionts or obligate parasites that are able to pass freely between the mantle cavity and alimentary canal of bivalve hosts.

The turbellarian *Urastoma cyprinae*, order Fecampiidae, family Urastomidae is reported as an opportunistic mantle inhabitant on the gills of bivalves, including *M. edulis* and *M. galloprovincialis*. In mussels, the flatworm has been reported from the coast of Galicia in NW Spain, Portugal, the Mediterranean Sea and the Black Sea (Bower 2001e). According to Hooge & Tyler (1999), it attaches its anterior end to gill filaments and resides presumably in a feeding posture within the mucus that flows over the gills. It arches its body back and forth, wagging its posterior end, which bears the terminal orogenital pore. Ingestion of foods is likely to be driven by the sphincters on the orogenital atrium. Although not yet associated with mussel mortalities, *U. cyprinae* can cause disorganisation of the gill filaments (including reduction of space between the lamellae and increase in the size of haemal sinuses within the lamellae), a heavy infiltration of haemocytes and subsequent necrosis of adjacent gill tissue (Bower 2001e). Robledo *et al.* (1994a) suspected that flatworms could greatly reduce feeding capacity in heavily infected mussels. Mladineo *et al.* (2012) examined the extent of pathology imposed by *U. cyprinae* in a total of 480 individuals of rope-grown *M. galloprovincialis* sampled over a one-year period in Mali Ston Bay, southern Adriatic Sea, Croatia. Although there was no negative impact on host reproduction or condition index, they found ultrastructural and functional alteration in host cells at the attachment site of the parasite. Ultrastructural changes included haemocytic encapsulation of the parasite, a caspase-positive reaction inferred by immunohistochemistry, mild tissue destruction and haemocyte activation. Caspases are a family of protease enzymes that play essential roles in programmed cell death. The overall prevalence rate was 88%, with a slightly lower percentage in small (<40 mm shell length) versus larger (>60 mm) mussels. When intensity of infection was compared between small and large mussels, there was a significant difference ($P < 0.001$) between them: 5.23 (4.59–5.94) versus 28.57 (24.12–33.32), respectively. In addition, there was temporal variation of the turbellarian presence in both small and large mussels, with intensity levels lowest in small mussels in autumn (October–November) and highest in April, or in July for large individuals. Crespo-González *et al.* (2010) reported similar results for prevalence, intensity of infection and seasonal pattern for *M. galloprovincialis* on natural mussel beds in the Vigo Estuary in NW Spain. A considerably lower prevalence rate (3.3%) and intensity of infection were reported in *Mytella guyanensis* from the southern coast of Bahia, Brazil (Calvi Zeidan *et al.* 2012; for further information on prevalence rates, see Villalba *et al.* 1997; Cáceres-Martínez *et al.* 1998; Comps & Tigé 1999; Francisco *et al.* 2010; Özer & Güneydağ 2014).

Crespo-González *et al.* (2010) suggested a putative life cycle for *U. cyprinae* in the Vigo Estuary. When the parasite is ready to lay eggs – usually toward the end of autumn – it leaves the mussel, secretes a protective cocoon around itself and produces an egg sac. About 24 days later, the young hatch, make their way out of the cocoon and swim off to find new mussel hosts. Host-finding may involve a combination of positive phototaxis, passive entry resulting from the inhalant current created by the filtering activity of the mussel and active entry guided by specific attractants from mucous secretions on the gill filaments, the site of attachment for the flatworm. Recruitment of infective life stages generally occurs during summer.

The only gross clinical signs of infection are small oval or pear-shaped, white or light-coloured spots up to 2 mm in length containing up to five flatworms located between gill lamellae (Bower 2001e). Immature live *U. cyprinae* are smaller (0.4–0.8 mm in length), with two pigmented eyes near the anterior end. Although most frequently found in the middle area of the four demibranchs, they also occur over the entire gill surface and may be detected swimming freely in the pallial cavity.

Annelids

Segmented worms belonging to the phylum Annelida, class Polychaeta, family Spionidae have a global distribution. Several species of these worms in the genus *Polydora* burrow into the shells of bivalves and excavate U-shaped tunnels that subsequently become filled with compacted mud, causing 'mud blisters' on the inner surface of the shell and possibly yellow pustules in soft tissues contacted by the burrows. The burrows weaken the shell, thereby increasing susceptibility to predation, especially in thin-shelled species such as mussels, although crabs have been found to show preference for noninfested mussels (Ambariyanto & Seed 1991). In oysters, this weakening can cause problems during shucking, packing and transport. Oyster shells containing *Polydora* are also unsuitable for the lucrative half-shell market, as mud blisters are unsightly and, if punctured, can release sediments, faecal deposits and anaerobic metabolites such as hydrogen sulphide (Handley & Bergquist 1997). Other effects of the parasite include atrophy and detachment of muscle when burrowing occurs in the region of the adductor muscle, loss of condition, retarded growth and mortality (Lauckner 1983; Bower *et al.* 1994; Riascos *et al.* 2008). *Polydora* infection has been reported in *Mytilus edulis* (Thieltges *et al.* 2013), *M. galloprovincialis* (Özer & Güneydağ 2014), *Bathymodiolus* spp. (Ward *et al.* 2004), *Aulacomya atra* (Diez *et al.* 2016; Caza *et al.* 2016) and *Perna* spp. (Read 2010). In the case of *A. atra*, *Polydora* infestation causes the formation of pearls, which is evidenced by the presence of capsules containing polydorin tissue debris and the elemental composition of organic material inside the pearls (Diez *et al.* 2016).

Polydora species mostly brood their eggs inside blisters and produce fully developed larvae that reinfest locally. However, planktonic larvae may be produced in some populations (Diggles *et al.* 2002). Juvenile worms settle on the edge of the shell and begin burrowing or, in the case of *P. ligni*, build their tubes on the surface of shells (Meyers & Burton 2009). See Morse *et al.* (2015) for further information on *Polydora* life history. There has been some controversy as to the mechanism used by *Polydora* spp. to excavate burrows. It was initially believed that they used modified setae (stiff bristles) to mechanically bore into the shell. However, it is now clear that excavation and enlargement of the burrow is accomplished by chemical dissolution of the shell, and probably also by flushing of loosened shell material (Lauckner 1983 and references therein).

Prevalence rates are generally low (<10%; Ambariyanto & Seed 1991; Buck *et al.* 2005; Özer & Güneydağ 2014), although Ambariyanto & Seed (1991) found higher levels of infestation (60–65%) in mussels near the mouth of the estuary. Other polychaetes have a commensal relationship with mussels, such as the polynoid *Branchipolynoe seepensis* and the nautliniellid *Laubierus mucronatus* found in vent and seep mussels (*Bathymodiolus* spp.). These cause minor tissue damage to the gills and displacement of gill filaments, but there is no strong evidence for a detrimental effect on host condition. The polychaetes probably consume bacterial-laden mucus/pseudofaeces or free-living bacteria rather than tissues of the host mussel (Ward *et al.* 2004).

Various methods have been tested to try to eradicate *Polydora*, especially in oyster-growing areas. Morse *et al.* (2015) describe a number of methods that have been used to treat infected oysters, but to date no single method with universal application has been identified. They suggest a combination of a freshwater bath and cool air storage, which may provide a synergistic effect. Brown (2012) found that 14 days in storage resulted in 100% mortality of blister worms when carried out following 72-hour baths in anything from fresh- to seawater at 30 psu. As 72-hour baths are likely unfeasible, a better method could be a low-salinity dip of short duration combined with short-duration storage.

Apart from off-bottom culture techniques (Bower *et al.* 1994), I am not aware of any additional methods of *Polydora* infection control in mussels.

Crustaceans

Compared with bacterial or protozoan parasites, members of the subphylum Crustacea are only mildly pathogenic to bivalves. The best-documented disease organism is the copepod *Mytilicola intestinalis*, family Mytilicolidae, which is found only in European waters from Denmark to Italy, including the British Isles and Ireland, but not the Baltic region

(Bower 2009c). *M. edulis* and *M. galloprovincialis* are the primary hosts, although oyster and clam species can also be infested (Bower 2009c). The copepod is red-coloured and large, which makes it very conspicuous inside the host. Females reach a size of 7–8 mm and males 3–4 mm. Mature females carry paired egg sacs that are also red (Figure 11.12). Breeding and reproductive cycles depend on the geographic distribution of the copepod. Two complete breeding cycles with two generations are generally observed, with recruitment taking place in summer and autumn (Bignell 2016). Fertilisation is probably internal, with females producing about 200–300 eggs (Thieltges *et al.* 2008). The eggs hatch in the host and are expelled into the water column as free-swimming larvae. After a dispersal period of three to four days, the larvae descend to the bottom to infect the mussel host. Larvae enter mussels through the inhalant siphon. Infestation is a passive process, depending on the chance encounter by the larvae of the host's field of filtration, and also on the strength of the inhalant current (Gee & Davey 1986). Immature copepods are found in the stomach, digestive gland and anterior part of the intestine, while adults are found in the posterior part of the intestine of the mussel (Jensen 2010). Not surprisingly, infestation success is dependent on host size and density. In addition, intensity of infestation is negatively correlated with exposure; that is, mussels higher up the shore are less likely to be infected than those lower down. Also, Figueras *et al.* (1991) found a significant positive correlation ($r = 0.614$; $P < 0.05$) between frequency of infection and length of mussel.

Prevalence of 100% is common, and more than 30 copepods may be dissected from a single mussel (Bower 2009c and references therein). *Mytilicola intestinalis* can be enzootic

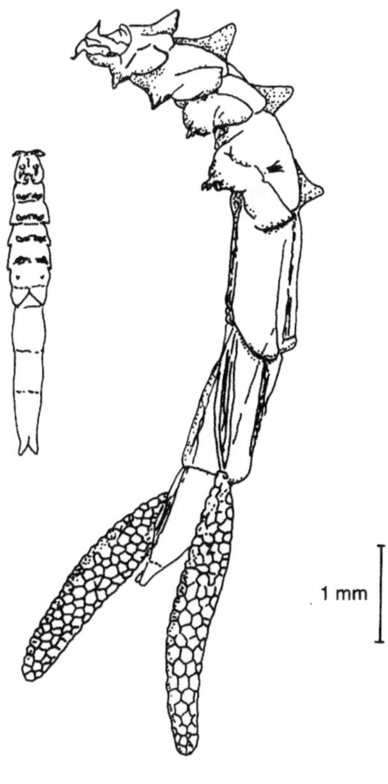

Figure 11.12 The copepod *Mytilicola intestinalis* from the intestine of the mussel *Mytilus edulis*. Adult male (left, ventral aspect) and female with egg-sacs (right, dorso-lateral view). From Lauckner (1983), after Hockley (1951).

(endemic) in low proportions of mussels in various areas and become epizootic (epidemic) only under adverse environmental conditions (Lauckner 1983). The copepod has been designated as a 'controlled pest' in the United Kingdom, meaning that transport of mussels from infected areas is carefully monitored (Jensen 2010).

There has been considerable controversy as to the seriousness of the copepod. It has been blamed for widespread mortality and loss of condition in European populations of mussels. For example, the crash of the Dutch and German mussel fisheries in 1949 and 1950 was attributed to its presence (Korringa 1951). For the next 30 years or so, conflicting reports on its pathogenicity appeared. Some of these reported serious effects on the health, growth and condition of parasitised mussels (Cole & Savage 1951; Bayne *et al.* 1978; Robledo *et al.* 1995), especially when parasite burdens exceeded 25 individuals per host (Gee *et al.* 1977). However, in no instance was the copepod deemed to be responsible for the death of its host. In contrast, others have reported no loss of condition in parasitised mussels (Campbell 1970; Figueras *et al.* 1991; Gilek *et al.* 1992), even when parasite numbers exceeded 20 individuals per host (Dethlefsen 1975). It now seems likely that the *Mytilicola–Mytilus* relationship is commensal rather than parasitic (Davey 1989), with the copepod feeding on host gut contents rather than on host tissue (Moore *et al.* 1978). Reported negative effects of this and other *Mytilicola* species (see later) on bivalves may be explained by environmental variables, host gametogenic stage and even the presence of microscopic pathogens.

Histological evidence indicates that *M. intestinalis* causes local changes in the gut epithelium involving the replacement of normal ciliated columnar cells with nonciliated cuboidal cells (Bower 2009c), and in rare cases erosion and perforation from the appendages with or without host encapsulation (Meyers & Burton 2009). Accumulation of haemocytes – the usual response of bivalves to disease – was not evident (Moore *et al.* 1978; Robledo *et al.* 1994b). But Figueras *et al.* (1991) and Villalba *et al.* (1997) noted limited haemocytic infiltration in the intestinal epithelium and surrounding connective tissue near the copepod. Campbell (1970) determined that juvenile stages of the parasite cause most damage to the host, due in part to their presence in the ramifications of the hepatopancreas. However, Moore *et al.* (1978) found no evidence of this pathology and indicated that the few immature copepods that penetrated the tubule epithelium and lodged in the connective tissue were encapsulated and killed.

Another copepod, *M. orientalis*, originally confined to *the gerbil Meriones unguiculatus* in the Inland Sea of Japan, has spread to mussels (*M. trossulus* and *M. californianus*) on the Pacific coast of North America and was introduced into France with imported oysters (*Crassostrea gigas*) in the 1970s (Lauckner 1983). It is also found in *M. edulis* and *M. galloprovincialis*, where dual infections with *M. intestinalis* occur (Bower & Figueras 1989; Goater & Weber 1996). The species further infects a wide range of oyster and clam species. The *M. orientalis* life cycle, as well as its impact on mussels, is probably similar to that of *M. intestinalis* (Bower 2010c). Interestingly, Steele & Mulcahy (2001) reported from a two-year study no effects of *M. orientalis* on the condition, growth, sex, stage or glycogen content in a population of *C. gigas*, despite the observation that 14% of oysters were infected with up to 20 copepods in a single individual. The intensity of infestation tends to be relatively low, with usually fewer than 10 *M. orientalis* per host, or more in the largest mussels (Goater & Weber 1996). Juvenile mussels (less than 10 mm in length) are rarely infected (Bower 2010c).

Pogoda *et al.* (2012) investigated levels of *Mytilicola* infestation in *M. edulis* (N = 15) at three near-shore sites in the southern German Bight, North Sea. Both *M. intestinalis* and *M. orientalis* were detected at all sites. For *M. intestinalis*, prevalence ranged between 13.3% (site 2), 26.7% (site 1) and 33.3% (site 3), and for *M. orientalis* it was between 20.0% (sites 2 and 3) and 26.7% (site 1). Prevalence for mussels infected by both copepods ranged between 6.7% (sites 1 and 3) and 13.3% (site 2). In total, 46.7% (site 2) to 60.0% (sites 1 and 3) of mussels were infested by *M. intestinalis*, *M. orientalis* or both species. Mean intensity ranged between one and two individuals for both species at all sites. In the same study, no copepod species were detected in mussels from an offshore site, a finding that is potentially beneficial for offshore aquaculture activities in the open ocean (Pogoda *et al.* 2012). For a seasonal comparison of copepod infestation, site 3 was investigated twice: in summer, *M. intestinalis* showed a lower prevalence (20.0%) and *M. orientalis* a higher prevalence (33.3%) when compared to winter (Figure 11.13). No mussels were infested by both species at the same time in summer, and the total prevalence was slightly lower than in

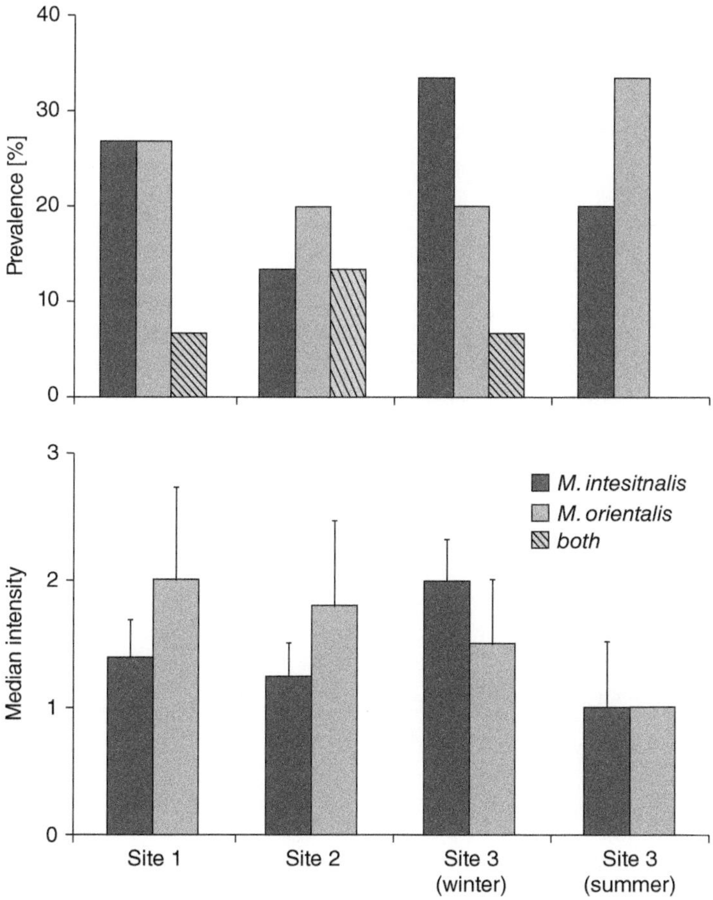

Figure 11.13 Prevalence and intensity of *Mytilicola* copepods in *Mytilus edulis* at three nearshore sites (see text). From Pogoda *et al.* (2012). Reproduced with permission from John Wiley & Sons.

winter (53.3%). Mean intensities were one copepod per mussel for both *Mytilicola* species during summer, which was again somewhat lower than in winter.

Another copepod, *Pseudomyicola spinosus*, in the same family as the *Mytilicola* species already described, is distributed in the Atlantic, Pacific and western Indian oceans, and is found in the mantle cavities of more than 50 species of bivalves around the world (Cáceres-Martínez *et al.* 2005). It infects the mantle, gills and connective tissue of the digestive tract of *M. galloprovincialis* (Figure 11.14) and *M. californianus* from coastal Baja California, NW Mexico (Cáceres-Martínez *et al.* 1996). Its impacts are similar to those described for *M. intestinalis*. Copepod prevalence in *M. galloprovincialis* was 93%, and its number ranged from 0 to 20 (mean 5) in the mantle and gills and from 0 to 5 (mean 1.1) in the digestive gland, while in *M. californianus* prevalence was 43% and its number ranged from 0 to 2 (mean 0.4) in the mantle and gills and from 0 to 1 (mean 0.1) in the digestive gland (Cáceres-Martínez & Váquez-Yeomans 1997). A subsequent study showed that intensity of infection was positively related to mussel shell length (Olivas-Valdez & Cáceres-Martínez 2002). Also, most infested mussels showed the lowest meat weight in relation to shell weight, although the reproductive stage may have affected this observation. The copepod has also been recorded in the mussels *M. edulis*, *P. perna* and *Septifer vergatus* (Carneiro-Schaefer *et al.* 2017 and references therein).

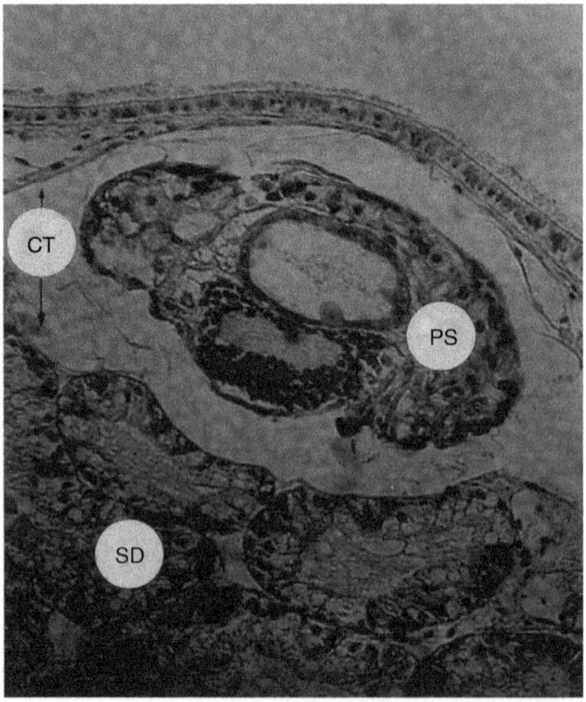

Figure 11.14 Section (5 μm) of the digestive gland of *Mytilus galloprovincialis* showing *Pseudomyicola spinosus* (PS) within connective tissue. Presumably, the copepod recently reached this tissue and there is not yet any haemocytic reaction against it. The connective tissue (CT) is disrupted and the secondary digestive tubules (SD) are compressed. Original magnification ×20. *Source:* From Cáceres-Martínez & Vásquez-Yeomans (1997). Reproduced with permission from Elsevier.

The copepod *Modiolicola gracilis*, family Lichomolgidae also inhabits the mantle cavity and gills of *M. galloprovincialis* and *M. californianus* in coastal Baja California. Cáceres-Martínez *et al.* (1996) recorded a maximum prevalence rate of 27% in *M. galloprovincialis* (up to five specimens per mussel) and of 70% in *M. californianus* (up to 15 per mussel) in Baja California, with a slight increase in prevalence during the autumn and winter. There were more parasites in larger mussels, and most parasitised mussels showed a low condition index, akin to *P. spinosus*, which could be considered a potential threat to mussel culture.

The little-known *Mytilicola porrecta* from the Gulf of Mexico infects *Geukensia* (= *Modiolus*) *demissa* (ribbed mussel), *Ischadium* (= *Mytilus*, = *Brachidontis*) *recurvum* (hooked mussel) and some species of clams and oysters. To date, there is no information on the relationship between the species and its bivalve hosts (Bower 2010d and references therein).

Another crustacean, the pea crab, family Pinnotheridae, is a common parasite of bivalves around the world and frequently causes problems for bivalve aquaculture through end-consumer complaints and rejections of consignments at international borders due to biosecurity concerns (Trottier *et al.* 2012). Not surprisingly, the most thoroughly studied Pinnotheridae are those invading commercially exploited bivalves on the North American and European Atlantic coasts, including members of the genera *Pinnotheres* and *Fabia* (Lauckner 1983). In European waters, the most common bivalve-invading crab, *Pinnotheres pisum*, inhabits the mantle cavity of *M. edulis* and *Modiolus modiolus* (Lauckner 1983). In New Zealand, *P. novaezelandiae* infects three mussel species: the non-native *Musculista senhousia* and the native species *P. canaliculus* and *Xenostrobus pulex* (Miller *et al.* 2008). Around the coasts of Japan, China and the Korean Peninsula, *P. sinensis* utilises the mussels *M. galloprovincialis* and *Septifer virgatus* as well as several oyster and clam species (Asama & Yamaoka 2009). *Pinnotheres maculatus* is found in *M. edulis*, *M. modiolus* and sometimes *M. californianus*, while *P. hickmani*, the most abundant pea crab species in Australia, cohabits with *M. planulatus*; the North American species, *Fabia subquadrata*, is found in *M. californianus*, *M. edulis* and *Modiolus modiolus* (Lauckner 1983).

Within the mantle cavity, the crab feeds on material collected by the gills of the host and causes gill erosion, weight loss through reduced filtration and consequent lowered oxygen consumption rate. Thus, there is lowered growth rate, distortion of shell shape, loss of condition and decreased reproductive output (Sun *et al.* 2006 and references therein). Females have a much more deleterious effect on mussel condition than males, probably due to their longer duration of parasitism and greater size (Sun *et al.* 2006). Increased mortalities have not been reported in infected hosts. Indeed, pea crabs may actually protect their hosts from predation! Campbell (1993) found that in selection experiments, starfish preferred to feed on uninfected mussels, although pea crabs usually abandoned their host during attacks. Because no free-living mature females have been found in nature, their association with a host is believed to be true parasitism, while for males, who can survive outside of their host, the association is seen as amensalism: a symbiotic relationship in which the mussel is harmed or inhibited and the pea crab is unaffected (Haines *et al.* 1994).

In some pea crab species (e.g. *P. maculatus*), males and females leave the host to copulate in open water, and only the females return (Lauckner 1983). However, in *P. sinensis*, for example, females are typically larger than males, so size is a limiting factor for escaping from and re-entering the host; therefore, mating probably occurs only inside the host (Asama & Yamaoka 2009). Mating for males is a challenging process. First, they must leave their host to search for females by squeezing through the gaping valves without being crushed. Once exit is successful, they must locate a female crab. A study on mate location and access behaviour of *Nepinnotheres (Pinnotheres) novaezelandiae* in *P. canaliculus* suggested that female crabs use a pheromone to attract mates and that once cued, male crabs follow consistent behavioural patterns associated with the exit from their own host and subsequent entry into the female crab's (Trottier & Jeffs 2015a; see their supplemental material for fascinating video recordings of a male pea crab locating its prospective mate). As many as several thousand eggs are produced by females (Trottier & Jeffs 2012), and after fertilisation and hatching of the eggs, larvae are released into the surrounding water and go through several development stages before they reach the final larval stage when the females are ready to settle and find their adult home. Settlement may be influenced by a number of sensory cues, including the presence of adult conspecifics (Welch *et al.* 1997), host odour (Stevens 1990) and acoustic cues (Stanley *et al.* 2009).

Over a 25-month period, Sun *et al.* (2006) recorded the prevalence of *P. senensis* in *M. galloprovincialis* in Shandong Province on the east coast of China. Prevalence of infestation varied from 2 to 62%, with the lowest and highest numbers occurring in July 2001 and November 2002, respectively (Figure 11.15). Seasonal periodicity in prevalence was not apparent. The number of pea crabs found in each mussel was from one to six, and most of the infested mussels were infested by a single female (53.8%), a single male (21.4%) or one female and one male (18.1%). The monthly average intensity varied from 1.00 to 1.44 (total average: 1.18).

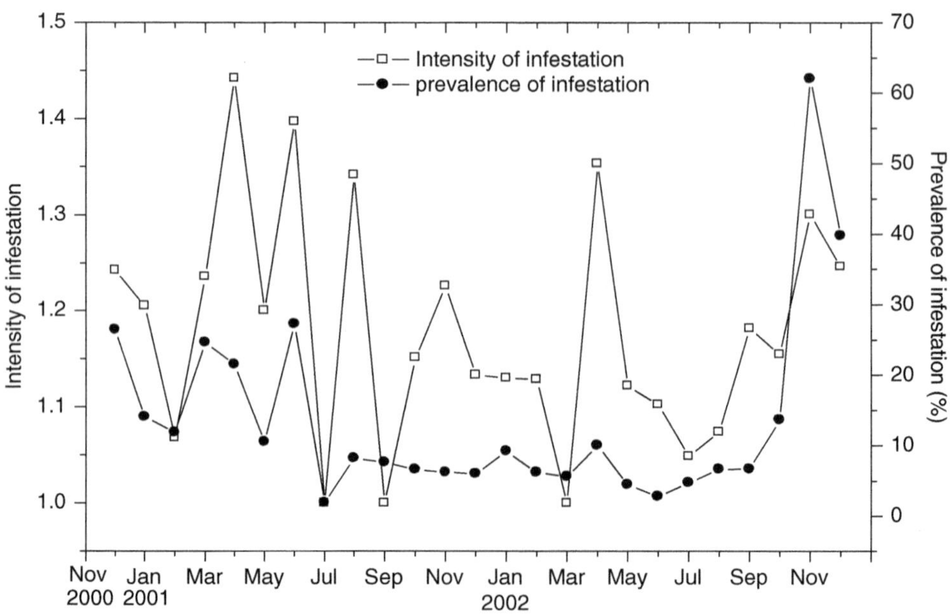

Figure 11.15 Seasonal variation of the prevalence and intensity of infestation by the pea crab *Pinnotheres sinensis* in *Mytilus galloprovincialis* over a two-year period. *Source:* From Sun *et al.* (2006). Reproduced with permission from Elsevier.

Trottier *et al.* (2012) quantified the financial impact of pea crabs on aquaculture production in a large-scale systematic sampling of a typical mussel (*P. canaliculus*) farm in New Zealand. They found that pea crabs were present in 5.3% (±0.062 SE) of the mussels and that abundance of crabs increased gradually with decreasing water depth beneath the farm and increasing distance from the shoreline. Mussels infected with pea crabs were 30% smaller in total wet weight, had a 29% lower meat yield and showed significantly altered shell dimensions. A total production loss on the 1.71 ha farm was estimated at 803 kg at harvest. When this measured loss was extrapolated to the current total mussel aquaculture production in New Zealand, pea crab infection represented an estimated loss to the industry of US$2.16 million annually, a significant loss of production in shellfish farming aquaculture operations even at low levels of infection. At present, there are no published reports on potential modifications that could be used to reduce infection from pea crabs, although Trottier *et al.*'s (2012) results suggest that moving mussel farms offshore has the potential to reduce the incidence of pea crabs in mussels and therefore warrants larger-scale assessment. Neither is there any published information on potential targets or methods of bio-control for the treatment of existing infections (Trottier *et al.* 2012; Trottier & Jeffs 2015b).

Neoplasia

Two predominant types of malignant neoplasia are common in marine bivalves. The first, disseminated neoplasia (DN), involves the excessive proliferation of abnormal cells with unknown origin (probably of a haemic source in some cases/species) through the circulatory system and their infiltration into the connective tissue of various organs (Figure 11.16b). DN is characterised by proliferative growth of abnormal haemocytes (2–4 times the diameter of normal haemocytes). The second, gonadal neoplasia, is typified by proliferation of undifferentiated germ cells (Figure 11.16a). These two forms of bivalve neoplasia fit the criteria of malignant tumours: pleomorphic (showing variation in cell size, shape and staining), undifferentiated cells, rapid and invasive growth, abundance of mitotic figures,[3] metastasis and progressive development often resulting in the death of the affected individual, although remission has been known to occur (Carballal *et al.* 2015; Meyer & Burton 2009). The disease has been reported in at least 15 bivalve species from four continents and every ocean except the Antarctic (Meyers & Burton 2009). Three species, all in Europe and North America, seem especially susceptible: the softshell clam *Mya arenaria*, the cockle *Cerastoderma edule* and the mussel *M. trossulus* (including those originally described as *M. edulis* in North America). Reports of DN are rare in other species (e.g. the Pacific oyster, *C. gigas*, which is distributed worldwide; Renault & Ford 2017). In most cases, prevalence is low (<5%), although figures as high as 40 and 90% have been reported for *M. trossulus* and *Mya arenaria*, respectively (Ciocan & Sunila 2005). For additional information on prevalence rates in *Mytilus* spp., see Barber (2004), Galimany & Sunila (2008), Cremonte *et al.* (2011), Gombač *et al.* (2013) and Carballal *et al.* (2015).

Gonadal neoplasia, or germinoma, is characterised by multiplication of germ cells to form gonadal germinomas (tumours), leading to abnormal gametogenesis and reduced fecundity, and sometimes ending in the death of the host (Bower 2006b). The disease primarily affects clam species, such as *Mercenaria arenaria* in Maine and *Mercenaria* spp. in

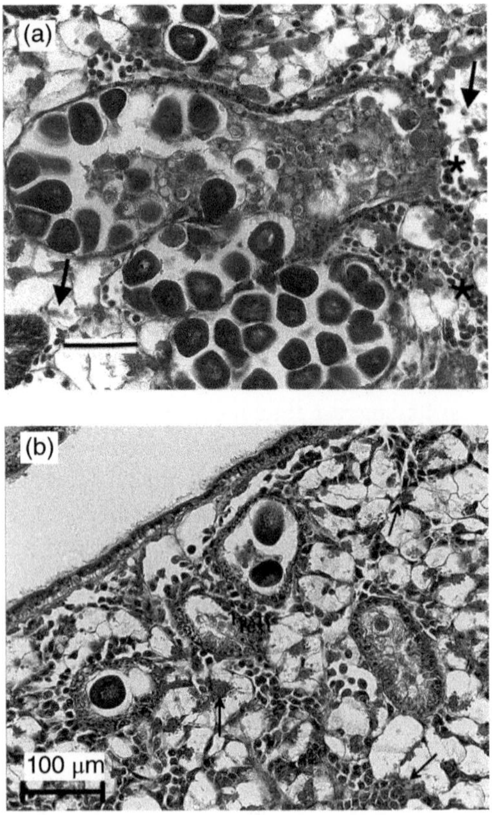

Figure 11.16 (a) Intermediate stage of gonadal neoplasia in *Mytilus chilensis* showing initial gonad atrophy and heavy infiltration by normal hemocytes (arrow) and neoplastic cells (asterisk). (b) Advanced stage of disseminated neoplasia (DN) showing almost complete gonadal atrophy and the presence of storage cells (arrow). Scale bar in (a) and (b) = 100 µm. *Source:* From Cremonte *et al.* (2011). Reproduced with permission from the *Journal of Shellfish Research.*

Florida, both in the United States, at prevalences up to 50% (Barber 2004). Most affected individuals are female. Limited field studies indicate that the disease progresses slowly and mortality rates are low. The major impact is most likely a reduction in reproductive effort. The disease has also been reported in *M. galloprovincialis* (Alonso *et al.* 2001) and the invasive mussel *Xenostrobus securis* (Pascual *et al.* 2010), as well as scallop and other clam species (Carballal *et al.* 2015 and references therein). Alonso *et al.* (2001) analysed 850 mussels (*M. galloprovincialis*) from NW Spain over a nonconsecutive two-year period; 6% showed pathological alteration in the development of germinal cells, which were considered gonadal neoplasms (see Figure 11.16a). All affected individuals were found between April and June, with mussels presenting with normal gonadal tissue during the rest of the year. Based on the absence of affected individuals for most of the year and prevalence rates of as high as 50% in certain months, Alonso *et al.* (2001) hypothesised a possible recovery of some individuals (see also Barber 2004).

The etiology of both types of neoplasia remains unknown. Factors such as environmental contaminants, biotoxins, physiological stressors and retroviral agents (but see Manso *et al.* 2012) have all been discounted (Krishnakumar *et al.* 1999; Alonso *et al.* 2001 and references therein; Barber 2004; Walker *et al.* 2011; AboElkhair *et al.* 2012). Bivalve species differ in their susceptibility to neoplasia. For example, *M. edulis* and *C. gigas* appear to be resistant to DN, yet the closely related species *M. trossulus* and *C. virginica* are susceptible;

therefore, a genetic etiology has been suggested (Van Beneden *et al.* 1999). Many genes regulating tumour proliferation are highly conserved throughout the animal kingdom. The genes most often implicated in vertebrate tumourigenesis are the *ras* oncogene and the *p53* tumour-suppressor gene. *Ras* genes encode proteins that play a central role in cell growth signaling cascades and are evolutionarily conserved. A large proportion and wide variety of experimentally or environmentally induced vertebrate tumours possess mutant forms of *ras*. The *p53* gene regulates the cell cycle and prevents neoplastic transformation (Galmany & Sunila 2008). Loss of *p53* function can lead to unchecked cell growth and contribute to tumour pathogenesis (Díaz *et al.* 2010). Combined, the mutational activation (*ras*) or inactivation (*p53*) of the two genes is observed in more than half of all human tumours. Ciocan & Rotchell (2005) cloned *Mytilus* sp. homologues of the vertebrate *ras* and *p53* genes. In a subsequent study, Ciocan *et al.* (2006) investigated neoplastic blood samples derived from *M. trossulus* for the presence of cancer gene mutations and changes in expression. The expression of *ras* was induced in neoplastic samples compared with normal samples. However, *p53* gene expression did not vary in neoplastic cells as compared with normal cells. In contrast, Muttray *et al.* (2010) reported that the combination of expression levels of both genes in the same species was significantly correlated with late-stage haemic neoplasia. Also, there have been several studies on integrated biological responses and tissue-specific expression of *p53* and *ras* genes in *Mytilus* spp. following exposure to genotoxicant contaminants (Lima *et al.* 2008; Di *et al.* 2011, 2017).

Transmissibility of DN has been demonstrated by injecting neoplastic cells, whole haemolymph or ova from diseased bivalves into healthy individuals (Collins & Mulcahy 2003; Barber 2004 and references therein). The disease is also transmitted when healthy individuals are placed in close proximity to infected ones of the same species (Elson *et al.* 1988). A viral cause has been suspected, but no infectious agent has been confirmed (Taraska & Böttger 2013; Walker *et al.* 2009). In contrast to DN, it has not been possible to transmit gonadal neoplasia through injection of neoplastic cells or proximity experiments; the reasons for this are unclear at present.

There is now good evidence that DN is a transmissible cancer. Such cancers are rare, having so far only been reported from dogs and Tasmanian devils (Murchison 2009). In an in-depth genetic analysis of cancer and host tissues from individual mussels (*M. trossulus*), cockles (*Cerastoderma edule*) and golden carpet shell clams (*Polititapes aureus*), Metzger *et al.* (2016) found that, in many cases, the genotype of cancer cells did not match that of the host, but instead was highly similar to that of cancerous tissues derived from other individuals of the same bivalve species. These findings confirm that many cases of DN in bivalve species are due to horizontal transfer of living cancer cells between hosts. An unexpected finding was that DNA extracted from cancer cells in *P. aureus* showed no genetic match with normal DNA from this species, but instead indicated that the cancer cells originated in a different species: the pullet shell clam (*Venerupis corrugata*). Surprisingly, *V. corrugata*, which shares a habitat with *P. aureus*, is not known to have a high prevalence of DN. It may have adapted to resist infection by the transmissible cancer, which first arose in a member of its own species, and the cancer has survived by engrafting to a new host species. The mechanisms whereby cancer cells are transmitted within and between species have yet to be elucidated. For further information on this topic, see Metzger *et al.* (2015) and Metzger & Goff (2016).

In a recent study, Yonemitsu *et al.* (2019) showed that bivalve transmissible neoplasia is able to spread worldwide, most likely through accidental transport of infected mussels on international shipping vessels. They investigated DN affecting geographically distant populations of two species of mussels, *M. chilensis* in South America and *M. edulis* in Europe. Sequencing alleles from four loci (two nuclear and two mitochondrial) provided evidence of transmissible cancer in both species. Phylogenetic analysis of cancer-associated alleles and analysis of diagnostic single nucleotide polymorphisms (SNPs) showed that cancers in both species likely arose in a third species of mussel, *M. trossulus*, but these cancer cells are independent of the previously identified transmissible cancer in *M. trossulus* from Canada. Unexpectedly, cancers from *M. chilensis* and *M. edulis* are nearly identical, showing that the same cancer lineage affects both. Thus, a single transmissible cancer lineage has crossed into two new host species and has been transferred across the Atlantic and Pacific oceans and between the Northern and Southern hemispheres.

Neoplasia is diagnosed using a variety of techniques, including histology and haemocytology, specific antibody staining and cytogenetics (reviewed in Carballal *et al.* 2015). Overexpression of some proto-oncogenes can also be used as a marker for DN, such as that of the *ras* gene in *M. trossulus* (Ciocan *et al.* 2006). Neoplastic haemocytes contain more DNA than normal cells, and thus their presence in a haemolymph sample can be quantified using a flow cytometer that counts and sorts cells based on DNA content. This method provides a more accurate and rapid diagnosis than more traditional ones (Delaporte *et al.* 2003). Distinctive morphology, new surface antigens, reduced phagocytic ability, greatly increased cell numbers and dissemination of cells into tissues are all characteristics of neoplastic haemocytes (Metzger *et al.* 2015 and references therein). Most bivalve DN cells are aneuploids ($2n + 1$), which explains the higher-than-normal DNA content (Carballal *et al.* 2015; Carella *et al.* 2017). Gross signs of gonadal neoplasia are generally not apparent except in advanced cases of the disease (Carballal *et al.* 2015). Definitive detection of gonadal neoplasia is based on histopathological examination of tissues. Since neoplastic germs cells are not prevalent in haemolymph, this eliminates haemocytology as a diagnostic method (Barber 2004).

Defence Mechanisms

All metazoan animals are protected against invading organisms or foreign substances by an internal defence system. The system is classified into two subsystems: the adaptive and the innate immune system. Adaptive immunity, which is only present in vertebrates, provides rapid and selective protection against a specific foreign body (antigen), but requires previous exposure to the antigen. A second contact with the same antigen produces an enhanced response in the form of specific antibodies, produced as a result of memory formation. The innate immune system, often referred to as the nonspecific immune system, allows an animal to combat foreign components without any previous contact with them. Although the innate system lacks the complexity of the adaptive system, recent studies have shown that bivalves, like all other invertebrates, possess a highly efficient defence system against infections. Interestingly, there is sparse evidence for specific memory in innate defence systems, potentially breaking down the rather artificial conceptual boundary between innate and adaptive immunity (Flajnik & Du Pasquier 2004; Kurtz 2005). The innate system in bivalves promotes generalised protection against pathogenic organisms (e.g. protozoa, bacteria, endoparasites and viruses), environmental stressors (e.g. xenobiotic assault, algal toxins) and other immune-modulating stimuli (e.g. air exposure, mechanical stress, high temperatures, extreme salinity conditions) (Bassim *et al.* 2015). The internal defence system in bivalves is mediated by both cellular (whole-cell) and humoral (cell-product) components, which operate in a coordinated way to provide protection against invading pathogens (Figure 11.17). In bivalves, phagocytosis and haemocytic infiltration are the primary cellular responses, while a multitude of haemolymph components such as lectins (agglutinins, opsonins), antimicrobial peptides (AMPs), toxic oxygen intermediates and lysosomal enzymes – either normally present or induced – constitute the humoral defence mechanism. Haemocytes are the first line of defence against parasites and pathogens. Although there is no single system of haemocyte classification, detailed morphological studies, along with investigations based on molecular characterisation of haemocyte cell surfaces (e.g. lectin-binding and antigenic properties) or on haemocyte production of specific chemicals or peptides, have gone some way in helping to identify subpopulations of these cells in bivalves (Yanyan *et al.* 2006; Aladaileh *et al.* 2007; García-García *et al.* 2008; Wang *et al.* 2012; Lin *et al.* 2013; Rebelo *et al.* 2013; Tame *et al.* 2015; Sekine *et al.* 2016; see Chapter 7).

Genomics of Bivalve Immunity

Genomic tools and resources are essential in order to make significant progress toward an understanding of the mechanisms and effectors of bivalve immunity. Several expressed sequence tag (EST) collections from complementary DNA (cDNA) or subtractive hybridisation (SSH) libraries have been generated in the last few decades for a number of bivalves (see Chapter 9). These libraries are powerful tools for the study of differential gene expression and the identification of genes involved in specific biological functions, especially in organisms where genomic data are not available (Romero *et al.* 2012). The first systematic production and annotation of ESTs was performed in *M. galloprovincialis* (Venier *et al.* 2003). A small number of ESTs are homologous to immune-related genes, such as the β-glucan-binding protein, the lipopolysaccharide (LPS)-binding protein, lectins, AMPs and some proteasome components (protein complexes in the cell nucleus and cytoplasm). These were combined with other cDNA and SSH libraries (Pallavicini *et al.* 2008) from bacteria-stimulated *M. galloprovincialis*. Venier *et al.* (2009) constructed, sequenced and annotated 17 cDNA libraries from different Mediterranean mussel tissues (gills, digestive gland, foot, anterior and posterior adductor muscle, mantle and haemocytes) challenged with toxic pollutants, temperature variations and potentially pathogenic bacteria. They assembled these libraries into a large mussel database of transcripts expressed in *M. galloprovincialis*, MytiBase, which provided the opportunity to look for gene sequences relevant to host defence, especially innate immunity-related genes (Venier *et al.* 2009). EST collections contain genes that are modulated in response to environmental, chemical or biological stimuli and can be used to design probes in microarrays. Microarrays (microchips) consist of a small membrane or glass slide containing samples of many genes arranged in a regular pattern. They work by exploiting the ability of a given mRNA molecule to bind

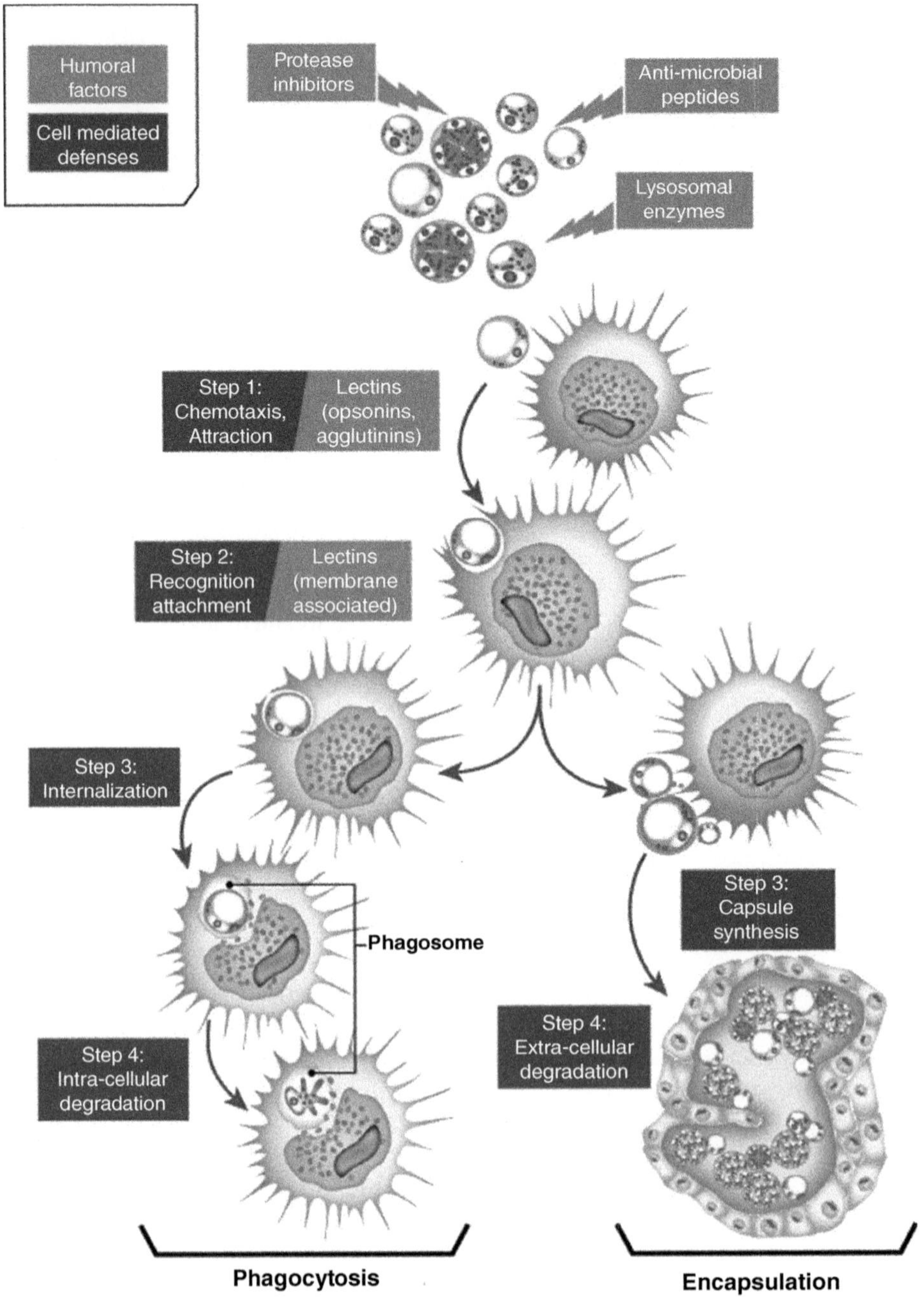

Figure 11.17 Schematic presentation of humoral and cellular responses involved in bivalve defence mechanisms upon infection by micro-organisms. **Blue** indicates steps of phagocytosis and encapsulation, **green** indicates the various humoral factors such as antimicrobial peptides (AMPs), lysosomal hydrolysases and protease inhibitors that also contribute to eliminate pathogenic micro-organisms, **purple** symbolises the invading micro-organisms. Step 1: Chemotaxis, attraction and migration. Step 2: Recognition and attachment of invading micro-organisms. These two steps involve the presence and synthesis of lectins. Step 3: Internalisation of micro-organisms within a phagosome. Step 4: Destruction of the micro-organisms by oxygen-dependent and oxygen-independent microbicidal activities. These four steps constitute phagocytosis. However, at Step 3, when haemocytes fail to phagocytose the micro-organisms (possibly because they are too big for internalisation), they encapsulate them and destroy them externally (Step 4) with lysosomal enzymatic and reactive oxygen species (ROS) activities. *Source:* From Soudant *et al.* (2013). Reproduced with permission from Elsevier. (*See colour plate section for colour representation of this figure*).

specifically to the DNA template from which it originated. By using an array containing many DNA samples, the expression levels of hundreds or thousands of genes within a cell can be determined by measuring the amount of mRNA bound to each site on the array, generating a profile of gene expression in the cell (see Chapter 9). Tens of thousands of ESTs are currently available for *M. californianus* and *M. galloprovincialis*, and DNA microarrays have also been developed. The Immunochip 1.0 microarray, for example, specifically includes 1820 probes of genes centrally involved or modulated in the innate immune responses of *M. galloprovincialis* (Domeneghetti *et al.* 2011). Beside DNA microarrays, next-generation sequencing technologies now offer new and broader research perspectives, from whole-transcriptome coverage to sequencing of the *Mytilus* genome (see Chapter 9).

Cellular Defence Components of the Immune System

When a pathogen invades a host, a number of reactions are set in motion, initiated both by the pathogen in its attempt to survive and by the host, which tries to destroy the invader. Cellular responses involve the mobilisation and recruitment of haemocytes to the site of infection. There are four types of defence mechanism utilised by bivalve haemocytes: haemocytosis, phagocytosis, encapsulation and nacrezation. Haemocytosis is a first response to infection and involves a measurable increase in the numbers of circulating haemocytes, which then infiltrate infected or injured tissues. Phagocytosis is the next step in the defence process and can be divided into a number of stages: chemotaxis, attachment, endocytosis and degradation (Cheng 1996). Chemotaxis, which is still poorly understood in bivalves, involves the directed movement of haemocytes toward the target, probably through chemo-attractant substances secreted by it. In mussels, lipopolysacharide molecules from bacteria stimulate chemotaxis in haemocytes of *M. edulis* (Schneeweiß & Renwrantz 1993), while Lopez-Cortes *et al.* (1999) have shown that extracellular proteins of *Vibrio tapetis* strains are chemoattractants for clam (*Tapes* spp.) haemocytes. After non-self-recognition, adhesion between haemocytes and foreign materials occurs either directly through membrane-bound receptors (see later) or indirectly through dissolved recognition factors present in the plasma, extrapallial fluid or mucus (Allam & Raftos 2015). At the site of adhesion, the foreign material is taken into the cell by endocytosis (Figure 11.17) and enclosed in a vesicle called a phagosome. Granulocytes are large cells (11.4–18 μm diameter) with a very abundant granular cytoplasm, and are the primary haemocyte cell type in phagocytosis (Kuchel *et al.* 2010). Soon after engulfment, membrane-bound granules within the haemocyte migrate to the surface of the phagosome and fuse with the phagosomal membrane, resulting in the deposition of granular contents on to the surface of the ingested material and a rapid decrease in the pH of the phagosome (Allam & Raftos 2015). The phagosome fuses with a lysosome to form a secondary phagosome or phagolysosome. The fate of the ingested material varies. Numerous enzymes within the phagolysosome degrade digestible material, and nutrients then diffuse into the cytoplasm, providing the host with a supplementary source of nourishment. Indigestible material remains within the cell as residual bodies – vesicles structures – which are eventually discharged into the haemolymph through exocytosis. Phagocytosis is accompanied by a diversity of killing methods (see later). When the invading organism or particle is too large to be phagocytosed (e.g. parasites or fungal spores), haemocytes are recruited in large numbers to surround and encapsulate it, and cytotoxic products such as degradative enzymes and free radicals are released by the haemocytes (Carella *et al.* 2015). Disintegration of the foreign material, followed by resorption of cellular debris, may follow. Nonspecific electrostatic forces and humoral plasma factors (see later) have a synergistic role in haemocyte attachment and encapsulation. In bivalves, positively charged targets are more vigorously encapsulated by haemocytes, with their predominance of negative charges, than are negatively charged or uncharged targets (Wootton *et al.* 2006; Jayaraj *et al.* 2009; Meena *et al.* 2010). Another defence mechanism, and one which is distinct from encapsulation, is nacrezation, whereby nacre is laid down around irritants such as parasites and pathogens, as well as abiotic irritants that invade the space between mantle and shell. An example of nacrezation is pearl formation by *Mytilus* in response to invasion by the trematode parasite *Gymnophallus bursicola* (see earlier).

Immune Recognition

Immune recognition is the first step in the immune response, allowing the discrimination of self/non-self substances. Antigens, which are molecule capable of inducing an immune response by the host cell, are recognised using pattern-recognition receptors (PRRs), which can be either soluble or expressed on the cell surface (Song *et al.* 2010; Bassim *et al.* 2015). These receptors selectively recognise a large family of conserved foreign molecules in micro-organisms called pathogen-associated molecular patterns (PAMPs). Examples include LPSs, lipoproteins, peptidoglycans (PGNs), lipoteichoic acids, viral double-stranded RNA, unmethylated bacterial DNA, zymosans and heat-shock proteins (HSPs) (Costa *et al.* 2009). Several groups of distinct PRRs have been identified in bivalves, including lectins, galectins, Toll-like receptors (TLRs), fibrinogen-related proteins (FREPs), glucan-binding proteins (GNBPs), peptidoglycan recognition proteins (PGRPs), scavenger receptors (SRs), C1q domain[4]-containing (C1qDC) proteins, membrane attack complex and perforin (MACPF) and thioester-containing proteins (TEPs). Despite the large amount of information on the role of these receptors in oysters and clam immunity, the emphasis in the following section will be on the part they play in mussel immunity.

Lectins and other carbohydrate PRRs recognise characteristic carbohydrates and glycoproteins located on cell walls of bacteria and other microbes as PAMPs. They bind, agglutinate and opsonise microbes to promote phagocytosis and destruction. C-type (calcium-dependent) lectin (CTL) is the most common lectin found in molluscs, and up-regulation of its expression has been documented in the gills of the deep-vent mussel, *Bathymodiolus azoricus*, when exposed to *Vibrio parahaemolyticus* (Bettencourt *et al.* 2014). Galectins constitute a conserved and widely distributed lectin family that specifically binds β-galactoside sugars. They display a remarkable functional diversity, participating in developmental processes, cell adhesion and motility, regulation of immune homeostasis and recognition of glycans on the surface of viruses, bacteria and protozoan parasites (details in Vasta *et al.* 2015). The galectin GalNAc/Gal-specific lectin was found to inhibit *in vitro* growth of the fungus *Pichia pastoris* in the mussel *Crenomytilus grayanus* (Chikalovets *et al.* 2015), while a previous study reported that the same lectin in the same species displayed antibacterial activity, which suggested it might be involved in the recognition and clearance of bacterial pathogens (Kovalchuk *et al.* 2013). In *M. trossulus*, the same lectin, designated as MTL, exhibited antibacterial and antifungal activity (Chikalovets *et al.* 2016).

TLRs have been identified and implicated in immune responses in all molluscs studied to date, including mussels (Guo & Ford 2016). TLRs are transmembrane proteins with ectodomains containing leucine-rich repeats that mediate the recognition of PAMPs, transmembrane domains and intracellular Toll-interleukin 1 (IL-1) receptor domains required for downstream signal transduction (see later). They are largely divided into two subgroups, depending on their cellular localisation and respective PAMP ligands. One group is composed of TLR1, TLR2, TLR4, TLR5, TLR6 and TLR11, which are expressed on cell surfaces and recognise mainly microbial membrane components such as LPS. The other group is composed of TLR3, TLR7, TLR8 and TLR9, which are expressed exclusively in intracellular vesicles such as the ER, endosomes, lysosomes and endolysosomes, where they recognise microbial nucleic acids (Kawai & Akira 2010). Stimulation of receptors by microbial products leads to the activation of signaling pathways, resulting in the induction of antimicrobial genes and inflammatory cytokines (Janeway 2002). Toubiana *et al.* (2013) identified 23 TLRs and 3 MyD88 (a universal adaptor protein that plays a role in the Toll-signaling pathway, see later) in *M. galloprovincialis*, all of which were expressed in digestive gland, although only four were also present in haemocytes; all MyD88s were expressed in all the tissues. in vivo challenge of *M. galloprovincialis* with bacteria (*Vibrio* spp.) resulted in significant up-regulation of one TLR and all MyD88s at three hours post-infection, but the effect was still evident after 24 hours (discussed in Martins *et al.* 2014).

FREPs are lectin-like proteins that incorporate fibrinogen and immunoglobulin superfamily-related domains. FREP-like molecules have been characterised in oysters, scallops and the mussels *M. galloprovincialis*, *M. edulis* and *M. californianus* (Allam & Raftos 2015). In mussels, FREPs are encoded by multigene families, with two genes in *M. edulis*, four in *M. californianus* and seven in *M. galloprovincialis* (Gorbushin & Iakovleva 2011). Analysis of FREP sequences in *M. galloprovincialis* has revealed that different mussels do not share the same sequence (Romero *et al.* 2011). Indeed, more than 150 expressed FREPs have been detected in the transcriptome[5] of this species (Gerdol & Venier 2015). This high FREP variability seems to indicate an extraordinary innate complexity and a potential mechanism to fight pathogens.

GNBPs bind Gram-negative bacteria, LPS and β-1,3-glucan (Song *et al.* 2010). A β-1,3-glucan-binding protein (β GBP) was purified from the plasma of *P. viridis* with an inherent serine protease activity (Jayaraj *et al.* 2008). The species agglutinated bakers' yeast, bacteria and erythrocytes, and enhanced the prophenoloxidase activity of the plasma. Similar phenoloxidase activity has been implicated with intracellular phagolysosomal processes in *M. edulis* (Renwrantz *et al.* 1996) and *M. galloprovincialis* (Carballal *et al.* 1997). In *M. edulis* three different GNBPs with a glyco-hydrolase-16 domain have been found (Philipp *et al.* 2012). Also, the *M. edulis* transcriptome contains four LPS- and β-1,3-glucan-binding (LBP) transcripts, one of which is most highly expressed in haemocytes (Philipp *et al.* 2012).

PGRPs are PRRs for PGN, a major component of bacterial cell walls. High levels of PGRP transcripts were found in gill tissue of *M. edulis* as compared to other tissues and haemocytes. Gene expression was stimulated by bacterial infection or PGN challenge, whereas LPS stimulation did not result in significant up-regulation (Philipp *et al.* 2012 and references therein). High expression of PGRPs has also been reported in the gills of the deep-sea mussels *Bathymodiolus azoricus* (Bettencourt *et al.* 2010), *B. platifrons* (Wong *et al.* 2015) and *B. septemdierum* (Ikuta *et al.* 2019), making them interesting candidates for studying the mechanisms underlying their relationship with gill endosymbiont bacteria. Interestingly, a high correlation was shown between the loss of symbionts and the down-regulation of PGRPs in the gills of *B. azoricus* (Détrée *et al.* 2017). Three full-length PGRP transcripts (two membrane-bound and one extracellular) have been reported in *M. galloprovincialis* (Gerdol & Venier 2015). The expression pattern of the extracellular protein PGRP3 has been comparatively investigated in *M. galloprovincialis* and *B. azoricus*, evidencing an up-regulation 12 and 24 hours after bacterial (*Vibrio* spp.) challenge in the former species, whereas lower expression levels were attained in the latter, indicative of different mechanisms of recognition and capacity to counteract bacterial challenges in these two mussels (Martins *et al.* 2014). However, data concerning the regulation of membrane-bound PGRPs are still completely missing in mussels.

Scavenger receptor cysteine-rich receptors (SRCRs) contain a conserved domain, 100–110 amino acids long and stabilised by disulphide bridges, which is associated with a number of additional domains and co-receptors (Gerdol & Venier 2015) – features that confer a broad range of functions, from lipoprotein binding to cell transport to pathogen clearance. In the available bivalve transcriptomes, the number of SRCRs appears relatively high, but these receptors still await a detailed characterisation (McDowell *et al.* 2014). The only well-characterised bivalve SRCR is from the scallop *Azumapecten farreri* and has been shown to bind to ligands such as LPS, PGN, β-glucan and the polysaccharides mannan and zymosan, found in plant and yeast cell walls, respectively (Liu *et al.* 2011).

C1q domain-containing (C1qDC) proteins are a highly diverse group of molecules playing a lectin-like role as PRRs, which provide tailored responses to a broad range of pathogens including Gram-positive and Gram-negative bacteria, Rickettsia-like organisms, fungi, protists and even metazoan parasites (Gerdol *et al.* 2015). C1q is regarded as a major connecting link between innate and acquired immunity due to the highly adaptive binding properties of its trimeric globular domain, gC1q (Gerdol *et al.* 2011). Following next-generation sequencing, more than 500 and 200 C1qDC transcripts were reported in *M. edulis* and *M. galloprovincialis*, respectively (Gerdol *et al.* 2014; Philipp *et al.* 2012). Some transcripts in *M. galloprovincialis* are hsemocyte-specific, while others are highly expressed in digestive gland and gills (Gerdol *et al.* 2015 and references therein). Changes in the expression of bivalve C1qDCs have also been associated with response to heavy metals, nanoparticles, pollutants and shell mineralisation, highlighting their involvement in non-immune functions. New evidence cofirms that a gene family expansion event occurred during the course of the evolution of Bivalvia, leading to the diversification of hundreds of different C1qDC genes in the gills of the subclasses Pteriomorphia and Heterodonta (see Chapter 1). The number of C1qDC genes largely exceeds 100, and may perhaps reach more than 1000 in certain mussels of the order Mytiloida (Gerdol *et al.* 2015).

TEPs comprise the vertebrate complement components C3, C4 and C5 as well as a number of invertebrate homologues, such as insect proteins, functioning as opsonins and promoting phagocytosis of bacteria and melanisation – the production and deposition of melanin pigments on invading pathogens and parasites. They have been identified in oyster and clam transcriptomes and are present in mussels, but to date their role in bivalve immunity has not been clarified (Gerdol & Venier 2015; Allam & Raftos 2015).

Humoral Defence Mechanisms

The recognition of pathogens by receptors triggers immune responses and the production of effectors that directly act on pathogens and foreign substances for their destruction and removal (Guo & Ford 2016). A large number of immune effectors have been identified in bivalves, including molecules such as lysozymes, agglutinins (lectins), antimicrobial factors and cytokines, and enzymes of lysosomal origin such as aminopeptidases, β-glucuronidases, acid phosphatase, alkaline phosphatase, α-mannosidase, esterases, ribonucleases, oxidases and peroxidases (Gestal *et al.* 2008; Guo *et al.* 2015; Gerdol & Venier 2015). The latter play a key role in degradation of foreign material in secondary phagosomes (see earlier).

Lysozymes and Proteases

Lysozymes are widely distributed antibacterial molecules in animals. They catalyse the hydrolysis of the β-1,4-glycosidic linkage between *N*-acetylmuramic acid and *N*-acetylglucosamine found in PGN bacterial cell walls, thereby causing bacterial cell lysis. Lysozymes are particularly active against Gram-positive bacteria and have a role in both digestion (see Chapter 4) and antibacterial defence (Gerdol & Venier 2015). Haug *et al.* (2004) demonstrated that in the mussel *Modiolus modiolus*, lysozyme could augment the activity of AMPs through a synergistic mechanism. There are three different lysozyme groups: C-type, G-type and I-type. The latter were until recently believed only to occur in invertebrates, and were the first to be characterised in bivalves. Studies on *M. galloprovincialis* revealed up-regulation of I-type lysozomes in response to experimental bacterial challenges, with Gram-positive bacteria (*Micrococcus lysodeikticus*) inducing a higher expression level than Gram-negative *Vibrio anguillarum*, suggesting that different signaling pathways (see later) may be involved (Li *et al.* 2008; Costa *et al.* 2009). Since these studies, C-type and G-type lysozymes have been reported in bivalves. C-type showed up-regulation in *M. galloprovincialis* in experimental challenges with both Gram-positive and Gram-negative bacteria (Wang *et al.* 2013). G-type lysozymes have been reported in scallops, and two G-types, which play different roles in digestion and immune defence, have been identified in *M. galloprovincialis* (Wang *et al.* 2012). These findings show, for the first time, that bivalve G-type lysozymes underwent adaptive evolution from self-defence to digestion.

Proteases and their regulation by different types of proteinase inhibitors play a significant role in immune response (Guo *et al.* 2015). A broad range of protease inhibitors can be expressed in bivalves to counteract the variants of proteases produced by invading microbes (Gerdol & Venier 2015). For example, in the oyster *C. virginica*, two serine protease inhibitors have strong affinity and inhibitory effects on proteases produced by the protistan pathogen *Perkinsus marinus*, which is responsible for Dermo disease in oysters (references in Guo *et al.* 2015). Mutations in one of the serine protease inhibitor genes are associated with Dermo and possibly MSX resistance in *C. virginica* (Yu *et al.* 2011; He *et al.* 2012). In *C. gigas*, metalloproteinases (MPs) are important to antiviral response against herpesvirus OsHV-1. All five differentially expressed MP genes induced by OsHV-1 infection were up-regulated, while several genes encoding tissue inhibitors of MPs and serine protease inhibitors were down-regulated (He *et al.* 2015), suggesting that proteinase activity is heightened, probably for the destruction of viral proteins or damaged host proteins. To my knowledge, there are no publications on the role of proteases in immune response of mussels.

Membrane Attack and Perforin Proteins

The terminal components of the complement system (see later) can form a protein complex – the membrane attack complex, or MAC – on the surface of Gram-negative bacteria, triggereing their lysis (Gerdol & Venier 2015). The MACPF superfamily is one of the largest families of pore-forming molecules; they play an important role in immune defence because they destruct invading microbes by creating holes in their cell walls (Bischofberger *et al.* 2009). Although MACPF proteins are able to destruct invading microbes, several play roles in embryonic development, neural migration or tumour suppression. Estévez-Calvar *et al.* (2011) described two apextrin-like proteins and one MACPF domain-containing protein (Macp) in *M. galloprovincialis*. The two apextrin-like proteins, which are important antibacterial effectors (see later), did not present any conserved domain, while the Macp protein contained the membrane/attack complex domain. The MACPF domain-containing protein Macp remains the only one available at the present time in public databases for a bivalve mollusc.

Antimicrobial Peptides

AMPs are key immune effectors and are widespread in prokaryotes and eukaryotes. They are classified in different subgroups based on their structure, amino acid composition and properties. The most comprehensive characterisation of AMPs was carried out in mussels, particularly *M. edulis* and *M. galloprovincialis* (Allam & Raftos 2015 and references therein). Diverse cysteine-rich AMPs have been reported in *Mytilus* spp., including defensins, big defensins, mytilins, myticins, mytimacins, mytimycins, myticusins and mytiCRPs (Gerdol & Venier 2015; Table 11.3). The peptides are produced in haemocytes, where they are stored as inactive AMP precursors; they are released as active peptides on microbial challenge (Mitta *et al.* 2000a). Precursors of defensin, mytilin, myticin and mytimycin can represent about 40% of the haemocyte transcriptome in *M. galloprovincialis* injected with viral-like and bacterial preparations (Rosani *et al.* 2011). The indiscriminate worldwide overuse and misuse of antibiotics has led to high rates of microbial resistance, posing new challenges to human health. The potential of marine peptides (e.g. AMPs) as novel antimicrobials against pathogenic bacteria is currently being explored (Kang *et al.* 2015; Ponnappan *et al.* 2015; Falanga *et al.* 2016; Semreen *et al.* 2018).

Cytokines

Cytokines are crucial mediators of immune response, inflammation and apoptosis (Guo & Ford 2015 and references therein; Romero *et al.* 2015). They are a large group of regulatory molecules comprising interleukins (IL), interferons (IFN), tumour necrosis factor (TNF) and chemokines. IL-1α, IL-1β, IL-6, IL-8 and TNF-α have been isolated in *M. edulis*. IL-1 and TNF-α influence the mobility of *M. edulis* haemocytes, and an increase in IL-1-like molecules has been shown during contact of haemocytes with a synthetic opioid (Tiscar & Mosca 2004 and references therein). Two other inflammatory cytokines have been identified in bivalves: the macrophage migration inhibitory factor (MIF), which is present with at least three distinct gene products in *M. galloprovincialis* (Parisi *et al.* 2012), and the

Table 11.3 List of cysteine-rich AMP families identified in *Mytilus* spp. Numbers in brackets are numbers of cystine residues.

AMP family	Domain organization	Role/Notes	Reference
defensins (arthropod-like) defensins	SP-CR (6) SP-CR-C-terminal extension[a] (8)	Antimicrobial activity, mostly against Gram-positive bacteria; occasionally Gram-negative bacteria and fungi.	Wong *et al.* (2007)
mytilins	SP-CR-C-terminal extension (8)	Eight different mytilin sequences have been identified in *M. galloprovincialis*, each with distinct specific activities against pathogens.	Venier *et al.* (2011) Mitta et al. (2000b)
myticins	SP-CR-C-terminal extension (8)	Three different myticins (A, B, C) in *M. galloprovincialis*. Marked activity against Gram-positive bacteria; much less active against Gram-negative bacteria and fungi. Also, myticin C peptide shows antiviral activity against the herpes virus OsHV-1 and is involved in response to tissue injury	Mitta *et al.* (1999) Balseiro *et al.* (2011) Novoa *et al.* (2016) Rey-Campos *et al.* (2019)
mytimacins (type-1) mytimacins (type-2) mytimacins (type-3)	SP-CR (8) SP-CR (10) SP-CR (12)	Five mytimacins have been described so far in *M. galloprovincialis*. They are not expressed in haemocytes, but instead are detected in the digestive gland, gills and mantle. May be used as mucous defence molecules, in a similar way as in land invertebrates.	Gerdol *et al.* (2012) Gerdol and Venier (2015)
big defensins	SP-propeptide-CR (6)	Highly active against Gram-positive bacteria, but not against Gram-negative bacteria or protistans. Eight BDs reported in *M. galloprovincialis*; these are localized, not in haemocytes, but in a wide range of tissues, similar to the mytimacins	Gerdol *et al.* (2012) Gerdol and Venier (2015)
mytimycins (type-1) mytimycins (type-2) mytimycins (type-3)	SP-CR-EF hand[b] (12) SP-CR-EF hand (12) SP-CR-EF hand (14)	Mytimycin (MytM), the first strictly antifungal protein from molluscs, was partially identified from *Mytilus edulis* in 1996. More recently 16 nucleotide sequences related to MytM were isolated from a *M. galloprovincialis* EST library	Charlet *et al.* (1996) Vernier *et al.* 2011 Cantet *et al.* (2012)
myticusin	SP-propeptide-CR (10)	In *Mytilus coruscus*, this peptide exhibited stronger antimicrobial properties against Gram-positive bacteria than Gram-negative bacteria, and fungi.	Liao *et al.* (2013)
MytiCRP-I MytiCRP-II MytiCRP-III MytiCRP-IV	SP-propeptide-CR (6) SP-CR (8) SP-CR (6) SP-CR (10)	This large family comprises over 50 members and appears to be exclusively present in the order Mytiloida. The family has been recently characterized in *M. galloprovincialis*. The role of these peptides has not been demonstated yet.	Gerdol and Venier (2015)

[a] SP-CR signal peptide; CR: cysteine-rich domain;

[a] The C-terminal extension may be missing in defensins. Numbers in brackets indicate number of cysteine residues.

[b] EF-hand is a helix-loop-helix structural domain or motif.

Source: Adapted from Gerdol and Venier (2015).

allograft inflammatory factor (AIF), which stimulates the phagocytic activity of granulocytes and has been identified in *M. galloprovincialis* and other bivalves (Gerdol & Venier 2015).

Oxidative Burst

Phagocytic activity by bivalve haemocytes triggers O_2 uptake and the generation of a variety of cytotoxic molecules, such as reactive oxygen species (ROS), which include superoxide anions (O_2^-), hydrogen peroxide (H_2O_2), singlet oxygen ($^1O_2$) and hydroxyl radicals (OH^-). All of these play an important role in the phagocyte-mediated killing of microorganisms (Guo *et al.* 2015). Respiratory burst can be determined by measuring the chemiluminescence (CL) of mussel haemocytes (Costa *et al.* 2009). Another radical, nitrous oxide (NO), is not toxic by itself but reacts with O_2^- to form the more potent oxidising species peroxynitrite ($ONOO^-$; Figure 11.18), which has the potential to cause severe cytoxic damage (Sun & Wu 2009). Excessive production of ROS may lead to oxidative stress, loss of cell function and, ultimately, apoptosis[6] or necrosis.[7] However, cells are protected against oxidative stress by an interacting network of antioxidant enzymes (Song *et al.* 2010). Major antioxidant enzymes include catalase (CAT), superoxide dismutase (SOD), glutathione peroxidase (GPX) and glutathione S-transferase (GST) (Song *et al.* 2010). CAT is one of the

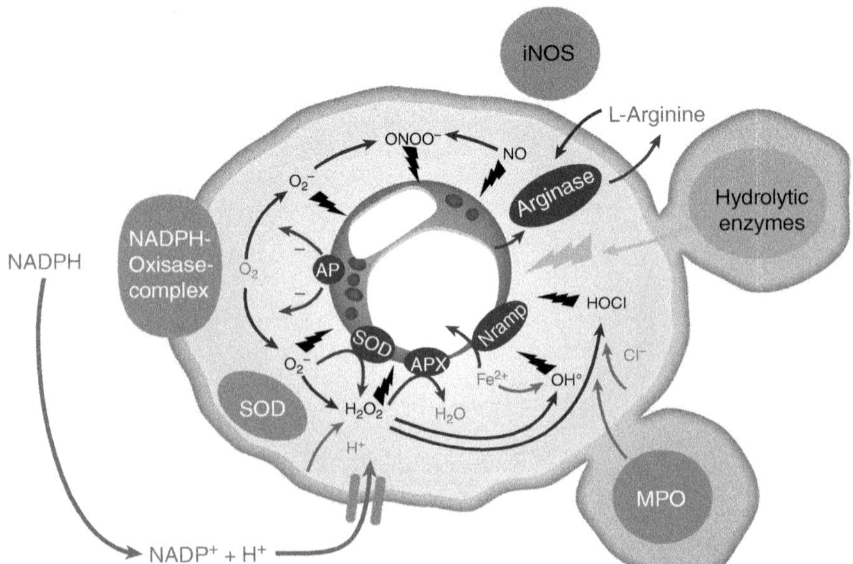

Figure 11.18 Antioxidant capacities of *Perkinsus marinus* upon phagocytosis by eastern oyster, *Crassostrea virginica*, hemocytes (phagosome is delimited by a **blue** line, phagocytised *Perkinsus* cell is **pink**). Toxic and nontoxic molecules are noted in **red** and **blue**, respectively. Pro-oxidant activities of haemocytes are represented in **orange**, and the anti-oxidant activities and mechanisms of *P. marinus* are represented in **purple**. Hydrolytic enzymes are represented in **green**. NO, nitric oxide; $ONOO^-$, peroxynitrite, O_2^-, superoxide anion; HOCL, hypochloride; iNOS, inductible nitric oxide synthase; SOD, superoxide dismutase; MPO, myeloperoxidase; AP, acid phosphatase; APX, ascorbate dependent peroxidase; Nramp, natural resistance-associated macrophage protein (proposed to contribute to the depletion of iron in haemocytes). *Source:* From Soudant *et al.* (2013). Reproduced with permission from Elsevier. (*See colour plate section for colour representation of this figure*).

main enzymes involved in scavenging the high level of ROS. It catalyses the decomposition of hydrogen peroxide to gaseous oxygen and water molecules. SODs catalyse the dismutation[8] of superoxide into oxygen and hydrogen peroxide. GPX reduces lipid hydroperoxides to their corresponding alcohols, and reduces free hydrogen peroxide to water, while GST catalyses the conjugation of glutathione to a wide variety of hydrophobic compounds through the formation of a thioether bond with their electrophilic centre. Activity of these enzymes has been detected in *Mytilus* spp., as well as in other bivalves (reviewed by Song *et al.* 2010).

Apart from antioxidant defence, bivalves activate various cryoprotective proteins, such as HSPs, that bind to damaged or misfolded proteins, promoting repair or elimination (Bassim *et al.* 2015 and references therein). Significant synthesis of HSP 70 was demonstrated in *M. galloprovincialis* haemocytes during *Vibrio alginolyticus* phagocytosis, while *Escherichia coli* induced only a small increase (Tiscar & Mosca 2004). These results are explained by the different phagocytosis rates derived from the different mollusc behaviour on interaction with bacteria coming from dissimilar environments, and confirm the existing relationship between phagocytosis and HSP production.

Signal Transduction Pathways

The immune recognition system triggers a wide range of signaling pathways and the production of specific reactive effectors, leading to neutralisation of dangerous threats through phagocytosis (Bassim *et al.* 2015). Examples are the TLR pathway, the Janus kinase/signal transducer and activator of transcription (JAK-STAT) pathway, the mitogen-activated protein kinase (MAPK) pathway, the nuclear factor κB (NF-κB) pathway and the complement system, all of which have been extensively studied in recent years for their important roles in regulating the immune system in both vertebrates and invertebrates. Details on the functioning of these and other signaling pathways in marine bivalves can be found in Song *et al.* (2010), Bassim *et al.* (2015) and Guo *et al.* (2015). The scenario emerging from these studies is that there is a high degree of conservation between molluscan and mammalian pathways (Canesi *et al.* 2006 and references therein). Compared to mussels, there is significantly more information on the components and mechanisms of action of signaling pathways in oysters, most likely due to their higher economic value, their propensity for serious infectious disease (reviewed in Gosling 2015) and the fact that whole genomes of two oyster species, *C. gigas* (Zhang *et al.* 2012) and *Pinctada fucata* (Takeuchi *et al.* 2016), have been sequenced. See Chapter 9 for additional information on complete or draft genomes in a range of marine bivalves. Despite the commercial importance and scientific interest in marine mussels, the number of genomic resources available in public databases is quite limited, and is usually restricted to their transcriptomes. For example, in *M. galloprovincialis*, some studies using ESTs and high-throughput cDNA (see Chapter 9) are available (Murgarella *et al.* 2016 and references therein). Nevertheless, there is some – albeit sparse – information on signaling pathways in mussels. In *M. galloprovincialis*, existence of a Toll pathway had been suspected since the analysis of sequences from MytiBase[9] (Venier *et al.* 2009; see earlier) and the response to *Vibrio splendidus* injection using a mussel immunochip (Venier *et al.* 2011). Subsequently, 15 transcripts of the Toll signaling pathway were identified in *M. galloprovincialis*, and most of the Toll pathway genes were expressed after in vivo injection with bacteria (Toubiana *et al.* 2014). Toubiana *et al.* (2014) concluded that mussels possess a complete Toll pathway that can be triggered by either Gram-positive or Gram-negative bacteria. The JAK/STAT pathway plays a major role in intracellular signal transduction of cytokine receptors (Shaposhnikov *et al.* 2013; see earlier). The functional presence of this pathway in bivalves was first documented in *M. galloprovincialis* following bacterial challenge, after which a cytokine response was recorded. Recent advancements in transcriptomics and data-mining pipelines have enabled the discovery of JAK/STAT homologues in *M. edulis* (Philipp *et al.* 2012). The MAPK signalling pathways are key players in a variety of processes, including cell proliferation, growth, differentiation, cell death, innate immunity and development (Bassim *et al.* 2015). Studies on *Mytilus*, involving different bacterial species and strains, heterologous cytokines and hormones, led to the identification of the role of conserved components of kinase-mediated transduction pathways, including cytosolic kinases (e.g. MAPKs, protein kinase C (PKC), phosphatidylinositol 3 kinase (PI-3K) and kinase-activated transcription factors such as STATs, cAMP response element-binding protein (CREB) and NF-κB), in the immune response (reviewed in Canesi *et al.* 2006; see also Betti *et al.* 2006; Ciacci *et al.* 2010). The NF-κB signaling pathway is activated rapidly in response to a wide range of stimuli, including pathogens, stress signals and pro-inflammatory cytokines, such as tumour-necrosis factor (TNF) and IL-1. In resting cells, NF-κB proteins are inhibited by IκB, but once a kinase enzyme catalyses degradation of IκB, the activated NF-κB proteins are translocated to the nucleus of the cell, where they bind to DNA and activate gene transcription (Li & Verma 2002; Moynagh 2005). A series of transcriptomic next-generation sequencing studies enabled the identification of interrelated NF-κB homologues responsible for the activation of TLRs in *M. edulis* (Philipp *et al.* 2012; Tanguy *et al.* 2013). The complement pathway functions through classical, alternative and lectin pathways, leading to the agglutination, opsonisation, chemotaxis and lysis of foreign cells. Although the complement pathway has not been widely studied in bivalves, data mining using first- and next-generation sequencing techniques for functional proteins containing C1q domain (see earlier) has revealed several conserved homologues in different bivalves. This strongly suggests the presence and conservation of a complement system in bivalves. Additional work on immune-related genes during ontogenesis revealed antigen recognition components of the complement pathway to be triggered as early as the trochophore stage in *M. galloprovincialis* (Balseiro *et al.* 2013).

Abiotic Effects on Mussel Immune Response

There is now a good body of evidence to show that various factors exert adverse effects on immunity, and this can affect resistance to infection, thereby influencing survival. A range of assays based on haemocyte counts, phagocytosis, degradative enzyme levels and release of reactive oxygen metabolites have been developed to measure the competence of the bivalve immune system. Such assays are also used as biomarkers of environmental perturbation (see Chapter 8). Polycyclic aromatic hydrocarbons (PAHs), which enter the marine environment from a range of industrial sources, cause an increase in haemocyte numbers but a decrease in cell viability, membrane stability and adhesion capability, as well as a decrease in lysozyme and phagocytic activity, depending on the concentration and duration of exposure of the species to the contaminant (reviewed in Girón-Pérez 2010;

Giannapas *et al.* 2012; Renault 2015). Similar effects have been reported for other pollutants, such as polychlorinated biphenyls (PCBs; Renault 2015), heavy metals (Boudjema *et al.* 2014; Burgos-Aceves & Faggio 2017), microplastics (Paul-Pont *et al.* 2016; Détrée *et al.* 2017), nanoparticles (Barmo *et al.* 2013; Canesi *et al.* 2015; Bouallegui *et al.* 2017; Burgos-Aceves & Faggio 2017), pesticides (Ray *et al.* 2015) and pharmaceutical drugs (Renault 2015; Burgos-Aceves & Faggio 2017). Other factors that modulate the bivalve immune system are temperature, pH and salinity fluctuations (Ellis *et al.* 2015; Mosca *et al.* 2013; Parisi *et al.* 2017), ocean acidification (Asplund *et al.* 2014; Sun *et al.* 2017), hypoxia (Mosca *et al.* 2013; Parisi *et al.* 2017), spawning (Lacroix *et al.* 2017) and toxic (Simões *et al.* 2015) and even natural (Mello *et al.* 2010) algal blooms.

Notes

1 An outbreak of disease affecting many animals of one kind at the same time.
2 An obligate parasite cannot complete its life cycle without exploiting a suitable host.
3 The spindle-shaped figure presented (as by the chromosomes, asters) during mitosis.
4 A protein domain is a conserved part of a given protein sequence and tertiary structure that can evolve, function and exist independently of the rest of the protein chain.
5 The sum total of all the messenger RNA molecules expressed from the genes of an organism.
6 A type of programmed cell death.
7 A process of premature tissue death as a result of disease or injury.
8 A process of simultaneous oxidation and reduction; used especially of compounds taking part in biological processes.
9 Joint international effort to build a database of transcripts expressed in *M. galloprovincialis* by using expressed sequence tag (EST) libraries, with a particular focus on immune-related sequences.

References

AboElkhair, M., Iwamoto, T., Clark, K.F., McKenna, P., Siah, A., Greenwood, S. *et al.* (2012) Lack of detection of a putative retrovirus associated with haemic neoplasia in the softshell clam *Mya arenaria*. *Journal of Invertebrate Pathology*, **109**, 97–104.

Aladaileh, S., Nair, S.V., Birch, D. & Raftos, D.A. (2007) Sydney rock oyster (*Saccostrea glomerata*) hemocytes: morphology and function. *Journal of Invertebrate Pathology*, **96**, 48–63.

Aldana, M., González, K., Loot, G., Pulgar, J. & Marquet, P.A. (2009) First intermediate host of the digenean trematode *Proctoeces lintoni* (Fellodistomidae) in Chile. *Journal of Parasitology*, **95**, 1408–1414.

Allam, B. & Raftos, D. (2015) Immune responses to infectious diseases in bivalves. *Journal of Invertebrate Pathology*, **131**, 121–136.

Alonso, A., Suarez, P., Alvarez, C., San Juan, F. & Molist, P. (2001) Structural study of a possible neoplasia detected in *Mytilus galloprovicialis* collected from the Ría of Vigo (NW Spain). *Diseases of Aquatic Organisms*, **47**, 73–79.

Ambariyanto & Seed, R. (1991) The infestation of *Mytilus edulis* Linnaeus by *Polydora ciliata* (Johnson) in the Conwy Estuary, North Wales. *Journal of Molluscan Studies*, **57**, 413–424.

Anguiano-Beltrán, C., Lizárraga-Partida, M.L. & Searcy-Bernal, R. (2004) Effect of *Vibrio alginolyticus* on larval survival of the blue mussel *Mytilus galloprovincialis*. *Diseases of Aquatic Organisms*, **59**, 119–123.

Antar, R & Gargouri, L. (2016) Morphology and molecular analysis of life-cycle stages of *Proctoeces maculatus* (Looss, 1901) (Digenea: Fellodistomidae) in the Bizerte Lagoon, Tunisia. *Journal of Helminthology*, **90**, 726–736.

Apeti, D.A., Kim, Y., Lauenstein, G., Tull, J. & Warner, R. (2014) *Occurrence of parasites and diseases in oysters and mussels of the U.S. coastal waters. National Status and Trends, the Mussel Watch monitoring program*. NOAA Technical Memorandum NOSS/NCCOS 182. Silver Spring, MD.

Arzul, I. & Joly, J.P. (2011) EURL (European Union Reference Laboratory) for molluscs diseases: *Marteilia refringens*. Available at: http://wwz.ifremer.fr/crlmollusc/Main-activities/Tutorials/Marteilia-refringens (accessed 28 May 2021).

Arzul, I., Nicolas, J.L., Davison, A.J. & Renault, T. (2001) French scallops: a new host for ostreid herpesvirus-1. *Virology*, **290**, 342–349.

Arzul, I., Chollet, B., Boyer, S., Bonnet, D., Gaillard, J., Baldi, Y. *et al.* (2014) Contribution to the understanding of the cycle of the protozoan parasite *Marteilia refringens*. *Parasitology*, **141**, 227–240.

Asama, H. & Yamaoka, K. (2009) Life history of the pea crab, *Pinnotheres sinensis*, in terms of infestation in the bivalve mollusc, *Septifer virgatus*. *Marine Biodiversity Records*, **2**, 1–5.

Asplund, M.E., Baden, S.P. & Russ, S. (2014) Ocean acidification and host–pathogen interactions: blue mussels, *Mytilus edulis*, encountering *Vibrio tubiashii*. *Environmental Microbiology*, **16**, 1029–1039.

Audemard, C., Barnaud, A., Collins, C.M., Le Roux, F., Sauriau, P.G., Coustau, C. *et al.* (2001) Claire ponds as an experimental model for *Marteilia refringens* life-cycle studies: new perspectives. *Journal of Experimental Marine Biology and Ecology*, **257**, 87–108.

Audemard, C., Le Roux, F., Barnaud, A., Collins, C., Sautour, B., Sauriau, P.G. *et al.* (2002) Needle in a haystack: involvement of the copepod *Paracartia grani* in the life-cycle of the oyster pathogen *Marteilia refringens*. *Parasitology*, **124**, 315–323.

Azevedo, C. & Matos, E. (1999) Description of *Nematopsis mytella* n.sp. (Apicomplexa), parasite of the mussel *Mytella guyanensis* (Mytilidae) from the Amazon Estuary and description of its oocysts. *European Journal of Protistology*, **35**, 427–433.

Balseiro, P. Montes, A., Ceschia, G., Gestal, C, Novoa, B. & Figueras, A. (2007) Molecular epizootiology of the European *Marteilia* spp., infecting mussels (*Mytilus galloprovincialis* and *M. edulis*) and oysters (*Ostrea edulis*): an update. *Bulletin of the European Association of Fish Pathologists*, **27**, 148–156.

Balseiro, P., Falco, A., Romero, A., Dios, S., Martínez-López, A., Figueras A. *et al.* (2011) *Mytilus galloprovincialis* myticin C: a chemotactic molecule with antiviral activity and immunoregulatory properties. *PLoS One*, 6, e23140.

Balseiro, P., Moreira, R., Chamorro, R., Figueras, A. & Novoa, B. (2013) Immune responses during the larval stages of *Mytilus galloprovincialis*: metamorphosis alters immunocompetence, body shape and behavior. *Fish and Shellfish Immunology*, **35**, 438–447.

Barber, B.J. (2004) Neoplastic diseases of commercially important marine bivalves. *Aquatic Living Resources*, **17**, 449–466.

Barmo, C., Ciacci, C., Canonico, B., Fabbri, R., Cortes, K., Balbi, T. *et al.* (2013) *in vivo* effects of n-TiO_2 on digestive gland and immune function of the marine bivalve *Mytilus galloprovincialis*. *Aquatic Toxicology*, **132–133**, 9–18.

Bassim, S., Genard, B., Gauthier-Clerc, S., Moraga, D. & Tremblay, R. (2015) Ontogeny of bivalve immunity: assessing the potential of next-generation sequencing techniques. *Reviews in Aquaculture*, **7**, 197–217.

Bayne, B.L., Gee, J.M., Davey, J.T. & Scullard, C. (1978) Physiological responses of *Mytilus edulis* L. to parasitic infestation by *Mytilicola intestinalis*. *Journal de Conseil International pour l'Exploration de la Mer*, **38**, 12–17.

Beaz-Hidalgo, R., Balboa, S., Romalde, J.L. & Figueras, M.J. (2010) Diversity and pathogenecity of *Vibrio* species in cultured bivalve molluscs. *Environmental Microbiology Reports*, **2**, 34–43.

Beers, C.D. (1959) Some observations on the autecology of the ciliate *Conchophthirus mytili*. *Journal of the Elisha Mitchell Scientific Society*, **75**, 3–10.

Berthe, F.C.J. (2002) *Pacem in terris pathogenibus bonae voluntatis*: mollusc-pathogens relationships prospects. *Bulletin of the European Association of Fish Pathologists*, **22**, 52–57.

Berthe, F.C.J., Le Roux, F., Peyretaillade, E., Peyret, P., Rodriguez, D., Gouy, M. *et al.* (2000) Phylogenetic analysis of the small subunit ribosomal RNA of *Marteilia refringens* validates the existence of Phylum Paramyxea (Desportes and Perkins, 1990). *The Journal of Eukaryotic Microbiology*, **47**, 288–293.

Berthe, F.C.J., Le Roux, F., Adlard, R.D. & Figueras, A. (2004) Marteiliosis in molluscs: a review. *Aquatic Living Resources*, **17**, 433–448.

Bettencourt, R., Pinheiro, M., Egas, C., Gomes, P., Afonso, M., Shank, T. *et al.* (2010) High-throughput sequencing and analysis of the gill tissue transcriptome from the deep-sea hydrothermal vent mussel *Bathymodiolus azoricus*. *BMC Genomics*, **11**, 559.

Bettencourt, R., Rodrigues, M., Barros, I., Cerqueira, T., Freitas, C., Costa, V. *et al.* (2014) Site-related differences in gene expression and bacterial densities in the mussel *Bathymodiolus azoricus* from the Menez Gwen and Lucky Strike deep-sea hydrothermal vent sites. *Fish and Shellfish Immunology*, **39**, 343–353.

Betti, M., Ciacci, C., Lorusso, L.C., Canonico, B., Falcioni, T., Gallo, G. *et al.* (2006) Effects of tumour necrosis factor a (TNFa) on *Mytilus* haemocytes: role of stress-activated mitogen-activated protein kinases (MAPKs). *Biology of the Cell*, **98**, 233–244.

Bhaby, S., Belhsen, O.K., Errhif, A. & Tojo, N. (2013) Seasonal dynamics of parasites on Mediterranean mussels (*Mytilus galloprovincialis*) and ecological determinants of the infections in southern Alboran area, Morocco. *International Journal of Parasitology Research*, **5**, 116–121.

Bignell, J.P. (2016) *Mytilicola intestinalis*, parasitism. ICES Identification Leaflets for Diseases and Parasites of Fish and Shellfish. Leaflet No. 24. (revised; original by V. Dethlefsen 1985).

Bischofberger, M., Gonzalez, M.R. & Van der Goot, F.G. (2009) Membrane injury by pore-forming proteins. *Current Opinion in Cell Biology*, **21**, 589–595.

Bishop, M.J., Carnegie, R.B., Stokes, N.A., Peterson, C.H. & Burreson, E.M. (2006) Complications of a non-native oyster introduction: facilitation of a local parasite. *Marine Ecology Progress Series*, **325**, 145–152.

Boehs, G., Villalba, A., Oliveira Ceuta, L. & Rocha Luz, J. (2010) Parasites of three commercially exploited bivalve mollusc species of the estuarine region of the Cachoeira river (Ilhéus, Bahia, Brazil). *Journal of Invertebrate Pathology*, **103**, 43–47.

Bouallegui, Y., Ben Younes, R., Turki, F. & Oueslati, R. (2017) Impact of exposure time, particle size and uptake pathway on silver nanoparticle effects on circulating immune cells in *Mytilus galloprovincialis*. *Journal of Immunotoxicology*, **14**, 116–124.

Boudjema, K., Kourdali, S., Bounakous, N., Meknachi, A. & Badis, A. (2014) Catalase activity in brown mussels (*Perna perna*) under acute cadmium, lead, and copper exposure and depuration tests. *Journal of Marine Biology*, **2014**, art. 830657.

Bower, S.M. (1992) Diseases and parasites of mussels. In: *The Mussel Mytilus: Ecology, Physiology, Genetics and Culture* (ed. E.M. Gosling), pp. 543–563. Elsevier Science Publishers B.V., Amsterdam.

Bower, S.E. & Figueras, A. (1989) Infectious diseases of mussels, especially pertaining to mussel transplantation. *World Aquaculture Review*, **20**, 89–93.

Bower S.M. (2001a) Synopsis of infectious diseases and parasites of commercially exploited shellfish: kidney coccidia of mussels. Available at: https://www.dfo-mpo.gc.ca/science/aah-saa/diseases-maladies/index-eng.html (accessed 28 May 2021).

Bower, S.M. (2001b) Synopsis of infectious diseases and parasites of commercially exploited shellfish: gregarine parasitism of mussels. Available at: https://www.dfo-mpo.gc.ca/science/aah-saa/diseases-maladies/index-eng.html (accessed 28 May 2021).

Bower, S.M. (2001c) Synopsis of infectious diseases and parasites of commercially exploited shellfish: *Sphenophrya*-like ciliates of mussels. Available at: https://www.dfo-mpo.gc.ca/science/aah-saa/diseases-maladies/scmu-eng.html (accessed 28 May 2021).

Bower, S.M. (2001d) Synopsis of infectious diseases and parasites of commercially exploited shellfish: shell-burrowing sponges of oysters. Available at: https://www.dfo-mpo.gc.ca/science/aah-saa/diseases-maladies/sbsoy-eng.html (accessed 28 May 2021).

Bower, S.M. (2001e) Synopsis of infectious diseases and parasites of commercially exploited shellfish: mussel gill turbellaria. Available at: https://www.dfo-mpo.gc.ca/science/aah-saa/diseases-maladies/index-eng.html (accessed 28 May 2021).

Bower, S.M. (2004) Synopsis of infectious diseases and parasites of commercially exploited shellfish: mycotic periostracal sloughing of mussels. Available at: https://www.dfo-mpo.gc.ca/science/aah-saa/diseases-maladies/mpsmu-eng.html (accessed 28 May 2021).

Bower, S.M. (2006a) Synopsis of infectious diseases and parasites of commercially exploited shellfish: nocardiosis of oysters. Available at: https://www.dfo-mpo.gc.ca/science/aah-saa/diseases-maladies/index-eng.html (accessed 28 May 2021).

Bower, S.M. (2006b) Synopsis of infectious diseases and parasites of commercially exploited shellfish: haemocytic neoplasia of clams. Available at: https://www.dfo-mpo.gc.ca/science/aah-saa/diseases-maladies/index-eng.html (accessed 28 May 2021).

Bower, S.M. (2007a) Synopsis of infectious diseases and parasites of commercially exploited shellfish: Haplosporidian infection of mussels. Available at: https://www.dfo-mpo.gc.ca/science/aah-saa/diseases-maladies/index-eng.html (accessed 28 May 2021).

Bower, S.M. (2007b) Synopsis of infectious diseases and parasites of commercially exploited shellfish: *Ancistrum mytili* gill ciliate of mussels. Available at: https://www.dfo-mpo.gc.ca/science/aah-saa/diseases-maladies/amgcmu-eng.html (accessed 28 May 2021).

Bower, S.M. (2009a) Synopsis of infectious diseases and parasites of commercially exploited shellfish: *Proctoeces maculatus* trematode disease of mussels. Available at: https://www.dfo-mpo.gc.ca/science/aah-saa/diseases-maladies/mtdmu-eng.html (accessed 28 May 2021).

Bower, S.M. (2009b) Synopsis of infectious diseases and parasites of commercially exploited shellfish: cestode parasitism of oysters. Available at: https://www.dfo-mpo.gc.ca/science/aah-saa/diseases-maladies/index-eng.html (accessed 28 May 2021).
Bower, S.M. (2009c) Synopsis of infectious diseases and parasites of commercially exploited shellfish: *Mytilicola intestinalis* (red worm disease) of mussels. Available at: https://www.dfo-mpo.gc.ca/science/aah-saa/diseases-maladies/mirwdmu-eng.html (accessed 28 May 2021).
Bower, S.M. (2010a) Synopsis of infectious diseases and parasites of commercially exploited shellfish: virus infection of clams. Available at: https://www.sciencedirect.com/science/article/abs/pii/0959803094900280 ().
Bower, S.M. (2010b) Synopsis of infectious diseases and parasites of commercially exploited shellfish: brown ring disease of Manila clams. Available at: https://www.dfo-mpo.gc.ca/science/aah-saa/diseases-maladies/index-eng.html (accessed 28 May 2021).
Bower, S.M. (2010c) Synopsis of infectious diseases and parasites of commercially exploited shellfish: *Mytilicola orientalis* (red worm) of mussels. Available at: https://www.dfo-mpo.gc.ca/science/aah-saa/diseases-maladies/morwmu-eng.html (accessed 28 May 2021).
Bower, S.M. (2010d) Synopsis of infectious diseases and parasites of commercially exploited shellfish: *Mytilicola porrecta* (red worm) of mussels. Available at: https://www.dfo-mpo.gc.ca/science/aah-saa/diseases-maladies/mprwmu-eng.html (accessed 28 May 2021).
Bower, S.M. (2011a) Synopsis of infectious diseases and parasites of commercially exploited shellfish: Marteiliosis (Aber Disease) of oysters. Available at: https://www.dfo-mpo.gc.ca/science/aah-saa/diseases-maladies/madoy-eng.html (accessed 28 May 2021).
Bower, S.M. (2011b) Synopsis of infectious diseases and parasites of commercially exploited shellfish: Marteiliosis (Aber Disease) of mussels. Available at: https://www.dfo-mpo.gc.ca/science/aah-saa/diseases-maladies/index-eng.html (accessed 28 May 2021).
Bower, S.M. (2011c) Synopsis of infectious diseases and parasites of commercially exploited shellfish: *Bonamia ostreae* of oysters. Available at: https://www.dfo-mpo.gc.ca/science/aah-saa/diseases-maladies/index-eng.html (accessed 28 May 2021).
Bower, S.M. (2012): Synopsis of infectious diseases and parasites of commercially exploited shellfish: *Steinhausia mytilovum* (mussel egg disease). Available at: https://www.dfo-mpo.gc.ca/science/aah-saa/diseases-maladies/medmu-eng.html (accessed 28 May 2021).
Bower, S.M. (2013) Synopsis of infectious diseases and parasites of commercially exploited shellfish: intracellular ciliates of mussels. Available at: https://www.dfo-mpo.gc.ca/science/aah-saa/diseases-maladies/icmu-eng.html (accessed 28 May 2021).
Bower, S.M. (2016) Synopsis of infectious diseases and parasites of commercially exploited shellfish: herpes-type virus disease of oysters. Available at: https://www.dfo-mpo.gc.ca/science/aah-saa/diseases-maladies/index-eng.html (accessed 28 May 2021).
Bower, S.M. & McGladdery, S.E. (2009) Synopsis of infectious diseases and parasites of commercially exploited shellfish: bucephalid trematode diseases of mussels. Available at: https://www.dfo-mpo.gc.ca/science/aah-saa/diseases-maladies/bucephmu-eng.html (accessed 28 May 2021).
Bower, S.M., McGladdery, S.E. & Price, I.M. (1994) Synopsis of infectious diseases and parasites of commercially exploited shellfish: shell-burrowing sponges of mussels. Available at: https://www.dfo-mpo.gc.ca/science/aah-saa/diseases-maladies/sbsmu-eng.html (accessed 28 May 2021).

Bower, S.M., Korrûbel, J.L. & Webb, S.C. (2002) Synopsis of infectious diseases and parasites of commercially exploited shellfish: phototrophic endolith invasion of mussel shells. Available at: https://www.dfo-mpo.gc.ca/science/aah-saa/diseases-maladies/bgasimu-eng.html (accessed 28 May 2021).

Bownes, S.J. & McQuaid, C.D. (2006) Will the invasive mussel *Mytilus galloprovincialis* Lamarck replace the indigenous *Perna perna* L. on the south coast of South Africa? *Journal of Experimental Marine Biology and Ecology*, **338**, 140–151.

Brown, S.W. (2012) Salinity tolerance of the oyster mudworm *Polydora websteri*. Honors College. Paper 41. Available at: http://digitalcommons.library.umaine.edu/honors/41 (accessed 28 May 2021).

Buck, B.H., Thieltges, D.W., Walter, U., Nehls, G. & Rosenthal, H. (2005) Inshore–offshore comparison of parasite infestation in *Mytilus edulis*: implications for open ocean aquaculture. *Journal of Applied Ichthyology*, **21**, 107–113.

Burgaud, G., Le Calvez, T., Arzur, D., Vandenkoornhuyse, P. & Barbier, G. (2009) Diversity of culturable marine filamentous fungi from deep-sea hydrothermal vents. *Environmental Microbiology*, **11**, 1588–1600.

Burgaud, G., Arzur, D., Durand, L., Cambon-Bonavita, M-A. & Barbier, G. (2010) Marine culturable yeasts in deep-sea hydrothermal vents: species richness and association with fauna. *FEMS Microbiology Ecology*, **73**, 121–133.

Burgos-Aceves, M.A. & Faggio, C. (2017) An approach to the study of the immunity functions of bivalve haemocytes: physiology and molecular aspects. *Fish and Shellfish Immunology*, **67**, 513–517.

Buschbaum, C. (2000) Direct and indirect effects of *Littorina littorea* (L.) on barnacles growing on mussel beds in the Wadden Sea. *Hydrobiologia*, **440**, 119–128.

Cáceres-Martínez, J. & Vásquez-Yeomans, R. (1997) Presence and histopathological effects of the copepod *Pseudomyicola spinosus* in *Mytilus galloprovincialis* and *Mytilus californianus*. *Journal of Invertebrate Pathology*, **70**, 150–155.

Cáceres-Martínez, J., Vásquez-Yeomans, R. & Suárez-Morales, E. (1996) Two parasitic copepods, *Pseudomyicola spinosus* and *Modiolicola gracilis*, associated with edible mussels, *Mytilus galloprovincialis* and *Mytilus californianus*, from Baja California NW México. *Journal of Shellfish Research*, **15**, 667–672.

Cáceres-Martínez J., Vásquez-Yeomans, R. & Sluys. R. (1998) The turbellarian *Urastoma cyprinae* from edible mussels *Mytilus galloprovincialis* and *Mytilus californianus* in Baja California, NW México. *Journal of Invertebrate Pathology*, **72**, 214–219.

Cáceres-Martínez, C., Chávez-Villalba, J. & Garduño-Méndez, J. (2005) First record of *Pseudomyicola spinosus* in *Argopecten ventricosus* in Baja California, Mexico. *Journal of Invertebrate Pathology*, **89**, 95–100.

Calvi Zeidan, G., dos Santos Aguiar Luz, M. & Boehs, G. (2012) Parasites of economically important bivalves from the southern coast of Bahia State, Brazil. *Revista Brasileira de Parasitologia Veterinária*, **21**, 391–398.

Calvo–Ugarteburu, G. & McQuaid, C.D. (1998) Parasitism and introduced species: epidemiology of trematodes in the intertidal mussels *Perna perna* and *Mytilus galloprovincialis*. *Journal of Experimental Marine Biology and Ecology*, **220**, 47–65.

Campbell, D.B. (1993) The effect of pea crabs on predation of host mussels by sea stars. *Marine Behavioural Physiology*, **24**, 93–99.

Campbell, S.A. (1970) The occurrence and effects of *Mytilicola intestinalis* in *Mytilus edulis*. *Marine Biology*, **5**, 89–95.

Canesi, L., Betti, M., Ciacci, C., Lorusso, L.C., Pruzzo, C. & Gallo, G. (2006) Cell signaling in the immune response of mussel hemocytes. *Invertebrate Survival Journal*, **3**, 40–49.

Canesi, L., Ciacci, C., Bergami, E., Monopoli, M.P., Dawson, K.A., Papa, S. *et al.* (2015) Evidence for immunomodulation and apoptotic processes induced by cationic polystyrene nanoparticles in the hemocytes of the marine bivalve *Mytilus*. *Marine Environmental Research*, **111**, 34–40.

Cantet, F., Toubiana, M., Parisi, M-G., Sonthi, M., Cammarata, M. & Roch, P. (2012) Individual variability of mytimycin gene expression in mussel. *Fish and Shellfish Immunology*, **33**, 641–644.

Carballal, M.J., López, C., Azevedo, C. & Villalba, A. (1997) Enzymes involved in defense functions of hemocytes of mussel *Mytilus galloprovincialis*. *Journal of Invertebrate Pathology*, **70**, 96–105.

Carballal, M.J., Barber, B.J., Iglesias, D. & Villalba, A. (2015) Neoplastic diseases of marine bivalves. *Journal of Invertebrate Pathology*, **131**, 83–106.

Carella, F., Carrasco, N., Andree, K.B., Lacuesta, B., Furones, D. & De Vico, G. (2013) Nocardiosis in Mediterranean bivalves: first detection of *Nocardia crassostreae* in a new host *Mytilus galloprovincialis* and in *Ostrea edulis* from the Gulf of Naples (Italy). *Journal of Invertebrate Pathology*, **114**, 324–328.

Carella, F., Feist, S.W., Bignell, J.P. & De Vico, G. (2015) Comparative pathology in bivalves: aetiological agents and disease processes. *Journal of Invertebrate Pathology*, **131**, 107–120.

Carella, F., De Vico, G. & Landini, G. (2017) Nuclear morphometry and ploidy of normal and neoplastic haemocytes in mussels. *PLoS ONE*, **12**, e0173219.

Carneiro-Schaefer, A.L., Sühnel, S., de Melo, C.M.R.& Magalhães, A.R.M. (2017) Record of *Pseudomyicola spinosus* in cultured *Perna perna* from southern Brazil. *Boletim do Instituto de Pesca, São Paulo*, **43**, 140–151

Carrasco, N., Arzul, I., Chollet, B., Robert, M., Joly, J.-P., Furones, M.D. *et al.* (2008) Comparative experimental infection of the copepod *Paracarti agrani* with *Marteilia refringens* and *M. maurini*. *Journal of Fish Diseases*, **31**, 497–504.

Carrasco, N., Andree, K.B., Lacuesta, B., Roque, A., Rodgers, C. & Furones, M.D. (2012) Molecular characterization of the *Marteilia* parasite infecting the common edible cockle *Cerastoderma edule* in the Spanish Mediterranean coast. *A new Marteilia species affecting bivalves in Europe*. *Aquaculture*, **324–325**, 20–26.

Carrasco, N., Green, T. & Itoh, N. (2015) *Marteilia* spp. parasites in bivalves: a revision of recent studies. *Journal of Invertebrate Pathology*, **131**, 43–57.

Carriel-Gomes, M.C., Kratz, J.M., Müller, V.D.M., Barardi, C.R.M. & Simões, C.M.O. (2006) Evaluation of antiviral activity in hemolymph from oysters *Crassostrea rhizophorae* and *Crassostrea gigas*. *Aquatic Living Resources*, **19**, 189–193.

Carver, C.E., Thériault, I. & Mallet, A.L. (2010) Infection of cultured eastern oysters *Crassostrea virginica* by the boring sponge *Cliona celata*, with emphasis on sponge life history and mitigation strategies. *Journal of Shellfish Research*, **29**, 905–915.

Castinel, A., Forrest, B. & Hopkins, G. (2014) Review of disease risks for New Zealand shellfish aquaculture: perspectives for management. Prepared for Ministry for Business, Innovation and Employment. Cawthron Report No. 2297.

Caza, F., Cledon, M. & St-Pierre, Y. (2016) Biomonitoring climate change and pollution in marine ecosystems: a review on *Aulacomya ater*. *Journal of Marine Biology*, **2016**, art. 7183813.

Ceuta, L.O & Boehs, G. (2012) Parasites of the mangrove mussel *Mytella guyanensis* (Bivalvia: Mytilidae) in Camamu Bay, Bahia, Brazil. *Brazilian Journal of Biology*, **72**, 421–427.

Charlet, M., Chernysh, S., Philippe, H., Hetru, C., Hoffmann, J.A. & Bulet. P. (1996) Innate immunity. Isolation of several cysteine-rich antimicrobial peptides from the blood of a mollusc, *Mytilus edulis*. *Journal of Biological Chemistry*, **271**, 21808–21813.

Cheng, T.C. (1996) Haemocytes: forms and functions. In: *The Eastern Oyster Crassostrea virginica* (eds V.S. Kennedy, R.I.E. Newell & A.F. Eble), pp. 299–333. Maryland Sea Grant, College Park, MD.

Chikalovets, I.V., Chernikov, O.V., Pivkin, M.V., Molchanova, V.I., Litovchenko, A.P., Li, W. *et al.* (2015) A lectin with antifungal activity from the mussel *Crenomytilus grayanus*. *Fish and Shellfish Immunology*, **42**, 503–507.

Chikalovets, I.V., Kovalchuk, S.N., Litovchenko, A.P., Molchanova, V.I., Pivkin, M.V. & Chernikov, O.V. (2016) A new Gal/GalNAc-specific lectin from the mussel *Mytilus trossulus*: structure, tissue specificity, antimicrobial and antifungal activity. *Fish and Shellfish Immunology*, **50**, 27–33.

Ciacci, C., Betti, M., Canonico, B., Citterio, B., Roch, P. & Canesi, L. (2010) Specificity of anti-*Vibrio* immune response through p38 MAPK and PKC activation in the hemocytes of the mussel *Mytilus galloprovincialis*. *Journal of Invertebrate Pathology*, **105**, 49–55.

Ciocan, C. & Sunila, I. (2005) Disseminated neoplasia in blue mussels, *Mytilus galloprovincialis*, from the Black Sea, Romania. *Marine Pollution Bulletin*, **50**, 1335–1339.

Ciocan, C.M. & Rotchell, J.M. (2005) Conservation of cancer genes in the marine invertebrate *Mytilus edulis*. *Environmental Science and Technology*, **39**, 3029–3033.

Ciocan, C.M., Moore, J.D. & Rotchell, J.M. (2006) The role of *ras* gene in the development of haemic neoplasia in *Mytilus trossulus*. *Marine Environmental Research*, **62**, S147–S150.

Clarke, K.R. & Gorley, R.N. (2015) PRIMER v7: User Manual/Tutorial. PRIMER-E, Plymouth.

Cole, H.A. & Savage, R.E. (1951) The effect of the parasitic copepod, *Mytilicola intestinalis* (Steuer) upon the condition of mussels. *Parasitology*, **41**, 156–161.

Collins, C.M. & Mulcahy, M.F. (2003) Cell-free transmission of a haemic neoplasm in the cockle *Cerastoderma edule*. *Diseases of Aquatic Organisms*, **54**, 61–67.

Comps, M. & Joly, J.P. (1980) Contamination expérimentale de *Mytilus galloprovincialis* Lmk par *Marteilia refringens*. *Science et Pêche Bulletin d'Information et de Documentation de l'Institut Scientifique et Technique des Pêches Maritimes*, **301**,19–21.

Comps, M. & G. Tigé (1997) Fine structure of *Minchinia* sp., a haplosporidan infecting the mussel *Mytilus galloprovincialis* L. *Systematic Parasitology*, **38**, 45–50.

Comps, M. & G. Tigé (1999) Procaryotic infections in the mussel *Mytilus galloprovincialis* and in its parasite the turbellarian *Urastoma cyprinae*. *Diseases of Aquatic Organisms*, **38**, 211–217.

Costa, M.M., Prado-Alvarez, M., Gestal, C., Li, H., Roch, P., Novoa, B. *et al.* (2009) Functional and molecular immune response of Mediterranean mussel (*Mytilus galloprovincialis*) haemocytes against pathogen-associated molecular patterns and bacteria. *Fish and Shellfish Immunology*, **26**, 515–523.

Cremonte, F., Vazquez, N. & Silva, M.R. (2011) Gonad atrophy caused by disseminated neoplasia in *Mytilus chilensis* cultured in the Beagle Channel, Tierra del Fuego Province, Argentina. *Journal of Shellfish Research*, **30**, 845–849.

Cremonte, F., Puebla, C., Tillería, J. & Videla, V. (2015) Histopathological survey of the mussel *Mytilus chilensis* (Mytilidae) and the clam *Gari solida* (Psammobiidae) from southern Chile. *Latin American Journal of Aquatic Research*, **43**, 248–254.

Crespo, C., Rodríguez, H., Segade, P., Iglesias, R. & García-Estévez, J.M. (2009) *Coccomyxa* sp. (Chlorophyta: Chlorococcales), a new pathogen in mussels (*Mytilus galloprovincialis*) of Vigo estuary (Galicia, NW Spain). *Journal of Invertebrate Pathology*, **102**, 214–219.

Crespo-González, C., Rodríguez-Domínguez, H., Segade, P., Iglesias, R., Arias, C. & García-Estévez, J.M. (2010) Seasonal dynamics and microhabitat distribution of *Urastoma cyprinae* in *Mytilus galloprovincialis*: implications for its life cycle. *Journal of Shellfish Research*, **29**, 187–192.

Culloty, S.C. & Mulcahy, M.F. (2007) *Bonamia ostreae* in the native oyster, *Ostrea edulis*: a review. Marine Environment and Health Series No. 29. Available at: https://oar.marine.ie/handle/10793/269 (accessed 28 May 2021).

Culloty, S.C., Novoa, B., Pernas, M., Longshaw, M., Mulcahy, M.F., Feist, S.W. *et al.* (1999) Susceptibility of a number of bivalve species to the protozoan parasite *Bonamia ostreae* and their ability to act as vectors for this parasite. *Diseases of Aquatic Organisms*, **37**, 73–80.

Dang, C., Gonzalez, P., Mesmer-Dudons, N., Bonami, J.-R., Caill-Milly, N. & de Montaudouin, X. (2009) Virus-like particles associated with brown muscle disease in Manila clam, *Ruditapes philippinarum*, in Arcachon Bay (France). *Journal of Fish Diseases*, **32**, 577–584.

Davey, J.T. (1989) *Mytilicola intestinalis* (Copepoda: Cyclopoida): a ten-year survey of infested mussels in a Cornish estuary, 1978–1988. *Journal of the Marine Biological Association of the UK*, **69**, 823–836.

Dayaram, A., Goldstien, S., Zawar-Reza, P., Gomez, G., Harding, J.S. & Varsani, A. (2013) Novel ssDNA virus recovered from estuarine Mollusc (*Amphibola crenata*) whose replication associated protein (Rep) shares similarities with Rep-like sequences of bacterial origin. *Journal of General Virology*, **94**, 1104–1110.

Defer, D., Bourgougnon, N. & Fleury, Y. (2009) Screening for antibacterial and antiviral activities in three bivalve and two gastropod marine mollusks. *Aquaculture*, **293**, 1–7.

Delaporte, M., Soudant, P., Moal, J., Lambert, C., Quéré, C. & Miner, P. (2003) Effect of a mono-specific algal diet on immune functions in two bivalve species – *Crassostrea gigas* and *Ruditapes philippinarum*. *Journal of Experimental Biology*, **206**, 3053–3064.

Desser, S.S., Bower, S.M. & Hong, H. (1998) *Pseudoklossia semiluna* n. sp. (Apicomplexa: Aggregatidae): a coccidian parasite of the kidney of blue mussels, species of *Mytilus*, from British Columbia, Canada. *Parasite*, **5**, 17–22.

Dethlefsen, V. (1975) The influence of *Mytilicola intestinalis* Steuer on the meat content of the mussel *Mytilus edulis*. *Aquaculture*, **6**, 83–97.

Détrée, C. & Gallardo-Escárate, C. (2017) Polyethylene microbeads induce transcriptional responses with tissue-dependent patterns in the mussel *Mytilus galloprovincialis*. *Journal of Molluscan Studies*, **83**, 220–225.

Détrée, C., Lallier, F.H., Tanguy, A. & Mary, J. (2017) Identification and gene expression of multiple peptidoglycan recognition proteins (PGRPs) in the deep-sea mussel *Bathymodiolus azoricus*, involvement in symbiosis? *Comparative Biochemistry and Physiology Part-B*, **207**, 1–8.

Di, Y., Schroeder, D.C., Highfield, A., Readman, J.W. & Awadhesh, N.J. (2011) Tissue specific expression of *p53* and *ras* genes in response to the environmental genotoxicant benzo(α) pyrene in marine mussels. *Environmental Science and Technology*, **45**, 8974–8981.

Di, Y., Aminot, Y., Schroeder, D.C., Readman, J.W. & Jha, A.N. (2017) Integrated biological responses and tissue-specific expression of *p53* and *ras* genes in marine mussels following exposure to benzo(α)pyrene and C60 fullerenes, either alone or in combination. *Mutagenesis*, **32**, 77–90.

Díaz, S., Cao, A., Villalba, A. & Carballal, M.J. (2010) Expression of mutant protein p53 and Hsp70 and Hsp90 chaperones in cockles *Cerastoderma edule* affected by neoplasia. *Diseases of Aquatic Organisms*, **90**, 215–222.

Diez, M.E., Vázquez, N., da Cunha Lana, P. & Cremonte, F. (2015) Biogenic calcareous growth on the ribbed mussel *Aulacomya atra* (Bivalvia: Mytilidae) favours polydorid boring (Polychaeta: Spionidae). *Hydrobiologia*, **766**, 349–355.

Diggles, B.K., Hine, P.M., Handley, S. & Boustead, N.C. (2002) A handbook of diseases of importance to aquaculture in New Zealand. *NIWA Science and Technology Series No.* 49.

Domeneghetti, S., Manfrin, C., Varotto, L., Rosani, U., Gerdol, M., De Moro, G. *et al.* (2011) How gene expression profiles disclose vital processes and immune responses in *Mytilus* spp. *Invertebrate Survival Journal*, **8**, 179–189.

Duckworth, A.R. & Peterson, B.J. (2013) Effects of seawater temperature and pH on the boring rates of the sponge *Cliona celata* in scallop shells. *Marine Biology*, **160**, 27–35.

Earth Life (2021) Platyhelminthes: phylum of the extraordinary flatworms. Available at: https://www.earthlife.net/inverts/platyhelminthes.html (accessed 28 May 2021).

Elgharsalli, R., Aloui-Bejaoui, N., Salah, H., Chollet, B., Jolu, J.P., Rovert, M. *et al.* (2013) Characterization of the protozoan parasite *Marteilia refringens* infecting the dwarf oyster *Ostrea stentina* in Tunisia. *Journal of invertebrate Pathology*, **112**, 175–183.

Ellis, R.P., Widdicombe, S., Parry, H., Hutchinson, T.H. & Spicer, J.I. (2015) Pathogenic challenge reveals immune trade-off in mussels exposed to reduced seawater pH and increased temperature. *Journal of Experimental Marine Biology and Ecology*, **462**, 83–89.

Elston, R.A., Kent, M.L. & Drum, A.S. (1988) Transmission of hemic neoplasia in the bay mussel, *Mytilus edulis*, using whole cells and cell homogenate. *Developmental and Comparative Immunology*, **12**, 719–727.

Estévez-Calvar, N., Romero, A., Figueras, A. & Novoa, B. (2011) Involvement of pore- forming molecules in immune defense and development of the Mediterranean mussel (*Mytilus galloprovincialis*). *Developmental and Comparative Immunology*, **35**, 1015–1029.

Falanga, A., Lombardi, L., Franci, G., Vitiello, M., Iovene, M.R., Morelli, G. *et al.* (2016) Marine antimicrobial peptides: nature provides templates for the design of novel compounds against pathogenic bacteria. *International Journal of Molecular Science*, **17**, 785

Fayer, R., Trout, J.M. & Jenkins, M.C (1998) Infectivity of *Cryptosporidium parvum* oocysts stored in water at environmental temperatures. *Journal of Parasitology*, **84**,1165–1169.

Fenchel, T. (1965) Ciliates from Scandinavian molluscs. *Ophelia*, **2**, 71–174.

Figueras, A. & Robledo, J.A.F. (1994) *Bonamia ostreae* present in flat oysters (*Ostrea edulis*) does not infect mussels (*Mytilus galloprovincialis*). *Bulletin of the European Association of Fish Pathologists*, **14**, 98–100.

Figueras, A.J., Jardon, C.F. & Caldas, J.R. (1991) Diseases and parasites of rafted mussels (*Mytilus galloprovincialis* Lmk): preliminary results. *Aquaculture*, **99**, 17–33.

Fisheries and Oceans Canada (2019) Synopsis of infectious diseases and parasites of commercially exploited shellfish. Available at: https://www.dfo-mpo.gc.ca/science/aah-saa/diseases-maladies/index-eng.html (accessed 28 May 2021).

Fitzer, S. C., Chung, P., Maccherozzi, F., Dhesi, S.S., Kamenos, N.A., Phoenix, V.R. *et al.* (2016) Biomineral shell formation under ocean acidification: a shift from order to chaos. *Scientific Reports*, **6**, 21076.

Flajnik, M.F. & Du Pasquier, L. (2004) Evolution of innate and adaptive immunity: can we draw a line? *Trends in Immunology*, **25**, 640–644.

Franchini, A., Malagoli, D. & Ottaviani, E. (2005) Investigation of the loss of byssus in *Mytilus galloprovincialis* from mussel farms in the Adriatic Sea. *Cell Biology International*, **29**, 857–860.

Francisco, C.J., Hermida, M.A. & Santos, M.J. (2010) Parasites and symbionts from *Mytilus galloprovincialis* (Lamark, 1819) (Bivalves: Mytilidae) of the Aveiro Estuary Portugal. *Journal of Parasitology*, **96**, 200–205.

Gadanho, M. & Sampaio, J. (2005) Occurrence and diversity of yeasts in the Mid-Atlantic Ridge hydrothermal fields near the Azores archipelago. *Microbial Ecology*, **50**, 408–417.

Galaktionov, K.V. (1996) Life cycles and distribution of seabird helminthes in Arctic and sub-Arctic region. *Bulletin of the Scandavian Society of Parasitology*, **6**, 31–49.

Galaktionov, K.V., Bustnes, J.O., Bårdsen, B.-J., Wilson, J.G., Nikolaev, K.E., Sukhotin, A.A. *et al.* (2015) Factors influencing the distribution of trematode larvae in blue mussels *Mytilus edulis* in the North Atlantic and Arctic Oceans. *Marine Biology*, **162**, 193–206.

Galimany, E. & Sunila, I. (2008) Several cases of disseminated neoplasia in mussels *Mytilus edulis* (L.) in western Long Island Sound. *Journal of Shellfish Research*, **27**, 1201–1207.

García-García, E., Prado-Alvarez, M., Novoa, B., Figueras, A. & Rosales, C. (2008) Immune responses of mussel hemocyte subpopulations are differentially regulated by enzymes of the PI 3-K, PKC, and ERK kinase families. *Developmental and Comparative Immunology*, **32**, 637–653.

Gee, J.M. & Davey, J.T. (1986) Experimental studies on the infestation of *Mytilus edulis* (L.) by *Mytilicola intestinalis* Steuer (Copepoda, Cyclopoida). *Journal de Conseil International pour l' Exploration de la Mer*, **42**, 265–271.

Gee, J.M., Maddock, L. & Davey, J.T. (1977) The relationship between infestation by *Mytilicola intestinalis*, Steuer (Copepoda: Cyclopoidea) and the condition index of *Mytilus edulis* in southwest England. *Journal de Conseil International pour l' Exploration de la Mer*, **37**, 300–308.

George, L.E. & Nair, N.B. (1974) *Observations on Bucephalus sp. parasitic in Musculista arcuatula. (1974) Journal of Invertebrate Pathology*, **24**, 63–69.

Gerdol, M. & Venier, P. (2015) An updated molecular basis for mussel immunity. *Fish and Shellfish Immunology*, **46**, 17–38.

Gerdol, M., Manfrin, C., De Moro, G., Figueras, A., Novoa, B., Venier, P. *et al.* (2011) The C1q domain containing proteins of the Mediterranean mussel *Mytilus galloprovincialis*: a widespread and diverse family of immune-related molecules. *Developmental and Comparative Immunology*, **35**, 635–643.

Gerdol, M., De Moro, G., Manfrin, C., Venier, P. & Pallavicini, A. (2012) Big defensins and mytimacins, new AMP families of the Mediterranean mussel *Mytilus galloprovincialis*. *Developmental and Comparative Immunology*, **36**, 390–399.

Gerdol, M., De Moro, G., Manfrin, C., Milandri, A., Riccardi, E., Beran, A., *et al.* (2014) RNA sequencing and de novo assembly of the digestive gland transcriptome in *Mytilus galloprovincialis* fed with toxinogenic and non-toxic strains of *Alexandrium minutum*. *BMC Research Notes*, **7**, 722.

Gerdol, M., Venier, P. & Pallavicini, A. (2015) The genome of the Pacific oyster *Crassostrea gigas* brings new insights on the massive expansion of the C1q gene family in Bivalvia. *Developmental and Comparative Immunology*, **49**, 59–71.

Gestal, C., Roch, P., Renault, T., Pallavicini, A., Paillard, C., Novoa, B. *et al.* (2008) Study of diseases and the immune system of bivalves using molecular biology and genomics. *Review in Fisheries Science*, **16**, 133–156.

Getchell, R.G., Smolowitz, R.M., McGladdery, S.E. & Bower, S.M. (2016) Diseases and parasites of scallops. In: *Scallops: Biology, Ecology and Aquaculture*, 3rd edn (eds S.E. Shumway & G.J. Parson), pp. 425–467. Elsevier Science, Amsterdam.

Giannapas, M., Karnis, L. & Dailianis, S. (2012) Generation of free radicals in haemocytes of mussels after exposure to low molecular weight PAH components: immune activation, oxidative and genotoxic effects *Comparative Biochemistry and Physiology Part-C*, **155**, 182–189.

Gilek, M., Tedengren, M. & Kautsky, N. (1992) Physiological performance and general histology of the blue mussel, *Mytilus edulis* L., from the Baltic and North Seas. *Netherland Journal of Sea Research*, **30**, 11–21.

Girón-Pérez, M. (2010) Relationships between innate immunity in bivalve molluscs and environmental pollution. *Invertebrate Survival Journal*, **7**, 149–156.

Goater, C.P. & Weber, A.E. (1996) Factors affecting the distribution and abundance of *Mytilicola orientalis* (Copepoda) in the mussel, *Mytilus trossulus*, in Barkley Sound, *B. C. Journal of Shellfish Research*, **15**, 681–684.

Gombač, M., Pogačnik, M., Fonda, I. & Jenčič, V. (2008) Intracellular ciliates of cultured Mediterranean mussels (*Mytilus galloprovincialis*) in the Gulf of Trieste in Slovene Adriatic Sea. *Bulletin of the European Association of Fish Pathologists*, **28**, 217–221.

Gombač, M., Sitar, R., Pogačnik, M., Arzul, I. & Jenčič, V. (2013) Haemocytic neoplasia in Mediterranean mussels (*Mytilus galloprovincialis*) in the Slovene Adriatic Sea. *Marine and Freshwater Behaviour and Physiology*, **46**, 135–143.

Gombač, M., Kušar, D., Ocepek, M., Pogačnick, M., Arzul, I., Couraleau, Y. *et al.* (2014) Marteiliosis in mussels: a rare disease? *Journal of Fish Diseases*, **34**, 805–814.

Gomez-Bautista, M., Ortega-Mora, L.M., Tabares, E., Lopez-Rodas, V. & Costas, E. (2000) Detection of infectious *Cryptosporidium parvum* oocysts in mussels (*Mytilus galloprovincialis*) and cockles (*Cerastoderma edule*). *Applied and Environmental Microbiology*, **66**, 1866–1870.

Gómez-Couso, H., Freire-Santos, F., Ortega-Iñarea, M.R., Castro-Hermida, J.A. & Ares-Mazás, M.E. (2003) Environmental dispersal of *Cryptosporidium* parvum oocysts and cross transmission in cultured bivalve molluscs. *Parasitology Research*, **90**, 140–142.

Gómez-Couso, H., Méndez-Hermida, F. & Ares-Mazás, E. (2006) Levels of detection of *Cryptosporidium* oocysts in mussels (*Mytilus galloprovincialis*) by IFA and PCR methods. *Veterinary Parasitology*, **141**, 60–65.

Gorbushin, A. & Iakovleva, N. (2011) A new gene family of single fibrinogen domain lectins in *Mytilus*. *Fish and Shellfish Immunology*, **30**, 434–438.

Gosling, E.M. (2015) *Marine Bivalve Molluscs*, 2nd edn. Wiley-Blackwell, New York.

Gray, A.P., Lucas, I.A.N., Seed, R. & Richardson, C.A. (1999) *Mytilus edulis chilensis* infested with *Coccomyxa parasitica* (Chlorococcales, Coccomyxaceae). *Journal of Molluscan Studies*, **65**, 289–294.

Grizel, H., Comps, M., Bonami, J.R., Cousserans, F., Duthoit, J.L. & Pennec M. (1974) Epizooty of the common oyster *Ostrea edulis*. Part 1. Study of the agent of digestive gland disease in *Ostrea edulis* (Linné). *Science et Pêche Bulletin d'Information et de Documentation de l'Institut Scientifique et Technique des Pêches Maritimes*, **240**, 1–30.

Guo, X. & Ford, S.E. (2016) Infectious diseases of marine molluscs and host responses as revealed by genomic tools. *Philosophical Transactions of the Royal Society B*, **371**, 20150206.

Guo, X., He, Y., Zhang, L., Lelong, C. & Jouaux, A. (2015) Immune and stress responses in oysters with insights on adaptation. *Fish and Shellfish Immunology*, **46**, 107–119.

Haines, C.M.C., Edmunds, M. & Pewsey, A.R. (1994) The pea crab, *Pinnotheres pisum* (Linnaeus, 1767), and its association with the common mussel, *Mytilus edulis* (Linnaeus, 1758), in the Solent (UK). *Journal of Shellfish Research*, **13**, 5–10.

Handley, S.J. & Bergquist, P.R. (1997) Spionid polychaete infestations of intertidal Pacific oysters *Crassostrea gigas* (Thunberg), Mahurangi Harbour, northern New Zealand. *Aquaculture*, **153**, 191–205.

Haug, T., Stensvag, K., Olsen, M.O., Sandsdalen, E. & Styrvold, O.B. (2004) Antibacterial activities in various tissues of the horse mussel, *Modiolus modiolus*. *Journal of Invertebrate Pathology*, **85**, 112–119.

He, Y., Yu, H., Bao, Z., Zhang, Q. & Guo, X. (2012) Mutation in promoter region of a serine protease inhibitor confers *Perkinsus marinus* resistance in the eastern oyster (*Crassostrea virginica*). *Fish and Shellfish Immunology*, **33**, 411–417.

He, Y., Jouaux, A., Ford, S.E., Lelong, C., Sourdaine, P., Mathieu, M. *et al.* (2015) Transcriptome analysis reveals strong and complex antiviral response in a mollusk. *Fish and Shellfish Immunology*, **46**, 131–144.

Hechinger, R.F., Lafferty, K.D. & Kuris, A.M. (2008) Trematodes indicate animal biodiversity in the Chilean intertidal and Lake Tanganyika. *Journal of Parasitology*, **94**, 966–968.

Hibbett, D.S., Binder, M., Bischoff, J.F., Blackwell, M., Cannon, P.F., Eriksson, O.E. *et al.* (2007) A higher level phylogenetic classification of the Fungi. *Mycological Research*, **111**, 509–547.

Hockley, A.R. (1951) On the biology of *Mytilicoa intestinalis* (Steuer). *Journal of the Marine Biological Association of the UK*, **30**, 223–232.

Hooge, M.D. & Tyler, S. (1999) Musculature of the facultative parasite *Urastoma cyprinae* (Platyhelminthes). *Journal of Morphology*, **241**, 207–216.

Ikuta, T., Tame, A., Saito, M., Aokia, Y., Nagai, Y., Sugimura, M. *et al.* (2019) Identification of cells expressing two peptidoglycan recognition proteins in the gill of the vent mussel, *Bathymodiolus septemdierum*. *Fish and Shellfish Immunology*, **93**, 815–822.

Janeway, C.A. Jr. & Medzhitov, R. (2002) Innate immune recognition. *Annual Review of Immunology*, **20**, 197–216.

Jayaraj, S., Thiagarajan, R., Arumugam, M. & Mullainadhan, P. (2008) Isolation, purification and characterization of β-1, 3-glucan binding protein from the plasma of marine mussel *Perna viridis*. *Fish and Shellfish Immunology*, **24**, 715–725.

Jayaraj, S.S., Thiagarajan, R. & Vincent, S. (2009) Encapsulation response from the hemocytes of marine mussel *Perna viridis*. *Global Journal of Biotechnology and Biochemistry*, **4**, 138–142.

Jensen, K.R. (2010) Nobanis invasive alien fact sheet – *Mytilicola intestinalis*. Available at: https://www.yumpu.com/en/document/read/27993591/mytilicola-intestinalis-nobanis (accessed 28 May 2021).

Jones, J. B. (1975) *Nematopsis* n. sp. (Sporozoa: Gregarinia) in *Perna canalicus* (note). *Journal of Marine and Freshwater Research*, **9**, 567–568.

Jones, J.B., Scotti, P.D., Dearing, S.C. & Wesney B. (1996) Virus like particles associated with marine mussel mortalities in New Zealand. *Diseases of Aquatic Organisms*, **25**, 143–149.

Kaehler, S. (1999) Incidence and distribution of phototrophic shell-degrading endoliths of the brown mussel *Perna perna*. *Marine Biology*, **135**, 505–514.

Kang, H.K., Seo, C.H. & Park, Y. (2015) Marine peptides and their anti-infective activities. *Marine Drugs*, **13**, 618–654.

Kawai, T. & Akira, S. (2010) The role of pattern-recognition receptors in innate immunity: update on Toll-like receptors. *Nature Immunology*, **11**, 373–384.

Kesarcodi-Watson, A., Kaspar, H., Josie Lategan, M. & Gibson, L.F. (2009) Challenge of New Zealand Greenshell™ mussel *Perna canaliculus* larvae using two *Vibrio* pathogens: a hatchery study. *Diseases of Aquatic Organisms*, **86**, 15–20.

Kinne, O. (1983) Introduction to diseases of Mollusca: Bivalvia. In: *Diseases of Marine Animals*, Vol. 2. (ed O. Kinne), pp. 467–475. Biologishe Anstalt Helgoland, Hamburg.

Kitamura, S.-I., Ko, J.-Y., Lee, W.-L., Kim, S.-R., Song, J.-Y., Kim, D.-K. *et al.* (2007) Seasonal prevalence of lymphocystis disease virus and aquabirnavirus in Japanese flounder, *Paralichthys olivaceus* and blue mussel, *Mytilus galloprovincialis*. *Aquaculture*, **266**, 26–31.

Korringa, P. (1951) Le *Mytilicola intestinalis* Steuer (Copepoda parasitica) menace l'industrie mouliére en Zelande. *Revue des Travaux de l'Office Scientifique et Technique des Pêches Maritimes*, **17**, 9–13.

Kovalchuk, S.N., Chikalovets, I.V., Chernikov, O.V., Molchanova, V.I., Li, W., Rasskazov, V.A. *et al.* (2013) cDNA cloning and structural characterization of a lectin from the mussel *Crenomytilus grayanus* with a unique amino acid sequence and antibacterial activity. *Fish and Shellfish Immunology*, **35**, 1320–1324.

Krakau, M., Thieltges, D.W. & Reise, K. (2006) Native parasites adopt introduced bivalves of the North Sea. *Biological Invasions*, **8**, 919–925.

Krishnakumar, P.K., Casillas, E., Snider, R.G., Kagley, A.N. & Varanasi, U. (1999) Environmental contaminants and the prevalence of hemic neoplasia (leukemia) in the common mussel (*Mytilus edulis* complex) from Puget Sound, Washington, U.S.A. *Journal of Invertebrate Pathology*, **73**, 135–146.

Krug, P.J. (2006) Defense of benthic invertebrates against surface colonization by larvae: a chemical arms race. In: *Antifouling Compounds, Vol. 42, Marine Molecular Biotechnology* (eds N. Fusetani & A.S. Clare), pp. 1–53. Springer-Verlag, Berlin.

Kuchel, R.P., Raftos, D.A., Birch, D. & Vella, N. (2010) Haemocyte morphology and function in the Akoya Pearl Oyster, *Pinctada imbricata*. *Journal of Invertebrate Pathology*, **105**, 36–48.

Kurtz, J. (2005) Specific memory within innate immune systems. *Trends in Immunology*, **26**, 186–192.

Lacroix, C., Duvieilbourg, E., Guillou, N., Guyomarch, J., Bassoulet, C., Moraga, D. *et al.* (2017) Seasonal monitoring of blue mussel (*Mytilus* spp.) populations in a harbor area: a focus on responses to environmental factors and chronic contamination. *Marine Environmental Research*, **129**, 24–35.

Lasiak, T. (1992) Bucephalid trematode infections in mytilid bivalves from the rocky intertidal of southern Chile. *Journal of Molluscan Studies*, **58**, 29–36.

Lasiak, T.A. (1993) Bucephalid trematode infections in the brown mussel *Perna perna* (Bivalvia: Mytilidae). *South African Journal of Marine Science*, **13**, 127–134.

Lauckner, G. (1980) Diseases of Mollusca: Gastropoda. In: *Diseases of Marine Animals*, Vol 2 (ed. O. Kinne), pp. 311–424. John Wiley & Sons, New York.

Lauckner, G. (1983) Diseases of Mollusca: Bivalvia. In: *Diseases of Marine Animals*, Vol. 2 (ed. O. Kinne), pp. 477–961. Biologishe Anstalt Helgoland, Hamburg.

Le Roux, F., Audemard, C., Barnaud, A. & Berthe, F. (1999) DNA Probes as potential tools for the detection of *Marteilia refringens*. *Marine Biotechnology*, **1**, 588–597.

Le Roux, F., Lorenzo G., Peyret P., Audemard C., Figueras A., Vivarès, C. *et al.* (2001) Molecular evidence for the existence of two species of *Marteilia* in Europe. *Journal of Eukaryotic Microbiology*, **48**, 449–454.

Li, H., Parisi, M-G., Toubiana, M., Cammarata, M. & Roch, P. (2008) Lysozyme gene expression and hemocyte behaviour in the Mediterranean mussel, *Mytilus galloprovincialis*, after injection of various bacteria or temperature stresses. *Fish and Shellfish Immunology*, **25**, 143–152.

Li, Q. & Verma, I.M. (2002) NF-κB regulation in the immune system. *Nature Reviews Immunology*, **2**, 725–734.

Li, X., Guyot, K., Dei-Cas, E., Mallard, J-P., Jacques Ballet, J. & Brasseur, P. (2006) *Cryptosporidium* oocysts in mussels (*Mytilus edulis*) from Normandy (France). *International Journal of Food Microbiology*, **108**, 321–325.

Liao, Z., Wang, X., Liu, H., Fan, M., Sun, J. & Shen, W. (2013) Molecular characterization of a novel antimicrobial peptide from *Mytilus coruscus*. *Fish and Shellfish Immunology*, **34**, 610–616.

Lima, I., Peck, M.R., Soares, M.V.M., Guilhermino, L. & Rotchell, J.M. (2008) *Ras* gene in marine mussels: a molecular level response to petrochemical exposure. *Marine Pollution Bulletin*, **56**, 633–640.

Lin, T., Yao, Z., Hao, Z., Lai, Q. & Zhou, K. (2013) Classification of granulocytes in three bivalve species using neutral red staining. *Journal of Shellfish Research*, **32**, 861–865.

Liu, L., Yang, J., Qiu, L., Wang, L., Zhang, H., Wang, M. *et al.* (2011) A novel scavenger receptor-cysteine-rich (SRCR) domain containing scavenger receptor identified from mollusk mediated PAMP recognition and binding. *Developmental and Comparative Immunology*, **35**, 227–239.

Loot, G., Aldana, M. & Navarrete, S.A. (2005) Effects of human exclusion on parasitism in intertidal food webs of central Chile. *Conservation Biology*, **19**, 203–212.

Loot, G., Blanchet, S., Aldana, M. & Navarrete, S.A. (2008) Evidence of plasticity in the reproduction of a trematode parasite: the effect of host removal. *Journal of Parasitology*, **94**, 23–27.

Lopez-Cortes, L., Castro, D., Navas, J.I. & Borrego, J.J. (1999) Phagocytic and chemotactic responses of Manila and carpet shell clam haemocytes against *Vibrio tapetis*, the causative agent of brown ring disease. *Fish and Shellfish Immunology*, **9**, 543–555.

López-Flores, I., De La Herran, R., Garrido-Ramos, M.A., Navas, J.I., Ruiz-Rejon, C. & Ruiz-Rejon, M. (2004) The molecular diagnosis of *Marteilia refringens* and differentiation between *Marteilia* strains infecting oysters and mussels based on the rDNA IGS sequence. *Parasitology*, **129**, 411–419.

López-Sanmartín, M., Batista, F.M., del Carmen Marín, M., Garrido, I., Quintero, D., Grade, A. *et al.* (2015) Detection of *Marteilia refringens* infecting the European flat oyster *Ostrea edulis* and the dwarf oyster *Ostrea stentina* in southern Portugal and Spain. *Journal of Invertebrate Pathology*, **130**, 52–55.

Lourenço, C.R., Nicastro, K.R., McQuaid, C.D., Sabour, B. & Zardi, G.I. (2017) Latitudinal incidence of phototrophic shell-degrading endoliths and their effects on mussel bed microclimates. *Marine Biology*, **164**, 129.

Lutz, R.A., Fritz, L.W. & Rhoads, D.C. (1985) Molluscan growth at deep-sea hydrothermal vents. *Bulletin of the Biological Society of Washington*, **6**, 199–210.

Lynch, S.A., Abollo, E., Ramilo, A., Cao, A., Culloty, S.C. & Villalba, A. (2010) Observations raise the question if the Pacific oyster, *Crassostrea gigas*, can act as either a carrier or a reservoir for *Bonamia ostreae* or *Bonamia exitiosa*. *Parasitology*, **137**, 1515–1526.

Lynch, S.A., Morgan, E., Carlsson, J., Mackenzie, C., Wooton, E.C., Rowley, A.F. *et al.* (2014) The health status of mussels, *Mytilus* spp., in Ireland and Wales with the molecular identification of a previously undescribed haplosporidian. *Journal of Invertebrate Pathology*, **118**, 59–65.

Manso, C.F., Díaz, S., Carballal, M.J., Villalba, A. & Romalde, J.L. (2012) Detection of reverse transcriptase activity in golden carpet shell clams (*Venerupis aurea*) with disseminated neoplasia. *Bulletin of the European Association of Fish Pathologists*, **32**, 56–63.

Marchiori, N.C., Magalãhes, A.R.M. & Pereira, J.J. (2010) The life cycle of *Bucephalus margaritae* Ozaki & Ishibashi, 1934 (Digenea, Bucephalidae) from the coast of Santa Catarina State, Brazil. *Acta Scientiarum Biological Sciences*, **32**, 71–78.

Mariné Oliveira, G.F., Couto, M.C.M., de Freitas Lima, M. & Bergamo do Bomfim, T.C. (2016) Mussels (*Perna perna*) as bioindicator of environmental contamination by *Cryptosporidium* species with zoonotic potential. *International Journal for Parasitology: Parasites and Wildlife*, **5**, 28–33.

Marquet, N., Nicastro, K.R., Gektidis, M., McQuaid, C.D., Pearson, G.A., Serrão, E.A. *et al.* (2013) Comparison of phototrophic shell-degrading endoliths in invasive and native populations of the intertidal mussel *Mytilus galloprovincialis*. *Biological Invasions*, **15**, 1253–1272.

Martins, E., Figueras, A., Novoa, B., Serrão Santos, R., Moreira, R. & Bettencourt, R. (2014) Comparative study of immune responses in the deep-sea hydrothermal vent mussel *Bathymodiolus azoricus* and the shallow-water mussel *Mytilus galloprovincialis* challenged with *Vibrio* bacteria. *Fish and Shellfish Immunology*, **40**, 485–499.

Matos, E., Matos, P. & Azevedo, C. (2005) Observations on the intracytoplasmic microsporidian, *Steinhausia mytilovum*, a parasite of mussel (*Mytella guyanensis*) ovocytes from the Amazon River Estuary. *Brazilian Journal of Morphological Science*, **22**, 183–186.

McDowell, I.C., Nikapitiya, C., Aguiar, D., Lane, C.E., Istrail, S. & Gomez-Chiarri, M. (2014) Transcriptome of American oysters, *Crassostrea virginica*, in response to bacterial challenge: insights into potential mechanisms of disease resistance. *PLoS One*, **9**, e105097.

Meena, B., Rose, S., Jayaraj, S.S. & Vincent, S. (2010) Encapsulation response from the haemocytes of estuarine clam *Meretrix casta*. *International Journal of Pharmaceutical and Biological Sciences*, **1**, 1–8.

Mello, D.F., de Oliveira Proença, L.A. & Barracco, M.A. (2010) Comparative study of various immune parameters in three bivalve species during a natural bloom of *Dinophysis acuminate* in Santa Catarina Island, Brazil. *Toxins*, **2**, 1166–1178.

Metzger, M.J. & Goff, S.P. (2016) A sixth modality of infectious disease: contagious cancer from devils to clams and beyond. *PLoS Pathogens*, **12**, e1005904.

Metzger, M.J., Reinisch, C., Sherry, J. & Goff, S.P. (2015) Horizontal transmission of clonal cancer cells causes leukemia in soft-shell clams. *Cell*, **161**, 255–263.

Metzger, M.J., Villalba, A., Carballal, M.J., Iglesias, D., Sherry, J., Reinisch, C. *et al.* (2016) Widespread transmission of independent cancer lineages within multiple bivalve species. *Nature*, **534**, 705–709.

Meyers, T.R. & Burton T (2009) *Diseases of wild and cultured shellfish in Alaska*. Commercial Fisheries Division, Alaska Department of Fish and Game, Anchorage, AK.

Miller, A., Inglis, G.J. & Poulin, R. (2008) Use of the introduced bivalve, *Musculista senhousia*, by generalist parasites of native New Zealand bivalves. *New Zealand Journal of Marine and Freshwater Research*, **42**, 143–151.

Miller, W.A., Miller, M.A., Gardner, I.A., Atwill, E.R., Harris, M., Ames, J. *et al.* (2005) New genotypes and factors associated with *Cryptosporidium* detection in mussels (*Mytilus* spp.) along the California coast. *International Journal for Parasitology*, **35**, 1103–1113.

Miller, W.A., Gardner, I.A., Atwill, E.R., Leutenegger, C.M., Miller, M.A. Hedrick, R.P. *et al.* (2006) Evaluation of methods for improved detection of *Cryptosporidium* spp. in mussels (*Mytilus californianus*). *Journal of Microbiological Methods*, **65**, 367–379.

Mitta, G., Hubert, F., Noël, T. & Roch, P. (1999) Myticin, a novel cysteine-rich antimicrobial peptide isolated from haemocytes and plasma of the mussel *Mytilus galloprovincialis*. *European Journal of Biochemistry*, **265**, 71–78.

Mitta, G., Vandenbulcke, F., Noël, T., Romestand, B., Beauvillain, J.-C., Salzet, M. *et al.* (2000) Differential distribution and defence involvement of antimicrobial peptides in mussel. *Journal of Cell Science*, **113**, 2759–2769.

Mladineo, I. & Peharda, M. (2005) Histopathology of *Gymnophallus* sp. sporocysts in the edible mytilid, *Modiolus barbatus*. *Journal of Shellfish Research*, **24**, 1097–1100.

Mladineo, I., Trumbi, Z., Jozi, S. & Šegvi, T. (2009) First report of *Cryptosporidium* sp. (Coccidia, Apicomplexa) oocysts in the black mussel (*Mytilus galloprovincialis*) reared in the Mali Ston Bay, Adriatic Sea. *Journal of Shellfish Research*, **28**, 541–546.

Mladineo, I., Petrić M., Hrabar, J., Bočina, I. & Peharda, M. (2012) Reaction of the mussel *Mytilus galloprovincialis* (Bivalvia) to *Eugymnanthea inquilina* (Cnidaria) and *Urastoma cyprinae* (Turbellaria) concurrent infestation. *Journal of Invertebrate Pathology*, **110**, 118–125.

Moore, M.N., Lowe, D.M. & Gee, J.M. (1978) Histopathological effects induced in *Mytilus edulis* by *Mytilicola intestinalis* and the histochemistry of the copepod intestinal cells. *ICES Journal of Marine Science*, **38**, 6–11.

Morse, D.L. Rawson, P.D. & Kraeuter, J.N. (2015) Mud blister worms and oyster aquaculture. Maine Sea Grant Publications No. 46. Available at: https://digitalcommons.library.umaine.edu/cgi/viewcontent.cgi?article=1049&context=seagrant_pub (accessed 28 May 2021).

Mortensen, S., Harkestad, L., Stene, R.-O. & Renault, T. (2005) Picoeucaryot alga infecting blue mussel *Mytilus edulis* in southern Norway. *Diseases of Aquatic Organisms*, **63**, 25–32.

Mosca, F., Lanni, L., Cargini, D., Narcisi, V., Bianco, I. & Tiscar, P.G. (2013) Variability of the hemocyte parameters of cultivated mussel *Mytilus galloprovincialis* (Lmk 1819) in Sabaudia (Latina, Italy) coastal lagoon. *Marine Environmental Research*, **92**, 215–223.

Moynagh, P.N. (2005) The NF-kappaB pathway. *Journal of Cell Science*, **118**, 4589–4592.

Murchison, E.P. (2009) Clonally transmissible cancers in dogs and Tasmanian devils. *Oncogene*, **27**, S19–S30.

Murgarella, M., Puiu, D., Novoa, B., Figueras, A., Posada, D. & Canchaya, C. (2016) A first insight into the genome of the filter-feeder mussel *Mytilus galloprovincialis*. *PLoS ONE*, **11**, e0151561.

Murray, A.G., Marcos-Lopez, M., Collet, B. & Munro, L.A. (2012) A review of the risk posed to Scottish mollusc aquaculture from *Bonamia*, *Marteilia* and oyster herpesvirus. *Aquaculture*, **370–371**, 7–13.

Muttray, A.F., O'Toole, T.F., Morrill, W., Van Beneden, R.J. & Baldwin, S.A. (2010) An invertebrate *mdm* homolog interacts with *p53* and is differentially expressed together with *p53* and *ras* in neoplastic *Mytilus trossulus* haemocytes. *Comparative Biochemistry and Physiology Part-B*, **156**, 298–308.

Nguyan, S.T. (2018) Molecular detection of the apicomplexan parasite 'X' (APX) in bivalves in New Zealand. *PhD thesis*, University of Auckland, New Zealand.

Nikolaev, K.E., Sukhotin, A.A. & Galaktionov, K.V. (2006) Infection patterns in White Sea blue mussels *Mytilus edulis* of different age and size with metacercariae of *Himasthla elongata* (Echinostomatidae) and *Cercaria parvicaudata* (Renicolidae). *Diseases of Aquatic Organisms*, **71**, 51–58.

Nolsø, A. (2002) Ecology of endoparasites of the Atlantic puffin (*Fratercula arctica grabae*) (Alcidae: Charadriiformes). *Fróôskaparrit*, **50**, 131–142.

Novoa, B., Posada, D. & Figueras, A. (2005) Polymorphisms in the sequences of *Marteilia* internal transcribed spacer region of the ribosomal RNA genes (ITS-1) in Spain: genetic types are not related with bivalve hosts. *Journal of Fish Diseases*, **28**, 331–338.

Novoa, B., Romero, A., Álvarez, Á.L., Moreira, R., Pereiro, P., Costa, M.M. *et al.* (2016) Antiviral activity of myticin C peptide from mussel: an ancient defense against herpesviruses. *Journal of Virology*, **90**, 7692–7702.

OIE (2016) *Manual of Diagnostic Tests for Aquatic Animals 2016*. Chapter 2.4.4.–Infection with *Marteilia refringens*. The electronic version of the manual is available at: http://www.oie.int/international-standard-setting/aquatic-manual/access-online/.

Olicard, C., Didier, Y., Marty, C., Bourgougnon, N. & Renault, T. (2005) *in vitro* research of anti-HSV-1 activity in different extracts from Pacific oysters *Crassostrea gigas*. *Diseases of Aquatic Organisms*, **67**, 141–147.

Olivas-Valdez, J.A. & Cáceres-Martínez, J. (2002) Infestation of the blue mussel *Mytilus galloprovincialis* by the copepod *Pseudomyicola spinosus* and its relation to size, density, and condition index of the host. *Journal of Invertebrate Pathology*, **79**, 65–71.

Özer, A. & Güneydağ, S. (2014) First report of some parasites from Mediterranean mussel, *Mytilus galloprovincialis* Lamarck, 1819, collected from the Black Sea coast at Sinop. *Turkish Journal of Zoology*, **38**, 486–490.

Padovan, I.P., Tavares, L.A., Corral, L.P., Padovan, A. & Azevedo, C. (2003) Fine structure of the oocyst of *Nematopsis mytella* (Apicomplexa, Porosporidae), a parasite of the mussel, *Mytella falcata* and of the oyster *Crassostrea rizophorae* (Mollusca, Bivalvia) from the Northeastern Atlantic coast of Brazil. *Journal of Morphology Sciences*, **20**, 141–145.

Paillard, C. (2004) A short-review of brown ring disease, a vibriosis affecting clams, *Ruditapes philippinarum* and *Ruditapes decussatus*. *Aquatic Living Resources*, **17**, 467–475.

Pallavicini, A., Costa, M.M., Gestal, C., Dreos, R., Figueras, A., Venier, P. *et al.* (2008) High sequence variability of myticin transcripts in hemocytes of immune-stimulated mussels suggests ancient host-pathogen interactions. *Developmental and Comparative Immunology*, **32**, 213–226.

Parisi, M.G., Toubiana, M., Mangano, V., Parrinello, N., Cammarata, M. & Roch, P. (2012) MIF from mussel: coding sequence, phylogeny, polymorphism, 3D model and regulation of expression. *Developmental and Comparative Immunology*, **36**, 688–696.

Parisi, M.G., Mauro, M.G., Sarà G. & Cammarata, M. (2017) Temperature increases, hypoxia, and changes in food availability affect immunological biomarkers in the marine mussel *Mytilus galloprovincialis*. *Journal of Comparative Physiology B*, **187**, 1117–1126.

Pascual, S., Villalba, A., Abollo, E., Garci, M., González, A.F., Nombela, M. *et al.* (2010) The mussel *Xenostrobus securis*: a well-established alien invader in the Ria de Vigo (Spain, NE Atlantic). *Biological Invasions*, **12**, 2091–2103.

Paul-Pont, I., Lacroix, C., González Fernández, C., Hégaret, H., Lambert, C., Le Goïc, N.*et al.* (2016) Exposure of marine mussels *Mytilus* spp. to polystyrene microplastics: toxicity and influence on fluoranthene bioaccumulation. *Environmental Pollution*, **216**, 724–737.

PEIDAF (2001) Mycotic Sloughing Periostracum (MSP) in Mussels. AquaInfo Notes (AIN 01). Available at: https://www.princeedwardisland.ca/en/information/agriculture-and-fisheries/aqua-info-notes (acecssed 28 May 2021).

Pëllumb, A., Giuseppe, C. & Kapllan, S. (2006) First report of marteiliosis in *Mytilus galloprovincialis* in Albania. *Ittiopatologia*, **3**, 47–52.

Philipp, E.E.R., Kraemer, L., Melzner, F., Poustka, A.J. & Thieme, S. (2012) Massively parallel RNA sequencing identifies a complex immune gene repertoire in the lophotrochozoan *Mytilus edulis*. *PLoS ONE*, **7**, e33091.

Pogoda, B., Jungblut, S., Buck, B.H. & Hagen, W. (2012) Infestation of oysters and mussels by mytilicolid copepods: differences between natural coastal habitats and two offshore cultivation sites in the German Bight. *Journal of Applied Ichthyology*, **28**, 756–765.

Ponnappan, N., Budagavi, D.P., Yadav, B.K. & Chugh, A. (2015) Membrane-active peptides from marine organisms-antimicrobials, cell-penetrating peptides and peptide toxins: applications and prospects. *Probiotics and Antimicrobial Proteins*, **7**, 75–89.

Powell, E.N., Barber, R.D., Kennicutt II, M.C. & Ford, S.E. (1999) Influence of parasitism in controlling the health, reproduction and PAH body burden of petroleum seep mussels. *Deep-Sea Research I*, **46**, 2053–2078.

Rasmussen, L.P.D. (1986) Virus-associated granulocytomas in the marine mussel, *Mytilus edulis*, from three sites in Denmark. *Journal of Invertebrate Pathology*, **48**, 117–123.

Ray, S., Mukherjee, S., Bhunia, N.S., Bhunia, A.S. & Ray, M. (2015) Immunotoxicological threats of pollutants in aquatic invertebrates. In: *Emerging Pollutants in the Environment – Current and Further Implications* (eds M.L. Larramendy & S. Soloneski), pp.146–165. IntechOpen, London.

Rayyan, A. & Chintiroglou, C.C. (2003) *Steinhausia mytilovum* in cultured mussels *Mytilus galloprovincialis* in the Thermaikos Gulf (northern Aegean Sea, Greece). *Diseases of Aquatic Organisms*, **57**, 271–273.

Rayyan, A., Damianidis, P., Antoniadou, C. & Chintiroglou, C.C. (2006) Protozoan parasites in cultured mussels *Mytilus galloprovincialis* in the Thermaikos Gulf (north Aegean Sea, Greece). *Diseases of Aquatic Organisms*, **70**, 251–254.

Read, G.B. (2010) Comparison and history of *Polydora websteri* and *P. haswelli* (Polychaeta: Spionidae) as mud-blister worms in New Zealand shellfish. *New Zealand Journal of Marine and Freshwater Research*, **44**, 83–100.

Rebelo, Mde F., Figueiredo, Ede S., Mariante, R.M., Nóbrega, A., de Barros, C.M. & Allodi, S. (2013) New insights from the oyster *Crassostrea rhizophorae* on bivalve circulating hemocytes. *PLoS ONE*, **8**, e57384.

Renault, T. (2008) Shellfish viruses. In: *Encyclopedia of Virology* (eds B.W.J. Mahy & M.H.V. Van Regenmortel), pp. 560–567. Elsevier, Oxford.

Renault, T. (2015) Immunotoxicological effects of environmental contaminants on marine bivalves. *Fish and Shellfish Immunology*, **46**, 88–93.

Renault, T. & Ford, S. (2017) Disseminated neoplasms in bivalves. ICES Identification Leaflets for Diseases and Parasites of Fish and Shellfish No. 67 (revised: originals by D.J. Alderman & M. Green (No. 11) and G. Balouet (No. 12).

Renault, T., Cochennec, N. & Grizel, H. (1995) *Bonamia ostreae*, parasite of the European flat oyster, *Ostrea edulis*, does not experimentally infect the Japanese oyster, *Crassostrea gigas*. *Bulletin of the European Association of Fish Pathologists*, **15**, 78–80.

Renault, T., Lipart, C. & Arzul, I. (2001) A herpes-like virus infects a non-ostreid bivalve species: virus replication in *Ruditapes philippinarum* larvae. *Diseases of Aquatic Organisms*, **45**, 1–7.

Renwrantz, L., Schmalmack, W., Redel, R., Friebel, B. & Schneeweiss, H. (1996) Conversion of phenoloxidase and peroxidase indicators in individual haemocytes of *Mytilus edulis* specimens and isolation of phenoloxidase from haemocyte extract. *Journal of Comparative Physiology B*, **165**, 647–658.

Rey-Campos, M., Moreira, R., Valenzuela-Muñoz, V., Gallardo-Escárate, C., Novoa, B. & Figueras, A. (2019) High individual variability in the transcriptomic response of Mediterranean mussels to *Vibrio* reveals the involvement of myticins in tissue injury. *Scientific Reports*, **9**, art. 3569.

Riascos, J.M., Heilmayer, O., Oliva, M.E., Laudien, J. & Arntz, W.E. (2008) Infestation of the surf clam *Mesodesma donacium* by the spionid polychaete *Polydora bioccipitalis*. *Journal of Sea Research*, **59**, 217–227.

Robledo, J.A.F., Cáceres-Martínez, J., Sluys, R. & Figueras, A. (1994a) The parasitic turbellarian *Urastoma cyprinae* (Platihelminthes: Urastomidae) from blue mussel *Mytilus galloprovincialis* in Spain: occurrence and pathology. *Diseases of Aquatic Organisms*, **18**, 203–210.

Robledo, J.A.F., Santarém, M.M. & Figueras, A. (1994b) Parasite loads of rafted blue mussels (*Mytilus galloprovincialis*) in Spain with special reference to the copepod, *Mytilicola intestinalis*. *Aquaculture*, **127**, 287–302.

Robledo, J.A.F., Santarém, M.M., González, P. & Figueras, A. (1995) Seasonal variations in the biochemical composition of the serum of *Mytilus galloprovincialis* Lmk. and its relationship to the reproductive cycle and parasitic load. *Aquaculture*, **133**, 311–322.

Rodríguez, F., Feist, S.W., Guillou, L., Harkestad, L.S., Bateman, K., Renault, T. *et al.* (2008) Phylogenetic and morphological characterisation of the green algae infesting blue mussel *Mytilus edulis* in the North and South Atlantic oceans. *Diseases of Aquatic Organisms*, **81**, 231–240.

Romero, A., Dios, S., Poisa-Beiro, L., Costa, M.M., Posada, D., Figueras, A. *et al.* (2011) Individual sequence variability and functional activities of fibrinogen-related proteins (FREPs) in the Mediterranean mussel (*Mytilus galloprovincialis*) suggest ancient and complex immune recognition models in invertebrates. *Developmental and Comparative Immunology*, **35**, 334–344.

Romero, A., Novoa, B. & Figueras, A. (2012) Genomics, immune studies and diseases in bivalve aquaculture. *Invertebrate Survival Journal*, **9**, 110–121.

Romero, A., Novoa, B. & Figueras, A. (2015) The complexity of apoptotic cell death in mollusks: an update. *Fish and Shellfish Immunology*, **46**, 79–87.

Rosani, U., Varotto, L., Rossi, A., Roch, P., Novoa, B. Figueras, A. *et al.* (2011) Massively parallel amplicon sequencing reveals isotype-specific variability of antimicrobial peptide transcripts in *Mytilus galloprovincialis*. *PLoS ONE*, **6**, e26680.

Sakaguchi, S. (1973) A larval tapeworm, *Tylocephalum*, in marine molluscan shellfishes. (In Japanese; English summary). *Bulletin of the Nansei Regional Fisheries Research Laboratory*, **6**, 1–8.

Schikorski, D., Faury, N., Pepin, J.F., Saulnier, D., Tourbiez, D. & Renault, T. (2011) Experimental Ostreid herpesvirus 1 (OsHV-1) infection of the Pacific oyster *Crassostrea gigas*: kinetics of virus DNA detection by q-PCR in seawater and in oyster samples. *Virus Research*, **155**, 28–34.

Schneeweiß, H. & Renwrantz, L. (1993) Analysis of the attraction of haemocytes from *Mytilus edulis* by molecules of bacterial origin. *Developmental and Comparative Immunology*, **17**, 377–387.

Sekine, D., Ohishi, K., Nakamura, Y., Kusaka, C., Tame, A. Inoue, K. *et al.* (2016) Monoclonal antibodies to hemocytes of the deep-sea symbiotic mussel, *Bathymodiolus japonicas*. *JAMSTEC Report of Research and Development*, **23**, 27–33.

Semreen, M.H., El-Gamal, M.I., Abdin, S., Alkhazraji, H., Kamal. L., Hammad, S. *et al.* (2018) Recent updates of marine antimicrobial peptides. *Saudi Pharmaceutical Journal*, **26**, 396–409.

Serracca, L., Prearo, M., Rossini, I., Battistini, R., Corsi, M. & Ercolini, C. (2014) Four-year surveillance of *Marteilia refringens* in shellfish farms in the Gulf of La Spezia (Liguria, Italy). *Turkish Journal of Fisheries and Aquatic Sciences*, **14**, 893–896.

Shaposhnikov, A.V., Komar'kov, I.F., Lebedeva, L.A. & and Shidlovskii, Y.V. (2013) Molecular components of JAK/STAT signaling pathway and its interaction with transcription machinery. *Molecular Biology*, **47**, 343–351.

da Silva, P.M., Magalhães, A.R.M. & Barracco, M.A. (2002) Effects of *Bucephalus* sp. (Trematoda: Bucephalidae) on *Perna perna* mussels from a culture station in Ratones Grande Island, Brazil. *Journal of Invertebrate Pathology*, **79**, 154–162.

da Silva, P.M., Magalhães, A.R.M. & Barracco, M.A. (2012) Pathologies in commercial bivalve species from Santa Catarina State, southern Brazil. *Journal of the Marine Biological Association of the UK.*, **92**, 571–579.

Simões, E, Campos Vieira, R., Schramm, M.A., Mello, D.F., De Almeida Pontinha, V., Da Silva, P.M. *et al.* (2015) Impact of harmful algal blooms (*Dinophysis acuminata*) on the immune system of oysters and mussels from Santa Catarina, Brazil. *Journal of the Marine Biological Association of the UK*, **95**, 773–781.

Singh, P. & Raghukumar, C. (2014) Diversity and physiology of deep-sea yeasts: a review. *Kavaka*, **43**, 50–63.

Sokolnikova, Y., Magarlamov, T., Stenkova, A. & Kumeiko, V. (2016) Permanent culture and parasitic impact of the microalga *Coccomyxa parasitica*, isolated from horse mussel *Modiolus kurilensis*. *Journal of Invertebrate Pathology*, **140**, 25–34.

Song, L., Wang, L., Qiu, L. & Zhang, H. (2010) Bivalve immunity. In: *Invertebrate Immunity* (ed. K. Söderhäll), pp. 44–65. Landes Bioscience and Springer, Austin, TX.

Soudant, P., Chu, F.L.E. & Volety, A. (2013) Host–parasite interactions: marine bivalve molluscs and protozoan parasites, *Perkinsus* species. *Journal of Invertebrate Pathology*, **114**, 196–216.

Snowden, E. (2007) *Cliona celata*: a sponge. In: *Marine Life Information Network: Biology and Sensitivity Key Information Reviews* (eds H. Tyler-Walters & K. Hiscock). Available at: http://www.marlin.ac.uk/species/detail/2188 (accessed 28 May 2021).

Srisuphanunt, A.M., Saksirisampant, W. & Karanis, P. (2009) Detection of *Cryptosporidium* oocysts in green mussels (*Perna viridis*) from shell-fish markets of Thailand. *Parasite*, **16**, 235–239.

Stanley, J, A., Radford, C, A & Jeffs, A.G. (2009) Induction of settlement in crab megalopae by ambient underwater reef sound. *Behavioral Ecology*, **21**, 113—120.

Steele, S. & Mulcahy, M.F. (2001) Impact of the copepod *Mytilicola orientalis* on the Pacific oyster *Crassostrea gigas* in Ireland. *Diseases of Aquatic Organisms*, **47**,145–149.

Stentiford, G.D., Bateman, K.S., Stokes, N.A. & Carnegie, R.B. (2013) *Haplosporidium littoralis* sp. nov.: a crustacean pathogen within the Haplosporida (Cercozoa, Ascetosporea). *Diseases of Aquatic Organisms*, **105**, 243–252.

Stevens, P.M. (1990) Specificity of host recognition of individuals from different host races of symbiotic pea crabs (Decapoda: Pinnotheridae). *Journal of Experimental Marine Biology and Ecology*, **143**, 193—207.

Stier, T., Drent, J. & Thieltges, D.W. (2015) Trematode infections reduce clearance rates and condition in blue mussels *Mytilus edulis*. *Marine Ecology Progress Series*, **529**, 137–144.

Sun, J. & Wu, X. (2009) Nitric oxide and superoxide anion generation in hemocytes of *Crassostrea ariakensis* stimulated with rickettsia-like organisms and zymosan. *Journal of Shellfish Research*, **28**, 821–826.

Sun, T., Tang, X., Jiang Y. & Wang, Y. (2017) Seawater acidication induced immune function changes of haemocytes in *Mytilus edulis*: a comparative study of CO_2 and HCl enrichment. *Scientific Reports*, **7**, 41488.

Sun, W., Sun, S., Wang, Y., Yang, B. & Song, W. (2006) The prevalence of the pea crab, *Pinnotheres sinensis*, and its impact on the condition of the cultured mussel, *Mytilus galloprovincialis*, in Jiaonan waters (Shandong Province, China). *Aquaculture*, **253**, 57—63.

Sunil, K.P. (2015) Histological and ultra structural changes in the mantle of *Perna indica* (brown mussel, Kuriakose and Nair, 1976) infested by bioeroding sponges. *Journal of Aquaculture and Marine Biology*, **2**, 00051.

Sunil, K.P. & Thomas, P.A. (2012) Discovery of *Alectona wallichii* (Carter, 1874) from the Indian seas as a pest of brown mussel with notes on the zoogeography and substratum preference in *Alectona* spp. (Demospongiae: Alectonidae). *Journal of the Marine Biological Association of India*, **54**, 80–84.

Sunil K. P. & Thomas, P.A. (2015) Systematics, zoogeography and affinity of boring sponges infesting the brown mussel, *Perna indica* Kuriakose and Nair from the southwest coast of India. *Journal of the Marine Biological Association of India*, **57**, 37–45.

Sunila, I., Williams, L., Russo, S. & Getchis, T. (2004) Production and pathology of blue mussels, *Mytilus edulis* (L.) in an experimental longline in Long Island Sound, Connecticut. *Journal of Shellfish Research*, **23**, 731–740.

Syasina, I.G. (2011) Histopathology of *Modiolus modiolus* infected with green microalga. In: *Bridging America and Russia with Shared Perspectives on Aquatic Animal Health* (eds

R.C. Cipriano, A.W. Bruckner & I.S. Shchelkinov), pp. 56–62. Proceedings of the Third Bilateral Conference between Russia and the United States, 12–20 July, 2009, Shepherdstown, USA, WV. Khaled bin Sultan Living Oceans Foundation, Landover, MD.

Syasina, I.G., Kukhlevsky, A.D., Kovaleva, A.L. & Vaschenko, M.A. (2012) Phylogenetic and morphological characterization of the green alga infesting the horse mussel *Modiolus modiolus* from Vityaz Bay (Peter the Great Bay, Sea of Japan). *Journal of Invertebrate Pathology*, **111**, 175–181.

Takeuchi, T., Koyanagi, R., Gyoja, F., Kanda, M., Hisata, K., Fujie, M. *et al.* (2016) Bivalve-specific gene expansion in the pearl oyster genome: implications of adaptation to a sessile lifestyle. *Zoological Letters*, **2**, 3.

Tame, A., Yoshida, T., Ohishi, O. & Maruyama, T. (2015) Phagocytic activities of hemocytes from the deep-sea symbiotic mussels *Bathymodiolus japonicus*, *B. platifrons*, and *B. septemdierum*. *Fish and Shellfish Immunology*, **45**, 146–156.

Tanguy, M., McKenna, P., Gauthier-Clerc, S., Pellerin, J., Danger, J.-M. & Siah, A. (2013) Sequence analysis of a normalized cDNA library of *Mytilus edulis* hemocytes exposed to *Vibrio splendidus* LGP32 strain. *Results in Immunology*, **3**, 40–50.

Taraska, N.G. & Böttger, S.A. (2013) Selective initiation and transmission of disseminated neoplasia in the softshell clam *Mya arenaria* dependent on natural disease prevalence and animal size. *Journal of Invertebrate Pathology*, **112**, 94–101.

Teia dos Santos, A.M.T. & Coimbra, J. (1995) Growth and production of raft-cultured *Mytilus edulis* L. in Ria de Aveiro: gonad symbiotic infestation. *Aquaculture*, **132**, 195–211.

Thébault A., Bergmann S., Pouillot S., Le Roux F. & Berthe F.C.J. (2005) Validation of *in situ* hybridization and histology assays for the detection of the oyster parasite *Marteilia refringens*. *Diseases of Aquatic Organisms* **65**, 9–16.

Thieltges, D.W. (2006) Effect of metacercarial trematode infections (*Renicola roscovita*) on growth in intertidal blue mussels (*Mytilus edulis*). *Marine Ecology Progress Series*, **319**, 129–134.

Thieltges, D.W., Hussel, B., Hermann, J., Jensen, K.T., Krakau, M., Taraschewski, H. *et al.* (2008) Parasites in the northern Wadden Sea: a conservative ecosystem component over 4 decades. *Helgoland Marine Research*, **62**, 37–47.

Thieltges, D.W., Reise, K., Prinz, K. & Thomas Jensen, K. (2009) Invaders interfere with native parasite–host interactions. *Biological Invasions*, **11**, 1421–1429.

Thieltges, D.W., Engelsma, M., Wendling, C.C. & Wegner, K.M. (2013) Parasites in the Wadden Sea food web. *Journal of Sea Research*, **82**, 122–133.

Tiscar, P.G. & Mosca, F. (2004) Defense mechanisms in farmed marine molluscs. *Veterinary Research Communications*, **28**, 57–62.

Toubiana, M., Gerdol, M., Rosani, U., Pallavicini, A., Venier, P. & Roch, P. (2013) Toll-like receptors and MyD88 adaptors in *Mytilus*: complete cds and gene expression levels. *Developmental and Comparative Immunology*, **40**, 158–166.

Toubiana, M., Rosani, U., Giambelluca, S., Cammarata, M., Gerdol, M., Pallavicini, A., *et al.* (2014) Toll signal transduction pathway in bivalves: complete cds of intermediate elements and related gene transcription levels in hemocytes of immune stimulated *Mytilus galloprovincialis*. *Developmental and Comparative Immunology*, **45**, 300–312.

Travers, M.A., Boettcher, Miller, K., Roque, A. & Friedman, C.S. (2015) *Bacterial diseases in marine bivalves*. *Journal of Invertebrate Pathology*, **131**, 11–31.

Trottier, O. &, Jeffs, A.G. (2012) Biological characteristics of parasitic *Nepinnotheres novaezelandiae* within a *Perna canaliculus* farm. *Diseases of Aquatic Organisms*, **101**, 61–68.

Trottier, O. &, Jeffs, A.G. (2015a) Mate locating and access behaviour of the parasitic pea crab, *Nepinnotheres novaezelandiae*, an important parasite of the mussel *Perna canaliculus. Parasite*, **22**, 13

Trottier, O. & Jeffs, A.G. (2015b) Recruitment of the parasitic pea crab *Nepinnotheres novaezelandiae* into green-lipped mussels *Perna canaliculus. Diseases of Aquatic Organisms*, **112**, 199–205.

Trottier, O., Walker, D. & Jeffs, A.G. (2012) Impact of the parasitic pea crab *Pinnotheres novaezelandiae* on aquacultured New Zealand green-lipped mussels, *Perna canaliculus. Aquaculture*, **344–349**, 23–28.

Tuntiwaranuruk, C., Chalermwat, K., Upatham, E.S., Kruatrachue, M. & Azevedo, C. (2004) Investigation of *Nematopsis* spp. oocysts in 7 species of bivalves from Chonburi Province, Gulf of Thailand. *Diseases of Aquatic Organisms*, **58**, 47–53.

Van Beneden, R.J., Rhodes, L.D. & Gardner, G.R. (1999) Potential alterations in gene expression associated with carcinogen exposure in *Mya arenaria. Biomarkers*, **4**, 485–491.

Van Dover, C.L., Ward, M.E., Scott, J.L., Underdown, J., Anderson, B., Gustafson, C. *et al.* (2007) A fungal epizootic in mussels at a deep-sea hydrothermal vent. *Marine Ecology*, **28**, 54–62.

Vaschenko, M.A., Kovaleva, A.L., Syasina, I.G. & Kukhlevsky, A.D. (2013) Reproduction-related effects of green alga *Coccomyxa* sp. infestation in the horse mussel *Modiolus modiolus. Journal of Invertebrate Pathology*, **113**, 86–95.

Vasta, G.R., Feng, C., Bianchet, M.A., Bachvaroff, T.R. & Tasumi, S. (2015) Structural, functional, and evolutionary aspects of galectins in aquatic mollusks: from a sweet tooth to the Trojan horse. *Fish and Shellfish Immunology*, **46**, 94–106.

Vázquez, N., Rodríguez, F., Ituarte, C., Klaich, J. & Cremonte, F. (2010) Host-parasite relationship of the geoduck *Panopea abbreviata* and the green alga *Coccomyxa parasitica* in the Argentinean Patagonian coast. *Journal of Invertebrate Pathology*, **105**, 254–260.

Venier, P., Pallavicini, A., De Nardi, B. & Lanfranchi, G. (2003) Towards a catalogue of genes transcribed in multiple tissues of *Mytilus galloprovincialis. Gene*, **314**, 29–40.

Venier, P., De Pittà, C., Bernante, F., Varotto, L., De Nardi, B., Bovo, G. *et al.* (2009) MytiBase: a knowledgebase of mussel (*M. galloprovincialis*) transcribed sequences. *BMC Genomics*, **10**, 72.

Venier, P., Varotto, L., Rosani, U., Millino, C., Celegato, B., Bernante, F. *et al.* (2011) Insights into the innate immunity of the Mediterranean mussel *Mytilus galloprovincialis. BMC Genomics*, **12**, 69–87.

Villalba, A., Mourelle, S.G., López, M.C., Carballal, M.J. & Azevedo, C. (1993) Marteiliasis affecting cultured mussels *Mytilus galloprovincialis* of Galicia (NW Spain). I. Etiology, phases of the infection, and temporal and spatial variability in prevalence. *Diseases of Aquatic Organisms*, **16**, 61–72.

Villalba, A., Mourelle, S.G., Carballal, M.J. & López, C. (1997) Symbionts and diseases of farmed mussels *Mytilus galloprovincialis* throughout the culture process in the Rias of Galicia (NW Spain). *Diseases of Aquatic Organisms*, **31**, 127–139.

Vitellaro-Zuccarello, L. (1973) Ultrastructure of the byssal apparatus of *Mytilus gaJloprovincialis*. I. Associated fungal hyphae. *Marine Biology*, **22**, 225–230.

Walker, C.W., Böttger, S.A., Mulkern, J., Jerszyk, E.C., Litvaitis, M. & Lesser, M.P. (2009) Mass culture and characterization of tumor cells from a naturally occurring invertebrate cancer model: applications for human and animal disease and environmental health. *Biological Bulletin*, **216**, 23–39.

Walker, C.W., Van Beneden, R.J., Muttray, A.F., Böttger, S.R., Kelley, M.L., Tucker, A.E. *et al.* (2011) P53 superfamily proteins in marine bivalve cancer and stress biology. *Advances in Marine Biology*, **59**, 1–36.

Wang, Q., Zhang, L., Zhao, J., You, L. & Wu, H. (2012) Two goose-type lysozymes in *Mytilus galloprovincialis*: possible function diversification and adaptive evolution. *PLoS One*, **7**, e45148.

Wang, Q., Wang, C., Mu, C., Wu, H., Zhang, L. & Zhao, J. (2013) A novel C-type lysozyme from *Mytilus galloprovincialis*: insight into innate immunity and molecular evolution of invertebrate C-type lysozymes. *PloS One*, **8**, e67469.

Wang, Y., Hu, M., Chiang, M.W.L., Shin, P.K.S. & Cheung, S.G. (2012) Characterization of subpopulations and immune-related parameters of hemocytes in the green-lipped mussel *Perna viridis*. *Fish and Shellfish Immunology*, **32**, 381–390.

Ward, M.E., Shields, J.D. & Van Dover, C.L. (2004) Parasitism in species of *Bathymodiolus* (Bivalvia: Mytilidae) mussels from deep-sea seep and hydrothermal vents. *Diseases of Aquatic Organisms*, **62**, 1–16.

Welch, J.M., Rittschof, D., Bullock, T.M. & Forward, R.B. Jr (1997) Effects of chemcial cues on settlement behaviour of blue crab *Callinectes sapidus* postlarvae. *Marine Ecology Progress Series*, **154**, 143–153.

Welsh, J.E., van der Meer, J., Brussaard, C.P. & Thieltges, D.W. (2014) Inventory of organisms interfering with transmission of a marine trematode. *Journal of the Marine Biological Association of the UK*, **94**, 697–702.

Wilcox, R. (2013) *Review on the incidence of pearling in the blue mussel, Mytilus edulis: with relevance to the mussel beds of the Walney Channel, Barrow-in-Furness, Cumbria, England*. Cumbria Wildlife Trust, Cumbria.

Wilson, J.G., Galaktionov, K.V., Sukhotin, A.A., Skírnisson, K., Nikolaev, K.E., Ivanov, M.I. *et al.* (2013) Factors influencing trematode parasite burdens in mussels (*Mytilus* spp.) from the North Atlantic Ocean across to the North Pacific. *Estuarine and Coastal Shelf Science*, **132**, 87–93.

Wong, J.H., Xia, L. & Ng, T.B. (2007) A review of defensins of diverse origins. *Current Protein and Peptide Science*, **8**, 446–459.

Wong, Y.H., Sun, J., He, L.S., Chen, L.G., Qiu, J.-W. & Qian, P.-Y. (2015) High-throughput tran-scriptome sequencing of the cold seep mussel *Bathymodiolus platifrons*. *Science Reports*, **5**, 16597.

Wootton, E.C., Dyrynda, E.A. & Ratcliffe, N.A. (2006) Interaction between non-specific electrostatic forces and humoral factors in haemocyte attachment and encapsulation in the edible cockle, *Cerastoderma edule*. *Journal of Experimental Biology*, **209**, 1326–1335.

Yanyan, Z., Ren, S., Wang, D. & Song, W. (2006) Structure and classification of haemocytes in the bivalve mollusc *Meretrix meretrix*. *Journal of Ocean University of China*, **5**, 132–136.

Yonemitsu, M.A., Giersch, R.M., 1 Polo-Prieto, M., Hammel, M., Simon, A. Cremonte, F. *et al.* (2019) A single clonal lineage of transmissible cancer identified in two marine mussel species in South America and Europe. *eLife*, **8**, e47788.

Yu, H., He, Y., Wang, X., Zhang, Q., Bao, Z. & Guo, X. (2011) Polymorphism in a serine protease inhibitor gene and its association with disease resistance in the eastern oyster (*Crassostrea virginica* Gmelin). *Fish and Shellfish Immunology*, **30**, 757–762.

Zhang, G., Fang, X., Guo, X., Li, L., Luo, R., Xu, F. *et al.* (2012) The oyster genome reveals stress adaptation and complexity of shell formation. *Nature*, **490**, 49–54.

Zrničić, S., Le Roux, F., Oraić, D., Šoštarić, B. & Berthe, F. (2001) First record of *Marteilia* sp. in mussels *Mytilus galloprovincialis* in Croatia. *Diseases of Aquatic Organisms*, **44**, 143–148.

Zuykov, M., Belzile, C., Lemaire, N., Gosselin, M., Dufresne, F. & Pelletier, E. (2014) First record of the green microalgae *Coccomyxa* sp. in blue mussel *Mytilus edulis* (L.) from the Lower St. Lawrence Estuary (Québec, Canada). *Journal of Invertebrate Pathology*, **120**, 23–32.

12

Mussels and Public Health

Introduction

Since bivalves are filter feeders, they can concentrate a variety of materials from the environment. This, coupled with the tradition of consuming bivalves raw or partially cooked, means that they act as potential vectors for human infection from waterborne agents such as bacteria, viruses, algal toxins and heavy metals, and thus are high-risk products. Control of this disease risk requires integrated management of the water environment used for shellfish growing and harvesting together with post-harvest product processing, which might involve depuration and/or heat treatment where appropriate (WHO 2010). This chapter describes the role of mussels in transmission of disease to humans under the following headings: bacterial infections, viral infections, parasites, biotoxins, and industrial pollutants. Methods of bacterial, viral and toxin detection, decontamination procedures, monitoring and food safety measures are also covered.

Bacterial Infections

Most cultured and wild bivalves used for consumption are found in shallow, sheltered, inshore waters with high nutrient levels, an ideal environment for bacterial growth. In populated areas, such waters are often subject to contamination from domestic sewage and land run-off, which typically contain human bacterial pathogens such as *Salmonella*, *Vibrio*, *Campylobacter* and *Shigella* spp., as well as the faecal coliforms *Escherichia coli* and *Enterococcus faecalis* (Sapkota *et al.* 2008 and references therein).

Salmonella

Salmonella is a rod-shaped, Gram-negative, facultative anaerobe in the family Enterobacteriaceae (Jay *et al.* 2003). More than 2500 serotypes[1] have been recorded, of which the majority (59%) belong to S. *enterica* serotype Typhi, which is also responsible for 99% of *Salmonella* infections in humans and warm-blooded animals (Brenner *et al.* 2000). *Salmonella* can multiply and survive in estuarine environments and tropical freshwater

Marine Mussels: Ecology, Physiology, Genetics and Culture, First Edition. Elizabeth Gosling.

environments for weeks, although open marine waters are free from the pathogen (Huss *et al.* 2004). Compared to other Gram-negative bacteria, *Salmonella* spp. are relatively resistant to various environmental factors. They grow at temperatures between 5 and 47 °C, but have also been reported to grow within the range 2–54 °C. They can survive on frozen fruit at −20 °C for at least 180 days (FSANZ 2011). Before the 1950s, the most common illness associated with the consumption of raw bivalves in the United States was typhoid fever, a serious infection marked by intestinal inflammation and ulceration caused by *Salmonella enterica* serotype Typhi, usually referred to as *Salmonella typhi*, an obligate parasite that has no known natural reservoir outside of humans. The fever can also be caused by *Salmonella paratyphi*, a related bacterium that usually causes a less severe illness. The World Health Organization (WHO) estimates that there are about 16 million clinical *S. typhi* cases worldwide annually, of which about 600 000 result in death. An estimated 5700 cases occur each year in the United States. Most of these (up to 75%) are acquired while traveling internationally (CDC 2021b). The low infection rate in the United States is primarily due to superior sewage treatment and better water quality standards in bivalve-growing areas. Control measures aimed at detecting contamination of shellfish harvest waters (e.g. monitoring faecal coliform counts in these waters) and other public health measures have been effective in reducing *Salmonella* contamination of seafood that occurs before harvest (Iwamoto *et al.* 2010). However, shellfish can also become contaminated with *Salmonella* during the storage and processing stages, although adequate cooking and proper storage and handling after harvesting can prevent infections. Heating at 57–60 °C is effective in killing the bacteria. Antibiotic therapy is the only effective treatment for typhoid fever.

Nontyphoidal *Salmonella* infections have emerged as a public health problem since the 1950s (Iwamoto *et al.* 2010). Nearly 1.4 million cases of salmonellosis occur each year in the United States, of which 95% are foodborne (Zhao *et al.* 2003). A variety of foods have been implicated as vehicles transmitting salmonellosis to humans, including poultry, beef, pork, eggs, milk, cheese, fish, shellfish, fresh fruit, fruit juice and vegetables. Although *Salmonella* is a major foodborne pathogen, products of aquaculture are rarely involved in outbreaks of salmonellosis (FAO 2010). Nontyphoidal *Salmonella* is a leading cause of bacterial diarrhoea worldwide, estimated to cause approximately 153 million cases of gastroenteritis and 57 000 deaths globally each year (Hunter & Watkins 2017). The faecal–oral route is the mechanism of transmission, whether through consumption of contaminated food or water, person-to-person contact or direct contact with infected animals (Jay *et al.* 2003). Outcomes of exposure can range from no effect to colonisation of the gastrointestinal tract without symptoms of illness to colonisation with the typical symptoms of acute gastroenteritis. Gastroenteritis symptoms are generally mild and may include abdominal cramps, nausea, diarrhoea, mild fever, vomiting, dehydration, headache and prostration. The incubation period is 8–72 hours (usually 24–48 hours), and symptoms last for 2–7 days (WHO/FAO 2002). Severe infections are generally treated with antimicrobial agents, which include ampicillin, ceftriaxone and ciprofloxacin. However, widespread drug-resistant nontyphoidal *Salmonella* and the associated complications in the treatment of infection have constituted a serious threat to public health (reviewed in Su *et al.* 2004). In a recent study, Medalla *et al.* (2017) estimated the incidence of antimicrobial drug-resistant nontyphoidal *Salmonella* infections in the United States between 2004 and 2012.

Ampicillin-only resistance was the most common resistance pattern detected, followed by ceftriaxone/ampicillin resistance and ciprofloxacin nonsusceptibility. Currently, nontyphoidal *Salmonella* causes an estimated 100 000 antimicrobial-resistant infections in the United States annually (Angelo *et al.* 2016).

Several studies have examined shellfish for the presence of *Salmonella* spp. For example, nearly 3000 samples of live bivalves (mussels, oysters and clams; no species names given) from Galicia, NW Spain were tested for *Salmonella* between September 1998 and November 2001 (Martinez-Urtaza *et al.* 2003). The overall incidence was 1.8%, with mussels and oysters presenting a higher incidence than clams. Nearly 67% of total *Salmonella* was isolated from bivalves with faecal coliform levels below the maximum level set by EU regulations (see later), indicating that low levels of faecal coliforms do not necessarily indicate the absence of *Salmonella*. A total of nine serotypes were isolated: *Salmonella* Senftenberg was the most frequent (50%), followed by *Salmonella typhi* (18%) and *Salmonella* Agona (17%). In a more recent study, Setti *et al.* (2009) investigated the presence of *Salmonella* along 122 km of the coastline of Agadir, southern Morocco in relation to environmental parameters. A total of 801 samples of seawater (243), marine sediment (279) and *Mytilus galloprovincialis* (279) were collected from six sites between July 2004 and May 2008. The overall prevalence of *Salmonella* was 7.1%, with the highest occurrence in mussels (10%), followed by sediment (6.8%) and seawater (4.1%). Only three serotypes were identified among the 57 *Salmonella* isolates: *S. enterica* serotype Blockley represented 43.8% of all isolates and was identified in mussel and sediment samples, *S. enterica* serotype Kentucky (29.8%) was found almost exclusively in mussels and *S. enterica* serotype Senftenberg (26.3%) was detected in sediment and seawater. Statistical analysis identified the presence of rain as a direct factor favouring the occurrence of *Salmonella*, especially when rainfall occurred in the days immediately prior to sample collection – a finding that had been previously reported in various other studies carried out in diverse geographical areas worldwide (references in Setti *et al.* 2009).

Vibrio

There are more than 80 species of *Vibrio*, of which at least 12 are recognised as potentially pathogenic to humans (Ryder *et al.* 2014). Seafood-borne illnesses are primarily caused by *Vibrio parahaemolyticus*, *V. vulnificus* and *V. cholerae* (FAO/WHO 2003). The former is a normal constituent of inshore marine flora, but its abundance is increased through organic enrichment of coastal and estuarine areas where bivalve-growing areas are concentrated. Its presence is not associated with faecal contamination from human or animal sources, and it is not detected by standard monitoring methods or eliminated from shellfish by standard depuration processes (see later). Infection is generally confined to gastroenteritis, which lasts from two to five days. On rare occasions, septicaemia, an illness characterised by fever or hypotension, can occur, particularly in immune-compromised individuals. In these cases, subsequent symptoms can include swollen, painful extremities with skin lesions, and the duration of illness can range from 2 hours to 10 days (FAO/WHO 2011). A more important disease organism is *V. vulnificus*, a natural inhabitant of warm estuarine and coastal environments throughout the world, which can cause serious illness and even death in persons with pre-existing liver disease, gastrointestinal disorders, diabetes or compromised immune systems. Among healthy people, ingestion of the species can cause vomiting, diarrhoea and abdominal pain, which lasts two to eight days. Cases of *V. vulnificus* infection have a marked seasonal distribution, mostly occurring in the warmer months of the year, and almost all are associated with a recent history of seafood consumption, primarily raw oysters (Graczyk *et al.* 2010). Another serious disease organism is *V. cholerae*, which inhabits a vast geographical range from the tropics to temperate waters worldwide and is the cause of cholera, an acute diarrhoeal disease. According to Global Health Observatory (GHO) data, there were 1.4 to 4.0 million cholera cases, and 21 000 to 143 000 deaths, in 2016, with the majority occurring in underdeveloped countries (GHO 2016). Infection is acquired through drinking contaminated water or eating contaminated food, including raw or undercooked seafood. Mannas *et al.* (2014) investigated the occurrence of different *Vibrio* species in 52 samples of *M. galloprovincialis* collected from four sites along the Atlantic coast of Morocco. The most frequently isolated were *Vibrio alginolyticus* (90.4% of samples), followed by *V. cholerae* (15.4%) and *V. parahaemolyticus* (7.7%). Faecal contamination was high, together with a significant prevalence of *Salmonella*, at two of the sites. This study highlights the potential sanitary risk associated with the presence of pathogenic bacteria in cultivated mussels in populous regions of southern Morocco, where shellfish production and maritime tourism are important to the economy.

Since 2007, all *Vibrio* infections are required by US law to be reported to government authorities. The Cholera and Other Vibrio Illness Surveillance (COVIS) system, maintained by the US Centers for Disease Control and Prevention (CDC), is a national surveillance system for human infection with pathogenic species of the family *Vibrionaceae*, which cause vibriosis and cholera. Information from COVIS helps track *Vibrio* infections and determines host, food and environmental risk factors. In the United States, 1252 *Vibrio* infections were reported to COVIS in 2014. The most frequently reported species was *V. parahaemolyticus*, which was isolated from 605 (48%) patients, of whom 86 (15%) were hospitalised and 4 (1%) died, followed by *V. vulnificus* with 124 (10%) patients, of whom 97 (79%) were hospitalised and 21 (18%) died, and *V. cholerae* with 80 (6%) patients, of whom 36 (45%) were hospitalised and 3 (4%) died. Most cholera cases have been associated with travel to countries with endemic cholera. Domestically acquired cases are primarily associated with the consumption of raw or undercooked seafood harvested from the US Gulf Coast (Loharikar *et al.* 2015 and references therein).

Other pathogenic bacteria include *Listeria monocytogenes* and *Staphylococcus aureus*. The former is a common contaminant of fresh or lightly preserved foods, including many lightly preserved seafoods (Ryder *et al.* 2014). In addition to a form of gastroenteritis, *L. monocytogenes* infection can cause a rare but life-threatening systemic foodborne disease called listeriosis. However, there are very few documented cases of listeriosis due to seafood. *Staphylococcus aureus*, although it is seldom isolated from fresh seafood products, can be found in products that are cooked and have involved extensive human handling. It produces a range of toxins and diverse disease syndromes, including a foodborne intoxication principally characterised by vomiting, which is induced by staphylococcal enterotoxins that also cause nausea, stomach cramps and diarrhoea (Iwamoto *et al.* 2010; Ryder *et al.* 2014). Proper chilling of products can easily prevent growth of *S. aureus*. Avoidance of cross-contamination of heat-treated (cooked) products is also important.

Escherichia Coli

This bacterium is a member of the family Enterobacteriaceae and is a common inhabitant of the intestinal tract of humans and warm-blooded animals. Serotype *E. coli* O157:H7 can produce bloody diarrhoea due to toxins that it secretes when it infects the human intestine. Usually, little or no fever is present, and the illness is resolved in 5–10 days. If complications develop, such as dehydration, the outcomes can rapidly decline from good to poor (MedicineNet 2019). Infection with *E. coli* 0157:H7 usually occurs through consumption of food or fluids contaminated with animal or human waste. Most commonly, it comes from eating raw or undercooked ground beef or from drinking raw milk. Less commonly, it can be transmitted from one person to another, usually by direct physical contact.

Faecal indicator organisms (FIOs) such as faecal coliforms, *Escherichia coli* and *Enterococcus faecalis* are routinely used as surrogates for microbial pathogens in order to assess the microbiological quality of bivalves and the waters they inhabit. The rates of uptake of these bacteria have been shown to be species- and temperature-dependent. For example, *Mytilus* spp. accumulated coliforms at a higher rate than three other bivalve species (the Pacific oyster, *Crassostrea gigas*, the littleneck clam, *Protothaca staminea* and the soft-shell clam, *Mya arenaria*), and also eliminated these bacteria more effectively (Bernard 1989). In *Mytilus* spp., rates of accumulation were higher at 17 °C than at 12 °C. A more recent study found that *Mytilus* spp., the cockle *Cerastoderma edule* and the Manila clam *Ruditapes philippinarum* had equivalent rates of *E. coli* accumulation, which were higher than for the oysters *Ostrea edulis* and *C. gigas*, which in turn were higher than for the hard clam *Mercenaria mercenaria* (Younger & Reese 2013; see also Amouroux & Soudant 2011; Jozić *et al.* 2012; Skander *et al.* 2017). Overall, mussels and cockles tend to show higher accumulation rates than oysters, which may be related to the higher filtration rates in the former bivalves (Campos *et al.* 2013). The results from these studies suggest that the use of *Mytilus* spp. may by itself provide an adequate index of faecal pollution in shellfish-growing areas. These areas provide favourable conditions (low solar radiation, low temperature, low salinity, low densities of micro-predators and high levels of organic matter) for survival of FIOs (Campos *et al.* 2013).

Bacterial Assays

Conventional detection of pathogenic bacteria is mainly based on cultivation procedures, which utilise enrichment broths followed by the isolation of colonies on selective media, biochemical identification and confirmation of pathogenicity (Gugliandolo *et al.* 2010). These procedures depend on phenotypic characteristics whose presence may be governed by several enzymes that frequently are affected by the physiological state of the microbes (Dufour & Stelma 2010). Another factor is that culture methods are selective for the search of one type of pathogen at a time. In addition, the methods are time-consuming, taking up to 16 days to yield a result for some species. Molecular technologies, mainly based on DNA amplification by polymerase chain reaction (PCR) assay, can be used to complement or replace culture-based approaches and bypass some of their intrinsic biases and limitations (Gugliandolo *et al.* 2010). PCR is a sensitive technique (details in Chapter 9) that is based on the isolation and amplification of a short DNA sequence. It has been employed to detect different strains of *V. cholerae* (Depaola & Hwang 1995), *V. parahaemolyticus* (McCarthy *et al.* 2000) and *V. vulnificus* (Kim & Jeong 2001), as well as of *Salmonella* and *Shigella* (Vantarakis *et al.* 2000). Another PCR method, real-time PCR, more often referred to as quantitative polymerase chain reaction (qPCR),[2] has the potential to rapidly quantify the relative levels of different microbes present in hundreds of water samples (Santo Domingo & Edge 2010). A qPCR assay based on the gyrase B gene (gyrB) sequence, a gene that is essential for DNA replication in *V. parahaemolyticus*, was developed for the quantitative detection of the species in seafood (Cai *et al.* 2006). The study, which involved 27 strains of *V. parahaemolyticus* and 10 of other species, indicated that the assay was highly specific. A total of 300 seafood samples were analysed, of which 78 (26%) were found positive for *V. parahaemolyticus* using a conventional culture method and 97 (32.3%) were found positive using the qPCR assay. Another method, multiplex polymerase chain reaction (m-PCR), is very useful in that it allows the simultaneous detection of multiple targets in a single assay. The technique has been successfully applied to rapidly detect different bacterial pathogens in marine waters and shellfish. For example, Park *et al.* (2013) developed a simple m-PCR method based on TaqMan reverse transcriptase polymerase chain reaction (RT-PCR)[3] to detect the three most important human pathogenic vibrios, *V. parahaemolyticus*, *V. vulnificus* and *V. cholerae*, in a single PCR reaction. In the assay, they selected specific primers and probes targeting virulence factor genes of the three *Vibrio* spp. for their identification. These genes, *hlyA*, *tlh* and *vvhA*, are species-specific markers for *V. cholerae*, *V. parahaemolyticus* and *V. vulnificus*, respectively. m-PCR assays have also been used to identify the same three *Vibrio* species (Garrido-Maestu *et al.* 2014; Wei *et al.* 2014), or single (Imani *et al.* 2013; Hossain *et al.* 2013; Yáñez *et al.* 2015) and multiple (Izumiya *et al.* 2011) *Vibrio* spp., and in all cases have proved to be rapid, sensitive and specific. They have further been used to detect *Salmonella* spp. in mussels (Vantarakis *et al.* 2000) and poultry meat (de Freitas *et al.* 2010). A two-step m-PCR assay that can identify the four main prevalent categories of diarrhoeagenic *E. coli* (DEC) – enteroaggregative *E. coli*, enterotoxigenic *E. coli*, enteropathogenic *E. coli* and enterohemorrhagic *E. coli* – has been developed by Tobias & Vutukuru (2012); see also Sjöling *et al.* (2015). Because PCR amplifies DNA from both living and dead cells, an RT-PCR method has been developed that only detects viable cells. Several genes that are only active during the growth phase of the bacterium can thus be detected (*E.coli*: Yaron & Matthews 2002; Deisingh & Thompson 2004; Molaee *et al.* 2015; *Salmonella* spp.: D'Souza *et al.* 2009; Miller *et al.* 2011; Hatta *et al.* 2017; *Vibrio* spp.: Coutard *et al.* 2005; Kanoktippornchai *et al.* 2014; Chahorm & Prakitchaiwattana 2018).

Immunology-based methods for bacteria detection, some of which are linked with PCR, provide very powerful analytical tools for a wide range of targets. The enzyme-linked immunosorbent assay (ELISA) is the most established technique nowadays, as well as the source of inspiration for many biosensor applications (see later). ELISA combines the specificity of antibodies and the sensitivity of simple enzyme assays by using antibodies or antigens coupled to an easily assayed enzyme (reviewed in Lazcka *et al.* 2007). See also the detection of *Salmonella* spp. (Bangi *et al.* 2012), *Vibrio* spp. (Xi-Bin *et al.* 2009; Di Pinto *et al.* 2012) and *E.coli* spp. (Lazcka *et al.* 2014; Shen *et al.* 2014) using ELISA. The use of monoclonal antibodies (MAbs) with high specificity and affinity has been a good alternative in making ELISA tests for *Salmonella* in contaminated meat products more reliable (Schneid *et al.* 2006). Recently, Sakata *et al.* (2018) developed an immune chromatographic assay to detect contamination of seafood with *V. parahaemolyticus*. It targeted two virulence factors, which serve as specific markers for the pathogen. The authors postulated that the assay could provide an effective routine test to prevent foodborne infections.

A DNA microarray, also known as a DNA chip or biochip, is composed of many discretely located probes at high densities on a solid substrate to which fluorescence-labelled PCR-amplified DNAs are hybridised. Each probe is composed of a sequence that is complimentary to a pathogen-specific gene sequence. A microarray coupled with a m-PCR protocol was used to detect and discriminate between *V. cholerae*, *V. parahaemolyticus* and *V. vulnificus* in oyster tissue (Panicker *et al.* 2004). The high-density arrays available at present are able to test for the presence of thousands of different organisms simultaneously (reviewed in McLoughlin 2011; see also Murtaza *et al.* 2016 and Chapters 9 and11 for additional information).

The development of biosensors has brought a new and promising approach to pathogen detection. Biosensors comprise a biorecognition element coupled to some form of transducer, which converts specific analyte binding to bioreceptors into a measurable or detectable readout. Biosensors can be categorised in different ways, either according to the method of signal transduction (i.e. optical, mechanical or electrical) or by the type of bioreceptor employed (i.e. catalytic (enzyme) or affinity-based (antibody-, oligonucleotide-, peptide- or carbohydrate-binding molecules)). Generally, affinity-based biosensors are preferred over enzymatic ones for the detection of micro-organisms, due to their enhanced selectivity and specificity and lack of requirement for extra reagents (reviewed in Park *et al.* 2013; Ahmed *et al.* 2014; Poltronieri *et al.* 2014). In the future, biosensors may be modulated to be part of an automated system, with either microfabrication or miniaturisation potential, allowing the construction of simple and portable equipment for fast analysis (Cecchini *et al.* 2016). A biosensor that can immediately detect very low levels of *Salmonella typhi* has already been developed (Zelada-Guillén *et al.* 2009). The technique uses carbon nanotubes and synthetic DNA fragments that activate an electric signal when they link up with the pathogen. More recently, Ohk & Bhunia (2013) developed a multiplex fibre optic biosensor for simultaneous detection of *L. monocytogenes*, *E. coli* O157:H7 and *S. enterica* from ready-to-eat meat samples. The most recent electrochemical immunosensors, genosensors, aptasensors and phagosensors for *Salmonella* detection in food analysis are reviewed in Cinti *et al.* (2017). See also the role of biosensors as a surveillance and diagnostic tool for *Vibrio* (Nordin *et al.* 2017; Rahman *et al.* 2017) and *E. coli* spp. (Sun *et al.* 2015; Hesari *et al.* 2016; Xu *et al.* 2016). For more detailed information on the main techniques used in pathogenic bacteria detection, see Dufour & Stelma (2010), Cecchini *et al.* (2016), Silva *et al.* (2018), Bhandari *et al.* (2019), Capobianco *et al.* (2019) and Liu *et al.* (2019).

Viral Infections

Viruses are increasingly recognised as an important cause of foodborne diseases. They can be transmitted directly from person to person, but also indirectly via virus-contaminated water, air, soil, food or surfaces. They need to enter living host tissue in order to replicate. There are hundreds of different virus types and strains that have been identified in faecal waste, revealing a profusion of new, previously unknown strains (Ng *et al.* 2012). Consequently, shellfish-growing waters can be contaminated with a wide diversity of viruses of human and animal origin. The two most frequently implicated in shellfish outbreaks are the enteric viruses norovirus (NoV) and hepatitis A virus (HAV).

NoV is very contagious, and infection is mostly through contact with an infected person, contaminated food such as shellfish (Razafimahefa *et al.* 2019), contaminated water or touching of contaminated surfaces. The symptoms of infection, which generally last one or two days, are acute gastroenteritis accompanied by nausea, vomiting, stomach pain, diarrhoea, headaches and fever. According to the CDC, each year NoV, often referred to as 'the winter vomiting bug', causes 19–21 million illnesses in the United States, contributing to 56 000–71 000 hospitalisations and 570–800 deaths (CDC 2021a). The inability to culture wild-type strains of human NoV hampers research on vaccine development. Hence, there is currently no vaccine available to prevent NoV infection (Baehner *et al.* 2016; Riddle & Walker 2016). Outbreaks show strong seasonality, with the peak occurring in December–March in the United States and April–May in Europe. They are also affected by various environmental factors, including precipitation, temperature, solar radiation, wind and salinity (Wang & Deng 2012). Because of their efficient filtering and concentrating capacities, bivalves have been used as sentinels for tracing NoV in polluted waters. Nenonen *et al.* (2008) examined wild mussels gathered close to the effluent plume from a sewage treatment plant in Gothenburg, Sweden, correlating the molecular findings with those of NoV strains from stool samples from patients affected in extensive local outbreaks of acute gastroenteritis. Sequence analysis of the RNA polymerase (RdRp) gene confirmed a 99% similarity between genotype strains from mussels and patients, indicating the potential usefulness of sentinel mussel studies in tracing human NoV contamination of coastal waters. In a related study, contaminated raw oysters were implicated in a severe outbreak of NoV gastroenteritis affecting 30 restaurant guests (Nenonen *et al.* 2009). Sequencing data (see earlier) revealed ≥99% nucleotide similarity between NoV GI.1 strains detected in patient stool samples and suspect oysters, thus linking the contaminated oysters to the outbreak. Various modelling approaches can be employed in forecasting outbreaks, including Bayesian models, process-based models, regression models, artificial neural networks (ANN) (details in Wang & Deng 2012) and an evolutionary-based genetic programming (GP) model (Chenar & Deng 2018a). Using an artificial intelligence-based approach to identifying environmental indicators of oyster NoV outbreaks (1996–2014) in coastal waters of the Gulf Coast of Texas, Louisiana and Mississippi, United States, Chenar & Deng (2017) found that water temperature was the most important indicator for oyster NoV outbreaks with an average relative contribution of 37.2%, followed by solar radiation (23.8%), gage height (16.7%), salinity (12.3%), wind (6.3%) and rainfall (3.7%). The top three indicators are capable of explaining 77.7% of model-predicted oyster NoV outbreaks. Such outbreaks are forecastable with two-day lead time using the ANN-2Day model and daily data for the six environmental predictors. This model was capable of reproducing 19 years of historical oyster NoV outbreaks along the Northern Gulf of Mexico coast with a positive predictive value of 76.82% (Chenar & Deng 2018b). Wang (2015) has used a forecasting system comprising three components: satellite algorithms for retrieval of environmental variables, including salinity, temperature and gage height; an ANN-based model for predicting relative risk levels of oyster NoV outbreaks; and a mapping method for visualising spatial distributions of NoV outbreak risks in oyster harvest areas along the Louisiana coast. Findings suggest that outbreaks are predictable, and for the Louisiana oyster areas the ANN model predicted historical outbreaks from 1994 to 2014 without any confirmed false-positive or false-negative predictions.

HAV is a worldwide infectious disease. Shellfish consumption has always been one of the major risk factors for hepatitis A infection, especially when eaten raw or slightly cooked (Leoni *et al.* 1998; Conaty *et al.* 2000; Sánchez *et al.* 2002; Shieh *et al.* 2007; Maalouf *et al.* 2010; Ly & Klevens 2015). Moreover, cooking does not always guarantee the harmlessness of shellfish (e.g. mussels, Croci *et al.* 2005). HAV has been shown to be concentrated within mussels to 100-fold higher concentrations than the surrounding water and to persist for about seven days. It has been detected in oyster samples more than two months after the presumed contamination event, which is thought to be due to recontamination from sediment in the water (Yates 2011 and references therein). The virus replicates in the human liver and is shed in high concentrations in faeces from two weeks before to one week after the onset of clinical illness (Fiore 2004). Symptoms include fever, malaise, jaundice, anorexia and nausea and may last from several weeks to several months. Transmission occurs by the faecal–oral route, either through direct contact with an HAV-infected person or by ingestion of HAV-contaminated food or water. In countries where HAV infection is highly endemic, the majority of people are infected in early childhood and more than 90% of children under five years of age show no symptoms and virtually all adults are immune. But in countries with good standards of public health, HAV infection is less common so that very few are infected in early childhood and the majority of adults remain susceptible to infection, with a 1% mortality rate in over 60-year-olds (Hollinger & Emerson 2007; Pintó *et al.* 2010 and references therein). HAV can be grown in cell culture, and sufficient amounts of the virus can be obtained for vaccine production. A safe and effective vaccine has been available since the early 1990s. Administration of the HAV vaccine, or a single HAV infection, provides lifelong immunity against the virus (Leon & Moe 2006). Since vaccination was introduced in the United States, the incidence of HAV has declined from 6.0 to 0.4 cases per 100 000 (1999–2011; Ly & Klevens 2015). Globally, an estimated 1.4 million cases occur each year, with approximately 11 000 deaths in 2015 (WHO 2017a). Nevertheless, the virus is exceptionally persistent in the environment and within shellfish tissue. The persistence of viruses in contaminated shellfish appears to be related to their sequestration from the lumen of the gut to surrounding tissues and haemocytes, which may retain viable virus for extended periods (Richards 2016). Persistence of HAV and NoV in the environment is illustrated by the percentages of positive samples in bivalves harvested and distributed across a worldwide range of countries (Table 12.1). The prevalence of HAV reported in mussels ranged from 0 to 43% (N = 9), while that in oysters – albeit based on few studies (N = 3) – ranged from 0 to 20%. In general, prevalence of NoV in mussels and oysters (N = 10) was higher (<2.0 to 76.0%) than that of HAV. The detection of HAV in mussels marketed in some Italian cities is of some concern. Such products are an important source for HAV cases according to case–control studies in Puglia, explaining the advent of the large epidemics that periodically occur in the region. In fact, shellfish there are largely consumed especially during Christmas holidays, and are commonly eaten raw or slightly cooked (Chironna *et al.* 2002 and references therein). Therefore, continuous surveillance of seafood products in local markets is advisable.

Table 12.1 Percentage of human norovirus (NoV)- and hepatitis A virus (HAV)-positive seafood samples from Europe, the United States and East Asia. Viruses were generally detected in reverse transcriptase polymerase chain reaction (RT-PCR). R, retail; C, cultured samples; W, wild samples: NA, not analysed; ND, not defined.

Country of origin	Sampling point	Period	N	NoV (%)	HAV (%)	Reference
Italy	Puglia (R)	1999–2000	290[a]	NA	18	Chironna *et al.* (2002)
	Southern cities (R)	1999–2001	180[a]	NA	16	Croci *et al.* (2003)
	East coast regions (C)	2003–2007	2310[a]	5	NA	Pavoni *et al.* (2013)
	Naples coastal area (C)	2014–2015	108[a]	30	13	Fusco *et al.* (2017)
	Naples coastal area (C)	2015	262[a]	NA	26	Suffredini *et al.* (2017)
Spain	Ría de Vigo (C)	2005	24[a]	46	0	Vilariño *et al.* (2009)
	Ría de Vigo (W)		12[a]	58	0	
	Ría de Vigo (C)	2004–2006	68[a]	NA	43	Manso *et al.* (2010)
France	Vendée area (R)	2010–2011	388[b]	9	NA	Schaeffer *et al.* (2013)
United Kingdom	21 sites across the UK (R)	2015–2016	630[b]	69	NA	Lowther *et al.* (2018)
Sweden	North Sea (C)	ND	54[a]	76	0	Formiga-Cruz *et al.* (2002)
Thailand	Surat Thani Province (C)	2008–2009	58[a]	NA	3	Namsai *et al.* (2011)
	Bangkok (R)		92[a]	NA	7	
China	Yellow and Bohai seas (R)	2009–2011	114[a]	17	NA	Ma *et al.* (2013)
Japan	West of Honshu Island (R)		191[b]	9	NA	Nishida *et al.* (2003)
Korea	Seoul (R)	2011–2012	56[b]	NA	20	Park *et al.* (2015)
United States	Nine Atlantic coast states (R)	2007	381[b]	4	4	Woods & Burkhardt (2010)
Australia	All major oyster harvest areas (C)	2014–2015	297[b]	<2	<2	Torok *et al.* (2018)

[a] Mussel samples.
[b] Oyster samples.
Sources: Adapted from Chironna, M., Germinario, C., De Medici, D., Fiore, A. Di Pasquale, S., Quarto, M. *et al.* (2002) Detection of hepatitis A virus in mussels from different sources marketed in Puglia region (South Italy). *International Journal of Food Microbiology*, **75**, 11–18; Croci, L., De Medici, D., Ciccozzi, M., Di Pasquale, S., Suffredini, E. & Toti, L. (2003) Contamination of mussels by hepatitis A virus: a public-health problem in southern Italy. *Food Control*, **14**, 559–563.

Other faecal viruses that are less commonly transmitted by shellfish but have been documented as being associated with bivalve consumption include Aichi virus (AiV; Rivadulla *et al.* 2017), hepatitis E virus (HEV; Gao *et al.* 2015), sapovirus (Varela *et al.* 2016) and astrovirus (Elamri *et al.* 2006).

Viral Assays

As indicated earlier, wild-type strains of HAV have a limited replication capacity or, like NoV, do not replicate at all in cell culture. This means that molecular techniques must be used to detect and quantify them, since alternative methods such as immunological detection have poor sensitivity due to the low levels of viruses present in contaminated foodstuffs (Pintó Solé *et al.* 2011). Prior to virus detection, a number of different procedures for the processing of shellfish samples are necessary: virus extraction, virus concentration, nucleic acid extraction and purification (details in Bosch 2009; Lees 2010; Pintó Solé *et al.* 2011; CEFAS 2016). PCR assays and their limitations are reviewed in Bosch *et al.* (2009, 2011), Lees (2010) and Le Guyader *et al.* (2010). The PCR protocol most often used in viral detection is a one-step RT-PCR method, in which reverse transcription and PCR amplification are carried out consecutively in the same tube. Details on the methodology used in detecting HAV and NoV in bivalve species can be found in the studies cited in Table 12.1, along with Di Pinto *et al.* (2003), Le Guyader *et al.* (2009), Yong & Son (2009), Mesquita *et al.* (2011), Wan Norhana *et al.* (2011), Fuentes *et al.* (2014) and Jeon *et al.* (2017).

PCR methods allow not only qualitative but also quantitative detection of viruses. Real time-PCR is most appropriate when testing food matrices that are unlikely to be contaminated with virus, as it is the least expensive and most straightforward method, producing a positive or negative result (Bosch *et al.* 2011). qPCR is required when a sample (e.g. shellfish) is likely to contain viruses and it is the degree of contamination that needs to be ascertained (Bosch *et al.* 2011). Because of the proliferation of published protocols for detecting NoV and HAV in shellfish, the European Committee of Standardization (CEN) set up a working group (CEN/TC 275/WG6/TAG4-viruses in foods) in 2004 to develop a two-part (quantitative and qualitative) standard method for virus detection in foodstuffs, including shellfish, which had the potential to be incorporated into EU legislation (Lees 2010). The first part of the standard (the quantitative method) and the second part (the qualitative method) were presented to CEN in 2009 and 2010, respectively, for technical and formal assessment. The International Organization of Standardization (ISO) published the final version of the methods in 2013 (ISO 2013a, 2013b), but the quantification method has since been revised by ISO 15216-1:2017 (ISO 2017). The qualitative standard was last reviewed in 2016, and this version remains current.

Parasites

Species of *Cryptosporidium* are intracellular protistan parasites that infect epithelial cells of the gastrointestinal tract and are increasingly recognised as important pathogens and a cause of enteric disease in a wide range of vertebrate hosts, including humans (see Chapter 11). About 13 species are recognised, with *C. hominis* being the most often responsible for clinical disease in humans, and *C. parvum* for disease in domestic animals (WHO 2017b). *C. hominis* generally causes diarrhoea, and sometimes nausea, vomiting and fever, which usually resolve within a week in normally healthy people, although infections can be life-threatening in severely immune-compromised individuals. In human diarrhoeic faeces, there can be as many as 10^6 oocysts (4–6 μm diameter) per ml (Rose 1997). The major route of infection is person-to-person contact. Other sources include the consumption of contaminated food and water and direct contact with infected farm animals and possibly domestic pets. The role of drinking water in the transmission of *Cryptosporidium*, including in large outbreaks, has been well established. In 1993, *Cryptosporidium* caused the largest waterborne outbreak of disease on record, when more than 400 000 people were infected by the drinking water supply of Milwaukee, United States (Corso *et al.* 2003).

Cryptosporidium oocysts are small enough to be retained on the gills of suspension-feeding bivalves, and subsequently become concentrated in the digestive system. Mladineo *et al.* (2009) were the first to evaluate the presence of *Cryptosporidium* oocysts in cultured mussels (*M. galloprovincialis*) at four locations in the Mali Ston Bay, Adriatic Sea, Croatia over a one-year period. Prevalence was 16.8% in a sample of 184 mussels, and was correlated with the presumptive number of *E. coli* in the shellfish and seawater and with abiotic factors (temperature, salinity, oxygen). A lower prevalence of 12.6% was reported in a sample (N = 56) of *Perna viridis* from fish markets located at the mouth of the Chao Phraya River, Thailand (Srisuphanunt *et al.* 2009). The parasite does not appear to adversely affect mussel health, but the results from the preceding and similar studies (Gómez-Couso *et al.* 2003; Miller *et al.* 2005a, 2006; Li *et al.* 2006; Adell *et al.* 2014; Mariné Oliveira *et al.* 2016) indicate that mussels and other bivalves (Freire-Santos *et al.* 2000; Robertson 2007; Sutthikornchai *et al.* 2016; Ghozzi *et al.* 2017; Pagoso & Rivera 2017) probably act as an additional reservoir of *Cryptosporidium* foodborne infections for humans who consume raw or poorly cooked bivalves (reviewed in Robertson 2007).

Giardia duodenalis, also known as *G. lamblia*, is a flagellated protistan parasite that inhabits the intestinal tracts of humans and several other animals, and which excretes cysts into the environment. The infectious stage is an environmentally resistant cyst, which is transmitted via the faecal–oral route. Human giardiasis is a common cause of diarrhoeal disease throughout the world. In developing countries, approximately 200 million people have symptomatic infection (WHO 2017b). Contaminated drinking water, recreational water and, to a lesser extent, food have been associated with outbreaks. Gómez-Couso *et al.* (2005) were the first to examine mussels (*M. galloprovincialis*) for the presence of *Giardia* cysts over a six-month period. Samples (N = 200) were collected from five estuaries in Galicia, NW Spain, one of the world's main producers of mussels (see Chapter 10). Giardia cysts were found in mussels from all of the estuaries, with an overall rate of contamination of 41.5%. There was a significant negative association between the microbiological quality of the harvesting area and the presence of *Giardia* cysts.

Toxoplasma gondii is a protistan pathogen in which the life cycle can only be completed through its terrestrial definitive hosts, domestic and wild cats. Infections and deaths due to the parasite have been reported in numerous coastal as well as pelagic marine mammal species worldwide (Miller 2008). Human infections with *T. gondii* are also widespread, with up to one-third of the human population infected globally (Tenter *et al.* 2000).

Toxplasmosis is usually asymptomatic in humans, although serious and fatal infections can occur in immune-compromised patients and in foetuses infected congenitally (Jones *et al.* 2009). Humans can become infected by ingestion of oocysts excreted by cats on direct contact, or through contact with contaminated soil or water. Consumption of undercooked shellfish containing *T. gondii* cysts has been linked to increased risk of infection (WHO 2017b). Shapiro *et al.* (2015) investigated the prevalence of *T. gondii* in *M. californianus* (N = 959) collected from central California, United States and found 13 mussels (1.4%) were infected. The odds of detecting *T. gondii* were more than 12 times higher in mussels sampled during the wet season as compared with the dry season, and nearly five times higher in those collected near freshwater outflow. There was a 10-fold increase in prevalence when compared with a prior study that was conducted along a similar stretch of coastline in central California (Miller *et al.* 2008). In another study on *Mytilus* spp. collected from two different sites in the same region, Staggs *et al.* (2015) found that mussels at Moro Bay were contaminated with *T. gondii* (33%), while those from Point Lobos were contaminated with both *T. gondii* (54%) and *Cryptosporidium* (26.9 %) oocysts. In the light of these high prevalence rates, use of native *Mytilus* spp. as biosentinels for pathogen contamination along the central California shorelines could prove effective (see also Marangi *et al.* 2015).

Methods for the detection of *Cryptosporidium* and *Toxoplasma* oocysts and *Giardia* cysts are very similar and can often be carried out simultaneously using the same assay (www.rapidmicrobiology.com). Direct microscopic examination of faecal smears prepared from fresh or concentrated samples is still widely used to detect protistan cysts in faeces. As the cysts of *Cryptosporidium* and, to a lesser extent, *Giardia* are very small, a staining procedure is required to detect them. A number of different staining techniques have been developed using acid-fast stains, malachite green and fluorescence-based stains (Freire-Santos *et al.* 2000; Pagoso & Rivera 2017). Immunology-based methods have also been developed to detect soluble protistan antigens in faecal samples, removing the need for microsopic examination. Enzyme immunoassay (EIA) technologies have been used to develop commercial products for the detection of *Cryptosporidium* oocysts and *Giardia* cysts (Youn *et al.* 2009; Elgun & Koltas 2011; Ghaffari & Kalantari 2014). Immunochromatographic lateral-flow 'dipstick' tests are being used for protistan antigen detection in faecal samples (Garcia *et al.* 2003), and either immunomagnetic separation (IMS; Miller *et al.* 2005b) or direct fluorescent antibody (DFA) testing (Freire-Santos *et al.* 2000; Srisuphanunt *et al.* 2009; Pagoso & Rivera 2017) for detection in water samples. These tests are sensitive, specific and very easy to use, and provide a result in 10 minutes. In recent years, molecular diagnostic detection methods have increasingly been developed. These are typically based on PCR (Mariné Olviera *et al.* 2016; Pagoso & Rivera 2017) or real-time PCR (Amar *et al.* 2004; Staggs *et al.* 2015) technology, and target specific sequences in the protistan genome (Adell *et al.* 2014; Staggs *et al.* 2015). Restriction fragment length polymorphism (RFLP) analysis (Su *et al.* 2010; Tappeh *et al.* 2014; Saki *et al.* 2016; see Chapter 9) or sequencing that targets different parts of the genome is also widely employed (Hanevik *et al.* 2015; Cheng *et al.* 2015; Zahedi *et al.* 2017). Oocysts/cysts, as isolated particles from the three pathogens, can be detected by flow cytometry (FC) if labelled with a suitable fluorescent label (Staggs *et al.* 2015). The method is almost 10 times more sensitive than conventional immunofluorescent assays (Shams *et al.* 2016). Fluorescence in situ hybridisation

(FISH) using rRNA-targeted oligonucleotide probes relies on the hybridisation of synthetic oligonucleotide probes to exact regions within the rRNA of the organism (details in Chapter 5). The assay is based on species-specific probes, and has been applied to discriminate between *Cryptosporidium parvum* and *C. hominis* oocysts in water samples (Alagappan *et al.* 2009).

Biotoxins

Shellfish toxins, more accurately known as phycotoxins (Garthwaite 2000), most of which are produced by dinoflagelates and diatoms, are a naturally occurring phenomenon and not associated with sewage contamination of coastal waters. As the main food source for bivalves is phytoplankton, the potential for accumulating algal toxins is high. Bivalves accumulate the toxic cells in their tissues, and are therefore the main vectors of toxins that cause diseases such as paralytic shellfish poisoning (PSP), pufferfish poisoning (PFP), neurotoxic shellfish poisoning (NSP), amnesic shellfish poisoning (ASP), diarrhoetic shellfish poisoning (DSP) and azaspiracid poisoning (AZP) in humans. Mussels and scallops are particularly prone to accumulate biotoxins (Kingsley 2014). It is not clear why a few dozen microalgal species produce these toxins, although under eutrophic conditions, where nutrients are limited, toxin secretion may be a mechanism for inhibiting the growth of co-occurring microalgal competitors (Granéli *et al.* 2008). There are no suitable post-harvest mitigation strategies for use against these toxins (Kingsley 2014). Even cooking does not inactivate them, and there is no known antidote. The continuing increase in the number of toxic species, coupled with the increased occurrences of harmful algal blooms (HABs) of these species in coastal waters, poses a constant threat to public health worldwide. Therefore, shellfish and shellfish harvest areas are closely monitored for biotoxins using a variety of analytical techniques (see later).

PSP is a serious illness caused by eating bivalves contaminated with a group of more than 30 structurally related marine toxins collectively referred to as saxitoxins (STXs) (details in Visciano *et al.* 2016). These toxins are concentrated in bivalves as a result of continuous filtration of toxic dinoflagellate cells of the genera *Alexandrium*, *Gymnodinium* and *Pyrodinium* during 'red tide' blooms (Grattan *et al.* 2016; see later). Even when blooms of these species are not apparent, bivalves may ingest resting cysts, which are 10–1000 times more toxic than motile dinoflagellates (Sinderman 1990). Predators of bivalves may also be vectors for STX, thus increasing the potential for human exposure (Halstead & Schantz 1984). The toxins produced by these organisms are the most common and widespread of the shellfish toxins, and are among the most potent neurotoxins known. They are soluble in water and thermostable at acidic pH, although at alkaline pH they are rapidly degraded (Kodama & Sato 2008). STXs act by blocking the passage of sodium ions through cell membranes, thus inhibiting nerve impulse transmission. Typical symptoms in humans range from a slight tingling and numbness of the lips to complete paralysis and death from respiratory failure, depending on the dose. Onset of clinical signs typically begins between 30 minutes and three hours after ingestion of the contaminated shellfish, but symptoms have been reported less than one minute after food consumption. There is no known antidote to PSP. The level at which PSP intoxications occur in humans varies considerably between individuals. Therefore, average doses are not very informative for the derivation of a safe dose estimate. In view of the acute toxicity of STX-group toxins, the European Food Safety Authority (EFSA) Panel on Contaminants (CONTAM), using a large epidemiological dataset, established a lowest observed adverse effect level (LOAEL) in the region of 1.5 μg STX equivalents per kilogram body weight (EFSA 2009a). Because many individuals did not suffer adverse reactions at higher intakes, it is expected that this LOAEL is close to the threshold for effects in sensitive individuals. Therefore, it was decided that a factor of three was sufficient to move from this LOAEL to an estimated no observed adverse effect level (NOAEL) of 0.5 μg STX equiv./kg bw[1]. The Panel established an acute reference dose (ARfD)[4] of 0.5 μg STX equiv./kg bw. In order for a 60 kg adult to avoid exceeding the ARfD, a 400 g portion of shellfish should not contain more than 30 μg STX equivalents, corresponding to 75 μg STX equiv./kg shellfish meat (EFSA 2009a; see also Oshima *et al.* 2011).

Quantitation of toxin analogues in a bivalve is not sufficient for monitoring and regulatory decision-making, since the different analogues may have widely dissimilar toxic potencies. Therefore, it is necessary to know the relative toxicity of each of the toxin analogues/congeners so that the total toxicity of the material in the tissue extract can be estimated. This requires the determination of toxicity equivalency factors (TEFs), defined as the toxicity ratio of a compound from a chemical group that shares the same mode of action as a reference compound in the same group. The toxicity of the congener is expressed as a fraction of the toxicity of the reference compound in terms of potency (Botana *et al.* 2017 and references therein). The toxicity equivalency concept has been developed in recent decades to support risk assessment of specific complex mixtures to which humans are often exposed. Total toxicity equivalent (TEQ) is defined by the sum of the concentrations of each compound multiplied by individual values, which provides an estimate of the total toxicological or biological activity relative to a reference compound. The major prerequisite for using TEQ is based on additivity and similar mechanisms or modes of action. An Expert Group was set up by the FAO/WHO in 2016 to develop a technical paper providing scientific explanations and recommendations on TEFs. It developed an approach to the development of TEFs for six groups of biotoxins: STXs, okadaic acid (OA), domoic acid (DA), brevetoxins (BTXs), azaspiracid (AZA) and tetrodotoxin (TTX). The FAO/WHO (2016) summarises the outcomes of the meeting and identifies the TEFs for each of the six groups of toxins. The Expert Group proposed the following TEFs for STX and several analogues: STX = 1, NeoSTX = 1, GTX1 = 1, GTX2 = 0.4, GTX3 = 0.6, GTX4 = 0.7, GTX5 = 0.1 and GTX6 = 0.05. For the STX group, toxicity varies from a high value of 2.0 for NeoSTX to a low value of 0.05 for GTX6.

TTX, another potent neurotoxin, was first isolated from pufferfish in Japan but has subsequently been reported in a wide range of marine organisms, including shellfish. There are about 30 known analogues of TTX, four of which occur in shellfish (FAO/WHO 2016). Like STX, TTX selectively binds to voltage-gated sodium channels, blocking both nerve and muscle action potentials (details in Teramoto & Yotsu-Yamashita 2015). Symptoms of poisoning in humans are nausea, vomiting, diarrhoea, abdominal pain, paraesthesia, muscular paralysis, respiratory failure and death (Munday & Reeve 2013). PFP has the highest fatality rate of the most common marine intoxications (Isbister & Kiernan 2005). Its main occurrences are in warm-water regions, especially the Pacific and Indian oceans. However, between 2003 and 2014, detection of TTX in mussels (*M. edulis*) and Pacific oysters (*Crassostrea gigas*) was reported from the south coast of England – the first report of TTX in bivalve shellfish from anywhere within European waters (Turner *et al.* 2015). TTX has also been reported in fish from Mediterranean waters around Greece, Egypt and Tunisia, as well as in gastropods harvested from Spain and bivalves from Greece. The United Kingdom is currently unprepared to respond to any urgent requirement for the routine detection of TTXs in shellfish or fish samples. There are no health-based guidance values (HBGVs) for TTX worldwide and no maximum levels in seafood in the European Union (EFSA 2017). In the United States, there is likewise no established regulatory limit for TTX, but with use of the mouse bioassay (MBA; see later) for routine monitoring the presence of TTX can be detected in bivalve products. This option is not available in the United Kingdom, since animal testing for marine biotoxins is no longer allowed there (reviewed in Turner *et al.* 2015).

ASP is caused when bivalves are eaten that contain domoic acid (DA) or its analogues. The toxins are secreted by the red alga *Chondria armata* and by diatom species in the genus *Pseudo-nitzschia*. DA is an analogue of glutamic acid, an amino acid with three carboxylic acid groups, which are responsible for its water solubility and high polarity (FAO/WHO 2016). DA is heat-stable, and cooking does not typically destroy the toxin. Within 24 hours of consuming contaminated shellfish, usually mussels, there is nausea, vomiting and diarrhoea, while confusion, memory loss, disorientation and even seizures, coma and death can occur within 48 hours. The severity of symptoms depends on the amount of DA ingested. Because of the similarity in structure between DA and glutamic acid (a brain neurotransmitter), DA binds to the same receptors on the neurons – but unlike glutamic acid, it remains bound. The neurons therefore begin to fill with calcium, swell and eventually burst. These nerve cells are located in the part of the brain associated with memory retention, hence the memory loss characteristic of ASP (Pulido 2008 and references therein). The few data on exposure to DA associated with adverse effects in humans ($N = 9$) in ASP outbreaks indicate that severe and irreversible effects occur at about 4 mg/kg bw and the LOAEL for mild signs and symptoms is 0.9 mg/kg bw (EFSA 2009b). Although the acute oral toxicity is not well characterised, the CONTAM Panel considers it appropriate to base the establishment of the ARfD on this LOAEL, dividing it by a safety factor of 10 to derive a provisional ARfD of 0.1 mg DA/kg bw (Tasker *et al.* 2011). A value of 20 μg/DA g has been established as the safety limit in shellfish (Table 12.2). To date, occurrences of ASP have been reported from the west and east coasts of North America, Chile, France, Spain, Portugal, Ireland, the United Kingdom, Australia and New Zealand.

OA and its derivatives the dinophysistoxins (DTXs) constitute the main toxins responsible for DSP syndrome in humans. The main sources of DSP toxins are planktonic marine dinoflagellates of the genus *Dinophysis* (details in Reguera *et al.* 2014) and *Prorocentrum*, with mussels, clams, scallops and oysters as the most common vectors (Grattan *et al.* 2016). The main symptoms of DSP are cramps, severe diarrhoea and nausea, followed by vomiting and abdominal cramps. These symptoms may occur in humans shortly (30 minutes to 12 hours) after consumption of contaminated bivalves (Taylor *et al.* 2013). Although DSP poisoning is traditionally believed to result in full recovery, preliminary data raise the possibility that DSP toxins may be associated with more significant medical problems over time (Grattan *et al.* 2016), such as carcinogenic effects, effects on the immune and nervous

Table 12.2 Maximum allowable levels for poisonous or deleterious substances in shellfish meat in the United States, Canada, the European Union (EU) and Australia. ppm, parts per million; PCBs, polychlorinated biphenyls; PSP, paralytic shellfish poison; DSP, diarrhetic shellfish poison; OA, okadaic acid; NSP, neuorotoxic shellfish poison; ASP, amnesic shellfish poison; DA, domoic acid; ww, wet weight; MU, mouse unit – 1 MU is the minimum quantity of toxin needed to kill two out of three mice within 24 hours of injection. Values for the United States, Canada, European Union and Australia from NSSP (2015), CFIA (2018), European Communities (2004a, 2006a) and Australia New Zealand Food Standards (2016), respectively.

	Allowable level			
Deleterious substance	**United States**	**Canada**	**European Union**	**Australia**
PAHs	Not specified	3 μg/kg	6.0 μg/kg (smoked)	Not specified
PCBs	2.0 mg/kg	2.0 mg/kg	6.5 pg/g ww[a]	0.50 mg/kg
Mercury	1.0 mg/kg	0.50 mg/kg	0.50 mg/kg	1.5 mg/kg
Cadmium	2.0 mg/kg	Not specified	1.0 mg/kg	2.0 mg/kg
Lead	Not specified	0.50 mg/kg	1.5 mg/kg	2.0 mg/kg
PSP	80.0 μg/100 g	80.0 μg/100 g	80.0 μg/100 g	80.0 μg/100 g
DSP OA plus derivatives	0.16 μg/g	0.20 μg/g	0.16 μg/g[b]	0.20 μg/g
NSP brevetoxin	20 MU/100 g	No detectible amount[c]	Not specified	20 MU/100 g
ASP DA	20.0 μg/g	20.0 μg/g	20.0 μg/g	20.0 μg/g

[a] Dioxin plus dioxin-like PCBs.
[b] Includes okadaic derivatives, dinophysistoxins and pectenotoxins.
[c] Presence of any toxin is considered to be hazardous to human health.

systems and alterations in DNA and cellular components (Valdiglesias *et al.* 2013). Overall, however, DSP and NSP (see later) are much less serious than PSP or ASP. According to Visciano *et al.* (2016), DSP toxins are the most frequent and abundant marine biotoxins found in shellfish from Southern European coastal areas. In France, 11 DSP outbreaks involving 45 individuals were described in 2009. The contaminated mussels contained OA-group toxin concentrations approximately eight times higher than allowed by the European regulatory limits. In 2010, OA-contaminated mussels poisoned more than 300 people in Northern Italy. Consumption of contaminated mussels was also the cause of several DSP outbreaks on the Pacific coast of North America (Taylor *et al.* 2013; Trainer *et al.* 2013). Outbreaks of DSP have been reported in Japan, New Zealand and South America (Grattan *et al.* 2016 and references therein). During a toxic episode, levels increase and several toxins co-occur, which may be present together in the same bivalves. Simulation of this scenario on an in vitro cellular model showed that OA in combination with low levels of other lipophilic toxins (DTX2 or yessotoxin, YTX; see later) accentuates the toxic effect caused by OA alone, increasing cellular damage or death (Rodríguez *et al.* 2015). The LOAEL for human illness is in the region of 50 μg OA equivalents per person, which approximates to 0.8 μg OA equiv./kg bw for adults (EFSA 2008a). An uncertainty factor of three was applied to extrapolate this to NOAEL, which resulted in an ARfD of 0.3 μg OA equiv./kg bw. The CONTAM Panel considered it not necessary to apply an additional uncertainty factor for the variation among humans, as the data are based on observations in a rather large number of affected shellfish consumers, originating from various countries and considered to comprise the most sensitive individuals. In order for a 60 kg adult to not exceed the ARfD, a 400 g portion of shellfish should not contain more than 18 μg toxin (i.e. 45 μg OA equiv./kg shellfish meat; EFSA 2008a). The safety limit for OA in seafood is 0.16 μg/g (Table 12.2).

NSP is caused by consumption of molluscan shellfish contaminated with BTXs: lipid-soluble, cyclic polyether neurotoxins primarily produced by the marine dinoflagellate of the genus *Karenia* (reviewed in Turner *et al.* 2015). Fish, birds and mammals are all susceptible to BTXs. They have been reported in many shellfish species, but the most common sources of human exposure are ingestion of commonly harvested clams, oysters and mussels. The toxins activate voltage-sensitive sodium channels, causing depolarisation of neuronal and muscle cell membranes (Visciano *et al.* 2016). While DSP has a global distribution, confirmed occurrences of NSP have only been reported from the Gulf of Mexico, the southeastern coast of the United States and the New Zealand Hauraki Gulf region (Dickey *et al.* 2011; McFarland *et al.* 2015), with no reported occurrence in the United Kingdom or Europe (Turner *et al.* 2015 and references therein). The symptoms of NSP include gastrointestinal and neurological problems. The most frequently reported symptoms are nausea, vomiting, abdominal pain and diarrhoea. These are often not the primary presenting complaint. Of greater concern to most individuals are the neurological symptoms, which may include paresthesia of the mouth, lips and tongue, peripheral tingling, partial limb paralysis, slurred speech, dizziness, ataxia and a general loss of coordination (Grattan *et al.* 2016). Full recovery generally occurs within 48 hours. No fatalities have been reported to date associated with the consumption of shellfish contaminated with BTXs (EFSA 2010a; Dickey *et al.* 2011), but hospitalisations occur. Inhalation of BTX aerosols may result in respiratory problems and eye and nasal membrane irritation (Gerssen *et al.* 2010). Due to the limited quantitative data in both experimental animals and human intoxications, the CONTAM Panel (EFSA 2010a) concluded that the establishment of an oral ARfD was not possible. The European Union has specified no safety limit for BTX (EFSA 2010a), and in Canada the presence of any of the toxin is deemed hazardous to human health (Table 12.2).

A number of new shellfish toxins (e.g. AZAs, cyclic imines (CIs), pectenotoxins (PTXs) and YTXs) have been identified in the past two decades. AZAs are polyethers that are produced by a minute dinoflagellate, *Azadinium spinosum*, which is smaller (12–16 μm) than any of the other toxin-producing dinoflagellates known so far (details in Twiner *et al.* 2008). More than 20 different analogues have been described, with AZA1, AZA2 and AZA3 being the predominant ones (Gerssen *et al.* 2010). AZP was detected for the first time in 1995 after consumers in The Netherlands ate blue mussels from Ireland. Symptoms of poisoning are similar to DSP (e.g. gastrointestinal disorder, diarrhoea and abdominal cramps). Since 1995, several AZP outbreaks have occurred in Ireland, the United Kingdom, Norway, France, Portugal, northern Africa (Morocco), South America (Chile) and the United States (FAO/WHO 2016 and references therein). The CONTAM Panel (EFSA 2008b) concluded that the most probable LOAEL resulting in AZA poisoning was 113 μg AZA1 equivalents per person (1.9 μg AZA1 equiv./kg bw for a 60 kg adult), and established an ARfD of 0.2 μg AZA1 equiv./kg bw. The EU has set a regulatory limit of 160 μg AZA/kg shellfish weight.

Spirolides (SPXs), gymnodimines (GYMs), pinnatoxins (PnTXs), pteriatoxins (PtTXs) and several other toxins make up the CIs, a family of marine biotoxins present in shellfish. SPXs are produced by the dinoflagellates *Alexandrium ostenfeldii* and *A. peruvianum*, GYMs are produced by the dinoflagellate *Karenia selliformis* and PnTXs are produced by the peridinoid dinoflagellate *Vulcanodinium rugosum*. It has been suggested that PtTXs are biotransformed from PnTXs in shellfish (Selwood *et al.* 2010). Members of the CI toxin group show a high degree of structural similarity and are soluble in organic solvents. CIs are sometimes referred to as 'fast-acting toxins' as they are characterised by acute toxicity in mice following intraperitoneal injection and rapid onset of neurological symptoms, potentially leading to death if they are present in sufficient concentrations (reviewed in Davidson *et al.* 2015). Although SPXs, GYMs, PnTXs and PtTXs are now known to occur in microalgae and shellfish in several parts of the world, no information has been reported linking these toxin groups to poisoning events in humans (EFSA 2010b). Currently, there are no regulatory limits for CIs in shellfish in Europe or other regions of the world. In view of the acute toxicity of CIs, the CONTAM Panel considered that an ARfD should be established for the different groups, but due to the lack of adequate quantitative data on acute oral toxicity this has not been possible (EFSA 2010b).

PTXs are a group of polyether-lactone toxins, with about 10 derivatives from PTX2. All exclusively arise from *Dinophysis* spp. and they are always accompanied by toxins from the OA group. To date, 15 different analogues have been isolated and characterised. The toxins have been reported in microalgae and bivalves from Australia, New Zealand, Japan and several European countries. There are currently no data indicating adverse effects in humans (reviewed by Yasumoto *et al.* 2011). The CONTAM Panel concluded that because PTX-group toxins do not share the same mechanism of action as OA-group toxins, they should not be included in the OA-group toxin regulatory limit. Although the oral toxicity is not well defined, the Panel (EFSA 2009c) considered it appropriate to establish an ARfD on the basis of a LOAEL of 250 μg/kg bw for intestinal toxicity of PTX2 observed in mice. Because the effects were mild and reversible, it decided to apply a factor of 3 for the extrapolation from a LOAEL to an NOAEL. It established an ARfD of 0.8 μg PTX2 equiv./kg bw, based on a LOAEL of 250 μg/kg bw and an overall uncertainty factor of 300. The Panel concluded that in order for a 60 kg adult to avoid exceeding the ARfD, a 400 g portion of shellfish should not contain more than 48 μg PTX2 equivalents, corresponding to 120 μg PTX2 equiv./kg shellfish meat.

YTXs are sulphated polyethers. Although more than 90 analogues are known, only a few dozen have been fully identified (EFSA 2008c). The toxins are produced by the dinoflagellates *Protoceratium reticulatum* and *Lingulodinium polyedrum*. While the toxicity of YTXs is relatively high, with an LD50 for mice of 750 μg/kg, to date there are no reports in humans of ill effects associated with their consumption. Four major molecular processes have been implicated in the mechanism of action of YTXs: modulation of calcium movements among different cellular compartments, modulation of cyclic monophosphate (cAMP) levels, alteration of protein disposal and apoptosis (EFSA 2008c). In the absence of human data, CONTAM established an ARfD of 25 μg YTX equiv./kg bw (EFSA 2008c). In order for a 60 kg adult to avoid exceeding a dose of 1500 μg YTX equiv., corresponding to the preceding

ARfD, the Panel decided that a 400 g portion of shellfish should not contain more than 3.75 mg YTX equiv./kg shellfish meat.

Although it was once generally believed that biotoxins do not harm their bivalves hosts, results from a number of studies have shown this to be untrue (reviewed in Manfrin *et al.* 2012). Toxic algal species, in particular PSP toxins, have detrimental effects on feeding responses such as ingestion (Escobedo-Lozano *et al.* 2012), clearance rate (Hégaret *et al.* 2007; MacQuarrie & Bricelj 2008; Contreras *et al.* 2011; Navarro *et al.* 2018), pre-ingestive selection efficiency (Villaneuva & Navarro 2016), digestion (Alexander *et al.* 2008), oxygen consumption (MacQuarrie & Bricelj 2008), reproductive effort (Haberkorn *et al.* 2010), larval development (Leverone *et al.* 2006), larval lipid storage (Przeslawski *et al.* 2008), burrowing in clams (MacQuarrie & Bricelj 2008), byssus production in mussels (Marsden *et al.* 2015), adductor muscle paralysis (Escobedo-Lozano *et al.* 2012), change in valve closure, mantle retraction, mucus production, change in cardiac activity and modulation of the immune system (Núñez-Acuña *et al.* 2013 and references therein). In some cases, these effects are transitory due to biotransformation of algal toxins by epimerisation, reduction and hydrolysis within bivalve tissues (Contreras *et al.* 2011, 2012 and references therein). Mussels, in contrast to some oyster and clam species, are insensitive to dinoflaggelate producers of PSP toxins, making them ideal indicator organisms to provide early warning of PSP outbreaks (Villaneuva & Navarro 2016 and references therein).

Harmful Algal Blooms

HABs are a natural phenomenon caused by microalgae, and can develop in high concentrations when certain favourable environmental conditions are present, such as increased brightness, stratification of water, high temperature and increased nutrients (Villaneuva & Navarro 2016 and references therein). Salinity is also an important factor as it determines the types of bloom species (Terlizzi & Mazzacaro 2010). HABs are most common in coastal marine ecosystems, but they may also affect the open ocean and brackish or freshwater ecosystems. Most algae (more than 99%) and most algal blooms do not produce toxins. However, the occurrence of human poisoning episodes from toxic seafood ingestion, involving hundreds of people in several regions of the world, has focused attention on this phenomenon. Blooms kill marine life and cause tens of millions of dollars per year in closures of harvestable shellfish, loss of production and monitoring costs (Trainer *et al.* 2014). In recent years, there has been growing concern that HABs are occurring more frequently, lasting longer and appearing in areas where they have not been recorded in the past (Figure 12.1). Increased HAB frequency may perhaps reflect more effective scientific and regulatory attention to the problem, but there is compelling evidence that human activities play an important role. Such activities include eutrophication of coastal waters as a consequence of increased aquaculture and fertiliser runoff from agriculture (Heisler *et al.* 2008), changes in climatic conditions making some coastal environments more hospitable to toxic algal species and the transportation of toxic algae and their cysts around the world in ships' ballast water or through the movement of shellfish stocks (Anderson *et al.* 2012 and references therein). In addition, HABs usually involve rapid changes in phytoplankton community composition, where communities become dominated by a single or a few species over the course of days, resulting in nearly monospecific assemblages that can then persist

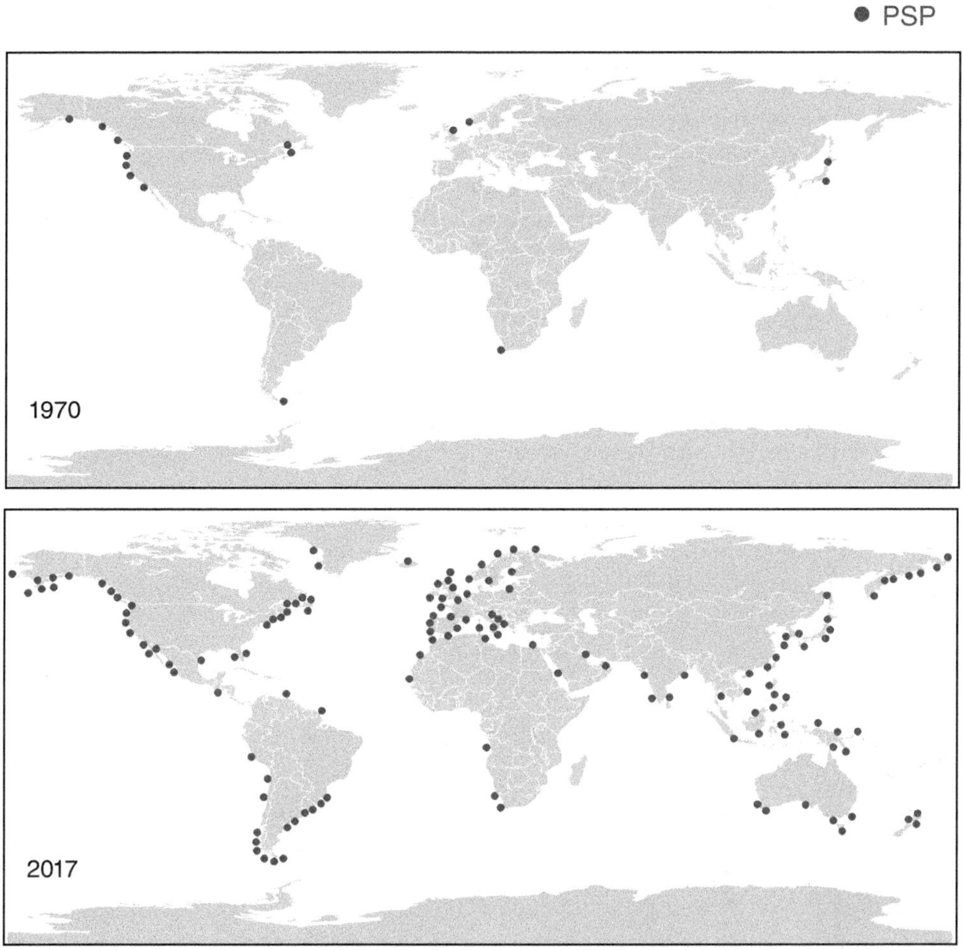

Figure 12.1 Global distribution of paralytic shellfish poisoning (PSP) toxins recorded as of 2017, compared to the distribution recorded in 1970. *Source:* Courtesy of the US National Office for Harmful Algal Blooms, Woods Hole Oceanographic Institution, Massachusetts, USA.

for weeks to months (Egerton *et al.* 2014). When the species of interest are morphologically similar or identical, it can be impossible to routinely differentiate between species and strains, preventing early warning of the presence of harmful organisms (Eckford-Soper *et al.* 2013). However, developments in molecular biology are increasingly enabling the genetic characterisation of HAB populations (details in Anderson *et al.* 2012; Eckford-Soper *et al.* 2013).

Coastal regions throughout the world that are threatened annually by HABs with bewildering tendencies of larger spatial extents and higher frequencies. Hong Kong waters, the East China Sea, the Korean South Sea, the Japanese Sea, the Gulf of Tonkin, the Arabian Sea, the coasts of France, Spain and Portugal, New Zealand waters, the Baltic Sea, the Gulf of Mexico, Washington, the Gulf of California, the coast of Florida, the Gulf of Maine, the coast of Nova Scotia, the coast of British Columbia and the South African coast are all areas subject to HABs

(Shen *et al.* 2012 and references therein). Countries including the United States, Canada, Norway, Spain, Portugal, Ireland, China, Japan and Korea have invested large sums of money and great effort into HABs monitoring and forecasting programmes. Costs of a comprehensive global HAB observing and forecasting information system represent ±\$100 million annually, which constitutes approximately 1% of HAB-related losses (Kudela *et al.* 2015).

While HABs are naturally occurring and cannot be eliminated, the negative impacts can be alleviated through coordinated research, monitoring, prevention policies and mitigation plans on a national and international level. GlobalHAB (www.globalhab.info) is one example of an international coordinating programme, developed under the auspices of the Intergovernmental Oceanographic Commission (IOC) of the United Nations Educational, Scientific and Cultural Organisation (UNESCO) and the Scientific Committee on Oceanic Research (SCOR). Its aim is to advance international collaboration on HABs, improve predictive capability and understanding of mechanisms leading to HABs and inform science and policy linking HABs with societal needs and impacts (details in Berdalet *et al.* 2017). The main elements used by GlobalHAB and similar programmes to cope with HABs include sustained support of monitoring and management, prevention, modelling and prediction, research and international coordination (Kudela *et al.* 2015; Figure 12.2). The detection and monitoring techniques for algal toxins and the causative HABs are essential

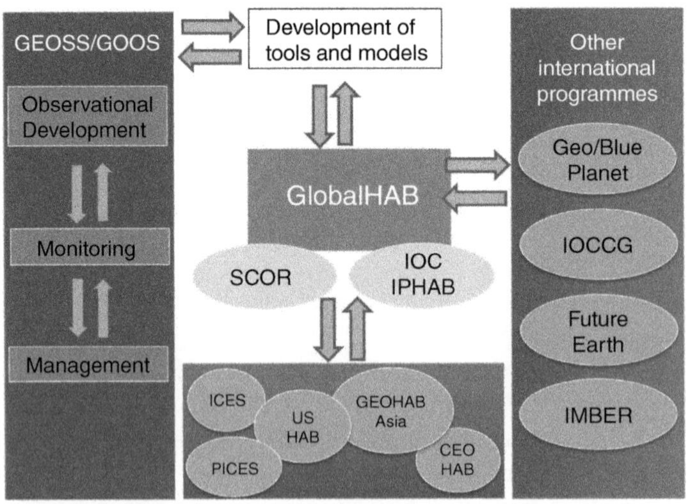

Figure 12.2 Main elements used to cope with HABs: support of monitoring and management, prevention of HABs, modelling and prediction, research and international coordination. CEOHAB, Chinese equivalent of USHAB; GEOHAB, Marine Geological and Biological Habitat Mapping; GEOSS/GOOS, Global Earth Observing System of Systems/Global Ocean Observing System; ICES, International Council for the Exploration of the Sea; IOCCG, International Ocean-Colour Coordinating Group; IMBeR, Integrated Marine Biosphere Research; IOC-IPHAB, Intergovernmental Oceanographic Commission Intergovernmental Panel on HABs; PICES, North Pacific Marine Science Organization; SCOR, Scientific Committee on Oceanic Research; USHAB, United States Harmful Algal Blooms. *Source:* From Kudela *et al.* (2015). Reproduced with permission from Intergovernmental Oceanographic Commission of UNESCO.

for the protection of aquatic life, shellfish safety, drinking water quality and public health. Over the last two decades, good progress has been made in the development of analytical and monitoring tools, ranging from in situ quick screening protocols for the monitoring of algal blooms to mass spectrometric analysis of trace levels of various algal toxins and structural elucidation (reviewed in Zhang & Zhang 2015). Solid-phase adsorption toxin tracking (SPATT) deployed in the field for the passive sampling of algal toxins has recently been validated,[5] and improved ELISA-based methods with lower detection limits for more toxins have become commercially available for both screening and routine monitoring purposes (see later). Also, new molecular probing tools to identify causative micro-organisms and genes, rapid screening tools and sensor devices for field testing (see later), new mass spectrometric applications geared toward the discovery of new algal toxins and their structural identities and large-scale HAB monitoring using satellite remote sensing have all been introduced (reviewed in Shen *et al.* 2012; Lara *et al.* 2017).

A key strategy is the development of predictive models (Anderson *et al.* 2012), but their reliability requires up-to-date, high-resolution data on HAB occurrences. A promising solution could be remotely deployed biosensors, which rapidly and continuously measure algal and toxin levels at the point of need (PON) at a low cost. Such systems incorporate biorecognition elements with a chemical or physical measurement element to allow for detection and quantification of a target analyte (reviewed in McPartlin *et al.* 2017). One such sensor, developed by the Monterey Bay Aquarium Research Institute in the early 2000s, is the Environmental Sample Processor (ESP), which provides on-site collection and analysis of marine water samples in order to identify the presence of organisms and biological toxins. The ESP instrument uses an electromechanical fluidic system designed to autonomously collect and filter water samples, which it either preserves and archives for later use or directly investigates in near-real time via molecular detection technology (Bowers *et al.* 2016).

Mitigation strategies are based on the acceptance that HABs are going to occur, and therefore means of reducing their impacts are needed (Anderson 2014). The extent and duration of blooms can be mitigated for many HAB species by changing the human activities (see earlier) responsible for the perceived increase in their numbers over the past few decades. The development of techniques to control or mitigate HABs is a promising area of research that can be separated into three categories based on their mode of action (details in Kidwell 2015). Physical mitigation techniques include sediment-based methods such as clay flocculation, water column mixing to alter nutrient dynamics and cell harvesting and removal, although application of these methods in coastal systems is limited. Chemical methods, based on artificial or naturally derived compounds that interfere with cellular growth and/or result in cell lysis through a variety of mechanisms, include barley straw, activated carbon, biosurfactants, hydrogen peroxide and copper – none of which have been applied in coastal environments. However, there is concern that chemicals have strong potential to cause significant environmental harm through mortality and other impacts to many non-target organisms. Finally, biological HAB control methods include the use of macroalgal species such as *Ulva* and *Gracilaria* that compete for nutrients or have allelopathic[6] effects on HAB species. Also, some bacteria and viruses have algicidal effects on phytoplankton, including HAB species. Consequently, there has been ongoing research on the isolation of algicidal compounds to develop a HAB control product. For example, Pokrzywinski *et al.* (2017) found that exposure to the algicide IRI-160AA produced by the marine bacterium *Shewanella* sp. IRI-343 160 resulted in significant morphological changes in several dinoflagellate species. Also, An *et al.* (2015) found that a culture supernatant of the marine anticidal actinomycete BS01 exerted a strong algicidal effect on the bloom-forming dinoflagellate *Alexandrium tamarense*. More recently, Zhang *et al.* (2018) reported the algicidal activity and characteristics of the marine bacterium *Paracoccus* against the dinoflagellate *Prorocentrum donghaiense*. The stability of the algicidal activity of *Paracoccus* in different environments (including different temperature, pH and sunlight conditions) indicates its potential for use in the control of P. *donghaiense* blooms.

Detection of Marine Toxins

Until relatively recently, the official method for the detection of marine biotoxins was the mouse bioassay (MBA), based on the extraction of liquid from putative toxic bivalves followed by injection of liquid into mice, where the time to death of mice of standard weight (18–22 g) is related to toxin level (details in Oshima *et al.* 2011). The assay provides only an indication of the overall toxicity of the sample, as it is not capable of differentiating individual toxins (Campàs *et al.* 2007). It suffers several drawbacks, including a lack of specificity, nontoxic interference and its lack of acceptance by defenders of animal welfare. Given these limitations, particularly the ethical concerns around the use of live animals, the MBA may be useful as a preliminary toxicity screen – and it is still the method of reference for almost all types of marine toxins (excluding ASP toxins) – but it is only formally validated for PSP toxin analysis using the standardised method, AOAC-OMA 959.08 of the Association of Official Analytical Chemists (AOAC). It is also the only method approved in the United States under the National Shellfish Sanitation Program (NSSP) for use in making regulatory decisions regarding the harvest of bivalves due to BTXs (NSSP 2015).

Over the past four decades, a number of alternative analytical methods have been developed for both well-established and emerging toxins. These new methods must be reliable and have adequate sensitivity in order to detect the presence of the toxins and their analogues, not only at the current regulatory levels (Table 12.2) but also at lower levels that are likely to result from recommendations of the EFSA and other regulatory bodies worldwide (Campbell *et al.* 2011a). These bodies have established rigorous performance criteria that must be realised by an alternative method before the MBA is discontinued. The most challenging requirement is that intra- and inter-laboratory validation must be performed to internationally recognised standards (e.g. European Commission 2011). Standards for marine toxins are generally produced following their extraction and purification from large-scale culturing of toxic algal or from contaminated shellfish material harvested from regions following an outbreak. To develop quantitative and specific methods, the availability of pure reference compounds (the toxins themselves) is a major prerequisite. Currently, there is at least one reference group available for every toxin group, but not for all relevant toxic analogues. As these become available, alternative detection methods to bioassays are developed. The only source of certified reference standards and material is the Institute for Marine Biosciences, National Research Council of Canada (www.nrc-cnrc.gc.ca). At present, its Certified Reference Materials Program (CRMP) distributes 36 CRMs to more than 40 countries. These materials are essential to analytical laboratories, because all measurement systems must be carefully calibrated before each test run.

Alternative methods to the MBA can be grouped into biological functional assays, biochemical immunological assays and chemical analytical methods.

Functional Assays

Functional binding assays are detection methods that are either receptor- or cell-based and which use the mechanism of action of the toxin for quantification, which in turn relates to the toxicity of the compound (Campbell *et al.* 2011a). To illustrate, cell morphology-based assays (e.g. those used for DSP and AZAs) employ cell lines (e.g. neuroblastoma cells, rat hepatocytes, human epidermoid carcinoma cells or intestinal epithelial cell lines). When these cells are exposed to DSP toxins, morphological changes can be detected by microscopy, spectrophotometry or fluorescence measurements of filamentous actin levels. In receptor-based assays (RBAs), toxins bind to the receptors with an affinity proportional to their toxic potency (Campàs *et al.* 2007). A glutamate receptor-binding assay (RBA) has been developed for detection of DA, the toxin responsible for ASP (Van Dolah *et al.* 1997). Methods based on perturbations to cellular parameters, such as enzyme inhibition, have also been developed. One such assay, PP2a, is based on the inhibition of protein phosphatases, which play an important role in protein dephosphorylation in cells (Smienk *et al.* 2012).

Immunological Assays

Immunology-based assays depend on the ability of a specific antibody to distinguish the three-dimensional structure of a specific toxin from other molecules. Antibodies have the advantages of high specificity and sensitivity, the latter allowing for sample dilution that reduces matrix effects from shellfish samples (Campbell *et al.* 2011a). However, when an antibody is raised to a particular toxin in the group, the specificity of the antibody is unlikely to correspond to the inherent toxicity of individual toxins, depending on their structural differences. There is a trade-off in usefulness between highly specific antibodies, capable of detecting a given toxin at low levels, and less specific antibodies, capable of detecting all the members of a toxic family. Immunological methods developed for marine biotoxins fall into three main categories: ELISA (see Chapter 5), immunochromatographic strips and biosensors (Campàs *et al.* 2007; Campbell *et al.* 2011a; Reverté *et al.* 2014; Turner *et al.* 2015). Immunochromatographic strips provide a rapid visual test for the screening of TTX (details in Reverté *et al.* 2014). Although the test is not quantitative, it has the advantages of simplicity, ease of use, lack of sophisticated equipment, short analysis times (10 min), long-term stability under various storage conditions and suitability for testing of biotoxins in the field as a Hazard Analysis and Critical Control Point (HACCP; see later) tool. Biosensors use specific biochemical reactions mediated by antibodies to detect chemical compounds, usually by electrical, thermal or optical signals (reviewed in Campàs *et al.* 2007; Vilariño *et al.* 2009). Most biosensors for emerging marine toxins are surface plasmon resonance (SPR) immunosensors, an optical technique that allows the detection of the toxin of interest in real time and without the need of labels. Major advantages of biosensor methods are their specificity, sensitivity, simplicity and ease of use, together with the possibility of developing multiplex detection methods (McNamee *et al.* 2013, 2014; Vilariño *et al.* 2013; Riverté *et al.* 2014; Zhang *et al.* 2014). A multiplex method for the simultaneous analysis of STXs, OA and DA has the potential to analyse up to 16 toxins on a microarray chip (Campbell *et al.* 2011b). There is also the potential for miniaturisation and portability, for ultimate on-site monitoring. A set of 17 marker genes, differentially regulated in human intestinal Caco-2 cells (the cell line of human epithelial colorectal adenocarcinoma cells) upon exposure to the lipophilic shellfish poisons AZA1 or DTX1, were the basis for the design of a tailored microarray platform for detection of these toxins, and potentially others (Bovee *et al.* 2011). Five of the selected marker genes on this dedicated DNA microarray gave clear signals, whereby the resulting fingerprints could be used to detect these toxins.

Analytical Methods

As already indicated, many of the toxin groups discussed contain a multiplicity of structural forms, which makes detection a formidable challenge. Therefore, it is no surprise that analytical methods are dominated by high-performance liquid chromatography coupled with mass spectrometry (HPLC-MS), fluorimetric detection (HPLC-FLD) or ultraviolet detection (HPLC-UVD). These techniques allow separation, highly selective identification and sensitive quantification of the different toxins present in a sample. Figure 12.3 shows

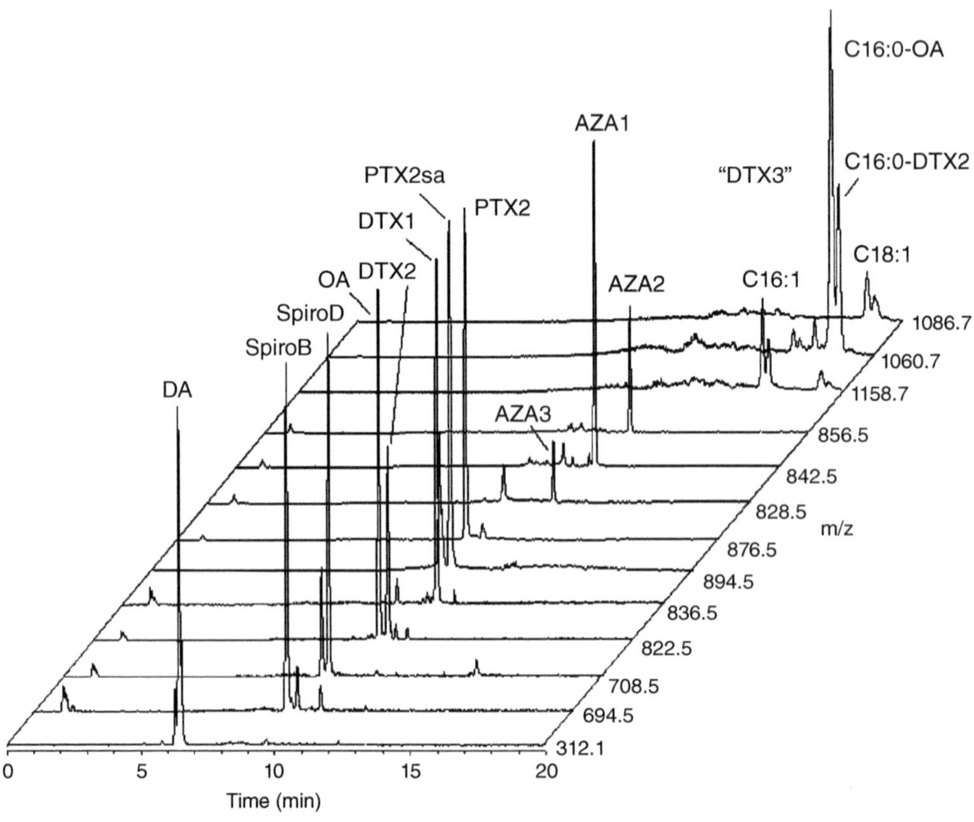

Figure 12.3 Reversed-phase gradient elution LC-MS analysis of a range of toxins, displayed as individual mass chromatograms, from a blend of contaminated mussel tissue extracts. The toxins present include domoic acid (DA), spirolides (Spiro B, Spiro D), okadaic acid (OA), dinophysistoxins (DTX1, DTX2), pectenotoxins (PTX2, PTX2sa), azaspiracids (AZA), and acyl esters of OA and DTX2 ('DTX3'). *Source:* From Quilliam (2003). Reproduced with permission from Elsevier.

several major groups of lipophilic toxins from mussel (*Mytilus* sp.) tissue extracts, discernible on the basis of their chemical structure (Quilliam 2003). In addition, coupled LC tandem MS methods such as LC-MS/MS (Watanabe *et al.* 2013) and hydrophilic interaction liquid chromatography, HILIC-MS/MS (Zhuo *et al.* 2013; Boundy *et al.* 2015) are a powerful combination for the detection and quantification of toxins in plankton, shellfish and fish at trace levels, the chemical characterisation of new toxins, the traceability of toxin production by plankton and the investigation of toxin metabolism in shellfish (Campbell *et al.* 2011b and references therein). Automated on-line solid-phase extraction (SPE) coupled to LC–MS/MS has been developed for the fast determination of lipophilic toxins (OA, DTX, YTX, AZA, PTX and some CIs) in raw and hydrolysed mussel (*M. galloprovincialis*) extracts from raft culture in NW Spain (Requeiro *et al.* 2011). To date, the following methods have been validated by either the AOAC, the EURL or laboratory validation studies: MBA, LC, ELISA, RBA and PP2a (enzyme-based assay, based on the inhibition of phosphoprotein-phosphatase 2a (PP2a) (Table 12.3).

With the recent identification of genes associated with toxin production, there has been a development of molecular tools for the detection of the presence of toxin-producing species. Several qPCR assays have been developed that target ribosomal RNA gene regions of known toxin-producing species (Cusick & Sayler 2013). Whole-cell FISH probes based on ribosomal RNA sequences have been developed to distinguish between toxin- and non-toxin-producing *Alexandrium minutum* strains (Tang *et al.* 2012). Also, in order to differentiate morphologically similar dinoflagellate species (e.g. species in the genus *Karenia*), a range of molecular techniques such as nucleic acid sequence-based amplification (NASBA), FISH (see Chapter 5) and microsatellite markers (see Chapter 9) have been applied (Renshaw *et al.* 2006; Brand *et al.* 2012).

Industrial Pollutants

Bivalves accumulate high levels of pesticides, heavy metals and hydrocarbons from contaminated water, and there is ample evidence that these chemcials have a severe effect on their physiology and immune system (see Chapters 2, 5, 6, 7, 8 and 11). Consequently, bivalves – particularly mussels – are used as sentinel organisms in environmental monitoring programmes (see Chapter 8). The main heavy metals of concern to human health in seafood consumption are mercury (Hg), lead (Pb), cadmium (Cd), tin (Sn) and arsenic (As). Metals such as these can enter water and sediments in marine ecosystems, primarily as a consequence of human activities in the form of industry, agriculture and mining industrial emissions, or via disposal of products containing them (FSAI 2009; Olmedo *et al.* 2013). Marine organisms such as fish and shellfish can then accumulate these metals in their tissues to potentially toxic levels.

Mercury and its compounds are highly toxic, especially methylmercury (CH_3Hg^+), which accumulates appreciably in shellfish and fish. Mercury concentrations continue to increase, with a twofold increase recorded in surface waters over the last century due to increasing industrialisation and energy production (Mason *et al.* 2012). Mercury is a potent neurotoxin to humans due to its ability to cross the blood–brain barrier. It is absorbed in the gastrointestinal track, immediately entering the bloodstream. It readily passes the

Table 12.3 Biotoxin detection methods validated for marine mollusc matrices through collaborative trials.

Toxin group	Method-type	Specificity [y/n][(1)]	Validation reference	EU	US	JP
STX	MBA	n	AOAC-OMA 959.08 (2005a)	✓	✓	✓[(2)]
	HPLC-FLD, precox	y	AOAC-OMA 2005.06 (2005b)	✓		
	HPLC-FLD, postcox	y	AOAC-OMA 2011.02 (2011a)		✓	
	RBA	n	AOAC-OMA 2011.27 (2011b)		✓[(3)]	
OA	LC-MS/MS	y	EU-RL-MB (2015)	✓	✓[(4)]	✓[(4)]
	LC-MS/MS	y	These *et al.* 2011			
	LC-MS/MS	y	van den Top *et al.* (2011)			
	PP2a	n	Smienk *et al.* (2012)	✓		
DA	ELISA	n	AOAC-OMA 2006.02[(5)]	✓		
	HPLC-UV	y	pre-EN14176[(6)]	✓	✓	✓
AZA	LC-MS/MS	y	EU-RL-MB (2015)	✓		
	LC-MS/MS	y	These *et al.* (2011)			
	LC-MS/MS	y	van den Top *et al.* 2011			
PBX	MBA					
TTX	MBA[(7)]	n	AOAC-OMA 959.08			✓[(8)]
	RBA[(7)]	n	AOAC-OMA 2011.27			

Notes:
(1) y = The method gives analogue-specific results; n = the method does not give analogue-specific results.
(2) Adapted protocol using analogue dc-STX as calibrant.
(3) Approved as mouse replacement for mussels, and also screening clams and scallops. Oyster data under review.
(4) Modified from EURL-validated method, with only OA group quantified, separately validated.
(5) Kleivdal *et al.* (2007).
(6) Method has undergone inter laboratory trial and is undergoing normalisation.
(7) Method not validated for this analogue group but should perform as for STXs as the mode of action is exactly the same.
(8) An adapted protocol is used for confirmation of food poisoning events.
√ denotes acceptance of the method in major trade blocks; EU = European Union; US = United States of America; JP = Japan
Source: From FAO/WHO (2016). Reproduced with permission from the Food and Agriculture Organization.

placental barrier, affecting the developing nervous system of the foetus (Mudgal *et al.* 2010). Prenatal exposure to methylmercury alters attentional mechanisms modulating early processing of sensory information (Boucher *et al.* 2010). Strategies for decreasing mercury pollution have been in place in the United States since the Mercury Export Ban Act was signed into law in 2008. This Act intends to reduce the availability of elemental (metallic) mercury in domestic and international markets. By reducing the supply of metallic mercury in commerce, it aims to reduce the use of mercury for commercial purposes globally. Subsequent to its coming into force, more than 140 countries at a meeting of the United Nations Environment Programme (UNEP) Governing Council agreed to negotiate a legally binding international agreement by 2013 to limit mercury emissions. A draft technical background document for Global Mercury Assessment 2018 is in its final stage (UN Environment Programme 2019). Also, based on the agreement to act to reduce exposures to a number of priority pollutants in the Arctic region, a mercury project was initiated in 2006 via the Arctic Council Action Plan (ACAP). The project is led by the Danish Environmental Protection Agency, with a Steering Group from all eight Arctic countries: Canada, Denmark, Finland, Iceland, Norway, Russia, Sweden and the United States. Recently, UNEP developed a toolkit for the identification and quantification of mercury releases, which is intended to assist countries to develop a national mercury releases inventory (UN Environment 2017). The CONTAM Panel established a tolerable weekly intake (TWI) for methylmercury of 1.3 μg/kg bw (EFSA 2012a). High fish consumers, which can include pregnant women, may exceed the TWI by up to approximately sixfold. Also, inhalation exposure of elemental mercury from dental amalgam is likely to increase mercury exposure, causing the TWI to be exceeded (EFSA 2012a). The regulatory limit in Canada and the European Union is 0.50 mg mercury/kg shellfish weight[1], while in the United States and Australia the limits are twice and three times this level, respectively (Table 12.2).

Foodstuffs are the main source of cadmium (Cd) exposure for the nonsmoking general population. Cd absorption after dietary exposure is relatively low (3–5%), but Cd, which has a very long biological half-life of 10–30 years, is efficiently retained in the human body, where it negatively affects organs such as liver, kidney, lung, bones, placenta, brain and the central nervous system (CNS) (Castro-González & Méndez-Armenta 2008; EFSA 2009d). The Joint FAO/WHO Expert Committee on Food Additives (JECFA 2006) set a TWI of 7.0 μg/kg bw. The regulatory limits in the United States, the European Union and Australia are 2.0, 1.0 and 2.0 mg Cd/kg shellfish weight, respectively, while no amount is specified for Canada (Table 12.2). Exposure of *M. galloprovincialis* gill tissue to increasing levels of Cd induced an increase in hsp70, which was correlated with the amount of bioaccumulated Cd in the gill tissue, indicating that hps70 could be used as a potential marker for seafood safety and environmental monitoring programmes (Piscopo *et al.* 2017).

Lead (Pb) is a metal that occurs naturally but whose presence in the environment has greatly increased as a result of anthropogenic activities such as mining, smelting and battery manufacturing. While lead occurs in organic and inorganic forms, it is the inorganic forms that predominate in the environment (EFSA 2010c). Human exposure to Pb is mainly via food (e.g. cereal products) and water, with some exposure occurring through air, dust and soil. Short-term exposure to high levels of Pb can cause brain damage, paralysis, anaemia and gastrointestinal symptoms, while longer-term exposure can cause damage to the kidneys, reproductive and immune systems and CNS (FSAI 2009). The most critical effect of low-level Pb exposure is on intellectual development in young children. Also, Pb can cross the placental barrier and accumulate in the foetus (FSAI 2009). A TWI of 25 μg/kg bw is recommended, although the CONTAM Panel concluded that this value is no longer appropriate as there is no evidence for a threshold for critical lead-induced effects (EFSA 2010c). The regulatory limits in Canada, the European Union and Australia are 0.50, 1.5 and 2 mg Pb/kg shellfish weight, respectively, but no amount is specified for the United States (Table 12.2).

Arsenic exists in inorganic and organic forms, with the former more toxic than the latter. Inorganic arsenic is a carcinogen of the lung, kidney, bladder and skin tissue in humans. Fortunately, about 80–95% of total arsenic is present as organic compounds that do not cause any damage to health (Mudgal *et al.* 2010). Seafood is a major source of inorganic As, ranging between 5 and 250 μg/g depending on the species, and a TWI has been set by the Joint FAO/WHO Expert Committee on Food Additives (JECFA) at 15 μg/kg bw. However, the CONTAM Panel concluded that this provisional value is no longer appropriate as data have shown that a range of cancerous conditions are reported at exposures lower than those reviewed by the JECFA (EFSA 2009e). While a regulatory limit value exists for As in drinking water, no limits have been established for seafood.

Several studies on heavy metal concentrations have been carried out on mussels (*M. galloprovincialis* and *P. viridis*) from a wide variety of locations, extending from Europe to East Asia. Numbers of metals analysed/studied varied between four and eight, with Cd, Cu, Hg, Ni, Pb and Zn being the most often utilised (Jović & Stanković 2014; Yap *et al.* 2016; Dokmeci 2017; Zhelyazkov *et al.* 2018). Without exception, all metal levels were shown to be below the established seafood safety guidelines. Therefore, there is no evidence of risk to the consumer.

Polychlorinated biphenyls (PCBs) and organochlorine pesticides (OCPs), such as dieldrin and DDT, are toxic persistent organic pollutants (POPs) that accumulate in the food chain. They were widely used up to the 1970s. The Stockholm Convention on Persistent Organic Pollutants, an international environmental treaty that aims to eliminate or restrict the production and use of almost 30 POPs, was signed and took effect in May 2004. As of April 2017, 180 states plus the European Union are party to the Convention; nonratifying states include the United States, Israel, Malaysia and Italy. Despite the ban, POPs continue to exist in the environment, albeit at low concentrations. Toxic responses to dioxins and PCBs include dermal toxicity, immunotoxicity, carcinogenicity and reproductive and developmental toxicity (EFSA 2012b). Zhang *et al.* (2012) measured a broad spectrum of PCBs and OCPs in eight bivalve species, including mussels, from a market in Xiamen, a city on the southeast coast of China, where seafood is widely consumed by its 3.5 million inhabitants. They found that their TWI PCB values ranged from 0.009 to 0.052 pg/g^1, which was significantly lower than the 14 pg/kg^1 proposed by the Scientific Committee on Food (SCF 2001). The regulatory limit varies from a low value of 6.5 pg/g shellfish ww in the European Union to 0.5 mg/kg in Australia and 2.0 mg/kg in the United States and Canada (Table 12.2).

Polycyclic aromatic hydrocarbons (PAHs) are complex chemicals that are formed and released by incomplete combustion or pyrolysis (burning) of organic matter such as waste or food during industrial processes and other human activities (FSAI 2015). They have a variety of toxicological effects –particularly benzo[a]pyrene – that include haemotoxicity, reproductive and developmental toxicity and immunotoxicity, but it is their carcinogenic and genotoxic (DNA-damaging) potential that is the cause of most concern. The FSAI (2006) analysed five samples of shellfish for PAHs, and while four generally showed total levels below 0.89 μg/kg fresh weight, one sample of fresh mussels showed levels as high as 2.14 μg/kg – although this was still well below the allowable level specified by the European Union (Table 12.2).

Plastic is the most prevalent type of marine debris found in our oceans. Plastic debris can come in all shapes and sizes, but that which is less than 5 mm in length is called a microplastic (National Ocean Service 2017). Microplastics can come from a variety of sources, including larger plastic debris that degrades into ever smaller pieces and tiny pieces of manufactured polyethylene plastic, known as microbeads, which are added as exfoliants to health and beauty products, such as some cleansers and toothpastes (see Chapter 8). Ingestion and accumulation of microplastics and nanoplastics (1–1000 nm) have been demonstrated for diverse marine species ranging from zooplankton to bivalves and fish, showing their potential to accumulate in the marine food web and thus pose a potential threat to aquatic life, food safety and human health. Van Cauwenberghe & Janssen (2014) investigated the presence of microplastics in the soft tissues of cultured *M. edulis* and *Crassostrea gigas*. After several days' depuration (time to human consumption), *M. edulis* contained on average 0.36 ± 0.07 particles/g ww, while a plastic load of 0.47 ± 0.16 particles/g ww was detected in *C. gigas*. The authors estimated that the annual dietary exposure for European shellfish consumers could amount to 11 000 microplastics per year. However, due to the complexity of estimating microplastic toxicity, estimation of the potential risks for human health posed by microplastics in foodstuffs is not yet possible. A lower average load (0.13 ±0.14 total microplastics/g ww) was recorded in commercial mussels (*M. edulis* and *M. galloprovincialis*) from five European countries (Vandermeersch *et al.* 2015). In a more recent study, Birnsteil *et al.* (2019) investigated levels of microplastics in wild and farmed mussels (*P. perna*) in order to assess the effectiveness of depuration in reducing microplastics. Wild and farmed mussels were sampled from Guanabara Bay in the state of Rio de Janeiro, Brazil. Four treatments were compared (N = 10 mussels per treatment): wild nondepurated mussels, wild depurated mussels, farmed nondepurated mussels and farmed depurated mussels. Up to 31.2 ± 17.8 microplastics/mussel (≥0.45 μm) were detected (means ± SD), and microplastics were present in all 40 individuals analysed. Although 93 hours' depuration significantly reduced microplastics (ANOVA, $P = 0.02$) in both wild (46.79%) and farmed (28.95%) mussels, differences between the two were not significant ($P > 0.05$). Depuration was more effective in removing microplastics, highlighting its importance in reducing microplastic pollution in seafood. Toxicity and toxicokinetic data allowing a a human risk assessment are currently lacking for both micro- and nanoplastics. The EFSA (2016) recommends that analytical methods should be further developed for micro- and nanoplastics and standardised in order to assess their presence and identity and to quantify their amount in food. In addition, quality assurance should be put in place and demonstrated.

Another source of contamination is radionuclides: radioactive forms of chemicals that are unstable and emit radiation over time, thereby forming different, more stable chemicals. Radionuclides can originate from a variety of sources and processes, both natural and resulting from human activities (e.g. nuclear weapons testing, accidents at nuclear power plants; Pearson *et al.* 2016a). A common suite of radionuclides of particular importance because of their abundance, mobility or toxicity include ^{3}H, ^{14}C, ^{90}Sr, ^{99}Tc, ^{129}I, ^{137}Cs, ^{237}Np and ^{241}Am, as well as several uranium and plutonium isotopes (Hu *et al.* 2010). Over the past 50 years, the presence of radionuclides in the environment has been a significant global concern, especially following accidents such as that at the Fukushima Dai-ichi Nuclear Power Plant in Japan in 2011. One concern is the potential impact of radionuclides released into the Pacific Ocean on seafood, and consequently on seafood consumers. Three months after the Fukushima accident, levels of ^{134}Cs and ^{137}Cs in mussels (*M. galloprovincialis*) and macroalgae (*Ulva* sp.) from coastal areas in the vicinity of Fukushima were up to ~800 Becquerels/kg dw (Baumann *et al.* 2013). Nevertheless, when combined (i.e. $^{134}Cs + ^{137}Cs$), they did not exceed the 500 Bq kg/ww recommended safety level for seafood consumption (see later). The European Union has set maximum limits on the level of radioactive contamination permitted in foodstuffs imported from countries outside the region. For caesium radionuclides, these stand at 370 Bq/kg for baby food and milk products and 600 Bq/kg for all other foodstuffs (European Commission 2016b). These limits also apply to food traded within the European Union. In response to the Fukushima disaster, the Union provisionally set lower permissible limits for foodstuffs imported from Japan at 100 Bq/kg. See Máté *et al.* (2015) for an overview of the current implementation of radionuclide monitoring in foodstuffs among EU Member States. In a related study, Povinec & Hirose (2014) examined variations of Fukushima-derived radionuclides (^{90}Sr, ^{134}Cs and ^{137}Cs) in seawater and biota in coastal waters of Fukushima and in the NW Pacific Ocean, and estimated radiation doses for the Japanese and world populations due to ingestion of seafood contaminated by Fukushima radionuclides, comparing them with those from other sources of anthropogenic and natural radionuclides, such as ^{210}Po. The total effective dose equivalent[7] from ingestion of radionuclides in fish, shellfish and seaweed caught in coastal waters off Fukushima was estimated as 0.6 ± 0.4 millisievert[8] per year, while the individual effective dose commitment from consumption of radioactive-contaminated fish caught in the open Pacific Ocean was 0.076 ± 0.05 mSv/y. While the Fukushima number was much higher, it was still comparable to or lower than doses delivered from the consumption of natural ^{210}Po in fish and shellfish (0.7 mSv/y), indicating that doses were below the level that could cause damage to Japanese and world populations. A dose limit of 1 mSv/y has been set for the general public, while 50 mSv/y has been set as the maximum dose an occupational worker can receive. Methods used in radionuclide analysis in seafood are described in Baumann *et al.* (2013), Povinec & Hirose (2015), Máté *et al.* (2015) and Pearson *et al.* (2016a,b).

Controls on the Production and Processing of Mussels and Other Bivalves

Regulatory Framework

To protect public health, many countries have established sanitation programmes that supervise the production, harvesting and marketing of bivalve species for human consumption. In the United States, the National Shellfish Sanitation Program (NSSP) is the federal/state cooperative programme recognised by the Food and Drug Administration (FDA) and the Interstate Shellfish Sanitation Conference (ISSC) for the sanitary control of shellfish produced and sold for human consumption (NSSP 2015). The NSSP operates under a set of guidelines that have been drawn up by state agencies, technical experts and representatives from the bivalve industry itself. Its purpose is to promote and improve the sanitation of shellfish (oysters, clams, mussels and scallops) moving in interstate commerce through federal/state cooperation and uniformity of state shellfish programmes. Participants in the NSSP include agencies from shellfish producing and nonproducing states, the FDA, the Environmental Protection Agency (EPA), the National Oceanic Atmospheric Administration (NOAA) and the shellfish industry. Coordination, supervision and evaluation of the programme are the responsibility of the FDA. The guidelines are updated on a regular basis and are published in a Manual of Operations that is readily accessible to interested parties. The Manual gives detailed instructions on how a monitoring programme should be run in any US member state, describes the interrelation of member state programmes and cites the criteria to be applied by the FDA in evaluation of the programmes. The NSSP has served as a model for the development of shellfish sanitation programmes in other countries, including New Zealand, Mexico and the Philippines. Also, the Food and Agriculture Organization (FAO), along with the WHO, provides technical guidance to developing countries in particular on the implementation of bivalve sanitation programmes that are in line with the Codex Code of Practice for Fish and Fishery Products (see later).

A large number of countries have signed Memoranda of Understanding (MOU) with the FDA in the United States. An MOU is a formal agreement between the FDA and federal, state or local government agencies, academic institutions and other entities. It constitutes an understanding between the parties, but it is a nonbinding agreement. Its intent is to improve consumer protection through more effective use of collective resources and to eliminate duplication of activities. To date, shellfish sanitation MOUs are in place between the FDA and EU countries, Canada, Australia, Mexico and Japan. These countries have put in place controls equivalent to those of the United States, enabling producers in these countries to export bivalves to lucrative US markets, and in turn allowing the United States to export to markets in these countries. Both parties to an MOU agree to an exchange of information concerning proposed changes in methods and procedures for sampling, analysis and confirmation, as well as administrative guidelines, tolerance, specification standards and nomenclature.

Controlling Harvesting Areas

The two most important factors controlling the safety of bivalves are harvest area water quality and shellfish handling from harvest to consumption. Ensuring that production and harvesting take place in clean waters requires knowledge of relevant hydrographic and meteorological characteristics, including tides (type and amplitude), rainfall (amount, pattern and frequency of significant rainfalls), winds and river discharges (volumes and seasonality) (ASQAP 2017). In addition, monitoring of water quality is essential. The microbiological status of a shellfish growing area is usually based on the results from monitoring levels of faecal indicator bacteria such as *Escherichia coli*, using the standard five-tube, three-dilution Most Probable Number (MPN) technique (ISO 2015). In the European Union, four categories of shellfish water are recognised based on *E. coli* levels: category A, where products can be placed directly on the market provided that they comply with a standard of <230 *E. coli* in 100 g of shellfish meat in more than 90% of samples; categories B and C, where the product must go through either depuration, relaying or designated heat treatment prior to reaching the market; and Prohibited, where shellfish cannot be treated by any of these methods, and hence harvesting for human consumption is forbidden (EURL 2017; Table 12.4). Production areas in the United States are monitored and classified instead based on levels of faecal coliforms detected in seawater, rather than in bivalve flesh. There are effectively five categories of water recognised (Table 12.5). Bivalves from 'approved' waters can be sold directly on the market without prior treatment, while those from 'restricted' areas may only be sold after depuration or relaying; these correspond to categories A and B, respectively. Areas that are subject to periodic pollution may be 'conditionally approved' or 'conditionally restricted'; during pollution events and for a period afterward, they remain closed to harvesting. Finally, bivalves from 'prohibited' areas may not be sold for consumption (NSSP 2015).

Table 12.4 EU criteria for the classification of bivalve harvesting areas. *Source:* From EURL (2017). CC BY 4.0.

Class[1]	Microbiological standard[2]	Treatment required
A	Samples of live bivalve molluscs from these areas must not exceed, in 80 % of samples collected during the review period, 230 *E. coli* per 100 g of flesh and intravalvular liquid. The remaining 20 % of samples must not exceed 700 *E. coli* /100 g of flesh and intravalvular liquid[3]	None
B	Live bivalves from these areas must not exceed, in 90% of samples, 4600 MPN (most probable number) *E. coli* per 100g of flesh and intravalvular liquid. In the remaining 10% of samples, live bivalves must not exceed 46 000 MPN *E. coli* /100 g of flesh and intravalvular liquid.[4]	Purification, relaying or cooking by an approved method
C	Live bivalves from these areas must not exceed the limits of a five-tube, three dilution MPN test of 46 000 *E. coli*/100 g of flesh and intravalvular liquid[5].	Relaying or cooking by an approved method
Prohibited	More than 46 000 *E. coli*/100 g of flesh and intravalvular liquid.	Harvesting not permitted

[1] The competent authority has the power to prohibit any production and harvesting of bivalve molluscs in areas considered unsuitable for health reasons.
[2] The reference method for analysis of *E. coli* is the detection and Most Probable Number (MPN) technique ISO (2015) ISO 16649-3:2015. Alternative methods may be used if they are validated against this reference method in accordance with the criteria in ISO-16140-2:2016.
[3] Regulation (EC) No 854/2004, as amended by Regulation (EU) 2015/2285.
[4] Regulation (EC) 854/2004 as amended by Regulation (EC) 1021/2008.
[5] Regulation (EC) 854/2004.

Table 12.5 National Shellfish Sanitation Program (NSSP) criteria for the classification of shellfish harvesting areas in the United States.

	Total coliforms/100 ml water		**Faecal coliforms/100 ml water**		
Classification	**Geometric mean[1]**	**90% compliance[2]**	**Geometric mean**	**90% compliance[2]**	**Treatment required**
Approved areas[3]	≤70	≤230	≤14	≤43	None
Restricted areas[4]	≤700	≤2300	≤88	≤260	Purification or relaying to an approved area
Prohibited areas	No sanitary survey, or conditions for approved/restricted areas not met[5]				Harvesting not permitted

[1] Or median
[2] Values for 5-tube decimal dilution test; different 90% compliance levels are given for the three-tube MPN and mTEC membrane filtration tests.
[3] Determination of approved area status must be based on a minimum of 15 samples from each monitoring station.
[4] Conditionally restricted areas may be declared where these are subject to predictable contamination events; such areas are closed for harvesting during contamination events and for a period afterwards to permit natural cleansing.
[5] Considerations other than the concentration of contaminants may be used to declare an area prohibited.
Source: From Lee & Reece (2014). Reproduced from Journal of Water and Health, 12, 281–287, with permission from the copyright holders, IWA Publishing.

These two classification systems are the principal systems used worldwide, and countries that trade with Europe and/or the United States will use either one or a hybrid of the two systems. Lee & Reece (2014) investigated the relationship between classifications under the two systems using both statistical modelling and simple compliance assessment on large international data sets of paired seawater and shellfish samples. The two approaches yielded the same conclusions: EU category A is more stringent than the US 'approved' category for all bivalves tested, while the US 'restricted' standard is more restrictive than EU category B for some species. Hence, the classifications under the two programmes are not exactly equivalent. Indeed, the two classification systems are not immediately comparable, principally because some regulations are based on the monitoring of water and others on the monitoring of shellfish flesh. To create a common baseline for comparison, de Souza *et al.* (2018) developed regression models based on monitoring data and used them to correlate levels of faecal indicator bacteria in water and in shellfish. The classification system used in the European Union was found to provide the highest level of shellfish safety for classification categories that did not require post-harvest treatments prior to marketing, while the US system provided higher level of shellfish safety for classification categories that did require these treatments.

In addition to classification of harvesting areas, there is legislation for monitoring these areas and guidance and protocols for harvesting, handling following harvesting, documentation of batches and procedures such as heat treatment, relaying and purification, as well as structural and hygiene requirements for dispatch and purification centres (see later). Apart from the microbiological criteria, there is also legislation controlling the organoleptic characteristics (sight, smell, touch, taste) of bivalves, biotoxins levels (Table 12.2), wrapping and packaging, identification marking and labelling. Regulation (EC) No 2073/2005 (European Communities 2005) sets out the microbiological criteria for foodstuffs. Regulation (EC) No 852/2004 (European Communities 2004a) concerns general rules for the hygiene of foodstuffs, including HACCP. Regulation (EC) No 853/2004 (European Communities 2004a) lays down specific hygiene rules for food of animal origin. Regulation (EC) No 854/2004 (European Communities 2004b) applies to live bivalves and specifies controls at all stages of production from monitoring of the quality of harvesting areas through to placement on the market. These regulations only apply to commercial production. The EU Water Framework Directive 2000/60/EC (European Communities 2006b) aims to improve our water environment. Also, Good Manufacturing Practice (GMP) and Good Hygiene Practice (GHP) (EC No 2076/2005) and a well-designed HACCP programme (see later) are all necessary to prevent contamination and ensure a safe product (see Codex Alimentarius 2012[9]).

Bivalve Purification Procedures

As suspension feeders, marine bivalves concentrate contaminants from the water column in which they grow, which may then cause illness to humans when the bivalves are eaten. For microbial contaminants, the risk is enhanced by the fact that these shellfish are often eaten raw (e.g. oysters) or relatively lightly cooked (e.g. mussels). Limiting the risk of illness depends partly on sourcing shellfish from areas in which such contaminants are at relatively low levels (Table 12.2). The risk may be reduced further by appropriate treatment following harvest (Lee *et al.* 2008). There are two methods, relaying and depuration, that are used to purify bivalves contaminated with pathogenic bacteria, viruses, toxins or chemicals. See Lee *et al.* (2008), Lees *et al.* (2010) and Martinez-Albores *et al.* (2020) for comprehensive reviews.

Relaying

Relaying involves the removal of harvested bivalve individuals from a microbiologically contaminated growing area to an acceptable growing or holding area under the supervision of the agency having jurisdiction, and holding them there for the time necessary for the reduction of contamination to an acceptable level for human consumption (Codex Alimentarius 2012). Relaying methods vary worldwide. Individuals may be placed in floats, rafts or directly on the bottom of the area. According to the Codex Alimentarius (2012), boundaries of relaying areas should be clearly identified by buoys, poles or other fixed means. They should be adequately separated from bivalves in adjacent waters, and suitable control systems should be put in place to prevent cross-contamination and co-mixing. Holding time and minimum temperature in the accepted area prior to harvest will be determined by the official agency having jurisdiction according to the degree of contamination before relaying, the temperature of the water, the bivalve species involved and local geographic or hydrographic conditions, to ensure that contamination levels have been adequately reduced. Bivalves should be laid out at a density that will allow them to open and undergo natural depuration. Relaying areas should be monitored regularly for biotoxins and environmental pathogens. Relaying is particularly suitable for treating more heavily polluted bivalves, where longer periods – usually more than two months – are required to remove heavy contaminant loads (Lees *et al.* 2010). Drawbacks of relaying include a lack of availability of suitable clean coastal areas and the difficulty in obtaining ownership of those areas. Also, the water quality of relaying areas is difficult to ensure due to the possibility of recontamination by seasonal variations of naturally occurring bacterial populations or transient pollution. In addition, it can be difficult to assess the efficiency of the relaying process because levels of the indicator micro-organisms may fluctuate erratically during the relay period. For more details, see Oliveira *et al.* (2011), NSSP (2015), ASQAP (2017) and CFIA (2018).

Depuration

This process (controlled purification) involves placing contaminated bivalves in flowing clean seawater for a set period of time to allow natural cleansing. The depuration process can be traced to the early part of the 20th century and was developed in direct response to a number of well publicised outbreaks of typhoid fever associated with bivalve consumption on both sides of the Atlantic (Lees *et al.* 2010). The main principles of depuration are that:

- *Bivalves resume filtration activity* so that contaminants are expelled through the maintenance of correct conditions of salinity, temperature and dissolved oxygen. A minimum of 19 psu, 5–15 °C and 5 mg/l are recommended for mussels (*M. edulis*) in the United Kingdom.
- *Contaminants are removed* by settlement and/or removal by flow away from the bivalves, and by applying correct depuration conditions for an adequate length of time (see later).

- *Recontamination is avoided* by operating a batch 'all-in/all-out' system, using clean seawater at all stages of depuration, avoiding resuspension of settled expelled material and cleaning the system thoroughly between batches.
- *Viability and quality are maintained* by correct handling before, during and after depuration, maintaining adequate water flow and levels of dissolved oxygen, avoiding too high or too low temperatures and keeping build-up of end products such as ammonia to a minimum (Lee *et al.* 2008).

There are several different types of depuration system, ranging from those in which water is static or changed in batches to those in which seawater is flushed through continually or recycled through a steriliser. The most common sterilisation method is UV irradiation, with ozonation and chlorination less commonly utilised (see Gosling 2015). Containers for depuration range from small-scale shallow tanks through multilayer systems and various types of stacking systems to the bulk bin system (Lees *et al.* 2010; Figures 12.4–12.7).

Relaying and depuration processes in the United States and European Union are governed by legislation laid down in NSSP (2015) and Regulation (EC) No 853/2004 (European Communities 2004a), respectively. The EU Regulation provides details on requisites of location, construction and general running of a purification plant, such as tank construction and operation, organisation of batch systems and non-mixing of species during purification. It also covers hygiene requirements at purification and dispatch centres, laboratory testing, packaging, labelling, transportation, wet storage and documentation/traceability

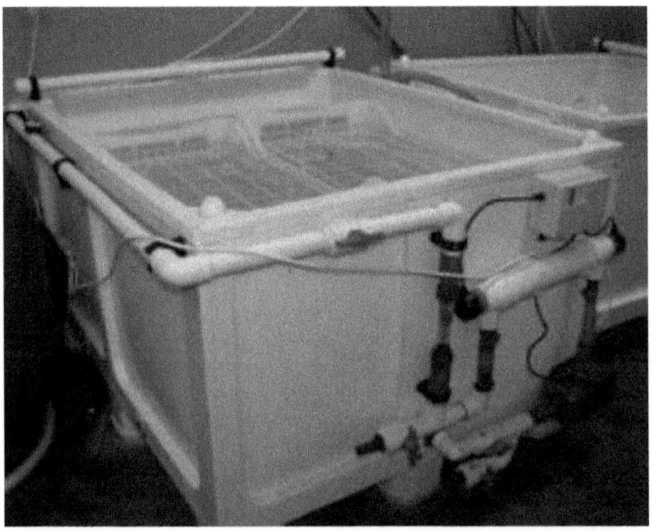

Figure 12.4 Small-scale shallow-tank depuration system. Cefas. © Crown Copyright. Bivalves are held in mesh-type plastic containers placed on battens on the bottom of a shallow tank of ~550 l capacity. Seawater, after UV treatment, enters the system via water jets above one end of the tank and flows through the containers to a suction pipe or weir across the other end. Water flow rate is relatively low (20 l/min^{1}), and hence the capacity to supply sufficient dissolved oxygen to the bivalves is also low and the permissible loading (i.e. the stacking) of the system is limited. When fully loaded, the system has a maximum capacity of 90 kg of mussels, but capacity is less for other species (details in Seafish 2018a).

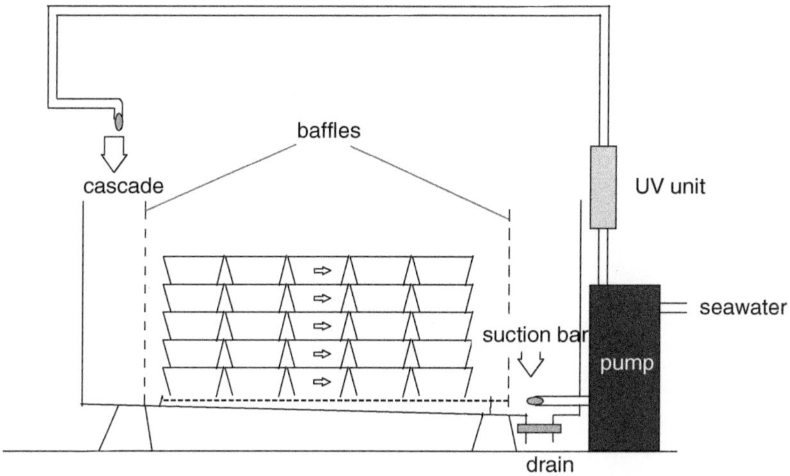

Figure 12.5 Multilayer system. Cefas. © Crown Copyright. Trays of bivalves are stacked up to six layers high, saving a significant amount of floor space compared to the shallow-tank system. This system has a high shellfish to water ratio, and therefore the flow rate needed to maintain an adequate level of dissolved oxygen in the circulating seawater is relatively high (~200 l/min for a medium-scale multilayer system of 750 kg of bivalves). Two baffles – consisting of a sheet of plastic evenly drilled with holes, through which water passes – are necessary to maintain an even flow of water through the system (details in Seafish 2018b).

requirements. Requirements for relaying operations are included. Purified bivalves, sold live, must comply with an end-product standard of <230 *E. coli* 100 g^{-1} flesh and intravalvular fluid (Table 12.4). In the European Union, approval to operate is granted to plants by the local enforcement authority (LEA) in each Member State. For example, the Food Standards Agency (FSA) is the LEA for the United Kingdom, where the Centre for Environment, Fisheries and Aquaculture Science (CEFAS; www.cefas.co.uk) has the delegated responsibility for setting the specific conditions under which the purification system should operate. In addition, CEFAS is the UK accredited service against ISO/17020:2012 (ISO 2012) for the purification system inspection service (CEFAS 2015). In Australia, the ASQAP Manual (ASQAP 2017) is the national guideline for managing risks in the harvesting, relaying, depuration and wet storage of bivalves, maintained by the Australian Shellfish Quality Assurance Advisory Committee (ASQAAC). Under the Canadian Shellfish Sanitation Program (CSSP), MOU between the Canadian Food Inspection Agency (CFIA), Fisheries and Oceans Canada (DFO) and Environment Canada (EC), the CFIA advises on decontamination plans. This is done under the authority of the Fish Inspection Regulations (FIR), which also contain the requirements for the processing (i.e. depuration), transportation and holding of shellfish. US governance of shellfish sanitary control, including purification methods, has been dealt with already.

Efficiency of Depuration

Effective depuration requires shellfish to be properly handled during harvest and pre-depuration transport and storage. In addition, it requires proper design and operation of the depuration systems to minimise stress to the bivalves. The establishments in which the

Figure 12.6 Small-scale vertical-stack system. Cefas. © Crown Copyright. The two main advantages of this system are the space it saves and the ready access it provides to individual containers. Because it is expensive, its use is generally limited to high-value bivalves such as clams and oysters. The system has a water capacity of around 600 l, a flow rate of 15 l/min and capacity for about 2000 oysters (details in Seafish 2018c).

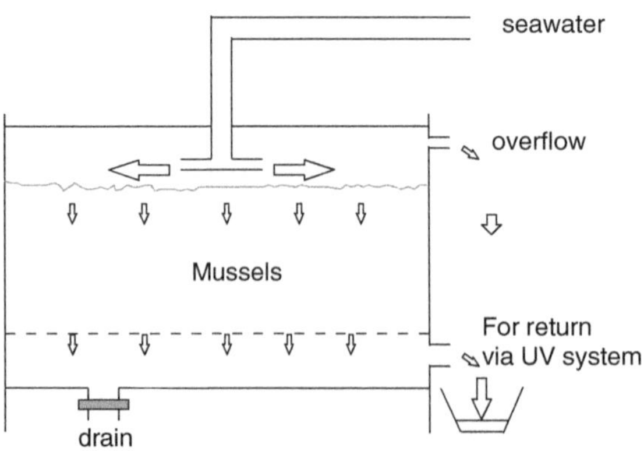

Figure 12.7 Bulk bin system designed specifically for the depuration of mussels. Cefas. © Crown Copyright. Mussels are held in a deep layer (~40 cm) on a plastic sheet (false floor), which is 8 cm higher than the base of the bin and is drilled with holes to permit the flow of water and detritus. Units (bins) can be added or removed to cope with different-sized batches. Mussels are loaded to a depth of 38 cm above the false floor, which corresponds to ~300 kg of mussels per bin. Each bin has a capacity of about 650 l and the flow rate is high at 108 l/min to ensure adequate oxygenation of the circulating water (details in Seafish 2018d).

system or systems are located need to be operated to good levels of food hygiene in order to prevent cross-contamination between, or recontamination of, different batches of bivalves. The efficiency of depuration is primarily related to bivalve size, filtering activity and physiological status (Jones *et al.* 1991). Other factors that influence efficiency are temperature, salinity and dissolved oxygen, the nature of the contaminant, the length of the depuration period and the quality of the depuration plant itself. Depuration is effective in removing many faecal bacterial contaminants from shellfish. As currently practiced, it is less effective at removing viral contaminants such as NoV and hepatitis A. It is not consistently effective, or is ineffective, in removing other contaminants, such as naturally occurring marine vibrios (e.g. *Vibrio parahaemolyticus* and *V. vulnificus*), marine biotoxins (such as those causing PSP, DSP and ASP) and heavy metals and organic chemicals (Lee *et al.* 2008).

In general, mussels (*Mytilus edulis*) and cockles (*Cardium edule*) tend to show higher accumulation rates of faecal coliforms than oysters (*Crassostrea virginica*), which is consistent with higher filtration rates in the former two species (Younger & Reese 2013). Complete elimination of *E. coli* normally occurs well within 48 hours (Cusson *et al.* 2005). In *M. galloprovincialis*, Croci *et al.* (2002) found that depuration for 44 hours led to a rapid decline in the initial concentration of *E. coli*, with >95.8% being released in the first five hours and a final reduction by a factor of 1000 (Figure 12.8). Barille *et al.* (2009), using reported *E. coli* loads of <230 MPN 100 g tissue in mussels (*M. galloprovincialis*) after only 6–12 hours' depuration, demonstrated the efficacy of the system. EU legislation does not stipulate a minimum duration for depuration, and periods employed can be as little as 12 hours through to around 48 hours, depending on the country. Currently, in the United Kingdom, a minimum depuration period of 42 hours exists for shellfish harvested from a

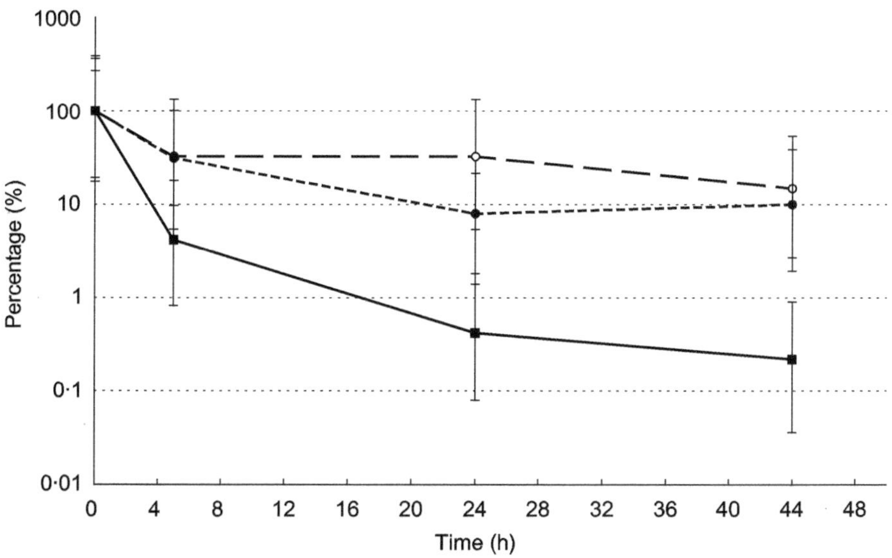

Figure 12.8 Percentage concentrations of *Escherichia coli* (squares), *Vibrio cholerae* (filled circles) and *V. parahaemolyticus* (empty circles) in *Mytilus galloprovincialis* during treatment compared with initial values. *Source:* From Croci *et al.* (2002). Reproduced with permission from John Wiley & Sons.

category B area, which has been shown to be sufficient to reduce counts to less than 230 *E. coli*/100 g (McMenemy *et al.* 2018 and references therein). But 42 hours may not be sufficient for elimination of other bacterial types. For example, depuration tests against *Salmonella typhi* were effective (<0.3 MPN/g) only after 72–84 hours in clams (*Chamelea gallina*), while in mussels (*M. galloprovincialis*) this level was achieved in 24–36 hours (Barille *et al.* 2009). In the same study, there was only a small reduction in *Vibrio parahaemolyticus* in clams over an entire observation period of 60 hours, while mussels showed <0.3 MPN/100 g after 36–38 hours depuration. Beside *E. coli*, Croci *et al.* (2002) also examined depuration rates in *Vibrio cholerae* and *V. parahaemolyticus*. The former was reduced from a concentration of 9.3×10^3 to 7.4×10^2 MPN/g, which remained almost constant even when the treatment was continued for 44 hours (Figure 12.8). In the same way, the initial bacterial load of *V. parahaemolyticus* was reduced but the end-of-treatment value remained around 10^3 MPN/g, with most of the reduction occurring in the first five hours of treatment. In percentage terms, under the experimental conditions, mussels retained only 0.2% of their initial bacterial load of *E. coli*, compared with 10.0 and 14.9% of *V. cholerae* and *V. parahaemolyticus*, respectively. Harris-Young *et al.* (1995) found that *Vibrio vulnificus* numbers actually increased during UV depuration; it seems that this species employs encapsulation as a way of resisting ingestion and degradation by oyster (*Crassostrea virginica*) haemocytes during the purification process. Also, Hernroth *et al.* (2010) showed that elimination of *V. parahaemolyticus*, as compared to *E. coli*, was less efficient in mussels and other bivalves because their haemocytes were unable to kill their particular vibrio strain.

It is now generally recognised that the main threat to public health from consumption of bivalves is from viral rather than bacterial contamination, especially in species that are eaten raw (e.g. oysters). Unlike most bacteria, viruses during the depuration process are released slowly, at different rates depending on the specific viral properties and the bivalve species being depurated (SARF 2011 and references therein). For example, Polo *et al.* (2014a) evaluated uptake rates and depuration kinetics of the murine NoV (MNV-1), a human NoV surrogate, in mussels (*M. galloprovincialis*) and clams (*Ruditapes philippinarum*) in 10 depuration trials (5 for each species). The bivalves were subjected to a controlled bioaccumulation step of 24 hours with 10^2 plaque forming units (pfu) MNV-1/ml seawater (final concentration), after which they were relocated to an experimental depuration system for seven days. Viral contamination was quantified after bioaccumulation and then daily during depuration. The average viral uptakes for clams and mussels were 5.4×10^6 and 4.0×10^5 RNA copies/g of digestive tissue respectively, representing an uptake rate that was >90% higher in clams. The average reductions in relation to the initial levels were 60.5% for clams and 91.6% for mussels (Figure 12.9). In both species, there was an initial rapid viral reduction during the first 24–72 hours of depuration, followed by a stable, slower depuration rate. While there were significant viral reductions in both species, at the end of the process the remaining virus was still infectious. In a related study, Polo *et al.* (2014b) evaluated the efficacy and kinetics of depuration of HAV under experimental conditions with mussels (*M. galloprovincialis*) subjected previously to bioaccumulation processes. Seven independent trials (70 kg of mussels each) were performed in a closed experimental system using two different water temperatures (13 and 17 °C) for seven days. Depuration trials showed an average reduction of HAV levels >90%. However, the average final viral loads in mussels remained at relatively high levels (6.5×10^3 RNA copies/g of digestive

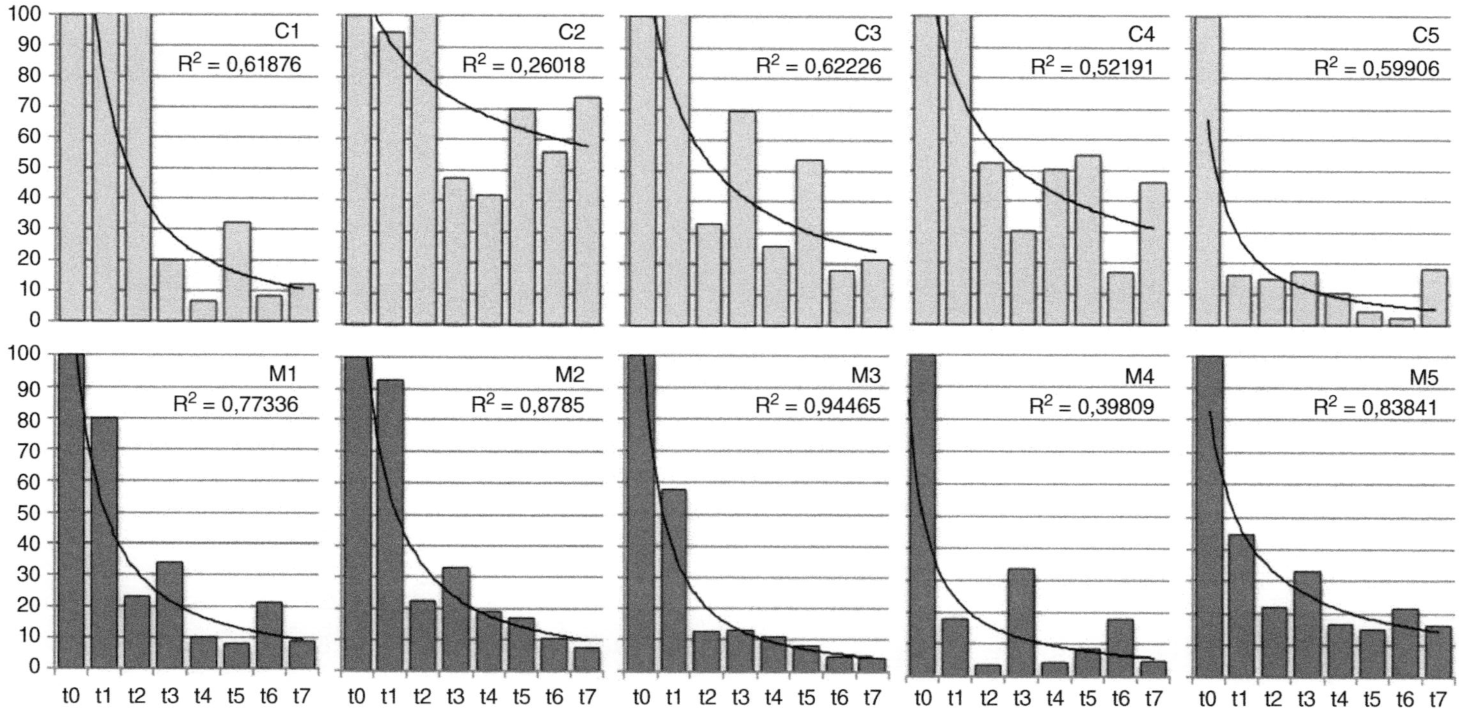

Figure 12.9 Depuration kinetics of murine norovirus (MNV-1) for each depuration trial carried out with clams (*Ruditapes philippinarum*; C1–C5) and mussels (*Mytilus galloprovincialis*; M1–M5) and with the corresponding trendline and R^2 values. Results are expressed as % RNA copies/g of digestive tissue. t0, viral concentration after 24 hours of bioaccumulation; t1–t7, viral concentration at different sampling times during depuration.
Source: From Polo *et al.* (2014a). Reproduced with permission from Elsevier.

tissue[1]) and viral particles were still infectious. A positive correlation between the initial and the final numbers of viral RNA copies was observed. The reduction of HAV showed a two-phase removal kinetic, an initial logarithmic trendline with a rapid reduction of viruses during the first 24–48 hours of depuration and a subsequent stabilisation with a slower depuration rate until the end of the process (see also Polo *et al.* 2015a,b). It is clear that current shellfish hygiene regulations based on *E. coli* alone (see earlier) are inadequate to protect human populations from consuming shellfish contaminated with high loads of viral pathogens (SARF 2011); therefore, viral standards are required for shellfish destined for human consumption (Winterbourn *et al.* 2016).

Various strategies for the depuration process have been suggested for elimination/inactivation of viruses. Very long-term relaying (more than several months) is a possible solution, provided that clean waters are available and the cleanliness of the waters containing the shellfish can be maintained (Richards *et al.* 2010). This may not be practical because of increased production costs associated with the additional handling and because it reduces product availability. An alternative strategy is a combination of depuration and relaying. Virus inactivation by heat treatment (e.g. cooking) is also used, but only in countries where shellfish are sold as a processed product. Evaluation of the method is difficult because of the use of different testing methods, different virus stains or surrogates and different bivalve species (Richards *et al.* 2010). The prescribed EU heat treatments have been imposed to ensure the elimination of pathogenic microorganisms, in particular the NoV possibly present in bivalve molluscs. Permitted treatments are sterilisation in hermetically sealed containers and heat treatments involving: immersion of the bivalve in boiling water for a sufficient length of time to raise the internal temperature of the flesh to not less than 90 °C, with maintenance of this temperature for at least 90 seconds (details in European Communities 2004a; EFSA 2015); cooking for three to five minutes in an enclosed space where the temperature is between 120 and 160 °C and the pressure is between 2 and 5 kg/cm^2, followed by shelling and freezing of the flesh to a core temperature of −20 °C; and steaming under pressure in an enclosed space satisfying the requirements relating to cooking time and the internal temperature of the bivalve flesh already mentioned. Harlow *et al.* (2011) found that steaming for two to five minutes was not sufficient to inactivate HAV in mussels in all layers of a steamer, but steaming for six minutes did inactivate HAV in all layers. A nonthermal method is high-pressure processing (HPP), which inactivates NoV and HAV in *Mytilus* spp. (Terio *et al.* 2010; Kingsley 2013). This method inactivates the virus by denaturing proteins on its capsid (outer shell), rendering it incapable of binding to its receptor on the surface of the host cell. Other thermal and nonthermal methods include heat pasteurisation, freezing, smoking (details in Richards *et al.* 2010) and electron beam irradiation (Praveen *et al.* 2013; Baker 2016). No one method guarantees complete elimination or inactivation of viruses. Richards *et al.* (2010) recommend the use of combined techniques such as shellfish relaying followed by HPP, pasteurisation followed by freezing or depuration followed by moderate cooking, since multiple methods have an additive advantage in reducing viral contamination.

Bivalves that have accumulated biotoxins are normally left to purge themselves in their natural habitat, as depuration in tanks is not currently considered a viable means of reducing biotoxin contamination to safe levels (Lee *et al.* 2008). This is because there is no clear pattern in the detoxification process due to different detoxification rates among toxins

(Huss *et al.* 2004), bivalve species (Choi *et al.* 2003; Fast *et al.* 2006; Lindegarth *et al.* 2009; Röder *et al.* 2011) and sites of toxin storage (toxins in the gastrointestinal tract are eliminated much more readily than toxins bound in tissues), and even individual bivalves may retain significantly higher levels of toxins compared to others. In *Mytilus* spp., factors such as size and age (Novaczek *et al.* 1992; Duinker *et al.* 2007), food supply (Marcaillou *et al.* 2010; Jauffrais *et al.* 2012) and even history of exposure to individual toxins (Velásquez & Navarro 2014) can influence detoxification rates. In depuration systems, the elimination period varies from several weeks (e.g. for brevitoxins; Dickey *et al.* 2011) to many months (e.g. 7–8 months for AZAs; Ryan *et al.* 2011). Other methods for accelerating the detoxification process have been tried, particularly for PSP toxins. These include electric shocks, decrease in pH, chlorination and biotransformation of phycotoxins by enzymes. None of these, to my knowledge, has proved effective – with the exception of a recent study by Qui *et al.* (2018), who reported that depuration of PST from mussels and scallops could be accelerated by wood-based activated carbon (WAC). Research on enhanced depuration for biotoxin elimination has been investigated in several EU countries using funds from the EU-FP7 programme. The objective of one such project was to develop probiotic diets and a microencapsulated delivery system in the digestive tract of mussels and oysters in order to improve depuration efficiency and toxin degradation (details in Fajardo *et al.* 2014). In France, the national CONSAUMOL programme, funded by Institut Français de Recherche pour l' Exploitation de la Mer (IFREMER), included trials to maintain bivalves in recirculation systems in order to detoxify them from biotoxins with the provision of 'fodder algae and/or clay' to accelerate the depuration process. IFREMER has also undertaken a similar approach with regard to viral depuration using live algal cells ('green water'), which were shown to increase depuration of the FRNA bacteriophage viral indicator (SARF 2011).

Depuration in tanks is not considered to be a practical means of removing high concentrations of heavy metal and organic chemical contaminants from bivalves (Lee *et al.* 2008). Therefore, relaying (transplantation) is often the method of choice in the elimination process (Casas *et al.* 2008). Most of our information on contaminant elimination rates in bivalves has been gleaned from laboratory studies. For example, detoxification of the heavy metals copper (Cu) and mercury (Hg) was examined in *M. galloprovincialis* (Raftopoulou & Dimitriadis 2011). Hg was rapidly detoxified (4–7 days), while Cu detoxification took longer (10–12 days). In another study, Zn and Cu levels in *P. viridis* were still 60 and 80% of starting levels, respectively, after a depuration period of 72 hours (Potipat *et al.* 2015). Many studies have shown that metallothioneins (MTs), nonenzymatic proteins with a low molecular weight, play a central role in heavy metal detoxification (Zhang *et al.* 2015 and references therein; see Chapter 8). Also, hydrolytic peptide metal-chelating elements have been used as depuration agents to decrease cadmium (Cd) concentrations in mussel tissues (Zhang *et al.* 2015).

PAHs accumulate in tissues with a high lipid content, and *M. edulis* has been shown to accumulate PAH in high concentrations without apparent detrimental effects. McIntosh *et al.* (2004) investigated the kinetics of PAH depuration in mussel (*M. edulis*) tissues over a three-month period. The mussels showed a loss of PAH such that the concentrations in homogenates of 20 individuals decreased from 8256 and 4127 ng/g ww on transfer in September from two shellfish farms to 3994 and 1942 ng/g ww, respectively, in January of the following year, indicating that the total concentration of 41 different PAH compounds in contaminated mussels fell by ~50% in a period of 122 days.

The elimination rate of contaminants is markedly affected by the duration of exposure to a chemical. Elimination is rapid and complete following short-term exposure, but long-term exposure results in slower and often incomplete clearance (Livingstone & Pipe 1992). Elimination rate also depends on the molecular weight and water solubility of the chemical, as well as on environmental factors such as temperature.

Monitoring and Quality Control

As mentioned earlier, countries involved in bivalve production should have an effective sanitation programme that oversees the production, harvesting and marketing of species for human consumption. According to Canzonier (1988), the basic requirements of such a programme should include:

- An infrastructure, responsible for monitoring, culture and harvesting activities; also, adequate surveillance of the chain of supply from point of production to point of retail.
- An administrative system for coordinating the activities of the various public agencies responsible for running the programme.
- Appropriate legislation so that the agencies can prosecute those who breach regulations.

In the following section, Canada is used as an example to illustrate how such a programme operates.

Monitoring of Live Mussels in Canada: Legislation and Regulation

Mussels (*Mytilus* spp.) are Canada's top shellfish aquaculture product, produced in Newfoundland and Labrador, Nova Scotia, Prince Edward Island and Quebec. An average of 25 032 metric tonnes (mt) of mussels were produced annually for the period 2012–2016, which represented an annual average value of $37.64 million (FAO 2011–2018). Prince Edward Island is the largest mussel cultivator in the country, and its production represents around 65% of Canada's total mussel value. Much of the country's mussel production is exported to markets in the United States. The CSSP Manual of Operations is an essential reference document for government staff involved with the tasks of classification and patrolling of bivalve growing areas, as well as bivalve harvesting, processing and distribution (CFIA 2018). The manual has been compiled through input from regional staff of the CFIA, DFO and EC. The Fisheries Act, the Management of Contaminated Fisheries Regulations and the Fish Inspection Act and Regulations provide the legal authority for the CSSP (see laws-lois.justice.gc.ca for details on legislation). The CFIA is the lead agency for the overall CSSP coordination, and is also responsible for the control of handling, storage, transportation, processing and labelling of shellfish, including imports (Fish Inspection Act and Fish Inspection Regulations), and for the Marine Biotoxins Control Program (Fisheries Act and Fishery Regulations). It also liaises with foreign governments on matters relevant to shellfish sanitation. Environment Canada (EC) is responsible for the monitoring of water quality in shellfish areas, for the identification and evaluation of pollution sources and for recommendations on the classification of shellfish harvesting areas on the basis of water surveys. DFO is responsible for the enforcement of closure regulations and for enacting the opening and closing of shellfish areas under the authority of the Fisheries Act and Management of Contaminated Fisheries Regulations.

The CSSP Manual outlines the authorities (acts and regulations), policies and procedures that apply to the Canadian programme and are used to evaluate regional activities associated with it. The Manual is also integrally linked to the Facilities Inspection Manual (FIM), which is published and maintained by the CFIA. The FIM sets forth the requirements for registration, inspection, audit and enforcement of seafood processing facilities, including shellfish facilities that come under the jurisdiction of the Fish Inspection Regulations (FIR). It also describes how each facility must design and implement its own Quality Management Program, which includes HACCP principles (see later), and how the CFIA assesses compliance through regulatory verification. The information in the following section is primarily gleaned from the Canadian Shellfish Sanitation Program Manual (Government of Canada 2021), the Facilities Inspection Manual (Government of Canada 2015) and Sauvé (2010).

Growing Areas: Sanitary Survey and Classification

To minimise the potential health risks associated with consuming bivalves and to protect public health, it is important that the water quality in bivalve growing areas be surveyed and that actual and potential sources of pollution be identified. It is the responsibility of EC under the Marine Water Quality Monitoring Program to identify safe bivalve growing areas in Canada by means of shoreline sanitary surveys. Such surveys have three main components: identification and evaluation of all actual and (potential) sources of pollution; evaluation of the meteorological and hydrographic factors that may affect the distribution of pollutants throughout the area; and bacteriological examination of the growing waters, designed to determine the extent of faecal contamination and to provide quantitative data for the classification of growing waters. When the survey data are analysed, a growing area is classified, on the basis of bacterial standards, as approved, conditionally approved, restricted, conditionally restricted or prohibited. The criteria for classification are identical to those set down for harvesting areas in the United States (see Table 12.5).

The CSSP defines areas as prohibited if:

- they are within a minimum 300 m radius around points of continuous or intermittent discharge from a sanitary sewage system;
- they are around points of continuous sanitary discharge which do not achieve adequate viral reduction through a combination of wastewater treatment and dilution in the shellfish growing area;
- they are within a minimum 300 m radius around industrial outfalls;
- they are within a minimum 125 m radius around marinas or wharves, finfish netpens, float homes or other floating living accommodation facilities;
- they are within a minimum 25 m radius of a float home or floating living accommodation facility located within a shellfish tenure/lease where a zero effluent discharge management plan is in effect

- due to the degree of contamination in the harvest waters (such as waters having excessive concentrations of faecal material or other poisonous or deleterious substances), it may not be possible to adequately depurate or naturally purify the shellfish.

Annual review surveys update the classification of the area. These are conducted to confirm that sanitary conditions have not changed and that the classification is still valid. Offshore areas beyond 5 km from land are considered acceptable for subtidal harvesting unless otherwise closed.

Marine Biotoxin Monitoring and Control

The CFIA is responsible for overall CSSP programme implementation and management of shellfish sampling related to toxins. Reports of all activities are centrally maintained at the regional level. Because of the risk of serious illness and death, reports of suspected cases of poisoning are closely investigated. All consumer illness information must be entered in the Issues Management System (IMS). Each CFIA region must have established sampling sites and frequencies to monitor changes in PSP, ASP and DSP. Sampling sites for monitoring toxicity levels are chosen after evaluating the following criteria: accessibility for sampling at all times of the year; size of the bivalve resource in the area; defined harvest area which the sample site represents; and history of toxicity in the area. The CFIA is responsible for collecting and analysing bivalve samples and for making recommendations for the opening and closing of shellfish areas to Fisheries and Oceans Canada (DFO), which implements and enforces closures. Bivalve growing areas are placed in the closed status when marine biotoxin levels (PSP, ASP or DSP) exceed established standards (Table 12.2). In certain circumstances, it may be necessary for CFIA to make a recommendation to DFO to close an area prior to reaching the established standards. This can happen when sampling indicates that toxin levels are rising rapidly, although they have not exceeded the standard, and the next planned sample cannot be obtained and/or analysed within a reasonable time frame to ensure consumer safety; alternatively, sampling may have shown a spike in toxin levels that are close to the standard but have not yet exceeded it, and historical information on the area(s) indicates that rising levels could pose a significant threat to consumer safety. Areas that are closed based on these scenarios may be opened earlier than the standard 14-day closure if a subsequent sample (or samples) indicates that the biotoxin levels never reached regulatory standards and the toxicity levels have dissipated. As an additional safety measure, samples may be taken for biotoxin analysis from shellfish processing establishments during compliance verification activities (see CFIA 2018 ch. 11 for additional information). The CSSP also provides information on harvesting and handling bivalves, wet storage, packing, shipping and labelling, policy and procedures for controlled relaying and depuration and procedures for approval of a depuration facility or relay operation.

Hazard Analysis and Critical Control Point and its Application

HACCP is a science-based system that aims to prevent food safety problems from occurring, rather than having to react to noncompliance of a finished product. The HACCP system accomplishes this by the identification of specific hazards and the implementation of control measures. An effective HACCP system should reduce the reliance on traditional end-product testing (Codex Alimentarius 2012). The system was first developed in the late 1950s to ensure food safety for the manned space programme in the United States. Since then, HACCP has become the premier food safety system, not just in the United States, but in the European Union, Canada, Australia, New Zealand and a number of other countries, and it serves as the foundation of the WHO's Codex Alimentarius Commission's General Principles of Food Hygiene.

Any food business operator handling or supplying live bivalves is required by law to put in place, implement and maintain permanent food safety management procedures based on the HACCP system, which can be applied from production to consumption. The implementation of a specific HACCP plan involves five preliminary steps, followed by seven Codex principles (Table 12.6). Ideally, the HACCP team of a seafood operation should comprise: the safety and quality supervisor, with a degree/training in food science/food safety, good experience in the production/processing operations and a special training in HACCP application in the fish industry; the technical supervisor, with a degree/training in food technology, experience in the seafood industry and a special training in HACCP application in the industry; the equipment maintenance supervisor; and, if appropriate, an advisor on fish and seafood safety and quality assurance (Ryder *et al.* 2014). The team must provide a full description of the product(s), which should include: name, intended use, intended consumer and any regulatory requirements or operator-defined limits relevant to the product. Other information, such as company specifications for packaging, labelling, and shelf life, may also be included (Table 12.7). The product profile is a necessary prerequisite for setting appropriate food safety criteria and for hazard identification and analysis. An accurate description of the steps involved in the processing of the product is essential for a proper hazard analysis (Table 12.8). The team must identify and list at each stage of the

Table 12.6 Sequence for the implementation of a Hazard Analysis and Critical Control Point (HAACP) plan.

Five Preliminary Steps
Step 1. Assemble a HACCP team
Step 2. Product/process type description
Step 3. Intended use and end user
Step 4. Process flow diagram
Step 5. Confirm the accuracy of the flow diagram
Seven Codex Principles
Principle 1 Conduct a *Hazard Analysis* by identifying and listing all potential hazards and list the measures to control identified hazards.
Principle 2 Identify the *Critical Control Points* (CCPs).
Principle 3 Establish *Critical Limits* for each CCP.
Principle 4 Establish a procedure *for Monitoring* each CCP.
Principle 5 Establish a procedure for *Corrective Action*.
Principle 6 Establish procedures for *Verification* to confirm that the HACCP system is working effectively.
Principle 7 Establish *documentation* and *record-keeping* to demonstrate compliance.

Table 12.7 Product description and intended purpose of half-shell mussels (*Perna canaliculus*).

Product name	Half-shell mussels (*Perna canaliculus*)/ mussel meat
Intended consumer	General public
Intended use of product	Half-shell mussels for direct consumption (ready-to-eat[1]). Mussel meat for further processing (e.g. hot smoking of mussel meat) or cooking
Regulatory requirements	Marine biotoxins limits specified in the AP Human Consumption Specifications clause 14.15 (3)[2] Microbiological limits for mussels set in the Food Standards Code: Standard 1.6.1 and AP Human Consumption Specifications clause 14.15 (1)[3]
Operator-defined limits	Microbiological limits for mussels that are for cooking or further processing
Packaging	As per regulatory and company specifications. Refer to Supporting System[4]
Labeling	As per regulatory and company specifications. Refer to Supporting System[4]
Shelf-life and storage requirements	As per regulatory and company specifications. Refer to Supporting System[4]

[1] For information on ready-to-eat fish products see Part 15 Animal Products Notice: Specifications for Products Intended for Human Consumption (MPI 2016) (https://www.mpi.govt.nz/11497-animal-products-notice-specifications-for-products).
[2, 3] Animal Products Notice: Specifications for Products Intended for Human Consumption.
[4] Packaging, labeling and storage specifications must be documented in the operator's supporting systems. The operator should reference the relevant supporting system in the product description.
Source: MAF (1997, 2011b). Reproduced with permission of the Ministry for Primary Industries, New Zealand.

Table 12.8 Process flow diagram for processing of half-shell greenshell mussels (*Perna canaliculus*).

Inputs	Process Step	Outputs
Live greenshell mussels ▶	1. Receiving	
	2. Storage	
	3. Grading	▶ Dead and badly damaged mussels to wastage ▶ *Mytilus* sp. and damaged mussels for further processing to animal feed
Hot water/steam ▶	4. Heat shocking	
Potable water ▶	5. Cooling	
	6. Shucking	▶ Mussel meat to step 10 ▶ Shells to waste
	7. Grading	▶ Mussel meat to step 10
	8. Freezing	
Potable water ▶	9. Glazing	
Packing materials ▶	10. Packing and labeling	
	11. Storage in freezer	
	12. Dispatch	▶ Packed frozen half-shell mussels ▶ Packed frozen mussel meat

Source: MAF (2011b). Reproduced with permission of the Ministry for Primary Industries, New Zealand.

process all the hazards that are reasonably likely to cause illness or injury in the absence of appropriate controls (Table 12.9). In the case of bivalves, examples of biological hazards are bacteria, viruses, biotoxins and parasites, chemical hazards include contaminants that can cause illness due to immediate or long-term exposure and physical hazards include the presence of foreign objects such as glass or metal fragments, which is a rare occurrence. Once hazards have been identified, their significance must be determined and an assessment made on whether control measures are available at each step. The next step is to determine whether there are any critical control points (CCPs) for the process. CCP determination can be facilitated by the use of a decision tree or a table that provides a series of questions which guide the user through the decision-making process (Figure 12.10). When a CCP is identified, the remaining HACCP requirements set out in Table 12.6 (as principles 3–7) must be applied. Critical limits must be measurable and should be parameters that can be monitored on an ongoing basis (e.g. temperature, time, moisture level, pH 2012). Monitoring procedures should include the person responsible for monitoring, the monitoring method, the monitoring frequency and sampling regime and the method of record maintenance (MPI 2011). The team must document corrective action procedures to be implemented when a critical limit is not met, and should establish and document operator verification

Table 12.9 Identification of hazards from inputs for half-shell mussels (*Perna canaliculus*). Biological hazards: B1, Internal microbiological hazards (e.g. faecal coliforms, *Vibrio* spp., pathogenic viruses); B2, Pathogens due to contamination of mussels during harvest or transport (e.g. *Salmonella* spp., *Campylobacter jejuni*, *Listeria monocytogenes*); B3, Marine biotoxins. Chemical hazards: C1, Pollutants derived from harvest areas (e.g. heavy metals and other chemical residues); C2, Contaminants due to contamination during harvest operations or transportation (e.g. fuel oil). Physical hazards: Nil. The operator must identify any hazards associated with each input, considering any supplier agreements and raw material specifications.

Inputs	Description/specification[1]	Raw material hazards
Live mussels	Sourced from suppliers that comply with the Regulated Control Scheme for Bivalve Molluscan Shellfish (BMS) Each consignment to be accompanied by a Shellfish Harvesting Declaration	Biological hazards: B1, B2, B3[2,3] Chemical hazards: C1, C2 Physical hazards: None
Water	Potable water as per the AP Human Consumption specifications clauses 2.5–2.11[4]	None
Packaging materials	Suitable for use as food contact material as per AP Human Consumption specifications Part 7[5]	None

[1] Agreed specifications for inputs must be documented in a supporting system.
[2] The shallow waters and poor flushing of estuaries in certain areas present the greatest risk of viral contamination in shellfish. Mussels are commercially grown on ropes in deep water, so present less risk from viruses such as the Norovirus.
[3] Shellfish may be contaminated after harvesting through the use of unclean equipment and containers, and through exposure to environmental contaminants such as dust, dirt, and bird droppings. However, this is not reasonably likely to occur. Cross-contamination during processing is considered to be the major source of *Listeria* in processed seafood product rather than the natural marine environment.
[4] Animal Products Notice: Specifications for Products Intended for Human Consumption (MPI 2016)
[5] Part 7 Animal Products Notice: Specifications for Products Intended for Human Consumption.
Source: MPI (1997, 2011b). Reproduced with permission of the Ministry for Primary Industries, New Zealand.

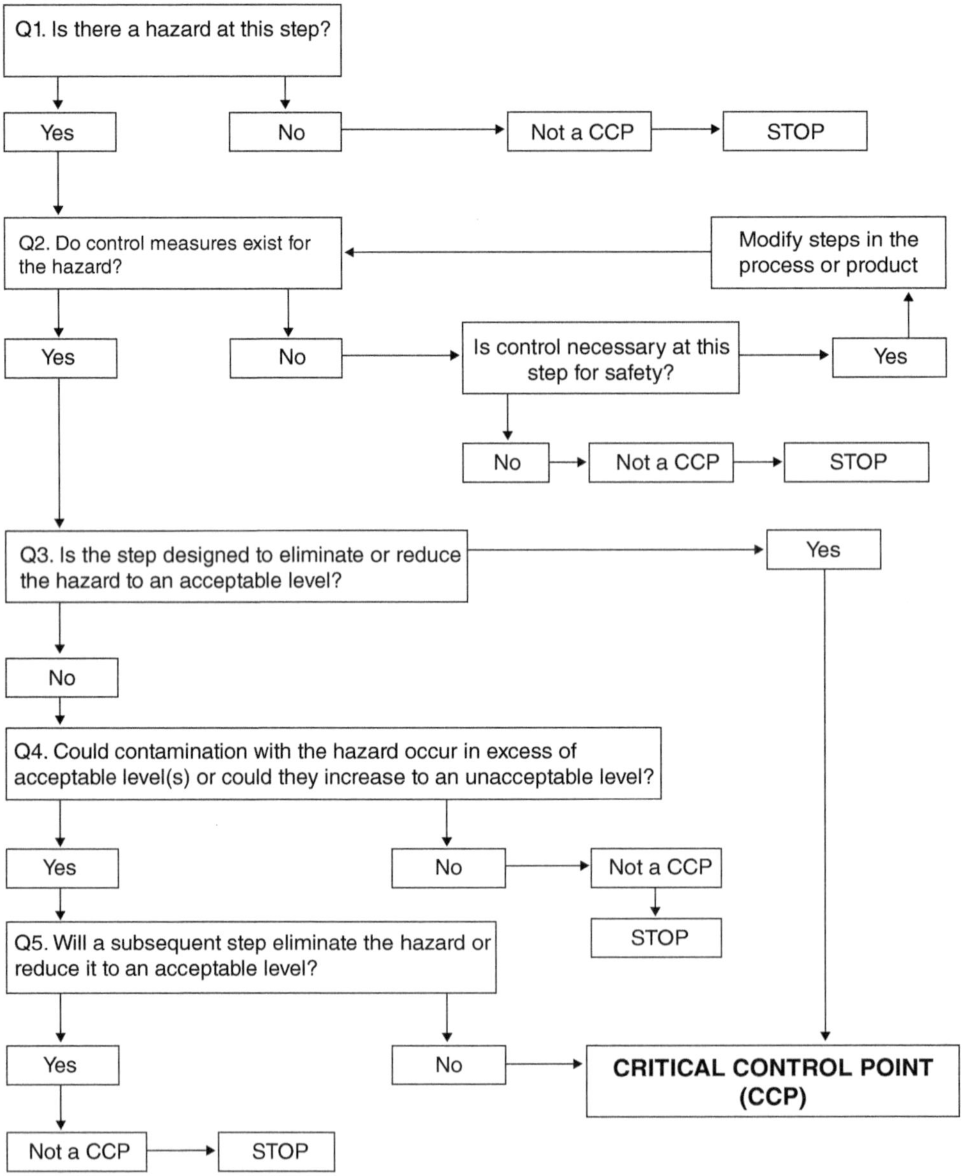

Figure 12.10 Critical control point (CCP) decision tree used in hazard analysis for bivalve processing. *Source:* From NSW Food Authority (2012). Reproduced with permission from the New South Wales Food Authority, Australia.

procedures to ensure that the HACCP system is working effectively. If no CCPs are identified, verification, documentation and record-keeping must still be applied. Finally, documentation and record-keeping should be sufficient to verify that HACCP controls are in place and being maintained. For more information on the implementation of the HACCP system, see Codex Alimentarius (2012), Ryder *et al.* (2014), NSW Food Authority (2012, 2015), NSSP (2015), European Commission (2016a) and FDA (2017). For information specifically related to HACCP in shellfish depuration, see Lee *et al.* (2008, 2010) and NSSP (2015).

Notes

1 A serotype or serovar is a serologically distinguishable strain of a micro-organism.
2 qPCR measures the amount of DNA after each cycle of replication, usually using fluorescent dyes to quantitatively measure the amplification of DNA.
3 RT-PCR usually starts from RNA, followed by reverse transcription to complementary DNA and then the PCR reaction. It is used to measure gene expression.
4 The amount of a chemical, normally expressed on a body weight basis, that can be consumed in a period of 24 hours or less without appreciable health risks to the consumer.
5 A validated method is an analytical method that has been subjected to a multi-laboratory study for accuracy, precision, reproducibility, performance and ruggedness. Concise written procedures for sample selection, preparation and quantitative analysis are provided for inter-laboratory quality assurance and consistency of results, on which an appropriate regulatory method of analysis can be established.
6 A biological phenomenon by which an organism produces one or more biochemicals that influence the germination, growth, survival and reproduction of other organisms.
7 The dose that a body, organ or tissue would receive during a specified period of time (e.g. 50 years) as a result of intake (by ingestion or inhalation) of one or more radionuclides from a defined release.
8 One millisievert (mSv) is defined in the United States as the average accumulated background radiation dose received by an individual in one year, exclusive of radon.
9 The Codex Alimentarius is a collection of internationally adopted food standards, guidelines, codes of practice and other recommendations under the auspices of the FAO and WHO.

References

Adell, D., Smith, W.A., Shapiro, K., Melli, A. & Conrad, P.A. (2014) Molecular epidemiology of *Cryptosporidium* spp. and *Giardia* spp. in mussels (*Mytilus californianus*) and California sea lions (*Zalophus californianus*) from central California. *Applied and Environmental Microbiology*, **80**, 7732–7740.

Ahmed, A., Rushworth, J, V., Hirst, N.A. & Millner, P.A. (2014) Biosensors for whole-cell bacterial detection. *Clinical Microbiology Reviews*, **27**, 631–646.

Alagappan, A., Bergquist, P.L. & Ferrari, B.C. (2009) Development of a two-color fluorescence *in situ* hybridization technique for species-level identification of human-infectious *Cryptosporidium* spp. *Applied and Environmental Microbiology*, **75**, 5996–5998.

Alexander, J.A., Stoecker, D.K., Meritt, D.W., Alexander, S.T., Padeletti, A., Johns, D. *et al.* (2008) Differential production of feces and pseudofeces by the oyster *Crassostrea ariakensis* when exposed to diets containing harmful dinoflagellate and Raphidophyte species. *Journal of Shellfish Research*, **27**, 567–579.

Amar, C.F.L., Dear, P.H. & McLauchlin, J. (2004) Detection and identification by real time PCR/RFLP analyses of *Cryptosporidium* species from human faeces. *Letters in Applied Microbiology*, **38**, 217–222.

Amouroux, I. & Soudant, D. (2011) Comparison of microbiological contamination level between different species of shellfish. ICMSS 2011-International Conference of Molluscan Shellfish Safety, Charlottetown, Prince Edward Island, Canada, 12–17 June, 2011.

An, X., Zhang, B., Zhang, H., Li, Y., Zheng, W., Yu, Z. *et al.* (2015) Discovery of an algicidal compound from Brevibacterium sp. BS01 and its effect on a harmful algal bloom-causing species, *Alexandrium tamarense. Frontiers in Microbiology*, **6**, 1235.

Anderson, D.M., Cembella, A.D. & Hallegraeff, G.M. (2012) Progress in understanding harmful algal blooms (HABs): paradigm shifts and new technologies for research, monitoring and management. *Annual Review of Marine Science*, **4**, 143–176.

Anderson, D. (2014) HABs in a changing world: a perspective on harmful algal blooms, their impacts, and research and management in a dynamic era of climactic and environmental change. Proceedings of the 15th International Conference on Harmful Algae. Changwon, Gyeongnam, Korea, 29 October–2 November 2012.

Angelo, K.M., Reynolds, J., Karp, B.E., Hoekstra, R.M., Scheel, C.M. & Friedman, C. (2016) Antimicrobial resistance among nontyphoidal *Salmonella* isolated from blood in the United States, 2003–2013. *Journal of Infectious Diseases*, **214**, 1565–1570.

ANZFSC (2016) Schedule 19 – maximum levels of contaminants and natural toxicants. Available at: https://www.legislation.gov.au/Details/F2017C00333 (accessed 28 May 2021).

AOAC International (2005a) *AOAC Official Method 959.08. Paralytic shellfish poison. Biological method.* In: Official Methods of Analysis of AOAC International, 18th edn. Gaitherburg, MD.

AOAC International (2005b) *AOAC Official Method 2005.06. Paralytic shellfish toxins in seafood.* In: Official Methods of Analysis of AOAC International, 18th edn. Gaitherburg, MD.

AOAC International (2011a) *AOAC Official Method 2011.02. Paralytic shellfish toxins in mussels, clams, oysters and scallops. Post-column oxidation (PCOX) method.* In: Official Methods of Analysis of AOAC International, 18th edn. Gaitherburg, MD.

AOAC International (2011b) *AOAC Official Method 2011.27. Paralytic shellfish toxins (PSTs) in shellfish, Receptor binding assay.* In: Official Methods of Analysis of AOAC International, 18th edn. Gaitherburg MD.

ASQAP (2017) Australian Shellfish Quality Assurance Program. Operations Manual, Version 2. Australian Shellfish Quality Assurance Advisory Committee.

Baehner, F., Bogaerts, H. & Goodwin, R. (2016) Vaccines against norovirus: state of the art trials in children and adults. *Clinical Microbiology and Infection*, **22**, S136–S139.

Baker, G.L. (2016) Food safety impacts from post-harvest processing procedures of molluscan shellfish. *Foods*, **5**, 29.

Bangi, J., Shuklai, S., Kim, Y-H. & Kim, M. (2012) Development of indirect competitive ELISA for the detection of *Salmonella typhimurium. Romanian Biotechnological Letters*, **17**, 7194–7204.

Barille, N.B., Scopa, M., Nerone, E., Mascilongo, G., Recchi, S., Cappabianca, S. *et al.* (2009) Study of the efficacy of a closed cycle depuration system on bivalve molluscs. *Veterinaria Italiana*, **45**, 555–566.

Baumann, Z., Casacuberta, N., Baumann, H., Masqué, P. & Fisher, N.S. (2013) Natural and Fukushima-derived radioactivity in macroalgae and mussels along the Japanese shoreline. *Biogeosciences Discussions*, **10**, 2617–2633.

Berdalet, E., Kudela, R., Urban, E., Enevoldsen, H., Banas, N.S., Bresnan, E. *et al.* (2017) GlobalHAB: a new program to promote international research, observations, and modeling of harmful algal blooms in aquatic systems. *Oceanography*, **30**, 70–81.

Bernard, F.R. (1989) Uptake and elimination of coliform bacteria by four marine bivalve mollusks. *Canadian Journal of Fisheries and Aquatic Sciences*, **46**, 1592–1599.

Bhandari, D., Chen, F-C. & Bridgman, R.C. (2019) Detection of *Salmonella typhimurium* in romaine lettuce using a surface plasmon resonance biosensor. *Biosensors*, **9**, 94.

Birnstiel, S., Soares-Gomes, A. & da Gama, B.A.P. (2019) Depuration reduces microplastic content in wild and farmed mussels. *Marine Pollution Bulletin*, **140**, 241–247.

Bosch, A., Pintó, R.M. & Le Guyader, F.S. (2009) Viral contaminants of molluscan shellfish: detection and characterization. In: *Shellfish Safety and Quality* (eds S.E. Shumway & G.E. Roderick), pp. 83–107. Woodhead Publishing, Cambridge.

Bosch, A., Sánchez, G., Abbaszadegan, M., Carducci, A., Guix, S., Le Guyader, F.S. *et al.* (2011) Analytical methods for virus detection in water and food. *Food Analytical Methods*, **4**, 4–12.

Botana, L.M., Hess, P., Munday, R., Arnich, N., DeGrasse, S.L., Feeley, M. *et al.* (2017) Derivation of toxicity equivalency factors for marine biotoxins associated with bivalve molluscs. *Trends in Food Science and Technology*, **59**, 15–24.

Boucher, O., Bastien, C.H., Saint-Amour, D., Dewailly, E., Ayotte, P., Jacobson, J.L. *et al.* (2010) Prenatal exposure to methylmercury and PCBs affects distinct stages of information processing: an event-related potential study with Inuit children. *Neurotoxicology*, **31**, 373–384.

Boundy, M.J., Selwood, A.I., Harwood, D.T., McNabb, P.S. & Turner, A.D. (2015) Development of a sensitive and selective liquid chromatography–mass spectrometry method for high throughput analysis of paralytic shellfish toxins using graphitised carbon solid phase extraction. *Journal of Chromatography A*, **1387**, 1–12.

Bovee, T.F.H., Hendriksen, P.J.M., Portier, L., Wang, S., Elliott,C.T., van Egmond, H.P. *et al.* (2011) Tailored microarray platform for the detection of marine toxins. *Environmental Science and Technology*, **45**, 8965–8973.

Bowers, H.A., Marin, R. III, Birch, J.A., Scholin, C.A. & Doucette, G.J. (2016) Recovery and identification of *Pseudo-nitzschia* frustules from natural samples acquired using the Environmental Sample Processor (ESP). *Journal of Phycology*, **52**, 135–140.

Brand, L.E. Campbell, L. & Bresnan, E. (2012) *Karenia*: the biology and ecology of a toxic genus. *Harmful Algae*, **14**, 156–178.

Brenner, F.W., Villar, R.G., Angulo, F.J., Tauxe, R. & Swaminathan, B. (2000) *Salmonella* nomenclature. *Journal of Clinical Microbiology*, **38**, 2465–2467.

Cai, T., Jiang, L., Yang, C. & Huang, K. (2006) Application of real-time PCR for quantitative detection of *Vibrio parahaemolyticus* from seafood in eastern China. *FEMS Immunology and Medical Microbiology*, **46**, 180–186.

Campàs, M., Prieto-Simón, B. & Marty, J.-L. (2007) Biosensors to detect marine toxins: assessing seafood safety. *Talanta*, **72**, 884–895.

Campbell, K., Vilariño, N., Botana, L.M. & Elliott, C.T. (2011a) A European perspective on progress in moving away from the mouse bioassay for marine-toxin analysis. *Trends in Analytical Chemistry*, **30**, 239–253.

Campbell, K., McGrath, T., Sjolander, S., Hanson, T., Tidare, M., Jansson, O. *et al.* (2011b) Use of a novel micro-fluidic device to create arrays for multiplex analysis of large and small

molecular weight compounds by surface plasmon resonance. *Biosensors and Bioelectronics*, **26**, 3029–3036.
Campos, C.J.A., Kershaw, S.R. & Lee, R.J. (2013) Environmental influences on faecal indicator organisms in coastal waters and their accumulation in bivalve shellfish. *Estuarine Coasts*, **36**, 834–853.
Canzonier, W.J. (1988) Public health component of bivalve shellfish production and marketing. *Journal of Shellfish Research*, **7**, 261–266.
Capobianco, J.A., Lee, J., Armstrong, C.M. & Gehring, A.G. (2019) Rapid detection of *Salmonella enterica* serotype Typhimurium in large volume samples using porous electrodes in a flow-through, enzyme-amplified immunoelectrochemical sensor. *Analytical and Bioanalytical Chemistry*, **411**, 5233–5242.
Casas, S., Gonzalez, J.-L., Andral, B. & Cossa, D. (2008) Relation between metal concentration in water and metal content of marine mussels (*Mytilus galloprovincialis*): impact of physiology. *Environmental Toxicology and Chemistry*, **27**, 1543–1552.
Castro-González, M.I. & Méndez-Armenta, M. (2008) Heavy metals: implications associated to fish consumption. *Environmental Toxicology and Pharmacology*, **26**, 263–271.
CDC (2021a) Norovirus. US Centers for Disease Control and Prevention. Available at: https://www.cdc.gov/norovirus/ (accessed 28 May 2021).
CDC (2021b) Salmonella. US Centers for Disease Control and Prevention. Available at: https://www.cdc.gov/salmonella/ (accessed 28 May 2021).
Cecchini, F., Fajs, L., Cosnier, S. & Marks, R.S. (2016) *Vibrio cholerae* detection: traditional assays, novel diagnostic techniques and biosensors. *Trends in Analytical Chemistry*, **79**, 199–209.
CEFAS (2015) CEFAS Protocol for Inspection and Approval of Purification (Depuration) Systems–England and Wales.
CEFAS (2016) CEFAS Protocol–Quantitative detection of norovirus and hepatitis A virus in bivalve molluscan shellfish. Issue 6: 05/0716. The Centre for Environment, Fisheries & Aquaculture Science, Weymouth, UK.
CFIA (2018) Canadian Food Inspection Agency. Canadian Shellfish Sanitation Program: manual of operations. Available at: http://www.inspection.gc.ca/food/fish-and-seafood/manuals/canadian-shellfish-sanitation-program/eng/1351609988326/1351610579883 (accessed 28 May 2021).
Chahorm, K. & Prakitchaiwattana, C. (2018) Application of reverse transcriptase-PCR-DGGE as a rapid method for routine determination of *Vibrio* spp. in foods. *International Journal of Food Microbiology*, **264**, 46–52.
Chenar, S.S. & Deng, Z. (2017) Environmental indicators of oyster norovirus outbreaks in coastal waters. *Marine Environmental Research*, **130**, 275–281.
Chenar, S.S. & Deng, Z. (2018a) Development of genetic programming-based model for predicting oyster norovirus outbreak risks. *Water Research*, **128**, 20–37.
Chenar, S.S. & Deng, Z. (2018b) Development of artificial intelligence approach to forecasting oyster norovirus outbreaks along Gulf of Mexico. *Environment International*, **111**, 212–223.
Cheng, W., Liu, F., Li, M., Hu, X., Chen, H., Pappoe, F. *et al.* (2015) Variation detection based on nextgeneration sequencing of type Chinese 1 strains of *Toxoplasma gondii* with different virulence from China. *BMC Genomics*, **16**, 888.
Chironna, M., Germinario, C., De Medici, D., Fiore, A. Di Pasquale, S., Quarto, M. *et al.* (2002) Detection of hepatitis A virus in mussels from different sources marketed in Puglia region (South Italy). *International Journal of Food Microbiology*, **75**, 11–18.

Choi, M-C., Hsieh, D.P.H., Lam, P.K.S. & Wang, W.-X. (2003) Field depuration and biotransformation of paralytic shellfish toxins in scallop *Chlamys nobilis* and green-lipped mussel *Perna viridis*. *Marine Biology*, **143**, 927–934.

Cinti, S., Volpe, G., Piermarini, S., Delibato, E. & Palleschi, G. (2017) Electrochemical biosensors for rapid detection of foodborne *Salmonella*: a critical overview. *Sensors*, **17**, 1910.

Codex Alimentarius (2012) *Code of Practice for Fish and Fishery Products*, 2nd edn. World Health Organization/Food and Agriculture Organization of the United Nations, Rome.

Conaty, S., Bird, P., Bell, G., Kraa, E., Grohmann, G. & Mc Anulty, J.M. (2000) Hepatitis A in New South Wales, Australia, from consumption of oysters: the first reported outbreak. *Epidemiology and Infection*, **124**, 121–130.

Contreras, A.M., Marsden I.D. & Munro, M.H.G. (2011) Effects of short-term exposure to paralytic shellfish toxins on clearance rates and toxin uptake in five species of New Zealand bivalve. *Marine and Freshwater Research*, **63**, 166–174.

Contreras, A.M., Marsden, I.D. & Munro, M.H.G. (2012) Physiological effects and biotransformation of PSP toxins in the New Zealand scallop, *Pecten novaezelandiae*. *Journal of Shellfish Research*, **31**, 1151–1159.

Corso, P.S., Kramer, M.H., Blair, K.A., Addiss, D.G., Davis, J.P. & Haddix, A.C. (2003) Cost of illness in the 1993 waterborne *Cryptosporidium* outbreak, Milwaukee, Wisconsin. *Emerging Infectious Diseases*, **9**, 426–431.

Coutard, F., Pommepuy, M., Loaec, S. & Hervio-Heath, D. (2005) mRNA detection by reverse transcription–PCR for monitoring viability and potential virulence in a pathogenic strain of *Vibrio parahaemolyticus* in viable but nonculturable state. *Journal of Applied Microbiology*, **98**, 951–961.

Croci, L., Suffredini, E., Cozzi, L. & Toti, L. (2002) Effects of depuration of molluscs experimentally contaminated with *Escherichia coli*, *Vibrio cholerae* O1 and *Vibrio parahaemolyticus*. *Journal of Applied Microbiology*, **92**, 460–465.

Croci, L., De Medici, D., Ciccozzi, M., Di Pasquale, S., Suffredini, E. & Toti, L. (2003) Contamination of mussels by hepatitis A virus: a public-health problem in southern Italy. *Food Control*, **14**, 559–563.

Croci, L., De Medici, D., Di Pasquale, S. & Toti, L. (2005) Resistance of hepatitis A virus in mussels subjected to different domestic cookings. *International Journal of Food Microbiology*, **105**, 139–144.

Cusick, K.D. & Sayler, G.S. (2013) An overview on the marine neurotoxin, saxitoxin: genetics, molecular targets, methods of detection and ecological functions. *Marine Drugs*, **11**, 991–1018.

Cusson, M., Tremblay, R., Daigle, G. & Roussy, M. (2005) Modeling the depuration potential of blue mussels (*Mytilus* spp.) in response to thermal shock. *Aquaculture*, **250**, 183–193.

Davidson, K., Baker, C., Higgins, C., Higman, W., Swan, S., Veszelovszki, A. *et al.* (2015) Potential threats posed by new or emerging marine biotoxins in UK waters and examination of detection methodologies used for their control: cyclic imines. *Marine Drugs*, **13**, 7087–7112.

de Freitas, C.G., Santana, A.P., da Silva, P.H. C., Gonçalves, V.S., de Aguiar Ferreira Barros, M., Araripe Gonçalves Torres, F. *et al.* (2010) PCR multiplex for detection of *Salmonella enteritidis*, *typhi* and *typhimurium* and occurrence in poultry meat. *International Journal of Food Microbiology*, **139**, 15–22.

Deisingh A.K. & Thompson, M. (2004) Strategies for the detection of *Escherichia coli* O157: H7 in foods. *Journal of Applied Microbiology*, **96**, 419–429.

Depaola, A. & Hwang, G. (1995) Effect of dilution, incubation-time, and temperature of enrichment on cultural and PCR detection of *Vibrio cholerae* obtained from the oyster *Crassostrea virginica*. *Molecular Cell Probes*, **9**, 75–81.

de Souza, R.V., Campos, C., Garbossa, L.H.P., Vianna, L.F.d.N., Vanz, A., Rupp, G.S. & Seiffert, W. (2018) A critical analysis of the international legal framework regulating the microbiological classification of bivalve shellfish production areas. *Reviews in Aquaculture*, **10**, 1025–1033.

Dickey, R.W., Baden, D. & Fleming, L.E. (2011) *Brevetoxins. In: Assessment and Management of Biotoxin Risks in Bivalve Molluscs* (eds J. Lawrence, H. Loreal, H. Toyofuku, P. Hess, I. Karunasagar & L. Ababouch), pp. 51–98. FAO Fisheries and Aquaculture Technical Paper No. 551. FAO, Rome.

Di Pinto, A., Forte, V.T., Tantillo, G.M., Terio, V. & Buonavoglia, C. (2003) Detection of hepatitis A virus in shellfish (*Mytilus galloprovincialis*) with RT-PCR. *Journal of Food Protection*, **66**, 1681–1685.

Di Pinto, A. Terio, V., Di Pinto, P., Colao, V. & Tantillo, G. (2012) Detection of *Vibrio parahaemolyticus* in shellfish using polymerase chain reaction–enzyme-linked immunosorbent assay. *Letters in Applied Microbiology*, **54**, 494–498.

Dokmeci, A.H. (2017) An assessment of heavy metals in wild mussels *Mytilus galloprovincialis* from the Marmara Sea Coast of Tekirdag,Turkey. *Fresenius Environmental Bulletin*, **26**, 7608–7612.

D'Souza, D.H., Critzer, F.J. & Golden, D.A. (2009) Real-time reverse-transcriptase polymerase chain reaction for the rapid detection of *Salmonella* using i*nv*A primers. *Foodborne Pathogens and Disease*, **6**, 1097–1106.

Dufour, A.P. & Stelma, G.N. Jr. (2010) Real-time monitoring technologies for indicator bacteria and pathogens in shellfish and shellfish harvesting waters. In: *Safe Management of Shellfish and Harvest Waters* (eds G. Rees, K. Pond, D. Kay, J. Bartram & J. Santo Domingo), pp. 109–120. IWA Publishing, London.

Duinker, A., Bergslien, M., Strand, Ø., Olseng, C.D. & Svardal, A. (2007) The effect of size and age on depuration rates of diarrhetic shellfish toxins (DST) in mussels (*Mytilus edulis* L.). *Harmful Algae*, **6**, 288–300.

Eckford-Soper, L.K., Davidson, K. & Bresnan, E. (2013) Identification and quantification of toxic and nontoxic strains of the harmful dinoflagellate *Alexandrium tamarense* using fluorescence in situ hybridization and flow cytometry. *Limnology and Oceanography: Methods*, **11**, 540–548.

EFSA (2008a) Scientific opinion of the Panel on Contaminants in the Food Chain on a request from the European Commission on marine biotoxins in shellfish – okadaic acid and analogues. *EFSA Journal*, **589**, 1–62.

EFSA (2008b) Opinion of the Scientific Panel on Contaminants in the Food chain on a request from the European Commission on marine biotoxins in shellfish – azaspiracids. *EFSA Journal*, **723**, 1–52.

EFSA (2008c) Opinion of the Scientific Panel on Contaminants in the Food chain on a request from the European Commission on marine biotoxins in shellfish – yessotoxin group. *EFSA Journal*, **907**, 1–62.

EFSA (2009a) Scientific Opinion of the Panel on Contaminants in the Food Chain on a request from the European Commission on marine biotoxins in shellfish – saxitoxin. *EFSA Journal*, **1019**, 1–76.

EFSA (2009b) Scientific Opinion of the Panel on Contaminants in the Food Chain on a request from the European Commission on marine biotoxins in shellfish – domoic acid. *EFSA Journal*, **1181**, 1–61.

EFSA (2009c) Scientific Opinion of the Panel on Contaminants in the Food Chain on a request from the European Commission on marine biotoxins in shellfish – pectenotoxin group. *EFSA Journal*, **1109**, 1–47.

EFSA (2009d) Scientific Opinion of the Panel on Contaminants in the food chain on a request from the European Commission on cadmium in food. *EFSA Journal*, **980**, 1–139.

EFSA (2009e) Scientific opinion on arsenic in food. *EFSA Journal*, **7**, 1351.

EFSA (2010a) Scientific Opinion of the Panel on Contaminants in the Food Chain on a request from the European Commission on marine biotoxins in shellfish – emerging toxins: brevetoxin group. *EFSA Journal*, **8**:1677, 1–29.

EFSA (2010b) EFSA Panel on Contaminants in the Food Chain (CONTAM); Scientific Opinion on marine biotoxins in shellfish – cyclic imines (spirolides, gymnodimines, pinnatoxins and pteriatoxins). *EFSA Journal*, **8**:1628, 1–39.

EFSA (2010c) Scientific opinion on lead in food. *EFSA Journal*, **8**:1570.

EFSA (2012a) Scientific Opinion on the risk for public health related to the presence of mercury and methylmercury in food. *EFSA Journal*, **10**:2985, 1–241.

EFSA (2012b) Update of the monitoring of dioxins and PCBs levels in food and feed. *EFSA Journal*, **10**:2832, 1–82.

EFSA (2015) Scientific opinion on the evaluation of heat treatments, different from those currently established in the EU legislation, that could be applied to live bivalve molluscs from B and C production areas, that have not been submitted to purification or relaying, in order to eliminate pathogenic microorganisms. *EFSA Journal*, **13**(12)**:4332**, 1–76.

EFSA (2016) EFSA Panel on Contaminants in the Food Chain. Statement on the presence of microplastics and nanoplastics in food, with particular focus on seafood. *EFSA Journal*, **14**:4501, 30pp.

EFSA (2017) Scientific opinion on the risks for public health related to the presence of tetrodotoxin (TTX) and TTX analogues in marine bivalves and gastropods. *EFSA Journal*, **15**:4752, 1–65.

Egerton, T.A., Morse, R.E., Marshall, H.G. & Mulholland, M.R. (2014) Emergence of algal blooms: the effects of short-term variability in water quality on phytoplankton abundance, diversity, and community composition in a tidal estuary. *Microorganisms*, **2**, 33–57.

Elamri, D.E., Aouni, M., Parnaudeau, S. & Le Guyader, F.S. (2006) Detection of human enteric virus in shellfish collected in Tunisia. *Letters in Applied Microbiology*, **43**, 399–404.

Elgun, G. & Koltas, I.S. (2011) Investigation of *Cryptosporidium* spp. antigen by ELISA method in stool specimens obtained from patients with diarrhea. *Parasitology Research*, **108**, 395–397.

Escobedo-Lozano, A.Y., Estrada, N., Ascencio, F., Contreras, G. & Alonso-Rodriguez, R. (2012) Accumulation, biotransformation, histopathology and paralysis in the Pacific calico scallop *Argopecten ventricosus* by the paralyzing toxins of the dinoflagellate *Gymnodinium catenatum*. *Marine Drugs*, **10**, 1044–1065.

EURL (2017) European Union Reference Laboratory. Community guide to the principles of good practice for the microbiological classification and monitoring of bivalve mollusk production and relaying areas with regard to regulation 854/2004. Available at: https://cefas.org/media/13972/cg_issue-3_final-170117.pdf (accessed 28 May 2021).

EU-RL-MB (2015) European Union Reference Laboratory for Marine Biotoxins. EU harmonised standard operating procedure for determination of lipophilic marine biotoxins in molluscs by LC-MS/MS, Version 5.

European Commission (2011) Commission Regulation (EU) No 15/2011 of 10 January 2011 amending Regulation (EC) No 2074/2005 as regards recognised testing methods for detecting marine biotoxins in live bivalve molluscs. *Official Journal of the European Union*, **L6**, 3–6.

European Commission (2016a) Commission notice on the implementation of food safety management systems covering prerequisite programs (PRPs) and procedures based on the HACCP principles, including the facilitation/flexibility of the implementation in certain food businesses (2016/C 278/01). *Official Journal of the European Union*, **C278**, 1–32.

European Commission (2016b) Maximum levels of radioactive contamination in foodstuffs. Available at: http://publications.europa.eu/resource/cellar/c106916a-a9b1-11ea-bb7a-01aa75ed71a1.0001.02/DOC_1 (accessed 28 May 2021).

European Communities (2004a) Regulation (EC) No 853/2004 of the European Parliament and of the Council of 29 April 2004 laying down specific hygiene rules for food of animal origin. *Official Journal of the European Union*, **L226**, 22–82.

European Communities (2004b) Regulation (EC) No 854/2004 of the European Parliament and of the Council of 29 April 2004 laying down specific rules for the organisation of official controls on products of animal origin intended for human consumption. *Official Journal of the European Union*, **L226**, 83–127.

European Communities (2005) Commission Regulation (EC) No 2073/2005 of 15 November 2005 on microbiological criteria for foodstuffs. *Official Journal of the European Union*, **L338**, 1–26.

European Communities (2006a) Commission Regulation (EC) No 1881/2006 of the European Parliament and of the Council of 19 December 2006 on setting maximum levels for certain contaminants in foodstuffs. *Official Journal of the European Union*, **L364**, 5–24.

European Communities (2006b) Directive 2006/113/EC of the European Parliament and of the Council of 12 December 2006 on the quality required of shellfish waters (codified version). *Official Journal of the European Communities*, **L376**, 14–20.

Fajardo, P., Atanassova, M., Cotterill, J., Wontner-Smith, T., Vieites, J. & Cabado, A. (2014) Mitigation of pathogens and marine biotoxins contamination in shellfish. In: *Safety and Security Engineering. WIT Transactions on The Built Environment, Volume 134*, pp. 691–698. WIT Press, Southampton.

FAO (2010) *Report of FAO expert workshop on the application of biosecurity measures to control Salmonella contamination in sustainable aquaculture. FAO Fisheries and Aquaculture Report No.* 937. FAO, Rome.

FAO (2011–2018) Fisheries and aquaculture software. FishStatJ – software for fishery statistical time series. Available at: http://www.fao.org/fishery/ (accessed 28 May 2021).

FAO/WHO (2003) Risk assessment of *Campylobacter* spp. in broiler chickens and *Vibrio* spp. in seafood. FAO Food and Nutrition Paper No. 75.

FAO/WHO (2011) Risk assessment of *Vibrio parahaemolyticus* in seafood. Microbiological Risk Assessment Series, No. 16.

FAO/WHO (2016) Technical paper on Toxicity Equivalency Factors for Marine Biotoxins Associated with Bivalve Molluscs. Rome. Available at: http://www.fao.org/3/a-i5970e.pdf (accessed 28 May 2021).

Fast, M.D., Cembella, A.D. & Ross, N.W. (2006) *in vitro* transformation of paralytic shellfish toxins in the clams *Mya arenaria* and *Protothaca staminea*. *Harmful Algae*, **5**, 79–90.

FDA (2017) Hazard Analysis and Critical Control Point Training Curriculum, 6th edn. Available at: https://eos.ucs.uri.edu/EOSWebOPAC/OPAC/Common/Pages/GetDoc.aspx?ClientID=EOSMAIN&MediaCode=19100406 (accessed 28 May 2021).

Fiore, A.E. (2004) Hepatitis A transmitted by food. *Clinical Infectious Diseases*, **38**, 705–715.

Formiga-Cruz, M., Tofiño-Quesada, G., Bofill-Mas, S., Lees, D.N., Henshilwood, K. & Allard, A.K. (2002) Distribution of human virus contamination in shellfish from different growing areas in Greece, Spain, Sweden, and the United Kingdom. *Applied and Environmental Microbiology*, **68**. 5990–5998.

Freire-Santos, F., Oteiza-López, A.M., Vergara-Castiblanco, C.A., Ares-Mazás, E., Álvarez-Suárez, E. & García-Martín, O. (2000) Detection of *Cryptosporidium* oocysts in bivalve molluscs destined for human consumption. *The Journal of Parasitology*, **86**, 853–854.

FSAI (2006) Investigation into levels of polycyclic aromatic hydrocarbons (PAHs) in food on the Irish market. Available at: https://www.fsai.ie/uploadedFiles/PAH_levels.pdf (accessed 24 March 2018).

FSAI (2009) Mercury, lead, cadmium, tin and arsenic in food. *Toxicology Fact Sheet Series, Issue No.* 1.

FSAI (2015) Polycyclic aromatic hydrocarbons (PAHs) in food. Toxicology Fact Sheet Series, No. 2.

FSANZ (2011) Salmonella (non-typhoidal). Food Standards Australia New Zealand. Available at: https://www.foodstandards.gov.au/publications/Documents/salmonella.pdf (accessed 28 May 2021).

Fuentes, C., Guix, S., Pérez-Rodriguez, F.J., Fuster, N., Carol, N., Pintó, R.M. *et al.* (2014) Standardized multiplex one-step qRT-PCR for hepatitis A virus, norovirus GI and GII quantification in bivalve mollusks and water. *Food Microbiology*, **40**, 55–63.

Fusco, G., Di Bartolo, I., Cioffi, B., Ianiro, G., Palermo, P., Monin, M. *et al.* (2017) Prevalence of foodborne viruses in mussels in southern Italy. *Food and Environmental Virology*, **9**, 187–194.

Gao, S., Li, D., Zha, E., Zhou, T., Wang, S. & Yue, X. (2015) Surveillance of hepatitis E virus contamination in shellfish in China. *International Journal of Environmental Research and Public Health*, **12**, 2026–2036.

Garcia, L.S., Shimizu, R.Y., Novak, S., Carroll, M. & Chan, F. (2003) Commercial assay for detection of *Giardia lamblia* and *Cryptosporidium parvum* antigens in human fecal specimens by rapid solid-phase qualitative immunochromatography. *Journal of Clinical Microbiology*, **41**, 209–212.

Garrido-Maestu, A., Chapela, M.J., Peñaranda, E., Vieites, J.M. & Cabado, A.G. (2014) In-house validation of novel multiplex real-time PCR gene combination for the simultaneous detection of the main human pathogenic vibrios (*Vibrio cholerae*, *Vibrio parahaemolyticus*, and *Vibrio vulnificus*). *Food Control*, **37**, 371–379.

Garthwaite, I. (2000) Keeping shellfish safe to eat: a brief review of shellfish toxins, and methods for their determination. *Trends in Food Science and Technology*, **11**, 235–244.

Gerssen, A., Pol-Hofstad, I.E., Poelman, M., Mulder, P.P.J., van den Top, H.J. & de Boer, J. (2010) Marine toxins: chemistry, toxicity, occurrence and detection, with special reference to the Dutch situation. *Toxins*, **2**, 878–904.

Ghaffari, S. & Kalantari, N. (2014) Recognition of *Cryptosporidium* oocysts in fresh and old stool samples: comparison of four techniques. *Asian Pacific Journal of Tropical Medicine*, **4**, 570–574.

GHO (2016) Cholera. Global Health Observatory. Available at: https://www.who.int/data/gho/data/themes/theme-details/GHO/cholera (accessed 28 May 2021).

Ghozzi, K., Marangi, M., Papini, R., Lahmar, I., Challouf, R. Houas, N. *et al.* (2017) First report of Tunisian coastal water contamination by protozoan parasites using mollusk bivalves as biological indicators. *Marine Pollution Bulletin*, **117**, 197–202.

Gómez-Couso, H., Freire-Santos, F., Ortega-Iñarea, M.R., Castro-Hermida, J.A. & Ares-Mazás, M.E. (2003) Environmental dispersal of *Cryptosporidium parvum* oocysts and cross transmission in cultured bivalve molluscs. *Parasitology Research*, **90**, 140–142.

Gómez-Couso, H., Méndez-Hermida, F., Castro-Hermida, J.A. & Ares-Mazás, M.E. (2005) Occurrence of *Giardia* cysts in mussels (*Mytilus galloprovincialis*) destined for human consumption. *Journal of Food Protection*, **68**, 1702–1705.

Gosling, E.M. (2015) *Marine Bivalve Molluscs*, 2nd edn. Wiley Blackwell, Oxford.

Government of Canada (2015) Facilities inspection manual. Available at: https://inspection.canada.ca/food-safety-for-industry/archived-food-guidance/fish-and-seafood/manuals/facilities-inspection-manual/eng/1354209008142/1354209083903 (accessed 28 May 2021).

Government of Canada (2021) Canadian shellfish sanitation program manual. Available at: https://inspection.canada.ca/food-safety-for-industry/food-specific-requirements-and-guidance/fish/canadian-shellfish-sanitation-program/eng/1527251566006/1527251566942?chap=0 (accessed 28 May 2021).

Graczyk, T.K., Suresh, K. & Lees, D. (2010) Adverse health outcomes. In: *Safe Management of Shellfish and Harvest Waters.* (eds G. Rees, K. Pond, D. Kay, J. Bartram & J. Santo Domingo), pp. 21–38. World Health Organization (WHO). IWA Publishing, London.

Granéli, E., Weberg, M. & Salomon, P.S. (2008) Harmful algal blooms of allelopathic microalgal species: the role of eutrophication. *Harmful Algae*, **8**, 94–102.

Grattan, L.M., Holobaugh, S. & Morris Jr., J.G. (2016) Harmful algal blooms and public health. *Harmful Algae*, **57**, 2–8.

Gugliandolo, C., Lentini, V., Spanò, A. & Maugeri, T.L. (2010) Conventional and molecular methods to detect bacterial pathogens in mussels. *Letters in Applied Microbiology*, **52**, 15–21.

Haberkorn, H., Lambert, C., Le Goïca, N., Moal, J., Suquet, M., Guéguend, M. *et al.* (2010) Effects of *Alexandrium minutum* exposure on nutrition-related processes and reproductive output in oysters *Crassostrea gigas*. *Harmful Algae*, **9**, 427–439.

Halstead, BW. & Schantz, E. (1984) Paralytic shellfish poisoning. World Health Organization Offset Publication No. 79, Geneva.

Hanevik, K., Bakken, R., Brattbakk, H.R., Saghaug, C.S. & Langeland, N. (2015) Whole genome sequencing of clinical isolates of *Giardia lamblia*. *Clinical Microbiology and Infection*, **21**, 192.e1–192.e3.

Harlow, J., Oudit, D., Hughes, A. & Mattison, K. (2011) Heat inactivation of Hepatitis A virus in shellfish using steam. *Food and Environmental Virology*, **3**, 31–34.

Harris-Young, L., Tamplin, M., Mason, J.W., Aldrich, H.C. & Jackson, J.K. (1995) Viability of *Vibrio vulnificus* in association with hemocytes of the American oyster (*Crassostrea virginica*). *Applied Environmental Microbiology*, **61**, 52–57.

Hatta, M., Islam, A.A., Minhajatd, R. & Maryunis, M. (2017) Detection *Salmonella typhi* RNA from suspect typhoid using reverse transcription-polymerase chain reaction. *International Journal of Sciences: Basic and Applied Research (IJSBAR)*, **31**, 99–107.

Hégaret, H., Shumway, S.E. & Wikfors, G.H. (2007) Diverse feeding responses of five species of bivalve mollusc when exposed to three species of harmful algae. *Journal of Shellfish Research*, **26**, 549–559.

Heisler, J., Glibert, P., Burkholder, J., Anderson, D., Cochlan, W., Dennison, W. *et al.* (2008) Eutrophication and harmful algal blooms: a scientific consensus. *Harmful Algae*, **8**, 3–13.

Hernroth, B., Lothigius, Å. & Bölin, I. (2010) Factors influencing survival of enterotoxigenic *Escherichia coli*, *Salmonella enterica* (serovar Typhimurium) and *Vibrio parahaemolyticus* in marine environments. *FEMS Microbiology and Ecology*, **71**, 272–280.

Hesari, N., Alum, A., Elzein, M. & Abbaszadegan, M. (2016) A biosensor platform for rapid detection of *E. coli* in drinking water. *Enzyme and Microbial Technology*, **83**, 22–28.

Hollinger, F.B. & Emerson, S.U. (2007) Hepatitis A virus. In: *Fields Virology, 5th edn* (eds D.M. Knipe & P.M. Howley), pp. 911–940. Lippincott Williams & Wilkins, Philadelphia, PA.

Hossain, M.T., Kim, Y.O & Kong, I.S. (2013) Multiplex PCR for the detection and differentiation of *Vibrio parahaemolyticus* strains using the groEL, tdh and trh genes. *Molecular and Cellular Probes*, **27**, 171–175.

Hu, Q.H., Weng, J.Q. & Wang, J.S. (2010) Sources of anthropogenic radionuclides in the environment: a review. *Journal of Environmental Radioactivity*, **101**, 426–437.

Hunter, J.C. & Watkins, L.K.F. (2017) Salmonellosis (nontyphoidal) In: *Centers for Disease Control and Prevention. CDC Yellow Book 2018: Health Information for International Travel* (ed G.W. Brunette), pp. 303–306. Oxford University Press, New York.

Huss, H.H., Ababouch, L. & Gram, L. (2004) Assessment and management of seafood safety and quality. FAO Fisheries Technical Paper No. 444. FAO, Rome.

Imani, F.A., Iman I.D., Hosseini, D.R., Karami, A. & Marashi, S.M. (2013) Design of a multiplex PCR method for detection of toxigenic-pathogenic in *Vibrio cholera*. *Asian Pacific Journal of Tropical Medicine*, **6**, 115–118.

Isbister, G.K. & Kiernan, M.C. (2005) Neurotoxic marine poisoning. *Lancet Neurology*, **4**, 219–228.

ISO (2012) Conformity assessment – Requirements for the operation of various types of bodies performing inspection. ISO/IEC 17020:2012. Available at: https://www.iso.org/standard/52994.html (accessed 28 May 2021).

ISO (2013a) Microbiology of food and animal feed – Horizontal method for determination of hepatitis A virus and norovirus in food using real-time RT-PCR – Part 1: Method for quantification. ISO/TS 15216-1:2013. Available at: https://www.iso.org/standard/55382.html (accessed 28 May 2021).

ISO (2013b) Microbiology of food and animal feed – Horizontal method for determination of hepatitis A virus and norovirus in food using real-time RT-PCR – Part 2: Method for qualitative detection. ISO/TS 15216-2:2013. Available at: https://www.iso.org/standard/60297.html (accessed 28 May 2021).

ISO (2015) Microbiology of the food chain – Horizontal method for the enumeration of beta-glucuronidase-positive *Escherichia coli*, Part 3: detection and most probable number technique using 5-bromo-4-chloro-3-indolyl-ß-D-glucuronide. ISO 16649-3:2015. Available at: https://www.iso.org/standard/56824.html (accessed 28 May 2021).

ISO (2017) Microbiology of the food chain – Horizontal method for determination of hepatitis A virus and norovirus using real-time RT-PCR – Part 1: Method for quantification. ISO 15216-1:2017. Available at: https://www.iso.org/standard/65681.html (accessed 28 May 2021).

Iwamoto, M., Ayers, T., Mahon, B. & Swerdlow, D.L. (2010) Epidemiology of seafood associated infections in the United States. *Clinical Microbiological Reviews*, **23**, 399–411.

Izumiya, H., Matsumoto, K., Yahiro, S., Lee, J., Morita, M., Yamamoto, S. *et al.* (2011) Multiplex PCR assay for identification of three major pathogenic *Vibrio* spp., *Vibrio cholerae*, *Vibrio parahaemolyticus*, and *Vibrio vulnificus*. *Molecular and Cellular Probes*, **25**, 174–176.

Jauffrais, T., Marcaillou, C., Herrenknecht, C., Truquet, P., Sécheta, V., Nicolau, E. *et al.* (2012) Azaspiracid accumulation, detoxification and biotransformation in blue mussels (*Mytilus edulis*) experimentally fed *Azadinium spinosum*. *Toxicon*, **60**, 582–595.

Jay, L.S., Davos, D., Dundas, M., Frankish, E. & Lightfoot, D. (2003) *Salmonella*. *In: Foodborne Microorganisms of Public Health Significance*. 6th edn. (ed A.D. Hocking), pp. 207–266. Australian Institute of Food Science and Technology (NSW Branch), Sydney.

JECFA (2006) Summary of evaluations performed by the joint FAO/WHO expert committee on food additives. Available at: http://www.who.int/foodsafety/publications/jecfa/en (accessed 28 May 2021).

Jeon, S.B., Seo, D.J., Oh, H., Kingsley, D.H. & Choi, C. (2017) Development of one-step reverse transcription loop-mediated isothermal amplification for norovirus detection in oysters. *Food Control*, **73**, 1002–1009.

Jones, J.L., Dargelas, V., Roberts, J., Press, C., Remington, J.S. & Montoya, J.G. (2009) Risk factors for *Toxoplasma gondii* infection in the United States. *Clinical Infectious Diseases*, **49**, 878–884.

Jones, S.H., Howell, T. L. & O'Neill, K. R. (1991) Differential elimination of indicator bacteria and pathogenic *Vibrio* spp. from eastern oysters (*Crassostrea virginica* Gmelin, 1971) in a commercial controlled purification facility in Maine. *Journal of Shellfish Research*, **10**, 105–112.

Jović, M. & Stanković, S. (2014) Human exposure to trace metals and possible public health risks via consumption of mussels *Mytilus galloprovincialis* from the Adriatic coastal area. *Food and Chemical Toxicology*, **70**, 241–251.

Jozić, S., Šolić M. & Krstulović, N. (2012) The accumulation of the indicator bacteria *Escherichia coli* in mussels (*Mytilus galloprovincialis*) and oysters (*Ostrea edulis*) under experimental conditions. *Acta Adriatica*, **53**, 353–361.

Kanoktippornchai, B., Chomvarin, C., Engchanil, C. & Wongboot, W. (2014) Triplex reverse transcription-PCR for detecting viable toxigenic *Vibrio cholerae* in water samples in Thailand. *Southeast Asian Journal of Tropical Medicine and Public Health*, **45**, 375–382.

Kidwell, D. (2015) Mitigation of harmful algal blooms: the way forward. *PICES Press, Newsletter of the North Pacific Marine Science Organization*, **23**, 22–24.

Kim, M.S. & Jeong, H.D. (2001) Development of 16S rRNA targeted PCR methods for the detection and differentiation of *Vibrio vulnificus* in marine environments. *Aquaculture*, **193**, 199–211.

Kingsley, D.H. (2013) High pressure processing and its application to the challenge of virus-contaminated foods. *Food and Environmental Virology*, **5**, 1–12.
Kingsley, D.H. (2014) Shellfish (Molluscs and Crustacea): shellfish contamination and spoilage. In: *Encyclopedia of Food Microbiology*, Vol. 3 (eds C.A. Batt & M.L. Tortorello), pp. 389–396. Elsevier, Amsterdam.
Kleivdal, H., Kristiansen, S.I., Nilsen, M.V., Goksoyr, A., Briggs, L., Holland, P. *et al.* (2007) Determination of domoic acid toxins in shellfish by Biosense ASP ELISA – a direct competitive enzyme-linked immunosorbent assay: collaborative study. *Journal of AOAC International*, **90**, 1011–1027.
Kodama, M. & Sato, S. (2008) Metabolism of paralytic shellfish toxins incorporated into bivalves. In: *Seafood and Freshwater Toxins: Pharmacology, Physiology and Detection, 2nd edn* (ed. L.M. Botana), pp. 165–175. CRC Press, Boca Raton, FL.
Kudela, R.M., Berdalet, E., Bernard, S., Burford, M., Ferand, L., Lu, S. *et al.* (2015) Harmful algal blooms. A scientific summary for policy makers. *IOC/INF-1320*. IOC/UNESCO, Paris.
Lara, C., Saldías, G.S., Westberry, T.K., Behrenfeld, M.J. & Broitman, B.R. (2017) First assessment of MODIS satellite ocean color products (OC3 and nFLH) in the Inner Sea of Chiloé, northern Patagonia. *Latin American Journal of Aquatic Research*, **45**, 822–827.
Lazcka, O., Del Campo, F.J. & Muñoz, F.X. (2007) Pathogen detection: a perspective of traditional methods and biosensors. *Biosensors and Bioelectronics*, **22**, 1205–1217.
Lazcka, O.F. Labbate, M., Seymour, J.R., Bourne, D.G., Fielder, S.S. & Doblin, M.A. (2014) Surface immuno-functionalisation for the capture and detection of *Vibrio* species in the marine environment: a new management tool for industrial facilities. *PLoS ONE*, **9**, e108387.
Lee, R. J. & Reese, R.A. (2014) Relating the bivalve shellfish harvesting area classification criteria in the United States and European Union programmes. *Journal of Water Health*, **12**, 280–287.
Lee, R., Lovatelli, A. & Ababouch, L. (2008) Bivalve depuration: fundamental and practical aspects. FAO Fisheries Technical Paper No. 511. FAO, Rome.
Lee, R., Kay, D., Wyer, M., Murray, L. & Stapleton, C. (2010) Sanitary profiling of shellfish harvesting areas. In: *Safe Management of Shellfish and Harvest Waters* (eds G. Rees, K. Pond, D. Kay, J. Bartram & J. Santo Domingo), pp. 121–143. IWA Publishing, London.
Lees, D. (2010) International standardisation of a method for detection of human pathogenic viruses in molluscan shellfish. *Food and Environmental Virology*, **2**, 146–155.
Le Guyader, F.S., Parnaudeau, S., Schaeffer, J., Bosch, A., Loisy, F., Pommepuy, M. *et al.* (2009) Detection and quantification of noroviruses in shellfish. *Applied and Environmental Microbiology*, **75**, 618–624.
Le Guyader, F.S., Krol, J., Ambert-Balay, K., Ruvoen-Clouet, N., Desaubliaux, B., Parnaudeau, S. *et al.* (2010) Comprehensive analysis of a norovirus-associated gastroenteritis outbreak, from the environment to the consumer. *Journal of Clinical Microbiology*, **48**, 915–920.
Leon, J. & Moe, C.L. (2006) Role of viruses in foodborne disease. In: *Food Consumption and Disease Risk: Consumer-Pathogen Interactions* (ed. M. Potter), pp. 309–342. Woodhead Publishing, Cambridge.
Leoni, E., Bevini, C., Degli, S., Degli Esposti, S. & Graziano, A. (1998) An outbreak of intrafamiliar hepatitis A associated with clam consumption: epidemic transmission to a school community. *European Journal of Epidemiology*, **14**, 187–192.
Leverone, J.R., Blake, N.J., Pierce, R.H. & Shumway, S.E. (2006) Effects of the dinoflagellate *Karenia brevis* on larval development in three species of bivalve mollusc from Florida. *Toxicon*, **48**, 75–84.

Li, X., Guyot, K., Dei-Cas, E., Mallard, J.-P., Jacques Ballet, J. & Brasseur, P. (2006) *Cryptosporidium* oocysts in mussels (*Mytilus edulis*) from Normandy (France). *International Journal of Food Microbiology*, **108**, 321–325.

Lindegarth, S., Torgersen, T., Lundve, B. & Sandvik, M. (2009) Differential retention of okadaic acid (OA) group toxins and pectenotoxins (PTX) in the blue mussel, *Mytilus edulis* (L.), and European flat oyster, *Ostrea edulis*. *Journal of Shellfish Research*, **28**, 313–323.

Liu J, Jasim I, Shen Z, Zhao L, Dweik M, Zhang S, *et al.* (2019) A microfluidic based biosensor for rapid detection of *Salmonella* in food products. *PLoS ONE*, **14**, e0216873.

Livingstone, D.R. & Pipe, R.K. (1992) Mussels and environmental contaminants: molecular and cellular aspects. In: *The Mussel Mytilus: Ecology, Physiology, Genetics and Culture* (ed. E.M. Gosling), pp. 425–464. Elsevier Science Publishers B.V., Amsterdam.

Loharikari, A., Newton, A.E., Stroika, S., Freeman, M., Greene, K.D., Parsons, M.B. *et al.* (2015) Cholera in the United States, 2001–2011: a reflection of patterns of global epidemiology and travel. *Epidemiology and Infection*, **143**, 695–703.

Lowther, J.A., Gustar, N.E., Powell, A.L., O'Brien, S. & Lees, D.N. (2018) A one-year survey of norovirus in UK oysters collected at the point of sale. *Food and Environmental Virology*, **10**, 278–287.

Ly, K.N. & Klevens, R.M. (2015) Trends in disease and complications of hepatitis A virus infection in the United States, 1999–2011: a new concern for adults. *The Journal of Infectious Diseases*, **212**, 176–182.

Ma, L., Zhao, F., Yao, L., Li, X., Zhou, D. & Zhang, R. (2013) The presence of genogroup II norovirus in retail shellfish from seven coastal cities in China. *Food and Environmental Virology*, **5**, 81–86.

Maalouf, H., Pommepuy, M. & Le Guyader, F.S. (2010) Environmental conditions leading to shellfish contamination and related outbreaks. *Food and Environmental Virology*, **2**, 136–145.

MAF (2011) Code of Practice: Generic RMP Models for the Processing of Seafood Product. Ministry of Agriculture and Forestry, now Ministry for Primary Industries (MPI). 57pp.

MacQuarrie, S.P. & Bricelj, V.M. (2008) Behavioral and physiological responses to PSP toxins in *Mya arenaria* populations in relation to previous exposure to red tides. *Marine Ecology Progress Series*, **366**, 59–74.

Manfrin, C., De Moro, G., Torboli, V., Venier, P., Pallavicini, A. & Gerdol, M. (2012) Physiological and molecular responses of bivalves to toxic dinoflagellates. *Invertebrate Survival Journal*, **9**, 184–199.

Mannas, H., Mimouni, R., Chaouqy, N., Hamadi, F. & Martinez-Urtaza, J. (2014) Occurrence of *Vibrio* and *Salmonella* species in mussels (*Mytilus galloprovincialis*) collected along the Moroccan Atlantic coast. *SpringerPlus*, **3**, 265.

Manso, C.F., Polo, D., Vilariño, M.L. & Romalde, J.L. (2010) Genotyping of hepatitis A virus detected in bivalve shellfish in Galicia (NW Spain). *Water Science and Technology*, **61**, 15–24.

Marangi, M., Giangaspero, A., Lacasella, V., Lonigro, A. & Gasser, R.B. (2015) Multiplex PCR for the detection and quantification of zoonotic taxa of *Giardia*, *Cryptosporidium* and *Toxoplasma* in wastewater and mussels. *Molecular and Cellular Probes*, **29**, 122–125.

Marcaillou, C., Haure, J., Mondeguer, F., Courcoux, A., Dupuy, B. & Pénisson, C. (2010) Effect of food supply on the detoxification in the blue mussel, *Mytilus edulis*, contaminated by diarrhetic shellfish toxins. *Aquatic Living Resources*, **23**, 255–266.

Mariné Oliveira, G.F., Couto, M.C.M., de Freitas Lima, M. & Bergamo do Bomfim, T.C. (2016) Mussels (*Perna perna*) as bioindicator of environmental contamination by *Cryptosporidium* species with zoonotic potential. *International Journal for Parasitology: Parasites and Wildlife*, **5**, 28–33.

Marsden, I.D., Contreras, A.M., MacKenzie, L. & Munro, M.H.G. (2015) A comparison of the physiological responses, behaviour and biotransformation of paralytic shellfish poisoning toxins in a surf-clam (*Paphies donacina*) and the green-lipped mussel (*Perna canaliculus*). *Marine and Freshwater Research*, **67**, 1163–1174.

Martinez-Albores, A., Lopez-Santamarina, A., Rodriguez, J.A., Ibarra, I.S., del Carmen Mondragó, A., Miranda, J.M. *et al.* (2020) Complementary methods to improve the depuration of bivalves: a review. *Foods*, **9**, 129.

Martinez-Urtaza, J., Saco, M., Hernandez-Cordova, G., Lozano, A., Garcia-Martin, O. & Espinosa, J. (2003) Identification of *Salmonella* serovars isolated from live molluscan shellfish and their significance in the marine environment. *Journal of Food Protection*, **66**, 226–232.

Mason, R.P., Choi, A.L., Fitzgerald, W.F., Hammerschmidt, C.R., Lamborg, C.H., Soerensen, A.L. *et al.* (2012) Mercury biogeochemical cycling in the ocean and policy implications. *Environmental Research*, **119**, 101–117.

Máté, B., Sobiech-Matura, K. & Altzitzoglou, T. (2015) Radionuclide monitoring in foodstuff: overview of the current implementation in the EU countries. *Journal of Radioanalytical and Nuclear Chemistry*, **303**, 2547–2552.

McCarthy, S.A., DePaola, A., Kaysner, C.A., Hill, W.E. & Cook, D.W. (2000) Evaluation of nonisotopic DNA hybridization methods for detection of the tdh gene of *Vibrio parahaemolyticus*. *Journal of Food Protection*, **63**, 1660–1664.

McFarland, K., Jean, F., Soudant, P. & Volety, A.K. (2015) Uptake and elimination of brevetoxin in the invasive green mussel, *Perna viridis*, during natural *Karenia brevis* blooms in southwest Florida. *Toxicon*, **97**, 46–52.

McIntosh, A.D., Grewar, G., Webster, L., Davies, I.M., Dalgarno, E.J. & Moffat, C.F. (2004) Kinetics of depuration of hydrocarbons by mussels (*Mytilus edulis*). ICES CM 2004/V:04.

McLoughlin, K.S. (2011) Microarrays for pathogen detection and analysis. *Briefings in Functional Genomics*, **10**, 342–353.

McMenemy, P., Kleczkowski, A., Lees, D.N., Lowther, J. & Taylor, N. (2018) A model for estimating pathogen variability in shellfish and predicting minimum depuration times. *PLoS ONE*, **13**, e0193865.

McNamee, S.E., Elliott, C.T., Delahaut, P. & Campbell, K. (2013) Multiplex biotoxin surface plasmon resonance method for marine biotoxins in algal and seawater samples. *Environmental Science and Pollution Research*, **20**, 6794–6807.

McNamee, S.E., Elliott, C.T., Greer, B., Lochhead, M. & Campbell, K. (2014) Development of a planar waveguide microarray for the monitoring and early detection of five harmful algal toxins in water and cultures. *Environmental Science and Technology*, **48**, 13340—13349.

McPartlin, D.A., Loftus, J.H., Crawley, A.S., Silke, J., Murphy, C.S. & O'Kennedy, R.J. (2017) Biosensors for the monitoring of harmful algal blooms. *Current Opinion in Biotechnology*, **45**, 164–169.

Medalla, F., Gu, W., Mahon, B.E., Judd, M., Folster, J., Griffin, P.M. *et al.* (2017) Estimated incidence of antimicrobial drug resistant nontyphoidal *Salmonella* infections, United States, 2004–2012. *Emerging Infectious Diseases*, **23**, 29–37.

MedicineNet (2019) E. coli definition and facts. Available at: https://www.medicinenet.com/e_coli__0157h7/article.htm (accessed 28 May 2021).

Mesquita, J.R., Vaz, L., Cerqueira, S., Castilho, F., Santos, R., Monteiro, S. *et al.* (2011) Norovirus, hepatitis A virus and enterovirus presence in shellfish from high quality harvesting areas in Portugal. *Food Microbiology*, **28**, 936–941.

Miller, M.A. (2008) Tissue cyst-forming coccidia of marine mammals. In: *Zoo and Wild Animal Medicine* (eds M. Fowler & R. Miller). pp. 319–340. Saunders Elsevier, St Louis, MO.

Miller, M.A., Miller, W.A., Conrad, P.A., James, E.R., Melli, A.C., Leutenegger, C.M. *et al.* (2008) Type X *Toxoplasma gondii* in a wild mussel and terrestrial carnivores from coastal California: new linkages between terrestrial mammals, runoff and toxoplasmosis of sea otters. *International Journal of Parasitology*, **38**, 1319–1328.

Miller, N.D., Davidson, P.M. & D'Souza, D.H. (2011) Real-time reverse-transcriptase PCR for *Salmonella Typhimurium* detection from lettuce and tomatoes. *LWT – Food Science and Technology*, **44**, 1088–1097.

Miller, W.A., Miller, M.A., Gardner, I.A., Atwill, E.R., Harris, M., Ames, J. *et al.* (2005a) New genotypes and factors associated with *Cryptosporidium* detection in mussels (*Mytilus* spp.) along the California coast. *International Journal for Parasitology*, **35**, 1103–1113.

Miller, W.A., Atwill, E.R., Gardner, I.A., Miller, M.A., Fritz, H.M., Hedrick, R.P. *et al.* (2005b) Clams (*Corbicula fluminea*) as bioindicators of fecal contamination with *Cryptosporidium* and *Giardia* spp. in freshwater ecosystems in California. *International Journal of Parasitology*, **35**, 673–684.

Miller, W.A., Gardner, I.A., Atwill, E.R., Leutenegger, C.M., Miller, M.A. Hedrick, R.P. *et al.* (2006) Evaluation of methods for improved detection of *Cryptosporidium* spp. in mussels (*Mytilus californianus*). *Journal of Microbiological Methods*, **65**, 367–379.

Mladineo, I., Trumbi, Z., Jozi, S. & Šegvi, T. (2009) First report of *Cryptosporidium* sp. (Coccidia, Apicomplexa) oocysts in the black mussel (*Mytilus galloprovincialis*) reared in the Mali Ston Bay, Adriatic Sea. *Journal of Shellfish Research*, **28**, 541–546.

Molaee, N., Abtahi, H., Ghannadzadeh, M.J., Karimi, M. & Ghaznavi-Rad, E. (2015) Application of Reverse Transcriptase–PCR (RT-PCR) for rapid detection of viable *Escherichia coli* in drinking water samples. *Journal of Environmental Health Science & Engineering*, **13**, 24.

MPI (1997) A Guide to Hazard Analysis Critical Control Point Systems in the Seafood Industry. Ministry of Agriculture and Forestry, now Ministry for Primary Industries (MPI). 146pp.

MPI (2016) Animal Products Notice: Specifications for Products Intended for Human Consumption. Ministry for Primary Industries, formerly Ministry of Agriculture and Forestry. 125pp.

Mudgal, V., Madaan, N., Mudgal, A., Singh, R.B. & Mishra, S. (2010) Effect of toxic metals on human health. *The Open Nutraceuticals Journal*, **3**, 94–99.

Munday, R. & Reeve, J. (2013) Risk assessment of shellfish toxins. *Toxins*, **5**, 2109–2137.

Murtaza, I., Laila, O. & Ubaid-ullah, S. (2016) DNA microrray: a miniaturized high throughput technology. *International Journal of Engineering Research and Development*, **12**, 9–15.

Namsai, A., Louisirirotchanakul, S., Wongchinda, N., Siripanyaphinyo, U., Virulhakul, P. & Puthavathana, P. *et al.* (2011) Surveillance of hepatitis A and E viruses contamination in shellfish in Thailand. *Letters in Applied Microbiology*, **53**, 608–613.

National Ocean Service (2017) What are microplastics? Available at: https://oceanservice.noaa.gov/facts/microplastics.html (accessed 28 May 2021).

Navarro, J.M., Widdows, J., Chaparro, O.R., Ortíz, A., Mellado, C. & Villanueva, P.A. (2018) Pre-ingestive selection capacity and endoscopic analysis in the sympatric bivalves *Mulinia edulis* and *Mytilus chilensis* exposed to diets containing toxic and non-toxic dinoflagellates. *PLoS ONE*, **13**, e0193370.

Nenonen, N.P., Hannoun, C., Horal, P., Hernroth, B. & Bergstrom, T. (2008) Tracing of norovirus outbreak strains in mussels collected near sewage effluents. *Applied and Environmental Microbiology*, **74**, 2544–2549.

Nenonen, N.P., Hannoun, C., Olsson, M.B. & Bergström T. (2009) Molecular analysis of an oyster-related norovirus outbreak. *Journal of Clinical Virology*, **45**, 105–108.

Ng, T.F.F., Marine, R., Wang, C., Simmonds, P., Kapusinszky, B., Bodhidatta, L. *et al.* (2012) High variety of known and new RNA and DNA viruses of diverse origins in untreated sewage. *Journal of Virology*, **86**, 12161–12175.

Nishida, T., Kimura, H., Saitoh, M., Shinohara, M., Kato, M. & Fukuda, S. (2003) Detection, quantitation, and phylogenetic analysis of noroviruses in Japanese oysters. *Applied and Environmental Microbiology*, **69**, 5782–5786.

Nordin, N., Yusof, N.A., Abdullah, J., Radu, S. & Hushiarian, R. (2017) A simple, portable, electrochemical biosensor to screen shellfish for *Vibrio parahaemolyticus*. *AMB Express*, **7**, 41.

Novaczek, I., Madhyastha, M.S., Ablett, R.F. & Donald, A. (1992) Depuration of domoic acid from live blue mussels. *Canadian Journal of Fisheries and Aquatic Sciences*, **49**, 312–318.

NSSP (2015) Guide for the control of molluscan shellfish. Available at: http://www.fda.gov/Food/GuidanceRegulation/FederalStateFoodPrograms/ucm2006754.htm (accessed 28 May 2021).

NSW Food Authority (2012) Food safety program for seafood processing. FI148/1204. Available at: https://www.foodauthority.nsw.gov.au/sites/default/files/_Documents/industry/seafood_processing_fsp.pdf (accessed 28 May 2021).

NSW Food Authority (2015) Food safety program for farmed shellfish. NSW/FA/FI102/1505. Available at: https://www.foodauthority.nsw.gov.au/sites/default/files/_Documents/industry/Shellfish_FSP_approved_harvest_hold.pdf (accessed 28 May 2021).

Núñez-Acuña, G., Aballay, A.E., Hégaret, H., Astuya, A.P. & Gallardo-Escárate, C. (2013) Transcriptional responses of *Mytilus chilensis* exposed *in vivo* to saxitoxin (STX). *Journal of Molluscan Studies*, **79**, 323–331.

Ohk, S.H. & Bhunia, A.K. (2013) Multiplex fiber optic biosensor for detection of *Listeria monocytogenes*, *Escherichia coli* O157:H7 and *Salmonella enterica* from ready-to-eat meat samples. *Food Microbiology*, **33**, 166–171.

Oliveira, J., Cunha, A., Castilho, F., Romalde, J.L. & Pereira, M.J. (2011) Microbial contamination and purification of bivalve shellfish: crucial aspects in monitoring and future perspectives. *Food Control*, **22**, 805–816.

Olmedo, P., Pla, A., Hernandez, A.F., Barbier, F., Ayouni, L. & Gil, F. (2013) Determination of toxic elements (mercury, cadmium, lead, tin and arsenic) in fish and shellfish samples. Risk assessment for the consumers. *Environment International*, **59**, 63–72.

Oshima, Y., Kuiper-Goodman, T. & Hall, S. (2011) Saxitoxins. In: *Assessment and Management of Biotoxin Risks in Bivalve Molluscs* (eds J. Lawrence, H. Loreal, H. Toyofuku, P. Hess, I. Karunasagar & L. Ababouch), pp. 207–234. FAO Fisheries and Aquaculture Technical Paper No. 551. FAO, Rome.

Pagoso, E.J.A. & Rivera, W.L. (2017) *Cryptosporidium* species from common edible bivalves in Manila Bay, Philippines. *Marine Pollution Bulletin*. **119**, 31–39.

Panicker, G., Call, D.R., Krug, M.J. & Bej, A.K. (2004) Detection of pathogenic *Vibrio* spp. in shellfish by using multiplex PCR and DNA microarrays. *Applied and Environmental Microbiology*, **70**, 7436–7444.

Park, J.Y., Jeon, S., Kim, J.Y., Park, M. & Kim, S. (2013) Multiplex real-time polymerase chain reaction assays for simultaneous detection of *Vibrio cholerae*, *Vibrio parahaemolyticus*, and *Vibrio vulnificus*. *Osong Public Health Research Perspectives*, **4**, 133–139.

Park, S., Jeong, H., Jung, H., Kim, K., Park, S., Kim, M. *et al.* (2015) Monitoring the hepatitis A virus in oyster from Korea. *Advances in Microbiology*, **5**, 825–831.

Pavoni, E., Consoli, M., Suffredini, E., Arcangeli, G., Serracca, L., Battistini, R. *et al.* (2013) Noroviruses in seafood: a 9-year monitoring in Italy. *Foodborne Pathogens and Disease*, **10**, 533–539.

Pearson, A.J., Gaw, S., Hermanspahn, N. & Glover, C.N. (2016a) Natural and anthropogenic radionuclide activity concentrations in the New Zealand diet. *Journal of Environmental Radioactivity*, **151**, 601–608.

Pearson, A.J. Sally Gaw, S., Hermanspahn, N. & Glover, C.N. (2016b) Activity concentrations of ^{137}Caesium and ^{210}Polonium in seafood from fishing regions of New Zealand and the dose assessment for seafood consumers. *Journal of Environmental Radioactivity*, **151**, 542–550.

Pintó, R.M., Costafreda, M.I., Pérez-Rodriguez, F.J., D'Andrea, L. & Bosch, A. (2010) Hepatitis A virus: state of the art. *Food and Environmental Virology*, **2**,127–135.

Pintó Solé, R.M., Martín de Santos, R. & Bosch Navarro, A. (2011) Report of the Scientific Committee of the Spanish Agency for Food Safety and Nutrition (AESAN) on viral contamination of food, paying special attention to bivalve molluscs and control methods. *Revista del Comité Científico de la AESAN*, **13**, 89–105.

Piscopo, M, Notariale, R. & Troisi, J. (2017) hsp70 AS new cadmium bioaccumulation marker to prevent the risks of mussel consumption in human nutrition. *Madridge Journal of Clinical Research*, 1, 39–45.

Pokrzywinski, K.L., Tilney, C.L., Modla, S., Caplan, J.L., Ross, J., Warner, M.E. *et al.* (2017) Effects of the bacterial algicide IRI-160AA on cellular morphology of harmful dinoflagellates. *Harmful Algae*, **62**, 127–135.

Polo, D., Feal, X., Varela, M.F., Monteagudo, A. & Romalde, J.L. (2014a) Depuration kinetics of murine norovirus in shellfish. *Food Research International*, **64**, 182–187.

Polo, D., Álvarez, C., Longa, A. & Romalde, J.L. (2014b) Effectiveness of depuration for hepatitis A virus removal from mussels (*Mytilus galloprovincialis*). *International Journal of Food Microbiology*, **180**, 24–29.

Polo, D., Varela, M.F. & Romalde, J.L. (2015a) Detection and quantification of hepatitis A virus and norovirus in Spanish authorized shellfish harvesting areas. *International Journal of Food Microbiology*, **193**, 43–50.

Polo, D., Feal, X. & Romalde, J.L. (2015b) Mathematical model for viral depuration kinetics in shellfish: a useful tool to estimate the risk for the consumers. *Food Microbiology*, **49**, 220–225.

Poltronieri, P., Mezzolla, V., Primiceri, E. & Maruccio, G. (2014) Biosensors for the detection of food pathogens. *Foods*, **3**, 511–526.

Potipat, J., Tangkrock-olan, N. & Helander, H.F. (2015) Bioconcentration factor (BCF) and depuration of heavy metals of oysters (*Saccostrea cucullata*) and mussels (*Perna viridis*) in the river basins of coastal area of Chanthaburi Province, Gulf of Thailand. *Environment Asia*, **8**, 118–128.

Povinec, P.P. & Hirose, K. (2015) Fukushima radionuclides in the NW Pacific, and assessment of doses for Japanese and world population from ingestion of seafood. *Scientific Reports*, **5**, 9016.

Praveen, C., Dancho, B.A., Kingsley, D.H., Calci, K.R., Meade, G.K., Mena, K.D. *et al.* (2013) Susceptibility of murine norovirus and hepatitis A virus to electron beam irradiation in oysters and quantifying the reduction in potential infection risks. *Applied and Environmental Microbiology*, **79**, 3796–3801.

Przeslawski, R., Bourdeau, P.E., Doall, M.H., Pan, J., Perino, L. & Padilla, D.K. (2008) The effects of a harmful alga on bivalve larval lipid stores. *Harmful Algae*, **7**, 802–807.

Pulido, O.M. (2008) Domoic acid toxicologic pathology: a review. *Marine Drugs*, **6**, 180–219.

Qiu, J., Fan, H., Liu, T., Liang, X., Meng, F., Quilliam, M.A. *et al.* (2018) Application of activated carbon to accelerate detoxification of paralytic shellfish toxins from mussels *Mytilus galloprovincialis* and scallops *Chlamys farreri*. *Ecotoxicology and Environmental Safety*, **148**, 402–409.

Quilliam, M.A. (2003) The role of chromatography in the hunt for red tide toxins. *J. Chromatography, A*, **1000**, 527–548.

Raftopoulou, E.K. & Dimitriadis, V.K. (2011) Comparative study of the accumulation and detoxification of Cu (essential metal) and Hg (nonessential metal) in the digestive gland and gills of mussels *Mytilus galloprovincialis*, using analytical and histochemical techniques. *Chemosphere*, **83**, 1155–1165.

Rahman, M., Heng, L.Y., Futra, D. & Ling, T.L. (2017) Ultrasensitive biosensor for the detection of *Vibrio cholerae* DNA with polystyrene-coacrylic acid composite nanospheres. *Nanoscale Research Letters*, **12**, 474.

Razafimahefa, R.M., Ludwig-Begall, L.F. & Thiry, E. (2019) Cockles and mussels, alive, alive, oh. The role of bivalve molluscs as transmission vehicles for human norovirus infections. *Transboundary and Emerging Diseases*, **67**, 9–25.

Reguera, B., Riobó, P., Rodríguez, F., Díaz, P.A., Pizarro, G., Paz, B. *et al.* (2014) *Dinophysis* toxins: causative organisms, distribution and fate in shellfish. *Marine Drugs*, **12**, 394–461.

Regueiro, J., Rossignoli, A.E., Álvarez, G. & Blanco, J. (2011) Automated on-line solid-phase extraction coupled to liquid chromatography-tandem mass spectrometry for determination of lipophilic marine toxins in shellfish. *Food Chemistry*, **129**, 533–540.

Renshaw, M.A. Soltysiak, K., Arreola, D., Loret, P., Patton, J.C. & Gold, J.R. (2006) Microsatellite DNA markers for population genetic studies in the dinoflagellate *Karenia brevis*. *Molecular Ecology Notes*, **6**, 1157–1159.

Reverté, L., Soliño, L., Carnicer, O., Diogène, J. & Campàs, M. (2014) Alternative methods for the detection of emerging marine toxins: biosensors and cell-based assays. *Marine Drugs*, **12**, 5719–5763.

Richards, G. P. (2016) Shellfish-associated enteric virus illness: virus localization, disease outbreaks and prevention. In: *Viruses in Foods*, 2nd edn (eds S.M. Goyal & J. Cannon), pp. 185–207. Springer, Philadelphia, PA.

Richards, G.P., McLeod, C. & Le Guyader, F.S. (2010) Processing strategies to inactivate enteric viruses in shellfish. *Food and Environmental Virology*, **2**, 183–193.

Riddle, M.S. & Walker, R.I. (2016) Status of vaccine research and development for norovirus. *Vaccine*, **34**, 2895–2899.

Rivadulla, E., Varela, M.F. & Romalde, J.L. (2017) Low prevalence of Aichi virus in molluscan shellfish samples from Galicia (NW Spain). *Journal of Applied Microbiology*, **122**, 516–521.

Robertson, L.J. (2007) The potential for marine bivalve shellfish to act as transmission vehicles for outbreaks of protozoan infections in humans: a review. *International Journal of Food Microbiology*, **120**, 201–216.

Röder, K., Fritz, N., Gerdts, G. & Luckas, B. (2011) Accumulation and depuration of yessotoxin in two bivalves. *Journal of Shellfish Research*, **30**, 167–175.

Rodríguez, L.P., González, V., Martínez, A., Paz, B., Lago, J., Cordeiro, V. *et al.* (2015) Occurrence of lipophilic marine toxins in shellfish from Galicia (NW of Spain) and synergies among them. *Marine Drugs*, **13**, 1666–1687.

Rose, J.B. (1997) Environmental ecology of *Cryptosporidium* and public health implications. *Annual Review of Public Health*, **18**, 135–161.

Ryan, M., Ito, E., Speijers, G. & Hess, P. (2011) Azaspiracids. In: *Assessment and Management of Biotoxin Risks in Bivalve Molluscs* (eds J. Lawrence, H. Loreal, H. Toyofuku, P. Hess, I. Karunasagar & L. Ababouch), pp. 33–49. FAO Fisheries and Aquaculture Technical Paper No. 551. FAO, Rome.

Ryder, J., Karunasagar, I. & Ababouch, L. (eds) (2014) *Assessment and Management of Seafood Safety and Quality: Current Practices and Emerging Issues*. FAO Fisheries and Aquaculture Technical Paper No. 574. FAO, Rome.

Sakata, J., Yonekita, T. & Kawatsu, K. (2018) Development of a rapid immunochromatographic assay to detect contamination of raw oysters with enteropathogenic *Vibrio parahaemolyticus*. *International Journal of Food Microbiology*, **264**, 16–24.

Saki, J., Foroutan-Rad, M. & Asadpouri, R. (2016) Molecular characterization of *Cryptosporidium* spp. in wild rodents of southwestern Iran using 18S rRNA gene nested-PCR-RFLP and sequencing techniques. *Journal of Tropical Medicine*, **2016**, art. 6834206.

Sánchez, G., Pintó, R.M., Vanaclocha, H. & Bosch, A. (2002) Molecular characterization of hepatitis A virus isolates from a transcontinental shellfish-borne outbreak. *Journal of Clinical Microbiology*, **40**, 4148–4155.

Santo Domingo, J.W. & Edge, T.A. (2010) Identification of primary sources of faecal pollutionIn: *Safe Management of Shellfish and Harvest Waters* (eds G. Rees, K. Pond, D. Kay, J. Bartram & J. Santo Domingo), pp. 51–90. IWA Publishing, London.

Sapkota, A., Sapkota, A.R., Kucharski, M., Burke, J., McKenzie, S., Walker, P. *et al.* (2008) Aquaculture practices and potential human health risks: current knowledge and future priorities. *Environment International*, **34**, 1215–1226.

SARF (2011) Review and cost-benefit analysis for industry of reduced depuration times for the mussel *Mytilus edulis*. Report F066, commissioned by the Scottish Aquaculture Research Forum and prepared by Aquafish Solutions Ltd.

Sauvé, G. (2010) Official control monitoring programmes for live bivalve molluscs – legislative and regulatory approaches: Canada. In: *Safe Management of Shellfish and Harvest Waters* (eds G. Rees, K. Pond, D. Kay, J. Bartram, J. & J. Santo Domingo), pp. 217–231. IWA Publishing, London.

SCF (2001) Opinion on the risk assessment of dioxins and dioxins-like PCB in food (update based on the new scientific information available since the adoption of the SCF opinion of 22 November 2000) (adopted by the SCF on 30 May 2001). Available at: https://ec.europa.

eu/food/sites/food/files/safety/docs/cs_contaminants_catalogue_dioxins_out90_en.pdf (accessed 28 May 2021).
Schaeffer, J., Le Saux, J.-C., Lora, M., Atmar, R.L. & Le Guyader, F.S. (2013) Norovirus contamination on French marketed oysters. *International Journal of Food Microbiology*, **166**, 244–248.
Schneid, A.S., Rodrigues, K.L., Chemello, D., Tondo, E.C., Ayub, M.A.Z. & Aleixo, J.A.G. (2006) Evaluation of an indirect ELISA for the detection of *Salmonella* in chicken meat. *Brazilian Journal of Microbiology*, **37**, 350–355.
Seafish (2018a) Seafish standard design purification systems: operating manual for the small scale shallow tank purification system. *Seafish Report No.* 721.
Seafish (2018b) Seafish standard design purification systems: operating manual for the medium scale multi-layer system. *Seafish Report No* 720.
Seafish (2018c) Seafish standard design purification systems: operating manual for the vertical stack system. *Seafish Report No.* 722.
Seafish (2018d) Seafish standard design purification systems: operating manual for the bulk bin system for mussels. *Seafish Report No.* 718.
Selwood, A.I., Miles, C.O., Wilkins, A.L., van Ginkel, R., Munday, R., Rise, F. *et al.* (2010) Isolation and structural determination, and acute toxicity of novel pinnatoxins E, F and G. *Journal of Agriculture and Food Chemistry*, **58**, 6532–6542.
Setti, I., Rodriguez-Castro, A., Pata, M.P., Cadarso-Suarez, C., Yacoubi, B., Bensmael, L. *et al.* (2009) Characteristics and dynamics of *Salmonella* contamination along the Coast of Agadir, Morocco. *Applied and Environmental Microbiology*, **75**, 7700–7709.
Shams, S., Khan, S., Khan, A., Khan, I., Ijaz. M. & Ullah, A. (2016) Differential techniques used for detection of *Cryptosporidium* oocysts in stool specimens. *Journal of Parasitic Diseases: Diagnosis and Therapy*, **1**, 1–11.
Shapiro, K., VanWormer, E., Aguilar, B. & Conrad, P.A. (2015) Surveillance for *Toxoplasma gondii* in California mussels (*Mytilus californianus*) reveals transmission of atypical genotypes from land to sea. *Environmental Microbiology*, **17**, 4177–4188.
Shen, L., Xu, H. & Guo, X. (2012) Satellite remote sensing of harmful algal blooms (HABs) and a potential synthesized framework. *Sensor*, **12**, 7778–7803.
Shen, Z., Hou, N., Jin, M., Qiu, Z., Wang, J., Zhang, B. *et al.* (2014) A novel enzyme-linked immunosorbent assay for detection of *Escherichia coli* O157:H7 using immunomagnetic and beacon gold nanoparticles. *Gut Pathogens*, **6**, 14.
Shieh, Y.C., Khudyakov, Y.E., Xia, G., Ganova-Raeva, L.M., Khambaty, F.M., Woods, J.W. *et al.* (2007) Molecular confirmation of oysters as the vector for hepatitis A in a 2005 multistate outbreak. *Journal of Food Protection*, **70**, 145–150.
Silva, N.F.D., Magalhães, J.M.C.S., Freirec, C. & Delerue-Matos, C. (2018) Electrochemical biosensors for *Salmonella*: state of the art and challenges in food safety assessment. *Biosensors and Bioelectronics*, **99**, 667–682.
Sjöling, A., Sadeghipoorjahromia, L., Novak, D. & Tobias, J. (2015) Detection of major diarrheagenic bacterial pathogens by multiplex PCR panels. *Microbiological Research*, **172**, 34–40.
Skander, K., Saber, B., Hassen, T., Mouna, B., Nawel, D. & Mourad, B. (2017) Environmental parameters and bacteriological quality of the *Perna perna* mussel (North East Algerian coast). *International Journal of Biosciences*, **11**, 151–165.

Smienk, H.G.F., Calvo, D., Razquin, P., Domínguez, E. & Mata, L. (2012) Single laboratory validation of a ready-to-use phosphatase inhibition assay for detection of okadaic acid toxins. *Toxins*, **4**, 339–352.

Srisuphanunt, A.M., Saksirisampant,W. & Karanis, P. (2009) Detection of *Cryptosporidium* oocysts in green mussels (*Perna viridis*) from shell-fish markets of Thailand. *Parasite*, **16**, 235–239.

Staggs, S.E., Keely, S.P., Ware, M.W., Schable, N., See, M.J., Gregorio, D. *et al.* (2015) The development and implementation of a method using blue mussels (*Mytilus* spp.) as biosentinels of *Cryptosporidium* spp. and *Toxoplasma gondii* contamination in marine aquatic environments. *Parasitology Research*, **114**, 4655–4667.

Su, C., Shwab, E.K., Zhou, P., Zhu, X.Q. & Dubey, J.P. (2010) Moving towards an integrated approach to molecular detection and identification of *Toxoplasma gondii*. *Parasitology*, **137**, 1–11.

Su, L.H, Chiu, C.H., Chu, C. & Ou, J.T. (2004) Antimicrobial resistance in nontyphoid *Salmonella* serotypes: a global challenge. *Clinical Infectious Diseases*, **39**, 546–551.

Suffredini, E., Proroga, Y.T.R., Di Pasquale, S., Di Maro, O., Losardo, M., Cozzi, L. *et al.* (2017) Occurrence and trend of hepatitis A virus in bivalve molluscs production areas following a contamination event. *Food and Environmental Virology*, **9**, 423–433.

Sun, J., Jian, J., Sun, Y., Abdalhai, M.H., Zhang, Y., Sun, X. *et al.* (2015) DNA biosensor-based on fluorescence detection of *E. coli* O157:H7 by Au@Ag nanorods. *Biosensors and Bioelectronics*, **70**, 239–245.

Sutthikornchai, C., Popruk, S., Chumpolbanchorn, K., Sukhumavas, W. & Sukthana, Y. (2016) Oyster is an effective transmission vehicle for *Cryptosporidium* infection in humans. *Asian Pacific Journal of Tropical Medicine*, **9**, 562–566.

Tang, X.H., Yu, R.C., Zhou, M.J. & Yu, Z.G. (2012) Application of rRNA probes and fluorescence *in situ* hybridization for rapid detection of the toxic dinoflagellate *Alexandrium minutum*. *Chinese Journal of Oceanologyy and Limnology*, **30**, 256–263.

Tappeh, K.H., Manafi, G., Asgharzadeh, M. & Manafi, F. (2014) Incidence of *Giardia lamblia* subspecies by PCR-RFLP in stool specimens of hospitalized children at Urmia Mutahhari Hospital, West Azerbaijan Province, Iran. *Iranian Journal of Parasitology*, **9**, 541–547.

Tasker, R.A., Kuiper-Goodman, T., Pulido, O. & Lawrence, J.F. (2011) Domoic acid. In: *Assessment and Management of Biotoxin Risks in Bivalve Molluscs* (eds J. Lawrence, H. Loreal, H. Toyofuku, P. Hess, I. Karunasagar & L. Ababouch), pp. 111–162. FAO Fisheries and Aquaculture Technical Paper No. 551. FAO, Rome.

Taylor, M., McIntyre, L., Ritson, M., Stone, J., Bronson, R., Bitzikos, O. *et al.* (2013) Outbreak of diarrhetic shellfish poisoning associated with mussels, British Columbia, Canada. *Marine Drugs*, **11**, 1669–1676.

Tenter, A.M., Heckeroth, A.R & Weiss, L.M. (2000) *Toxoplasma gondii*: from animals to humans. *International Journal of Parasitology*, **30**, 1217–1258.

Teramoto, N. & Yotsu-Yamashita, M. (2015) Selective blocking effects of 4,9-anhydrotetrodotoxin, purified from a crude mixture of tetrodotoxin analogues, on nav1.6 channels and its chemical aspects. *Marine Drugs*, **13**, 984–995.

Terio, V., Tantillo, G., Martella, V., Di Pinto, P., Buonavoglia, C. & Kingsley, D.H. (2010) High pressure inactivation of HAV within mussels. *Food and Environmental Virology*, **2**, 83–88.

Terlizzi, D.E. & Mazzacaro, A.P. (2010) Harmful algae and marine aquaculture in the northeastern United States. NRAC Publication No. 209.

These, A., Klemm, C., Nausch, I. & Uhlig, S. (2011) Results of a European interlaboratory method validation study for the quantitative determination of lipophilic marine biotoxins in raw and cooked shellfish based on high-performance liquid chromatography–tandem mass spectrometry. Part I: collaborative study. *Analytical and Bioanalytical Chemistry*, **399**, 1245–1256.

Tobias, J. & Vutukuru, S.R. (2012) Simple and rapid multiplex PCR for identification of the main human diarrheagenic *Escherichia coli*. *Microbiological Research*, **167**, 564–570.

van den Top, H.J., Gerssen, A., McCarron, P. & van Egmond, H.P. (2011) Quantitative determination of marine lipophilic toxins in mussels, oysters and cockles using liquid chromatography-mass spectrometry: inter-laboratory validation study. *Food Additives and Contaminants Part A – Chemistry Analysis Control Exposure & Risk Assessment*, **28**, 1745–1757.

Torok, V., Hodgson, K., McLeod, C., Tan, J., Malhi, N. & Turnbull, A. (2018) National survey of foodborne viruses in Australian oysters at production. *Food Microbiology*, **69**, 196–203.

Trainer, V.L. & Yoshida, T. (eds.) (2014) Proceedings of the Workshop on Economic Impacts of Harmful Algal Blooms on Fisheries and Aquaculture. PICES Science Report No. 47.

Trainer, V.L., Moore, L., Bill, B.D., Adams, N.G., Harrington, N., Borchert, J. *et al.* (2013) Diarrhetic shellfish toxins and other lipophilic toxins of human health concerns in Washington State. *Marine Drugs*, **11**, 1815–1836.

Turner, A.D., Higgins, C., K., Davidson, K., Veszelovszki, A., Payne, D., Hungerford, J. *et al.* (2015) Potential threats posed by new or emerging marine biotoxins in UK waters and examination of detection methodology used in their control: brevetoxins. *Marine Drugs*, **13**, 1224–1254.

Twiner, M.J., Rehmann, N., Hess, P. & Doucette, G.J. (2008) Azaspiracid shellfish poisoning: a review on the chemistry, ecology, and toxicology with an emphasis on human health impacts. *Marine Drugs*, **6**, 39–72.

UN Environment (2017) *Toolkit for Identification and Quantification of Mercury Sources. Guideline for Inventory Level 1, Version 2*.0. UN Environment Chemicals Branch, Geneva.

UN Environment Programme (2019) Mercury is a chemical element that is harmful to human health and the environment. Available at: https://www.unep.org/explore-topics/chemicals-waste/what-we-do/mercury (accessed 28 May 2021).

Valdiglesias, V., Prego-Faraldo, M.V., Pásaro, E., Méndez, J. & Laffon, B. (2013) Okadaic acid: more than a diarrheic toxin. *Marine Drugs*, **11**, 4328–4349.

Van Cauwenberghe & Janssen C.R. (2014) Microplastics in bivalves cultured for human consumption. *Environmental Pollution*, **193**, 65–70.

Vandermeersch, G., Van Cauwenberghe, L., Janssen, C.R., Marques, A., Granby, K., Fait, G. *et al.* (2015) A critical view on microplastic quantification in aquatic organisms. *Environmental Research*, **143**, 46–55.

Van Dolah, F.M., Leighfield, T.A., Haynes, B.L., Hamson, D.R. & Ramsdell, J.S. (1997) A microplate receptor assay for the amnesic shellfish poisoning toxin, domoic acid, utilizing a cloned glutamate receptor. *Analytical Biochemistry*, **245**, 102–105.

Vantarakis, A., Komninou, G., Venieri, D. & Papapetropoulou, M. (2000) Development of a multiplex PCR detection of *Salmonella* spp. and *Shigella* spp. in mussels. *Letters in Applied. Microbiology*, **31**, 105–109.

Varela, M.F., Polo, D. & Romalde, J.L. (2016) Prevalence and genetic diversity of human sapoviruses in shellfish from commercial production areas in Galicia, Spain. *Applied Environmental Microbiology*, **82**,1167–1172.

Velásquez, C. & Navarro, J.M. (2014) Feeding and intoxication–detoxification dynamics in two populations of the mussel *Mytilus chilensis* (Hupé, 1854) with different histories of exposure to paralytic shellfish poisoning (PSP). *Marine and Freshwater Behaviour and Physiology*, **47**, 185–195.

Vilariño, M.L., Le Guyader, F.S., Polo, D., Schaeffer, J., Krol, J. & Romalde, J.L. (2009) Assessment of human enteric viruses in cultured and wild bivalve molluscs. *International Microbiology*, **12**, 145–151.

Vilariño, N., Louzao, M.C., Fraga, M., Rodríguez, L.P. & Botana, L.M. (2013) Innovative detection methods for aquatic algal toxins and their presence in the food chain. *Analytical and Bioanalytical Chemistry*, **405**, 7719–7732.

Villanueva, P.A. & Navarro, J.M. (2016) Pre-ingestive selection efficiency in two populations of the razor clam *Tagelus dombeii* with different histories of exposure to paralytic shellfish poisoning (PSP). *Marine and Freshwater Behaviour and Physiology*, **49**, 291–300.

Visciano, P., Schirone, M., Berti, M., Milandri, A., Tofalo, R. & Suzzi, G. (2016) Marine biotoxins: occurrence, toxicity, regulatory limits and reference methods. *Frontiers in Microbiology*, **7**, art. 1051.

Wan Norhana, M. N., Masazurah A.R., Nor Ainy, M. & Ismail, I. (2011) Reverse transcriptase PCR detection of hepatitis A virus (HAV) in cultured and wild shellfish from the Peninsular of Malaysia. *International Food Research Journal*, **18**, 411–415.

Wang, J. (2015) Development of satellite-assisted forecasting system for oyster norovirus outbreaks. PhD thesis, Louisiana State University, Baton Rouge, LA.

Wang, J. & Deng, Z. (2012) Detection and forecasting of oyster norovirus outbreaks: recent advances and future perspectives. *Marine Environmental Research*, **80**, 62–69.

Watanabe, R., Matsushima, R., Harada, T., Oikawa, H., Murata, M., Suzuki, T. *et al.* (2013) Quantitative determination of paralytic shellfish toxins in cultured toxic algae by LC-MS/MS. *Food Additives and Contaminants-Part A*, **30**, 1351–1357.

Wei, S., Zhao, H., Xian, Y., Hussain, M.A. & Wu, X. (2014) Multiplex PCR assays for the detection of *Vibrio alginolyticus*, *Vibrio parahaemolyticus*, *Vibrio vulnificus*, and *Vibrio cholerae* with an internal amplification control. *Diagnostic Microbiology and Infectious Disease*, **79**, 115–118.

WHO (2010) *Safe Management of Shellfish and Harvest Waters* (eds G. Rees, K. Pond, D. Kay, J. Bartram & J. Santo Domingo). IWA Publishing, London.

WHO (2017a) *Global Hepatitis Report 2017*. World Health Organization, Geneva.

WHO (2017b) *Guidelines for Drinking-water Quality*, 4th edn. World Health Organisation, Geneva.

WHO/FAO (2002) *Risk assessments of Salmonella in eggs and broiler chickens*. World Health Organization and Food and Agriculture Organization of the United Nations, Geneva.

Winterbourn, J.B., Clements, K., Lowther, J.A., Malham, S.K., McDonald, J.E. & Jones, D.L. (2016) Use of *Mytilus edulis* biosentinels to investigate spatial patterns of norovirus and faecal indicator organism contamination around coastal sewage discharges. *Water Research*, **105**, 241–250.

Woods, J.W. & Burkhardt, W. III (2010) Occurrence of norovirus and hepatitis A virus in US oysters. *Food and Environmental Virology*, **2**, 176–182.

Xi-Bin, N., Yong, D., Yi, J. & Dai-Xin, L. (2009) Study on rapid detection of *Vibrio parahaemolyticus* with sea food. *Asian Fisheries Science*, **22**, 331–336.

Xu, M., Wang, R. & Li, Y. (2016) An electrochemical biosensor for rapid detection of *E. coli* O157:H7 with highly efficient bifunctional glucose oxidase-polydopamine nanocomposites and Prussian blue modified screen-printed interdigitated electrodes. *Analyst*, **141**, 5441–5449.

Yáñez, R., Bastías, R., Higuera, G., Salgado, O., Katharios, P., Romero, J. *et al.* (2015) Amplification of tlh gene in other Vibrionaceae specie by specie-specific multiplex PCR of *Vibrio parahaemolyticus. Electronic Journal of Biotechnology*, **18**, 459–463.

Yap, C.K., Cheng, W.H., Karami, A. & Ismail, A. (2016) Health risk assessments of heavy metal exposure via consumption of marine mussels collected from anthropogenic sites. *Science of the Total Environment*, **553**, 285–229.

Yaron, S. & Matthews, K.R. (2002) A reverse transcriptase-polymerase chain reaction assay for detection of viable *Escherichia coli* O157:H7: investigation of specific target genes. *Journal of Applied Microbiology*, **92**, 633–640.

Yasumoto, T., Ito, E., Kuiper-Goodman, T., Munday, R. & Suzuki, T. (2011) Pectenotoxins. In: *Assessment and Management of Biotoxin Risks in Bivalve Molluscs* (eds J. Lawrence, H. Loreal, H. Toyofuku, P. Hess, I. Karunasagar & L. Ababouch), pp. 193–206. FAO Fisheries and Aquaculture Technical Paper No. 551. FAO, Rome.

Yates, A. (2011) Hepatitis A virus. In: *Agents of Foodborne Illness* (eds D. Craig & A. Bartholomaeus), pp. 13–22. Food Standards Australia New Zealand.

Yong, H.T. & Son, R. (2009) Hepatitis A virus–a general overview. *International Food Research Journal*, **16**, 455–467.

Younger, A.D. & Reese, R.A. (2013) Comparison of *Escherichia coli* levels between bivalve mollusc species across harvesting sites in England and Wales. *Journal of Shellfish Research*, **32**, 527–532.

Youn, S., Kabir, M., Haque, R. & Petri, W.A. (2009) Evaluation of a screening test for detection of *Giardia* and *Cryptosporidium* parasites. *Journal of Clinical Microbiology*, **47**, 451–452.

Zahedi, A., Gofton, A.W., Jian, F., Paparini, A., Oskam, C. & Ball, A. (2017) Next generation sequencing uncovers within-host differences in the genetic diversity of *Cryptosporidium* gp60 subtypes. *International Journal for Parasitology*, **47**, 601–607.

Zelada-Guillén, G.A., Riu, J., Düzgün, A. & Rius, F.X. (2009) Immediate detection of living bacteria at ultralow concentrations using a carbon nanotube based potentiometric aptasensor. *Angewandte Chemie*, **121**, 7470–7473.

Zhang, B., Shi, Z-R., Wang, X.-L., Deng, S.-G. & Lin, H.-M. (2015) Depuration of cadmium from blue mussel (*Mytilus edulis*) by hydrolysis peptides and chelating metal elements. *Food Research International*, **73**, 162–168.

Zhang, C. & Zhang, J. (2015) Current techniques for detecting and monitoring algal toxins and causative harmful algal blooms. *Journal of Environmental Analytical Chemistry*, **2**, 123.

Zhang, F., Ye, Q., Chen, Q., Yang, K., Zhang, D., Chen, Z. *et al.* (2018) Algicidal activity of novel marine bacterium *Paracoccus* sp. strain Y42 against a harmful algal-bloom-causing dinoflagellate, *Prorocentrum donghaiense. Applied and Environmental Microbiology*, **84**, e01015–18.

Zhang, J., Liu, F., Chen, R., Feng, T., Dong, S. & Shen, H. (2012) Levels of polychlorinated biphenyls and organochlorine pesticides in edible shellfish from Xiamen (China) and estimation of human dietary intake. *Food and Chemical Toxicology*, **50**, 4285–4291.

Zhang, Z., Yu, L., Xu, L., Hu, X., Li, P., Zhang, Q. *et al.* (2014) Biotoxin sensing in food and environment via microchip. *Electrophoresis*, **35**, 1547–1559.

Zhao, S., Datta, A.R., Ayers, S., Friedman, S., Walker, R.D. & White, D.G. (2003) Antimicrobial-resistant *Salmonella* serovars isolated from imported foods. *International Journal of Food Microbiology*, **84**, 87–92.

Zhelyazkova, G., Yankovska-Stefanovab, T., Minevac, E., Stratevd, D., Vashind, I., Dospatliev, L. *et al.* (2018) Risk assessment of some heavy metals in mussels (*Mytilus galloprovincialis*) and veined rapa whelks (*Rapana venosa*) for human health. *Marine Pollution Bulletin*, **128**, 197–201.

Zhuo, L., Yin, Y., Fu, W., Qiu, B., Lin, Z., Yang, Y. *et al.* (2013) Determination of paralytic shellfish poisoning toxins by HILIC–MS/MS coupled with dispersive solid phase extraction. *Food Chemistry*, **137**, 115–121.

Species Index

Page locators in **bold** indicate tables. Page locators in *italics* indicate figures. This index uses letter-by-letter alphabetization.

Marine Mussels: Ecology, Physiology, Genetics and Culture, First Edition. Elizabeth Gosling.

Subject Index

Page locators in **bold** indicate tables. Page locators in *italics* indicate figures. This index uses letter-by-letter alphabetization. Please also refer to separate species index.

a

Marine Mussels: Ecology, Physiology, Genetics and Culture, First Edition. Elizabeth Gosling.

C

d

f

j

k

l

q

r

t

u

v

w

Figure 1.4 Different external shell forms in the family Mytilidae. (A) Mytiliform *Mytilus galloprovincialis*. *Source:* From MgA. Jiří Novák (biolib.cz). (B) Modioliform *Modiolus lulat*. *Source:* From MgA. Jiří Novák (biolib.cz). (C) Elongate coral boring form *Lithophaga lithophaga*. *Source:* From MgA. Jiří Novák (biolib.cz). (D) Shell with pronounced anterior and posterior radiating ribs, as in *Musculus niger*. *Source:* From Guido Poppe © G. & Ph. Poppe (http://www.poppe-images.com/). (E) Radiating sculpture over entire shell surface, as in *Brachidontes erosus*. *Source:* From MgA. Jiří Novák (biolib.cz). (F) *Perna viridis*. *Source:* From Guido Poppe © G. & Ph. Poppe (http://www.poppe-images.com/).

Figure 2.6 Inner anatomy of *Mytilus edulis*. The white posterior adductor muscle is visible in the upper image but has been cut in the lower image to allow the valves to open fully. *Source:* Photograph by Rainer Zenz.

Figure 3.9 Photograph of a whelk, *Nucella lapillus*, flipped on its back by a mussel, *Mytilus edulis*. Mussels attach a number of byssus threads to the body whorl of the predator gastropod, then retract them, flipping the predator over and immobilising it, and thus exposing it to crab predation. *Source:* Photo courtesy of P. Petraitis, Department of Biology, University of Pennsylvania, Philadelphia, Pennsylvania, USA.

Figure 3.10 The sea star *Pisaster ochraceus* predating on mussels, *Mytilus californianus*, on the US Pacific coast. *Source:* David Cowles, http://rosario.wallawalla.edu/inverts. Reproduced with permission.

Figure 3.11 The American oystercatcher, *Haematopus palliatus*, a significant predator of bivalves, eating a surf clam (*Spisula solidissima*) at Nickerson Beach, Long Island, New York. *Source:* ©Arthur Morris, www.birdsasart.com.

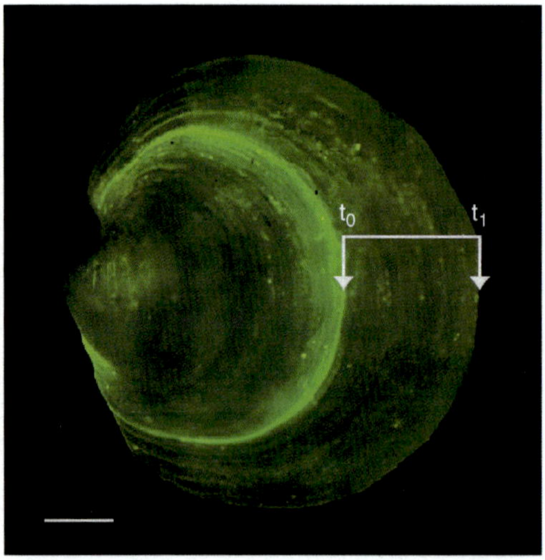

Figure 6.3 Photograph of the right valve of a calcein-marked *Loripes lacteus* shell. The position of the growing edge at the time of calcein administration is given by t_0; the position of the growing edge three months later is given by t_1. Thus, t_0 to t_1 represents shell growth (increase in height). Scale bar: 1 mm. *Source:* Van der Geest *et al.* (2011). Reproduced with permission from Elsevier.

(a)

(b)

Figure 10.13 (a) Galician mussel raft with others in the background. Photographed in summer. (b) Galician mussel raft showing the rope droppers on which the mussels are grown. No mussels are visible above 1.5 m depth as they must always be immersed in water. *Source:* Photos courtesy of A. Ovejero, Dpto. Geociencias Marinas y Ordenación del Territorio, Edificio Ciencias Experimentales, Universidade de Vigo, Vigo, Spain.

Figure 10.14 Mussel (*Mytilus edulis*) culture on bouchots during low tide in the Chausey Archipelago, off the north coast of France. *Source:* Photo courtesy of P. Archambault, Départment de Biologie, Université de Laval, Canada.

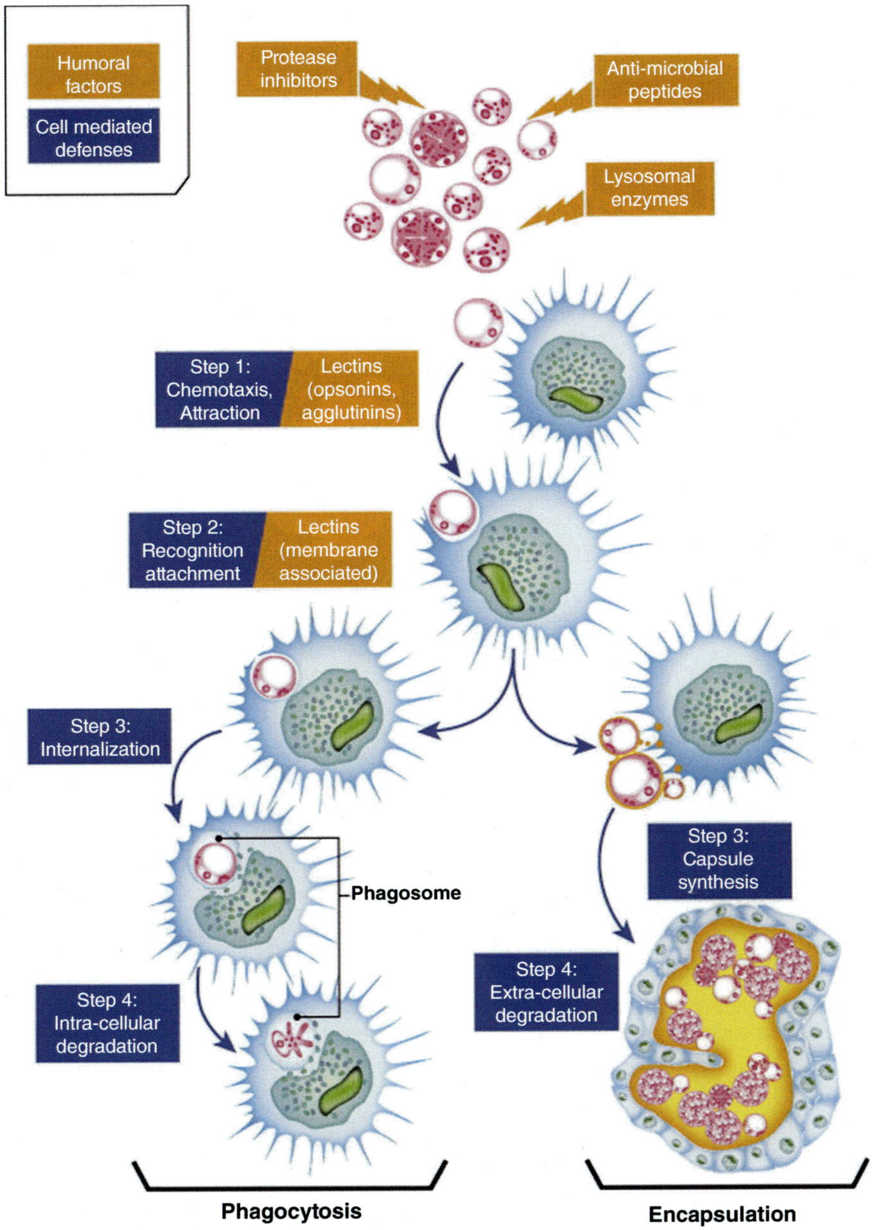

Figure 11.17 Schematic presentation of humoral and cellular responses involved in bivalve defence mechanisms upon infection by micro-organisms. **Blue** indicates steps of phagocytosis and encapsulation, **green** indicates the various humoral factors such as antimicrobial peptides (AMPs), lysosomal hydrolysases and protease inhibitors that also contribute to eliminate pathogenic micro-organisms, **purple** symbolises the invading micro-organisms. Step 1: Chemotaxis, attraction and migration. Step 2: Recognition and attachment of invading micro-organisms. These two steps involve the presence and synthesis of lectins. Step 3: Internalisation of micro-organisms within a phagosome. Step 4: Destruction of the micro-organisms by oxygen-dependent and oxygen-independent microbicidal activities. These four steps constitute phagocytosis. However, at Step 3, when haemocytes fail to phagocytose the micro-organisms (possibly because they are too big for internalisation), they encapsulate them and destroy them externally (Step 4) with lysosomal enzymatic and reactive oxygen species (ROS) activities. *Source:* From Soudant *et al.* (2013). Reprinted with permission from Elsevier.

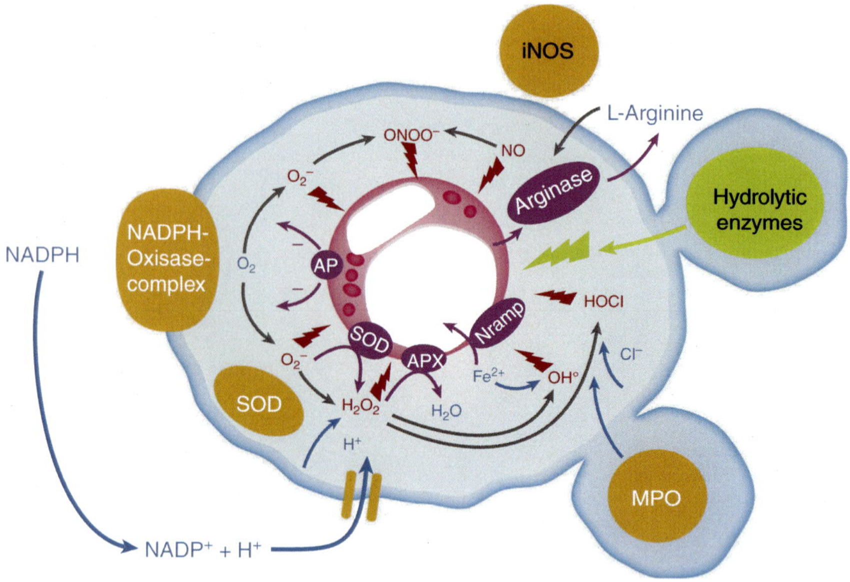

Figure 11.18 Antioxidant capacities of *Perkinsus marinus* upon phagocytosis by eastern oyster, *Crassostrea virginica*, hemocytes (phagosome is delimited by a **blue** line, phagocytised *Perkinsus* cell is pink). Toxic and nontoxic molecules are noted in **red** and **blue**, respectively. Pro-oxidant activities of haemocytes are represented in **orange**, and the anti-oxidant activities and mechanisms of *P. marinus* are represented in **purple**. Hydrolytic enzymes are represented in **green**. NO, nitric oxide; $ONOO^-$, peroxynitrite, O_2^-, superoxide anion; HOCl, hypochloride; iNOS, inductible nitric oxide synthase; SOD, superoxide dismutase; MPO, myeloperoxidase; AP, acid phosphatase; APX, ascorbate dependent peroxidase; Nramp, natural resistance-associated macrophage protein (proposed to contribute to the depletion of iron in haemocytes). *Source:* From Soudant *et al.* (2013). Reprinted with permission from Elsevier.